Electrical Networks

About the Author

Ravish R Singh is presently Vice-Principal and Head, Department of Electronics and Telecommunication Engineering at Thakur College of Engineering and Technology, Mumbai. He obtained his BE degree from the University of Mumbai in 1991 and completed his M.Tech from IIT Bombay in 2001. Currently, he is pursuing a doctorate degree from the Faculty of Technology of the University of Mumbai.

Ravish R Singh has over fifteen years of teaching experience in the field of Electronics and Telecommunication Engineering. His fields of interest include Networks and Circuits, Signals and Systems and Digital Signal Processing.

Electrical Networks

RAVISH R SINGH

Vice-Principal
Thakur College of Engineering and Technology
Mumbai

Tata McGraw-Hill Publishing Company Limited
NEW DELHI

McGraw-Hill Offices

New Delhi New York St Louis San Francisco Auckland Bogotá Caracas
Kuala Lumpur Lisbon London Madrid Mexico City Milan Montreal
San Juan Santiago Singapore Sydney Tokyo Toronto

Tata McGraw-Hill

Published by the Tata McGraw-Hill Publishing Company Limited,
7 West Patel Nagar, New Delhi 110 008.

ISBN 13: 978-0-07-069977-9
ISBN 10: 0-07-069977-1

Managing Director: *Ajay Shukla*

General Manager: Publishing—SEM & Tech Ed: *Vibha Mahajan*
Sponsoring Editor: *Shukti Mukherjee*
Editorial Executive: *Suman Sen*
Executive—Editorial Services: *Sohini Mukherjee*
Senior Production Manager: *P L Pandita*

General Manager: Marketing—Higher Education & School: *Michael J Cruz*
Product Manager: SEM & Tech Ed: *Biju Ganesan*

Controller—Production: *Rajender P Ghansela*
Asst. General Manager—Production: *B L Dogra*

Published by the Tata McGraw-Hill Publishing Company Limited, 7 West Patel Nagar, New Delhi 110 008, typeset at Bukprint India, B-180A, Guru Nanak Pura, Laxmi Nagar-110 092 and printed at Adarsh Printers, C-50/51, Panchsheel Garden, Naveen Shahdara, Delhi 110 032

Cover: SDR Printers

RAXCRRBFDLXCY

The McGraw-Hill Companies

To

My Parents

Contents

Preface

Electrical Networks is one of the core subjects for students of Electrical Engineering, Electronics Engineering, Electronics and Telecommunication Engineering, Instrumentation Engineering, Biomedical Engineering and related branches of Engineering in the third/fourth semester course of almost all universities in India. This subject is also one of the important topics of competitive examinations such as IAS, IES, etc., and examinations conducted by various public-sector undertakings in this field. As per the course requirement, five major topics that are mandatory to be covered by any book on this subject are dc and ac circuits, transient analysis, network functions, two-port networks and network synthesis.

Generally, numerical problems are expected in university examinations in this subject. The weightage given to problems is more than 70%–80% in examinations. This book attempts to cover almost all the topics and solved problems on these important topics. Objective-type questions from various papers of competitive examinations are included in each chapter. This will help the students in sharpening their knowledge about core concepts. This text attempts to provide a simple explanation about the concepts of Electrical Networks with brief theory and a large number of problems. Numerous examples and exercise problems have been included to help the reader develop an intuitive grasp of the contents. It covers both analysis and synthesis of networks.

The salient features of the book are

- Complete coverage of dc circuits with dependent and independent sources covered
- Chapter on graph theory included
- Over 500 solved examples
- Hundreds of additional practice problems with answers
- Network analysis as well as synthesis covered

The first chapter of the book covers basic circuit elements and basic laws pertaining to networks. Next, dc network theorems are covered in chapters 2 and 3. Chapter 4 deals with the analysis of ac circuits. Chapter 5 covers ac network theorems. Three-phase circuits are covered in Chapter 6. Chapter 7 deals with graph theory which is useful for solving complex network problems. Chapters 8 and 9 discuss transient analysis in time domain and frequency domain respectively. Chapters 10 and 11 cover network functions and two-port networks. Network synthesis is covered in the last chapter.

Appendices on Fourier Series, Network Filters, Bode Plots and Attenuators can be accessed at http://www.mhhe.com/ravish/en.

I would like to express my gratitude to my colleagues Ameya Kadam, Sanjeev Ghosh, Payel Saha, Kalawati Patil, Aradhana Manekar, Vipul Gohil, Archana Deshpande and Deepak Jain in Thakur College of Engineering and Technology for helping me in proofreading. I would also like to thank my staff Sanjay Rawat in preparing the manuscript. I am grateful to the following reviewers who took out time to review the manuscript and gave me noteworthy suggestions.

T L Singal	*Chitkara Institute of Engineering and Technology, Rajpura, Punjab*
Manoj Kumar	*DAV Engineering College, Jalandhar, Punjab*
P Purkait	*Haldia Institute of Technology, Haldia, West Bengal*
Anirudha Mukherjee	*Hooghly Engineering College and Technology, Hooghly, West Bengal*
A B Nanda	*Institute of Technical Education and Research, Bhubaneswar, Orissa*
Sonali Sawant	*Excelsior Education Society's K C College of Engineering, Thane, Maharashtra*
Ujwal Chaudhari	*M H Saboo Siddik College of Engineering, Mumbai, Maharashtra*
Sanjay Singh Thakur	*Don Bosco Institute of Technology, Mumbai , Maharashtra*
K Manivanan	*Pondicherry Engineering College, Pondicherry*
P Murugan	*Kalasalingam University, Madurai, Tamil Nadu*
Divya Prabha	*Sri Siddhartha Institute of Technology, Tumkur, Karnataka*
Gopal Rao	*R V College of Engineering, Bangalore, Karnataka*

My sincere thanks are due to my friend Mukul Bhatt for always inspiring and encouraging me in this project. I would also like to thank all the staff at McGraw-Hill Education India, especially Suman Sen, Sagar Divekar, Sohini Mukherjee and P L Pandita for coordinating with me during the editorial, copyediting and production stages of this book. Finally, I thank my wife, Nitu; son, Aman; and daughter, Aditri, for supporting me throughout the preparation of this project.

Any suggestions for the improvement of the book are welcome. Please feel free to write to me at ravishrsingh@yahoo.com.

RAVISH R SINGH

Walkthrough

1 Basic Circuit Concepts

Chapter introduction gives the concept of the topic that will be discussed in the chapter.

1.1 INTRODUCTION

We know that like charges repel each other whereas unlike charges attract each other. To overcome this force of attraction, an electromagnetic force (EMF) must be applied. When the charges are separated, it is said that a potential difference exists and the work or energy per unit charge utilized in this process is known as voltage or potential difference. When W joules of energy are supplied to Q coulombs of charge, the voltage is given by

$$V = \frac{W}{Q} \text{ volts (V)}$$

The phenomenon of transfer of charge from one point to another is termed current. Current (I) is defined as the rate of flow of electrons in a conductor. It is measured by the number of electrons that flow in unit time.

$$I = \frac{Q}{t} \text{ amperes (A)}$$

Energy is the total work done in the electric circuit. The rate at which the work is done in an electric circuit is called electic power.

$$\text{Power} = \frac{\text{Energy}}{\text{Time}} = \frac{W}{t}$$

Definitions along with illustrations are used exhaustively to provide an understanding of the points.

7.3 DEFINITIONS ASSOCIATED WITH A GRAPH

1. Planar graph A graph drawn on a two-dimensional plane is said to be planar if two branches do not intersect or cross at a point which is other than a node. Figure 7.2 shows such graphs.

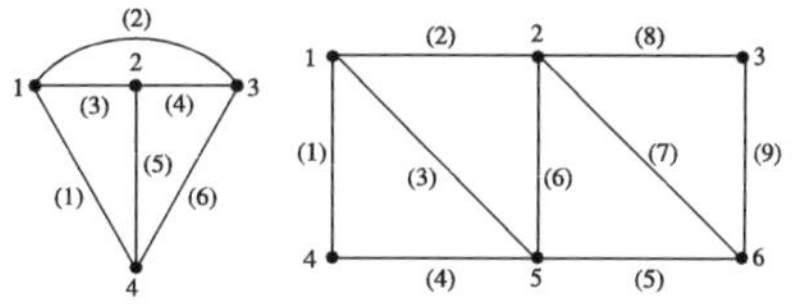

Fig. 7.2 *Planar graphs*

2. Non-planar graph A graph drawn on a two-dimensional plane is said to be non-planar if there is intersection of two or more branches at another point which is not a node. Figure 7.3 shows non-planar graphs.

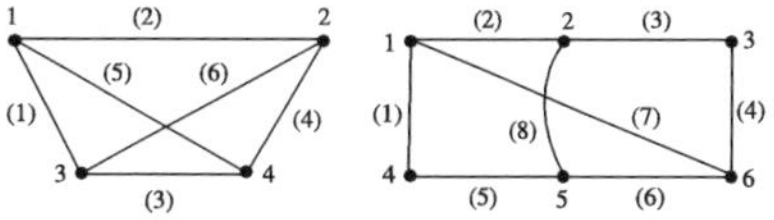

Fig. 7.3 *Non-planar graphs*

3. Sub-graph It is a subset of branches and nodes of a graph. It is a proper sub-graph if it contains branches and nodes less than those on a graph. A sub-graph can be just a node or only one branch of the graph. Figure 7.4 shows a graph and its proper sub-graph.

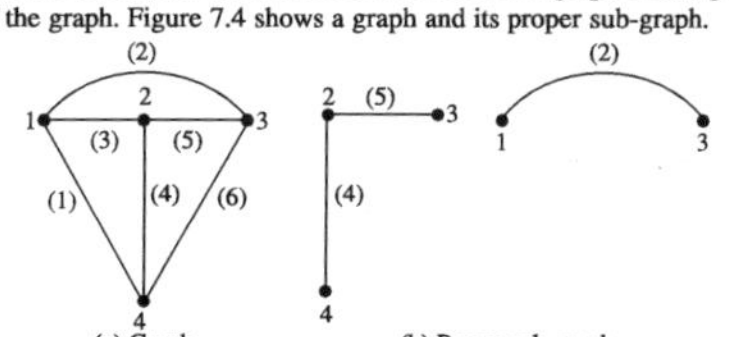

(a) Graph (b) Proper sub-graph

Fig. 7.4

11.2 OPEN-CIRCUIT IMPEDANCE PARAMETERS (Z PARAMETERS)

The Z parameters of a two-port network may be defined by expressing two-port voltages V_1 and V_2 in terms of two-port currents I_1 and I_2.

$$(V_1, V_2) = f(I_1, I_2) \quad \ldots(11.1)$$

$$V_1 = Z_{11} I_1 + Z_{12} I_2 \quad \ldots(11.2)$$

$$V_2 = Z_{21} I_1 + Z_{22} I_2 \quad \ldots(11.3)$$

In matrix form, we can write

$$\begin{bmatrix} V_1 \\ V_2 \end{bmatrix} = \begin{bmatrix} Z_{11} & Z_{12} \\ Z_{21} & Z_{22} \end{bmatrix} \begin{bmatrix} I_1 \\ I_2 \end{bmatrix}$$

$$[V] = [Z]\,[I] \quad \ldots(11.4)$$

The individual Z parameters for a given network can be defined by setting each of the port currents equal to zero.

Case 1 When the output port is open circuited,

i.e. $I_2 = 0$

$$Z_{11} = \left.\frac{V_1}{I_1}\right|_{I_2=0} \quad \ldots(11.5)$$

where Z_{11} is the driving-point impedance with the output port open circuited. It is also called *open circuit input impedance*.

Similarly,

$$Z_{21} = \left.\frac{V_2}{I_1}\right|_{I_2=0} \quad \ldots(11.6)$$

where Z_{21} is the transfer impedance with the output port open circuited. It is also called *open circuit forward transfer impedance*.

Case 2: When input port is open circuited,

Preliminary discussions following definitions consist of **derivations or methods** which provide a systematic approach to problem solving.

Example 3.4 *Find the voltage V_{AB}.*

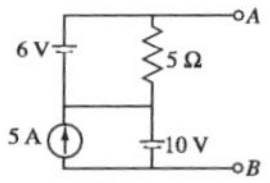

Fig. 3.24

Step I: *When the 6-V source is acting alone*

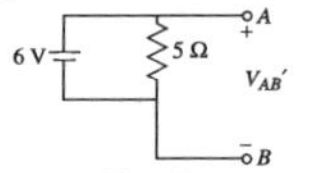

Fig. 3.25

From Fig. 3.25,

$$V_{AB}' = 6 \text{ V}$$

Solved numericals are used exhaustively to illustrate the concepts and methods described.

Step II: *Calculation of R_N*

Replacing all voltage sources by short circuits,

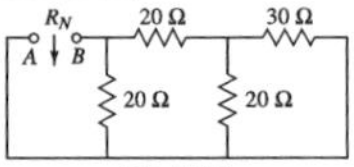

Fig. 3.125

$$R_N = [(20 \parallel 30) + 20] \parallel 20 = 12.3\ \Omega$$

Step III: *Calculation of I_L*

$$I_L = 0.81 \times \frac{12.3}{12.3 + 10} = 0.45 \text{ A}$$

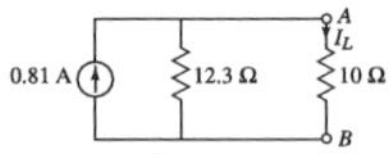

Fig. 3.126

Example 3.28 Obtain Norton's equivalent network as seen by R_L.

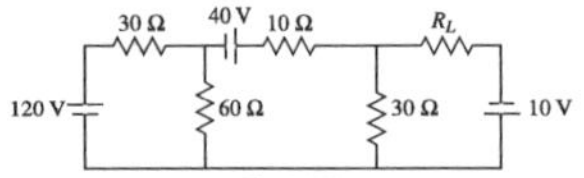

Fig. 3.127

Step I: *Calculation of I_{SC}*

Applying KVL to Mesh 1,

$$120 - 30I_1 - 60\,(I_1 - I_2) = 0$$

$$90I_1 - 60I_2 = 120 \quad \ldots(1)$$

Applying KVL to Mesh 2,

$$-60\,(I_2 - I_1) + 40 - 10I_2 - 30\,(I_2 - I_3) = 0$$

$$-60I_1 + 100I_2 - 30I_3 = 40 \quad \ldots(2)$$

Applying KVL to Mesh 3,

$$-30\,(I_3 - I_2) + 10 = 0$$

$$30I_2 - 30I_3 = -10 \quad \ldots(3)$$

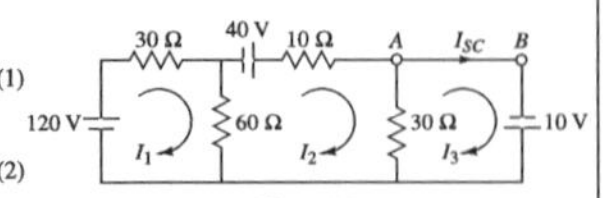

Fig. 3.128

Problem-solving approach from simpler to complex ones gives the reader an in-depth understanding of the topics.

3.60 *Electrical Networks*

Exercises

SUPERPOSITION THEOREM

1. Find the current through the 10-Ω resistor.

Fig. 3.226

[0.37 A]

2. Find the current through the 8-Ω resistor.

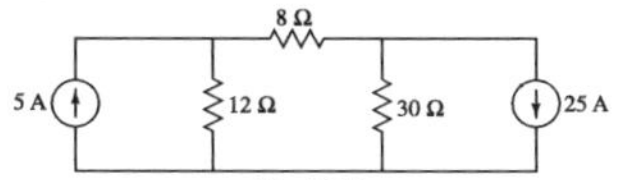

Fig. 3.227

[16.2 A]

3. Find the potential across the 3-Ω resistor.

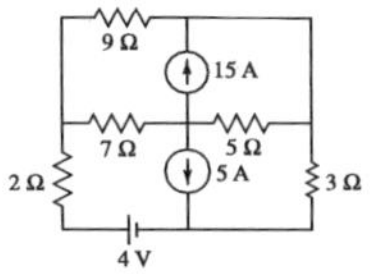

Exercises will help readers review their understanding of the concepts discussed in the chapter.

Network Theorems-II 3.61

5. Find the current through the 1-Ω resistor.

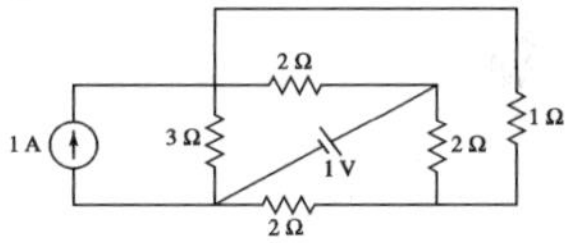

Fig. 3.230

[0.41 A]

6. Find the current through the 4-Ω resistor.

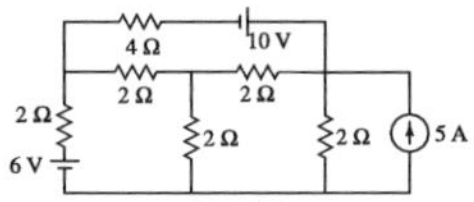

Fig. 3.231

[1.33 A]

7. Find the current I_x.

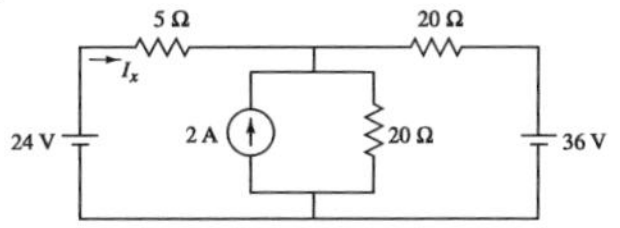

Fig. 3.232

[–1.143 A]

Answers have been provided along with each unsolved numerical.

7.46 *Electrical Networks*

Objective-Type Questions

1. The number of independent loops for a network with n nodes and b branches is
   (a) $n - 1$ (b) $b - n$
   (c) $b - n + 1$ (d) independent of the number of nodes
2. A network has 7 nodes and 5 independent loops. The number of branched in the network is
   (a) 13 (b) 12 (c) 11 (d) 10
3. Identify which of the following is NOT a true of the graph shown in Fig. 7.90.
   (a) *begh* (b) *defg*
   (c) *adfg* (d) *aegh*

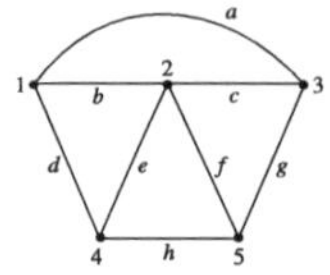

Fig. 7.90

4. The minimum number of equations required to analyze the circuit shown in Fig. 7.91 is
   (a) 3 (b) 4
   (c) 6 (d) 7

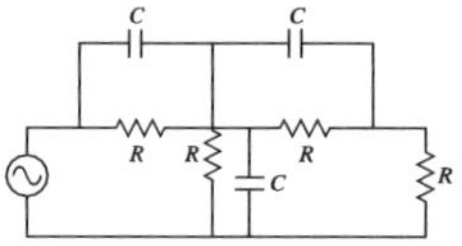

Fig. 7.91

Objective-type questions provided at the end of each chapter enable the reader to prepare for competitive examinations.

6.36 *Electrical Networks*

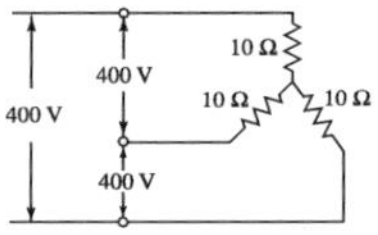

Fig. 6.25

10. If a balanced delta load has an impedance of $(6 + j9)$ ohms per phase, then the impedance of each phase if equivalent star load is
    (a) $(6 + j9)$ ohms (b) $(2 + j3)$ ohms (c) $12 + j8$) ohms (d) $(3 + j4.5)$ ohms
11. In two wattmeter method of measurement, if one of the wattmeters reads zero, then power factor will be
    (a) zero (b) unity (c) 0.5 (d) 0.866

Answers to the objective questions are provided for verification.

Answers to Objective-Type Questions

1. (d) 2. (c) 3. (d) 4. (b) 5. (c) 6. (b)
7. (d) 8. (b) 9. (a) 10. (b) 11. (c)

1 Basic Circuit Concepts

1.1 INTRODUCTION

We know that like charges repel each other whereas unlike charges attract each other. To overcome this force of attraction, an electromagnetic force (EMF) must be applied. When the charges are separated, it is said that a potential difference exists and the work or energy per unit charge utilized in this process is known as voltage or potential difference. When W joules of energy are supplied to Q coulombs of charge, the voltage is given by

$$V = \frac{W}{Q} \text{ volts (V)}$$

The phenomenon of transfer of charge from one point to another is termed current. Current (I) is defined as the rate of flow of electrons in a conductor. It is measured by the number of electrons that flow in unit time.

$$I = \frac{Q}{t} \text{ amperes (A)}$$

Energy is the total work done in the electric circuit. The rate at which the work is done in an electric circuit is called electic power.

$$\text{Power} = \frac{\text{Energy}}{\text{Time}} = \frac{W}{t}$$

We can also write

$$P = \frac{W}{Q} \cdot \frac{Q}{t}$$

$$= V \, . \, I \quad \text{watts (W)}$$

Energy is measured in joules (J) and power in watts (W).

1.2 RESISTANCE

Resistance is the property of a substance due to which it opposes the flow of electric current through it.

Certain substances offer very little opposition to the flow of electric current and are called conductors, e.g., metals, acids and salt solutions. Certain substances offer very high resistance to the flow of electric current and are called insulators, e.g., mica, glass, rubber, bakelite, etc.

The practical unit of resistance is **ohm** and is represented by the symbol Ω. A conductor is said to have resistance of one ohm if a potential difference of one volt across its terminals causes a current of one ampere to flow through it.

The resistance of a conductor depends on the following factors.

(i) It is directly proportional to its length.
(ii) It is inversely proportional to the area of cross section of the conductor.
(iii) It depends on the nature of the material.
(iv) It also depends on the temperature of the conductor.

Hence we can say that

$$R \propto \frac{l}{A}$$

$$R = \rho \frac{l}{A}$$

where l is length of the conductor, A is the cross-sectional area and ρ is a constant known as specific resistance or resistivity of the material.

Power dissipated in a resistor We know that $v = R \cdot i$

When current flows through any resistor, power is absorbed by the resistor which is given by

$$p = v \cdot i$$

The power dissipated in the resistor is converted to heat which is given by

$$W = \int_0^t v \cdot i \, dt$$

$$= \int_0^t R \cdot i \cdot i \, dt$$

$$= i^2 R t$$

1.3 INDUCTANCE

Inductance is the property of a coil that opposes any change in the amount of current flowing through it. If the current in the coil is increasing, the self-induced emf is set up in such a direction so as to oppose the rise of current. Similarly, if the current in the coil is decreasing, the self-induced emf will be in the same direction as the applied voltage.

Inductance is defined as the ratio of flux linkage to the current flowing through the coil. The practical unit of inductance is **henry** and is represented by the symbol H. A coil is said to have an inductance of one henry if a current of one ampere when flowing through it produces flux linkages of one weber-turn in it.

The inductance of an inductor depends on the following factors.

(i) It is directly proportional to the square of the number of turns.
(ii) It is directly proportional to the area of cross section.
(iii) It is inversely proportional to the length.
(iv) It depends on the absolute permeability of the magnetic material.

Hence we can say that

$$L \propto \frac{N^2 A}{l}$$

$$L = \mu \frac{N^2 A}{l}$$

where l is the mean length, A is the cross-sectional area and μ is the absolute permeability of the magnetic material.

Current–voltage relationships in an inductor We know that

$$v = L \frac{di}{dt}$$

Expressing inductor current as a fuction of voltage,

$$di = \frac{1}{L} v\, dt$$

Integrating both the sides,

$$\int_{i(0)}^{i(t)} di = \frac{1}{L} \int_0^t v\, dt$$

$$i(t) = \frac{1}{L} \int_0^t v\, dt + i(0)$$

The quantity $i(0)$ denotes the initial current through the inductor. When there is no initial current through the inductor,

$$i(t) = \frac{1}{L} \int_0^t v\, dt$$

Energy stored in an inductor Consider a coil of inductance L carrying a changing current I. When the current is changed from zero to a maximum value I, every change is opposed by the self-induced emf produced. To overcome this opposition, some energy is needed and this energy is stored in the magnetic field. The voltage v is given by

$$v = L \frac{di}{dt}$$

Energy supplied to the inductor during interval dt is given by

$$\begin{aligned} dW &= v \,.\, i \,.\, dt \\ &= L \frac{di}{dt} \,.\, i \,.\, dt \\ &= L \,.\, i \,.\, di \end{aligned}$$

Hence, total energy supplied to the inductor when current is increased from 0 to I amperes is

$$E = \int_0^I L.\,i\,\mathrm{d}i = \frac{1}{2}LI^2$$

1.4 CAPACITANCE

Capacitance is the property of a capacitor to store an electric charge when its plates are at different potentials. If Q coulombs of charge is given to one of the plates of a capacitor and if a potential difference of V volts is applied between the two plates then its capacitance is given by

$$C = \frac{Q}{V}$$

The practical unit of capacitance is **farad** and is represented by the symbol F. A capacitor is said to have capacitance of one farad if a charge of one coulomb is required to establish a potential difference of one volt between its plates.

The capacitance of a capacitor depends on the following factors.

(i) It is directly proportional to the area of the plates.

(ii) It is inversely proportional to the distance between two plates.

(iii) It depends on the absolute permittivity of the medium between the plates.

Hence we can say that

$$C \propto \frac{A}{d}$$

$$C = \varepsilon\,\frac{A}{d}$$

where d is the distance between two plates, A is the cross-sectional area of the plates and ε is absolute permittivity of the medium between the plates.

Current–voltage relationships in a capacitor The charge on a capacitor is given by

$$q = C\,v$$

where q denotes the charge and v is the potential difference across the plates at any instant.

We know that

$$i = \frac{\mathrm{d}q}{\mathrm{d}t} = \frac{d}{\mathrm{d}t}\,Cv = C\frac{\mathrm{d}v}{\mathrm{d}t}$$

Expressing capacitor voltage as a function of current,

$$\mathrm{d}v = \frac{1}{C}\,i\,\mathrm{d}t$$

Integrating both the sides,

$$\int_{v(0)}^{v(t)} \mathrm{d}v = \frac{1}{C}\int_0^t i\,\mathrm{d}t$$

$$v(t) = \frac{1}{C}\int_0^t i\,\mathrm{d}t + v\,(0)$$

The quantity v (0) denotes the initial voltage across the capacitor. When there is no initial voltage on the capacitor,

$$v(t) = \frac{1}{C}\int_0^t i \, dt$$

Energy stored in a capacitor Let a capacitor of capacitance C farads be charged from a source of V volts. Then current i is given by

$$i = C\frac{dv}{dt}$$

Energy supplied to the capacitor during interval dt is given by

$$dW = v \, . \, i \, . \, dt$$

$$= v \, . \, C\frac{dv}{dt} \, . \, dt$$

Hence, total energy supplied to the capacitor when potential difference is increased from 0 to V volts is

$$E = \int_0^V C \, . \, v \, dv$$

$$= \frac{1}{2}CV^2$$

1.5 SOURCES

Source is a basic network element which supplies energy to the networks. There are two classes of sources, namely,

(1) Independent sources
(2) Dependent sources

1.5.1 Independent Sources

Output characteristics of an independent source are not dependent on any network variable such as a current or voltage. Its characteristics, however, may be time-varying. There are two types of independent sources:

(1) Independent voltage source
(2) Independent current source

Independent voltage source An independent voltage source is a two-terminal network element that establishes a specified voltage across its terminals. The value of this voltage at any instant is independent of the value or direction of the current that flows through it. The symbols for such voltage sources are shown in Fig. 1.1.

The terminal voltage may be a constant, or it may be some specified function of time.

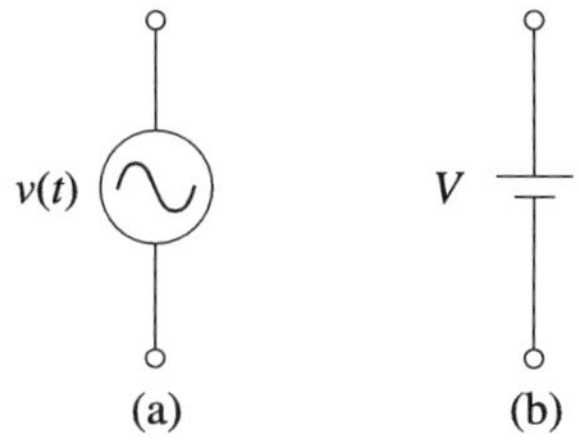

Fig. 1.1

Independent current source An independent current source is a two-terminal network element which produces a specified current. The value and direction of this current at any instant is independent of the value or direction of the voltage that appears across the terminals of the source. The symbols for such current sources are shown in Fig. 1.2.

The output current may be a constant or it may be a function of time.

1.5.2 Dependent Sources

If the voltage or current of a source depends in turn upon some other voltage or current, it is called as dependent or controlled source. The dependent sources are of four kinds, depending on whether the control variable is voltage or current and the controlled source is a voltage source or current source.

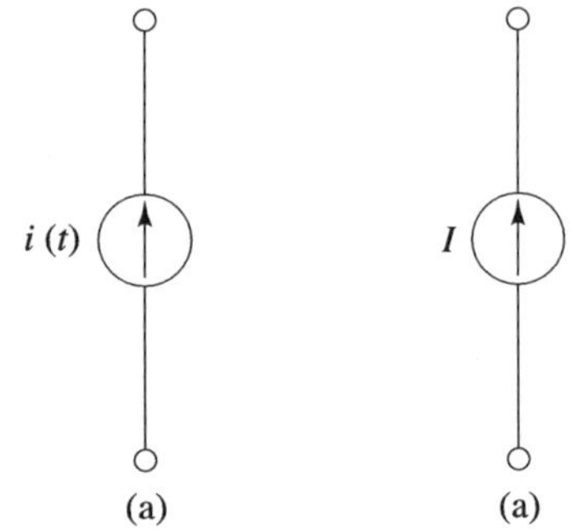

Fig. 1.2

Voltage-controlled voltage source (VCVS) A voltage-controlled voltage source is a four-terminal network component that establishes a voltage v_{cd} between two points c and d in the circuit that is proportional to a voltage v_{ab} between two points a and b.

The symbol for such a source is shown in Fig. 1.3.

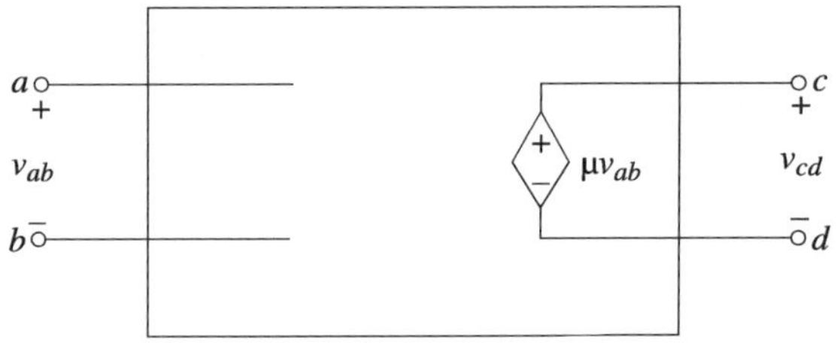

Fig. 1.3

The (+) and (–) sign inside the diamond of the component symbol identifies the component as a voltage source.

$$v_{cd} = \mu\, v_{ab}$$

The voltage v_{cd} depends upon the control voltage v_{ab} and the constant μ, a dimensionless constant called voltage gain.

Voltage-controlled current source (VCCS) A voltage-controlled current source is a four-terminal network component that establishes a current i_{cd} in a branch of the circuit that is proportional to the voltage v_{ab} between two points a and b.

The symbol for such a source is shown in Fig. 1.4.

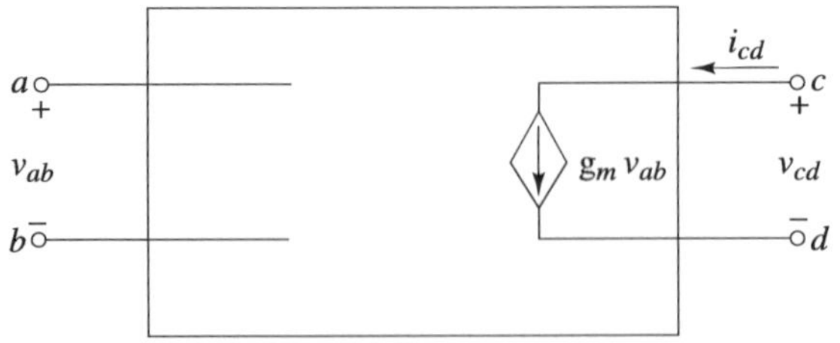

Fig. 1.4

The arrow inside the diamond of the component symbol identifies the component as a current source.

$$i_{cd} = g_m\, v_{ab}$$

The current i_{cd} depends only on the control voltage v_{ab} and the constant g_m, called the transconductance or mutual conductance. The constant g_m has dimension of ampere per volt or siemens (S).

Current-controlled voltage source (CCVS) A current-controlled voltage source is a four-terminal network component that establishes a voltage v_{cd} between two points c and d in the circuit that is proportional to the current i_{ab} in some branch of the circuit.

The symbol for such a source is shown in Fig. 1.5.

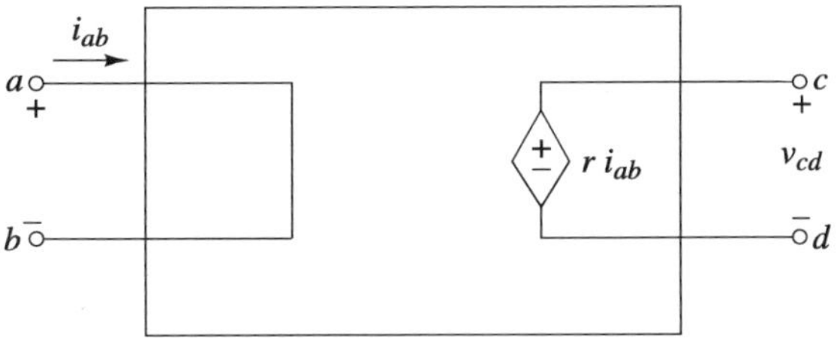

Fig. 1.5

$$v_{cd} = r\, i_{ab}$$

The voltage v_{cd} depends only on the control current i_{ab} and the constant r called the transresistance or mutual resistance. The constant r has dimension of volt per ampere or ohm (Ω).

Current-controlled current source (CCCS) A current-controlled current source is a four-terminal network component that establishes a current i_{cd} in one branch of a circuit that is proportional to the current i_{ab} in some branch of the network.

The symbol for such a source is shown in Fig. 1.6.

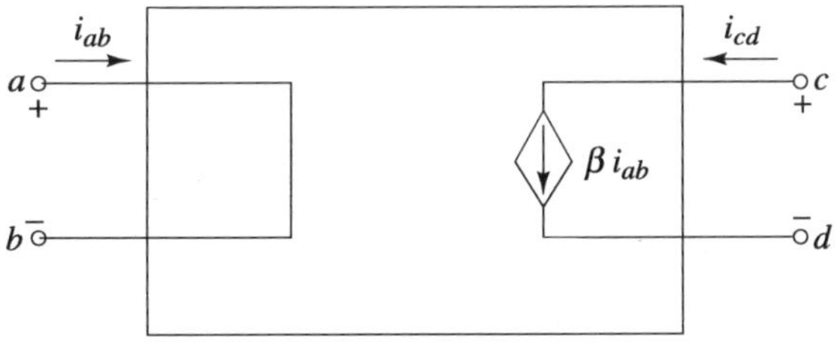

Fig. 1.6

$$i_{cd} = \beta\, i_{ab}$$

The current i_{cd} depends only on the control current i_{ab} and the dimensionless constant β, called the current gain.

1.6 SOME DEFINITIONS

1. Network and circuit The interconnection of two or more circuit elements (viz., voltage sources, resistors, inductors and capacitors) is called an *electric network*. If the network contains at least one closed path, it is called an electric circuit. Every circuit is a network, but all networks are not circuits. Figure 1.7(a) shows a network which is not a circuit and Fig. 1.7(b) shows a network which is a circuit.

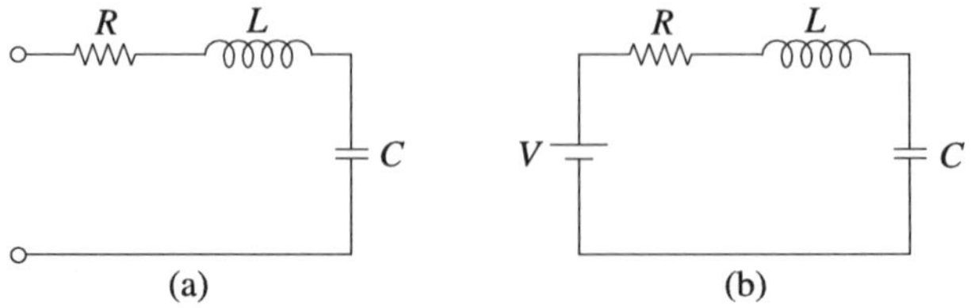

Fig. 1.7

2. *Linear and non-linear elements* If the resistance, inductance or capacitance offered by an element does not change linearly with the change in applied voltage or circuit current, the element is termed as *linear element*. Such an element shows a linear relation between voltage and current as shown in Fig. 1.8. Ordinary resistors, capacitors and inductors are examples of linear elements.

A non-linear circuit element is one in which the current does not change linearly with the change in applied voltage. A semiconductor diode operating in the curved region of characteristics as shown in Fig. 1.8 is common example of non-linear element.

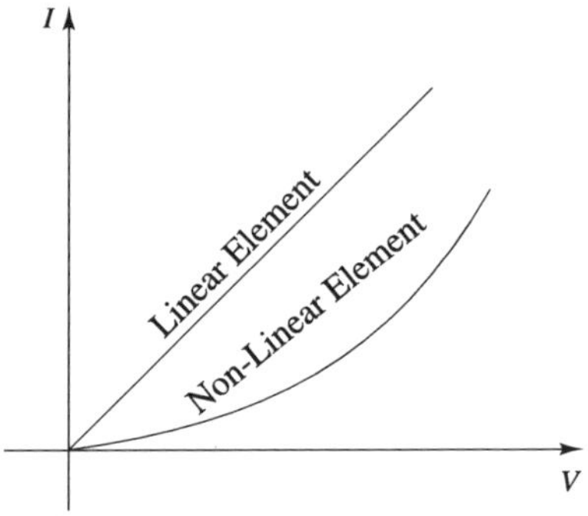

Fig. 1.8

Other examples of non-linear elements are voltage-dependent resistor (VDR), voltage-dependent capacitor (varactor), temperature-dependent resistor (thermistor), light-dependent resistor (LDR), etc. Linear elements obey Ohm's law whereas non-linear elements do not obey Ohm's law.

3. *Active and passive elements* An element which is a source of electrical signal or which is capable of increasing the level of signal energy is termed as *active element*. Batteries, BJTs, FETs or OP-AMPs are treated as active elements because these can be used for the amplification or generation of signals. All other circuit elements, such as resistors, capacitors, inductors, VDR, LDR, thermistors, etc., are termed *passive elements*. The behaviour of active elements cannot be described by Ohm's law.

4. *Unilateral and bilateral elements* If the magnitude of current flowing through a circuit element is affected when the polarity of the applied voltage is changed, the element is termed *unilateral element*. Consider the example of a semiconductor diode. Current flows through the diode only in one direction. Hence, it is called an unilateral element. Next, consider the example of a resistor. When the voltage is applied, current starts to flow. If we change the polarity of the applied voltage, the direction of the current is changed but its magnitude is not affected. Such an element is called a *bilateral element*.

5. *Active and passive networks* A network which contains at least one active element such as an independent voltage or current source is an active network. A network which does not contain any active element is a passive network.

1.7 KIRCHHOFF'S LAWS

The entire electric circuit analysis is based on Kirchhoff's laws only. But before discussing this, it is essential to familiarise ourselves with the following terms:

(a) Node A node is a junction where two or more circuit elements are connected together.

(b) Branch An element or number of elements connected between two nodes constitutes a branch.

(c) Loop A loop is any closed part of the circuit.

(d) Mesh A mesh is the most elementary form of a loop and cannot be further divided into other loops. All the meshes are loops but all the loops are not meshes.

1. Kirchhoff's current law (KCL) *The algebraic sum of currents meeting at a junction or node in an electric circuit is zero.*

Consider five conductors, carrying currents I_1, I_2, I_3, I_4 and I_5 meeting at a point O as shown in Fig. 1.9. Assuming the incoming currents to be positive and outgoing currents negative, we have

$$I_1 + (-I_2) + I_3 + (-I_4) + I_5 = 0$$
$$I_1 - I_2 + I_3 - I_4 + I_5 = 0$$
$$I_1 + I_3 + I_5 = I_2 + I_4$$

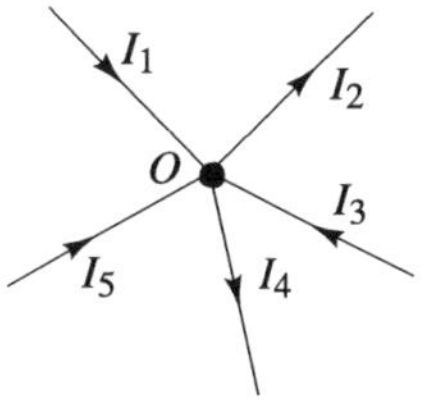

Fig. 1.9

Thus, above law can also be stated as the sum of currents flowing towards any junction in an electric circuit is equal to the sum of the currents flowing away from that junction.

2. Kirchhoff's voltage law (KVL) *The algebraic sum of all the voltages in any closed circuit or mesh or loop is zero.*

If we start from any point in a closed circuit and go back to that point, after going round the circuit, there is no increase or decrease in potential at that point. This means that the sum of emfs and the sum of voltage drops or rises meeting on the way is zero.

Determination of sign A rise in potential can be assumed to be positive while a fall in potential can be considered negative. The reverse is also possible and both conventions will give the same result.

(a) If we go from the positive terminal of the battery or source to the negative terminal, there is a fall in potential and so the emf should be assigned a negative sign. If we go from the negative terminal of the battery or source to the positive terminal, there is a rise in potential and so the emf should be given a positive sign.

(a) Rise in voltage (b) Fall in voltage

Fig. 1.10

(b) When current flows through a resistor, there is a voltage drop across it. If we go through the resistance in the same direction as the current, there is a fall in the potential and so the sign of this voltage drop is negative. If we go opposite to the direction of the current flow, there is a rise in potential and hence, this voltage drop should be given a positive sign.

(a) Fall in voltage (b) Rise in voltage

Fig. 1.11

Example 1.1 *The voltage drop across the 15-Ω resistance is 30 V, having the polarity indicated. Find the value of R.*

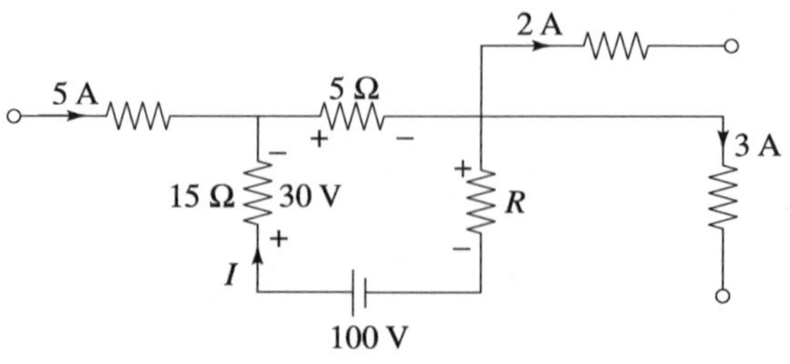

Fig. 1.12

Solution Current through the 15-Ω resistor is given by

$$I = \frac{30}{15} = 2\text{ A}$$

Current through the 5-Ω resistor is given by

$$= 5 + 2 = 7\text{ A}$$

Applying KVL to the closed path,

$$-5\,(7) - R(I) + 100 - 30 = 0$$
$$-35 - 2R + 100 - 30 = 0$$
$$R = 17.5\ \Omega$$

Example 1.2 *Determine the currents I_1, I_2 and I_3.*

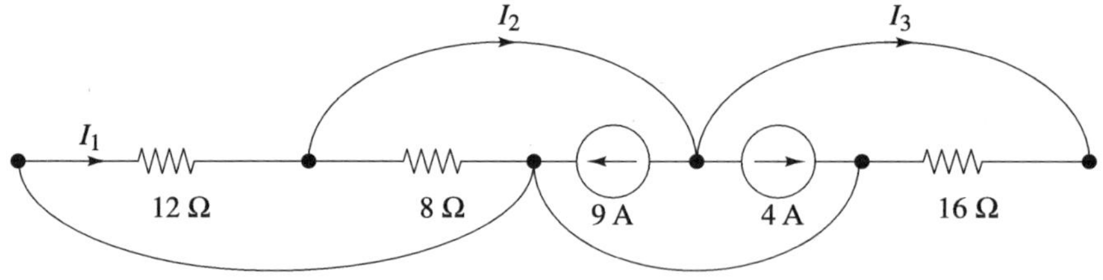

Fig. 1.13

Solution Assigning currents to all the branches,

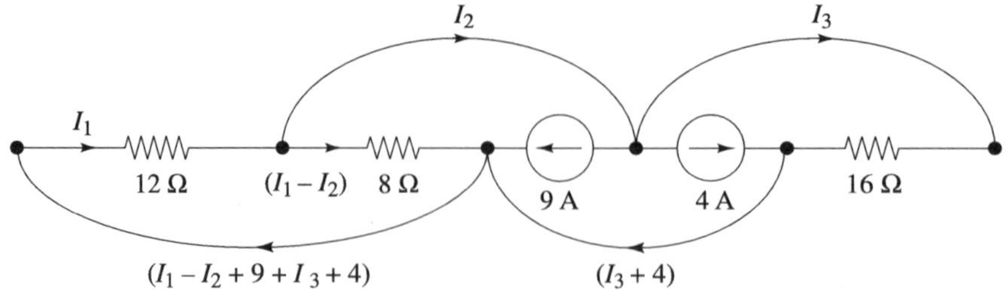

Fig. 1.14

From Fig. 1.14,

$$I_1 = I_1 - I_2 + 9 + I_3 + 4$$
$$I_2 - I_3 = 13 \qquad \text{...(i)}$$

Also, $-12I_1 - 8(I_1 - I_2) = 0$

$-20I_1 + 8I_2 = 0$ ...(ii)

Also, $-12I_1 - 16I_3 = 0$ ...(iii)

Solving Eqs (i), (ii) and (iii),

$I_1 = 4$ A

$I_2 = 10$ A

$I_3 = -3$ A

Example 1.3 *Find currents in all the branches of the network shown.*

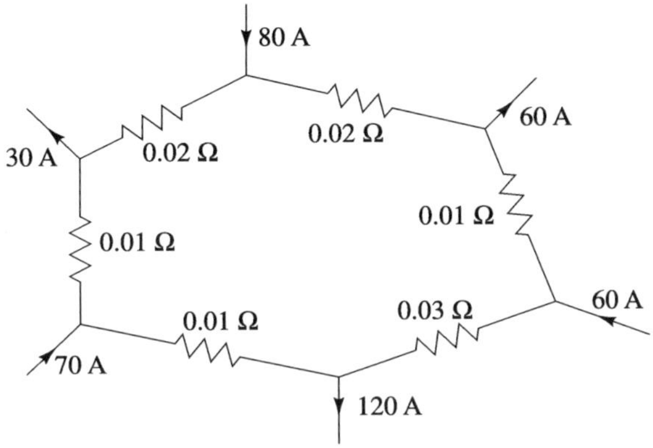

Fig. 1.15

Solution Let

$I_{AF} = x$

then $I_{FE} = x - 30$

$I_{ED} = x + 40$

$I_{DC} = x - 80$

$I_{CB} = x - 20$

$I_{BA} = x - 80$

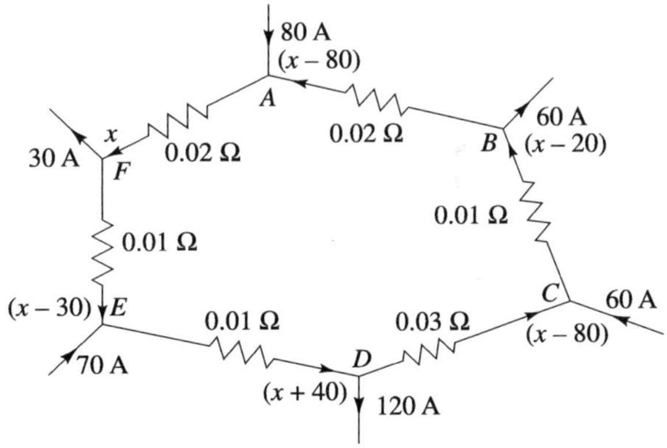

Fig. 1.16

Applying KVL to the closed path *AFEDCBA*,

$$-0.02\,x - 0.01\,(x - 30) - 0.01\,(x + 40) - 0.03\,(x - 80) - 0.01\,(x - 20) - 0.02\,(x - 80) = 0$$

$x = 41$

$I_{AF} = 41$ A

$I_{BA} = 41 - 80 = -39$ A

$I_{AB} = 39$ A

$I_{CB} = 41 - 20 = 21$ A

$I_{DC} = 41 - 80 = -39$ A

$I_{CD} = 39$ A

$I_{ED} = 41 + 40 = 81$ A

$I_{FE} = 41 - 30 = 11$ A

Example 1.4 *Find currents in all the branches of the network shown.*

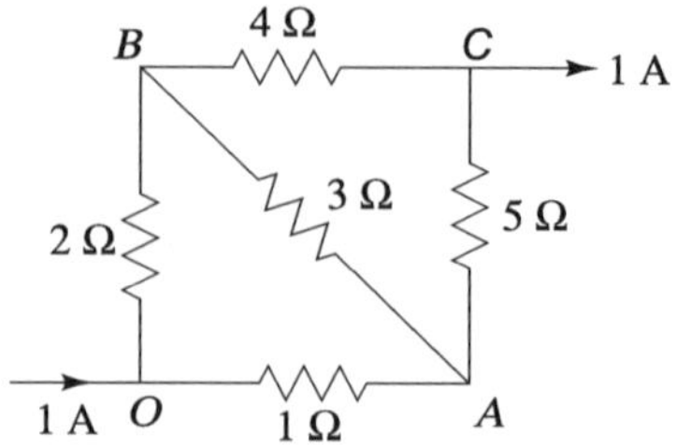

Fig. 1.17

Solution Assigning currents to all the branches,

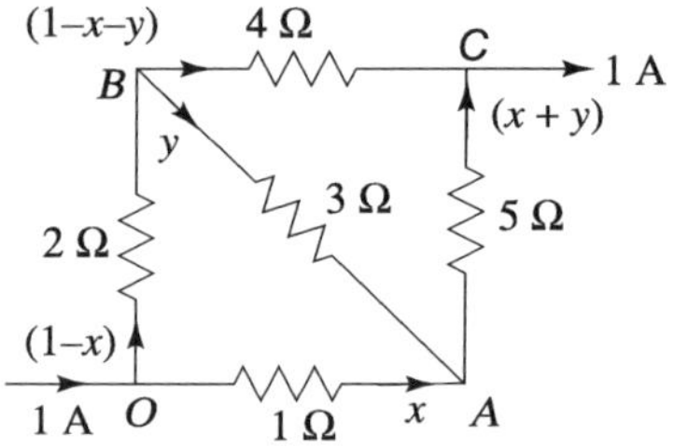

Fig. 1.18

Applying KVL to the closed path *OBAO*,

$$-2(1-x) - 3y + x = 0$$

$$3x - 3y = 2 \qquad \text{...(i)}$$

Applying KVL to the closed path *ABCA*,

$$3y - 4(1-x-y) + 5(x+y) = 0$$

$$9x + 12y = 4 \qquad \text{...(ii)}$$

Solving Eqs (i) and (ii),

$$x = 0.57 \text{ A}$$

$$y = -0.095 \text{ A}$$

$$I_{OA} = 0.57 \text{ A}$$

$$I_{OB} = 1 - 0.57 = 0.43 \text{ A}$$

$$I_{AB} = 0.095 \text{ A}$$

$$I_{AC} = 0.57 - 0.095 = 0.475 \text{ A}$$

$$I_{BC} = 1 - 0.57 + 0.095 = 0.525 \text{ A}$$

Example 1.5 *What is the p.d. between points x and y in the network shown?*

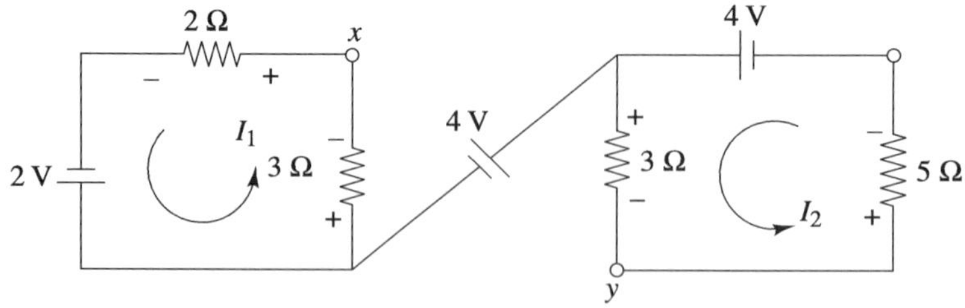

Fig. 1.19

Solution

$$I_1 = \frac{2}{5} = 0.4 \text{ A}$$

$$I_2 = \frac{4}{8} = 0.5 \text{ A}$$

Potential difference between points x and $y = V_{xy} = V_x - V_y$

Writing KVL equation for the path x to y,

$$V_x + 3I_1 + 4 - 3I_2 - V_y = 0$$

$$V_x + 3(0.4) + 4 - 3(0.5) - V_y = 0$$

$$V_x - V_y = -3.7$$

$$V_{xy} = -3.7 \text{ V}$$

Example 1.6 *In the network shown, find the voltage between points A and B.*

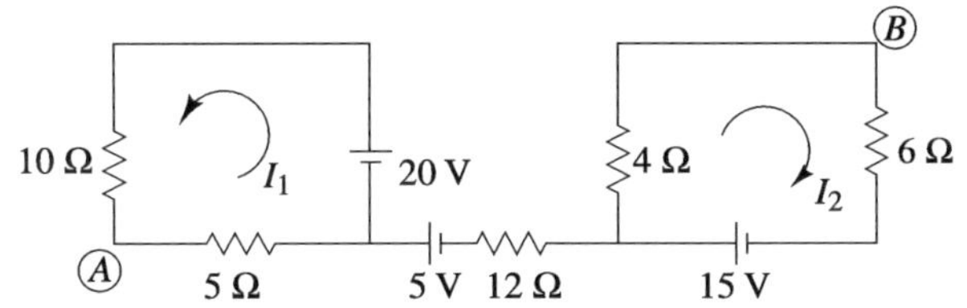

Fig. 1.20

Solution

$$I_1 = \frac{20}{15} = 1.33 \text{ A}$$

$$I_2 = \frac{15}{10} = 1.5 \text{ A}$$

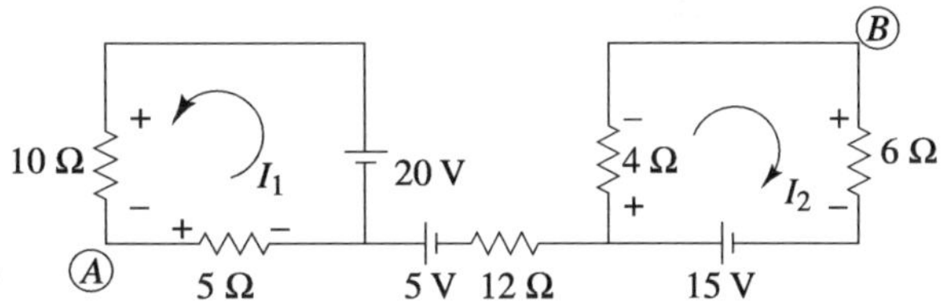

Fig. 1.21

Voltage between points A and $B = V_{AB} = V_A - V_B$

Writing KVL equation for the path A to B,

$$V_A - 5I_1 - 5 - 15 + 6I_2 - V_B = 0$$

$$V_A - 5(1.33) - 5 - 15 + 6(1.5) - V_B = 0$$

$$V_A - V_B = 17.65$$

$$V_{AB} = 17.65 \text{ V}$$

Example 1.7 *Determine the potential difference V_{AB} for the given network.*

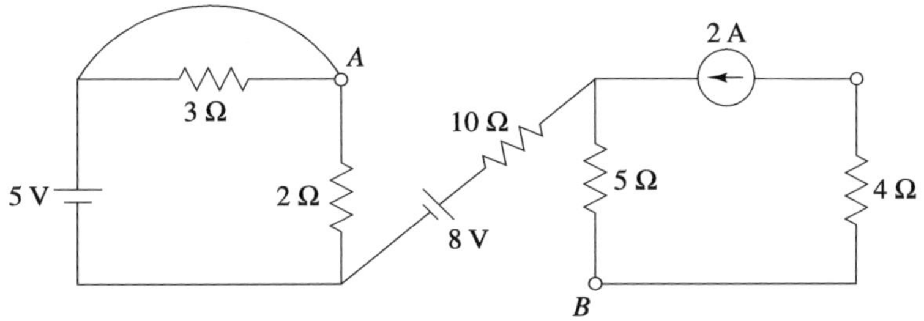

Fig. 1.22

Solution The resistance of 3 Ω is connected across a short circuit. Hence, it gets shorted.

$$I_1 = \frac{5}{2} = 2.5 \text{ A}$$

$$I_2 = 2 \text{ A}$$

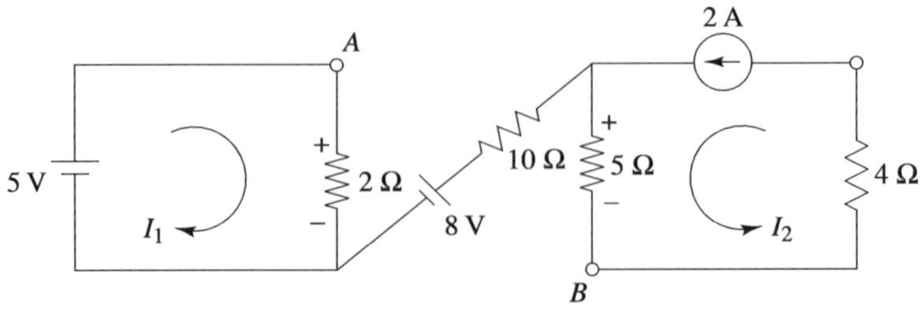

Fig. 1.23

Potential difference $V_{AB} = V_A - V_B$

Writing KVL equation for the path A to B,

$$V_A - 2I_1 + 8 - 5I_2 - V_B = 0$$

$$V_A - 2(2.5) + 8 - 5(2) - V_B = 0$$

$$V_A - V_B = 7$$

$$V_{AB} = 7 \text{ V}$$

Example 1.8 *Find the voltage of the point A w.r.t. B.*

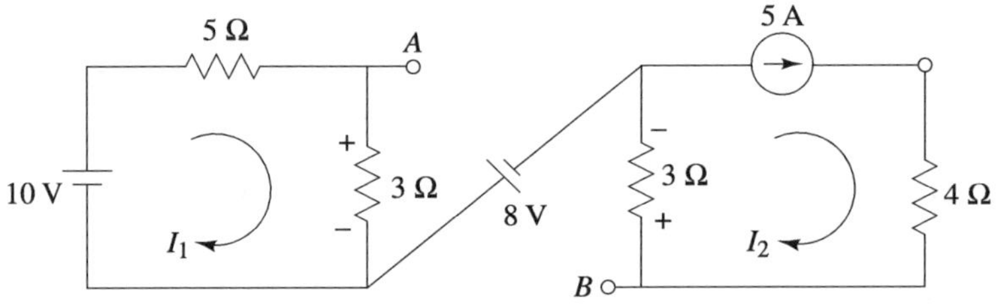

Fig. 1.24

Solution

$$I_1 = \frac{10}{8} = 1.25 \text{ A}$$

$$I_2 = 5 \text{ A}$$

Applying KVL to the path from A to B,

$$V_A - 3I_1 - 8 + 3I_2 - V_B = 0$$

$$V_A - 3(1.25) - 8 + 3(5) - V_B = 0$$

$$V_A - V_B = -3.25$$

$$V_{AB} = -3.25 \text{ V}$$

1.8 SOURCE TRANSFORMATION

A voltage source with a series resistance can be converted into an equivalent current source with a parallel resistance. Conversely, a current source with a parallel resistance can be converted into voltage source with a series resistance.

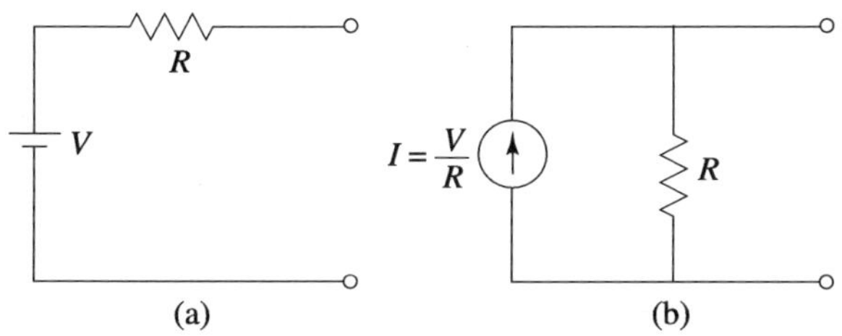

Fig. 1.25

Source transformation can be applied to controlled sources as well. The controlling variable, however, must not be tampered with any way since the operation of the controlled source depends on it.

Example 1.9 *Replace the circuit between A and B with voltage source in series with a single resistor.*

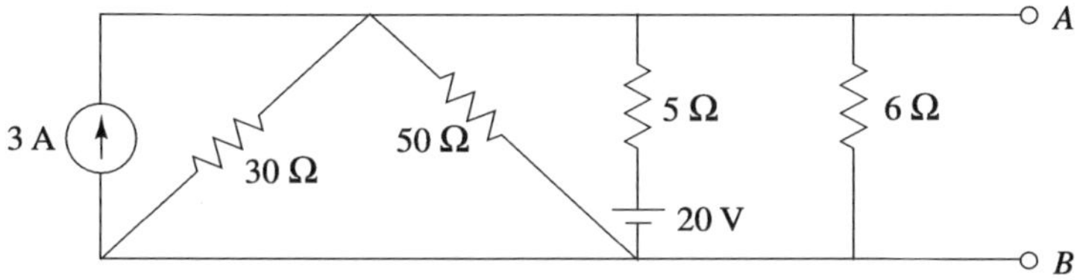

Fig. 1.26

Solution Converting the series combination of voltage source of 20 V and a resistance of 5 Ω into equivalent parallel combination of current source and resistance, we have

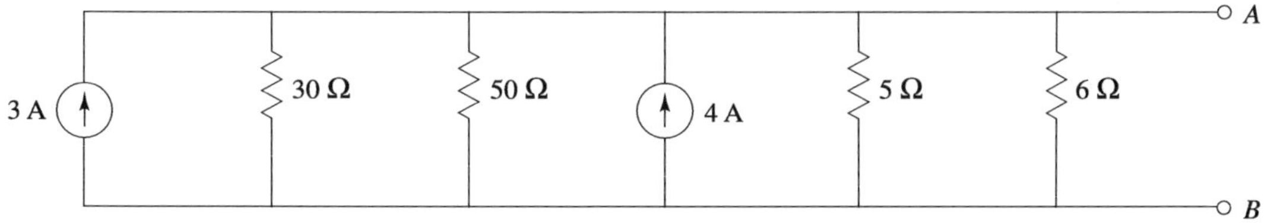

Fig. 1.27

Adding the two current sources and simplifying the circuit,

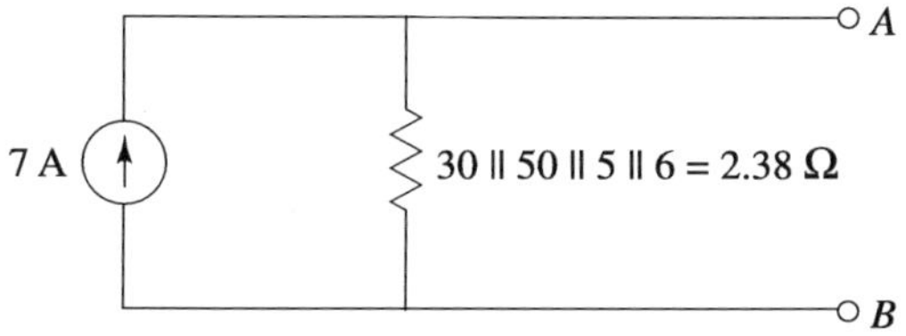

Fig. 1.28

By source conversion,

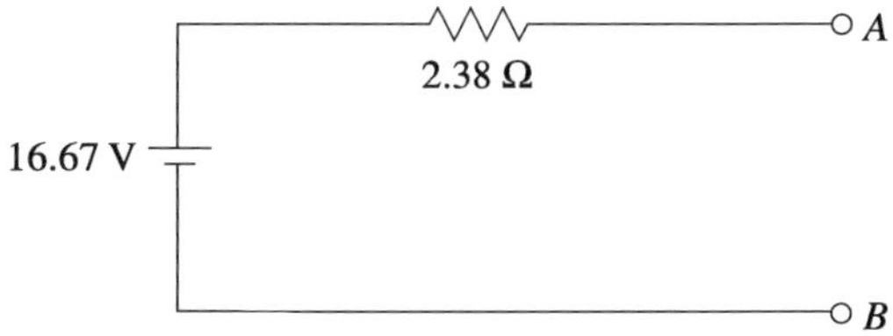

Fig. 1.29

Example 1.10 *Replace the given network with a single current source and a resistor.*

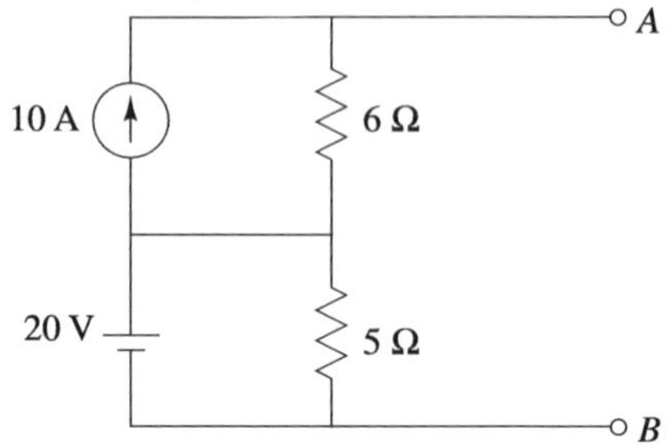

Fig. 1.30

Solution Since the resistance of 5 Ω is connected in parallel with the voltage source of 20 V, it becomes redundant. Converting parallel combination of current source and resistor into equivalent voltage source and resistor, we have

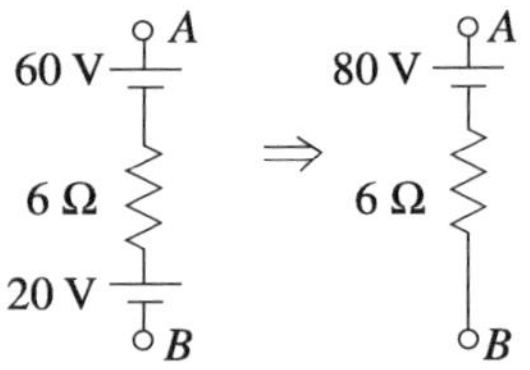

Fig. 1.31

By source conversion,

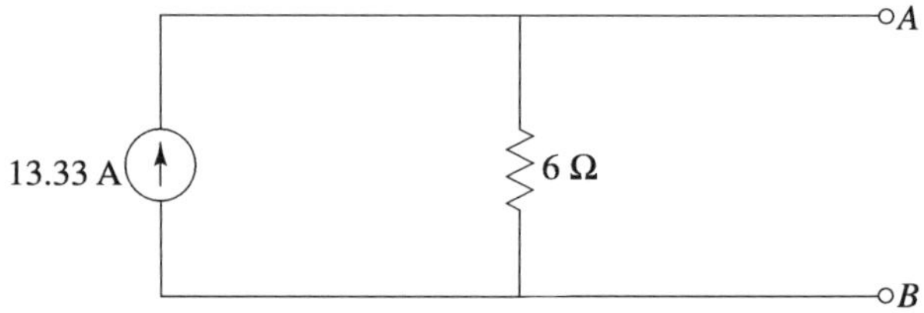

Fig. 1.32

Example 1.11 *Reduce network shown into a single source and a single resistor between terminals A and B.*

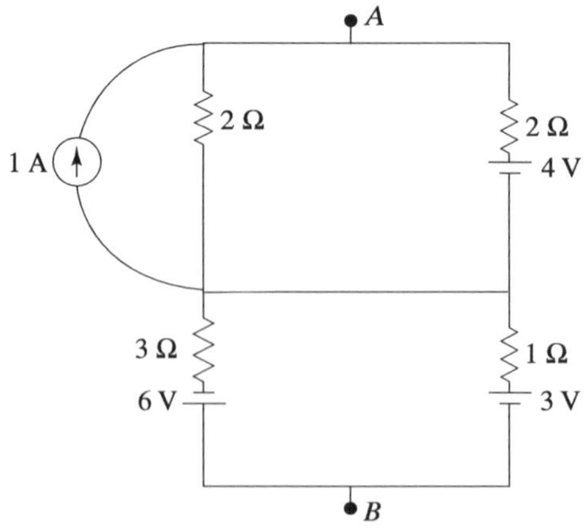

Fig. 1.33

Solution Converting all voltage sources into equivalent current sources,

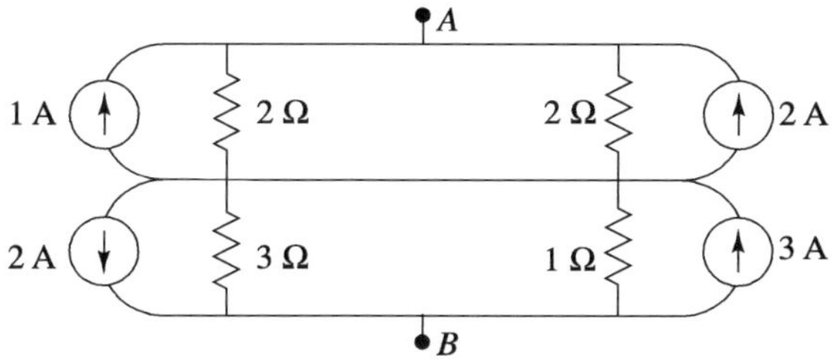

Fig. 1.34

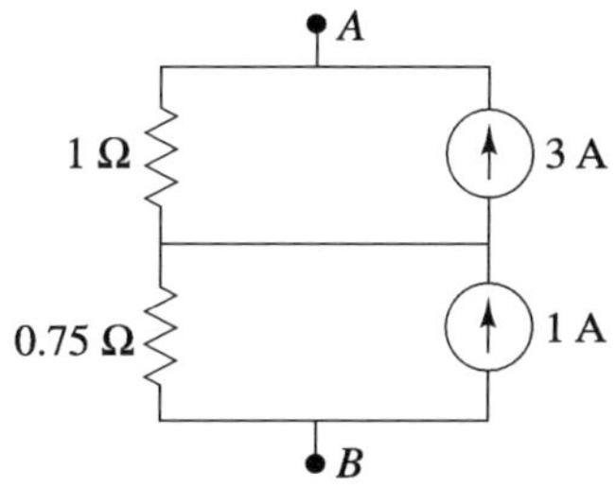

Fig. 1.35

Converting the current sources into equivalent voltage sources,

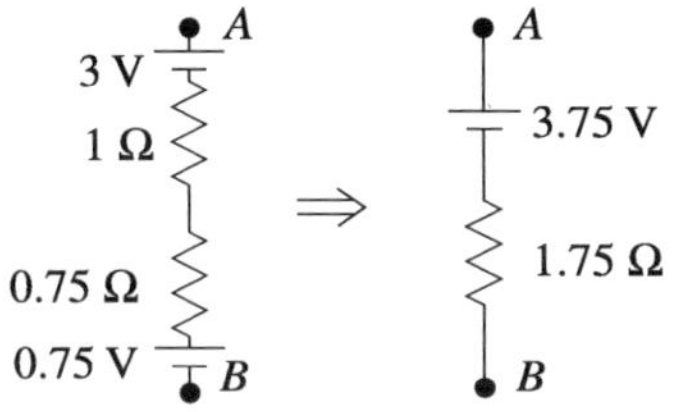

Fig. 1.36

Example 1.12 *Find the power delivered by the 50-V source in the circuit.*

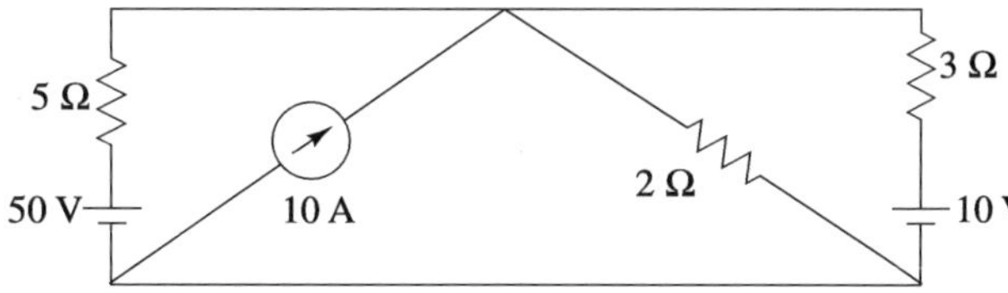

Fig. 1.37

Solution Converting the series combination of voltage source of 10 V and resistor of 3 Ω into equivalent current source and resistor, we have

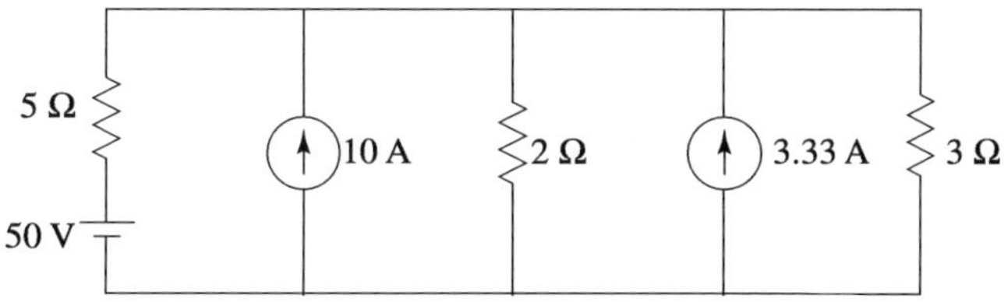

Fig. 1.38

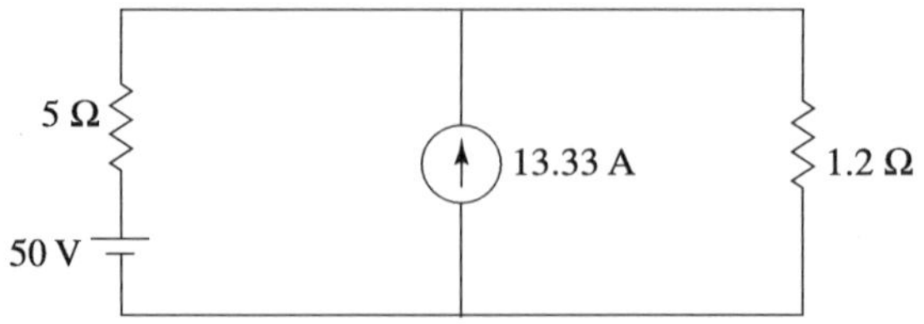

Fig. 1.39

By source conversion,

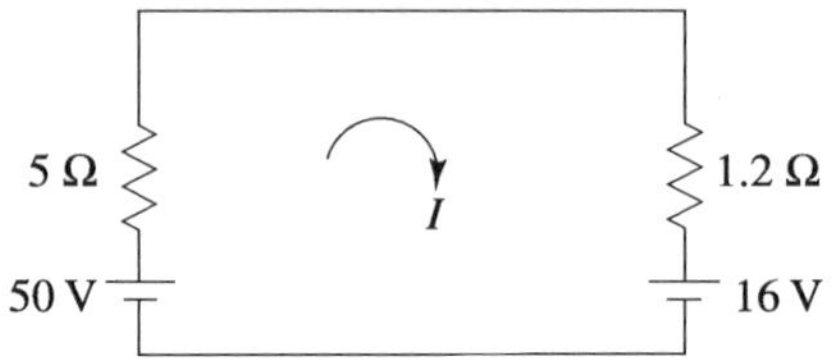

Fig. 1.40

Applying KVL to the circuit,

$$50 - 5I - 1.2I - 16 = 0$$

$$I = 5.48 \text{ A}$$

$$\text{Power delivered by the 50-V source} = 50 \times 5.48$$

$$= 274 \text{ W}$$

Example 1.13 *Find the current in the 4-Ω resistor.*

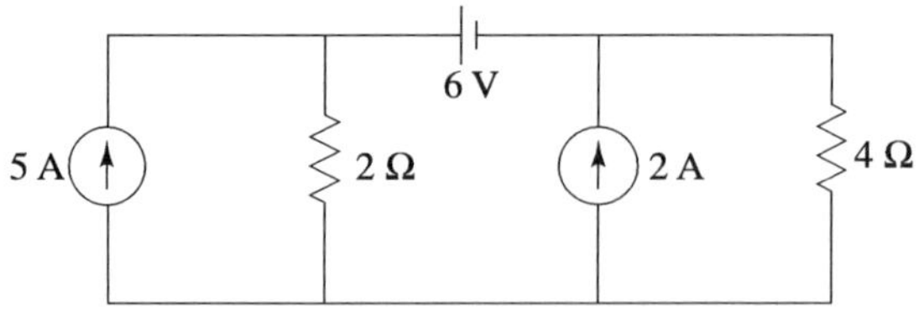

Fig. 1.41

Solution Converting the parallel combination of the current source of 5 A and the resistor of 2 Ω into an equivalent series combination of voltage source and resistor, we have

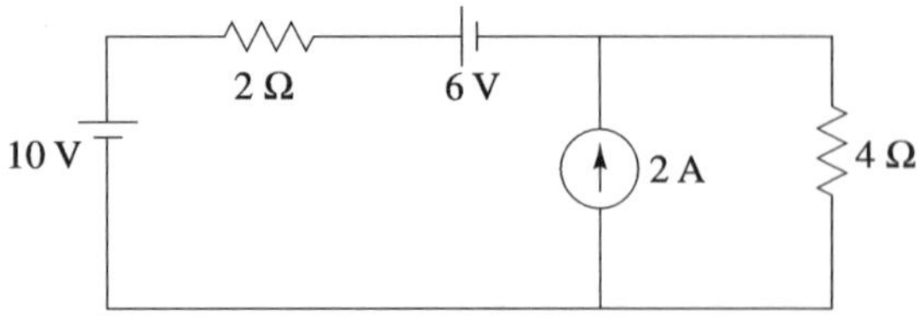

Fig. 1.42

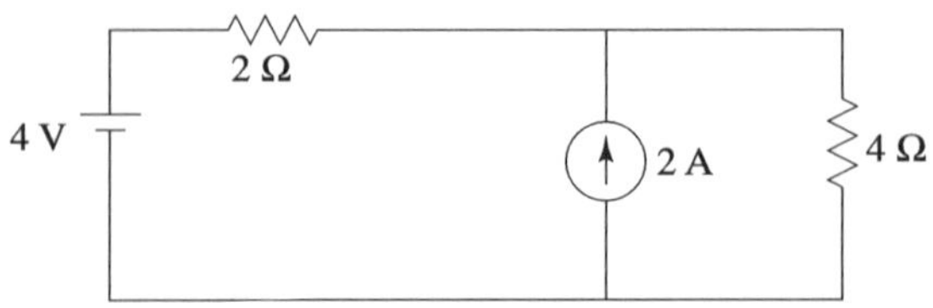

Fig. 1.43

Again by source conversion,

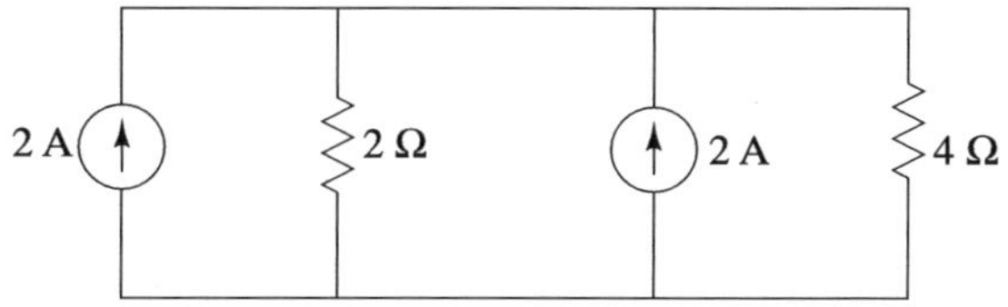

Fig. 1.44

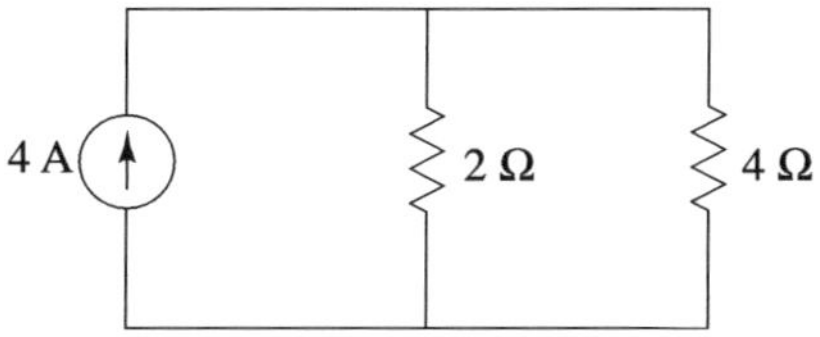

Fig. 1.45

By current-division formula,

$$I_{4\Omega} = 4 \times \frac{2}{2+4} = 1.33 \text{ A}$$

Example 1.14 *Find the voltage at Node 2.*

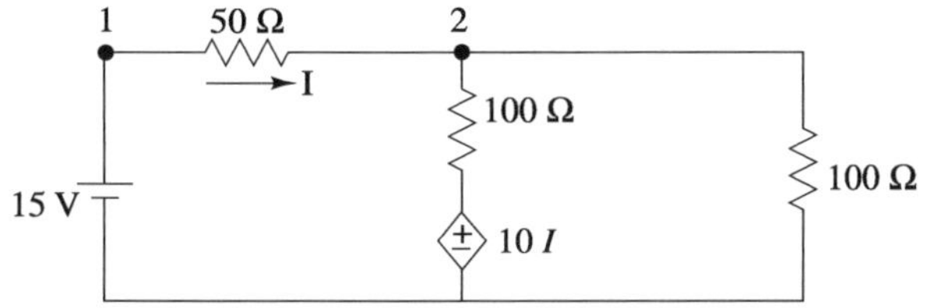

Fig. 1.46

Solution We cannot change the network between nodes 1 and 2 since the controlling current *I*, for the controlled source, is in the resistor between these nodes. Applying source transformation to series combination of controlled source and the 100-Ω resistor, we get

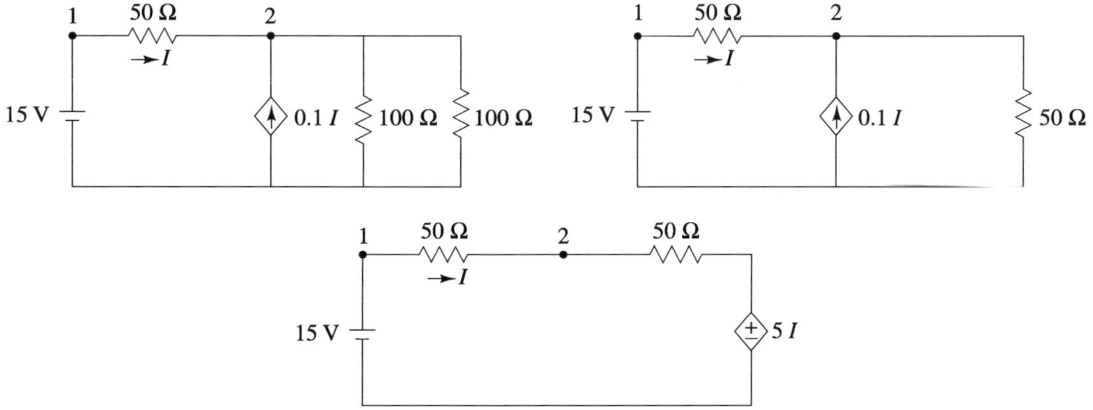

Fig. 1.47

Applying KVL to the mesh,

$$15 - 50I - 50I - 5I = 0$$

$$I = \frac{15}{105} = 0.143 \text{ A}$$

$$\text{Voltage at Node 2} = 15 - 50I$$

$$= 15 - 50 \times 0.143 = 7.86 \text{ V}$$

1.9 SOURCE SHIFTING

Source shifting is the simplification technique used when there is no resistor in series with a voltage source or a resistor in parallel with a current source.

Example 1.15 *Calculate the voltage across the 6-Ω resistor using source-shifting technique.*

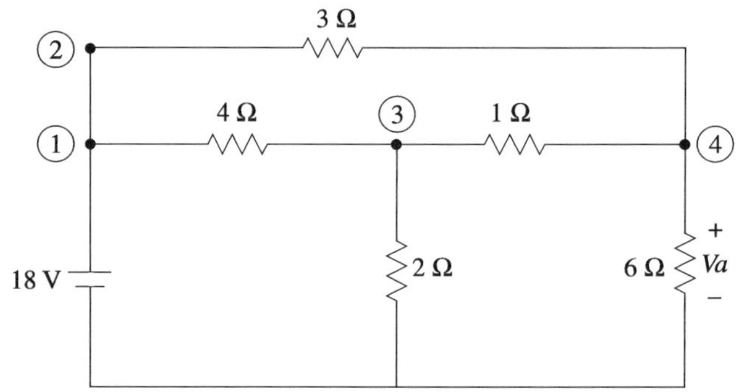

Fig. 1.48

Solution Adding a voltage source of 18 V to the network and connecting to Node 2, we have

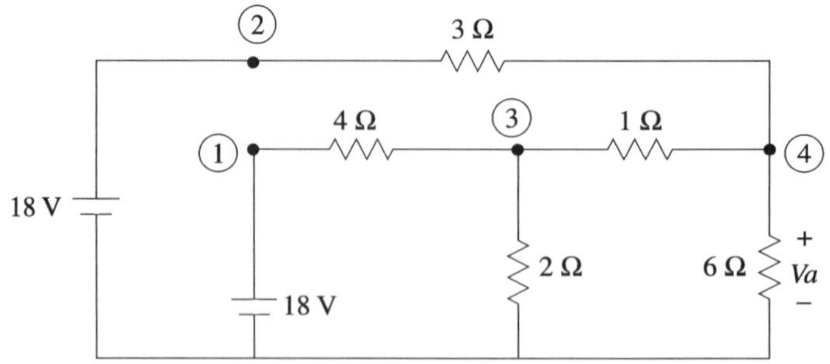

Fig. 1.49

Since nodes 1 and 2 are maintained at the same voltage by the sources, the connection between nodes 1 and 2 is removed. Now the two voltage sources have resistors in series and source transformation can be applied.

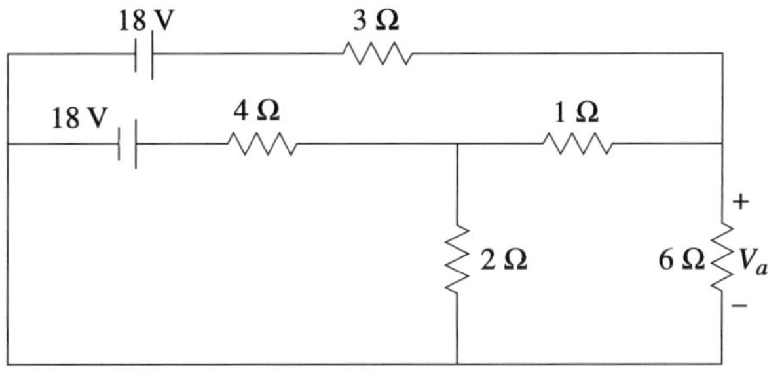

Fig. 1.50

The steps are shown in Fig. 1.51 to simplify the network.

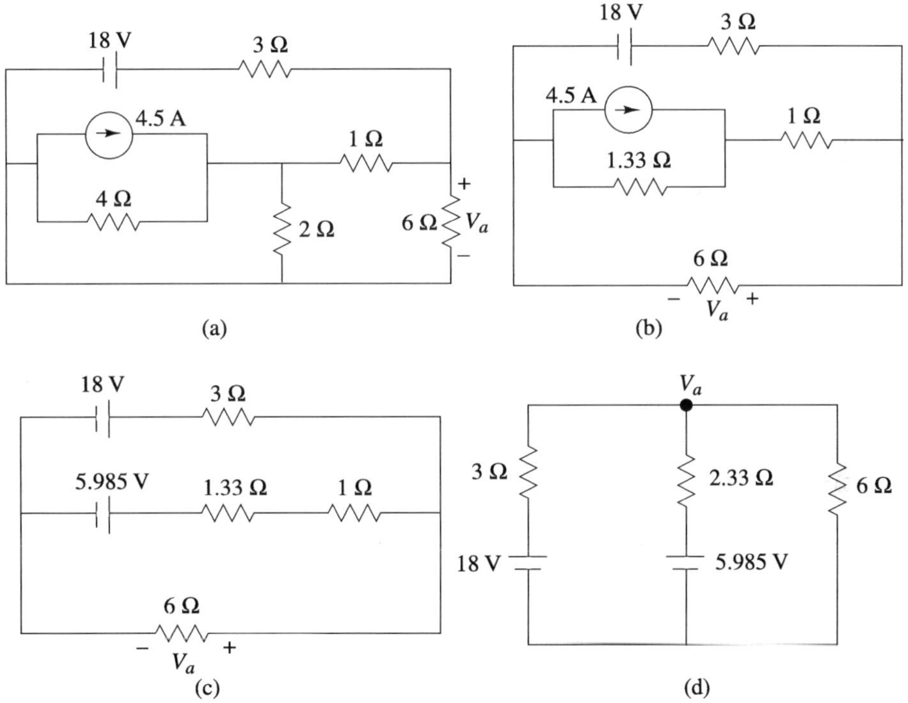

Fig. 1.51

Applying KCL at the node,

$$\frac{V_a - 18}{3} + \frac{V_a - 5.985}{2.33} + \frac{V_a}{6} = 0$$

Solving the equation, we get,

$$V_a = 9.23 \text{ V}$$

Exercises

KIRCHHOFF'S LAWS

1. Replace the network of sources shown below with (i) $V_{aa'}$, and (ii) $I_{bb'}$.

(i)

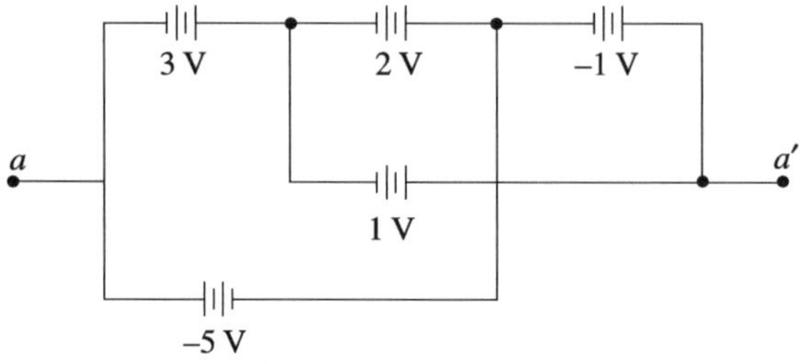

Fig. 1.52

[−4 V]

(ii)

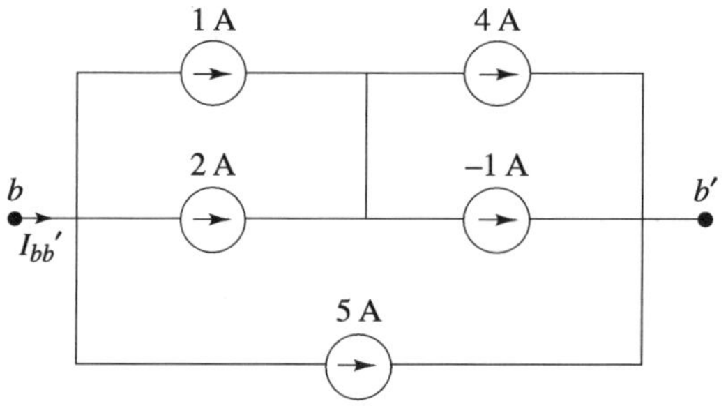

Fig. 1.53

[8 A]

2. For the network shown, find V_1/V_o and V_2/V_o.

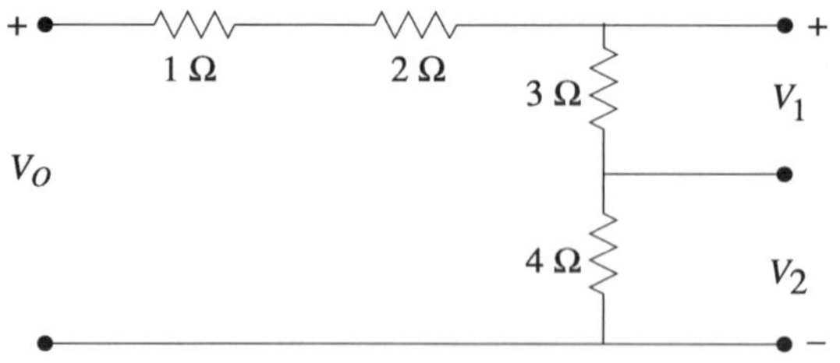

Fig. 1.54

[0.3, 0.4]

3. What is the value of R_{AB} in the circuit shown? Each side of the cube is R ohms.

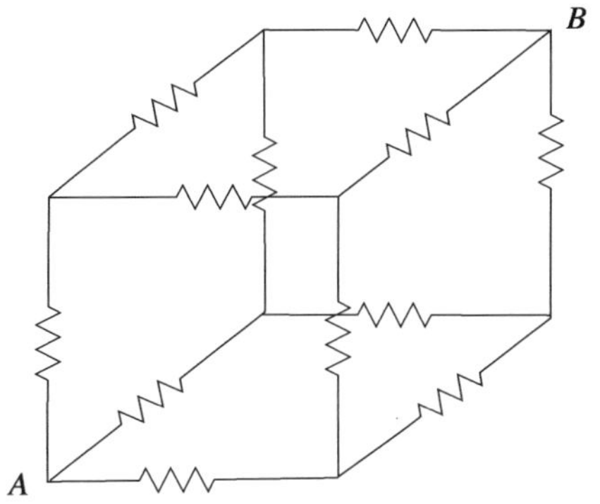

Fig. 1.55

$\left[\frac{5}{6} R\ \Omega\right]$

SOURCE TRANSFORMATION

4. Replace the given network with single voltage source and a resistor.

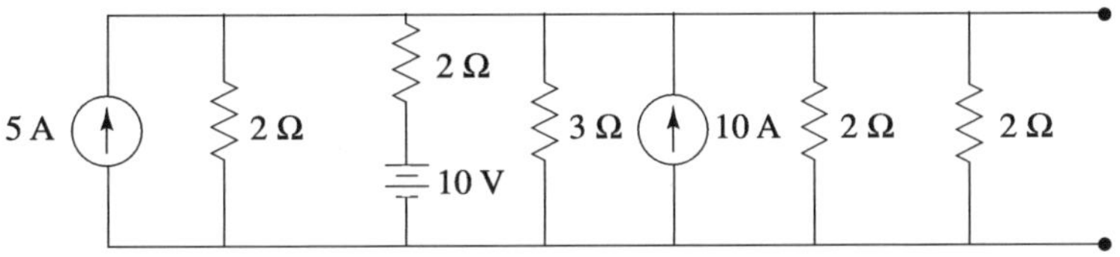

Fig. 1.56

[8.6 V, 0.43 Ω]

5. Use source transformation to simplify the network until two elements remain to the left of terminals *a* and *b*.

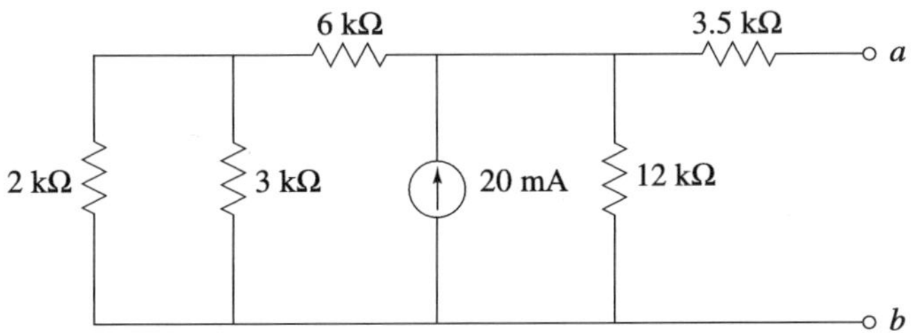

Fig. 1.57

[88.42 V, 7.92 kΩ]

6. Determine the voltage V_x by source-shifting technique.

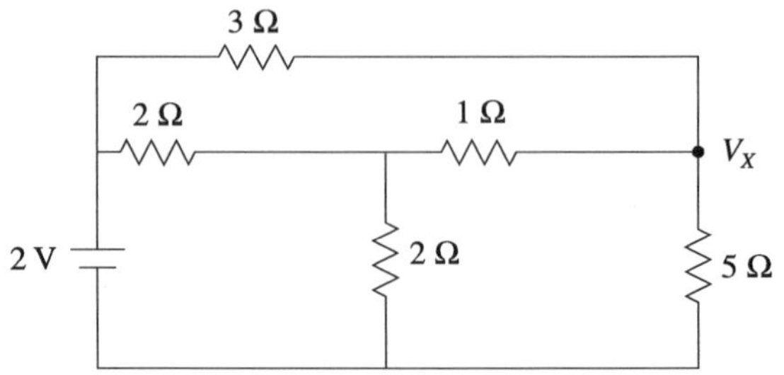

Fig. 1.58

[1.129 V]

Objective-Type Questions

1. A network contains linear resistors and ideal voltage sources. If values of all the resistors are doubled then the voltage across each resistor is
   (a) halved (b) doubled
   (c) increased by four times (d) not changed
2. The current I_4 in the circuit of Fig. 1.59 is equal to
   (a) 12 A (b) –12 A (c) 4 A (d) none of the above

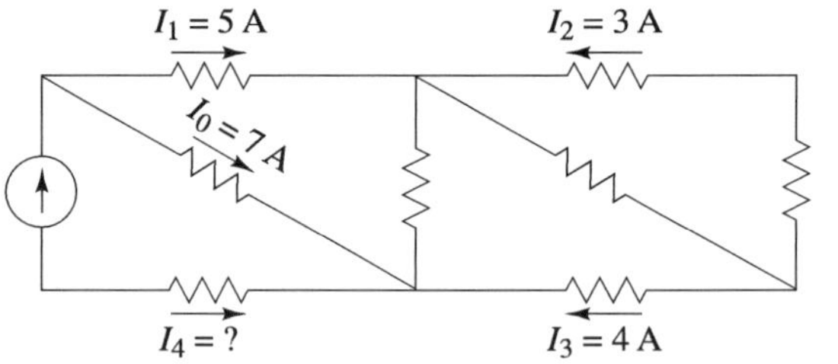

Fig. 1.59

3. The voltage V in Fig. 1.60 is equal to
   (a) 3 V (b) –3 V (c) 5 V (d) none of the above

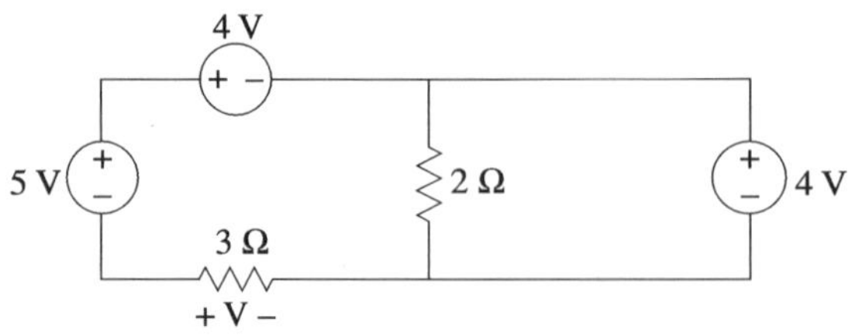

Fig. 1.60

4. The voltage V in Fig. 1.61 is equal to

(a) 9 V (b) 5 V (c) 1 V (d) none of the above

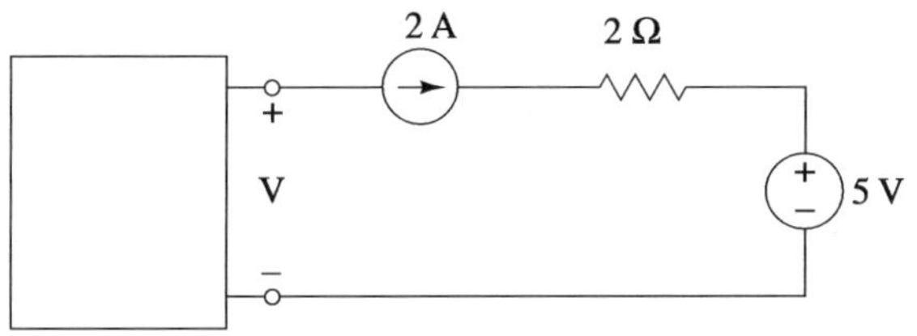

Fig. 1.61

5. The voltage V in Fig. 1.62 is

(a) 10 V (b) 15 V (c) 5 V (d) none of the above

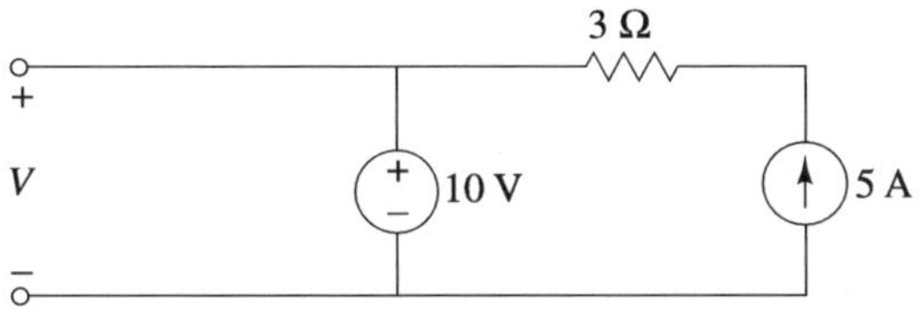

Fig. 1.62

6. In the circuit of Fig. 1.63, the value of the voltage source E is

(a) −16 V (b) 4 V (c) −6 V (d) 16 V

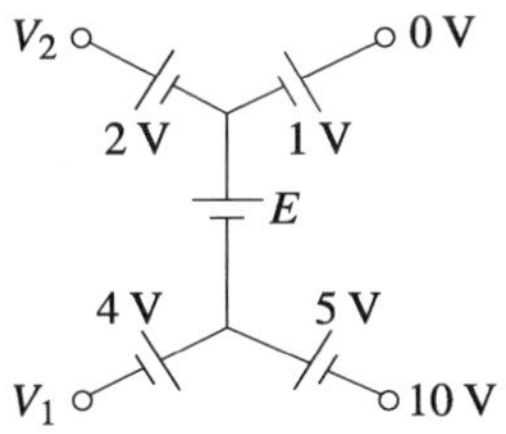

Fig. 1.63

7. The voltage V_0 in Fig. 1.64 is

(a) 2 V (b) $\frac{4}{3}$ V (c) 4 V (d) 8 V

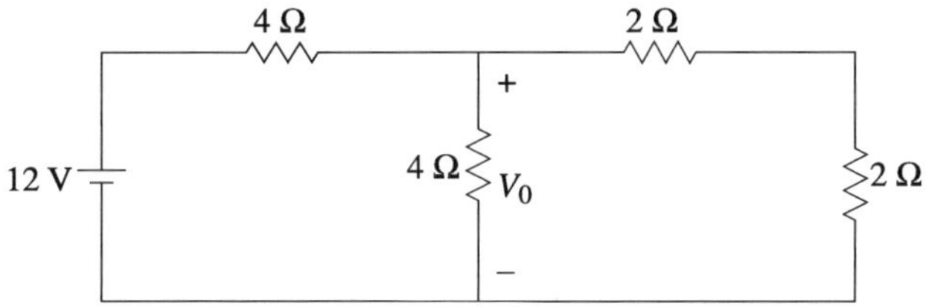

Fig. 1.64

8. If $R_1 = R_2 = R_4 = R$ and $R_3 = 1.1\ R$ in the bridge circuit shown in Fig. 1.65, then the reading in the ideal voltmeter connected between a and b is

(a) 0.238 V (b) 0.138 V (c) – 0.238 V (d) 1 V

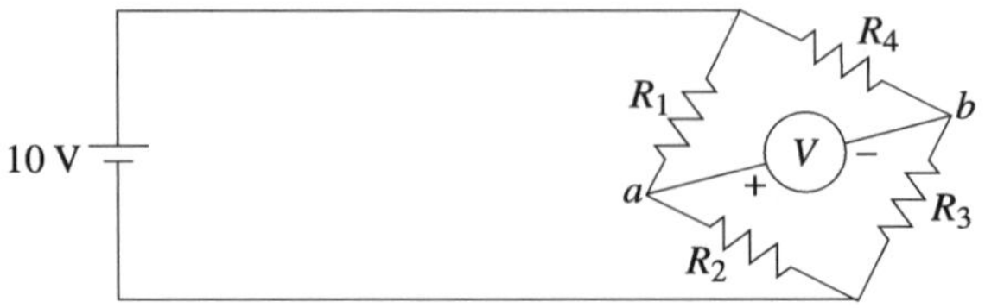

Fig. 1.65

9. A 10 V battery with an internal resistance of 1 Ω is connected across a nonlinear load whose V-I characteristic is given by $7\,I = V^2 + 2\,V$. The current delivered by the battery is

(a) 0 (b) 10 A (c) 5 A (d) 8 A

10. If the length of a wire of resistance R is uniformly stretched to n times its original value, its new resistance is

(a) nR (b) $\frac{R}{n}$ (c) n^2R (d) $\frac{R}{n^2}$

11. All the resistances in Fig. 1.66 are 1 Ω each. The value of I will be

(a) $\frac{1}{15}$ A (b) $\frac{2}{15}$ A (c) $\frac{4}{15}$ A (d) $\frac{8}{15}$ A

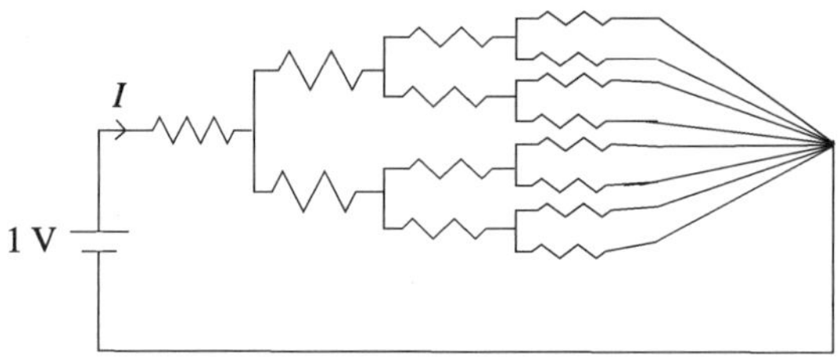

Fig. 1.66

12. The current waveform in a pure resistor at 10 Ω is shown in Fig. 1.67. Power dissipated in the resistor is

(a) 7.29 W (b) 52.4 W (c) 135 W (d) 270 W

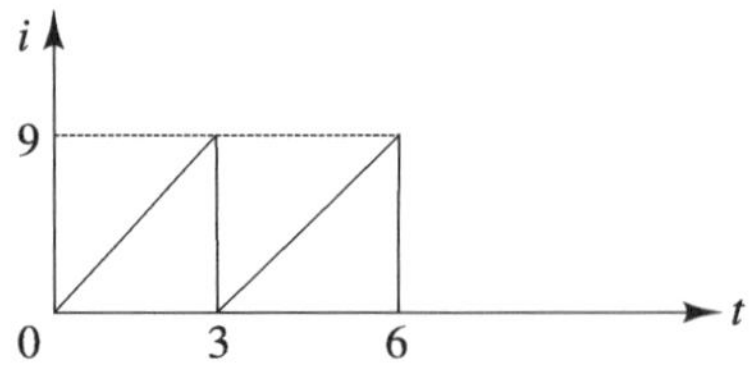

Fig. 1.67

13. Two wires *A* and *B* of the same material and length *L* and 2*L* have radius *r* and 2*r* respectively. The ratio of their specific resistance will be

(a) 1 : 1 (b) 1 : 2 (c) 1 : 4 (d) 1 : 8

Answers to Objective-Type Questions

1. (d)	**2.** (b)	**3.** (a)	**4.** (d)	**5.** (a)	**6.** (a)
7. (c)	**8.** (c)	**9.** (c)	**10.** (c)	**11.** (d)	**12.** (d)
13. (b)					

2 Network Theorems-I

2.1 INTRODUCTION

In Chapter 1, we have studied basic circuit concepts. In network analysis, we have to find currents and voltages in various parts of the circuit. In this chapter, we will study elementary network theorems like star-delta transformation, mesh analysis and node analysis. These methods are applicable to all types of circuits. The first step in analysing circuits by mesh analysis and node analysis is to apply Ohm's law and Kirchoff's laws. The second step is the solving of these equations by mathematical tools.

2.2 STAR–DELTA TRANSFORMATION

When a circuit cannot be simplified by normal series–parallel reduction technique, the star-delta transformation can be used.

Figure 2.1(a) shows three resistances R_A, R_B and R_C connected in delta.

Figure 2.1(b) shows three resistances R_1, R_2 and R_3 connected in star.

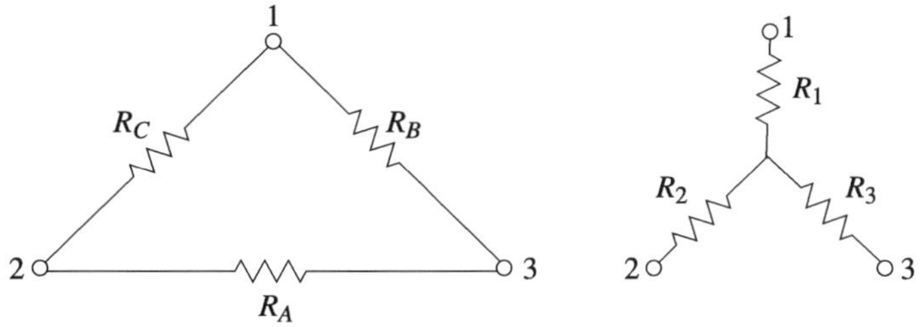

(a) (b)

Fig. 2.1

These two networks will be electrically equivalent if the resistance as measured between any pair of terminals is the same in both the arrangements.

2.2.1 Delta to Star Transformation

Referring to delta network shown in Fig. 2.1(a),

The resistance between terminals 1 and 2 = $R_C \,||\, (R_A + R_B)$

$$= \frac{R_C(R_A + R_B)}{R_A + R_B + R_C} \quad ...(2.1)$$

Referring to the star network shown in Fig. 2.1(b),

The resistance between terminals 1 and 2 = $R_1 + R_2$ ...(2.2)

Since the two networks are electrically equivalent,

$$R_1 + R_2 = \frac{R_C(R_A + R_B)}{R_A + R_B + R_C} \quad ...(2.3)$$

Similarly,

$$R_2 + R_3 = \frac{R_A(R_B + R_C)}{R_A + R_B + R_C} \quad ...(2.4)$$

and

$$R_3 + R_1 = \frac{R_B(R_A + R_C)}{R_A + R_B + R_C} \quad ...(2.5)$$

Subtracting Eq. (2.4) from Eq. (2.3),

$$R_1 - R_3 = \frac{R_B R_C - R_A R_B}{R_A + R_B + R_C} \quad ...(2.6)$$

Adding Eq. (2.6) and Eq. (2.5),

$$R_1 = \frac{R_B R_C}{R_A + R_B + R_C} \quad ...(2.7)$$

Similarly,

$$R_2 = \frac{R_A R_C}{R_A + R_B + R_C} \quad ...(2.8)$$

$$R_3 = \frac{R_A R_B}{R_A + R_B + R_C} \quad ...(2.9)$$

Thus, star resistance connected to terminal is equal to the product of the two delta resistances connected to the same terminal divided by the sum of the delta resistances.

2.2.2 Star to Delta Transformation

Multiplying the above equations,

$$R_1\, R_2 = \frac{R_A R_B R_C^2}{(R_A + R_B + R_C)^2} \quad ...(2.10)$$

$$R_2\, R_3 = \frac{R_A^2 R_B R_C}{(R_A + R_B + R_C)^2} \quad ...(2.11)$$

$$R_3\, R_1 = \frac{R_A R_B^2 R_C}{(R_A + R_B + R_C)^2} \quad ...(2.12)$$

Adding Eqs (2.10), (2.11) and (2.12),

$$R_1 R_2 + R_2 R_3 + R_3 R_1 = \frac{R_A R_B R_C (R_A + R_B + R_C)}{(R_A + R_B + R_C)^2}$$

$$= \frac{R_A R_B R_C}{R_A + R_B + R_C} = R_A R_1$$

$$= R_B R_2$$

$$= R_C R_3$$

Hence

$$R_A = \frac{R_1 R_2 + R_2 R_3 + R_3 R_1}{R_1}$$

$$= R_2 + R_3 + \frac{R_2 R_3}{R_1}$$

$$R_B = \frac{R_1 R_2 + R_2 R_3 + R_3 R_1}{R_2}$$

$$= R_1 + R_3 + \frac{R_3 R_1}{R_2}$$

$$R_C = \frac{R_1 R_2 + R_2 R_3 + R_3 R_1}{R_3}$$

$$= R_1 + R_2 + \frac{R_1 R_2}{R_3}$$

Thus, delta resistance between the two terminals is the sum of two star resistances connected to the same terminals plus the product of the two resistances divided by the remaining third star resistance.

Note: When three equal resistances are connected in delta, the equivalent star resistance is given by

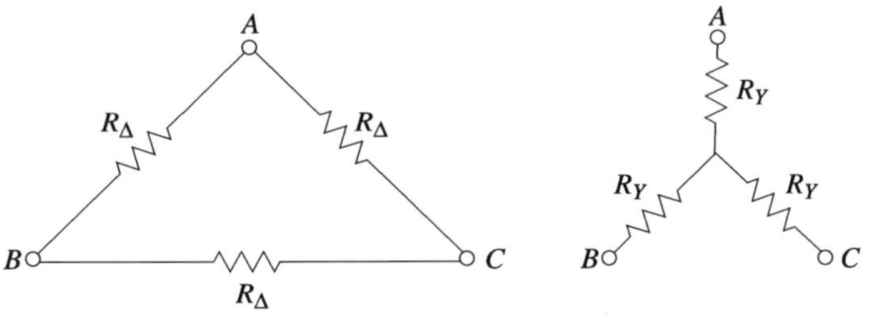

Fig. 2.2

$$R_Y = \frac{R_\Delta R_\Delta}{R_\Delta + R_\Delta + R_\Delta} = \frac{R_\Delta}{3}$$

or

$$R_\Delta = 3R_Y$$

Example 2.1 *Find an equivalent resistance between A and B.*

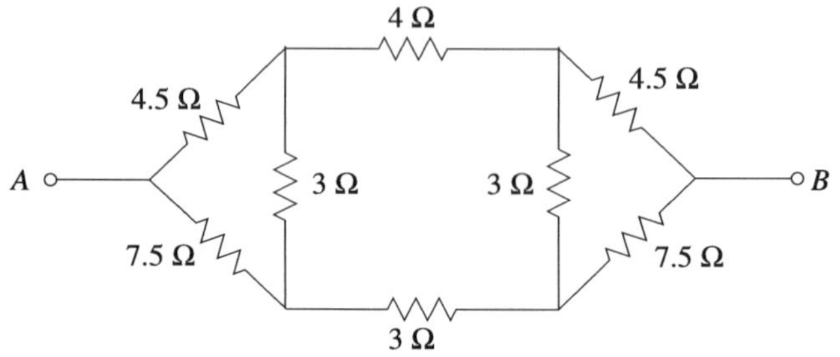

Fig. 2.3

Solution Converting the two delta networks formed by resistors 4.5 Ω, 3 Ω and 7.5 Ω into equivalent star networks, we have

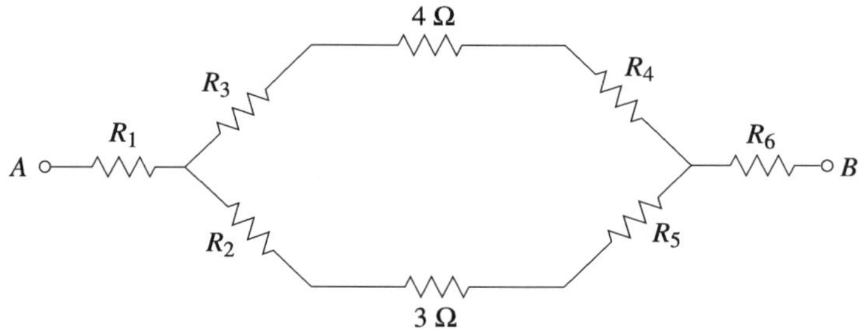

Fig. 2.4

$$R_1 = R_6 = \frac{4.5 \times 7.5}{4.5 + 7.5 + 3} = 2.25\ \Omega$$

$$R_2 = R_5 = \frac{7.5 \times 3}{4.5 + 7.5 + 3} = 1.5\ \Omega$$

$$R_3 = R_4 = \frac{4.5 \times 3}{4.5 + 7.5 + 3} = 0.9\ \Omega$$

The simplified network is shown in Fig. 2.5.

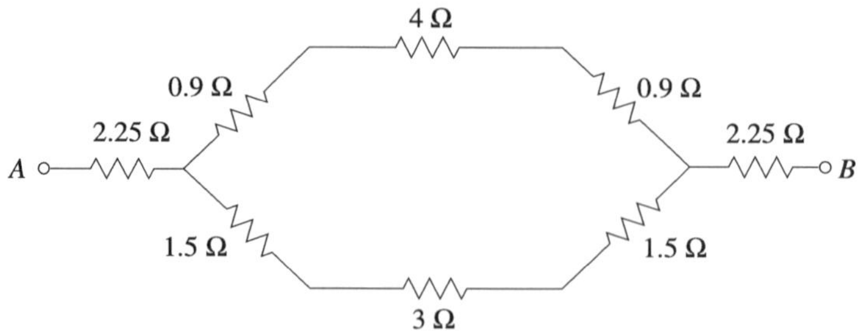

Fig. 2.5

The network can be simplified as follows:

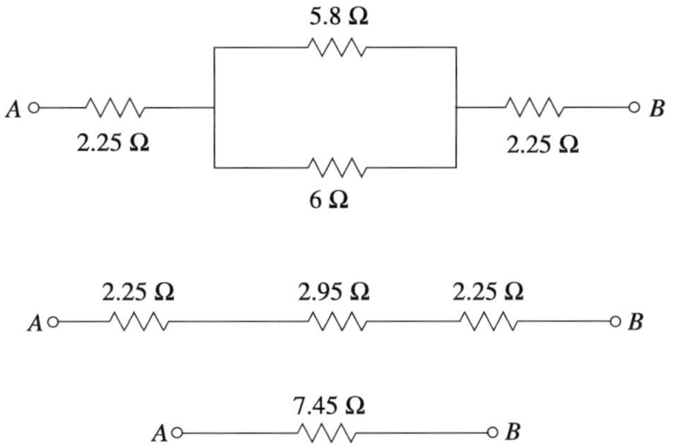

Fig. 2.6

$R_{AB} = 7.45\ \Omega$

Example 2.2 *Find an equivalent resistance between A and B.*

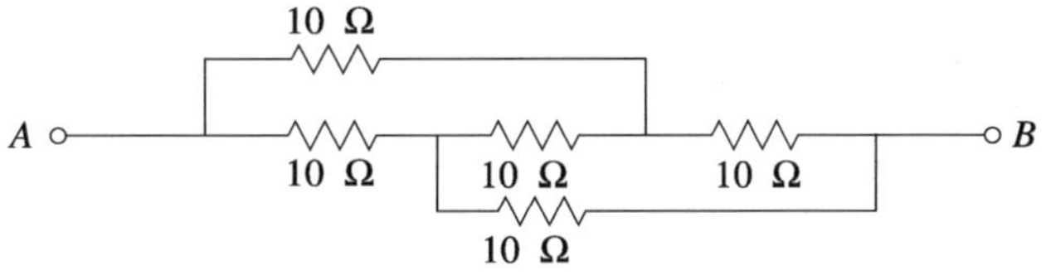

Fig. 2.7

Solution Redrawing the network, we have

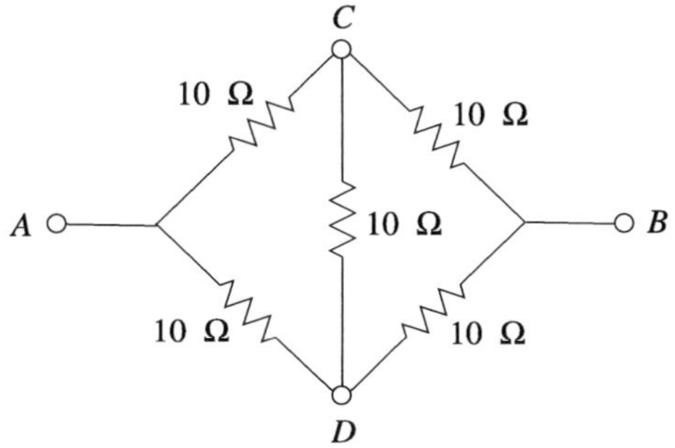

Fig. 2.8

Converting the delta formed by three resistors of 10 Ω into an equivalent star network,

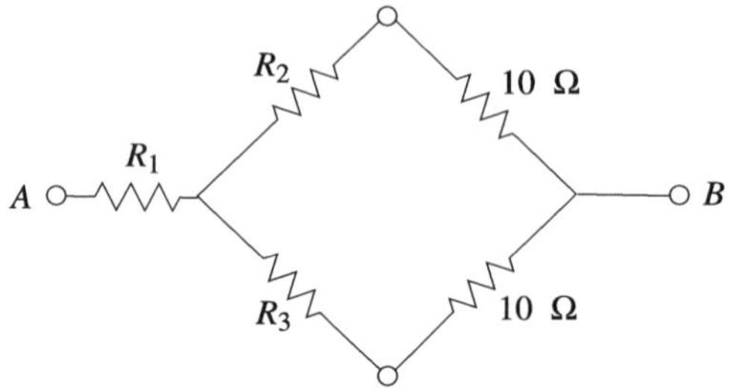

Fig. 2.9

$$R_1 = R_2 = R_3 = \frac{10 \times 10}{10 + 10 + 10} = \frac{10}{3}\ \Omega$$

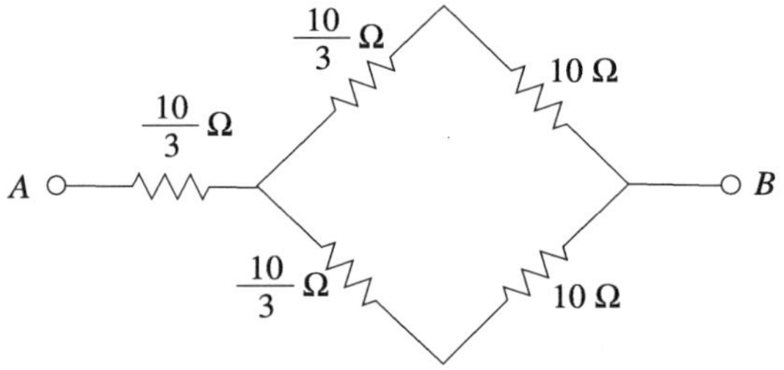

Fig. 2.10

The network can be simplified as follows:

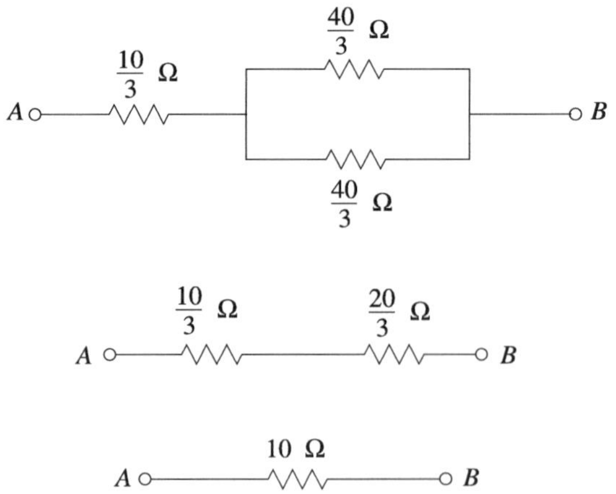

Fig. 2.11

$$R_{AB} = 10\ \Omega$$

Example 2.3 *Find an equivalent resistance between A and B.*

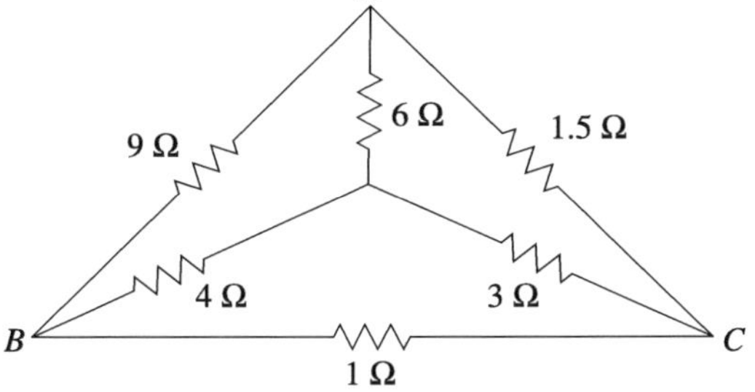

Fig. 2.12

Solution Converting the star network formed by resistors 3 Ω, 4 Ω and 6 Ω into an equivalent delta network,

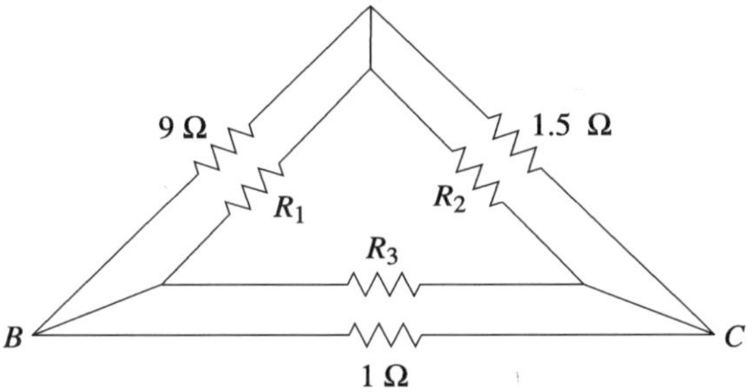

Fig. 2.13

$$R_1 = 6 + 4 + \frac{6\times 4}{3} = 18\ \Omega$$

$$R_2 = 6 + 3 + \frac{6\times 3}{4} = 13.5\ \Omega$$

$$R_3 = 4 + 3 + \frac{4\times 3}{6} = 9\ \Omega$$

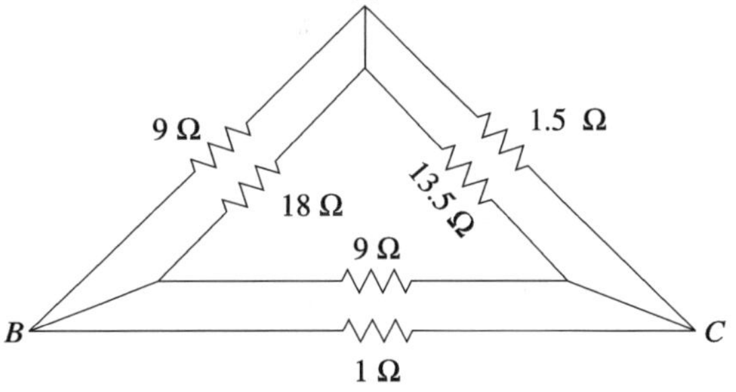

Fig. 2.14

The network can be simplified as shown in Fig. 2.15.

$$R_{AB} = 6 \,||\, (1.35 + 0.9)$$

$$= 6 \,||\, 2.25$$

$$= 1.64\ \Omega$$

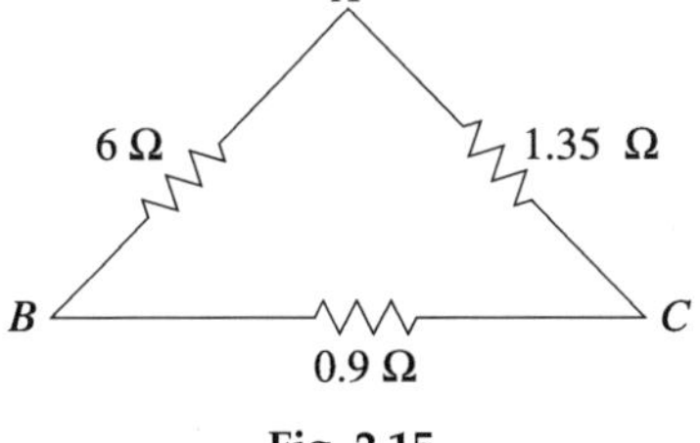

Fig. 2.15

Example 2.4 *Find an equivalent resistance between A and B.*

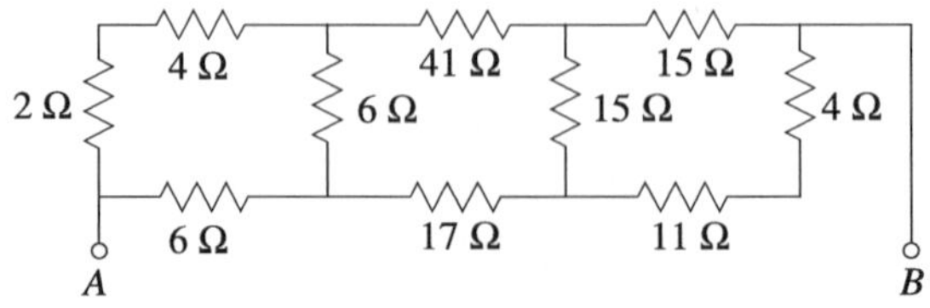

Fig. 2.16

Solution The resistances of 2 Ω and 4 Ω and the resistances of 4 Ω and 11 Ω are in series.

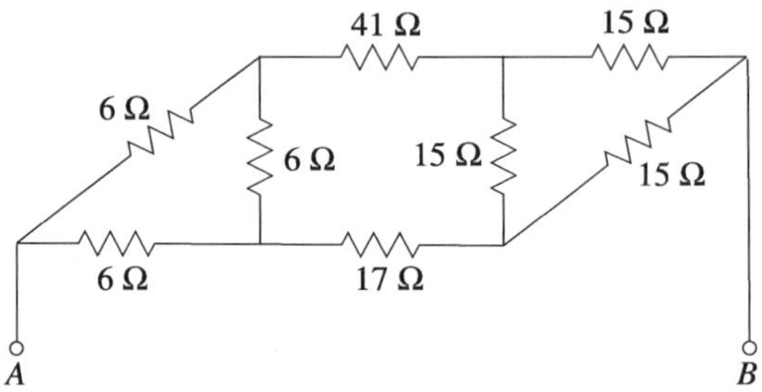

Fig. 2.17

Converting the two outer delta networks into equivalent star networks,

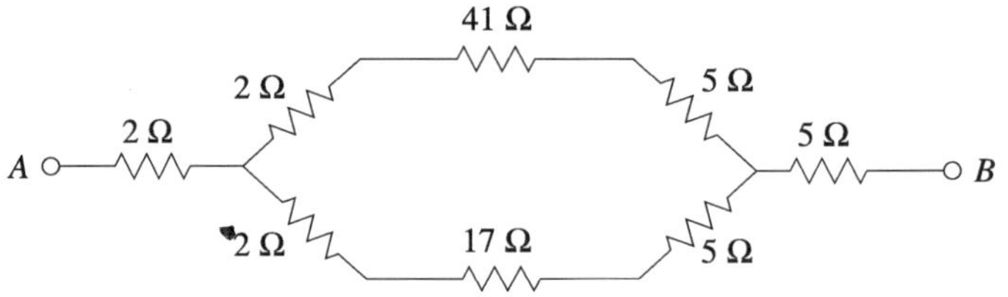

Fig. 2.18

The network can be further simplified as follows:

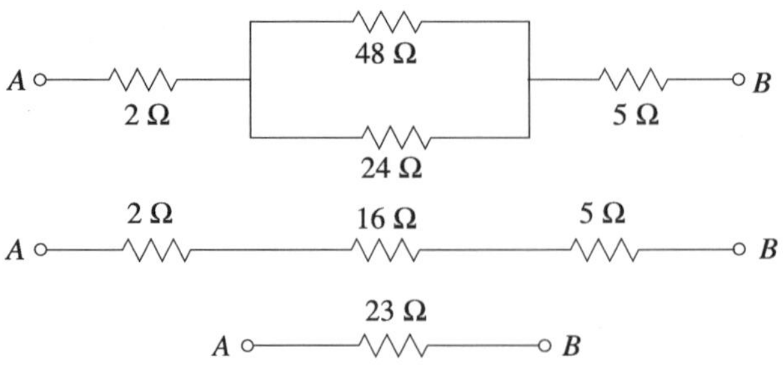

Fig. 2.19

$R_{AB} = 23\ \Omega$

Example 2.5 *Find an equivalent resistance between A and B.*

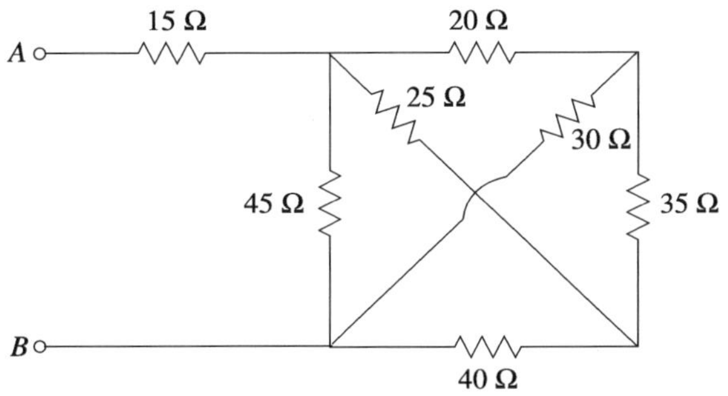

Fig. 2.20

Solution Drawing the resistance of 30 Ω from outside,

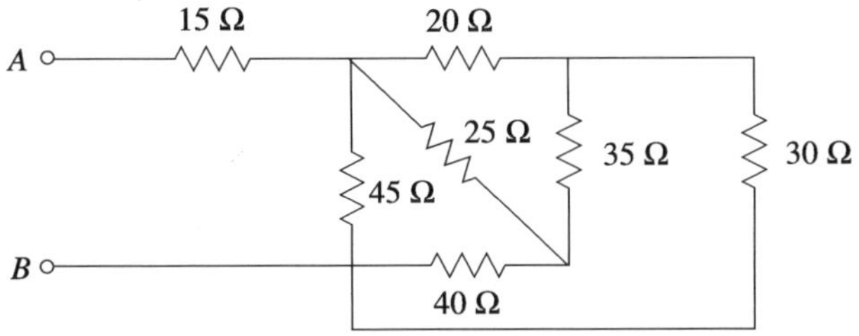

Fig. 2.21

Converting the delta network formed by resistors 20 Ω, 25 Ω and 35 Ω into equivalent star network,

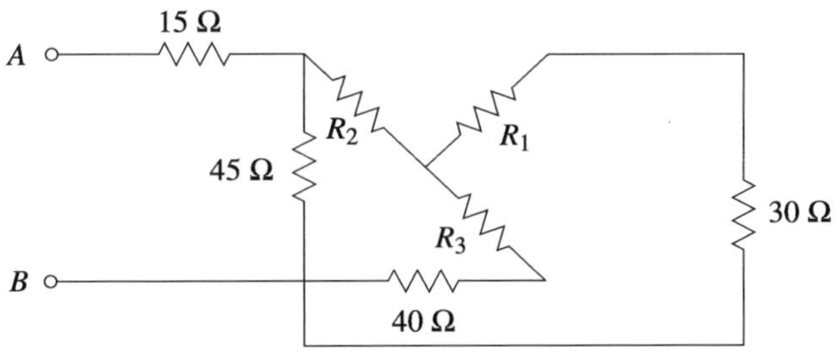

Fig. 2.22

$$R_1 = \frac{20 \times 35}{20 + 35 + 25} = 8.75\ \Omega$$

$$R_2 = \frac{20 \times 25}{20 + 35 + 25} = 6.25\ \Omega$$

$$R_3 = \frac{35 \times 25}{20 + 35 + 25} = 10.94\ \Omega$$

The network can be redrawn as follows:

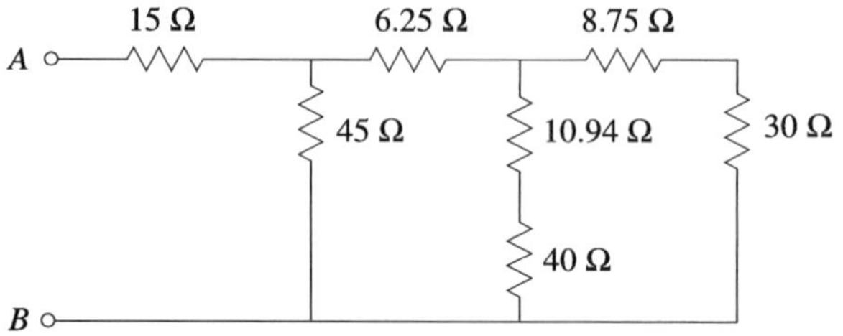

Fig. 2.23

By series–parallel reduction technique,

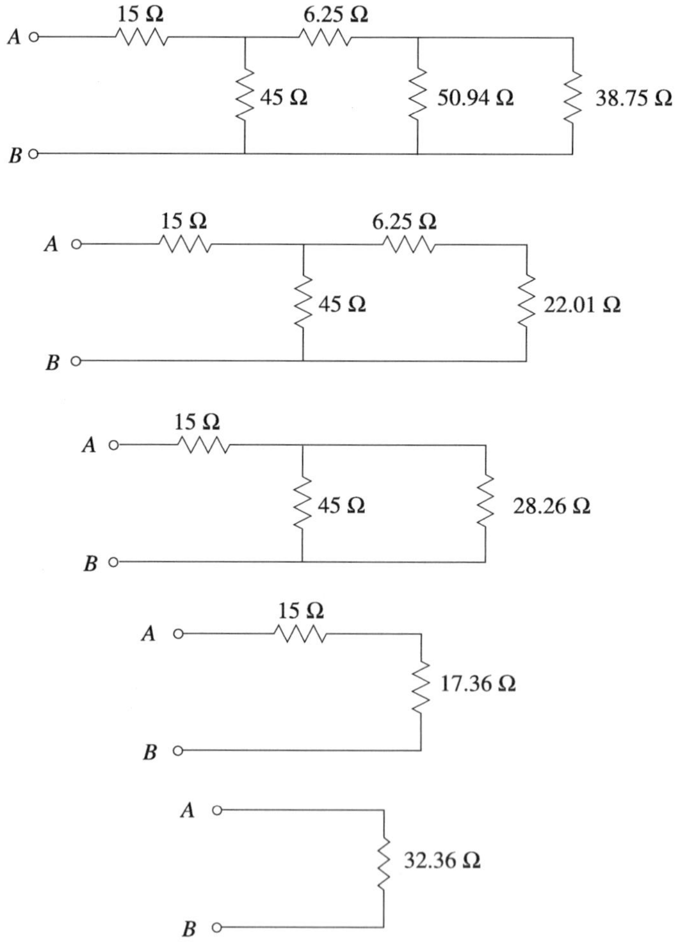

Fig. 2.24

$R_{AB} = 32.36\ \Omega$

Example 2.6 *Find the equivalent resistance between A and B.*

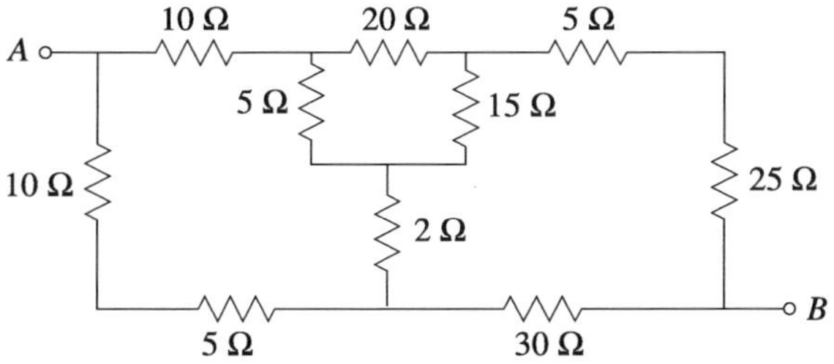

Fig. 2.25

Solution The resistances of 5 Ω and 25 Ω and the resistances of 10 Ω and 5 Ω are in series.

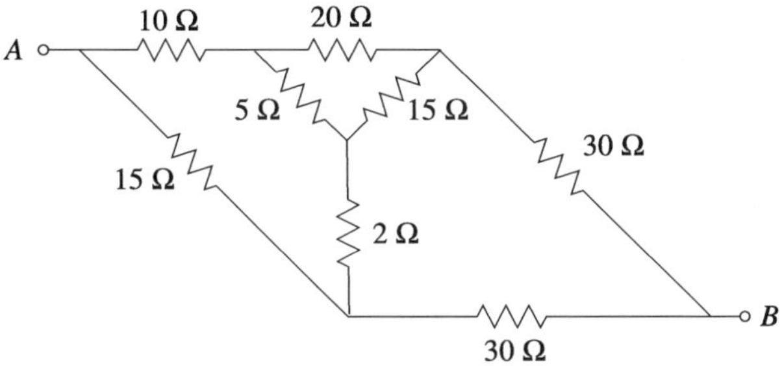

Fig. 2.26

Converting the delta network formed by resistance of 20 Ω, 5 Ω and 15 Ω into equivalent star network,

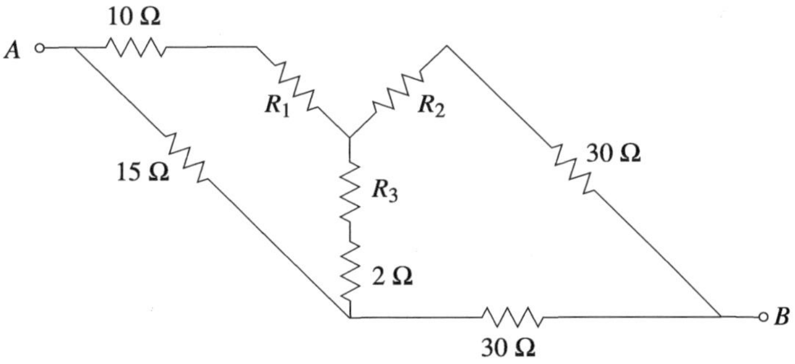

Fig. 2.27

$$R_1 = \frac{20 \times 5}{20+5+15} = 2.5\ \Omega$$

$$R_2 = \frac{20 \times 15}{20+5+15} = 7.5\ \Omega$$

$$R_3 = \frac{5 \times 15}{20+5+15} = 1.875\ \Omega$$

The network can be redrawn as follows:

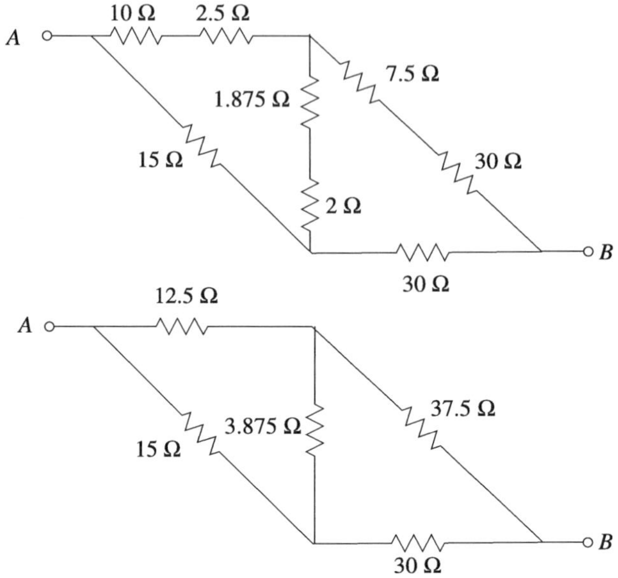

Fig. 2.28

Converting the delta network formed by resistance of 3.875 Ω, 37.5 Ω and 30 Ω into equivalent star network, we have

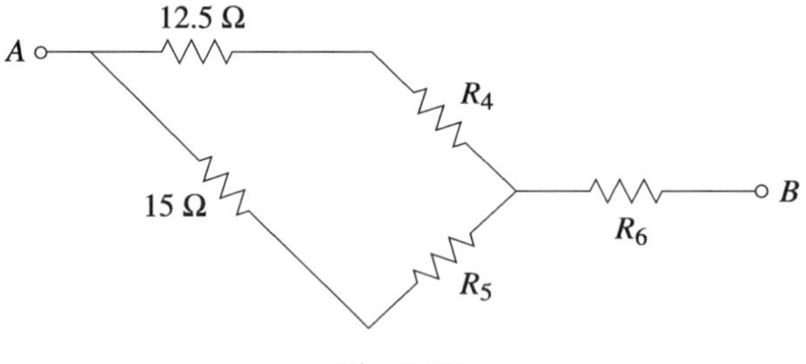

Fig. 2.29

$$R_4 = \frac{3.875 \times 37.5}{3.875 + 37.5 + 30} = 2.04\ \Omega$$

$$R_5 = \frac{3.875 \times 30}{3.875 + 37.5 + 30} = 1.63\ \Omega$$

$$R_6 = \frac{37.5 \times 30}{3.875 + 37.5 + 30} = 15.76\ \Omega$$

The network can be simplified as follows:

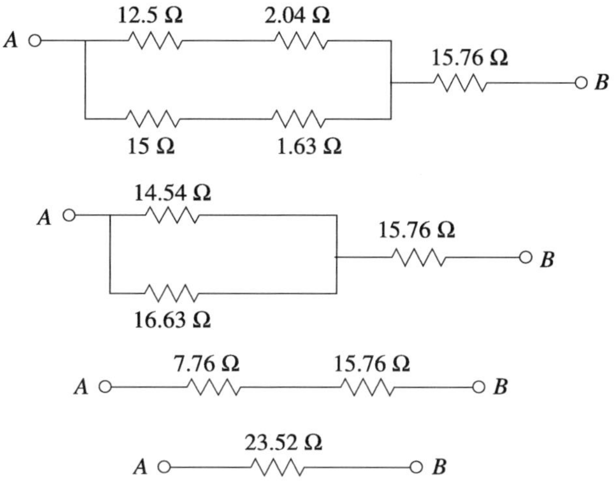

Fig. 2.30

$R_{AB} = 23.52\ \Omega$

Example 2.7 *Find an equivalent resistance between A and B.*

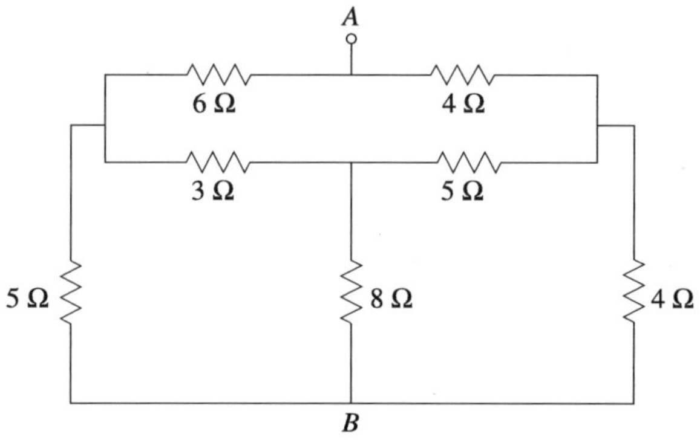

Fig. 2.31

Solution Converting the star network formed by resistances of 3 Ω, 5 Ω and 8 Ω into an equivalent delta network,

$$R_1 = 3 + 5 + \frac{3 \times 5}{8}$$

$$= 9.875\ \Omega$$

$$R_2 = 3 + 8 + \frac{3 \times 8}{5}$$

$$= 15.8\ \Omega$$

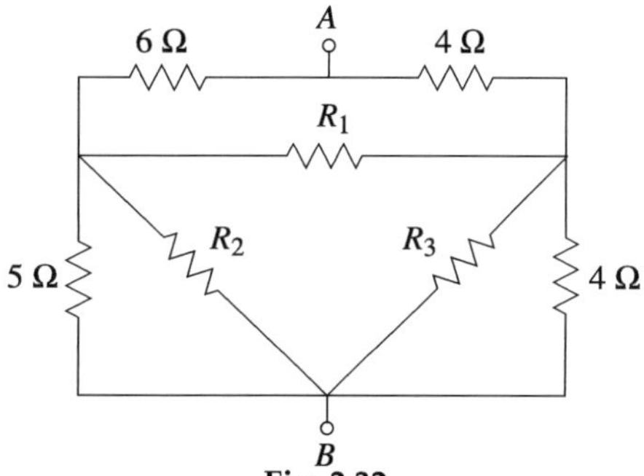

Fig. 2.32

$$R_3 = 5 + 8 + \frac{5 \times 8}{3}$$

$$= 26.33\ \Omega$$

The network can be redrawn as follows:

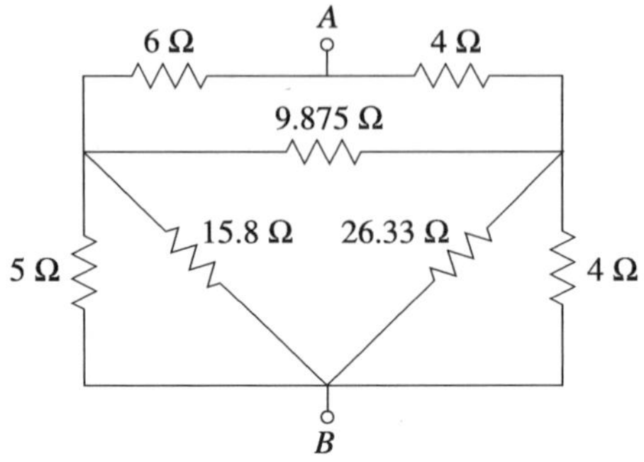

Fig. 2.33

The resistances of 15.8 Ω and 5 Ω and resistances of 26.33 Ω and 4 Ω are in parallel.

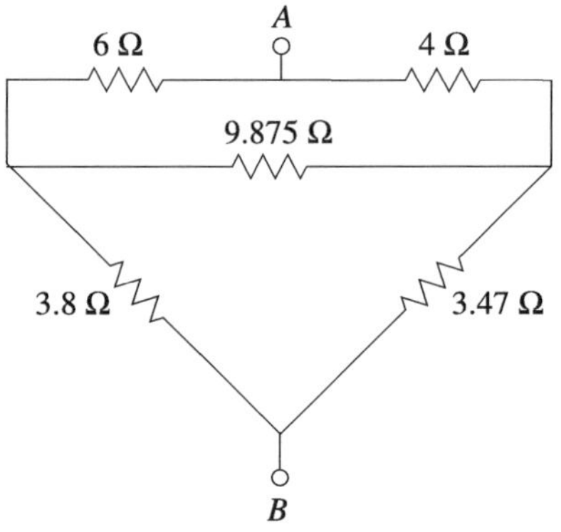

Fig. 2.34

Converting the delta network into star network,

$$R_4 = \frac{3.8 \times 9.875}{3.8 + 9.875 + 3.47}$$

$$= 2.19\ \Omega$$

$$R_5 = \frac{3.8 \times 3.47}{3.8 + 9.875 + 3.47}$$

$$= 0.77\ \Omega$$

$$R_6 = \frac{3.47 \times 9.875}{3.8 + 9.875 + 3.47}$$

$$= 2\ \Omega$$

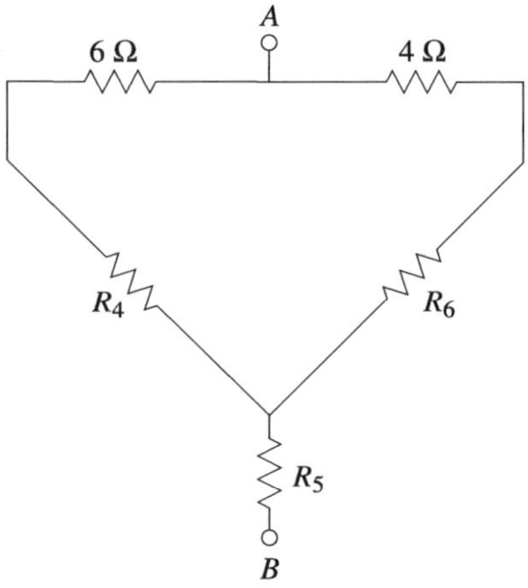

Fig. 2.35

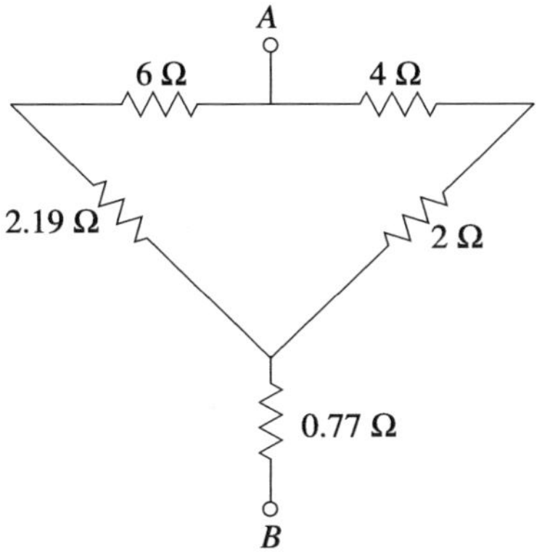

Fig. 2.36

The network can be simplified as follows:

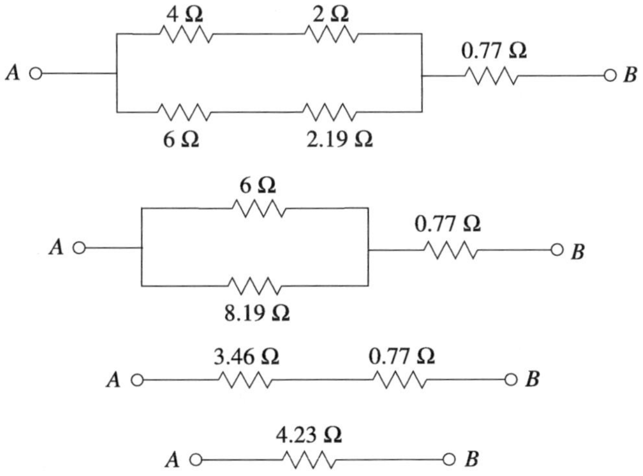

Fig. 2.37

$$R_{AB} = 4.23\ \Omega$$

Example 2.8 *Find an equivalent resistance between A and B.*

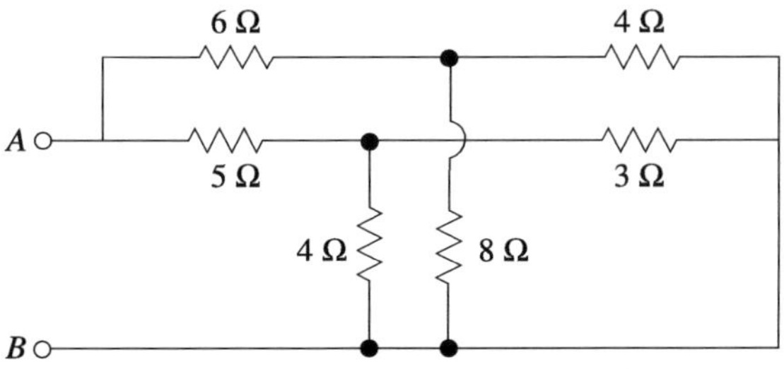

Fig. 2.38

Solution Converting the star network formed by resistors of 3 Ω, 4 Ω and 5 Ω into equivalent delta network,

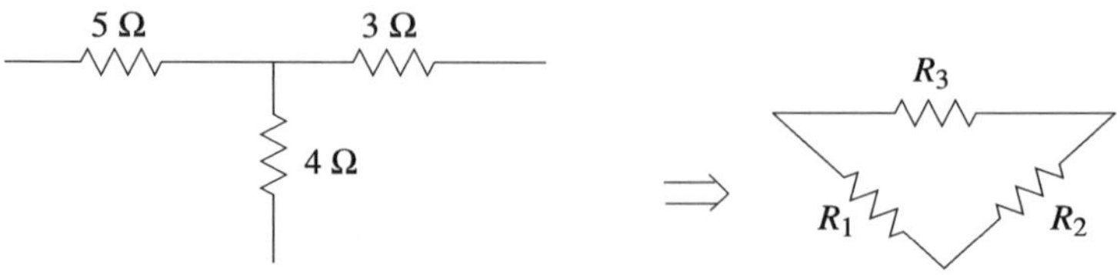

Fig. 2.39

$$R_1 = 5 + 4 + \frac{5 \times 4}{3} = 15.67\ \Omega$$

$$R_2 = 3 + 4 + \frac{3 \times 4}{5} = 9.4\ \Omega$$

$$R_3 = 5 + 3 + \frac{5 \times 3}{4} = 11.75\ \Omega$$

Similarly, converting the star network formed by resistors of 4 Ω, 6 Ω and 8 Ω into equivalent delta network,

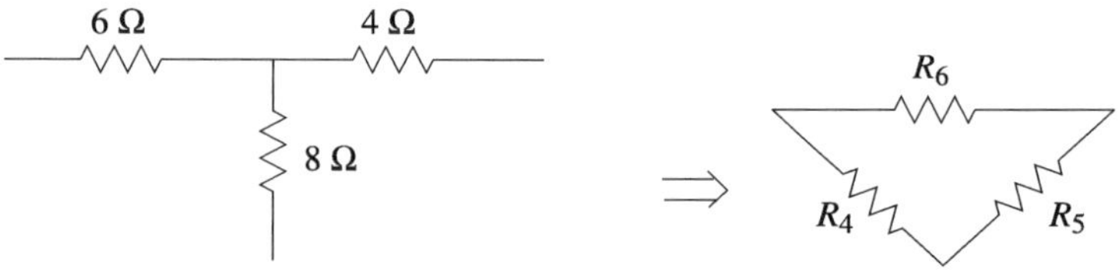

Fig. 2.40

$$R_4 = 6 + 8 + \frac{6 \times 8}{4} = 26\ \Omega$$

$$R_5 = 4 + 8 + \frac{4 \times 8}{6} = 17.33\ \Omega$$

$$R_6 = 6 + 4 + \frac{6 \times 4}{8} = 13\ \Omega$$

These two delta networks are connected in parallel between points A and B.

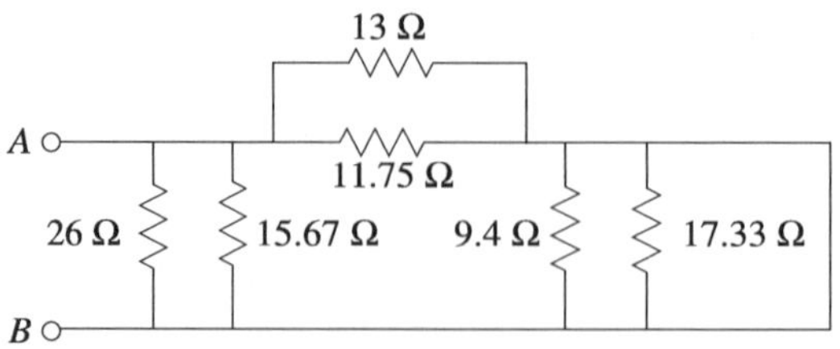

Fig. 2.41

The resistances of 9.4 Ω and 17.33 Ω are in parallel with a short. Hence, equivalent resistance of this combination becomes zero. Simplifying the parallel networks, we get

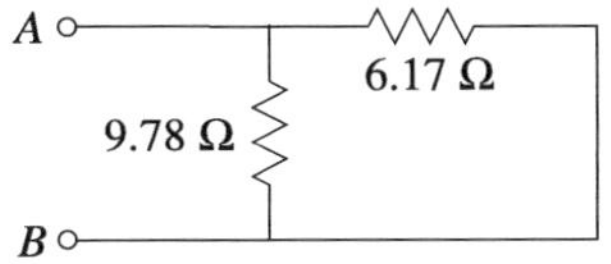

Fig. 2.42

$$R_{AB} = 6.17 \mid\mid 9.78 = 3.78\ \Omega$$

2.3 MESH ANALYSIS

A mesh is defined as a loop which does not contain any other loops within it. Mesh analysis is applicable only for planar networks. A network is said to be planar if it can be drawn on a plane surface without crossovers. In this method, the currents in different meshes are assigned continuous paths so that they do not split at a junction into branch currents. If a network has a large number of voltage sources, it is useful to use mesh analysis. Basically, this analysis consists of writing mesh equations by Kirchoff's voltage law in terms of unknown mesh currents.

2.3.1 Steps to be followed in Mesh Analysis

1. Identify the mesh, assign a direction to it and assign an unknown current in each mesh.
2. Assign the polarities for voltage across the branches.
3. Apply KVL around the mesh and use Ohm's law to express the branch voltages in terms of unknown mesh currents and the resistance.
4. Solve the simultaneous equations for unknown mesh currents.

Consider the network shown in Fig. 2.43 which has three meshes. Let the mesh currents for the three meshes be I_1, I_2 and I_3 and all the three mesh currents may be assumed to flow in the clockwise direction. The choice of direction for any mesh current is arbitrary.

Applying KVL to Mesh 1,

$$V_1 - R_1 (I_1 - I_2) - R_2 (I_1 - I_3) = 0$$

$$(R_1 + R_2) I_1 - R_1 I_2 - R_2 I_3 = V_1 \qquad ...(2.13)$$

Applying KVL to Mesh 2,

$$V_2 - R_3 I_2 - R_4 (I_2 - I_3) - R_1 (I_2 - I_1) = 0$$

$$-R_1 I_1 + (R_1 + R_3 + R_4) I_2 - R_4 I_3 = V_2 \qquad ...(2.14)$$

Applying KVL to Mesh 3,

$$-R_2 (I_3 - I_1) - R_4 (I_3 - I_2) - R_5 I_3 + V_3 = 0$$

$$-R_2 I_1 - R_4 I_2 + (R_2 + R_4 + R_5) I_3 = V_3 \qquad ...(2.15)$$

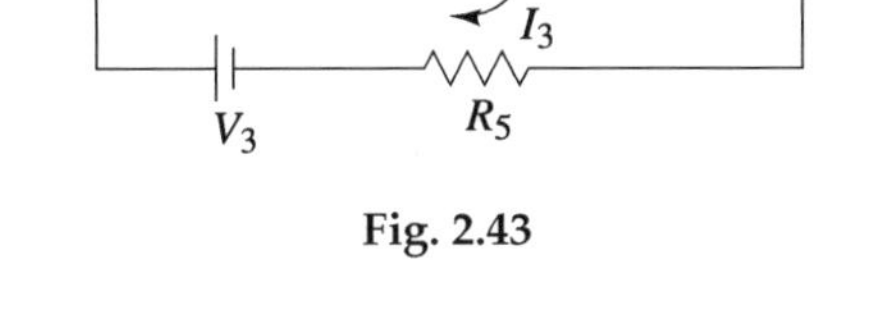

Fig. 2.43

Writing Eqs (2.13), (2.14) and (2.15) in matrix form,

$$\begin{bmatrix} R_1 + R_2 & -R_1 & -R_2 \\ -R_1 & R_1 + R_3 + R_4 & -R_4 \\ -R_2 & -R_4 & R_2 + R_4 + R_5 \end{bmatrix} \begin{bmatrix} I_1 \\ I_2 \\ I_3 \end{bmatrix} = \begin{bmatrix} V_1 \\ V_2 \\ V_3 \end{bmatrix}$$

In general,

$$\begin{bmatrix} R_{11} & R_{12} & R_{13} \\ R_{21} & R_{22} & R_{23} \\ R_{31} & R_{32} & R_{33} \end{bmatrix} \begin{bmatrix} I_1 \\ I_2 \\ I_3 \end{bmatrix} = \begin{bmatrix} V_1 \\ V_2 \\ V_3 \end{bmatrix}$$

where,

R_{11} = Self-resistance or sum of all the resistances of Mesh 1

$R_{12} = R_{21}$ = Mutual resistance or sum of all the resistances common to Meshes 1 and 2

$R_{13} = R_{31}$ = Mutual resistance or sum of all the resistances common to Meshes 1 and 3

R_{22} = Self-resistance or sum of all the resistances of Mesh 2

$R_{23} = R_{32}$ = Mutual resistance or sum of all the resistances common to Meshes 2 and 3

R_{33} = Self-resistance or sum of all the resistances of Mesh 3

If the directions of the currents passing through the common resistance are the same, the mutual resistance will have a positive sign, and if the direction of the currents passing through common resistance are opposite then the mutual resistance will have a negative sign. If each mesh currents are assumed to flow in the clockwise direction then all self-resistances will be always positive and all mutual resistances will always be negative.

The voltages V_1, V_2 and V_3 represent the algebraic sum of all the voltages in meshes 1, 2 and 3 respectively. While going along the current, if we go from negative terminal of the battery to the positive terminal then its emf is taken as positive. Otherwise, it is taken as negative.

Example 2.9 *Find the current through the 5-Ω resistor.*

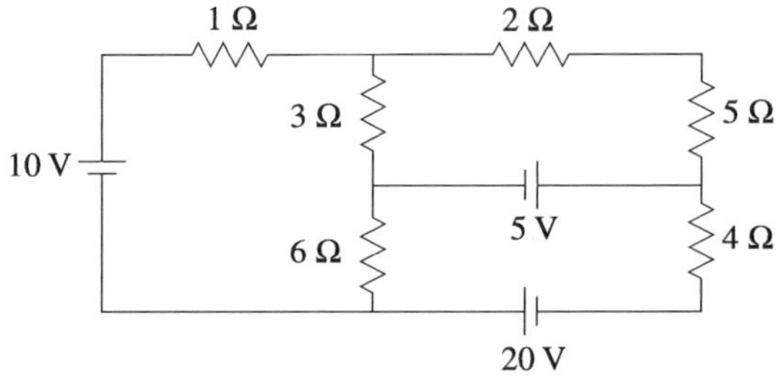

Fig. 2.44

Solution Assigning clockwise currents in three meshes,

Applying KVL to Mesh 1,

$$10 - I_1 - 3(I_1 - I_2) - 6(I_1 - I_3) = 0$$

$$10I_1 - 3I_2 - 6I_3 = 10 \quad \text{...(i)}$$

Applying KVL to Mesh 2,

$$-3\,(I_2 - I_1) - 2I_2 - 5I_2 - 5 = 0$$

$$-3I_1 + 10I_2 = -5 \quad \text{...(ii)}$$

Applying KVL to Mesh 3,

$$-6\,(I_3 - I_1) + 5 - 4I_3 + 20 = 0$$

$$-6I_1 + 10I_3 = 25 \quad \text{...(iii)}$$

Writing equations in matrix form,

$$\begin{bmatrix} 10 & -3 & -6 \\ -3 & 10 & 0 \\ -6 & 0 & 10 \end{bmatrix} \begin{bmatrix} I_1 \\ I_2 \\ I_3 \end{bmatrix} = \begin{bmatrix} 10 \\ -5 \\ 25 \end{bmatrix}$$

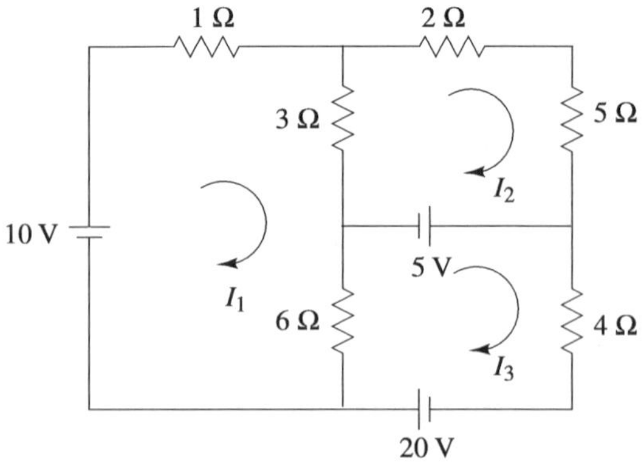

Fig. 2.45

We can write matrix equation directly from Fig. 2.44.

$$\begin{bmatrix} R_{11} & R_{12} & R_{13} \\ R_{21} & R_{22} & R_{23} \\ R_{31} & R_{32} & R_{33} \end{bmatrix} \begin{bmatrix} I_1 \\ I_2 \\ I_3 \end{bmatrix} = \begin{bmatrix} V_1 \\ V_2 \\ V_3 \end{bmatrix}$$

where R_{11} = Self-resistance of Mesh 1 = 1 + 3 + 6 = 10 Ω

R_{12} = Mutual resistance common to Meshes 1 and 2 = –3 Ω

Here, negative sign indicates that the currents through common resistance are in opposite direction.

R_{13} = Mutual resistance common to Meshes 1 and 3 = –6 Ω

Similarly,

$R_{21} = -3\ \Omega$

$R_{22} = 3 + 2 + 5 = 10\ \Omega$

$R_{23} = 0$

$R_{31} = -6\ \Omega$

$R_{32} = 0$

$R_{33} = 6 + 4 = 10\ \Omega$

For voltage matrix,

$V_1 = 10$ V

$V_2 = -5$ V

V_3 = algebraic sum of all the voltages in Mesh 3 = 5 + 20 = 25 V

Solving equations (i), (ii) and (iii),

$I_1 = 4.27$ A

$I_2 = 0.78$ A

$I_3 = 5.06$ A

$I_{5\Omega} = 0.78$ A

Example 2.10 *Find the current through the 2-Ω resistor.*

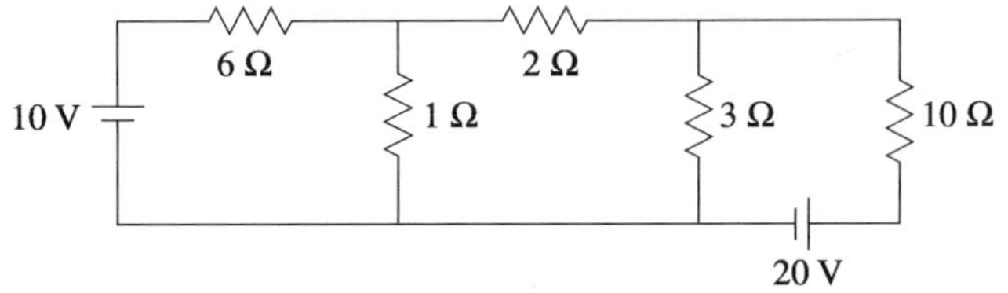

Fig. 2.46

Solution Assigning clockwise currents in three meshes,

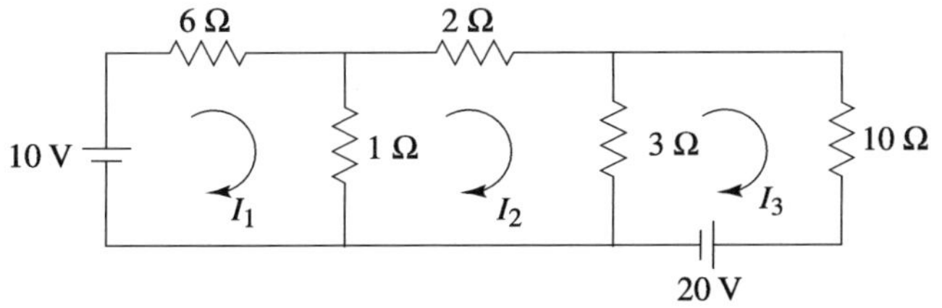

Fig. 2.47

Applying KVL to Mesh 1,

$$10 - 6I_1 - 1(I_1 - I_2) = 0$$

$$7I_1 - I_2 = 10 \qquad \text{...(i)}$$

Applying KVL to Mesh 2,

$$-(I_2 - I_1) - 2I_2 - 3(I_2 - I_3) = 0$$

$$-I_1 + 6I_2 - 3I_3 = 0 \qquad \text{...(ii)}$$

Applying KVL to Mesh 3,

$$-3(I_3 - I_2) - 10I_3 - 20 = 0$$

$$-3I_2 + 13I_3 = -20 \qquad \text{...(iii)}$$

Writing equations in matrix form,

$$\begin{bmatrix} 7 & -1 & 0 \\ -1 & 6 & -3 \\ 0 & -3 & 13 \end{bmatrix} \begin{bmatrix} I_1 \\ I_2 \\ I_3 \end{bmatrix} = \begin{bmatrix} 10 \\ 0 \\ -20 \end{bmatrix}$$

Solving equations (i), (ii) and (iii),

$$I_1 = 1.34 \text{ A}$$

$$I_2 = -0.62 \text{ A}$$

$$I_3 = -1.68 \text{ A}$$

$$I_{2\,\Omega} = -0.62 \text{ A}$$

Example 2.11 *Determine the current through the 5-Ω resistor.*

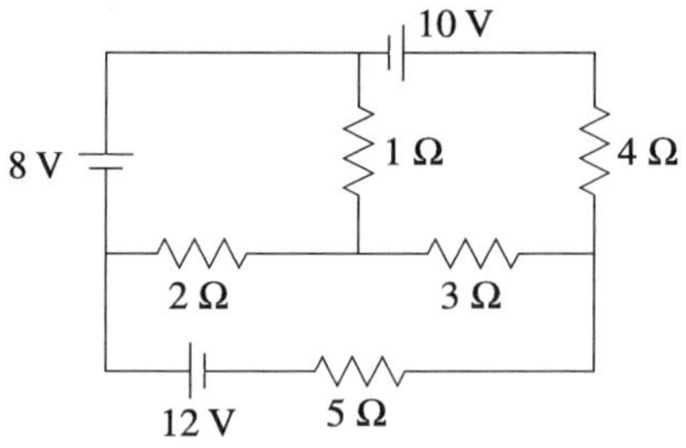

Fig. 2.48

Solution Assigning clockwise currents in three meshes,

Applying KVL to Mesh 1,

$$8 - 1(I_1 - I_2) - 2(I_1 - I_3) = 0$$

$$3I_1 - I_2 - 2I_3 = 8 \qquad \text{...(i)}$$

Applying KVL to Mesh 2,

$$10 - 4I_2 - 3(I_2 - I_3) - 1(I_2 - I_1) = 0$$

$$-I_1 + 8I_2 - 3I_3 = 10 \qquad \text{...(ii)}$$

Applying KVL to Mesh 3,

$$-2(I_3 - I_1) - 3(I_3 - I_2) - 5I_3 + 12 = 0$$

$$-2I_1 - 3I_2 + 10I_3 = 12 \qquad \text{...(iii)}$$

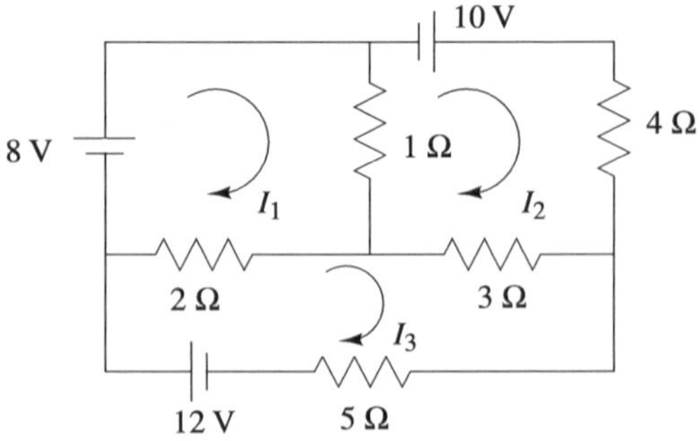

Fig. 2.49

Writing equations in matrix form,

$$\begin{bmatrix} 3 & -1 & -2 \\ -1 & 8 & -3 \\ -2 & -3 & 10 \end{bmatrix} \begin{bmatrix} I_1 \\ I_2 \\ I_3 \end{bmatrix} = \begin{bmatrix} 8 \\ 10 \\ 12 \end{bmatrix}$$

Solving equations (i), (ii) and (iii),

$$I_1 = 6.01 \text{ A}$$
$$I_2 = 3.27 \text{ A}$$
$$I_3 = 3.38 \text{ A}$$
$$I_{5\,\Omega} = 3.38 \text{ A}$$

Example 2.12 *Find the current supplied by the battery.*

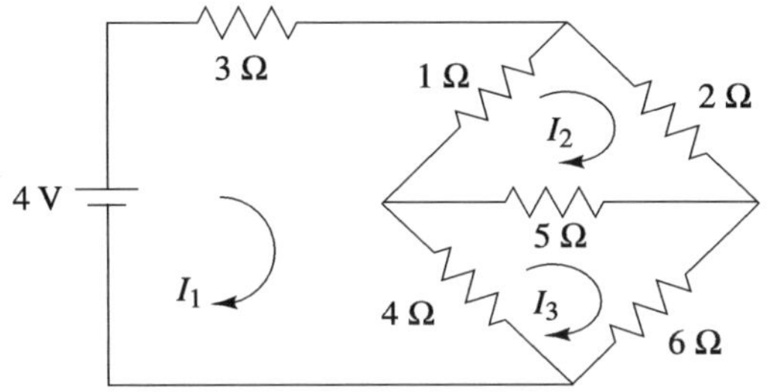

Fig. 2.50

Solution Applying KVL to Mesh 1,

$$4 - 3I_1 - 1(I_1 - I_2) - 4(I_1 - I_3) = 0$$
$$8I_1 - I_2 - 4I_3 = 4 \qquad \text{...(i)}$$

Applying KVL to Mesh 2,

$$-2I_2 - 5(I_2 - I_3) - 1(I_2 - I_1) = 0$$
$$-I_1 + 8I_2 - 5I_3 = 0 \qquad \text{...(ii)}$$

Applying KVL to Mesh 3,

$$-6I_3 - 4\,(I_3 - I_1) - 5(I_3 - I_2) = 0$$
$$-4I_1 - 5I_2 + 15I_3 = 0 \qquad \text{...(iii)}$$

Writing equations in matrix form,

$$\begin{bmatrix} 8 & -1 & -4 \\ -1 & 8 & -5 \\ -4 & -5 & 15 \end{bmatrix} \begin{bmatrix} I_1 \\ I_2 \\ I_3 \end{bmatrix} = \begin{bmatrix} 4 \\ 0 \\ 0 \end{bmatrix}$$

Solving equations (i), (ii) and (iii),

$$I_1 = 0.66 \text{ A}$$
$$I_2 = 0.24 \text{ A}$$
$$I_3 = 0.26 \text{ A}$$

Current supplied by the battery = 0.66 A.

Example 2.13 *Determine the voltage V which causes the current I_1 to be zero.*

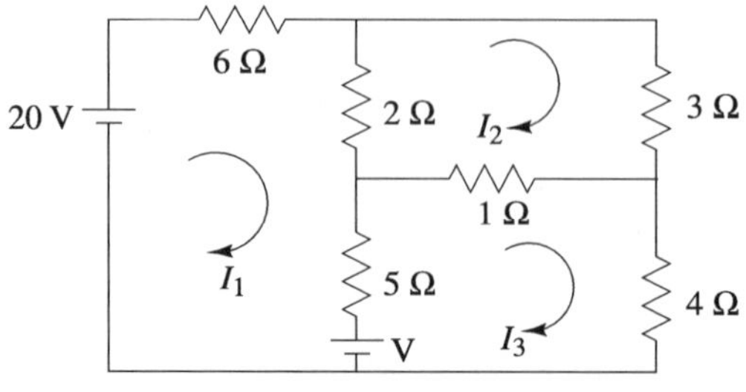

Fig. 2.51

Solution Applying KVL to Mesh 1,

$$20 - 6I_1 - 2(I_1 - I_2) - 5(I_1 - I_3) - V = 0$$

$$V + 13I_1 - 2I_2 - 5I_3 = 20 \qquad \text{...(i)}$$

Applying KVL to Mesh 2,

$$-2(I_2 - I_1) - 3I_2 - 1(I_2 - I_3) = 0$$

$$2I_1 - 6I_2 + I_3 = 0 \qquad \text{...(ii)}$$

Applying KVL to Mesh 3,

$$-1(I_3 - I_2) - 4I_3 + V - 5(I_3 - I_1) = 0$$

$$V + 5I_1 + I_2 - 10I_3 = 0 \qquad \text{...(iii)}$$

Putting $I_1 = 0$ in equations (i), (ii) and (iii),

$$V - 2I_2 - 5I_3 = 20$$

$$-6I_2 + I_3 = 0$$

$$V + I_2 - 10I_3 = 0$$

Writing equations in matrix form,

$$\begin{bmatrix} 1 & -2 & -5 \\ 0 & -6 & 1 \\ 1 & 1 & -10 \end{bmatrix} \begin{bmatrix} V \\ I_2 \\ I_3 \end{bmatrix} = \begin{bmatrix} 20 \\ 0 \\ 0 \end{bmatrix}$$

Solving equations (i), (ii) and (iii),

$$V = 43.7 \text{ V}$$

Example 2.14 *Find the current through the 2-Ω resistor.*

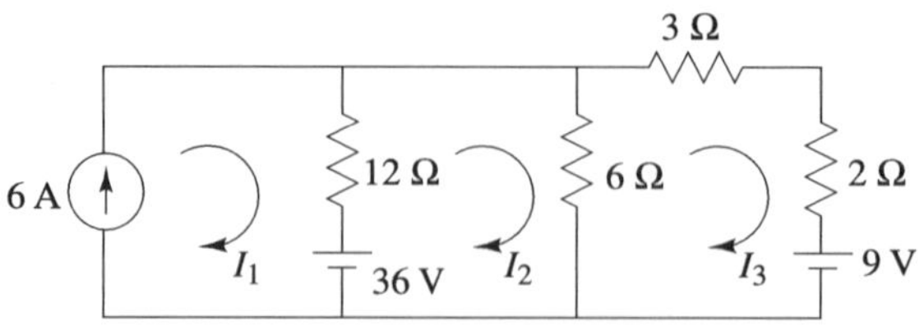

Fig. 2.52

Solution Mesh 1 contains a current source of 6 A. Hence, we cannot write KVL equation for Mesh 1. Since direction of current source and mesh current I_1 are same,

$$I_1 = 6\text{ A} \qquad \text{...(i)}$$

Applying KVL to Mesh 2,

$$36 - 12(I_2 - I_1) - 6(I_2 - I_3) = 0$$
$$36 - 12(I_2 - 6) - 6I_2 + 6I_3 = 0$$
$$18I_2 - 6I_3 = 108 \qquad \text{...(ii)}$$

Applying KVL to Mesh 3,

$$-6(I_3 - I_2) - 3I_3 - 2I_3 - 9 = 0$$
$$6I_2 - 11I_3 = 9 \qquad \text{...(iii)}$$

Solving equations (ii) and (iii),

$$I_3 = 3\text{ A}$$
$$I_{2\Omega} = 3\text{ A}$$

Example 2.15 *Obtain the branch currents.*

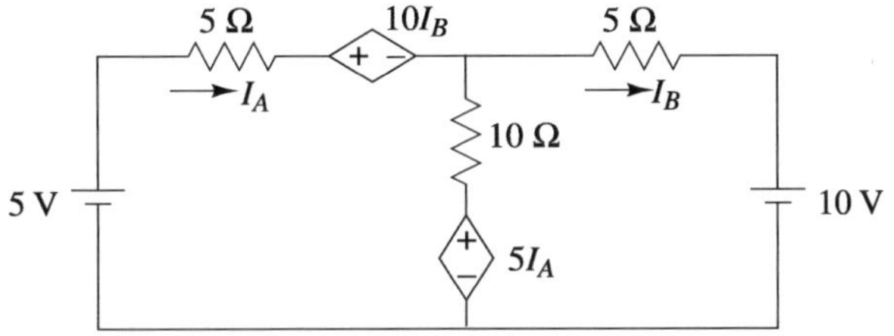

Fig. 2.53

Solution Assigning clockwise currents in two meshes,

From the figure,

$$I_A = I_1 \qquad \text{...(i)}$$
$$I_B = I_2 \qquad \text{...(ii)}$$

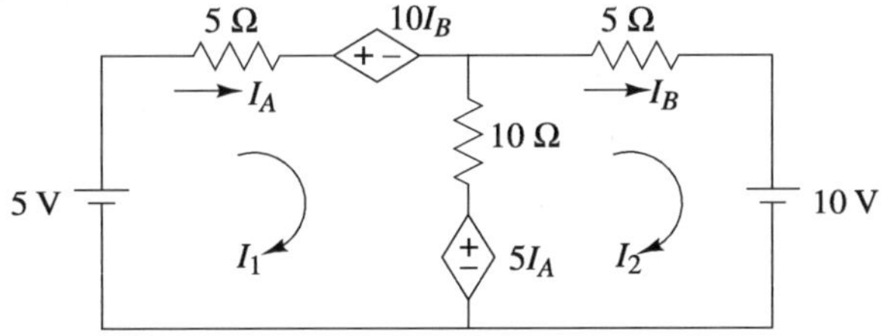

Fig. 2.54

Applying KVL to Mesh 1,

$$5 - 5I_1 - 10I_B - 10(I_1 - I_2) - 5I_A = 0$$
$$5 - 5I_1 - 10I_2 - 10I_1 + 10I_2 - 5I_1 = 0$$
$$-20I_1 = -5$$
$$I_1 = \frac{1}{4} = 0.25\text{ A} \qquad \text{...(iii)}$$

Applying KVL to Mesh 2,

$$5I_A - 10(I_2 - I_1) - 5I_2 - 10 = 0$$
$$5I_1 - 10I_2 + 10I_1 - 5I_2 = 10$$
$$15I_1 - 15I_2 = 10 \qquad \text{...(iv)}$$

Putting $I_1 = 0.25$ A in equation (iv),

$$15(0.25) - 15I_2 = 10$$
$$3.75 - 15I_2 = 10$$
$$-6.25 = 15I_2$$
$$I_2 = -0.416\text{ A}$$

Example 2.16 *Find currents in the three meshes.*

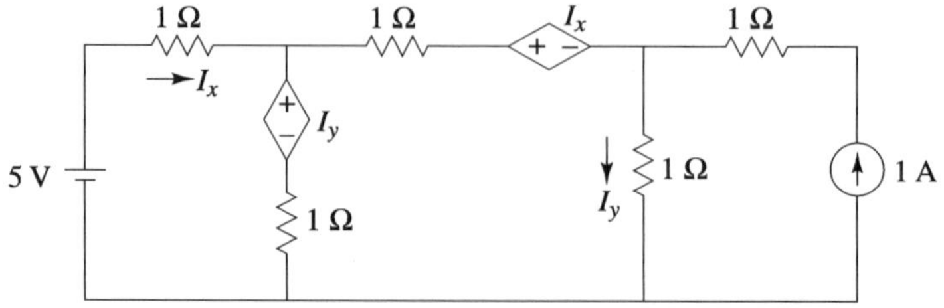

Fig. 2.55

Solution Assigning clockwise currents in three meshes,

From the figure,

$$I_x = I_1$$
$$I_y = I_2 - I_3$$

But
$$I_3 = 1\text{ A}$$
$$I_y = I_2 + 1$$

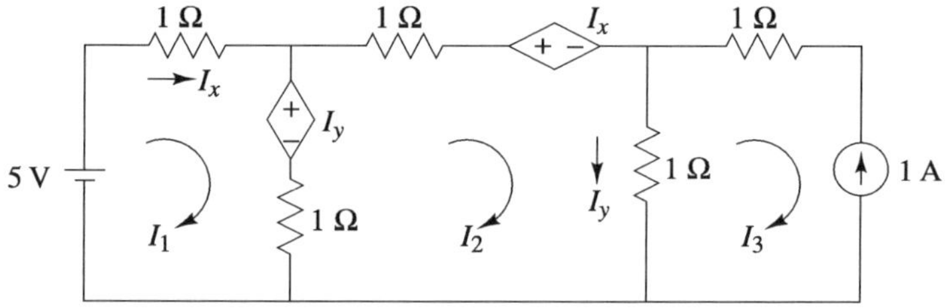

Fig. 2.56

Applying KVL to Mesh 1,

$$5 - I_1 - I_y - (I_1 - I_2) = 0$$
$$5 - I_1 - (I_2 + 1) - (I_1 - I_2) = 0$$
$$-I_1 - I_2 - I_1 + I_2 = -5 + 1$$
$$-2I_1 = -4$$
$$I_1 = 2\text{ A}$$

Applying KVL to Mesh 2,

$$-(I_2 - I_1) + I_y - I_2 - I_x - (I_2 - I_3) = 0$$
$$-I_2 + I_1 + (I_2 + 1) - I_2 - I_1 - I_2 + I_3 = 0$$
$$-2I_2 + 1 - 1 = 0$$
$$I_2 = 0$$

Example 2.17 *Find the mesh currents in the network shown.*

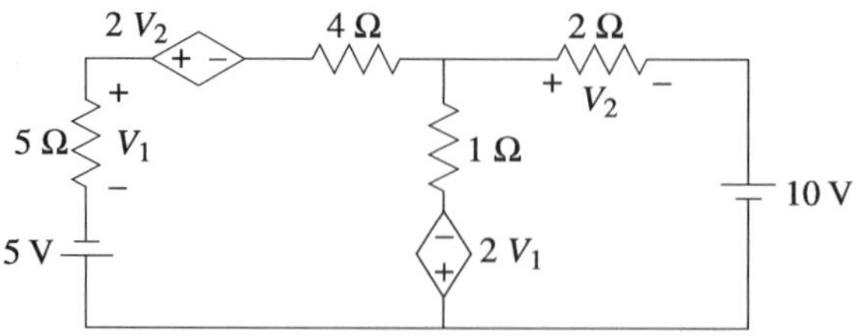

Fig. 2.57

Solution Assigning clockwise currents in two meshes,

From the figure,

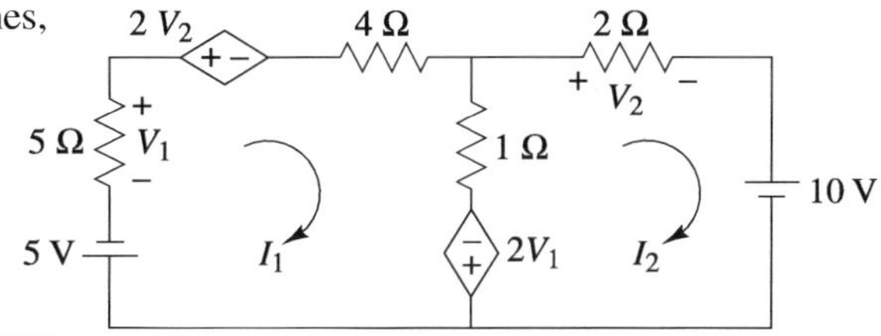

Fig. 2.58

$$V_1 = -5I_1 \quad \ldots\text{(i)}$$
$$V_2 = 2I_2 \quad \ldots\text{(ii)}$$

Applying KVL to Mesh 1,

$$-5 - 5I_1 - 2V_2 - 4I_1 - (I_1 - I_2) + 2V_1 = 0$$
$$-5 - 5I_1 - 2(2I_2) - 4I_1 - I_1 + I_2 + 2(-5I_1) = 0$$
$$20I_1 + 3I_2 = -5 \quad \ldots\text{(iii)}$$

Applying KVL to Mesh 2,

$$-2V_1 - (I_2 - I_1) - 2I_2 - 10 = 0$$
$$-2(-5I_1) - I_2 + I_1 - 2I_2 = 10$$
$$11I_1 - 3I_2 = 10 \quad \ldots\text{(iv)}$$

Solving equations (iii) and (iv),

$$I_1 = 0.161 \text{ A}$$
$$I_2 = -2.742 \text{ A}$$

Example 2.18 *Find currents I_x and I_y.*

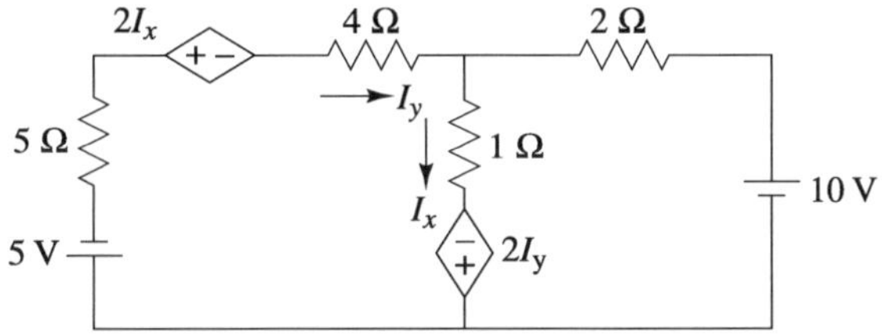

Fig. 2.59

Solution Assigning clockwise currents in two meshes,

From the figure,

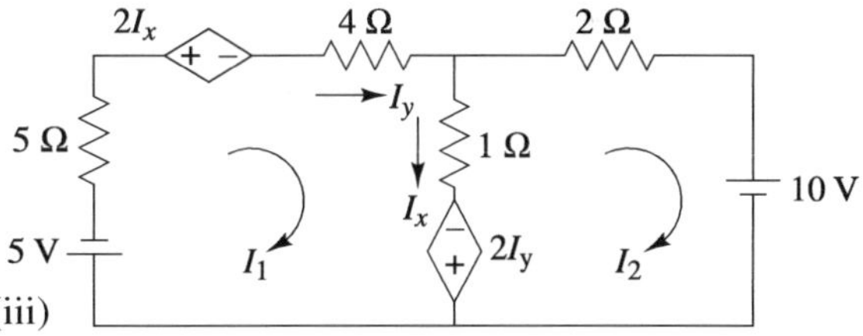

Fig. 2.60

$$I_y = I_1 \quad \ldots\text{(i)}$$
$$I_x = I_1 - I_2 \ldots\text{(ii)}$$

Applying KVL to Mesh 1,

$$-5 - 5I_1 - 2I_x - 4I_1 - (I_1 - I_2) + 2I_y = 0$$
$$-5 - 5I_1 - 2(I_1 - I_2) - 4I_1 - I_1 + I_2 + 2I_1 = 0$$
$$-5 - 5I_1 - 2I_1 + 2I_2 - 4I_1 - I_1 + I_2 + 2I_1 = 0$$
$$-10I_1 + 3I_2 = 5 \quad \ldots\text{(iii)}$$

Applying KVL to Mesh 2,

$$-2I_y - (I_2 - I_1) - 2I_2 - 10 = 0$$
$$-2I_1 - I_2 + I_1 - 2I_2 = 10$$
$$-I_1 - 3I_2 = 10 \quad \ldots\text{(iv)}$$

Solving equations (iii) and (iv),

$$I_1 = -\frac{15}{11} = -1.364 \text{ A}$$
$$I_2 = -2.878 \text{ A}$$
$$I_y = -1.364 \text{ A}$$
$$I_x = I_1 - I_2$$
$$= -1.364 + 2.878 = 1.514 \text{ A}$$

2.4 SUPERMESH ANALYSIS

Meshes that share a current source with other meshes, none of which contains a current source in the outer loop, form a supermesh. A path around a supermesh doesn't pass through a current source. A path around each mesh contained within a supermesh passes through a current source. The total number of equations required for a supermesh is equal to the number of meshes contained in the supermesh. A supermesh requires one mesh current equation, that is, a KVL equation. The remaining mesh current equations are KCL equations.

Example 2.19 *Find the current through the 10-Ω resistor.*

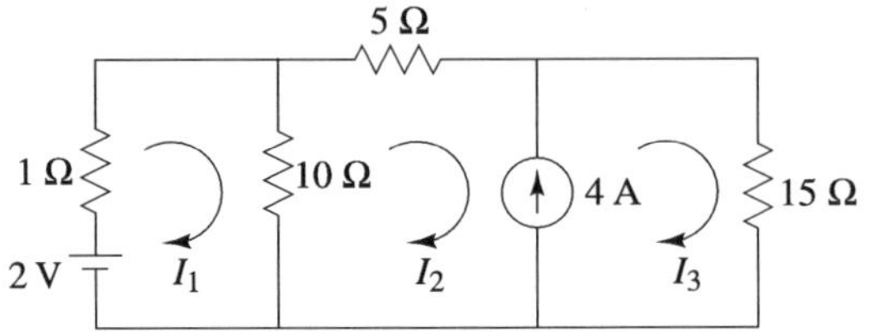

Fig. 2.61

Solution Applying KVL to Mesh 1,

$$2 - I_1 - 10(I_1 - I_2) = 0$$

$$11I_1 - 10I_2 = 2 \quad \text{...(i)}$$

Since meshes 2 and 3 contain a current source of 4 A, these two meshes will form a supermesh. A supermesh is formed by two adjacent meshes that have a common current source. The direction of the current source of 4 A and current $(I_3 - I_2)$ will be same, i.e., in the upward direction.

Writing current equation to supermesh,

$$I_3 - I_2 = 4 \quad \text{...(ii)}$$

Applying KVL to outer path of supermesh,

$$-10(I_2 - I_1) - 5I_2 - 15I_3 = 0$$

$$10I_1 - 15I_2 - 15I_3 = 0$$

$$2I_1 - 3I_2 - 3I_3 = 0 \quad \text{...(iii)}$$

Solving equations (i), (ii) and (iii),

$$I_1 = -2.35 \text{ A}$$

$$I_2 = -2.78 \text{ A}$$

$$I_3 = 1.22 \text{ A}$$

$$\text{Current through the 10-}\Omega \text{ resistor} = I_1 - I_2$$

$$= -(2.35) - (-2.78) = 0.43 \text{ A}$$

Example 2.20 *Find the current in the 3-Ω resistor.*

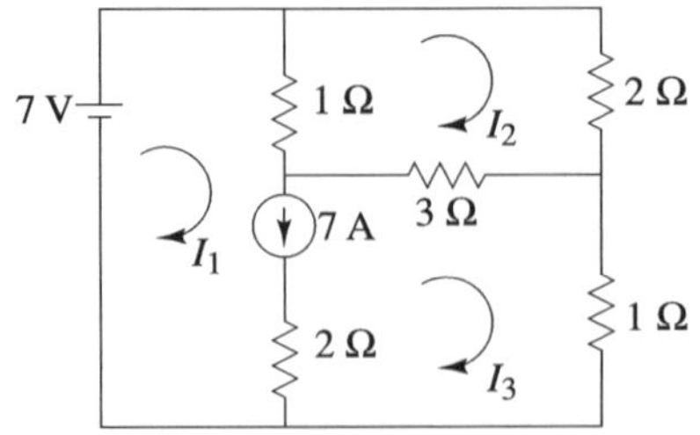

Fig. 2.62

Solution Meshes 1 and 3 will form a supermesh.

Writing current equation for supermesh,

$$I_1 - I_3 = 7 \quad \text{...(i)}$$

Applying KVL to the outer path of the supermesh,

$$7 - 1(I_1 - I_2) - 3(I_3 - I_2) - I_3 = 0$$
$$-I_1 + 4I_2 - 4I_3 = -7 \quad \text{...(ii)}$$

Applying KVL to Mesh 2,

$$-1(I_2 - I_1) - 2I_2 - 3(I_2 - I_3) = 0$$
$$I_1 - 6I_2 + 3I_3 = 0 \quad \text{...(iii)}$$

Solving equations (i), (ii) and (iii),

$$I_1 = 9 \text{ A}$$
$$I_2 = 2.5 \text{ A}$$
$$I_3 = 2 \text{ A}$$

Current through the 3-Ω resistor $= I_2 - I_3$

$= 2.5 - 2 = 0.5$ A

Example 2.21 *Find the current in the 5-Ω resistor.*

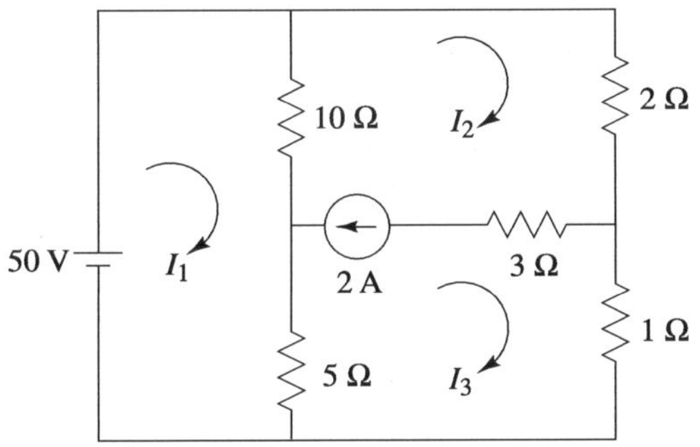

Fig. 2.63

Solution Applying KVL to Mesh 1,

$$50 - 10(I_1 - I_2) - 5(I_1 - I_3) = 0$$
$$15I_1 - 10I_2 - 5I_3 = 50 \quad \text{...(i)}$$

Meshes 2 and 3 will form a supermesh as these two meshes share a common current source of 2 A.

Writing current equation for the supermesh,

$$I_2 - I_3 = 2 \text{ A} \quad \text{...(ii)}$$

Applying KVL to the outer path of the supermesh,

$$-10(I_2 - I_1) - 2I_2 - I_3 - 5(I_3 - I_1) = 0$$
$$-15I_1 + 12I_2 + 6I_3 = 0 \quad \text{...(iii)}$$

Solving equations (i), (ii) and (iii),

$$I_1 = 20 \text{ A}$$
$$I_2 = 17.33 \text{ A}$$
$$I_3 = 15.33 \text{ A}$$

Current through the 5-Ω resistor $= I_1 - I_3$

$= 20 - 15.33 = 4.67$ A

Example 2.22 *Find currents I_1, I_2, I_3 and I_4.*

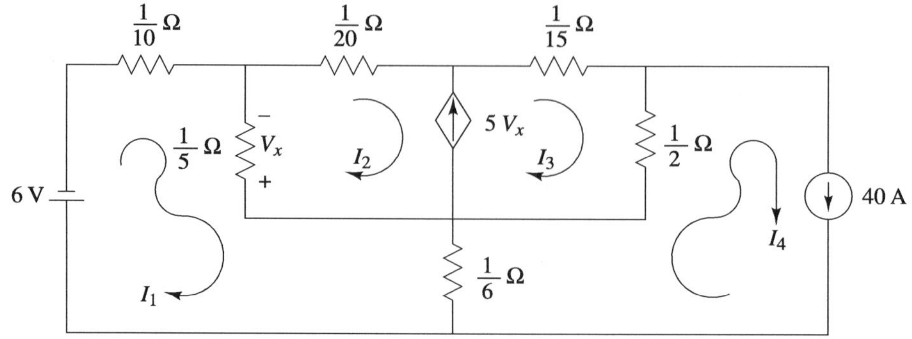

Fig. 2.64

Solution From the figure,

$$I_4 = 40\text{ A} \quad \dots \text{(i)}$$

Meshes 2 and 3 form a supermesh. Writing current equation for supermesh,

$$I_3 - I_2 = 5V_x$$

But

$$V_x = \frac{1}{5}(I_2 - I_1)$$

$$I_3 = 2I_2 - I_1 \quad \dots \text{(ii)}$$

Applying KVL to supermesh,

$$-\frac{1}{5}(I_2 - I_1) - \frac{1}{20}I_2 - \frac{1}{15}I_3 - \frac{1}{2}(I_3 - I_4) = 0 \quad \dots \text{(iii)}$$

Applying KVL to Mesh 1,

$$-6 - \frac{1}{10}I_1 - \frac{1}{5}(I_1 - I_2) - \frac{1}{6}(I_1 - I_4) = 0 \quad \dots \text{(iv)}$$

Solving Eqs (i), (ii), (iii) and (iv), we get

$$I_1 = 10\text{ A}$$
$$I_2 = 20\text{ A}$$
$$I_3 = 2\,(20) - 10 = 30\text{ A}$$
$$I_4 = 40\text{ A}$$

Example 2.23 *Find the currents I_1 and I_2.*

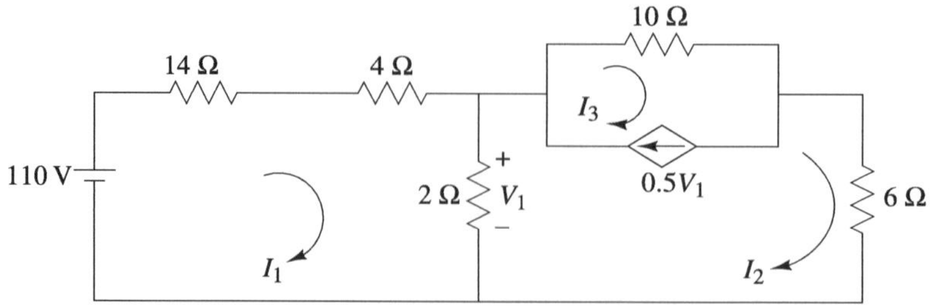

Fig. 2.65

Solution Meshes 2 and 3 form a supermesh.

Writing current equation for supermesh,

$$I_3 - I_2 = 0.5\ V_1$$

But

$$V_1 = 2(I_1 - I_2)$$

$$I_3 - I_2 = 0.5 \times 2(I_1 - I_2)$$

$$= I_1 - I_2$$

$$I_3 = I_1$$

Applying KVL to supermesh,

$$-2(I_2 - I_1) - 10I_3 - 6I_2 = 0$$

$$-2I_2 + 2I_1 - 10I_1 - 6I_2 = 0$$

$$I_1 = -I_2$$

Applying KVL to Mesh 1,

$$110 - 14I_1 - 4I_1 - 2(I_1 - I_2) = 0$$

$$110 - 20I_1 + 2I_2 = 0$$

$$110 + 20I_2 + 2I_2 = 0$$

$$I_2 = -5\ \text{A}$$

$$I_1 = -I_2 = 5\ \text{A}$$

2.5 NODAL ANALYSIS

Nodal analysis is based on Kirchhoff's current law which states that the algebraic sum of currents meeting at a point is zero. Every junction where two or more branches meet is regarded as a node. One of the nodes in the network is taken as *reference node* or *datum node*. If there are n nodes in any network, the number of simultaneous equations to be solved will be $(n - 1)$.

2.5.1 Steps to be followed in Nodal Analysis

1. Assuming that a network has n nodes, assign a reference node and the reference directions, and assign a current and a voltage name for each branch and node respectively.
2. Apply KCL at each node except for the reference node and apply Ohm's law to the branch currents.
3. Solve the simultaneous equations for the unknown node voltages.
4. Using these voltages, find any branch currents required.

Example 2.24 *Find voltage at nodes 1 and 2.*

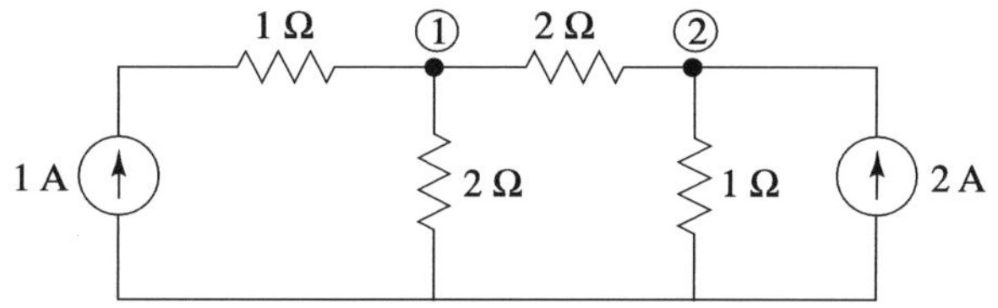

Fig. 2.66

Solution Assigning voltages V_1 and V_2 at nodes 1 and 2 respectively,

Assume that the currents are moving away from the nodes.

Applying KCL at Node 1,

$$1 = \frac{V_1}{2} + \frac{V_1 - V_2}{2}$$

$$2V_1 - V_2 = 2 \qquad \text{...(i)}$$

Applying KCL at Node 2,

$$2 = \frac{V_2}{1} + \frac{V_2 - V_1}{2}$$

$$3V_2 - V_1 = 4 \qquad \text{...(ii)}$$

Solving Eqs (i) and (ii),

$$V_1 = 2 \text{ V}$$
$$V_2 = 2 \text{ V}$$

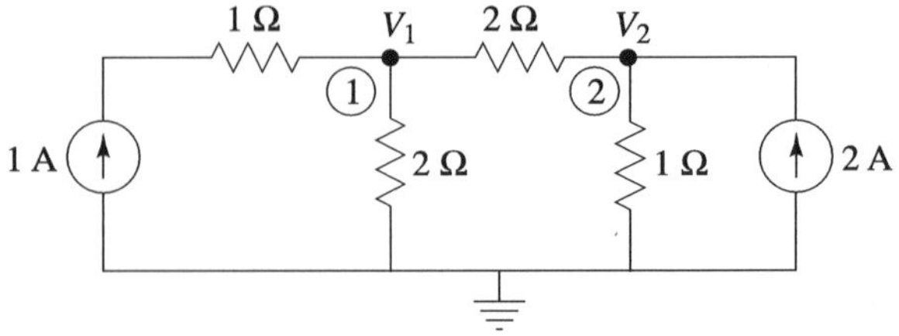

Fig. 2.67

Example 2.25 *Find V_A and V_B.*

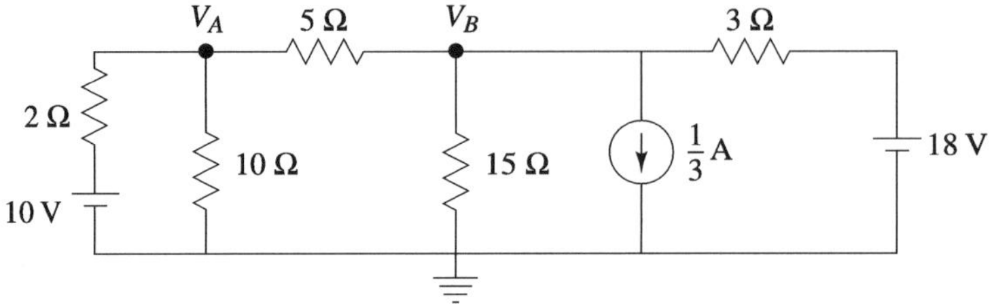

Fig. 2.68

Solution Assume that the currents are moving away from the nodes.

Applying KCL at Node A,

$$\frac{V_A - 10}{2} + \frac{V_A}{10} + \frac{V_A - V_B}{5} = 0$$

$$\frac{5V_A - 50 + V_A + 2V_A - 2V_B}{10} = 0$$

$$8V_A - 2V_B = 50 \qquad \text{...(i)}$$

Applying KCL at Node B,

$$\frac{V_B - V_A}{5} + \frac{V_B}{15} + \frac{1}{3} + \frac{V_B - 18}{3} = 0$$

$$\frac{3V_B - 3V_A + V_B + 5 + 5V_B - 90}{15} = 0$$

$$-3V_A + 9V_B = 85 \qquad \text{...(ii)}$$

Solving Eqs (i) and (ii),

$$V_A = 9.39 \text{ V}$$
$$V_B = 12.58 \text{ V}$$

Example 2.26 *Calculate the current through the 5-Ω resistor.*

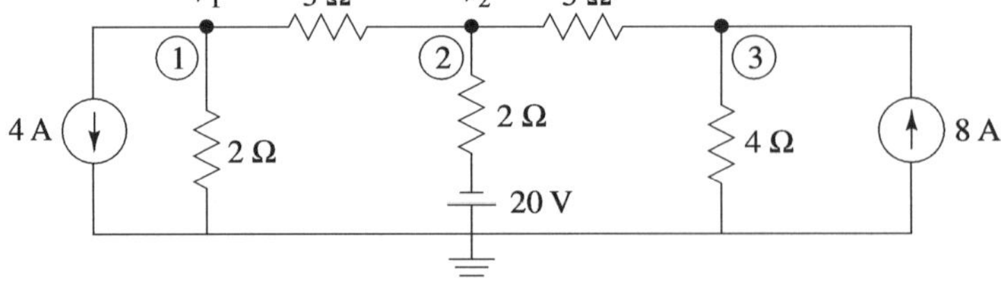

Fig. 2.69

Solution Assume that the currents are moving away from the nodes.

Applying KCL at Node 1,

$$4 + \frac{V_1}{2} + \frac{V_1 - V_2}{3} = 0$$

$$\frac{24 + 3V_1 + 2V_1 - 2V_2}{6} = 0$$

$$5V_1 - 2V_2 = -24 \qquad \text{...(i)}$$

Applying KCL at Node 2,

$$\frac{V_2 - V_1}{3} + \frac{V_2 - (-20)}{2} + \frac{V_2 - V_3}{5} = 0$$

$$\frac{10V_2 - 10V_1 + 15V_2 + 300 + 6V_2 - 6V_3}{30} = 0$$

$$10V_1 - 31V_2 + 6V_3 = 300 \qquad \text{...(ii)}$$

Applying KCL at Node 3,

$$\frac{V_3 - V_2}{5} + \frac{V_3}{4} = 8$$

$$4V_3 - 4V_2 + 5V_3 = 160$$

$$-4V_2 + 9V_3 = 160 \qquad \text{...(iii)}$$

Solving Eqs (i), (ii) and (iii),

$$V_1 = -8.77 \text{ V}$$

$$V_2 = -9.92 \text{ V}$$

$$V_3 = 13.37 \text{ V}$$

$$\text{Current through the 5-}\Omega\text{ resistor} = \frac{V_3 - V_2}{5}$$

$$= \frac{13.37 - (-9.92)}{5} = 4.66 \text{ A}$$

Example 2.27 *Find the current in the 100-Ω resistor.*

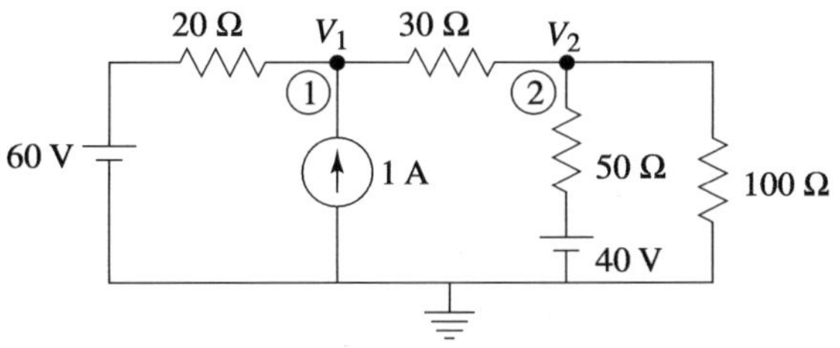

Fig. 2.70

Solution Assume that the currents are moving away from the nodes.

Applying KCL at Node 1,

$$\frac{V_1 - 60}{20} + \frac{V_1 - V_2}{30} = 1$$

$$\frac{30V_1 - 1800 + 20V_1 - 20V_2}{600} = 1$$

$$50V_1 - 20V_2 = 2400 \qquad \text{...(i)}$$

Applying KCL at Node 2,

$$\frac{V_2 - V_1}{30} + \frac{V_2 - 40}{50} + \frac{V_2}{100} = 0$$

$$\frac{10V_2 - 10V_1 + 6V_2 - 240 + 3V_2}{300} = 0$$

$$-10V_1 + 19V_2 = 240 \qquad \text{...(ii)}$$

Solving Eqs (i) and (ii),

$$V_1 = 67.2 \text{ V}$$
$$V_2 = 48 \text{ V}$$

$$\text{Current through the 100-}\Omega\text{ resistor} = \frac{V_2}{100} = \frac{48}{100} = 0.48 \text{ A}$$

Example 2.28 *Find V_A and V_B.*

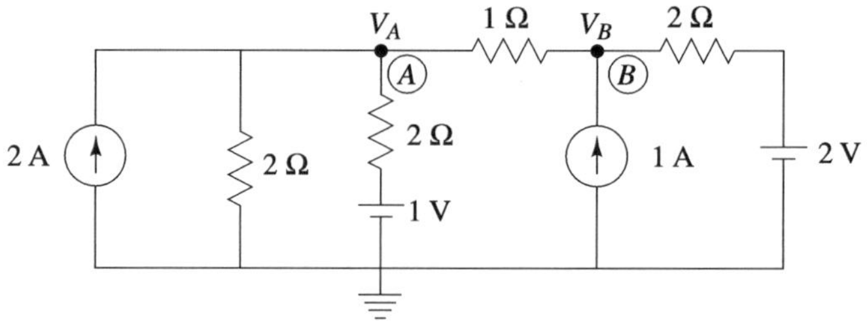

Fig. 2.71

Solution Assume that the currents are moving away from the nodes.

Applying KCL at Node *A*,

$$2 = \frac{V_A}{2} + \frac{V_A - 1}{2} + \frac{V_A - V_B}{1}$$

$$2 = \frac{V_A + V_A - 1 + 2V_A - 2V_B}{2}$$

$$4V_A - 2V_B = 5 \qquad \text{...(i)}$$

Applying KCL at Node *B*,

$$\frac{V_B - V_A}{1} + \frac{V_B - 2}{2} = 1$$

$$\frac{2V_B - 2V_A + V_B - 2}{2} = 1$$

$$-2V_A + 3V_B = 4 \qquad \text{...(ii)}$$

Solving Eqs (i) and (ii),

$$V_A = 2.88 \text{ V}$$

$$V_B = 3.25 \text{ V}$$

Example 2.29 *Find the voltage across the 5-Ω resistor.*

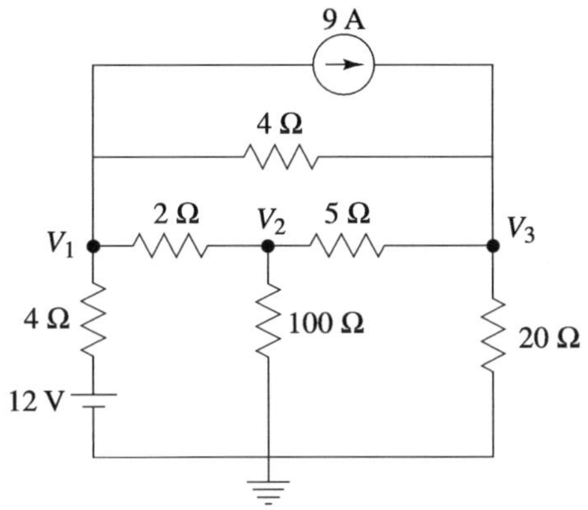

Fig. 2.72

Solution Assume that the currents are moving away from the node.

Applying KCL at Node 1,

$$\frac{V_1 - 12}{4} + \frac{V_1 - V_2}{2} + \frac{V_1 - V_3}{4} + 9 = 0$$

$$\frac{V_1 - 12 + 2V_1 - 2V_2 + V_1 - V_3 + 36}{4} = 0$$

$$4V_1 - 2V_2 - V_3 = -24 \qquad \text{...(i)}$$

Applying KCL at Node 2,

$$\frac{V_2 - V_1}{2} + \frac{V_2}{100} + \frac{V_2 - V_3}{5} = 0$$

$$\frac{50V_2 - 50V_1 + V_2 + 20V_2 - 20V_3}{100} = 0$$

$$-50V_1 + 71V_2 - 20V_3 = 0 \qquad \text{...(ii)}$$

Applying KCL at Node 3,

$$\frac{V_3 - V_2}{5} + \frac{V_3}{20} + \frac{V_3 - V_1}{4} = 9$$

$$\frac{4V_3 - 4V_2 + V_3 + 5V_3 - 5V_1}{20} = 9$$

$$-5V_1 - 4V_2 + 10V_3 = 180 \qquad \text{...(iii)}$$

Solving Eqs (i), (ii) and (iii),

$$V_1 = 6.35 \text{ V}$$
$$V_2 = 11.76 \text{ V}$$
$$V_3 = 25.88 \text{ V}$$

Voltage across the 5-Ω resistor $= V_3 - V_2$

$$= 25.88 - 11.76 = 14.12 \text{ V}$$

Example 2.30 *Find currents I_1, I_2 and I_3.*

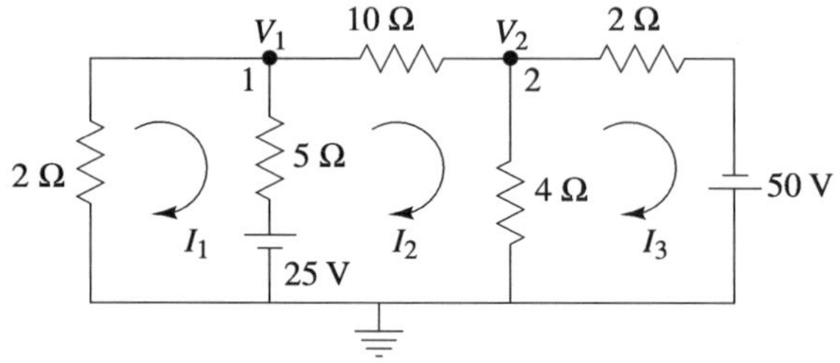

Fig. 2.73

Solution Assume that the currents are moving away from the nodes.

Applying KCL at Node 1,

$$\frac{V_1}{2} + \frac{V_1 - 25}{5} + \frac{V_1 - V_2}{10} = 0$$

$$\frac{5V_1 + 2V_1 - 50 + V_1 - V_2}{10} = 0$$

$$8V_1 - V_2 = 50 \qquad \text{...(i)}$$

Applying KCL at Node 2,

$$\frac{V_2 - V_1}{10} + \frac{V_2}{4} + \frac{V_2 - (-50)}{2} = 0$$

$$\frac{2V_2 - 2V_1 + 5V_2 + 10V_2 + 500}{20} = 0$$

$$-2V_1 + 17V_2 = -500 \qquad \text{...(ii)}$$

Solving Eqs (i) and (ii),

$$V_1 = 2.61 \text{ V}$$
$$V_2 = -29.1 \text{ V}$$

$$I_1 = -\frac{V_1}{2} = \frac{-2.61}{2} = -1.31 \text{ A}$$

$$I_2 = \frac{V_1 - V_2}{10} = \frac{2.61 - (-29.1)}{10} = 3.17 \text{ A}$$

$$I_3 = \frac{V_2 + 50}{2} = \frac{-29.1 + 50}{2} = 10.45 \text{ A}$$

Example 2.31 *Find currents I_1, I_2 and I_3 and voltages V_a and V_b.*

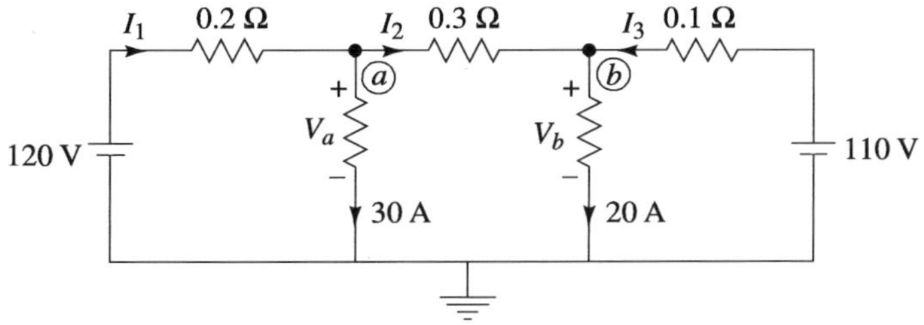

Fig. 2.74

Applying KCL at Node a,

$$I_1 = 30 + I_2$$

$$\frac{120 - V_a}{0.2} = 30 + \frac{V_a - V_b}{0.3}$$

$$36 - 0.3V_a = 1.8 + 0.2V_a - 0.2V_b$$

$$0.5V_a - 0.2V_b = 34.2 \quad \text{...(i)}$$

Applying KCL at Node b,

$$I_2 + I_3 = 20$$

$$\frac{V_a - V_b}{0.3} + \frac{110 - V_b}{0.1} = 20$$

$$\frac{0.1V_a - 0.1V_b + 33 - 0.3V_b}{0.03} = 20$$

$$0.1V_a - 0.4V_b = -32.4 \quad \text{...(ii)}$$

Solving Eqs (i) and (ii),

$$V_a = 112 \text{ V}$$

$$V_b = 109 \text{ V}$$

$$I_1 = \frac{120 - V_a}{0.2} = \frac{120 - 112}{0.2} = 40 \text{ A}$$

$$I_2 = \frac{V_a - V_b}{0.3} = \frac{112 - 109}{0.3} = 10 \text{ A}$$

$$I_3 = \frac{110 - V_b}{0.1} = \frac{110 - 109}{0.1} = 10 \text{ A}$$

Example 2.32 *Find the current in the 10-Ω resistor.*

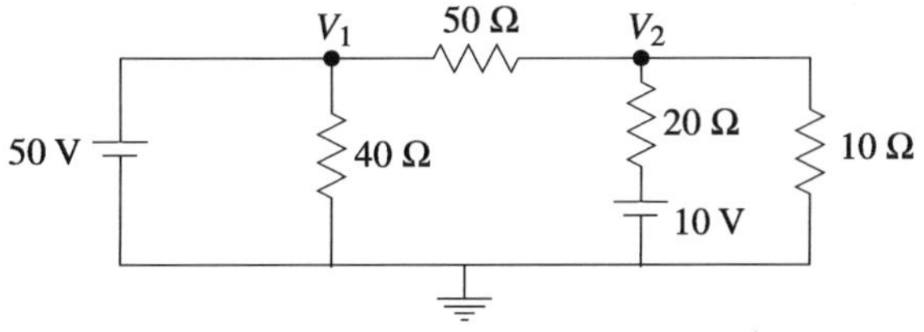

Fig. 2.75

Solution Node 1 is directly connected to a voltage source of 50 V. Hence, we cannot write KCL equation at Node 1.

At Node 1,

$$V_1 = 50 \qquad \text{...(i)}$$

Assume that the current are moving away from the node.

Applying KCL at Node 2,

$$\frac{V_2 - V_1}{50} + \frac{V_2 - 10}{20} + \frac{V_2}{10} = 0$$

$$\frac{2V_2 - 2V_1 + 5V_2 - 50 + 10V_2}{100} = 0$$

$$-2V_1 + 17V_2 = 50 \qquad \text{...(ii)}$$

Solving Eqs (i) and (ii),

$$V_1 = 50 \text{ V}$$
$$V_2 = 8.82 \text{ V}$$

$$\text{Current in the 10-}\Omega\text{ resistor} = \frac{V_2}{10}$$

$$= \frac{8.82}{10} = 0.88 \text{ A}$$

Example 2.33 *Find V_1 and V_2.*

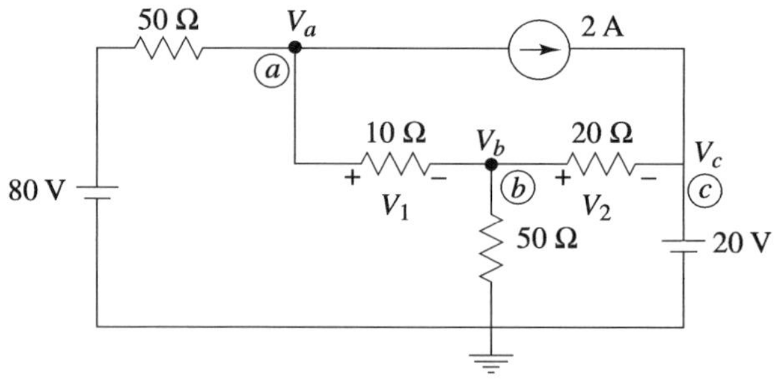

Fig. 2.76

Solution Assume that the currents are moving away from the nodes.

Applying KCL at Node *a*,

$$\frac{V_a - 80}{50} + \frac{V_a - V_b}{10} + 2 = 0$$

$$\frac{V_a - 80 + 5V_a - 5V_b + 100}{50} = 0$$

$$6V_a - 5V_b = -20 \qquad \text{...(i)}$$

Applying KCL at Node b,

$$\frac{V_b - V_a}{10} + \frac{V_b}{50} + \frac{V_b - V_c}{20} = 0$$

$$\frac{10V_b - 10V_a + 2V_b + 5V_b - 5V_c}{100} = 0$$

$$-10V_a + 17V_b - 5V_c = 0 \qquad \text{...(ii)}$$

Node c is directly connected to a voltage source of 20 V. Hence, we cannot write KCL equation at Node c,

At Node c,

$$V_c = 20 \qquad \text{...(iii)}$$

Solving Eqs (i), (ii) and (iii),

$$V_a = 3.08 \text{ V}$$

$$V_b = 7.69 \text{ V}$$

$$V_1 = V_a - V_b = 3.08 - 7.69 = -4.61 \text{ V}$$

$$V_2 = V_b - V_c = 7.69 - 20 = -12.31 \text{ V}$$

Example 2.34 *Find the voltage across the 100-Ω resistor.*

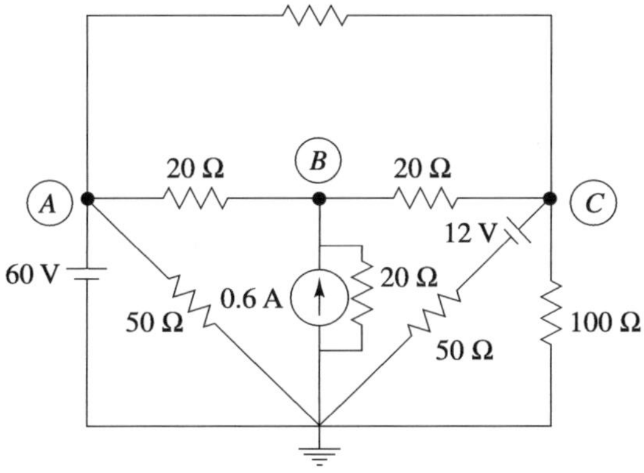

Fig. 2.77

Solution Node A is directly connected to a voltage source of 20 V. Hence, we cannot write KCL equation at Node A.

At Node A,

$$V_A = 60 \qquad \text{...(i)}$$

Assume that the currents are moving away from the nodes.

Applying KCL at Node B,

$$\frac{V_B - V_A}{20} + \frac{V_B - V_C}{20} + \frac{V_B}{20} = 0.6$$

$$-V_A + 3V_B - V_C = 12 \qquad \text{...(ii)}$$

Applying KCL at Node C,

$$\frac{V_C - V_A}{50} + \frac{V_C - V_B}{20} + \frac{V_C - 12}{50} + \frac{V_C}{100} = 0$$

$$2V_C - 2V_A + 5V_C - 5V_B + 2V_C - 24 + V_C = 0$$

$$-2V_A - 5V_B + 10V_C = 24 \quad \text{...(iii)}$$

Solving equations (i), (ii) and (iii),

$$V_C = 31.68 \text{ V}$$

$$\text{Voltage across the 100-}\Omega\text{ resistor} = 31.68 \text{ V}$$

Example 2.35 *Find voltages V_1 and V_2.*

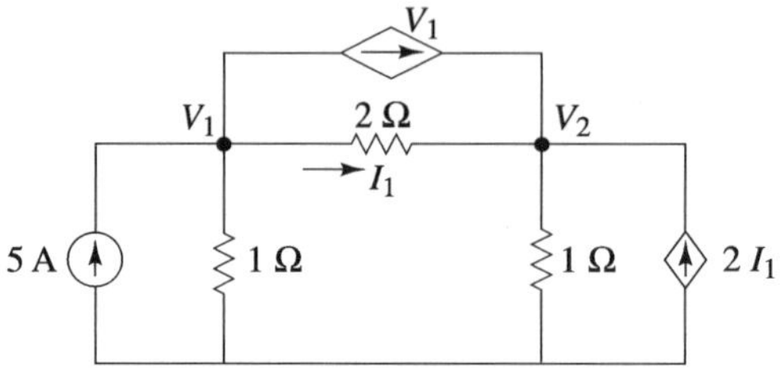

Fig. 2.78

Solution From the figure,

$$I_1 = \frac{V_1 - V_2}{2} \quad \text{... (i)}$$

Assume that the currents are moving away from the node.
Applying KCL at Node 1,

$$5 = \frac{V_1}{1} + \frac{V_1 - V_2}{2} + V_1$$

$$2.5V_1 - 0.5V_2 = 5 \quad \text{... (ii)}$$

Applying KCL at Node 2,

$$\frac{V_1 - V_2}{2} + V_1 + 2I_1 = \frac{V_2}{1}$$

$$\frac{V_1 - V_2}{2} + V_1 + 2\left(\frac{V_1 - V_2}{2}\right) = V_2$$

$$\frac{V_1}{2} - \frac{V_2}{2} + V_1 + V_1 - V_2 - V_2 = 0$$

$$2.5V_1 - 2.5V_2 = 0$$

$$V_1 = V_2 \quad \text{... (iii)}$$

Solving Eqs (ii) and (iii),

$$V_1 = 2.5 \text{ V}$$

$$V_2 = 2.5 \text{ V}$$

Example 2.36 *Find the voltage across the 5-Ω resistor.*

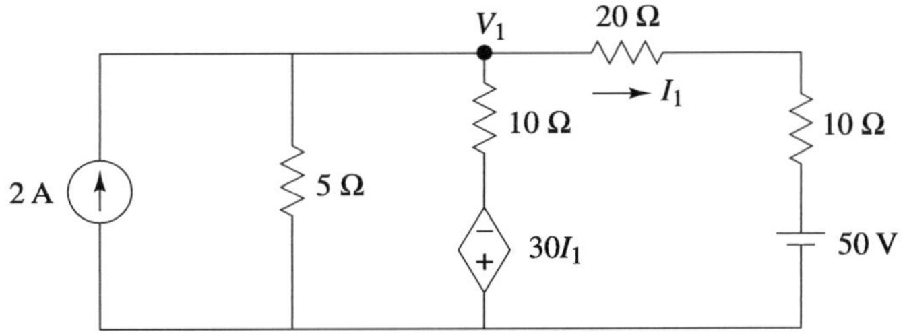

Fig. 2.79

Solution From the figure,

$$I_1 = \frac{V_1 - 50}{20 + 10} = \frac{V_1 - 50}{30} \quad \text{... (i)}$$

Assume that the currents are moving away from the node.
Applying KCL at Node 1,

$$2 = \frac{V_1}{5} + \frac{V_1 + 30I_1}{10} + \frac{V_1 - 50}{30}$$

$$2 = \frac{V_1}{5} + \frac{V_1 + 30\left(\frac{V_1 - 50}{30}\right)}{10} + \frac{V_1 - 50}{30}$$

$$2 = \frac{V_1}{5} + \frac{2V_1 - 50}{10} + \frac{V_1 - 50}{30} \quad \text{... (ii)}$$

Solving Eq. (ii),

$$V_1 = 20 \text{ V}$$

Voltage across the 5-Ω resistor = 20 V

Example 2.37 *Find the nodal voltages V_1 and V_2.*

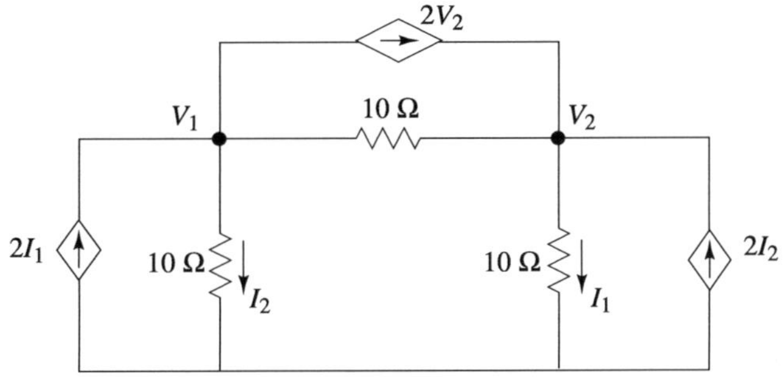

Fig. 2.80

Solution From the figure,

$$I_1 = \frac{V_2}{10} \quad \text{... (i)}$$

$$I_2 = \frac{V_1}{10} \quad \text{... (ii)}$$

Assume that the currents are moving away from the node.
Applying KCL at Node 1,

$$2I_1 = \frac{V_1}{10} + \frac{V_1 - V_2}{10} + 2V_2$$

$$2\left(\frac{V_2}{10}\right) = \frac{V_1}{10} + \frac{V_1}{10} - \frac{V_2}{10} + 2V_2$$

$$2V_1 + 17V_2 = 0 \quad \text{... (iii)}$$

Applying KCL at Node 2,

$$2I_2 + 2V_2 = \frac{V_2 - V_1}{10} + \frac{V_2}{10}$$

$$2\left(\frac{V_1}{10}\right) + 2V_2 = \frac{V_2}{10} - \frac{V_1}{10} + \frac{V_2}{10}$$

$$3V_1 + 18V_2 = 0$$

$$V_1 + 6V_2 = 0 \quad \text{... (iv)}$$

Solving Eqs (iii) and (iv),

$$V_1 = V_2 = 0$$

Example 2.38 Find voltages V_a, V_b and V_c.

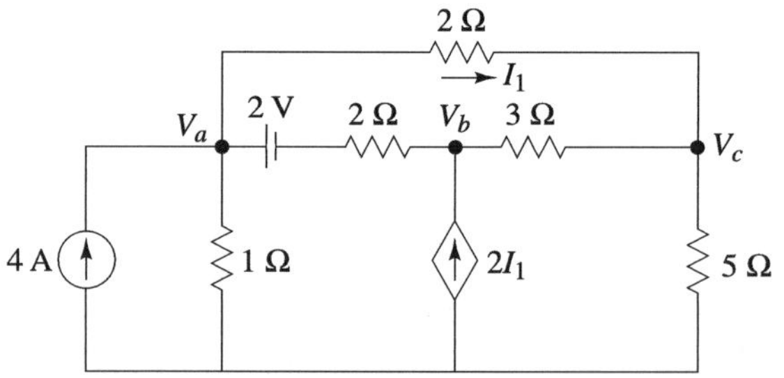

Fig. 2.81

Solution From the figure,

$$I_1 = \frac{V_a - V_c}{2}$$

Assume that the currents are moving away from the node.
Applying KCL at Node a,

$$4 = \frac{V_a}{1} + \frac{V_a - V_c}{2} + \frac{V_a - 2 - V_b}{2} \quad \text{...(i)}$$

Applying KCL at Node b,

$$\frac{V_b + 2 - V_a}{2} + \frac{V_b - V_c}{3} = 2I_1$$

$$\frac{V_b + 2 - V_a}{2} + \frac{V_b - V_c}{3} = 2\left(\frac{V_a - V_c}{2}\right) \qquad \ldots\text{(ii)}$$

Applying KCL at Node c,

$$\frac{V_c - V_b}{3} + \frac{V_c}{5} = I_1$$

$$\frac{V_c - V_b}{3} + \frac{V_c}{5} = \frac{V_a - V_c}{2} \qquad \ldots\text{(iii)}$$

Solving Eqs (i), (ii) and (iii),

$$V_a = 4.303 \text{ V}$$
$$V_b = 3.87 \text{ V}$$
$$V_c = 3.33 \text{ V}$$

2.6 SUPERNODE ANALYSIS

Nodes that are connected to each other by voltage sources, but not to the reference node by a path of voltage sources, form a *supernode*. A supernode requires one node voltage equation, that is, a KCL equation. The remaining node voltage equations are KVL equations.

Example 2.39 *Find the nodal voltages in the circuit.*

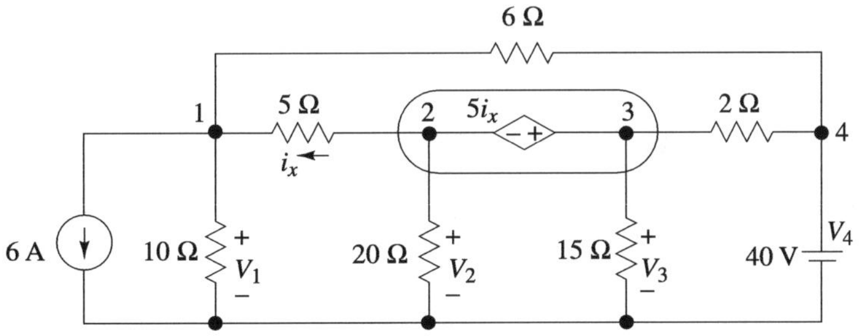

Fig. 2.82

Solution From the figure,

$$V_4 = 40 \text{ V} \qquad \ldots\text{(i)}$$

Nodes 2 and 3 form a supernode.

$$V_3 = 5i_x + V_2 = 5\left[\left(\frac{V_2 - V_1}{5}\right)\right] + V_2 = 2V_2 - V_1 \qquad \ldots\text{(ii)}$$

Applying KCL at Node 1,

$$6 + \frac{V_1}{10} + \frac{V_1 - V_2}{5} + \frac{V_1 - V_4}{6} = 0$$

$$6+\frac{V_1}{10}+\frac{V_1-V_2}{5}+\frac{V_1-40}{6}=0$$

$$\frac{7}{15}V_1-\frac{1}{5}V_2=\frac{2}{3} \qquad \text{... (iii)}$$

Applying KCL for the supernode,

$$\frac{V_2-V_1}{5}+\frac{V_2}{20}+\frac{V_3}{15}+\frac{V_3-V_4}{2}=0$$

$$\frac{V_2-V_1}{5}+\frac{V_2}{20}+\frac{(2V_2-V_1)}{15}+\frac{(2V_2-V_1)-40}{2}=0$$

$$-\frac{23}{30}V_1+\frac{83}{60}V_2=20 \qquad \text{... (iv)}$$

Solving Eqs (iii) and (iv),

$$V_1 = 10 \text{ V}$$
$$V_2 = 20 \text{ V}$$
$$V_3 = 2V_2 - V_1 = 40 - 10 = 30 \text{ V}$$

Example 2.40 *Find the nodal voltages in the circuit.*

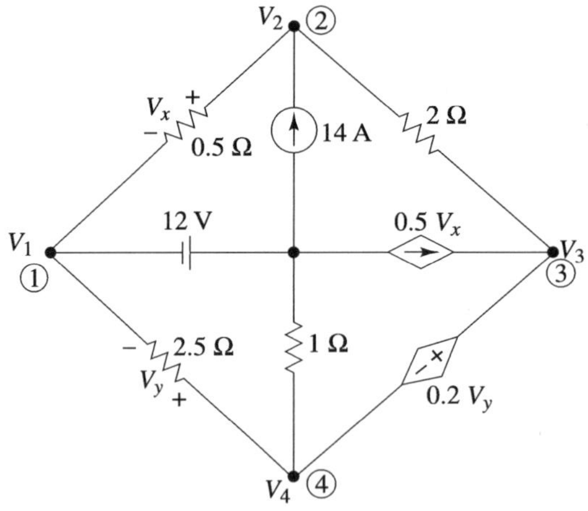

Fig. 2.83

Solution Selecting the central node as reference node,

$$V_1 = -12 \text{ V} \qquad \text{...(i)}$$

Applying KCL at Node 2,

$$\frac{V_2-V_1}{0.5}+\frac{V_2-V_3}{2}=14$$

$$-2V_1+2.5V_2-0.5V_3=14 \qquad \text{...(ii)}$$

Nodes 3 and 4 form a supernode,

$$V_3 - V_4 = 0.2V_y = 0.2(V_4 - V_1)$$

$$0.2V_1 + V_3 - 1.2V_4 = 0 \qquad \ldots \text{(iii)}$$

Applying KCL to the supernode,

$$\frac{V_3 - V_2}{2} - 0.5V_x + \frac{V_4}{1} + \frac{V_4 - V_1}{2.5} = 0$$

$$\frac{V_3 - V_2}{2} - 0.5\,(V_2 - V_1) + V_4 + \frac{V_4 - V_1}{2.5} = 0$$

$$0.1V_1 - V_2 + 0.5V_3 + 1.4V_4 = 0 \qquad \ldots \text{(iv)}$$

Solving Eqs (i), (ii), (iii) and (iv),

$$V_1 = -12 \text{ V}$$
$$V_2 = -4 \text{ V}$$
$$V_3 = 0$$
$$V_4 = -2 \text{ V}$$

Exercises

STAR–DELTA TRANSFORMATION

1. Find an equivalent resistance between A and B.

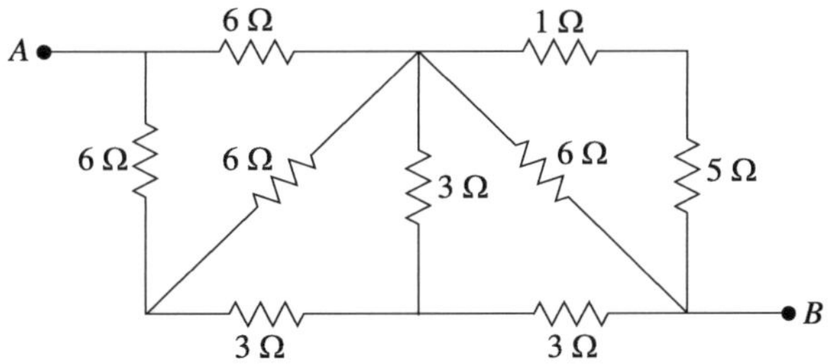

Fig. 2.84

[5 Ω]

2. Find an equivalent resistance between A and B.

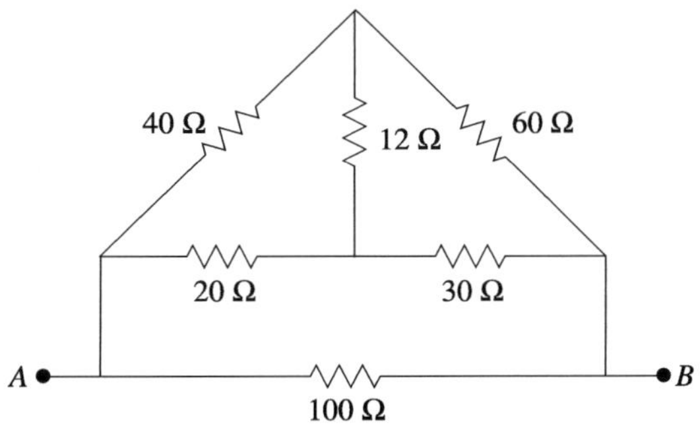

Fig. 2.85

[25 Ω]

3. Find an equivalent resistance between A and B.

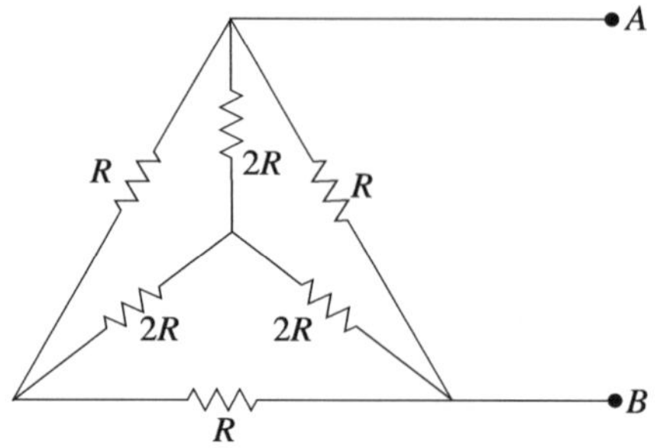

Fig. 2.86

[4/7 R]

4. Find an equivalent resistance between A and B.

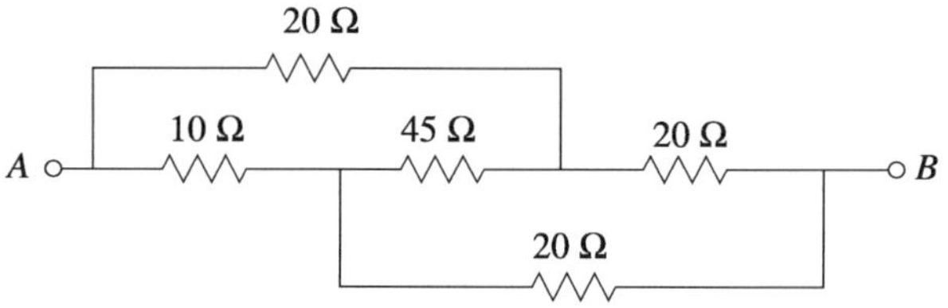

Fig. 2.87

[17 Ω]

5. Find R_{AB} by solving the outer delta (X–B–Y) only.

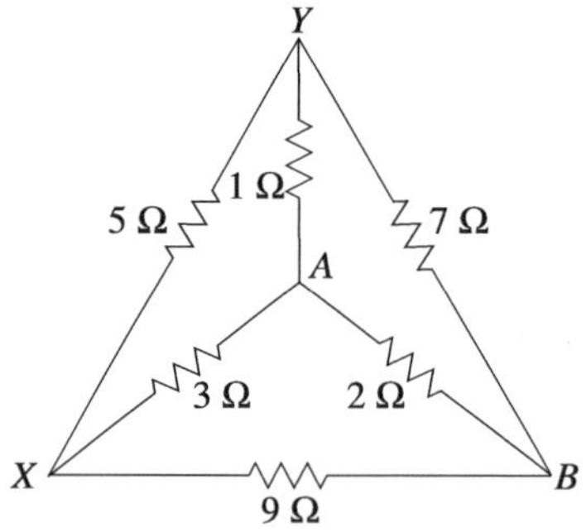

Fig. 2.88

[1.41 Ω]

6. Find an equivalent resistance between A and B.

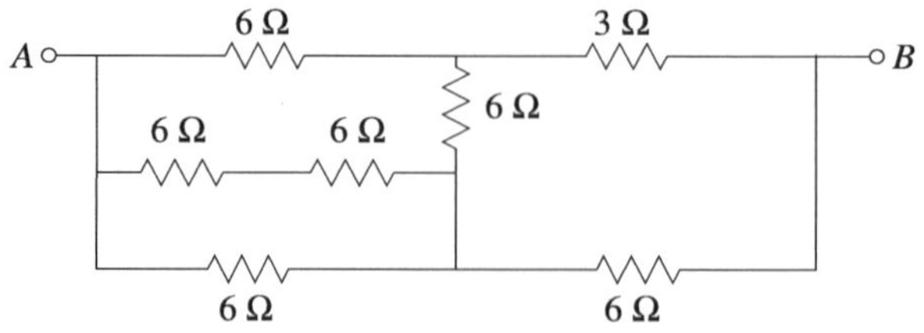

Fig. 2.89

[4.59 Ω]

7. Find the voltage between terminals A and B.

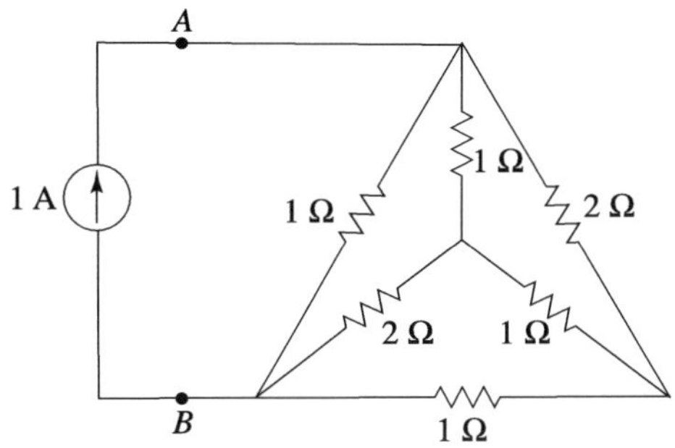

Fig. 2.90

[0.559 V]

8. Determine the power supplied to the network.

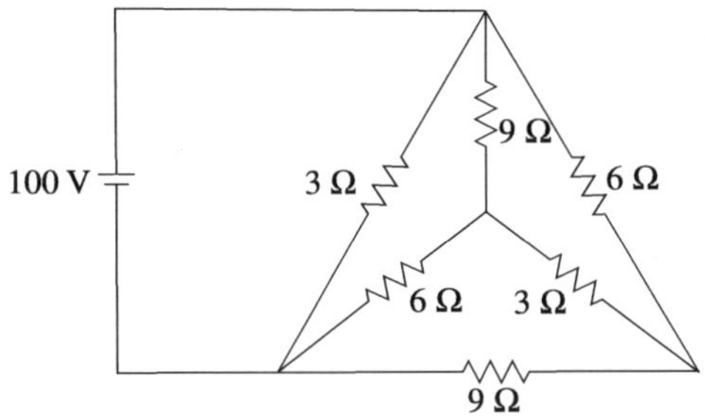

Fig. 2.91

[4705.88 Ω]

9. Find the current I.

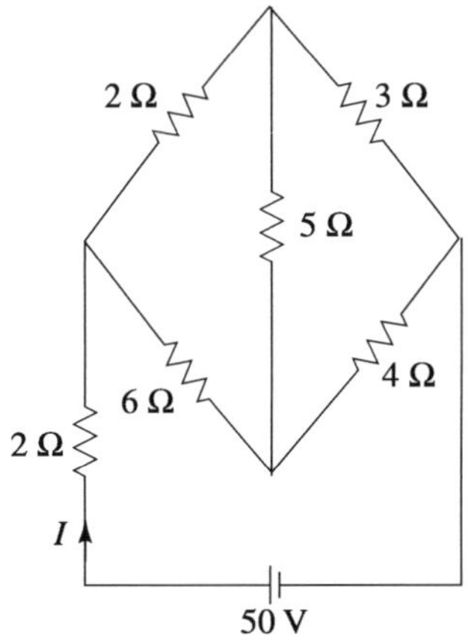

Fig. 2.92

[9.465 A]

10. Determine the current through the 10-Ω resistor.

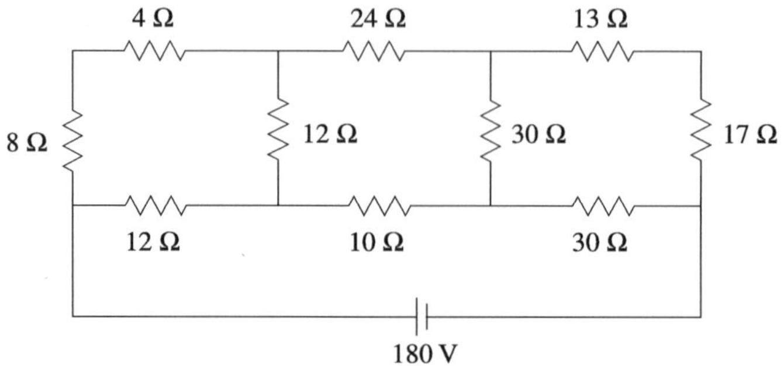

Fig. 2.93

[3.843 A]

MESH ANALYSIS

11. Find the current through the 1-Ω resistor.

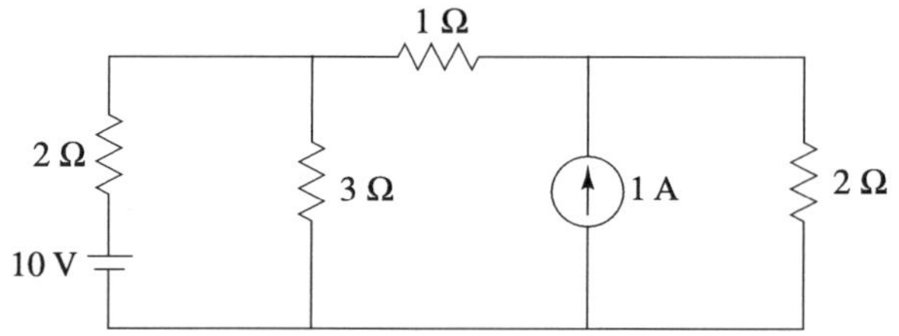

Fig. 2.94

[0.95 A]

12. Find the current through the 4-Ω resistor.

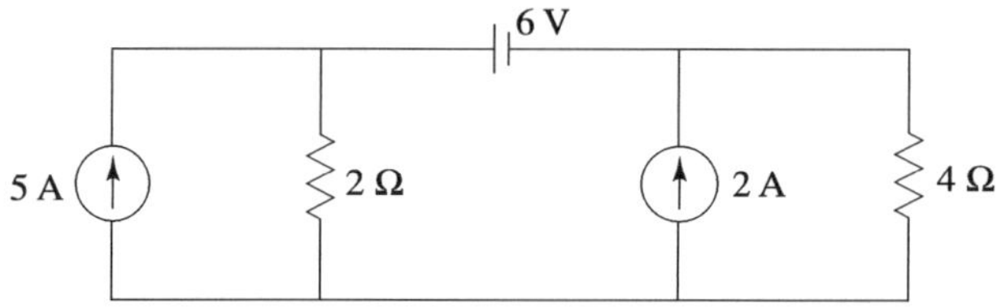

Fig. 2.95

[1.33 A]

13. Find the potential across the 3-Ω resistor.

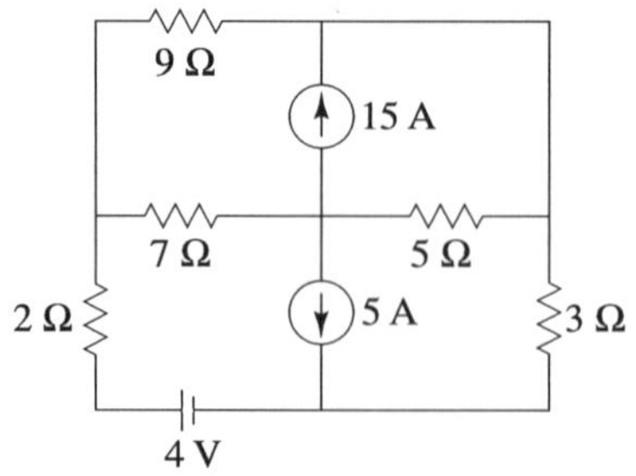

Fig. 2.96

[3.3 V]

14. Find the current I_1.

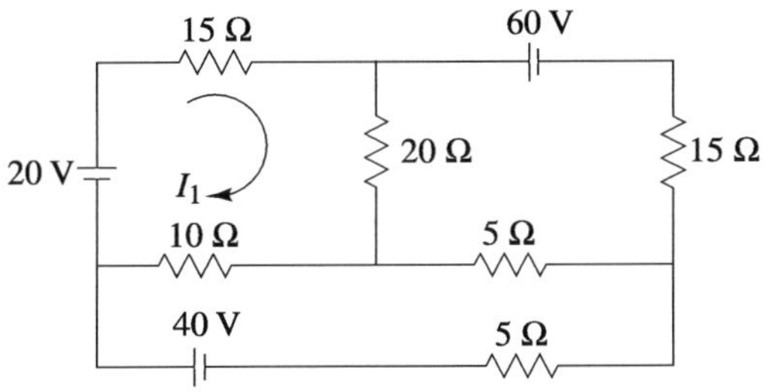

Fig. 2.97

[0.41 A]

15. Find currents I_x and I_y.

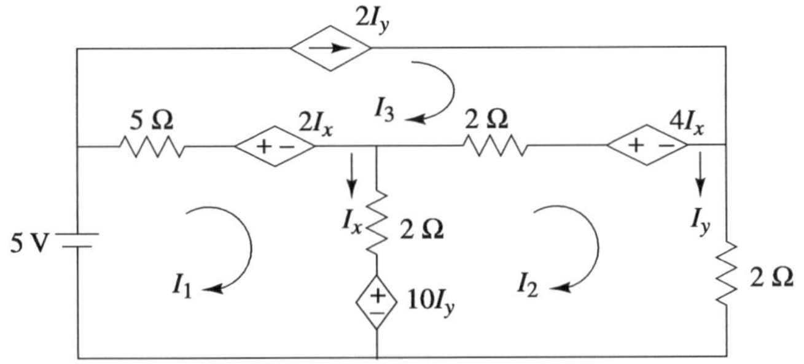

Fig. 2.98

(0.5 A, 0.1 A)

16. Use mesh analysis to find V_3 if element A is a

(a) short circuit
(b) 5-Ω resistor
(c) 20 V independent voltage source, positive reference on the right
(d) dependent voltage source of $1.5i_1$, with positive reference on the right
(e) dependent current source $5i_1$, arrow directed to the right

(69.4 V, 72.38 V, 73.68 V, 70.71 V, 97.39 V)

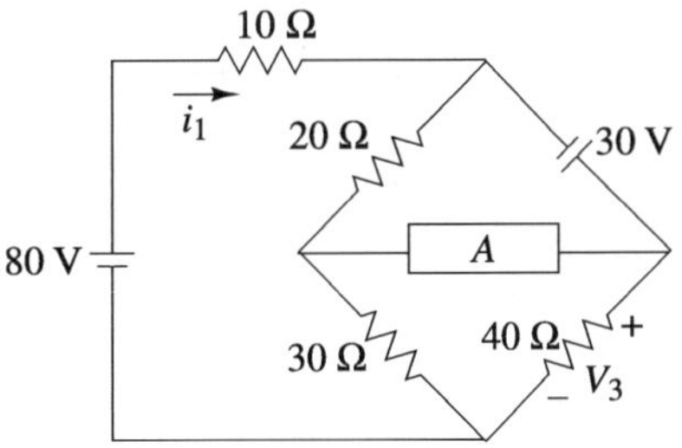

Fig. 2.99

17. Find I_1 if the dependent voltage is labelled (i) $2V_2$, and (ii) $1.5V_3$.

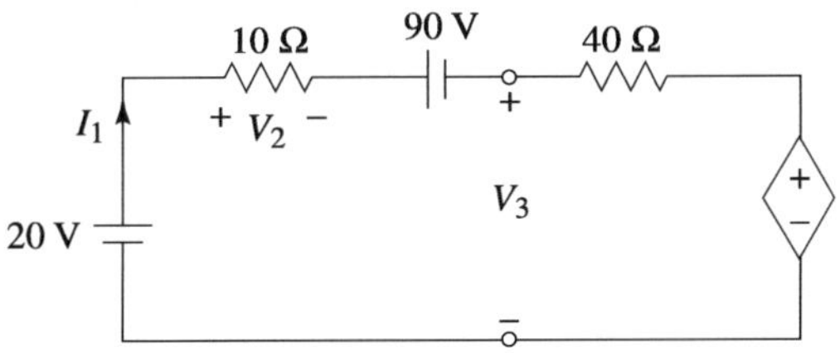

Fig. 2.100

(–1 A, 1 A)

18. Find currents I_1, I_2 and I_3.

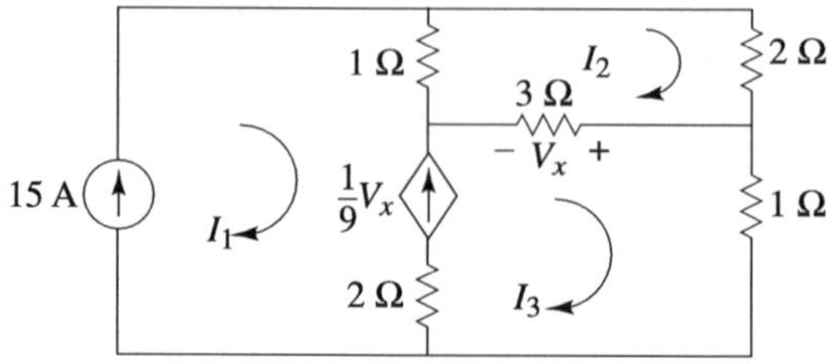

Fig. 2.101

(15 A, 11 A, 17 A)

19. Find the current I_x.

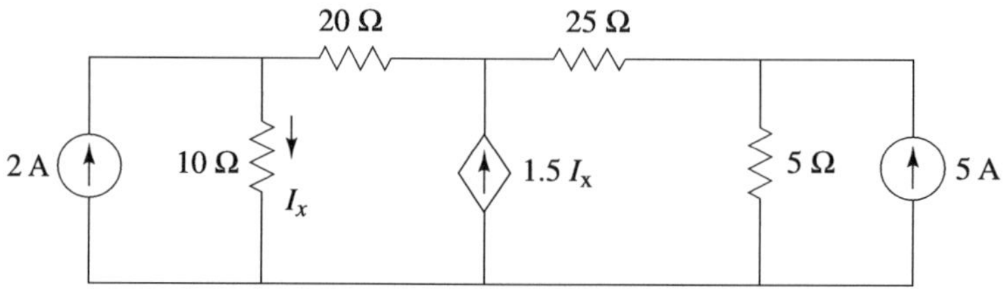

Fig. 2.102

(8.33 A)

20. Find the current I_1.

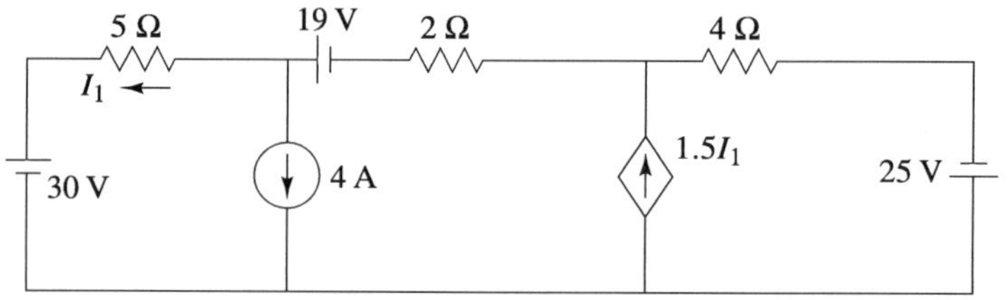

Fig. 2.103

(–12 A)

NODAL ANALYSIS

21. Determine the current through the 5-Ω resistor.

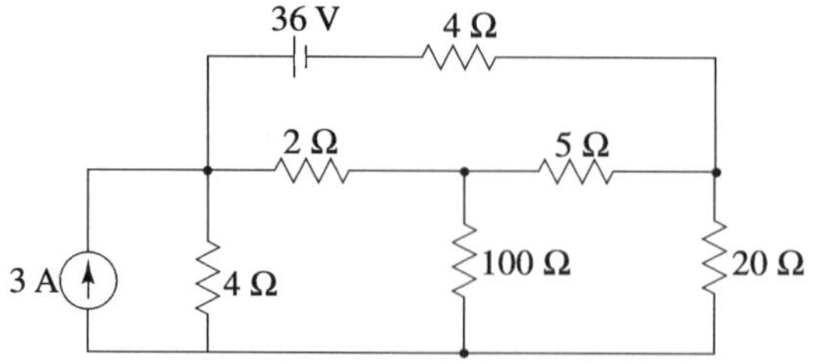

Fig. 2.104

[3.11 A]

22. Find the current through the 4-Ω resistor.

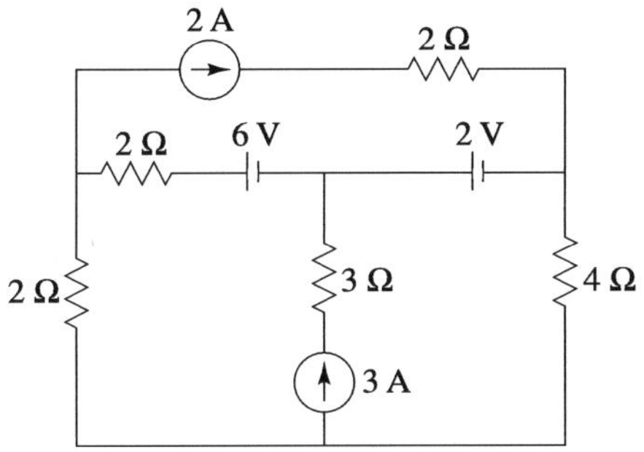

Fig. 2.105

[1 A]

23. Find the voltage V_x.

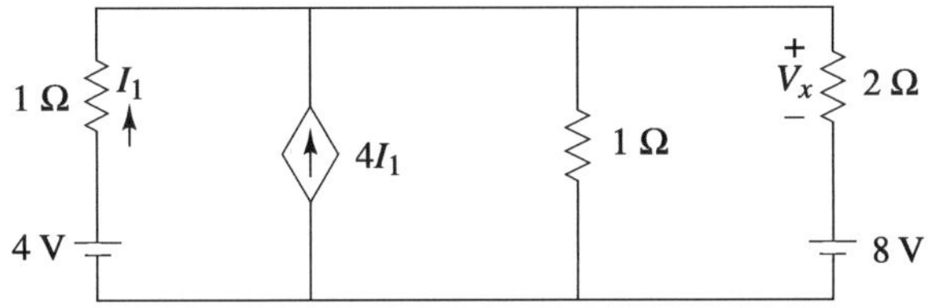

Fig. 2.106

(–4.31 V)

24. Find the voltage V_x.

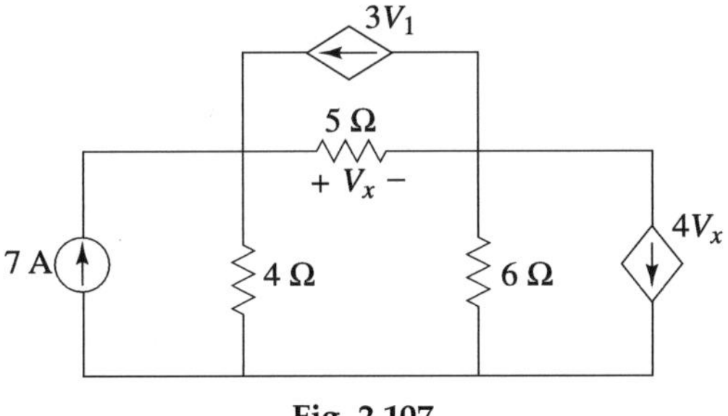

Fig. 2.107

(2.09 V)

25. Find the voltage V_x.

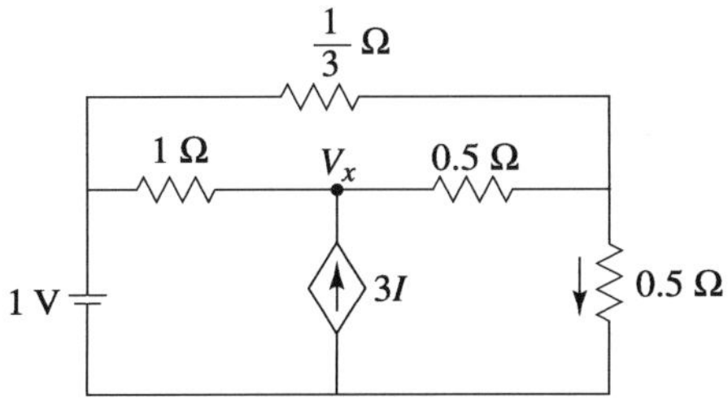

Fig. 2.108

(6.2 V)

26. Determine V_1.

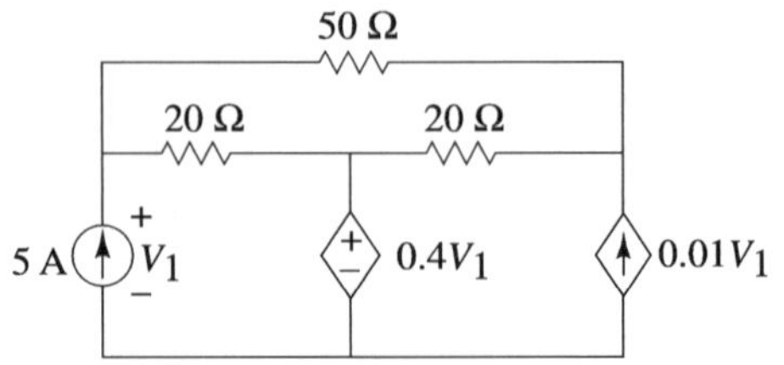

Fig. 2.109

(140 V)

27. Find the voltage V_y.

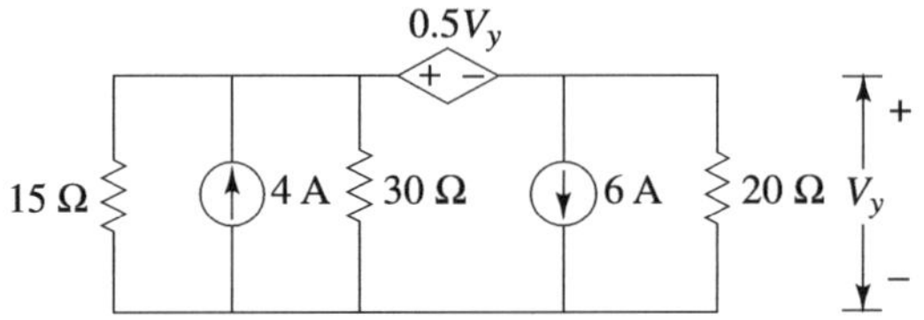

Fig. 2.110

(–10 V)

28. Find the voltage V_2.

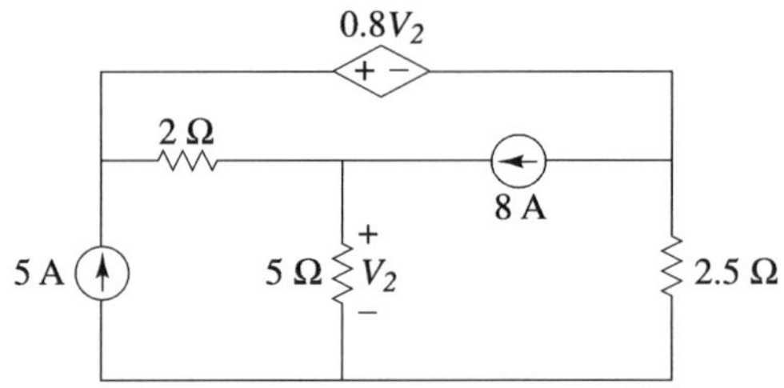

Fig. 2.111

(25.9 V)

Objective-Type Questions

1. Two electrical sub-networks N_1 and N_2 are connected through three resistors as shown in Fig. 2.112. The voltages across the 5-Ω resistor and 1-Ω resistor are given to be 10 V and 5 V respectively. Then the voltage across the 15-Ω resistor is

(a) –105 V (b) 105 V
(c) –15 V (d) 15 V

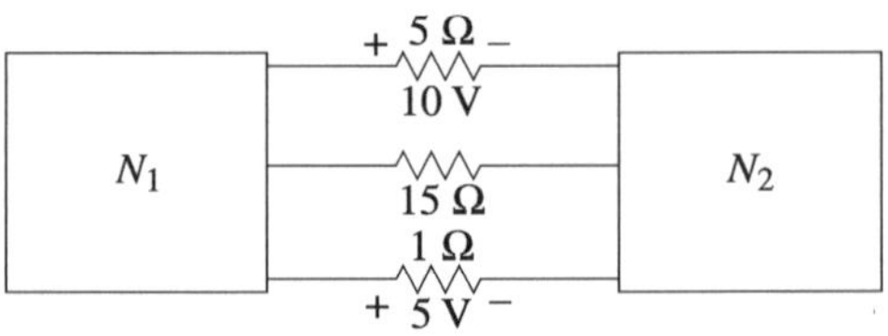

Fig. 2.112

2. The nodal method of circuit analysis is based on
   (a) KVL and Ohm's law
   (b) KCL and Ohm's law
   (c) KCL and KVL
   (d) KCL, KVL and Ohm's law

3. The voltage across terminals a and b in Fig. 2.113 is
   (a) 0.5 V
   (b) 3 V
   (c) 3.5 V
   (d) 4 V

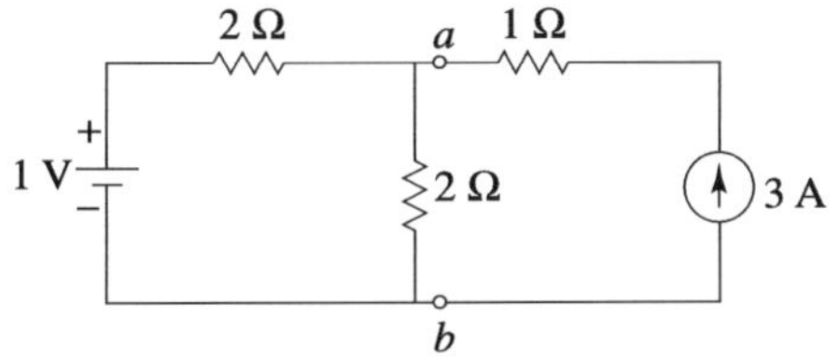

Fig. 2.113

4. A delta-connected network with its wye-equivalent is shown in Fig. 2.114. The resistances R_1, R_2 and R_3 (in ohms) are respectively
   (a) 1.5, 3 and 9
   (b) 3, 9 and 1.5
   (c) 9, 3 and 1.5
   (d) 3, 1.5 and 9

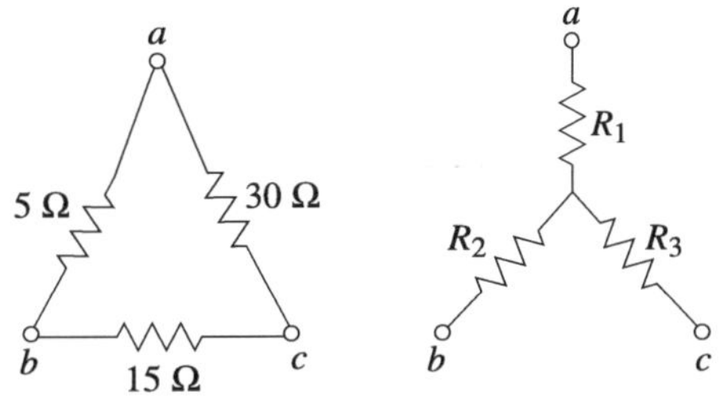

Fig. 2.114

5. If each branch of a delta circuit has resistance $\sqrt{3}\,R$, then each branch of the equivatent wye circuit has resistance
   (a) $\dfrac{R}{\sqrt{3}}$
   (b) $3R$
   (c) $3\sqrt{3}\,R$
   (d) $\dfrac{R}{3}$

6. The voltage V_o in Fig. 2.115 is
   (a) 48 V
   (b) 24 V
   (c) 36 V
   (d) 28 V

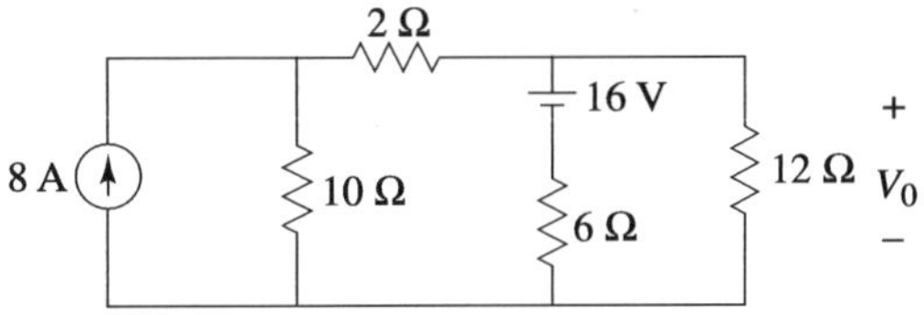

Fig. 2.115

7. The dependent current source shown in Fig. 2.116.
   (a) delivers 80 W
   (b) absorbs 80 W
   (c) delivers 40 W
   (d) absorbs 40 W

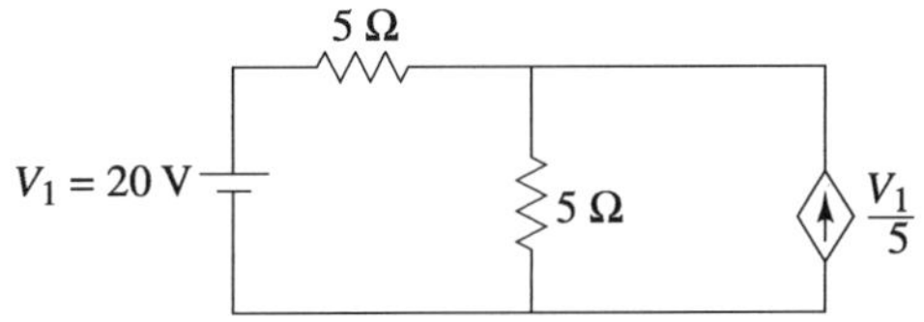

Fig. 2.116

8. If $V = 4$ in Fig. 2.117, the value of I_S is given by
   (a) 6 A
   (b) 2.5 A
   (c) 12 A
   (d) none of these

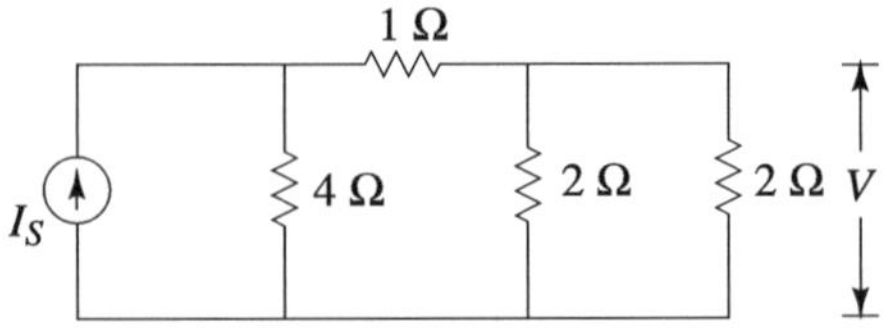

Fig. 2.117

9. The value of V_x, V_y and V_z in Fig. 2.118 shown are
   (a) –6, 3, –3
   (b) –6, –3, 1
   (c) 6, 3, 3
   (d) 6, 1, 3

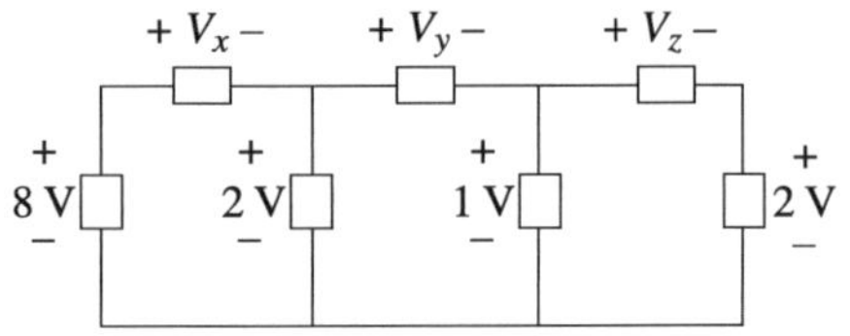

Fig. 2.118

10. Viewed from the terminal AB, the following circuit can be reduced to an equivalent circuit of a single voltage source in series with a single resistor with the following parameters
   (a) 5-Volt source in series with a 10-Ω resistor
   (b) 1-Volt source in series with a 2.4-Ω resistor
   (c) 15-Volt source in series with a 2.4-Ω resistor
   (d) 1-Volt source in series with a 10-Ω resistor

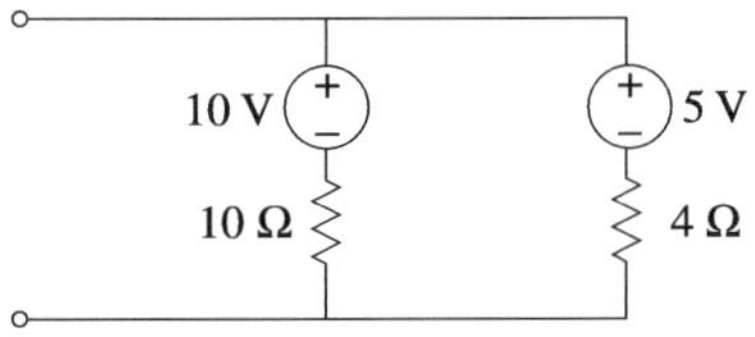

Fig. 2.119

11. The circuit shown in Fig. 2.120 is equivalent to a load of
   (a) $\frac{4}{3}$ Ω (b) $\frac{8}{3}$ Ω
   (c) 4 Ω (d) 2 Ω

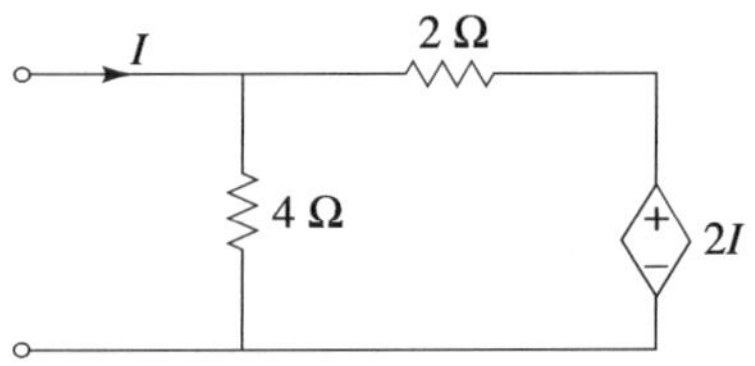

Fig. 2.120

12. In the network shown in Fig. 2.121 the effective resistance faced by the voltage source is
   (a) 4 Ω (b) 3 Ω
   (c) 2 Ω (d) 1 Ω

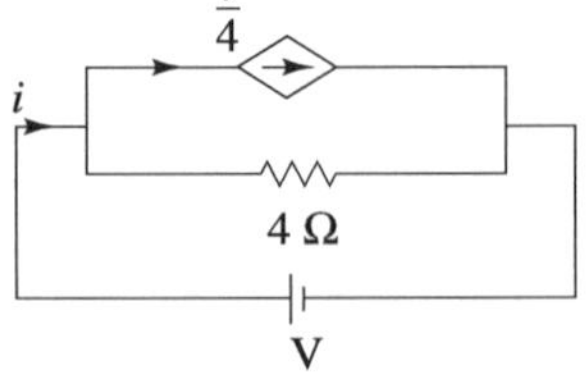

Fig. 2.121

13. A network contains only an independent current source and resistors. If the values of all resistors are doubled, the value of the node voltages will
   (a) become half (b) remain unchanged (c) become double (d) none of these

14. Consider the star network shown in Fig. 2.122. The resistance between terminals A and B with C open is 6 Ω, between terminals B and C with A open is 11 Ω and between terminals C and A with B open is

(a) $R_A = 4\ \Omega$, $R_B = 2\ \Omega$, $R_C = 5\ \Omega$
(b) $R_A = 2\ \Omega$, $R_B = 4\ \Omega$, $R_C = 7\ \Omega$
(c) $R_A = 3\ \Omega$, $R_B = 3\ \Omega$, $R_C = 4\ \Omega$
(d) $R_A = 5\ \Omega$, $R_B = 1\ \Omega$, $R_C = 10\ \Omega$

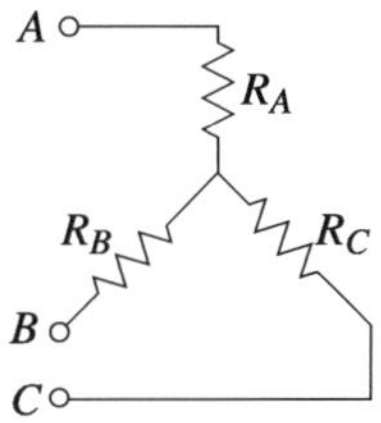

Fig. 2.122

Answers to Objective-Type Questions

1. (a)	**2.** (b)	**3.** (c)	**4.** (d)	**5.** (a)	**6.** (d)
7. (a)	**8.** (a)	**9.** (d)	**10.** (b)	**11.** (b)	**12.** (b)
13. (c)	**14.** (b)				

3 Network Theorems-II

3.1 INTRODUCTION

In Chapter 2, we have studied elementary network theorems like star–delta transformation, mesh analysis and node analysis. There are some other methods also to analyse circuits. In this chapter, we will study superposition theorem, Thevenin's theorem, Norton's theorem and maximum power transfer theorem. We can find currents and voltages in various parts of the circuits with these methods.

3.2 SUPERPOSITION THEOREM

It states that *'In a linear network containing more than one independent sources and dependent sources, the resultant current in any element is the algebraic sum of the currents that would be produced by each independent source acting alone, all the other independent sources being represented meanwhile by their respective internal resistances.'*

The independent voltage sources are represented by their internal resistance if given or simply with zero resistance, i.e., short circuits if internal resistances are not mentioned. The independent current sources are represented by infinite resistance, i.e., open circuits.

The dependent sources are not sources but dissipative components—hence they are active at all the times. A dependent source has zero value only when its control voltage or current is zero.

A linear network is one whose parameters are constant, i.e., they do not change with voltage and current.

Explanation Consider the circuit shown in Fig. 3.1.

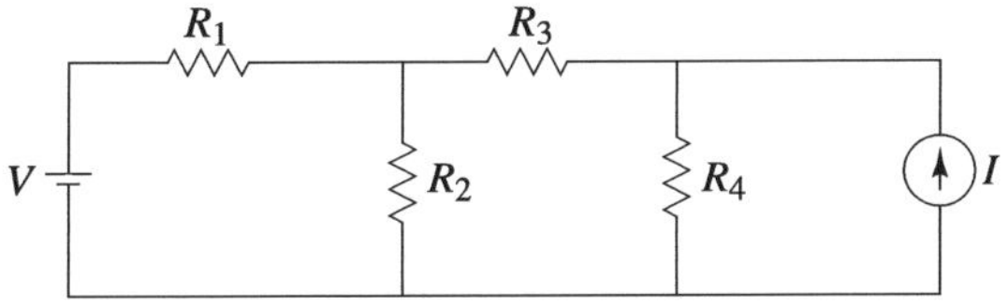

Fig. 3.1

Steps to be followed in superposition theorem

1. The current flowing through R_4 due to constant voltage source V is found to be say I_4' (with proper direction), representing constant current source with infinite resistance, i.e., open circuit.
2. The current flowing through R_4 due to constant current source of I amp is found to be say I_4'' (with proper direction), representing the constant voltage source with zero resistance or short circuit.
3. The resultant current I_4 through R_4 is found by superposition theorem.

$$I_4 = I_4' + I_4''$$

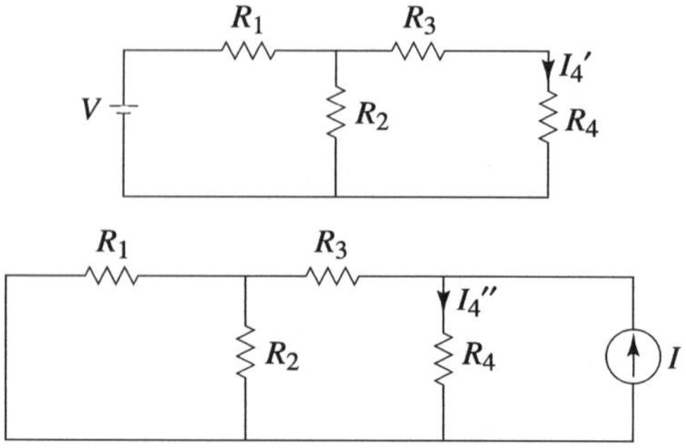

Fig. 3.2

Example 3.1 *Determine the current in the 10-Ω resistor.*

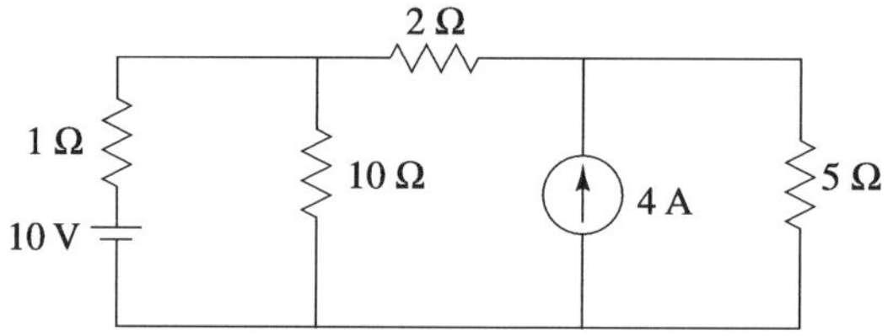

Fig. 3.3

Solution

Step I: *When the 10-V source is acting alone*

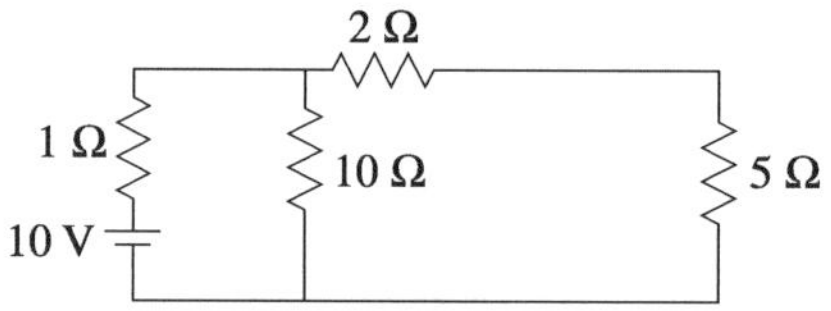

Fig. 3.4

By source transformation,

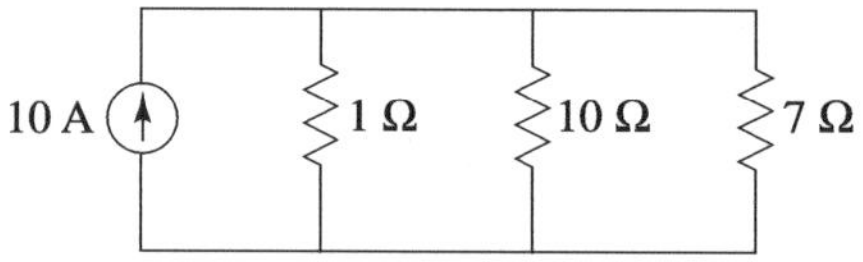

Fig. 3.5

Since we have to find current through the 10-Ω resistor, the parallel combination of resistances of 1 Ω and 7 Ω is combined into an equivalent resistance of 0.875 Ω.

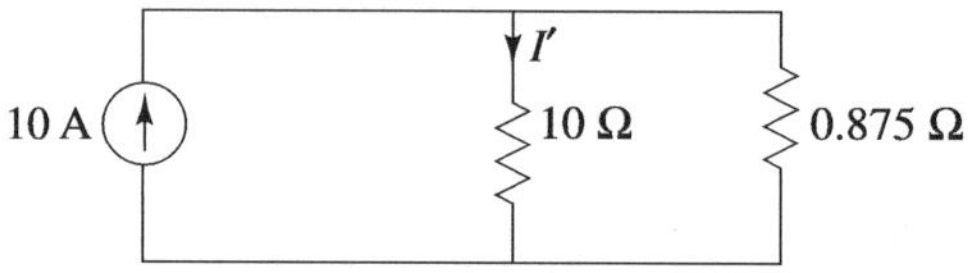

Fig. 3.6

By current-division formula,

$$I' = 10 \times \frac{0.875}{10+0.875} = 0.8 \text{ A } (\downarrow)$$

Step II: *When the 4-A source is acting alone*

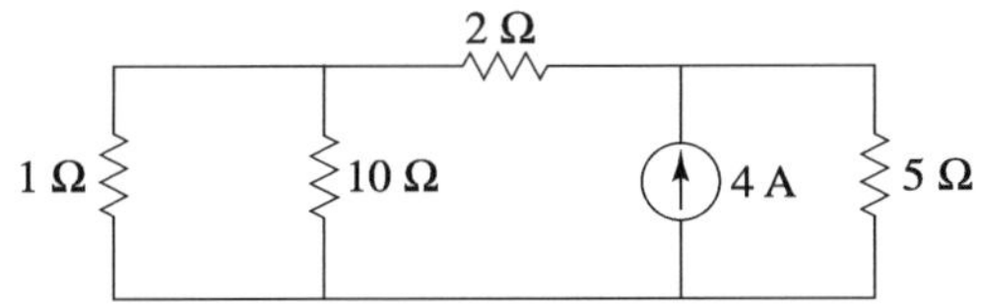

Fig. 3.7

By source transformation,

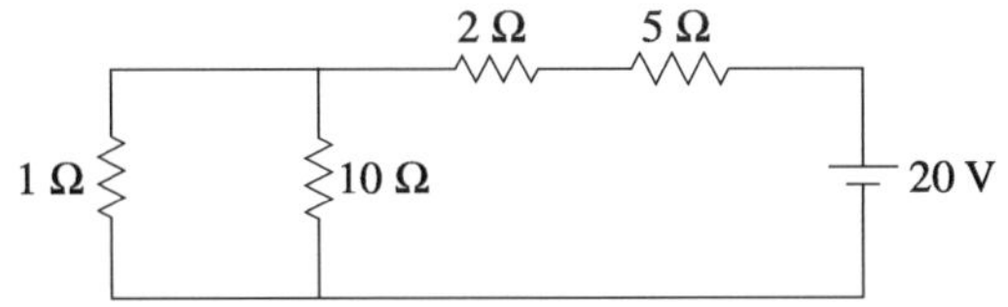

Fig. 3.8

Again by source transformation,

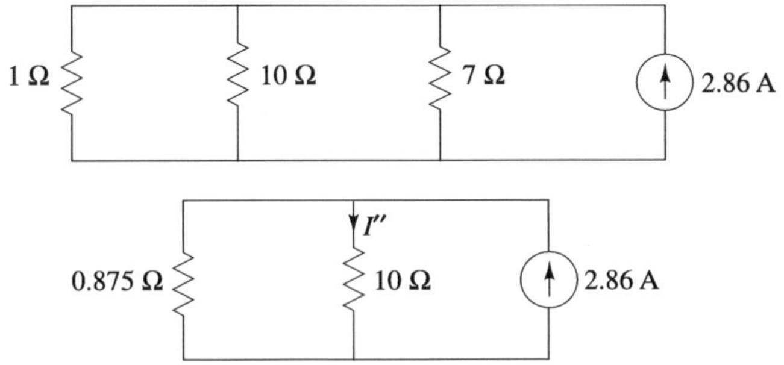

Fig. 3.9

By current-division formula,

$$I'' = 2.86 \times \frac{0.875}{10+0.875} = 0.23 \text{ A } (\downarrow)$$

Step III: *By superposition theorem,*

$$I = I' + I''$$
$$= 0.8 + 0.23 = 1.03 \text{ A } (\downarrow)$$

Example 3.2 *Find the current through the 6-Ω resistor.*

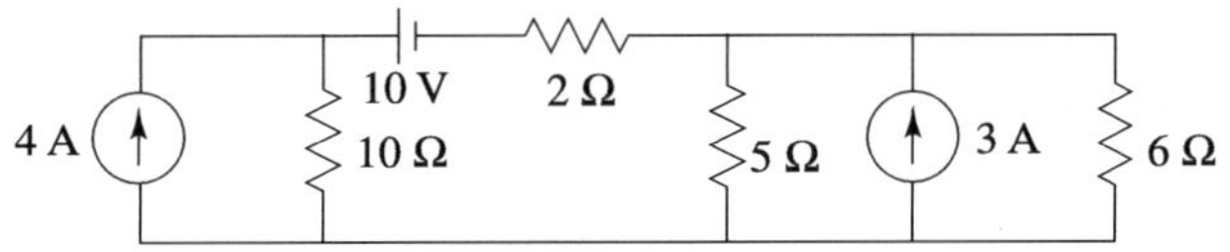

Fig. 3.10

Solution

Step I: *When the 4-A source is acting alone*

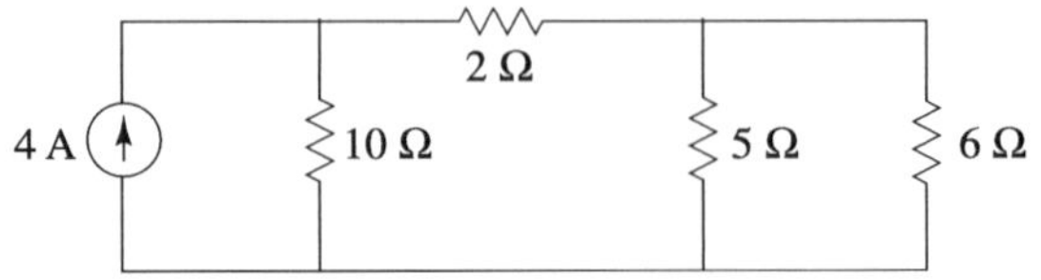

Fig. 3.11

By source transformation,

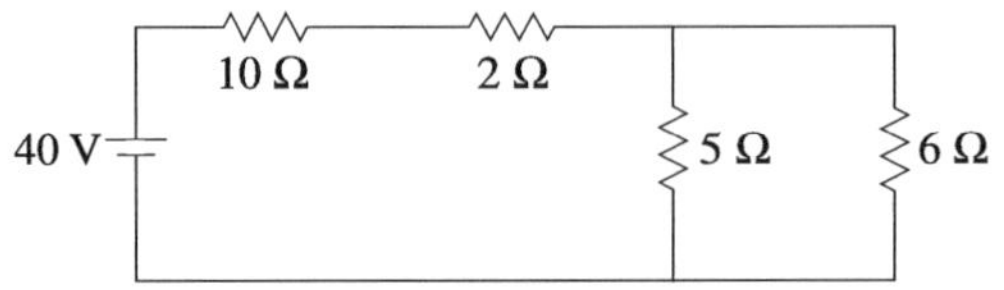

Fig. 3.12

Again by source transformation,

Fig. 3.13

By series–parallel reduction technique,

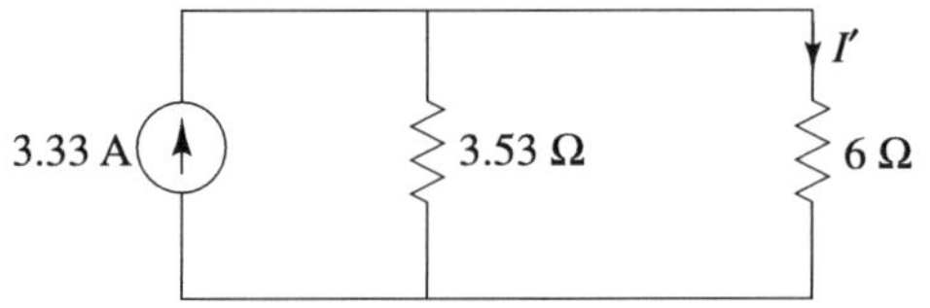

Fig. 3.14

By current-division formula,

$$I' = 3.33 \times \frac{3.53}{6 + 3.53} = 1.23 \text{ A } (\downarrow)$$

Step II: *When the 10-V source is acting alone*

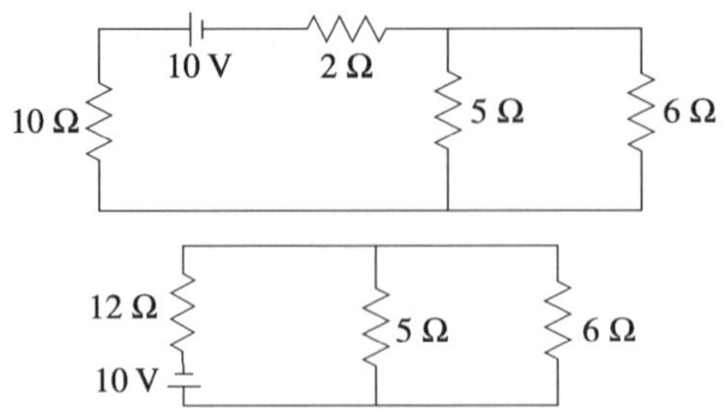

Fig. 3.15

By source transformation,

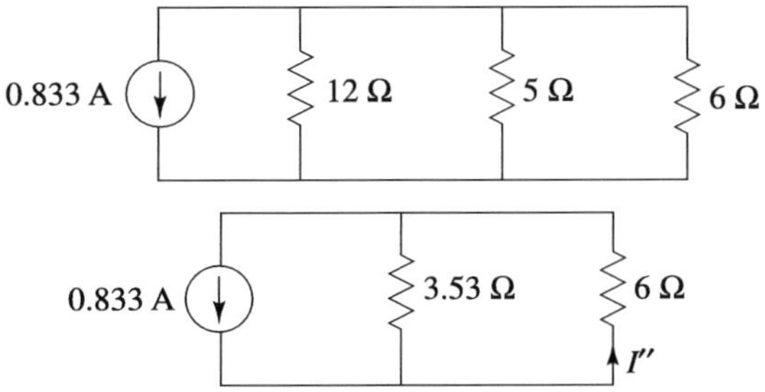

Fig. 3.16

By current-division formula,

$$I'' = 0.833 \times \frac{3.53}{6+3.53}$$
$$= 0.31\text{ A }(\uparrow) = -0.31\text{ A }(\downarrow)$$

Step III: *When the 3-A source is acting alone*

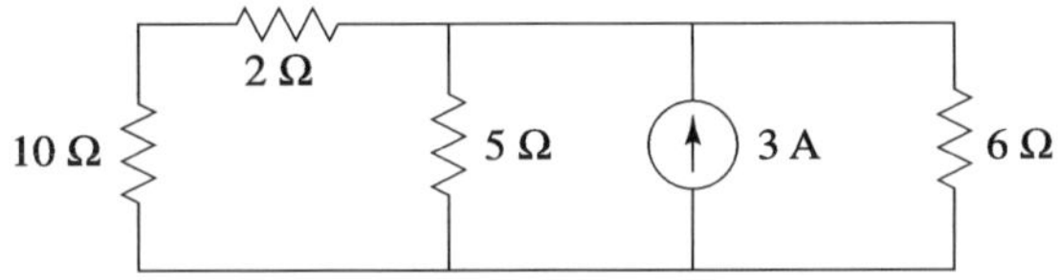

Fig. 3.17

By series–parallel reduction technique to the left of the 3-A source,

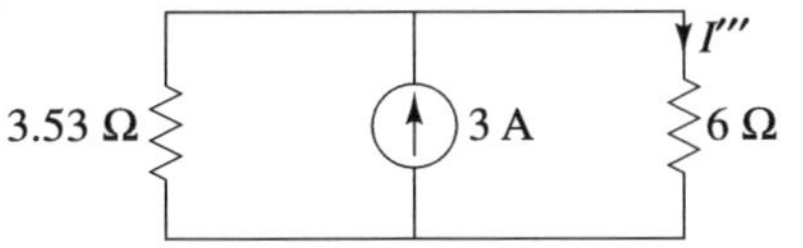

Fig. 3.18

$$I''' = 3 \times \frac{3.53}{6+3.53} = 1.11\text{ A }(\downarrow)$$

Step IV: *By superposition theorem,*

$$I = I' + I'' + I'''$$
$$= 1.23 - 0.31 + 1.11 = 2.03\text{ A }(\downarrow)$$

Example 3.3 *Find the current in the 1-Ω resistor.*

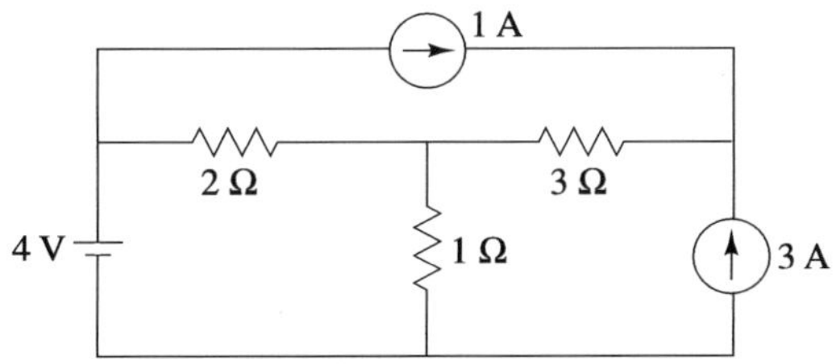

Fig. 3.19

Step I: *When the 4-V source is acting alone*

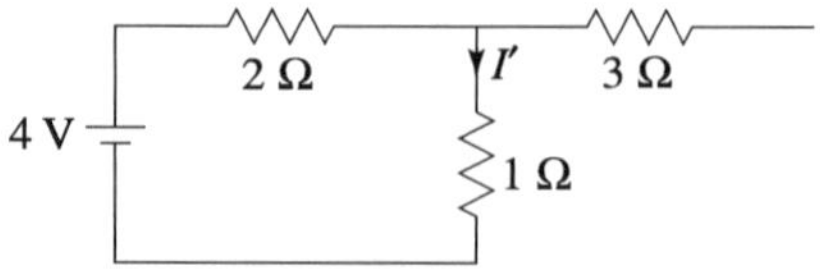

Fig. 3.20

$$I' = \frac{4}{2+1} = 1.33 \text{ A } (\downarrow)$$

Step II: *When the 3-A source is acting alone*

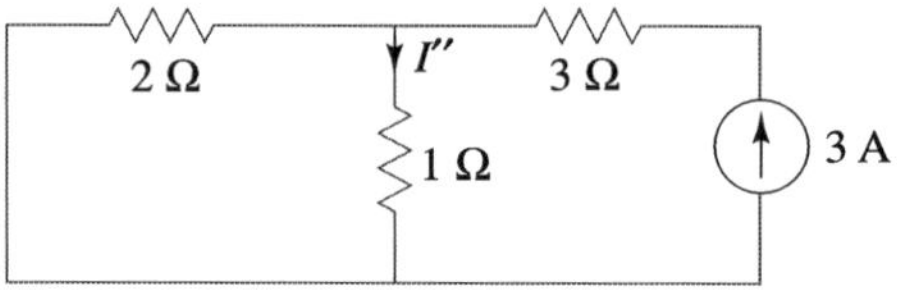

Fig. 3.21

By current-division formula,

$$I'' = 3 \times \frac{2}{1+2} = 2 \text{ A } (\downarrow)$$

Step III: *When the 1-A source is acting alone*

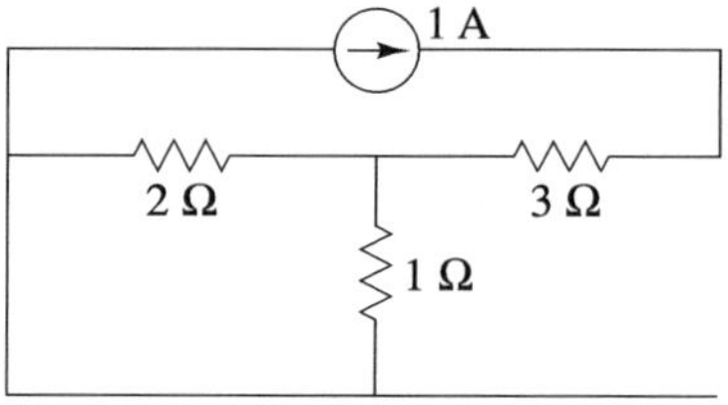

Fig. 3.22

The circuit can be redrawn as shown:

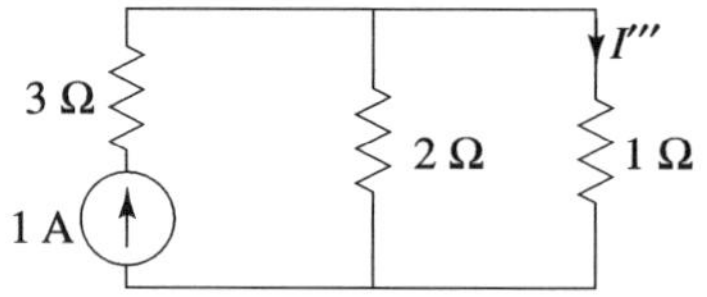

Fig. 3.23

By current-division formula,

$$I''' = 1 \times \frac{2}{2+1} = 0.66 \text{ A } (\downarrow)$$

Step IV: *By superposition theorem,*

$$I = I' + I'' + I'''$$
$$= 1.33 + 2 + 0.66 = 4 \text{ A } (\downarrow)$$

Example 3.4 *Find the voltage V_{AB}.*

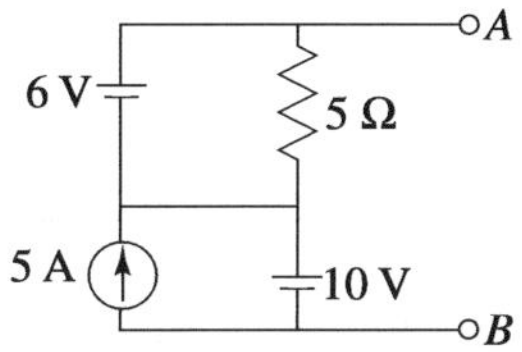

Fig. 3.24

Step I: *When the 6-V source is acting alone*

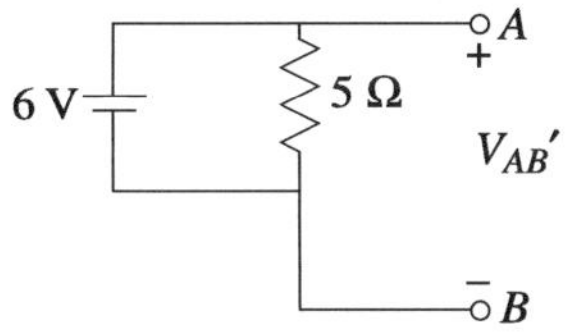

Fig. 3.25

From Fig. 3.25,

$$V_{AB}' = 6 \text{ V}$$

Step II: *When the 10-V source is acting alone*

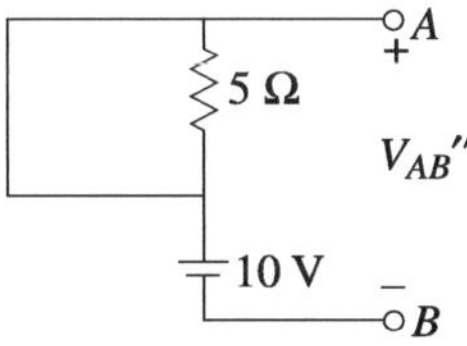

Fig. 3.26

Since the resistor of 5 Ω is shorted, the voltage across it is zero.

$$V_{AB}'' = 10 \text{ V}$$

Step III: *When the 5-A source is acting alone*

Due to short circuit in both the parts,

$$V_{AB}''' = 0 \text{ V}$$

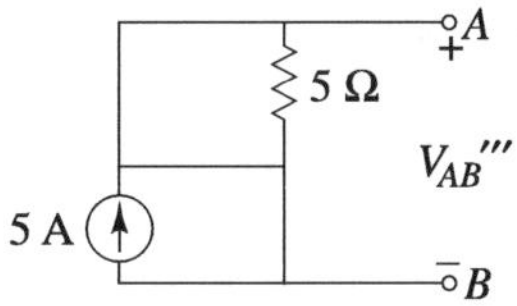

Fig. 3.27

Step IV: *By superposition theorem,*

$$V_{AB} = V_{AB}' + V_{AB}'' + V_{AB}'''$$
$$= 6 + 10 + 0 = 16 \text{ V}$$

Example 3.5 *Find the current through the 4-Ω resistor.*

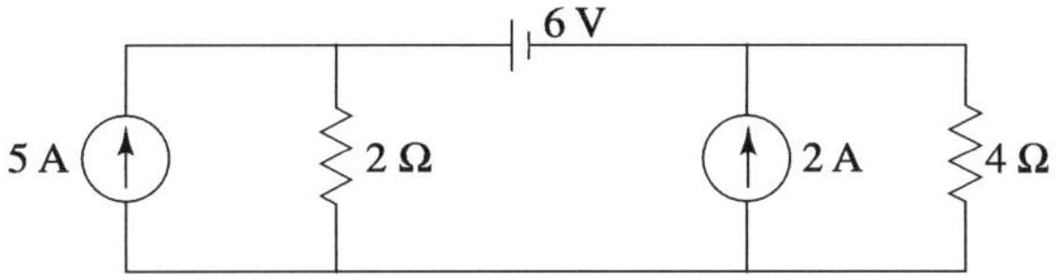

Fig. 3.28

Step I: *When the 5-A source is acting alone*

By current-division formula,

$$I' = 5 \times \frac{2}{2+4} = 1.67 \text{ A } (\downarrow)$$

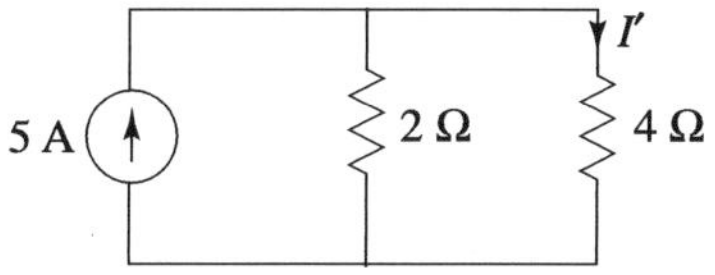

Fig. 3.29

Step II: *When the 2-A source is acting alone*
By current-division formula,

$$I'' = 2 \times \frac{2}{2+4} = 0.67 \text{ A } (\downarrow)$$

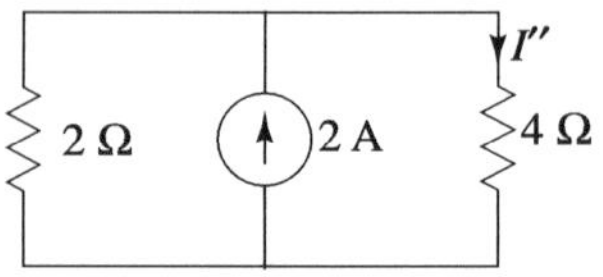

Fig. 3.30

Step III: *When the 6-V source is acting alone*
Applying KVL to the mesh,

$$-2I''' - 6 - 4I''' = 0$$
$$I''' = -1 \text{ A } (\downarrow)$$

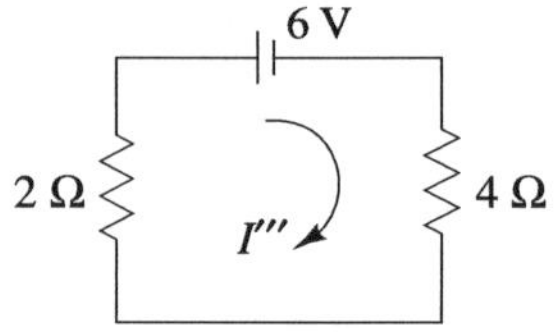

Fig. 3.31

Step IV: *By superposition theorem,*

$$I = I' + I'' + I'''$$
$$= 1.67 + 0.67 - 1 = 1.34 \text{ A } (\downarrow)$$

Example 3.6 *Find the current through the 6-Ω resistor.*

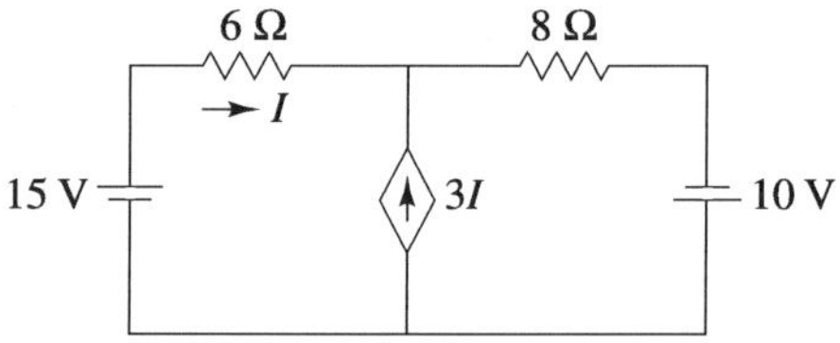

Fig. 3.32

Step I: *When the 15-V source is acting alone*
From the figure,

$$I' = \frac{15 - V_1}{6}$$
$$V_1 = 15 - 6I'$$

Applying KCL at Node 1,

$$I' + 3I' = \frac{V_1}{8} = \frac{15 - 6I'}{8}$$
$$32I' = 15 - 6I'$$
$$38I' = 15$$
$$I' = \frac{15}{38} \text{ A } (\rightarrow)$$

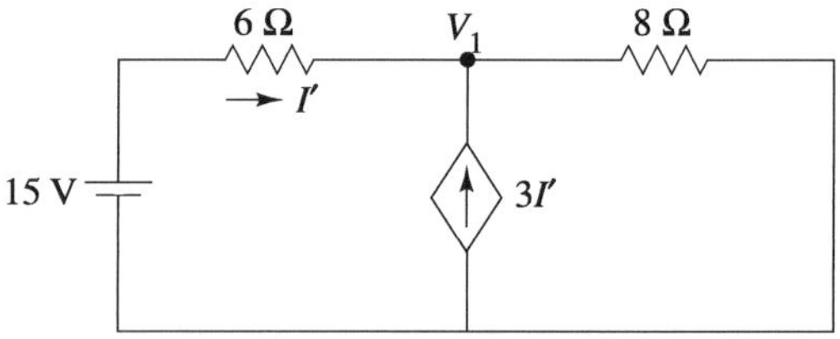

Fig. 3.33

Step II: *When the 10-V source is acting alone*

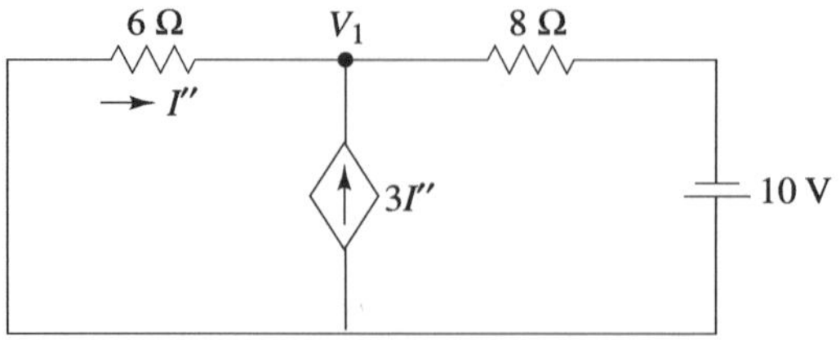

Fig. 3.34

From the figure,

$$I'' = -\frac{V_1}{6}$$

Applying KCL at Node 1,

$$I'' + 3I'' = \frac{V_1 + 10}{8}$$

$$8\,(4I'') = -6I'' + 10$$

$$38I'' = 10$$

$$I'' = \frac{10}{38}\text{ A } (\rightarrow)$$

Step III: *By superposition theorem,*

$$I = I' + I''$$

$$= \frac{15}{38} + \frac{10}{38} = \frac{25}{38}\text{ A } (\rightarrow)$$

Example 3.7 *Find the current I_x.*

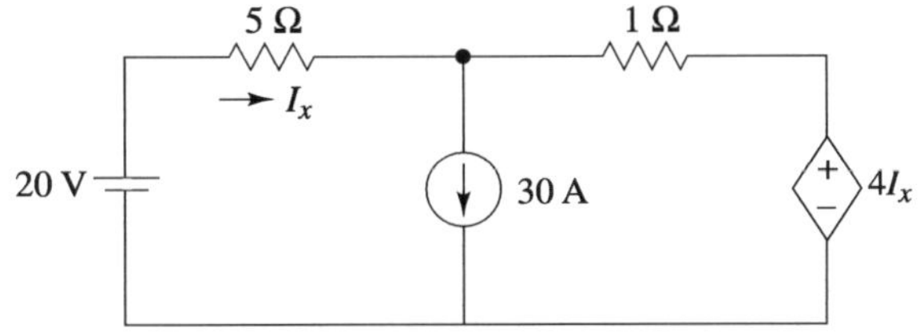

Fig. 3.35

Step I: *When the 30-A source is acting alone*

From the figure,

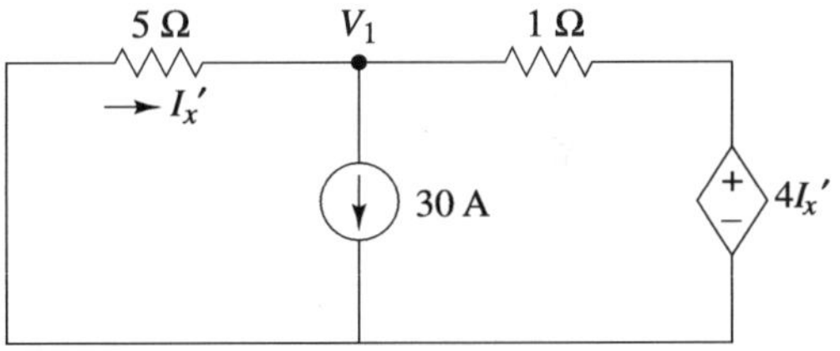

Fig. 3.36

$$I_x' = -\frac{V_1}{5}$$

Applying KCL at Node 1,

$$I_x' = 30 + \frac{V_1 - 4I_x'}{1}$$

$$I_x' = 30 + V_1 - 4I_x'$$

$$I_x' + 4I_x' = 30 - 5I_x'$$

$$10I_x' = 30$$

$$I_x' = 3\text{ A } (\rightarrow)$$

Step II: *When the 20-V source is acting alone*

Applying KVL to the mesh,

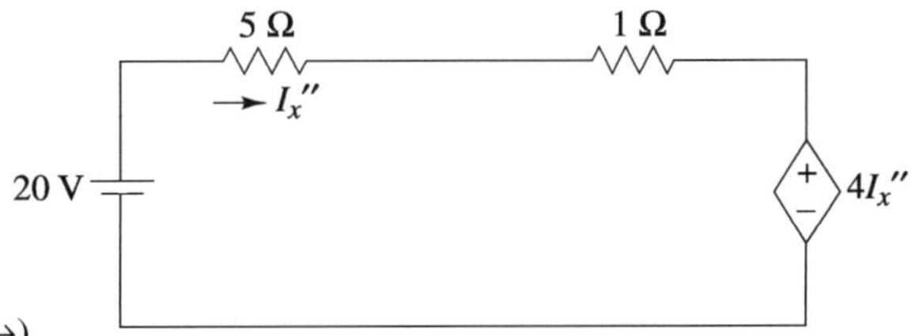

Fig. 3.37

$$20 - 5I_x'' - 1I_x'' - 4I_x'' = 0$$

$$20 = 10I_x''$$

$$I_x'' = 2\text{ A } (\rightarrow)$$

Step III: *By superposition theorem,*

$$I_x = I_x' + I_x''$$

$$= 3 + 2 = 5\text{ A } (\rightarrow)$$

Example 3.8 *Find the current I_1.*

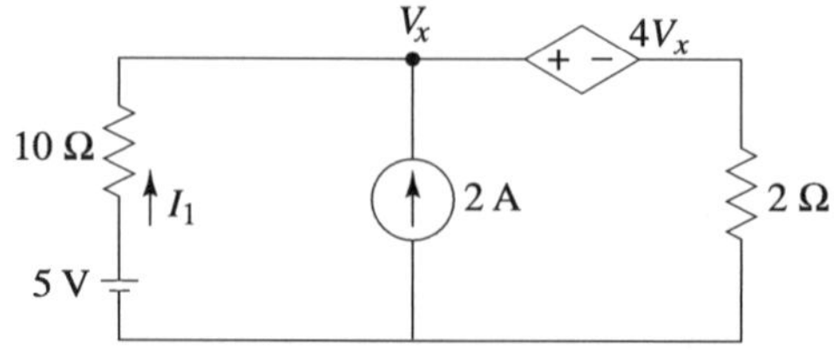

Fig. 3.38

Step I: *When the 5-V source is acting alone*

From the figure,

$$V_x = 5 - 10I_1'$$

Applying KVL to the mesh,

$$5 - 10I_1' - 4V_x - 2I_1' = 0$$
$$5 - 10I_1' - 4(5 - 10I_1') - 2I_1' = 0$$
$$5 - 10I_1' - 20 + 40I_1' - 2I_1' = 0$$
$$28I_1' = 15$$
$$I_1' = \frac{15}{28} = 0.535 \text{ A } (\uparrow)$$

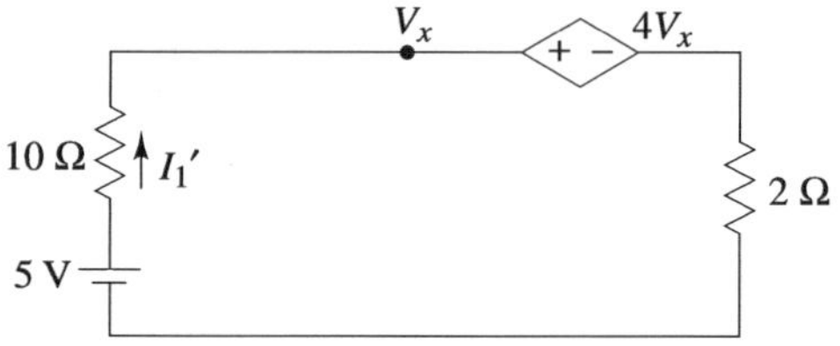

Fig. 3.39

Step II: *When the 2-A source is acting alone*

From the figure,

$$I_1'' = -\frac{V_x}{10}$$

Applying KCL at Node x,

$$-\frac{V_x}{10} + 2 = \frac{V_x - 4V_x}{2}$$
$$-\frac{V_x}{10} + \frac{3V_x}{2} = -2$$
$$14V_x = -20$$
$$V_x = -\frac{20}{14} = -\frac{10}{7} \text{ V}$$
$$I_1'' = -\frac{V_x}{10} = \frac{1}{7} = 0.1428 \text{ A } (\uparrow)$$

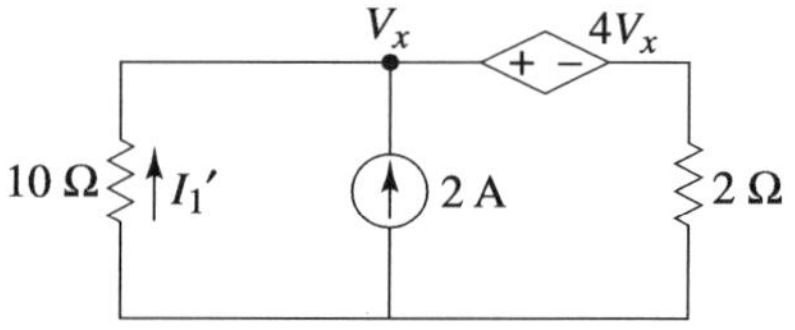

Fig. 3.40

Step III: *By superposition theorem,*

$$I_1 = I_1' + I_1''$$
$$= 0.535 + 0.1428 = 0.678 \text{ A } (\uparrow)$$

Example 3.9 *Determine the current through the 10-Ω resistor.*

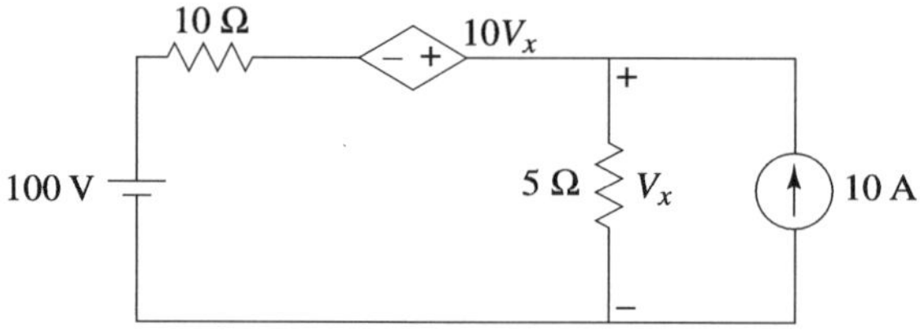

Fig. 3.41

Step I: *When the 100-V source is acting alone*

From the figure,

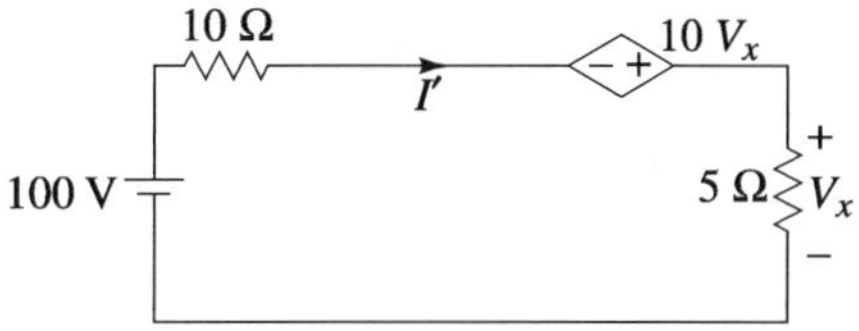

Fig. 3.42

$$V_x = 5I'$$

Applying KVL to the mesh,

$$100 - 10I' + 10V_x - 5I' = 0$$

$$100 - 10I' + 10(5I') - 5I' = 0$$

$$I' = -\frac{100}{35} = -\frac{20}{7}\text{ A } (\rightarrow) = \frac{20}{7}\text{ A } (\leftarrow)$$

Step II: *When the 10-A source is acting alone*

From the figure,

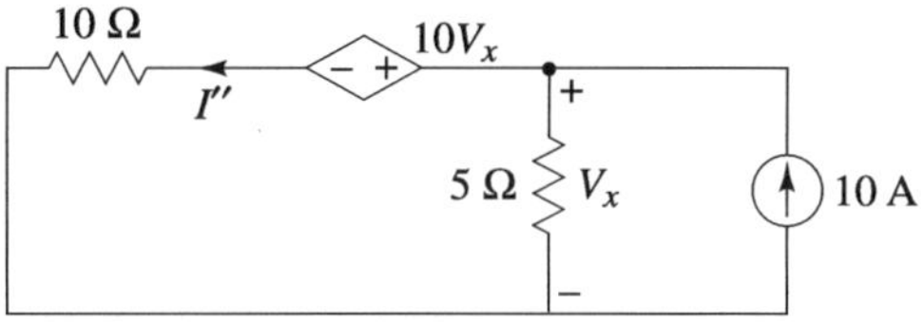

Fig. 3.43

$$I'' = \frac{V_x - 10V_x}{10}$$

Applying KCL at Node 1,

$$10 = \frac{V_x}{5} + I''$$

$$10 = \frac{V_x}{5} + \frac{V_x - 10V_x}{10} = \frac{2V_x}{10} - \frac{9V_x}{10}$$

$$10 = -\frac{7V_x}{10}$$

$$100 = -7V_x$$

$$V_x = -\frac{100}{7}\ V$$

$$I'' = -\frac{9V_x}{10} = -\frac{9}{10} \times \left(-\frac{100}{7}\right)$$

$$= \frac{90}{7}\text{ A } (\leftarrow)$$

Step III: *By superposition theorem*

$$I = I' + I''$$

$$= \frac{20}{7} + \frac{90}{7}$$

$$= \frac{110}{7}\text{ A } (\leftarrow)$$

Example 3.10 *Find the current I in the circuit.*

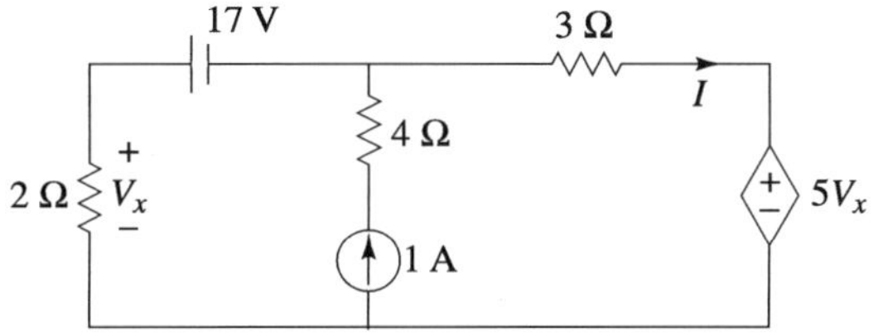

Fig. 3.44

Step I: *When the 17-V source is acting alone*

From the figure,

$$V_x = -2I'$$

Applying KVL to the mesh,

$$-2I' - 17 - 3I' - 5V_x = 0$$
$$-2I' - 17 - 3I' - 5(-2I') = 0$$
$$5I' = 17$$
$$I' = \frac{17}{5}\text{ A } (\rightarrow)$$

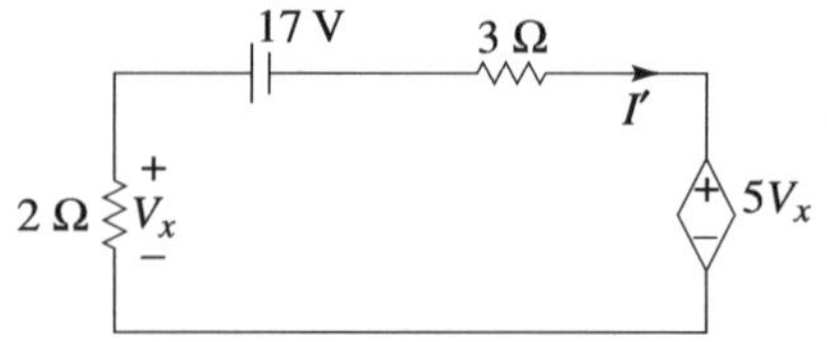

Fig. 3.45

Step II: *When the 1-A source is acting alone*

Applying KCL at Node *x*,

$$\frac{V_x}{2} + \frac{V_x - 5V_x}{3} = 1$$
$$\frac{V_x}{2} - \frac{4V_x}{3} = 1$$
$$-\frac{5V_x}{6} = 1$$
$$V_x = -\frac{6}{5}$$
$$I'' = -\frac{4V_x}{3}$$
$$= -\frac{4}{3}\times\left(-\frac{6}{5}\right) = \frac{8}{5}\text{ A } (\rightarrow)$$

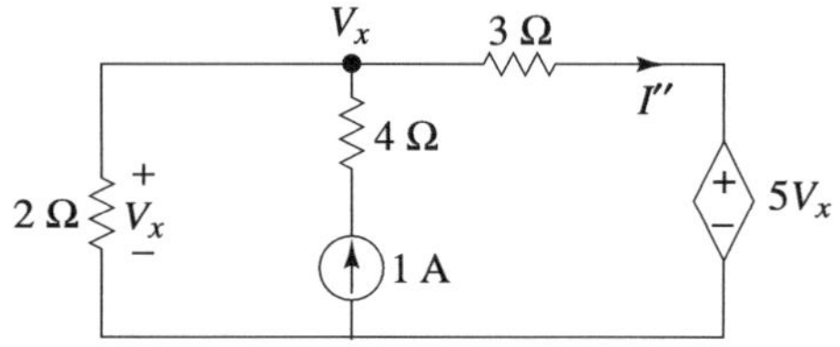

Fig. 3.46

Step III: *By superposition theorem*

$$I = I' + I''$$
$$= \frac{17}{5} + \frac{8}{5} = \frac{25}{5} = 5\text{ A } (\rightarrow)$$

Example 3.11 *Find the voltage* V_1.

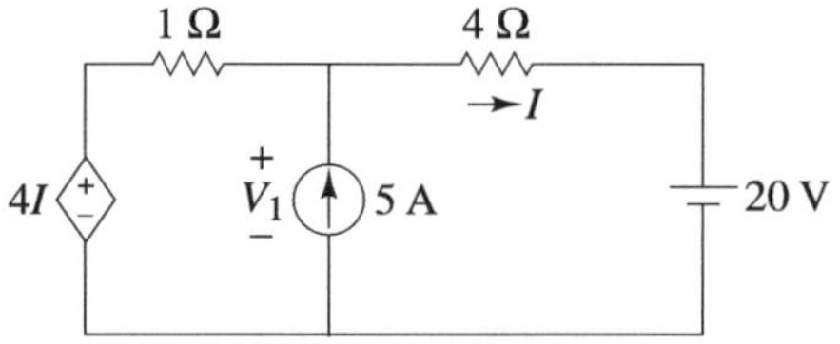

Fig. 3.47

Step I: *When the 5-A source is acting alone*

From the figure,

$$I = \frac{V_1'}{4}$$

Applying KCL at Node 1,

$$\frac{V_1' - 4I}{1} + \frac{V_1'}{4} = 5$$

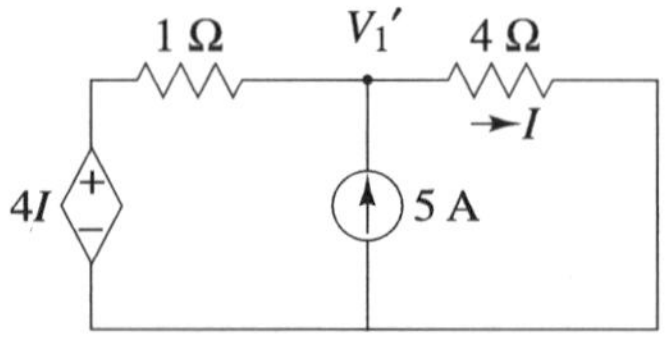

Fig. 3.48

$$V_1' - 4 \times \frac{V_1'}{4} + \frac{V_1'}{4} = 5$$

$$\frac{V_1'}{4} = 5$$

$$V_1' = 20 \text{ V}$$

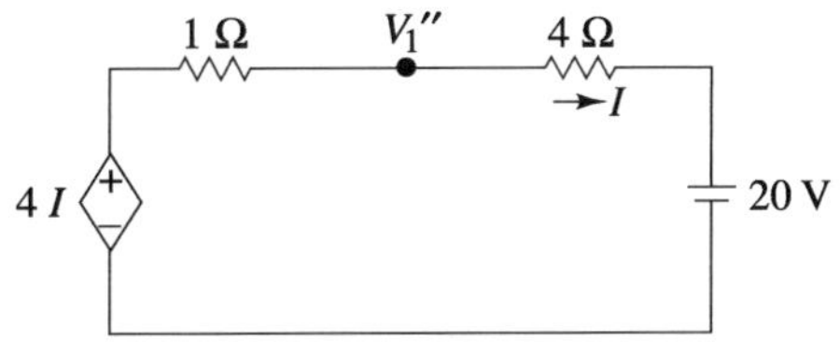

Fig. 3.49

Step II: *When the 20-V source is acting alone*

Applying KVL to the mesh,

$$4I - I - 4I - 20 = 0$$

$$I = -20 \text{ A}$$

$$V_1'' = 4I - 1(I) = 3I$$

$$= 3(-20) = -60 \text{ V}$$

Step III: *By superposition theorem,*

$$V_1 = V_1' + V_1''$$

$$= 20 - 60 = -40 \text{ V}$$

Example 3.12 *Find the current in the 6-Ω resistor.*

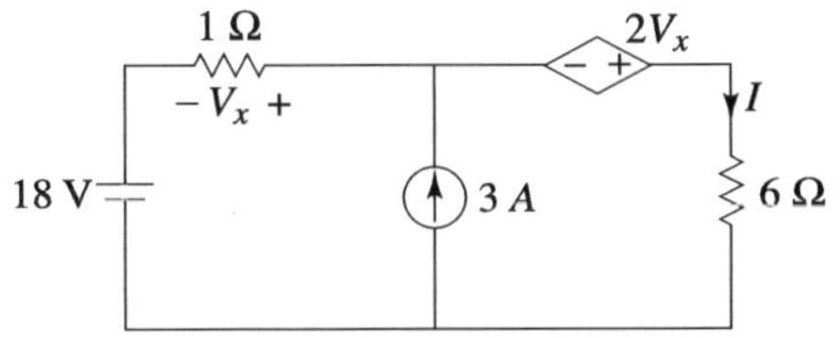

Fig. 3.50

Step I: *When the 18-V source is acting alone*

From Fig. 3.51,

$$V_x = -I'$$

Applying KVL to the mesh,

$$18 + V_x + 2V_x - 6I' = 0$$

$$18 - I' - 2I' - 6I' = 0$$

$$I' = 2 \text{ A } (\downarrow)$$

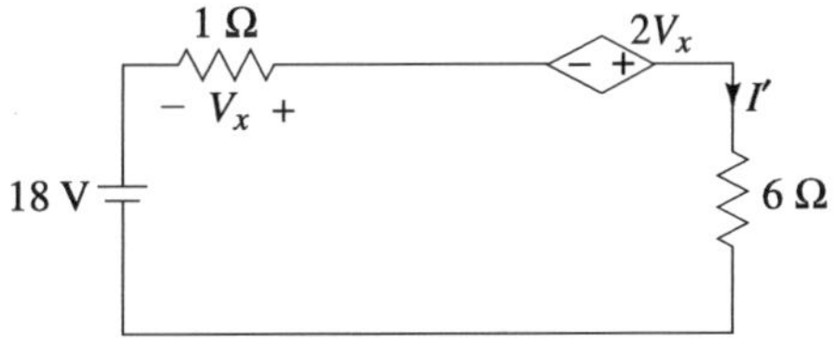

Fig. 3.51

Step II: *When the 3-A source is acting alone*

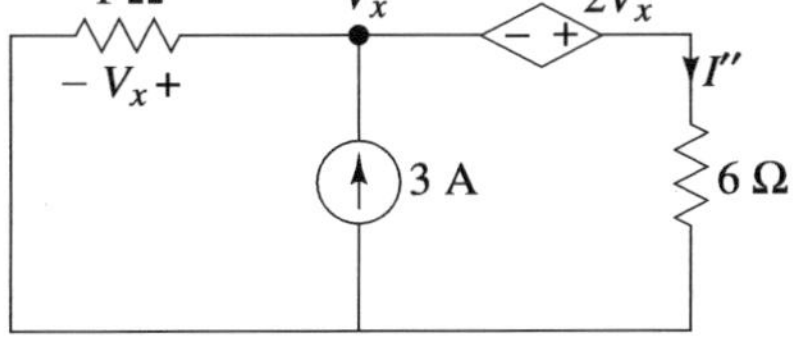

Fig. 3.52

Applying KCL at Node 1,

$$3 = \frac{V_x}{1} + \frac{V_x + 2V_x}{6}$$

$$V_x = 2 \text{ V}$$

$$I'' = \frac{V_x + 2V_x}{6}$$

$$= \frac{2+4}{6} = 1\text{ A }(\downarrow)$$

Step III: *By superposition theorem,*

$$I_{6\,\Omega} = I' + I''$$
$$= 2 + 1 = 3\text{ A }(\downarrow)$$

Example 3.13 *Find the current I_y.*

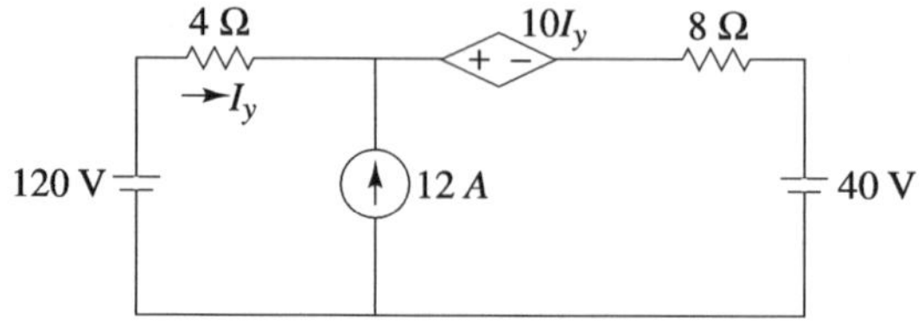

Fig. 3.53

Step I: *When the 120-V source is acting alone*

Applying KVL to the mesh,

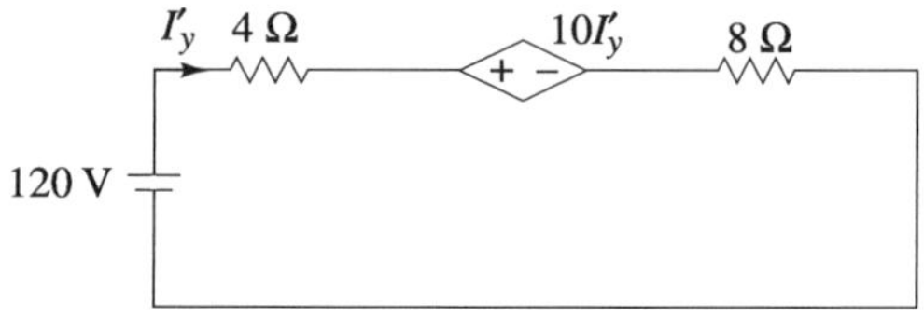

Fig. 3.54

$$120 - 4I_y' - 10I_y' - 8I_y' = 0$$
$$120 = 22I_y'$$
$$I_y' = 5.45\text{ A }(\rightarrow)$$

Step II: *When the 12-A source is acting alone*

From the figure,

$$I_y'' = -\frac{V_1}{4}$$

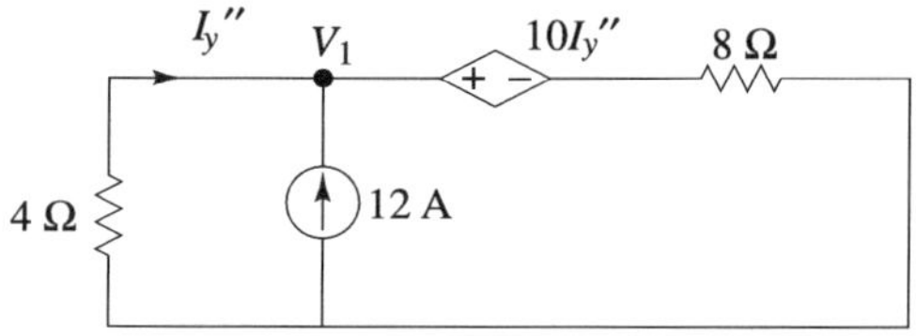

Fig. 3.55

Applying KCL at Node 1,

$$I_y'' + 12 = \frac{V_1 - 10I_y''}{8}$$
$$I_y'' + 12 = \frac{-4I_y'' - 10I_y''}{8}$$
$$I_y'' = -\frac{12}{2.75} = -4.36\text{ A }(\rightarrow)$$

Step III: *When the 40-V source is acting alone*

Applying KVL to the mesh,

$$-4\,I_y''' - 10I_y''' - 8I_y''' - 40 = 0$$
$$-22I_y''' = 40$$
$$I_y''' = -\frac{40}{22}$$
$$= -1.82\text{ A }(\rightarrow)$$

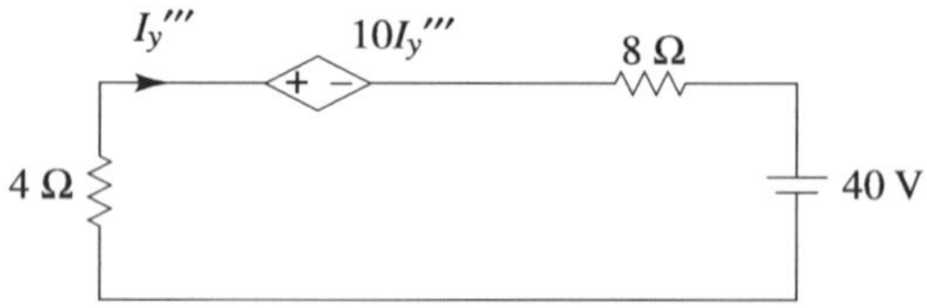

Fig. 3.56

Step IV: *By the superposition theorem,*

$$I_y = I_y' + I_y'' + I_y'''$$
$$= 5.45 - 4.36 - 1.82$$
$$= -0.73\text{ A }(\rightarrow)$$

Example 3.14 *Find the voltage V_x.*

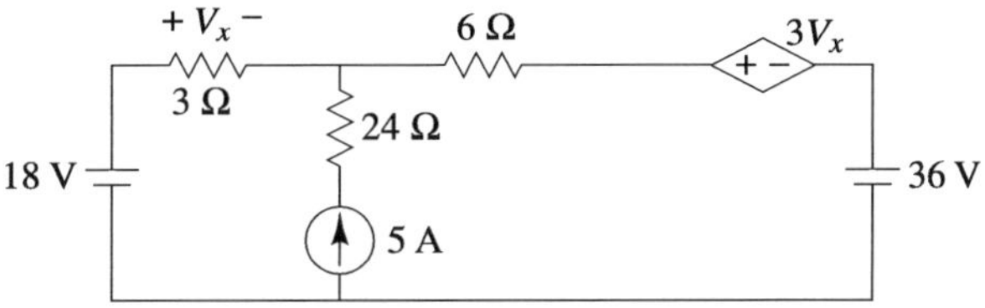

Fig. 3.57

Step I: *When the 18-V source is acting alone*

From the figure,

$$V_x' = 3I$$

Applying KVL to the mesh,

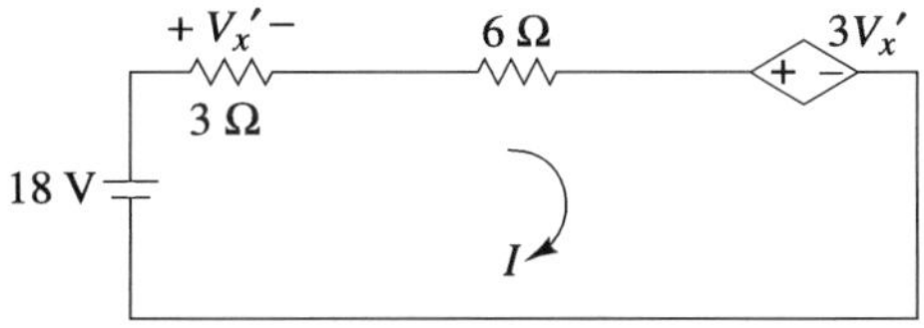

Fig. 3.58

$$18 - 3I - 6I - 3V_x' = 0$$
$$18 - 3I - 6I - 3\,(3I) = 0$$
$$18 = 18I$$
$$I = 1 \text{ A}$$
$$V_x' = 3 \text{ V}$$

Step II: *When the 5-A source is acting alone*

From the figure,

$$V_1 = -V_x''$$

Applying KCL at Node 1,

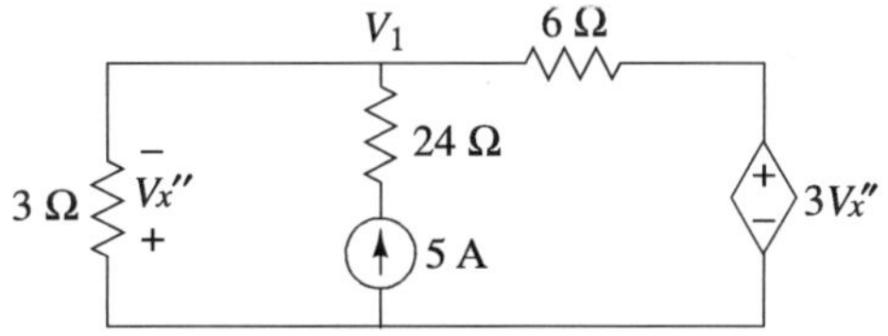

Fig. 3.59

$$\frac{V_1}{3} + \frac{V_1 - 3V_x''}{6} = 5$$
$$\frac{V_1}{3} + \frac{V_1 + 3V_1}{6} = 5$$
$$\frac{V_1}{3} + \frac{4V_1}{6} = 5$$
$$V_1 = 5 \text{ V}$$
$$V_x'' = -5 \text{ V}$$

Step III: *When the 36-V source is acting alone,*

From the figure,

$$V_x''' = -3I$$

Applying KVL to the mesh,

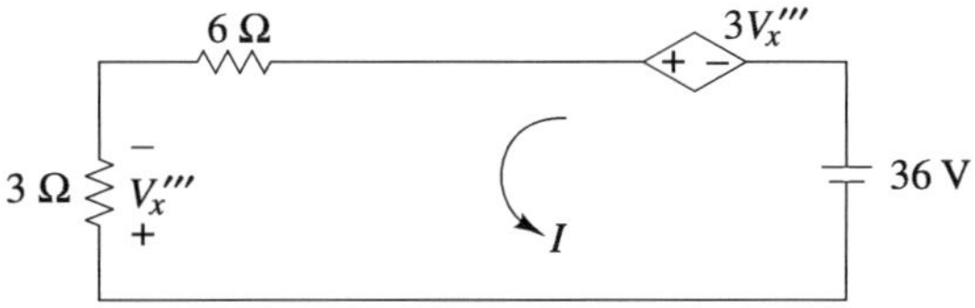

Fig. 3.60

$$36 + 3V_x''' - 6I - 3I = 0$$
$$36 + 3V_x''' - 6\left(\frac{-V_x'''}{3}\right) - 3\left(\frac{-V_x'''}{3}\right) = 0$$
$$36 + 3V_x''' + 2V_x''' + V_x''' = 0$$
$$36 = -6V_x'''$$
$$V_x''' = -6 \text{ V}$$

Step IV: *By superposition theorem,*

$$V_x = V_x' + V_x'' + V_x'''$$
$$= 3 - 5 - 6 = -8 \text{ V}$$

3.3 THEVENIN'S THEOREM

It states that *'Any two terminals of a network can be replaced by an equivalent voltage source and an equivalent series resistance. The voltage source is the voltage across the two terminals with load, if any, removed. The series resistance is the resistance of the network measured between two terminals with load removed and constant voltage source being replaced by its internal resistance (or if it is not given with zero resistance, i.e., short circuit) and constant current source replaced by infinite resistance, i.e., open circuit.'*

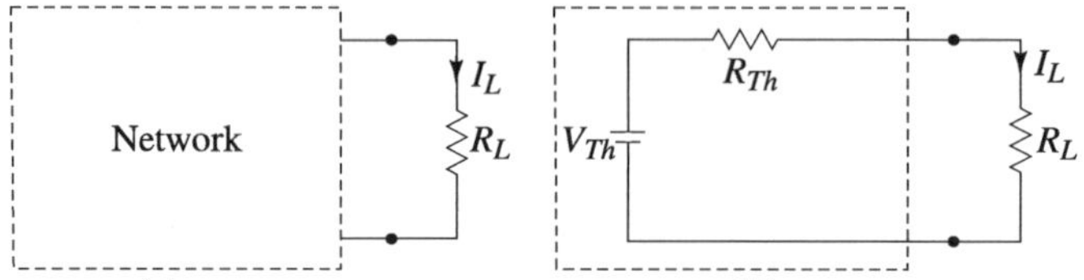

Fig. 3.61

Explanation The above method of determining the load current through a given load resistance can be explained with the help of following circuit.

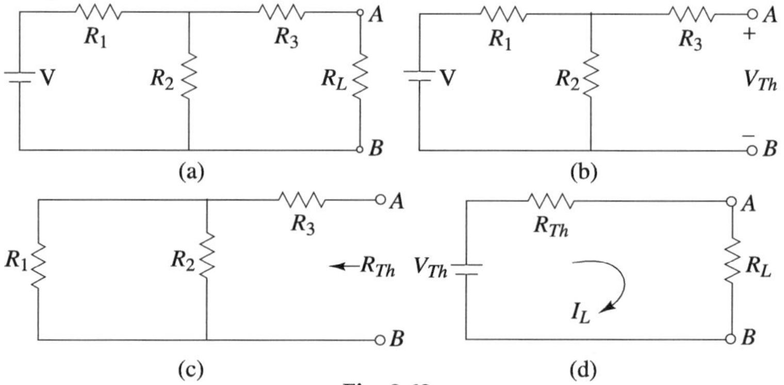

Fig. 3.62

Steps to be followed in Thevenin's theorem

1. Remove the load resistance R_L.
2. Find the open circuit voltage V_{Th} across points A and B.
3. Find the resistance R_{Th} as seen from points A and B with the voltage source V replaced by a short circuit.
4. Replace the network by a voltage source V_{Th} in series with resistance R_{Th}.
5. Find the current through R_L using Ohm's law.

$$I_L = \frac{V_{Th}}{R_{Th} + R_L}$$

Example 3.15 *Find the current through the 10-Ω resistor.*

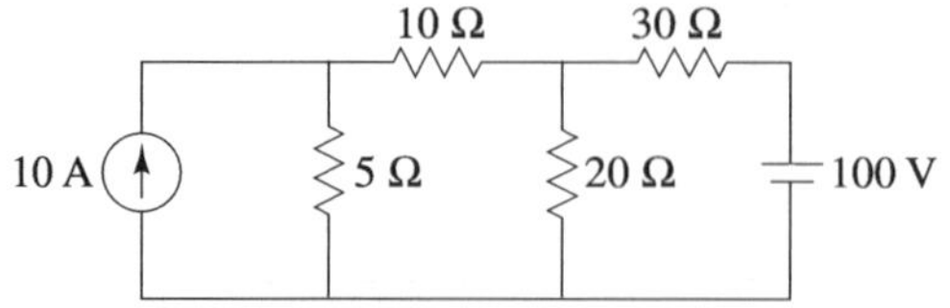

Fig. 3.63

Step I: *Calculation of* V_{Th}

Removing the 10-Ω resistor from the circuit,

For Mesh 1,

$$I_1 = 10 \text{ A}$$

Applying KVL to Mesh 2,

$$100 - 30I_2 - 20I_2 = 0$$

$$I_2 = 2 \text{ A}$$

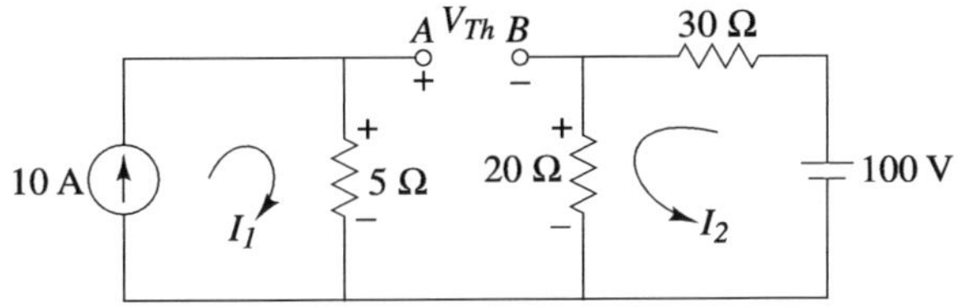

Fig. 3.64

Writing V_{Th} equation,

$$5I_1 - V_{Th} - 20I_2 = 0$$

$$V_{Th} = 5I_1 - 20I_2$$

$$= 5(10) - 20(2) = 10 \text{ V}$$

Step II: *Calculation of* R_{Th}

Replacing the current source of 10 A with an open circuit and the voltage source of 100 V with a short circuit,

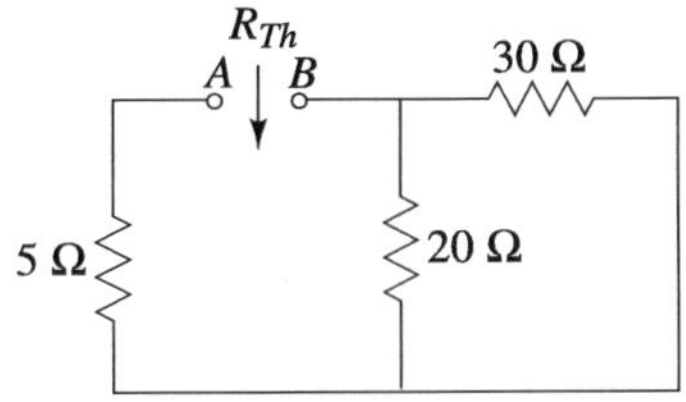

Fig. 3.65

$$R_{Th} = 5 + (20 \,||\, 30) = 17 \,\Omega$$

Step III: *Calculation of* I_L

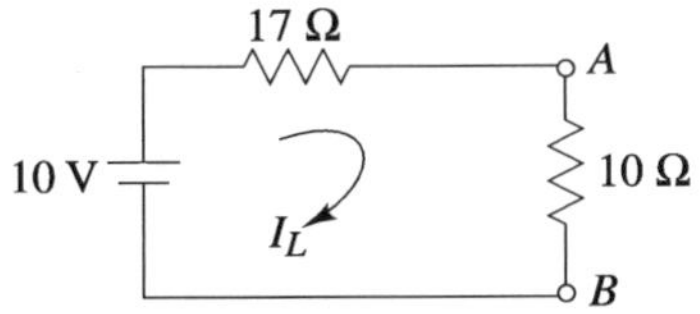

Fig. 3.66

$$I_L = \frac{10}{17+10} = 0.37 \text{ A}$$

Example 3.16 Determine the current through the 24-Ω resistor.

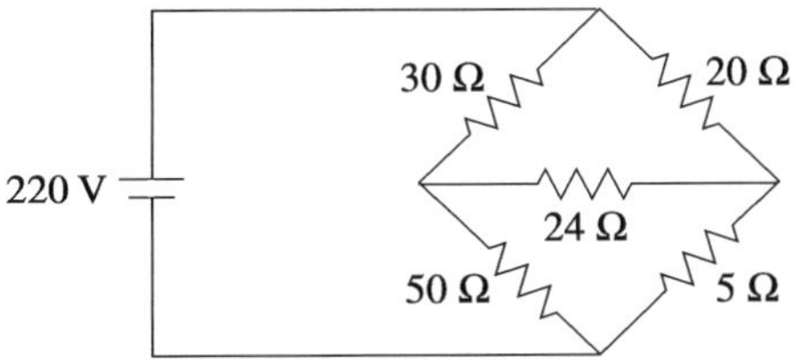

Fig. 3.67

Step I: *Calculation of* V_{Th}

Removing the 24-Ω resistor from the network,

$$I_1 = \frac{220}{30+50} = 2.75 \text{ A}$$

$$I_2 = \frac{220}{20+5} = 8.8 \text{ A}$$

Writing V_{Th} equation,

$$V_{Th} + 30I_1 - 20I_2 = 0$$

$$V_{Th} = 20I_2 - 30I_1$$

$$= 20\,(8.8) - 30(2.75) = 93.5 \text{ V}$$

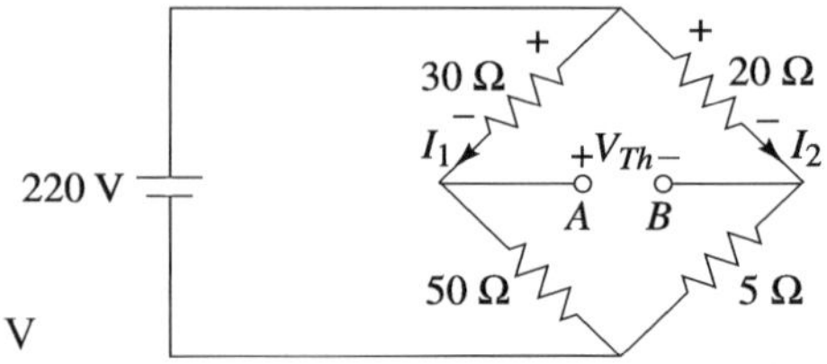

Fig. 3.68

Step II: *Calculation of R_{Th}*

Replacing the 220-V source with short circuit,

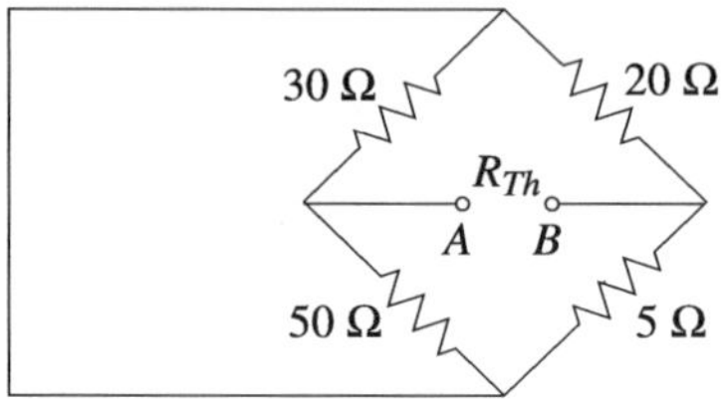

Fig. 3.69

The circuit can be redrawn as shown:

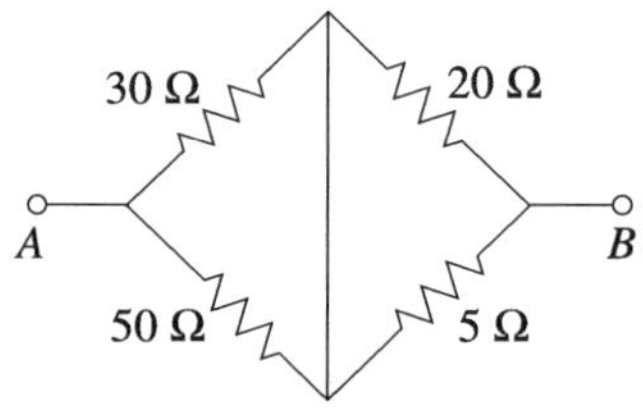

Fig. 3.70

$$R_{Th} = (30 \,||\, 50) + (20 \,||\, 5) = 22.75\ \Omega$$

Step III: *Calculation of I_L*

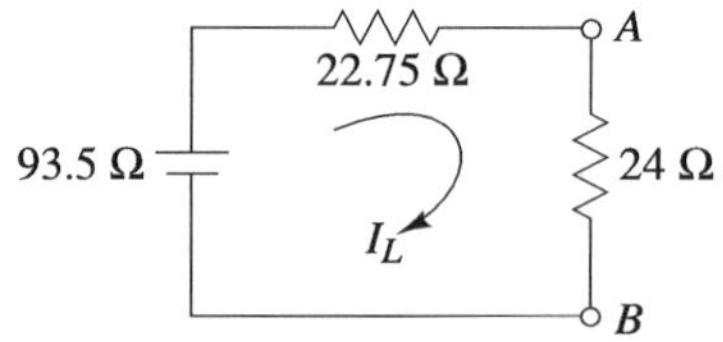

Fig. 3.71

$$I_L = \frac{93.5}{22.75+24} = 2 \text{ A}$$

Example 3.17 *Find the current through the 3-Ω resistor.*

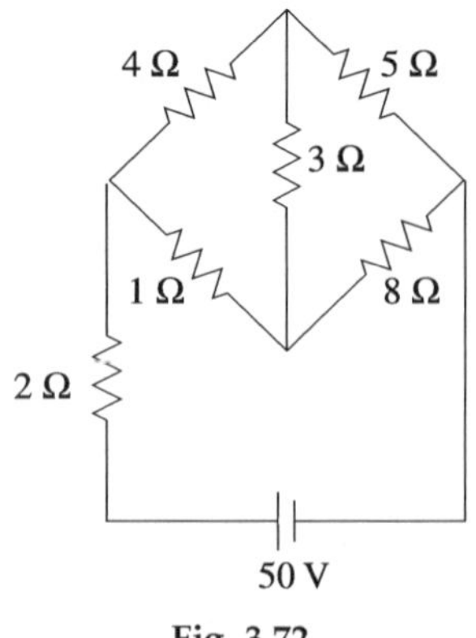

Fig. 3.72

Step I: *Calculation of* V_{Th}

Removing the 3-Ω resistor from the network,

Applying KVL to Mesh 1,

$$50 - 2I_1 - 1(I_1 - I_2) - 8(I_1 - I_2) = 0$$

$$11I_1 - 9I_2 = 50 \qquad ...(1)$$

Applying KVL to Mesh 2,

$$-4I_2 - 5I_2 - 8(I_2 - I_1) - (I_2 - I_1) = 0$$

$$-9I_1 + 18I_2 = 0 \qquad ...(2)$$

Solving Eqs (1) and (2),

$$I_1 = 7.69 \text{ A}$$

$$I_2 = 3.85 \text{ A}$$

Writing V_{Th} equation,

$$V_{Th} - 5I_2 - 8(I_2 - I_1) = 0$$

$$V_{Th} = 5I_2 + 8(I_2 - I_1)$$

$$= 5(3.85) + 8(3.85 - 7.69) = -11.47 \text{ V}$$

$$= 11.47 \text{ V (the terminal } B \text{ is positive w.r.t. } A)$$

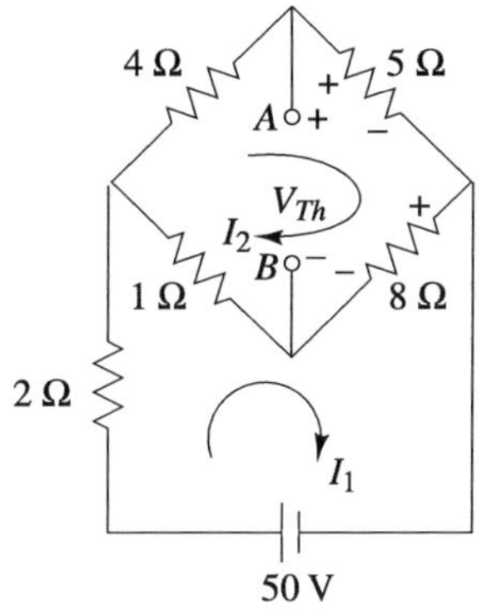

Fig. 3.73

Step II: *Calculation of* R_{Th}

Replacing the voltage source of 50 V with a short circuit,

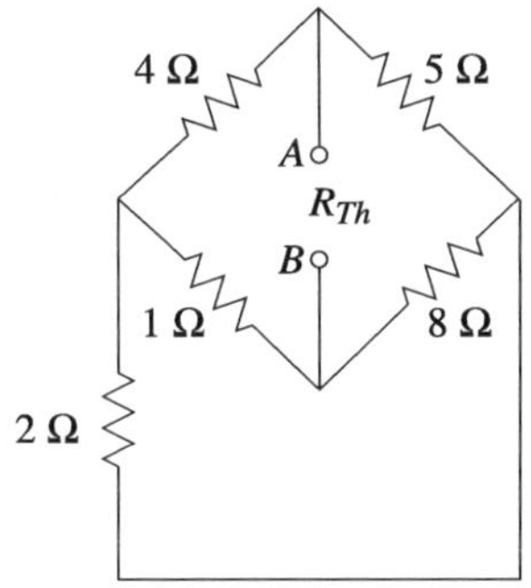

Fig. 3.74

The circuit can be redrawn as follows:

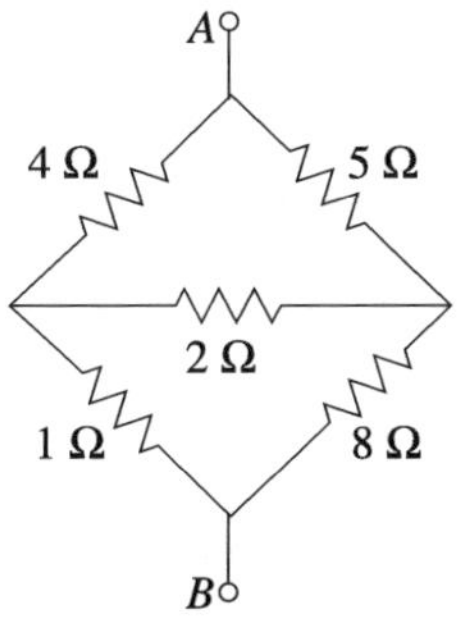

Fig. 3.75

Converting the upper delta into equivalent star network,

$$R_1 = \frac{4 \times 2}{4+2+5} = 0.73\ \Omega$$

$$R_2 = \frac{4 \times 5}{4+2+5} = 1.82\ \Omega$$

$$R_3 = \frac{5 \times 2}{4+2+5} = 0.91\ \Omega$$

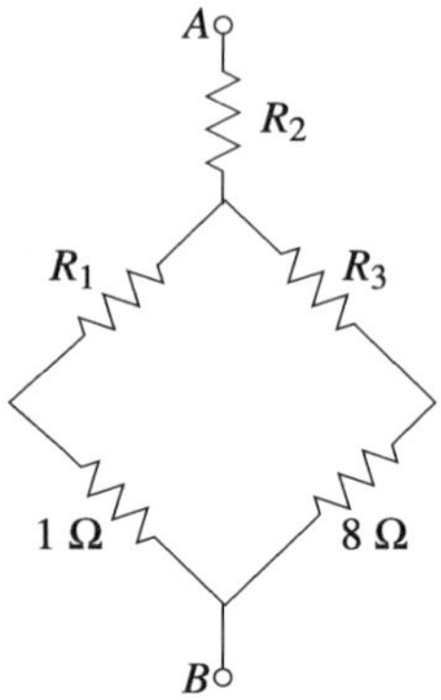

Fig. 3.76

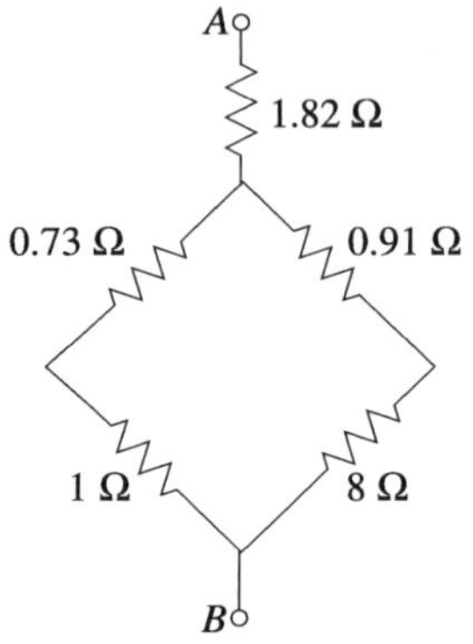

Fig. 3.77

The simplified network is drawn as follows:

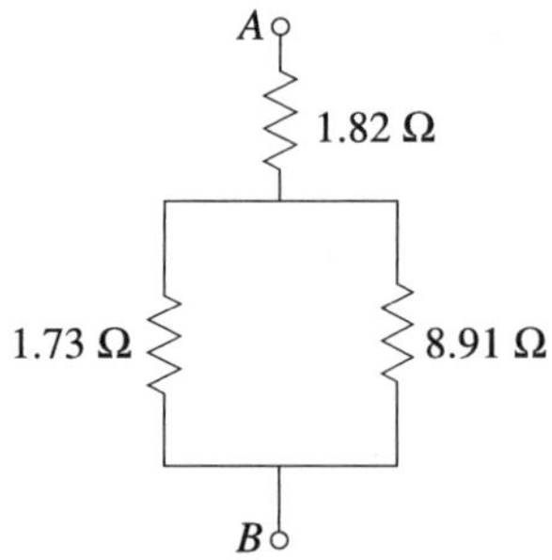

Fig. 3.78

$$R_{Th} = 1.82 + (1.73 \parallel 8.91) = 3.27\ \Omega$$

Step III: *Calculation of* I_L

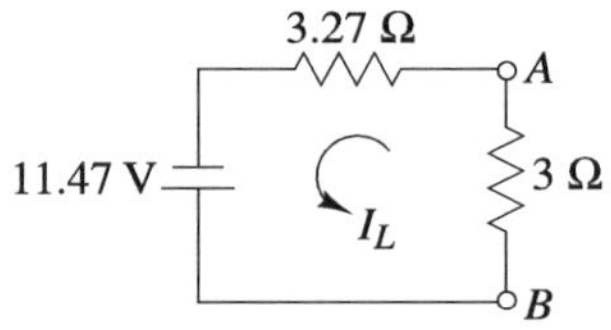

Fig. 3.79

$$I_L = \frac{11.47}{3.27+3} = 1.83\ \text{A}\ (\uparrow)$$

Example 3.18 *Find the current through the 20-Ω resistor.*

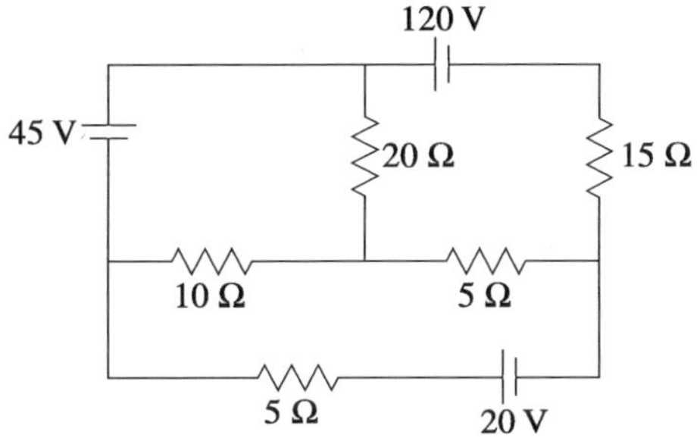

Fig. 3.80

Step I: *Calculation of* V_{Th}

Removing the 20-Ω resistor from the network,

Applying KVL to Mesh 1,

$$45 - 120 - 15I_1 - 5(I_1 - I_2) - 10(I_1 - I_2) = 0$$

$$30I_1 - 15I_2 = -75 \quad ...(1)$$

Applying KVL to Mesh 2,

$$20 - 5I_2 - 10(I_2 - I_1) - 5(I_2 - I_1) = 0$$

$$-15I_1 + 20I_2 = 20 \quad ...(2)$$

Solving equations (1) and (2),

$$I_1 = -3.2 \text{ A}$$

$$I_2 = -1.4 \text{ A}$$

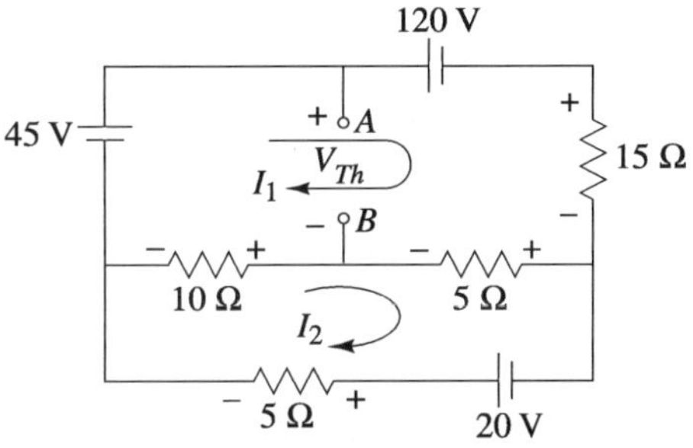

Fig. 3.81

Writing V_{Th} equation,

$$45 - V_{Th} - 10\,(I_1 - I_2) = 0$$

$$V_{Th} = 45 - 10\,(I_1 - I_2)$$

$$= 45 - 10\,[-3.2 - (-1.4)] = 63 \text{ V}$$

Step II: *Calculation of* R_{Th}

Replacing all voltage sources with short circuit,

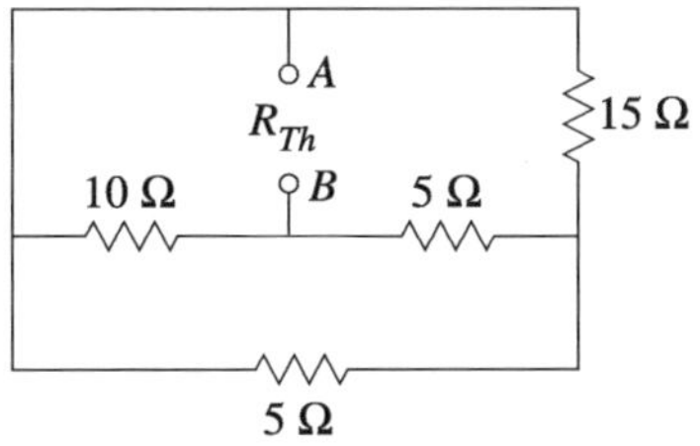

Fig. 3.82

Converting the delta formed by resistances of 10 Ω, 5 Ω and 5 Ω into equivalent star network,

$$R_1 = \frac{10 \times 5}{20} = 2.5\ \Omega$$

$$R_2 = \frac{10 \times 5}{20} = 2.5\ \Omega$$

$$R_3 = \frac{5 \times 5}{20} = 1.25\ \Omega$$

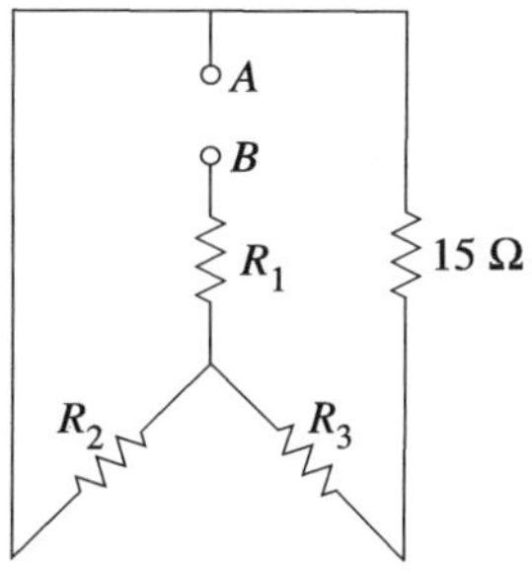

Fig. 3.83

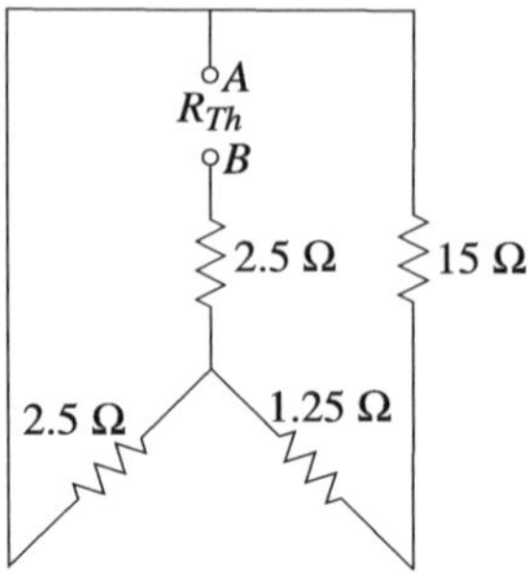

Fig. 3.84

The circuit can be simplified as follows:

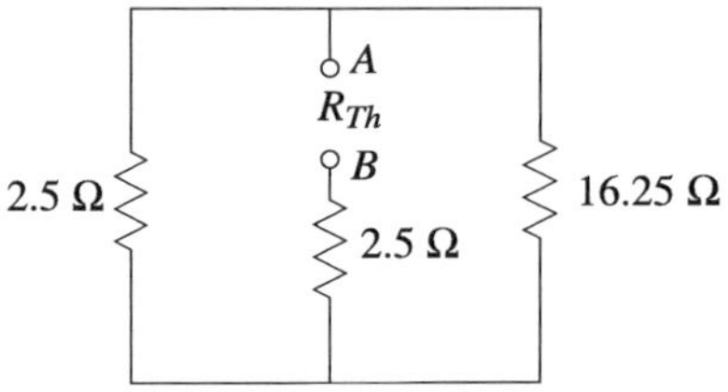

Fig. 3.85

$$R_{\text{Th}} = (16.25 \,||\, 2.5) + 2.5 = 4.67\ \Omega$$

Step III: *Calculation of I_L*

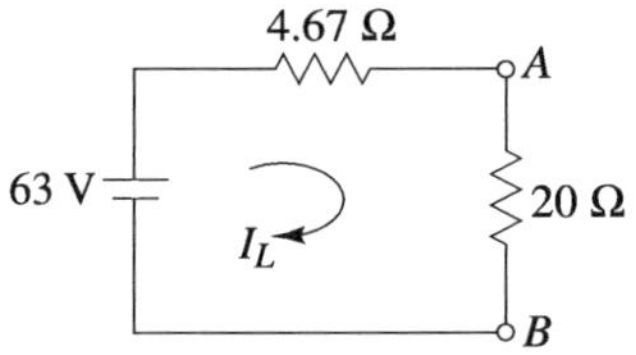

Fig. 3.86

$$I_L = \frac{63}{4.67 + 20} = 2.55\ \text{A}$$

Example 3.19 *Find the current through the 3-Ω resistor.*

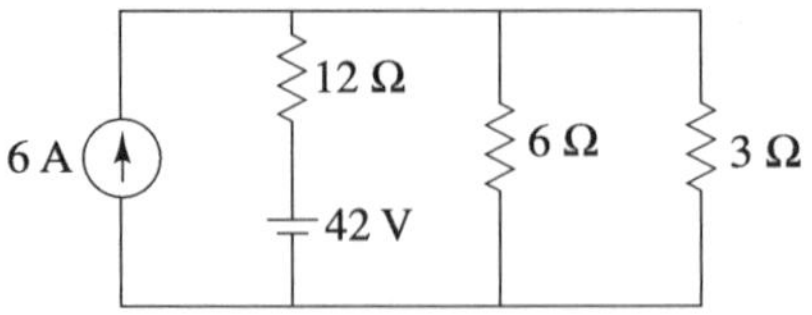

Fig. 3.87

Step I: *Calculation of V_{Th}*

Removing the 3-Ω resistor from the network,

Writing equation for Mesh 1,

$$I_1 = 6 \quad ...(1)$$

Applying KVL to Mesh 2,

$$42 - 12(I_2 - I_1) - 6\,I_2 = 0$$

$$-12\,I_1 + 18\,I_2 = 42 \quad ...(2)$$

Substituting value of I_1 in Eq. (2),

$$I_2 = 6.33 \text{ A}$$

Writing V_{Th} equation,

$$V_{Th} = 6\,I_2 = 38 \text{ V}$$

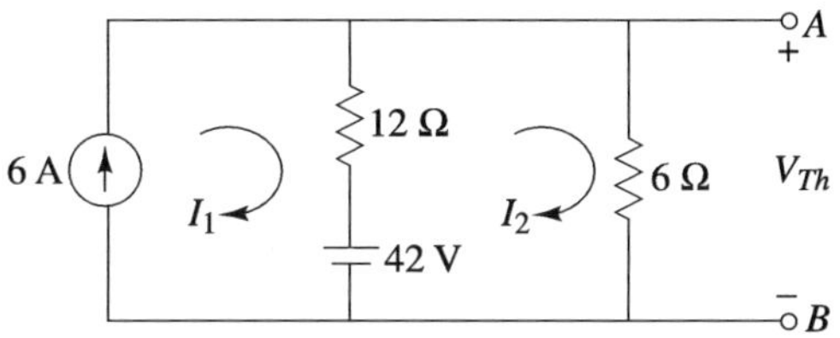

Fig. 3.88

Step II: *Calculation of R_{Th}*

Replacing voltage source by short circuit and current source by open circuit,

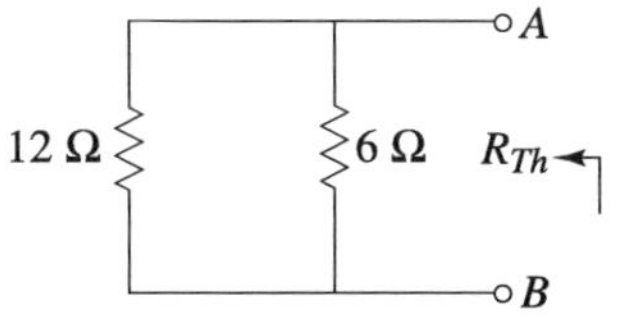

Fig. 3.89

$$R_{Th} = 6 \parallel 12 = 4\ \Omega$$

Step III: *Calculation of I_L*

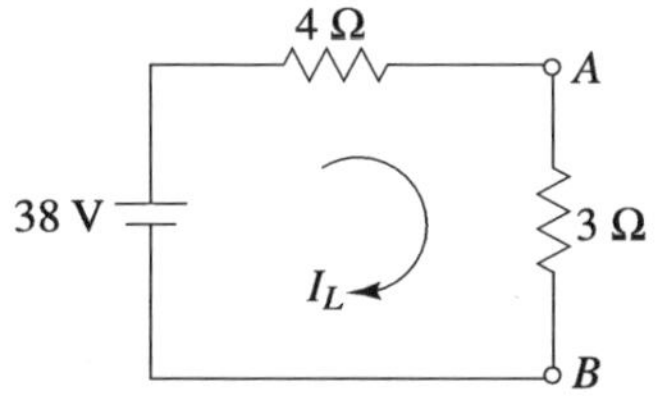

Fig. 3.90

$$I_L = \frac{38}{4+3} = 5.43 \text{ A}$$

Example 3.20 *Find the current through the 30-Ω resistor.*

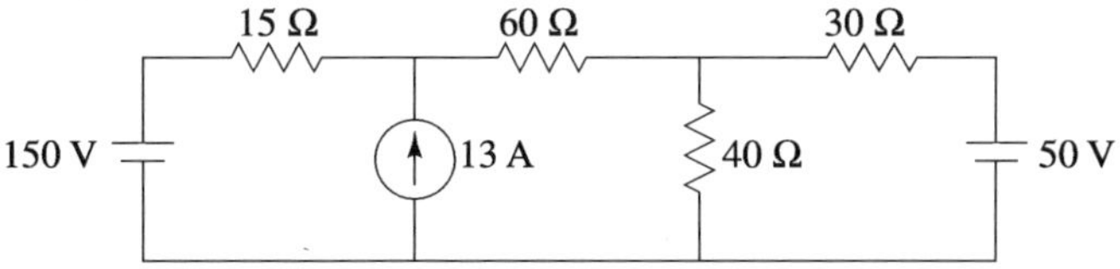

Fig. 3.91

Step I: *Calculation of V_{Th}*

Removing the 30-Ω resistor from the network,

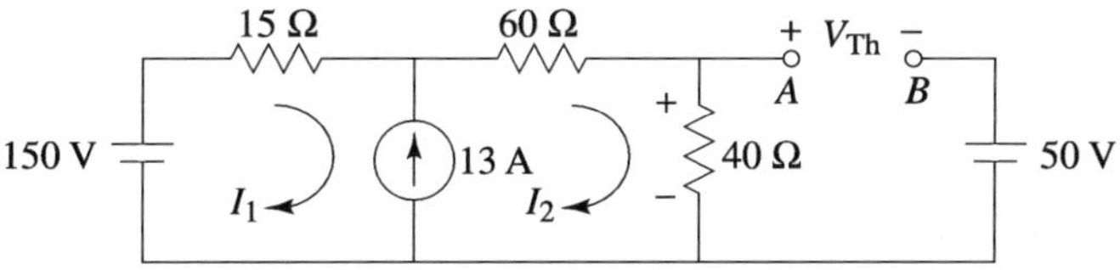

Fig. 3.92

Meshes 1 and 2 form a supermesh.
Writing current equation for supermesh,

$$I_2 - I_1 = 13 \qquad \text{...(1)}$$

Writing voltage equation for supermesh,

$$150 - 15I_1 - 60I_2 - 40I_2 = 0$$

$$15I_1 + 100I_2 = 150 \qquad \text{...(2)}$$

Solving Eqs (1) and (2),

$$I_1 = -10 \text{ A}$$

$$I_2 = 3 \text{ A}$$

Writing V_{Th} equation,

$$40I_2 - V_{Th} - 50 = 0$$

$$V_{Th} = 40I_2 - 50 = 40(3) - 50 = 70 \text{ V}$$

Step II: *Calculation of R_{Th}*
Replacing the voltage sources by short circuits and the current source by an open circuit,

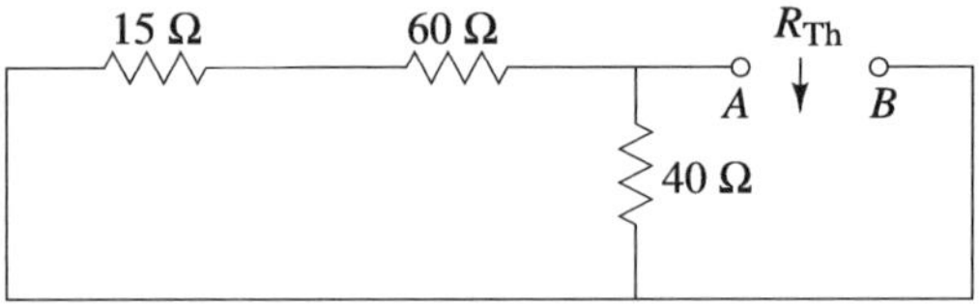

Fig. 3.93

$$R_{Th} = 75 \parallel 40 = 26.09\ \Omega$$

Step III: *Calculation of I_L*

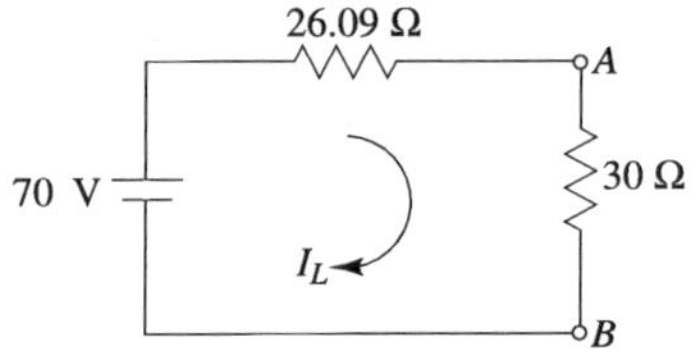

Fig. 3.94

$$I_L = \frac{70}{26.09 + 30} = 1.25 \text{ A}$$

Example 3.21 *Find the current through the 20-Ω resistor.*

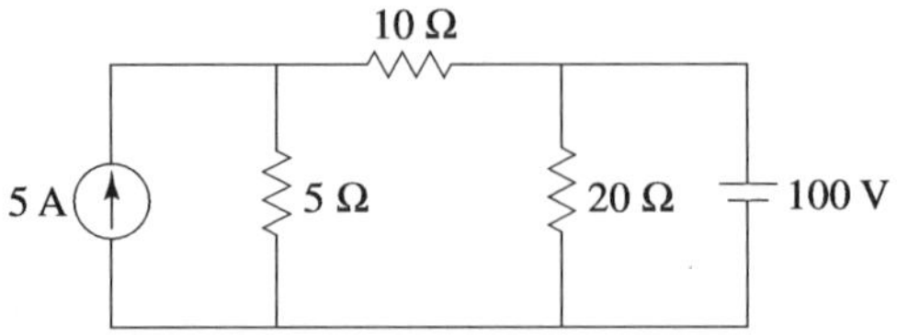

Fig. 3.95

Step I: *Calculation of V_{Th}*
Removing the 20-Ω resistor from the network,

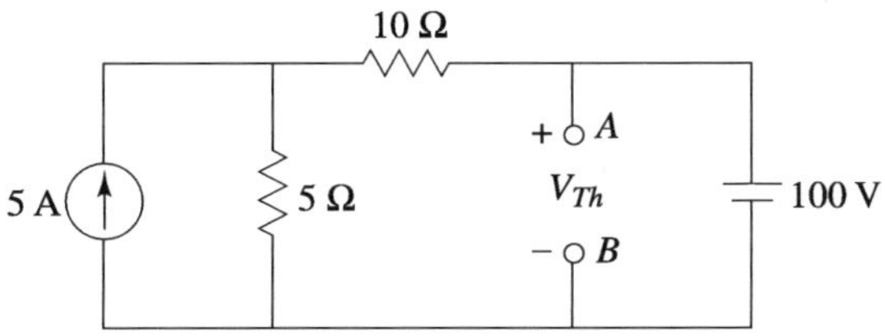

Fig. 3.96

From Fig. 3.96

$$V_{Th} = 100 \text{ V}$$

Step II: *Calculation of R_{Th}*

Replacing the voltage source by a short circuit and the current source by an open circuit,

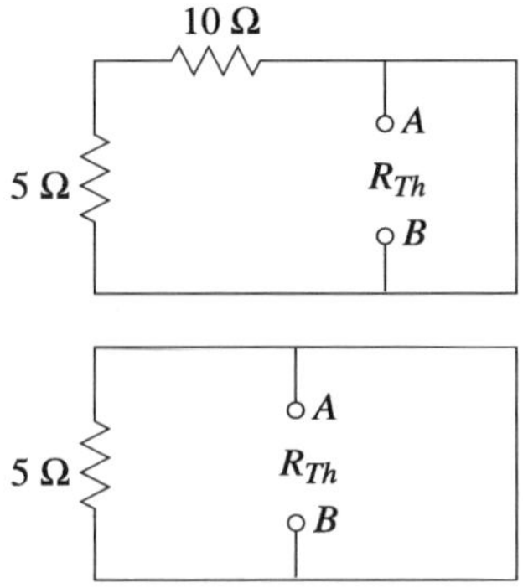

Fig. 3.97

$$R_{\text{Th}} = 0$$

Step III: *Calculation of I_L*

$$I_L = \frac{100}{20} = 5 \text{ A}$$

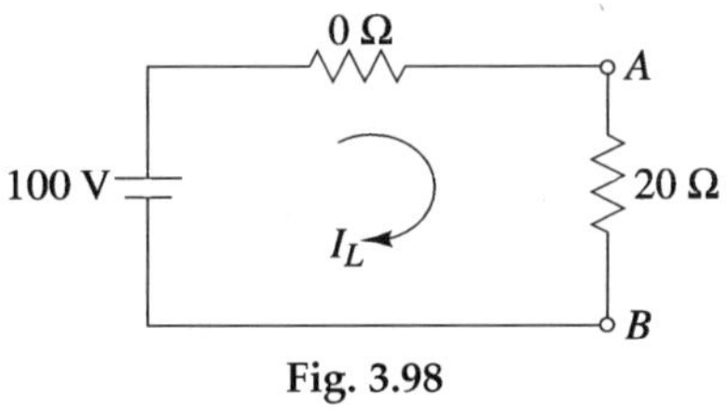

Fig. 3.98

Example 3.22 *Find the current through the 10-Ω resistor.*

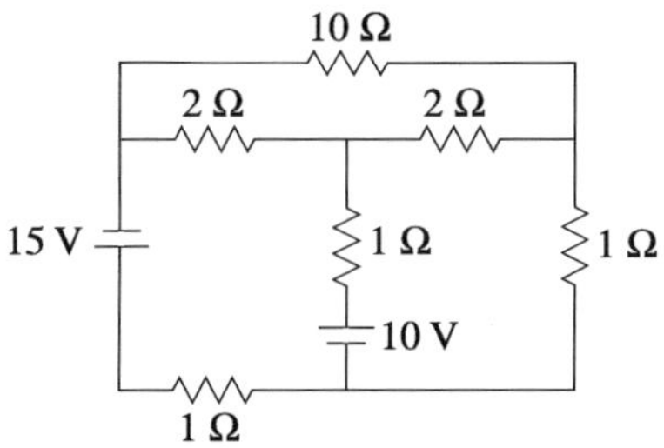

Fig. 3.99

Step I: *Calculation of V_{Th}*

Removing the 10-Ω resistor from the network,

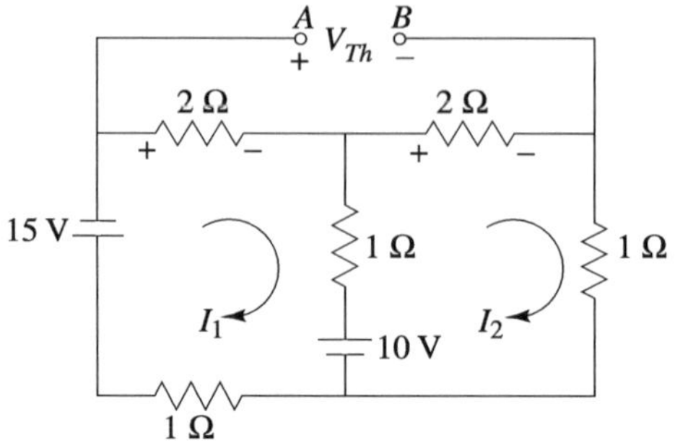

Fig. 3.100

Applying KVL to Mesh 1,

$$-15 - 2I_1 - 1(I_1 - I_2) - 10 - 1I_1 = 0$$

$$4I_1 - I_2 = -25 \quad ...(1)$$

Applying KVL to Mesh 2,

$$10 - (I_2 - I_1) - 2I_2 - I_2 = 0$$

$$-I_1 + 4I_2 = 10 \quad ...(2)$$

Solving Eqs (1) and (2),

$$I_1 = -6 \text{ A}$$

$$I_2 = 1 \text{ A}$$

Writing V_{Th} equation,

$$-V_{Th} + 2I_2 + 2I_1 = 0$$

$$V_{Th} = 2I_1 + 2I_2$$

$$= 2(-6) + 2(1) = -10 \text{ V}$$

$$= 10 \text{ V (the terminal } B \text{ is positive w.r.t. } A)$$

Step II: *Calculation of R_{Th}*

Replacing voltage sources by a short circuit,

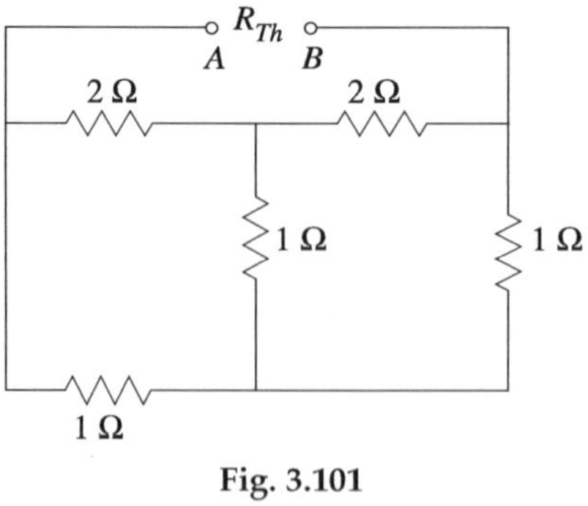

Fig. 3.101

Converting the star network formed by resistances of 2 Ω, 2 Ω and 1 Ω into an equivalent delta network.

$$R_1 = 2 + 2 + \frac{2\times 2}{1} = 8\ \Omega$$

$$R_2 = 2 + 1 + \frac{2\times 1}{2} = 4\ \Omega$$

$$R_3 = 2 + 1 + \frac{2\times 1}{2} = 4\ \Omega$$

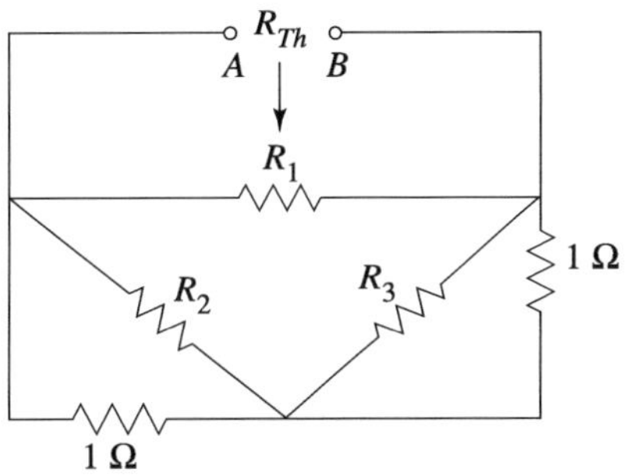

Fig. 3.102

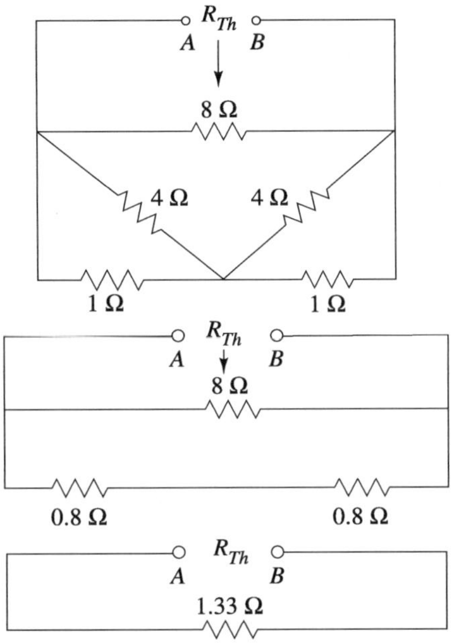

Fig. 3.103

Step III: *Calculation of I_L*

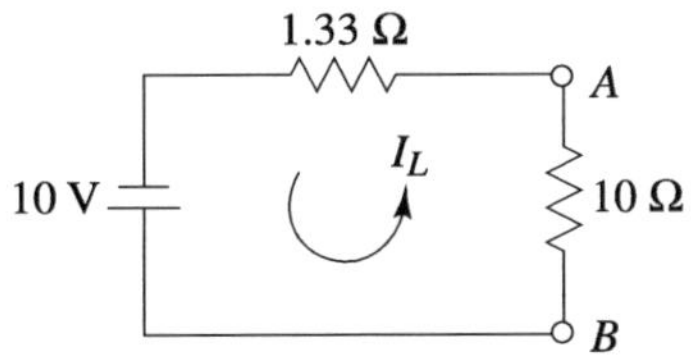

Fig. 3.104

$$I_L = \frac{10}{1.33+10} = 0.88\ \text{A}\ (\uparrow)$$

Example 3.23 *Find the current through the 1-Ω resistor.*

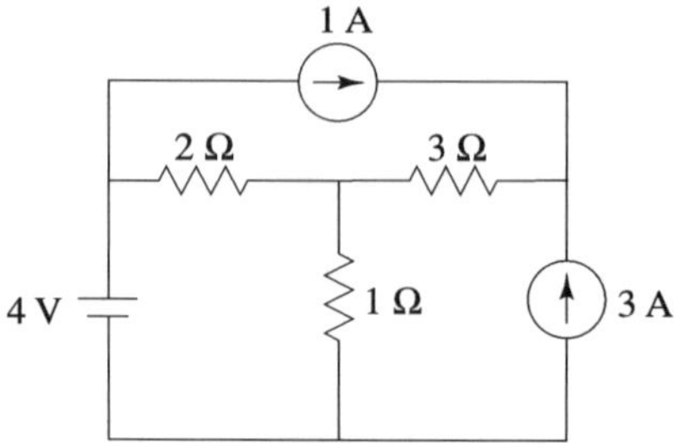

Fig. 3.105

Step I: *Calculation of V_{Th}*

Removing the 1-Ω resistor from the network,

Writing the current equation for Meshes 1 and 2,

$$I_1 = -3 \text{ A}$$
$$I_2 = 1 \text{ A}$$

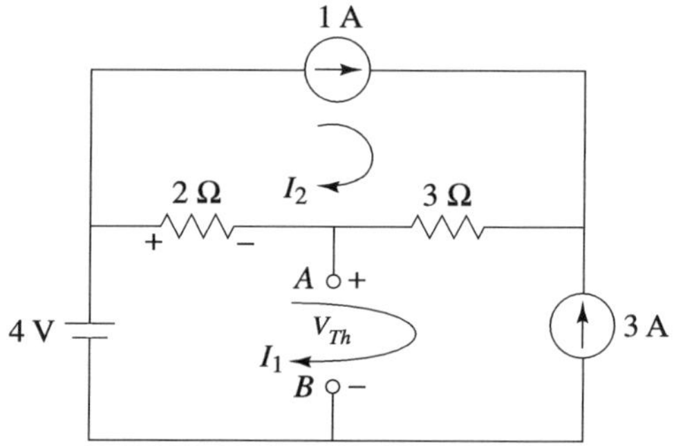

Fig. 3.106

Writing V_{Th} equation,

$$4 - 2(I_1 - I_2) - V_{Th} = 0$$
$$V_{Th} = 4 - 2(-3 - 1)$$
$$= 4 - 2(-4) = 12 \text{ V}$$

Step II: *Calculation of R_{Th}*

Replacing the voltage source by a short circuit and the current source by an open circuit,

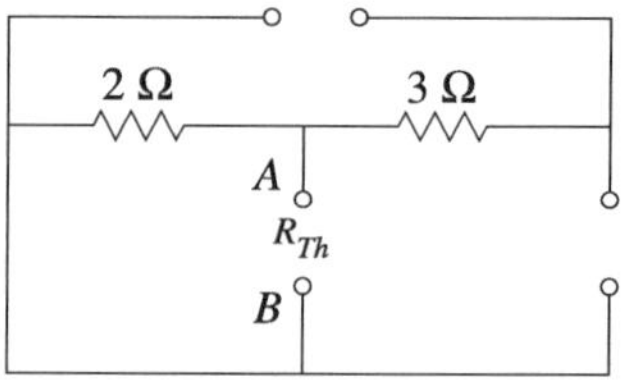

Fig. 3.107

$$R_{Th} = 2 \ \Omega$$

Step III: *Calculation of I_L*

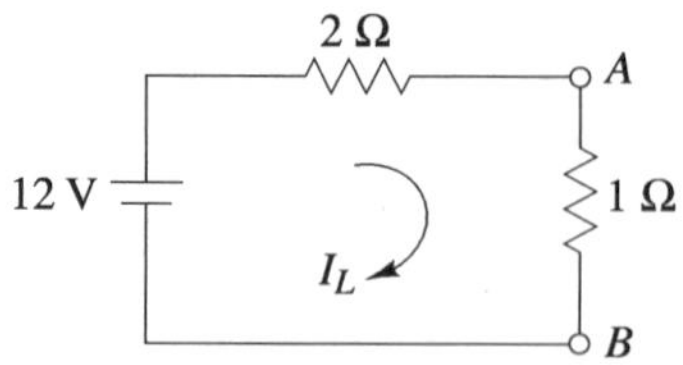

Fig. 3.108

$$I_L = \frac{12}{2+1} = 4 \text{ A}$$

3.4 NORTON'S THEOREM

It states that *'Any two terminals of a network can be replaced by an equivalent current source and an equivalent parallel resistance.'* The constant current is equal to the current which would flow in a short circuit placed across the terminals. The parallel resistance is the resistance of the network when viewed from these open-circuited terminals after all voltage and current sources have been removed and replaced by internal resistances.

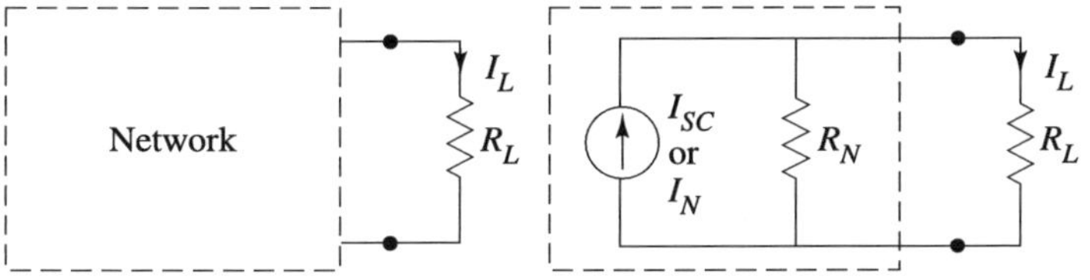

Fig. 3.109

Explanation: The method of determining the load current through a given load resistance can be explained with the help of the following circuit.

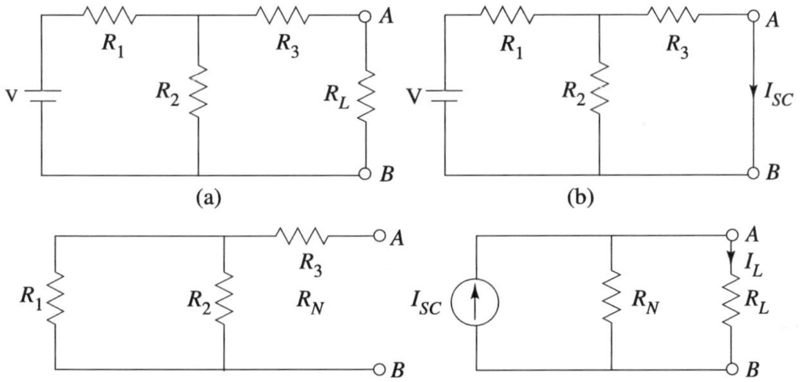

Fig. 3.110

Steps to be followed in Norton's theorem

1. Remove the load resistance R_L and put a short circuit across the terminals.
2. Find the short-circuit current I_{SC} or I_N.
3. Find the resistance R_N as seen from points A and B by replacing the voltage source by a short circuit.
4. Replace the network by a current source I_{SC} in parallel with resistance R_N.
5. Find current through R_L by current–division formula.

$$I_L = \frac{I_{SC} R_N}{R_N + R_L}$$

Example 3.24 *Find the current through the 10-Ω resistor.*

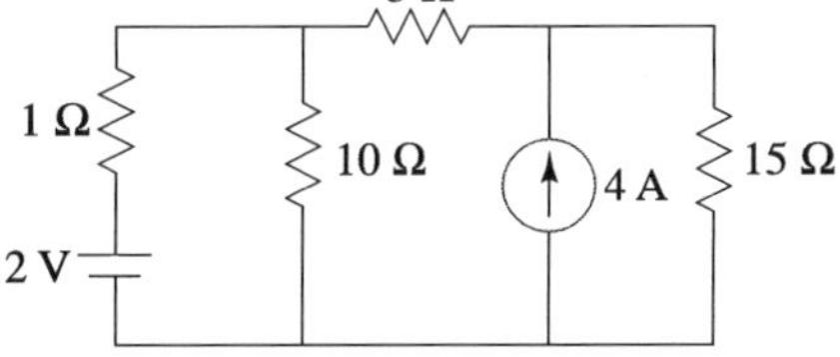

Fig. 3.111

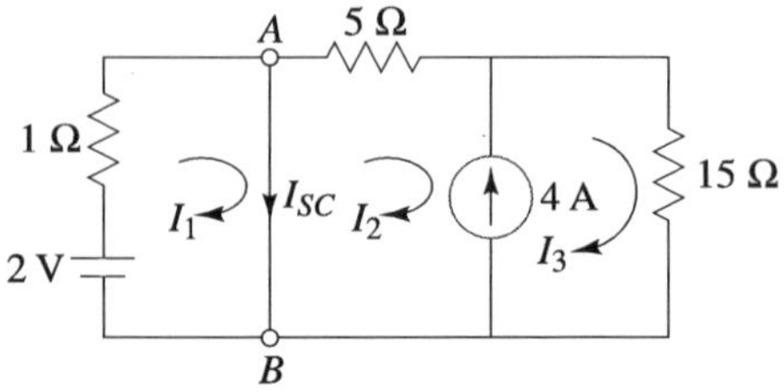

Fig. 3.112

Step I: *Calculation of I_{SC}*

Applying KVL to Mesh 1,

$$2 - I_1 = 0 \quad \text{...(1)}$$

$$I_1 = 2$$

Meshes 2 and 3 will form a supermesh.

Writing current equation for the supermesh,

$$I_3 - I_2 = 4 \quad \text{...(2)}$$

Applying KVL to the supermesh,

$$-5I_2 - 15I_3 = 0 \quad \text{...(3)}$$

Solving Eqs (1), (2) and (3),

$$I_1 = 2 \text{ A}$$

$$I_2 = -3 \text{ A}$$

$$I_3 = 1 \text{ A}$$

$$I_{SC} = I_1 - I_2 = 2 - (-3) = 5 \text{ A}$$

Step II: *Calculation of R_N*

Replacing the voltage source by a short circuit and current source by an open circuit,

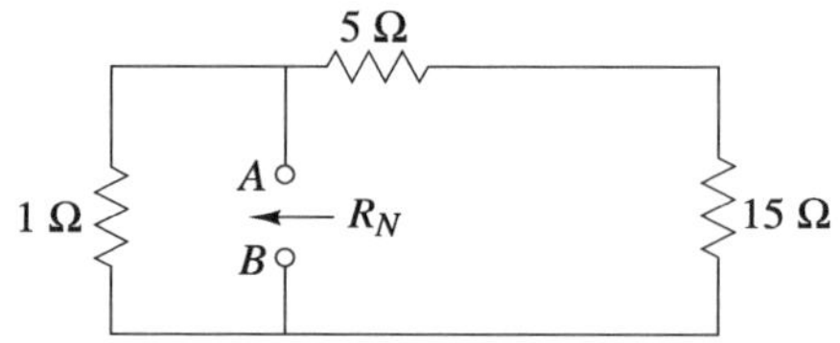

Fig. 3.113

$$R_N = 1 \parallel (5 + 15) = 0.95 \ \Omega$$

Step III: *Calculation of I_L*

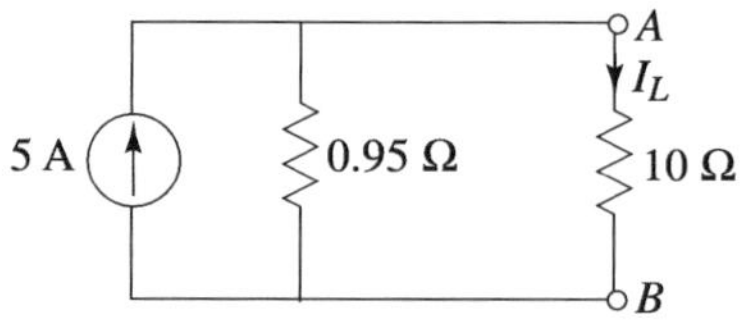

Fig. 3.114

$$I_L = 5 \times \frac{0.95}{10 + 0.95} = 0.43 \text{ A}$$

Example 3.25 *Find the current through the 10-Ω resistor.*

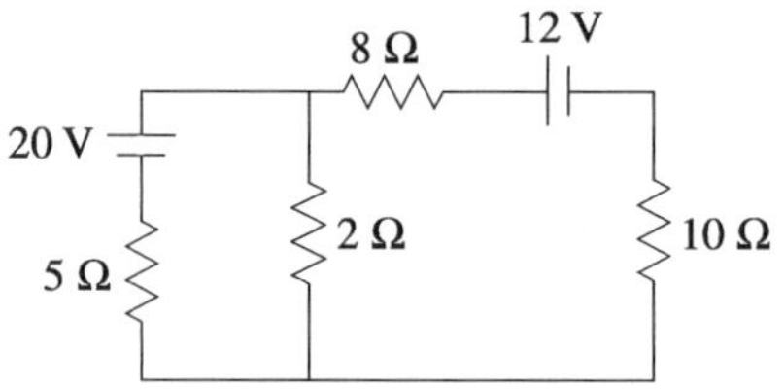

Fig. 3.115

Step I: *Calculation of* I_{SC}

Applying KVL to Mesh 1,

$$-5I_1 + 20 - 2(I_1 - I_2) = 0$$

$$7I_1 - 2I_2 = 20 \quad ...(1)$$

Applying KVL to Mesh 2,

$$-2(I_2 - I_1) - 8I_2 - 12 = 0$$

$$-2I_1 + 10I_2 = -12 \quad ...(2)$$

Solving Eqs (1) and (2),

$$I_2 = -0.67 \text{ A}$$

$$I_{SC} = I_2 = -0.67 \text{ A}$$

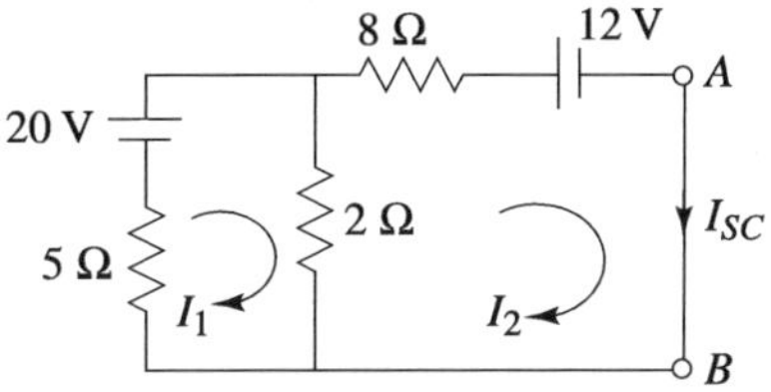

Fig. 3.116

Step II: *Calculation of* R_N

Replacing voltage sources with short circuits,

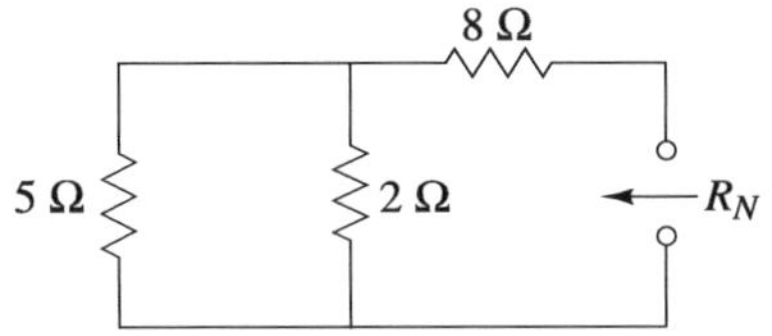

Fig. 3.117

$$R_N = (5 \,||\, 2) + 8 = 9.43\ \Omega$$

Step III: *Calculation of* I_L

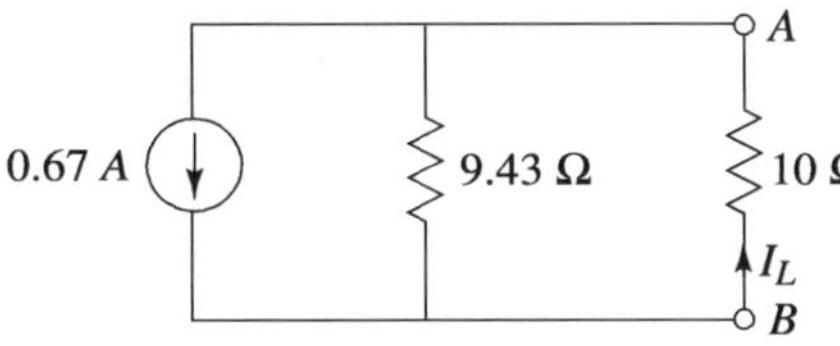

Fig. 3.118

$$I_L = 0.67 \times \frac{9.43}{9.43 + 10}$$

$$= 0.33 \text{ A } (\uparrow)$$

Example 3.26 *Find the current through the 10-Ω resistor in Fig. 3.119.*

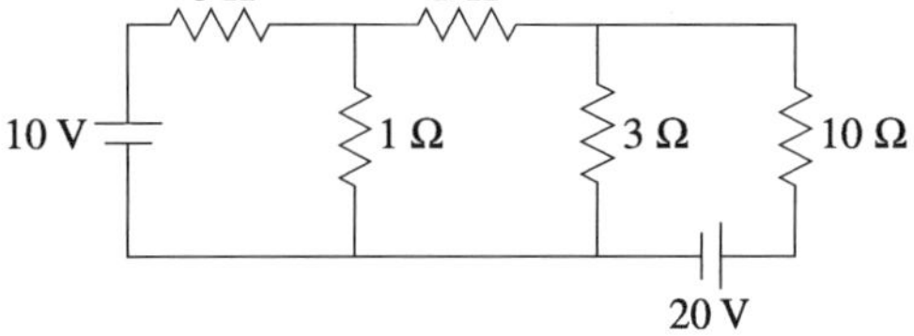

Fig. 3.119

Step I: *Calculation of* I_{SC}

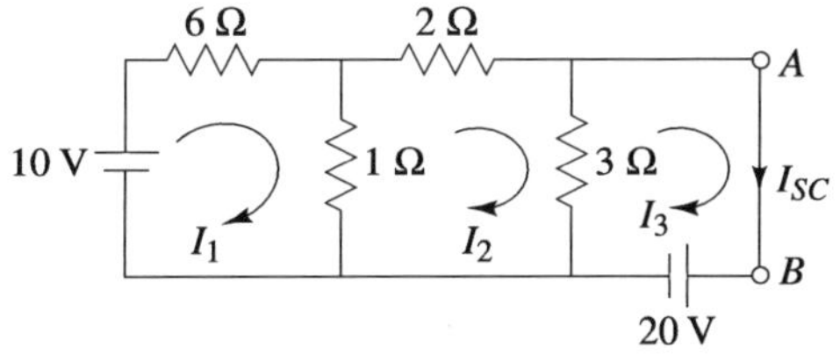

Fig. 3.120

Applying KVL to Mesh 1,

$$10 - 6I_1 - 1\,(I_1 - I_2) = 0$$

$$7I_1 - I_2 = 10 \qquad \text{...(1)}$$

Applying KVL to Mesh 2,

$$-1\,(I_2 - I_1) - 2I_2 - 3(I_2 - I_3) = 0$$

$$-I_1 + 6I_2 - 3I_3 = 0 \qquad \text{...(2)}$$

Applying KVL to Mesh 3,

$$-3\,(I_3 - I_2) - 20 = 0$$

$$3I_2 - 3I_3 = 20 \qquad \text{...(3)}$$

Solving Eqs (1), (2) and (3),

$$I_3 = -13.17 \text{ A}$$

$$I_{SC} = I_3 = -13.17 \text{ A}$$

Step II: *Calculation of R_N*

Replacing voltage sources with short circuits,

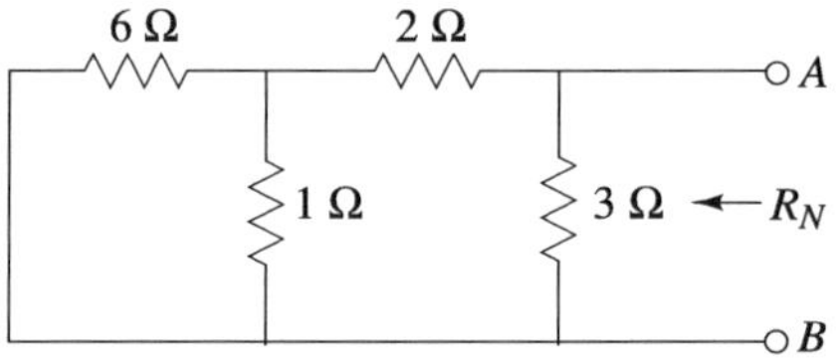

Fig. 3.121

$$R_N = [(6 \parallel 1) + 2] \parallel 3 = 1.46\ \Omega$$

Step III: *Calculation of I_{SC}*

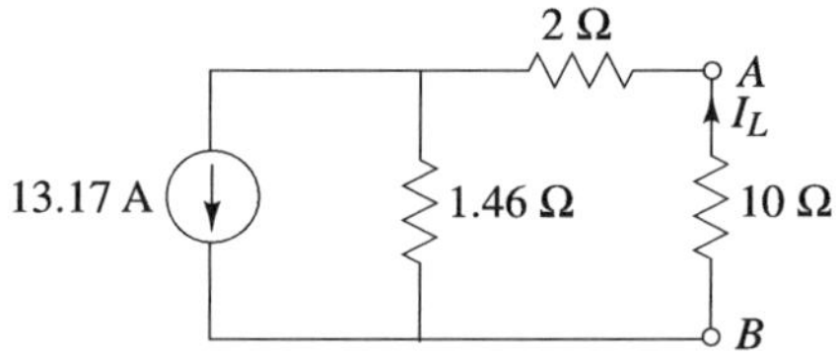

Fig. 3.122

$$I_L = 13.17 \times \frac{1.46}{1.46 + 10} = 1.68 \text{ A } (\uparrow)$$

Example 3.27 *Find the current through the 10-Ω resistor.*

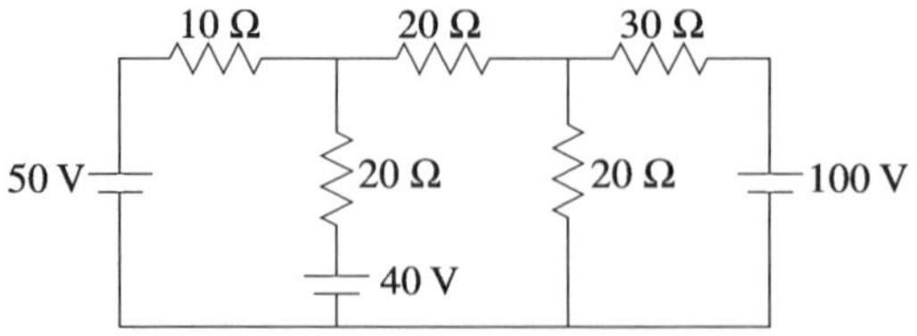

Fig. 3.123

Step I: *Calculation of* I_{SC}

Applying KVL to Mesh 1,

$$50 - 20(I_1 - I_2) - 40 = 0$$
$$20I_1 - 20I_2 = 10 \quad ...(1)$$

Applying KVL to Mesh 2,

$$40 - 20(I_2 - I_1) - 20I_2 - 20(I_2 - I_3) = 0$$
$$-20I_1 + 60I_2 - 20I_3 = 40 \quad ...(2)$$

Applying KVL to Mesh 3,

$$-20(I_3 - I_2) - 30I_3 - 100 = 0$$
$$-20I_2 + 50I_3 = -100 \quad ...(3)$$

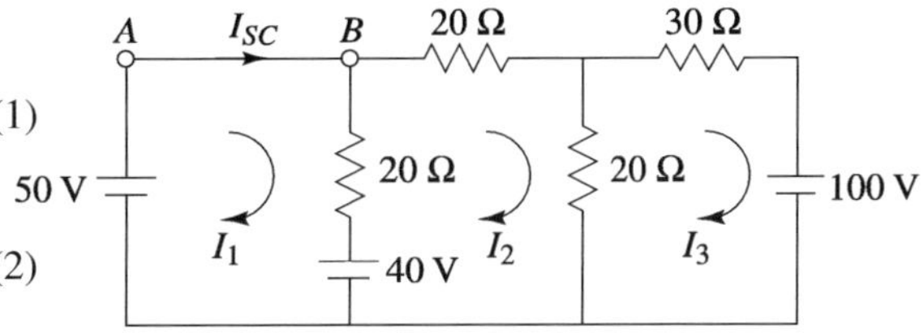

Fig. 3.124

Solving Eqs (1), (2) and (3),

$$I_1 = 0.81 \text{ A}$$
$$I_{SC} = I_1 = 0.81 \text{ A}$$

Step II: *Calculation of* R_N

Replacing all voltage sources by short circuits,

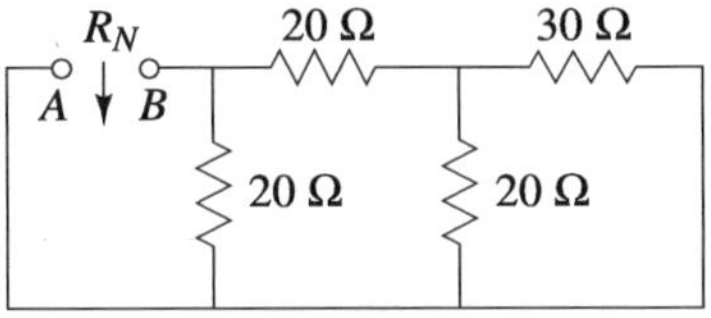

Fig. 3.125

$$R_N = [(20 \parallel 30) + 20] \parallel 20 = 12.3\ \Omega$$

Step III: *Calculation of* I_L

$$I_L = 0.81 \times \frac{12.3}{12.3 + 10} = 0.45 \text{ A}$$

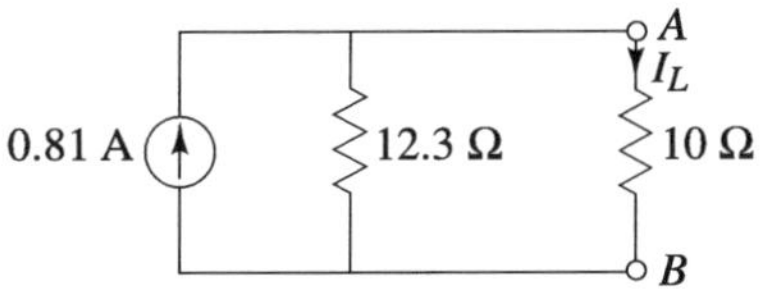

Fig. 3.126

Example 3.28 Obtain Norton's equivalent network as seen by R_L.

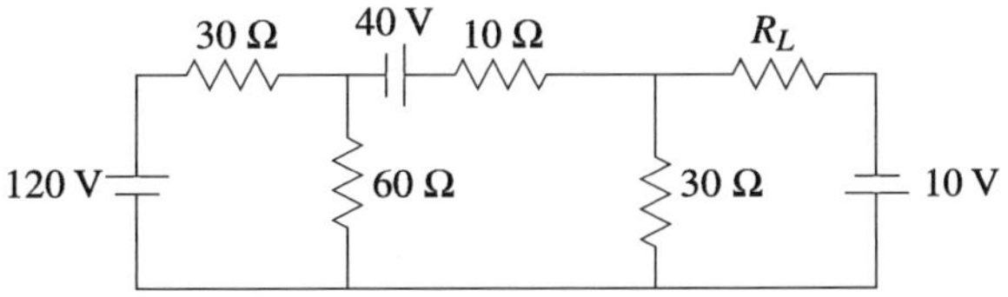

Fig. 3.127

Step I: *Calculation of* I_{SC}

Applying KVL to Mesh 1,

$$120 - 30I_1 - 60(I_1 - I_2) = 0$$
$$90I_1 - 60I_2 = 120 \quad ...(1)$$

Applying KVL to Mesh 2,

$$-60(I_2 - I_1) + 40 - 10I_2 - 30(I_2 - I_3) = 0$$
$$-60I_1 + 100I_2 - 30I_3 = 40 \quad ...(2)$$

Applying KVL to Mesh 3,

$$-30(I_3 - I_2) + 10 = 0$$
$$30I_2 - 30I_3 = -10 \quad ...(3)$$

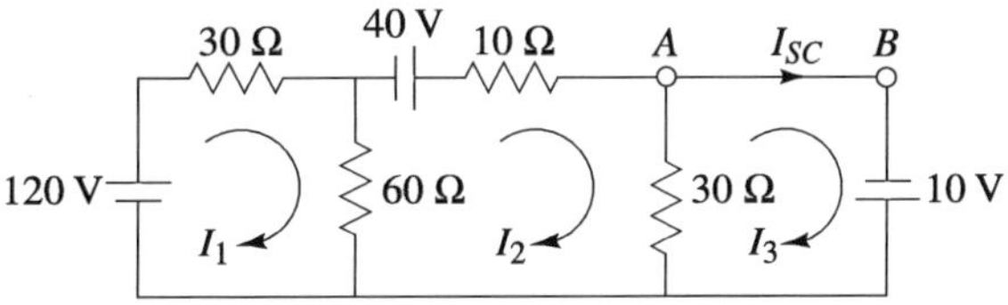

Fig. 3.128

Solving Eqs (1), (2) and (3),

$$I_3 = 4.67 \text{ A}$$
$$I_{SC} = I_3 = 4.67 \text{ A}$$

Step II: *Calculation of R_N*

Replacing voltage sources by short circuits,

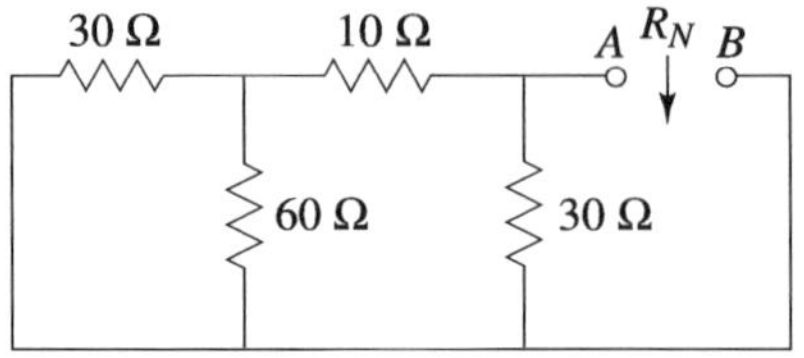

Fig. 3.129

$$R_N = [(30 \parallel 60) + 10] \parallel 30 = 15\ \Omega$$

Step III: *Norton's equivalent Network*

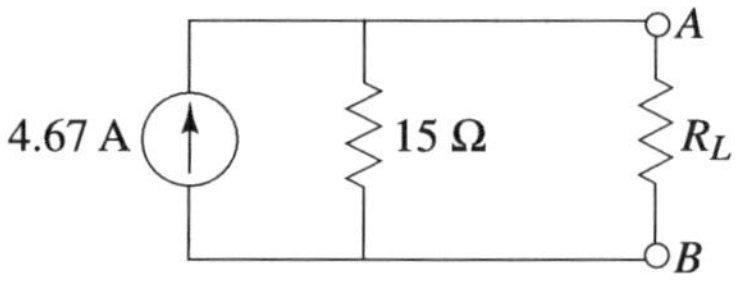

Fig. 3.130

Example 3.29 *Find the current through the 8-Ω resistor.*

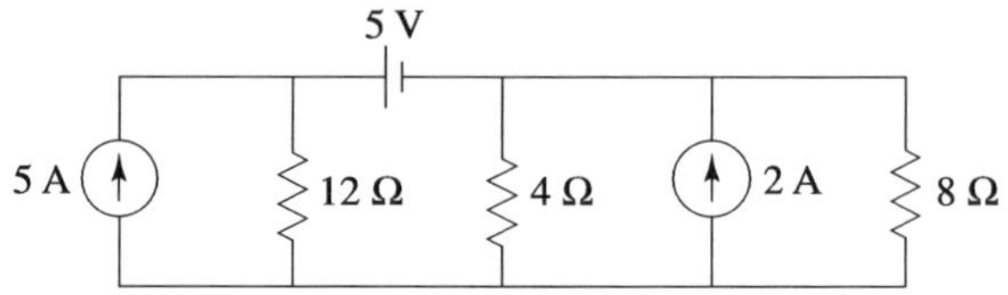

Fig. 3.131

Step I: *Calculation of I_{SC}*

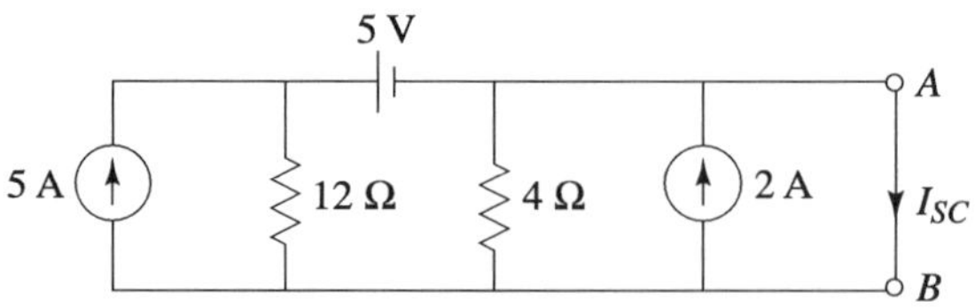

Fig. 3.132

The resistor of the 4-Ω source gets shorted as it is in parallel with the short circuit. Simplifying the network by source transformation,

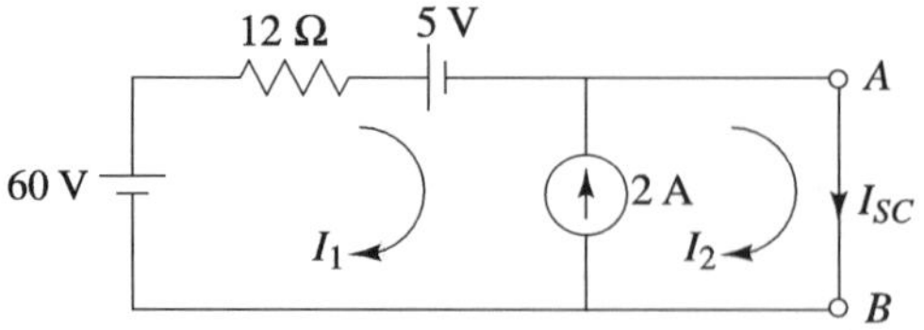

Fig. 3.133

Meshes 1 and 2 will form a supermesh.

Writing current equation for the supermesh,

$$I_2 - I_1 = 2 \quad \text{...(1)}$$

Applying KVL to the supermesh,

$$60 - 12I_1 - 5 = 0$$

$$12I_1 = 55 \quad \text{...(2)}$$

Solving Eqs (1) and (2),

$$I_1 = 4.58 \text{ A}$$

$$I_2 = 6.58 \text{ A}$$

$$I_{SC} = I_2 = 6.58 \text{ A}$$

Step II: *Calculation of R_N*

Replacing the voltage source by a short circuit and the current source by an open circuit,

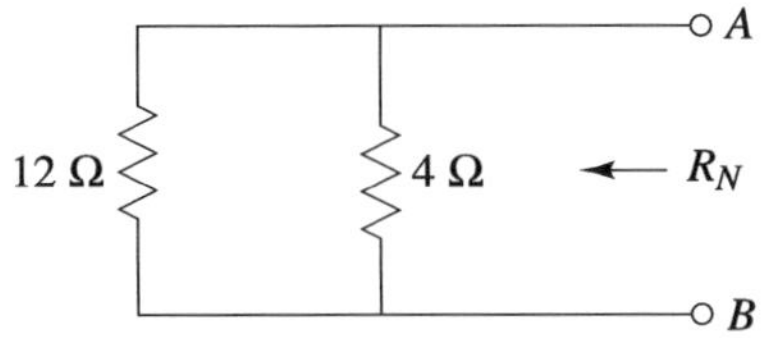

Fig. 3.134

$$R_N = 12 \parallel 4 = 3\ \Omega$$

Step III: *Calculation of I_L*

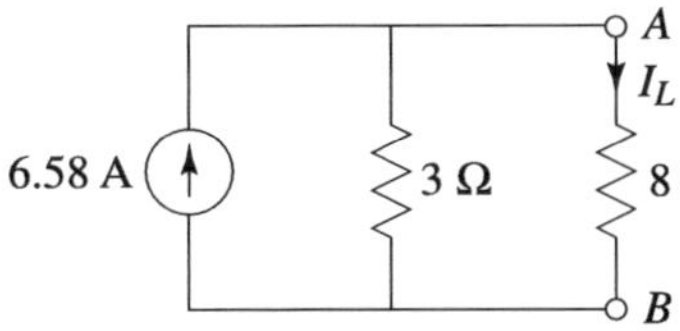

Fig. 3.135

$$I_L = 6.58 \times \frac{3}{3+8} = 1.79 \text{ A}$$

Example 3.30 *Find current through the 1-Ω resistor.*

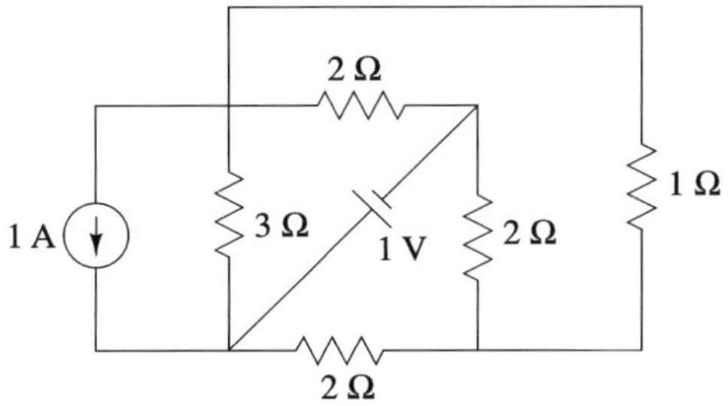

Fig. 3.136

Step I: *Calculation of* I_{SC}

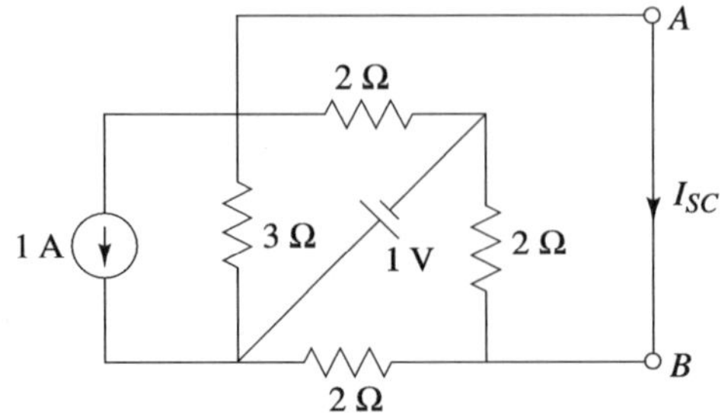

Fig. 3.137

By source transformation,

Applying KVL to Mesh 1,

$$-3 - 3I_1 - 2(I_1 - I_3) + 1 = 0$$

$$5I_1 - 2I_3 = -2 \qquad \text{...(1)}$$

Applying KVL to Mesh 2,

$$-1 - 2\,(I_2 - I_3) - 2\,I_2 = 0$$

$$4I_2 - 2I_3 = -1 \qquad \text{...(2)}$$

Applying KVL to Mesh 3,

$$-2\,(I_3 - I_1) - 2\,(I_3 - I_2) = 0$$

$$-2I_1 - 2I_2 + 4I_3 = 0 \qquad \text{...(3)}$$

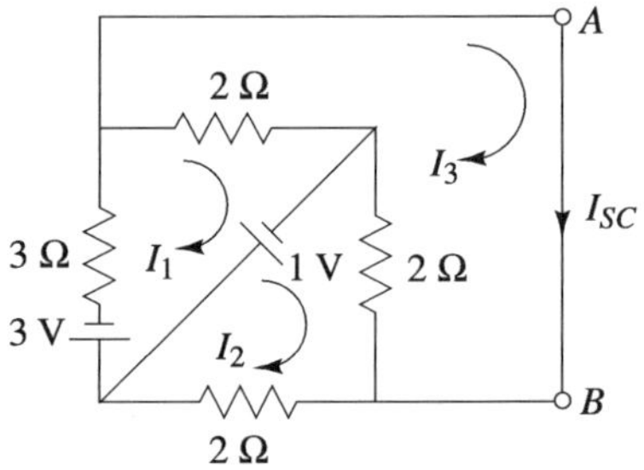

Fig. 3.138

Solving Eqs (1), (2) and (3),

$$I_1 = -0.64 \text{ A}$$

$$I_2 = -0.55 \text{ A}$$

$$I_3 = -0.59 \text{ A}$$

$$I_{SC} = I_3 = -0.59 \text{ A}$$

Step II: *Calculation of* R_N

Replacing the voltage source by a short circuit and the current source by an open circuit,

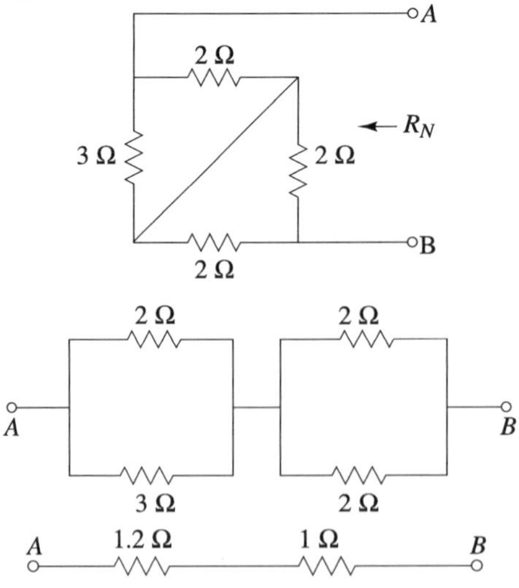

Fig. 3.139

$$R_N = 2.2\ \Omega$$

Step III: *Calculation of I_L*

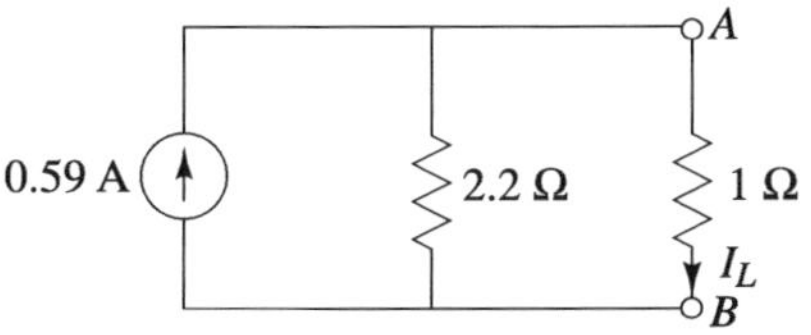

Fig. 3.140

$$I_L = 0.59 \times \frac{2.2}{2.2+1} = 0.41 \text{ A}$$

3.5 THEVENIN'S AND NORTON'S THEOREM WITH DEPENDENT SOURCES

In a resistive circuit containing dependent and independent sources, we shall often find it more convenient to determine either the Thevenin or Norton equivalent by finding both the open-circuit voltage and short-circuit current and then determining the value of R_{Th} as,

$$R_{Th} = \frac{V_{Th}}{I_{SC}}$$

Dependent sources are active at all the times. These have zero value only when its control voltage or current is zero. R_{Th} may be negative in some cases which indicates negative resistance region of the device, i.e., as voltage increases, current decreases in this region.

Thevenin's theorem and Norton's theorem are the dual of each other. If we apply source transformation to one network, we will obtain the other network. For example, if we transform the Norton equivalent network, we obtain a voltage source $R_{Th}\, I_{SC}$ in series with resistance R_{Th}. This gives the Thevenin equivalent network.

$$V_{Th} = R_{Th}\, I_{SC}$$

Example 3.31 *Obtain the Thevenin equivalent network for the given network at terminals A and B.*

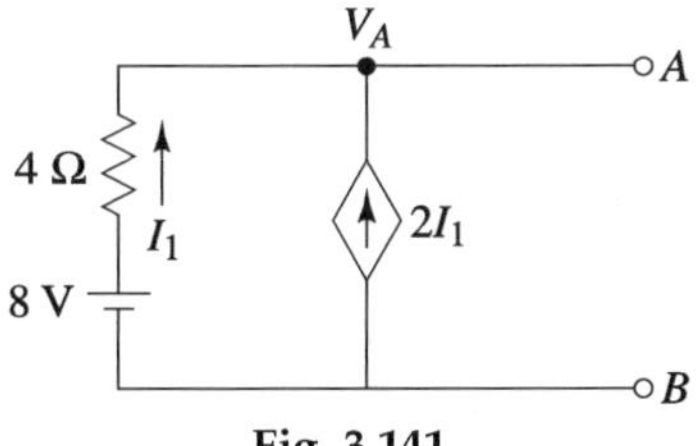

Fig. 3.141

Step I: *Calculation of V_{Th}*

From the figure,

$$I_1 = \frac{8 - V_A}{4}$$

Applying KCL at Node A,

$$I_1 + 2I_1 = 0$$

$$3I_1 = 0$$

$$3\left(\frac{8 - V_A}{4}\right) = 0$$

$$V_A = 8 \text{ V}$$

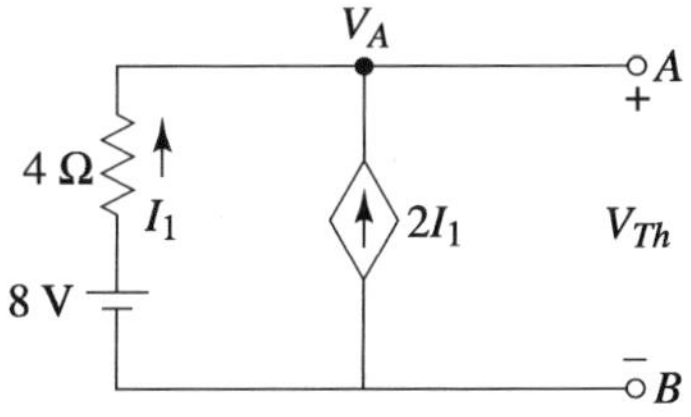

Fig. 3.142

Writing V_{Th} equation,

$$V_{Th} = V_A = 8 \text{ V}$$

Step II: *Calculation of I_{SC}*

From the figure,

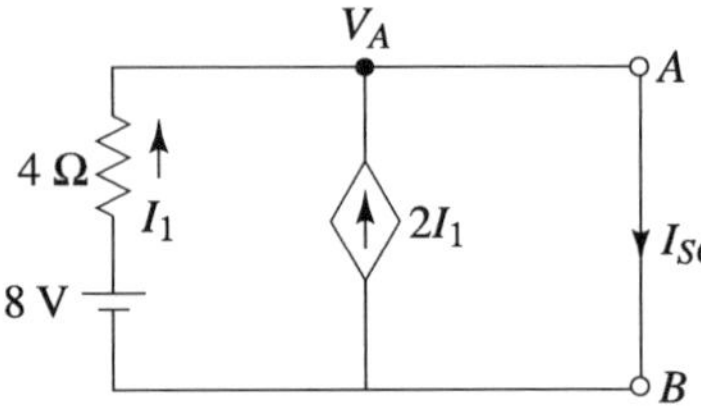

Fig. 3.143

$$I_1 = \frac{8 - V_A}{4}$$

Applying KCL at Node A,

$$I_1 + 2I_1 = I_{SC}$$
$$I_{SC} = 3I_1$$
$$= 3\left(\frac{8 - V_A}{4}\right) = 6 \text{ A} \qquad (\because V_A = 0)$$

Step III: *Calculation of R_{Th}*

$$R_{Th} = \frac{V_{Th}}{I_{SC}}$$
$$= \frac{8}{6} = 1.33\ \Omega$$

Example 3.32 *Obtain the Thevenin equivalent network for the load R_L in the network.*

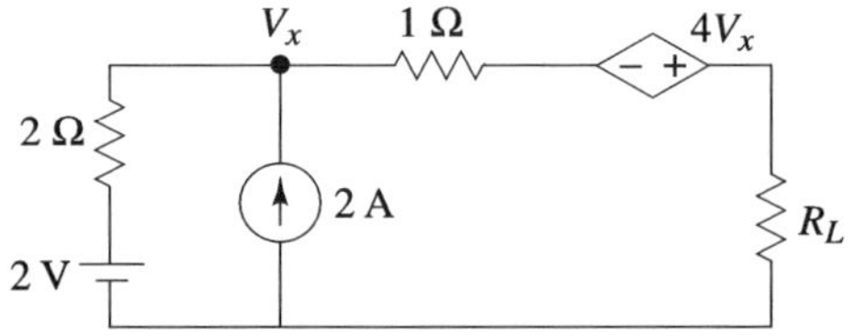

Fig. 3.144

Step I: *Calculation of V_{Th}*

Applying KCL at Node x,

$$\frac{V_x - 2}{2} = 2$$
$$V_x - 2 = 4$$
$$V_x = 6 \text{ V}$$

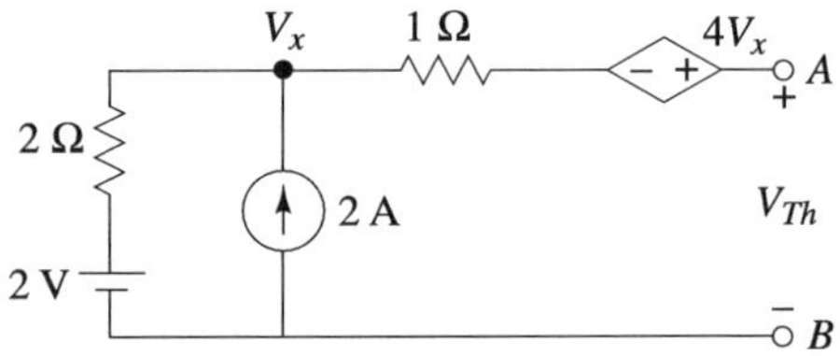

Fig. 3.145

Writing V_{Th} equation,

$$V_{Th} = V_x + 4V_x$$
$$= 5V_x$$
$$= 5 \times 6 = 30 \text{ V}$$

Step II: *Calculation of I_{SC}*

Applying KCL at Node x,

$$\frac{V_x - 2}{2} + \frac{V_x + 4V_x}{1} = 2$$
$$\frac{V_x}{2} - 1 + 5V_x = 2$$
$$5.5V_x = 3$$
$$V_x = 0.545 \text{ V}$$

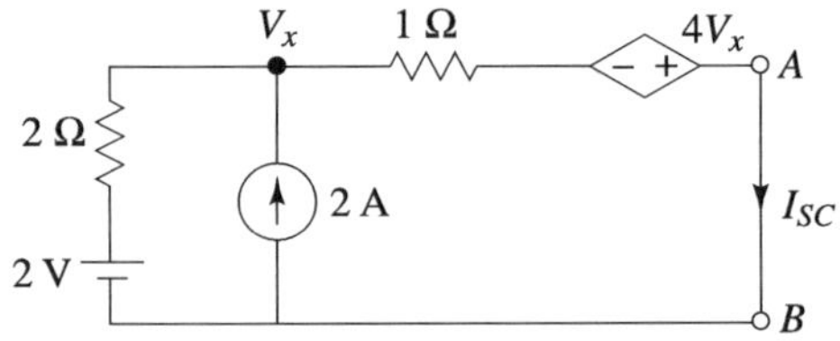

Fig. 3.146

$$I_{SC} = \frac{5V_x}{1} = \frac{5 \times 0.545}{1} = 2.73 \text{ A}$$

Step III: *Calculation of R_{Th}*

$$R_{Th} = \frac{V_{Th}}{I_{SC}} = \frac{30}{2.73} = 10.98\ \Omega$$

Example 3.33 *Find the Thevenin equivalent network for the terminals A and B.*

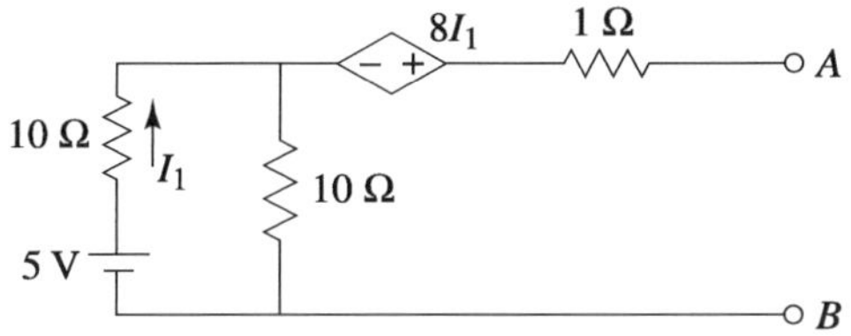

Fig. 3.147

Step I: *Calculation of V_{Th}*

Applying KVL to the Mesh,

$$5 - 10I_1 - 10I_1 = 0$$

$$I_1 = \frac{5}{20} = 0.25\ \text{A}$$

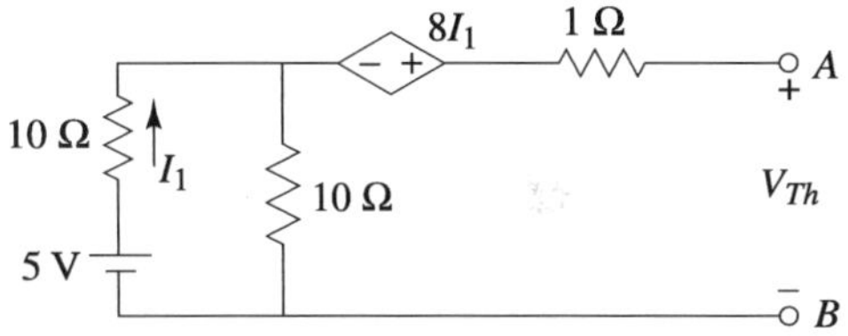

Fig. 3.148

Writing V_{Th} equation,

$$V_{Th} = 10I_1 + 8I_1$$
$$= 18I_1$$
$$= 18 \times 0.25 = 4.5\ \text{V}$$

Step II: *Calculation of I_{SC}*

Applying KVL to Mesh 1,

$$5 = 10I_1 + 10(I_1 - I_2)$$
$$5 = 20I_1 - 10I_2$$
$$1 = 4I_1 - 2I_2 \quad \ldots(1)$$

Applying KVL to Mesh 2,

$$8I_1 - I_2 - 10(I_2 - I_1) = 0$$
$$18I_1 = 11I_2$$
$$I_2 = 1.61I_1$$
$$I_1 = 0.611I_2 \quad \ldots(2)$$

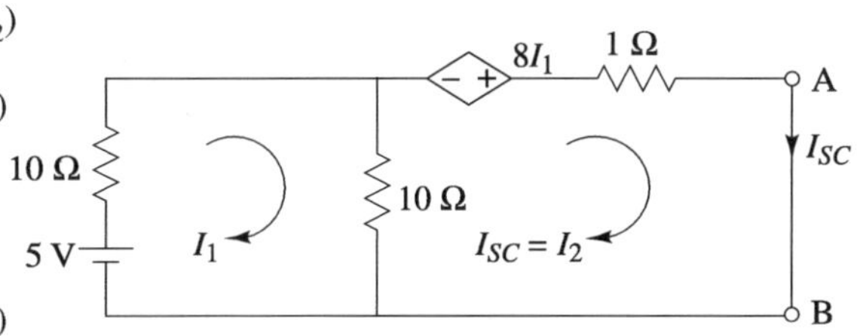

Fig. 3.149

Solving Eqs (1) and (2),

$$I_2 = 2.25\ \text{A}$$
$$I_{SC} = 2.25\ \text{A}$$

Step III: *Calculation of R_{Th}*

$$R_{Th} = \frac{V_{Th}}{I_{SC}} = \frac{4.5}{2.25} = 2\ \Omega$$

Example 3.34 *Find R_{Th} and V_{Th} between A and B.*

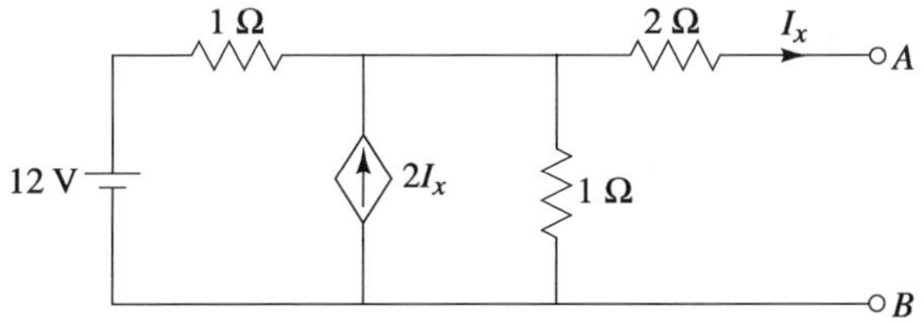

Fig. 3.150

Step I: *Calculation of V_{Th}*

$$I_x = 0$$

The dependent source $2I_x$ depends on the controlling variable I_x. When $I_x = 0$, the dependent source vanishes, i.e., $2I_x = 0$

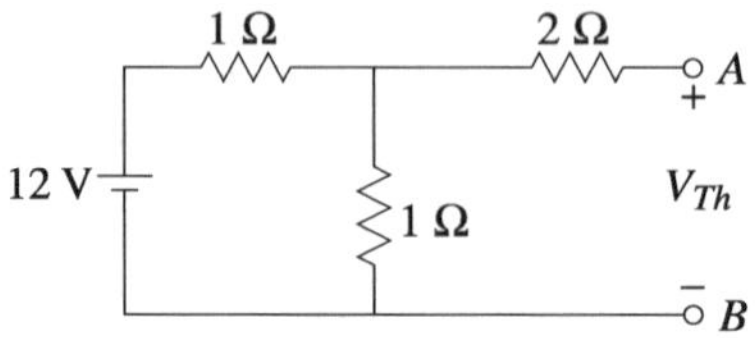

Fig. 3.151

$$V_{Th} = 12 \times \frac{1}{1+1} = 6 \text{ V}$$

Step II: *Calculation of I_{SC}*

From the figure,

$$I_x = I_{SC}$$

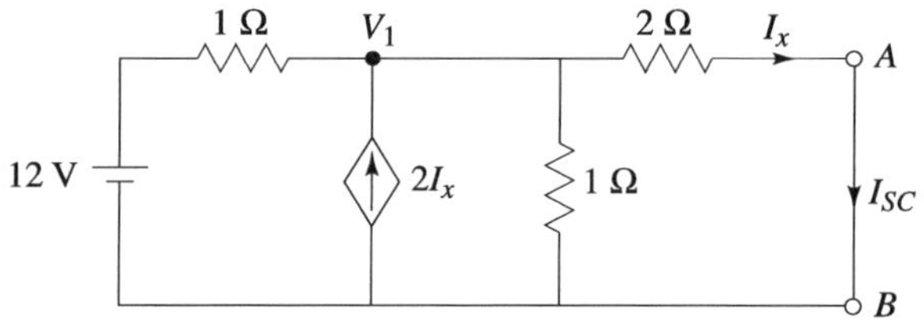

Fig. 3.152

Applying KCL at Node 1,

$$\frac{V_1 - 12}{1} + \frac{V_1}{1} + \frac{V_1}{2} = 2I_x$$

$$V_1 + V_1 + \frac{V_1}{2} - 12 = 2\left(\frac{V_1}{2}\right)$$

$$1.5V_1 = 12$$

$$V_1 = 8 \text{ V}$$

$$I_{SC} = \frac{V_1}{2} = \frac{8}{2} = 4 \text{ A}$$

Step III: *Calculation of R_{Th}*

$$R_{Th} = \frac{V_{Th}}{I_{SC}}$$

$$= \frac{6}{4} = 1.5 \ \Omega$$

Example 3.35 *Find the current in the 9-Ω resistor.*

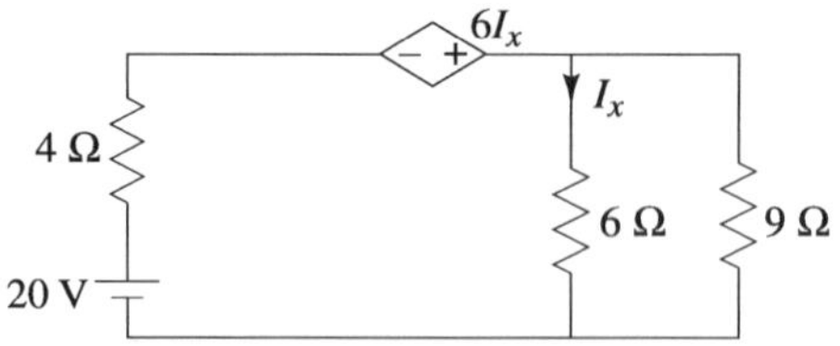

Fig. 3.153

Step I: *Calculation of V_{Th}*

Applying KVL to the Mesh,

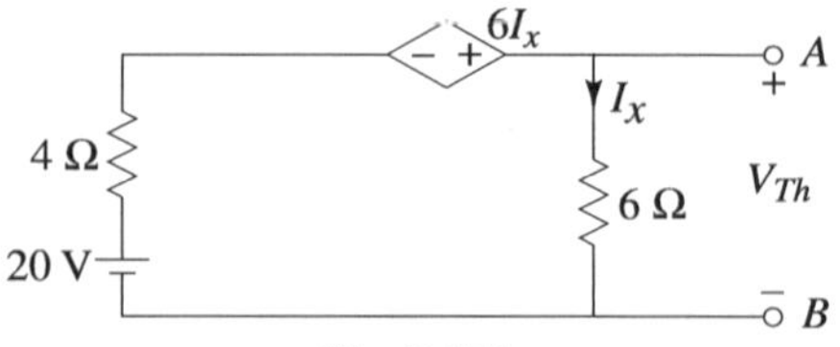

Fig. 3.154

$$20 - 4I_x + 6I_x - 6I_x = 0$$

$$20 = 4I_x$$

$$I_x = 5 \text{ A}$$

$$V_{Th} = 6I_x$$

$$= 6 \times 5 = 30 \text{ V}$$

Step II: *Calculation of I_{SC}*

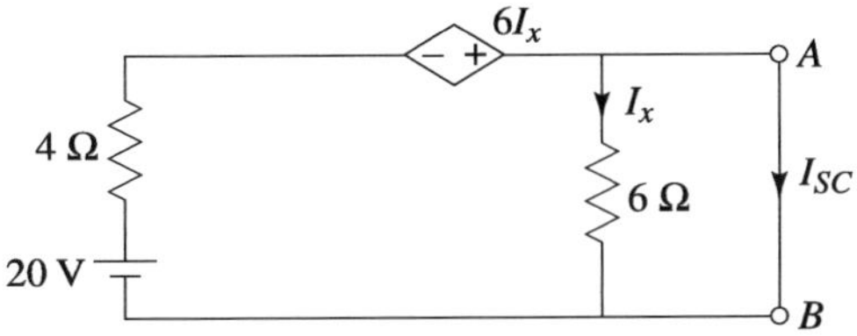

Fig. 3.155

From the figure,

$$I_x = 0$$

The dependent source $6I_x$ depends on the controlling variable I_x. When $I_x = 0$, the dependent source vanishes, i.e., $6I_x = 0$

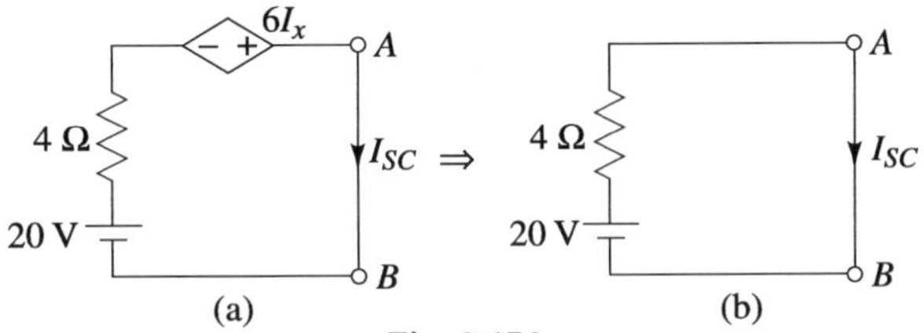

Fig. 3.156

$$I_{SC} = \frac{20}{4} = 5 \text{ A}$$

Step III: *Calculation of R_{Th}*

$$R_{Th} = \frac{V_{Th}}{I_{SC}} = \frac{30}{5} = 6\ \Omega$$

Step IV: *Calculation of I_L*

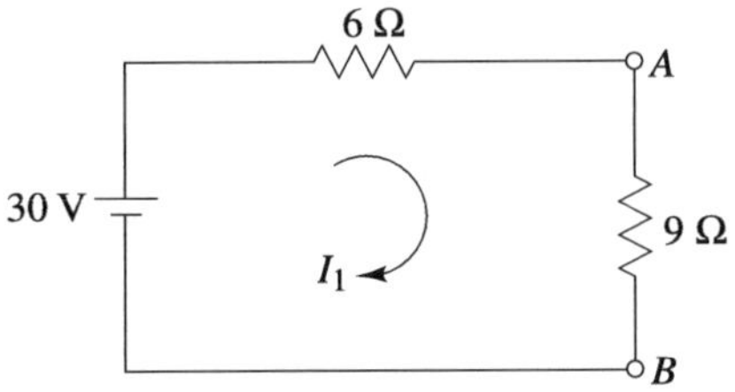

Fig. 3.157

$$I_L = \frac{30}{6+9} = 2 \text{ A}$$

Example 3.36 *Find the current in the 10-Ω resistor .*

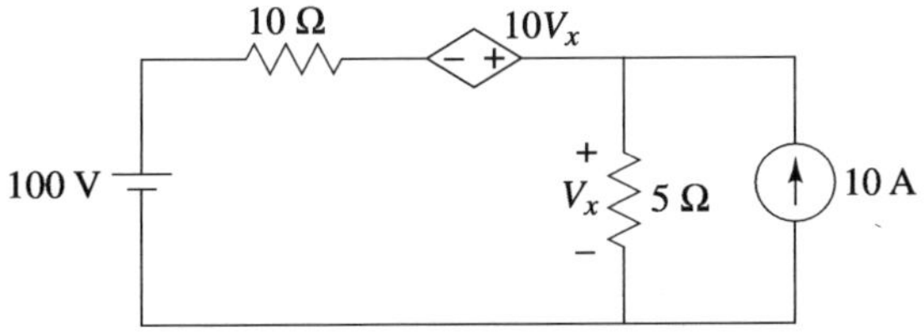

Fig. 3.158

Step I: *Calculation of V_{Th}*

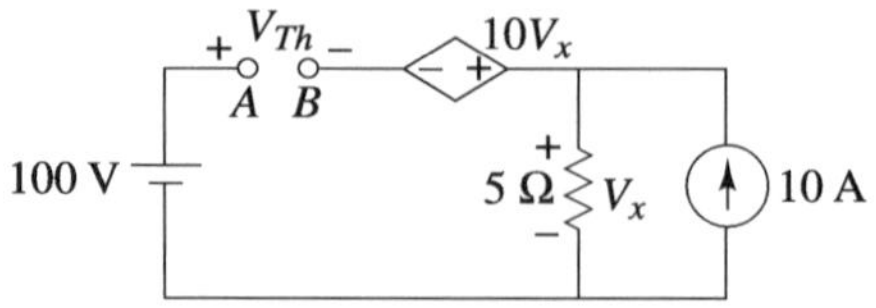

Fig. 3.159

From the figure,

$$V_x = 10 \times 5 = 50 \text{ V}$$
$$V_A = 100 \text{ V}$$
$$V_B = -10V_x + V_x = -9V_x = -9 \times 50 = -450 \text{ V}$$

Writing V_{Th} equation,

$$V_{Th} = V_A - V_B = 100 - (-450) = 550 \text{ V}$$

Step II: *Calculation of I_{SC}*

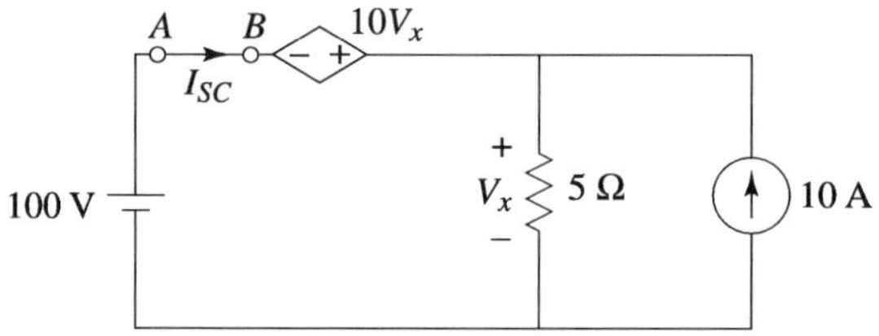

Fig. 3.160

$$100 + 10V_x - V_x = 0$$
$$V_x = -\frac{100}{9} \text{ V}$$

Also,
$$V_x = 5(I_{SC} + 10)$$
$$-\frac{100}{9} = 5I_{SC} + 50$$
$$I_{SC} = -\frac{550}{45} \text{ A}$$

Step III: *Calculation of R_{Th}*

$$R_{Th} = \frac{550}{-\frac{550}{45}} = -45\ \Omega$$

Step IV: *Calculation of I_L*

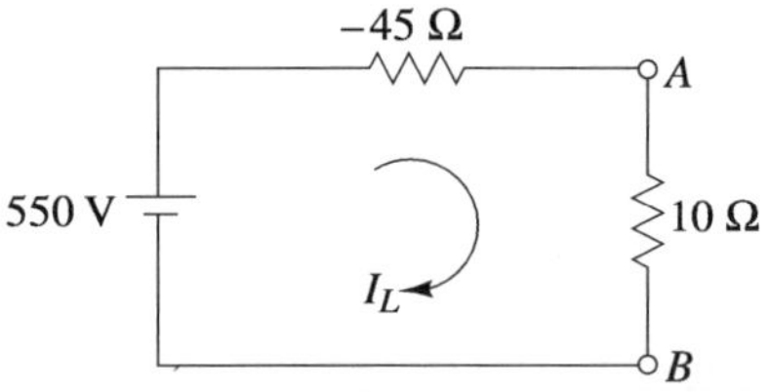

Fig. 3.161

$$I_L = \frac{550}{-45+10} = -\frac{550}{35} = -\frac{110}{7} \text{ A}$$

Example 3.37 *Determine the current in the 16-Ω resistor .*

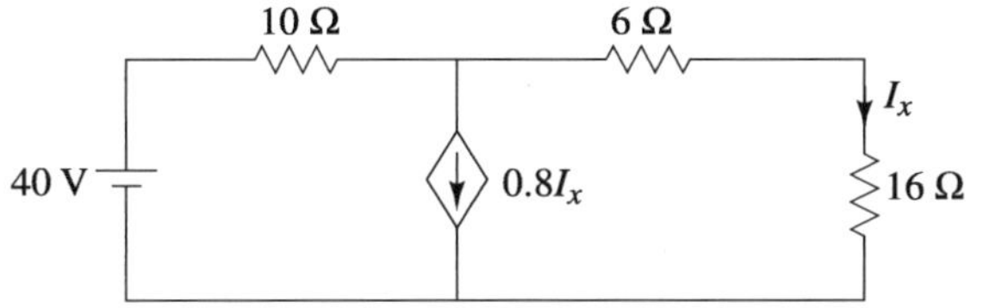

Fig. 3.162

Step I: *Calculation of V_{Th}*

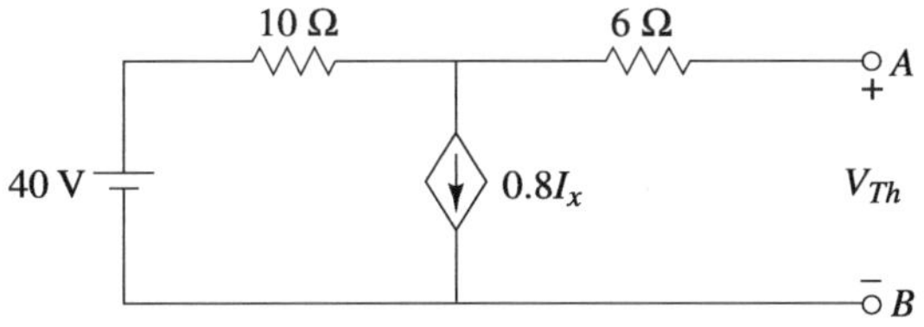

Fig. 3.163

From the figure,

$$I_x = 0$$

The dependent source $0.8I_x$ depends on the controlling variable I_x. When $I_x = 0$, the dependent source vanishes,

i.e., $$0.8I_x = 0$$

$$V_{Th} = 40 \text{ V}$$

Step II: *Calculation of I_{SC}*

From the figure,

$$I_x = \frac{V_1}{6}$$

Applying KCL at Node 1,

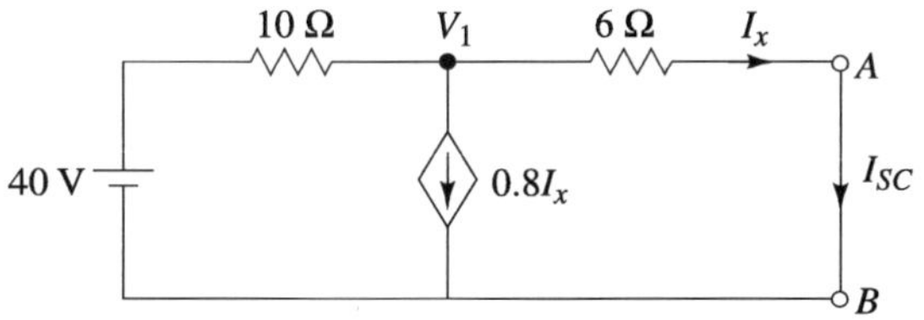

Fig. 3.164

$$\frac{V_1 - 40}{10} + 0.8I_x + I_x = 0$$

$$\frac{V_1}{10} - 4 + 1.8I_x = 0$$

$$\frac{V_1}{10} + 1.8\frac{V_1}{6} = 4$$

$$0.1V_1 + 0.3V_1 = 4$$

$$V_1 = \frac{4}{0.4} = 10 \text{ V}$$

$$I_{SC} = I_x = \frac{V_1}{6} = \frac{10}{6} \text{ A}$$

Step III: *Calculation of R_{Th}*

$$R_{Th} = \frac{V_{Th}}{I_{SC}} = \frac{40}{\frac{10}{6}} = 24 \ \Omega$$

Step IV: *Calculation of I_L*

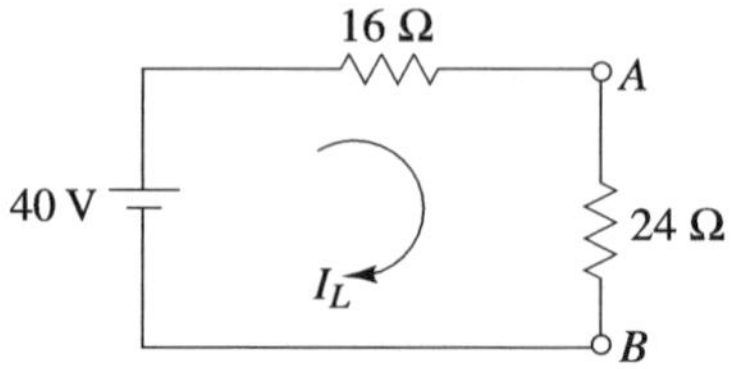

Fig. 3.165

$$I_L = \frac{40}{16+24} = 1 \text{ A}$$

Example 3.38 *Find the current in the 6-Ω resistor.*

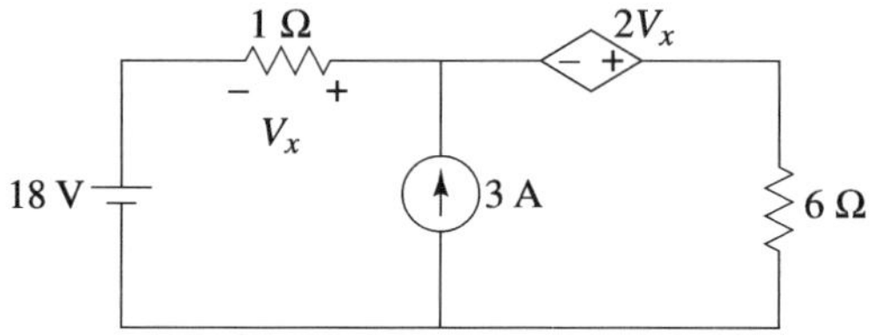

Fig. 3.166

Step I: *Calculation of V_{Th}*
Applying KCL at Node 1,

$$\frac{V_1 - 18}{1} = 3$$

$$V_1 = 21$$

Writing V_{Th} equation,

$$\begin{aligned} V_{Th} &= V_1 + 2V_x \\ &= V_1 + 2\,(V_1 - 18) \\ &= V_1 + 2V_1 - 36 = 3V_1 - 36 \\ &= 63 - 36 = 27 \text{ V} \end{aligned}$$

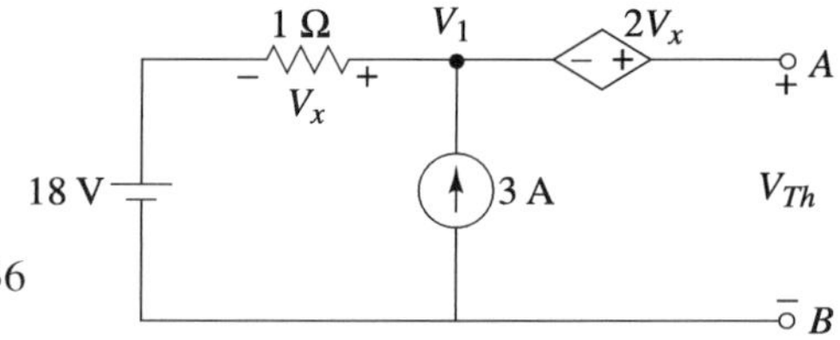

Fig. 3.167

Step II: *Calculation of I_{SC}*
From the figure,

$$\begin{aligned} I_2 &= I_{SC} \\ V_x &= -1\,(I_1) \\ I_1 &= -V_x \\ I_2 - I_1 &= 3 \\ I_{SC} + V_x &= 3 \\ I_{SC} &= 3 - V_x \end{aligned}$$

Also

$$\begin{aligned} 18 + V_x + 2V_x &= 0 \\ 18 + 3V_x &= 0 \\ V_x &= -6 \text{ V} \\ I_{SC} &= 3 - (-6) = 9 \text{ A} \end{aligned}$$

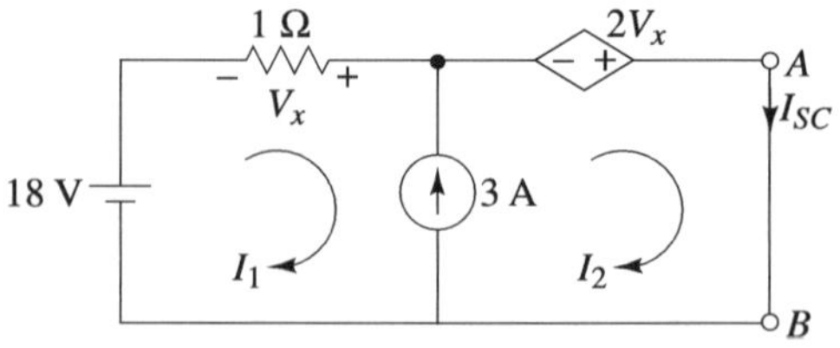

Fig. 3.168

Step III: *Calculation of R_{Th}*

$$R_{Th} = \frac{V_{Th}}{I_{SC}} = \frac{27}{9} = 3\ \Omega$$

Step IV: *Calculation of I_L*

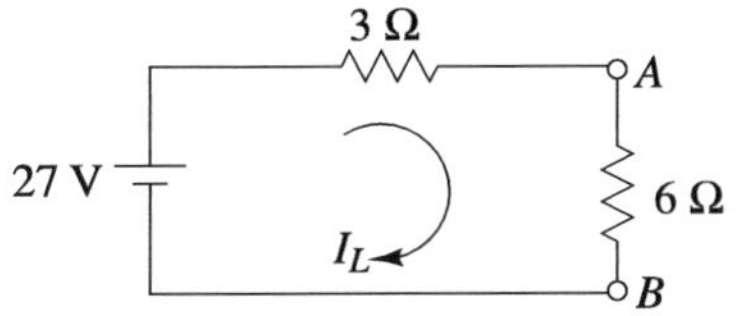

Fig. 3.169

$$I_L = \frac{27}{3+6} = 3 \text{ A}$$

Example 3.39 *Obtain the Thevenin equivalent network for the given network at terminals a and b.*

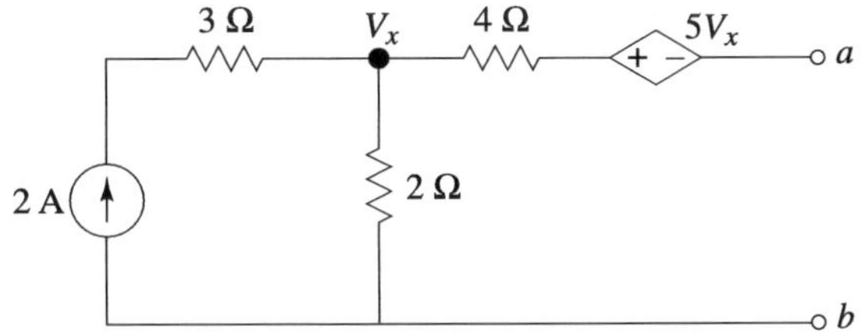

Fig. 3.170

Step I: *Calculation of V_{Th}*
Applying KCL at Node x,

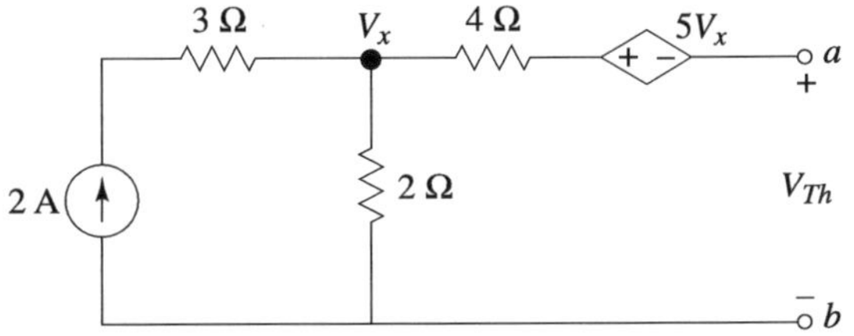

Fig. 3.171

$$2 = \frac{V_x}{2}$$

$$V_x = 4 \text{ V}$$

Writing V_{Th} equation,

$$V_{Th} = V_x - 5V_x = -4V_x$$

$$= -16 \text{ V}$$

(the terminal a is negative w.r.t. b)

Step II: *Calculation of I_{SC}*
Applying KCL at Node x,

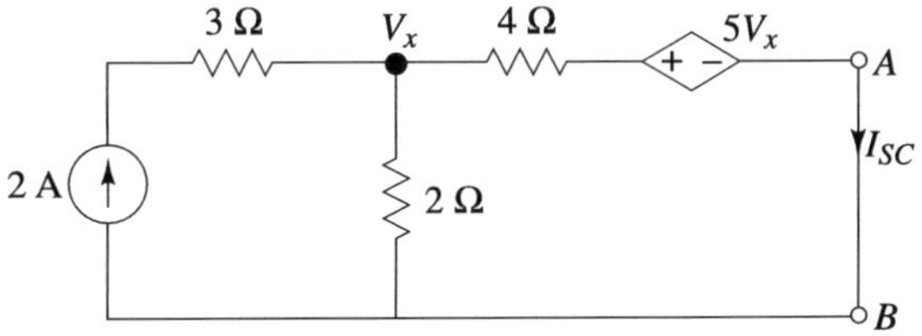

Fig. 3.172

$$2 = \frac{V_x}{2} + \frac{V_x - 5V_x}{4}$$

$$2 = \frac{V_x}{2} - V_x = -\frac{V_x}{2}$$

$$V_x = -4 \text{ V}$$

$$I_{SC} = \frac{V_x - 5V_x}{4}$$

$= -V_x = -4$ (I_{SC} is flowing towards Node x)

Step II: *Calculation of R_{Th}*

$$R_{Th} = \frac{V_{Th}}{I_{SC}}$$

$$= \frac{16}{-4} = -4\ \Omega$$

Example 3.40 *Obtain the Thevenin equivalent network for the given network.*

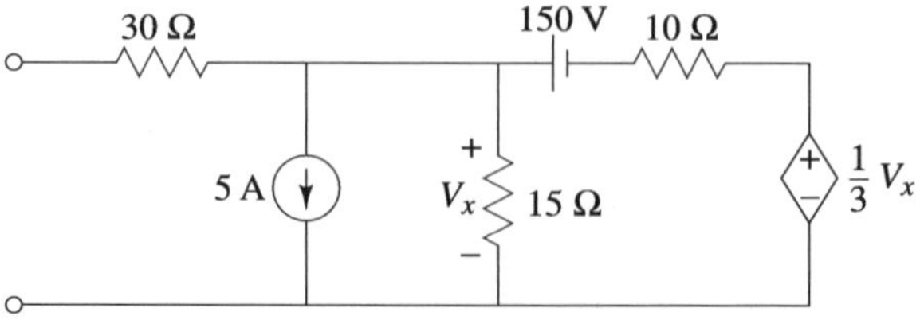

Fig. 3.173

Step I: *Calculation of V_{Th}*

Applying KCL at the node,

$$\frac{V_x - 150 - \frac{1}{3}V_x}{10} + \frac{V_x}{15} + 5 = 0$$

$$V_x = 75 \text{ V}$$

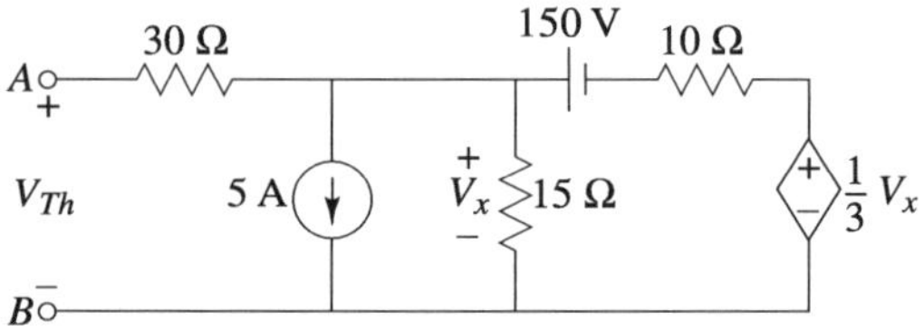

Fig. 3.174

From the figure,

$$V_x = V_{Th}$$

$$V_{Th} = 75 \text{ V}$$

Step II: *Calculation of I_{SC}*

Applying KCL at Node x,

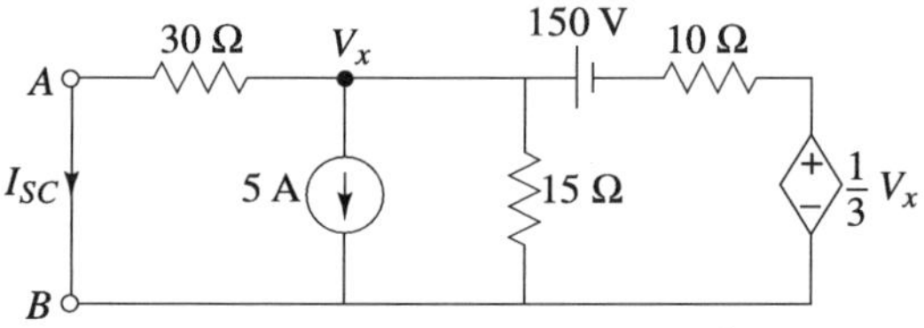

Fig. 3.175

$$\frac{V_x}{30} + 5 + \frac{V_x}{15} + \frac{V_x - 150 - \frac{1}{3}V_x}{10} = 0$$

$$\frac{V_x}{30} + \frac{V_x}{15} + \frac{V_x}{10} - \frac{V_x}{30} = 15 - 5$$

$$5V_x = 300$$

$$V_x = 60 \text{ V}$$

$$I_{SC} = \frac{V_x}{30} = \frac{60}{30} = 2 \text{ A}$$

Step III: Calculation of R_{Th}

$$R_{Th} = \frac{V_{Th}}{I_{SC}} = \frac{75}{2} \; \Omega$$

3.6 MAXIMUM POWER TRANSFER THEOREM

It states that the maximum power is delivered from a source to a load when the load resistance is equal to the source resistance.

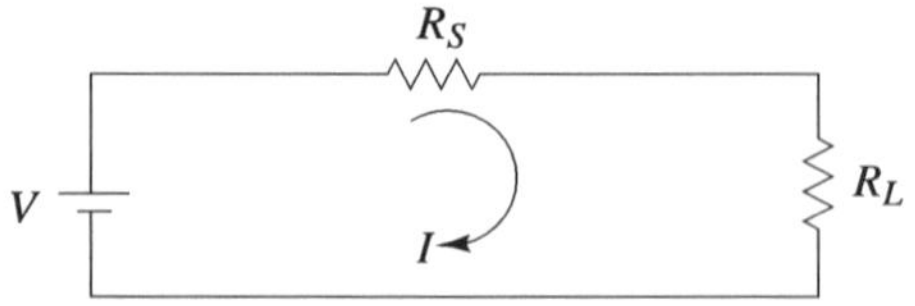

Fig. 3.176

$$I = \frac{V}{R_S + R_L}$$

Power delivered to the load $R_L = P = I^2 R_L = \dfrac{V^2 R_L}{(R_S + R_L)^2}$

To determine the value of R_L for maximum power to be transferred to the load,

$$\frac{dP}{dR_L} = 0$$

$$\frac{dP}{dR_L} = \frac{d}{dR_L}\frac{V^2}{(R_S + R_L)^2}R_L$$

$$= \frac{V^2[(R_S + R_L)^2 - (2R_L)(R_S + R_L)]}{(R_S + R_L)^4}$$

$$(R_S + R_L)^2 - 2R_L(R_S + R_L) = 0$$

$$R_S^2 + R_L^2 + 2R_S R_L - 2R_L R_S - 2R_L^2 = 0$$

$$R_S = R_L$$

Hence, the maximum power will be transferred to the load when load resistance is equal to the source resistance.

Steps to be followed in maximum power transfer theorem

1. Remove the variable load resistor R_L.
2. Find the open circuit voltage V_{Th} across points A and B.
3. Find the resistance R_{Th} as seen from points A and B with voltage source and current source replaced by internal resistance.
4. Find the resistance R_L for maximum power transfer.

$$R_L = R_{Th}$$

5. Find the maximum power.

$$I_L = \frac{V_{Th}}{R_{Th} + R_L} = \frac{V_{Th}}{2R_{Th}}$$

$$P_{\max} = I_L^2 R_L$$

$$= \frac{V_{Th}^2}{4R_{Th}^2} \times R_{Th} = \frac{V_{Th}^2}{4R_{Th}}$$

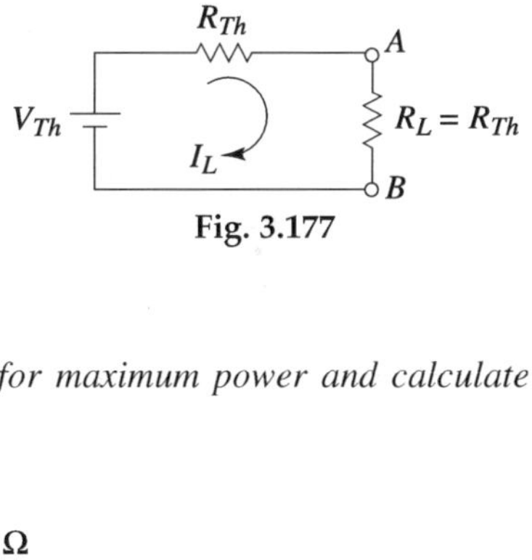

Fig. 3.177

Example 3.41 *For the circuit shown, find value of resistance R_L for maximum power and calculate maximum power.*

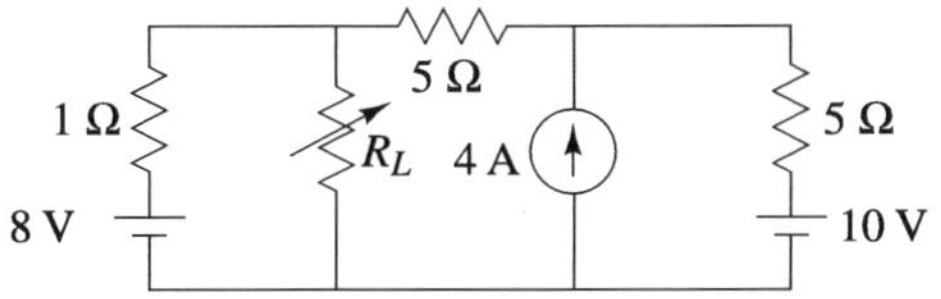

Fig. 3.178

Step I: *Calculation of V_{Th}*

Removing the variable resistor R_L from the network,

$$I_2 - I_1 = 4 \qquad \text{...(1)}$$

Applying KVL to the outer path,

$$8 - I_1 - 5I_1 - 5I_2 - 10 = 0$$

$$-6I_1 - 5I_2 = 2 \qquad \text{...(2)}$$

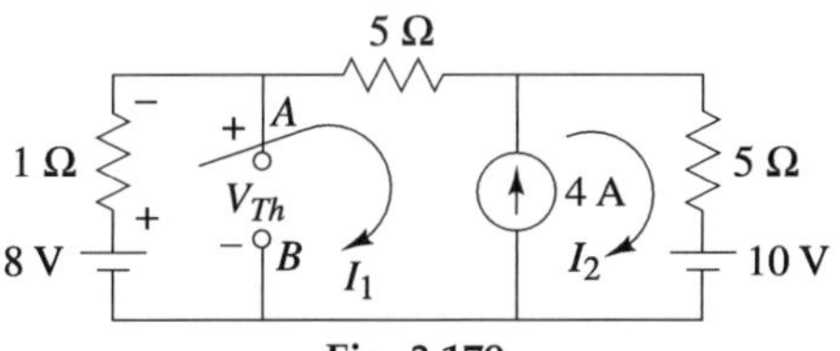

Fig. 3.179

Solving Eqs (1) and (2),

$$I_1 = -2 \text{ A}$$
$$I_2 = 2 \text{ A}$$

Writing V_{Th} equation,

$$8 - 1(I_1) - V_{Th} = 0$$
$$V_{Th} = 8 - I_1 = 8 - (-2) = 10 \text{ V}$$

Step II: *Calculation of R_{Th}*

Replacing the voltage sources by short circuits and current source by an open circuit,

$$R_{Th} = 10\ \Omega \parallel 1\ \Omega = 0.91\ \Omega$$

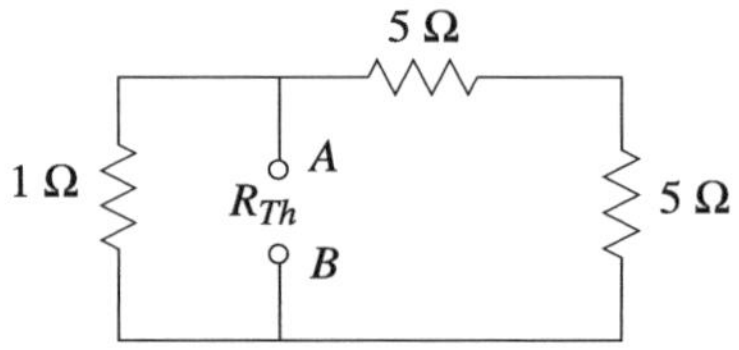

Fig. 3.180

Step III: *Value of R_L*

For maximum power transfer,

$$R_L = R_{Th} = 0.91\ \Omega$$

Step IV: *Calculation of P_{max}*

$$P_{\max} = \frac{V_{Th}^2}{4R_{Th}}$$

$$= \frac{(10)^2}{4 \times 0.91} = 27.47 \text{ W}$$

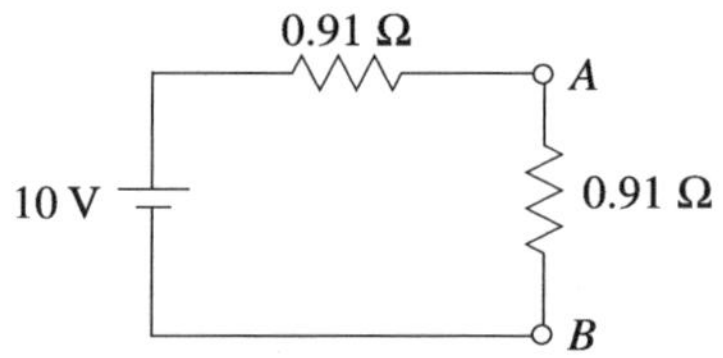

Fig. 3.181

Example 3.42 *For the circuit shown, find the value of the resistance R_L for maximum power and calculate the maximum power.*

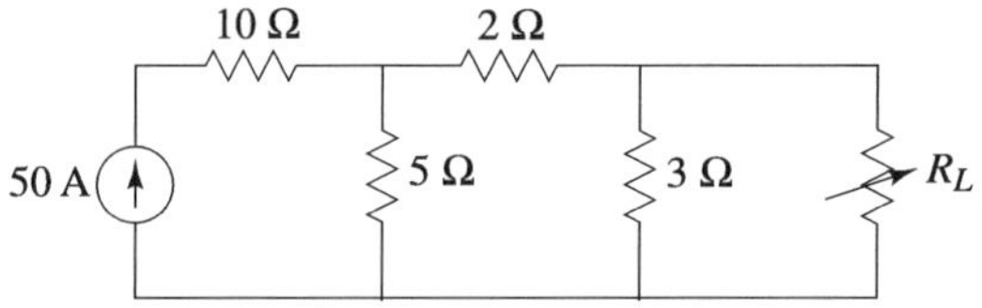

Fig. 3.182

Step I: *Calculation of V_{Th}*

Removing the variable resistor R_L from the circuit,

For Mesh 1,

$$I_1 = 50 \text{ A}$$

Applying KVL to Mesh 2,

$$-5(I_2 - I_1) - 2I_2 - 3I_2 = 0$$
$$5I_1 - 10I_2 = 0$$
$$I_1 = 2I_2$$
$$I_2 = 25 \text{ A}$$
$$V_{Th} = 3I_2 = 3(25) = 75 \text{ V}$$

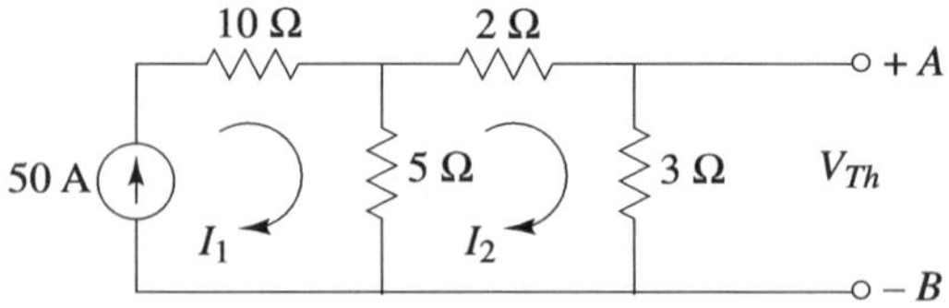

Fig. 3.183

Step II: *Calculation of R_{Th}*

Replacing the current source of 50 A with an open circuit,

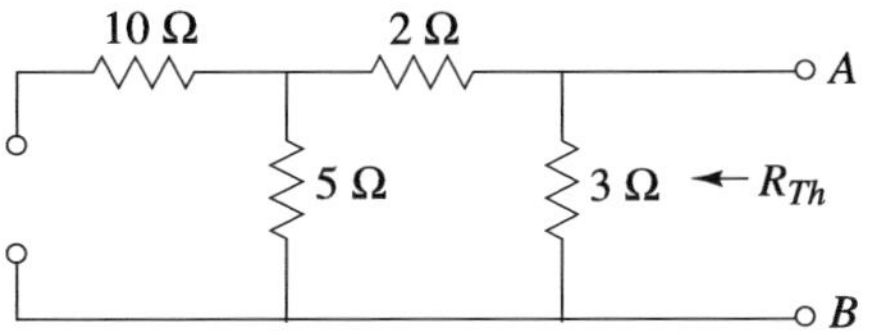

Fig. 3.184

$$R_{Th} = 7 \parallel 3 = 2.1\ \Omega$$

Step III: *Value of R_L*

For maximum power transfer,

$$R_L = R_{Th} = 2.1\ \Omega$$

Step IV: *Calculation of P_{max}*

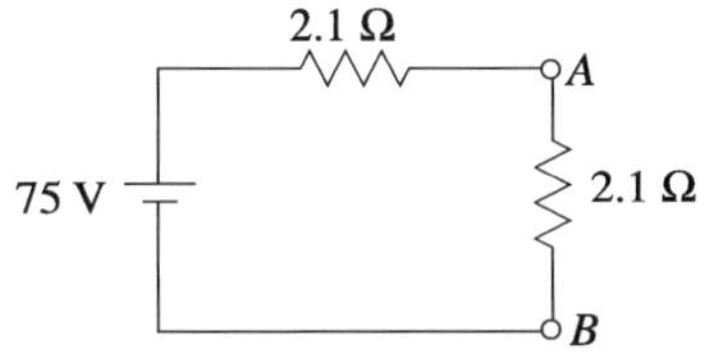

Fig. 3.185

$$P_{\max} = \frac{V_{Th}^2}{4R_{Th}} = \frac{(75)^2}{4 \times 2.1} = 669.64\ \text{W}$$

Example 3.43 *For the circuit shown, find value of resistance R_L for maximum power and calculate maximum power.*

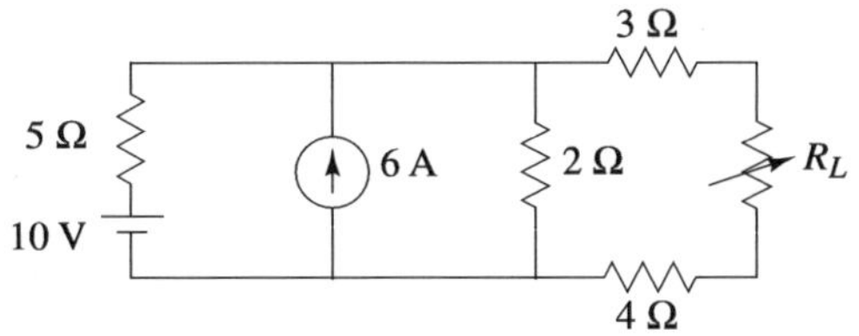

Fig. 3.186

Step I: *Calculation of V_{Th}*

Removing the variable resistor R_L from the circuit,

Writing the current equation for the supermesh,

$$I_2 - I_1 = 6 \qquad \text{...(1)}$$

Applying KVL to the supermesh,

$$10 - 5I_1 - 2I_2 = 0$$

$$5I_1 + 2I_2 = 10 \qquad \text{...(2)}$$

Solving equations (1) and (2),

$$I_1 = -0.29\ \text{A}$$

$$I_2 = 5.71\ \text{A}$$

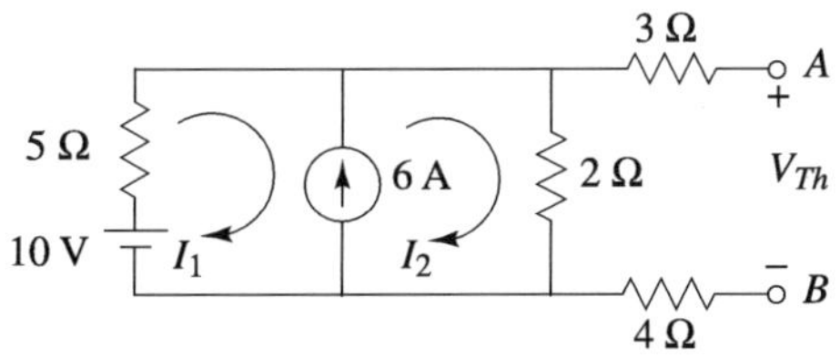

Fig. 3.187

Writing V_{Th} equation,

$$V_{Th} = 2I_2 = 11.42\ \text{V}$$

Step II: *Calculation of R_{Th}*

Replacing the voltage source by a short circuit and the current source by an open circuit,

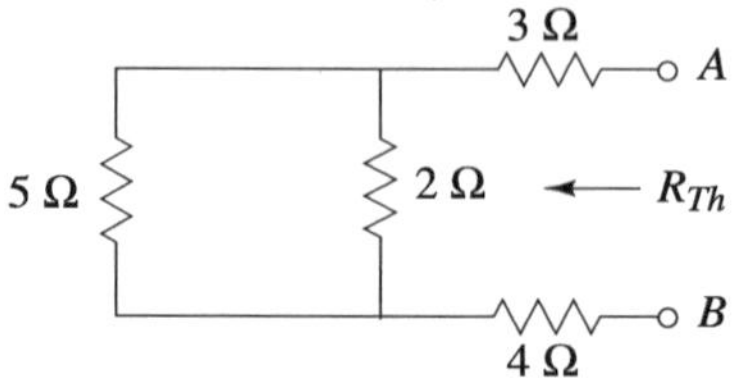

Fig. 3.188

$$R_{Th} = (5 \parallel 2) + 3 + 4 = 8.43\ \Omega$$

Step III: *Calculation of R_L*

For maximum power transfer

$$R_L = R_{Th} = 8.43\ \Omega$$

Step IV: *Calculation of P_{max}*

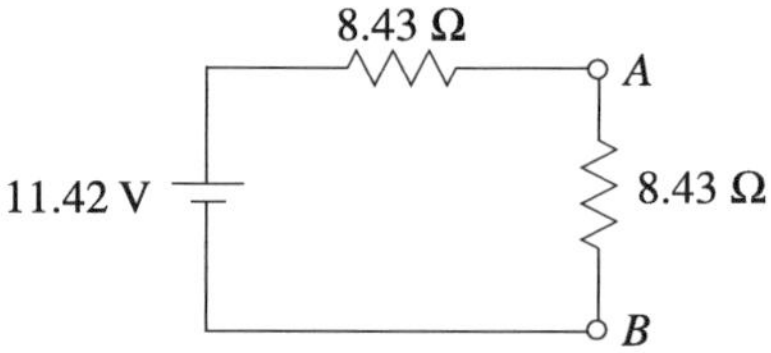

Fig. 3.189

$$P_{\max} = \frac{V_{Th}^2}{4R_{Th}} = \frac{(11.42)^2}{4 \times 8.43} = 3.87\ \text{W}$$

Example 3.44 *For the circuit shown, find the value of the resistance R_L for maximum power and calculate the maximum power.*

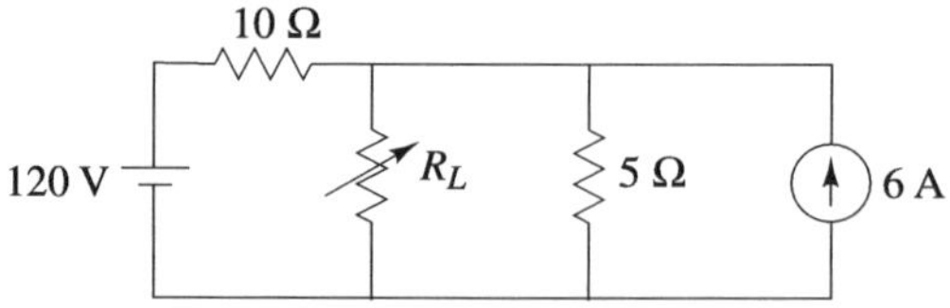

Fig. 3.190

Step I: *Calculation of V_{Th}*

Removing the variable resistor R_L from the circuit,

Applying KVL to Mesh 1,

$$120 - 10I_1 - 5(I_1 - I_2) = 0$$

$$15I_1 - 5I_2 = 120 \qquad \text{...(1)}$$

Writing current equation for Mesh 2,

$$I_2 = -6\ \text{A} \qquad \text{...(2)}$$

Solving Eqs (1) and (2),

$$I_1 = 6\ \text{A}$$

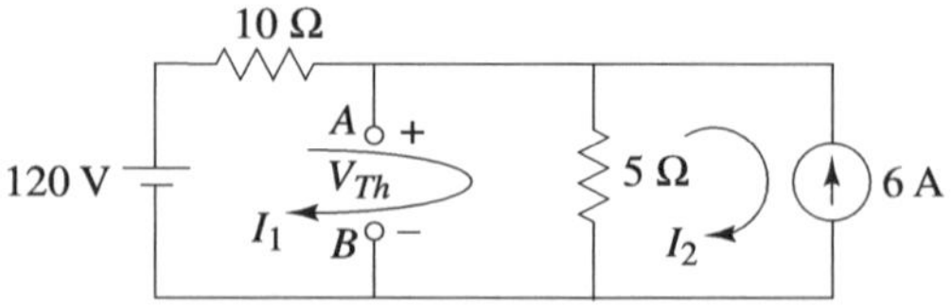

Fig. 3.191

Writing V_{Th} equation,

$$120 - 10I_1 - V_{Th} = 0$$

$$V_{Th} = 120 - 10\,(6)$$

$$= 60 \text{ V}$$

Step II: *Calculation of R_{Th}*

Replacing the voltage source by a short circuit and the current source by an open circuit,

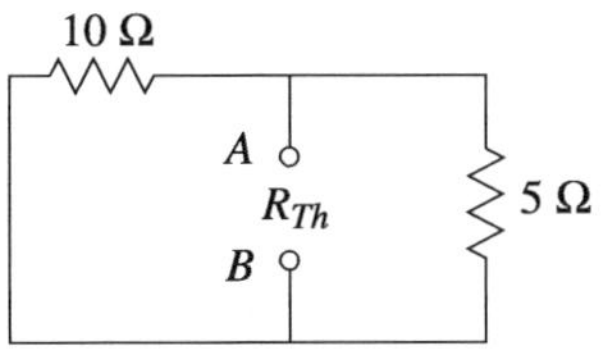

Fig. 3.192

$$R_{Th} = 10 \,||\, 5 = 3.33\ \Omega$$

Step III: *Calculation of R_L*

For maximum power transfer

$$R_L = R_{Th} = 3.33\ \Omega$$

Step IV: *Calculation of P_{max}*

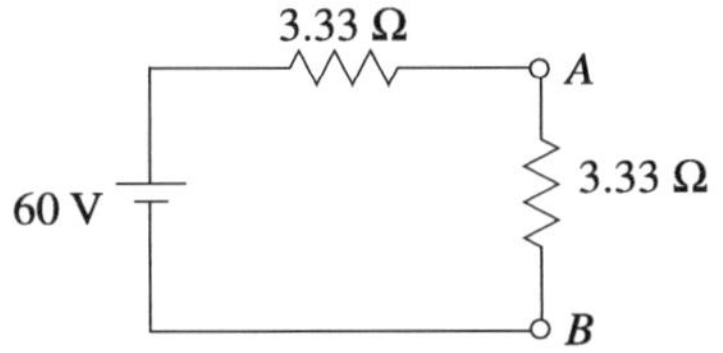

Fig. 3.193

$$P_{\max} = \frac{V_{Th}^2}{4R_{Th}} = \frac{(60)^2}{4\times 3.33} = 270.27 \text{ W}$$

Example 3.45 *For the circuit shown, find the value of the resistance R_L for maximum power and calculate the maximum power.*

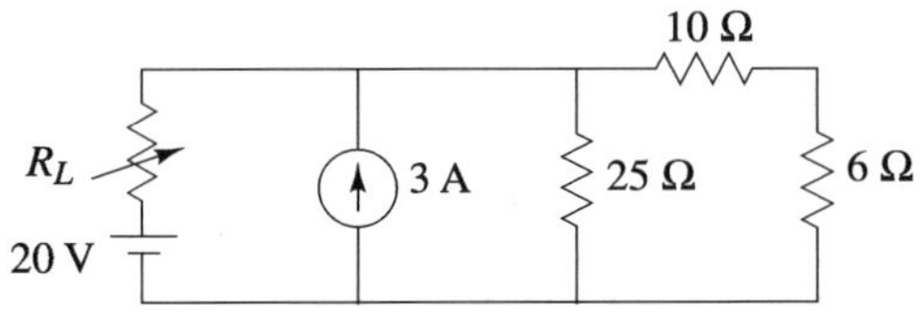

Fig. 3.194

Step I: *Calculation of V_{Th}*

Removing the variable resistor R_L from the circuit,

$$I_1 = 3 \text{ A} \quad ...(1)$$

Applying KVL to Mesh 2,

$$-25(I_2 - I_1) - 10I_2 - 6I_2 = 0$$

$$-25I_1 + 41I_2 = 0 \quad ...(2)$$

Solving Eqs (1) and (2),

$$I_2 = 1.83 \text{ A}$$

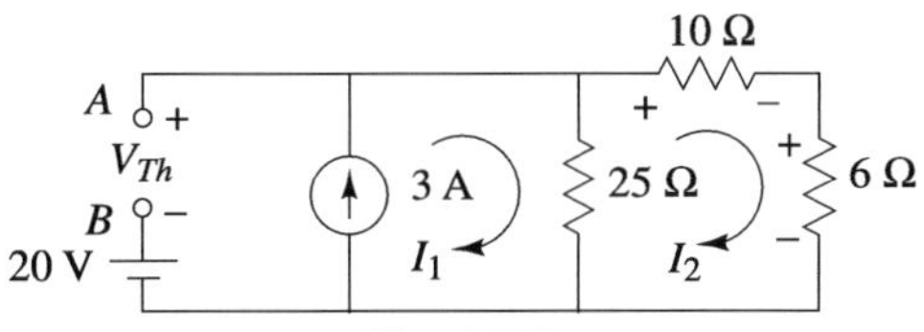

Fig. 3.195

Writing V_{Th} equation,

$$20 + V_{Th} - 10I_2 - 6I_2 = 0$$

$$V_{Th} = -20 + 10\,(1.83) + 6\,(1.83) = 9.28\text{ V}$$

Step II: *Calculation of R_{Th}*

Replacing the voltage source by a short circuit and the current source by an open circuit,

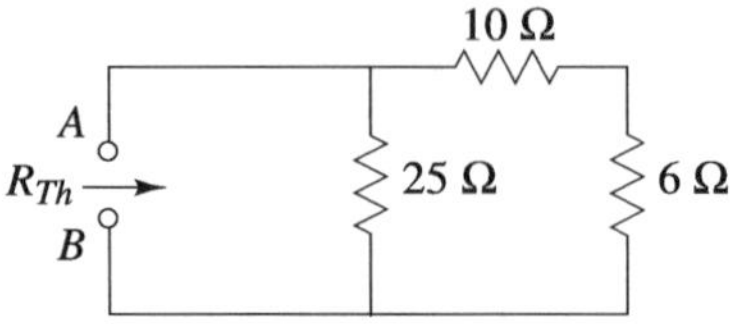

Fig. 3.196

$$R_{Th} = 25 \parallel 16 = 9.76\ \Omega$$

Step III: *Calculation of R_L*

For maximum power transfer

$$R_L = R_{Th} = 9.76\ \Omega$$

Step IV: *Calculation of P_{max}*

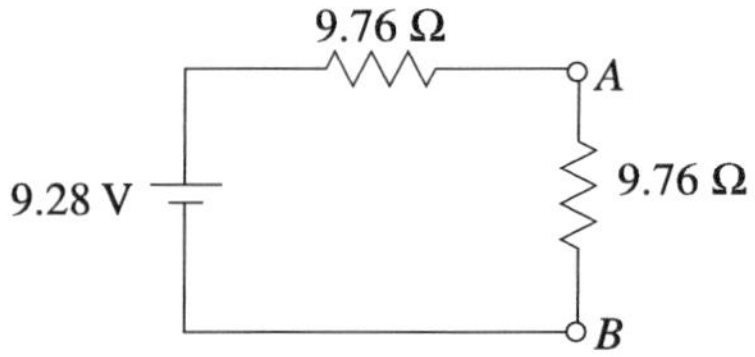

Fig. 3.197

$$P_{\max} = \frac{V_{Th}^2}{4R_{Th}} = \frac{(9.28)^2}{4\times 9.76} = 2.21\text{ W}$$

Example 3.46 *For the circuit shown, find the value of the resistance R_L for maximum power and calculate the maximum power.*

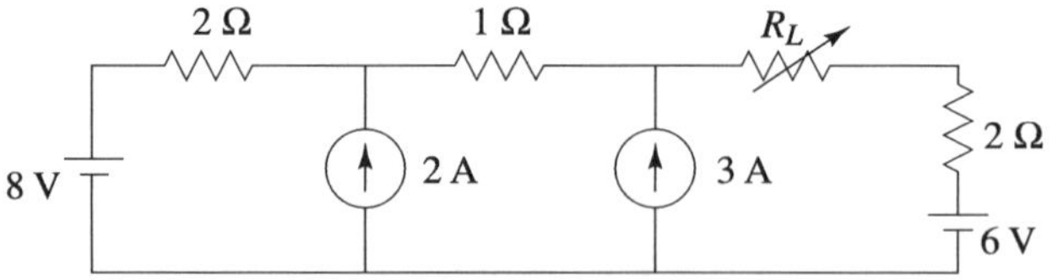

Fig. 3.198

Step I: *Calculation of V_{Th}*

Removing the variable resistor R_L from the circuit,

$$I_2 - I_1 = 2 \qquad \text{...(1)}$$

$$I_2 = -3\text{ A} \qquad \text{...(2)}$$

Solving Eqs (1) and (2),

$$I_1 = -5\text{ A}$$

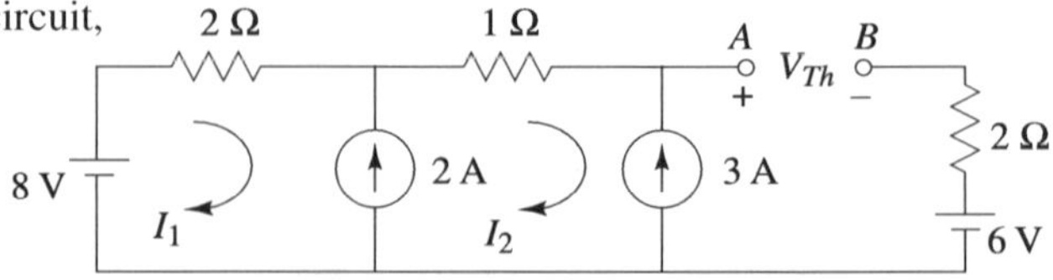

Fig. 3.199

Writing V_{Th} equation,

$$8 - 2I_1 - I_2 - V_{Th} - 6 = 0$$

$$V_{Th} = 8 - 2\,(-5) - (-3) - 6 = 15\text{ V}$$

Step II: *Calculation of R_{Th}*

Replacing the voltage source by a short circuit and the current source by an open circuit,

$$R_{Th} = 5\ \Omega$$

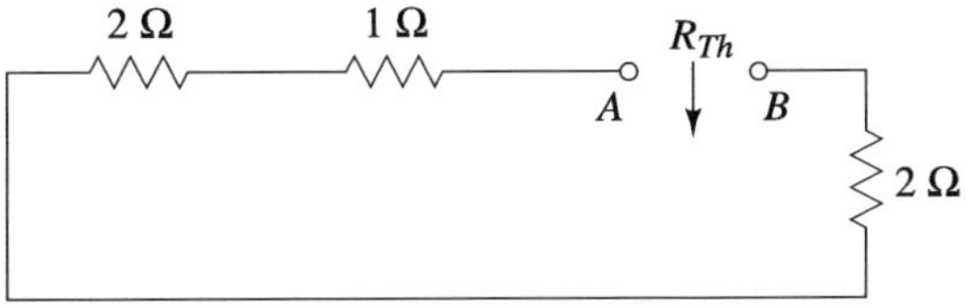

Fig. 3.200

Step III: *Calculation of R_L*

For maximum power transfer

$$R_L = R_{Th} = 5\ \Omega$$

Step IV: *Calculation of P_{max}*

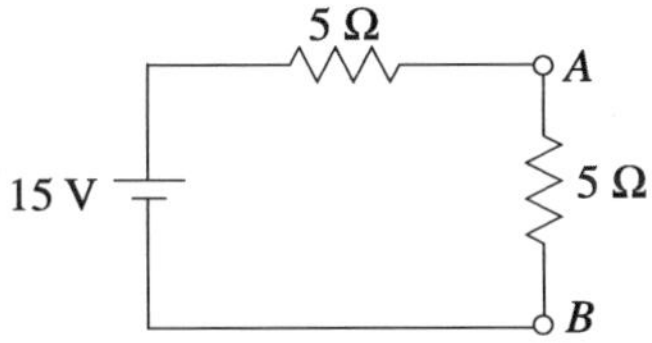

Fig. 3.201

$$P_{\max} = \frac{V_{Th}^2}{4R_{Th}} = \frac{(15)^2}{4 \times 5} = 11.25\ \text{W}$$

Example 3.47 *For the circuit shown, find the value of resistance the R_L for maximum power and calculate the maximum power.*

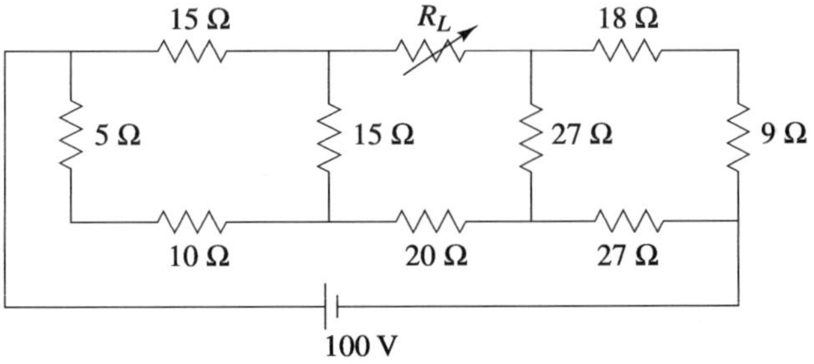

Fig. 3.202

Step I: *Calculation of V_{Th}*

Removing the variable resistor R_L from the circuit,

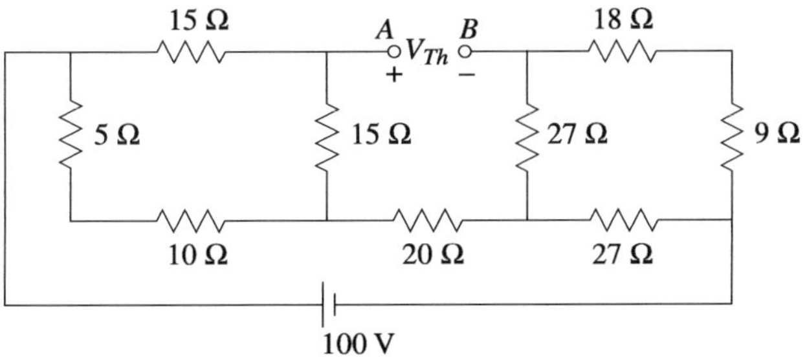

Fig. 3.203

By star–delta transformation,

$$I = \frac{100}{5+5+20+9+9} = 2.08 \text{ A}$$

Writing V_{Th} equation,

$$100 - 5I - V_{Th} - 9I = 0$$

$$V_{Th} = 100 - 14I$$

$$= 100 - 14(2.08) = 70.88 \text{ V}$$

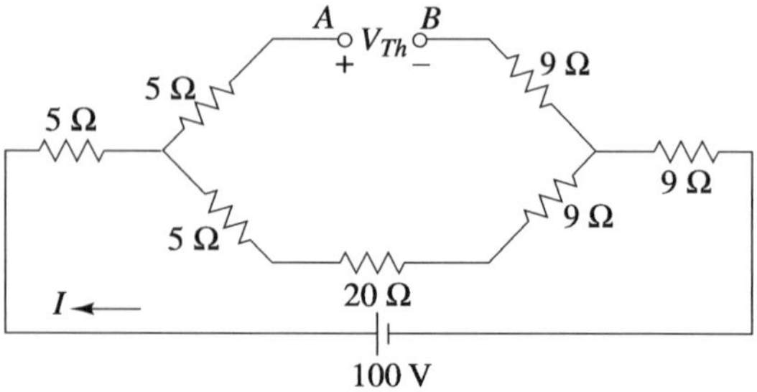

Fig. 3.204

Step II: *Calculation of R_{Th}*

Replacing the voltage source by a short circuit,

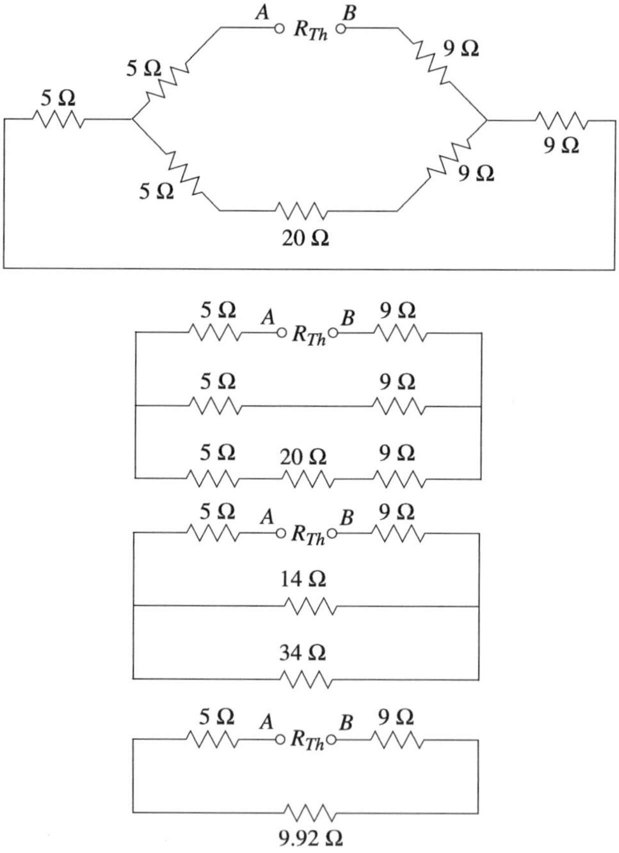

Fig. 3.205

$$R_{Th} = 23.92 \ \Omega$$

Step III: *Calculation of R_L*

For maximum power transfer,

$$R_L = R_{Th} = 23.92 \ \Omega$$

Step IV: *Calculation of P_{max}*

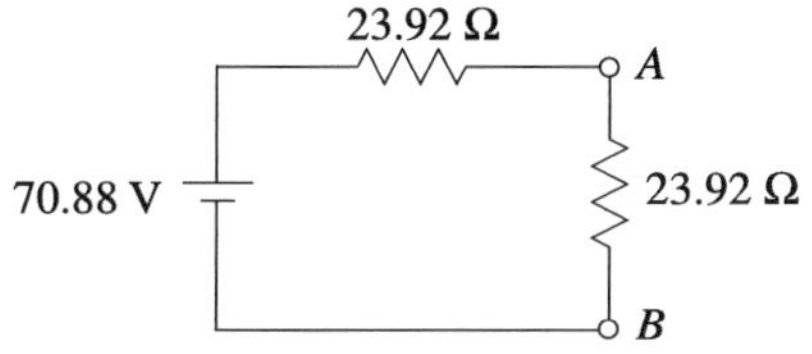

Fig. 3.206

$$P_{max} = \frac{V_{Th}^2}{4R_{Th}} = \frac{(70.88)^2}{4 \times 23.92} = 52.51 \text{ W}$$

Example 3.48 *For the circuit shown, find the value of the resistance R_L for maximum power and calculate the maximum power.*

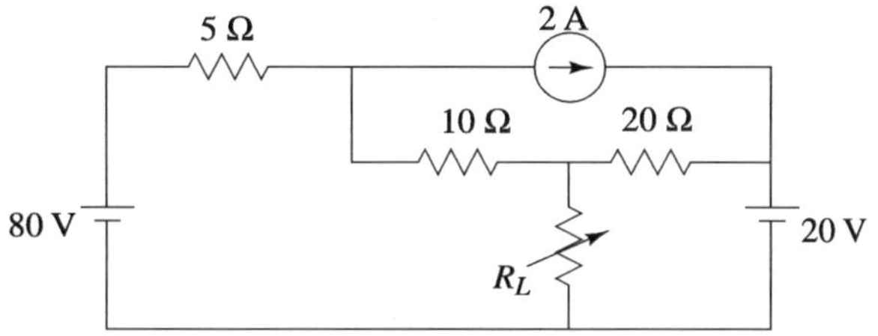

Fig. 3.207

Step I: *Calculation of V_{Th}*

Removing the variable resistor R_L from the circuit,

Applying KVL to Mesh 1,

$$80 - 5I_1 - 10(I_1 - I_2) - 20(I_1 - I_2) - 20 = 0$$

$$35I_1 - 30I_2 = 60 \quad \text{...(1)}$$

Writing the current equation for Mesh 2,

$$I_2 = 2 \quad \text{...(2)}$$

Solving Eqs (1) and (2),

$$I_1 = 3.43 \text{ A}$$

Writing V_{Th} equation,

$$V_{Th} - 20\,(I_1 - I_2) - 20 = 0$$

$$V_{Th} = 20(3.43 - 2) + 20 = 48.6 \text{ V}$$

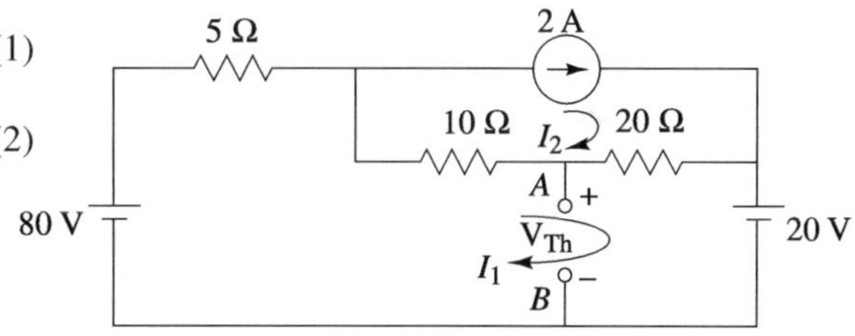

Fig. 3.208

Step II: *Calculation of R_{Th}*

Replacing the voltage sources by short circuits and the current source by an open circuit,

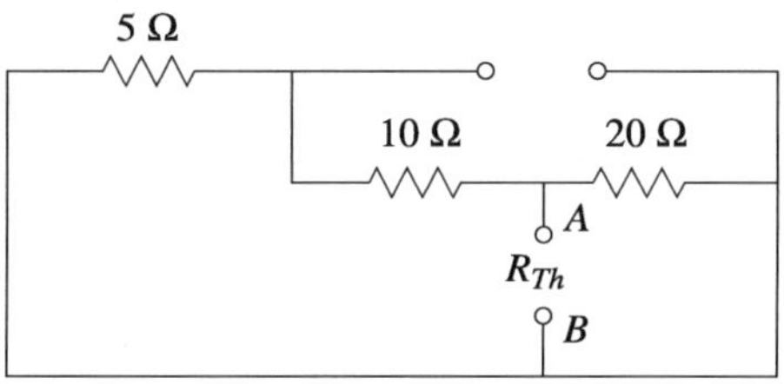

Fig. 3.209

$$R_{Th} = 15 \parallel 20 = 8.57\ \Omega$$

Step III: *Calculation of R_L*

For maximum power transfer,

$$R_L = R_{Th} = 8.57\ \Omega$$

Step IV: *Calculation of P_{max}*

$$P_{\max} = \frac{V_{Th}^2}{4R_{Th}} = \frac{(48.6)^2}{4\times 8.57} = 68.9\ \text{W}$$

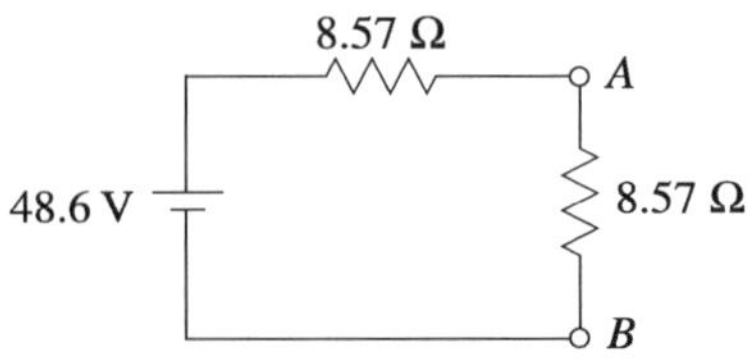

Fig. 3.210

Example 3.49 *For the circuit shown, find the value of the resistance R_L for maximum power and calculate the maximum power.*

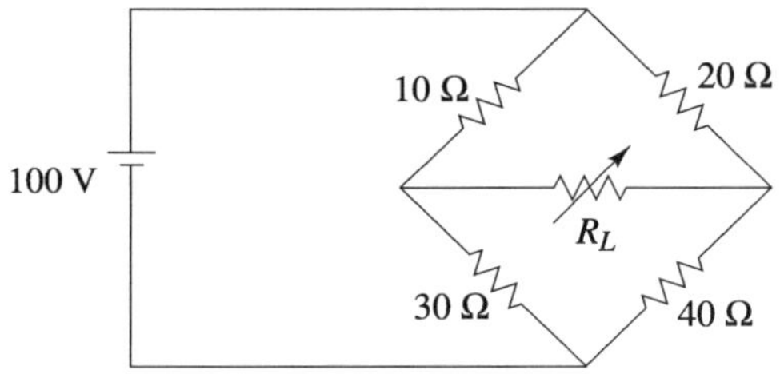

Fig. 3.211

Step I: *Calculation of V_{Th}*

Removing the variable resistor R_L from the network,

$$I_1 = \frac{100}{10+30} = 2.5\ \text{A}$$

$$I_2 = \frac{100}{20+40} = 1.66\ \text{A}$$

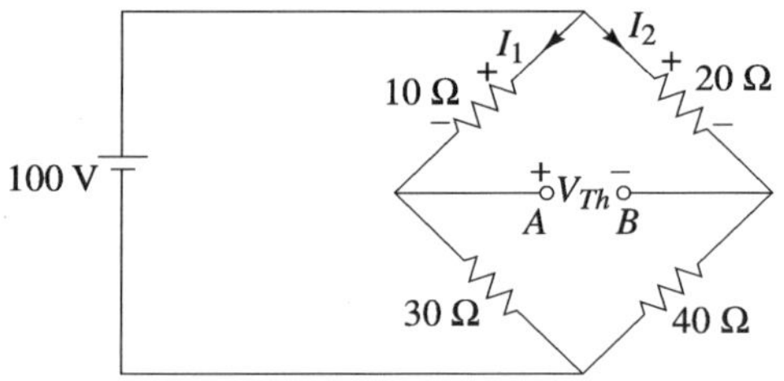

Fig. 3.212

Writing V_{Th} equation,

$$V_{Th} + 10I_1 - 20I_2 = 0$$

$$V_{Th} = 20I_2 - 10I_1$$

$$= 20(1.66) - 10(2.5) = 8.2\ \text{V}$$

Step II: *Calculation of R_{Th}*

Replacing the voltage source of 100 V with short circuits,

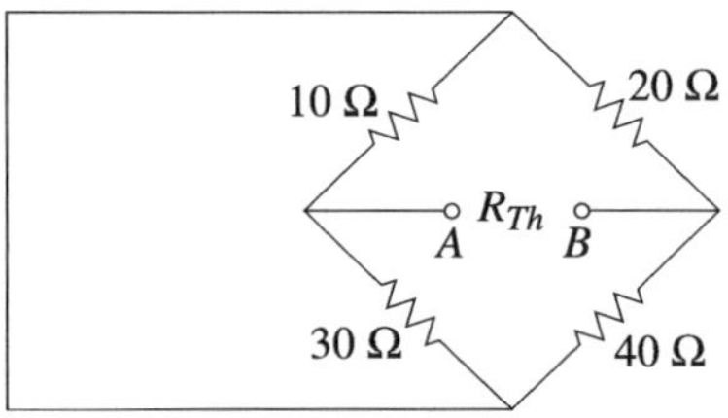

Fig. 3.213

The above circuit can be redrawn as shown:

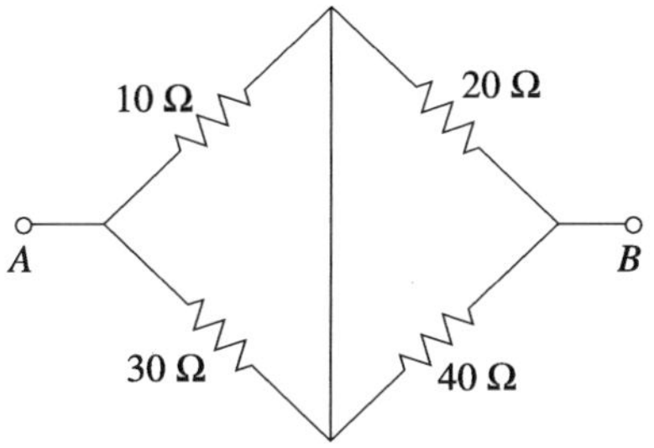

Fig. 3.214

$$R_{Th} = (10 \parallel 30) + (20 \parallel 40) = 20.83\ \Omega$$

Step III: *Value of R_L*

For maximum power transfer,

$$R_L = R_{Th} = 20.83\ \Omega$$

Step IV: *Calculation of P_{max}*

$$P_{\max} = \frac{V_{Th}^2}{4R_{Th}} = \frac{(8.2)^2}{4 \times 20.83} = 0.81\ \text{W}$$

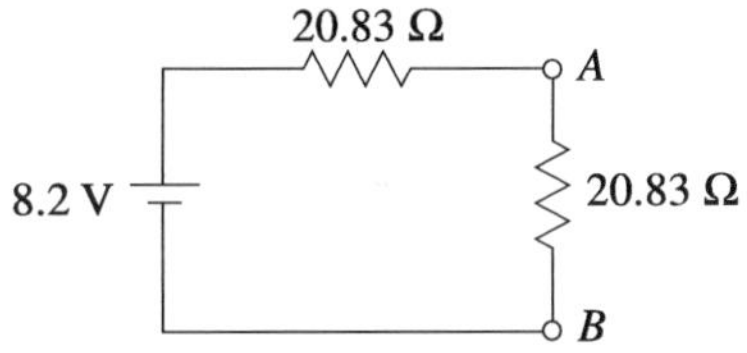

Fig. 3.215

Example 3.50 *For the circuit shown, find the value of the resistance R_L for maximum power and calculate the maximum power.*

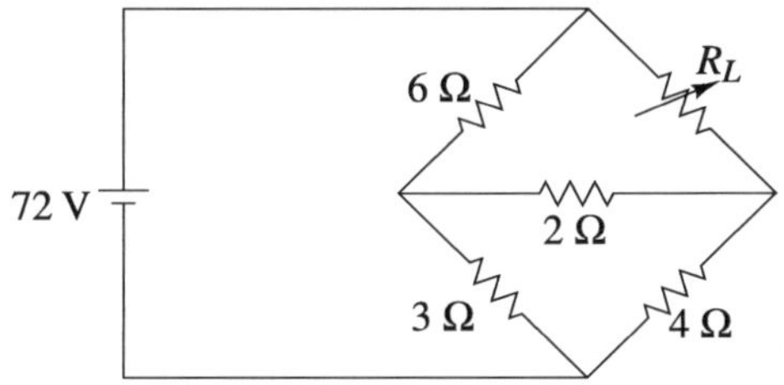

Fig. 3.216

Step I: *Calculation of V_{Th}*

Removing the variable resistor R_L from the circuit,

Applying KVL to Mesh 1,

$$72 - 6I_1 - 3(I_1 - I_2) = 0$$

$$9I_1 - 3I_2 = 72 \quad ...(1)$$

Applying KVL to Mesh 2,

$$-3(I_2 - I_1) - 2I_2 - 4I_2 = 0$$

$$-3I_1 + 9I_2 = 0 \quad ...(2)$$

Solving equations (1) and (2),

$$I_1 = 9\ \text{A}$$

$$I_2 = 3\ \text{A}$$

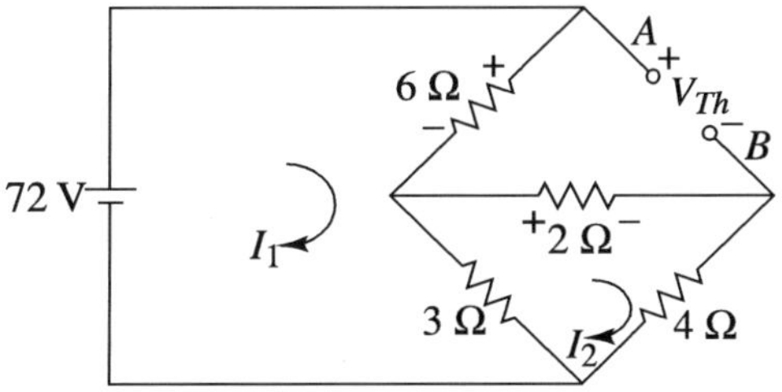

Fig. 3.217

Writing V_{Th} equation

$$V_{Th} - 6I_1 - 2I_2 = 0$$

$$V_{Th} = 6I_1 + 2I_2$$

$$= 6(9) + 2(3) = 60\ \text{V}$$

Step II: *Calculation of* R_{Th}

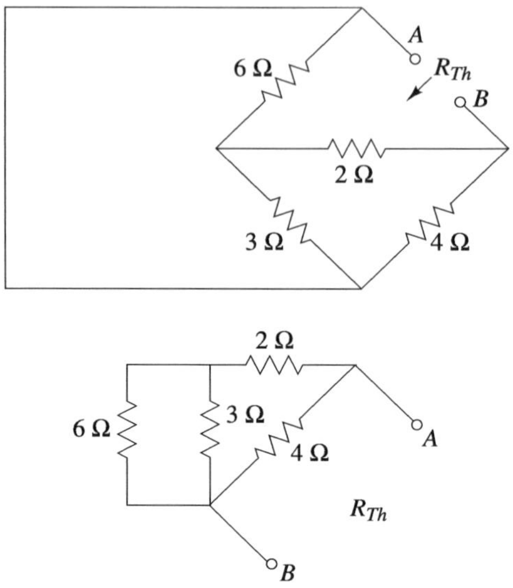

Fig. 3.218

$$R_{Th} = \{(6 \parallel 3) + 2\} \parallel 4 = 2\ \Omega$$

Step III: *Calculation of* R_L

For maximum power transfer

$$R_L = R_{Th} = 2\ \Omega$$

Step IV: *Calculation of* P_{max}

$$P_{\max} = \frac{V_{Th}^{2}}{4R_{Th}}$$

$$= \frac{(60)^2}{4 \times 2} = 450\ \text{W}$$

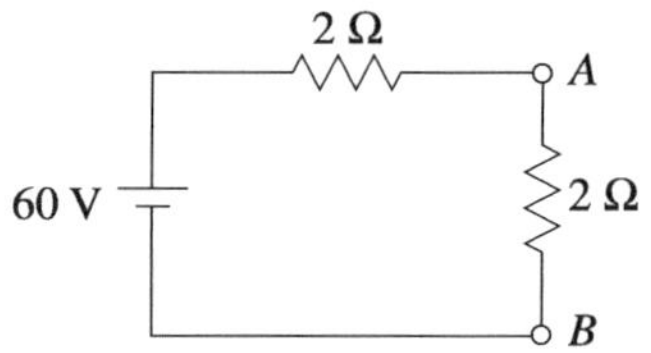

Fig. 3.219

Example 3.51 *What will be the value of* R_L *to get maximum power delivered to it? What is the value of this power?*

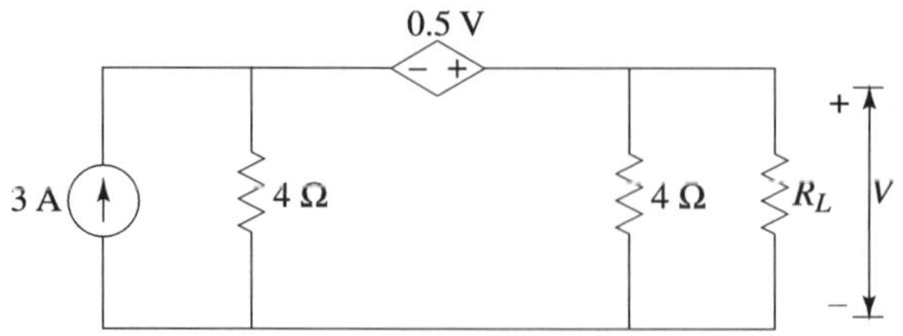

Fig. 3.220

Step I: *Calculation of* V_{Th}

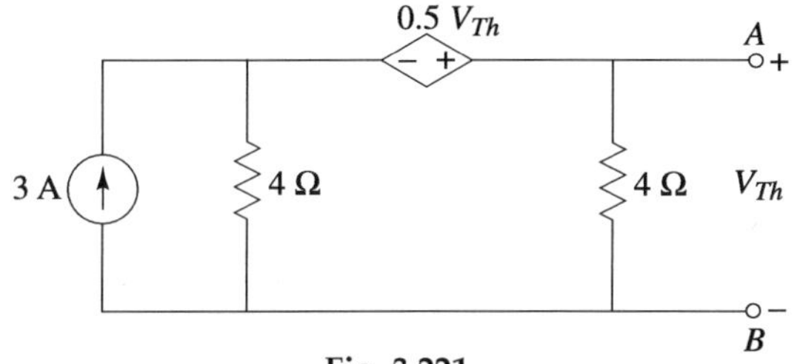

Fig. 3.221

By source transformation,
From the figure,

$$V_{Th} = 4I$$

Applying KVL to the mesh,

$$12 - 4I + 0.5V_{Th} - 4I = 0$$
$$12 - V_{Th} + 0.5V_{Th} - V_{Th} = 0$$
$$V_{Th} = 8 \text{ V}$$

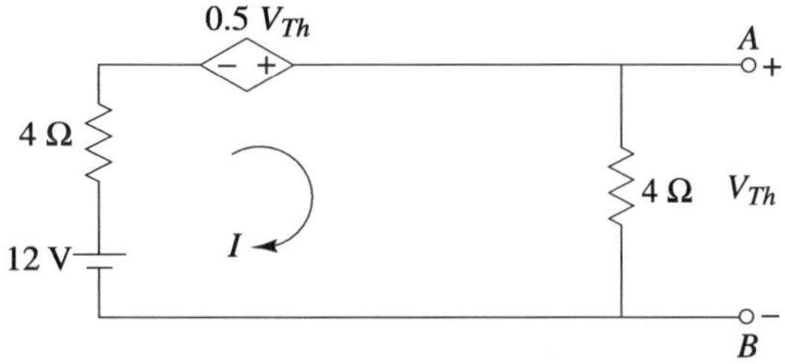

Fig. 3.222

Step II: *Calculation of* I_{SC}

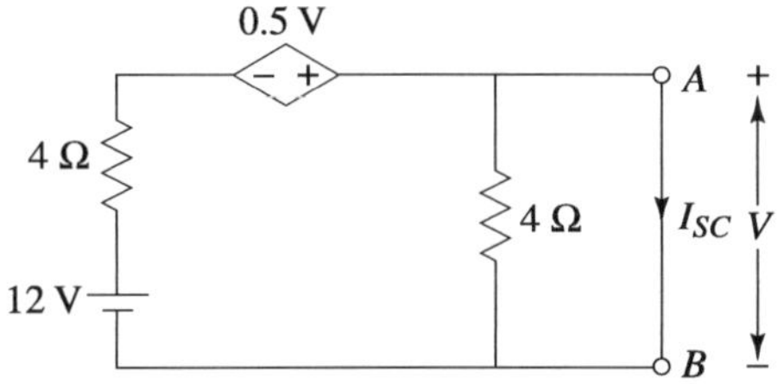

Fig. 3.223

If two terminals *A* and *B* are shorted, the 4-Ω resistor gets shorted.

$$V = 0$$

Dependent source 0.5 V depends on the controlling variable *V*. When *V* = 0, the dependent source vanishes.

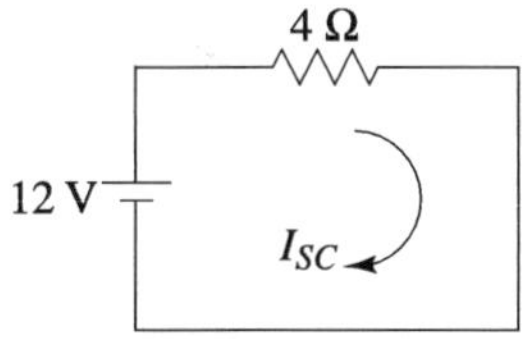

Fig. 3.224

$$I_{SC} = \frac{12}{4} = 3 \text{ A}$$

Step III: *Calculation of* R_{Th}

$$R_{Th} = \frac{V_{Th}}{I_{SC}} = \frac{8}{3} = 2.667\ \Omega.$$

Step IV: *Calculation of* R_L

For maximum power transfer

$$R_L = R_{Th} = 2.667\ \Omega.$$

Step V: *Calculation of* P_{max}

$$I_L = \frac{8}{2.667 + 2.667} = 1.5 \text{ A}$$

$$P_{\max} = (1.5)^2 \times 2.667 = 6 \text{ W}$$

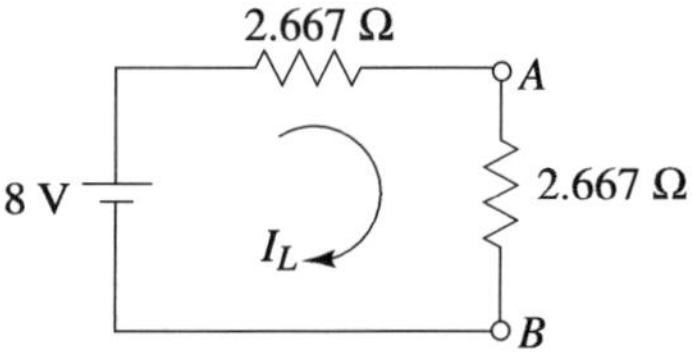

Fig. 3.225

Exercises

SUPERPOSITION THEOREM

1. Find the current through the 10-Ω resistor.

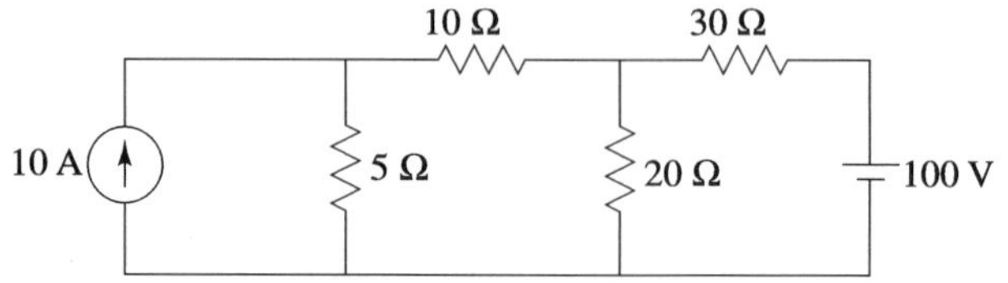

Fig. 3.226

[0.37 A]

2. Find the current through the 8-Ω resistor.

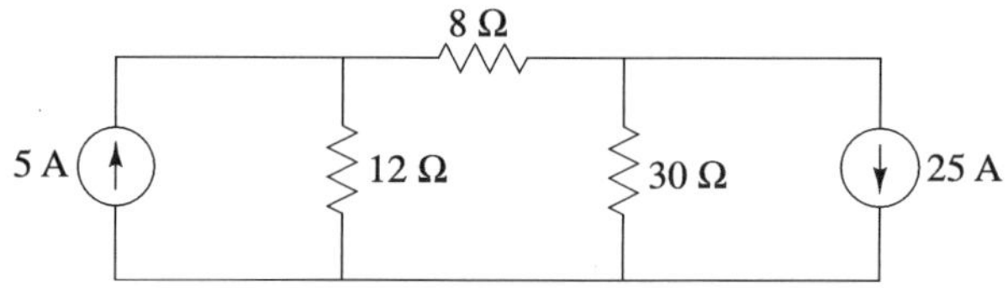

Fig. 3.227

[16.2 A]

3. Find the potential across the 3-Ω resistor.

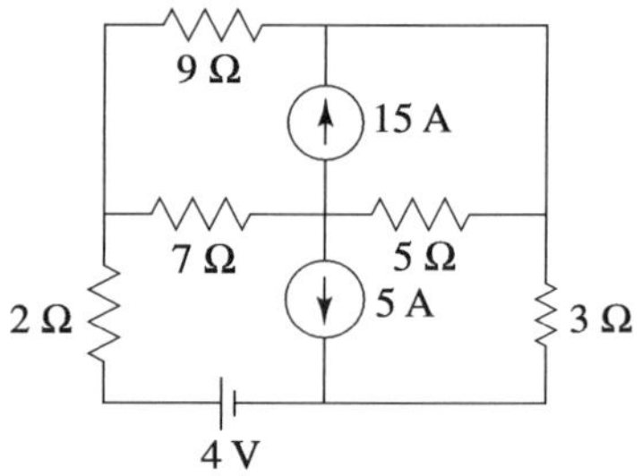

Fig. 3.228

[3.3 V]

4. Calculate the current through the 10-Ω resistor.

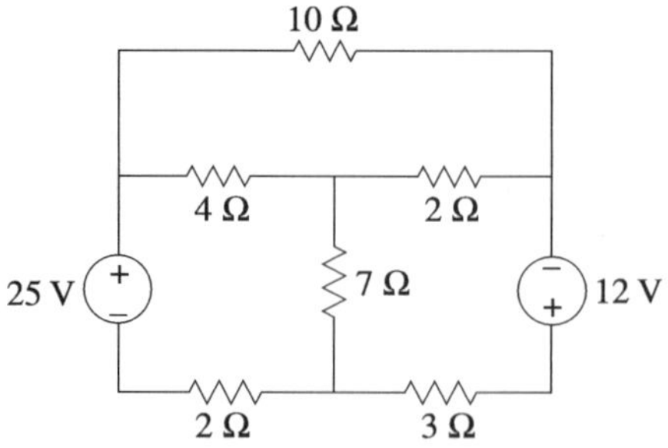

Fig. 3.229

[1.62A]

5. Find the current through the 1-Ω resistor.

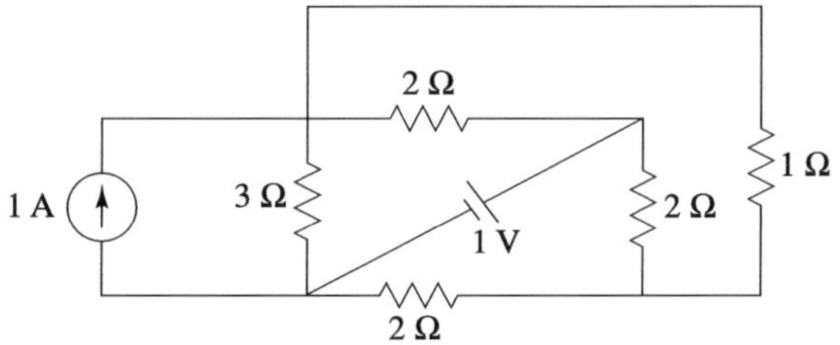

Fig. 3.230

[0.41 A]

6. Find the current through the 4-Ω resistor.

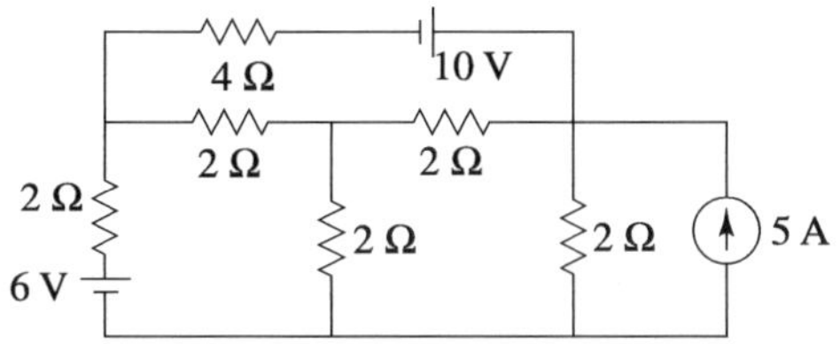

Fig. 3.231

[1.33 A]

7. Find the current I_x.

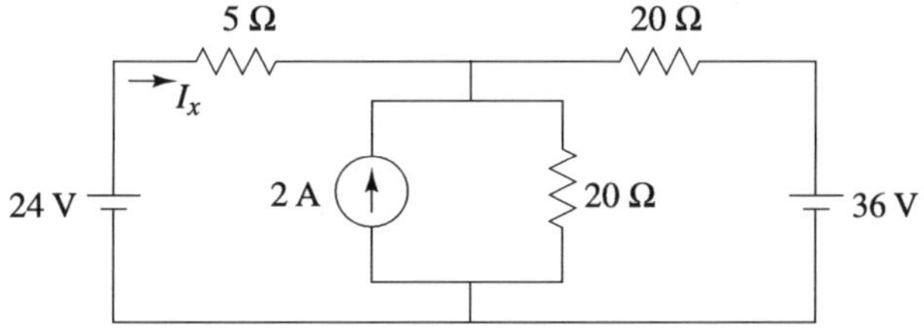

Fig. 3.232

[–1.143 A]

8. Find the voltage V_x.

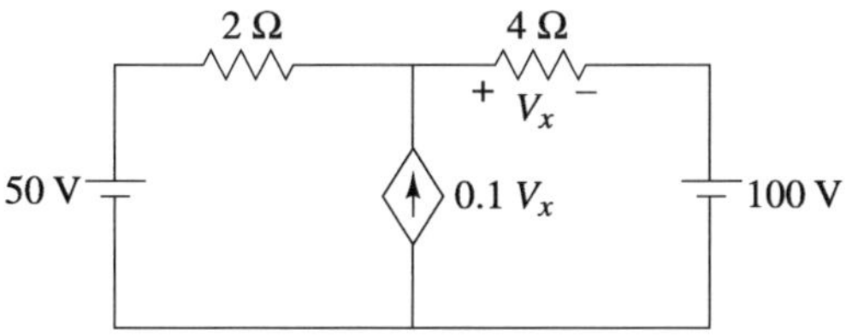

Fig. 3.233

[–38.5 V]

9. Determine the voltages V_1 and V_2.

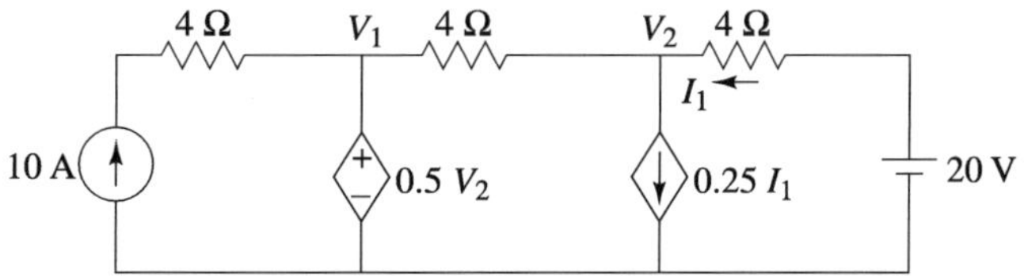

Fig. 3.234

[6 V, 12 V]

10. Find the voltage V_x.

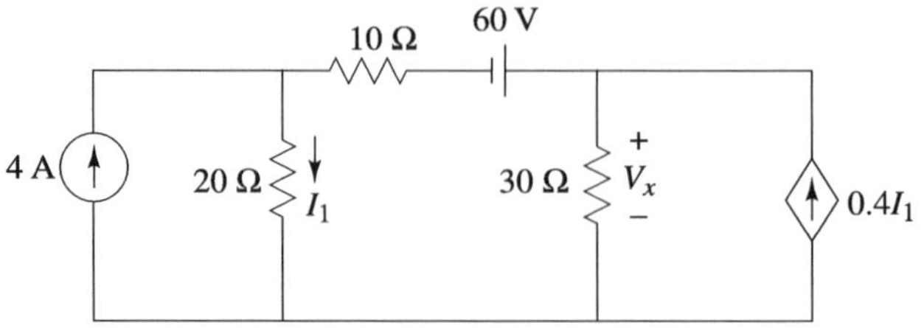

Fig. 3.235

[82.5 V]

THEVENIN'S THEOREM

11. Find the current through the 5-Ω resistor.

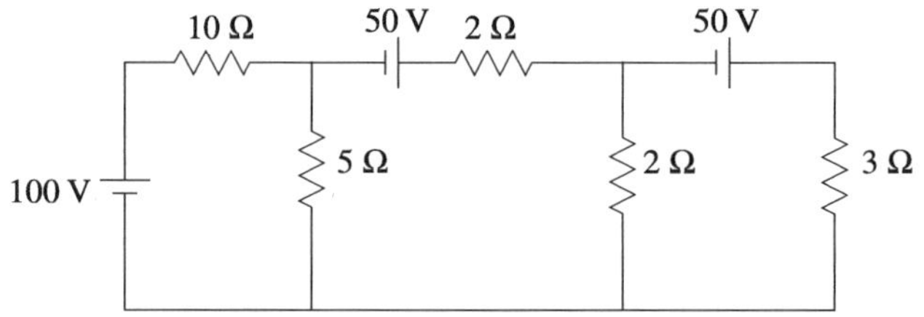

Fig. 3.236

[3.87 A]

12. Find the current through the 6-Ω resistor.

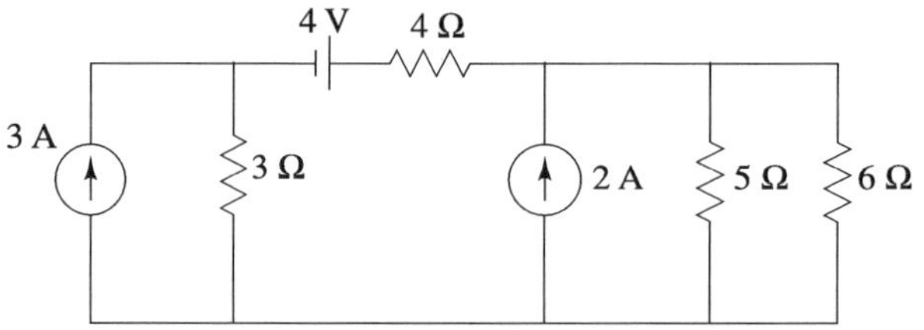

Fig. 3.237

[1.26 A]

13. Find the current through the 6-Ω resistor.

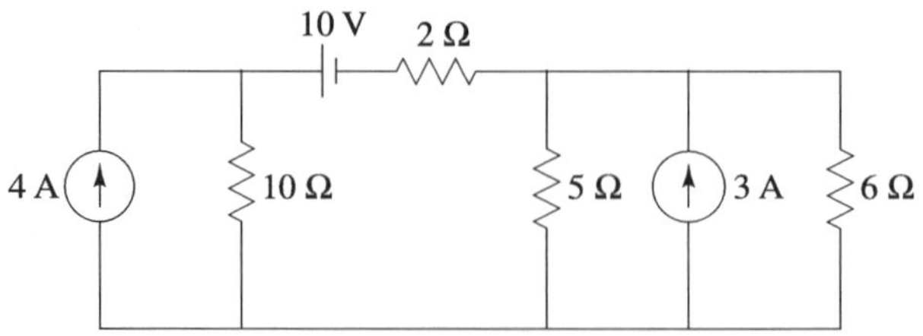

Fig. 3.238

[2.04A]

14. Find the current through the 2-Ω resistor connected between terminals *A* and *B*.

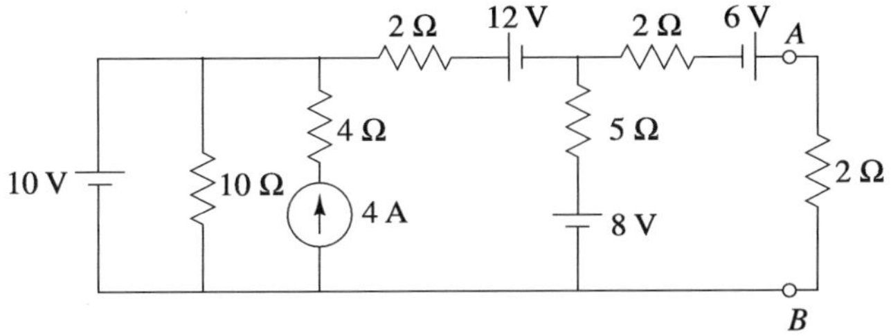

Fig. 3.239

[1.26A]

15. Find the current through the 5-Ω resistor.

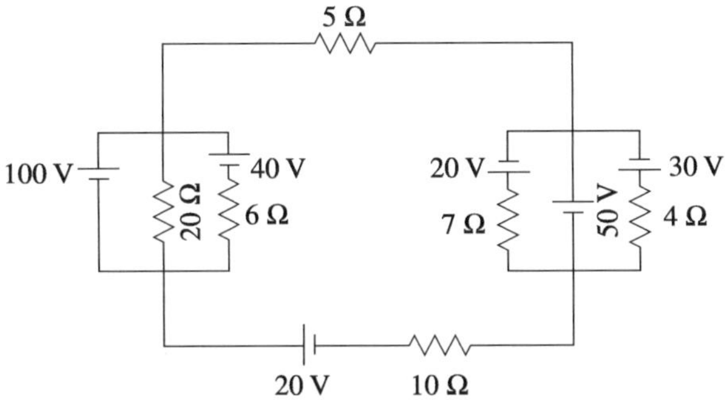

Fig. 3.240

[4.67A]

16. Find the current through the 20-Ω resistor.

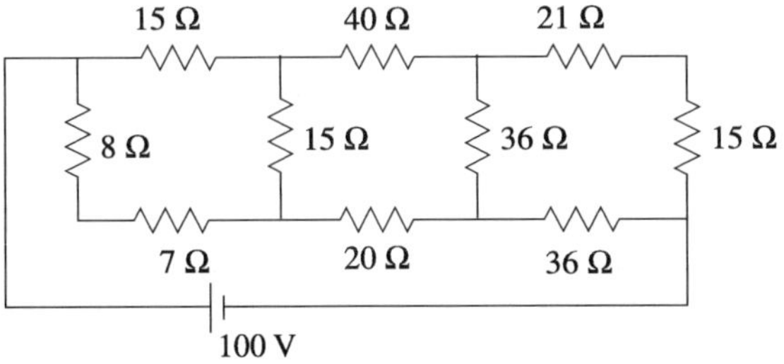

Fig. 3.241

[1.54 A]

17. Calculate the current through the 10-Ω resistor.

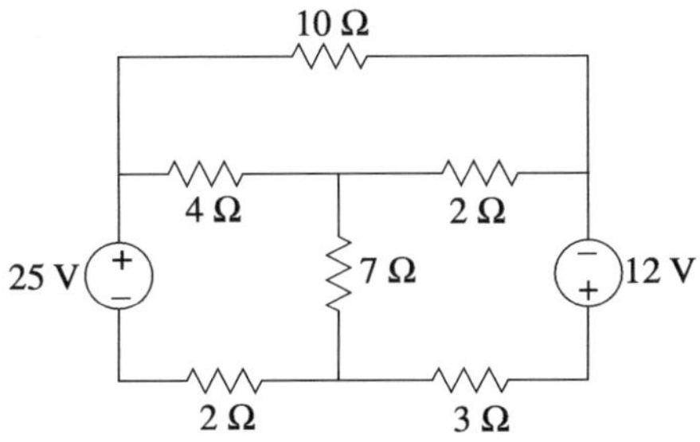

Fig. 3.242

[1.62 A]

NORTON'S THEOREM

18. Find the current through the 10-Ω resistor.

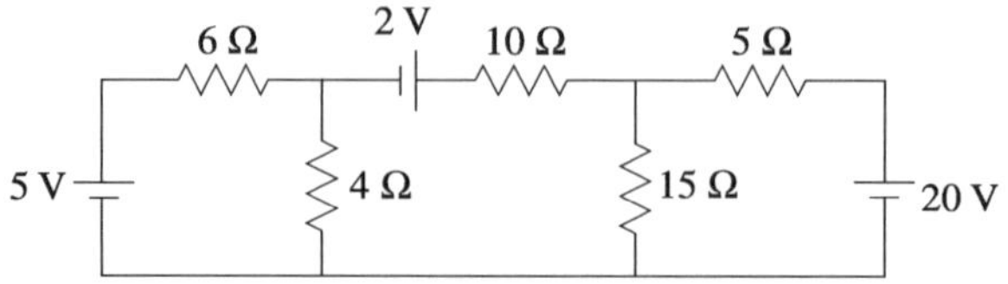

Fig. 3.243

[0.68 A]

19. Find the current through the 20-Ω resistor.

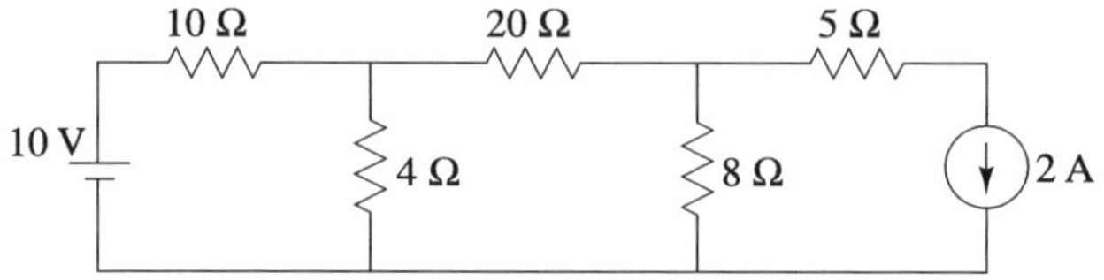

Fig. 3.244

[0.61 A]

20. Find the current through the 2-Ω resistor.

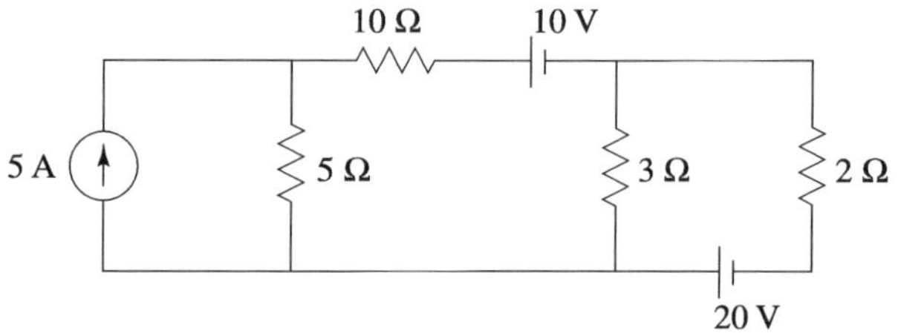

Fig. 3.245

[5 A]

21. Find the current through the 5-Ω resistor.

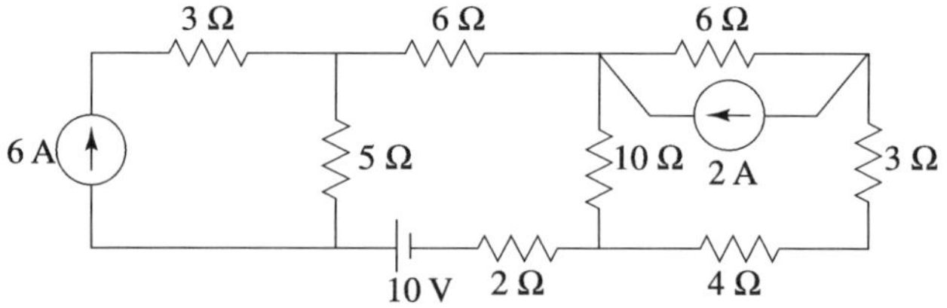

Fig. 3.246

[4.13 A]

22. Find Norton's equivalent circuit for the portion of network shown in Fig. 3.247 to the left of *ab*. Hence obtain the current in the 10-Ω resistor.

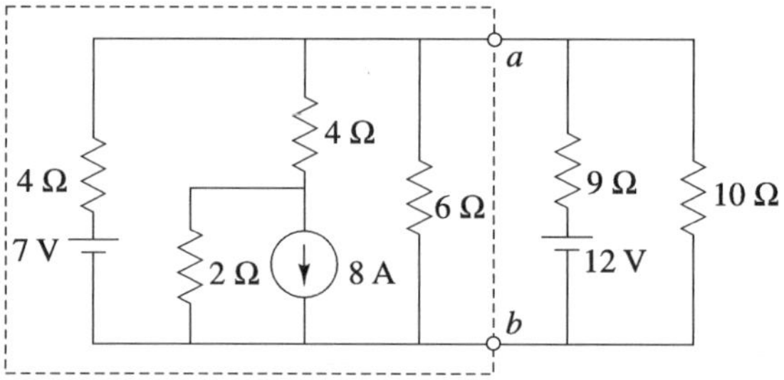

Fig. 3.247

[0.053 A]

THEVENIN'S AND NORTON'S THEOREMS WITH DEPENDENT SOURCES

23. Determine Thevenin's equivalent network for figures shown below.

(a)

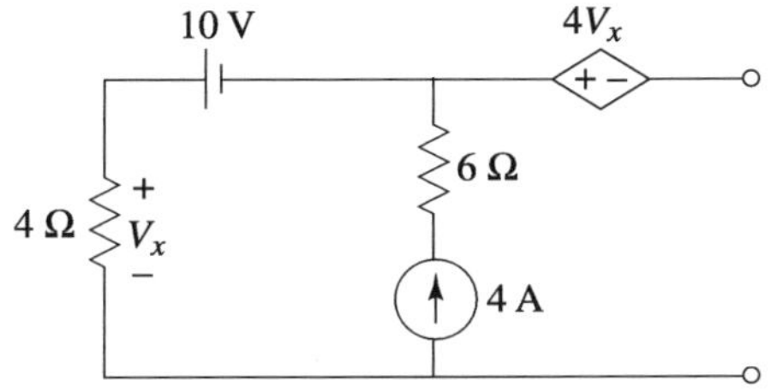

Fig. 3.248

[–58 V, 12 Ω]

(b)

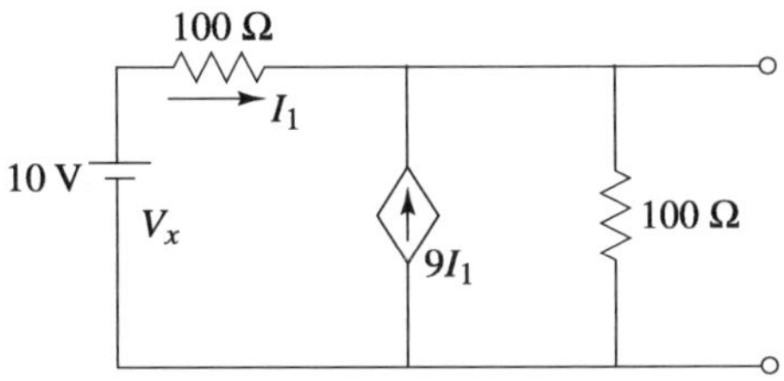

Fig. 3.249

[9.09 V, 9.09 Ω]

(c)

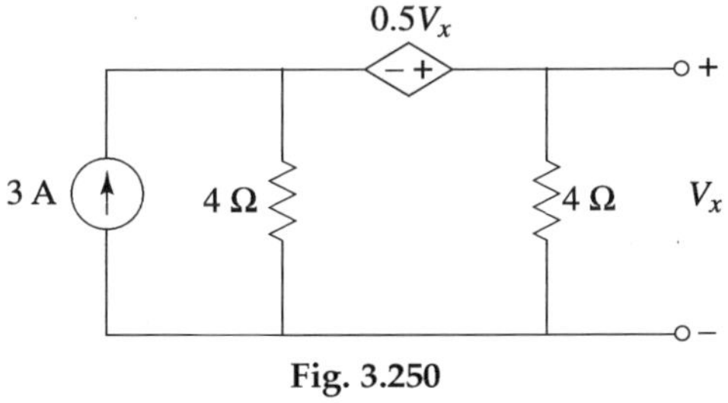

Fig. 3.250

[8 V, 2.66 Ω]

(d)

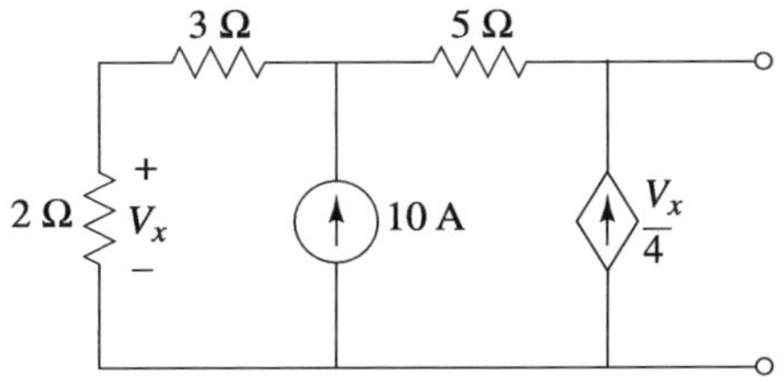

Fig. 3.251

[150 V, 20 Ω]

(e)

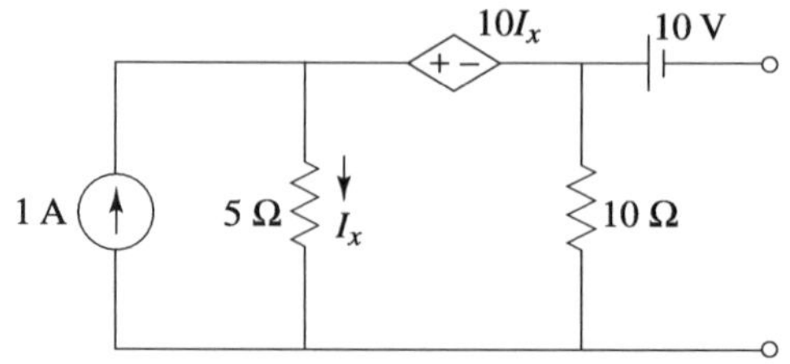

Fig. 3.252

[–20 V, –10 Ω]

24. Find Norton's equivalent network and hence find the current in the 10-Ω resistor.

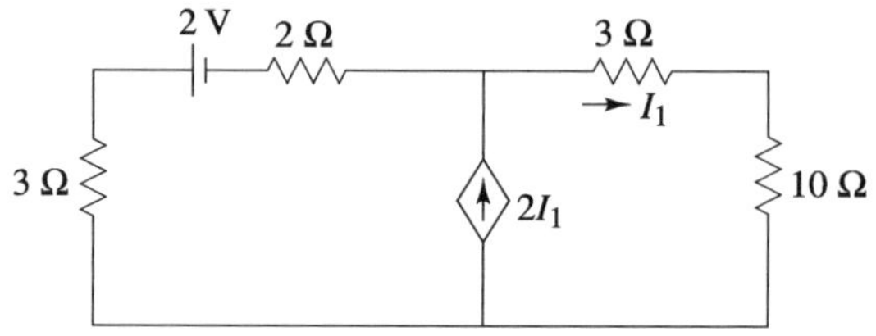

Fig. 3.253

[0.25 A]

25. Find the current I_x.

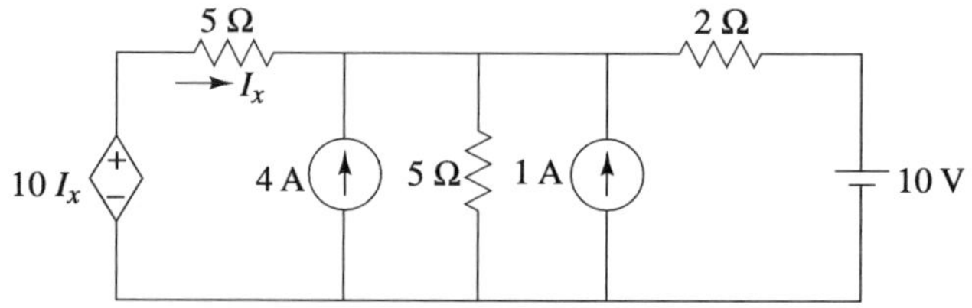

Fig. 3.254

[4 A]

26. Find the current in the 24-Ω resistor.

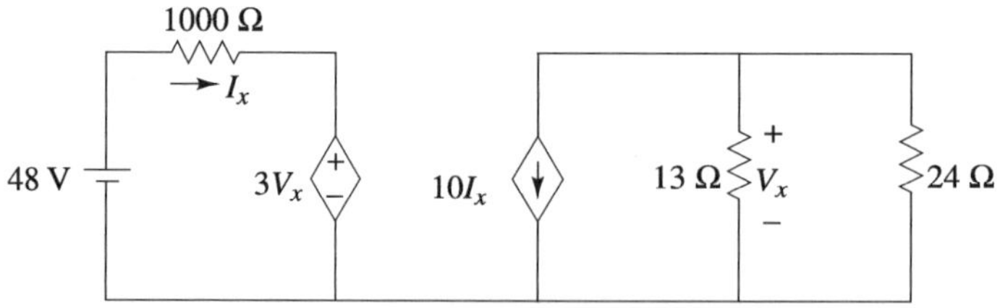

Fig. 3.255

[0.225 A]

27. Find Norton's equivalent network.

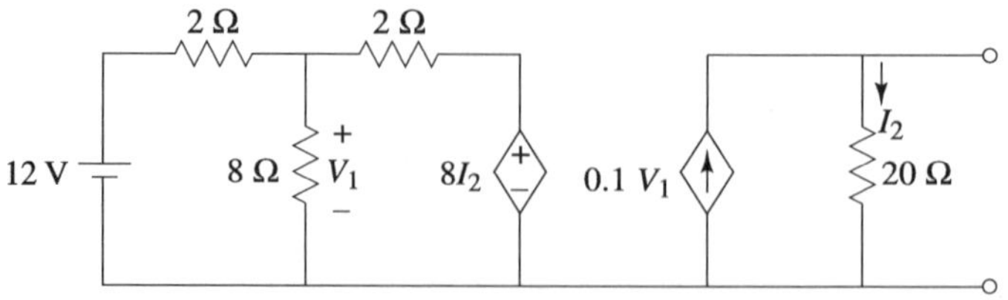

Fig. 3.256

[0.533 A, 31 Ω]

MAXIMUM POWER TRANSFER THEOREM

28. Find the value of the resistance R_L for maximum power transfer and calculate the maximum power.

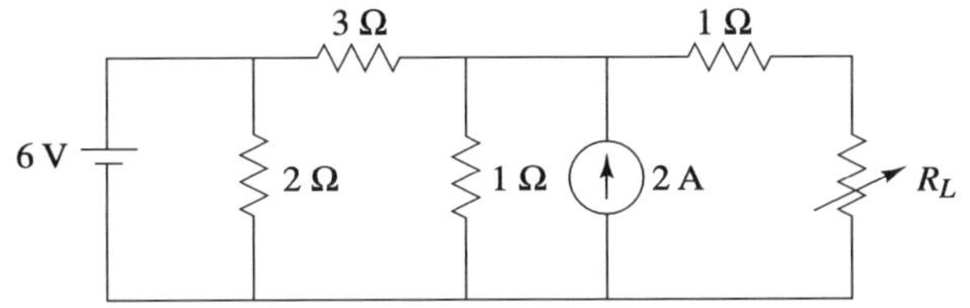

Fig. 3.257

[1.75 Ω, 1.29 W]

29. Find the value of the resistance R_L for maximum power transfer and calculate the maximum power.

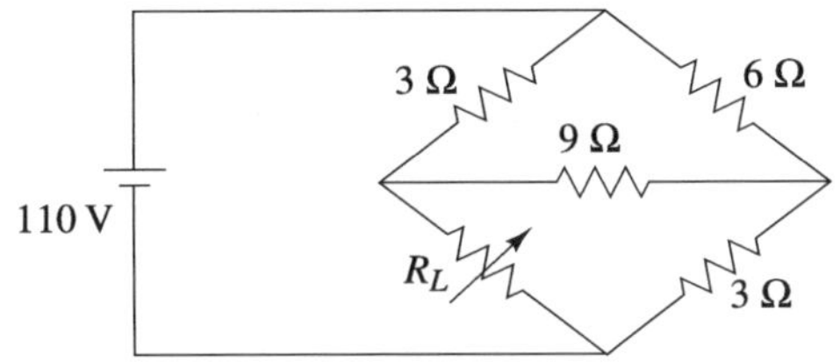

Fig. 3.258

[2.36 Ω, 940 W]

30. Find the value of the resistance R_L for maximum power transfer and calculate the maximum power.

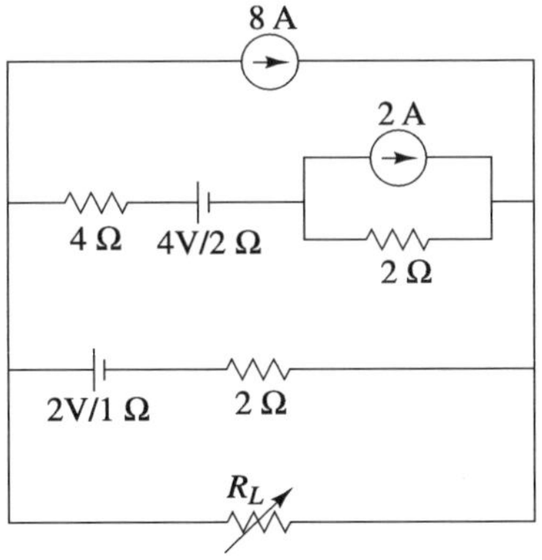

Fig. 3.259

[2.18 Ω, 29.35 W]

31. Find the value of the resistance R_L for maximum power transfer and calculate the maximum power.

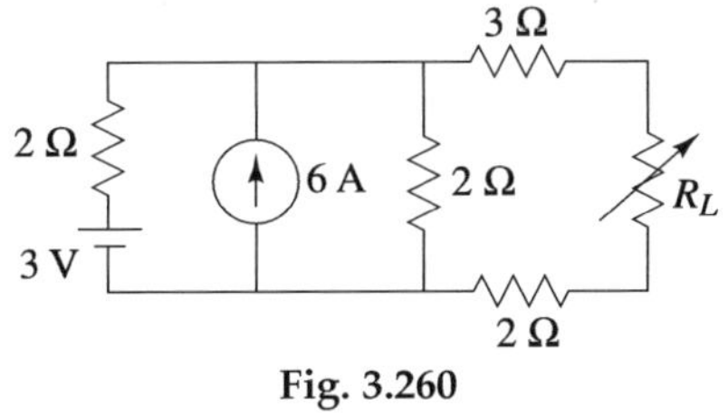

Fig. 3.260

[3 Ω, 2.52 W]

32. Find the value of the resistance R_L for maximum power transfer and calculate the maximum power.

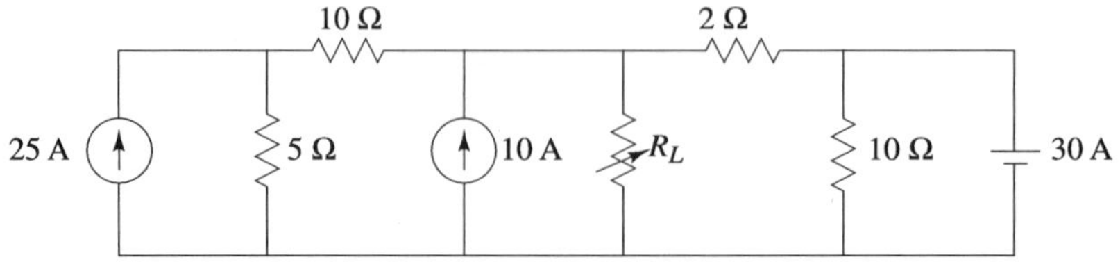

Fig. 3.261

[1.76 Ω, 490.187 W]

Objective-Type Questions

1. The value of the resistance R connected across the terminals A and B, which will absorb the maximum power is
(a) 4 kΩ (b) 4.11 kΩ
(c) 8 kΩ (d) 9 k Ω

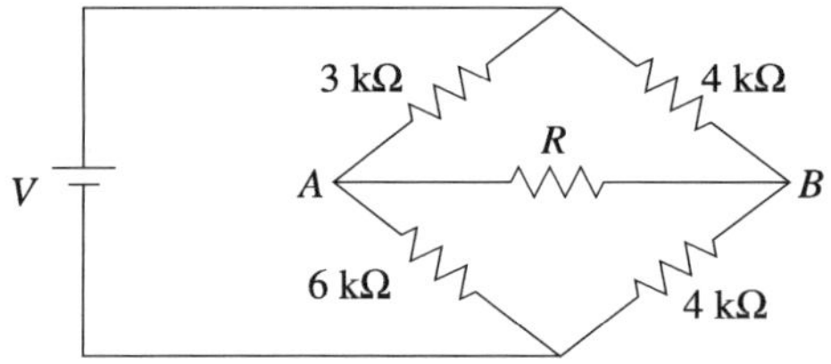

Fig. 3.262

2. Superposition theorem is not applicable to networks containing
(a) nonlinear elements (b) dependent voltage source
(c) dependent current source (d) transformers

3. The value of R required for maximum power transfer in the network shown in Fig. 3.263 is
(a) 2 Ω (b) 4 Ω
(c) 8 Ω (d) 16 Ω

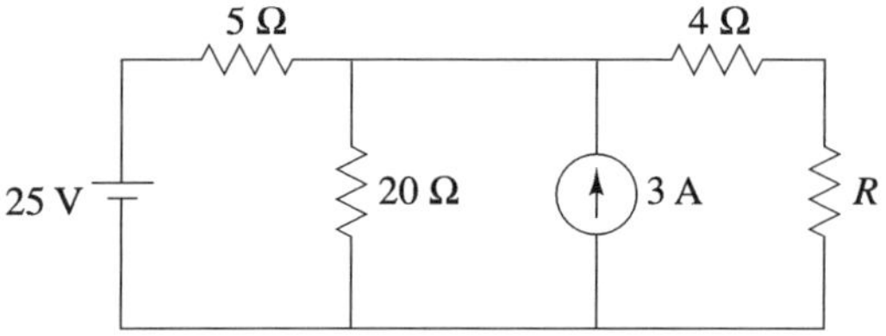

Fig. 3.263

4. In the network of Fig. 3.264 the maximum power is delivered to R_L if its value is

(a) 16 Ω (b) $\frac{40}{3}$ Ω

(c) 60 Ω (d) 20 Ω

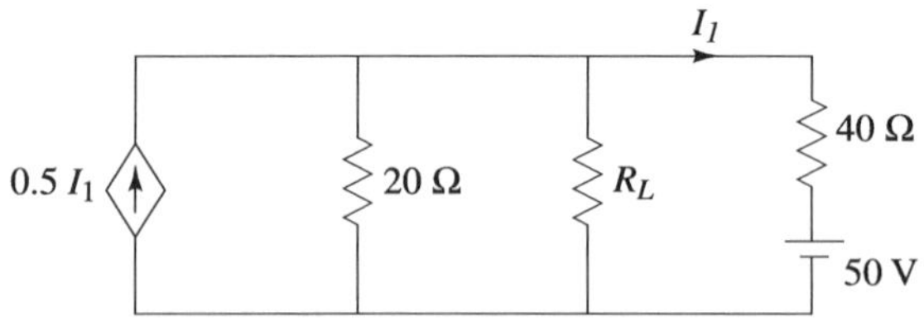

Fig. 3.264

5. The maximum power that can be transferred to the load R_L from the voltage source in Fig. 3.265 is
(a) 1 W (b) 10 W
(c) 0.25 W (d) 0.5 W

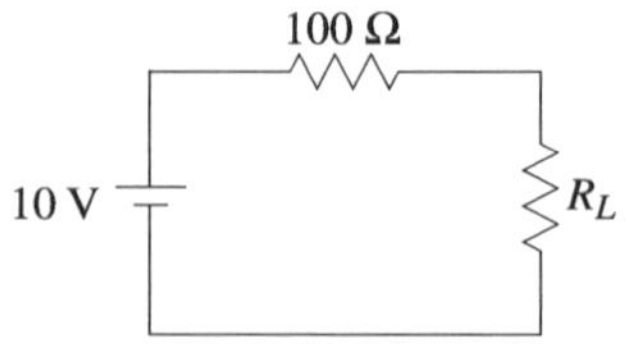

Fig. 3.265

6. For the circuit shown in Fig. 3.266. Thevenin's voltage and Thevenin's equivalent resistance at terminals *a-b* is

(a) 5 V and 2 Ω (b) 7.5 V and 2.5 Ω
(c) 4 V and 2 Ω (d) 3 V and 2.5 Ω

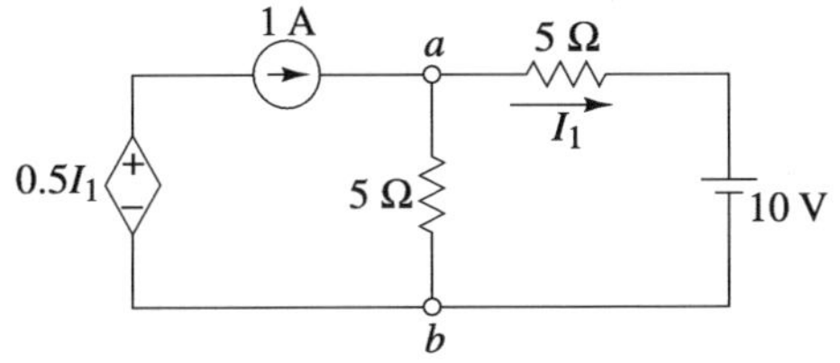

Fig. 3.266

7. The value of R_L for maximum power transfer is

(a) 3 Ω (b) 1.125 Ω
(c) 4.1785 Ω (d) none of these

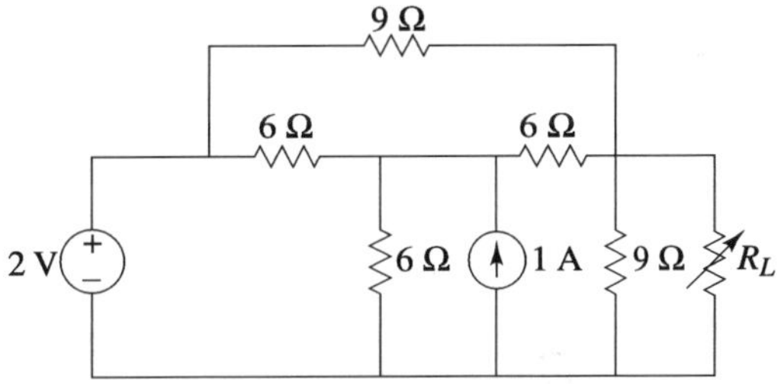

Fig. 3.267

Answers to Objective-Type Questions

1. (a) **2.** (a) **3.** (c) **4.** (a) **5.** (c) **6.** (b)
7. (a)

4 AC Circuits

4.1 INTRODUCTION

An alternating waveform changes its magnitude and direction periodically. Figure 4.1 shows various ac waveforms.

Many times alternating voltages and currents are represented by a sinusoidal waveform.

A sinusoidal voltage can be represented as

$$\begin{aligned} v &= V_m \sin \theta \\ &= V_m \sin \omega t \\ &= V_m \sin 2\pi f t \\ &= V_m \sin \frac{2\pi}{T} t \end{aligned}$$

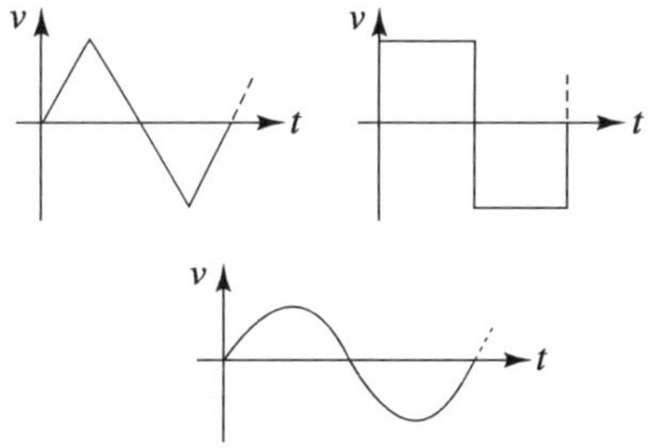

Fig. 4.1

4.2 TERMS RELATED WITH ALTERNATING QUANTITY

1. Waveform A waveform is a graph in which the instantaneous value of any quantity is plotted against time. Figure 4.1 shows few waveforms.

2. Cycle One complete set of positive and negative values of an alternating quantity is termed as cycle.

3. Frequency The number of cycles per second of an alternating quantity is known as frequency. It is denoted by f and is expressed in hertz (Hz) or cycles per second (c/s).

4. Time period The time taken by an alternating quantity to complete one cycle is called time period. It is denoted by T and is expressed in seconds.

$$T = \frac{1}{f}$$

5. Amplitude The maximum positive or negative value of an alternating quantity is called the amplitude.

6. Phase The phase of an alternating quantity is the time that has elapsed since the quantity has last passed through zero point of reference.

7. Phase difference This term is used to compare the phases of two alternating quantities. Two alternating quantities are said to be in phase when they reach their maximum and zero values at the same time. Their maximum value may be different in magnitude.

A leading alternating quantity is one which reaches its maximum or zero value earlier as compared to the other quantity.

A lagging alternating quantity is one which attains its maximum or zero value later than the other quantity.

A plus (+) sign when used in connection with the phase difference denotes 'lead' whereas a minus (–) sign denotes 'lag'.

$$v_A = V_m \sin \omega t$$
$$v_B = V_m \sin (\omega t + \phi)$$

Here, quantity B leads A by a phase angle ϕ.

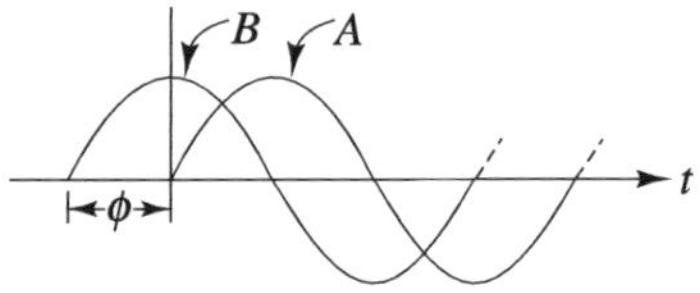

Fig. 4.2

4.3 ROOT MEAN SQUARE (RMS) OR EFFECTIVE VALUE

Normally, the current is measured by the amount of work it will do or the amount of heat it will produce. Hence, rms or effective value of alternating current is defined as that value of steady current (direct current) which will do the same amount of work in the same time or would produce the same heating effect as when the alternating current is applied for the same time.

Figure 4.3 shows the positive half cycle of a non-sinusoidal alternating current waveform. The waveform is divided in m equal intervals with the instantaneous currents, these intervals being $i_1, i_2, ..., i_m$. This waveform is applied to a circuit consisting of a resistance of R ohms. Then work done in different intervals will be

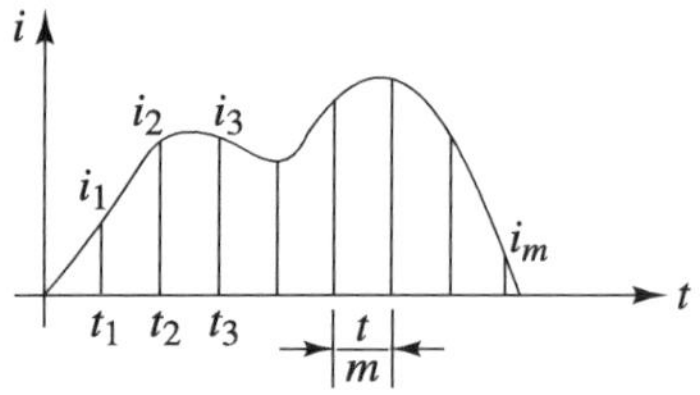

Fig. 4.3

$$\left(i_1^2 R \times \frac{t}{m}\right), \left(i_2^2 R \times \frac{t}{m}\right), ..., \left(i_m^2 R \times \frac{t}{m}\right) \text{joules.}$$

Thus, the total work done in t seconds on applying alternating current waveform to a resistance

$$R = \frac{i_1^2 + i_2^2 + \ldots + i_m^2}{m} \times Rt \text{ joules}$$

Let I be the value of the direct current that while flowing through the same resistance does the same amount of work in the same time t. Then

$$I^2 Rt = \frac{i_1^2 + i_2^2 + \ldots + i_m^2}{m} \times Rt$$

$$I^2 = \frac{i_1^2 + i_2^2 + \ldots + i_m^2}{m}$$

Hence, rms value of alternating current is given by

$$I_{rms} = \sqrt{\frac{i_1^2 + i_2^2 + \ldots + i_m^2}{m}} = \sqrt{\text{Mean value of}\,(i)^2}$$

RMS value of any current $i\ (t)$ over the specified interval t_1 and t_2 is expressed mathematically as

$$I_{rms} = \sqrt{\frac{1}{t_2 - t_1}\int_{t_1}^{t_2} i^2(t)\,dt}$$

The rms value of an alternating current is of considerable importance in practice because the ammeters and voltmeters record the rms value of alternating current and voltage respectively.

4.3.1 RMS Value of Sinusoidal Waveform

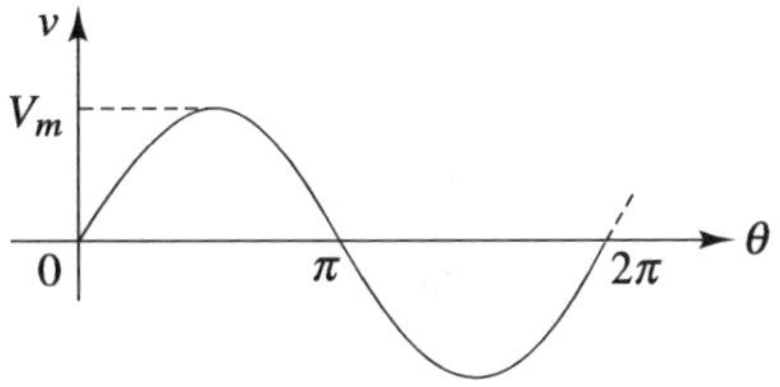

Fig. 4.4

$$v = V_m \sin\theta \qquad 0 < \theta < 2\pi$$

$$V_{rms} = \sqrt{\frac{1}{2\pi}\int_0^{2\pi} v^2(\theta)\,d\theta}$$

$$= \sqrt{\frac{1}{2\pi}\int_0^{2\pi} V_m^2 \sin^2\theta\,d\theta} = \sqrt{\frac{V_m^2}{2\pi}\int_0^{2\pi} \sin^2\theta\,d\theta}$$

$$= \sqrt{\frac{V_m^2}{2\pi}\int_0^{2\pi}\left(\frac{1-\cos 2\theta}{2}\right)d\theta} = \sqrt{\frac{V_m^2}{2\pi}\left[\frac{\theta}{2} - \frac{\sin 2\theta}{4}\right]_0^{2\pi}}$$

$$= \sqrt{\frac{V_m^2}{2\pi}\left[\frac{2\pi}{2} - 0 - 0 + 0\right]}$$

$$= \sqrt{\frac{V_m^2}{2}} = \frac{V_m}{\sqrt{2}} = 0.707\ V_m$$

Crest or peak or amplitude factor It is defined as the ratio of maximum value to rms value of the given quantity.

$$\text{Peak factor}\ (k_p) = \frac{\text{Maximum value}}{\text{RMS value}}$$

4.4 AVERAGE VALUE

The average value of an alternating quantity is defined as the arithmetic mean of all the values over one complete cycle.

In case of symmetrical alternating waveform (whether sinusoidal or non-sinusoidal), the average value over a complete cycle is zero. Hence, in such a case, the average value is obtained over half the cycle only.

Referring to Fig. 4.3, the average value of the current is given by

$$I_{avg} = \frac{i_1 + i_2 + \ldots + i_m}{m}$$

The average value of any current $i(t)$ over the specified interval t_1 and t_2 is expressed mathematically as

$$I_{avg} = \frac{1}{t_2 - t_1}\int_{t_1}^{t_2} i(t)\,dt$$

4.4.1 Average Value of Sinusoidal Waveform

$$v = V_m \sin\theta \qquad 0 < \theta < 2\pi$$

Since this is a symmetrical waveform, the average value is calculated over half the cycle.

$$V_{avg} = \frac{1}{\pi}\int_0^{\pi} v(\theta)\,d\theta = \frac{1}{\pi}\int_0^{\pi} V_m \sin\theta\; d\theta$$

$$= \frac{V_m}{\pi}\int_0^{\pi}\sin\theta\; d\theta = \frac{V_m}{\pi}\left[-\cos\theta\right]_0^{\pi}$$

$$= \frac{V_m}{\pi}[1+1] = \frac{2V_m}{\pi} = 0.637\,V_m$$

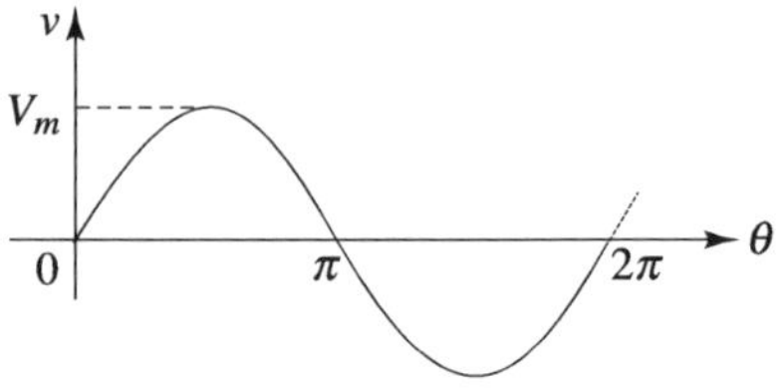

Fig. 4.5

Form factor It is defined as the ratio of rms value to the average value of the given quantity.

$$\text{Form factor } (k_f) = \frac{\text{RMS value}}{\text{Average value}}.$$

Example 4.1 *An alternating current takes 3.375 ms to reach 15 A for the first time after becoming instantaneously zero. The frequency of the current is 40 Hz. Find the maximum value of the alternating current.*

Solution

Data $\quad i = 15\text{ A} \qquad t = 3.375\text{ ms} \qquad f = 40\text{ Hz}$

$$i = I_m \sin 2\pi f t$$

$$15 = I_m \sin(2\pi \times 40 \times 3.375 \times 10^{-3})$$

$$15 = I_m \times 0.75$$

$$I_m = 20\text{ A}$$

Example 4.2 *An alternating current of frequency 50 c/s has a maximum value of 100 A. (a) Calculate its value* $\frac{1}{600}$ *second and after the instant the current is zero. (b) In how many seconds after the zero value will the current attain the value of 86.6 A?*

Solution

Data $\quad f = 50\text{ c/s} \qquad I_m = 100\text{ A}$

(a)

$$i = I_m \sin 2\pi f t = 100\sin\left(2\pi \times 50 \times \frac{1}{600}\right)$$

$$= 100\sin(30^\circ) = 50\text{ A}$$

(b)

$$i = I_m \sin 2\pi f t$$

$$86.6 = 100\sin(2\pi \times 50 \times t)$$

$$\sin(100\,\pi t) = 0.866$$

$$100\,\pi t = 60^\circ$$

$$t = \frac{60}{100 \times 180} = \frac{1}{300}\text{ second}$$

Example 4.3 *An alternating current varying sinusoidally with a frequency of 50 c/s has an rms value of 20 A. Write down the equation for the instantaneous value and find this value at (a) 0.0025 s (b) 0.0125 s after passing through zero and increasing positively. (c) At what time, measured from zero, will the value of the instantaneous current be 14.14 A?*

Solution

Data

$$f = 50 \text{ c/s}$$
$$I = 20 \text{ A}$$
$$I_m = I \times \sqrt{2} = 20\sqrt{2} = 28.28 \text{ A}$$

Equation of current,

$$i = I_m \sin 2\pi ft$$
$$= 28.28 \sin (2\pi \times 50 \times t) = 28.28 \sin 100\pi t$$

(a) At

$$t = 0.0025 \text{ second}$$
$$i = 28.28 \sin (100\pi \times 0.0025)$$
$$= 28.28 \sin (45^\circ) = 20 \text{ A}$$

(b) At

$$t = 0.0125 \text{ second}$$
$$i = 28.28 \sin (100\pi \times 0.0125)$$
$$= 28.28 \sin (225^\circ) = -20 \text{ A}$$

(c)

$$i = 28.28 \sin 100\pi t$$
$$14.14 = 28.28 \sin 100\pi t$$
$$\sin 100\pi t = 0.5$$
$$100\pi t = 30^\circ$$
$$t = 1.66 \times 10^{-3} \text{ second}$$

Example 4.4 *Find the following parameters of a voltage v = 200 sin 314 t: (i) frequency, (ii) form factor, and (iii) crest factor.*

Solution

Data

$$v = 200 \sin 314\, t$$
$$v = V_m \sin 2\pi ft$$
$$2\pi f = 314$$
$$f = \frac{314}{2\pi} = 50 \text{ Hz}$$

For a sinusoidal waveform,

$$V_{\text{avg}} = \frac{2V_m}{\pi}$$
$$V_{\text{rms}} = \frac{V_m}{\sqrt{2}}$$

$$\text{Form factor} = \frac{V_{\text{rms}}}{V_{\text{avg}}} = \frac{\dfrac{V_m}{\sqrt{2}}}{\dfrac{2V_m}{\pi}} = 1.11$$

$$\text{Crest factor} = \frac{V_m}{V_{\text{rms}}} = \frac{V_m}{\dfrac{V_m}{\sqrt{2}}} = 1.414$$

Example 4.5 *A non-sinusoidal voltage is having a form factor of 1.2 and peak factor of 1.5. If the average value of the voltage is 10 V, calculate (i) rms value, and (ii) maximum value.*

Solution

Data

$$k_f = 1.2$$
$$k_p = 1.5$$
$$V_{avg} = 10$$

$$\text{Form factor } k_f = \frac{V_{rms}}{V_{avg}}$$

$$1.2 = \frac{V_{rms}}{10}$$

$$V_{rms} = 12 \text{ V}$$

$$\text{Peak factor } k_p = \frac{V_m}{V_{rms}}$$

$$1.5 = \frac{V_m}{12}$$

$$V_m = 18 \text{ V}$$

Example 4.6 *Find the average value and rms value of the waveform shown in Fig. 4.6.*

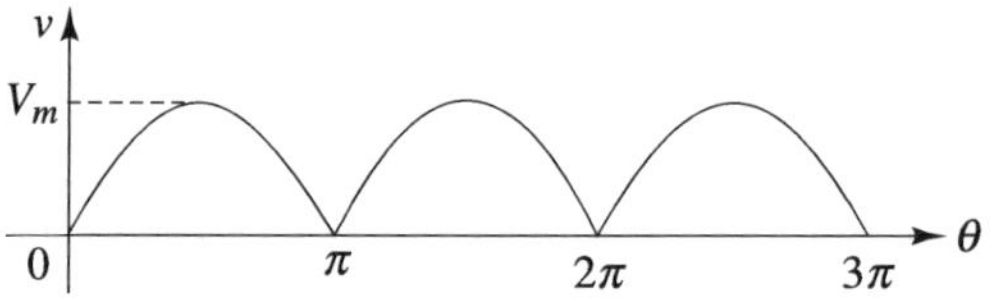

Fig. 4.6

Solution

$$v = V_m \sin\theta \qquad 0 < \theta < \pi$$

$$V_{avg} = \frac{1}{\pi}\int_0^{\pi} v(\theta)\, d\theta = \frac{1}{\pi}\int_0^{\pi} V_m \sin\theta\, d\theta$$

$$= \frac{V_m}{\pi}\left[-\cos\theta\right]_0^{\pi} = \frac{V_m}{\pi}[1+1]$$

$$= \frac{2V_m}{\pi} = 0.637\, V_m$$

$$V_{rms} = \sqrt{\frac{1}{\pi}\int_0^{\pi} v^2(\theta)\, d\theta}$$

$$= \sqrt{\frac{1}{\pi}\int_0^{\pi} V_m^2 \sin^2\theta\, d\theta} = \sqrt{\frac{V_m^2}{\pi}\int_0^{\pi} \sin^2\theta\, d\theta}$$

$$= \sqrt{\frac{V_m^2}{\pi}\int_0^{\pi}\left(\frac{1-\cos 2\theta}{2}\right)d\theta} = \sqrt{\frac{V_m^2}{\pi}\left[\frac{\theta}{2} - \frac{\sin 2\theta}{4}\right]_0^{\pi}}$$

$$= \sqrt{\frac{V_m^2}{\pi}\left[\frac{\pi}{2} - \frac{\sin 2\pi}{4} - 0 + \frac{\sin 0}{4}\right]} = \sqrt{\frac{V_m^2}{2}}$$

$$= \frac{V_m}{\sqrt{2}} = 0.707\ V_m$$

Example 4.7 *Find the average and rms value of the waveform shown in Fig. 4.7.*

Solution

$$v = V_m \sin\theta \qquad 0 < \theta < \pi$$
$$= 0 \qquad \pi < \theta < 2\pi$$

$$V_{avg} = \frac{1}{2\pi}\int_0^{2\pi} v(\theta)\, d\theta$$

$$= \frac{1}{2\pi}\left[\int_0^{\pi} V_m \sin\theta\ d\theta + \int_{\pi}^{2\pi} 0\, d\theta\right]$$

$$= \frac{1}{2\pi}\int_0^{\pi} V_m \sin\theta\ d\theta$$

$$= \frac{V_m}{2\pi}[-\cos\theta]_0^{\pi}$$

$$= \frac{V_m}{2\pi}[1+1] = \frac{V_m}{\pi} = 0.318\ V_m$$

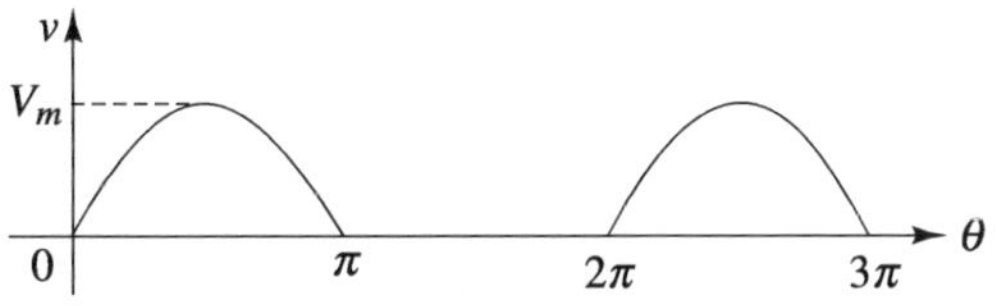

Fig. 4.7

$$V_{rms} = \sqrt{\frac{1}{2\pi}\int_0^{2\pi} v^2(\theta)\, d\theta}$$

$$= \sqrt{\frac{1}{2\pi}\left[\int_0^{\pi} V_m^2 \sin^2\theta\ d\theta + \int_{\pi}^{2\pi} 0\, d\theta\right]}$$

$$= \sqrt{\frac{1}{2\pi}\int_0^{\pi} V_m^2 \sin^2\theta\ d\theta} = \sqrt{\frac{V_m^2}{2\pi}\int_0^{\pi} \sin^2\theta\ d\theta}$$

$$= \sqrt{\frac{V_m^2}{2\pi}\int_0^{\pi}\left(\frac{1-\cos 2\theta}{2}\right) d\theta} = \sqrt{\frac{V_m^2}{2\pi}\left[\frac{\theta}{2} - \frac{\sin 2\theta}{4}\right]_0^{\pi}}$$

$$= \sqrt{\frac{V_m^2}{2\pi}\left[\frac{\pi}{2} - \frac{\sin 2\pi}{4} - 0 + \frac{\sin 0}{4}\right]}$$

$$= \sqrt{\frac{V_m^2}{4}} = \frac{V_m}{2} = 0.5\ V_m$$

Example 4.8 *Find the average value and rms value of the waveform shown in Fig. 4.8.*

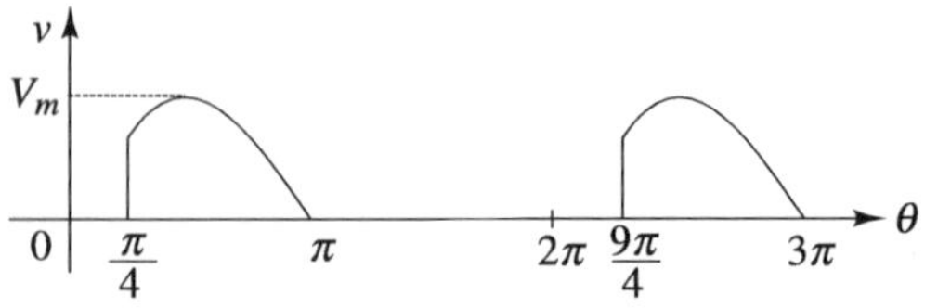

Fig. 4.8

Solution

$$v = 0 \qquad 0 < \theta < \pi/4$$
$$= V_m \sin\theta \qquad \pi/4 < \theta < \pi$$
$$= 0 \qquad \pi < \theta < 2\pi$$

$$V_{avg} = \frac{1}{2\pi}\int_0^{2\pi} v(\theta)\,d\theta = \frac{1}{2\pi}\int_{\pi/4}^{\pi} V_m \sin\theta\,d\theta$$

$$= \frac{V_m}{2\pi}[-\cos\theta]_{\pi/4}^{\pi} = \frac{V_m}{2\pi} = [1 + 0.707] = 0.272\ V_m$$

$$V_{rms} = \sqrt{\frac{1}{2\pi}\int_0^{2\pi} v^2(\theta)\,d\theta}$$

$$= \sqrt{\frac{1}{2\pi}\int_{\pi/4}^{\pi} V_m^2 \sin^2\theta\,d\theta} = \sqrt{\frac{V_m^2}{2\pi}\int_{\pi/4}^{\pi} \sin^2\theta\,d\theta}$$

$$= \sqrt{\frac{V_m^2}{2\pi}\int_{\pi/4}^{\pi}\left(\frac{1-\cos 2\theta}{2}\right)d\theta} = \sqrt{\frac{V_m^2}{2\pi}\left[\frac{\theta}{2} - \frac{\sin 2\theta}{4}\right]_{\pi/4}^{\pi}}$$

$$= \sqrt{\frac{V_m^2}{2\pi}\left[\frac{\pi}{2} - \frac{\sin 2\pi}{4} - \frac{\pi}{8} + \frac{\sin \pi/2}{4}\right]}$$

$$= \sqrt{0.227 V_m^2}$$
$$= 0.476\ V_m$$

Example 4.9 *A full-wave rectified wave is clipped at 70.7% of its maximum value as shown in Fig. 4.9. Find its average and rms value.*

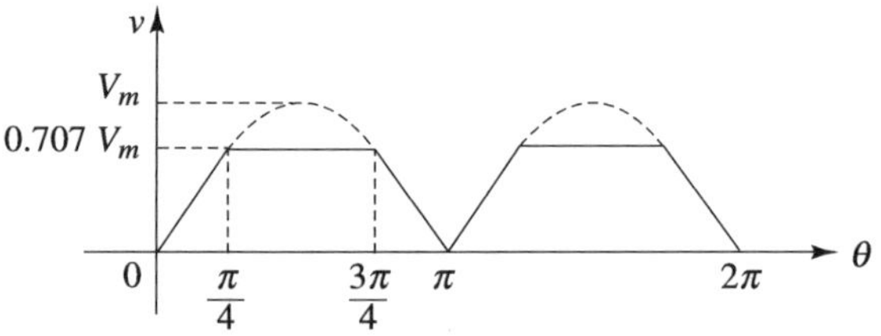

Fig. 4.9

Solution

$$v = V_m \sin\theta \qquad 0 < \theta < \pi/4$$
$$= 0.707\ V_m \qquad \pi/4 < \theta < 3\pi/4$$
$$= V_m \sin\theta \qquad 3\pi/4 < \theta < \pi$$

$$V_{avg} = \frac{1}{\pi}\int_0^{\pi} v(\theta)\,d\theta$$

$$= \frac{1}{\pi}\left[\int_0^{\pi/4} V_m \sin\theta\,d\theta + \int_{\pi/4}^{3\pi/4} 0.707 V_m\,d\theta + \int_{3\pi/4}^{\pi} V_m \sin\theta\,d\theta\right]$$

$$= \frac{V_m}{\pi}\left\{[-\cos\theta]_0^{\pi/4} + 0.707[\theta]_{\pi/4}^{3\pi/4} + [-\cos\theta]_{3\pi/4}^{\pi}\right\}$$

$$= \frac{V_m}{\pi}(0.293 + 1.11 + 0.293) = 0.54\ V_m$$

$$V_{rms} = \sqrt{\frac{1}{\pi}\int_0^{\pi} v^2(\theta)\,d\theta}$$

$$= \sqrt{\frac{1}{\pi}\left[\int_0^{\pi/4} V_m^2 \sin^2\theta\,d\theta + \int_{\pi/4}^{3\pi/4} (0.707V_m)^2 d\theta + \int_{3\pi/4}^{\pi} V_m^2 \sin^2\theta\,d\theta\right]}$$

$$= \sqrt{\frac{V_m^2}{\pi}\left\{\left[\frac{\theta}{2} - \frac{\sin 2\theta}{4}\right]_0^{\pi/4} + 0.499[\theta]_{\pi/4}^{3\pi/4} + \left[\frac{\theta}{2} - \frac{\sin 2\theta}{4}\right]_{3\pi/4}^{\pi}\right\}}$$

$$= \sqrt{0.341\,V_m^2} = 0.584\,V_m$$

Example 4.10 *Find the rms value of the waveform shown in Fig. 4.10.*

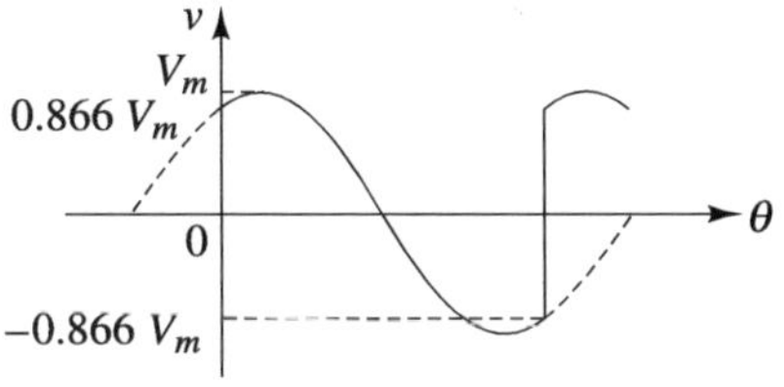

Fig. 4.10

Solution The equation of the waveform is given by
$v = V_m \sin(\theta + \phi)$ where ϕ is the phase difference.

When $\theta = 0$, $v = 0.866\,V_m$.

$$0.866\,V_m = V_m \sin(0 + \phi)$$

$$\phi = \sin^{-1}(0.866) = \frac{\pi}{3}$$

$$v = V_m \sin\left(\theta + \frac{\pi}{3}\right)$$

The time period of a complete sine wave is always 2π. Since some part of the waveform is chopped from both the sides,

$$\text{time period} = 2\pi - \frac{\pi}{3} - \frac{\pi}{3} = \frac{4\pi}{3}$$

$$V_{rms} = \sqrt{\frac{1}{4\pi/3}\int_0^{4\pi/3} V_m^2 \sin^2\left(\theta + \frac{\pi}{3}\right)d\theta}$$

$$= \sqrt{\frac{3}{4\pi}\int_0^{4\pi/3} V_m^2 \sin^2\left(\theta + \frac{\pi}{3}\right)d\theta}$$

$$= \sqrt{\frac{3V_m^2}{4\pi}\int_0^{4\pi/3}\left[\frac{1 - \cos 2(\theta + \pi/3)}{2}\right]d\theta}$$

$$= \sqrt{\frac{3V_m^2}{4\pi}\left[\frac{\theta}{2} - \frac{\sin 2(\theta + \pi/3)}{4}\right]_0^{4\pi/3}}$$

$$= \sqrt{0.6031\, V_m^2} = 0.776\, V_m$$

Example 4.11 *Find the rms and average value of the waveform shown in Fig. 4.11.*

Solution

$$v = V_m \qquad 0 < t < T/2$$
$$= 0 \qquad T/2 < t < T$$

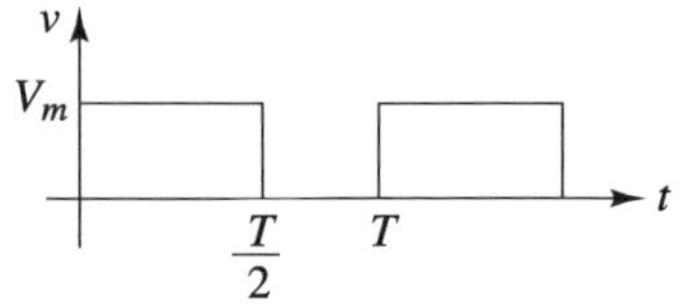

Fig. 4.11

$$V_{avg} = \frac{1}{T}\int_0^T v(t)\,dt$$

$$= \frac{1}{T}\left[\int_0^{T/2} V_m\,dt + \int_{T/2}^{T} 0\,dt\right] = \frac{1}{T}\int_0^{T/2} V_m\,dt$$

$$= \frac{V_m}{T}[t]_0^{T/2} = \frac{V_m}{T}\cdot\frac{T}{2} = 0.5\,V_m$$

$$V_{rms} = \sqrt{\frac{1}{T}\int_0^T v^2(t)\,dt} = \sqrt{\frac{1}{T}\int_0^{T/2} V_m^2\,dt}$$

$$= \sqrt{\frac{V_m^2}{T}[t]_0^{T/2}} = \sqrt{\frac{V_m^2}{T}\cdot\frac{T}{2}}$$

$$= \sqrt{\frac{V_m^2}{2}} = 0.707\,V_m$$

Example 4.12 *Find the rms and average value of the waveform shown in Fig. 4.12.*

Solution

$$v = \frac{V_m}{T}t \qquad 0 < t < T$$

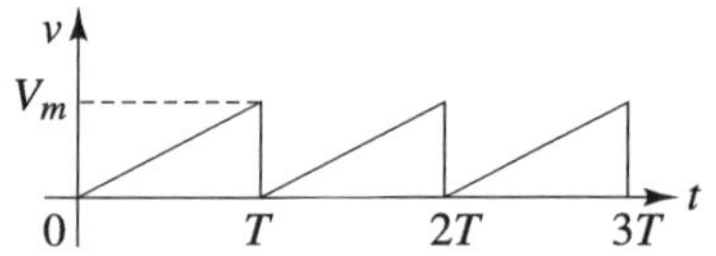

Fig. 4.12

$$V_{avg} = \frac{1}{T}\int_0^T v(t)\,dt = \frac{1}{T}\int_0^T \frac{V_m}{T}t\,dt$$

$$= \frac{V_m}{T^2}\left[\frac{t^2}{2}\right]_0^T = \frac{V_m}{T^2}\cdot\frac{T^2}{2} = 0.5\,V_m$$

$$V_{rms} = \sqrt{\frac{1}{T}\int_0^T v^2(t)\,dt}$$

$$= \sqrt{\frac{1}{T}\int_0^T \frac{V_m^2}{T^2}.t^2\,dt} = \sqrt{\frac{V_m^2}{T^3}\left[\frac{t^3}{3}\right]_0^T}$$

$$= \sqrt{\frac{V_m^2}{T^3}\left[\frac{T^3}{3}\right]} = \sqrt{\frac{V_m^2}{3}} = 0.577\,V_m$$

Example 4.13 *Find the rms and average value of the waveform shown in Fig. 4.13.*

Solution

$$v = t \qquad 0 < t < 1$$
$$= 1 \qquad 1 < t < 2$$

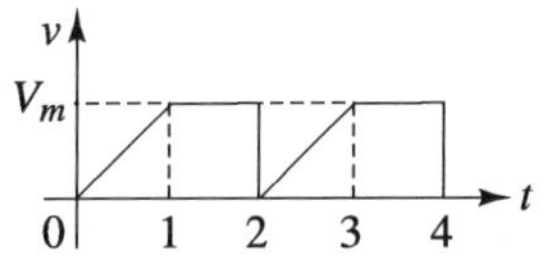

Fig. 4.13

$$V_{avg} = \frac{1}{T}\int_0^T v(t)\,dt = \frac{1}{2}\left[\int_0^1 t\,dt + \int_1^2 1\,dt\right]$$
$$= \frac{1}{2}\left\{\left[\frac{t^2}{2}\right]_0^1 + [t]_1^2\right\}$$
$$= \frac{1}{2}\left[\frac{1}{2} - 0 + 2 - 1\right] = \frac{3}{4} = 0.75\text{ V}$$
$$V_{rms} = \sqrt{\frac{1}{T}\int_0^T v^2(t)\,dt}$$
$$= \sqrt{\frac{1}{2}\left[\int_0^1 t^2 dt + \int_1^2 (1)^2\,dt\right]} = \sqrt{\frac{1}{2}\left\{\left[\frac{t^3}{3}\right]_0^1 + [t]_1^2\right\}}$$
$$= \sqrt{\frac{1}{2}\left[\frac{1}{3} - 0 + 2 - 1\right]} = \sqrt{\frac{4}{6}} = 0.816\text{ V}$$

Example 4.14 *Find the rms and average value of the waveform shown in Fig. 4.14.*

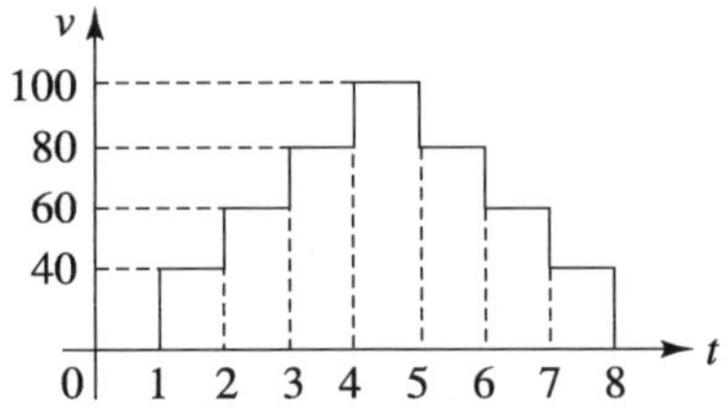

Fig. 4.14

Solution

$$V_{avg} = \frac{0 + 40 + 60 + 80 + 100 + 80 + 60 + 40}{8} = 57.5\text{ V}$$
$$V_{rms} = \sqrt{\frac{0^2 + (40)^2 + (60)^2 + (80)^2 + (100)^2 + (80)^2 + (60)^2 + (40)^2}{8}} = 64.42\text{ V}$$

Example 4.15 *Find the rms and average value of the waveform shown in Fig. 4.15.*

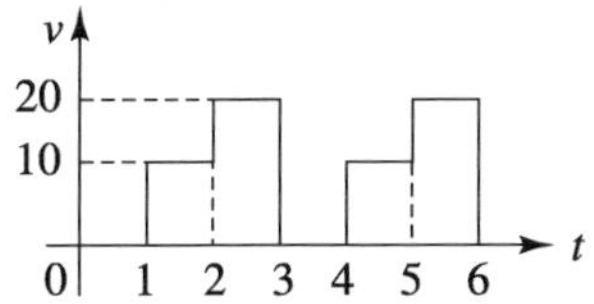

Fig. 4.15

Solution

$$V_{avg} = \frac{0+10+20}{3} = 10 \text{ V}$$

$$V_{rms} = \sqrt{\frac{0^2+(10)^2+(20)^2}{3}} = 12.9 \text{ V}$$

Example 4.16 *Find the effective value of the resultant current which carries simultaneously a direct current of 10 A and a sinusoidally alternating current with a peak value of 10 A.*

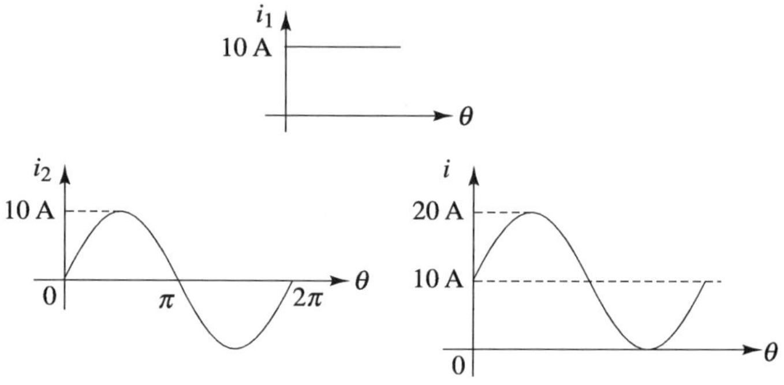

Fig. 4.16

Solution $i = 10 + 10 \sin \theta$

$$I_{eff} = I_{rms} = \sqrt{\frac{1}{2\pi}\int_0^{2\pi} i^2(\theta)\,d\theta}$$

$$= \sqrt{\frac{1}{2\pi}\int_0^{2\pi} (10+10\sin\theta)^2 d\theta} = \sqrt{\frac{1}{2\pi}\int_0^{2\pi} (100+200\sin\theta+100\sin^2\theta)\,d\theta}$$

$$= \sqrt{\frac{100}{2\pi}\int_0^{2\pi} (1+2\sin\theta+\sin^2\theta)\,d\theta}$$

$$= \sqrt{\frac{100}{2\pi}\int_0^{2\pi} \left[1+2\sin\theta+\left(\frac{1-\cos 2\theta}{2}\right)\right]d\theta}$$

$$= \sqrt{\frac{100}{2\pi}\left[\theta-2\cos\theta+\frac{\theta}{2}-\frac{\sin 2\theta}{4}\right]_0^{2\pi}}$$

$$= \sqrt{\frac{100}{2\pi}\left[2\pi-2\cos 2\pi+\frac{2\pi}{2}-\frac{\sin 4\pi}{4}-0+2\cos 0-0+\frac{\sin 0}{4}\right]}$$

$$= \sqrt{\frac{100}{2\pi}\left[2\pi-2+\frac{2\pi}{2}+2\right]} = \sqrt{\frac{100}{2\pi}\times 3\pi}$$

$$= \sqrt{150} = 12.25 \text{ A}$$

Example 4.17 *Find the relative heating effects of two current waves of equal peak value, one sinusoidal and the other, rectangular in shape.*

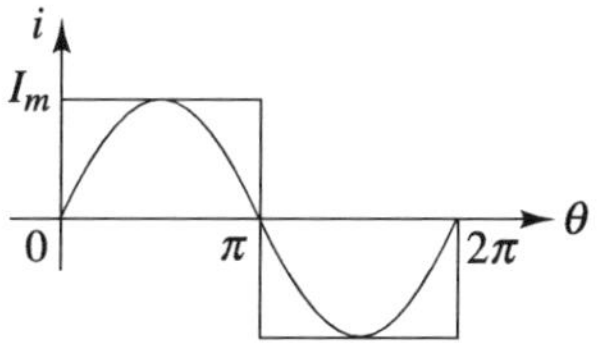

Fig. 4.17

Solution RMS value of the rectangular wave $= I_m$

$$\text{RMS value of sinusoidal current wave} = \frac{I_m}{\sqrt{2}}$$

$$\text{Heating effect due to rectangular current wave} = (I_m)^2 RT$$

$$\text{Heating effect due to sinusoidal current wave} = \left(\frac{I_m}{\sqrt{2}}\right)^2 RT = \frac{(I_m)^2}{2} RT$$

$$\text{Relative heating effects} = \frac{(I_m)^2}{2} RT : (I_m)^2 RT$$

$$= \frac{1}{2} : 1 = 1 : 2$$

4.5 PHASOR REPRESENTATION OF ALTERNATING QUANTITIES

The alternating quantities are represented by phasors. A phasor is a line of definite length rotating in an anticlockwise direction at a constant angular velocity ω. The length of a phasor is equal to the maximum value of the alternating quantity and the angular velocity is equal to the angular velocity of alternating quantity.

As shown in Fig. 4.18(a), consider a phasor $OP = I_m$, where I_m is the maximum value of the alternating current. Let this phasor rotate in an anticlockwise direction at a uniform angular velocity of ω rad/second. The projection of the phasor OP on the Y-axis at any instant gives the instantaneous value of that alternating current.

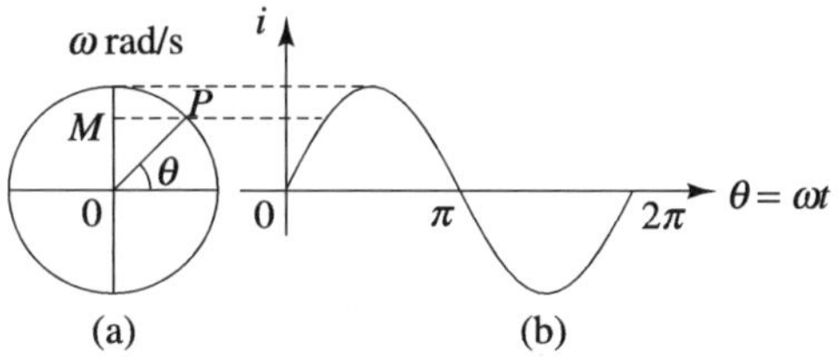

Fig. 4.18

$$OM = OP \sin \omega t$$
$$= I_m \sin \omega t = i$$

Thus, if we plot the projections of the phasor on the Y-axis versus its angular position point by point, a sinusoidal alternating current waveform is obtained.

Phasor diagram using rms values A sinusoidal alternating current and voltages can be represented by phasors. The electrical measuring instruments like ammeter and voltmeter are calibrated to read the rms value of ac quantities. Hence, instead of using maximum value, it is more convenient to draw phasor diagrams using rms values of alternating quantities. However, such a phasor diagram will not generate a sine wave of proper amplitude unless the length of the phasor is multiplied by $\sqrt{2}$.

4.6 MATHEMATICAL REPRESENTATION OF PHASORS

A phasor can be represented in four forms.

(i) Rectangular form

$$\overline{V} = X \pm jY$$

$$\text{Magnitude of phasor, } V = \sqrt{X^2 + Y^2}$$

$$\text{Phase angle } \phi = \tan^{-1}\left(\frac{Y}{X}\right)$$

(ii) Trigonometric form

$$\bar{V} = V(\cos\phi \pm j\sin\phi)$$

(iii) Exponential form

$$\bar{V} = Ve^{\pm j\phi}$$

(iv) Polar form

$$\bar{V} = V\angle\pm\phi$$

Significance of operator j The operator j is used in rectangular form. It is used to indicate anticlockwise rotation of a phasor through 90°. Mathematically,

$$j = \sqrt{-1}$$

Whenever a phasor is multiplied by j, the phasor is rotated once in anticlockwise direction through 90°. The power of j represents the number of times the phasor should be rotated through 90° in anticlockwise direction.

Example 4.18 *Two sinusoidal currents are given as*

$$i_1 = 10\sqrt{2}\ \sin\omega t,\ i_2 = 20\sqrt{2}\ \sin(\omega t + 60°).$$

Find the expression for the sum of these currents.

Solution

Data

$$i_1 = 10\sqrt{2}\ \sin\omega t$$

$$i_2 = 20\sqrt{2}\ \sin(\omega t + 60°)$$

Writing currents i_1 and i_2 in the phasor form,

$$\bar{I}_1 = \frac{10\sqrt{2}}{\sqrt{2}}\angle 0° = 10\angle 0°$$

$$\bar{I}_2 = \frac{20\sqrt{2}}{\sqrt{2}}\angle 60° = 20\angle 60°$$

$$\bar{I} = \bar{I}_1 + \bar{I}_2$$

$$= 10\angle 0° + 20\angle 60° = 26.46\ \angle 40.89°$$

$$i = 26.46\sqrt{2}\sin(\omega t + 40.89°) = 37.42\sin(\omega t + 40.89°)$$

Example 4.19 *The following three sinusoidal currents flow into the junction* $i_1 = 3\sqrt{2}\ \sin\omega t$, $i_2 = 5\sqrt{2}\ \sin(\omega t + 30°)$ *and* $i_3 = 6\sqrt{2}\ \sin(\omega t - 120°)$. *Find the expression for the resultant current which leaves the junction.*

Solution

Data

$$i_1 = 3\sqrt{2}\sin\omega t$$

$$i_2 = 5\sqrt{2}\sin(\omega t + 30°)$$

$$i_3 = 6\sqrt{2}\sin(\omega t - 120°)$$

Writing currents i_1, i_2 and i_3 in the phasor form,

$$\bar{I}_1 = \frac{3\sqrt{2}}{\sqrt{2}}\angle 0° = 3\angle 0°$$

$$\bar{I}_2 = \frac{5\sqrt{2}}{\sqrt{2}}\angle 30° = 5\angle 30°$$

$$\bar{I}_3 = \frac{6\sqrt{2}}{\sqrt{2}}\angle{-120°} = 6\angle{-120°}$$

The resultant current which leaves the junction is given by

$$\bar{I} = \bar{I}_1 + \bar{I}_2 + \bar{I}_3 = 3\angle 0° + 5\angle 30° + 6\angle{-120°}$$
$$= 5.1\ \angle{-31.9°}$$
$$i = 5.1\sqrt{2}\sin(\omega t - 31.9°) = 7.21\sin(\omega t - 31.9°)$$

Example 4.20 *In a circuit, four currents are meeting at a point. Find the resultant current.*

$$i_1 = 5\sin\omega t, \qquad i_2 = 10\sin(\omega t - 30°)$$
$$i_3 = 5\cos(\omega t - 30°) \qquad i_4 = -10\sin(\omega t + 45°)$$

Solution

Data

$$i_1 = 5\sin\omega t$$
$$i_2 = 10\sin(\omega t - 30°)$$
$$i_3 = 5\cos(\omega t - 30°) = 5\sin(\omega t + 60°)$$
$$i_4 = -10\sin(\omega t + 45°) = 10\sin(\omega t + 225°)$$

Writing currents i_1, i_2, i_3 and i_4 in the phasor form,

$$\bar{I}_1 = \frac{5}{\sqrt{2}}\angle 0° = 3.54\angle 0°$$

$$\bar{I}_2 = \frac{10}{\sqrt{2}}\angle{-30°} = 7.07\angle{-30°}$$

$$\bar{I}_3 = \frac{5}{\sqrt{2}}\angle 60° = 3.54\angle 60°$$

$$\bar{I}_4 = \frac{10}{\sqrt{2}}\angle 225° = 7.07\angle 225°$$

$$\text{Resultant current} = \bar{I}_1 + \bar{I}_2 + \bar{I}_3 + \bar{I}_4$$
$$= 3.54\angle 0° + 7.07\angle{-30°} + 3.54\angle 60° + 7.07\angle 225° = 8.44\angle{-40.36°}$$
$$i = 8.44\sqrt{2}\sin(\omega t - 40.36°) = 11.94\sin(\omega t - 40.36°)$$

Example 4.21 *Find the resultant voltage and its equation for the given voltages.*

$$e_1 = 20\sin\omega t, \qquad e_2 = 30\sin\left(\omega t - \frac{\pi}{4}\right), \qquad e_3 = 40\cos\left(\omega t + \frac{\pi}{6}\right)$$

Solution

Data

$$e_1 = 20\sin\omega t$$

$$e_2 = 30\sin\left(\omega t - \frac{\pi}{4}\right) = 30\sin(\omega t - 45°)$$

$$e_3 = 40\cos\left(\omega t + \frac{\pi}{6}\right) = 40\sin(\omega t + 120°)$$

Writing voltages e_1, e_2 and e_3 in the phasor form,

$$\bar{E}1 = \frac{20}{\sqrt{2}}\angle 0° = 14.14\angle 0°$$

$$\bar{E}2 = \frac{30}{\sqrt{2}}\angle{-45°} = 21.21\angle{-45°}$$

$$\bar{E}_3 = \frac{40}{\sqrt{2}} \angle 120^\circ = 28.28\angle 120^\circ$$

$$\text{Resultant voltage } \bar{E} = \bar{E}_1 + \bar{E}_2 + \bar{E}_3$$

$$= 14.14\angle 0^\circ + 21.21\angle -45^\circ + 28.28\angle 120^\circ = 17.75\angle 32.33^\circ$$

$$e = 17.75\sqrt{2}\sin(\omega t + 32.33^\circ) = 25.1\sin(\omega t + 32.33^\circ)$$

Example 4.22 *Obtain the sum of the three voltages.*

$$v_1 = 147.3\cos(\omega t + 98.1^\circ)$$
$$v_2 = 294.6\cos(\omega t - 45^\circ)$$
$$v_3 = 88.4\sin(\omega t + 135^\circ)$$

Solution

Data

$$v_1 = 147.3\cos(\omega t + 98.1^\circ) = 147.3\sin(\omega t + 188.1^\circ)$$
$$v_2 = 294.6\cos(\omega t - 45^\circ) = 294.6\sin(\omega t + 45^\circ)$$
$$v_3 = 88.4\sin(\omega t + 135^\circ)$$

Writing the voltages v_1, v_2 and v_3 in the phasor form,

$$\bar{V}_1 = \frac{147.3}{\sqrt{2}} \angle 188.1^\circ = 104.16\angle 188.1^\circ$$

$$\bar{V}_2 = \frac{294.6}{\sqrt{2}} \angle 45^\circ = 208.31\angle 45^\circ$$

$$\bar{V}_3 = \frac{88.4}{\sqrt{2}} \angle 135^\circ = 62.51\angle 135^\circ$$

$$\text{Resultant voltage } \bar{V} = \bar{V}_1 + \bar{V}_2 + \bar{V}_3$$

$$= 104.16\angle 188.1^\circ + 208.31\angle 45^\circ + 62.51\angle 135^\circ = 176.82\angle 90^\circ$$

$$v = 176.82\sqrt{2}\sin(\omega t + 90^\circ) = 250.06\sin(\omega t + 90^\circ)$$

Example 4.23 *Find vectorially the resultant of the following four voltages.*

$$e_1 = 25\sin\omega t, \qquad e_2 = 30\sin\left(\omega t + \frac{\pi}{6}\right),$$

$$e_3 = 30\cos\omega t, \qquad e_4 = 20\sin\left(\omega t - \frac{\pi}{6}\right),$$

Obtain the answer in similar form.

Solution

Data

$$e_1 = 25\sin\omega t$$

$$e_2 = 30\sin\left(\omega t + \frac{\pi}{6}\right) = 30\sin(\omega t + 30^\circ)$$

$$e_3 = 30\cos\omega t = 30\sin(\omega t + 90^\circ)$$

$$e_4 = 20\sin\left(\omega t - \frac{\pi}{6}\right) = 20\sin(\omega t - 30^\circ)$$

Writing voltages e_1, e_2, e_3 and e_4 in the phasor form,

$$\bar{E}_1 = \frac{25}{\sqrt{2}} \angle 0^\circ = 17.68\angle 0^\circ$$

$$\overline{E}_2 = \frac{30}{\sqrt{2}}\angle 30° = 21.21\angle 30°$$

$$\overline{E}_3 = \frac{30}{\sqrt{2}}\angle 90° = 21.21\angle 90°$$

$$\overline{E}_4 = \frac{20}{\sqrt{2}}\angle -30° = 14.14\angle -30°$$

Resultant voltage $\overline{E} = \overline{E}_1 + \overline{E}_2 + \overline{E}_3 + \overline{E}_4$

$$= 17.68\angle 0° + 21.21\angle 30° + 21.21\angle 90° + 14.14\angle -30° = 54.26\angle 27.13°$$

$$e = 54.26\sqrt{2}\sin(\omega t + 27.13°) = 76.74\sin(\omega t + 27.13°)$$

Example 4.24 *Two currents are represented by* $i_1 = 15\sin\left(\omega t + \frac{\pi}{3}\right)$ *and* $i_2 = 25\sin\left(\omega t + \frac{\pi}{4}\right)$. *These currents are fed into a common conductor. Find the total current in the form* $i = I_m \sin(\omega t + \phi)$. *If the conductor has a resistance of 10 Ω, what will be the energy loss in 24 hours?*

Solution

Data

$$i_1 = 15\sin\left(\omega t + \frac{\pi}{3}\right)$$

$$i_2 = 25\sin\left(\omega t + \frac{\pi}{4}\right)$$

$$R = 10\ \Omega$$

$$t = 24 \text{ hours} = 86400 \text{ seconds}$$

Writing currents i_1 and i_2 in phasor form,

$$\overline{I}_1 = \frac{15}{\sqrt{2}}\angle 60° = 10.61\angle 60°$$

$$\overline{I}_2 = \frac{25}{\sqrt{2}}\angle 45° = 17.68\angle 45°$$

Total current $\overline{I} = \overline{I}_1 + \overline{I}_2 = 10.61\angle 60° + 17.68\angle 45° = 28.06\angle 50.62°$

$$i = 28.06\sqrt{2}\sin(\omega t + 50.62°) = 39.68\sin(\omega t + 50.62°)$$

Energy loss in 24 hours, $E = I^2Rt$ where I is rms value of current

$$E = (28.06)^2 \times 10 \times 86400 = 6.8 \times 10^8 \text{ J}$$

Example 4.25 *The voltage drops across four series connected impedances are given:*

$$v_1 = 60\sin\left(\omega t + \frac{\pi}{6}\right) \qquad v_2 = 75\sin\left(\omega t - \frac{5\pi}{6}\right),$$

$$v_3 = 100\cos\left(\omega t + \frac{\pi}{4}\right), \qquad v_4 = V_{4m}\sin(\omega t + \phi_4)$$

Calculate the values of V_{4m} *and* ϕ_4 *if the voltage applied across the series circuit is* $140\sin\left(\omega t + \frac{3\pi}{5}\right)$.

Solution

Data

$$v_1 = 60 \sin\left(\omega t + \frac{\pi}{6}\right) = 60 \sin(\omega t + 30°)$$

$$v_2 = 75 \sin\left(\omega t - \frac{5\pi}{6}\right) = 75 \sin(\omega t - 150°)$$

$$v_3 = 100 \cos\left(\omega t + \frac{\pi}{4}\right) = 100 \sin(\omega t + 135°)$$

$$v = 140 \sin\left(\omega t + \frac{3\pi}{5}\right) = 140 \sin(\omega t + 108°)$$

Writing voltages v_1, v_2, v_3 and v in the phasor form,

$$\bar{V}_1 = \frac{60}{\sqrt{2}}\angle 30° = 42.43\angle 30°$$

$$\bar{V}_2 = \frac{75}{\sqrt{2}}\angle -150° = 53.03\angle -150°$$

$$\bar{V}_3 = \frac{100}{\sqrt{2}}\angle 135° = 70.71\angle 135°$$

$$\bar{V} = \frac{140}{\sqrt{2}}\angle 108° = 98.99\angle 108°$$

For series-connected impedances,

$$\bar{V} = \bar{V}_1 + \bar{V}_2 + \bar{V}_3 + \bar{V}_4$$

$$\bar{V}_4 = \bar{V} - \bar{V}_1 - \bar{V}_2 - \bar{V}_3$$

$$= 98.99\angle 108° - 42.43\angle 30° - 53.03\angle -150° - 70.71\angle 135° = 57.13\angle 59.96°$$

$$v_4 = 57.13\sqrt{2} \sin(\omega t + 59.96°) = 80.79 \sin(\omega t + 59.96°)$$

$$V_{4m} = 80.79 \text{ V}$$

$$\phi_4 = 59.96°$$

Example 4.26 *Two voltages having rms values of 50 V and 75 V have a phase difference of 60°. Find the resultant sum of these two voltages.*

Solution

Data

$$V_1 = 50 \text{ V}$$

$$V_2 = 75 \text{ V}$$

$$\phi = 60°$$

Let

$$\bar{V}_1 = 50\angle 0° \text{ V}$$

$$\bar{V}_2 = 75\angle -60° \text{ V}$$

Resultant voltage $\bar{V} = \bar{V}_1 + \bar{V}_2$

$$= 50\angle 0° + 75\angle -60° = 108.97\angle -36.58° \text{ V}$$

Example 4.27 *Two single-phase alternators supply 300 A and 400 A respectively at a phase difference of 20° to a common load. Find the resultant current and its phase relation to its component.*

Solution

Data

$$I_1 = 300 \text{ A}$$
$$I_2 = 400 \text{ A}$$
$$\phi = 20°$$

Let

$$\bar{I}_1 = 300 \angle 0° \text{ A}$$
$$\bar{I}_2 = 400 \angle -20° \text{ A}$$

$$\text{Resultant current } \bar{I} = \bar{I}_1 + \bar{I}_2$$
$$= 300\angle 0° + 400\angle -20° = 689.59 \angle -11.44° \text{ A}$$

Example 4.28 *Two voltage sources have equal emfs and a phase difference α. When they are connected in series, the voltage is 200 V. When one source is reversed, the voltage is 15 V. Find their emfs and phase angle α.*

Solution

Data

$$\bar{E}_1 = E \angle 0°$$
$$\bar{E}_2 = E \angle \alpha°$$
$$E_1 = E_2 = E$$

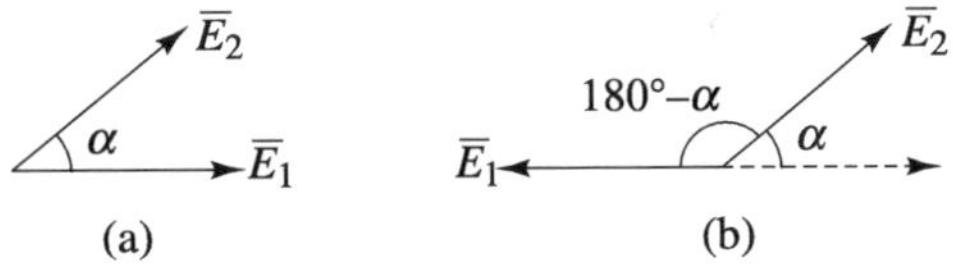

Fig. 4.19

When two sources are connected in series,

$$\sqrt{E_1^2 + E_2^2 + 2E_1E_2\cos\alpha} = 200$$

$$\sqrt{E^2 + E^2 + 2E^2\cos\alpha} = 200$$

$$2E^2 + 2E^2 \cos \alpha = 40000 \quad \text{...(i)}$$

When one source is reversed,

$$\sqrt{E_1^2 + E_2^2 - 2E_1E_2\cos\alpha} = 15$$

$$\sqrt{E^2 + E^2 - 2E^2\cos\alpha} = 15$$

$$2E^2 - 2E^2 \cos \alpha = 225 \quad \text{...(ii)}$$

Adding Eqs (i) and (ii),

$$4E^2 = 40225$$
$$E^2 = 10056.25$$
$$E = 100.28 \text{ V}$$
$$2E^2 + 2E^2 \cos \alpha = 40000$$
$$20112.5 + 20112.5 \cos \alpha = 40000$$
$$\cos \alpha = 0.988$$
$$\alpha = 8.58°$$

Example 4.29 *Two sinusoidal sources of emf have rms values E_1 and E_2 and a phase difference α. When connected in series, the resultant voltage is 41.1 V. When one of the sources is reversed, the resultant emf is 17.52 V. When phase displacement is made zero, the resultant emf is 42.5 V. Calculate E_1, E_2 and α.*

Solution

Data

$$\bar{E}_1 = E_1 \angle 0°$$

$$\overline{E}_2 = E_2 \angle \alpha°$$

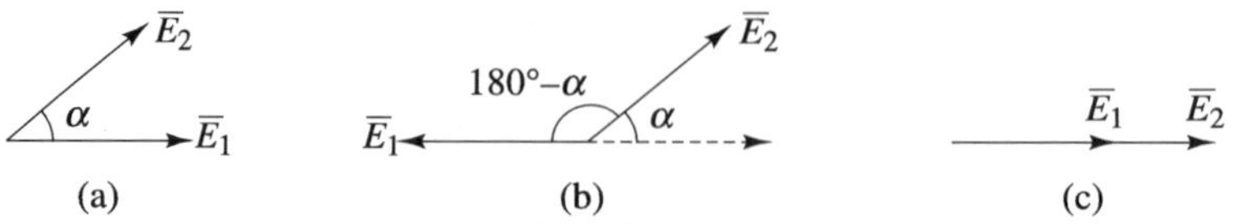

Fig. 4.20

When two sources are connected in series,

$$\sqrt{E_1^2 + E_2^2 + 2E_1E_2\cos\alpha} = 41.1$$

$$E_1^2 + E_2^2 + 2E_1E_2 \cos \alpha = 1689.21 \qquad \text{...(i)}$$

When one of the source is reversed,

$$\sqrt{E_1^2 + E_2^2 - 2E_1E_2\cos\alpha} = 17.52$$

$$E_1^2 + E_2^2 - 2E_1E_2 \cos \alpha = 306.95 \qquad \text{...(ii)}$$

When phase displacement is made zero,

$$\sqrt{E_1^2 + E_1^2 + 2E_1E_2\cos 0°} = 42.5$$

$$E_1 + E_2 = 42.5 \qquad \text{...(iii)}$$

Adding Eqs (i) and (ii),

$$2(E_1^2 + E_2^2) = 1996.16$$

$$E_1^2 + E_2^2 = 998.08$$

$$(42.5 - E_2)^2 + E_2^2 = 998.08$$

$$1806.25 - 85E_2 + E_2^2 + E_2^2 = 998.08$$

$$E_2^2 - 42.5E_2 + 404.09 = 0 \qquad \text{...(iv)}$$

Solving Eq. (iv),

$$E_2 = 28.14 \text{ V} \quad \text{or} \quad E_2 = 14.36 \text{ V}$$

$$E_1 = 14.36 \text{ V} \quad \text{or} \quad E_1 = 28.14 \text{ V}$$

Subtracting Eq. (ii) from Eq. (i),

$$4E_1E_2 \cos \alpha = 1382.26$$

$$4 \times 14.37 \times 28.14 \cos \alpha = 1382.26$$

$$\cos \alpha = 0.855$$

$$\alpha = 31.24°$$

4.7 BEHAVIOUR OF A PURE RESISTOR IN AN AC CIRCUIT

Consider a pure resistor R connected across an alternating voltage source v as shown in Fig. 4.21. Let the alternating voltage $v = V_m \sin \omega t$.

The alternating current i is given by

$$i = \frac{v}{R} = \frac{V_m}{R} = \sin \omega t = I_m \sin \omega t \qquad \ldots \left(I_m = \frac{V_m}{R} \right)$$

where I_m is the maximum value of the alternating current. From the voltage and current equation, it is clear that the current is in phase with the voltage in a pure resistive circuit.

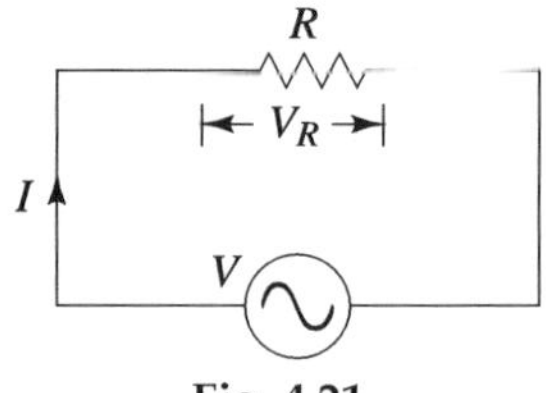

Fig. 4.21

Waveforms

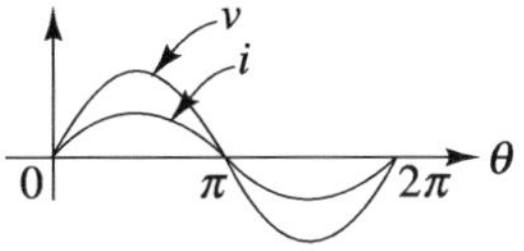

Fig. 4.22

Phasor diagram

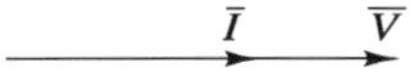

Fig. 4.23

Impedance It is the resistance offered to the flow of current in an ac circuit. In a pure resistive circuit,

$$Z = \frac{V}{I} = \frac{V_m}{I_m} = \frac{V_m}{V_m / R} = R$$

Phase difference Since the voltage and current are in phase with each other, the phase difference is zero.

$$\phi = 0°$$

Power factor It is defined as the cosine of the angle between voltage and current phasor.

$$\text{Power factor} = \cos \phi = \cos (0°) = 1$$

Power Instantaneous power $p = vi$

$$= V_m \sin \omega t \,.\, I_m \sin \omega t = V_m I_m \sin^2 \omega t$$

$$= \frac{V_m I_m}{2}(1 - \cos 2\omega t)$$

$$= \frac{V_m I_m}{2} - \frac{V_m I_m}{2} \cos 2\omega t$$

The power consists of a constant part $\frac{V_m I_m}{2}$ and a fluctuating part $\frac{V_m I_m}{2}$ cos $2\omega t$. The frequency of the fluctuating power is twice the applied voltage frequency and its average value over one complete cycle is zero.

$$\text{Average power } P = \frac{V_m I_m}{2} = \frac{V_m}{\sqrt{2}} \cdot \frac{I_m}{\sqrt{2}} = VI$$

Thus, power in a pure resistive circuit is equal to the product of rms values of voltage and current.

4.8 BEHAVIOUR OF A PURE INDUCTOR IN AN AC CIRCUIT

Consider a pure inductor L connected across an alternating voltage v as shown in Fig. 4.24. Let the alternating voltage $v = V_m \sin \omega t$.

The voltage across the inductor is given by

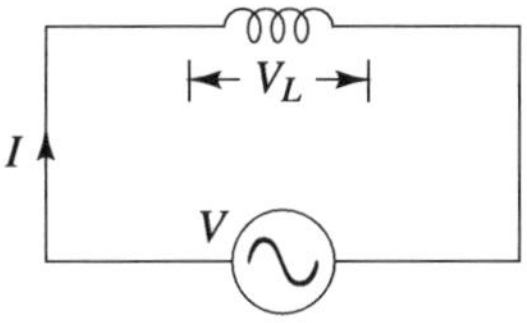

Fig. 4.24

$$v = L \frac{di}{dt}$$

$$i = \frac{1}{L}\int v \, dt = \frac{1}{L}\int V_m \sin \omega t \, dt$$

$$= \frac{V_m}{\omega L}(-\cos \omega t) = -\frac{V_m}{\omega L} \cos \omega t$$

$$= \frac{V_m}{\omega L} \sin\left(\omega t - \frac{\pi}{2}\right) = I_m \sin\left(\omega t - \frac{\pi}{2}\right) \quad \ldots \left(I_m = \frac{V_m}{\omega L}\right)$$

where I_m is the maximum value of the alternating current. From the voltage and current equation, it is clear that the current lags behind the voltage by 90° in a pure inductive circuit.

Waveform

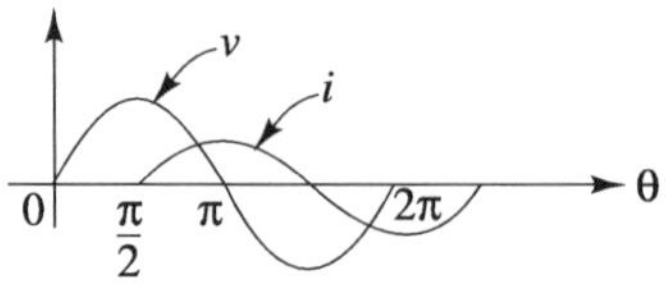

Fig. 4.25

Phasor diagram

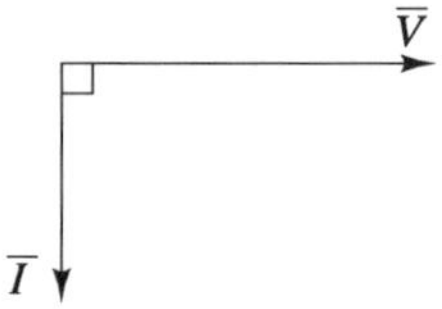

Fig. 4.26

Impedance

$$Z = \frac{V}{I} = \frac{V_m}{I_m} = \frac{V_m}{V_m/\omega L} = \omega L$$

The quantity ωL is called inductive reactance, is denoted by X_L and is expressed in ohms.

For a dc supply, $f = 0 \quad \therefore \quad X_L = 0$

Thus, an inductor acts as a short circuit for a dc supply.

Phase difference It is the angle between the voltage and current phasor.

$$\phi = 90°$$

Power factor It is defined as the cosine of the angle between the voltage and current phasor.

$$pf = \cos\phi = \cos(90°) = 0$$

Power Instantaneous power,

$$\begin{aligned} p &= vi \\ &= V_m \sin\omega t \,.\, I_m \sin\left(\omega t - \frac{\pi}{2}\right) \\ &= -V_m I_m \sin\omega t \cos\omega t \\ &= -\frac{V_m I_m}{2}\sin 2\omega t \end{aligned}$$

The average power for one complete cycle, $P = 0$.

Hence, power consumed by a pure inductive circuit is zero.

4.9 BEHAVIOUR OF A PURE CAPACITOR IN AN AC CIRCUIT

Consider a pure capacitor C connected across an alternating voltage v as shown in Fig. 4.27. Let the alternating voltage $v = V_m \sin\omega t$.

The current through capacitor is given by,

$$i = C\frac{dv}{dt}$$

$$= C\frac{d}{dt}(V_m \sin \omega t)$$
$$= \omega C\, V_m \cos \omega t$$
$$= \omega C\, V_m \sin (\omega t + 90^\circ)$$
$$= I_m \sin (\omega t + 90^\circ) \quad ...(I_m = \omega C\, V_m)$$

where I_m is the maximum value of the alternating current. From the voltage and current equation, it is clear that current leads voltage by 90° in pure capacitive circuit.

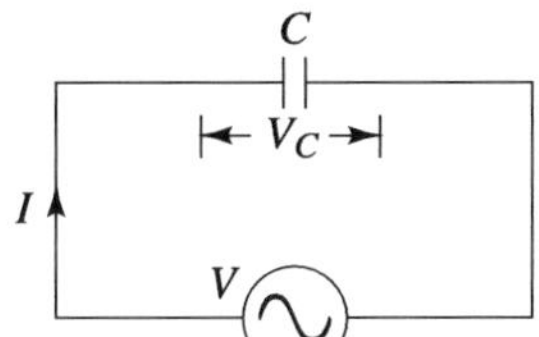

Fig. 4.27

Waveform

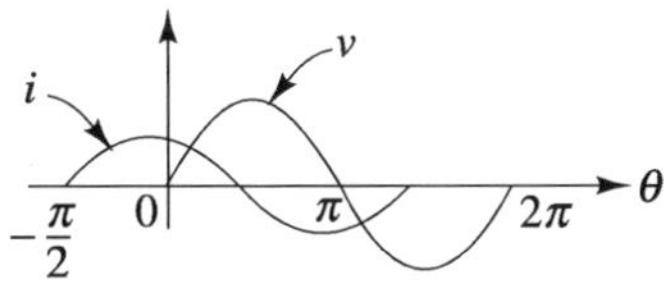

Fig. 4.28

Phasor diagram

Fig. 4.29

Impedance

$$Z = \frac{V}{I} = \frac{V_m}{I_m} = \frac{V_m}{\omega C V_m} = \frac{1}{\omega C}$$

The quantity $\frac{1}{\omega C}$ is called capacitive reactance and is denoted by X_C and is expressed in ohms.

For a dc supply, $f = 0 \quad \therefore \quad X_C = \infty$

Thus, the capacitor acts as an open circuit for dc supply.

Phase difference

$$\phi = 90^\circ$$

Power factor

$$\text{pf} = \cos \phi = \cos (90^\circ) = 0$$

Power Instantaneous power,

$$p = vi$$
$$= V_m \sin \omega t \,.\, I_m \sin (\omega t + 90^\circ)$$
$$= V_m I_m \sin \omega t \cos \omega t$$
$$= \frac{V_m I_m}{2} \sin 2\omega t$$

The average power for one complete cycle, $P = 0$.

Hence, power consumed by a pure capacitive circuit is zero.

4.10 SERIES R-L CIRCUIT

Figure 4.30 shows a pure resistor R connected in series with a pure inductor L across an alternating voltage v.

Let V and I be the rms values of applied voltage and current.

Potential difference across the resistor = $V_R = R \,.\, I$

Potential difference across the inductor = $V_L = X_L \,.\, I$

The voltage $\overline{V_R}$ is in phase with current $\overline{I}$ whereas voltage $\overline{V_L}$ leads current $\overline{I}$ by 90°.

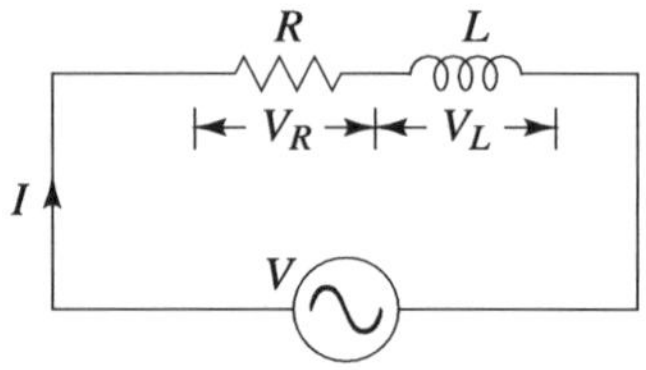

Fig. 4.30

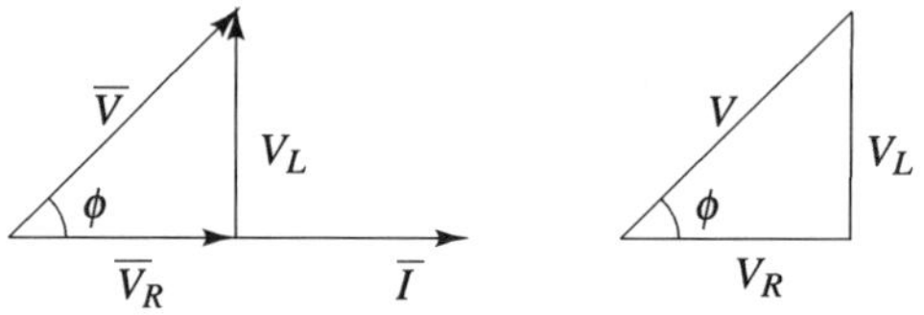

(a) Phasor diagram (b) Voltage triangle

Fig. 4.31

$$\cos\phi = \frac{V_R}{V}$$

Impedance

$$\overline{V} = \overline{V_R} + \overline{V_L} = R\overline{I} + jX_L\,\overline{I}$$

$$= (R + jX_L)\,\overline{I}$$

$$\frac{\overline{V}}{\overline{I}} = R + jX_L = \overline{Z}$$

$$\overline{Z} = Z\angle\phi$$

$$Z = \sqrt{R^2 + X_L^2} = \sqrt{R^2 + \omega^2 L^2}$$

$$\phi = \tan^{-1}\left(\frac{X_L}{R}\right) = \tan^{-1}\left(\frac{\omega L}{R}\right)$$

The quantity Z is called complex impedance of the R-L circuit.

Impedance triangle

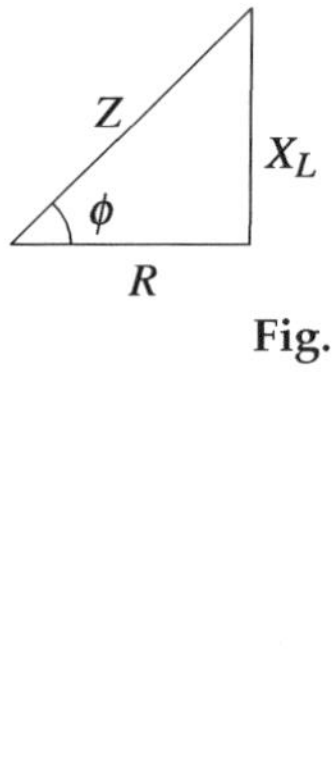

Fig. 4.32

$$\cos\phi = \frac{R}{Z}$$

Current From the phasor diagram, it is clear that the current I lags behind voltage V by an angle ϕ. If the applied voltage is given by $v = V_m \sin \omega t$, then the current equation will be

$$i = I_m \sin(\omega t - \phi)$$

where
$$I_m = \frac{V_m}{Z}$$

and
$$\phi = \tan^{-1}\left(\frac{\omega L}{R}\right)$$

Waveforms

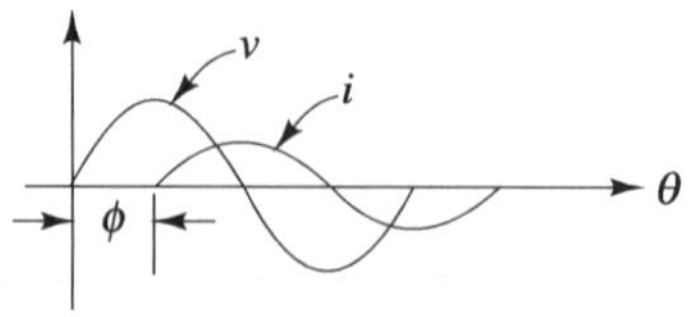

Fig. 4.33

Power Instantaneous power

$$
\begin{aligned}
p &= v \,.\, i \\
&= V_m \sin \omega t \,.\, I_m \sin (\omega t - \phi) \\
&= V_m I_m \sin \omega t \,.\, \sin (\omega t - \phi) \\
&= V_m I_m \left[\frac{\cos\phi - \cos(2\omega t - \phi)}{2} \right] \\
&= \frac{V_m I_m}{2} \cos \phi - \frac{V_m I_m}{2} \cos (2\omega t - \phi)
\end{aligned}
$$

Thus, power consists of a constant part $\frac{V_m I_m}{2} \cos \phi$ and a fluctuating part $\frac{V_m I_m}{2} \cos (2\omega t - \phi)$. The frequency of the fluctuating part is twice the applied voltage frequency and its average value over one complete cycle is zero.

$$\text{Average power } P = \frac{V_m I_m}{2} \cos \phi = \frac{V_m}{\sqrt{2}} \frac{I_m}{\sqrt{2}} \cos \phi = VI \cos \phi$$

Thus, power is dependent upon the in-phase component of the current. The average power is also called active power and is measured in watts.

We know that pure a inductor and capacitor consume no power because all the power received from the source in a half cycle is returned to the source in the next half cycle. This circulating power is called *reactive power*. It is a product of the voltage and reactive component of the current, i.e., $I \sin \phi$ and is measured in VAR (volt–ampere-reactive).

$$\text{Reactive power } Q = VI \sin \phi.$$

The product of voltage and current is known as apparent power (S) and is measured in volt–ampere (VA).

$$S = \sqrt{P^2 + Q^2}$$

Power triangle In terms of circuit components,

$$\cos \phi = \frac{R}{Z}$$

and

$$
\begin{aligned}
V &= Z \,.\, I \\
P &= VI \cos \phi = Z \,.\, I \,.\, I \frac{R}{Z} = I^2 R \text{ (W)} \\
Q &= VI \sin \phi = Z \,.\, I \,.\, I \frac{X_L}{Z} = I^2 X_L \text{ (VAR)} \\
S &= VI = Z \,.\, I \,.\, I = I^2 Z \text{ (VA)}
\end{aligned}
$$

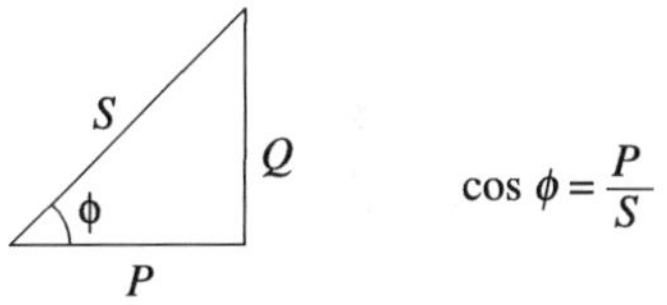

Fig. 4.34

Power factor It is defined as the cosine of the angle between the voltage and current phasor.

$$\text{pf} = \cos \phi$$

From voltage triangle, $\text{pf} = \frac{V_R}{V}$

From impedance triangle, $\text{pf} = \frac{R}{Z}$

From power triangle, $\text{pf} = \frac{P}{S}$

In case of *R-L* series circuit, the power factor is lagging in nature.

Example 4.30 *An alternating voltage of 80 + j60 V is applied to a circuit and the current flowing is 4 – j2 A. Find the (a) impedance, (b) power consumed, (c) phase angle, and (d) power factor.*

Solution

Data

$$\bar{V} = 80 + j60 \text{ V}$$

$$\bar{I} = 4 - j2 \text{ A}$$

$$\bar{Z} = \frac{\bar{V}}{\bar{I}} = \frac{80 + j60}{4 - j2} = \frac{100\angle 36.87^\circ}{4.47\angle -26.56^\circ} = 22.37\ \angle 63.43^\circ\ \Omega$$

Impedance $Z = 22.37\ \Omega$

Phase angle $\phi = 63.43^\circ$

Power factor $\text{pf} = \cos\phi = \cos(63.43^\circ) = 0.447$ (lagging)

Power consumed

$$P = VI\cos\phi$$

$$= 100 \times 4.47 \times 0.447 = 199.81 \text{ W}$$

Example 4.31 *The voltage and current in a circuit are given by $\bar{V} = 150\ \angle 30^\circ$ V and $\bar{I} = 2\ \angle -15^\circ$ A. If the circuit works on a 50-Hz supply, determine the power factor, power loss, impedance, resistance, and reactance considering the circuit as a simple series circuit.*

Solution

Data

$$\bar{V} = 150\ \angle 30^\circ \text{ V}$$

$$\bar{I} = 2\angle -15^\circ \text{ A}$$

$$f = 50 \text{ Hz}$$

$$\bar{Z} = \frac{\bar{V}}{\bar{I}} = \frac{150\angle 30^\circ}{2\angle -15^\circ}$$

$$= 75\ \angle 45^\circ\ \Omega = 53.03 + j53.03\ \Omega$$

Impedance $Z = 75\ \Omega$

Resistance $R = 53.03\ \Omega$

Reactance $X = 53.03\ \Omega$

Power factor $\text{pf} = \cos\phi = \cos(45^\circ) = 0.707$ (lagging)

Power loss

$$P = VI\cos\phi$$

$$= 150 \times 2 \times 0.707 = 212.1 \text{ W}$$

Example 4.32 *An rms voltage of 100 $\angle 0^\circ$ V is applied to a series combination of Z_1 and Z_2 when $Z_1 = 20\ \angle 30^\circ\ \Omega$. The effective voltage drop across Z_1 is known to be 40 $\angle -30^\circ$ V. Find the reactive component of Z_2.*

Solution

Data

$$\bar{V} = 100\ \angle 0^\circ \text{ V}$$

$$\bar{Z}_1 = 20\ \angle 30^\circ\ \Omega$$

$$\bar{V}_1 = 40\ \angle -30^\circ \text{ V}$$

$$\bar{I} = \frac{\bar{V}_1}{\bar{Z}_1} = \frac{40\angle -30^\circ}{20\angle 30^\circ} = 2\ \angle -60^\circ \text{ A}$$

Total impedance

$$\bar{Z} = \frac{\bar{V}}{\bar{I}} = \frac{100\angle 0^\circ}{2\angle -60^\circ} = 50\ \angle 60^\circ = 25 + j43.3\ \Omega$$

$$\bar{Z}_1 = 20\ \angle 30^\circ = 17.32 + j10\ \Omega$$

$$\bar{Z} = \bar{Z}_1 + \bar{Z}_2$$

$$\sqrt{2} = \bar{Z} - \bar{Z}_1$$

$$= 25 + j43.3 - 17.32 - j10 = 7.68 + j33.3\ \Omega$$

Reactive component of $\bar{Z}_2 = 33.3\ \Omega$

Example 4.33 *A voltage v (t) = 177 sin (314t + 10°) is applied to a circuit. It causes a steady state current to flow, which is described by i (t) = 14.14 sin (314t – 20°). Determine the power factor an average power delivered to the circuit.*

Solution

Data $v(t) = 177 \sin (314t + 10^\circ)$

$i(t) = 14.14 \sin (314t - 20^\circ)$

Current $i(t)$ lags behind voltage $v(t)$ by 30°.

$\phi = 30^\circ$

Power factor pf = cos (30°) = 0.866 (lagging)

Power consumed $P = VI \cos \phi$

$$= \frac{177}{\sqrt{2}} \times \frac{14.14}{\sqrt{2}} \times 0.866 = 1083.7\ \text{W}$$

Example 4.34 *When a sinusoidal voltage 120 V (rms) is applied to a series R-L circuit, it is found that there occurs a power dissipation of 1200 W and a current flow given by i (t) = 28.3 sin (314 t – φ). Find the circuit resistance and inductance.*

Solution

Data $V = 120$ V

$i(t) = 28.3 \sin (314t - \phi)$

$P = 1200$ W

$$I = \frac{28.3}{\sqrt{2}} = 20.01\ \text{A}$$

$$P = VI \cos \phi$$

$$1200 = 120 \times 20.01 \times \cos \phi$$

$$\cos \phi = 0.499$$

$$\phi = 60.02^\circ$$

$$Z = \frac{V}{I} = \frac{120}{20.01} = 6\ \Omega$$

$$\bar{Z} = Z\angle\phi = 6\ \angle 60.02^\circ = 3 + j5.2\ \Omega$$

Resistance $R = 3\ \Omega$

Reactance $X_L = 5.2\ \Omega$

$X_L = \omega L$

$5.195 = 314 \times L$

Inductance $L = 0.0165$ H

Example 4.35 *In a series circuit containing resistance and inductance, the current and voltage are expressed as* $i(t) = 5 \sin\left(314t + \frac{2\pi}{3}\right)$ *and* $v(t) = 20 \sin\left(314t + \frac{5\pi}{6}\right)$. *(a) What is the impedance of the circuit? (b) What are the values of resistance, inductance and power factor? (c) What is the average power drawn by the circuit?*

Solution

Data

$$i(t) = 5 \sin\left(314t + \frac{2\pi}{3}\right)$$

$$v(t) = 20 \sin\left(314t + \frac{5\pi}{6}\right)$$

$$I = \frac{5}{\sqrt{2}} \qquad V = \frac{20}{\sqrt{2}}$$

$$\text{Impedance } Z = \frac{V}{I} = \frac{20/\sqrt{2}}{5/\sqrt{2}} = 4\ \Omega$$

Current $i(t)$ lags behind voltage $v(t)$ by an angle $\phi = 150° - 120° = 30°$

Power factor $\quad \text{pf} = \cos\phi = \cos(30°) = 0.866$ (lagging)

$$\bar{Z} = 4\angle 30° = 3.464 + j2\ \Omega$$

Resistance $\quad R = 3.464\ \Omega$

Reactance $\quad X_L = 2$

$$X_L = \omega L$$

$$2 = 314 \times L$$

Inductance $\quad L = 6.37$ mH

Average power $\quad P = VI\cos\phi$

$$= \frac{20}{\sqrt{2}} \times \frac{5}{\sqrt{2}} \times 0.866 = 43.3 \text{ W}$$

Example 4.36 *A series circuit consists of non-inductive resistance of 6 Ω and an inductive reactance of 10 Ω. When connected to a single phase ac supply, it draws a current* $i(t) = 27.89 \sin(628t - 45°)$. *Calculate (i) the voltage applied to the series circuit in the form* $V_m \sin(\omega t \pm \phi)$, *(ii) inductance, and (iii) power drawn by the circuit.*

Solution

Data

$$R = 6\ \Omega$$

$$X_L = 10\ \Omega$$

$$i(t) = 27.89 \sin(628t - 45°)$$

$$\bar{Z} = R + jX_L = 6 + j10 = 11.66\angle 59.04°\ \Omega$$

$$\bar{I} = \frac{27.89}{\sqrt{2}}\angle -45° = 19.72\angle -45° \text{ A}$$

$$\bar{V} = \bar{Z}\,.\,\bar{I} = (11.66\angle 59.04°)(19.72\angle -45°) = 229.95\angle 14.04° \text{ V}$$

$$v = 229.95\sqrt{2}\sin(\omega t + 14.04°) = 325.2\sin(\omega t + 14.04°)$$

$$X_L = \omega L$$

$$10 = 628 \times L$$

Inductance $L = 0.0159\text{ H} = 15.9\text{ mH}$

Power $P = VI\cos\phi$

$$= 229.95 \times 19.72 \times \cos(59.04°)$$
$$= 2332.78\text{ W}$$

Example 4.37 *A choke coil is connected in series with a fixed resistor. A 240-V, 50-Hz supply is applied and current of 2.5 A flows. If the voltage drop across the coil and fixed resistor are 140 V and 160 V respectively, calculate the resistance and inductance of the coil, and the value of the fixed resistor and power drawn by the coil.*

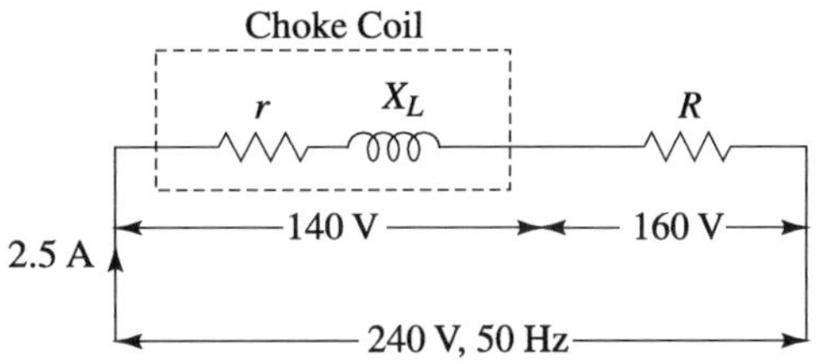

Fig. 4.35

Solution

Resistance of fixed resistor $R = \dfrac{160}{2.5} = 64\ \Omega$

Impedance of choke coil $Z_{coil} = \dfrac{140}{2.5} = 56\ \Omega$

$$Z_{coil} = \sqrt{r^2 + X_L^2} = 56$$
$$r^2 + X_L^2 = 3136 \quad \text{...(i)}$$

Total impedance $Z = \dfrac{240}{2.5} = 96\ \Omega$

$$\bar{Z} = (R + r) + jX_L$$
$$Z = \sqrt{(R+r)^2 + X_L^2}$$
$$96 = \sqrt{(r+64)^2 + X_L^2}$$
$$(r + 64)^2 + X_L^2 = 9216 \quad \text{...(ii)}$$

Subtracting Eq. (ii) from (i),

$$(r + 64)^2 - r^2 = 6080$$
$$r^2 + 128\,r + 4096 - r^2 = 6080$$
$$128\,r = 1984$$
$$r = 15.5\ \Omega$$

Resistance of coil $r = 15.5\ \Omega$

Substituting the value of r in the Eq. (i),

$$(15.5)^2 + X_L^2 = 3136$$
$$X_L^2 = 2895.75$$
$$X_L = 53.81\ \Omega$$
$$X_L = 2\pi f L$$
$$53.81 = 2\pi \times 50 \times L$$

Inductance of coil $L = 0.17\text{ H}$

Power drawn by the coil $= I^2 r$

$$= (2.5)^2 \times 15.5 = 96.875\text{ W}$$

Example 4.38 *A 100-Ω resistance is connected in series with a choke coil. When a 400-V, 50-Hz supply is applied to this combination, the voltage across the resistance and the choke coil are 200 V and 300 V respectively. Find the power consumed by the choke coil. Also, calculate the power factor of the choke coil and the power factor of the circuit.*

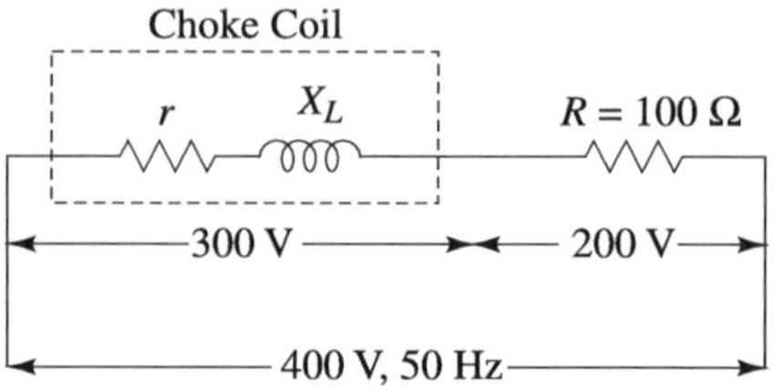

Fig. 4.36

Solution

$$\text{Current } I = \frac{200}{100} = 2 \text{ A}$$

$$\text{Impedance of choke coil} \quad Z_{coil} = \frac{300}{2} = 150 \ \Omega$$

$$\sqrt{r^2 + X_L^2} = 150$$

$$r^2 + X_L^2 = 22500 \qquad \text{...(i)}$$

$$\text{Total impedance} \quad Z = \frac{400}{2} = 200 \ \Omega$$

$$\bar{Z} = (R + r) + jX_L$$

$$Z = \sqrt{(R+r)^2 + X_L^2} = 200$$

$$(100 + r)^2 + X_L^2 = 40000 \qquad \text{...(ii)}$$

Subtracting the Eq. (i) from (ii),

$$(100 + r)^2 - r^2 = 17500$$

$$10000 + 200r + r^2 - r^2 = 17500$$

$$200r = 7500$$

$$r = 37.5 \ \Omega$$

Substituting the value of r in the Eq. (i),

$$(37.5)^2 + X_L^2 = 22500$$

$$X_L^2 = 21093.75$$

$$X_L = 145.24 \ \Omega$$

$$\text{Power consumed by choke coil} = I^2 r$$

$$= (2)^2 \times 37.5 = 150 \text{ W}$$

$$\text{Power factor of choke coil} = \frac{r}{Z_{coil}} = \frac{37.5}{150} = 0.25 \text{ (lagging)}$$

$$\text{Power factor of circuit} = \frac{R + r}{Z} = \frac{100 + 37.5}{200} = 0.6875 \text{ (lagging)}$$

Example 4.39 *A resistance of 25 Ω is connected in series with a choke coil. The series combination when connected across a 250-V, 50-Hz supply, draws a current of 4-A which lags behind the voltage by 65°. Calculate (i) total power, (ii) power consumed by resistance, (iii) power consumed by choke coil, and (iv) resistance and inductance of the coil.*

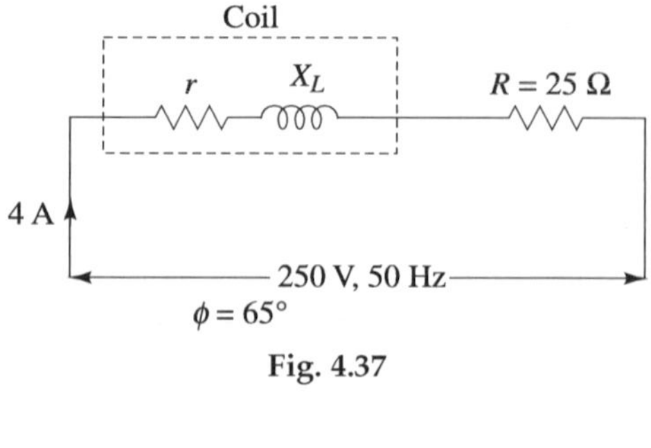

Fig. 4.37

Solution

Total impedance $Z = \dfrac{250}{4} = 62.5\ \Omega$

$\bar{Z} = Z\angle\phi = 62.5\ \angle 65° = 26.41 + j56.64\ \Omega$

But $\bar{Z} = (R + r) + jX_L$

$X_L = 56.64\ \Omega$

$R + r = 26.41$

Resistance of coil $r = 26.41 - 25 = 1.41\ \Omega$

Total power $P = I^2 (R + r) = (4)^2 \times 26.41 = 422.56\ \text{W}$

Power consumed by resistance $P_R = I^2R = (4)^2 \times 25 = 400\ \text{W}$

Power consumed by choke coil $P_{coil} = I^2r = (4)^2 \times 1.41 = 22.56\ \text{W}$

$X_L = 2\pi fL$

$56.64 = 2\pi \times 50 \times L$

Inductance of coil $L = 0.18\ \text{H}$

Example 4.40 *When a resistor and a coil in series are connected to a 240-V supply, a current of 3-A flows, lagging 37° behind the supply voltage. The voltage across the coil is 171 volts. Find the resistance of the resistor and the resistance and reactance of the coil.*

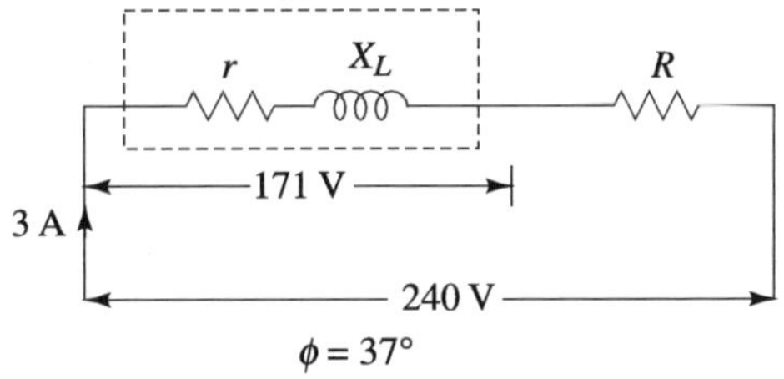

Fig. 4.38

Solution

Impedance of the coil $Z_{coil} = \dfrac{171}{3} = 57\ \Omega$

$\sqrt{r^2 + X_L^2} = 57$

$r^2 + X_L^2 = 3249$

Total impedance $Z = \dfrac{240}{3} = 80\ \Omega$

$\bar{Z} = Z\angle\phi = 80\ \angle 37° = 63.89 + j48.15$

But $\bar{Z} = (R + r) + jX_L$

Reactance of coil $X_L = 48.15\ \Omega$

$r^2 + X_L^2 = 3249$

$r^2 + (48.145)^2 = 3249$

$r^2 = 931.04$

Resistance of coil $r = 30.51$

$(R + r) = 63.89$

$R + 30.51 = 63.89$

Resistance of resistor $R = 33.38\ \Omega$

Example 4.41 *A choke coil and a resistor are connected in series across a 230-V, 50-Hz ac supply. The circuit draws a current of 2 A at 0.866 lagging pf. The voltage drop across the resistor is 100 V. Calculate the power factor of the choke coil.*

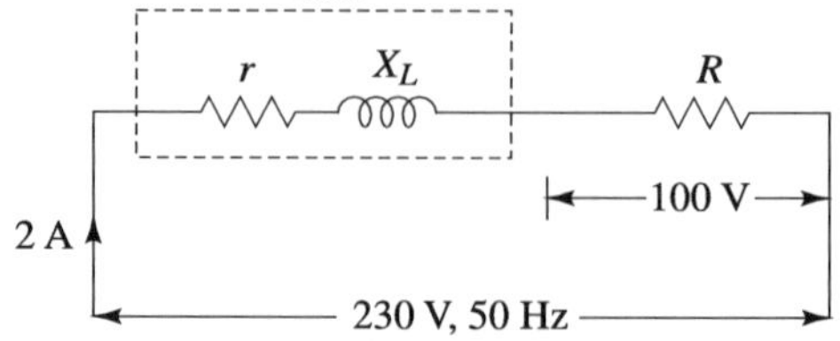

pf = 0.866 (lagging)

Fig. 4.39

Solution

Resistance $$R = \frac{100}{2} = 50\ \Omega$$

Total impedance $$Z = \frac{230}{2} = 115\ \Omega$$

$$\text{pf} = 0.866 \text{ (lagging)}$$
$$\phi = \cos^{-1}(0.866) = 30^\circ$$
$$\bar{Z} = Z\angle\phi = 115\angle 30^\circ = 99.59 + j57.5\ \Omega$$
$$R + r = 99.59$$
$$50 + r = 99.59$$
$$r = 49.59\ \Omega$$
$$X_L = 57.5\ \Omega$$

Impedance of choke coil $$Z_{\text{coil}} = \sqrt{r^2 + X_L^2}$$
$$= \sqrt{(49.59)^2 + (57.5)^2} = 75.93\ \Omega$$

Power factor of choke coil $$= \frac{r}{Z_{\text{coil}}} = \frac{49.59}{75.93} = 0.653 \text{ (lagging)}$$

Example 4.42 *A circuit consists of a pure resistance and coil in series. Power dissipated in the resistance and in the coil are 1000 W and 250 W respectively. The voltage drops across the resistance and the coil are 200 V and 300 V respectively. Determine (i) value of pure resistance, (ii) resistance and reactance of the coil, (iii) coil impedance, (iv) combined resistance of the circuit, (v) combined impedance, and (vi) supply voltage.*

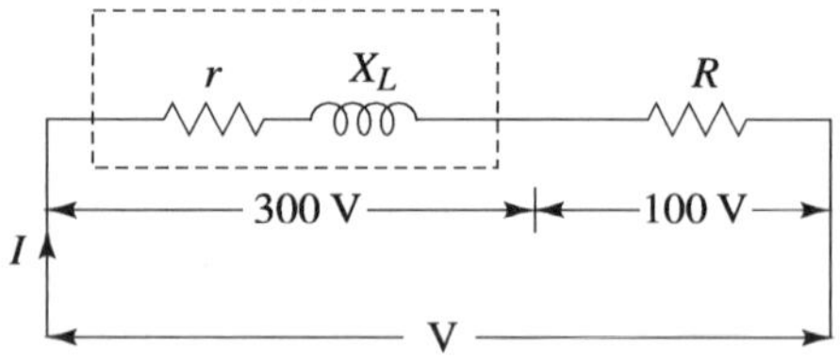

Fig. 4.40

Solution

$$P_R = 1000 \text{ W} \qquad V_R = 200 \text{ V}$$
$$P_{\text{coil}} = 250 \text{ W} \qquad V_{\text{coil}} = 300 \text{ V}$$
$$P_R = \frac{V_R^2}{R}$$

$$1000 = \frac{(200)^2}{R}$$
$$R = 40\ \Omega$$
$$V_R = RI$$
$$200 = 40I$$
$$I = 5\text{ A}$$
$$P_{coil} = I^2 r$$
$$250 = (5)^2 \times r$$

Resistance of coil $r = 10\ \Omega$

Impedance of coil $Z_{coil} = \frac{V_{coil}}{I} = \frac{300}{5} = 60\ \Omega$

Reactance of coil $X_L = \sqrt{Z_{coil}^2 - r^2}$

$$= \sqrt{(60)^2 - (10)^2} = 59.2\ \Omega$$

Combined resistance $R_T = R + r = 40 + 10 = 50\ \Omega$

Combined impedance $Z_T = \sqrt{(R+r)^2 + X_L^2}$

$$= \sqrt{(50)^2 + (59.2)^2} = 77.5\ \Omega$$

Supply voltage $V = Z_T \,.\, I = 77.5 \times 5 = 387.5\text{ V}$

Example 4.43 *A coil A takes 2 A at a power factor of 0.8 lagging with an applied p.d. of 10 V. A second coil B takes 2 A with a power factor of 0.7 lagging with an applied voltage of 5 V. What voltage will be required to produce a total current of 2 A with coils A and B in series? Find the power factor in this case.*

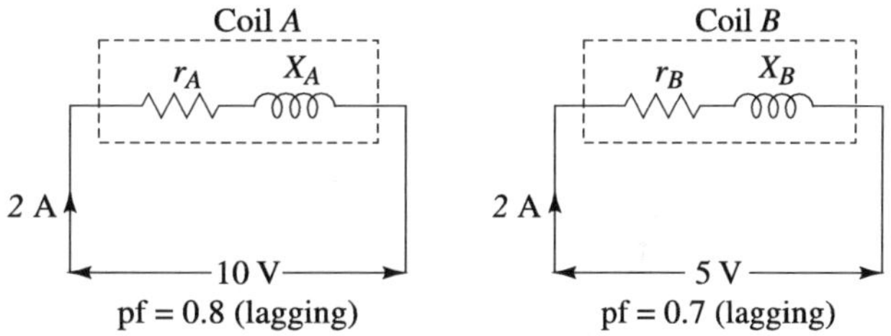

Fig. 4.41

Solution

For coil A,

$$\phi_A = \cos^{-1}(0.8) = 36.87°$$
$$Z_A = \frac{10}{2} = 5\ \Omega$$
$$\bar{Z}_A = 5\ \angle 36.87°$$
$$= 4 + j3\ \Omega$$
$$r_A = 4\ \Omega$$
$$X_A = 3\ \Omega$$

For coil B,

$$\phi_B = \cos^{-1}(0.7) = 45.57°$$
$$Z_B = \frac{5}{2} = 2.5\ \Omega$$
$$\bar{Z}_B = 2.5\ \angle 45.57°$$

$$\bar{Z}_B = 1.75 + j1.78\ \Omega$$
$$r_B = 1.75\ \Omega$$
$$X_B = 1.78\ \Omega$$

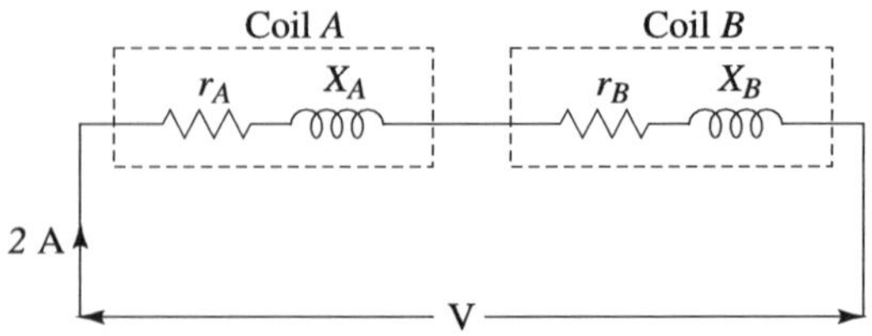

Fig. 4.42

When coils A and B are connected in series,

$$\bar{Z} = r_A + jX_A + r_B + jX_B = 4 + j3 + 1.75 + j1.78$$
$$= 5.75 + j4.78 = 7.48\ \angle 39.74°\ \Omega$$
$$Z = 7.48\ \Omega$$
$$\phi = 39.74°$$
$$V = Z\,.\,I = 7.48 \times 2 = 14.96\text{ V}$$
$$\text{pf} = \cos\phi = \cos(39.74°) = 0.77\text{ (lagging)}$$

Example 4.44 *When a voltage of 100 V is applied to a coil A, the current taken is 8 A and the power is 120 W. When applied to a coil B, the current is 10 A and the power is 500 W. What current and power will be taken when 100 V is applied to the two coils connected in series?*

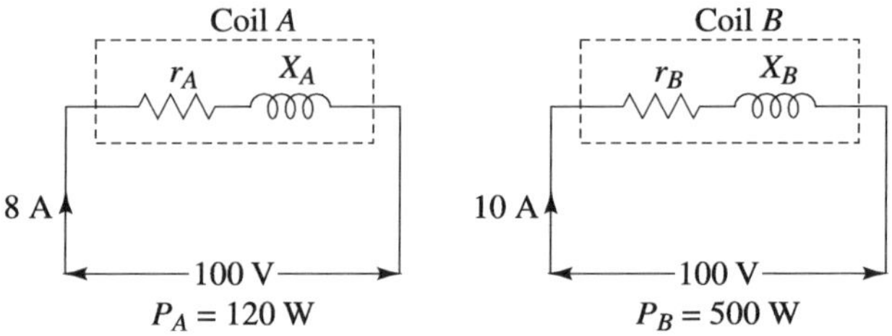

Fig. 4.43

Solution

For the coil A,

$$Z_A = \frac{100}{8} = 12.5\ \Omega$$
$$P_A = I_A^2\, r_A$$
$$120 = (8)^2 \times r_A$$
$$r_A = 1.875\ \Omega$$
$$X_A = \sqrt{Z_A^2 - r_A^2} = \sqrt{(12.5)^2 - (1.875)^2} = 12.36\ \Omega$$

For the coil B,

$$Z_B = \frac{100}{10} = 10\ \Omega$$
$$P_B = I_B^2\, r_B$$
$$500 = (10)^2 \times r_B$$
$$r_B = 5\ \Omega$$
$$X_B = \sqrt{Z_B^2 - r_B^2} = \sqrt{(10)^2 - (5)^2} = 8.66\ \Omega$$

When coils A and B are connected in series,

$$\bar{Z} = r_A + jX_A + r_B + jX_B$$
$$= 1.875 + j12.36 + 5 + j8.66$$
$$= 6.875 + j21.02$$
$$= 22.11\ \angle 71.89°\ \Omega$$
$$Z = 22.11\ \Omega$$
$$\phi = 71.89°$$

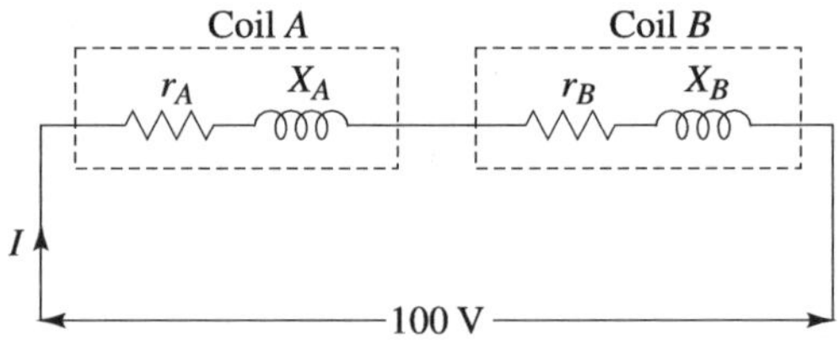

Fig. 4.44

$$I = \frac{V}{Z} = \frac{100}{22.11} = 4.52 \text{ A}$$

$$P = I^2 (r_A + r_B) = (4.52)^2 \times (6.875) = 140.64 \text{ W}$$

Example 4.45 *In a particular circuit a voltage of 10 V at 25 Hz produces 100 mA, while the same voltage at 75 Hz produces 60 mA. Find the values of components of the circuit.*

Solution

Data $V_1 = 10$ V, $f_1 = 25$ Hz, $I_1 = 10$ mA

$V_2 = 10$ V, $f_2 = 75$ Hz, $I_2 = 60$ mA

Case (i) $V_1 = 10$ V, $f_1 = 25$ Hz, $I_1 = 100$ mA

$$Z_1 = \frac{V_1}{I_1} = \frac{10}{100 \times 10^{-3}} = 100\ \Omega$$

Case (ii) $V_2 = 10$ V, $f_1 = 75$ Hz, $I_2 = 60$ mA

$$Z_2 = \frac{V_2}{I_2} = \frac{10}{60 \times 10^{-3}} = 166.67\ \Omega$$

As frequency increases, impedance of the circuit increases. In a series *R-L* circuit, inductive reactance X_L increases with frequency. Hence impedance increases.

Hence, circuit consists of a resistance *R* and inductance *L*.

$$Z_1 = \sqrt{R^2 + X_{L_1}^2} = \sqrt{R^2 + (2\pi \times 25 \times L)^2} = 100$$

$$R^2 + (50\pi L)^2 = 10000 \quad \text{...(i)}$$

$$Z_2 = \sqrt{R^2 + X_{L_2}^2} = \sqrt{R^2 + (2\pi \times 75 \times L)^2} = 166.67$$

$$R^2 + (150\pi L)^2 = 27778.89 \quad \text{...(ii)}$$

Solving Eqs (i) and (ii),

$$R = 88.1\ \Omega$$

$$L = 0.3 \text{ H}$$

Example 4.46 *When 1 A is passed through three coils A, B and C in series, the voltage across them are 6 V, 3 V and 8 V respectively on a dc supply and 7 V, 5 V and 10 V respectively on an ac supply. Find the power factor and the power dissipated in each coil and the power factor of the whole circuit.*

Solution

Data $I = 1$ A

On dc supply, $V_A = 6$ V, $V_B = 3$ V, $V_C = 8$ V

On ac supply, $V_A = 7$ V, $V_B = 5$ V, $V_C = 10$ V

For dc supply, $f = 0$

$$X_L = 2\pi f L = 0$$

The coil behaves as a pure resistor.

$$R_A = \frac{6}{1} = 6\ \Omega \qquad R_B = \frac{3}{1} = 3\ \Omega \qquad R_C = \frac{8}{1} = 8\ \Omega$$

For ac supply,

$$Z_A = \frac{7}{1} = 7\ \Omega \qquad Z_B = \frac{5}{1} = 5\ \Omega \qquad Z_C = \frac{10}{1} = 10\ \Omega$$

$$X_A = \sqrt{Z_A^2 - R_A^2} = \sqrt{(7)^2 - (6)^2} = 3.6\ \Omega$$

$$X_B = \sqrt{Z_B^2 - R_B^2} = \sqrt{(5)^2 - (3)^2} = 4\ \Omega$$

$$X_C = \sqrt{Z_C^2 - R_C^2} = \sqrt{(10)^2 - (8)^2} = 6\ \Omega$$

Power factor of coil $A = \dfrac{R_A}{Z_A} = \dfrac{6}{7} = 0.857$ (lagging)

Power factor of coil $B = \dfrac{R_B}{Z_B} = \dfrac{3}{5} = 0.6$ (lagging)

Power factor of coil $C = \dfrac{R_C}{Z_C} = \dfrac{8}{10} = 0.8$ (lagging)

Power dissipated in coil $A = I^2R_A = (1)^2 \times 6 = 6$ W

Power dissipated in coil $B = I^2R_B = (1)^2 \times 3 = 3$ W

Power dissipated in coil $C = I^2R_C = (1)^2 \times 8 = 8$ W

Total impedance

$$\begin{aligned}\bar{Z} &= R_A + jX_A + R_B + jX_B + R_C + jX_C \\ &= 6 + j3.6 + 3 + j4 + 8 + j6 \\ &= 17 + j13.6 = 21.77 \angle 38.68^\circ\ \Omega\end{aligned}$$

Power factor of the whole circuit = cos (38.68°) = 0.78 (lagging)

Example 4.47 *An air-cored coil takes 5 A current and consumes 600 W power when connected across a 200-V, 50-Hz ac supply. Calculate the value of the current drawn by the coil if the supply frequency increases to 60 Hz.*

Solution

Data

$$I = 5\text{ A}$$
$$V = 200\text{ V}$$
$$P = 600\text{ W}$$

For $f = 50$ Hz,

$$Z = \frac{V}{I} = \frac{200}{5} = 40\ \Omega$$

$$P = I^2r$$
$$600 = (5)^2 \times r$$
$$r = 24\ \Omega$$

$$X_L = \sqrt{Z^2 - r^2} = \sqrt{(40)^2 - (24)^2} = 32\ \Omega$$

$$X_L = 2\pi f L$$
$$32 = 2\pi \times 50 \times L$$
$$L = 0.1019\text{ H}$$

For $f = 60$ Hz

$$X_L = 2\pi \times 60 \times 0.1019 = 38.4\ \Omega$$
$$r = 24\ \Omega$$

$$Z = \sqrt{r^2 + X_L^2} = \sqrt{(24)^2 + (38.4)^2} = 45.28\ \Omega$$

$$I = \frac{V}{Z} = \frac{200}{45.28} = 4.417\text{ A}$$

Example 4.48 *When an iron-cored choking coil is connected to a 12-V dc supply, it draws a current of 2.5 A and when it is connected to a 230-V, 50-Hz supply, it draws a 2-A current and consumes 50 W of power. Determine for this value of current (i) power loss in the iron core, (ii) inductance of the coil, (iii) power factor, and the (iv) value of the series resistance which is equivalent to the effect of iron loss.*

Solution

Data For dc $V = 12$ V, $I = 2.5$ A

For ac $V = 230$ V, $I = 2$ A, $P = 50$ W

In an iron-cored coil, there are two types of losses.

(i) Losses in core known as core or iron loss

(ii) Losses in winding known as copper loss

$$P = I^2R + Pi$$

$$\frac{P}{I^2} = R + \frac{Pi}{I^2}$$

$$R_T = R + \frac{Pi}{I^2}$$

where R is the resistance of the coil and $\frac{Pi}{I^2}$ is the resistance which is equivalent to the effect of iron loss.

For dc supply,
$$f = 0$$
$$X_L = 0$$
$$R = \frac{12}{2.5} = 4.8\ \Omega$$

For ac supply,
$$Z = \frac{230}{2} = 115\ \Omega$$

Iron loss
$$Pi = P - I^2R = 50 - (2)^2 \times 4.8 = 30.8\ \text{W}$$
$$R_T = \frac{P}{I^2} = \frac{50}{(2)^2} = 12.5\ \Omega$$
$$X_L = \sqrt{Z^2 - R_T^2} = \sqrt{(115)^2 - (12.5)^2} = 114.3\ \Omega$$
$$X_L = 2\pi fL$$
$$114.3 = 2\pi \times 50 \times L$$

Inductance
$$L = 0.363\ H$$

Power factor
$$= \frac{R_T}{Z} = \frac{12.5}{115} = 0.108\ \text{(lagging)}$$

The series resistance equivalent to the effect of iron loss $= \frac{Pi}{I^2} = \frac{30.8}{(2)^2} = 7.7\ \Omega$

Example 4.49 *An iron-cored coil takes 4 A at a power factor of 0.5 when connected to a 200-V, 50-Hz supply. When the iron core is removed and the voltage is reduced to 40 V, the current rises to 5 A at a pf of 0.8. Find the iron loss in the core and inductance in each case.*

Solution

Data With iron core $I = 4$ A, pf = 0.5, $V = 200$ V

Without iron core $I = 5$ A, pf = 0.8, $V = 40$ V

When the iron core is removed,

$$Z = \frac{V}{I} = \frac{40}{5} = 8\ \Omega$$

$$\text{pf} = \frac{R}{Z}$$

$$0.8 = \frac{R}{8}$$

$$R = 6.4\ \Omega$$

$$X_L = \sqrt{Z^2 - R^2} = \sqrt{(8)^2 - (6.4)^2} = 4.8\ \Omega$$

$$X_L = 2\pi f L$$

$$4.8 = 2\pi \times 50 \times L$$

Inductance $L = 0.0153$ H

With iron core,

$$Z = \frac{200}{4} = 50\ \Omega$$

$$\text{pf} = \frac{R_T}{Z}$$

$$0.5 = \frac{R_T}{50}$$

$$R_T = 25\ \Omega$$

$$X_L = \sqrt{Z^2 - R_T^2} = \sqrt{(50)^2 - (25)^2} = 43.30\ \Omega$$

$$X_L = 2\pi f L$$

$$43.3 = 2\pi \times 50 \times L$$

Inductance $L = 0.1378$ H

Iron loss $Pi = P - I^2R$

$$= VI\cos\phi - I^2R$$

$$= 200 \times 4 \times 0.5 - (4)^2 \times 6.4 = 297.6\text{ W}$$

4.11 SERIES R-C CIRCUIT

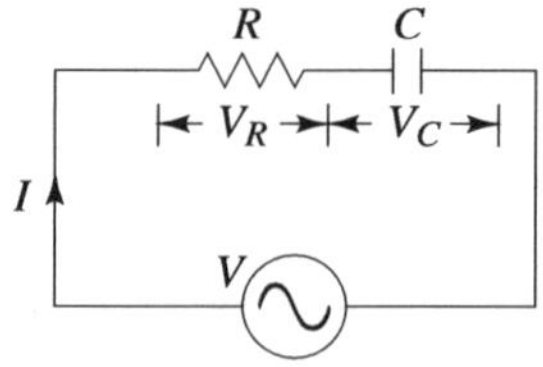

Fig. 4.45

Figure 4.45 shows a pure resistor R connected in series with a pure capacitor C across an alternating voltage v.

Let V and I be the rms values of applied voltage and current.

Potential difference across the resistor = $V_R = R \,.\, I$

Potential difference across the capacitor = $V_C = X_C \,.\, I$

The voltage $\bar{V}_R$ is in phase with current $\bar{I}$ whereas voltage $\bar{V}_C$ lags behind current $\bar{I}$ by 90°.

$$\text{Applied voltage } \bar{V} = \bar{V}_R + \bar{V}_C$$

Phasor diagram Since the same current flows through R and C, current I is taken as reference phasor.

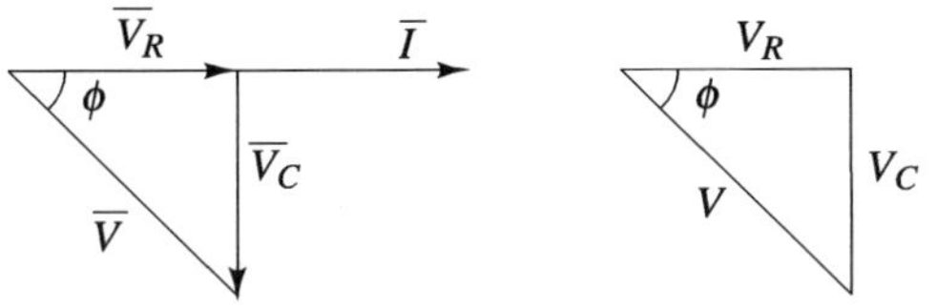

(a) Phasor Diagram (b) Voltage Triangle

$$\cos\phi = \frac{V_R}{V}$$

Fig. 4.46

Impedance

$$\begin{aligned}
\overline{V} &= \overline{V}_R + \overline{V}_C \\
&= R\overline{I} - jX_C\overline{I} \\
&= (R - jX_C)\ \overline{I} \\
\frac{\overline{V}}{\overline{I}} &= R - jX_C = \overline{Z} \\
\overline{Z} &= Z\angle -\phi \\
Z &= \sqrt{R^2 + X_C^2} = \sqrt{R^2 + \frac{1}{\omega^2 C^2}} \\
\phi &= \tan^{-1}\left(\frac{X_C}{R}\right) = \tan^{-1}\left(\frac{1}{\omega RC}\right)
\end{aligned}$$

The quantity $\overline{Z}$ is called complex impedance of the *R-C* circuit.

Impedance triangle

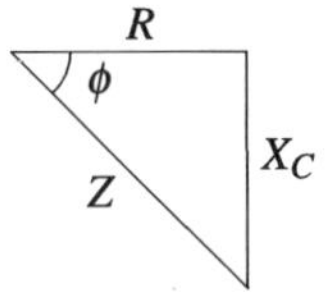

$$\cos\phi = \frac{R}{Z}$$

Fig. 4.47

Current From the phasor diagram, it is clear that the current *I* leads voltage *V* by an angle ϕ. If the applied voltage is given by $v = V_m \sin \omega t$ then the current equation will be

$$i = I_m \sin(\omega t + \phi)$$

where $$I_m = \frac{V_m}{Z}$$

and $$\phi = \tan^{-1}\left(\frac{X_C}{R}\right) = \tan^{-1}\left(\frac{1}{\omega RC}\right)$$

Waveforms

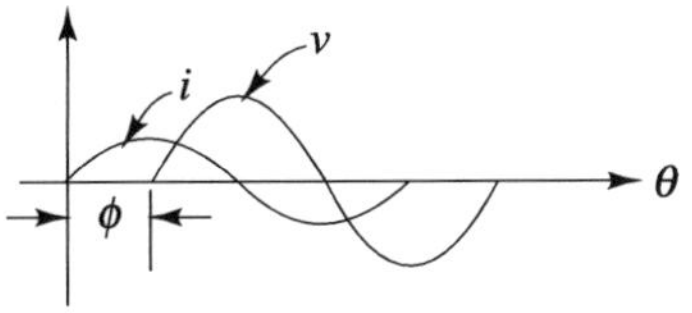

Fig. 4.48

Power

Active power $P = VI\cos\phi = I^2R$

Reactive power $Q = VI\sin\phi = I^2X_C$

Apparent power $S = VI = I^2Z$

Power triangle

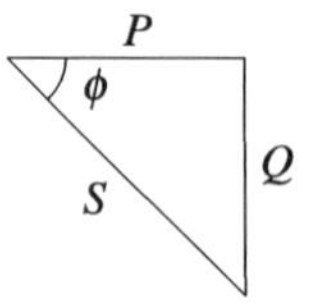

$$\cos\phi = \frac{P}{S}$$

Fig. 4.49

Power factor It is defined as the cosine of the angle between voltage and current phasor.

$$\text{pf} = \cos\phi$$

From voltage triangle, $\text{pf} = \dfrac{V_R}{V}$

From impedance triangle, $\text{pf} = \dfrac{R}{Z}$

From power triangle, $\text{pf} = \dfrac{P}{S}$

In case of an *R-C* series circuit, the power factor is leading in nature.

Example 4.50 *The voltage applied to a circuit is e = 100 sin (ωt + 30°) and the current flowing in the circuit is i = 15 sin (ωt + 60°). Determine impedance, resistance, reactance, power and power factor.*

Solution

Data

$$e = 100\sin(\omega t + 30°)$$
$$i = 15\sin(\omega t + 60°)$$
$$\bar{E} = \frac{100}{\sqrt{2}}\angle 30°\ \text{V}$$
$$\bar{I} = \frac{15}{\sqrt{2}}\angle 60°\ \text{A}$$
$$\bar{Z} = \frac{\bar{E}}{\bar{I}} = \frac{\frac{100}{\sqrt{2}}\angle 30°}{\frac{15}{\sqrt{2}}\angle 60°} = 6.67\angle -30° = 5.77 - j3.33\ \Omega$$

Impedance $Z = 6.67\ \Omega$

Resistance $R = 5.77\ \Omega$

Reactance $X_C = 3.33\ \Omega$

Power factor $= \cos\phi = \cos(30°) = 0.866$ (leading)

Power $P = VI\cos\phi = \dfrac{100}{\sqrt{2}} \times \dfrac{15}{\sqrt{2}} \times 0.866 = 649.5\ \text{W}$

Example 4.51 *A series circuit consumes 2000 W at 0.5 leading power factor, when connected to 230 V, 50 Hz ac supply. Calculate (i) kVA, (ii) kVAR, and (iii) current.*

Solution

Data

$$P = 2000\ \text{W}$$
$$\text{pf} = 0.5\ \text{(leading)}$$

$$V = 230 \text{ V}$$
$$P = VI \cos \phi$$
$$2000 = 230 \times I \times 0.5$$
$$I = 17.39 \text{ A}$$
$$S = VI = \frac{P}{\cos \phi} = \frac{2000}{0.5} = 4000 \text{ VA} = 4 \text{ kVA}$$
$$\phi = \cos^{-1}(0.5) = 60°$$
$$Q = VI \sin \phi = 230 \times 17.39 \times \sin(60°) = 3.464 \text{ kVAR}$$

Example 4.52 *A resistor R in series with a capacitance C is connected to a 240-V, 50-Hz ac supply. Find the value of C so that R absorbs 300 W at 100 V. Find also the maximum charge and maximum stored energy in C.*

Solution

Data $V = 240 \text{ V}$, $V_R = 100 \text{ V}$, $P = 300 \text{ W}$, $f = 50 \text{ Hz}$

$$P = \frac{V_R^2}{R}$$
$$300 = \frac{(100)^2}{R}$$
$$R = 33.33 \ \Omega$$
$$P = I^2 R$$
$$300 = I^2 \times 33.33$$
$$I = 3 \text{ A}$$
$$Z = \frac{V}{I} = \frac{240}{3} = 80 \ \Omega$$
$$X_C = \sqrt{Z^2 - R^2} = \sqrt{(80)^2 - (33.33)^2} = 72.72 \ \Omega$$
$$X_C = \frac{1}{2\pi fC}$$
$$72.72 = \frac{1}{2\pi \times 50 \times C}$$
$$C = 43.77 \ \mu\text{F}$$

Voltage across capacitor $V_C = \sqrt{V^2 - V_R^2} = \sqrt{(240)^2 - (100)^2} = 218.17 \text{ V}$

Maximum value of $V_C = 218.17 \times \sqrt{2} = 308.54 \text{ V}$

Maximum charge $Q_{max} = CV_{Cmax} = 43.77 \times 10^{-6} \times 308.54 = 0.0135 \text{ C}$

Maximum stored energy $E_{max} = \frac{1}{2} C (V_{Cmax})^2$

$$= \frac{1}{2} \times 43.77 \times 10^{-6} \times (308.54)^2 = 2.08 \text{ J}$$

Example 4.53 *A capacitor of 35 μF is connected in series with a variable resistor. The circuit is connected across 50-Hz mains. Find the value of the resistor for a condition when the voltage across the capacitor is half the supply voltage.*

Solution

Data $C = 35\ \mu F \qquad f = 50$ Hz

$$V_C = \frac{1}{2} V$$

$$X_C = \frac{1}{2\pi fC} = \frac{1}{2\pi \times 50 \times 35 \times 10^{-6}} = 90.946\ \Omega$$

$$V_C = \frac{1}{2} V$$

$$X_C \cdot I = \frac{1}{2} Z \cdot I$$

$$X_C = \frac{1}{2} Z$$

$$Z = 2X_C$$

$$Z = \sqrt{R^2 + X_C^2}$$

$$(2X_C)^2 = R^2 + X_C^2$$

$$3X_C^2 = R^2$$

$$R^2 = 3 \times (90.946)^2 = 24813.35$$

$$R = 157.5\ \Omega$$

Example 4.54 *A voltage of 125 V at 50 Hz is applied across a non-inductive resistor connected in series with a capacitor. The current is 2.2 A. The power loss in the resistor is 96.8 W. Calculate the resistance and capacitance.*

Solution

Data $V = 125$ V $\qquad P = 96.8$ W

$I = 2.2$ A $\qquad f = 50$ Hz

$$Z = \frac{V}{I} = \frac{125}{2.2} = 56.82\ A$$

$$P = I^2 R$$

$$96.8 = (2.2)^2 \times R$$

$$R = 20\ \Omega$$

$$X_C = \sqrt{Z^2 - R^2} = \sqrt{(56.82)^2 - (20)^2} = 53.18\ \Omega$$

$$X_C = \frac{1}{2\pi fC}$$

$$53.18 = \frac{1}{2\pi \times 50 \times C}$$

$$C = 59.85\ \mu F$$

Example 4.55 *A resistor and a capacitor are connected across a 250-V supply. When the supply frequency is 50 Hz, the current drawn is 5 A. When the frequency is increased to 60 Hz, it draws 5.8 A. Find the value of R and C and power drawn in the second case.*

Solution

Data $V = 250$ V

$f_1 = 50$ Hz, $I_1 = 5$ A

$f_2 = 60$ Hz, $I_2 = 5.8$ A

For $f_1 = 50$ Hz,

$$Z_1 = \frac{250}{5} = 50 = \sqrt{R^2 + \left(\frac{1}{2\pi f_1 C}\right)^2} = \sqrt{R^2 + \left(\frac{1}{100\pi C}\right)^2}$$

$$R^2 + \left(\frac{1}{100\pi C}\right)^2 = 2500 \quad \text{... (i)}$$

For $f_2 = 60$ Hz,

$$Z_2 = \frac{250}{5.8} = 43.1\ \Omega$$

$$Z_2 = \sqrt{R^2 + \left(\frac{1}{2\pi f_2 C}\right)^2} = \sqrt{R^2 + \left(\frac{1}{120\pi C}\right)^2}$$

$$R^2 + \left(\frac{1}{120\pi C}\right)^2 = 1857.9\ \Omega \quad \text{...(ii)}$$

Solving Eqs (i) and (ii),

$R = 19.96\ \Omega$

$C = 69.4\ \mu$F

Power drawn in the second case= $I^2R = (5.8)^2 \times 19.96 = 671.45$ W

4.12 SERIES R-L-C CIRCUIT

Figure 4.50 shows a pure resistor R, pure inductor L and pure capacitor C connected in series across an alternating voltage v.

Let V and I be the rms values of applied voltage and current.

Potential difference across the resistor = $V_R = R \cdot I$

Potential difference across the inductor = $V_L = X_L \cdot I$

Potential difference across the capacitor = $V_C = X_C \cdot I$

The voltage $\overline{V_R}$ is in phase with current $\overline{I}$, voltage $\overline{V_L}$ leads current $\overline{I}$ by 90° and voltage $\overline{V_C}$ lags behind current $\overline{I}$ by 90°.

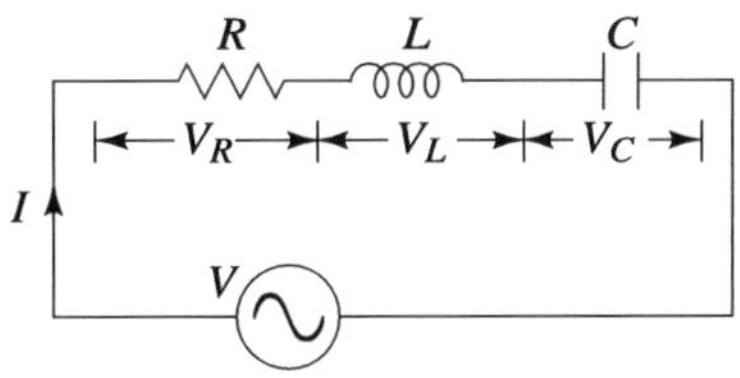

Fig. 4.50

$$\text{Applied voltage } \overline{V} = \overline{V_R} + \overline{V_L} + \overline{V_C}$$

Phasor diagram Since the same current flows through R, L and C, current I is taken as reference phasor.

Case (i) $X_L > X_C$

The reactance X will be inductive in nature and circuit will behave like an R-L circuit.

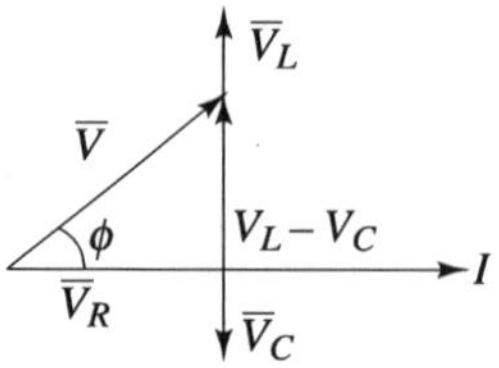

Fig. 4.51

Case (ii) $X_C > X_L$

The reactance X will be capacitive in nature and the circuit will behave like an R-C circuit.

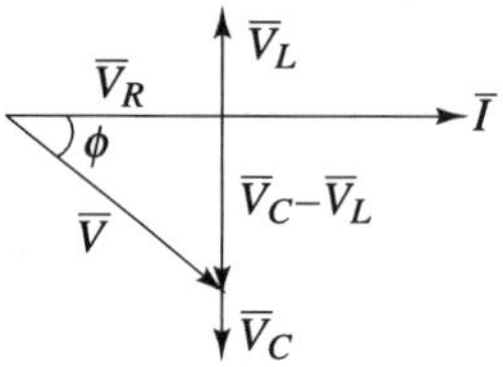

Fig. 4.52

Impedance

$$\overline{V} = \overline{V_R} + \overline{V_L} + \overline{V_C} = R\overline{I} + jX_L I - jX_C \overline{I} = [R + j(X_L - X_C)]\overline{I}$$

$$\frac{\overline{V}}{\overline{I}} = R + j(X_L - X_C) = \overline{Z}$$

$$\overline{Z} = Z\angle\phi$$

$$Z = \sqrt{R^2 + (X_L - X_C)^2}$$

$$\phi = \tan^{-1}\left(\frac{X_L - X_C}{R}\right)$$

Impedance triangle

Case (i) $X_L > X_C$

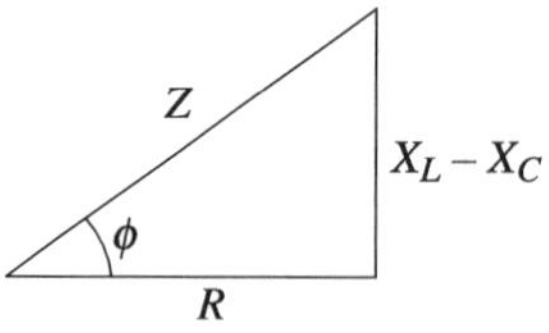

Fig. 4.53

Case (ii) $X_C > X_L$

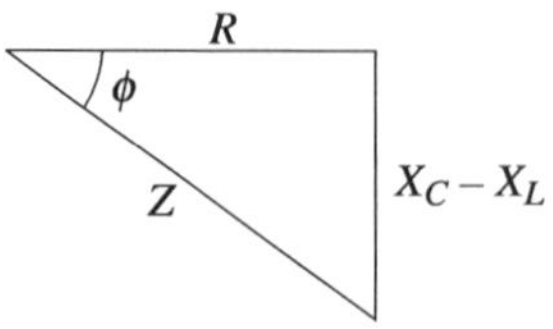

Fig. 4.54

Current equation If the applied voltage is given by $v = V_m \sin \omega t$ then current equation will be

$$i = I_m \sin (\omega t \pm \phi)$$

'–' sign is used when $X_L > X_C$
'+' sign is used when $X_C > X_L$

Waveforms

Case (i) $X_L > X_C$

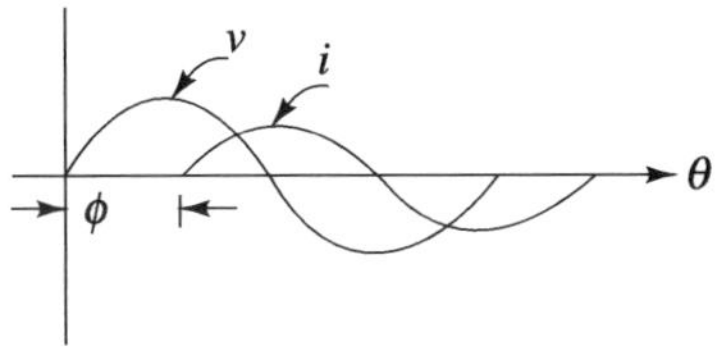

Fig. 4.55

Case (ii) $X_C > X_L$

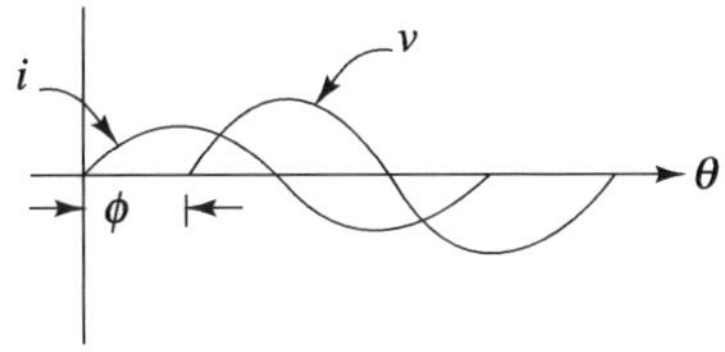

Fig. 4.56

Power

Average power	$P = VI \cos \phi = I^2R$
Reactive power	$Q = VI \sin \phi = I^2X$
Apparent power	$S = VI = I^2Z$

Power triangle

Case (i) $X_L > X_C$

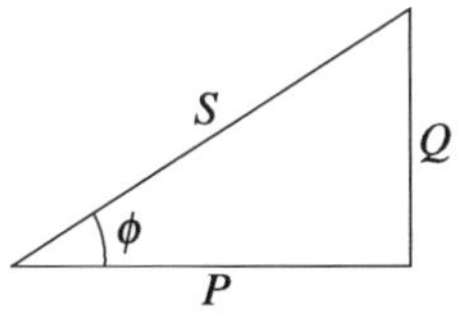

Fig. 4.57

Case (ii) $X_C > X_L$

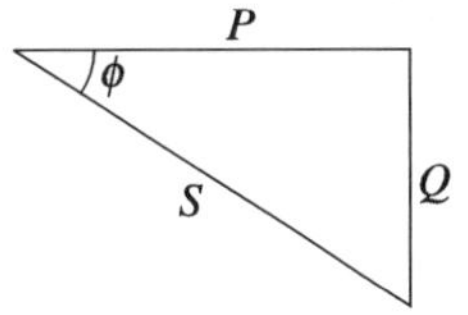

Fig. 4.58

Power factor It is defined as the cosine of the angle between voltage and current phasor.

$$\text{pf} = \cos \phi$$

$$\text{pf} = \frac{V_R}{V} = \frac{R}{Z} = \frac{P}{S}$$

Example 4.56 *Two impedances Z_1 and Z_2 having the same numerical value are connected in series. If Z_1 is having a pf of 0.866 lagging and Z_2 is having a pf of 0.8 leading, calculate the pf of the series combination.*

Solution

Data

$\text{pf}_1 = 0.866$ (lagging)
$\text{pf}_2 = 0.8$ (leading)
$Z_1 = Z_2 = Z$
$\phi_1 = \cos^{-1} (0.866) = 30°$
$\phi_2 = \cos^{-1} (0.8) = 36.87°$

$$\bar{Z}_1 = Z\angle\phi_1 = Z\angle 30° = 0.866\,Z + j0.5\,Z$$

$$\bar{Z}_2 = Z\angle -\phi_2 = Z\angle -36.87° = 0.8\,Z - j0.6\,Z$$

For a series combination,

$$\bar{Z} = \bar{Z}_1 + \bar{Z}_2 = 0.866\,Z + j0.5\,Z + 0.8\,Z - j0.6\,Z$$

$$= 1.6666\,Z - j0.1\,Z = Z\,(1.666 - j0.1) = 1.668\,Z\angle -3.43°$$

$$\text{pf} = \cos(3.43°) = 0.9982 \text{ (leading)}$$

Example 4.57 *A coil of resistance 3 Ω and an inductance of 0.22 H is connected in series with an imperfect capacitor. When such a series circuit is connected across a 200-V, 50-Hz supply, it has been observed that their combined impedance is (3.8 + j6.4) Ω. Calculate the resistance and capacitance of imperfect capacitor.*

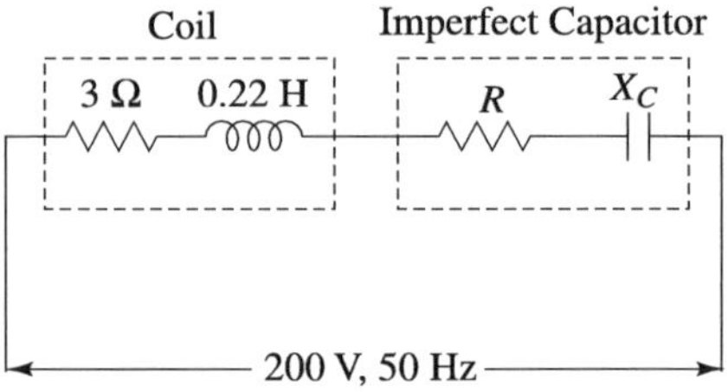

Fig. 4.59

Solution

Data

$$\bar{Z} = 3.8 + j6.4\ \Omega$$

$$X_L = 2\pi fL = 2\pi \times 50 \times 0.22 = 69.12\ \Omega$$

Total impedance

$$\bar{Z} = 3 + j69.12 + R - jX_C$$

$$= (3 + R) + j\,(69.12 - X_C)$$

$$3 + R = 3.8$$

$$R = 0.8\ \Omega$$

$$69.12 - X_C = 6.4$$

$$X_C = 62.72\ \Omega$$

$$X_C = \frac{1}{2\pi fC}$$

$$62.72 = \frac{1}{2\pi \times 50 \times C}$$

$$C = 50.75\ \mu\text{F}$$

Example 4.58 *An R-L-C series circuit has a current which lags the applied voltage by 45°. The voltage across the inductance has maximum value equal to twice the maximum value of voltage across the capacitor. Voltage across the inductance is 300 sin (1000t) and R = 20 Ω. Find the value of inductance and capacitance.*

Solution

Data

$$v_L = 300 \sin(1000t)$$

$$R = 20\ \Omega$$

$$\phi = 45°$$

$$V_{L(\max)} = 2V_{C(\max)}$$

$$\sqrt{2}\,V_L = 2\sqrt{2}\,V_C$$

$$I \times X_L = 2I \times X_C$$

$$X_L = 2X_C$$

$$\cos\phi = \frac{R}{Z}$$

$$\cos(45°) = \frac{20}{Z}$$

$$Z = 28.28\ \Omega$$

For a series *R-L-C* circuit,

$$Z = \sqrt{R^2 + (X_L - X_C)^2}$$

$$(28.28)^2 = (20)^2 + (2X_C - X_C)^2$$

$$799.76 = 400 + X_C^2$$

$$X_C = 20\ \Omega$$

$$X_L = 2X_C = 40\ \Omega$$

$$X_L = \omega L$$

$$40 = 1000 \times L$$

$$L = \frac{40}{1000} = 0.04\ \text{H}$$

$$X_C = \frac{1}{\omega C}$$

$$20 = \frac{1}{1000 \times C}$$

$$C = 50\ \mu\text{F}$$

Example 4.59 *A coil having a power factor of 0.5 is in series with a 79.57 μF capacitor and when connected across a 50-Hz supply, the p.d. across the coil is equal to the p.d. across the capacitor. Find the resistance and inductance of the coil.*

Solution

Data

$$\text{pf} = 0.5$$

$$C = 79.57\ \mu\text{F}$$

$$f = 50\ \text{Hz}$$

$$V_{\text{coil}} = V_C$$

$$X_C = \frac{1}{2\pi fC} = \frac{1}{2\pi \times 50 \times 79.57 \times 10^{-6}} = 40\ \Omega$$

$$V_{\text{coil}} = V_C$$

$$I \,.\, Z_{\text{coil}} = I \,.\, X_C$$

$$Z_{\text{coil}} = X_C = 40\ \Omega$$

$$\text{pf of coil} = \cos\phi = \frac{R}{Z_{\text{coil}}}$$

$$0.5 = \frac{R}{40}$$

Resistance of coil $R = 20\ \Omega$

$$X_L = \sqrt{Z_{\text{coil}}^2 - R^2} = \sqrt{(40)^2 - (20)^2} = 34.64\ \Omega$$

$$X_L = 2\pi fL$$

$$34.64 = 2\pi \times 50 \times L$$

Inductance of coil = 0.11 H

Example 4.60 *A 250-V, 50-Hz voltage is applied to a coil having resistance of 5 Ω and an inductance of 9.55 H in series with a capacitor C. If the voltage across the coil is 300 V, find the value of C.*

Solution

Data

$$V = 250 \text{ V}$$
$$R = 5\ \Omega$$
$$L = 9.55 \text{ H}$$
$$V_{\text{coil}} = 300 \text{ V}$$
$$X_L = 2\pi f L = 2\pi \times 50 \times 9.55 = 3000\ \Omega$$
$$Z_{\text{coil}} = \sqrt{R^2 + X_L^2} = \sqrt{(5)^2 + (3000)^2} = 3000\ \Omega$$
$$I = \frac{V_{\text{coil}}}{Z_{\text{coil}}} = \frac{300}{3000} = 0.1 \text{ A}$$

Total impedance
$$Z = \frac{V}{I} = \frac{250}{0.1} = 2500\ \Omega$$

When $X_L > X_C$,

$$Z = \sqrt{R^2 + (X_L - X_C)^2}$$
$$(2500)^2 = (5)^2 + (3000 - X_C)^2$$
$$(3000 - X_C) = 2500$$
$$X_C = 500$$
$$C = \frac{1}{2\pi f X_C} = \frac{1}{2\pi \times 50 \times 500} = 6.37\ \mu\text{F}$$

When $X_C > X_L$,

$$Z = \sqrt{R^2 + (X_C - X_L)^2}$$
$$(2500)^2 = (5)^2 + (X_C - 3000)^2$$
$$2500 = X_C - 3000$$
$$X_C = 5500$$
$$C = \frac{1}{2\pi \times 50 \times 5500} = 0.58\ \mu\text{F}.$$

Example 4.61 *Draw the phasor diagram for the series circuit shown in Fig. 4.60 when the current in the circuit is 2 A. Find the values of V_1 and V_2 and show these voltages on the phasor diagram.*

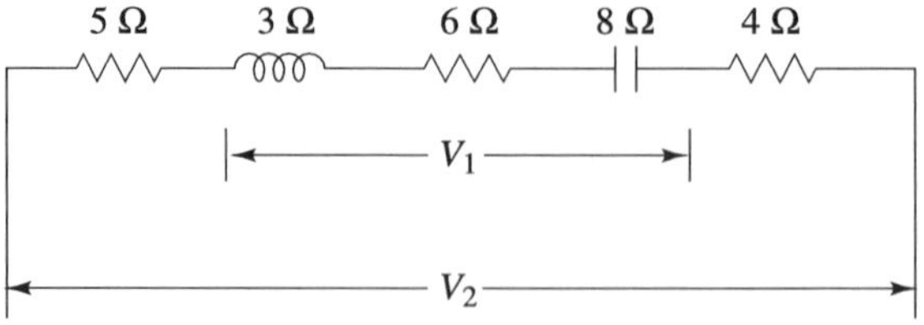

Fig. 4.60

Solution

$$\bar{Z}_1 = j3 + 6 - j8 = 6 - j5 = 7.81\ \angle{-39.8°}\ \Omega$$
$$\bar{Z}_2 = 5 + j3 + 6 - j8 + 4 = 15 - j5 = 15.81\ \angle{-18.43°}\ \Omega$$

$$I = 2 \text{ A}$$

Let $\bar{I} = 2\angle 0° \text{ A}$

$$\bar{V}_1 = \bar{Z}_1 \cdot \bar{I} = (7.81\angle -39.8°)(2\angle 0°)$$
$$= 15.62\angle -39.8° \text{ V}$$
$$\bar{V}_2 = \bar{Z}_2 \cdot \bar{I} = (15.81\angle -18.43°)(2\angle 0°)$$
$$= 31.62\angle -18.43° \text{V}$$

Phasor diagram

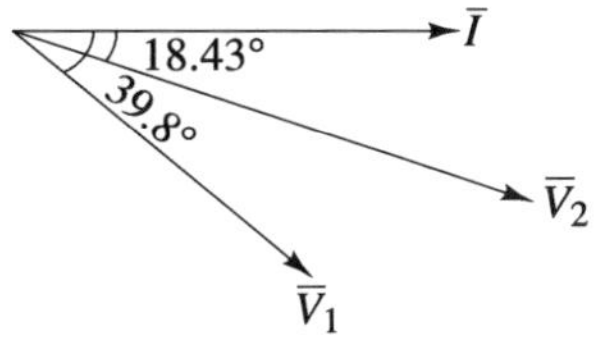

Fig. 4.61

Example 4.62 *Draw a vector diagram for the circuit shown in Fig. 4.62 indicating terminal voltages V_1 and V_2 and the current. Find the value of (a) current, (b) V_1 and V_2, and (c) power factor.*

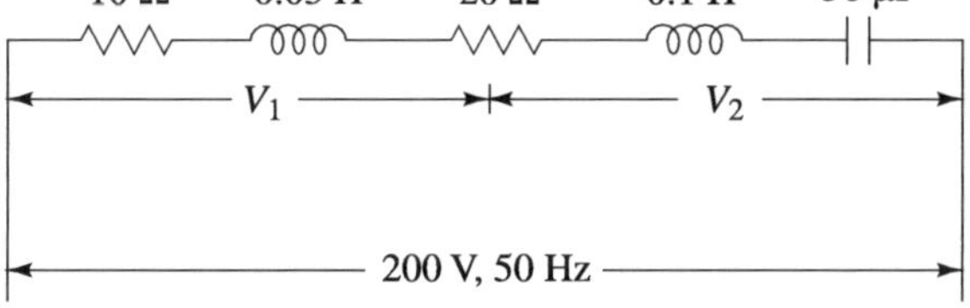

Fig. 4.62

Solution

$$X_{L_1} = 2\pi f L = 2\pi \times 50 \times 0.05 = 15.71 \ \Omega$$

$$X_{L_2} = 2\pi f L = 2\pi \times 50 \times 0.1 = 31.42 \ \Omega$$

$$X_C = \frac{1}{2\pi f C} = \frac{1}{2\pi \times 50 \times 50 \times 10^{-6}} = 63.66 \ \Omega$$

Total impedance

$$\bar{Z} = 10 + j15.71 + 20 + j31.42 - j63.66$$
$$= 30 - j16.53 = 34.25\angle -28.85° \ \Omega$$

$$I = \frac{V}{Z} = \frac{200}{34.25} = 5.84 \text{ A}$$

Let $\bar{I} = 5.84\angle 0° \text{ A}$

$$\bar{Z}_1 = 10 + j15.71 = 18.62\angle 57.52 \ \Omega$$
$$\bar{V}_1 = \bar{Z}_1 \cdot \bar{I} = (18.62\angle 57.52°)(5.84\angle 0°) = 108.74\angle 57.52° \text{ V}$$
$$\bar{Z}_2 = 20 + j31.42 - j63.66$$
$$= 20 - j32.24 = 37.94\angle -58.19° \ \Omega$$
$$\bar{V}_2 = \bar{Z}_2 \cdot \bar{I} = (37.94\angle -58.19°)(5.84\angle 0°) = 221.57\angle -58.19° \text{ V}$$
$$\text{pf} = \cos\phi = \cos(28.85°) = 0.875 \text{ (leading)}$$

Vector diagram

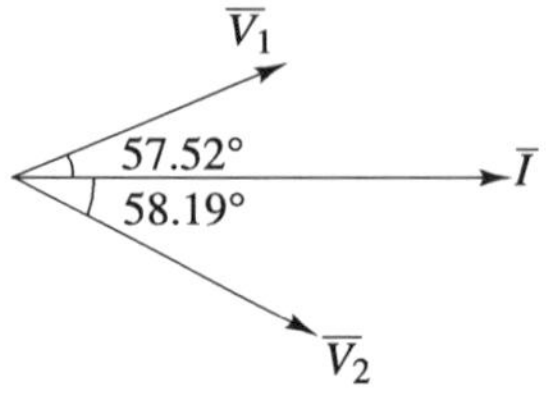

Fig. 4.63

Example 4.63 *Find the values of R and C so that $V_x = 3V_y$. V_x and V_y are in quadrature.*

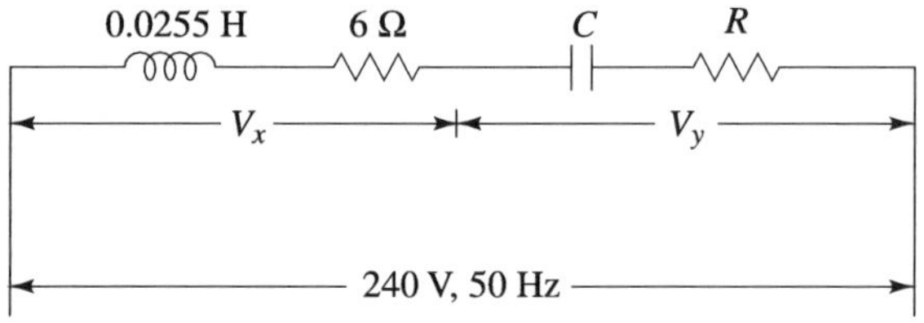

Fig. 4.64

Solution

$$X_L = 2\pi f L = 2\pi \times 50 \times 0.0255 = 8\ \Omega$$

$$\overline{Z}_x = 6 + j8 = 10\ \angle 53.13°\ \Omega$$

$$V_x = 3V_y$$

$$I \times Z_x = 3 \times I \times Z_y$$

$$Z_x = 3Z_y$$

V_x and V_y are in quadrature, i.e., phase angle between V_x and V_y is 90°. Hence, the angle between Z_x and Z_y will be 90°. The impedance Z_y is capacitive in nature.

$$\overline{Z}_y = Z_y\ \angle -\phi$$

$$\overline{Z}_y = \frac{10}{3} \angle (53.13 - 90)° = 3.33\ \angle -36.87° = 2.66 - j2\ \Omega$$

$$R = 2.66\ \Omega$$

$$X_C = 2\ \Omega$$

$$X_C = \frac{1}{2\pi f C}$$

$$C = \frac{1}{2\pi \times 50 \times 2} = 1.59\ \text{mF}$$

4.13 ADMITTANCE AND ITS COMPONENTS

The admittance Y of a circuit is defined as a reciprocal of impedance. The impedance $\overline{Z}$ can be written as

$$\overline{Z} = R + jX$$

The real part of impedance is called resistance and the imaginary part is called reactance. Similarly, admittance also can be expressed in terms of real part and imaginary part.

$$\frac{1}{\bar{Z}} = \frac{1}{R + jX} = \frac{R - jX}{(R + jX)(R - jX)} = \frac{R - jX}{R^2 + X^2}$$

$$\bar{Y} = \frac{R}{R^2 + X^2} - j\frac{X}{R^2 + X^2}$$

The real part of admittance is called conductance and the imaginary part is called susceptance and are measured in mho (℧) or siemens (S).

In general,

$$\bar{Y} = G \pm jB$$

If $\bar{Z} = R + jX_L$ $\qquad \bar{Y} = \frac{1}{\bar{Z}} = G - jB_L$

If $\bar{Z} = R - jX_C$ $\qquad \bar{Y} = \frac{1}{\bar{Z}} = G + jB_C$

Example 4.64 *Two circuits, the impedances of which are given by $Z_1 = (6 + j8)\ \Omega$ and $Z_2 = (8 - j6)\ \Omega$ are connected in parallel. If the applied voltage to the combination is 100 V, find (i) current and pf of each branch, (ii) overall current and pf of the combination, and (iii) power consumed by each impedance.*

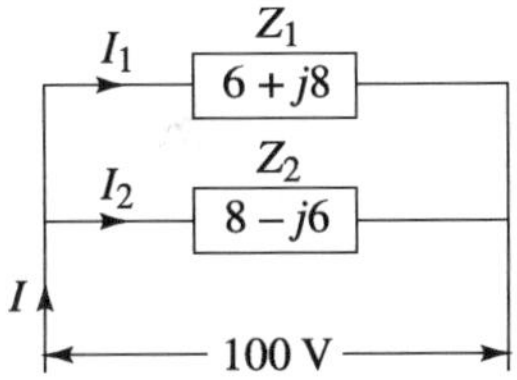

Fig. 4.65

Solution

Data $\bar{Z}_1 = 6 + j8\ \Omega \quad \bar{Z}_2 = 8 - j6\ \Omega$

$$V = 100\ V$$

$$\bar{V} = 100\ \angle 0°\ \text{V}$$

$$\bar{I}_1 = \frac{\bar{V}}{\bar{Z}_1} = \frac{100\angle 0°}{6 + j8} = 10\ \angle{-53.13°}\ \text{A}$$

$$\bar{I}_2 = \frac{\bar{V}}{\bar{Z}_1} = \frac{100\angle 0°}{8 - j6} = 10\ \angle 36.9°\ \text{A}$$

$$\cos\phi_1 = \cos(53.13°) = 0.6\ \text{(lagging)}$$

$$\cos\phi_2 = \cos(36.9°) = 0.8\ \text{(leading)}$$

$$\bar{I} = \bar{I}_1 + \bar{I}_2 = 10\ \angle{-53.13°} + 10\ \angle 36.9°$$

$$= 14.14\ \angle{-8.13°}\ \text{A}$$

$$\text{pf} = \cos\phi = \cos(8.13°) = 0.989\ \text{(lagging)}$$

$$P_1 = I_1^2 R_1 = (10)^2 \times (6) = 600\ \text{W}$$

$$P_2 = I^2 R_2 = (10)^2 \times (8) = 800\ \text{W}$$

Example 4.65 *An impedance of $(7 + j5)\ \Omega$ is connected in parallel with another impedance of $(10 - j8)\ \Omega$ across a 230-V, 50-Hz supply. Calculate (i) admittance, conductance and susceptance of the combined circuit, and (ii) total current and power factor.*

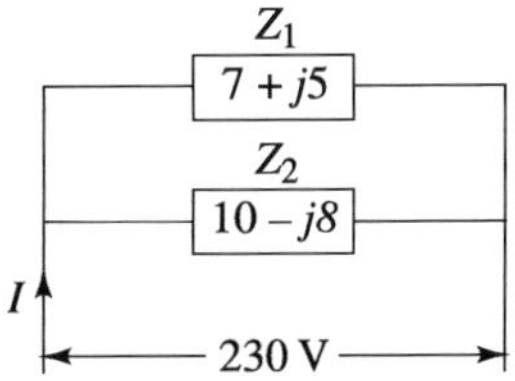

Fig. 4.66

Solution

Data $\bar{Z}_1 = (7 + j5)\ \Omega$

$$\bar{Z}_2 = (10 - j8)\ \Omega$$
$$V = 230\text{ V}$$
$$\bar{Y}_1 = \frac{1}{\bar{Z}_1} = \frac{1}{7 + j5} = 0.12\angle{-35.54^\circ}\ \mho$$
$$\bar{Y}_2 = \frac{1}{\bar{Z}_2} = \frac{1}{10 - j8} = 0.08\angle 38.66^\circ\ \mho$$
$$\begin{aligned}\bar{Y} &= \bar{Y}_1 + \bar{Y}_2\\ &= 0.12\angle{-35.54^\circ} + 0.08\angle 38.66^\circ\\ &= 0.16\ \angle{-7.04^\circ}\ \mho\\ &= 0.16 - j0.02\ \mho\end{aligned}$$

Admittance $Y = 0.16\ \mho$

Conductance $G = 0.16\ \mho$

Susceptance $B = 0.02\ \mho$

$$\begin{aligned}\bar{I} &= \bar{V}\cdot\bar{Y}\\ &= (230\ \angle 0^\circ)\ (0.16\ \angle{-7.04^\circ}) = 36.8\ \angle{-7.04^\circ}\text{ A}\end{aligned}$$
$$I = 36.8\text{ A}$$
$$\text{pf} = \cos\phi = \cos\ (7.04^\circ) = 0.99\ \text{(lagging)}$$

Example 4.66 *Two impedances Z_1 and Z_2 are connected in parallel across a 200-V, 50-Hz ac supply. The current drawn by impedance Z_1 is 4 A at 0.8 lagging pf. The total current drawn from the supply is 5 A at unity pf. Calculate the impedance Z_2.*

Solution

Data $V = 200$ V

$I_1 = 4$ A at 0.8 lagging pf

$I = 5$ A at unity pf

$$\bar{I}_1 = 4\ \angle{-\cos^{-1}}\ (0.8) = 4\ \angle{-36.87^\circ}\text{ A}$$
$$\bar{I} = 5\ \angle\cos^{-1}\ (1) = 5\ \angle 0^\circ\text{ A}$$
$$\bar{I} = \bar{I}_1 + \bar{I}_2$$
$$\bar{I}_2 = \bar{I} - \bar{I}_1 = 5\angle 0^\circ - 4\ \angle{-36.87^\circ} = 3\ \angle 53.13^\circ\text{ A}$$
$$\bar{Z}_2 = \frac{\bar{V}}{\bar{I}_2} = \frac{200\angle 0^\circ}{3\angle 53.13^\circ} = 66.67\ \angle{-53.13^\circ}\ \Omega$$

Example 4.67 *Compute Z_{eq} and Y_{eq} for the given circuit.*

Fig. 4.67

Solution

Data

$$\bar{Z}_1 = j5\,\Omega$$

$$\bar{Z}_2 = 5 + j8.66\ \Omega$$

$$\bar{Z}_3 = 15\ \Omega$$

$$\bar{Z}_4 = -j10\ \Omega$$

$$\bar{Y}_{eq} = \bar{Y}_1 + \bar{Y}_2 + \bar{Y}_2 + \bar{Y}_4$$

$$= \frac{1}{\bar{Z}_1} + \frac{1}{\bar{Z}_2} + \frac{1}{\bar{Z}_3} + \frac{1}{\bar{Z}_4}$$

$$= \frac{1}{j5} + \frac{1}{5 + j8.66} + \frac{1}{15} + \frac{1}{-j10}$$

$$= 0.22\ \angle{-57.99^\circ}\ \mho$$

$$\bar{Z}_{eq} = \bar{Y}_{eq} = \frac{1}{0.22\angle{-57.99^\circ}} = 4.54\ \angle 57.99^\circ\ \Omega$$

Example 4.68 *Find currents I_1 and I_2.*

Solution

Data

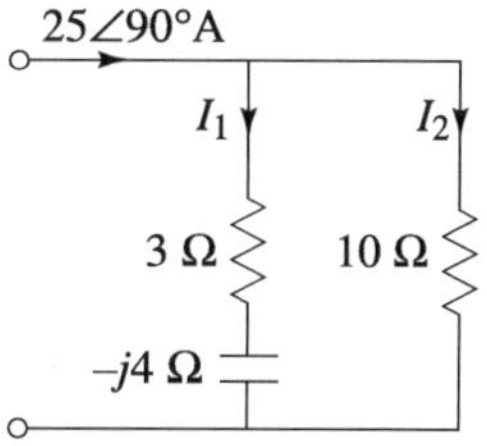

Fig. 4.68

$$\bar{Z}_1 = 3 - j4\ \Omega$$

$$\bar{Z}_2 = 10\ \Omega$$

By current–division formula,

$$\bar{I}_1 = \bar{I} \cdot \frac{\bar{Z}_2}{\bar{Z}_1 + \bar{Z}_2} = (25\ \angle 90^\circ) \cdot \frac{10}{3 - j4 + 10} = 18.38\ \angle 107.1^\circ\ \text{A}$$

$$\bar{I}_2 = \bar{I} - \bar{I}_1 = 25\ \angle 90^\circ - 18.38\ \angle 107.1^\circ = 9.19\ \angle 54^\circ\ \text{A}$$

Example 4.69 *Three impedances of 25 ∠53.1° Ω, 5 ∠–53.1° Ω and 10 ∠36.9° Ω are connected in parallel. The combination is in series with another impedance of 14.14 ∠45° Ω. Calculate the equivalent impedance of the circuit.*

Solution

Data

$$\bar{Z}_1 = 25\ \angle 53.1^\circ\ \Omega$$

$$\bar{Z}_2 = 5\ \angle{-53.1^\circ}\ \Omega$$

$$\bar{Z}_3 = 10\ \angle 36.9^\circ\ \Omega$$

$$\bar{Z}_4 = 14.14\ \angle 45^\circ\ \Omega$$

$$\bar{Y} = \frac{1}{\bar{Z}} = \frac{1}{\bar{Z}_1} + \frac{1}{\bar{Z}_2} + \frac{1}{\bar{Z}_3}$$

$$= \frac{1}{25\angle 53.1^\circ} + \frac{1}{5\angle -53.1^\circ} + \frac{1}{10\angle 36.9^\circ} = 0.23\ \angle 16.86^\circ\ \mho$$

$$\bar{Z} = 4.27\ \angle -16.86^\circ\ \Omega$$

$$\bar{Z}_{eq} = \bar{Z} + \bar{Z}_4$$

$$= 4.27\ \angle -16.86^\circ + 14.14\ \angle 45^\circ = 16.58\ \angle 31.87^\circ\ \Omega$$

Example 4.70 *A voltage of 200 ∠25° V is applied to a circuit composed of two parallel branches. If the branch currents are 10 ∠40° A and 20 ∠–30° A, determine the kVA, kVAR, kW in each branch. Also, calculate the pf of the combined load.*

Solution

Data $\bar{I}_1 = 10\ \angle 40^\circ$ A $\quad \bar{I}_2 = 20\ \angle -30^\circ$ A

$$\bar{V} = 200\ \angle 25^\circ \text{ V}$$

Phase difference between V and I_1, $\phi_1 = 40^\circ - 25^\circ = 15^\circ$

Phase difference between V and I_2, $\phi_2 = 25^\circ - (-30^\circ) = 55^\circ$

$$\cos\phi_1 = \cos(15^\circ) = 0.97 \text{ (leading)}$$

$$\cos\phi_2 = \cos(55^\circ) = 0.57 \text{ (lagging)}$$

For the branch current of 10 ∠40° A,

$$P_1 = VI_1 \cos\phi_1 = 200 \times 10 \times 0.97 = 1.94 \text{ kW}$$

$$Q_1 = VI_1 \sin\phi_1 = 200 \times 10 \times \sin(15^\circ) = 0.52 \text{ kVAR}$$

$$S_1 = VI_1 = 200 \times 10 = 2 \text{ kVA}$$

For the branch current of 20 ∠–30° A,

$$P_2 = VI_2 \cos\phi_2 = 200 \times 20 \times 0.57 = 2.28 \text{ kW}$$

$$Q_2 = VI_2 \sin\phi_2 = 200 \times 20 \times \sin(55^\circ) = 3.28 \text{ kVAR}$$

$$S_2 = VI_2 = 200 \times 20 = 4 \text{ kVA}$$

For the combined load,

$$\bar{I}_1 = 10\ \angle 40^\circ \text{ A}$$

$$\bar{I}_2 = 20\ \angle -30^\circ \text{ A}$$

$$\bar{I} = \bar{I}_1 + \bar{I}_2$$

$$= 10\ \angle 40^\circ + 20\ \angle -30^\circ = 25.24\ \angle -8.14^\circ \text{ A}$$

$$\text{Phase difference} = 25^\circ - (-8.14^\circ)$$

$$= 33.14^\circ$$

$$\text{pf} = \cos(33.13^\circ) = 0.84 \text{ (lagging)}$$

Example 4.71 *Two circuits have the same numerical value of impedance. The pf of one is 0.8 and that of the other is 0.6. What is the pf of combination if they are connected in parallel?*

Solution

Data

$$\text{pf}_1 = 0.8$$

$$\text{pf}_2 = 0.6$$

$$\bar{Z}_1 = Z\ \angle \cos^{-1}(0.8) = Z\ \angle 36.87^\circ\ \Omega$$

$$\bar{Z}_2 = Z\ \angle \cos^{-1}(0.6) = Z\ \angle 53.13^\circ\ \Omega$$

For parallel combination,

$$\overline{Z} = \frac{\overline{Z_1}\cdot\overline{Z_2}}{\overline{Z_1}+\overline{Z_2}} = \frac{(Z\angle 36.87^\circ)(Z\angle 53.13^\circ)}{Z\angle 36.87^\circ + Z\angle 53.13^\circ}$$

$$= \frac{Z^2\angle 90^\circ}{Z(1.4 + j1.4)}$$

$$= \frac{Z^2\angle 90^\circ}{1.98Z\angle 45^\circ} = 0.505\ Z\ \angle 45^\circ$$

$$\text{pf} = \cos(45^\circ) = 0.707$$

Example 4.72 *When a 240-V, 50-Hz supply is fed to a 15-Ω resistor in parallel with an inductor, the total current is 22.1 A. What value must the frequency have for the total current to be 34 A?*

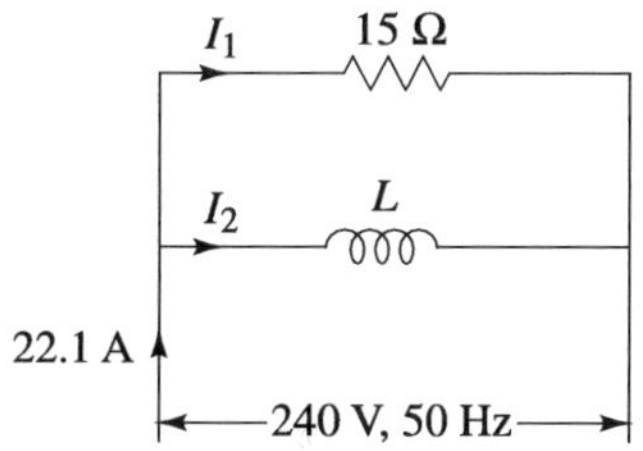

Fig. 4.69

Solution

Data

$$V = 240\text{ V}$$

$$R = 15\ \Omega$$

$$I = 22.1\text{ A}$$

$$\overline{I_1} = \frac{240\angle 0^\circ}{15\angle 0^\circ}\ 16\ \angle 0^\circ = 16\text{ A}$$

$$\overline{I_2} = \frac{240\angle 0^\circ}{X_L\angle 90} = \frac{240}{X_L}\angle{-90^\circ} = -j\,\frac{240}{X_L}\text{ A}$$

$$\overline{I} = 16 - j\,\frac{240}{X_L}$$

$$\sqrt{(16)^2 + \left(\frac{240}{X_L}\right)^2} = 22.1$$

$$256 + \frac{57600}{X_L^2} = 488.41$$

$$X_L = 15.74\ \Omega$$

$$X_L = 2\pi f L$$

$$15.74 = 2\pi \times 50 \times L$$

$$L = 0.05\text{ H}$$

Let the new frequency be f. Then

$$\sqrt{(16)^2 + \left(\frac{240}{2\pi f \times 0.05}\right)^2} = 34$$

$$256 + \frac{57600}{0.0987 f^2} = 1156$$

$$f = 25.47\text{ Hz}$$

Example 4.73 *Determine the current in the circuit of Fig. 4.70. Also, find the power consumed as well as pf.*

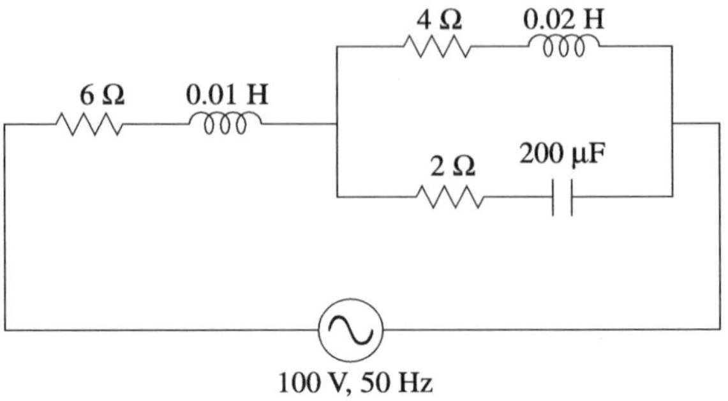

Fig. 4.70

Solution

$$X_{L_1} = 2\pi \times 50 \times 0.01 = 3.14\ \Omega$$

$$X_{L_2} = 2\pi \times 50 \times 0.02 = 6.28\ \Omega$$

$$X_C = \frac{1}{2\pi \times 50 \times 200 \times 10^{-6}} = 15.92\ \Omega$$

$$\overline{Z_1} = 6 + j3.14\ \Omega$$

$$\overline{Z_2} = 4 + j6.28\ \Omega$$

$$\overline{Z_3} = 2 - j15.92\ \Omega$$

$$\overline{Z} = \overline{Z_1} + \frac{\overline{Z_2}.\overline{Z_3}}{\overline{Z_2} + \overline{Z_3}}$$

$$= (6 + j3.14) + \frac{(4 + j6.28)(2 - j15.92)}{(4 + j6.28) + (2 - j15.92)} = 17.27\ \angle 30.75^\circ\ \Omega$$

$$\overline{I} = \frac{\overline{V}}{\overline{Z}} = \frac{100\angle 0^\circ}{17.27\angle 30.75^\circ} = 5.79\ \angle -30.75^\circ\ \text{A}$$

$$P = VI\cos\phi = 100 \times 5.79 \times \cos(30.75^\circ)$$

$$= 497.94\ \text{W}$$

$$\text{pf} = \cos\phi = \cos(30.75^\circ) = 0.86\ \text{(lagging)}$$

Example 4.74 *Find the applied voltage V_{AB} so that 10 A current may flow through the capacitor.*

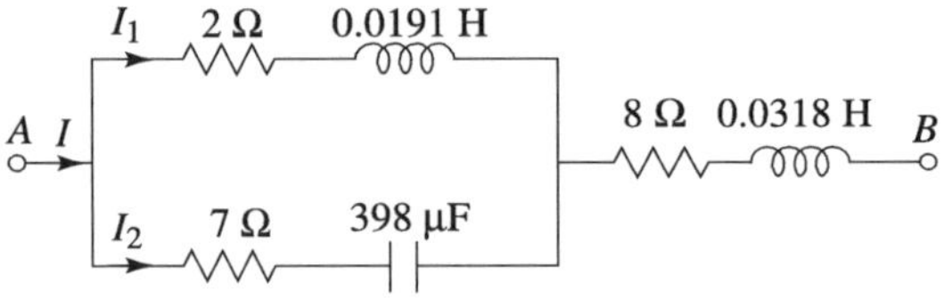

Fig. 4.71

Solution

$$X_{L_1} = 2\pi \times 50 \times 0.0191 = 6\ \Omega$$
$$X_{L_2} = 2\pi \times 50 \times 0.0318 = 10\ \Omega$$
$$X_C = \frac{1}{2\pi \times 50 \times 398 \times 10^{-6}} = 8\ \Omega$$
$$\overline{Z_1} = 2 + j6 = 6.32\ \angle 71.56°\ \Omega$$
$$\overline{Z_2} = 7 - j8 = 10.63\ \angle -48.8°\ \Omega$$
$$\overline{Z_3} = 8 + j10 = 12.8\ \angle 51.34°\ \Omega$$
$$\overline{Z} = \frac{\overline{Z_1}\,\overline{Z_2}}{\overline{Z_1} + \overline{Z_2}} + \overline{Z_3}$$
$$= \frac{(2+j6)(7-j8)}{(2+j6+7-j8)} + (8 + j10) = 19.91\ \angle 45.53°\ \Omega$$

Let

$$\overline{I_2} = 10\ \angle 0°\ \text{A}$$
$$\overline{V_2} = \overline{Z_2}\,.\,\overline{I_2} = (10.63\ \angle -48.8°)\ (10\ \angle 0°) = 106.3\ \angle -48.8°\ \text{V}$$
$$\overline{V_1} = \overline{V_2} = 106.3\ \angle -48.8°\ \text{V}$$
$$\overline{I_1} = \frac{\overline{V_1}}{\overline{Z_1}} = \frac{106.3\angle -48.8°}{6.32\angle 71.56°} = 16.82\ \angle -120.36°\ \text{A}$$
$$\overline{I} = \overline{I_1} + \overline{I_2} = 16.82\ \angle -120.36° - 10\angle 0° = 14.58\ \angle -84.09°\ \text{A}$$
$$\overline{V_{AB}} = \overline{Z}\,.\,\overline{I} = (19.91\ \angle 45.53°)\ (14.58\ \angle -84.09°)$$
$$= 290.28\ \angle -38.56°\ \text{V}$$

Example 4.75 *If a voltage of 150 V applied between terminals A and B produces a current of 32 A for the circuit shown in Fig. 4.72, calculate the value of resistance R and pf of the circuit.*

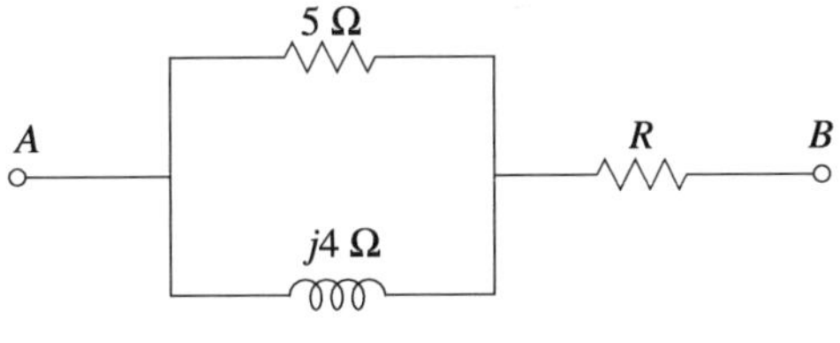

Fig. 4.72

Solution

Data

$$V = 150\ \text{V}$$
$$I = 32\ \text{A}$$
$$Z = \frac{150}{32} = 4.687\ \Omega$$
$$\overline{Z} = \frac{(5)(j4)}{5 + j4} + R = \frac{(5)(4\angle 90°)}{64\angle 38.66°} + R$$
$$= 3.125\ \angle 51.34° + R = 1.95 + j2.44 + R$$
$$\sqrt{(1.95 + R)^2 + (2.44)^2} = 4.687$$
$$(1.95 + R)^2 + (2.44)^2 = (4.687)^2$$
$$(1.95 + R)^2 = (4.687)^2 - (2.44)^2$$
$$1.95 + R = 4$$
$$R = 2.05\ \Omega$$

$$\text{pf} = \frac{\text{Total resistance}}{\text{Total impedance}} = \frac{1.95 + 2.05}{4.687} = 0.853 \text{ (lagging)}$$

Example 4.76 *An impedance of R + jX ohm is connected in parallel with another impedance of –j5 ohm. The combination is then connected in series with a pure resistance of 2 Ω. When connected across a 100-V, 50-Hz ac supply, the total current drawn by the circuit is 20 A and the total power consumed by the circuit is 2 kW. Calculate (i) R and L, and (ii) currents through parallel branches.*

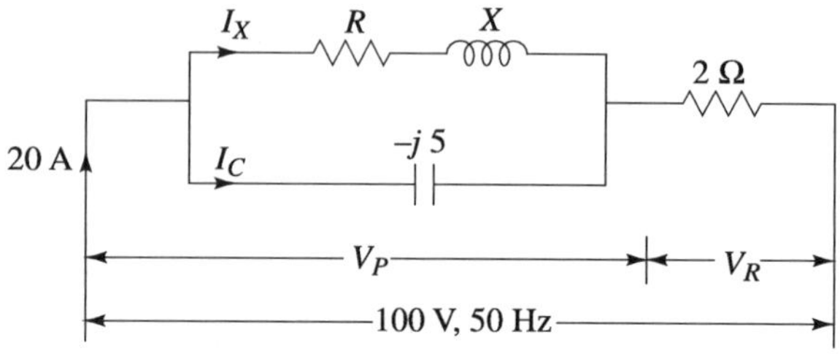

Fig. 4.73

Solution

Data

$$P = 2 \text{ kW}$$
$$V = 100 \text{ V}$$
$$I = 20 \text{ A}$$
$$P = VI \cos \phi$$
$$2000 = 100 \times 20 \times \cos \phi$$
$$\cos \phi = 1$$
$$\phi = 0$$
$$\bar{V} = 100 \angle 0° \text{ V}$$
$$\bar{I} = 20 \angle 0° \text{ A}$$
$$\overline{V_R} = 2 \times 20° \angle 0° = 40 \angle 0° \text{ V}$$
$$\overline{V_P} = \bar{V} - \overline{V_R}$$
$$= 100 \angle 0° - 40 \angle 0° = 60 \angle 0° \text{ V}$$
$$\overline{I_C} = \frac{\overline{V_P}}{Z_C} = \frac{60\angle 0°}{5\angle -90°} = 12 \angle 90° \text{ A}$$
$$\overline{I_X} = \bar{I} - \overline{I_C}$$
$$= 20 \angle 0° - 12 \angle 90° = 23.32 \angle -30.96° \text{ A}$$
$$\overline{Z_X} = \frac{60\angle 0°}{23.32\angle -30.96°} = 2.57 \angle 30.96° \ \Omega = 2.2 + j1.32 \ \Omega$$
$$R = 2.2 \ \Omega$$
$$X_L = 1.32 \ \Omega$$
$$X_L = 2\pi f L$$
$$1.32 = 2\pi \times 50 \times L$$
$$L = 4.2 \text{ mH}$$

Example 4.77 *The circuit of Fig. 4.74 takes 12 A at a lagging power factor and dissipates 1800 W. The reading of the voltmeter is 200 V. Find R_1, X_1 and X_2.*

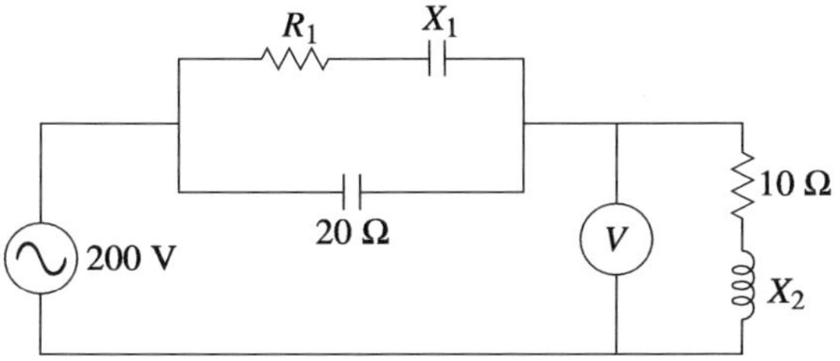

Fig. 4.74

Solution

Data

$$I = 12\text{ A}$$
$$P = 1800\text{ W}$$

Let

$$\bar{I} = 12\angle 0^\circ \text{ A}$$
$$Z_2 = \frac{200}{12} = 16.667\ \Omega$$
$$Z_2 = \sqrt{R_2^2 + X_2^2}$$
$$16.67 = \sqrt{(10)^2 + X_2^2}$$
$$(16.67)^2 = 10^2 + X_2^2$$
$$277.88 = 100 + X_2^2$$
$$X_2 = 13.33\ \Omega$$
$$\overline{V_2} = (12\angle 0^\circ)(10 + j13.33)$$
$$= (12\angle 0^\circ)(16.666\angle 53.13^\circ) = 200\angle 53.13^\circ \text{ V}$$
$$P = VI\cos\phi$$
$$1800 = 200 \times 12 \times \cos\phi$$
$$\cos\phi = 0.75$$
$$\phi = 41.41^\circ$$

Applied voltage $\overline{V_{eq}} = 200\angle 41.41^\circ$ V

Voltage across parallel branches = $200\angle 41.41^\circ - 200\angle 53.13^\circ = 40.84\angle -42.73^\circ$ V

$$\text{Current through capacitor} = \frac{40.84\angle -42.73^\circ}{20\angle -90^\circ} = 2.04\angle 47.27^\circ \text{ A}$$

Current through R_1 and X_1 = $12\angle 0^\circ - 2.04\angle 47.27^\circ = 10.72\angle -8.03^\circ$ A

$$\overline{Z_1} = \frac{40.84\angle -42.73^\circ}{10.72\angle -8.03^\circ}$$
$$= 3.81\angle -34.7^\circ\ \Omega = 3.13 - j2.17\ \Omega$$
$$R_1 = 3.13\ \Omega$$
$$X_1 = 2.17\ \Omega$$

Example 4.78 *For the circuit shown, calculate (i) total admittance, total conductance and total susceptance, (ii) total current and total pf, and (iii) value of pure capacitance to be connected in parallel with the above combination to make the total pf unity.*

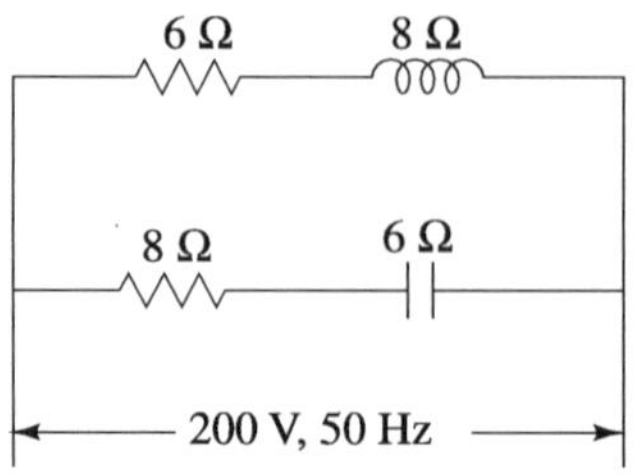

Fig. 4.75

Solution

(i)

$$\overline{Z_1} = 6 + j8 = 10\angle 53.13°\ \Omega$$

$$\overline{Z_2} = 8 - j6 = 10\angle -36.87°\ \Omega$$

$$\overline{Y_1} = \frac{1}{\overline{Z_1}} = \frac{1}{10\angle 53.13°} = 0.1\angle -53.13°\ \mho$$

$$\overline{Y_2} = \frac{1}{\overline{Z_2}} = \frac{1}{10\angle -36.87°} = 0.1\angle 36.87°\ \mho$$

$$\overline{Y} = \overline{Y_1} + \overline{Y_2}$$
$$= 0.1\angle -53.13° + 0.1\angle 36.87°$$
$$= 0.14 - j0.02\ \mho = 0.14\angle -8.13°\ \mho$$

Total admittance $Y = 0.14\ \mho$

Total conductance $G = 0.14\ \mho$

Total susceptance $B = 0.02\ \mho$

Let $\overline{V} = 200\angle 0°\ \text{V}$

(ii)

$$I = \overline{V}\cdot\overline{Y}$$
$$= (200\angle 0°)(0.14\angle -8.13°) = 28\angle -8.13°\ \text{A}$$

Total pf = cos (8.13°) = 0.989 (lagging)

(iii) Since the current lags behind voltage, the circuit is inductive in nature. In order to make the total pf unity, a pure capacitor is connected in parallel so that pf becomes unity and imaginary part of $\overline{Y}_{\text{req}}$ becomes zero.

$$\overline{Y_{\text{req}}} = 0.14 - j0.02 + j0.02 = 0.14$$

$$\frac{1}{X_C} = 0.02$$

$$X_C = 50\ \Omega$$

$$C = \frac{1}{2\pi \times 50 \times 50} = 63.66\ \mu\text{F}$$

4.14 SERIES RESONANCE

A circuit containing reactance is said to be in resonance if the voltage across the circuit is in phase with the current through it. At resonance, the circuit thus behaves as a pure resistance and the net reactance is zero.

Consider the series *R-L-C* circuit as shown in Fig. 4.76. The impedance of the circuit is

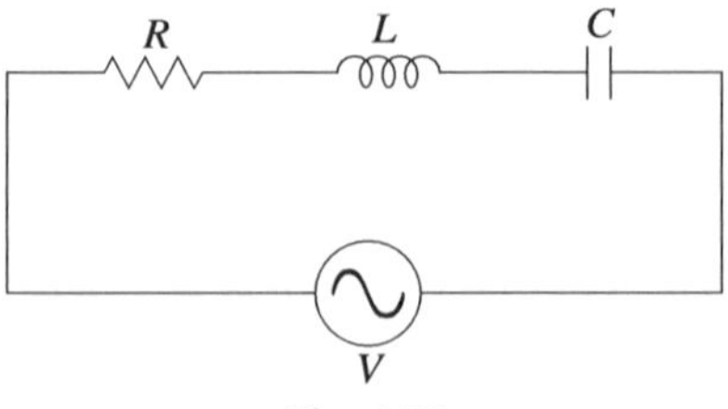

Fig. 4.76

$$\bar{Z} = R + jX_L - jX_C$$
$$= R + j\omega L - \frac{j}{\omega C}$$
$$= R + j\left(\omega L - \frac{1}{\omega C}\right)$$

(i) At resonance, Z must be resistive. Therefore the condition for resonance is

$$\omega L - \frac{1}{\omega C} = 0$$

$$\omega = \omega_0 = \frac{1}{\sqrt{LC}}$$

$$f = f_0 = \frac{1}{2\pi\sqrt{LC}}$$

where f_0 is called the resonant frequency of the circuit.

(ii) Power factor

$$\text{Power factor} = \cos\phi = \frac{R}{Z} = \frac{R}{\sqrt{R^2 + \left(\omega L - \frac{1}{\omega C}\right)^2}}$$

But at resonance
$$\omega L = \frac{1}{\omega C}$$
$$\text{Power factor} = \frac{R}{R} = 1$$

(iii) Current Since impedance is minimum, the current is maximum at resonance. Thus, the circuit accepts more current and as such, an *R-L-C* circuit under resonance is called an *acceptor circuit.*

$$I_0 = \frac{V}{Z} = \frac{V}{R}$$

(iv) Voltage At resonance,

$$\omega_0 L = \frac{1}{\omega_0 C}$$

$$\omega_0 L\, I_0 = \frac{1}{\omega_0 C} I_0$$

$$V_{L_0} = V_{C_0}$$

Thus, potential difference across inductance equal to potential difference across capacitance being equal and opposite cancel each other. Also, since I_0 is maximum, V_{L_0} and V_{C_0} will also be maximum. Thus, voltage magnification takes place during resonance. Hence, it is also referred to as voltage magnification circuit.

(v) Phasor diagram

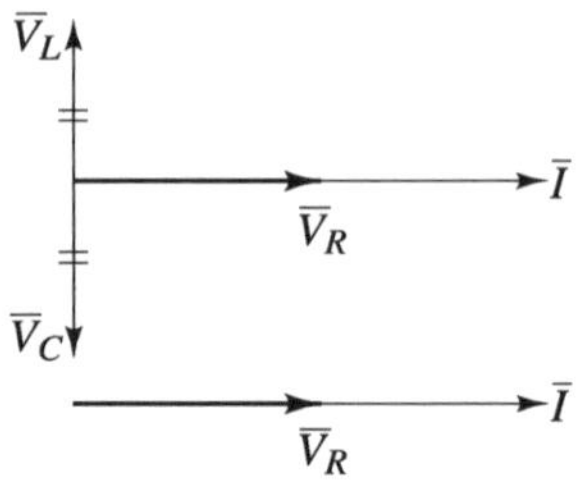

Fig. 4.77

(vi) Behaviour of R, L and C with change in frequency

Resistance remains constant with the change in frequencies. Inductive reactance X_L is directly proportional to frequency f. It can be drawn as a straight line passing through the origin. Capacitive reactance X_C is inversely proportional to the frequency f. It can be drawn as a rectangular hyperbola in the fourth quadrant.

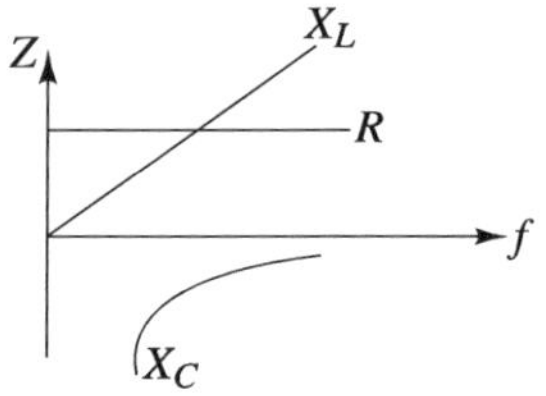

Fig. 4.78

Total impedance

$$\overline{Z} = R + j\,(X_L - X_C)$$

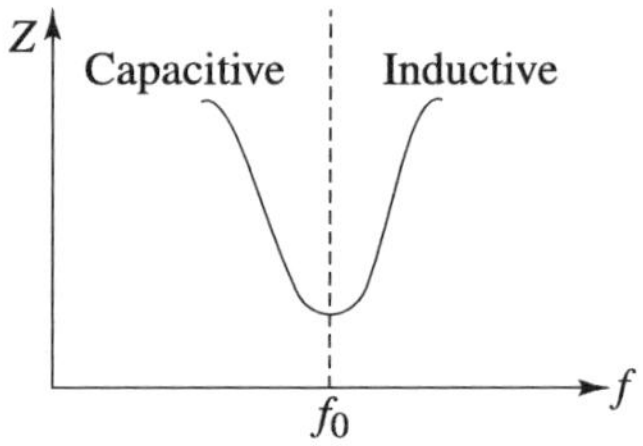

Fig. 4.79

(a) When $f < f_0$, impedance is capacitive and decreases up to f_0. The power factor is leading in nature.
(b) At $f = f_0$, impedance is resistive. Power factor is unity.
(c) When $f > f_0$, impedance is inductive and goes on increasing beyond f_0. Power factor is lagging in nature.

(vii) Bandwidth

For the series *R-L-C* circuit, bandwidth is defined as the range of frequencies for which the power delivered to R is greater than or equal to $\frac{P_0}{2}$ where P_0 is the power delivered to R at resonance. From the shape of the resonance curve, it is clear that there are two frequencies for which the power delivered to R is half the power at resonance. For this reason, these frequencies are referred as those corresponding to the half-power points. The magnitude of the current at each half-power point is the same.

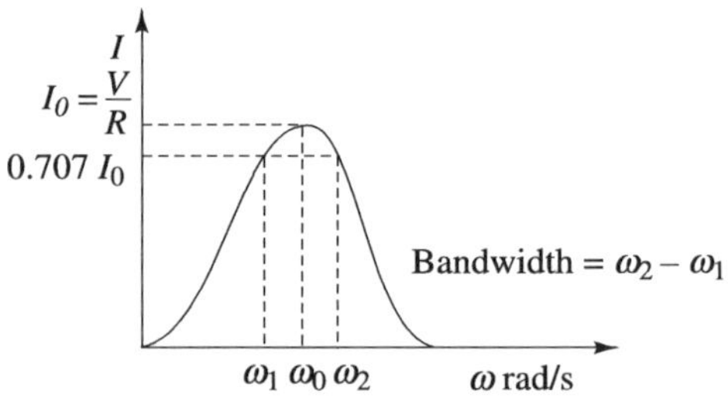

Fig. 4.80

Hence $$I_1^2 R = \frac{1}{2} I_0^2 R = I_2^2 R$$

where the subscript 1 denotes the lower half point and the subscript 2, the higher half point. It follows then that

$$I_1 = I_2 = \frac{I_0}{\sqrt{2}} = 0.707 I_0$$

Accordingly, the bandwidth may be identified on the resonance curve as the range of frequencies over which the magnitude of the current is equal to or greater than 0.707 of the current at resonance. In Fig. 4.80, the bandwidth is $\omega_2 - \omega_1$.

Expression for the bandwidth Generally, at any frequency ω,

$$I = \frac{V}{Z} = \frac{V}{\sqrt{R^2 + (X_L - X_C)^2}} = \frac{V}{\sqrt{R^2 + \left(\omega L - \frac{1}{\omega C}\right)^2}} \quad \text{... (4.1)}$$

At half-power points,

$$I = \frac{I_0}{\sqrt{2}}$$

But $$I_0 = \frac{V}{R}$$

$$I = \frac{V}{\sqrt{2} R} \quad \text{... (4.2)}$$

From Eqs (4.1) and (4.2), we get

$$\frac{V}{\sqrt{R^2 + \left(\omega L - \frac{1}{\omega C}\right)^2}} = \frac{V}{\sqrt{2} R}$$

$$\frac{1}{\sqrt{R^2 + \left(\omega L - \frac{1}{\omega C}\right)^2}} = \frac{1}{\sqrt{2} R}$$

Squaring both sides we get,

$$R^2 + \left(\omega L - \frac{1}{\omega C}\right)^2 = 2R^2$$

$$\left(\omega L - \frac{1}{\omega C}\right)^2 = R^2$$

$$\omega L - \frac{1}{\omega C} \pm R = 0$$

$$\omega^2 \pm \frac{1}{2} \omega - \frac{1}{LC} = 0$$

$$\omega = \pm \frac{R}{2L} \pm \sqrt{\frac{R^2}{4L^2} + \frac{1}{LC}}$$

For low values of R, the term $\left(\frac{R^2}{4L^2}\right)$ can be neglected in comparison with the term $\frac{1}{LC}$.

Then ω is given by, $$\omega = \pm \frac{R}{2L} \pm \sqrt{\frac{1}{LC}} = \pm \frac{R}{2L} \pm \frac{1}{\sqrt{LC}}$$

The resonant frequency for this circuit is given by

$$f_0 = \frac{1}{2\pi\sqrt{LC}}$$

$$\omega_0 = \frac{1}{\sqrt{LC}}$$

$$\omega = \pm\frac{R}{2L} + \omega_0 \qquad \text{(considering only +ve sign of } \omega_0\text{)}$$

$$\omega_1 = \omega_0 - \frac{R}{2L}$$

and

$$\omega_2 = \omega_0 + \frac{R}{2L}$$

$$\text{Bandwidth} = \omega_2 - \omega_1 = \frac{R}{L}\ \text{rad/s}$$

or

$$\text{Bandwidth} = f_2 - f_1 = \text{c/s}$$

(viii) Quality factor Q_0 of the R-L-C circuit It is the ratio of the resonant frequency to the bandwidth. It is a measure of the selectivity or sharpness of tuning of the series *R-L-C* circuit.

$$Q_0 = \frac{\omega_0}{\text{Bandwidth}}$$

$$Q_0 = \frac{\omega_0}{R/L} = \frac{\omega_0 L}{R} = \frac{X_{L0}}{R} = \frac{1}{\omega_0 RC}$$

$$Q_0 = \frac{1/\sqrt{LC}}{R/L} = \frac{1}{R}\sqrt{\frac{L}{C}}$$

$$Q_0 = \frac{V_{L_0}}{V} = \frac{V_{C_0}}{V}$$

where V_{L_0} and V_{C_0} are both measured at resonance. Hence, Q_0 is also called *voltage magnification factor.*

Example 4.79 *A series R-L-C circuit has the following parameter values: R = 10 Ω, L = 0.01 H, C = 100 μF. Compute the resonant frequency, quality factor of the circuit, bandwidth, and lower and upper frequency of the bandwidth.*

Solution

Data

$$R = 10\ \Omega$$
$$L = 0.01\ \text{H}$$
$$C = 100\ \mu\text{F}$$

Resonant frequency

$$f_0 = \frac{1}{2\pi\sqrt{LC}} = \frac{1}{2\pi\sqrt{0.01\times 100\times 10^{-6}}} = 159.15\ \text{Hz}$$

Bandwidth

$$BW = \frac{R}{2\pi L} = \frac{10}{2\pi\times 0.01} = 159.15\ \text{Hz}$$

Lower frequency of bandwidth $f_1 = f_0 - \dfrac{BW}{2}$

$$= 159.15 - \frac{159.15}{2} = 79.58\ \text{Hz}$$

Higher frequency of bandwidth $f_2 = f_0 + \dfrac{BW}{2} = 159.15 + \dfrac{159.15}{2} = 238.73\ \text{Hz}$

Example 4.80 *An R-L-C series circuit with a resistance of 10 Ω, inductance of 0.2 H and a capacitance of 40 μF is supplied with a 100-V supply at variable frequency. Find the following w.r.t. series resonant circuit:*

(i) the frequency at which resonance takes place
(ii) at resonance, find the current
(iii) power
(iv) power factor
(v) voltage across *R-L-C* at that time
(vi) quality factor
(vii) half-power points
(viii) resonance and phasor diagram.

Solution

Data

$$R = 10\ \Omega$$
$$L = 0.2\ \text{H}$$
$$C = 40\ \mu\text{F}$$
$$V = 100\ \text{V}$$

(i) Resonant frequency $f_0 = \dfrac{1}{2\pi\sqrt{LC}} = \dfrac{1}{2\pi\sqrt{0.2 \times 40 \times 10^{-6}}} = 56.3$ Hz

(ii) Current $I_0 = \dfrac{V}{R} = \dfrac{100}{10} = 10$ A

(iii) Power $P_0 = I_0^2\, R = (10)^2 \times 10 = 1000$ W

(iv) Power factor pf = 1

(v) Voltage across $R = R\,.\,I = 10 \times 10 = 100$ V

Voltage across $L = X_L\,.\,I = 2\pi \times 56.3 \times 0.2 \times 10 = 707.5$ V

Voltage across $C = X_C\,.\,I = \dfrac{1}{2\pi \times 56.3 \times 40 \times 10^{-6}} \times 10 = 707.5$ V

(vi) Quality factor $Q = \dfrac{1}{R} \times \sqrt{\dfrac{L}{C}} = \dfrac{1}{10}\sqrt{\dfrac{0.2}{40 \times 10^{-6}}} = 7.07$

(vii) Half-power points

$$f_1 = f_0 - \frac{R}{4\pi L} = 56.3 - \frac{10}{4\pi(0.2)} = 52.32\ \text{Hz}$$

$$f_2 = f_0 + \frac{R}{4\pi L} = 56.3 + \frac{10}{4\pi(0.2)} = 60.3\ \text{Hz}$$

(viii) *Resonance and phasor diagram*

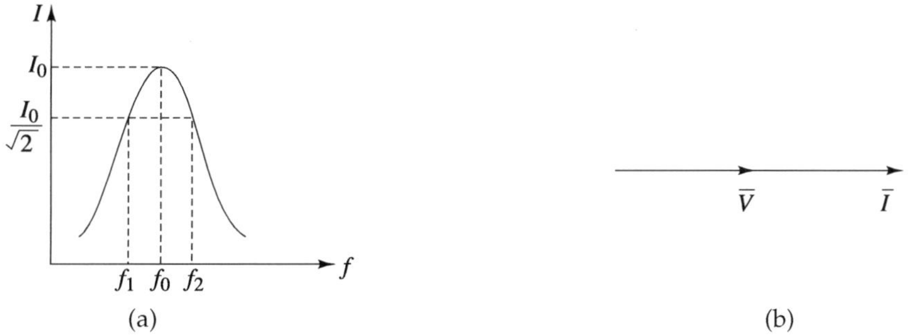

Fig. 4.81

Example 4.81 *A series R-L-C circuit is connected to a 200-V ac supply. The current drawn by the circuit at resonance is 20 A. The voltage drop across the capacitor is 5000 V at series resonance. Calculate resistance and inductance if capacitance is 4 μF. Also, calculate the resonant frequency.*

Solution

Data $V = 200$ V, $I_0 = 20$ A, $V_C = 5000$ V, $C = 4\ \mu$F

Resistance $R = \frac{V}{I_0} = \frac{200}{20} = 10\ \Omega$

$$X_{C_0} = \frac{V_{C0}}{I_0} = \frac{5000}{20} = 250\ \Omega$$

$$X_{C_0} = \frac{1}{2\pi f_0 C}$$

$$f_0 = \frac{1}{2\pi X_{C0} C} = \frac{1}{2\pi \times 250 \times 4 \times 10^{-6}} = 159.15 \text{ Hz}$$

At resonance $X_{C_0} = X_{L_0}$

$$X_{L_0} = 250\ \Omega$$

$$X_{L_0} = 2\pi f_0 L$$

$$250 = 2\pi \times 159.15 \times L$$

Inductance $L = 0.25$ H

Example 4.82 *A resistor and a capacitor are connected in series with a variable inductor. When the circuit is connected to a 230-V, 50-Hz supply, the maximum current obtained by varying the inductance is 2 A. The voltage across the capacitor is 500 V. Calculate the resistance, inductance and capacitance of the circuit.*

Solution

Data $V = 230$ V, $f_0 = 50$ Hz, $I_0 = 2$ A, $V_{C_0} = 500$ V

Resistance $R = \frac{V}{I_0} = \frac{230}{2} = 115\ \Omega$

$$X_{C_0} = \frac{V_{C0}}{I_0} = \frac{500}{2} = 250\ \Omega$$

$$X_{C_0} = \frac{1}{2\pi f_0 C}$$

$$250 = \frac{1}{2\pi \times 50 \times C}$$

Capacitance $C = 12.73\ \mu$F

At resonance $X_{C_0} = X_{L_0}$

$$X_{L_0} = 250\ \Omega$$

$$X_{L_0} = 2\pi f_0 L$$

$$250 = 2\pi \times 50 \times L$$

Inductance $L = 0.795$ H

Example 4.83 *A coil of 2 Ω resistance and 0.01-H inductance is connected in series with a capacitor across 200-V mains. What must be the capacitance in order that maximum current occurs at a frequency of 50 Hz? Find also the current and voltage across the capacitor.*

Solution

Data

$$R = 2\ \Omega$$
$$L = 0.01\text{ H}$$
$$V = 200\text{ V}$$
$$f_0 = 50\text{ Hz}$$
$$f_0 = \frac{1}{2\pi\sqrt{LC}}$$
$$50 = \frac{1}{2\pi\sqrt{0.01 \times C}}$$

Capacitance $$C = 1013.2\ \mu\text{F}$$

Current $$I_0 = \frac{V}{R} = \frac{200}{2} = 100\text{ A}$$

Voltage across capacitor $$V_{C_0} = I_0\ X_{C_0}$$
$$= 100 \times \frac{1}{2\pi \times 50 \times 1013.2 \times 10^{-6}} = 314.16\text{ V}$$

Example 4.84 *A voltage v(t) = 10 sin ωt is applied to a series R-L-C circuit. At the resonant frequency of the circuit, the voltage across the capacitor is found to be 500 V. The bandwidth of the circuit is known to be 400 rad/s and the impedance of the circuit at resonance is 100 Ω. Determine resonant frequency, upper and lower cut-off frequencies, inductance and capacitance.*

Solution

Data

$$v(t) = 10 \sin \omega t \qquad V_{C_0} = 5000\text{ V}$$
$$BW = 400\text{ rad/s} \qquad R = 100\ \Omega$$
$$V = \frac{10}{\sqrt{2}} = 7.07\text{ V}$$
$$I_0 = \frac{V}{R} = \frac{7.07}{100} = 0.0707\text{ A}$$
$$BW = \frac{R}{L}$$
$$400 = \frac{100}{L}$$
$$L = 0.25\text{ H}$$
$$Q_0 = \frac{V_{C0}}{V} = \frac{500}{7.07} = 70.72$$
$$Q_0 = \frac{1}{R}\sqrt{\frac{L}{C}}$$
$$70.72 = \frac{1}{100}\sqrt{\frac{0.25}{C}}$$

$$C = 4.99 \times 10^{-9}\ \text{F}$$

$$f_0 = \frac{1}{2\pi\sqrt{LC}} = \frac{1}{2\pi \times \sqrt{0.25 \times 4.99 \times 10^{-9}}} = 4506.09\ \text{Hz}$$

Lower cut-off frequency $\quad f_1 = f_0 - \dfrac{R}{4\pi L} = 4506.09 - \dfrac{100}{4\pi \times 0.25} = 4474.26\ \text{Hz}$

Upper cut-off frequency $\quad f_2 = f_0 + \dfrac{R}{4\pi L} = 4506.09 + \dfrac{100}{4\pi \times 0.25} = 4537.92\ \text{Hz}$

Example 4.85 *A series resonant circuit has impedance of 500 Ω at resonant frequency. Cut-off frequencies are 10 kHz and 100 Hz. Determine (i) resonant frequency, (ii) value of R-L-C, and (iii) quality factor at resonant frequency.*

Solution

Data

$$R = 500\ \Omega$$

$$f_1 = 100\ \text{Hz}$$

$$f_2 = 10\ \text{kHz}$$

$$BW = f_2 - f_1 = 10 \times 10^3 - 100 = 9900\ \text{Hz}$$

$$f_2 = f_0 - \frac{BW}{2} \qquad \text{...(i)}$$

$$f_2 = f_0 + \frac{BW}{2} \qquad \text{...(ii)}$$

Adding Eqs (i) and (ii),

$$f_1 + f_2 = 2f_0$$

$$f_0 = \frac{f_1 + f_2}{2} = \frac{10000 + 100}{2} = 5050\ \text{Hz}$$

$$BW = \frac{R}{2\pi L}$$

$$9900 = \frac{500}{2\pi L}$$

$$L = 8.038\ \text{mH}$$

$$X_{L_0} = 2\pi f_0 L = 2\pi \times 5050 \times 8.038 \times 10^{-3} = 255.05\ \Omega$$

At resonance $\quad X_{L_0} = X_{C_0} = 255.05\ \Omega$

$$X_{C_0} = \frac{1}{2\pi f_0 C}$$

$$255.05 = \frac{1}{2\pi \times 5050 \times C}$$

$$C = 0.12\ \mu\text{F}$$

$$Q_0 = \frac{1}{R}\sqrt{\frac{L}{C}} = \frac{1}{500}\sqrt{\frac{8.038 \times 10^{-3}}{0.12 \times 10^{-6}}} = 0.5176$$

Example 4.86 *Impedance of a circuit is observed to be capacitive and decreasing from 1 Hz to 100 Hz. Beyond 100 Hz, the impedance starts increasing. Find the values of circuit elements if the power drawn by this circuit is 100 W at 100 Hz, when the current is 1 A. The power factor of the circuit at 70 Hz is 0.707.*

Solution

Data $f_0 = 100$ Hz $P_0 = 100$ W

$I_0 = 1$ A $(\text{pf})_{70\text{ Hz}} = 0.707$

The impedance of the circuit is capacitive and decreasing from 1 Hz to 100 Hz. Beyond 100 Hz, the impedance starts increasing.

$$f_0 = 100 \text{ Hz}$$

$$P_0 = I_0^2 R$$

$$100 = (1)^2 \times R$$

$$R = 100\ \Omega$$

$$f_0 = \frac{1}{2\pi\sqrt{LC}}$$

$$100 = \frac{1}{2\pi\sqrt{LC}}$$

$$LC = 2.53 \times 10^{-6} \quad \text{... (i)}$$

Power factor at 70 Hz is 0.707.

$$\frac{R}{Z} = 0.707$$

$$Z = \frac{100}{0.707} = 141.44$$

$$\text{Impedance at 70 Hz} = \sqrt{R^2 + (X_C - X_L)^2}$$

$$141.44 = \sqrt{(100)^2 + \left(\frac{1}{2\pi \times 70 \times C} - 2\pi \times 70 \times L\right)^2}$$

$$\frac{2.27 \times 10^{-3}}{C} - 439.82\ L = 100.02 \quad \text{... (ii)}$$

Solving Eqs (i) and (ii),

$$L = 0.2187 \text{ H}$$

$$C = 11.58\ \mu\text{F}$$

Example 4.87 *A constant voltage at a frequency of 1 MHz is applied to an inductor in series with a variable capacitor. When the capacitor is set to 500 pF, the current has its maximum value while it is reduced to one-half when the capacitance is 600 pF. Find resistance, inductance and Q-factor of inductor.*

Solution

Data $f_0 = 1$ MHz

$C_1 = 500$ pF

$C_2 = 600$ pF

At resonance $C = 500 \text{ pF} = 500 \times 10^{-12}$ F

$$f_0 = \frac{1}{2\pi\sqrt{LC}}$$

$$10^6 = \frac{1}{2\pi\sqrt{L \times 500 \times 10^{-12}}}$$

$$L = 0.05 \text{ mH}$$

$$X_L = 2\pi f_0 L = 2\pi \times 10^6 \times 0.05 \times 10^{-3} = 314.16\ \Omega$$

When capacitance is 600 pF, the current reduces to one-half of the current at resonance,

$$X_C = \frac{1}{2\pi fC}$$

$$= \frac{1}{2\pi \times 10^6 \times 600 \times 10^{-12}} = 265.26\ \Omega$$

$$I = \frac{1}{2} I_0$$

$$\frac{V}{Z} = \frac{1}{2}\frac{V}{R}$$

$$Z = 2R$$

$$\sqrt{R^2 + (X_L - X_C)^2} = 2R$$

$$R^2 + (314.16 - 265.26)^2 = 4R^2$$

$$3R^2 = 2391.21$$

Resistance of inductor, $R = 28.23\ \Omega$

Quality factor $Q_0 = \frac{1}{R}\sqrt{\frac{L}{C}}$

$$= \frac{1}{28.23}\sqrt{\frac{0.05 \times 10^{-3}}{500 \times 10^{-12}}} = 11.2$$

4.15 PARALLEL RESONANCE

Consider a parallel circuit consisting of a coil and a capacitor as shown in Fig. 4.82. The impedance of two branches are

$$\bar{Z}_1 = R + jX_L$$

$$\bar{Z}_2 = -jX_C$$

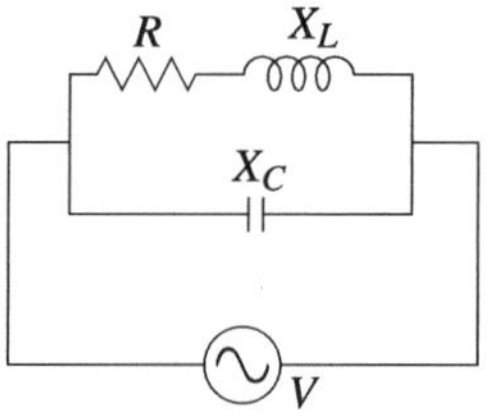

Fig. 4.82

$$\bar{Y}_1 = \frac{1}{\bar{Z}_1} = \frac{1}{R + jX_L}$$

$$= \frac{R - jX_L}{R^2 + X_L^2}$$

$$\bar{Y}_2 = \frac{1}{\bar{Z}_2} = \frac{1}{-jX_C} = \frac{j}{X_C}$$

Admittance of the circuit $\bar{Y} = \bar{Y}_1 + \bar{Y}_2$

$$= \frac{R - jX_L}{R^2 + X_L^2} + j\frac{1}{X_C}$$

$$= \frac{R}{R^2 + X_L^2} - j\left(\frac{X_L}{R^2 + X_L^2} - \frac{1}{X_C}\right)$$

At resonance, the circuit is purely resistive. Therefore the condition for resonance is

$$\frac{X_L}{R^2 + X_L^2} - \frac{1}{X_C} = 0$$

$$\frac{X_L}{R^2 + X_L^2} = \frac{1}{X_C}$$

$$X_L \, . \, X_C = R^2 + X_L^2$$

$$\omega_0 L \, . \, \frac{1}{\omega_0 C} = R^2 + \omega_0^2 L^2$$

$$\omega_0^2 L^2 = \frac{L}{C} - R^2$$

$$\omega_0^2 = \frac{1}{L^2}\left(\frac{L}{C} - R^2\right) = \frac{1}{LC} - \frac{R^2}{L^2}$$

$$\omega_0 = \sqrt{\frac{1}{LC} - \frac{R^2}{L^2}}$$

$$f_0 = \frac{1}{2\pi}\sqrt{\frac{1}{LC} - \frac{R^2}{L^2}}$$

where f_0 is called the resonant frequency of the circuit.

If R is very small as compared to L, then

$$f_0 = \frac{1}{2\pi\sqrt{LC}}$$

Dynamic Impedance of Parallel Circuit At resonance, the circuit is purely resistive. The real part of admittance is $\frac{R}{R^2 \times X_L^2}$. Hence, the dynamic impedance at resonance is given by

$$Z = \frac{R^2 + X_L^2}{R}$$

At resonance,

$$R^2 + X_L^2 = X_L \, . \, X_C = \frac{L}{C}$$

$$Z = \frac{L}{CR}$$

Table 4.1 *Comparison of Series and Parallel Resonant Circuits*

Parameter	*Series Circuit*	*Parallel Circuit*
Current at resonance	$I = \frac{V}{R}$ and is maximum	$I = \frac{V}{(L/CR)}$ and is minimum
Impedance at resonance	$Z = R$ and is minimum	$Z = \frac{L}{CR}$ and is maximum
Power factor at resonance	Unity	Unity
Resonant frequency	$f_0 = \frac{1}{2\pi\sqrt{LC}}$	$f_0 = \frac{1}{2\pi}\sqrt{\frac{1}{LC} - \frac{R^2}{L^2}}$
Q-factor	$Q = \frac{2\pi fL}{R}$	$Q = \frac{2\pi fL}{R}$
It magnifies	Voltage across L and C	Current through L and C

Example 4.88 *A coil of 20-Ω resistance has an inductance of 0.2 H and is connected in parallel with a condenser of 100 μF capacitance. Calculate the frequency at which this circuit will behave as a non-inductive resistance. Find also the value of dynamic resistance.*

Solution

Data $R = 20\ \Omega$ $L = 0.2$ H

$C = 100\ \mu F$

$$f_0 = \frac{1}{2\pi}\sqrt{\frac{1}{LC} - \frac{R^2}{L^2}}$$

$$= \frac{1}{2\pi}\sqrt{\frac{1}{0.2 \times 100 \times 10^{-6}} - \left(\frac{20}{0.2}\right)^2} = 31.83 \text{ Hz}$$

Dynamic resistance $= \dfrac{L}{CR}$

$$= \frac{0.2}{100 \times 10^{-6} \times 20} = 100\ \Omega$$

Example 4.89 *A coil having a resistance of 20 Ω and an inductance of 200 μH is connected in parallel with a variable capacitor. This parallel combination is connected in series with a resistance of 8000 Ω. A voltage of 230 V at a frequency of 10^6 Hz is applied across the circuit. Calculate (a) the value of capacitance at resosnance, (b) Q factor of the circuit, (c) dynamic impedance of the cirucit, and (d) total circuit current.*

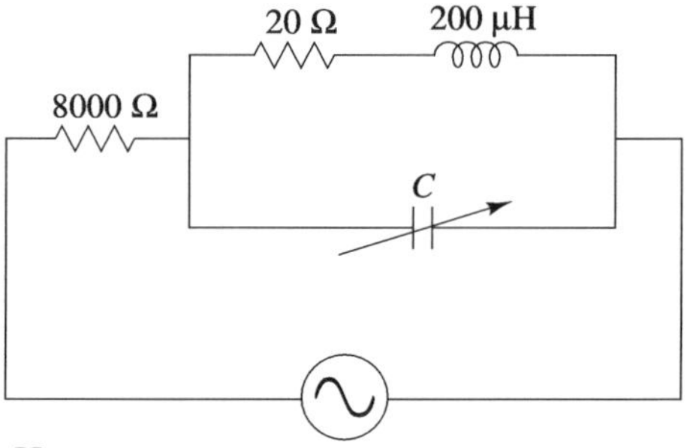

Fig. 4.83

Solution

Data $R = 20\ \Omega$ $L = 200\ \mu H$

$f = 10^6$ Hz $V = 230$ V

$R_S = 8000\ \Omega$

$X_L = 2\pi f L = 2 \times \pi \times 10^6 \times 200 \times 10^{-6} = 1256.6\ \Omega$

$$f_0 = \frac{1}{2\pi}\sqrt{\frac{1}{LC} - \frac{R^2}{L^2}}$$

$$10^6 = \frac{1}{2\pi}\sqrt{\frac{1}{200 \times 10^{-6} \times C} - \frac{(20)^2}{(200 \times 10^{-6})^2}}$$

$C = 126.65 \times 10^{-12}$ F $= 126.65$ pF

Quality Factor $Q_0 = \dfrac{2\pi fL}{R}$

$$= \frac{2\pi \times 10^6 \times 200 \times 10^{-6}}{20} = 62.83$$

Dynamic Impedance $Z = \dfrac{L}{CR}$

$$= \frac{200 \times 10^{-6}}{126.65 \times 10^{-12} \times 20} = 78958\ \Omega$$

Total equivalent impedance of the circuit at resonance

$$= 78958 + 8000 = 86958\ \Omega$$

$$\text{Total circuit current} = \frac{230}{86958}$$

$$= 2.645 \times 10^{-3}\ \text{A}$$

$$= 2.65\ \text{mA}$$

Exercises

1. Find the average value and rms value of the following waveforms:

(i)

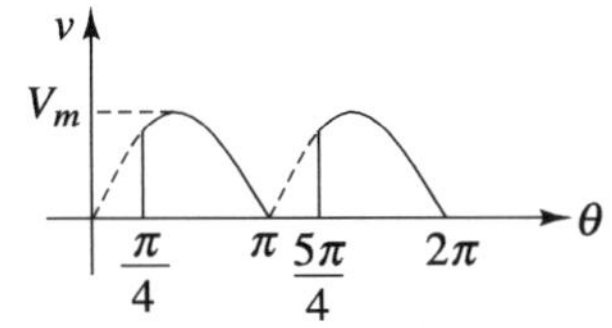

Fig. 4.84

[0.543 V_m, 0.674 V_m]

(ii)

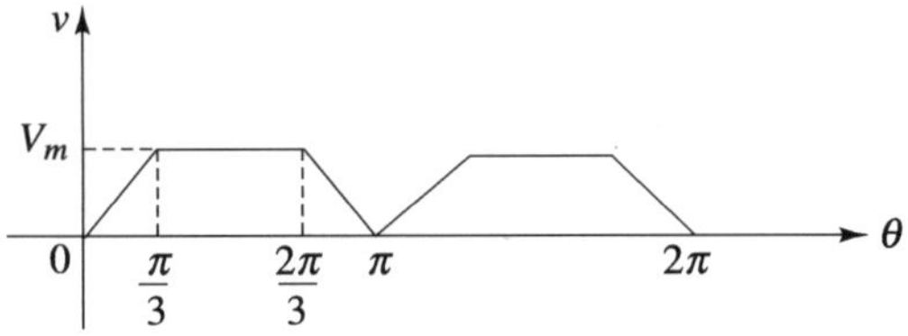

Fig. 4.85

[0.67 V_m, 0.745 V_m]

(iii)

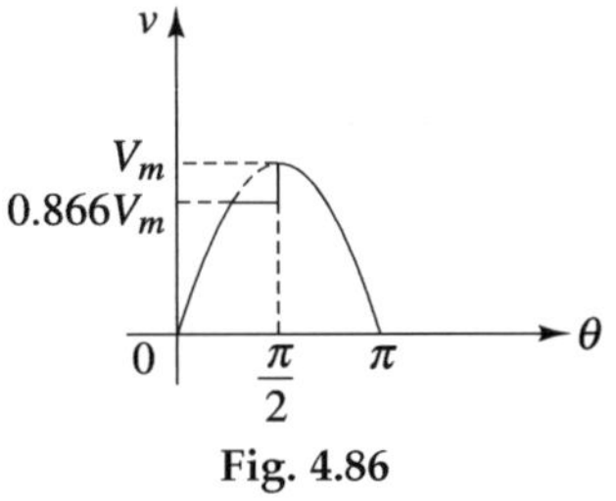

Fig. 4.86

[0.622 V_m, 0.687 V_m]

(iv)

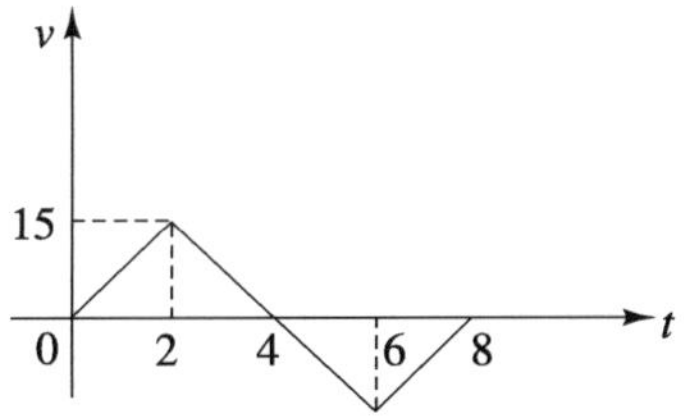

Fig. 4.87

[7.5 V, 8.66 V]

2. Find the rms value of resultant current in a circuit which simultaneously carries a direct current of 5 A and a sinusoidal alternating current with a peak value of 10 A. [8.66 A]
3. Find the resultant of the following:

$$e_1 = 25 \sin \omega t$$
$$e_2 = 10 \sin (\omega t + \pi/6)$$
$$e_3 = 30 \cos \omega t$$
$$e_4 = 20 \sin (\omega t - \pi/4)$$

Draw all phasors. [52.14 sin (ωt + 23.57°)]
4. A current of 5 A flows through a non-inductive resistance in series with a choking coil when supplied at 250 V, 50 Hz. If the voltage drop across the coil and fixed resistor are 200 V and 125 V respectively, calculate the resistance and inductance of the coil, value of fixed resistor and power drawn by the coil. Draw the phasor diagram. [5.5 Ω, 0.126 H, 25 Ω, 137.5 W]
5. For Fig. 4.88 shown, find *R* and *L*.

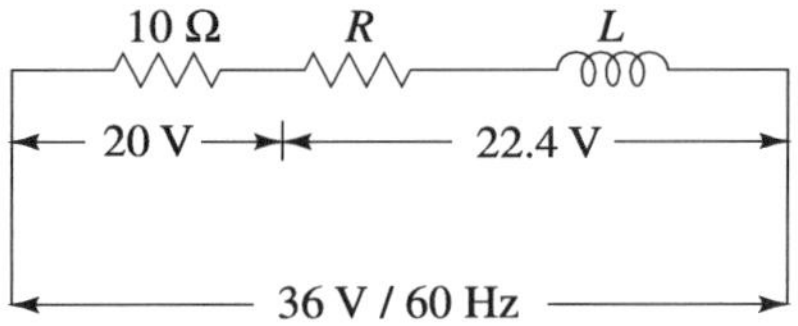

Fig. 4.88

[4.928 Ω, 0.0266 H]
6. Two coils *A* and *B* are connected in series across a 200-V, 50-Hz ac supply. The power input to the circuit is 2 kW and 1.15 kVAR. If the resistance and reactance of the coil *A* are 5 Ω and 8 Ω respectively, calculate the resistance and reactance of the coil *B*. Also, calculate the power consumed by coils *A* and *B*. [10.03 Ω, 0.64 Ω, 665.3 W, 1334.7 W]
7. Two impedances 10 ∠30° Ω and 20 ∠–45° Ω are connected in series. Calculate the power factor of the series combination. [0.9281 (lagging)]
8. A capacitive load takes 10 kVA and 5 kVAR, when connected to 200 V, 50 Hz ac supply. Calculate (i) resistance, (ii) capacitance, (iii) active power, and (iv) pf. [3.464 Ω, 1.59×10^{-3} F, 8.66 kW, 0.866 (leading)]
9. Two impedances Z_1 and Z_2 are connected in series across a 230-V, 50-Hz ac supply. The total current drawn by a series combination is 2.3 A. The pf of Z_1 is 0.8 lagging. The voltage drop across Z_1 is twice the voltage drop across Z_2 and it is 90° out of phase with it. Determine the value of Z_2. [44.719 Ω]
10. In Fig 4.89., find the applied voltage, the frequency and loss in the iron-cored inductor *L*.

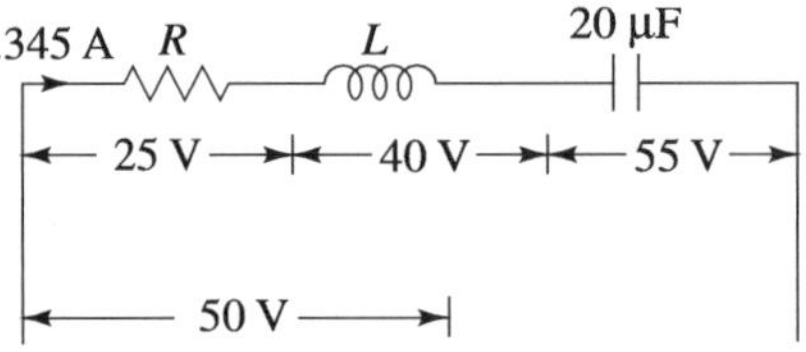

Fig. 4.89

[34.2 ∠–26.76° V, 50 Hz, 1.9 W]
11. A non-inductive 10-ohm resistor is in series with a coil of 1.3-ohm resistance and 0.018-H inductance. If a voltage of maximum value of 100 V at a frequency of 100 Hz is applied to this circuit, what will be the voltage across the resistor? [62.54 V]
12. A load consisting of a capacitor in series with a resistor has an impedance of 50 Ω and a pf of 0.707 leading. The load is connected in series with a 40-Ω resistor across ac supply and the resulting current is 3 A. Determine the supply voltage and the overall phase angle. [249.69 V, 25.13°]

13. Two impedances $\bar{Z}_1 = 10 - j15$ ohms and $\bar{Z}_2 = 4 + j8$ ohms are connected in parallel. The supply voltage is 100 V, 25 Hz. Calculate (i) the admittance, conductance and susceptance of the combined circuit, and (ii) total current drawn and pf. [0.097 ℧, 0.081 ℧, 0.054 ℧, 9.7 A, 0.83 (lagging)]

14. A voltage of 200 ∠53.13° V is applied across two impedances in parallel. The values of the impedances are (12 + *j*16) Ω and (10 – *j*20) Ω. Determine kVA, kVAR and kW in each branch and pf of whole circuit. [2 kVA, 1.2 kW, 1.6 kVAR, 1.788 kVA, 0.8 kW, 1.6 kVAR, unity pf]

15. For the circuit shown, evaluate the current through and voltage across each element.

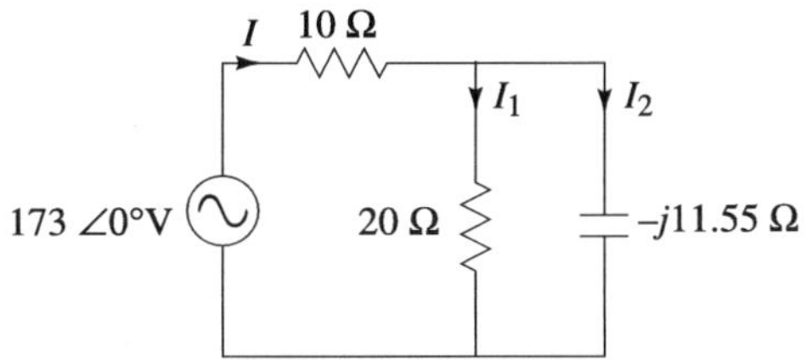

Fig. 4.90

[$\bar{I}_1 = 5\angle{-30°}$ A, $\bar{I}_2 = 8.66\angle 60°$ A, $\bar{V}_1 = 100\angle{-30°}$ V, $\bar{V}_2 = 100\angle{-30°}$ V, $\bar{V} = 100\angle 30°$ V]

16. Two impedances Z_1 and Z_2 are connected in parallel. The first branch takes a leading current of 16 A and has a resistance of 5 Ω, while the second branch takes a lagging current at pf 0.8. The total power supplied is 5 kW, the applied voltage being (100 + *j*200) V. Determine branch currents and total current. [16 ∠132.46° A, 20.8 ∠26.57° A, 22.49 A]

17. For the parallel branch shown, find the value of R_2 when the overall power factor is 0.92 lag.

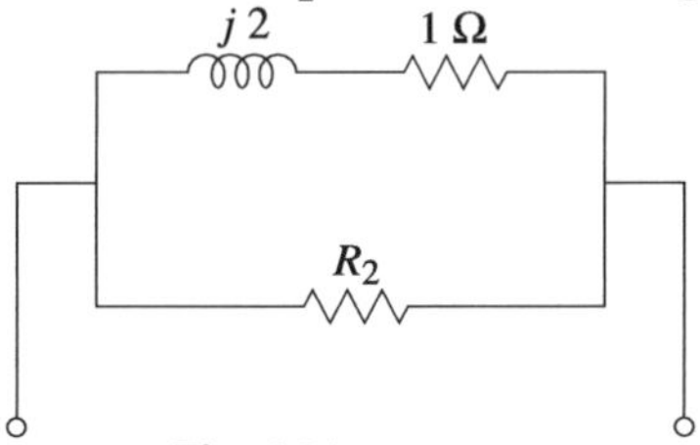

Fig. 4.91

[1.35 Ω]

18. In a series–parallel circuit, two parallel branches *A* and *B* are in series with *C*. The impedances are $Z_A = (10 + j8)$ Ω, $Z_B = (9 - j6)$ Ω and $Z_C = (3 + j2)$ Ω. If the voltage across Z_C is 100 ∠0° V, determine the values I_A and I_B. [15.7 ∠–73.39° A, 18.59 ∠–1.04° A]

19. Find the total impedance, supply current and pf of the entire circuit.

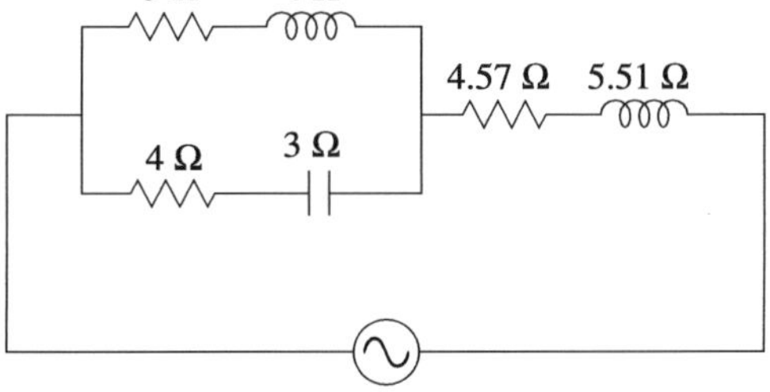

Fig. 4.92

[10.06 ∠36.68° Ω, 19.88 ∠–36.68° A, 0.801 lagging]

20. The power dissipated in the coil *A* is 300 W and in the coil *B*, is 400 W. Each coil takes a current of 5 A when connected to a 110-V, 50-Hz supply. Find the current drawn when the coils are connected in parallel. [9.93 ∠–50.11° A]

21. A capacitor is placed in parallel with two inductive loads. The current through the first inductor is 20 A at 30° lag and the current through the second is 40 A at 60° lag. What must be the current in the capacitor so that the currnet in the external circuit is of unity power factor. [44.64 ∠90° A]

22. A circuit of resistance 15 Ω and inductive reactance 12 Ω is connected in parallel with another circuit consisting of a resistor of 25 Ω in series with a capacitive reactance of 17 Ω. This combination is energised from a 200-V, 40-Hz mains. Find the branch currents, total current and power factor of the circuit. It is desired to raise the power factor of this circuit to unity by connecting a capacitor in parallel. Determine the value of the capacitance of the capacitor.
[10.42 ∠–38.56° A, 6.61 ∠34.21° A, 13.95 ∠–11.56° A, 0.98 lagging 54.9 μF]

23. A resistor of 30 Ω and a capacitor of unknown value are connected in parallel across a 110-V, 50-Hz supply. The combination draws a current of 5 A from the supply. Find the value of the unknown capacitance of the capacitor. This combination is again connected across a 110-V supply of unknown frequency. It is observed that total current drawn from the mains falls to 4 A. Determine the frequency of the supply. [98.58 μF, 23.68 Hz]

24. Two reactive circuits have an impedance of 20 Ω each. One of them has a lagging power factor of 0.8 and other has a leading power factor of 0.6. Find (a) voltage necessary to send a current of 10 A through the two in series, and (b) current drawn from 200 V supply if the two are connected in parallel. Draw the phasor diagram in each case. [282.8 V, 14.14 A]

25. Inductive loads of 0.8 kW and 1.2 kW at lagging power factors of 0.8 and 0.6 respectively are connected across a 200-V, 50-Hz supply. Find the total current, power factor and the value of capacitor to be put in parallel to both to raise the overall power factor of 0.9 lagging. [14.87 A, 0.673 lagging, 98 μF]

Objective-Type Questions

1. In a series *R-L-C* high *Q* circuit, the current peaks at a frequency
(a) equal to the resonant frequency (b) greater than the resonant frequency
(c) less than the resonant frequency (d) none of the above

2. In Fig. 4.93 A_1, A_2 and A_3 are ideal ammeters. If A_1 reads 5 A, A_2 reads 12 A, then A_3 should read
(a) 7 A (b) 12 A
(c) 13 A (d) 17 A

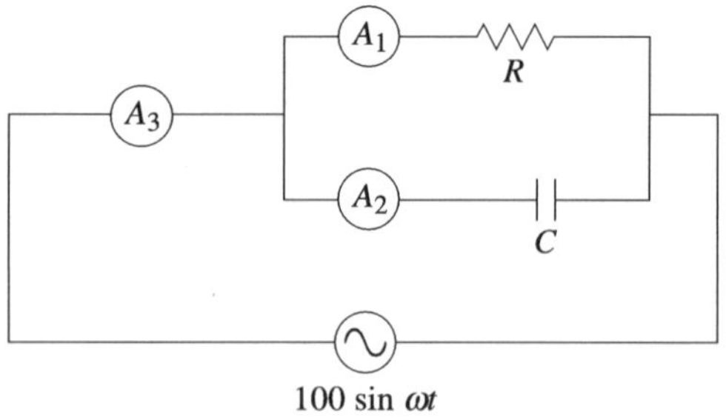

Fig. 4.93

3. A series *R-L-C* circuit consisting of $R = 10\ \Omega$, $X_L = 20\ \Omega$ and $X_C = 20\ \Omega$ is connected across an ac supply of 200 V rms. The rms voltage across the capacitor is
(a) 200 ∠–90° V (b) 200 ∠90° V (c) 400 ∠–90° V (d) 400 ∠90° V

4. The current $i(t)$ through a 10-Ω resistor in series with an inductance is given by

$$i(t) = 3 + 4 \sin(100t + 45°) + 4 \sin(300t + 60°)$$

The rms value of the current and the power dissipated in the circuit are

(a) $\sqrt{41}$ A_1, 410 W (b) $\sqrt{35}$ A, 350 W (c) 5 A, 250 W (d) 11 A, 1210 W

5. A series *R-L-C* circuit has a Q of 100 and an impedance of $(100 + j0)$ Ω at its resonant angular frequency of 10^7 rad/sec. The values of R and L are

(a) 100 Ω, 10^3 H (b) 100 Ω, 10^{-3} H (c) 10 Ω, 10 H (d) 10 Ω, 0.1 H

6. The parallel *R-L-C* circuit shown in Fig. 4.94 is in resonance. In this circuit

(a) $|I_R| < 1$ mA (b) $|I_R + I_L| > 1$ mA
(c) $|I_R + I_C| < 1$ mA (d) $|I_R + I_C| > 1$ mA

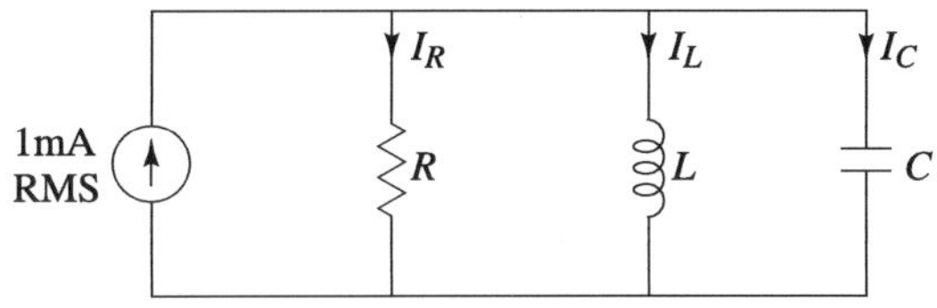

Fig. 4.94

7. A series *R-L-C* circuit has a resonance frequency of 1 kHz and a qualify factor Q of 100. If each of R, L and C is doubled from its original value, the new Q of the circuit is

(a) 25 (b) 50 (c) 100 (d) 200

8. An input voltage $v(t) = 10\sqrt{2}\cos(t + 10°) + 10\sqrt{3}\cos(2t + 10°)$ is applied to a series combination of $R = 1$ Ω and $L = 1$ H. The resulting steady state current $i(t)$ in ampere is

(a) $10\cos(t + 55°) + 10\cos(2t + 10° + \tan^{-1} 2)$ (b) $10\cos(t + 55°) + 10\sqrt{\frac{3}{2}}\cos(2t + 55°)$

(c) $10\cos(t - 35°) + 10\cos(2t + 10° - \tan^{-1} 2)$ (d) $10\cos(t - 35°) + 10\sqrt{\frac{3}{2}}\cos(2t - 35°)$

9. The circuit shown in Fig. 4.95 with $R = \frac{1}{3}\Omega$, $L = \frac{1}{4}\text{H}$, $C = 3$ F has input voltage $v(t) = \sin 2t$.

The resulting current $i(t)$ is

(a) $5\sin(2t + 53.1°)$ (b) $5\sin(2t - 53.1°)$
(c) $25\sin(2t + 53.1°)$ (d) $25\sin(2t - 53.1°)$

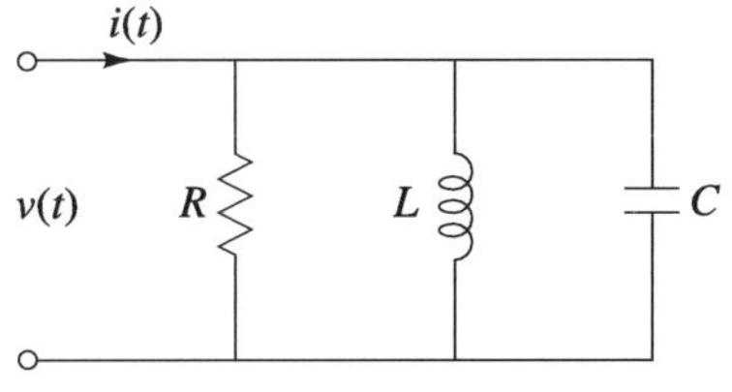

Fig. 4.95

10. In a series *R-L-C* circuit, $R = 2$ kΩ, $L = 1$ H, $C = \frac{1}{400}\mu\text{F}$. The resonant frequency is

(a) 2×10^4 Hz (b) $\frac{1}{\pi} \times 10^4$ Hz (c) 10^4 Hz (d) $2\pi \times 10^4$ Hz

11. For a series resonant circuit at low frequency, circuit impedance is __________ and at high frequency circuit impedance is __________ .

(a) capacitive, inductive (b) inductive, capacitive
(c) resistive, inductive (d) capacitive, resistive

12. In Fig. 4.96, A_1, A_2 and A_3 are ideal ammeters. If A_1 and A_3 read 5 A and 13 A respectively, the reading of A_2 will be
(a) 8 A (b) 12 A
(c) 18 A (d) 10 A

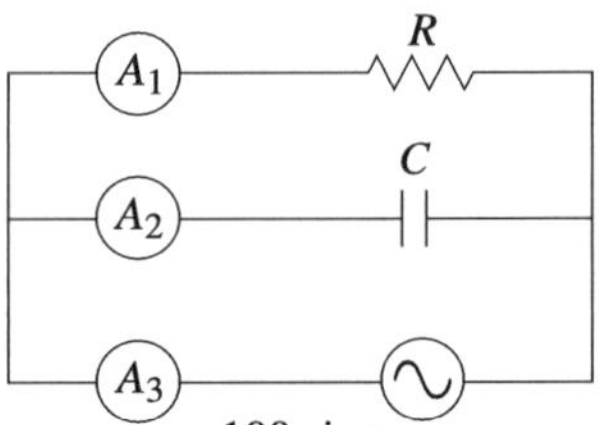

Fig. 4.96

13. In Fig. 4.97 $i(t)$ under steady state is
(a) zero (b) 5
(c) 7.07 sin t (d) 7.07 sin (t – 45°)

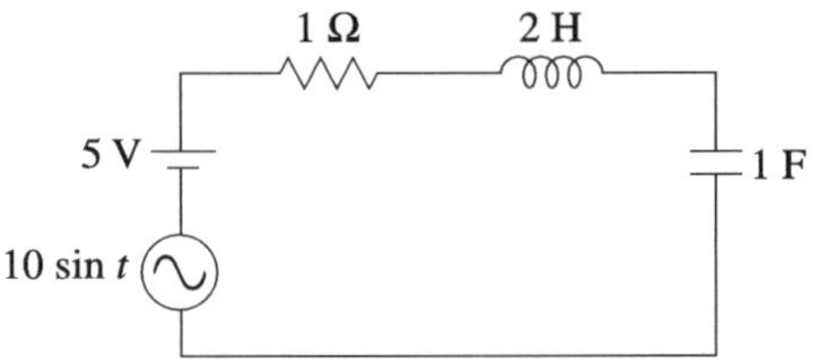

Fig. 4.97

14. The following circuit in Fig. 4.98 resonates at
(a) all frequencies (b) 0.5 rad/s
(b) 0.5 rad/s (d) 1 rad/s

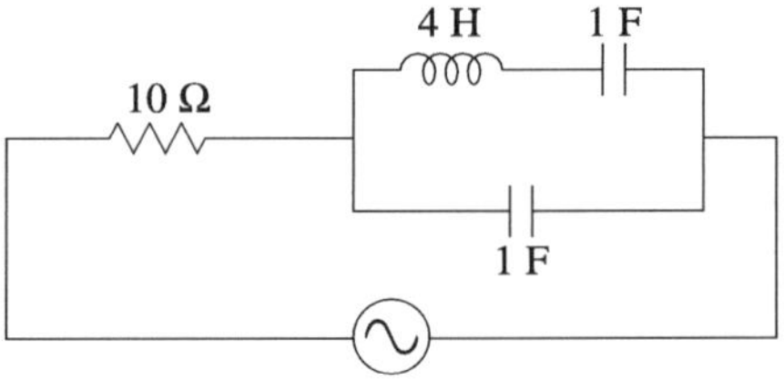

Fig. 4.98

15. At resonance, the parallel circuit shown in Fig. 4.99 behaves like
(a) an open circuit
(b) a short circuit
(c) a pure resistor of value R
(d) a pure resistor of value much higher than R

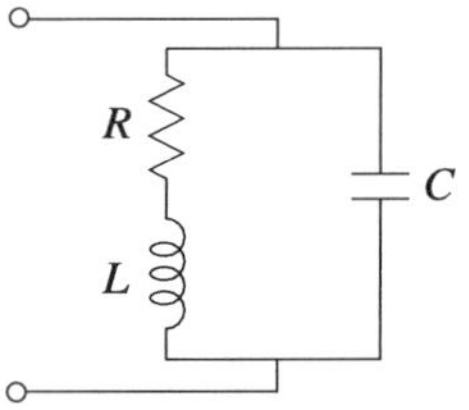

Fig. 4.99

16. A circuit with a resistor, inductor and capacitor in series is resonant of fo Hz. If all the component values are now doubled, the new resonant frequency is
(a) $2f_o$ (b) f_o (c) $f_o/4$ (d) $f_o/2$

17. In a series R-L-C circuit at resonance, the magnitude of the voltage developed across the capacitor
(a) is always zero.
(b) can never be greater than the input voltage.
(c) can be greater than the input voltage, however it is 90° out of phase with the input voltage.
(d) can be greater than the input voltage and is in phase with the input voltage.

18. A 240-V, single-phase ac source is connected to a load with an impedance of $10\angle 60°\ \Omega$. A capacitor is connected in parallel with the load. If the capacitor supplies 1250 VAR, the real power supplied by the source is
(a) 3600 W (b) 2880 W (c) 2400 W (d) 1200 W

19. The power in a series *R-L-C* circuit will be half of that at resonance when the magnitude of the current is equal to

(a) $\dfrac{V}{2R}$ (b) $\dfrac{V}{\sqrt{3}R}$ (c) $\dfrac{V}{\sqrt{2}R}$ (d) $\dfrac{\sqrt{2}\,V}{R}$

20. If a network has an impedance of $(1 - j)$ at a specific frequency, the circuit would consists of series
1. *R* and *C* 2. *R* and *L* 3. *R*, *L* and *C*
Which of these statements are correct?
(a) 1 and 2 (b) 1 and 3 (c) 1, 2 and 3 (d) 2 and 3

Answers to Objective-Type Questions

1. (a)	**2.** (c)	**3.** (c)	**4.** (c)	**5.** (b)	**6.** (c)
7. (b)	**8.** (c)	**9.** (a)	**10.** (b)	**11.** (a)	**12.** (b)
13. (d)	**14.** (b)	**15.** (d)	**16.** (d)	**17.** (c)	**18.** (b)
19. (c)	**20.** (b)				

5 Steady-State AC Analysis

5.1 INTRODUCTION

We have discussed the network theorems with reference to resistive load and dc sources. Now, all the theorems will be discussed when a network consists of ac sources, resistors, inductors and capacitors. All the theorems are also valid for ac sources.

5.2 MESH ANALYSIS

Mesh analysis is useful if a network has a large number of voltage sources. In this method, currents are assigned in each mesh. We can write mesh equations by Kirchhoff's voltage law in terms of unknown mesh currents,

Example 5.1 *Find mesh currents I_1 and I_2 in the network of Fig. 5.1.*

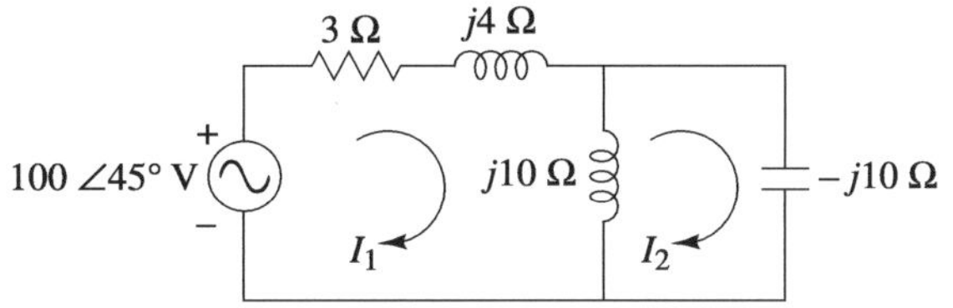

Fig. 5.1

Solution Applying KVL to Mesh 1,

$$100 \angle 45^\circ - (3 + j4)\, I_1 - j10(I_1 - I_2) = 0$$

$$(3 + j14)\, I_1 - j10 I_2 = 100 \angle 45^\circ \qquad \text{...(i)}$$

Applying KVL to Mesh 2,

$$-j10(I_2 - I_1) + j10(I_2) = 0$$

$$j10 I_1 = 0 \qquad \text{...(ii)}$$

$$I_1 = 0$$

Substituting I_1 in Eq. (i), we get

$$-j10 I_2 = 100 \angle 45^\circ$$

$$I_2 = \frac{100\angle 45^\circ}{-j10} = \frac{100\angle 45^\circ}{10\angle -90^\circ} = 10 \angle 135^\circ \text{ A}$$

Example 5.2 *Find mesh current I_1, I_2 and I_3 in the network of Fig. 5.2.*

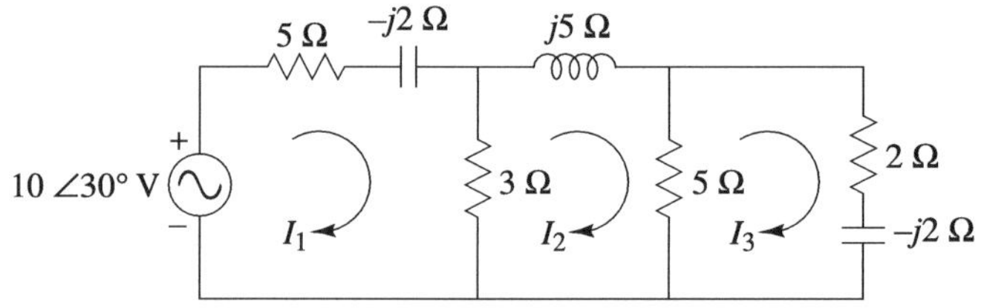

Fig. 5.2

Solution Applying KVL to Mesh 1,

$$10 \angle 30^\circ - (5 - j2) I_1 - 3 (I_1 - I_2) = 0$$

$$(8 - j2) I_1 - 3I_2 = 10 \angle 30^\circ \qquad \text{... (i)}$$

Applying KVL to Mesh 2,

$$-3(I_2 - I_1) - j5I_2 - 5(I_2 - I_3) = 0$$

$$-3I_1 + (8 + j5)I_2 - 5I_3 = 0 \qquad \text{... (ii)}$$

Applying KVL to Mesh 3,

$$-5(I_3 - I_2) - (2 - j2) I_3 = 0$$

$$-5I_2 + (7 - j2) I_3 = 0 \qquad \text{... (iii)}$$

Writing these equations in matrix form,

$$\begin{bmatrix} 8 - j2 & -3 & 0 \\ -3 & 8 + j5 & -5 \\ 0 & -5 & 7 - j2 \end{bmatrix} \begin{bmatrix} I_1 \\ I_2 \\ I_3 \end{bmatrix} = \begin{bmatrix} 10\angle 30^\circ \\ 0 \\ 0 \end{bmatrix}$$

By Cramer's rule,

$$I_1 = \frac{\begin{vmatrix} 10\angle 30^\circ & -3 & 0 \\ -3 & 8 + j5 & -5 \\ 0 & -5 & 7 - j2 \end{vmatrix}}{\begin{vmatrix} 8 - j2 & -3 & 0 \\ -3 & 8 + j5 & -5 \\ 0 & -5 & 7 - j2 \end{vmatrix}} = 1.43 \angle 38.7^\circ \text{ A}$$

$$I_2 = \frac{\begin{vmatrix} 8-j2 & 10\angle 30^\circ & 0 \\ -3 & 0 & -5 \\ 0 & 0 & 7-j2 \end{vmatrix}}{\begin{vmatrix} 8-j2 & -3 & 0 \\ -3 & 8+j5 & -5 \\ 0 & -5 & 7-j2 \end{vmatrix}} = 0.693 \angle -2.2^\circ \text{ A}$$

$$I_3 = \frac{\begin{vmatrix} 8-j2 & -3 & 10\angle 30^\circ \\ -3 & 8+j5 & 0 \\ 0 & -5 & 0 \end{vmatrix}}{\begin{vmatrix} 8-j2 & -3 & 0 \\ -3 & 8+j5 & -5 \\ 0 & -5 & 7-j2 \end{vmatrix}} = 0.476 \angle 13.8^\circ \text{ A}$$

Example 5.3 *In the network of Fig. 5.3, find the value of V_2 so that the current through (2 + j3) ohm impedance is zero.*

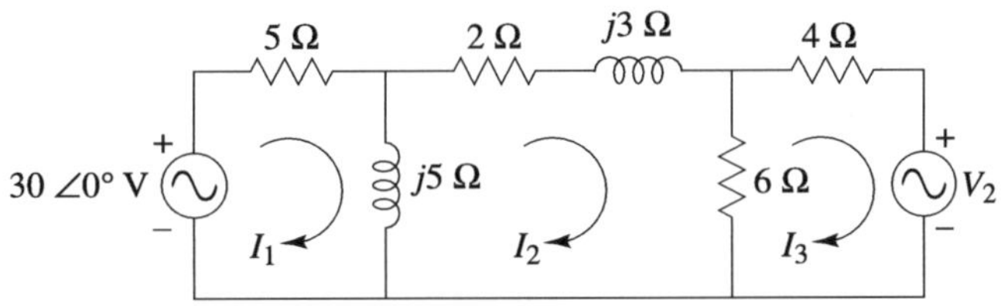

Fig. 5.3

Solution Applying KVL to Mesh 1,

$$30 \angle 0^\circ - 5I_1 - j5(I_1 - I_2) = 0$$

$$(5 + j5)\, I_1 - j5I_2 = 30 \angle 0^\circ \qquad \text{...(i)}$$

Applying KVL to Mesh 2,

$$-j5(I_2 - I_1) - (2 + j3)\, I_2 - 6(I_2 - I_3) = 0$$

$$-j5I_1 + (8 + j8)\, I_2 - 6I_3 = 0 \qquad \text{...(ii)}$$

Applying KVL to Mesh 3,

$$-6(I_3 - I_2) - 4I_3 - V_2 = 0$$

$$-6I_2 + 10I_3 = -V_2 \qquad \text{...(iii)}$$

Writing equations in matrix form,

$$\begin{bmatrix} 5+j5 & -j5 & 0 \\ -j5 & 8+j8 & -6 \\ 0 & -6 & 10 \end{bmatrix} \begin{bmatrix} I_1 \\ I_2 \\ I_3 \end{bmatrix} = \begin{bmatrix} 30\angle 0^\circ \\ 0 \\ -V_2 \end{bmatrix}$$

By Cramer's rule,

$$I_2 = \frac{\begin{vmatrix} 5+j5 & 30\angle 0^\circ & 0 \\ -j5 & 0 & -6 \\ 0 & -V_2 & 10 \end{vmatrix}}{\begin{vmatrix} 5+j5 & -j5 & 0 \\ -j5 & 8+j8 & -6 \\ 0 & -6 & 10 \end{vmatrix}} = 0$$

$$(5 + j5)(-6V_2) - (30)(-j50) = 0$$

$$V_2 = \frac{j1500}{30 + j30} = 35.36 \angle 45° \text{ V}$$

Example 5.4 *Find the value of the current I_3 in the network shown in Fig. 5.4.*

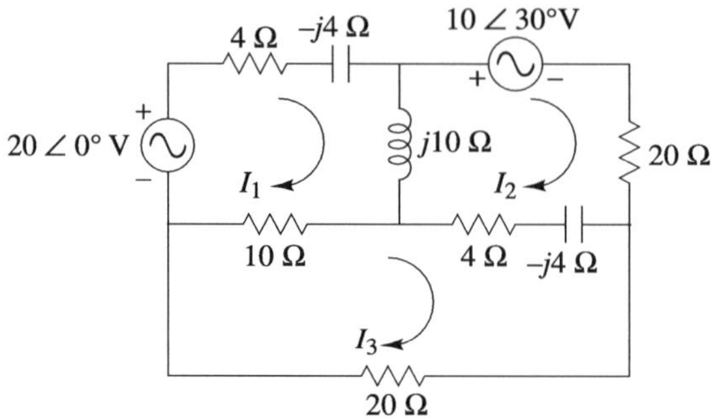

Fig. 5.4

Solution Applying KVL to Mesh 1,

$$20 \angle 0° - (4 - j4) I_1 - j10 (I_1 - I_2) - 10 (I_1 - I_3) = 0$$

$$(14 + j6) I_1 - j10 I_2 - 10I_3 = 20 \angle 0^{\circ} \qquad \text{... (i)}$$

Applying KVL to Mesh 2,

$$-j10 (I_2 - I_1) - 10 \angle 30^{\circ} - 20I_2 - (4 - j4)(I_2 - I_3) = 0$$

$$-j10I_1 + (24 + j6) i_2 - (4 - j4) I_3 = -10 \angle 30^{\circ} \qquad \text{... (ii)}$$

Applying KVL to Mesh 3,

$$-10(I_3 - I_1) - (4 - j4)(I_3 - I_2) - 20I_3 = 0$$

$$-10I_1 - (4 - j4) I_2 + (34 - j4) I_3 = 0 \qquad \text{... (iii)}$$

Writing equations in matrix form,

$$\begin{bmatrix} 14 + j6 & -j10 & -10 \\ -j10 & 24 + j6 & -(4 - j4) \\ -10 & -(4 - j4) & (34 - j4) \end{bmatrix} \begin{bmatrix} I_1 \\ I_2 \\ I_3 \end{bmatrix} = \begin{bmatrix} 20\angle 0° \\ -10\angle 30° \\ 0 \end{bmatrix}$$

By Cramer's rule,

$$I_3 = \frac{\begin{vmatrix} 14 + j6 & -j10 & 20\angle 0° \\ -j10 & 24 + j6 & -10\angle 30° \\ -10 & -(4 - j4) & 0 \end{vmatrix}}{\begin{vmatrix} 14 + j6 & -10 & -10 \\ -j10 & 24 + j6 & -(4 - j4) \\ -10 & -(4 - j4) & 34 - j4 \end{vmatrix}} = 0.44 \angle -14° \text{ A}$$

Example 5.5 *Find voltage V_{AB} in the network of Fig. 5.5.*

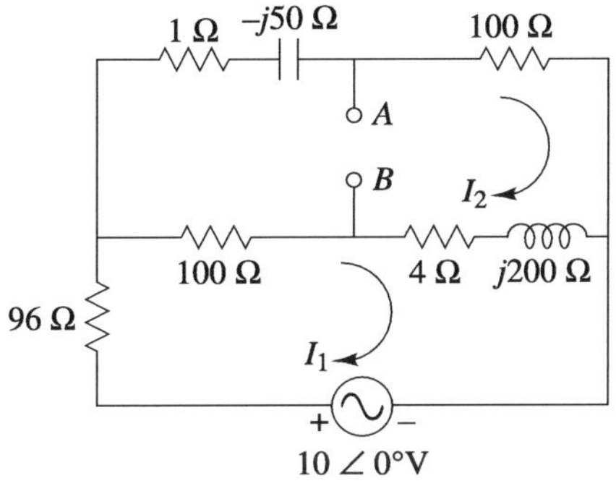

Fig. 5.5

Solution Applying KVL to Mesh 1,

$$-96I_1 - (100 + 4 + j200)(I_1 - I_2) + 10\angle 0° = 0$$

$$(200 + j200)\, I_1 - (104 + j200)\, I_2 = 10\angle 0° \quad \ldots\text{(i)}$$

Applying KVL to Mesh 2,

$$-(1 - j50 - 100)\, I_2 - (100 + 4 + j200)(I_2 - I_1) = 0$$

$$-(104 + j200)\, I_1 + (205 + j150)\, I_2 = 0 \quad \ldots\text{(ii)}$$

Writing equations in matrix form,

$$\begin{bmatrix} 200 + j200 & -(104 + j200) \\ -(104 + j200) & 205 + j150 \end{bmatrix} \begin{bmatrix} I_1 \\ I_2 \end{bmatrix} = \begin{bmatrix} 10\angle 0° \\ 0 \end{bmatrix}$$

By Cramer's rule,

$$I_1 = \frac{\begin{vmatrix} 10\angle 0° & -(104 + j200) \\ 0 & 205 + j150 \end{vmatrix}}{\begin{vmatrix} 200 + j200 & -(104 + j200) \\ -(104 + j200) & 205 + j150 \end{vmatrix}} = 5.05 \times 10^{-2} \angle -0.074° \text{ A}$$

$$I_2 = \frac{\begin{vmatrix} 200 + j200 & 10\angle 0° \\ -(104 + j200) & 0 \end{vmatrix}}{\begin{vmatrix} 200 + j200 & -(104 + j200) \\ -(104 + j200) & 205 + j150 \end{vmatrix}} = 4.48 \times 10^{-2} \angle 25.6° \text{ A}$$

$$V_{AB} = 100 I_2 - (4 + j200)(I_1 - I_2)$$

$$= 100\,(4.48 \times 10^{-2} \angle 25.6°) - (4 + j200)\,(5.05 \times 10^{-2} \angle 0.074° - 4.48 \times 10^{-2} \angle 25.6°) = 0$$

5.3 NODAL ANALYSIS

Nodal analysis uses Kirchhoff's current law for finding currents and voltage in a network. For ac networks, Kirchhoff's current law states that the phasor sum of currents meeting at a point is equal to zero.

Example 5.6 *In the network shown, determine V_a and V_b.*

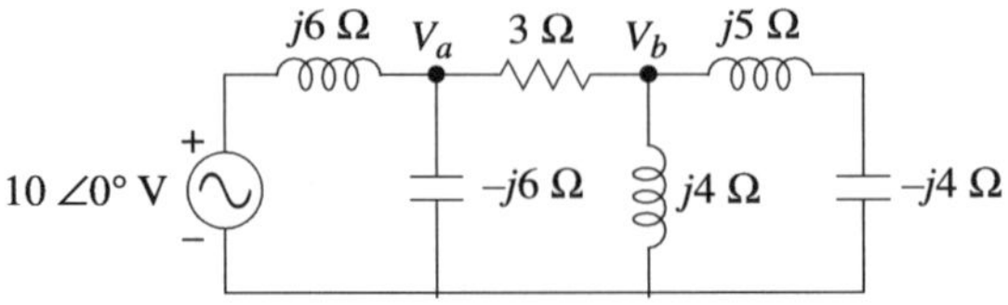

Fig. 5.6

Solution Applying KCL at Node *a*,

$$\frac{V_a - 10\angle 0°}{j6} + \frac{V_a}{-j6} + \frac{V_a - V_b}{3} = 0$$

$$\left(\frac{1}{j6} - \frac{1}{j6} + \frac{1}{3}\right) V_a - \frac{1}{3} V_b = \frac{10\angle 0°}{j6}$$

$$0.33\ V_a - 0.33\ V_b = 1.67\ \angle -90° \qquad \text{...(i)}$$

Applying KCL at Node *b*,

$$\frac{V_b - V_a}{3} + \frac{V_b}{j4} + \frac{V_b}{j1} = 0$$

$$-\frac{1}{3} V_a + \left(\frac{1}{3} + \frac{1}{j4} + \frac{1}{j1}\right) V_b = 0$$

$$-0.33 V_a + (0.33 - j1.25)\ V_b = 0 \qquad \text{...(ii)}$$

Adding Eqs (i) and (ii), we get

$$-j1.25\ V_b = 1.67\ \angle -90°$$

$$V_b = \frac{1.67\angle -90°}{-j1.25}$$

$$= 1.34\ \angle 0°\ \text{V}$$

Substituting V_b in Eq. (i), we get

$$0.33 V_a - 0.33\ (1.34\ \angle 0°) = 1.67\ \angle -90°$$

$$V_a = \frac{1.73\angle -75.17°}{0.33}$$

$$= 5.24\ \angle -75.17°\ \text{V}$$

Example 5.7 *For the network shown, find voltages V_1 and V_2.*

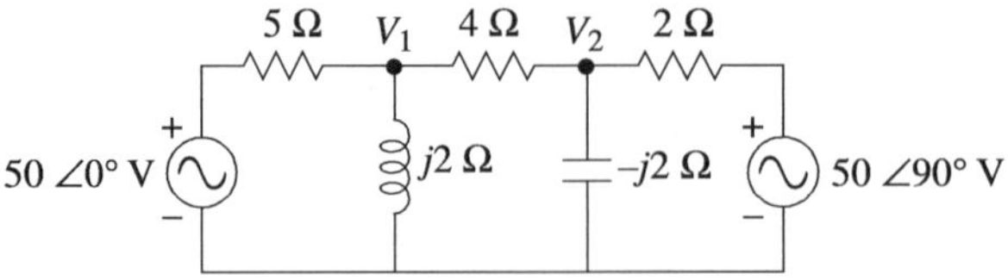

Fig. 5.7

Applying KCL to Node 1,

$$\frac{V_1 - 50\angle 0°}{5} + \frac{V_1}{j2} + \frac{V_1 - V_2}{4} = 0$$

$$\left(\frac{1}{5} + \frac{1}{j2} + \frac{1}{4}\right) V_1 - \frac{1}{4} V_2 = 10\ \angle 0°$$

$$(0.45 - j0.5)\ V_1 - 0.25\ V_2 = 10\ \angle\ 0° \qquad \text{...(i)}$$

Applying KCL at Node 2,

$$\frac{V_2 - V_1}{4} + \frac{V_2}{-j2} + \frac{V_2 - 50\angle 90^\circ}{2} = 0$$

$$-\frac{1}{4}V_1 + \left(\frac{1}{4} + \frac{1}{-j2} + \frac{1}{2}\right)V_2 = 25\ \angle 90^\circ$$

$$-0.25\ V_1 + (0.75 + j0.5)\ V_2 = 25\ \angle 90^\circ \qquad \text{...(ii)}$$

Writing Eqs (i) and (ii) in matrix form,

$$\begin{bmatrix} 0.45 - j0.5 & -0.25 \\ -0.25 & 0.75 + j0.5 \end{bmatrix} \begin{bmatrix} V_1 \\ V_2 \end{bmatrix} = \begin{bmatrix} 10\angle 0^\circ \\ 25\angle 90^\circ \end{bmatrix}$$

By Cramer's rule,

$$V_1 = \frac{\begin{vmatrix} 10\angle 0^\circ & -0.25 \\ j25 & 0.75 + j0.5 \end{vmatrix}}{\begin{vmatrix} 0.45 - j0.5 & -0.25 \\ -0.25 & 0.75 + j0.5 \end{vmatrix}} = \frac{13.5\angle 56.3^\circ}{0.55\angle -15.95^\circ} = 24.7\ \angle 72.25^\circ \text{ V}$$

$$V_2 = \frac{\begin{vmatrix} 0.45 - j0.5 & 10\angle 0^\circ \\ -0.25 & 25\angle 90^\circ \end{vmatrix}}{\begin{vmatrix} 0.45 - j0.5 & -0.25 \\ -0.25 & 0.75 + j0.5 \end{vmatrix}} = \frac{18.75\angle 36.87^\circ}{0.55\angle -15.95^\circ} = 34.34\ \angle 52.82^\circ \text{ V}$$

Example 5.8 *Find the voltage V_{AB} in the network of Fig. 5.8.*

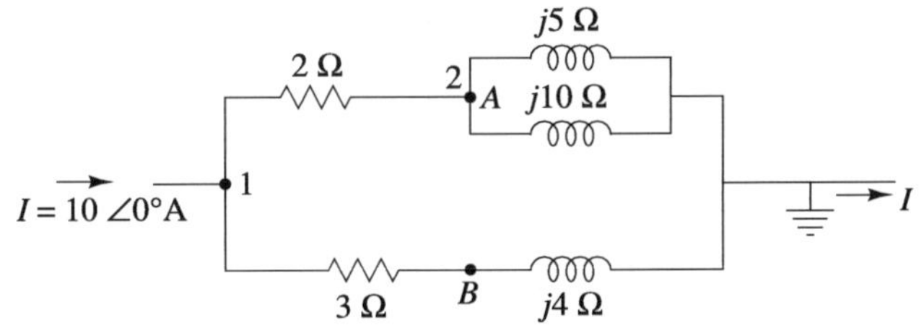

Fig. 5.8

Solution Applying KCL at Node 1,

$$10\ \angle 0^\circ = \frac{V_1 - V_2}{2} + \frac{V_1}{3 + j4}$$

$$\left(\frac{1}{2} + \frac{1}{3 + j4}\right)V_1 - \frac{1}{2}V_2 = 10\ \angle 0^\circ$$

$$(0.62 - j0.16)\ V_1 - 0.5V_2 = 10\ \angle 0^\circ \qquad \text{...(i)}$$

Applying KCL at Node 2,

$$\frac{V_2 - V_1}{2} + \frac{V_2}{j5} + \frac{V_2}{j10} = 0$$

$$-\frac{1}{2}V_1 + \left(\frac{1}{2} + \frac{1}{j5} + \frac{1}{j10}\right)V_2 = 0$$

$$-0.5V_1 + (0.5 - j0.3)\ V_2 = 0 \qquad \text{...(ii)}$$

Writing Eqs (i) and (ii) in matrix form,

$$\begin{bmatrix} 0.62 - j0.16 & -0.5 \\ -0.5 & 0.5 - j0.3 \end{bmatrix} \begin{bmatrix} V_1 \\ V_2 \end{bmatrix} = \begin{bmatrix} 10\angle 0^\circ \\ 0 \end{bmatrix}$$

By Cramer's rule,

$$V_1 = \frac{\begin{vmatrix} 10\angle 0^\circ & -0.5 \\ 0 & 0.5 - j0.3 \end{vmatrix}}{\begin{vmatrix} 0.62 - j0.16 & -0.5 \\ -0.5 & 0.5 - j0.3 \end{vmatrix}} = \frac{5.83\angle -31^\circ}{0.267\angle -87.42^\circ} = 21.8\ \angle 56.42^\circ \text{ V}$$

$$V_2 = \frac{\begin{vmatrix} 0.62 - j0.16 & 10\angle 0^\circ \\ -0.5 & 0 \end{vmatrix}}{\begin{vmatrix} 0.62 - j0.16 & -0.5 \\ -0.5 & 0.5 - j0.3 \end{vmatrix}} = \frac{5\angle 0^\circ}{0.267\angle -87.42^\circ} = 18.7\ \angle 87.42^\circ \text{ V}$$

$$V_A = V_2$$

$$V_B = \frac{V_1}{3 + j4}(j4)$$

$$= \frac{21.8\angle 56.42^\circ}{(3 + j4)}(j4) = 17.45\ \angle 93.32^\circ \text{ V}$$

$$V_{AB} = V_A - V_B$$

$$= (18.7\ \angle 87.42^\circ) - (17.45\ \angle 93.32^\circ) = 2.23\ \angle 34.1^\circ \text{ V}$$

5.4 SUPERPOSITION THEOREM

The superposition theorem can be used to analyse an ac network containing more than one source. The superposition theorem states that *in a network containing more than one voltage source or current source, the total current or voltage in any branch of the network is the phasor sum of currents or voltages produced in that branch by each source acting separately*. As each source is considered, all of the other sources are replaced by their internal impedances. This theorem is valid only for linear systems.

Example 5.9 *Find current through the 3 + j4 ohm impedance.*

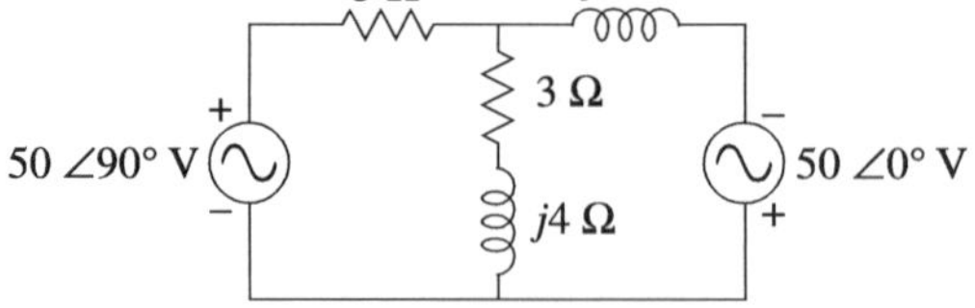

Fig. 5.9

Solution

Step I *When the 50 ∠90° V source is acting alone*

$$Z_T = 5 + \frac{(3 + j4)(j5)}{3 + j9}$$

$$= 5.83 + j2.5 = 6.35\ \angle\ 23.2^\circ\ \Omega$$

$$I_T = \frac{50\angle 90°}{6.35\angle 23.2°} = 7.87\ \angle 66.8°\ \text{A}$$

$$I' = (7.87\ \angle 66.8°)\left(\frac{j5}{3+j9}\right) = 4.15\ \angle 85.3°\ \text{A}\ (\downarrow)$$

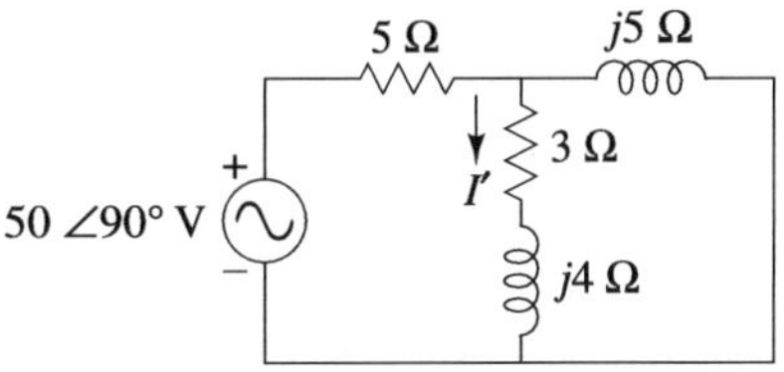

Fig. 5.10

Step II *When the 50 ∠0° V source is acting alone*

$$Z_T = j5 + \frac{5(3+j4)}{8+j4}$$

$$= 2.5 + j6.25 = 6.74\ \angle 68.2°\ \Omega$$

$$I_T = \frac{50\angle 0°}{6.74\angle 68.2°} = 7.42\ \angle -68.2°\ \text{A}$$

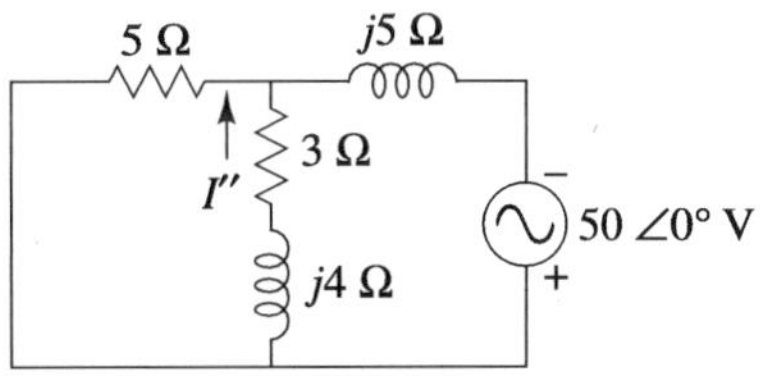

Fig. 5.11

$$I'' = (7.42\ \angle -68.2°)\left(\frac{5}{8+j4}\right)$$

$$= 4.15\ \angle -94.77°\ (\uparrow) = 4.15\ \angle 85.3°\text{A}\ (\downarrow)$$

Step III *By superposition theorem*

$$I = I' + I''$$

$$= 4.15\ \angle 85.3° + 4.15\ \angle 85.3° = 8.31\ \angle 85.3°\ \text{A}$$

Example 5.10 *Determine the voltage across the (2 + j5) ohm impedance for the network shown in Fig. 5.12.*

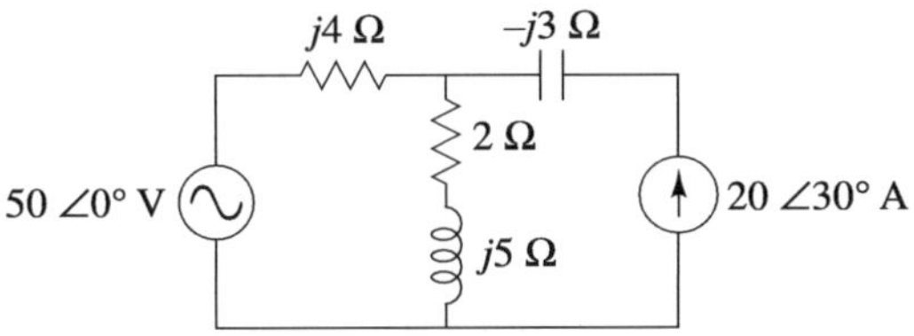

Fig. 5.12

Solution

Step I *When the 50 ∠0° source is acting alone*

$$I = \frac{50\angle 0°}{2 + j4 + j5} = \frac{50\angle 0°}{2 + j9}$$

$$= \frac{50\angle 0°}{9.22\angle 77.47°} = 5.42\ \angle -77.47°\ \text{A}$$

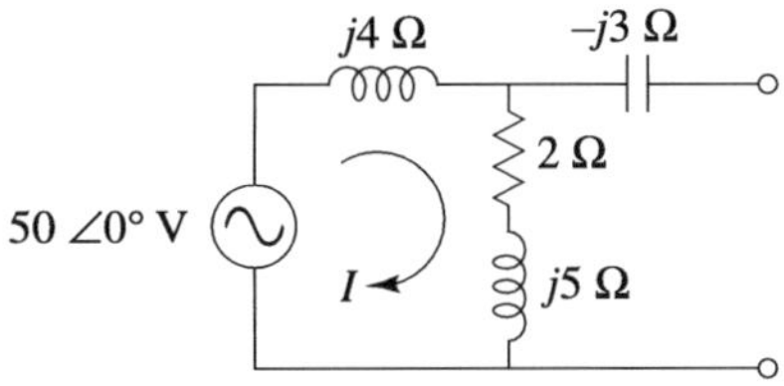

Fig. 5.13

Voltage cross (2 + j5) Ω impedance

$$V' = (5.42\ \angle -77.47°)\ (2 + j5)$$

$$= 29.16\ \angle -9.28°\ \text{V}$$

Step II *When the 20 ∠30° A source is acting alone*

$$I = (20\ \angle 30°)\left(\frac{j4}{2+j9}\right)$$

$$= \frac{20\angle 30° \times 4\angle 90°}{9.22\angle 77.47°} = 8.68\ \angle 42.53°\ \text{A}$$

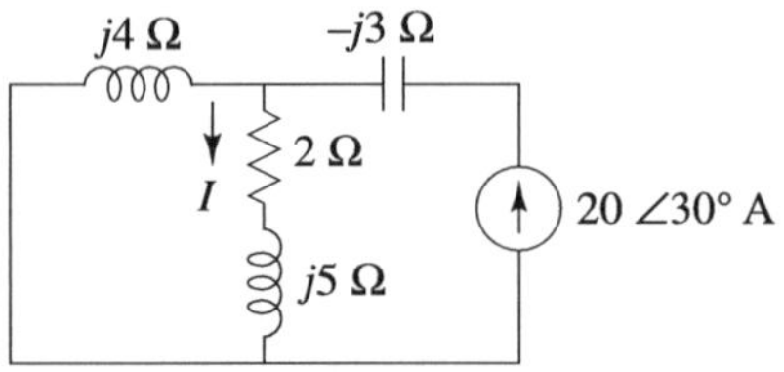

Fig. 5.14

Voltage across (2 + *j*5) Ω impedance

$$V'' = (8.68\ \angle 42.53°)\ (2+ j5) = 46.69\ \angle 110.72°\ \text{V}$$

Step III *By superposition theorem*

$$\begin{aligned} V &= V' + V'' \\ &= 29.16\ \angle -9.28° + 46.69\ \angle 110.72° \\ &= 28.78 - j4.7 - 16.52 + j43.67 \\ &= 12.26 + j38.97 = 40.85\ \angle\ 72.53°\ \text{V} \end{aligned}$$

Example 5.11 *Determine voltage V_{AB} for the network shown in Fig. 5.15.*

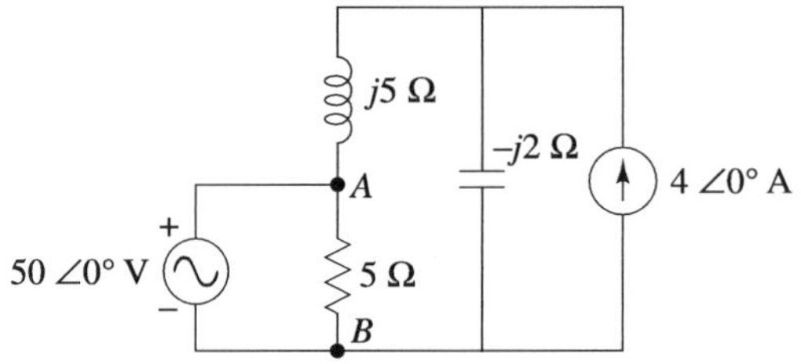

Fig. 5.15

Solution

Step I *When the 50 ∠0° V source is acting alone*

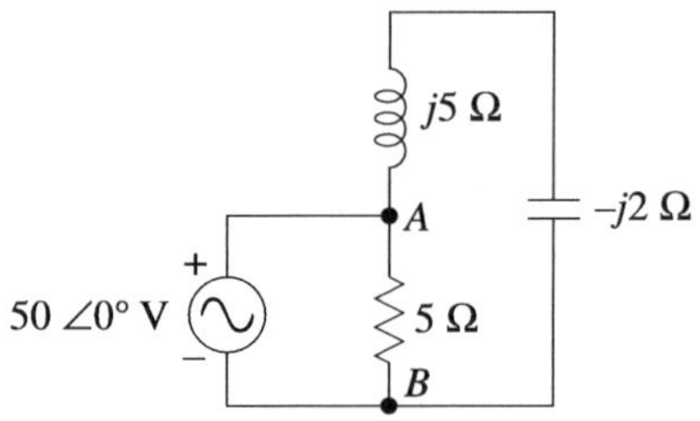

Fig. 5.16

$$V_{AB}' = 50\ \angle 0°\ \text{V}$$

Step II *When the 4 ∠0° A source is acting alone*

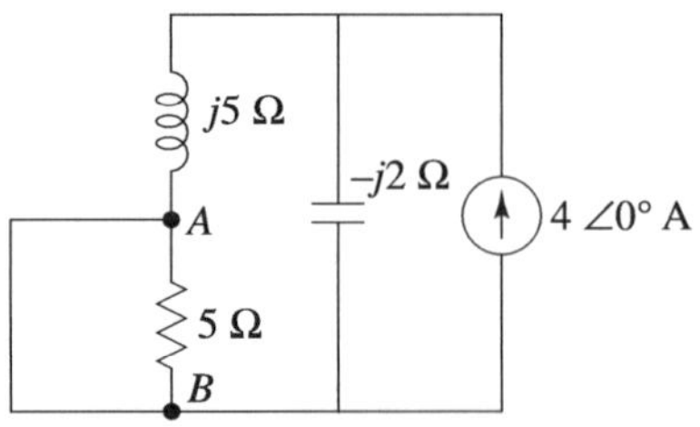

Fig. 5.17

$$V_{AB}'' = 0$$

Step III *By superposition theorem*

$$V_{AB} = V_{AB}' + V_{AB}''$$
$$= 50\ \angle 0° + 0 = 50\ \angle 0°\ \text{V}$$

Example 5.12 *Find the current I in the network shown in Fig. 5.18.*

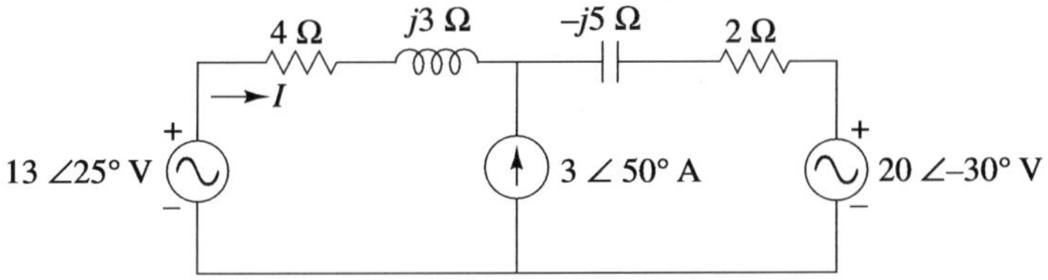

Fig. 5.18

Solution

Step I *When the 13 ∠25° V source is acting alone*

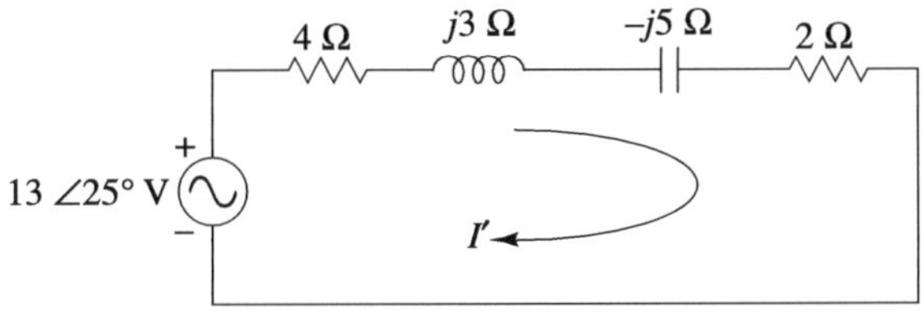

Fig. 5.19

$$I' = \frac{13\angle 25°}{6 - j2} = \frac{13\angle 25°}{6.32\angle -18.43°} = 2.057\ \angle 43.43°\ (\rightarrow)$$

Step II *When the 20 ∠–30° V source is acting alone*

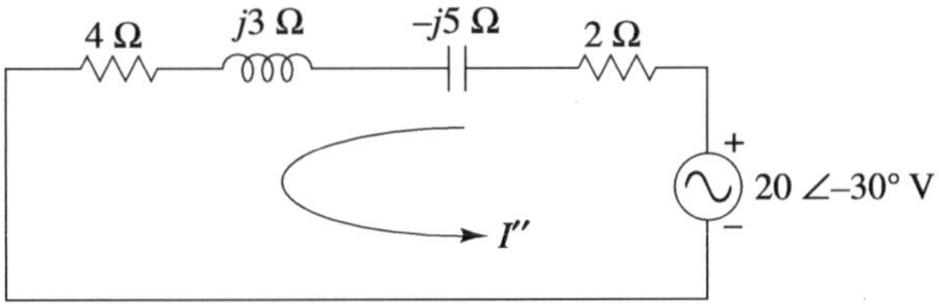

Fig. 5.20

$$I'' = \frac{20\angle -30°\ \text{V}}{6 - j2} = \frac{20\angle -30°}{6.32\angle -18.43°} = 3.16\angle -11.57°\ \text{A}\ (\leftarrow)$$
$$= 3.16\ \angle 168.43°\ \text{A}\ (\rightarrow)$$

Step III *When the 3 ∠50° A source is acting alone*

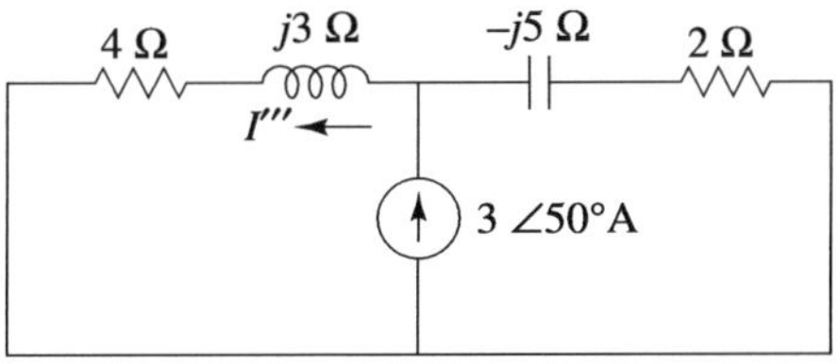

Fig. 5.21

$$I''' = 3\ \angle 50^\circ \times \frac{2 - j5}{6 - j2}$$

$$= 3\ \angle 50^\circ \times \frac{5.39\angle -68.2^\circ}{6.32\angle -18.43^\circ} = 2.56\ \angle 0.23^\circ \text{ A } (\leftarrow)$$

$$= 2.56\ \angle -179.77^\circ \text{ A } (\rightarrow)$$

Step IV *By superposition theorem*

$$I = I' + I'' + I'''$$

$$= 2.057\ \angle 43.13^\circ + 3.16\ \angle 168.43^\circ + 2.56\ \angle -179.77^\circ \text{ A}$$

$$= 4.62\ \angle 153.99^\circ \text{ A } (\rightarrow)$$

5.5 THEVENIN'S AND NORTON'S THEOREMS

Thevenin's and Norton's theorems give us a method for simplifying a network. In Thevenin's theorem, any linear network can be replaced by a voltage source V_{Th} in series with an impedance Z_{Th}. In Norton's theorem, any linear network can be replaced by a current source I_{SC} in parallel with an impedance Z_N where I_{SC} is the current flowing through short-circuited path placed across the terminals.

Example 5.13 *Obtain Thevenin's equivalent network for the terminals A and B in Fig. 5.22.*

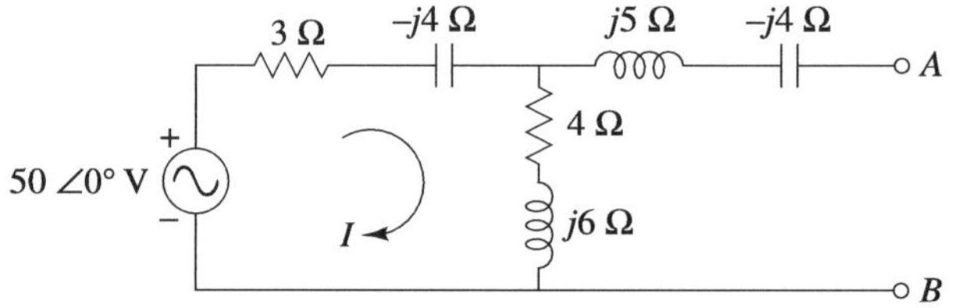

Fig. 5.22

Solution

Step I *Calculation of V_{Th}*

Applying KVL to the Mesh,

$$50\ \angle 0^\circ - (3 - j4)\,I - (4 + j6)\,I = 0$$

$$I = \frac{50\angle 0^\circ}{(3 - j4) + (4 + j6)}$$

$$= \frac{50\angle 0^\circ}{7 + j2} = \frac{50\angle 0^\circ}{7.28\angle 15.95^\circ} = 6.87\ \angle -15.95^\circ \text{ A}$$

$$V_{Th} = (4 + j6)\,I$$

$$= (4 + j6)\,(6.87\ \angle -15.95^\circ)$$

$$= (7.21\ \angle 56.3^\circ)\,(6.87\ \angle -15.95^\circ) = 49.5\ \angle 40.35^\circ \text{ V}$$

Step II *Calculation of Z_{Th}*

Replacing the voltage source by a short circuit,

The impedance seen from terminals A and B is

$$Z_{Th} = (j5 - j4) + \frac{(3 - j4)(4 + j6)}{(3 - j4) + (4 + j6)}$$

$$= j1 + \frac{5\angle -53.13^\circ \times 7.21\angle 56.3^\circ}{7.28\angle 15.95^\circ}$$

$$= j1 + 4.95\ \angle -12.78^\circ = j1 + 4.83 - j1.095$$

$$= 4.83 - j0.095 = 4.83\ \angle -1.13^\circ\ \Omega$$

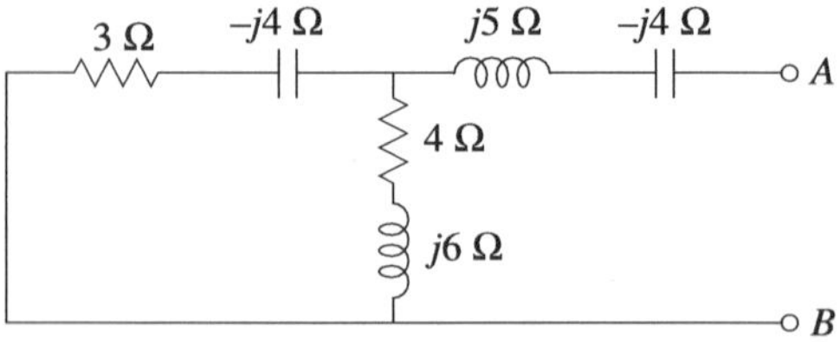

Fig. 5.23

Thevenin's equivalent network is shown in Fig. 5.24.

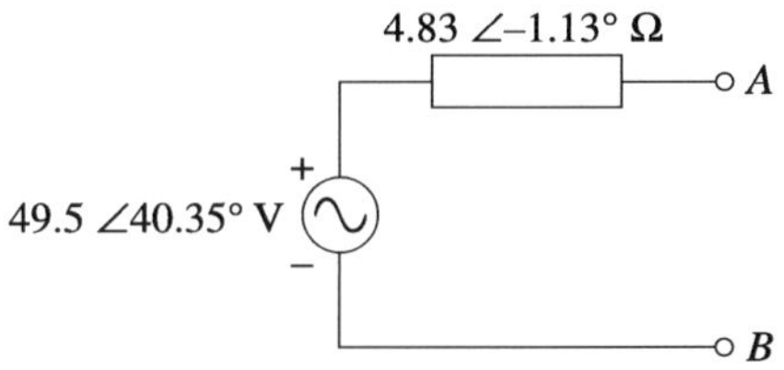

Fig. 5.24

Example 5.14 *Find Thevenin's equivalent network for Fig. 5.25 shown.*

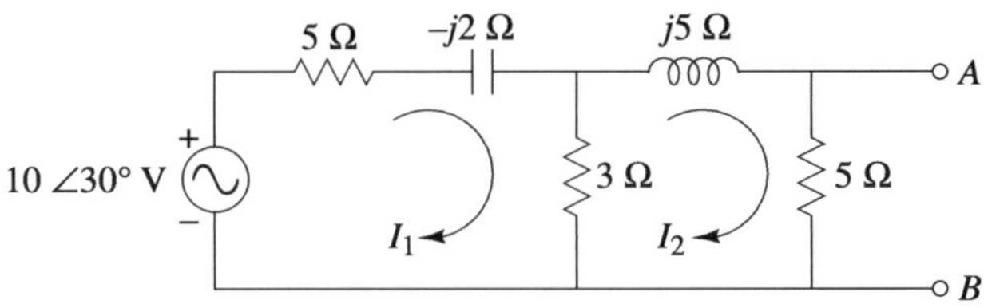

Fig. 5.25

Solution

Step I *Calculation of V_{Th}*

Applying KVL to Mesh 1,

$$10\angle 30^\circ - (5 - j2)\, I_1 - 3(I_1 - I_2) = 0$$

$$(8 - j2)\, I_1 - 3I_2 = 10\angle 30^\circ \quad \ldots\text{(i)}$$

Applying KVL to Mesh 2,

$$-3\,(I_2 - I_1) - j5I_2 - 5I_2 = 0$$

$$-3I_1 + (8 + j5)\, I_2 = 0 \quad \ldots\text{(ii)}$$

Solving Eqs (i) and (ii) by Cramer's rule,

$$I_2 = \frac{\begin{vmatrix} 8 - j2 & 10\angle 30^\circ \\ -3 & 0 \end{vmatrix}}{\begin{vmatrix} 8 - j2 & -3 \\ -3 & 8 + j5 \end{vmatrix}} = \frac{30\angle 30^\circ}{69.25\angle 20.3^\circ} = 0.433\angle 9.7^\circ \text{ A}$$

$$V_{Th} = V_{AB} = 5I_2$$

$$= 5\,(0.433\angle 9.7^\circ) = 2.16\angle 9.7^\circ \text{ V}$$

Step II *Calculation of Z_{Th}*

The impedance seen from terminals *A* and *B* with short-circuiting voltage source is

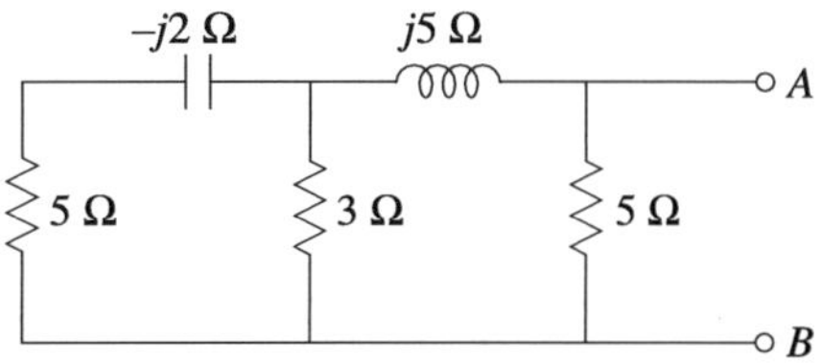

Fig. 5.26

$$Z_{Th} = \left[\left\{\frac{(5 - j2)3}{5 - j2 + 3}\right\} + j5\right] \parallel 5$$

$$= [1.94 - j0.265 + j5] \parallel 5 = (1.94 + j4.735) \parallel 5$$

$$= \frac{(1.94 + j4.735)5}{6.94 + j4.735} = 3.04\angle 33.4^\circ\ \Omega$$

Thevenin's equivalent network is shown in Fig. 5.27.

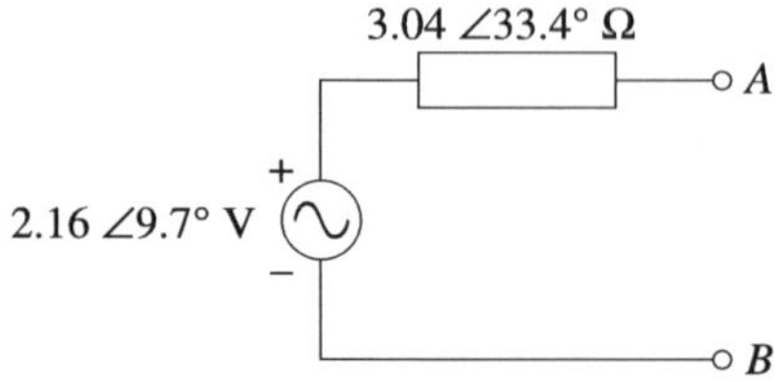

Fig. 5.27

Example 5.15 *Obtain Thevenin's equivalent network for Fig. 5.28 shown.*

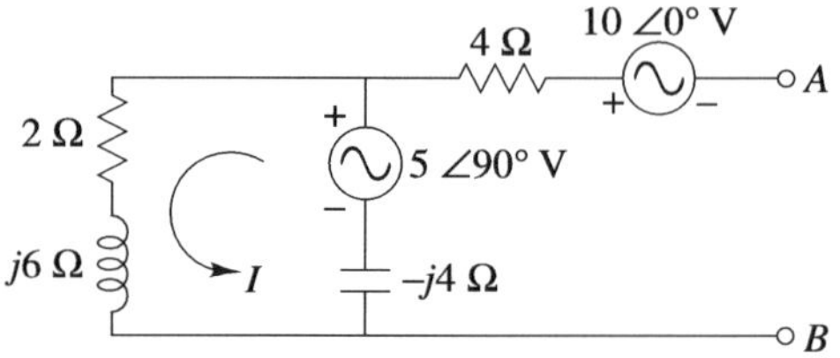

Fig. 5.28

Solution

Step I *Calculation of V_{Th}*

$$I = \frac{5\angle 90^\circ}{2 + j2}$$

$$= \frac{5\angle 90^\circ}{2.83\angle 45^\circ} = 1.77\ \angle 45^\circ\ \text{A}$$

$$V_{Th} = (-j4)\,I + 5\ \angle 90^\circ - 10\ \angle\ 0^\circ$$

$$= (4\ \angle -90^\circ)\ (1.77\ \angle 45^\circ) + 5\ \angle 90^\circ - 10\ \angle 0^\circ$$

$$= -15 + j10 = 18\ \angle 146.31^\circ\ \text{V}$$

Step II *Calculation of Z_{Th}*

The impedance seen from terminals A and B, when voltage sources are replaced by short circuits, is

$$Z_{Th} = 4 + \frac{(2 + j6)(-j4)}{2 + j2}$$

$$= 4 + \frac{6.32\angle 71.56^\circ \times 4\angle -90^\circ}{2.83\angle 45^\circ}$$

$$= 4 + 8.93\ \angle -63.44^\circ = 4 + 4 - j7.98$$

$$= 8 - j7.98 = 11.3\ \angle -44.93^\circ\ \Omega$$

Thevenin's equivalent network is shown in Fig. 5.29.

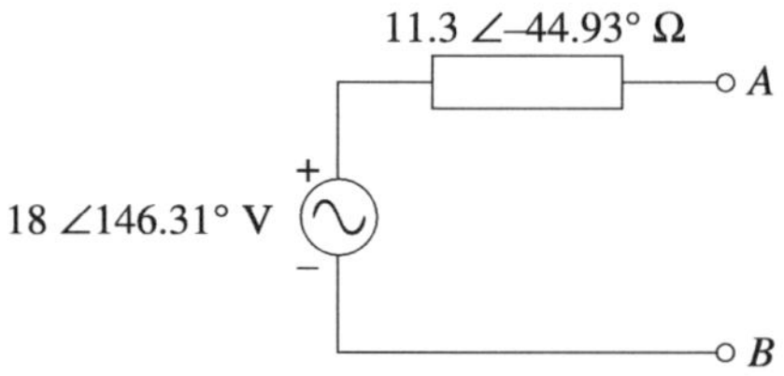

Fig. 5.29

Example 5.16 *Obtain Thevenin's equivalent network for Fig. 5.30 shown.*

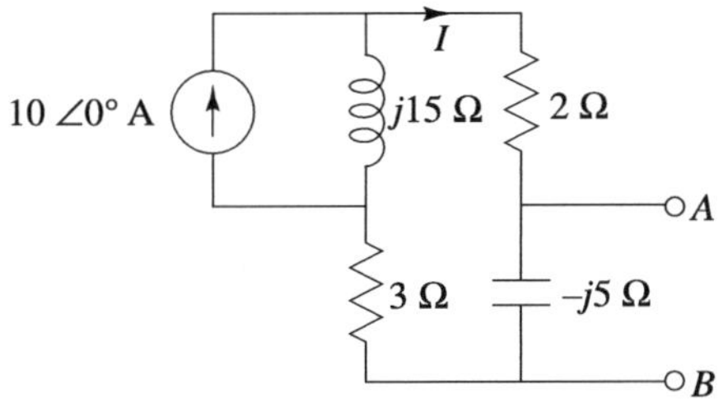

Fig. 5.30

Solution

Step I *Calculation of V_{Th}*

By current-division formula,

$$I = \frac{(10\angle 0^\circ)(j15)}{5 - j5 + j15}$$

$$= \frac{150\angle 90^\circ}{5 + j10} = \frac{150\angle 90^\circ}{11.18\angle 63.43^\circ}$$

$$= 13.42\ \angle 26.57^\circ \text{ A}$$

$$V_{Th} = V_{AB} = (-j5)\, I$$

$$= (5\ \angle -90^\circ)\ (13.42\ \angle 26.57^\circ) = 67.08\ \angle -63.43^\circ \text{ V}$$

Step II *Calculation of Z_{Th}*

The impedance seen from terminals A and B, with current source open circuited is,

$$Z_{Th} = \frac{(-j5)(5 + j15)}{-j5 + 5 + j15} = 7.07\ \angle -81.86^\circ\ \Omega$$

Thevenin's equivalent network is shown in Fig. 5.31.

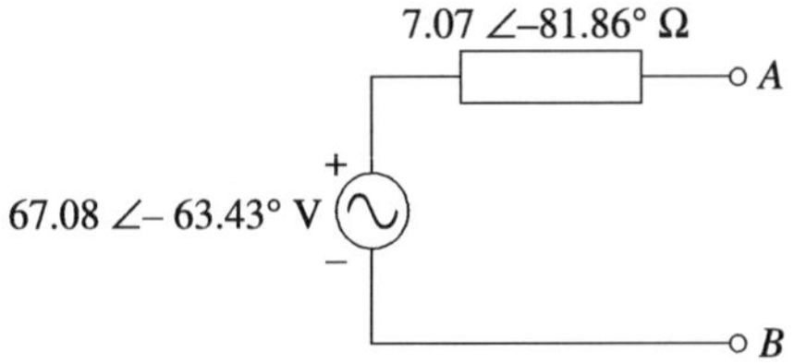

Fig. 5.31

Example 5.17 *Obtain Thevenin's equivalent network for Fig. 5.32 shown.*

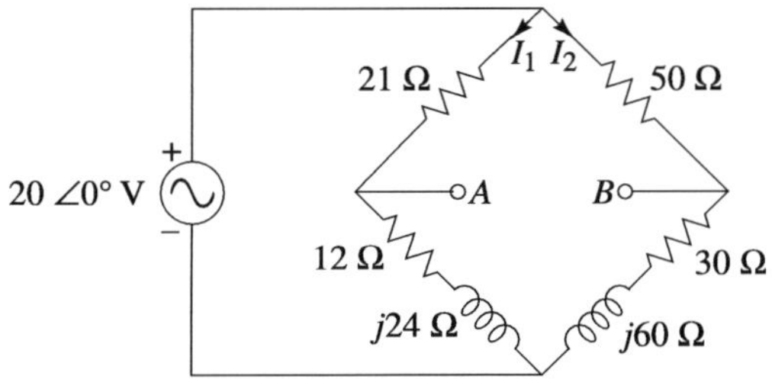

Fig. 5.32

Solution

Step I *Calculation of* V_{Th}

$$I_1 = \frac{20\angle 0^\circ}{21+12+j24} = \frac{20\angle 0^\circ}{33+j24} = \frac{20\angle 0^\circ}{40.8\angle 36.02^\circ} = 0.49\angle -36.02^\circ \text{ A}$$

$$I_2 = \frac{20\angle 0^\circ}{80+j60} = \frac{20\angle 0^\circ}{100\angle 36.86^\circ} = 0.2\angle -36.86^\circ \text{ A}$$

$$V_{Th} = V_{AB} = (12+j24)\,I_1 - (30+j60)\,I_2$$
$$= (26.83\angle 63.43^\circ)(0.49\angle -36.02^\circ) - (67.08\angle 63.43^\circ)(0.2\angle -36.86^\circ)$$
$$= 0.33\angle 171.12^\circ \text{ V}$$

Step II *Calculation of* Z_{Th}

The impedance seen from terminals *A* and *B* with short-circuiting voltage source is shown below:

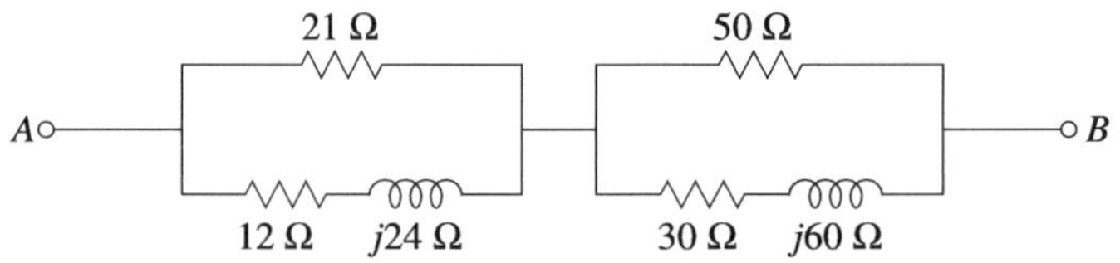

Fig. 5.33

$$Z_{Th} = \frac{21(12+j24)}{33+j24} + \frac{50(30+j60)}{80+j60} = 47.4\angle 26.8^\circ\ \Omega$$

Thevenin's equivalent network is shown in Fig. 5.34

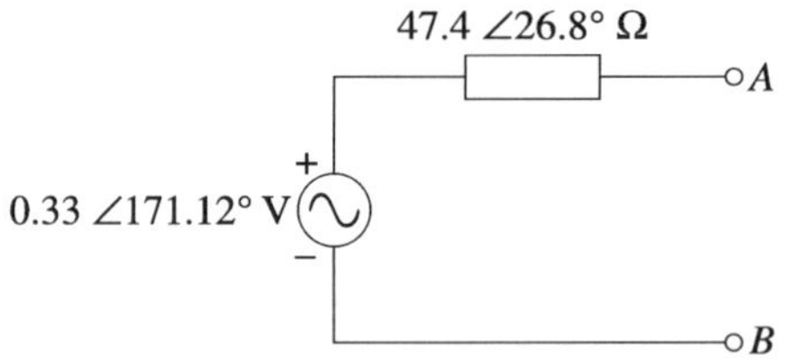

Fig. 5.34

Example 5.18 *Obtain Norton's equivalent network between terminals A and B as shown in Fig. 5.35.*

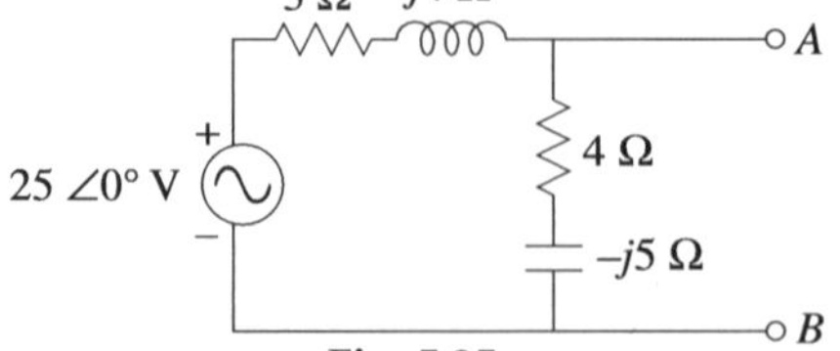

Fig. 5.35

Solution

Step I *Calculation of* I_{SC}

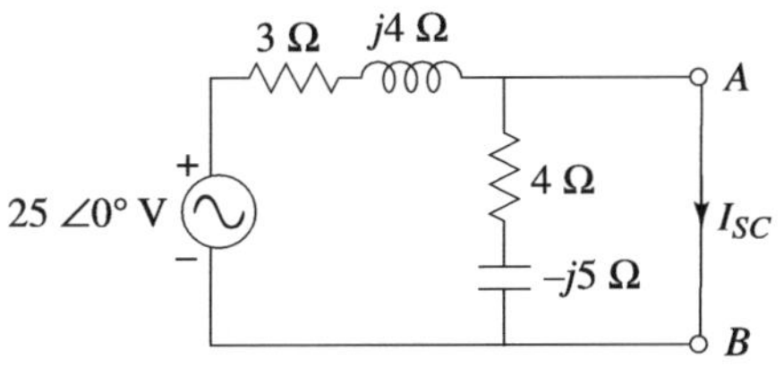

Fig. 5.36

$$I_{SC} = \frac{25\angle 0^\circ}{3 + j4} = \frac{25\angle 0^\circ}{5\angle 53.13^\circ} = 5\ \angle{-53.13^\circ}\ \text{A}$$

Step II *Calculation of* Z_N

The impedance seen from terminals A and B is, when valtage source is short circuited, is

$$Z_N = \frac{(3 + j4)(4 - j5)}{3 + j4 + 4 - j5}$$

$$= \frac{5\angle 53.13^\circ \times 6.4\angle -51.34^\circ}{7.07\angle -8.13^\circ} = 4.53\ \angle 9.92^\circ\ \Omega$$

Norton's equivalent network is shown in Fig. 5.37.

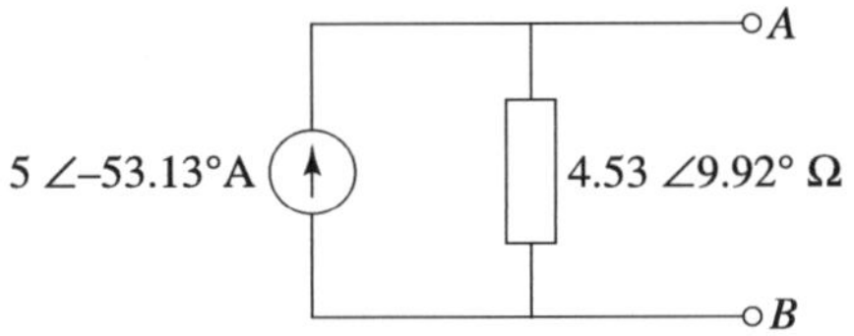

Fig. 5.37

5.6 MAXIMUM POWER TRANSFER THEOREM

The maximum power transfer theorem states that the maximum power is delivered from a source to the load when the load resistance is equal to the source resistance. This theorem can be applied to complex impedance circuits. If the source impedance is complex then the maximum power transfer occurs when the load impedance is the complex conjugate of the source impedance.

Example 5.19 *Find the impedance* Z_L *so that maximum power can be transferred to it in the network of Fig. 5.38. Find maximum power.*

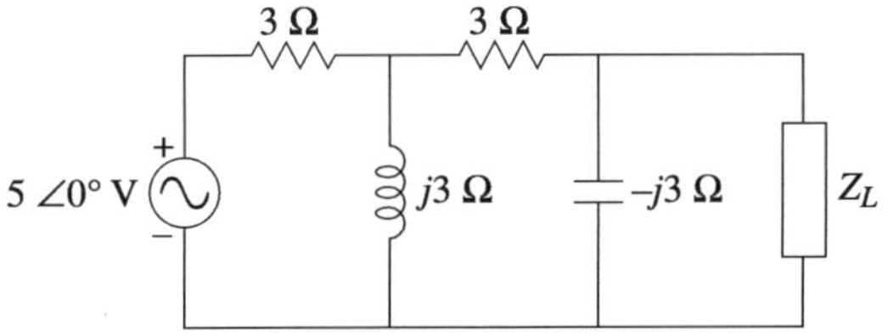

Fig. 5.38

Solution

Step I *Calculation of* V_{eq}

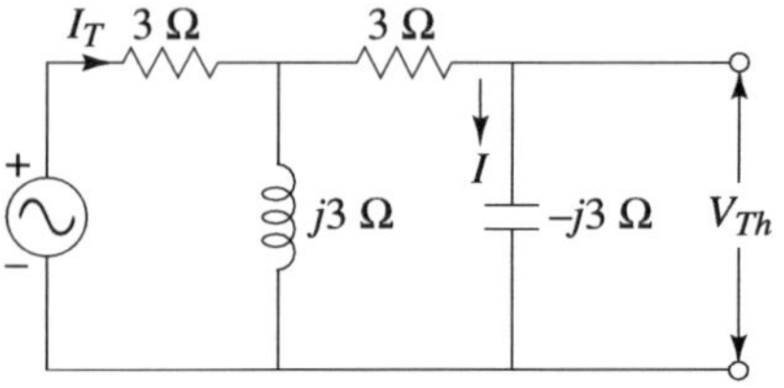

Fig. 5.39

$$Z_{eq} = 3 + \frac{j3(3-j3)}{3+j3-j3}$$

$$= 3 + \frac{9+j9}{3} = 3 + 3 + j3$$

$$= 6 + j3 = 6.71 \angle 26.57° \ \Omega$$

$$I_T = \frac{5\angle 0°}{6.71\angle 26.57°} = 0.75 \angle{-26.57°} \text{ A}$$

$$I = 0.75 \angle{-26.57°} \times \frac{j3}{3+j3-j3} = 0.75 \angle 63.43° \text{ A}$$

$$V_{Th} = (-j3)(0.75 \angle 63.43°)$$

$$= 2.24 \angle -26.57° \text{ V}$$

Step II *Calculation of* Z_{Th}

The impedance seen from the open terminal after the short-circuiting voltage source is shown below:

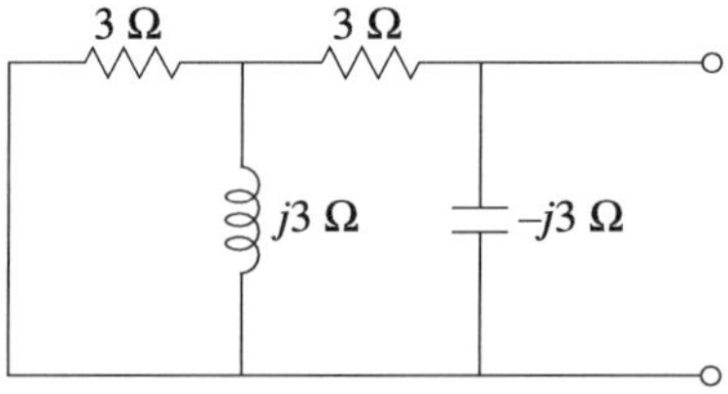

Fig. 5.40

$$Z_{Th} = [(3 \,||\, j3) + 3] \,||\, (-j3)$$

$$= 3 \angle{-53.12°} \ \Omega = 1.8 - j2.4 \ \Omega$$

Step III *Calculation of* Z_L

For maximum power transfer, the load impedance should be a complex conjugate of the source impedance.

$$Z_L = 1.8 + j2.4 \ \Omega$$

Step IV *Calculation of* P_{max}

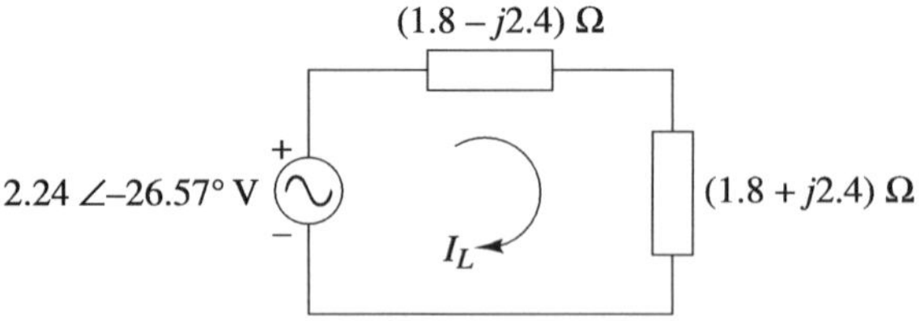

Fig. 5.41

$$I_L = \frac{2.24\angle -26.57°}{1.8 - j2.4 + 1.8 + j2.4} = 0.621 \angle{-26.57°} \text{ A}$$

$$P_{\max} = I_L^2 R_L$$

$$= (0.621)^2 \times 1.8 = 0.694 \text{ W}$$

Example 5.20 *Find the value of Z_L for maximum power transfer in the network shown and find maximum power.*

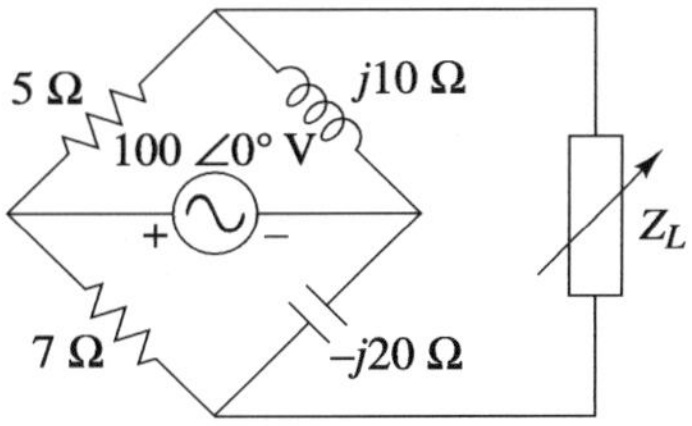

Fig. 5.42

Solution

Step I *Calculation of V_{Th}*

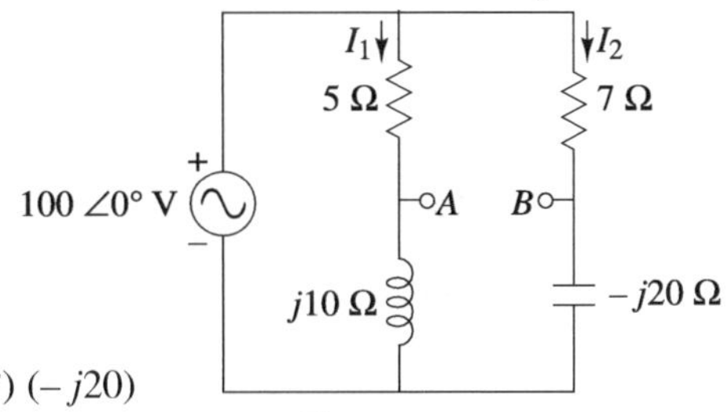

Fig. 5.43

$$I_1 = \frac{100\angle 0^\circ}{5 + j10} = \frac{100\angle 0^\circ}{11.18\angle 63.43^\circ} = 8.94\ \angle -63.43^\circ \text{ A}$$

$$I_2 = \frac{100\angle 0^\circ}{7 - j20} = \frac{100\angle 0^\circ}{21.19\angle -70.7^\circ} = 4.72\ \angle 70.7^\circ \text{ A}$$

$$V_{Th} = V_A - V_B = (8.94\ \angle -63.43^\circ)\,(j10) - (4.72\ \angle 70.7^\circ)\,(-j20)$$

$$= 89.4\ \angle 26.57^\circ - 94.4\ \angle -19.3^\circ$$

$$= 79.96 + j39.98 - 89.09 + j31.2 = -9.13 + j71.18 = 71.76\ \angle 97.3^\circ \text{ V}$$

Step II *Calculation of Z_{Th}*

The impedance seen from the terminals *A* and *B* with short-circuiting voltage source is shown below:

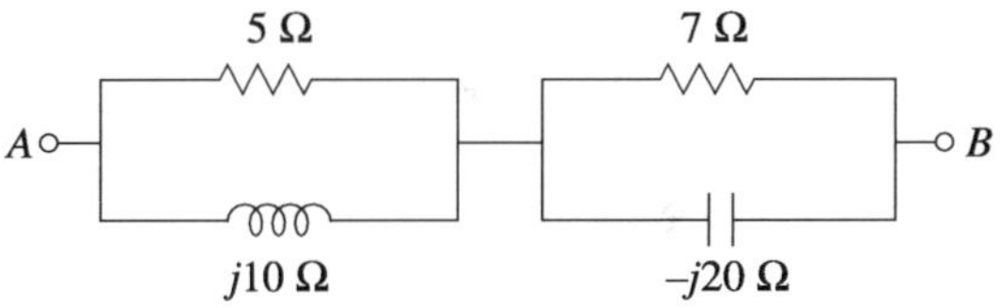

Fig. 5.44

$$Z_{Th} = \frac{5(j10)}{5 + j10} + \frac{7(-j20)}{7 - j20}$$

$$= \frac{50\angle 90^\circ}{11.18\angle 63.43^\circ} + \frac{140\angle -90^\circ}{21.19\angle -70.7^\circ}$$

$$= 4.47\ \angle 26.57^\circ + 6.6\ \angle -19.3^\circ$$

$$= 4 + j2 + 6.23 - j2.18 = 10.23 - j0.18\ \Omega$$

Step III *For maximum power transfer, the load impedance should be complex conjugate of the source impedance.*

$$Z_L = 10.23 + j0.18\ \Omega$$

Step IV *Calculation of P_{max}*

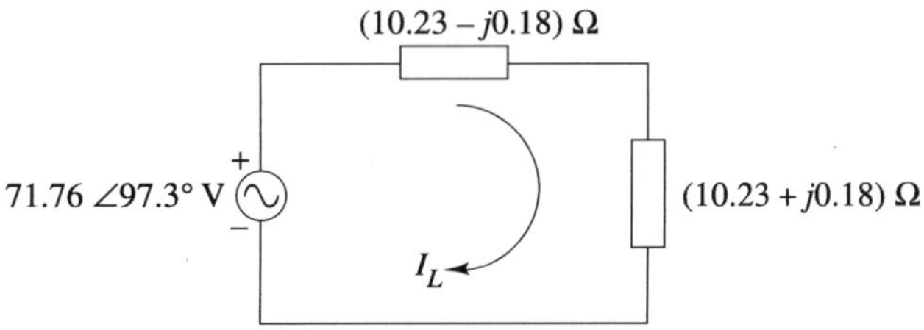

Fig. 5.45

$$I_L = \frac{71.76\angle 97.3^\circ}{10.23 - j0.18 + 10.23 + j0.18}$$

$$= \frac{71.76\angle 97.3^\circ}{20.46} = 3.51\ \angle 97.3^\circ \text{ A}$$

$$P_{\max} = I_L^2 R_L$$

$$= (3.51)^2 \times 10.23 = 125.91 \text{ W}$$

Example 5.21 *Find the value of load impedance Z_L so that maximum power can be transferred to it in the network of Fig. 5.46. Find maximum power.*

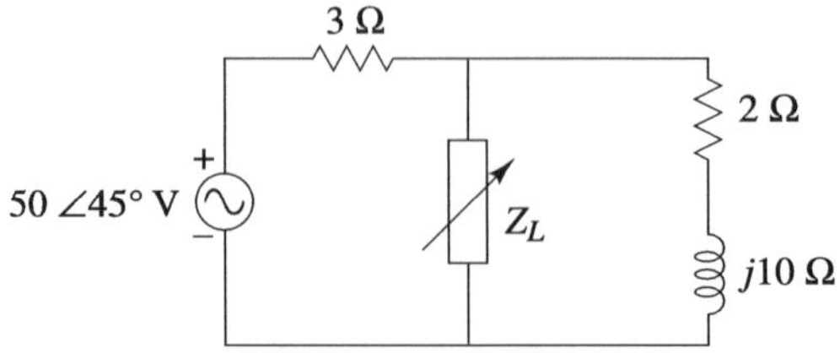

Fig. 5.46

Solution

Step I *Calculation of V_{Th}*

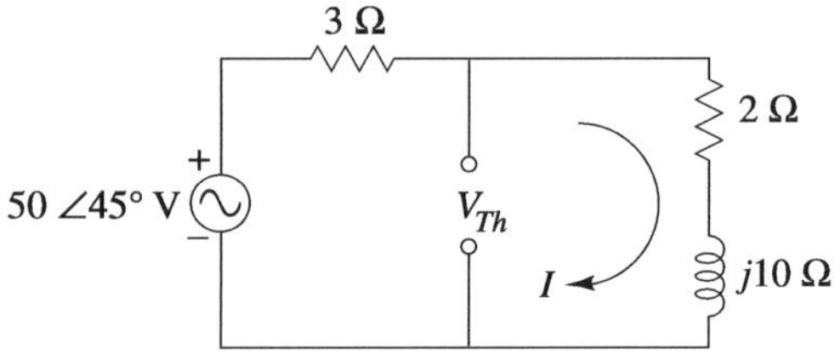

Fig. 5.47

$$I = \frac{50\angle 45^\circ}{3+2+j10} = \frac{50\angle 45^\circ}{5+j10}$$

$$= \frac{50\angle 45^\circ}{11.18\angle 63.43^\circ} = 4.47\angle -18.43^\circ \text{ A}$$

$$V_{Th} = (2+j10)(4.47\angle -18.43^\circ)$$

$$= (10\angle 78.69^\circ)(4.47\angle -18.43^\circ) = 45.6\angle 60.26^\circ \text{ V}$$

Step II *Calculation of Z_{Th}*

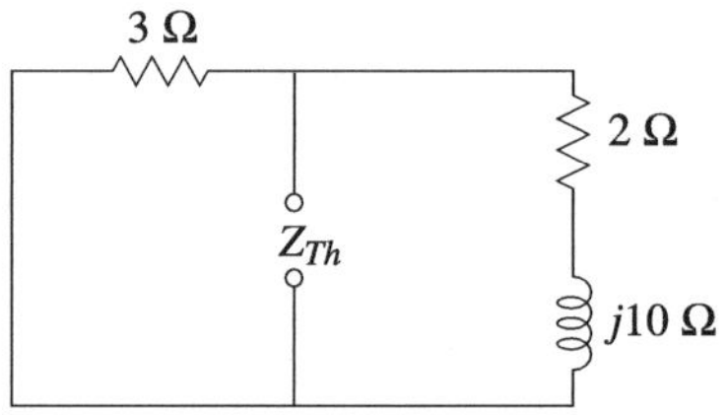

Fig. 5.48

$$Z_{Th} = \frac{3(2+j10)}{3+2+j10} = (2.64 + j0.72)\ \Omega$$

Step III *Calculation of Z_L*

For maximum power transfer, the load impedance should be complex conjugate of the source impedance.

$$Z_L = 2.64 - j0.72\ \Omega$$

Step IV *Calculation of P_{max}*

$$I_L = \frac{45.6\angle 60.26^\circ}{2.64 + j0.72 + 2.64 - j0.72}$$

$$= \frac{45.6\angle 60.26^\circ}{5.28} = 8.64\angle 60.26^\circ \text{ A}$$

$$P_{\max} = I_L^2 R_L$$

$$= (8.64)^2 \times 2.64 = 197.07 \text{ W}$$

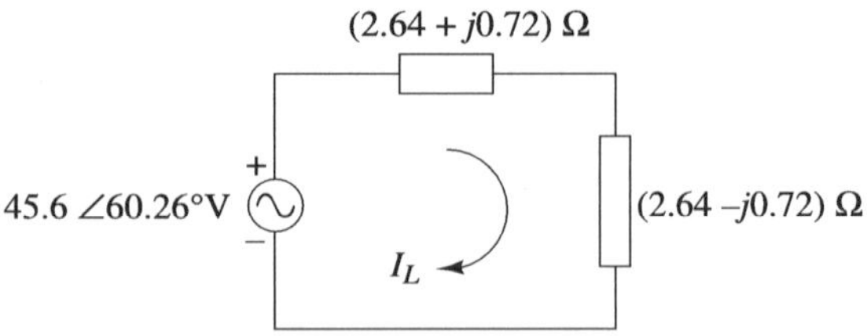

Fig. 5.49

Example 5.22 *Determine the load Z_L required to be connected in the network of Fig. 5.50 for maximum power transfer. Determine the maximum power drawn.*

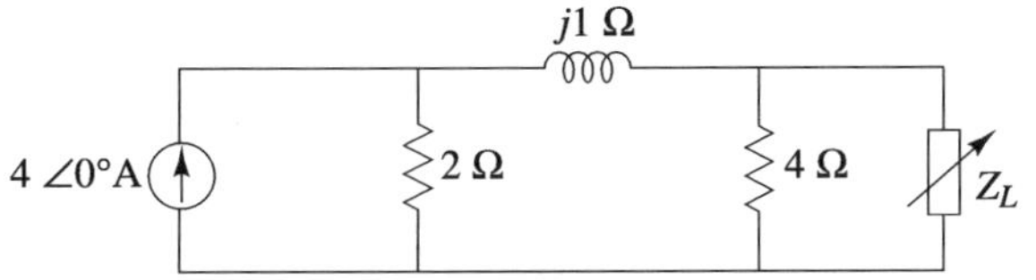

Fig. 5.50

Solution

Step I *Calculation of V_{Th}*

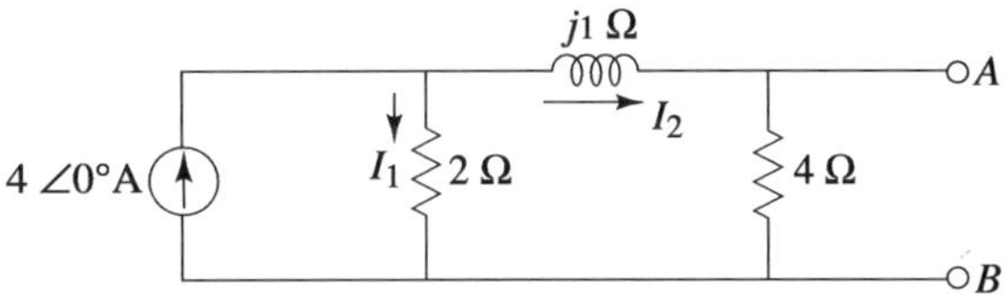

Fig. 5.51

$$I_2 = 4\angle 0^\circ \times \frac{2}{6+j1} = \frac{8\angle 0^\circ}{6+j1}$$

$$= \frac{8\angle 0^\circ}{6.08\angle 9.46} = 1.315\angle -9.46^\circ \text{ A}$$

$$V_{Th} = V_{AB} = 4I_2 = 4(1.315\angle -9.46^\circ) = 5.26\angle -9.46^\circ \text{ V}$$

Step II *Calculation of Z_{Th}*

The impedance Z_{Th} seen from the terminals A and B, when current source is open circuited, is

$$Z_{Th} = \frac{4(2+j1)}{4+2+j1}$$

$$= \frac{8+j4}{6+j1} = \frac{8.94\angle 26.56^\circ}{6.08\angle 9.46^\circ}$$

$$= 1.47\angle 17.1^\circ = 1.41 + j0.43\ \Omega$$

Step III *Calculation of Z_L*

For maximum power transfer, the load impedance should be the complex conjugate of the source impedance.

$$Z_L = 1.41 - j0.43\ \Omega$$

Step IV *Calculation of P_{max}*

$$I_L = \frac{5.26\angle -9.46^\circ}{1.41 + j0.43 + 1.41 - j0.43}$$

$$= \frac{5.26\angle -9.46^\circ}{2.82}$$

$$I_L = 1.86\angle -9.46^\circ \text{ A}$$

$$P_{max} = I_L^2 R_L$$

$$= (1.86)^2 \times 1.41 = 4.88 \text{ W}$$

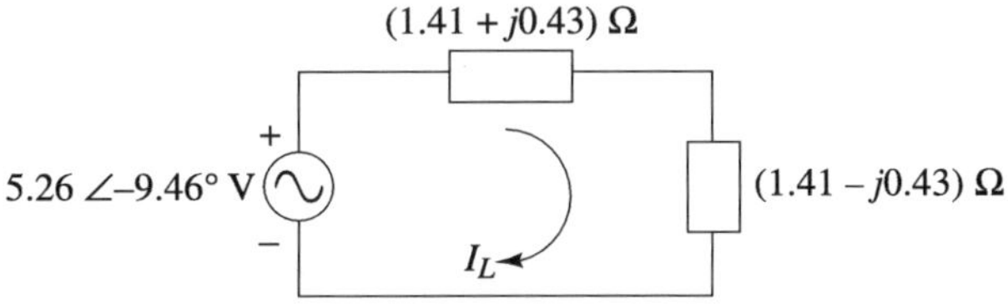

Fig. 5.52

5.7 COUPLED CIRCUITS

Consider two coils located physically close to one another as shown in Fig. 5.53.

When current I_1 flows in the first coil and $I_2 = 0$ in the second coil, flux ϕ_1 is produced in the coil. A fraction of this flux also links the second coil and induces a voltage in this coil. The voltage V_1 induced in the first coil is then,

$$V_1 = L_1 \left. \frac{dI_1}{dt} \right|_{I_2 = 0}$$

The voltage V_2 induced in the second coil is

$$V_2 = M \left. \frac{dI_1}{dt} \right|_{I_2 = 0}$$

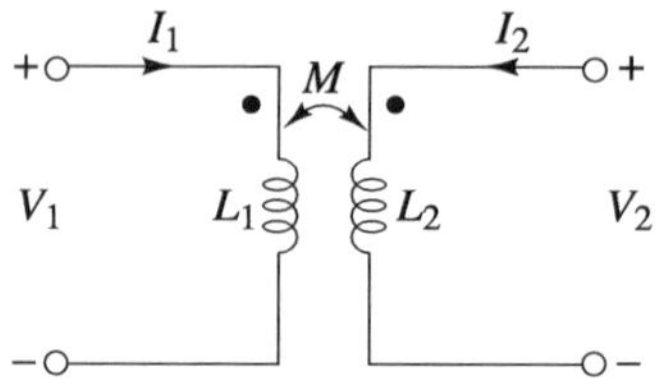

Fig. 5.53

The polarity of the voltage induced in the second coil depends on the way the coils are wound and it is usually indicated by dots. The dots signify that the induced voltages in the two coils (due to single current) have the same polarities at the dotted ends of the coils. Thus, due to I_1, the induced voltage V_1 must be positive at the dotted end of Coil 1. The voltage V_2 is also positive at the dotted end in Coil 2.

The same reasoning applies if a current I_2 flows in Coil 2 and $I_1 = 0$ in Coil 1. The induced voltages V_2 and V_1 are

$$V_2 = L_2 \left. \frac{dI_2}{dt} \right|_{I_1 = 0}$$

and

$$V_1 = M \left. \frac{dI_2}{dt} \right|_{I_1 = 0}$$

The polarities of V_1 and V_2 follow the dot convention. The voltage polarity is positive at the doted end of inductor L_2 when the current direction for I_2 is as shown in Fig. 5.53. Therefore, the voltage induced in Coil 1 must be positive at the dotted end also.

Now if both currents I_1 and I_2 are present, by using superposition principle, we can write

$$V_1 = L_1 \frac{dI_1}{dt} + M \frac{dI_2}{dt}$$

$$V_2 = M \frac{dI_1}{dt} + L_2 \frac{dI_2}{dt}$$

This can be represented in terms of dependent sources, as shown in Fig. 5.54.

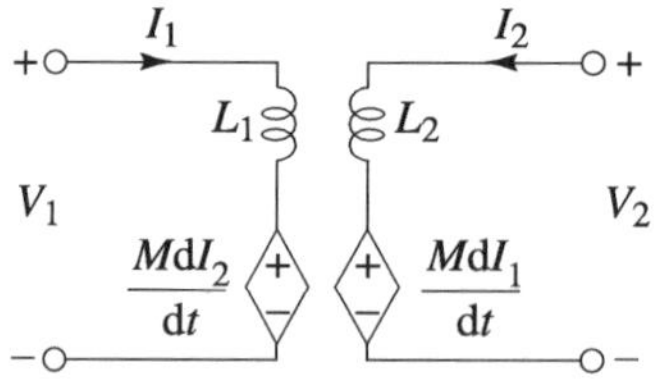

Fig. 5.54

Now consider the case when the dots are placed at the opposite ends in the two coils, as shown in Fig. 5.55.

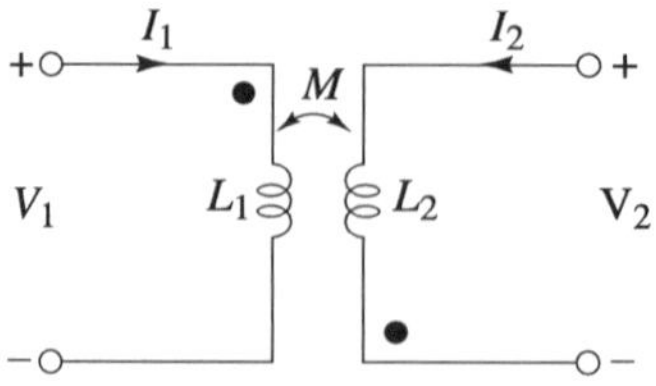

Fig. 5.55

Due to I_1, with $I_2 = 0$, the dotted end in Coil 1 is positive, so the induced voltage in Coil 2 is positive at the dot, which is the reverse of the designated polarity for V_2. Similarly, due to I_2, with $I_1 = 0$, the dotted ends have negative polarities for the induced voltages. The mutually induced voltages in both cases have polarities that are the reverse of terminal voltages and the equations are

$$V_1 = L_1 \frac{dI_1}{dt} - M \frac{dI_2}{dt}$$

$$V_2 = -M \frac{dI_1}{dt} + L_2 \frac{dI_2}{dt}$$

This can be repressed in terms of dependent sources as shown in Fig. 5.56.

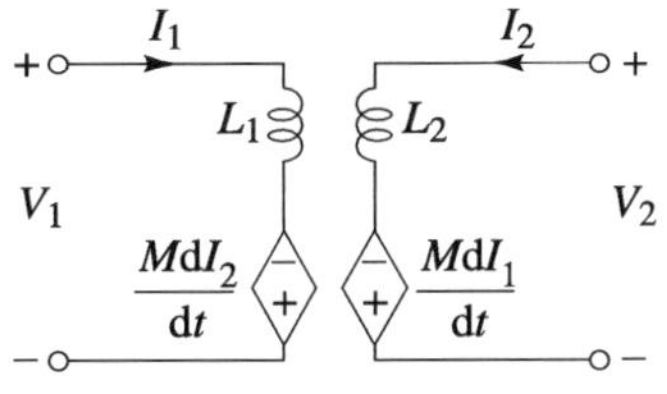

Fig. 5.56

Hence various cases can be summarised as follows:

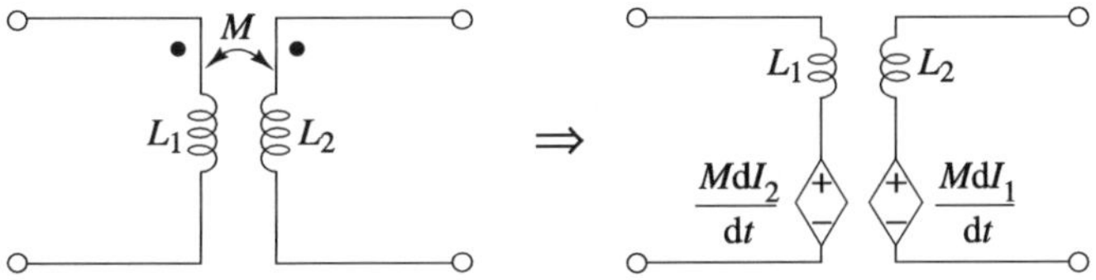

Fig. 5.57

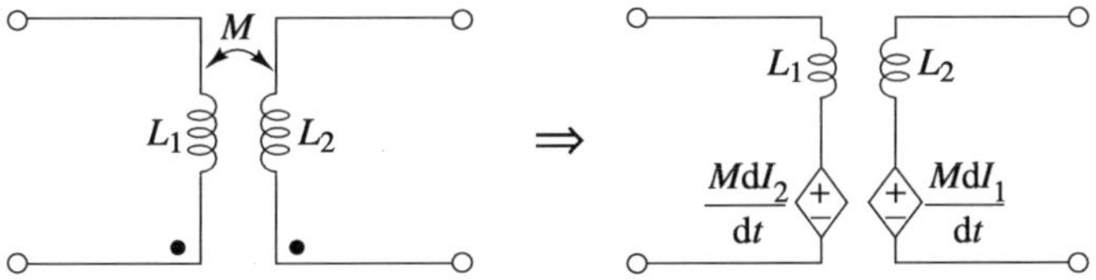

Fig. 5.58

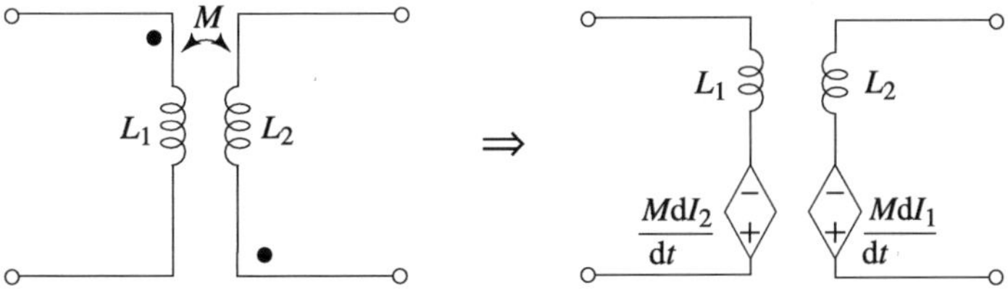

Fig. 5.59

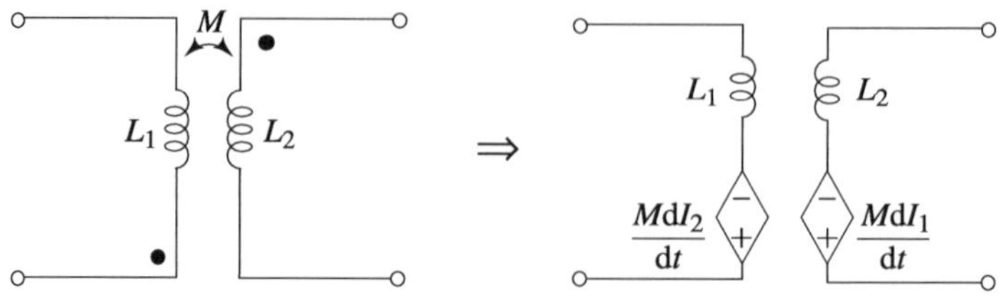

Fig. 5.60

Example 5.23 *Find voltage across the 5-Ω resistor using mesh analysis.*

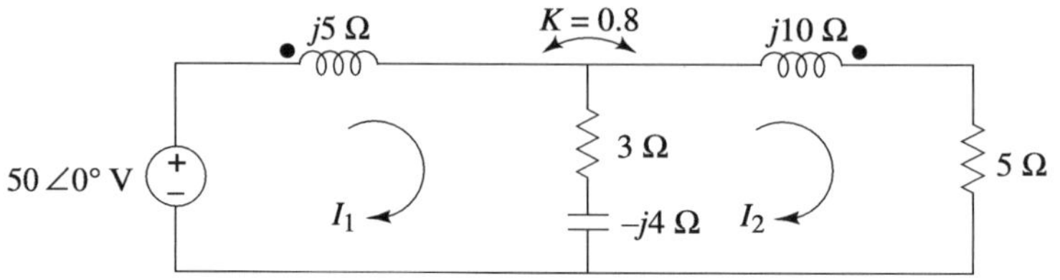

Fig. 5.61

Solution For a magnetically coupled circuit,

$$M = K\sqrt{5(10)}$$

$$X_m = K\sqrt{X_{L_1}X_{L_2}}$$

$$= 0.8\sqrt{5(10)} = 5.66\ \Omega$$

The equivalent circuit in terms of dependent sources can be drawn as

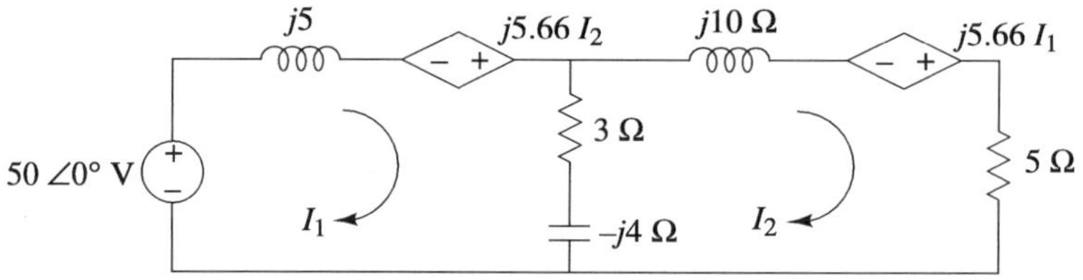

Fig. 5.62

Applying KVL to Mesh 1,

$$50\angle 0° - j5I_1 - 3(I_1 - I_2) + j4(I_1 - I_2) + j5.66I_2 = 0$$

$$50\angle 0° = (3 + j1)I_1 - (3 + j1.66)I_2$$

$$(3 + j1)\,\mathrm{I}_2 + (-3 - j1.66)I_2 = 50\angle 0° \qquad \text{...(i)}$$

Applying KVL to Mesh 2,

$$j4(I_2 - I_1) - 3(I_2 - I_1) - j10I_2 + j5.66I_1 - 5I_2 = 0$$

$$j4I_2 - j4I_1 - 3I_2 + 3I_1 - j10I_2 + j5.66I_1 - 5I_2 = 0$$

$$-j4I_2 + j4I_1 + 3I_2 - 3I_1 + j10I_2 - j5.66I_1 + 5I_2 = 0$$

$$(-3 - j1.66)I_1 + (8 + j6)I_2 = 0 \qquad \text{...(ii)}$$

By Cramer's rule,

$$I_2 = \frac{\begin{vmatrix} 3 + j1 & 50\angle 0° \\ -3 - j1.66 & 0 \end{vmatrix}}{\begin{vmatrix} 3 + j1 & -3 - j1.66 \\ -3 - j1.66 & 8 + j6 \end{vmatrix}} = 8.62\angle -24.79°\ \text{A}$$

$$V = 5I_2 = 5(8.62\angle -24.79°) = 43.1\angle -24.79°\ \text{V}$$

Example 5.24 *Find the voltage across the 5-Ω resistor using mesh analysis.*

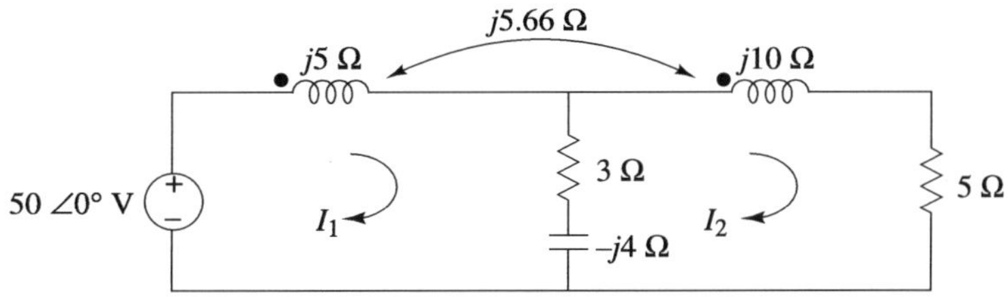

Fig. 5.63

Solution The equivalent circuit in terms of dependent sources can be drawn as shown below:

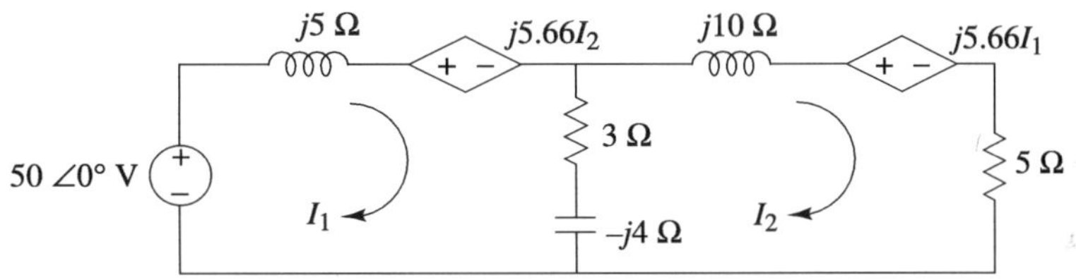

Fig. 5.64

Applying KVL to Mesh 1,

$$50\angle 0^\circ - j5I_1 - 3(I_1 - I_2) + j4(I_1 - I_2) - j5.66I_2 = 0$$

$$50\angle 0^\circ = (3 + j1)I_1 - (3 - j9.66)I_2$$

$$(3 + j1)I_2 + (-3 + j9.66)I_2 = 50\angle 0^\circ \qquad \dots (1)$$

Applying KVL to Mesh 2,

$$j4(I_2 - I_1) - 3(I_2 - I_1) - j10I_2 - j5.66I_1 - 5I_2 = 0$$

$$j4I_2 - j4I_1 - 3I_2 + 3I_1 - j10I_2 - j5.66I_1 - 5I_2 = 0$$

$$-j4I_2 + j4I_1 + 3I_2 - 3I_1 + j10I_2 + j5.66I_1 + 5I_2 = 0$$

$$(-3 + j9.66)I_1 + (8 + j6)I_2 = 0 \qquad \dots (2)$$

By Cramer's rule,

$$I_2 = \frac{\begin{vmatrix} 3 + j1 & 50\angle 0^\circ \\ -3 + j9.66 & 0 \end{vmatrix}}{\begin{vmatrix} 3 + j1 & -3 + j9.66 \\ -3 + j9.66 & 8 + j6 \end{vmatrix}} = 3.82\angle -112.14^\circ \text{ A}$$

$$V = 5I_2 = 5(3.82\angle -112.14^\circ) = 19.1\angle -112.14 \text{ V}$$

Example 5.25 *Find the current I_2 using mesh analysis.*

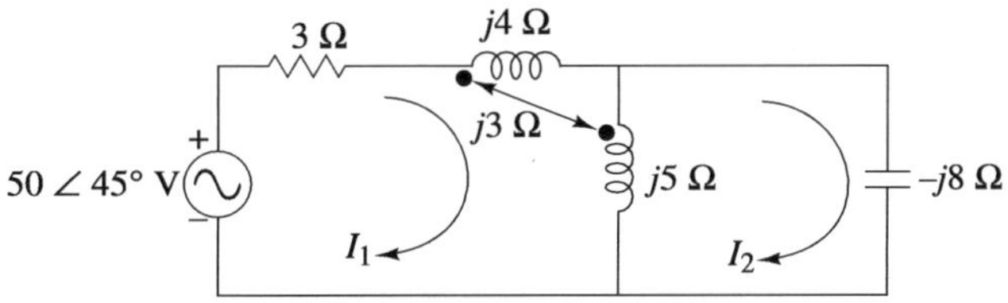

Fig. 5.65

Solution The equivalent circuit in terms of dependent sources can be drawn as shown,

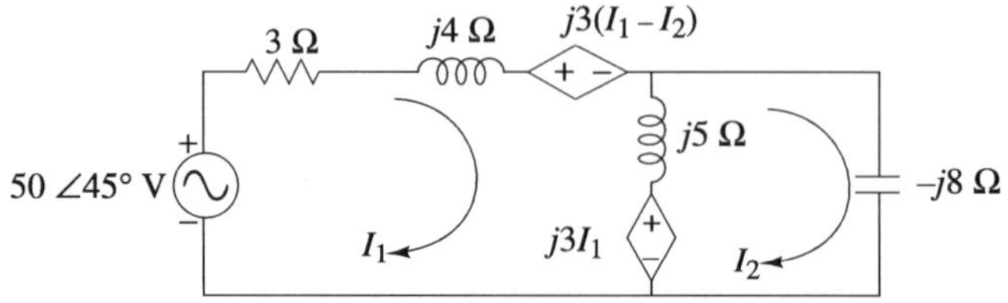

Fig. 5.66

Applying KVL to Mesh 1,

$$50\ \angle 45^\circ - 3I_1 - j4I_1 - j3\ (I_1 - I_2) - j3\ I_1 - j5\ (I_1 - I_2) = 0$$

$$50\ \angle 45^\circ = (3 + j4 + j5 + j3 + j3)\ I_1 - (\ j5 + j3)\ I_2 = (3 + j15)\ I_1 - j8I_2$$

Applying KVL to Mesh 2,

$$j3I_1 - j5\ (I_2 - I_1) + j8I_2 = 0$$

$$(\ j3 + j5)\ I_1 + (\ j8 - j5)\ I_2 = 0$$

By Cramer's rule,

$$I_1 = \frac{\begin{vmatrix} 50\angle 45^\circ & -j8 \\ 0 & j3 \end{vmatrix}}{\begin{vmatrix} 3+j15 & -j8 \\ j8 & j3 \end{vmatrix}} = \frac{150\angle 135^\circ}{109\angle 175.33^\circ} = 1.37\ \angle{-40.33^\circ}\ \text{A}$$

$$I_2 = \frac{\begin{vmatrix} 3+j15 & 50\angle 45^\circ \\ j8 & 0 \end{vmatrix}}{\begin{vmatrix} 3+j15 & -j8 \\ j8 & j3 \end{vmatrix}} = \frac{400\angle -135^\circ}{109\angle 175.33^\circ} = 3.66\ \angle{-310.33}\ \text{A}$$

Exercises

MESH ANALYSIS

1. Find the current through the 3 + *j*4 Ω impedance.

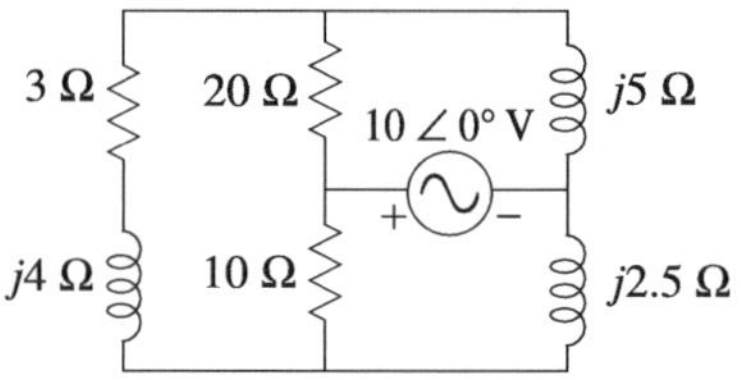

Fig. 5.67

[0]

2. In the network of Fig. 5.68, find V_0.

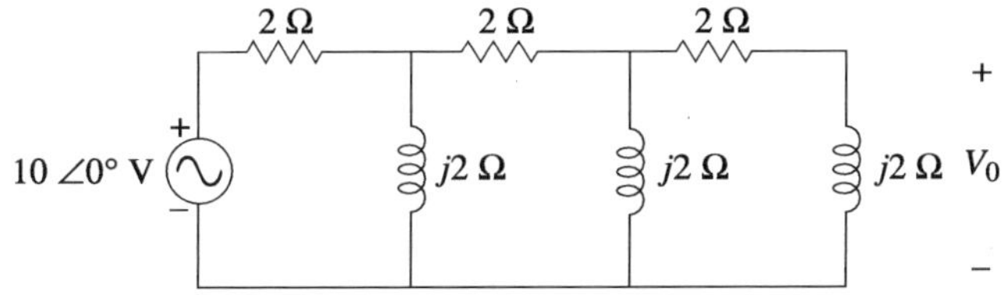

Fig. 5.68

[1.56 ∠128.7° V]

3. Find the current I_3 in the network of Fig. 5.69.

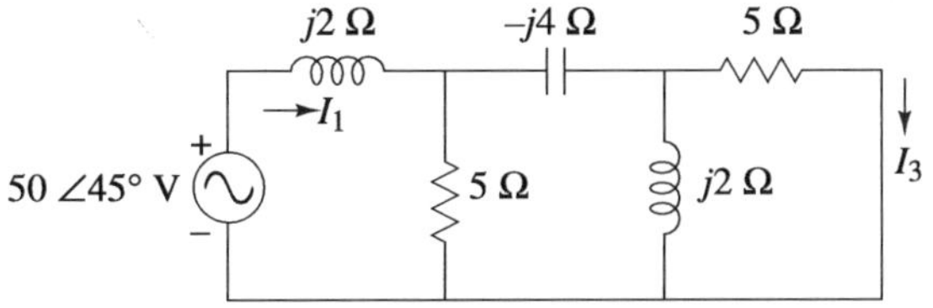

Fig. 5.69

[11.6 ∠113.2° A]

4. In the network of Fig. 5.70, find V_2 which results in zero current through the 4-Ω resistor.

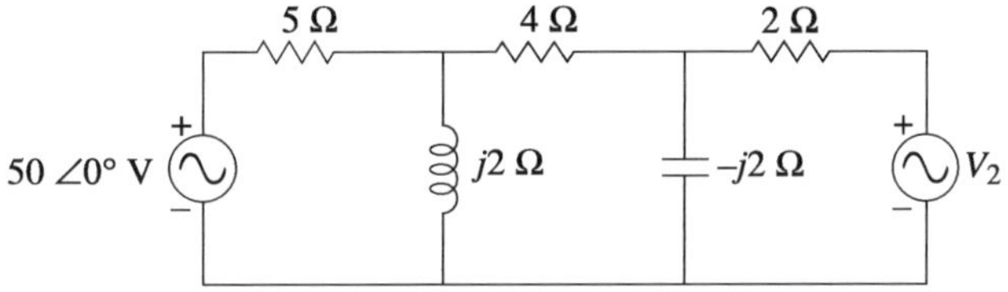

Fig. 5.70

[26.3 ∠113.2° V]

NODAL ANALYSIS

5. For the network shown in Fig. 5.71, find the voltage V_{AB}.

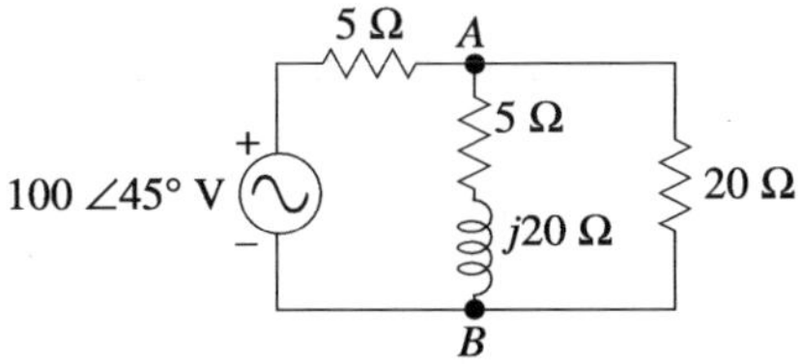

Fig. 5.71

[75.4 ∠55.2° V]

6. Find the voltages at nodes 1 and 2 in the network of Fig. 5.72.

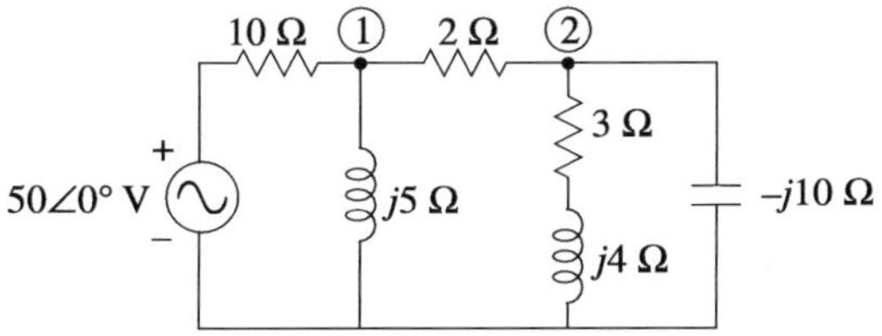

Fig. 5.72

[15.95 ∠49.94° V, 12.9 ∠55.5° V]

7. In the network of Fig. 5.73, find the current in the 10 ∠30° V source.

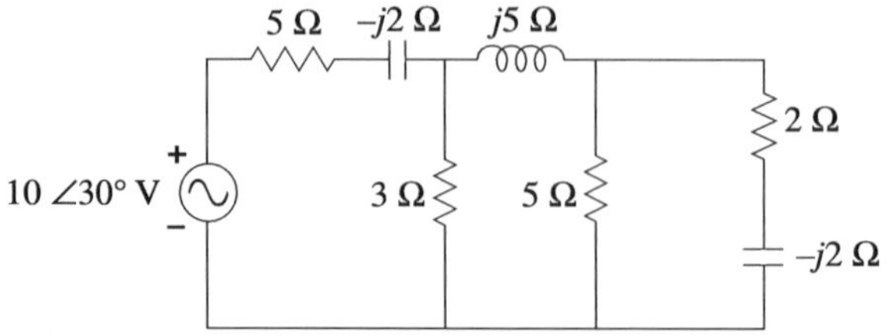

Fig. 5.73

[1.44 ∠38.8° A]

SUPERPOSITION THEOREM

8. For the network shown, find the current in the 10-Ω resistor.

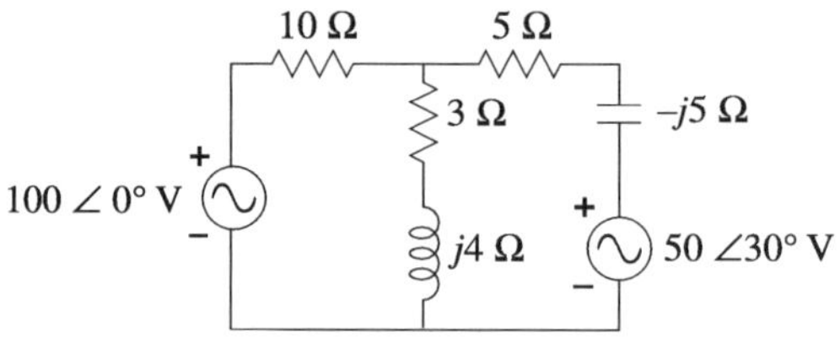

Fig. 5.74

[73.4 ∠–21.84° A]

9. In the network of Fig. 5.75, find the current through capacitance.

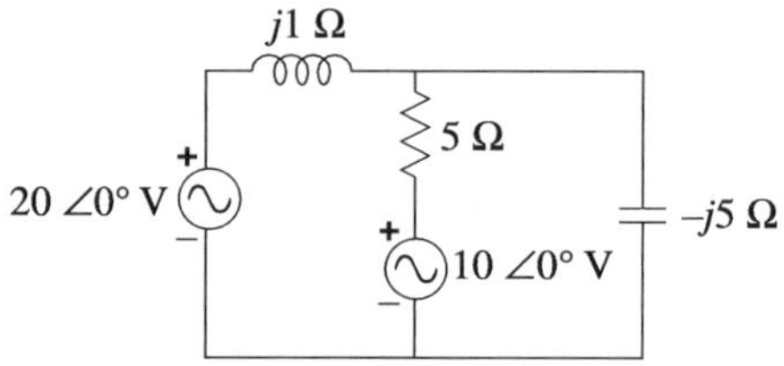

Fig. 5.75

[4.86 ∠80.8° A]

THEVENIN'S AND NORTON'S THEOREM

10. Obtain Thevenin's equivalent network.

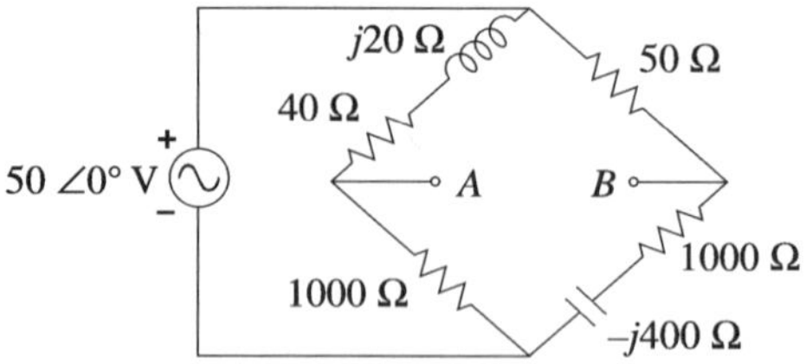

Fig. 5.76

[0.192 ∠–43.4° V, 88.7 ∠11.55° Ω]

11. Obtain Thevenin's equivalent network for Fig. 5.77 shown.

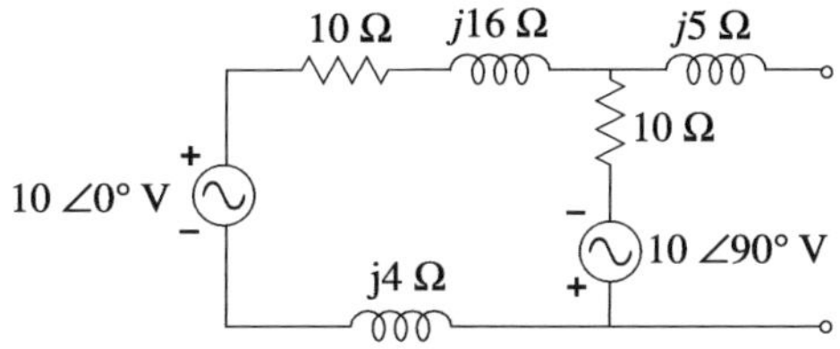

Fig. 5.77

[11.17 ∠–63.4° V, 10.6 ∠45° Ω]

12. Find Norton's equivalent network for Fig. 5.78 shown.

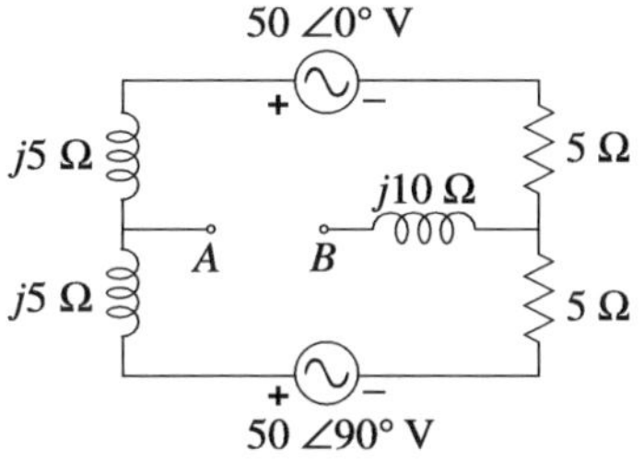

Fig. 5.78

[2.77 ∠–33.7° A, 2.5 + j12.5 Ω]

13. Find the current through the 3 + j4 Ω impedance.

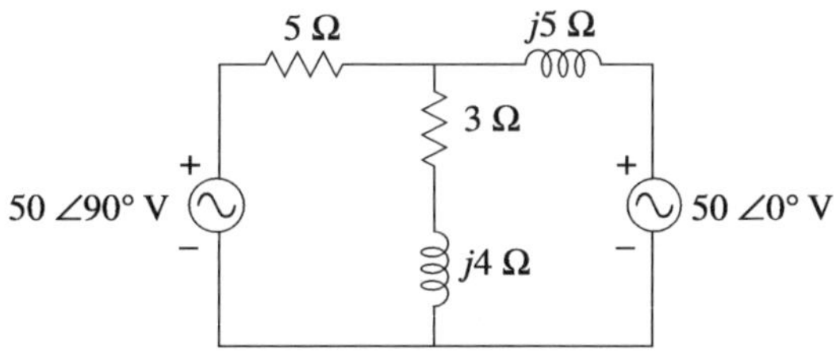

Fig. 5.79

[8.3 ∠85.2° A]

MAXIMUM POWER TRANSFER THEOREM

14. Determine the maximum power delivered to the load in the network shown in Fig. 5.80.

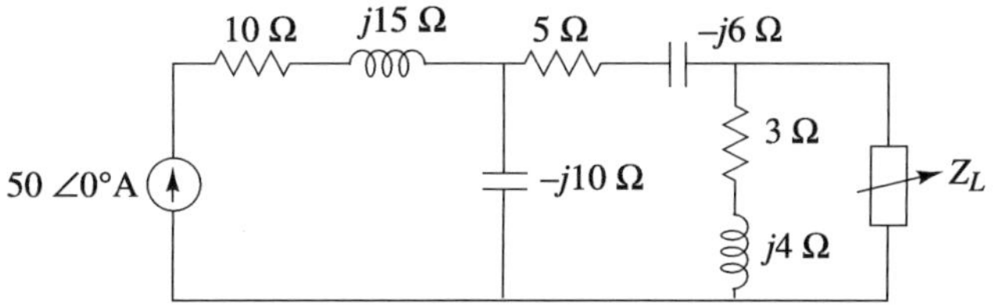

Fig. 5.80

[1032.35 W]

15. For the network shown, find the value of Z_L that will receive the maximum power. Determine also this power.

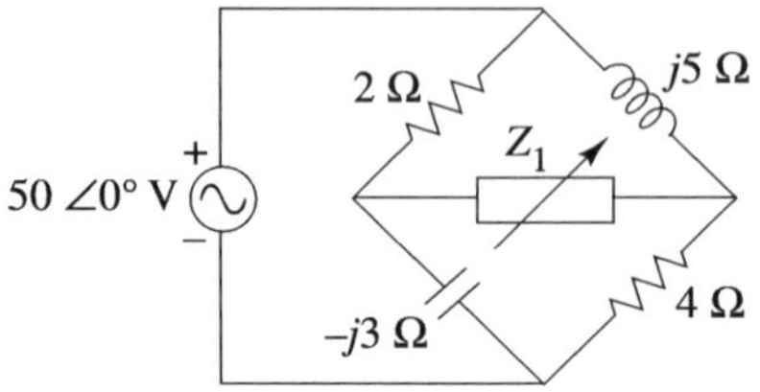

Fig. 5.81

[3.82 – j1.03 Ω, 54.5 W]

Objective-Type Questions

1. In the Fig. 5.82 shown, the equivalent impedance seen across terminals *a*, *b*, is

   (a) $\frac{16}{3}$ Ω (b) $\frac{8}{3}$ Ω

   (c) $\left(\frac{8}{3} + j12\right)$ Ω (d) none of the above

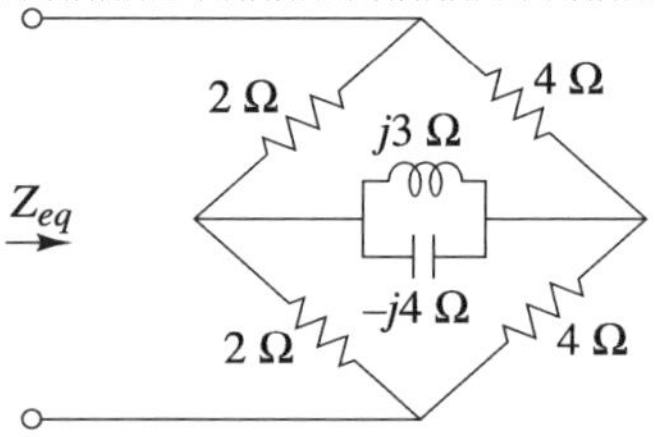

Fig. 5.82

2. The Thevenin equivalent voltage V_{Th} appearing between the terminals *A* and *B* of the network shown in Fig. 5.83 is given by

   (a) $j16(3 - j4)$ (b) $j16(3 + j4)$

   (c) $16(3 + j4)$ (d) $16(3 - j4)$

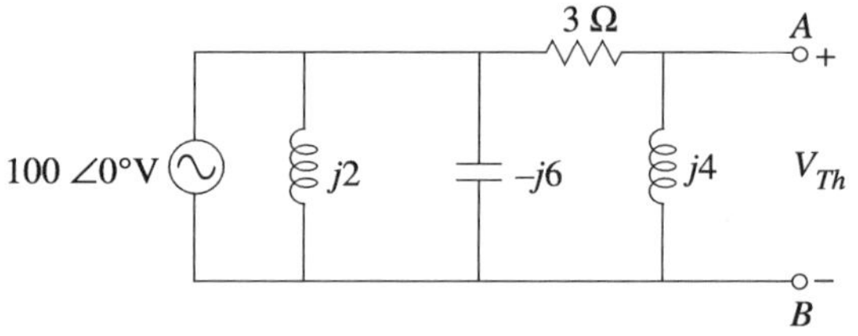

Fig. 5.83

3. A source of angular frequency of 1 rad/s has a source impedance consisting of a 1-Ω resistance in series with a 1-H inductance. The load that will obtain the maximum power fransfer is

   (a) 1-Ω resistance

   (b) 1-Ω resistance in parallel with 1-H inductance

   (c) 1-Ω resistance in series with 1-F capacitance

   (d) 1-Ω resistance in parallel with 1-F capacitance

4. The equivalent inductance measured between the terminals 1 and 2 for circuit shown in Fig. 5.84 is

   (a) $L_1 + L_2 + M$ (b) $L_1 + L_2 - M$

   (c) $L_1 + L_2 + 2M$ (d) $L_1 + L_2 - 2M$

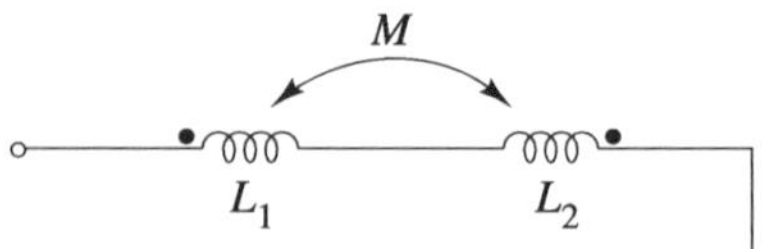

Fig. 5.84

5. For the circut shown, the instantaneous current $i(t)$ is

(a) $\frac{10\sqrt{3}}{2}\ \angle 90°$ A

(b) $\frac{10\sqrt{3}}{2}\ \angle -90°$ A

(c) $5\ \angle 60°$ A

(d) $5\angle -60°$ A

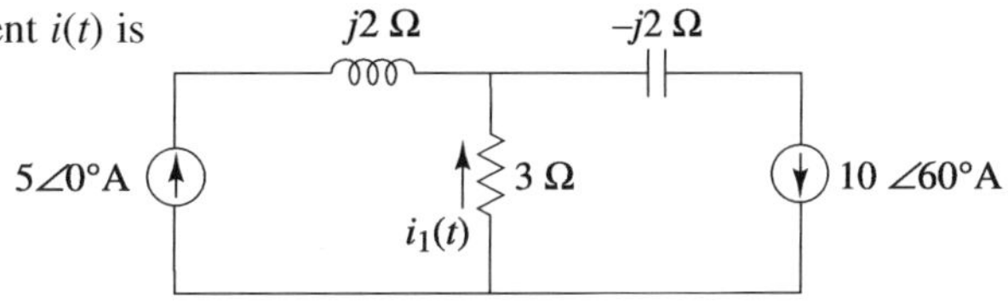

Fig. 5.85

6. Impedance Z as shown in given Fig. 5.86 is

(a) $j29\ \Omega$ (b) $j9\ \Omega$

(c) $j19\ \Omega$ (d) $j39\ \Omega$

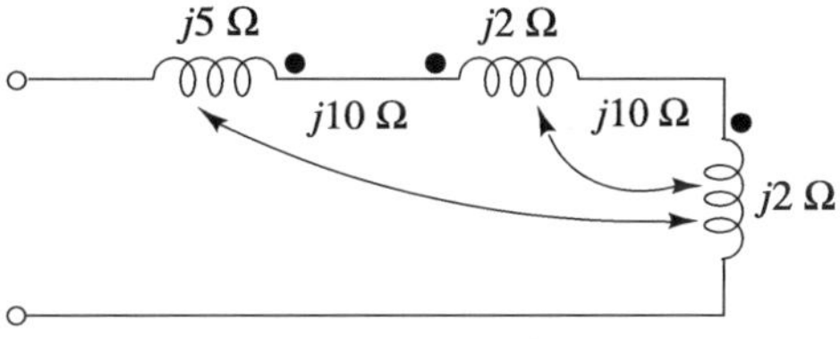

Fig. 5.86

7. In the circuit shown in the Fig. 5.87 the current supplied by the sinusoidal current source I is

(a) 28 A (b) 4 A

(c) 20 A (d) cannot be determined

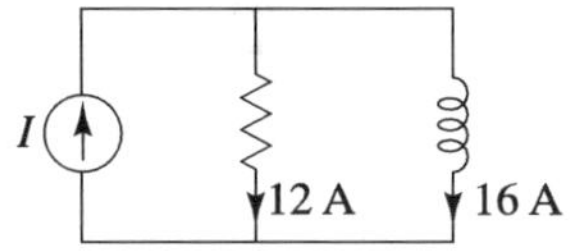

Fig. 5.87

8. In the circuit of Fig. 5.88 the magnitudes of V_L and V_C are twice that of V_R. The inductance of the coil is

(a) 2.14 mH (b) 5.3 H

(c) 31.8 mH (d) 1.32 H

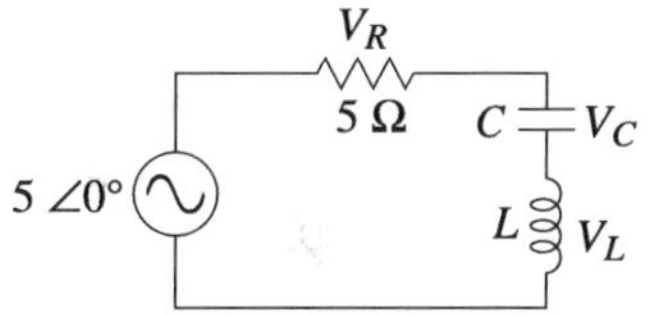

Fig. 5.88

9. Phase angle of the current I with respect to the voltage V_1 in the circuit shown in Fig. 5.89 is

(a) 0° (b) 45°

(c) –45° (d) –90°

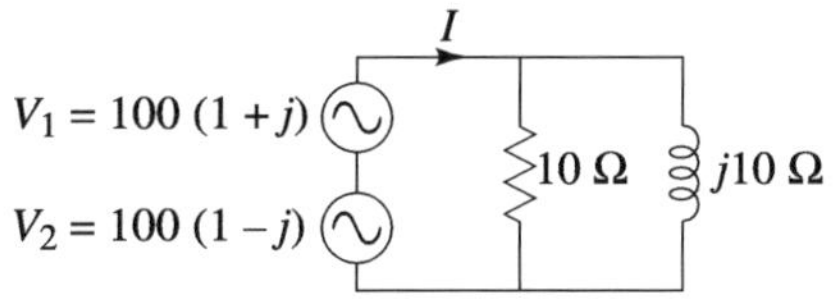

Fig. 5.89

Answers to Objective-Type Questions

1. (b) **2.** (d) **3.** (c) **4.** (d) **5.** (a) **6.** (b)

7. (c) **8.** (c) **9.** (d)

6 Three-Phase Circuits

6.1 INTRODUCTION

A system which generates a single alternating voltage and current is termed a *single-phase system*. It utilizes only one winding. A *polyphase system* utilizes more than one winding. It will produce as many induced voltages as the number of windings.

A three-phase system consists of three separate but identical windings that are displaced by 120 electrical degrees from each other. When these three windings are rotated in an anticlockwise direction with constant angular velocity in a uniform magnetic field, the emfs are induced in each winding which have the same magnitude and frequency but displaced 120° from one another.

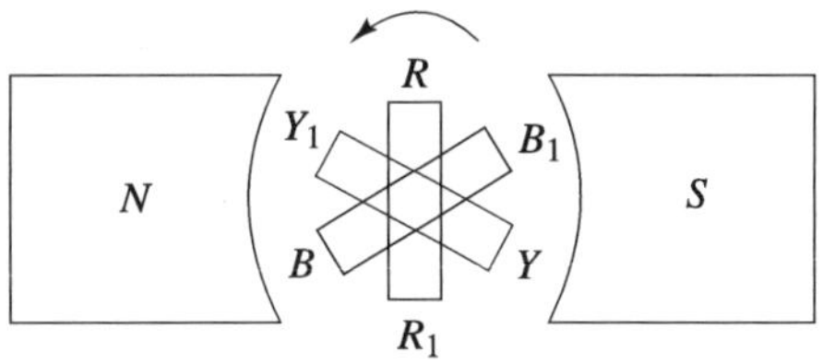

Fig. 6.1

The instantaneous values of generated voltage in windings RR_1, YY_1 and BB_1 are given by

$$e_R = E_m \sin \theta$$

$$e_Y = E_m \sin (\theta - 120°)$$

$$e_B = E_m \sin (\theta - 240°)$$

where E_m is the maximum value of the induced voltage in each winding. The waveforms of these three voltages are shown in Fig. 6.2.

Figure 6.3 shows the phasor diagram of these three induced voltages.

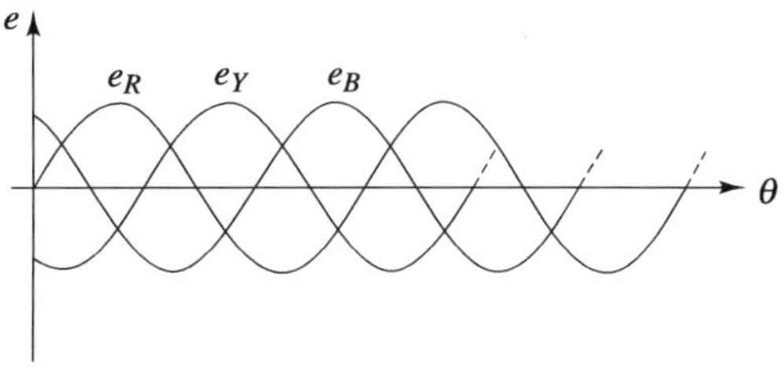

Fig. 6.2

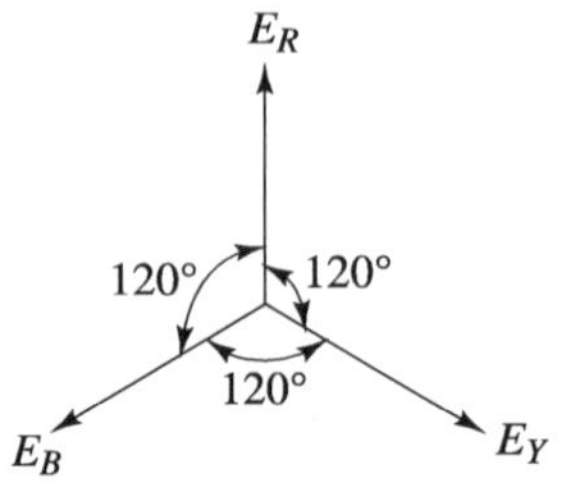

Fig. 6.3

6.2 ADVANTAGES OF A THREE-PHASE SYSTEM

1. In a single-phase system, the instantaneous power is fluctuating in nature. However, in a three-phase system, it is constant at all times.
2. The output of a three-phase system is greater than that of a single-phase system.
3. Transmission and distribution of a three-phase system is cheaper than that of a single-phase system.
4. Three-phase motors are more efficient and have higher power factor than single-phase motors of the same frequency.
5. Three-phase motors are self-starting whereas single-phase motors are not self-starting.

6.3 SOME DEFINITIONS

1. Phase Sequence The sequence in which the voltages in the three phases reach maximum positive value is called the *phase sequence* or *phase order*. From the phasor diagram of a three-phase system, it is clear that the voltage in the coil R attains maximum positive value first, next in the coil Y and then in the coil B. Hence, the phase sequence is R-Y-B.

2. Phase Voltage The voltage induced in each winding is called the *phase voltage*.

3. Phase Current The current flowing through each winding is called the *phase current*.

4. Line Voltage The voltage available between any pair of terminals or lines is called the *line voltage*.

5. Line Current The current flowing through each line is called the *line current*.

6. Balanced System A three-phase system is said to be balanced if the

(a) voltages in the three phases are equal in magnitude and differ in phase from one another by 120°
(b) currents in the three phases are equal in magnitude and differ in phase from one another by 120°
(c) loads connected across the three phases are identical, i.e., all the loads have the same magnitude and power factor

6.4 INTERCONNECTION OF THREE PHASES

In a three-phase system, there are three windings. Each winding has two terminals, viz., 'start' and 'finish'. If a separate load is connected across each winding as shown in Fig. 6.4, six conductors are required to transmit and distribute power. This will make the system complicated and expensive.

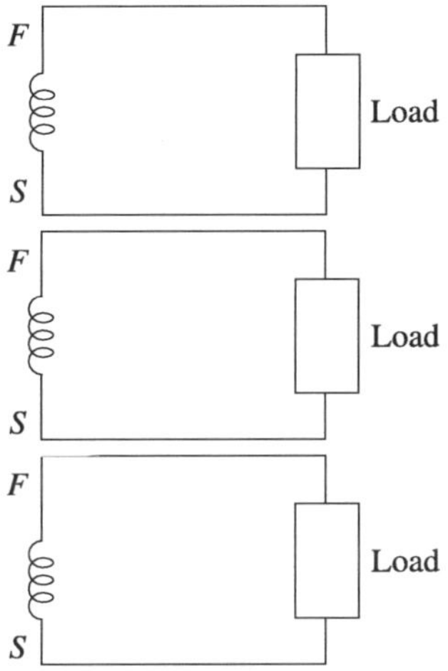

Fig. 6.4

In order to reduce the number of conductors, the three windings are connected in the following two ways :

1. Star or Wye Connection
2. Delta or Mesh Connection.

6.5 STAR OR WYE CONNECTION

In this method, similar terminals (start or finish) of the three windings are joined together as shown in Fig. 6.5. The common point is called *star* or *neutral point.*

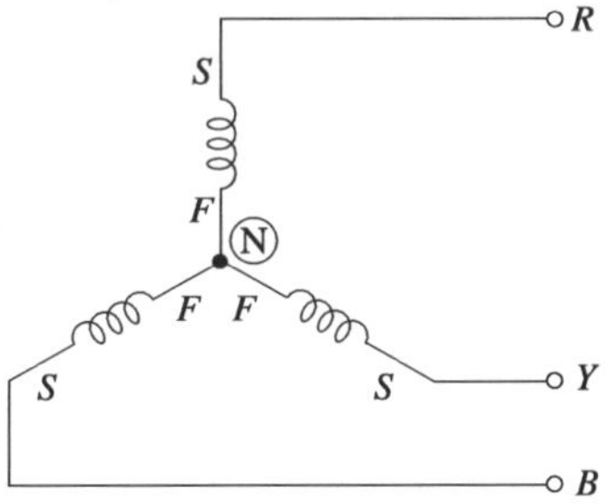

Fig. 6.5

Figure 6.6 shows a three-phase system in star connection.

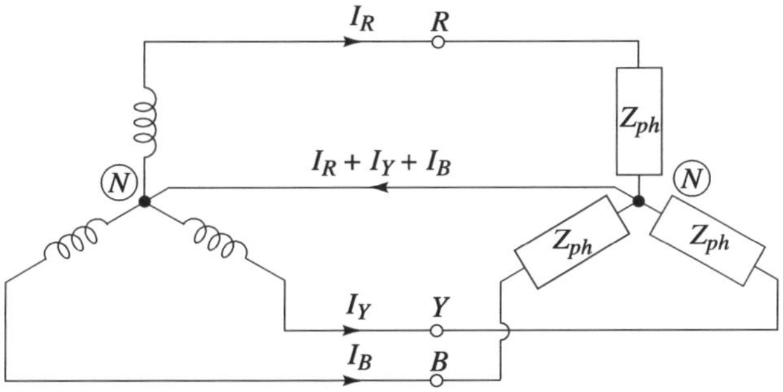

Fig. 6.6

This system is called a three-phase, four-wire system. If three identical loads are connected to each phase, the current flowing through the neutral wire is the sum of the three currents I_R, I_Y and I_B. Since the impedances are identical, the three currents are equal in magnitude but differ in phase from one another by 120°.

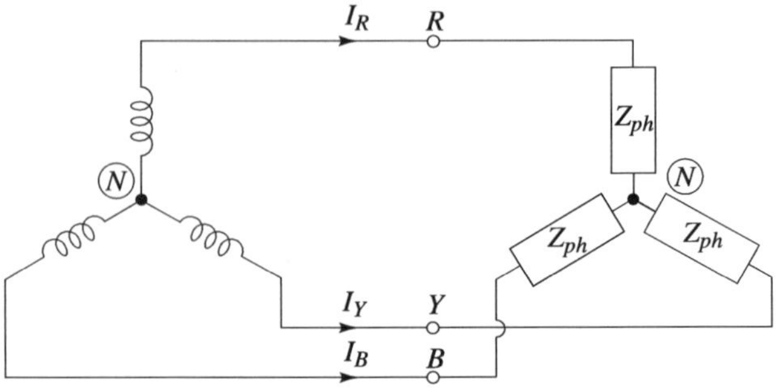

Fig. 6.7

$$i_R = I_m \sin \theta$$
$$i_Y = I_m \sin (\theta - 120°)$$
$$i_B = I_m \sin (\theta - 240°)$$
$$i_R + i_Y + i_B = I_m \sin \theta + I_m \sin (\theta - 120°) + I_m \sin (\theta - 240°) = 0$$

Therefore, the neutral wire can be removed without any way affecting the voltages or currents in the circuit as shown in Fig. 6.7. This constitutes a three-phase, three-wire system. If the load is not balanced, the neutral wire carries some current.

6.6 DELTA OR MESH CONNECTION

In this method, dissimilar terminals of the three windings are joined together, i.e., the 'finish' terminal of one winding is connected to the 'start' terminal of the other winding, and so on, as shown in Fig. 6.8. This system is also called three-phase, three-wire system.

For a balanced system, the sum of the three phase voltages round the closed mesh is zero. The three emfs are equal in magnitude but differ in phase from one another by 120°.

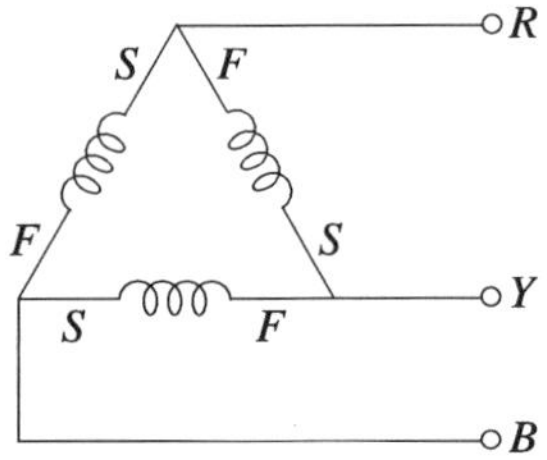

Fig. 6.8

$$e_R = E_m \sin \theta$$
$$e_Y = E_m \sin (\theta - 120°)$$
$$e_B = E_m \sin (\theta - 240°)$$
$$e_R + e_Y + e_B = E_m \sin \theta + E_m \sin (\theta - 120°) + E_m \sin (\theta - 240°) = 0$$

6.7 VOLTAGE, CURRENT AND POWER RELATIONS IN A BALANCED STAR-CONNECTED LOAD

6.7.1 Relation Between Line Voltage and Phase Voltage

Since the system is balanced, the three-phase voltages V_{RN}, V_{YN} and V_{BN} are equal in magnitude and differ in phase from one another by 120°.

Let $\qquad V_{RN} = V_{YN} = V_{BN} = V_{ph}$

where V_{ph} indicates the rms value of phase voltage.

$$\overline{V_{RN}} = V_{ph} \angle 0°$$
$$\overline{V_{YN}} = V_{ph} \angle -120°$$
$$\overline{V_{BN}} = V_{ph} \angle -240°$$

Let $\qquad V_{RY} = V_{YB} = V_{BR} = V_L$

where V_L indicates the rms value of line voltage.

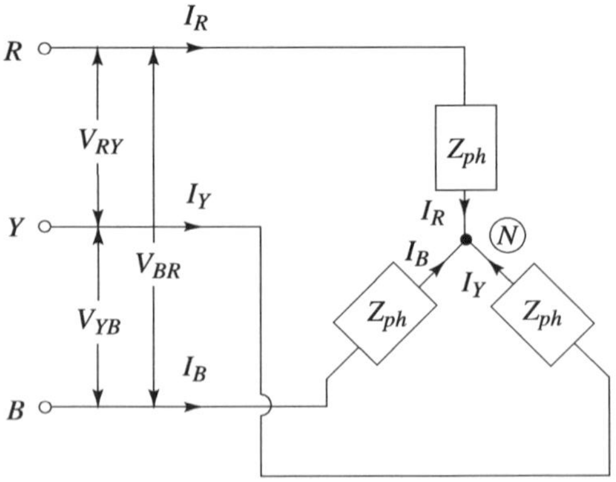

Fig. 6.9

Applying Kirchhoff's voltage law,

$$\begin{aligned}\overline{V_{RY}} &= \overline{V_{RN}} + \overline{V_{NY}} = \overline{V_{RN}} - \overline{V_{NY}}\\ &= V_{ph}\angle 0^\circ - V_{ph}\angle -120^\circ\\ &= V_{ph} + j0 + 0.5\,V_{ph} + j0.866\,V_{ph}\\ &= 1.5\,V_{ph} + j0.866\,V_{ph}\\ &= \sqrt{3}\,V_{ph}\angle 30^\circ\end{aligned}$$

Similarly,

$$\begin{aligned}\overline{V_{YB}} &= \overline{V_{YN}} + \overline{V_{NB}} = \sqrt{3}\,V_{ph}\angle 30^\circ\\ \overline{V_{BR}} &= \overline{V_{BN}} + \overline{V_{NR}} = \sqrt{3}\,V_{ph}\angle 30^\circ\end{aligned}$$

Thus in a star-connected, three-phase system, $V_L = \sqrt{3}\,V_{ph}$ and line voltages lead respective phase voltages by 30°.

6.7.2 Phasor Diagram

(Lagging power factor)

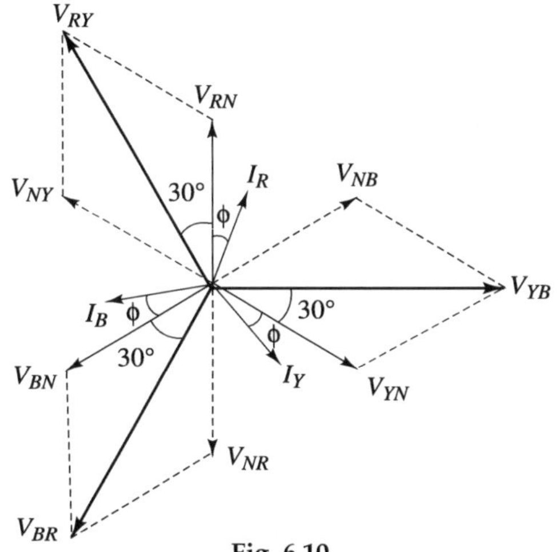

Fig. 6.10

6.7.3 Relation Between Line Current and Phase Current

From Fig. 6.9, it is clear that line current is equal to the phase current.

$$I_L = I_{ph}$$

6.7.4 Power

The total power in a three-phase system is the sum of powers in the three phases. For a balanced load, the power consumed in each load phase is the same.

Total power $P = 3 \times$ power in each phase $= 3\,V_{ph}\,I_{ph}\cos\phi$

In a star-connected, three-phase system,

$$V_{ph} = \frac{V_L}{\sqrt{3}}$$

$$I_{ph} = I_L$$

$$P = 3 \times \frac{V_L}{\sqrt{3}} \times I_L \cos\phi = \sqrt{3}\,V_L\,I_L\cos\phi$$

where ϕ is the phase difference between phase voltage and corresponding phase current.

Similarly, total reactive power $Q = 3\ V_{ph}\ I_{ph} \sin \phi$

$$= \sqrt{3}\ V_L\ I_L \sin \phi$$

Total apparent power $S = 3\ V_{ph}\ I_{ph} = \sqrt{3}\ V_L\ I_L$

6.8 VOLTAGE, CURRENT AND POWER RELATIONS IN A BALANCED DELTA-CONNECTED LOAD

6.8.1 Relation Between Line Voltage and Phase Voltage

From Fig. 6.11, it is clear that line voltage is equal to phase voltage.

$$V_L = V_{ph}$$

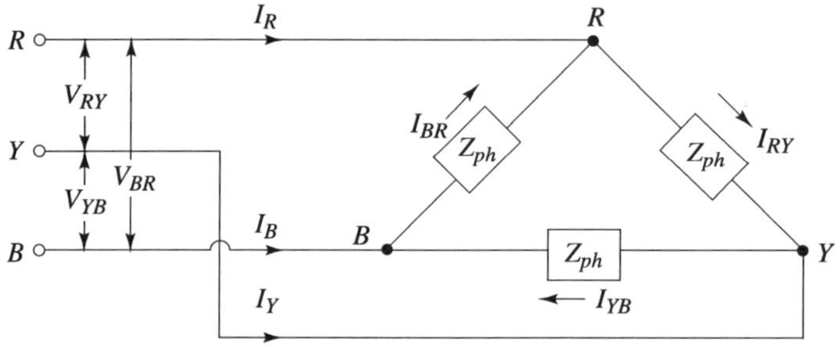

Fig. 6.11

6.8.2 Relation Between Line Current and Phase Current

Since the system is balanced, the three-phase currents I_{RY}, I_{YB} and I_{BR} are equal in magnitude but differ in phase from one another by 120°.

Let $I_{RY} = I_{YB} = I_{BR} = I_{ph}$

where I_{ph} indicates rms value of the phase current.

$$\overline{I_{RY}} = I_{ph} \angle 0°$$

$$\overline{I_{YB}} = I_{ph} \angle -120°$$

$$\overline{I_{BR}} = I_{ph} \angle -240°$$

Let $I_R = I_Y = I_B = I_L$

where I_L indicates rms value of the line current.

Applying Kirchhoff's current law,

$$\overline{I_R} + \overline{I_{BR}} = \overline{I_{RY}}$$

$$\overline{I_R} = \overline{I_{RY}} - \overline{I_{BR}} = I_{ph} \angle 0° - I_{ph} \angle -240°$$

$$= I_{ph} + j0 + 0.5\ I_{ph} - j0.866\ I_{ph}$$

$$= \sqrt{3}\ I_{ph} \angle -30°$$

Similarly,

$$\overline{I_Y} = \overline{I_{YB}} - \overline{I_{RY}} = \sqrt{3}\ I_{ph} \angle -30°$$

$$\overline{I_B} = \overline{I_{BR}} - \overline{I_{YB}} = \sqrt{3}\ I_{ph} \angle -30°$$

Thus in a delta-connected, three-phase system, $I_L = \sqrt{3}\ I_{ph}$ and line currents are 30° behind the respective phase currents.

6.8.3 Phasor Diagram

(Lagging power factor)

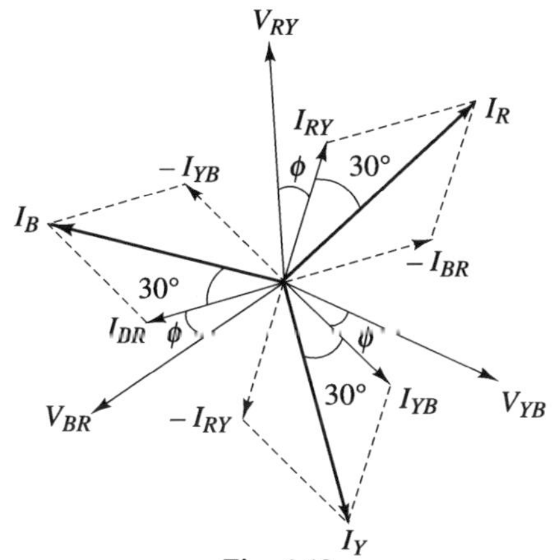

Fig. 6.12

6.8.4 Power

Total power $P = 3\, V_{ph}\, I_{ph} \cos\phi$

In a delta-connected, three-phase system,

$$V_{ph} = V_L$$

$$I_{ph} = \frac{I_L}{\sqrt{3}}$$

$$P = 3 \times V_L \times \frac{I_L}{\sqrt{3}} \cos\phi = \sqrt{3}\, V_L\, I_L \cos\phi$$

Total reactive power $\quad Q = 3\, V_{ph}\, I_{ph} \sin\phi = \sqrt{3}\, V_L\, I_L \sin\phi$

Total apparent power $\quad S = 3\, V_{ph}\, I_{ph} = \sqrt{3}\, V_L\, I_L$

6.9 BALANCED Y/Δ AND Δ/Y CONVERSIONS

Any balanced star-connected system can be completely converted into the equivalent delta-connected system and vice versa.

For a balanced star-connected load,

$$\text{Line voltage} = V_L$$

$$\text{Line current} = I_L$$

$$\text{Impedance/phase} = Z_Y$$

$$V_{ph} = \frac{V_L}{\sqrt{3}}$$

$$I_{ph} = I_L$$

$$Z_Y = \frac{V_{ph}}{I_{ph}} = \frac{V_L}{\sqrt{3} I_L}$$

For an equivalent delta-connected system, the line voltages and currents must have the same values as in the star-connected system, i.e.,

$$\text{Line voltage} = V_L$$

$$\text{Line current} = I_L$$

$$\text{Impedance/phase} = Z_\Delta$$

$$V_{ph} = V_L$$

$$I_{ph} = \frac{I_L}{\sqrt{3}}$$

$$Z_\Delta = \frac{V_{ph}}{I_{ph}} = \frac{V_L}{\frac{I_L}{\sqrt{3}}} = \sqrt{3}\frac{V_L}{I_L} = 3\ Z_Y$$

$$Z_Y = \frac{Z_\Delta}{3}$$

6.10 RELATION BETWEEN POWER IN DELTA AND STAR SYSTEM

Let a balanced load be connected in star having impedance per phase as Z_{ph}.

For a star-connected load,

$$V_{ph} = \frac{V_L}{\sqrt{3}}$$

$$I_{ph} = \frac{V_{ph}}{Z_{ph}} = \frac{V_L}{\sqrt{3}Z_{ph}}$$

$$I_L = I_{ph} = \frac{V_L}{\sqrt{3}Z_{ph}}$$

Now

$$P_Y = \sqrt{3}\ V_L\ I_L \cos\phi$$

$$= \sqrt{3} \times V_L \times \frac{V_L}{\sqrt{3}Z_{ph}} \cos\phi = \frac{V_L^2}{Z_{ph}} \cos\phi$$

For a delta-connected load,

$$V_{ph} = V_L$$

$$I_{ph} = \frac{V_{ph}}{Z_{ph}} = \frac{V_L}{Z_{ph}}$$

$$I_L = \sqrt{3}\ I_{ph} = \sqrt{3}\frac{V_L}{Z_{ph}}$$

Now

$$P_\Delta = \sqrt{3}\ V_L\ I_L \cos\phi$$

$$= \sqrt{3} \times V_L \times \sqrt{3}\ \frac{V_L}{Z_{ph}} \cos\phi$$

$$= 3\frac{V_L^2}{Z_{ph}} \cos\phi = 3P_Y$$

Thus, power consumed by a balanced, star-connected load is one-third of that in the case of a delta-connected load.

6.11 COMPARISON BETWEEN STAR AND DELTA CONNECTION

Star Connection	*Delta Connection*
1. $V_L = \sqrt{3}\, V_{ph}$	1. $V_L = V_{ph}$
2. $I_L = I_{ph}$	2. $I_L = \sqrt{3}\, I_{ph}$
3. Line voltage leads the respective phase voltage by 30°.	3. Line current lags behind the respective phase current by 30°.
4. Power in star connection is one-third of power in delta connection.	4. Power in delta connection is 3 times of the power in star connection.
5. Three-phase, three-wire and three-phase, four-wire systems are possible.	5. Only three-phase, three-wire system is possible.
6. The phasor sum of all the phase currents is zero.	6. The phasor sum of all the phase voltages is zero.

Example 6.1 *Three equal impedances, each of 8 + j10 ohms are connected in star. This is further connected to a 440-V, 50-Hz, three-phase supply. Calculate the active and reactive power and line and phase currents.*

Solution

Data

$$\bar{Z}_{ph} = 8 + j10\ \Omega$$
$$V_L = 440\ \text{V}$$
$$f = 50\ \text{Hz}$$

For a star-connected load,

$$V_{ph} = \frac{V_L}{\sqrt{3}} = \frac{440}{\sqrt{3}} = 254.03\ \text{V}$$
$$\bar{Z}_{ph} = 8 + j10 = 12.81\ \angle 51.34°\ \Omega$$
$$Z_{ph} = 12.81\ \Omega$$
$$\phi = 51.34°$$
$$I_{ph} = \frac{V_{ph}}{Z_{ph}} = \frac{254.03}{12.81} = 19.83\ \text{A}$$
$$I_L = I_{ph} = 19.83\ \text{A}$$
$$P = \sqrt{3}\, V_L I_L \cos\phi = \sqrt{3} \times 440 \times 19.83 \times \cos(51.34°) = 9.44\ \text{kW}$$
$$Q = \sqrt{3}\, V_L I_L \sin\phi = \sqrt{3} \times 440 \times 19.83 \times \sin(51.34°) = 11.81\ \text{kVAR.}$$

Example 6.2 *A balanced delta-connected load of impedance (8 – j6) ohm per phase is connected to a three-phase, 230-V, 50-Hz supply. Calculate (i) line current, (ii) power factor, and (iii) reactive power.*

Solution

Data

$$\bar{Z}_{ph} = 8 - j6\ \Omega$$
$$V_L = 230\ \text{V}$$
$$f = 50\ \text{Hz}$$

For delta-connected load,

$$V_{ph} = V_L = 230\ \text{V}$$

$$\bar{Z}_{ph} = 8 - j6 = 10 \angle -36.87° \ \Omega$$

$$Z_{ph} = 10 \ \Omega$$

$$\phi = 36.87°$$

$$\text{Power factor} = \cos(36.87°) = 0.8 \text{ (leading)}$$

$$I_{ph} = \frac{V_{ph}}{Z_{ph}} = \frac{230}{10} = 23 \text{ A}$$

$$I_L = \sqrt{3}\, I_{ph} = \sqrt{3} \times 23 = 39.84 \text{ A}$$

$$Q = \sqrt{3}\ V_L I_L \sin\phi = \sqrt{3} \times 230 \times 39.84 \times \sin(36.87°) = 9.52 \text{ kVAR}$$

Example 6.3 *Three coils, each having resistance and inductance of 8-Ω and 0.02-H respectively, are connected in star across a three-phase, 230-V, 50-Hz supply. Find the line current, power factor, power, reactive voltamperes and total voltamperes.*

Solution

Data

$$R = 8 \ \Omega$$

$$L = 0.02 \text{ H}$$

$$V_L = 230 \text{ V}$$

$$f = 50 \text{ Hz}$$

For a star-connected load,

$$V_{ph} = \frac{V_L}{\sqrt{3}} = \frac{230}{\sqrt{3}} = 132.79 \text{ V}$$

$$X_L = 2\pi f L = 2\pi \times 50 \times 0.02 = 6.28 \ \Omega$$

$$\bar{Z}_{ph} = R + jX_L$$

$$= 8 + j6.28 = 10.17 \angle 38.13° \ \Omega$$

$$Z_{ph} = 10.17 \ \Omega$$

$$\phi = 38.13°$$

$$\text{Power factor} = \cos(38.13°) = 0.786 \text{ (lagging)}$$

$$I_{ph} = \frac{V_{ph}}{Z_{ph}} = \frac{132.79}{10.17} = 13.05 \text{ A}$$

$$I_L = I_{ph} = 13.05 \text{ A}$$

$$P = \sqrt{3}\ V_L I_L \cos\phi = \sqrt{3} \times 230 \times 13.05 \times 0.786 = 4.088 \text{ kW}$$

$$Q = \sqrt{3}\ V_L I_L \sin\phi$$

$$= \sqrt{3} \times 230 \times 13.05 \times \sin(38.13°) = 3.21 \text{ kVAR}$$

$$S = \sqrt{3}\, V_L I_L$$

$$= \sqrt{3} \times 230 \times 13.05 = 5.198 \text{ kVA}$$

Example 6.4 *Three coils each having a resistance of 8-Ω and inductance of 0.02-H are connected in delta to a three-phase, 400-V, 50-Hz supply. Calculate the line current and power absorbed.*

Solution

Data

$$R = 8 \ \Omega$$

$$L = 0.02 \text{ H}$$

$$V_L = 400 \text{ V}$$

$$f = 50 \text{ Hz}$$

For a delta-connected load,

$$V_L = V_{ph} = 400 \text{ V}$$
$$X_L = 2\pi f L = 2\pi \times 50 \times 0.02 = 6.28\ \Omega$$
$$\bar{Z}_{ph} = R + jX_L = 8 + j6.28 = 10.17\ \angle 38.13^\circ\ \Omega$$
$$Z_{ph} = 10.17\ \Omega$$
$$\phi = 38.13^\circ$$
$$I_{ph} = \frac{V_{ph}}{Z_{ph}} = \frac{400}{10.17} = 39.33 \text{ A}$$
$$I_L = \sqrt{3}\ I_{ph} = \sqrt{3} \times 39.33 = 68.12 \text{ A}$$
$$P = \sqrt{3}\ V_L I_L \cos\phi$$
$$= \sqrt{3} \times 400 \times 68.12 \times \cos(38.13^\circ) = 37.12 \text{ kW}$$

Example 6.5 *Three similar coils A, B, and C are available. Each coil has a 9-Ω resistance and a 12-Ω reactance. They are connected in delta to a three-phase, 440-V, 50-Hz supply. Calculate for this load the (i) phase current, (ii) line current, (iii) power factor, (iv) total kVA, (v) active power, and (vi) reactive power. If these coils are connected in star across the same supply, calculate all the above quantities.*

Solution

Data $R = 9\ \Omega$ $\quad f = 50$ Hz

$X_L = 12\ \Omega$ $\quad V_L = 440$ V

For a delta-connected load,

$$V_L = V_{ph} = 440 \text{ V}$$
$$\bar{Z}_{ph} = 9 + j12 = 15\ \angle 53.13^\circ\ \Omega$$
$$Z_{ph} = 15\ \Omega$$
$$\phi = 53.13^\circ$$
$$I_{ph} = \frac{V_{ph}}{Z_{ph}} = \frac{440}{15} = 29.33 \text{ A}$$
$$I_L = \sqrt{3}\ I_{ph} = \sqrt{3} \times 29.33 = 50.8 \text{ A}$$
$$\text{Power factor} = \cos\phi = \cos(53.13^\circ) = 0.6 \text{ (lagging)}$$
$$S = \sqrt{3}\ V_L I_L$$
$$= \sqrt{3} \times 440 \times 50.8 = 38.71 \text{ kVA}$$
$$P = \sqrt{3}\ V_L I_L \cos\phi$$
$$= \sqrt{3} \times 440 \times 50.8 \times 0.6 = 23.23 \text{ kW}$$
$$Q = \sqrt{3}\ V_L I_L \sin\phi$$
$$= \sqrt{3} \times 440 \times 50.8 \times \sin(53.13^\circ) = 30.97 \text{ kVAR}$$

If these coils are connected in star across the same supply,

$$V_{ph} = \frac{V_L}{\sqrt{3}} = \frac{440}{\sqrt{3}} = 254.03 \text{ V}$$
$$Z_{ph} = 15\ \Omega$$

$$I_{ph} = \frac{V_{ph}}{Z_{ph}} = \frac{254.03}{15} = 16.94 \text{ A}$$

$$I_L = I_{ph} = 16.94 \text{ A}$$

$$\text{Power factor} = 0.6 \text{ (lagging)}$$

$$S = \sqrt{3}\ V_L I_L = \sqrt{3} \times 440 \times 16.94 = 12.91 \text{ kVA}$$

$$P = \sqrt{3}\ V_L I_L \cos\phi = \sqrt{3} \times 440 \times 16.94 \times 0.6 = 7.74 \text{ kW}$$

$$Q = \sqrt{3}\ V_L I_L \sin\phi$$

$$= \sqrt{3} \times 440 \times 16.94 \times 0.8 = 12.33 \text{ kVAR}$$

Example 6.6 *A 415-V, 50-Hz, three-phase voltage is applied to three star-connected identical impedances. Each impedance consists of a resistance of 15 Ω, a capacitance of 177 μF and an inductance of 0.1 henry in series. Find the (i) phase current, (ii) line current, (iii) power factor, (iv) active power, (v) reactive power, and (vi) total VA. Draw a neat phasor diagram. If the same impedances are connected in delta, find the (i) line current, and (ii) power consumed.*

Solution

Data $V_L = 415\ \Omega$ $\quad f = 50$ Hz

$R = 15\ \Omega$ $\quad C = 177\ \mu$F

$L = 0.1$ H

For a star-connected load,

$$V_{ph} = \frac{V_L}{\sqrt{3}} = \frac{415}{\sqrt{3}} = 239.6 \text{ V}$$

$$X_L = 2\pi f L = 2\pi \times 50 \times 0.1 = 31.42\ \Omega$$

$$X_C = \frac{1}{2\pi f C} = \frac{1}{2\pi \times 50 \times 177 \times 10^{-6}} = 17.98\ \Omega$$

$$\bar{Z}_{ph} = R + jX_L - jX_C = 15 + j31.42 - j17.98$$

$$= 15 + j13.44 = 20.14 \angle 41.86^\circ\ \Omega$$

$$Z_{ph} = 20.14\ \Omega$$

$$\phi = 41.86^\circ$$

$$\text{Power factor} = \cos\phi = \cos(41.86^\circ) = 0.744 \text{ (lagging)}$$

$$I_{ph} = \frac{V_{ph}}{Z_{ph}} = \frac{239.6}{20.14} = 11.9 \text{ A}$$

$$I_L = I_{ph} = 11.9 \text{ A}$$

$$P = \sqrt{3}\ V_L I_L \cos\phi$$

$$= \sqrt{3} \times 415 \times 11.9 \times 0.744 = 6.36 \text{ kW}$$

$$Q = \sqrt{3}\ V_L I_L \sin\phi$$

$$= \sqrt{3} \times 415 \times 11.9 \times \sin(41.86^\circ) = 5.71 \text{ kVAR}$$

$$S = \sqrt{3}\ V_L I_L$$

$$= \sqrt{3} \times 415 \times 11.9 = 8.55 \text{ kVA}$$

Phasor Diagram

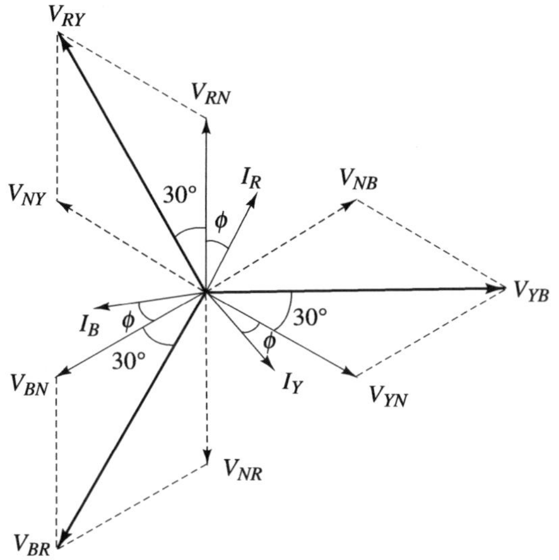

$\phi = 41.86°$

Fig. 6.13

If the same impedances are connected in delta,

$$V_L = V_{ph} = 415 \text{ V}$$
$$Z_{ph} = 20.14\ \Omega$$
$$I_{ph} = \frac{V_{ph}}{Z_{ph}} = \frac{415}{20.14} = 20.61 \text{ A}$$
$$I_L = \sqrt{3}\ I_{ph} = \sqrt{3} \times 20.61 = 35.69 \text{ A}$$
$$P = \sqrt{3}\ V_L I_L \cos \phi$$
$$= \sqrt{3} \times 415 \times 35.69 \times 0.744 = 19.09 \text{ kW}$$

Example 6.7 *Each phase of a delta-connected load consists of a 50-mH inductor in series with a parallel combination of a 50-Ω resistor and a 50-μF capacitor. The load is connected to a three-phase, 550-V, 800-rad/s ac supply. Find the (i) phase current, (ii) line current , (iii) power drawn, (iv) power factor, (v) reactive power, and (vi) kVA rating of the load.*

Solution

Data

$$L = 50 \text{ mH}$$
$$R = 50\ \Omega$$
$$C = 50\ \mu\text{F}$$
$$V_L = 550 \text{ V}$$
$$\omega = 800 \text{ rad/s}$$

For a delta-connected load,

$$V_L = V_{ph} = 550 \text{ V}$$
$$X_L = \omega L = 800 \times 50 \times 10^{-3} = 40\ \Omega$$

$$X_C = \frac{1}{\omega C} = \frac{1}{800 \times 50 \times 10^{-6}} = 25\ \Omega$$

$$\bar{Z}_{ph} = jX_L + \frac{R(-jX_C)}{R - jX_C}$$

$$= j40 + \frac{50(-j25)}{50 - j25}$$

$$= 10 + j20 = 22.36\ \angle 63.43°\ \Omega$$

$$Z_{ph} = 22.36\ \Omega$$

$$\phi = 63.43°$$

$$I_{ph} = \frac{V_{ph}}{Z_{ph}} = \frac{550}{22.36} = 24.6\ \text{A}$$

$$I_L = \sqrt{3}\, I_{ph} = \sqrt{3} \times 24.6 = 42.61\ \text{A}$$

$$\text{Power factor} = \cos\phi$$

$$= \cos(63.43°) = 0.447\ \text{(lagging)}$$

$$P = \sqrt{3}\ V_L I_L \cos\phi$$

$$= \sqrt{3} \times 550 \times 42.61 \times 0.447 = 18.14\ \text{kW}$$

$$Q = \sqrt{3}\ V_L I_L \sin\phi$$

$$= \sqrt{3} \times 550 \times 42.61 \times \sin(63.43°) = 36.3\ \text{kVAR}$$

$$S = \sqrt{3}\ V_L I_L$$

$$= \sqrt{3} \times 550 \times 42.61 = 40.59\ \text{kVA}$$

Example 6.8 *Three identical coils connected in delta to a 440-V, three-phase supply take a total power of 50 kW and a line current of 90 A. Find the (i) phase current, (ii) power factor, and (iii) total apparent power taken by the coils.*

Solution

Data

$$V_L = 440\ \text{V}$$

$$P = 50\ \text{kW}$$

$$I_L = 90\ \text{A}$$

For a delta-connected load,

$$V_L = V_{ph} = 440\ \text{V}$$

$$I_{ph} = \frac{I_L}{\sqrt{3}} = \frac{90}{\sqrt{3}} = 51.96\ \text{A}$$

$$P = \sqrt{3}\, V_L I_L \cos\phi$$

$$50 \times 10^3 = \sqrt{3} \times 440 \times 90 \times \cos\phi$$

$$\cos\phi = 0.73\ \text{(lagging)}$$

$$S = \sqrt{3}\, V_L I_L = \sqrt{3} \times 440 \times 90 = 68.59\ \text{kVA}$$

Example 6.9 *Three similar choke coils are connected in star to a three-phase supply. If the line current is 15 A, the total power consumed is 11 kW and the volt-ampere input is 15 kVA, find the line and phase voltages, the VAR input and the reactance and resistance of each coil. If these coils are now connected in delta to the same supply, calculate phase and line currents, active and reactive power.*

Solution

Data $I_L = 15\text{ A}$ $\quad P = 11\text{ kW}$

$S = 15\text{ kVA}$

For a star-connected load,

$$S = \sqrt{3}\, V_L I_L$$

$$15 \times 10^3 = \sqrt{3} \times V_L \times 15$$

$$V_L = 577.35\text{ V}$$

$$V_{ph} = \frac{V_L}{\sqrt{3}} = \frac{577.35}{\sqrt{3}} = 333.33\text{ V}$$

$$\cos\phi = \frac{P}{S} = \frac{11\times10^3}{15\times10^3} = 0.733$$

$$\phi = 42.86°$$

$$Q = \sqrt{3}\, V_L I_L \sin\phi = \sqrt{3} \times 577.35 \times 15 \times \sin(42.86°) = 10.2\text{ kVAR}$$

$$I_{ph} = I_L = 15\text{ A}$$

$$Z_{ph} = \frac{V_{ph}}{I_{ph}} = \frac{333.33}{15} = 22.22\ \Omega$$

$$R = Z_{ph}\cos\phi = 22.22 \times 0.733 = 16.29\ \Omega$$

$$X_L = Z_{ph}\sin\phi$$

$$= 22.22 \times \sin(42.86°) = 15.11\ \Omega$$

If these coils are now connected in delta,

$$V_{ph} = V_L = 577.35\text{ V}$$

$$Z_{ph} = 22.22\ \Omega$$

$$I_{ph} = \frac{V_{ph}}{Z_{ph}} = \frac{577.35}{22.22} = 25.98\text{ A}$$

$$I_L = \sqrt{3}\, I_{ph}$$

$$= \sqrt{3} \times 25.98 = 45\text{ A}$$

$$P = \sqrt{3}\, V_L I_L \cos\phi$$

$$= \sqrt{3} \times 577.35 \times 45 \times 0.733 = 32.98\text{ kW}$$

$$Q = \sqrt{3}\, V_L I_L \sin\phi$$

$$= \sqrt{3} \times 577.35 \times 45 \times \sin(42.86°) = 30.61\text{ kVAR}$$

Example 6.10 *A three-phase, star-connected source feeds 1500 kW at 0.85 power factor lag to a balanced mesh-connected load. Calculate the current, its active and reactive components in each phase of the source and the load. The line voltage is 2.2 kV.*

Solution

Data $P = 1500$ kW $\quad pf = 0.85$ (lagging)

$V_L = 2.2$ kV

For a delta-connected load,

$$P = \sqrt{3}\ V_L I_L \cos\phi$$

$$1500 \times 10^3 = \sqrt{3} \times 2.2 \times 10^3 \times I_L \times 0.85$$

$$I_L = 463.12\ \text{A}$$

$$I_{ph} = \frac{I_L}{\sqrt{3}} = \frac{463.12}{\sqrt{3}} = 267.38\ \text{A}$$

Active component in each phase of the load $= I_{ph} \cos\phi$

$= 267.38 \times 0.85 = 227.27$ A

Reactive component in each phase of the load $= I_{ph} \sin\phi$

$= 267.38 \times \sin(\cos^{-1} 0.85)$

$= 267.38 \times 0.526 = 140.85$ A

For a star-connected source, the phase current in the source will be the same as the line current drawn by the load.

Active component of this current in each phase of the source

$= 463.12 \times 0.85 = 393.65$ A

Reactive component of this current in each phase of the source

$= 463.12 \times 0.526 = 243.6$ A

Example 6.11 *A three-phase, 208-volt generator supplies a total of 1800 W at a line current of 10 A when three identical impedances are arranged in a Wye connection across the line terminals of the generator. Compute the resistive and reactive components of each phase impedance.*

Solution

Data $V_L = 208$ V $\quad P = 1800$ W

$I_L = 10$ A

For a Wye-connected load,

$$V_{ph} = \frac{V_L}{\sqrt{3}} = \frac{208}{\sqrt{3}} = 120.09\ \text{V}$$

$$I_{ph} = I_L = 10\ \text{A}$$

$$Z_{ph} = \frac{V_{ph}}{I_{ph}} = \frac{120.09}{10} = 12.09\ \Omega$$

$$P = \sqrt{3}\ V_L I_L \cos\phi$$

$$1800 = \sqrt{3} \times 208 \times 10 \times \cos\phi$$

$$\cos\phi = 0.5$$

$$\phi = 60°$$

$$R_{ph} = Z_{ph} \cos\phi = 12.09 \times 0.5 = 6.05\ \Omega$$

$$X_{ph} = Z_{ph} \sin\phi = 12.09 \times \sin(60°) = 10.47\ \Omega$$

Example 6.12 *A balanced, three-phase, star-connected load of 100 kW takes a leading current of 80 A, when connected across a three-phase, 1100-V, 50-Hz supply. Find the circuit constants of the load per phase.*

Solution

Data

$$P = 100 \text{ kW}$$
$$I_L = 80 \text{ A}$$
$$V_L = 1100 \text{ V}$$
$$f = 50 \text{ Hz}$$

For a star-connected load,

$$V_{ph} = \frac{V_L}{\sqrt{3}} = \frac{1100}{\sqrt{3}} = 635.08 \text{ V}$$
$$I_{ph} = I_L = 80 \text{ A}$$
$$Z_{ph} = \frac{V_{ph}}{I_{ph}} = \frac{635.08}{80} = 7.94 \ \Omega$$
$$P = \sqrt{3}\ V_L I_L \cos\phi$$
$$100 \times 10^3 = \sqrt{3} \times 1100 \times 80 \times \cos\phi$$
$$\cos\phi = 0.656 \text{ (leading)}$$
$$\phi = 49°$$
$$R_{ph} = Z_{ph}\cos\phi = 7.94 \times 0.656 = 5.21 \ \Omega$$
$$X_{ph} = Z_{ph}\sin\phi = 7.94 \times \sin(49°) = 6 \ \Omega$$

This reactance will be capacitive in nature as the current is leading.

$$X_C = \frac{1}{2\pi fC} = 6$$
$$C = \frac{1}{2\pi f X_C} = \frac{1}{2\pi \times 50 \times 6} = 530.52 \ \mu\text{F}$$

Example 6.13 *Three identical impedances are connected in delta to a three-phase supply of 400 V. The line current is 34.65 A, and the total power taken from the supply is 14.4 kW. Calculate the resistance and reactance values of each impedance.*

Solution

Data

$$V_L = 400 \text{ V}$$
$$I_L = 34.65 \text{ A}$$
$$P = 14.4 \text{ kW}$$

For a delta-connected load,

$$V_L = V_{ph} = 400 \text{ V}$$
$$I_{ph} = \frac{I_L}{\sqrt{3}} = \frac{34.65}{\sqrt{3}} = 20 \text{ A}$$
$$Z_{ph} = \frac{V_{ph}}{I_{ph}} = \frac{400}{20} = 20 \ \Omega$$
$$P = \sqrt{3}\ V_L I_L \cos\phi$$
$$14.4 \times 10^3 = \sqrt{3} \times 400 \times 34.65 \times \cos\phi$$

$$\cos \phi = 0.6$$
$$\phi = 53.13°$$
$$R_{ph} = Z_{ph} \cos \phi$$
$$= 20 \times 0.6 = 12\ \Omega$$
$$X_{ph} = Z_{ph} \sin \phi$$
$$= 20 \times \sin (53.13°) = 16\ \Omega$$

Example 4.14 *A balanced, three-phase load connected in delta draws a power of 10.44 kW at 200 V at a power factor of 0.5 lead. Find the values of the circuit elements and the reactive voltamperes drawn.*

Solution

Data $P = 10.44$ kW $\quad V_L = 200$ V

$pf = 0.5$ (leading)

For a delta-connected load,

$$V_L = V_{ph} = 200 \text{ V}$$
$$P = \sqrt{3}\, V_L I_L \cos \phi$$
$$10.44 \times 10^3 = \sqrt{3} \times 200 \times I_L \times 0.5$$
$$I_L = 60.28 \text{ A}$$
$$I_{ph} = \frac{I_L}{\sqrt{3}} = \frac{60.28}{\sqrt{3}} = 34.8 \text{ A}$$
$$Z_{ph} = \frac{V_{ph}}{I_{ph}} = \frac{200}{34.8} = 5.75\ \Omega$$
$$R_{ph} = Z_{ph} \cos \phi = 5.75 \times 0.5 = 2.875\ \Omega$$
$$X_{ph} = Z_{ph} \sin \phi = 5.75 \times \sin (\cos^{-1} 0.5)$$
$$= 5.747 \times 0.866 = 4.98\ \Omega$$
$$Q = \sqrt{3}\ V_L I_L \sin \phi$$
$$= \sqrt{3} \times 200 \times 60.28 \times 0.866 = 18.08 \text{ kVAR}$$

Example 6.15 *For a balanced, three-phase Wye-connected load, the phase voltage V_R is $100 \angle{-45°}$ V and it draws a line current I_y of $5 \angle 180°$A. (i) Find the complex impedance per phase. (ii) Draw a power triangle and identify all its sides with magnitudes and appropriate units. Assume phase sequence R-Y-B.*

Solution

Data $V_R = V_{ph} = 100 \angle{-45°}$ V

$I_y = 5 \angle 180°$ A

For a Wye-connected load,

$$\overline{V_R} = 100 \angle{-45°} \text{ V}$$

The current I_R leads current I_y by angle 120°.

$$\overline{I_R} = 5 \angle{-60°} \text{ A}$$
$$\overline{Z_{ph}} = \frac{\overline{V_R}}{\overline{I_R}} = \frac{100\angle -45°}{5\angle -60°}$$
$$= 20 \angle 15°\ \Omega = 19.32 + j5.18\ \Omega$$

Active power $P = 3\, V_{ph} I_{ph} \cos \phi$

$$= 3 \times 100 \times 5 \times \cos (15°) = 1.45 \text{ kW}$$

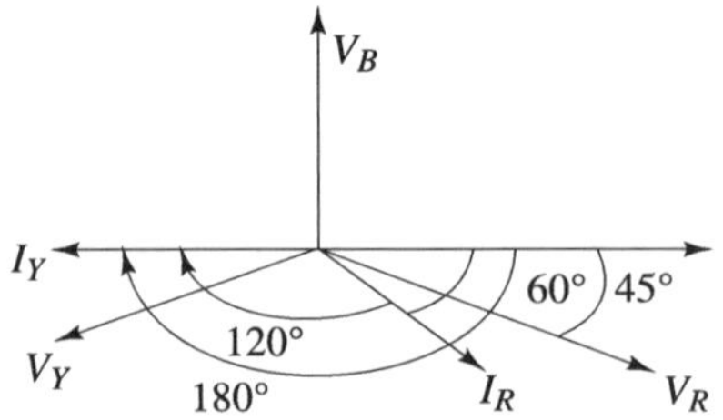

$I_R = 5 \angle{-60°}$

Fig. 6.14

Reactive power $Q = 3\, V_{ph}\, I_{ph} \sin \phi$

$= 3 \times 100 \times 5 \times \sin(15°) = 0.39$ kVAR

Apparent power $S = 3\, V_{ph}\, I_{ph}$

$= 3 \times 100 \times 5 = 1.5$ kVA

Power Triangle

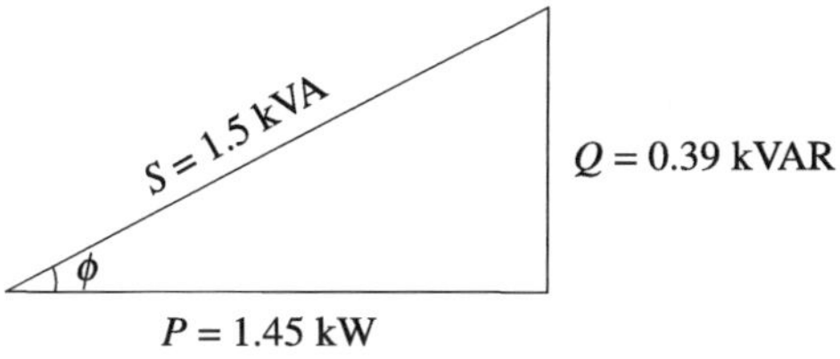

Fig. 6.15

Example 6.16 *Each leg of a balanced, delta-connected load consists of a 7-Ω resistance in series with a 4-Ω inductive reactance. The line-to-line voltages are*

$$E_{ab} = 2360\ \angle 0°\ \text{V}$$
$$E_{bc} = 2360\ \angle -120°\ \text{V}$$
$$E_{ca} = 2360\ \angle 120°\ \text{V}$$

Determine (i) phase current I_{ab}, I_{bc} and I_{ca} (both magnitude and phase),

(ii) each line current and its associated phase angle,

(iii) the load power factor, and

(iv) find the impedance per phase that draws the same power at the same power factor.

Solution

Data

$$R = 7\ \Omega$$
$$X_L = 4\ \Omega$$
$$V_L = 2360\ \text{V}$$

For a delta-connected load,

$$V_{ph} = V_L = 2360\ \text{V}$$
$$\bar{Z}_{ph} = 7 + j4 = 8.06\ \angle 29.74°\ \Omega$$

Phase current

$$\bar{I}_{ab} = \frac{\bar{E}_{ab}}{\bar{Z}_{ph}}$$
$$= \frac{2360\angle 0°}{8.06\angle 29.74°} = 292.8\ \angle -29.74°\ \text{A}$$
$$\bar{I}_{bc} = \frac{2360\angle -120°}{8.06\angle 29.74°} = 292.8\ \angle -149.71°\ \text{A}$$
$$\bar{I}_{ca} = \frac{2360\angle 120°}{8.06\angle 29.74°} = 292.8\ \angle 90.26°\ \text{A}$$

In a delta-connected, three-phase system, line currents lag behind respective phase currents by 30°.

$$I_L = \sqrt{3}\, I_{ph}$$
$$= \sqrt{3} \times 292.8 = 507.14\ \text{A}$$

$$I_{La} = 507.14 \angle -59.71° \text{ A}$$
$$I_{Lb} = 507.14 \angle -179.71° \text{ A}$$
$$I_{Lc} = 507.14 \angle 60.26° \text{ A}$$
$$\text{Load power factor} = \cos (29.74°) = 0.868 \text{ (lagging)}$$

Assuming that impedances are now connected in star, the power per phase and power factor remains the same.

For a delta-connected load,

$$\text{Power per phase} = V_{ph} I_{ph} \cos \phi = 2360 \times 292.8 \times 0.868 = 599.79 \text{ kW}$$

For a star-connected load,

$$V_{ph} = \frac{V_L}{\sqrt{3}} = \frac{2360}{\sqrt{3}} = 1362.55 \text{ V}$$
$$\text{Power per phase} = V_{ph} I_{ph} \cos \phi$$
$$599.79 \times 10^3 = 1362.55 \times I_{ph} \times 0.868$$
$$I_{ph} = 507.14 \text{ A}$$
$$Z_{ph} = \frac{V_{ph}}{I_{ph}} = \frac{1362.55}{507.14} = 2.69 \ \Omega$$

Example 6.17 *A three-phase, 200-kW, 50-Hz, delta-connected induction motor is supplied from a three-phase, 440-V, 50-Hz supply system. The efficiency and power factor of the three-phase induction motor are 91% and 0.86 respectively. Calculate (i) line currents, (ii) currents in each phase of the motor, and (iii) active and reactive components of phase current.*

Solution

Data

$$P_o = 200 \text{ kW}$$
$$f = 50 \text{ Hz}$$
$$V_L = 440 \text{ V}$$
$$\eta = 91\%$$
$$pf = 0.86$$

For a delta-connected load (induction motor),

$$V_{ph} = V_L = 440 \text{ V}$$

Efficiency

$$\eta = \frac{\text{Output power}}{\text{Input power}}$$
$$0.91 = \frac{200 \times 10^3}{\text{Input power}}$$

Input power

$$P_i = 219.78 \text{ kW}$$
$$P_i = \sqrt{3} V_L I_L \cos \phi$$
$$219.78 \times 10^3 = \sqrt{3} \times 440 \times I_L \times 0.86$$
$$I_L = 335.3 \text{ A}$$
$$I_{ph} = \frac{I_L}{\sqrt{3}} = \frac{335.3}{\sqrt{3}} = 193.6 \text{ A}$$
$$\text{Active component of phase current} = I_{ph} \cos \phi = 193.6 \times 0.86 = 166.5 \text{ A}$$
$$\text{Reactive component of phase current} = I_{ph} \sin \phi = 193.6 \times 0.51 = 98.7 \text{ A}$$

Example 6.18 *A three-phase, 400-V, star-connected alternator supplies a three-phase, 112-kW, mesh-connected induction motor of efficiency and power factor 0.88 and 0.86 respectively. Find the (i) current in each motor phase, (ii) current in each alternator phase, and (iii) active and reactive components of current in each case.*

Solution

Data $V_L = 400$ V

$P_o = 112$ kW

$pf = 0.86$

$\eta = 0.88$

For a mesh-connected load (induction motor),

$$V_{ph} = V_L = 400 \text{ V}$$

$$\eta = \frac{\text{Output power}}{\text{Input power}}$$

$$0.88 = \frac{112 \times 10^3}{\text{Input power}}$$

Input power $P_i = 127.27$ kW

$$P_i = \sqrt{3}\, V_L\, I_L \cos\phi$$

$$127.27 \times 10^3 = \sqrt{3} \times 400 \times I_L \times 0.86$$

$$I_L = 213.6 \text{ A}$$

$$I_{ph} = \frac{I_L}{\sqrt{3}} = \frac{213.6}{\sqrt{3}} = 123.32 \text{ A}$$

Current in a star-connected alternator phase will be same as the line current drawn by the motor.

Therefore, current in each alternator phase = 213.6 A

Active component of current in each phase of motor

$$= I_{ph} \cos\phi$$

$$= 123.32 \times 0.86 = 105.06 \text{ A}$$

Reactive component of current in each phase of the motor

$$= I_{ph} \sin\phi$$

$$= 123.32 \times 0.51 = 62.89 \text{ A}$$

Active component of current in each alternator phase

$$= 213.6 \times 0.86 = 183.7 \text{ A}$$

Reactive component of current in each alternator phase

$$= 213.6 \times 0.51 = 108.94 \text{ A}$$

Example 6.19 *Three similar resistors are connected in star across 400-V, three-phase lines. The line current is 5 A. Calculate the value of each resistor. To what value should the line voltage be changed to obtain the same line current with the resistors connected in delta?*

Solution

Data $V_L = 400$ V

$I_L = 5$ A

For a star-connected load,

$$V_{ph} = \frac{V_L}{\sqrt{3}} = \frac{400}{\sqrt{3}} = 230.94 \text{ V}$$

$$I_{ph} = I_L = 5 \text{ A}$$

$$Z_{ph} = R_{ph} = \frac{V_{ph}}{I_{ph}} = \frac{230.94}{5} = 46.19\ \Omega$$

For a delta-connected load,

$$I_L = 5\text{ A}$$
$$R_{ph} = 46.19\ \Omega$$
$$I_{ph} = \frac{I_L}{\sqrt{3}} = \frac{5}{\sqrt{3}}\text{ A}$$
$$V_{ph} = I_{ph} R_{ph}$$
$$= \frac{5}{\sqrt{3}} \times 46.19 = 133.33\text{ V}$$

Voltage needed is one-third of the star value.

Example 6.20 *Three 100-Ω, non-inductive resistances are connected in (a) star, and (b) delta across a 400-V, 50-Hz, three-phase supply. Calculate the power taken from the supply in each case. If one of the resistances is open circuited, what would be the value of total power taken from the mains in each of the two cases?*

Solution

Data

$$V_L = 400\text{ V}$$
$$Z_{ph} = 100\ \Omega$$

For a star-connected load,

$$V_{ph} = \frac{V_L}{\sqrt{3}} = \frac{400}{\sqrt{3}} = 230.94\text{ V}$$
$$I_{ph} = \frac{V_{ph}}{Z_{ph}} = \frac{230.94}{100} = 2.31\text{ A}$$
$$I_L = I_{ph} = 2.31\text{ A}$$
$$\cos\phi = 1$$
$$P = \sqrt{3}\, V_L I_L \cos\phi$$
$$= \sqrt{3} \times 400 \times 2.31 \times 1 = 1600.41\text{ W}$$

For a delta-connected load,

$$V_{ph} = V_L = 400\text{ V}$$
$$I_{ph} = \frac{V_{ph}}{Z_{ph}} = \frac{400}{100} = 4\text{ A}$$
$$I_L = \sqrt{3}\, I_{ph}$$
$$= \sqrt{3} \times 4 = 6.93\text{ A}$$
$$P = \sqrt{3}\, V_L I_L \cos\phi$$
$$= \sqrt{3} \times 400 \times 6.93 \times 1 = 4801.24\text{ W}$$

When one of the resistors is open circuited

(i) Star connection

The circuit consists of two 100-Ω resistors in series across a 400-V supply.

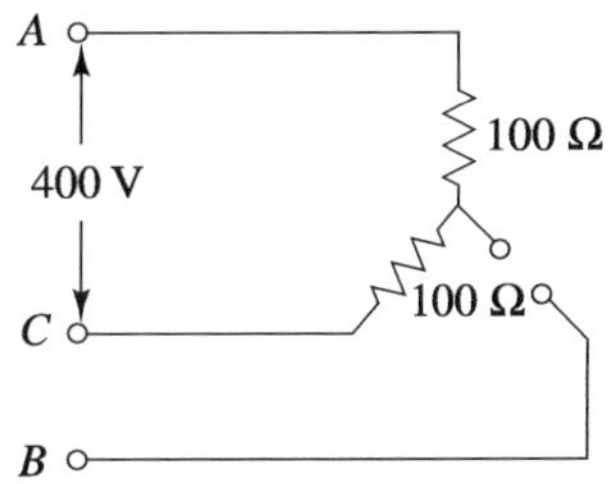

Fig. 6.16

$$\text{Currents in lines } A \text{ and } C = \frac{400}{200} = 2 \text{ A}$$

$$\text{Power taken from the mains} = 400 \times 2 = 800 \text{ W}$$

Hence, when one resistor is open circuited, the power consumption is reduced by half.

(ii) Delta connection

In this case, currents in A and C remain as usual 120° out of phase with each other.

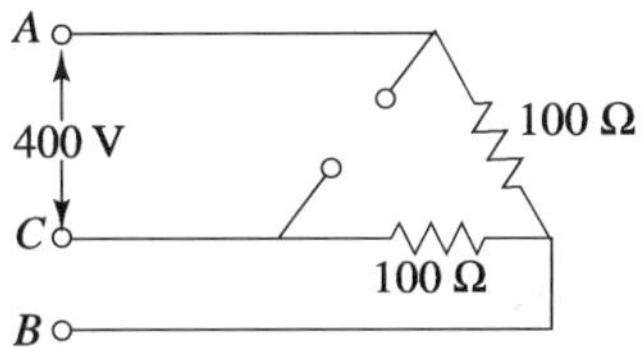

Fig. 6.17

$$\text{Current in each phase} = \frac{400}{100} = 4 \text{ A}$$

$$\text{Power taken from the mains} = 2 \times 4 \times 400 = 3200 \text{ W}$$

Hence when one resistor is open circuited, the power consumption is reduced by one-third.?

Example 6.21 *Three identical impedances of 10 ∠30° Ω each are connected star and another set of three identical impedances of 18 ∠60° Ω are connected in delta. If both the sets of impedances are connected across a balanced, three-phase 400 V supply, find the line current, total voltamperes, active power and reactive power.*

Solution

Data

$$\bar{Z}_Y = 10 \angle 30^\circ \ \Omega$$

$$\bar{Z}_\Delta = 18 \angle 60^\circ \ \Omega$$

$$V_L = 400 \text{ V}$$

Three identical delta impedances can be converted into equivalent star impedances.

$$\bar{Z}'_Y = \frac{\bar{Z}_\Delta}{3} = \frac{18\angle 60^\circ}{3} = 6 \angle 60^\circ \ \Omega$$

Now two star-connected impedances 10 ∠30° Ω and 6 ∠60° Ω are in parallel across a three-phase supply.

$$\bar{Z}_{eq} = \frac{(10\angle 30^\circ)(6\angle 60^\circ)}{10\angle 30^\circ + 6\angle 60^\circ} = 3.87 \angle 48.83^\circ \ \Omega$$

For a star-connected load,

$$V_{ph} = \frac{V_L}{\sqrt{3}} = \frac{400}{\sqrt{3}} = 230.94 \text{ V}$$

$$I_{ph} = \frac{V_{ph}}{Z_{ph}} = \frac{V_{ph}}{Z_{eq}} = \frac{230.94}{3.87} = 59.67 \text{ A}$$

$$I_L = I_{ph} = 59.67 \text{ A}$$

$$S = \sqrt{3}\, V_L I_L = \sqrt{3} \times 400 \times 59.67 = 41.34 \text{ kVA}$$

$$P = \sqrt{3}\, V_L I_L \cos\phi = \sqrt{3} \times 400 \times 59.67 \times \cos(48.83^\circ) = 27.21 \text{ kW}$$

$$Q = \sqrt{3}\, V_L I_L \sin\phi = \sqrt{3} \times 400 \times 59.67 \times \sin(48.83^\circ) = 31.12 \text{ kVAR}$$

Example 6.22 *Three star-connected impedances $Z_Y = (20 + j37.7)\ \Omega$ per phase are connected in parallel with three delta-connected impedances $Z_\Delta = (30 - j159.3)\ \Omega$ per phase. The line voltage is 398 V. Find the line current, pf, active and reactive power taken by the combination.*

Solution

Data

$$\bar{Z}_Y = 20 + j37.7 = 42.68 \angle 62.05^\circ\ \Omega$$

$$\bar{Z}_\Delta = 30 - j159.3 = 162.1 \angle -79.33^\circ\ \Omega$$

$$V_L = 398\ \text{V}$$

Three identical delta-connected impedances can be converted by an equivalent star impedances.

$$\bar{Z}'_Y = \frac{162.1 \angle -79.3^\circ}{3} = 54.03 \angle -79.3^\circ\ \Omega$$

Now two star-connected impedances of 42.68 ∠62.05° Ω and 54.03 ∠–79.3° Ω are in parallel across the three-phase supply.

$$\bar{Z}_{eq} \frac{(42.68 \angle 62.05^\circ)(54.03 \angle -79.3^\circ)}{42.68 \angle 62.05^\circ + 54.03 \angle -79.3^\circ} = 68.33 \angle 9.88^\circ\ \Omega$$

For a star-connected load,

$$V_{ph} = \frac{V_L}{\sqrt{3}} = \frac{398}{\sqrt{3}} = 229.79\ \text{V}$$

$$I_{ph} = \frac{V_{ph}}{Z_{ph}} = \frac{V_{ph}}{Z_{eq}}$$

$$= \frac{229.79}{68.33} = 3.36\ \text{A}$$

$$I_L = I_{ph} = 3.36\ \text{A}$$

$$pf = \cos\phi = \cos(9.88^\circ) = 0.99\ \text{(lagging)}$$

$$P = \sqrt{3}\, V_L I_L \cos\phi = \sqrt{3} \times 398 \times 3.36 \times 0.99 = 2.29\ \text{kW}$$

$$Q = \sqrt{3}\, V_L I_L \sin\phi = \sqrt{3} \times 398 \times 3.36 \times \sin(9.88^\circ) = 397.43\ \text{VAR}$$

Example 6.23 *A balanced, delta-connected load having an impedance $Z_L = (300 + j210)$ ohm in each phase is supplied from a 400-V, three-phase supply through a three-phase line having an impedance of $Z_S = (4 + j8)$ ohm in each phase. Find current and voltage in each phase of the load.*

Solution

Data

$$\bar{Z}_L = 300 + j210\ \Omega$$

$$\bar{Z}_S = 4 + j8\ \Omega$$

$$V_L = 400\ \text{V}$$

Three identical delta impedances can be converted into equivalent star impedances.

$$\bar{Z}'_L = \frac{\bar{Z}_L}{3} = \frac{300 + j210}{3} = 100 + j70\ \Omega$$

The circuit can be drawn as

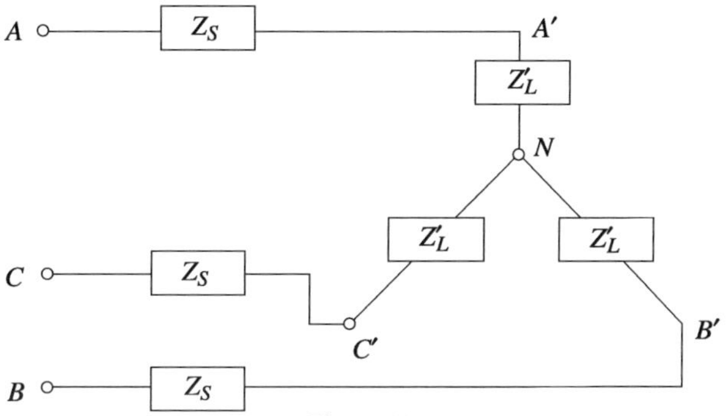

Fig. 6.18

$$\bar{Z}_{AN} = \bar{Z}_S + \bar{Z}_L'$$
$$= 4 + j8 + 100 + j70$$
$$= 104 + j78\ \Omega = 130\ \angle 36.87°\ \Omega$$

For a star-connected load,

$$V_{AN} = \frac{400}{\sqrt{3}} = 230.94\ \text{V}$$

$$\bar{I}_{AN} = \frac{230.94\angle 0°}{130\angle 36.87°} = 1.78\ \angle -36.87°\ \text{A}$$

Voltage across $\bar{Z}_S = \bar{I}_{AN} \cdot \bar{Z}_S$

$$= (1.78\ \angle -36.87°) \times (4 + j8) = 15.92\ \angle 26.56°\ \text{V}$$

Voltage across $Z_L' = 230.94\ \angle 0° - 15.92\ \angle 26.56°$

$$= 216.82\ \angle -1.88°\ \text{V}$$

Voltage in each phase of the load $= 216.82$ V

Current in each phase of the load $= 1.78$ A

Example 6.24 *Three coils each having a resistance of 20 Ω and a reactance of 15 Ω are connected in star to a 400-V, three-phase, 50-Hz supply. Calculate (i) line current, (ii) power supplied, and (iii) power factor. If three capacitors, each of same capacitance, are connected in delta to the same supply so as to form parallel circuit with the above coils, calculate the capacitance of each capacitor to obtain a resultant power factor of 0.95 lagging.*

Solution

Data

$$R_{ph} = 20\ \Omega$$
$$X_{ph} = 15\ \Omega$$
$$V_L = 400\ \text{V}$$

For a star-connected load,

$$\bar{Z}_{ph} = R_{ph} + jX_{ph}$$
$$= 20 + j15 = 25\ \angle 36.87°\ \Omega$$

$$V_{ph} = \frac{V_L}{\sqrt{3}} = \frac{400}{\sqrt{3}} = 230.94\ \text{V}$$

$$I_{ph} = \frac{V_{ph}}{Z_{ph}} = \frac{230.94}{25} = 9.24 \text{ A}$$

$$I_L = I_{ph} = 9.24 \text{ A}$$

$$P_1 = \sqrt{3}\, V_L I_L \cos \phi_1$$

$$= \sqrt{3} \times 400 \times 9.24 \times \cos (36.87°) = 5.12 \text{ kW}$$

$$pf = \cos (36.87°) = 0.8 \text{ (lagging)}$$

$$Q_1 = \sqrt{3}\, V_L I_L \sin \phi_1$$

$$= \sqrt{3} \times 400 \times 9.24 \times \sin (36.87°) = 3.84 \text{ kVAR}$$

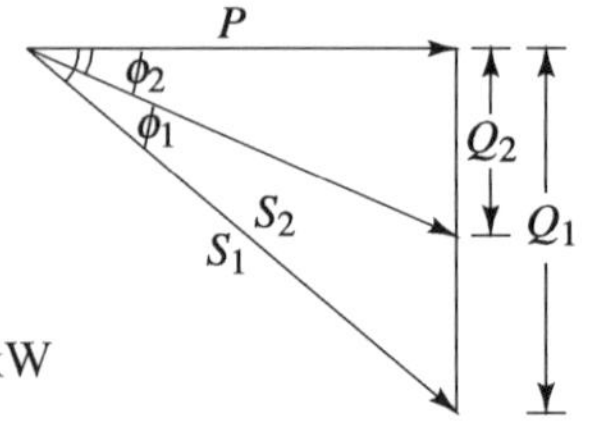

Fig. 6.19

When capacitors are connected in delta to the same supply.

$$pf = 0.95$$

$$\phi_2 = \cos^{-1} (0.95) = 18.19°$$

$$\tan \phi_2 = \tan (18.19°) = 0.33$$

Since capacitors do not absorb any power, power remains the same even when capacitors are connected. But reactive power changes.

$$P_2 = 5.12 \text{ kW}$$

$$Q_2 = P_2 \tan \phi_2$$

$$= 5.12 \times 0.33 = 1.69 \text{ kVAR}$$

Difference in reactive power is supplied by three capacitors.

$$Q = Q_1 - Q_2$$

$$= 3.84 - 1.69 = 2.15 \text{ kVAR}$$

$$Q = \sqrt{3}\, V_L I_L \sin \phi$$

$$2.15 \times 10^3 = \sqrt{3} \times 400 \times I_L \times \sin (90°)$$

$$I_L = 3.1 \text{ A}$$

$$I_{ph} = \frac{I_L}{\sqrt{3}} = 1.79 \text{ A}$$

$$I_{ph} = \frac{V_{ph}}{X_C} = V_{ph} \times 2\pi f C$$

$$C = \frac{I_{ph}}{V_{ph} \times 2\pi f} = \frac{1.79}{400 \times 2\pi \times 50} = 14.24 \text{ µF}$$

6.12 MEASUREMENT OF THREE-PHASE POWER

In a three-phase system, total power is the sum of powers in three phases. The power is measured by wattmeter. It consists of two coils. (i) current coil, and (ii) voltage coil. The current coil is connected in series with the load and it senses current. Voltage coil is connected across supply terminals and it senses voltages.

6.12.1 Two-Wattmeter Method

This method is used for balanced as well as unbalanced load. The current coils of the two wattmeters are inserted in any two lines and the voltage coil of each wattmeter is joined to a third line. The load may be star or delta connected. The sum of the two wattmeter readings gives three-phase power.

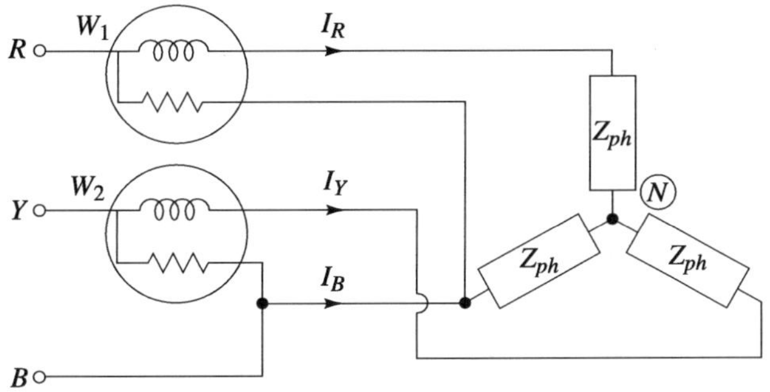

(a)

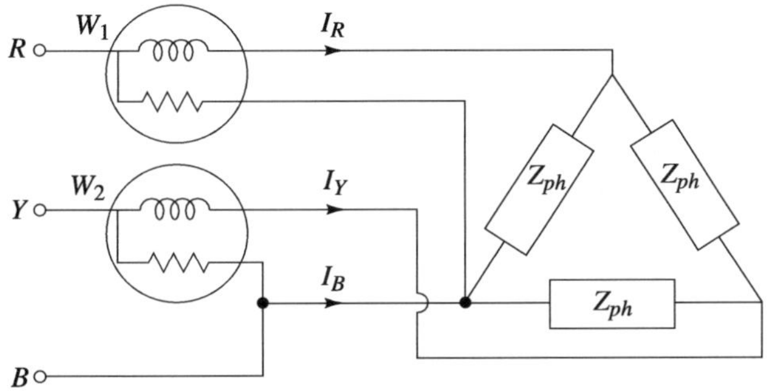

(b)

Fig. 6.20

Total power $P = W_1 + W_2$

Measurement of power Figure 6.20 shows a balanced, star-connected load, the load may be assumed to be inductive. Let V_{RN}, V_{YN} and V_{BN} be the three phase voltages. I_R, I_Y and I_B be the phase currents. The phase currents will lag behind their respective phase voltages by angle ϕ.

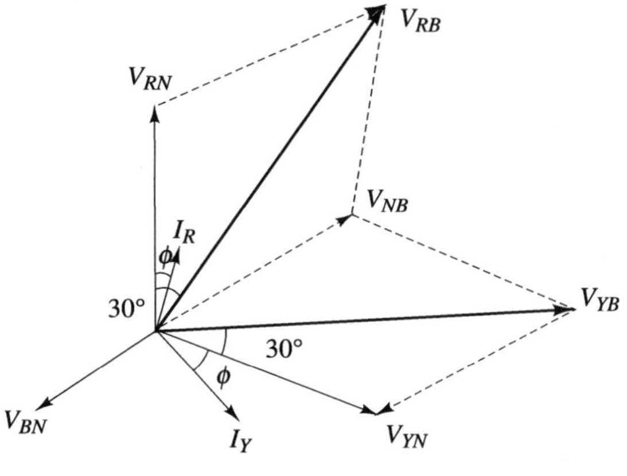

Fig. 6.21

Current through current coil of $W_1 = I_R$

Voltage across voltage coil of $W_1 = V_{RB} = V_{RN} + V_{NB} = V_{RN} - V_{BN}$

From phasor diagram, it is clear that the phase angle between V_{RB} and I_R is $(30° - \phi)$

$$W_1 = V_{RB}\, I_R \cos(30° - \phi)$$

Current through current coil of $W_2 = I_Y$

Voltage across voltage coil of $W_2 = V_{YB} = V_{YN} + V_{NB} = V_{YN} - V_{BN}$

From phasor diagram it is clear that phase angle between V_{YB} and I_Y is $(30° + \phi)$

$$W_2 = V_{YB}\, I_Y \cos(30° + \phi)$$

But

$$I_R = I_Y = I_L$$

$$V_{RB} = V_{YB} = V_L$$

$$W_1 = V_L I_L \cos(30° - \phi)$$

$$W_2 = V_L I_L \cos(30° + \phi)$$

$$W_1 + W_2 = V_L I_L [\cos(30° + \phi) + \cos(30° - \phi)]$$

$$= V_L I_L (2 \cos 30° \cos \phi) = \sqrt{3}\, V_L I_L \cos \phi$$

Thus, the sum of two wattmeter readings gives three-phase power.

Measurement of power factor

(i) Lagging power factor

$$W_1 = V_L I_L \cos(30° - \phi)$$

$$W_2 = V_L I_L \cos(30° + \phi)$$

$$W_1 + W_2 = \sqrt{3}\, V_L I_L \cos \phi$$

$$W_1 - W_2 = V_L I_L [\cos(30° - \phi) - \cos(30° + \phi)] = V_L I_L \sin \phi$$

$$\frac{W_1 - W_2}{W_1 + W_2} = \frac{V_L I_L \sin \phi}{\sqrt{3} V_L I_L \cos \phi}$$

$$\tan \phi = \sqrt{3}\,\frac{W_1 - W_2}{W_1 + W_2}$$

$$\phi = \tan^{-1}\left(\sqrt{3}\,\frac{W_1 - W_2}{W_1 + W_2}\right)$$

$$pf = \cos \phi = \cos\left\{\tan^{-1}\left(\sqrt{3}\,\frac{W_1 - W_2}{W_1 + W_2}\right)\right\}$$

(ii) Leading power factor

Phasor Diagram

$$W_1 = V_L I_L \cos(30° + \phi)$$

$$W_2 = V_L I_L \cos(30° - \phi)$$

$$W_1 + W_2 = \sqrt{3}\, V_L I_L \cos \phi$$

$$W_1 - W_2 = -V_L I_L \sin \phi$$

$$\tan \phi = -\sqrt{3}\,\frac{(W_1 - W_2)}{(W_1 + W_2)}$$

$$\phi = \tan^{-1}\left\{-\sqrt{3}\,\frac{(W_1 - W_2)}{(W_1 + W_2)}\right\}$$

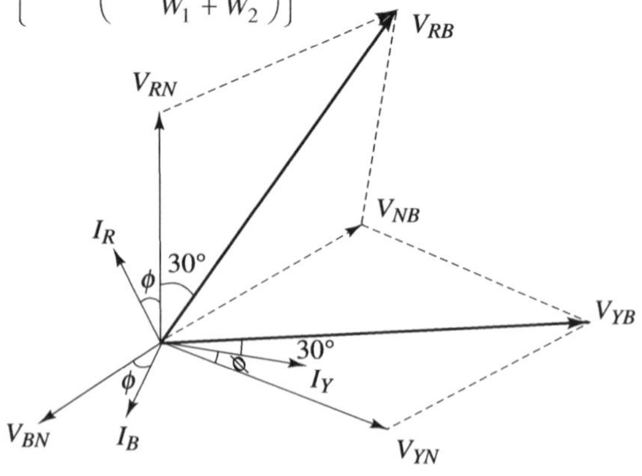

Fig. 6.22

$$pf = \cos\phi = \cos\left\{\tan^{-1}\left(-\sqrt{3}\,\frac{W_1 - W_2}{W_1 + W_2}\right)\right\}$$

Example 6.25 *Three coils each with a resistance of 10 Ω and reactance of 10 Ω are connected in star across a three phase, 50-Hz, 400-V supply. Calculate (i) line current, and (ii) readings on the two wattmeters connected to measure the power.*

Solution

Data $R = 10\ \Omega \qquad X_L = 10\ \Omega$

$V_L = 400\ \text{V}$

For a star-connected load,

$$V_{ph} = \frac{V_L}{\sqrt{3}} = \frac{400}{\sqrt{3}} = 230.94\ \text{V}$$

$$\bar{Z}_{ph} = R + jX_L = 10 + j10 = 14.14\ \angle 45°\ \Omega$$

$$Z_{ph} = 14.14\ \Omega$$

$$\phi = 45°$$

$$\text{Power factor} = \cos\phi = \cos(45°) = 0.707\ \text{(lagging)}$$

$$I_{ph} = \frac{V_{ph}}{Z_{ph}} = \frac{230.94}{14.14} = 16.33\ \text{A}$$

$$I_L = I_{ph} = 16.33\ \text{A}$$

$$P = \sqrt{3}\, V_L I_L \cos\phi = \sqrt{3} \times 400 \times 16.33 \times 0.707 = 7998.83\ \text{W}$$

$$W_1 + W_2 = 7998.83\ \text{W} \qquad \text{...(i)}$$

Also,

$$\tan\phi = \sqrt{3}\,\frac{W_1 - W_2}{W_1 + W_2}$$

$$\tan 45° = \sqrt{3}\,\frac{W_1 - W_2}{7998.83}$$

$$W_1 - W_2 = 4618.13\ \text{W} \qquad \text{...(ii)}$$

Solving Eqs (i) and (ii),

$$W_1 = 6308.48\ \text{W}$$

$$W_2 = 1690.35\ \text{W}$$

Example 6.26 *Three coils each having a resistance of 20 Ω and reactance of 15 Ω are connected in (i) star, and (ii) delta, across a three-phase, 400-V, 50-Hz supply. Calculate in each case, the readings on two wattmeters connected to measure the power input.*

Solution

Data $R = 20\ \Omega \qquad X_L = 15\ \Omega \qquad V_L = 400\ \text{V}$

(i) For a star-connected load,

$$V_{ph} = \frac{V_L}{\sqrt{3}} = \frac{400}{\sqrt{3}} = 230.94\ \text{V}$$

$$\bar{Z}_{ph} = 20 + j15 = 25\ \angle 36.87°\ \Omega$$

$$Z_{ph} = 25\ \Omega$$

$$\phi = 36.87°$$

$$I_{ph} = \frac{V_{ph}}{Z_{ph}} = \frac{230.94}{25} = 9.24 \text{ A}$$

$$I_L = I_{ph} = 9.24 \text{ A}$$

$$W_1 = V_L I_L \cos(30° - \phi)$$
$$= 400 \times 9.24 \times \cos(30° - 36.87°) = 3669.46 \text{ W}$$

$$W_2 = V_L I_L \cos(30° + \phi)$$
$$= 400 \times 9.24 \times \cos(30° + 36.87°)$$
$$= 1451.86 \text{ W}$$

(ii) For a delta-connected load,

$$V_{ph} = V_L = 400 \text{ V}$$
$$Z_{ph} = 25\ \Omega$$
$$\phi = 36.87°$$

$$I_{ph} = \frac{V_{ph}}{Z_{ph}} = \frac{400}{25} = 16 \text{ A}$$

$$I_L = \sqrt{3}\, I_{ph} = \sqrt{3} \times 16 = 27.72 \text{ A}$$

$$W_1 = V_L I_L \cos(30° - \phi)$$
$$= 400 \times 27.72 \times \cos(30° - 36.87°) = 11008.39 \text{ W}$$

$$W_2 = V_L I_L \cos(30° + \phi)$$
$$= 400 \times 27.72 \times \cos(30° + 36.87°) = 4355.57 \text{ W}$$

Example 6.27 *Two wattmeters connected to measure the input to a balanced, three-phase circuit indicate 2000 W and 500 W respectively. Find the power factor of the circuit (i) when both readings are positive and (ii) when the latter is obtained after reversing the connection to the current coil of one instrument.*

Solution

Data

$$W_1 = 2000 \text{ W}$$
$$W_2 = 500 \text{ W}$$

(i) When both readings are positive,

$$W_1 = 2000 \text{ W}$$
$$W_2 = 500 \text{ W}$$

$$\tan\phi = \sqrt{3}\frac{W_1 - W_2}{W_1 + W_2} = \sqrt{3}\frac{(2000 - 500)}{(2000 + 500)} = 1.039$$

$$\phi = 46.102°$$

$$\text{Power factor} = \cos\phi = \cos(46.102°) = 0.693$$

(ii) When the latter reading is obtained after reversing the connection to the current coil of one instrument,

$$W_1 = 2000 \text{ W}$$
$$W_2 = -500 \text{ W}$$

$$\tan\phi = \sqrt{3}\frac{W_1 - W_2}{W_1 + W_2}$$

$$= \sqrt{3}\frac{(2000 + 500)}{(2000 - 500)} = 2.887$$

$$\phi = 70.89°$$

$$\text{Power factor} = \cos\phi = \cos(70.89°) = 0.33$$

Example 6.28 *Find the power and power factor of the balanced circuit in which the wattmeter readings are 5 kW and 0.5 kW, the latter being obtained after the reversal of the current coil terminals of the wattmeter.*

Solution

Data

$$W_1 = 5 \text{ kW}$$
$$W_2 = 0.5 \text{ kW}$$

When the latter readings are obtained after the reversal of the current coil terminals of the wattmeter,

$$W_1 = 5 \text{ kW}$$
$$W_2 = -0.5 \text{ kW}$$
$$\text{Power} = W_1 + W_2 = 5 + (-0.5) = 4.5 \text{ kW}$$

$$\tan\phi = \sqrt{3}\,\frac{W_1 - W_2}{W_1 + W_2} = \sqrt{3}\,\frac{(5+0.5)}{(5-0.5)} = 2.12$$

$$\phi = 64.72°$$

$$\text{Power factor} = \cos\phi = \cos(64.72°) = 0.43$$

Example 6.29 *A three-phase, 10-kVA load has a power factor of 0.342. The power is measured by the two-wattmeter method. Find the reading of each wattmeter when the (i) power factor is leading, and the (ii) power factor is lagging.*

Solution

Data $S = 10$ kVA $\quad pf = 0.342$

$$S = \sqrt{3}\, V_L I_L$$
$$10 \times 10^3 = \sqrt{3}\, V_L I_L$$
$$V_L I_L = 5.77 \text{ kVA}$$
$$\cos\phi = 0.342$$
$$\phi = 72°$$

(i) When the power factor is leading,

$$W_1 = V_L I_L \cos(30° + \phi) = 5.77 \cos(30° + 70°) = -1 \text{ kW}$$
$$W_2 = V_L I_L \cos(30° - \phi) = 5.77 \cos(30° - 70°) = 4.42 \text{ kW}$$

(ii) When the power factor is lagging

$$W_1 = V_L I_L \cos(30° - \phi) = 4.42 \text{ kW}$$
$$W_2 = V_L I_L \cos(30° + \phi) = -1 \text{ kW}$$

Example 6.30 *The power input to a 2000-V, 50-Hz, three-phase motor running on full load at an efficiency of 90% is measured by two wattmeters which indicate 300 kW and 100 kW respectively. Calculate the (i) input power, (ii) power factor, and (iii) line current.*

Solution

Data

$$V_L = 2000 \text{ V}$$
$$\eta = 0.9$$

$$W_1 = 300 \text{ kW}$$
$$W_2 = 100 \text{ kW}$$
$$\text{Input power } P = W_1 + W_2 = 300 + 100 = 400 \text{ kW}$$
$$\tan\phi = \sqrt{3}\,\frac{W_1 - W_2}{W_1 + W_2} = \sqrt{3}\,\frac{(300-100)}{(300+100)} = 0.866$$
$$\phi = 40.89°$$
$$\text{Power factor} = \cos\phi = \cos(40.89°) = 0.76$$
$$P = \sqrt{3}\, V_L I_L \cos\phi$$
$$400 \times 10^3 = \sqrt{3} \times 2000 \times I_L \times 0.76$$
$$I_L = 151.93 \text{ A}$$

Example 6.31 *A three-phase, 220-V, 50-Hz, 11.2-kW induction motor has a full load efficiency of 88 per cent and draws a line current of 38 A under full load, when connected to a three-phase, 220-V supply. Find the reading on two wattmeters connected in the circuit to measure the input to the motor. Determine power factor at which the motor is operating.*

Solution

Data

$$V_L = 220 \text{ V}$$
$$P_o = 11.2 \text{ kW}$$
$$\eta = 88\%$$
$$I_L = 38 \text{ A}$$
$$\eta = \frac{P_o}{P_i}$$
$$0.88 = \frac{11.2 \times 10^3}{P_i}$$
$$P_i = 12.73 \text{ kW}$$

But

$$P_i = \sqrt{3}\, V_L I_L \cos\phi$$
$$12.73 \times 10^3 = \sqrt{3} \times 220 \times 38 \times \cos\phi$$
$$\cos\phi = 0.88 \text{ lagging}$$
$$\phi = 28.36°$$
$$W_1 = V_L I_L \cos(30° - \phi) = 220 \times 38 \times \cos(30° - 28.36°)$$
$$= 8356.58 \text{ W}$$
$$W_2 = V_L I_L \cos(30° + \phi) = 220 \times 38 \times \cos(30° + 28.36°)$$
$$= 4385.49 \text{ W}$$

Example 6.32 *What will be the relation between readings on the wattmeter connected to measure power in a three-phase balanced circuit with (i) unity power factor, (ii) zero power factor, and (iii) power factor = 0.5.*

Solution

Data (i) $pf = 1$ (ii) $pf = 0$ (iii) $pf = 0.5$

(i)

$$\text{Power factor} = 1$$
$$\cos\phi = 1$$
$$\phi = 0$$
$$\tan\phi = \tan 0° = 0$$

$$\tan \phi = \sqrt{3}\,\frac{W_1 - W_2}{W_1 + W_2}$$

$$0 = \sqrt{3}\,\frac{W_1 - W_2}{W_1 + W_2}$$

$$W_1 = W_2$$

(ii) Power factor = 0

$$\cos \phi = 0$$

$$\phi = 90°$$

$$\tan \phi = \tan 90° = \infty$$

$$\tan \phi = \sqrt{3}\,\frac{W_1 - W_2}{W_1 + W_2}$$

$$W_1 + W_2 = 0$$

$$W_1 = -W_2$$

(iii) Power factor = 0.5

$$\cos \phi = 60°$$

$$\tan \phi = \tan 60° = 1.732$$

$$\tan \phi = \sqrt{3}\,\frac{W_1 - W_2}{W_1 + W_2}$$

$$1.732 = \sqrt{3}\,\frac{W_1 - W_2}{W_1 + W_2}$$

$$W_1 - W_2 = W_1 + W_2$$

$$W_2 = 0$$

Example 6.33 *Two wattmeters are used to measure power in a three-phase balanced load. Find the power factor if (i) two readings are equal, and (ii) two readings are equal and opposite.*

Solution

Data (i) $W_1 = W_2$

(ii) $W_1 = -W_2$

(i) If two readings are equal,

$$W_1 = W_2$$

$$\tan \phi = \sqrt{3}\,\frac{W_1 - W_2}{W_1 + W_2}$$

$$\tan \phi = \sqrt{3}\,(0) = 0$$

$$\phi = 0°$$

$$\text{Power factor} = \cos \phi = \cos 0° = 1$$

(ii) If two readings are equal and opposite,

$$W_1 = -W_2$$

$$\tan \phi = \sqrt{3}\,\frac{W_1 - W_2}{W_1 + W_2} = \infty$$

$$\phi = 90°$$

$$\text{Power factor} = \cos \phi = \cos (90°) = 0$$

Exercises

1. Three inductors each with a resistance of 5 Ω and an inductive resistance of 6 Ω are connected in closed delta and supplied from a 440-V, three-phase system. Calculate the line and phase currents, the power factor of the system and the intake in watts. [97.58 A, 56.33 A, 0.64 lagging, 47.61 kW]
2. Three coils each having a resistance of 10 Ω and an inductance of 0.02 H are connected (i) in star, (ii) in delta to a three-phase, 50-Hz supply, the line voltage being 500 volts. Calculate for each case the line current and the total power taken from the supply.
[(i) Star: 24.46 A, 17.94 kW, (ii) Delta: 73.39 A, 53.83 kW]
3. A balanced, three-phase, star-connected load of 150 kW takes a leading current of 100 A with a line voltage of 1100 V, 50 Hz. Find the circuit constant of the load per phase. [5 Ω, 813 μF]
4. Three pure elements connected in star draw *x* kVAR. What will be the value of elements that will draw the same kVAR, when connected in delta across the same supply? [$Z_\Delta = 3Z_Y$]
5. Three coils each having an impedance of (4 + *j*3) ohm are connected in star to a 440-V, three-phase, 50-Hz balanced supply. Calculate the line current and active power. Now, if three pure capacitors each of *C* farads connected in delta, are connected across the same supply, it is found that the total power factor of the circuit becomes 0.96 lag. Find the value of *C*. Also find the total line current.
[50.8 A, 30.976 kW, 77.75 μF, 42.34 A]
6. A three-phase, 500-V motor load has a power factor of 0.4. Two wattmeters connected to measure power show the input to be 30 kW. Find the reading on each instrument. [35 kW, –5 kW]
7. The power in a three-phase circuit is measured by two wattmeters. If the total power is 100 kW and the power factor is 0.66 leading, what will be the reading of each wattmeter. [17.26 kW, 82.74 kW]
8. Two wattmeters are connected to measure the input to a 400 V, three-phase connected motor outputting 24.4 kW at a power factor of 0.4 lag and 80% efficiency. Calculate the (i) resistance and reactance of motor per phase, and the (ii) reading of each wattmeter. [2.55 Ω, 5.58 Ω, 34915 W, –4850 W]
9. In a balanced, three-phase, 400-V circuit, the line current is 115.5 A. When the power is measured by two-wattmeter method, one meter reads 40 kW and the other, zero. What is the power factor of the load? If the power factor were unity and the line current the same, what would be the reading of each wattmeter? [0.5, 40 kW, 40 kW]
10. A 440-V, three-phase, delta-connected induction motor has an output of 14.92 kW at pf of 0.82 and efficiency of 85%. Calculate the readings on each of the two wattmeters connected to measure the input. If another star-connected load of 10 kW at 0.85 *pf* lagging is added in parallel to the motor, what will be the current drawn from the line and the power taken from the line?
[12.35 kW, 5.26 kW, 43.56 A, 27.6 kW]

Objective-Type Questions

1. If the 3-phase balanced source in Fig. 6.23 delivers 1500 W at a leading power factor of 0.844, then the value of Z_L (in ohm) is approximately
(a) 90 ∠32.44° (b) 80 ∠32.44° (c) 80 ∠–32.44° (d) 90 ∠–32.44°

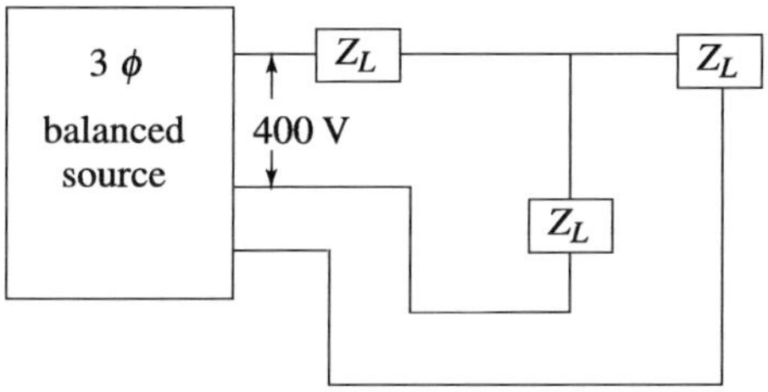

Fig. 6.23

2. v_{RN}, v_{YN} and v_{BN} are the instantaneous line to neutral voltages and i_R, i_Y and i_B are instantaneous line currents in a balanced three phase circuit, the computation v_{RN} $(i_Y - i_B) - (v_{YN} - v_{BN})$ i_R will yield a quantity proportional to

(a) active power (b) power factor (c) reactive power (d) complex power

3. In a 3-phase system, $\bar{V}_{YN} = 100 \angle{-120°}$V and $\bar{V}_{BN} = 100 \angle 120°$V. Then $\bar{V}_{YB}$ will be

(a) 170 < 90° (b) 173 < – 90° V (c) 200 < 60° (d) none of the above

4. A 3 phase load is balanced if all the three phases have the same.

(a) impedance (b) power factor
(c) impedance and power factor (d) none of the above

5. The power consumed in the star connected load shown in Fig. 6.24 is 690 W. The line current is

(a) 2.5 A (b) 1 A
(c) 1.725 A (d) none of the above

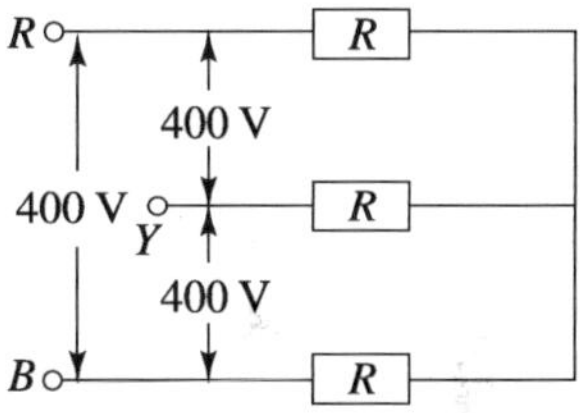

Fig. 6.24

6. Three identical resistances connected in star carry a line current of 12A. If the same resistances are connected in delta across the same supply, the line current will be

(a) 12 A (b) 4 A (c) 8 A (d) 36 A

7. Three delta-connected resistors absorb 60 kW when connected to a 3-phase line. If the resistors are connected in star, the power absorbed is

(a) 60 kW (b) 20 kW (c) 40 kW (d) 180 kW

8. If one of the resistors in Fig. 6.25 is open circuited, power consumed in the circuit is

(a) 8000 W (b) 4000 W (c) 16000 W (d) none of the above

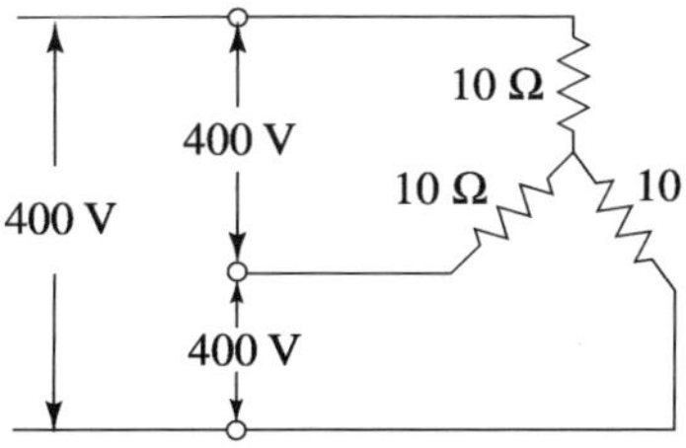

Fig. 6.25

9. If a balanced delta load has an impedance of $(6 + j9)$ ohms per phase, then the impedance of each phase if equivalent star load is

(a) $(6 + j9)$ ohms (b) $(2 + j3)$ ohms (c) $12 + j8)$ ohms (d) $(3 + j4.5)$ ohms

10. In two wattmeter method of measurement, if one of the wattmeters reads zero, then power factor will be

(a) zero (b) unity (c) 0.5 (d) 0.866

Answers to Objective-Type Questions

1. (d)	**2.** (c)	**3.** (d)	**4.** (c)	**5.** (b)	**6.** (d)
7. (b)	**8.** (a)	**9.** (b)	**10.** (c)		

7 Graph Theory

7.1 INTRODUCTION

The purpose of network analysis is to find voltage across and currents through all the elements. When the network is complicated and having a large number of nodes and closed paths, network analysis can be done conveniently by using 'Graph Theory'. This theory does not make any distinction between different types of physical elements of the network but makes the study based on a geometric pattern of the network. The basic elements of this theory are branches, nodes, loops and meshes.

Node It is defined as a point at which two or more elements have a common connection.

Branch It is a line connecting a pair of nodes, the line representing a single element.

Loop Whenever there is more than one path between two nodes, there is a circuit or loop.

Mesh It is a loop which does not contain any other loops within it.

7.2 GRAPH OF A NETWORK

A linear graph is a collection of nodes and branches. The nodes are joined together by branches.

The graph of a network is drawn by first marking the nodes and then joining these nodes by lines which correspond to the network elements of each branch. All the voltage and current sources are replaced by their internal impedances. The voltage sources are replaced by short circuits as their internal impedances are zero whereas current sources are replaced by open circuits as their internal impedances are infinite. Nodes and branches are numbered. Figure 7.1 shows a network and its associated graphs.

Each branch of a graph may be given an orientation or a direction with the help of an arrow head which represents the assigned reference direction for current. Such a graph is then referred to as a directed or oriented graph.

Branches whose ends fall on a node are said to be incident at that node. Branches 2, 3 and 4 are incident at node 2 in Fig. 7.1(c).

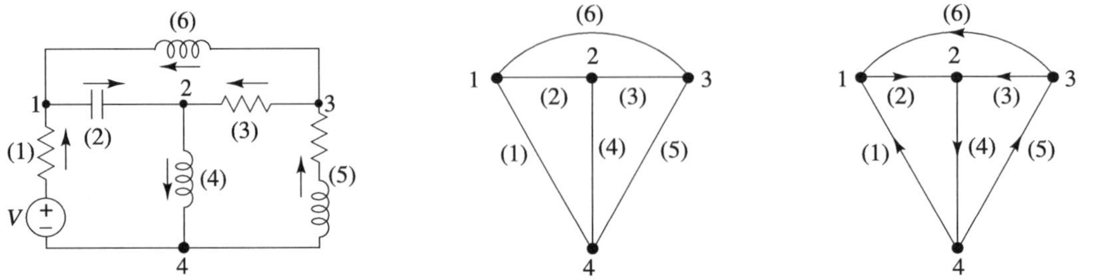

(a) Network (b) Undirected graph (c) Directed or oriented graph

Fig. 7.1

7.3 DEFINITIONS ASSOCIATED WITH A GRAPH

1. Planar graph A graph drawn on a two-dimensional plane is said to be planar if two branches do not intersect or cross at a point which is other than a node. Figure 7.2 shows such graphs.

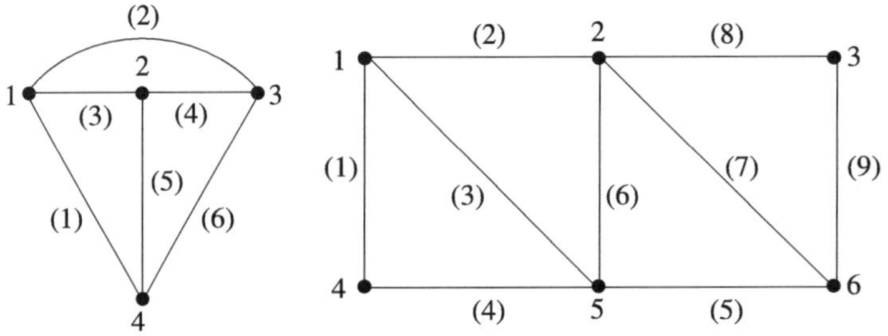

Fig. 7.2 *Planar graphs*

2. Non-planar graph A graph drawn on a two-dimensional plane is said to be non-planar if there is intersection of two or more branches at another point which is not a node. Figure 7.3 shows non-planar graphs.

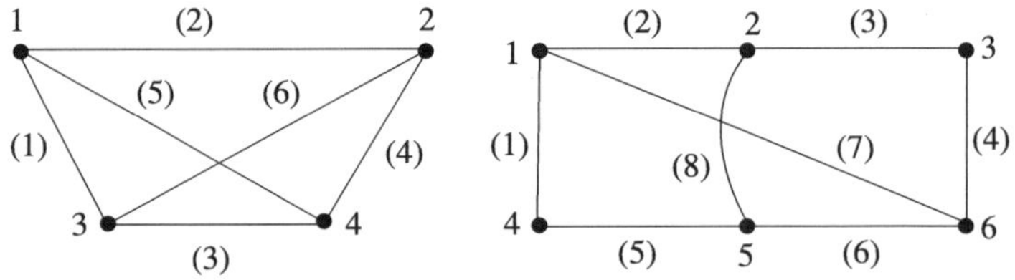

Fig. 7.3 *Non-planar graphs*

3. Sub-graph It is a subset of branches and nodes of a graph. It is a proper sub-graph if it contains branches and nodes less than those on a graph. A sub-graph can be just a node or only one branch of the graph. Figure 7.4 shows a graph and its proper sub-graph.

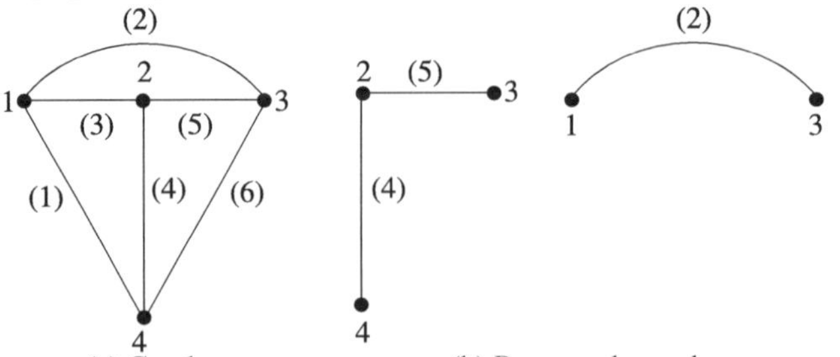

(a) Graph (b) Proper sub-graph

Fig. 7.4

4. Path It is an improper sub-graph having the following properties:

(a) At two of its nodes called terminal nodes, there is incident only one branch of sub-graph.

(b) At all remaining nodes called internal nodes, there are incident two branches of a graph.

In Fig. 7.5, branches 2, 5 and 6 together with all the four nodes, constitute a path.

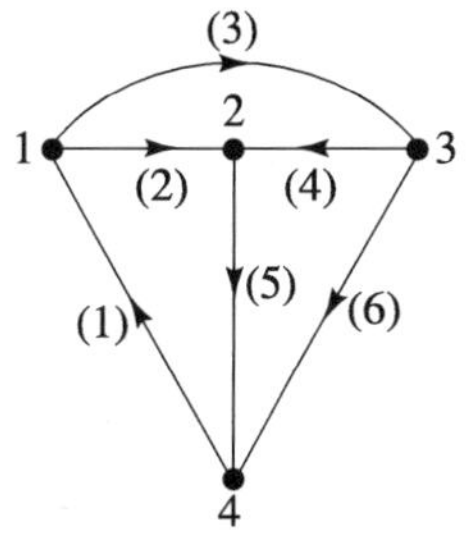

Fig. 7.5

5. Connected graph A graph is said to be connected if there exists a path between any pair of nodes. Otherwise, the graph is disconnected.

6. Rank of a graph If there are n nodes in a graph, the rank of the graph is $(n - 1)$.

7. Loop or circuit A loop is a connected sub-graph of a connected graph at each node of which are incident exactly two branches. If two terminals of a path are made to coincide, it will result in a loop or circuit.

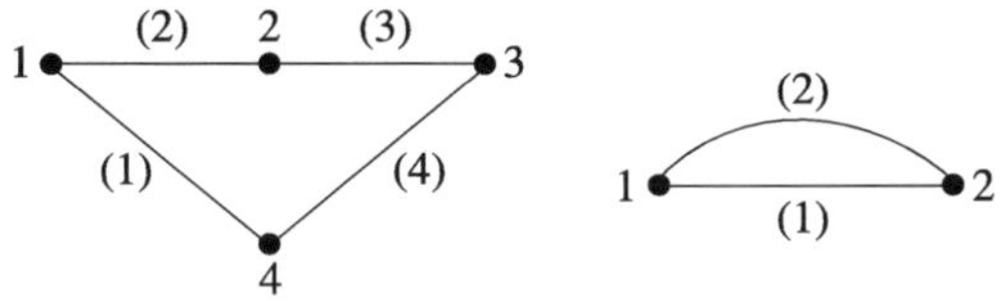

Loops: {1, 2, 3, 4} {1, 2}

Fig. 7.6

Loops of a graph have the following properties:

(1) There are at least two branches in a loop.

(2) There are exactly two paths between any pair of nodes in a circuit.

(3) The maximum number of possible branches is equal to the number of nodes.

8. Tree A tree is a set of branches with every node connected to every other node in such a way that removal of any branch destroys this property.

Alternately, a tree is defined as a connected sub-graph of a connected graph containing all the nodes of the graph but not containing any loops.

Branches of a tree are called twigs. A tree contains $(n - 1)$ twigs where n is the number of nodes in the graph.

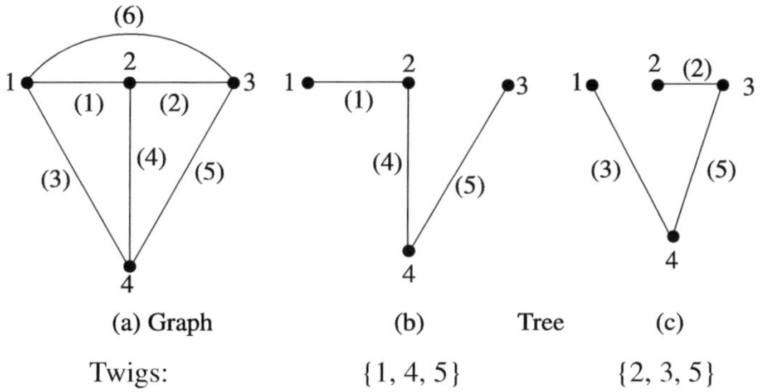

Twigs: {1, 4, 5} {2, 3, 5}

Fig. 7.7

9. Co-tree Branches which are not on a tree are called links or chords. All links of a tree together constitute the compliment of the corresponding tree and is called the co-tree.

A co-tree contains $b - (n - 1)$ links where b is the number of branches of the graph.

In Fig. 7.7(b) and (c) the links are {2, 3, 6} and {1, 4, 6} respectively.

Trees have the following properties:

(1) There exists only one path between any pair of nodes in a tree.
(2) A tree contains all nodes of the graph.
(3) If n is the number of nodes of the graph, there are $(n - 1)$ branches in the tree.
(4) Trees do not contain any loops.
(5) Every connected graph has at least one tree.
(6) The minimum terminal nodes in a tree are two.

7.4 INCIDENCE MATRIX

A linear graph is made up of nodes and branches. When a graph is given, it is possible to tell which branches are incident at which nodes and what are its orientations relative to the nodes.

7.4.1 Complete Incidence Matrix (A_a)

For a graph with n nodes and b branches, the complete incidence matrix is a rectangular matrix of order $n \times b$.

Elements of this matrix have the following values:

a_{ij} = 1, if branch j is incident at node i and is oriented away from node i.
= –1, if branch j is incident at node i and is oriented towards node i.
= 0, if branch j is not incident at node i.

For a graph shown in Fig. 7.8, the complete incidence matrix is as given below:

Nodes ↓	Branches → 1	2	3	4	5	6
1	1	1	0	0	0	1
2	0	–1	1	–1	0	0
3	0	0	0	1	1	–1
4	–1	0	–1	0	–1	0

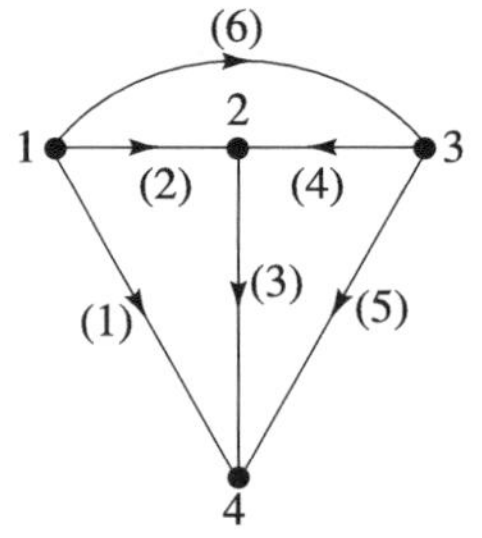

Fig. 7.8

$$A_a = \begin{bmatrix} 1 & 1 & 0 & 0 & 0 & 1 \\ 0 & -1 & 1 & -1 & 0 & 0 \\ 0 & 0 & 0 & 1 & 1 & -1 \\ -1 & 0 & -1 & 0 & -1 & 0 \end{bmatrix}$$

It is seen from the matrix A_a that the sum of the elements in any column is zero. Hence, any one row of the complete incidence matrix can be obtained by the algebraic manipulation of other rows.

7.4.2 Reduced Incidence Matrix (*A*)

The reduced incidence matrix A is obtained from the complete incidence matrix A_a by eliminating one of the rows. It is also called *incidence matrix*. It is of order $(n-1) \times b$.

Eliminating the third row of matrix A_a, we get

$$A = \begin{bmatrix} 1 & 1 & 0 & 0 & 0 & 1 \\ 0 & -1 & 1 & -1 & 0 & 0 \\ -1 & 0 & -1 & 0 & -1 & 0 \end{bmatrix}$$

When a tree is selected for the graph as shown in Fig. 7.9, the incidence matrix is obtained by arranging a column such that the first $(n-1)$ column corresponds to twigs of the tree and the last $b-(n-1)$ branches corresponds to the links of the selected tree.

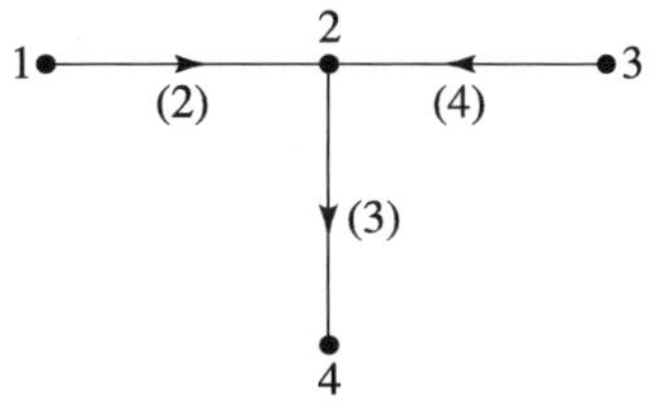

Twigs: {2, 3, 4}
Links: {1, 5, 6}

Fig. 7.9

$$A = \begin{array}{c} \quad \text{Twigs} \qquad \text{Links} \\ \begin{array}{cccccc} 2 & 3 & 4 & 1 & 5 & 6 \end{array} \\ \begin{bmatrix} 1 & 0 & 0 & 1 & 0 & 1 \\ -1 & 1 & -1 & 0 & 0 & 0 \\ 0 & -1 & 0 & -1 & -1 & 0 \end{bmatrix} \end{array}$$

The matrix A can be subdivided into submatrices A_t and A_l.

$$A = [A_t : A_l]$$

where A_t the is twig matrix whereas A_l is the link matrix.

7.4.3 Number of Possible Trees of a Graph

Let the transpose of the reduced incidence matrix A be A^T. It can be shown that the number of possible trees of a graph will be given by

$$\text{Number of possible trees} = \left| AA^T \right|$$

For the graphs shown in Fig. 7.8, the reduced incidence matrix is given by

$$A = \begin{bmatrix} 1 & 1 & 0 & 0 & 0 & 1 \\ 0 & -1 & 1 & -1 & 0 & 0 \\ -1 & 0 & -1 & 0 & -1 & 0 \end{bmatrix}$$

Then transpose of this matrix will be

$$A^T = \begin{bmatrix} 1 & 0 & -1 \\ 1 & -1 & 0 \\ 0 & 1 & -1 \\ 0 & -1 & 0 \\ 0 & 0 & -1 \\ 1 & 0 & 0 \end{bmatrix}$$

Hence, number of all possible trees of the graph

$$AA^T = \begin{bmatrix} 1 & 1 & 0 & 0 & 0 & 1 \\ 0 & -1 & 1 & -1 & 0 & 0 \\ -1 & 0 & -1 & 0 & -1 & 0 \end{bmatrix} \begin{bmatrix} 1 & 0 & -1 \\ 1 & -1 & 0 \\ 0 & 1 & -1 \\ 0 & -1 & 0 \\ 0 & 0 & -1 \\ 1 & 0 & 0 \end{bmatrix} = \begin{bmatrix} 3 & -1 & -1 \\ -1 & 3 & -1 \\ -1 & -1 & 3 \end{bmatrix}$$

$$\left|AA^T\right| = \begin{vmatrix} 3 & -1 & -1 \\ -1 & 3 & -1 \\ -1 & -1 & 3 \end{vmatrix} = 3\,(9-1) + (1)\,(-3-1) - 1\,(1+3) = 16$$

Thus, 16 different trees can be drawn.

7.5 LOOP MATRIX OR CIRCUIT MATRIX

When a graph is given, it is possible to tell which branches constitute which loop or circuit. Alternately, if a loop matrix or circuit matrix is given, we can reconstruct the graph.

For a graph having n nodes and b branches, the loop matrix B_a is a rectangular matrix of order b columns and as many rows as there are loops.

Its elements have the following values:

b_{ij} = 1, if branch j is in loop i and their orientations coincide.

= –1, if branch j is in loop i and their orientations do not coincide.

= 0, if branch j is not in loop i.

A graph and its loops are shown in Fig. 7.10.

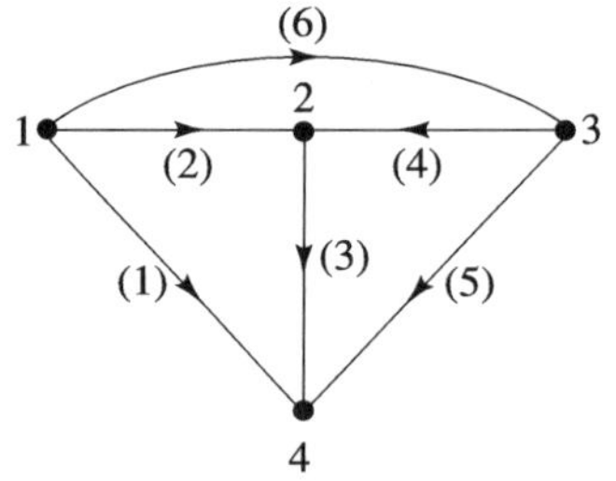

Loop 1: {1, 2, 3}
Loop 2: {3, 4, 5}
Loop 3: {2, 4, 6}
Loop 4: {1, 2, 4, 5}
Loop 5: {1, 5, 6}
Loop 6: {2, 3, 5, 6}
Loop 7: {1, 3, 4, 6}

Fig. 7.10

All the loop currents are assumed to be flowing in a clockwise direction.

Loops ↓	Branches → 1	2	3	4	5	6
1	–1	1	1	0	0	0
2	0	0	–1	–1	1	0
3	0	–1	0	1	0	1
4	–1	1	0	–1	1	0
5	–1	0	0	0	1	1
6	0	–1	–1	0	1	1
7	–1	0	1	1	0	1

$$B_a = \begin{bmatrix} -1 & 1 & 1 & 0 & 0 & 0 \\ 0 & 0 & -1 & -1 & 1 & 0 \\ 0 & -1 & 0 & 1 & 0 & 1 \\ -1 & 1 & 0 & -1 & 1 & 0 \\ -1 & 0 & 0 & 0 & 1 & 1 \\ 0 & -1 & -1 & 0 & 1 & 1 \\ -1 & 0 & 1 & 1 & 0 & 1 \end{bmatrix}$$

7.5.1 Fundamental Circuit (Tieset) and Fundamental Circuit Matrix

When a graph is given, first select a tree and remove all the links. When a link is replaced, a closed loop or circuit is formed. Circuits formed in this way are called fundamental circuits or f-circuits or tiesets.

Orientation of an *f*-circuit is given by the orientation of the connecting link. The number of f-circuits is same as the number of links for a graph. In a graph having b branches and n nodes, the number of f-circuits or tiesets will be $(b - n + 1)$. Figure 7.11 shows a tree and f-circuits (tiesets) for the graph shown in Fig. 7.10.

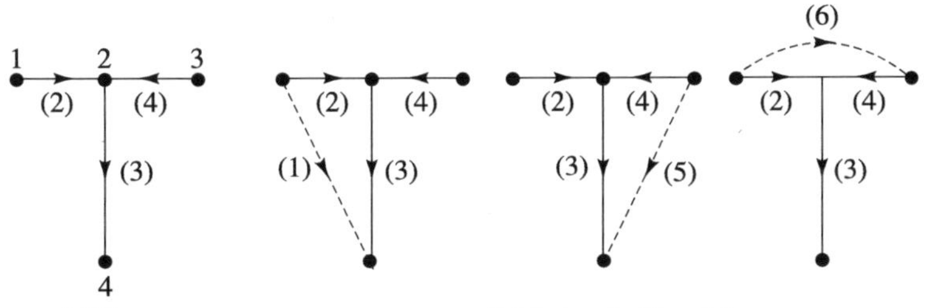

(a) Tree (b) f-circuits (tiesets)

Fig. 7.11

Here, $b = 6$ and $n = 4$.

$$\text{Number of tiesets} = b - n + 1 = 6 - 4 + 1 = 3$$

f-circuits are shown in Fig. 7.11. The orientation of each *f*-circuit is given by the orientation of the corresponding connecting link. Writing the tieset, with the link as the first entry and other branches in sequence, we have

tieset 1: {1, 2, 3}
tieset 5: {5, 3, 4}
tieset 6: {6, 2, 4}

Then, the tieset schedule will be written as

Tiesets ↓	Branches → 1	2	3	4	5	6
1	1	−1	−1	0	0	0
5	0	0	−1	−1	1	0
6	0	−1	0	1	0	1

Hence, an *f*-circuit matrix or tieset matrix will be given as

$$B = \begin{bmatrix} 1 & -1 & -1 & 0 & 0 & 0 \\ 0 & 0 & -1 & -1 & 1 & 0 \\ 0 & -1 & 0 & 1 & 0 & 1 \end{bmatrix}$$

Usually, the *f*-circuit matrix B is rearranged so that the first $(n-1)$ columns correspond to the twigs and $b-(n-1)$ columns to the links of the selected tree.

$$B = \begin{array}{c} \text{Twigs} \qquad\quad \text{Links} \\ \begin{array}{cccccc} 2 & 3 & 4 & 1 & 5 & 6 \end{array} \\ \begin{bmatrix} -1 & -1 & 0 & 1 & 0 & 0 \\ 0 & -1 & -1 & 0 & 1 & 0 \\ -1 & 0 & 1 & 0 & 0 & 1 \end{bmatrix} \end{array}$$

The matrix B can be partitioned into two matrices B_t and B_l.

$$B = [\, B_t : B_l \,]$$
$$= [\, B_t : U \,]$$

where B_t is the twig matrix, B_l is the link matrix and U is the unit matrix.

7.5.2 Orthogonal Relationship between Matrix *A* and Matrix *B*

For a linear graph, if the columns of the two matrices A_a and B_a are arranged in the same order, it can be shown that

$$A_a B_a^T = 0$$

or

$$B_a A_a^T = 0$$

The above equations describe the orthogonal relationship between the matrices A_a and B_a.

If the reduced incidence matrix A and the *f*-circuit matrix B are written for the same tree, it can be shown that

$$A\,B^T = 0$$

or

$$B\,A^T = 0$$

These two equations show the orthogonal relationship between matrices A and B.

7.6 CUTSET MATRIX

Consider a linear graph. By removing a set of branches without affecting the nodes, two connected sub-graphs are obtained and the original graph becomes unconnected. The removal of this set of branches which results in cutting the graph into two parts are known as a *cutset*. The cutset separates the nodes of the graph into two groups, each being in one of the two groups.

Figure 7.12 shows a graph.

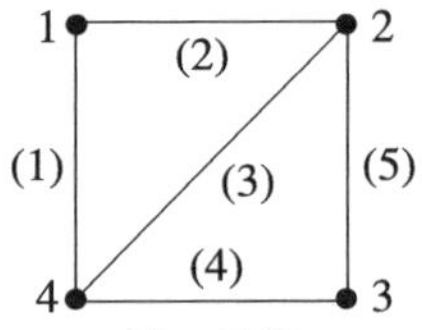

Fig. 7.12

Branches 1, 3 and 4 will form a cutset. This set of branches separates the graph into two parts. One having an isolated Node 4 and other part having branches 2 and 5 and nodes 1, 2 and 3.

Similarly, branches 1 and 2 will form a cutset. Each branch of the cutset has one of its terminals incident at a node in one part and its other end incident at other nodes in the other parts. The orientation of a cutset is made to coincide with orientation of defining branch.

For a graph having n nodes and b branches, the cutset matrix Q_a is a rectangular matrix of order b columns and as many rows as there are cutsets. Its elements have the following values:

q_{ij} = 1, if branch j is in the cutset i and the orientations coincide.

= –1, if branch j is in the cutset i and the orientations do not coincide.

= 0, if branch j is not in the cutset i.

Figure 7.13 shows a directed graph and its cutsets.

Cutsets ↓	1	2	3	4	5	6
1	1	1	0	0	0	1
2	0	1	–1	1	0	0
3	1	0	1	0	1	0
4	0	0	0	1	1	–1
5	0	–1	1	0	1	–1
6	1	0	1	–1	0	1

(Branches →)

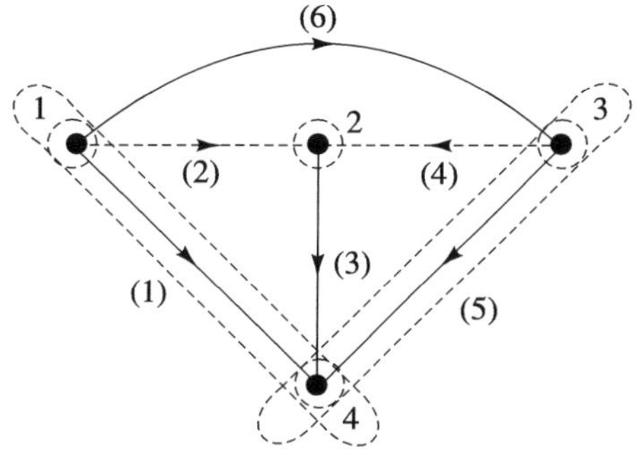

Cutset 1: {1, 2, 6}
Cutset 2: {2, 3, 4}
Cutset 3: {3, 1, 5}
Cutset 4: {4, 5, 6}
Cutset 5: {5, 2, 3, 6}
Cutset 6: {6, 1, 3, 4}

Fig. 7.13

For the cutset 2, which cuts the branches 2, 3 and 4 and is shown by a dotted circle, the entry in the cutset schedule for the branch (2) is 1, since the orientation of this cutset is given by the orientation of the branch 2 and hence it coincides. The entry for branch 3 is –1 as orientation of branch 3 is opposite to that of cutset 2 i.e. branch 2 goes into cutset while branch 3 goes out of cutset. The entry for the branch 4 is 1 as the branch 2 and the branch 4 go into the cutset. Thus their orientations coincide.

Hence, the cutset matrix Q_a will be given as

$$Q_a = \begin{bmatrix} 1 & 1 & 0 & 0 & 0 & 1 \\ 0 & 1 & -1 & 1 & 0 & 0 \\ 1 & 0 & 1 & 0 & 1 & 0 \\ 0 & 0 & 0 & 1 & 1 & -1 \\ 0 & -1 & 1 & 0 & 1 & -1 \\ 1 & 0 & 1 & -1 & 0 & 1 \end{bmatrix}$$

7.6.1 Fundamental Cutset and Fundamental Cutset Matrix

When a graph is given, first select a tree and note down its twigs. When a twig is removed from the tree, it separates a tree into two parts (one of the separated part may be an isolated node). Now, all the branches connecting one part of the disconnected tree to the other along with the twig removed constitutes a cutset. This set of branches is called a fundamental cutset or *f*-cutset. A matrix formed by these *f*-cutsets is called an *f*-cutset matrix. The orientation of the *f*-cutset is made to coincide with the orientation of the defining branch, i.e., twig. The number of *f*-cutsets is the same as the number of twigs for a graph.

Figure 7.14 shows a graph, selected tree and *f*-cutsets corresponding to the selected tree.

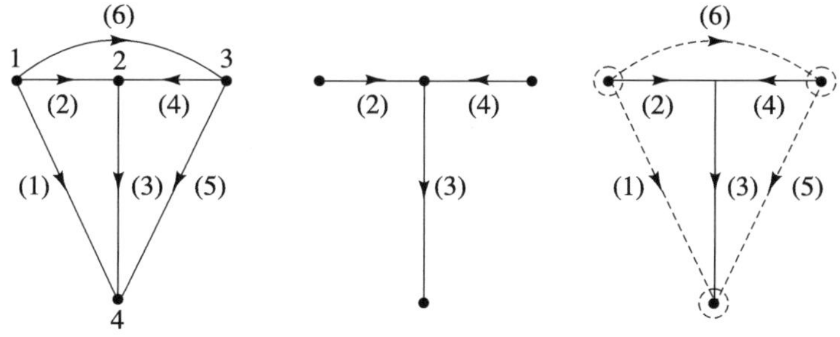

(a) Graph (a) Tree (a) *f*-cutsets

Fig. 7.14

f-cutset 2: {2, 1, 6}
f-cutset 3: {3, 1, 5}
f-cutset 4: {4, 5, 6}

The cutset schedule is written as below:

f-cutsets	Branches →					
↓	1	2	3	4	5	6
2	1	1	0	0	0	1
3	1	0	1	0	1	0
4	0	0	0	1	1	–1

Hence, the *f*-cutset matrix Q is given by

$$Q = \begin{bmatrix} 1 & 1 & 0 & 0 & 0 & 1 \\ 1 & 0 & 1 & 0 & 1 & 0 \\ 0 & 0 & 0 & 1 & 1 & -1 \end{bmatrix}$$

The *f*-cutset matrix Q is rearranged so that the first $(n - 1)$ columns correspond to twigs and $b - (n - 1)$ columns to links of the selected tree.

$$\begin{matrix} & \text{Twigs} & & & \text{Links} & \\ 2 & 3 & 4 & 1 & 5 & 6 \end{matrix}$$

$$Q = \begin{bmatrix} 1 & 0 & 0 & 1 & 0 & 1 \\ 0 & 1 & 0 & 1 & 1 & 0 \\ 0 & 0 & 1 & 0 & 1 & -1 \end{bmatrix}$$

The matrix Q can be subdivided into matrices Q_t and Q_l.

$$\begin{aligned} Q &= [\, Q_t : Q_l \,] \\ &= [\, U : Q_l \,] \end{aligned}$$

where Q_t is the twig matrix, Q_l is the link matrix and U is the unit matrix.

7.6.2 Orthogonal Relationship between Matrix *B* and Matrix *Q*

For a linear graph, if the columns of two matrices B_a and Q_a are arranged in the same order, it can be shown that

$$Q_a B_a^T = 0$$

or

$$B_a Q_a^T = 0$$

If the *f*-circuit matrix B and the *f*-cutset matrix Q are written for the same selected tree, it can be shown that

$$B\, Q^T = 0$$

or

$$Q\, B^T = 0$$

These two equations show the orthogonal relationship between matrices *A* and *B*.

7.7 RELATIONSHIP AMONG SUBMATRICES OF *A*, *B* AND *Q*

Arranging the columns of matrices *A*, *B* and *Q* with twigs for a given tree first and then the links, we get the partitioned forms as

$$\begin{aligned} A &= [\, A_t : A_l \,] \\ B &= [\, B_t : B_l \,] = [\, B_t : U] \\ Q &= [\, Q_t : Q_l \,] = [\, U : Q_l \,] \end{aligned}$$

From the orthogonal relation, $AB^T = 0$, we get

$$AB^T = [\,A_t : A_l\,]\begin{bmatrix} B_t^T \\ \cdots \\ B_l^T \end{bmatrix}$$

$$A_t B_t^T + A_l B_l^T = 0$$

$$A_t B_t^T = -A_l B_l^T$$

Since A_t is non-singular, i.e., $|A| \neq 0$,

A_t^{-1} exists

Premultiplying with A_t^{-1}, we get

$$B_t^T = -A_t^{-1} A_l B_l^T$$

$$B_t = -B_l (A_t^{-1} \cdot A_l)^T$$

Since B_l is a unit matrix

$$B_t = -(A_t^{-1} \cdot A_l)^T$$

Hence, matrix B is written as

$$B = [-(A_t^{-1} \cdot A_l)^T : U] \qquad \ldots(7.1)$$

We know that $AB^T = 0$

$$A_l B_l^T = -A_t B_t^T$$

Postmultiplying with $(B_l^T)^{-1}$

$$A_l = -A_t B_t^T (B_l^T)^{-1}$$

$$= -A_t B_t^T (B_l^{-1})^T = -A_t (B_l^{-1} \cdot B_t)^T$$

Hence matrix A can be written as

$$A = [\,A_t : -A_t (B_l^{-1} \cdot B_t)^T\,]$$

$$= A_t [\,U : -(B_l^{-1} \cdot B_t)^T\,] \qquad \ldots(7.2)$$

Similarly we can prove that

$$Q = [\,U : -(B_l^{-1} \cdot B_t)^T\,] \qquad \ldots(7.3)$$

From Eqs (7.2) and (7.3), we can write

$$A = A_t Q \qquad \ldots(7.4)$$

$$Q = A_t^{-1} A \qquad \ldots(7.5)$$

$$= A_t^{-1} [\,A_t : A_l\,]$$

$$= [\,U : A_t^{-1} A_l\,] \qquad \ldots(7.6)$$

We have shown that $B_t = -(A_t^{-1} \cdot A_l)^T$

$$B_t^T = -(A_t^{-1} \cdot A_l)$$

Hence Q can be written as

$$Q = [\,U : -B_t^T\,]$$

$$Q_l = -B_t^T \qquad \ldots(7.7)$$

Example 7.1 *For the circuit shown in Fig. 7.15, draw the oriented graph and write the (i) incidence matrix, (ii) f-cutset matrix, and (iii) tieset matrix.*

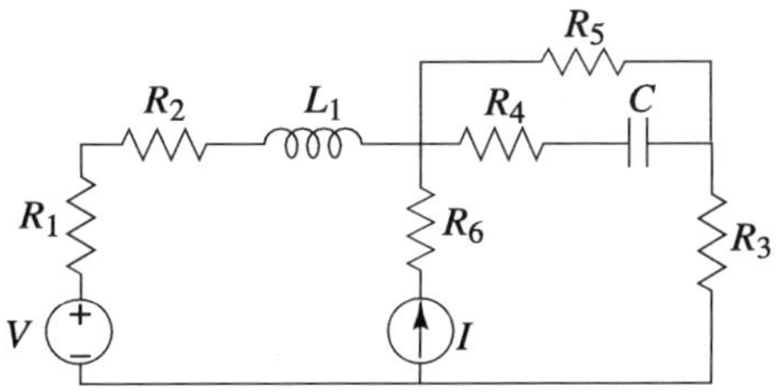

Fig. 7.15

Solution For drawing the oriented graph,

(1) replace all resistors, inductors and capacitors by line segments,
(2) replace the voltage source by short circuit and the current source by an open circuit,
(3) assume the directions of branch currents arbitrarily, and
(4) number all the nodes and branches.

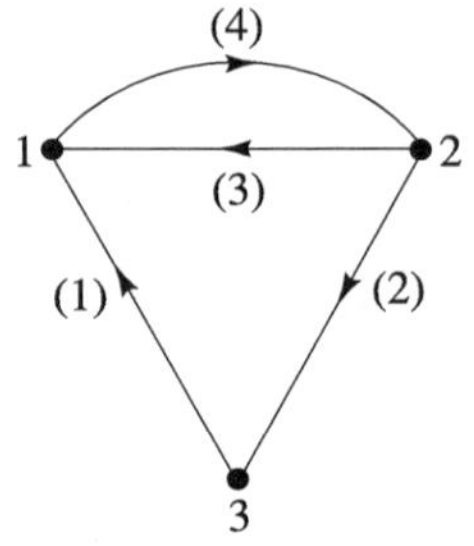

Fig. 7.16

Complete Incidence Matrix (A_a)

Nodes ↓ \ Branches →	1	2	3	4
1	–1	0	–1	1
2	0	1	1	–1
3	1	–1	0	0

$$A_a = \begin{bmatrix} -1 & 0 & -1 & 1 \\ 0 & 1 & 1 & -1 \\ 1 & -1 & 0 & 0 \end{bmatrix}$$

Eliminating the third row from the matrix A_a, we get the incidence matrix A.

$$A = \begin{bmatrix} -1 & 0 & -1 & 1 \\ 0 & 1 & 1 & -1 \end{bmatrix}$$

Tieset Matrix (*B*)

$$B = \begin{matrix} & \begin{matrix} 1 & 2 & 3 & 4 \end{matrix} \\ \begin{matrix} 3 \\ 4 \end{matrix} & \begin{bmatrix} -1 & -1 & 1 & 0 \\ 1 & 1 & 0 & 1 \end{bmatrix} \end{matrix}$$

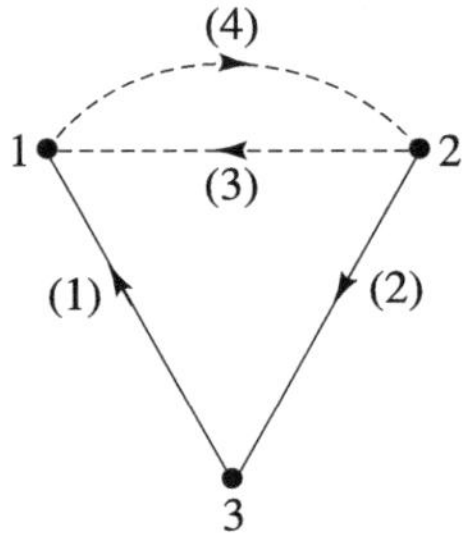

Fig. 7.17

Twigs: {1, 2}
Links: {3, 4}
Tieset 3: {3, 1, 2}
Tieset 4: {4, 1, 2}

f-cutset Matrix (*Q*)

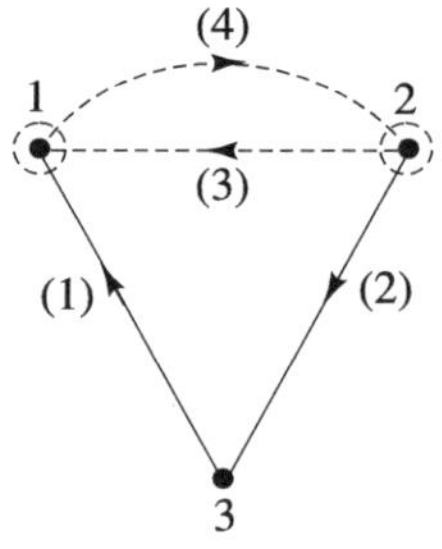

Fig. 7.18

f-cutset 1: {1, 3, 4}
f-cutset 2: {2, 3, 4}

$$Q = \begin{matrix} & \begin{matrix} 1 & 2 & 3 & 4 \end{matrix} \\ \begin{matrix} 1 \\ 2 \end{matrix} & \begin{bmatrix} 1 & 0 & 1 & -1 \\ 0 & 1 & 1 & -1 \end{bmatrix} \end{matrix}$$

Example 7.2 *For the network shown in Fig. 7.19, draw the oriented graph and write the (i) incidence matrix, (ii) tieset matrix, and (iii) f-cutset matrix.*

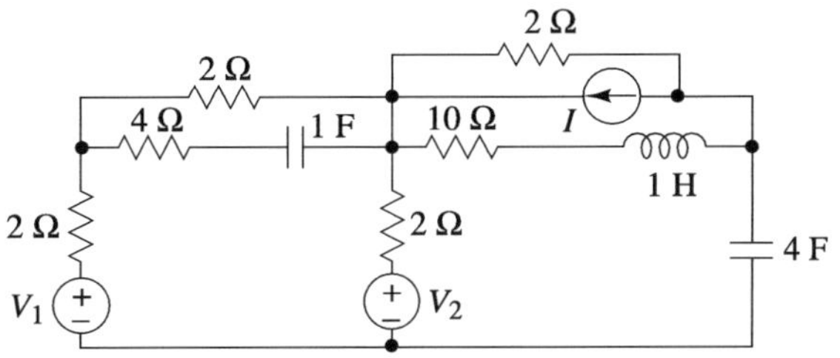

Fig. 7.19

Solution For drawing the oriented graph,

(1) replace all resistors, inductors and capacitors by line segments,

(2) replace all voltage sources by short circuits and current source by an open circuit,

(3) assume directions of branch currents arbitrarily, and

(4) number all the nodes and branches.

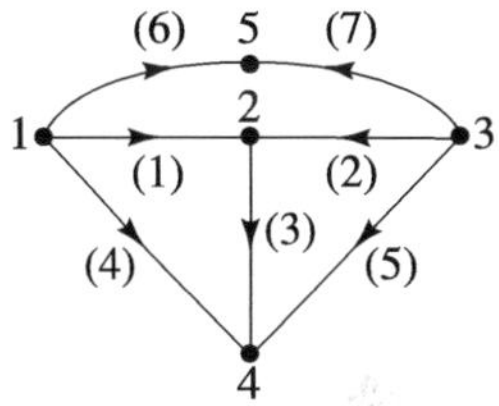

Fig. 7.20

Complete Incidence Matrix (A_a)

Nodes ↓ \ Branches →	1	2	3	4	5	6	7
1	1	0	0	1	0	1	0
2	–1	–1	1	0	0	0	0
3	0	1	0	0	1	0	1
4	0	0	–1	–1	–1	0	0
5	0	0	0	0	0	–1	–1

$$A_a = \begin{bmatrix} 1 & 0 & 0 & 1 & 0 & 1 & 0 \\ -1 & -1 & 1 & 0 & 0 & 0 & 0 \\ 0 & 1 & 0 & 0 & 1 & 0 & 1 \\ 0 & 0 & -1 & -1 & -1 & 0 & 0 \\ 0 & 0 & 0 & 0 & 0 & -1 & -1 \end{bmatrix}$$

Eliminating the last row from the matrix A_a, we get the incidence matrix A.

$$A = \begin{bmatrix} 1 & 0 & 0 & 1 & 0 & 1 & 0 \\ -1 & -1 & 1 & 0 & 0 & 0 & 0 \\ 0 & 1 & 0 & 0 & 1 & 0 & 1 \\ 0 & 0 & -1 & -1 & -1 & 0 & 0 \end{bmatrix}$$

Tieset Matrix (B)

$$B = \begin{array}{c} \\ 2 \\ 4 \\ 7 \end{array}\!\!\begin{array}{c} \begin{array}{ccccccc} 1 & 2 & 3 & 4 & 5 & 6 & 7 \end{array} \\ \begin{bmatrix} 0 & 1 & 1 & 0 & -1 & 0 & 0 \\ -1 & 0 & -1 & 1 & 0 & 0 & 0 \\ 1 & 0 & 1 & 0 & -1 & -1 & 1 \end{bmatrix} \end{array}$$

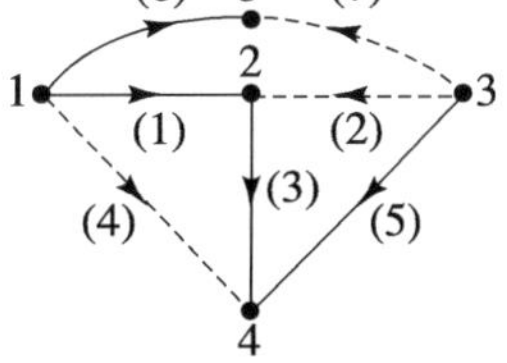

Fig. 7.21

Twigs: {1, 3, 5, 6}
Links: {2, 4, 7}
Tieset 2: {2, 3, 5}
Tieset 4: {4, 1, 3}
Tieset 7: {7, 6, 1, 3, 5}

f-cutset Matrix (Q)

$$Q = \begin{array}{c} \\ 1 \\ 3 \\ 5 \\ 6 \end{array}\begin{array}{c} \begin{matrix} 1 & 2 & 3 & 4 & 5 & 6 & 7 \end{matrix} \\ \begin{bmatrix} 1 & 0 & 0 & 1 & 0 & 0 & -1 \\ 0 & -1 & 1 & 1 & 0 & 0 & -1 \\ 0 & 1 & 0 & 0 & 1 & 0 & 1 \\ 0 & 0 & 0 & 0 & 0 & 1 & 1 \end{bmatrix} \end{array}$$

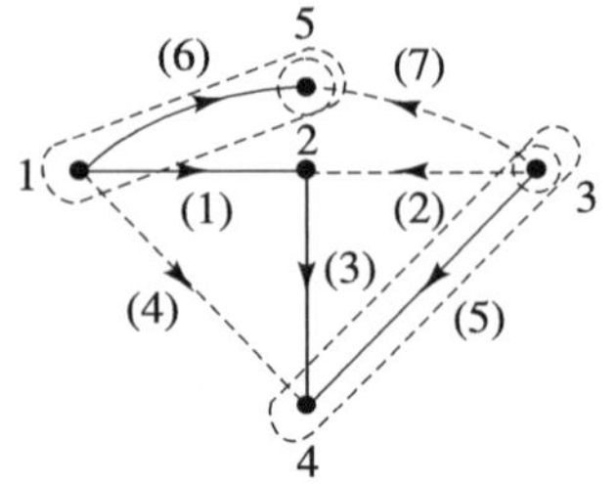

f-cutset 1: {1, 4, 7}
f-cutset 3: {3, 4, 7, 2}
f-cutset 5: {5, 2, 7}
f-cutset 6: {6, 7}

Fig. 7.22

Example 7.3 *For the circuit shown in Fig. 7.23, (i) draw its graph, (ii) draw its tree, and (iii) write the fundamental cutset matrix.*

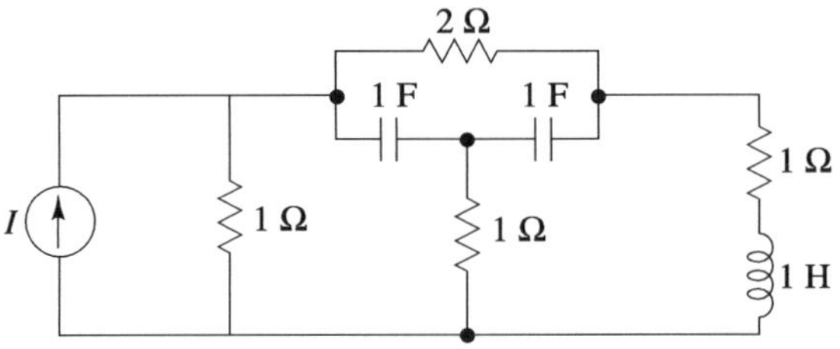

Fig. 7.23

Solution For drawing the oriented graph,

(1) replace all resistors, inductors and capacitors by line segments,
(2) replace the current source by an open circuit,
(3) assume directions of branch currents, and
(4) number all the nodes and branches.

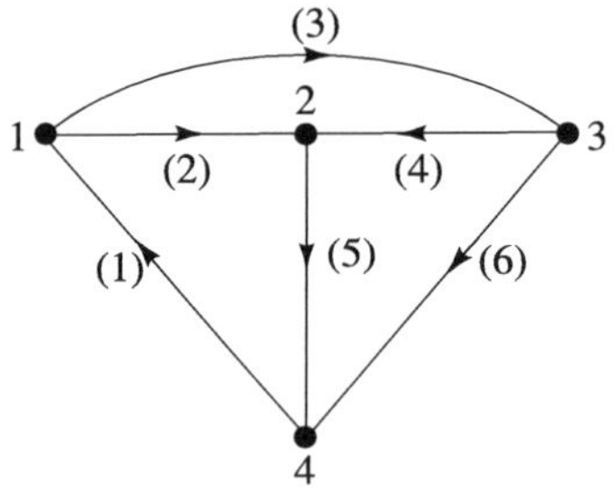

Fig. 7.24

Fundamental Cutset Matrix (Q)

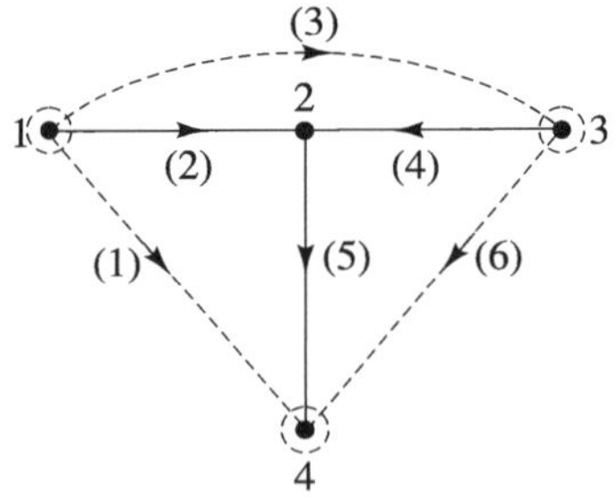

Fig. 7.25

Twigs: {2, 4, 5}
Links: {1, 3, 6}
f-cutset 2: {2, 1, 3}
f-cutset 4: {4, 3, 6}
f-cutset 5: {5, 1, 6}

$$Q = \begin{array}{c} \\ 2 \\ 4 \\ 5 \end{array}\begin{array}{c} \begin{matrix} 1 & 2 & 3 & 4 & 5 & 6 \end{matrix} \\ \begin{bmatrix} 1 & 1 & 1 & 0 & 0 & 0 \\ 0 & 0 & -1 & 1 & 0 & 1 \\ 1 & 0 & 0 & 0 & 1 & 1 \end{bmatrix} \end{array}$$

Example 7.4 *For the circuit shown in Fig. 7.26, draw the oriented graph and write (i) incidence matrix, (ii) tieset matrix, and (iii) cutset matrix.*

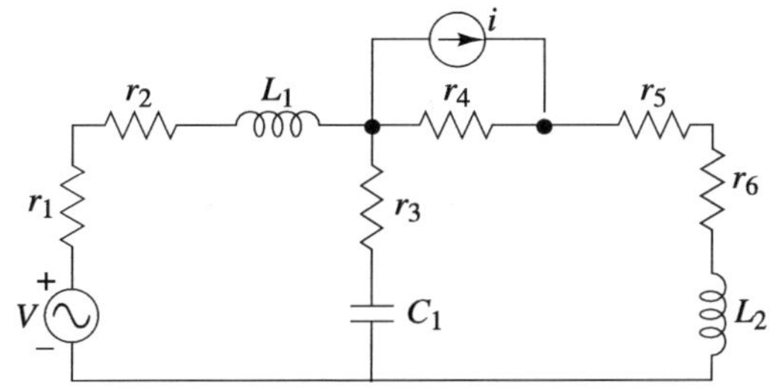

Fig. 7.26

Solution For drawing the oriented graph,

(1) replace all resistors, inductors and capacitors by line segments,
(2) replace voltage source by short circuit and current source by an open circuit,
(3) assume directions of branch currents arbitrarily, and
(4) number the nodes and branches.

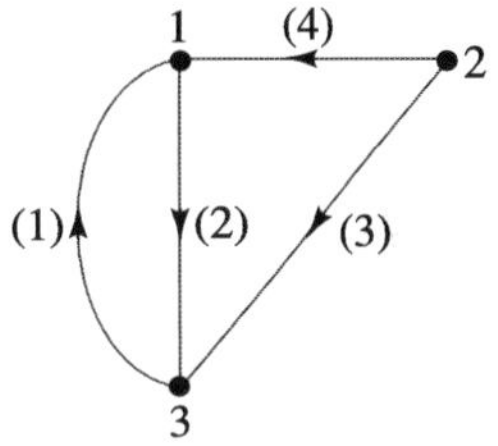

Fig. 7.27

Complete Incidence Matrix (A_a)

Nodes ↓	Branches → 1	2	3	4
1	−1	1	0	−1
2	0	0	1	1
3	1	−1	−1	0

$$A_a = \begin{bmatrix} -1 & 1 & 0 & -1 \\ 0 & 0 & 1 & 1 \\ 1 & -1 & -1 & 0 \end{bmatrix}$$

Eliminating the third row from the matrix A_a, we get the incidence matrix A.

$$A = \begin{bmatrix} -1 & 1 & 0 & -1 \\ 0 & 0 & 1 & 1 \end{bmatrix}$$

Tieset Matrix (B)

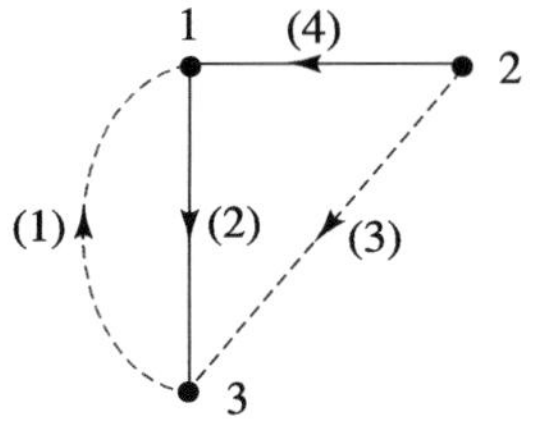

Twigs: {2, 4}
Links: {1, 3}
Tieset 1: {1, 2}
Tieset 3: {3, 2, 4}

Fig. 7.28

$$B = \begin{matrix} & \begin{matrix} 1 & 2 & 3 & 4 \end{matrix} \\ \begin{matrix} 1 \\ 3 \end{matrix} & \begin{bmatrix} 1 & 1 & 0 & 0 \\ 0 & -1 & 1 & -1 \end{bmatrix} \end{matrix}$$

f-cutset Matrix (Q)

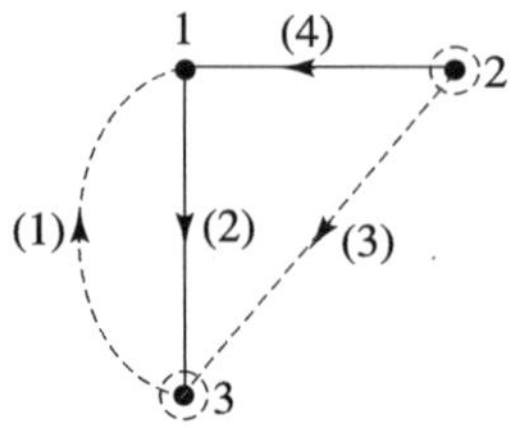

f-cutset 2: {2, 1, 3}
f-cutset 4: {4, 3}

Fig. 7.29

$$Q = \begin{array}{c} \\ 2 \\ 4 \end{array}\begin{array}{c} \begin{array}{cccc} 1 & 2 & 3 & 4 \end{array} \\ \begin{bmatrix} -1 & 1 & 1 & 0 \\ 0 & 0 & 1 & 1 \end{bmatrix} \end{array}$$

Example 7.5 *The graph of a network is shown in Fig. 7.30. Write the (i) incidence matrix, (ii) f-cutset matrix, and (iii) f-circuit matrix.*

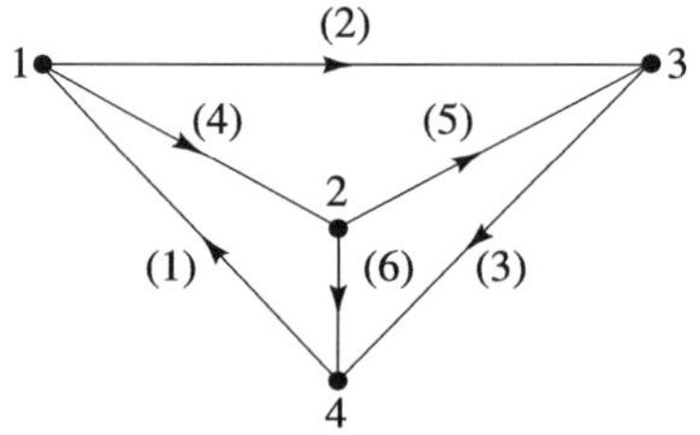

Fig. 7.30

Solution

Complete Incidence Matrix (A_a)

$$A_a = \begin{array}{c} \\ 1 \\ 2 \\ 3 \\ 4 \end{array}\begin{array}{c} \begin{array}{cccccc} 1 & 2 & 3 & 4 & 5 & 6 \end{array} \\ \begin{bmatrix} -1 & 1 & 0 & 1 & 0 & 0 \\ 0 & 0 & 0 & -1 & 1 & 1 \\ 0 & -1 & 1 & 0 & -1 & 0 \\ 1 & 0 & -1 & 0 & 0 & -1 \end{bmatrix} \end{array}$$

The incidence matrix A is obtained by eliminating any row from the matrix A_a.

$$A = \begin{bmatrix} -1 & 1 & 0 & 1 & 0 & 0 \\ 0 & 0 & 0 & -1 & 1 & 1 \\ 0 & -1 & 1 & 0 & -1 & 0 \end{bmatrix}$$

Tieset Matrix (B)

$$B = \begin{array}{c} \\ 1 \\ 2 \\ 3 \end{array}\begin{array}{c} \begin{array}{cccccc} 1 & 2 & 3 & 4 & 5 & 6 \end{array} \\ \begin{bmatrix} 1 & 0 & 0 & 1 & 0 & 1 \\ 0 & 1 & 0 & -1 & -1 & 0 \\ 0 & 0 & 1 & 0 & 1 & -1 \end{bmatrix} \end{array}$$

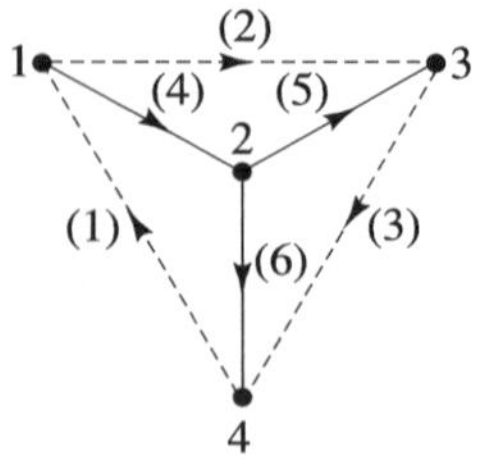

Twigs: {4, 5, 6}
Links: {1, 2, 3}
Tieset 1: {1, 4, 6}
Tieset 2: {2, 4, 5}
Tieset 3: {3, 5, 6}

Fig. 7.31

f-cutset Matrix (Q)

$$Q = \begin{matrix} & \begin{matrix} 1 & 2 & 3 & 4 & 5 & 6 \end{matrix} \\ \begin{matrix} 4 \\ 5 \\ 6 \end{matrix} & \begin{bmatrix} -1 & 1 & 0 & 1 & 0 & 0 \\ 0 & 1 & -1 & 0 & 1 & 0 \\ -1 & 0 & 1 & 0 & 0 & 1 \end{bmatrix} \end{matrix}$$

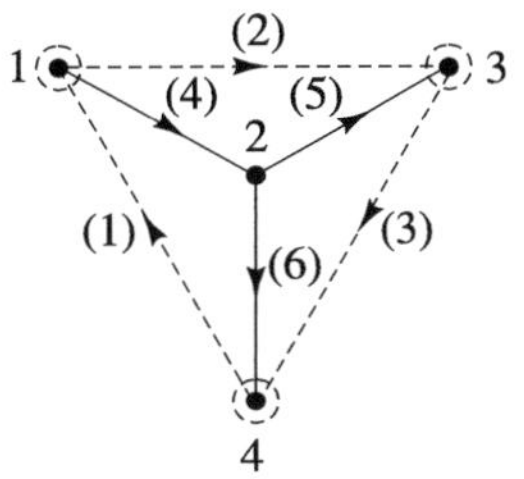

Fig. 7.32

f-cutset 4: {4, 1, 2}
f-cutset 5: {5, 2, 3}
f-cutset 6: {6, 1, 3}

Example 7.6 *For the graph shown, write the incidence matrix, tieset matrix and f-cutset matrix.*

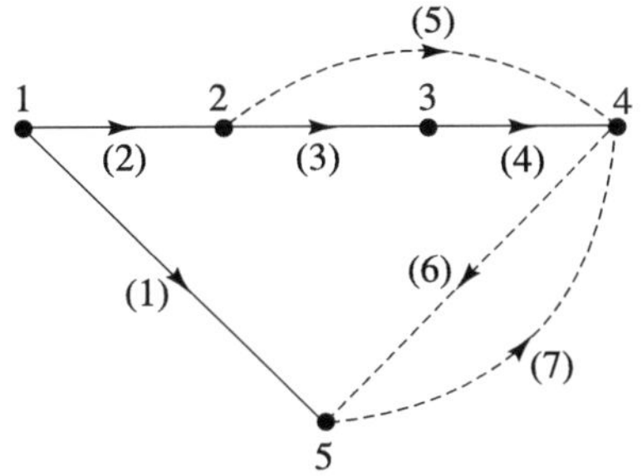

Fig. 7.33

Solution

Complete Incidence Matrix (A_a)

$$A_a = \begin{matrix} & \begin{matrix} 1 & 2 & 3 & 4 & 5 & 6 & 7 \end{matrix} \\ \begin{matrix} 1 \\ 2 \\ 3 \\ 4 \\ 5 \end{matrix} & \begin{bmatrix} 1 & 1 & 0 & 0 & 0 & 0 & 0 \\ 0 & -1 & 1 & 0 & 1 & 0 & 0 \\ 0 & 0 & -1 & 1 & 0 & 0 & 0 \\ 0 & 0 & 0 & -1 & -1 & 1 & -1 \\ -1 & 0 & 0 & 0 & 0 & -1 & 1 \end{bmatrix} \end{matrix}$$

The incidence matrix A is obtained by eliminating any row from the matrix A_a.

$$A = \begin{bmatrix} 1 & 1 & 0 & 0 & 0 & 0 & 0 \\ 0 & -1 & 1 & 0 & 1 & 0 & 0 \\ 0 & 0 & -1 & 1 & 0 & 0 & 0 \\ 0 & 0 & 0 & -1 & -1 & 1 & -1 \end{bmatrix}$$

Tieset Matrix (B)

Twigs: {1, 2, 3, 4}
Links: {5, 6, 7}
Tieset 5: {5, 3, 4}
Tieset 6: {6, 1, 2, 3, 4}
Tieset 7: {7, 1, 2, 3, 4}

$$B = \begin{matrix} & \begin{matrix} 1 & 2 & 3 & 4 & 5 & 6 & 7 \end{matrix} \\ \begin{matrix} 5 \\ 6 \\ 7 \end{matrix} & \begin{bmatrix} 0 & 0 & -1 & -1 & 1 & 0 & 0 \\ -1 & 1 & 1 & 1 & 0 & 1 & 0 \\ 1 & -1 & -1 & -1 & 0 & 0 & 1 \end{bmatrix} \end{matrix}$$

f-cutset Matrix (Q)

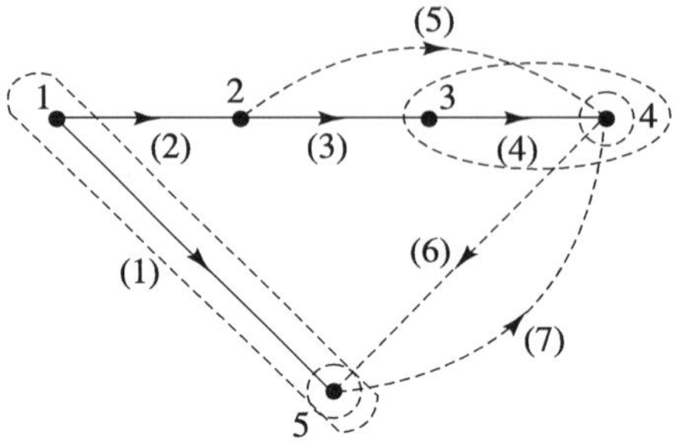

Fig. 7.34

f-cutset 1: {1, 6, 7}
f-cutset 2: {2, 6, 7}
f-cutset 3: {3, 5, 6, 7}
f-cutset 4: {3, 5, 6, 7}

$$Q = \begin{array}{c} \\ 1 \\ 2 \\ 3 \\ 4 \end{array} \begin{array}{c} \begin{array}{ccccccc} 1 & 2 & 3 & 4 & 5 & 6 & 7 \end{array} \\ \begin{bmatrix} 1 & 0 & 0 & 0 & 0 & 1 & -1 \\ 0 & 1 & 0 & 0 & 0 & -1 & 1 \\ 0 & 0 & 1 & 0 & 1 & -1 & 1 \\ 0 & 0 & 0 & 1 & 1 & -1 & 1 \end{bmatrix} \end{array}$$

Example 7.7 *For the graph shown, write the incidence matrix,tieset matrix and f-cutset matrix.*

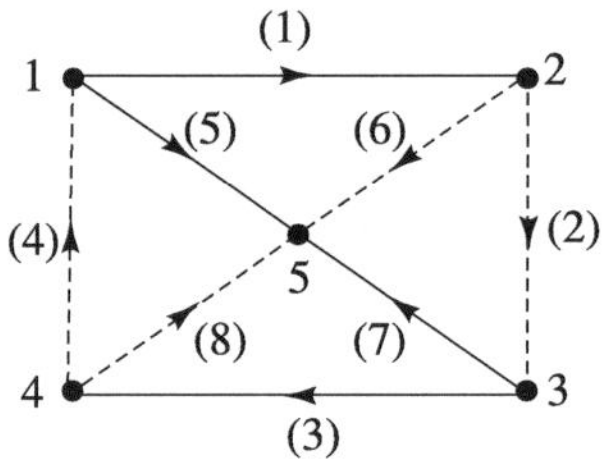

Fig. 7.35

Solution

Complete Incidence Matrix (A_a)

$$A_a = \begin{array}{c} \\ 1 \\ 2 \\ 3 \\ 4 \\ 5 \end{array} \begin{array}{c} \begin{array}{cccccccc} 1 & 2 & 3 & 4 & 5 & 6 & 7 & 8 \end{array} \\ \begin{bmatrix} 1 & 0 & 0 & -1 & 1 & 0 & 0 & 0 \\ -1 & 1 & 0 & 0 & 0 & 1 & 0 & 0 \\ 0 & -1 & 1 & 0 & 0 & 0 & 1 & 0 \\ 0 & 0 & -1 & 1 & 0 & 0 & 0 & 1 \\ 0 & 0 & 0 & 0 & -1 & -1 & -1 & -1 \end{bmatrix} \end{array}$$

The incidence matrix is obtained by eliminating any one row.

$$A = \begin{bmatrix} 1 & 0 & 0 & -1 & 1 & 0 & 0 & 0 \\ -1 & 1 & 0 & 0 & 0 & 1 & 0 & 0 \\ 0 & -1 & 1 & 0 & 0 & 0 & 1 & 0 \\ 0 & 0 & -1 & 1 & 0 & 0 & 0 & 1 \end{bmatrix}$$

Tieset Matrix (*B*)

Twigs: {1, 3, 5, 7}
Links: {2, 4, 6, 8}
Tieset 2: {2, 7, 5, 1}
Tieset 4: {4, 5, 7, 3}
Tieset 6: {6, 5, 1}
Tieset 8: {8, 7, 3}

$$B = \begin{array}{c} \\ 2 \\ 4 \\ 6 \\ 8 \end{array}\begin{array}{c} \begin{array}{cccccccc} 1 & 2 & 3 & 4 & 5 & 6 & 7 & 8 \end{array} \\ \begin{bmatrix} 1 & 1 & 0 & 0 & -1 & 0 & 1 & 0 \\ 0 & 0 & 1 & 1 & 1 & 0 & -1 & 0 \\ 1 & 0 & 0 & 0 & -1 & 1 & 0 & 0 \\ 0 & 0 & 1 & 0 & 0 & 0 & -1 & 1 \end{bmatrix} \end{array}$$

f-cutset Matrix (*Q*)

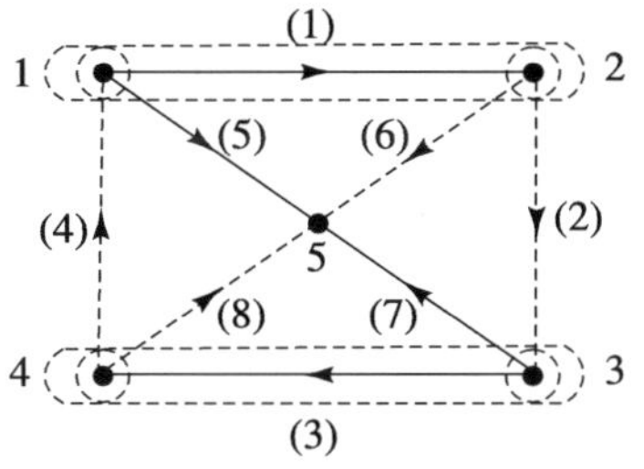

Fig. 7.36

f-cutset 1: {1, 6, 2}
f-cutset 3: {3, 4, 8}
f-cutset 5: {5, 4, 6, 2}
f-cutset 7: {7, 2, 8, 4}

$$Q = \begin{array}{c} \\ 1 \\ 3 \\ 5 \\ 7 \end{array}\begin{array}{c} \begin{array}{cccccccc} 1 & 2 & 3 & 4 & 5 & 6 & 7 & 8 \end{array} \\ \begin{bmatrix} 1 & -1 & 0 & 0 & 0 & -1 & 0 & 0 \\ 0 & 0 & 1 & -1 & 0 & 0 & 0 & -1 \\ 0 & 1 & 0 & -1 & 1 & 1 & 0 & 0 \\ 0 & -1 & 0 & 1 & 0 & 0 & 1 & 1 \end{bmatrix} \end{array}$$

Example 7.8 *How many trees are possible for the graph of the network of Fig. 7.37?*

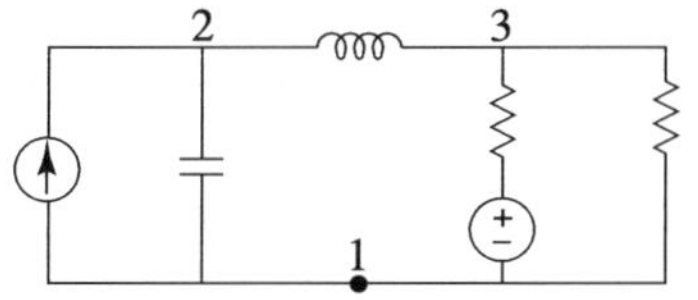

Fig. 7.37

Solution To draw the graph,

(1) replace all resistors, inductors and capacitors by line segments,
(2) replace voltage source by short circuit and current source by an open circuit,
(3) assume directions of branch currents arbitrarily, and
(4) number all the nodes and branches.

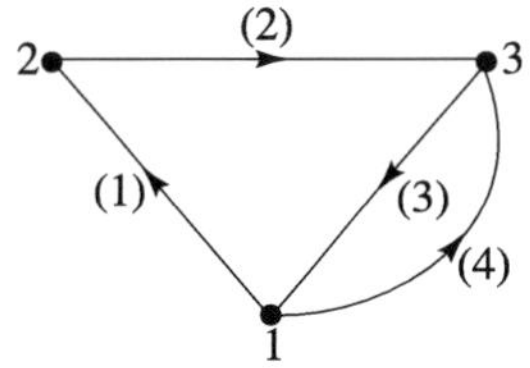

Fig. 7.38

Complete Incidence Matrix (A_a)

$$A_a = \begin{array}{c} \\ 1 \\ 2 \\ 3 \end{array}\begin{array}{c} \begin{array}{cccc} 1 & 2 & 3 & 4 \end{array} \\ \begin{bmatrix} 1 & 0 & -1 & 1 \\ -1 & 1 & 0 & 0 \\ 0 & -1 & 1 & -1 \end{bmatrix} \end{array}$$

The reduced incidence matrix A is obtained by eliminating the last row from matrix A_a.

$$A = \begin{bmatrix} 1 & 0 & -1 & 1 \\ -1 & 1 & 0 & 0 \end{bmatrix}$$

The number of possible trees = $|AA^T|$.

$$AA^T = \begin{bmatrix} 1 & 0 & -1 & 1 \\ -1 & 1 & 0 & 0 \end{bmatrix}\begin{bmatrix} 1 & -1 \\ 0 & 1 \\ -1 & 0 \\ 1 & 0 \end{bmatrix} = \begin{bmatrix} 3 & -1 \\ -1 & 2 \end{bmatrix}$$

$$|AA^T| = \begin{vmatrix} 3 & -1 \\ -1 & 2 \end{vmatrix} = 6 - 1 = 5.$$

Example 7.9 *Draw the oriented graph from the complete incidence matrix given below:*

Nodes ↓	Branches → 1	2	3	4	5	6	7	8
1	1	0	0	0	1	0	0	1
2	0	1	0	0	–1	1	0	0
3	0	0	1	0	0	–1	1	–1
4	0	0	0	1	0	0	–1	0
5	–1	–1	–1	–1	0	0	0	0

Solution First, note down the nodes 1, 2, 3, 4, 5 as shown in Fig. 7.39. From the complete incidence matrix, it is clear that the branch number 1 is between nodes 1 and 5 and it is going away from Node 1 and towards Node 5 as the entry against Node 1 is 1 and that against 5 is –1. Hence, connect the nodes 1 and 5 by a line, point the arrow towards 5 and call it Branch 1 as shown in Fig. 7.39. Similarly, draw the other oriented branches.

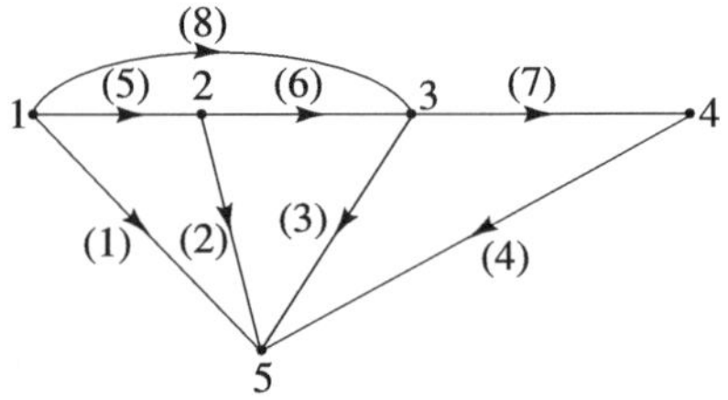

Fig. 7.39

Example 7.10 *The reduced incidence matrix of an oriented graph is given below. Draw the graph.*

$$A = \begin{bmatrix} 0 & -1 & 1 & 1 & 0 \\ 0 & 0 & -1 & -1 & -1 \\ -1 & 0 & 0 & 0 & 1 \end{bmatrix}$$

Solution First, writing the complete incidence matrix from the matrix A such that the sum of all entries in each column of A_a will be zero,we have

$$A_a = \begin{array}{c} \\ 1 \\ 2 \\ 3 \\ 4 \end{array}\begin{array}{c} \begin{matrix} 1 & 2 & 3 & 4 & 5 \end{matrix} \\ \begin{bmatrix} 0 & -1 & 1 & 1 & 0 \\ 0 & 0 & -1 & -1 & -1 \\ -1 & 0 & 0 & 0 & 1 \\ 1 & 1 & 0 & 0 & 0 \end{bmatrix} \end{array}$$

Now, the oriented graph can be drawn with matrix A_a.

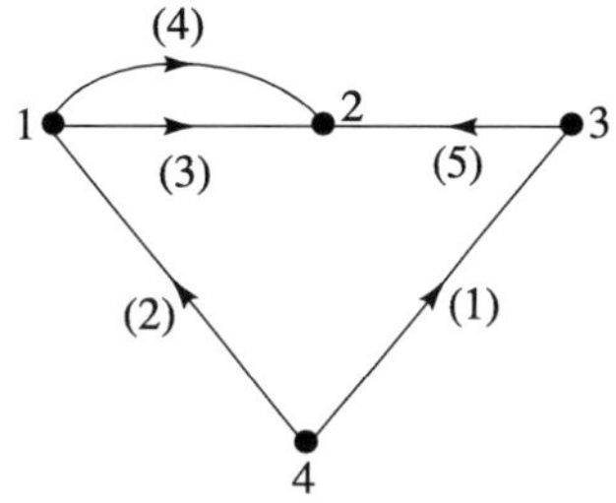

Fig. 7.40

Example 7.11 *The reduced incidence matrix of an oriented graph is*

$$A = \begin{bmatrix} 0 & -1 & 1 & 0 & 0 \\ 0 & 0 & -1 & -1 & -1 \\ -1 & 0 & 0 & 0 & 1 \end{bmatrix}$$

(i) Draw the graph. (ii) How many trees are possible for this graph? (iii) Write the tieset and cutset matrices.

Solution First, writing the complete incidence matrix A_a such that the sum of all the entries in each column of A_a is zero, we have

$$A_a = \begin{array}{c} \\ 1 \\ 2 \\ 3 \\ 4 \end{array}\begin{array}{c} \begin{matrix} 1 & 2 & 3 & 4 & 5 \end{matrix} \\ \begin{bmatrix} 0 & -1 & 1 & 0 & 0 \\ 0 & 0 & -1 & -1 & -1 \\ -1 & 0 & 0 & 0 & 1 \\ 1 & 1 & 0 & 1 & 0 \end{bmatrix} \end{array}$$

Now, the oriented graph can be drawn with the matrix A_a.

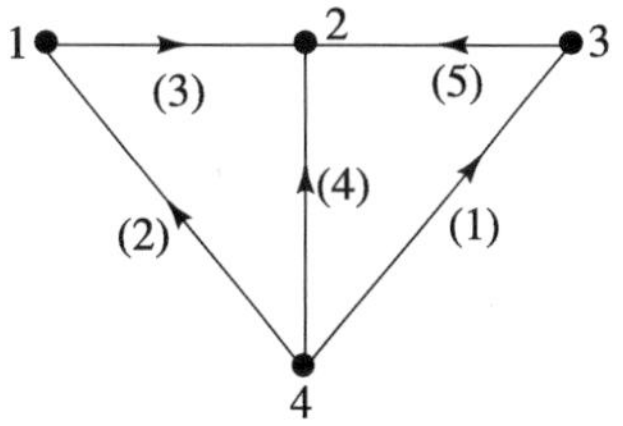

Fig. 7.41

The number of possible trees = $\left|AA^T\right|$

$$AA^T = \begin{bmatrix} 0 & -1 & 1 & 0 & 0 \\ 0 & 0 & -1 & -1 & -1 \\ -1 & 0 & 0 & 0 & 1 \end{bmatrix}\begin{bmatrix} 0 & 0 & -1 \\ -1 & 0 & 0 \\ 1 & -1 & 0 \\ 0 & -1 & 0 \\ 0 & -1 & 1 \end{bmatrix} = \begin{bmatrix} 2 & -1 & 0 \\ -1 & 3 & -1 \\ 0 & -1 & 2 \end{bmatrix}$$

$$\left|AA^T\right| = \begin{vmatrix} 2 & -1 & 0 \\ -1 & 3 & -1 \\ 0 & -1 & 2 \end{vmatrix} = 2\,(6-1) + 1\,(-2) = 8$$

The number of possible trees = 8.

Tieset Matrix (*B*)

$$B = \begin{array}{c} \\ 1 \\ 2 \end{array}\begin{array}{c} \begin{matrix} 1 & 2 & 3 & 4 & 5 \end{matrix} \\ \begin{bmatrix} 1 & 0 & 0 & -1 & 1 \\ 0 & 1 & 1 & -1 & 0 \end{bmatrix} \end{array}$$

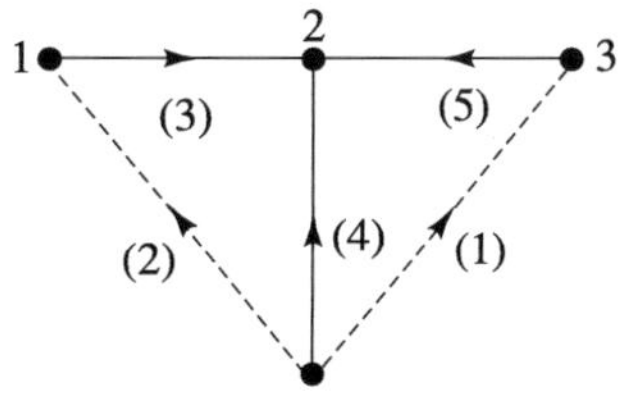

Fig. 7.42

Twigs: {3, 4, 5}
Links: {1, 2}
Tieset 1: {1, 4, 5}
Tieset 2: {2, 3, 4}

f-cutset Matrix (Q)

$$Q = \begin{array}{c} \\ 3 \\ 4 \\ 5 \end{array}\begin{array}{c} \begin{array}{ccccc} 1 & 2 & 3 & 4 & 5 \end{array} \\ \begin{bmatrix} 0 & -1 & 1 & 0 & 0 \\ 1 & 1 & 0 & 1 & 0 \\ -1 & 0 & 0 & 0 & 1 \end{bmatrix} \end{array}$$

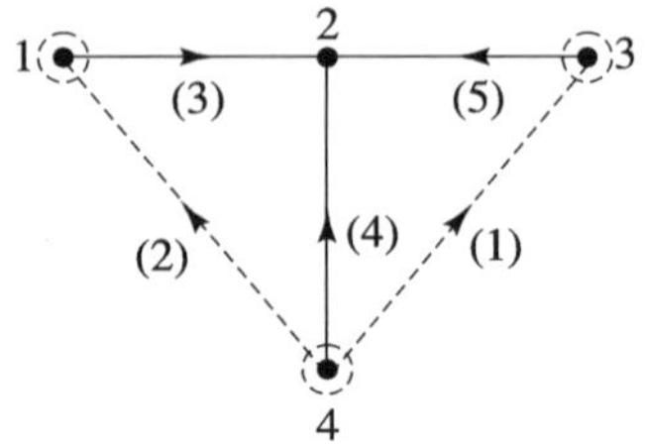

Fig. 7.43

f-cutset 3: {3, 2}
f-cutset 4: {4, 2, 1}
f-cutset 5: {5, 1}

Example 7.12 *The fundamental cutset matrix of a network is given as follows:*

Twigs			*Links*		
a	*c*	*e*	*b*	*d*	*f*
1	0	0	1	0	1
0	1	0	0	1	1
0	0	1	1	1	1

Draw the oriented graph.

Solution

No. of links $l = b - n + 1$

No. of nodes $n = b - l + 1$

$= 6 - 3 + 1 = 4$

f-cutsets are written as,

f-cutset a: $\{a, b, f\}$

f-cutset c: $\{c, d, f\}$

f-cutset e: $\{e, b, d, f\}$

The oriented graph is drawn as

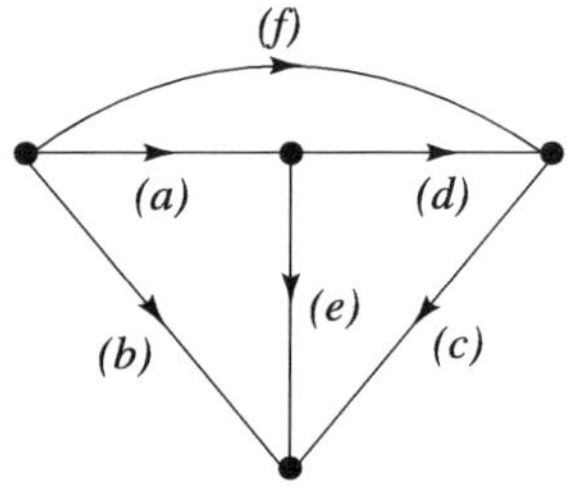

Fig. 7.44

Twigs: $\{a, c, e\}$
Links: $\{b, d, f\}$

Example 7.13 *Draw the oriented graph of a network with the f-cutset matrix as shown below:*

Twigs				*Links*		
1	2	3	4	5	6	7
1	0	0	0	–1	0	0
0	1	0	0	1	0	1
0	0	1	0	0	1	1
0	0	0	1	0	1	0

Solution

No. of links $l = b - n + 1$

No. of nodes $n = b - l + 1 = 7 - 3 + 1 = 5$

f-cutsets are written as

f-cutset 1: {1, 5}

f-cutset 2: {2, 5, 7}

f-cutset 3: {3, 6, 7}

f-cutset 4: {4, 6}

Then oriented graph can be drawn as

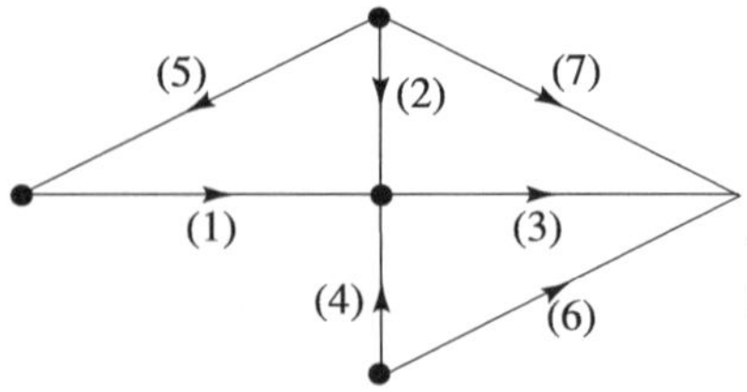

Fig. 7.45

Twigs: {1, 2, 3, 4}
Links: {5, 6, 7}

7.8 KIRCHHOFF'S VOLTAGE LAW

KVL states that if v_k is the voltage drop across the kth branch, then

$$\Sigma\, v_k = 0 \qquad \ldots(7.8)$$

the sum being taken over all the branches in a given loop. If l is the number of loops or f-circuits, then there will be l number of KVL equations, one for each loop. The KVL equation for the f-circuit or loop 'i' can be written as

$$\sum_{k=1}^{b} b_{ik}\, v_k = 0 \qquad (h = 1, 2, \ldots, l)$$

where b_{ik} is the elements of the tieset matrix B, b being the number of branches. The set of l KVL equations can be written in matrix form.

$$B\, V_b = 0 \qquad \ldots(7.9)$$

where $V_b = \begin{bmatrix} v_1 \\ v_2 \\ \vdots \\ v_b \end{bmatrix}$ is a column vector of branch voltages.

and B is the fundamental circuit matrix.

7.9 KIRCHHOFF'S CURRENT LAW

KCL states that if i_k is the current in the kth branch then at a given node

$$\Sigma i_k = 0 \qquad \ldots(7.10)$$

the sum being taken over all the branches incident at a given node. If there are 'n' nodes, there will 'n' such equations, one for each nodes

$$\sum_{k=1}^{b} a_{ik}\, i_k = 0 \qquad (h = 1, 2, \ldots, n)$$

so that set of n equations can be written in matrix form.

$$A_a\, I_b = 0 \qquad \ldots(7.11)$$

where $I_b = \begin{bmatrix} i_1 \\ i_2 \\ \vdots \\ i_b \end{bmatrix}$ is a column vector of branch currents.

and A_a is the complete incidence matrix.

If one node is taken as reference node or datum node, we can write the equation (7.11) as,

$$A\, I_b = 0 \qquad \ldots(7.12)$$

where A is the incidence matrix of order $(n - 1) \times b$.

We know that $\quad A = A_t\, Q$

Equation (7.12) can be written as

$$A_t\, Q \cdot I_b = 0$$

Premultiplying with A_t^{-1}, we get

$$A_t^{-1}\, A_t\, Q\, I_b = A_t^{-1} \cdot 0$$

$$I\, Q\, I_b = 0$$

$$Q\, I_b = 0 \qquad \ldots(7.13)$$

where Q is the f-cutset matrix.

7.10 RELATION BETWEEN BRANCH VOLTAGE MATRIX V_b, TWIG VOLTAGE MATRIX V_t AND NODE VOLTAGE MATRIX V_n

We know that $$B\,V_b = 0$$

$$[B_t : B_l]\begin{bmatrix} V_t \\ \cdots \\ V_l \end{bmatrix} = 0$$

$$B_t V_t + B_l V_l = 0$$

$$B_l V_l = -B_t V_t$$

Premultiplying with B_l^{-1}.

$$V_l = -B_l^{-1} B_t V_t$$

$$= -(B_l^{-1} B_t)\,V_t \quad \text{...(7.14)}$$

Now

$$V_b = \begin{bmatrix} V_t \\ \cdots \\ V_l \end{bmatrix}$$

$$= \begin{bmatrix} V_t \\ \cdots\cdots\cdots \\ -\left(B_l^{-1}B_t\right)V_t \end{bmatrix} = \begin{bmatrix} U \\ \cdots\cdots\cdots \\ -\left(B_l^{-1}B_t\right) \end{bmatrix} \cdot V_t$$

$$V_b = Q^T V_t \quad \text{...(7.15)}$$

Also,

$$Q = A_t^{-1} A$$

$$Q^T = A^T (A_t^{-1})^T$$

$$= A^T (A_t^T)^{-1}$$

Hence the Eq. (7.15) can be written as

$$V_b = A^T (A_t^T)^{-1} V_t$$

$$= A^T \{(A_t^T)^{-1} V_t\}$$

$$= A^T V_n \quad \text{...(7.16)}$$

where $V_n = (A_t^T)^{-1} V_n$ is node voltage matrix.

7.11 RELATION BETWEEN BRANCH CURRENT MATRIX I_b AND LOOP CURRENT MATRIX I_l

We know that, $$A\,I_b = 0$$

$$[A_t : A_l]\begin{bmatrix} I_t \\ \cdots \\ I_l \end{bmatrix} = 0$$

$$A_t I_t + A_l I_l = 0$$

$$A_t I_t = -A_l I_l$$

Premultiplying with A_t^{-1}.

$$I_t = -A_t^{-1} A_l I_l$$

$$= -(A_t^{-1} A_l)\,I_l \quad \text{...(7.17)}$$

Now

$$I_b = \begin{bmatrix} I_t \\ \cdots \\ I_l \end{bmatrix}$$

$$= \begin{bmatrix} -(A_t^{-1}A_l)I_l \\ \cdots\cdots\cdots\cdots \\ I_l \end{bmatrix} = \begin{bmatrix} -(A_t^{-1}A_l) \\ \cdots\cdots\cdots\cdots \\ U \end{bmatrix} \cdot I_l$$

$$I_b = B^T I_l \qquad \ldots(7.18)$$

7.12 NETWORK EQUILIBRIUM EQUATION

7.12.1 KVL Equation

(1) If there is a voltage source v_{sk} in the branch k having impedance z_k and carrying current i_k then

$$v_k = z_k i_k - v_{sk} \qquad (k = 1, 2, \ldots, b) \quad \ldots(7.19)$$

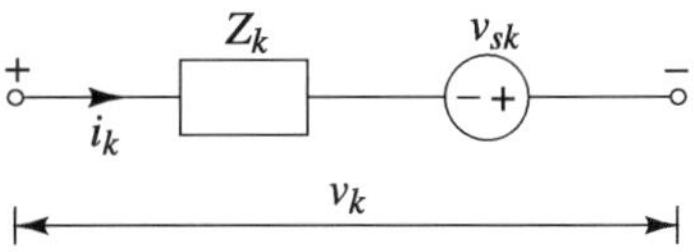

Fig. 7.46

In matrix form,

$$V_b = Z_b I_b - V_s \qquad \ldots(7.20)$$

where Z_b is the branch impedance matrix, I_b is the column vector of branch currents and V_s is the column vector of source voltages. Hence, KVL equation can be written as

$$B V_b = 0$$
$$B (Z_b I_b - V_s) = 0$$
$$B Z_b I_b = BV_s$$

Also, $\quad I_b = B^T I_l$

$$B Z_b B^T I_l = B V_s \qquad \ldots(7.21)$$
$$Z I_l = E$$

where $\quad E = B V_s$

and $\quad Z = B Z_b B^T$

The matrix Z is called *loop impedance matrix.*

(2) If there is a voltage source in series with an impedance and a current source in parallel with the combination as shown in Fig. 7.47,

$$i_k = \frac{(v_k + v_{sk})}{zk} - i_{sk}$$

$$v_k = z_k i_k + z_k i_{sk} - v_{sk} \qquad \ldots(7.22)$$

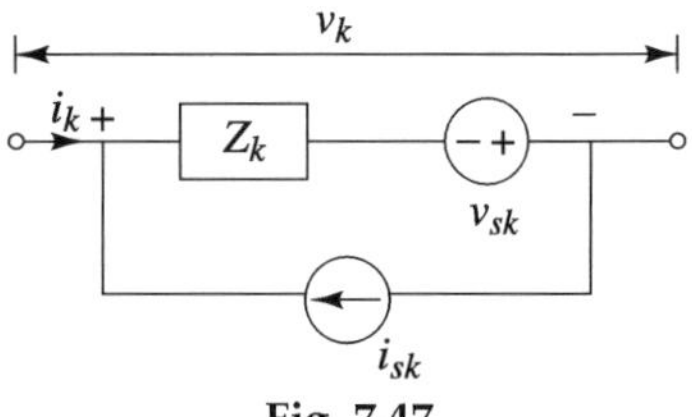

Fig. 7.47

In matrix form, we can write,

$$V_b = Z_b I_b + Z_b I_s - V_s \qquad \ldots(7.23)$$

KVL equation is $B V_b = 0$

$$BV_b = B (Z_b I_b + Z_b I_s - V_s) = 0$$
$$B Z_b I_b = B V_s - B Z_b I_s$$

Now $\quad I_b = B^T I_l$

$$B Z_b B^T I_l = B V_s - B Z_b I_s \qquad \ldots(7.24)$$
$$Z I_l = B V_s - B Z_b I_s$$

where Z = $B Z_b B^T$ is the loop impedance matrix. This is the generalised KVL equation.

7.12.2 KCL Equation

(1) If the branch k contains an input current source i_{sk} and an admittance y_k, then

$$i_k = y_k v_k - i_{sk} \qquad (k = 1, 2, \ldots, b) \qquad \ldots(7.25)$$

In the matrix form,

$$I_b = Y_b V_b - I_s \qquad \ldots(7.26)$$

where Y_b is the branch admittance matrix.

Hence KCL equation is given by,

$$A\, I_b = 0$$

$$A\, (Y_b\, V_b - I_s) = 0$$

$$A\, Y_b\, V_b = A\, I_s$$

Also $$V_b = A^T\, V_n$$

$$A\, Y_b\, A^T\, V_n = A\, I_s \qquad \ldots(7.27)$$

$$Y\, V_n = I$$

where $$Y = A\, Y_b\, A^T$$

and $$I = A\, I_s$$

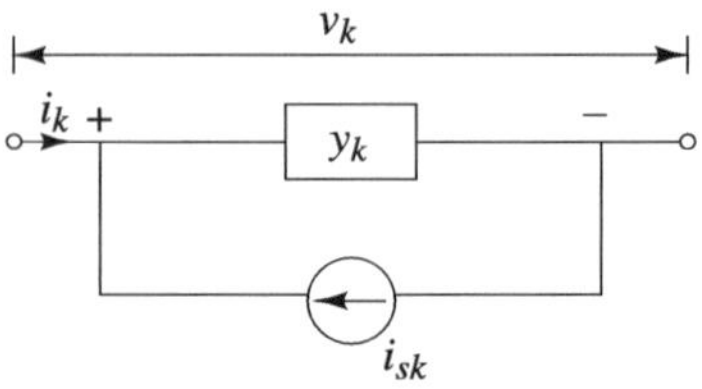

Fig. 7.48

The matrix Y is called admittance matrix. This is the KCL equation in matrix form.

In terms of f-cutset matrix, the KCL equation can be written as

$$Q\, I_b = 0$$

$$Q\, (Y_b\, V_b - V_s) = 0$$

$$Q\, Y_b\, V_b = Q\, I_s$$

Also $$V_b = Q^T\, V_t$$

$$Q\, Y_b\, Q^T\, V_t = Q\, I_s \qquad \ldots(7.28)$$

$$Y\, V_t = I$$

where $$Y = Q\, Y_b\, Q^T$$

and $$I = Q\, I_s$$

This is the KCL equation in matrix form.

(2) If there is a voltage source in series with an impedance and a current source in parallel with the combination as shown in Fig. 7.49,

$$y_k = \frac{1}{z_k}$$

$$i_k = y_k\, v_k + y_k\, v_{sk} - i_{sk} \qquad \ldots(7.29)$$

In matrix form,

$$I_b = Y_b\, V_b + Y_b\, V_s - I_s \qquad \ldots(7.30)$$

KCL equation will be given by,

$$A\, I_b = 0$$

$$A\, (Y_b\, I_b + Y_b\, V_s - I_s) = 0$$

$$A\, Y_b\, V_b = A\, I_s - A\, Y_b\, V_s$$

Also $$V_b = A^T\, V_n$$

$$A\, Y_b\, A^T\, V_n = A\, I_s - A\, Y_b\, V_s \qquad \ldots(7.31)$$

$$Y\, V_n = A\, I_s - A\, V_b\, V_s$$

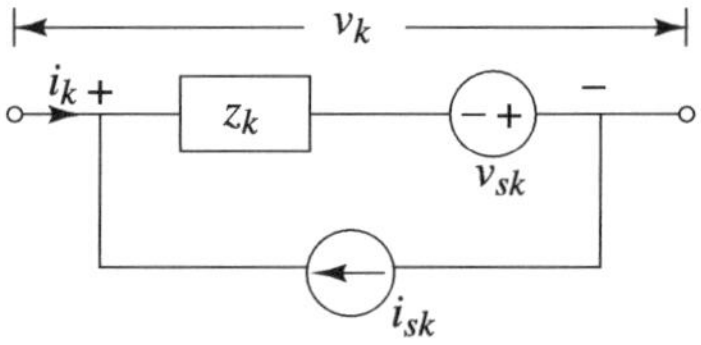

Fig. 7.49

where $Y = A\, Y_b\, A^T$ is the node admittance matrix. This is a generalised KCL equation.

In terms of f-cutset matrix, the KCL equation can be written as

$$Q\, I_b = 0$$

$$Q\, (Y_b\, V_b + Y_b\, V_s - I_s) = 0$$

$$Q\, Y_b\, V_b = Q\, I_s - Q\, Y_b\, V_s$$

Also $$V_b = Q^T\, V_t$$

$$Q\, Y_b\, Q^T\, V_t = Q\, I_s - Q\, Y_b\, V_s$$

$$Y\, V_t = Q\, I_s - Q\, Y_b\, V_s \qquad \ldots(7.32)$$

This is a generalised KCL equation.

Example 7.14 *For the network shown in Fig. 7.50, write down the tieset matrix and obtain the network equilibrium equation in matrix form using KVL. Calculate the loop currents and branch currents.*

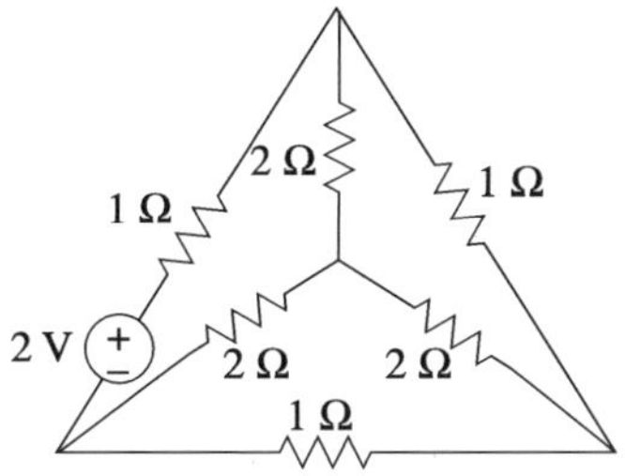

Fig. 7.50

Solution The oriented graph and one of its trees are shown in Fig. 7.51.

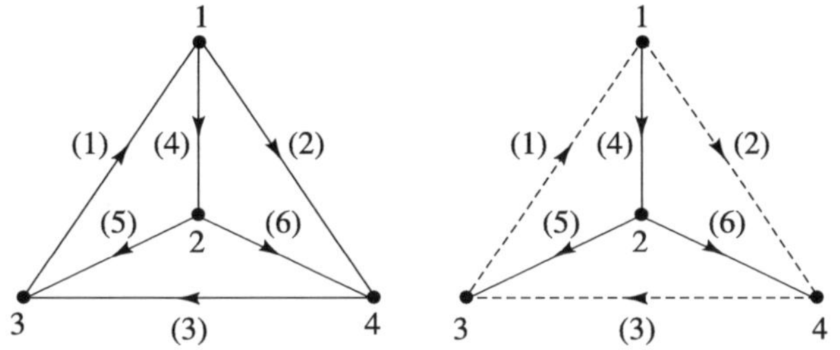

Fig. 7.51

Tieset1: {1, 4, 5}
Tieset2: {2, 4, 6}
Tieset3: {3, 5, 6}

Tieset Matrix (*B*)

$$B = \begin{array}{c} \\ 1 \\ 2 \\ 3 \end{array}\begin{array}{c} \begin{matrix} 1 & 2 & 3 & 4 & 5 & 6 \end{matrix} \\ \begin{bmatrix} 1 & 0 & 0 & 1 & 1 & 0 \\ 0 & 1 & 0 & -1 & 0 & -1 \\ 0 & 0 & 1 & 0 & -1 & 1 \end{bmatrix} \end{array}$$

The KVL equation in matrix form is given by

$$B\,Z_b\,B^T\,I_l = B\,V_s - B\,Z_b\,I_s$$

Here, $I_s = 0$,

$$B\,Z_b\,B^T\,I_l = B\,V_s$$

$$Z_b = \begin{bmatrix} 1 & 0 & 0 & 0 & 0 & 0 \\ 0 & 1 & 0 & 0 & 0 & 0 \\ 0 & 0 & 1 & 0 & 0 & 0 \\ 0 & 0 & 0 & 2 & 0 & 0 \\ 0 & 0 & 0 & 0 & 2 & 0 \\ 0 & 0 & 0 & 0 & 0 & 2 \end{bmatrix} \qquad B^T = \begin{bmatrix} 1 & 0 & 0 \\ 0 & 1 & 0 \\ 0 & 0 & 1 \\ 1 & -1 & 0 \\ 1 & 0 & -1 \\ 0 & -1 & 1 \end{bmatrix} \qquad V_s = \begin{bmatrix} 2 \\ 0 \\ 0 \\ 0 \\ 0 \\ 0 \end{bmatrix}$$

$$B\,Z_b = \begin{bmatrix} 1 & 0 & 0 & 1 & 1 & 0 \\ 0 & 1 & 0 & -1 & 0 & -1 \\ 0 & 0 & 1 & 0 & -1 & 1 \end{bmatrix}\begin{bmatrix} 1 & 0 & 0 & 0 & 0 & 0 \\ 0 & 1 & 0 & 0 & 0 & 0 \\ 0 & 0 & 1 & 0 & 0 & 0 \\ 0 & 0 & 0 & 2 & 0 & 0 \\ 0 & 0 & 0 & 0 & 2 & 0 \\ 0 & 0 & 0 & 0 & 0 & 2 \end{bmatrix} = \begin{bmatrix} 1 & 0 & 0 & 2 & 2 & 0 \\ 0 & 1 & 0 & -2 & 0 & -2 \\ 0 & 0 & 1 & 0 & -2 & 2 \end{bmatrix}$$

$$B\,Z_b\,B^T = \begin{bmatrix} 1 & 0 & 0 & 2 & 2 & 0 \\ 0 & 1 & 0 & -2 & 0 & -2 \\ 0 & 0 & 1 & 0 & -2 & 2 \end{bmatrix} \begin{bmatrix} 1 & 0 & 0 \\ 0 & 1 & 0 \\ 0 & 0 & 1 \\ 1 & -1 & 0 \\ 1 & 0 & -1 \\ 0 & -1 & 1 \end{bmatrix} = \begin{bmatrix} 5 & -2 & -2 \\ -2 & 5 & -2 \\ -2 & -2 & 5 \end{bmatrix}$$

$$B\,V_s = \begin{bmatrix} 1 & 0 & 0 & 1 & 1 & 0 \\ 0 & 1 & 0 & -1 & 0 & -1 \\ 0 & 0 & 1 & 0 & -1 & 1 \end{bmatrix} \begin{bmatrix} 2 \\ 0 \\ 0 \\ 0 \\ 0 \\ 0 \end{bmatrix} = \begin{bmatrix} 2 \\ 0 \\ 0 \end{bmatrix}$$

The KVL equation in matrix form is given by

$$\begin{bmatrix} 5 & -2 & -2 \\ -2 & 5 & -2 \\ -2 & -2 & 5 \end{bmatrix} \begin{bmatrix} I_{l_1} \\ I_{l_2} \\ I_{l_3} \end{bmatrix} = \begin{bmatrix} 2 \\ 0 \\ 0 \end{bmatrix}$$

Solving this matrix equation, we get

$$I_{l_1} = \frac{6}{7}\text{ A}$$

$$I_{l_2} = \frac{4}{7}\text{ A}$$

$$I_{l_3} = \frac{4}{7}\text{ A}$$

The branch currents are given by

$$I_b = B^T I_l$$

$$\begin{bmatrix} i_1 \\ i_2 \\ i_3 \\ i_4 \\ i_5 \\ i_6 \end{bmatrix} = \begin{bmatrix} 1 & 0 & 0 \\ 0 & 1 & 0 \\ 0 & 0 & 1 \\ 1 & -1 & 0 \\ 1 & 0 & -1 \\ 0 & -1 & 1 \end{bmatrix} \begin{bmatrix} 6/7 \\ 4/7 \\ 4/7 \end{bmatrix} = \begin{bmatrix} 6/7 \\ 4/7 \\ 4/7 \\ 2/7 \\ 2/7 \\ 0 \end{bmatrix}$$

Example 7.15 *For the network shown in Fig. 7.52, write down the tieset matrix and obtain the network equilibrium equation in matrix form using KVL. Calculate loop currents.*

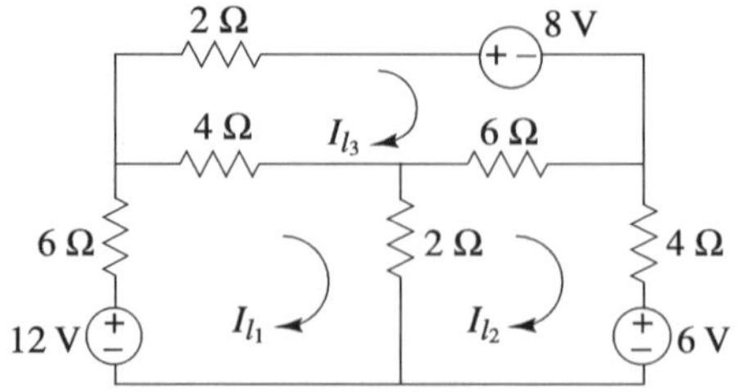

Fig. 7.52

Solution The oriented graph and its selected tree are shown in Fig. 7.53.

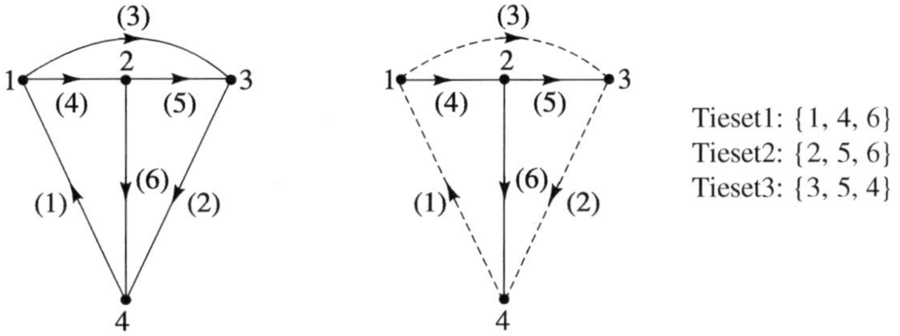

Fig. 7.53

Tieset Matrix (*B*)

$$B = \begin{array}{c} \\ 1 \\ 2 \\ 3 \end{array}\begin{array}{c} \begin{array}{cccccc} 1 & 2 & 3 & 4 & 5 & 6 \end{array} \\ \begin{bmatrix} 1 & 0 & 0 & 1 & 0 & 1 \\ 0 & 1 & 0 & 0 & 1 & -1 \\ 0 & 0 & 1 & -1 & -1 & 0 \end{bmatrix} \end{array}$$

The KVL equation in matrix form is given by

$$B\,Z_b\,B^T\,I_l = B\,V_s - B\,Z_b\,I_s$$

Here, $I_s = 0$,

$$B\,Z_b\,B^T\,I_l = B\,V_s$$

$$Z_b = \begin{bmatrix} 6 & 0 & 0 & 0 & 0 & 0 \\ 0 & 4 & 0 & 0 & 0 & 0 \\ 0 & 0 & 2 & 0 & 0 & 0 \\ 0 & 0 & 0 & 4 & 0 & 0 \\ 0 & 0 & 0 & 0 & 6 & 0 \\ 0 & 0 & 0 & 0 & 0 & 2 \end{bmatrix} \quad B^T = \begin{bmatrix} 1 & 0 & 0 \\ 0 & 1 & 0 \\ 0 & 0 & 1 \\ 1 & 0 & -1 \\ 0 & 1 & -1 \\ 1 & -1 & 0 \end{bmatrix} \quad V_s = \begin{bmatrix} 12 \\ -6 \\ -8 \\ 0 \\ 0 \\ 0 \end{bmatrix}$$

$$B\,Z_b = \begin{bmatrix} 1 & 0 & 0 & 1 & 0 & 1 \\ 0 & 1 & 0 & 0 & 1 & -1 \\ 0 & 0 & 1 & -1 & -1 & 0 \end{bmatrix}\begin{bmatrix} 6 & 0 & 0 & 0 & 0 & 0 \\ 0 & 4 & 0 & 0 & 0 & 0 \\ 0 & 0 & 2 & 0 & 0 & 0 \\ 0 & 0 & 0 & 4 & 0 & 0 \\ 0 & 0 & 0 & 0 & 6 & 0 \\ 0 & 0 & 0 & 0 & 0 & 2 \end{bmatrix} = \begin{bmatrix} 6 & 0 & 0 & 4 & 0 & 2 \\ 0 & 4 & 0 & 0 & 6 & -2 \\ 0 & 0 & 2 & -4 & -6 & 0 \end{bmatrix}$$

$$B\,Z_b\,B^T = \begin{bmatrix} 6 & 0 & 0 & 4 & 0 & 2 \\ 0 & 4 & 0 & 0 & 6 & -2 \\ 0 & 0 & 2 & -4 & -6 & 0 \end{bmatrix}\begin{bmatrix} 1 & 0 & 0 \\ 0 & 1 & 0 \\ 0 & 0 & 1 \\ 1 & 0 & -1 \\ 0 & 1 & -1 \\ 1 & -1 & 0 \end{bmatrix} = \begin{bmatrix} 12 & -2 & -4 \\ -2 & 12 & -6 \\ -4 & -6 & 12 \end{bmatrix}$$

$$B\,V_s = \begin{bmatrix} 1 & 0 & 0 & 1 & 0 & 1 \\ 0 & 1 & 0 & 0 & 1 & -1 \\ 0 & 0 & 1 & -1 & -1 & 0 \end{bmatrix} \begin{bmatrix} 12 \\ -6 \\ -8 \\ 0 \\ 0 \\ 0 \end{bmatrix} = \begin{bmatrix} 12 \\ -6 \\ -8 \end{bmatrix}$$

Hence, the KVL equation in matrix form is given by

$$\begin{bmatrix} 12 & -2 & -4 \\ -2 & 12 & -6 \\ -4 & -6 & 12 \end{bmatrix} \begin{bmatrix} I_{l_1} \\ I_{l_2} \\ I_{l_3} \end{bmatrix} = \begin{bmatrix} 12 \\ -6 \\ -8 \end{bmatrix}$$

Solving this matrix equation, we get

$$I_{l_1} = 0.55 \text{ A}$$
$$I_{l_2} = -0.866 \text{ A}$$
$$I_{l_3} = -0.916 \text{ A}$$

Example 7.16 *For the network shown in Fig. 7.54, write down the tieset matrix and obtain the network equilibrium equation in matrix form using KVL.*

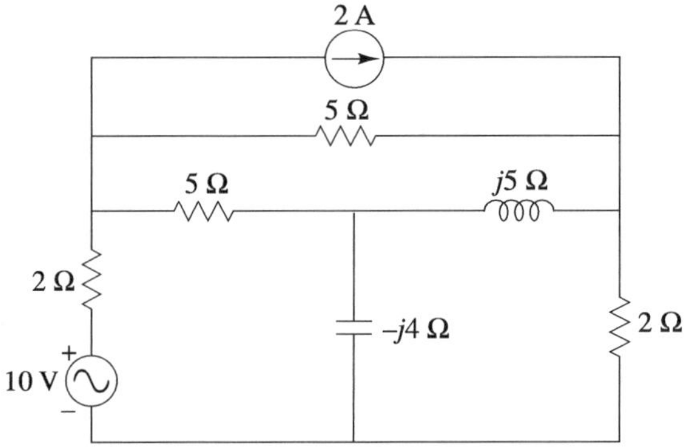

Fig. 7.54

Solution The oriented graph and its selected tree are shown in Fig. 7.55.

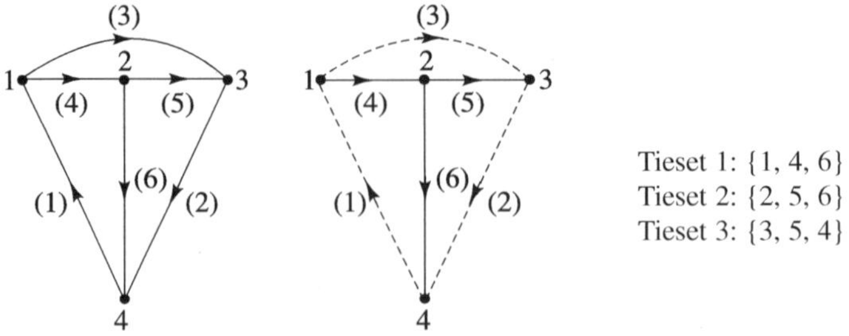

Fig. 7.55

Tieset Matrix (*B*)

$$B = \begin{matrix} & \begin{matrix} 1 & 2 & 3 & 4 & 5 & 6 \end{matrix} \\ \begin{matrix} 1 \\ 2 \\ 3 \end{matrix} & \begin{bmatrix} 1 & 0 & 0 & 1 & 0 & 1 \\ 0 & 1 & 0 & 0 & 1 & -1 \\ 0 & 0 & 1 & -1 & -1 & 0 \end{bmatrix} \end{matrix}$$

The KVL equation in matrix form is given by

$$B\,Z_b\,B^T\,I_l = B\,V_s - B\,Z_b\,I_s$$

$$Z_b = \begin{bmatrix} 2 & 0 & 0 & 0 & 0 & 0 \\ 0 & 2 & 0 & 0 & 0 & 0 \\ 0 & 0 & 5 & 0 & 0 & 0 \\ 0 & 0 & 0 & 5 & 0 & 0 \\ 0 & 0 & 0 & 0 & j5 & 0 \\ 0 & 0 & 0 & 0 & 0 & -j4 \end{bmatrix} \quad B^T = \begin{bmatrix} 1 & 0 & 0 \\ 0 & 1 & 0 \\ 0 & 0 & 1 \\ 1 & 0 & -1 \\ 0 & 1 & -1 \\ 1 & -1 & 0 \end{bmatrix} \quad V_s = \begin{bmatrix} 10 \\ 0 \\ 0 \\ 0 \\ 0 \\ 0 \end{bmatrix} \quad I_s = \begin{bmatrix} 0 \\ 0 \\ -2 \\ 0 \\ 0 \\ 0 \end{bmatrix}$$

$$B\,Z_b = \begin{bmatrix} 1 & 0 & 0 & 1 & 0 & 1 \\ 0 & 1 & 0 & 0 & 1 & -1 \\ 0 & 0 & 1 & -1 & -1 & 0 \end{bmatrix} \begin{bmatrix} 2 & 0 & 0 & 0 & 0 & 0 \\ 0 & 2 & 0 & 0 & 0 & 0 \\ 0 & 0 & 5 & 0 & 0 & 0 \\ 0 & 0 & 0 & 5 & 0 & 0 \\ 0 & 0 & 0 & 0 & j5 & 0 \\ 0 & 0 & 0 & 0 & 0 & -j4 \end{bmatrix} = \begin{bmatrix} 2 & 0 & 0 & 5 & 0 & -j4 \\ 0 & 2 & 0 & 0 & j5 & j4 \\ 0 & 0 & 5 & -5 & -j5 & 0 \end{bmatrix}$$

$$B\,Z_b\,B^T = \begin{bmatrix} 2 & 0 & 0 & 5 & 0 & -j4 \\ 0 & 2 & 0 & 0 & j5 & j4 \\ 0 & 0 & 5 & -5 & -j5 & 0 \end{bmatrix} \begin{bmatrix} 1 & 0 & 0 \\ 0 & 1 & 0 \\ 0 & 0 & 1 \\ 1 & 0 & -1 \\ 0 & 1 & -1 \\ 1 & -1 & 0 \end{bmatrix} = \begin{bmatrix} 7-j4 & j4 & -5 \\ j4 & 2+j1 & -j5 \\ -5 & -j5 & 10+j5 \end{bmatrix}$$

$$B\,V_s = \begin{bmatrix} 1 & 0 & 0 & 1 & 0 & 1 \\ 0 & 1 & 0 & 0 & 1 & -1 \\ 0 & 0 & 1 & -1 & -1 & 0 \end{bmatrix} \begin{bmatrix} 10 \\ 0 \\ 0 \\ 0 \\ 0 \\ 0 \end{bmatrix} = \begin{bmatrix} 10 \\ 0 \\ 0 \end{bmatrix}$$

$$B\,Z_b\,I_s = \begin{bmatrix} 2 & 0 & 0 & 5 & 0 & -j4 \\ 0 & 2 & 0 & 0 & j5 & j4 \\ 0 & 0 & 5 & -5 & -j5 & 0 \end{bmatrix} \begin{bmatrix} 0 \\ 0 \\ -2 \\ 0 \\ 0 \\ 0 \end{bmatrix} = \begin{bmatrix} 0 \\ 0 \\ -10 \end{bmatrix}$$

Hence, the KCL equation in matrix form is given by

$$B\,Z_b\,B^T\,I_l = B\,V_s - B\,Z_b\,I_s$$

$$\begin{bmatrix} 7-j4 & j4 & -5 \\ j4 & 2+j1 & -j5 \\ -5 & -j5 & 10+j5 \end{bmatrix}\begin{bmatrix} I_{l_1} \\ I_{l_2} \\ I_{l_3} \end{bmatrix} = \begin{bmatrix} 10 \\ 0 \\ 0 \end{bmatrix} - \begin{bmatrix} 0 \\ 0 \\ -10 \end{bmatrix} = \begin{bmatrix} 10 \\ 0 \\ 10 \end{bmatrix}$$

Example 7.17 *For the network shown in Fig. 7.56, write down the tieset matrix and obtain the network equilibrium equation in matrix form using KVL.*

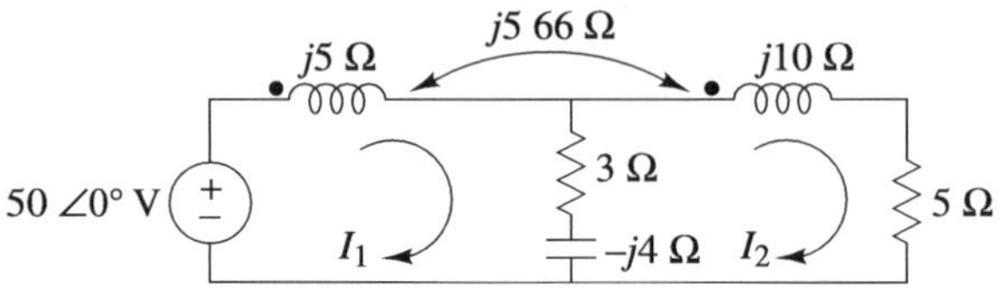

Fig. 7.56

Solution The branch currents are so chosen that they assume the direction out of the dotted terminals. Because of this choice of current direction, the mutual inductance is positive. The oriented graph and its selected tree are shown in Fig. 7.57.

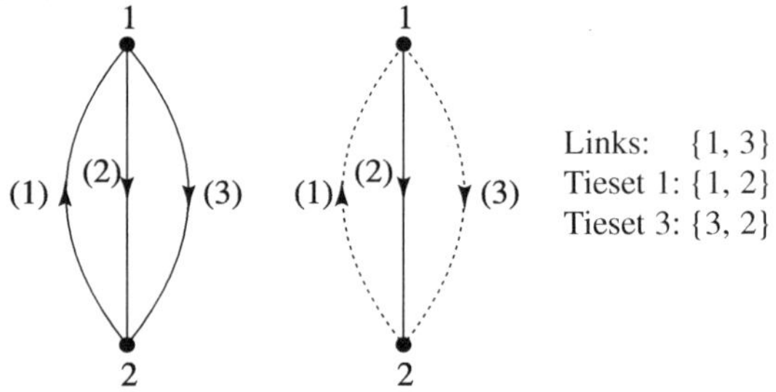

Fig. 7.57

Tieset Matrix (*B*)

$$B = \begin{array}{c} \begin{matrix} 1 & 2 & 3 \end{matrix} \\ \begin{bmatrix} 1 & 1 & 0 \\ 0 & -1 & 1 \end{bmatrix} \end{array}$$

The KVL equation in matrix form is given by

$$B\,Z_b\,B^T\,I_l = B\,V_s - B\,Z_b\,I_s$$

Here $I_s = 0,$

$$B\,Z_b\,B^T\,I_l = B\,V_s$$

$$Z_b = \begin{bmatrix} j5 & 0 & j5.66 \\ 0 & 3-j4 & 0 \\ j5.66 & 0 & 5+j10 \end{bmatrix} \qquad B^T = \begin{bmatrix} 1 & 0 \\ 1 & -1 \\ 0 & 1 \end{bmatrix} \qquad V_s = \begin{bmatrix} 50\angle 0^\circ \\ 0 \\ 0 \end{bmatrix}$$

$$B\,Z_b = \begin{bmatrix} 1 & 1 & 0 \\ 0 & -1 & 1 \end{bmatrix}\begin{bmatrix} j5 & 0 & j5.66 \\ 0 & 3-j4 & 0 \\ j5.66 & 0 & 5+j10 \end{bmatrix} = \begin{bmatrix} j5 & 3-j4 & j5.66 \\ j5.66 & -3+j4 & 5+j10 \end{bmatrix}$$

$$B\,Z_b\,B^T = \begin{bmatrix} j5 & 3-j4 & j5.66 \\ j5.66 & -3+j4 & 5+j10 \end{bmatrix} = \begin{bmatrix} 1 & 0 \\ 1 & -1 \\ 0 & 1 \end{bmatrix} = \begin{bmatrix} 3+j1 & -3+j9.66 \\ -3+j9.66 & 8+j6 \end{bmatrix}$$

$$B\,V_s = \begin{bmatrix} 1 & 1 & 0 \\ 0 & -1 & 1 \end{bmatrix}\begin{bmatrix} 50\angle 0^\circ \\ 0 \\ 0 \end{bmatrix}\begin{bmatrix} 50\angle 0^\circ \\ 0 \end{bmatrix}$$

Hence, the KVL equation in matrix form is given by

$$\begin{bmatrix} 3+j1 & -3+j9.66 \\ -3+j9.66 & 8+j6 \end{bmatrix} \begin{bmatrix} I_{l_1} \\ I_{l_2} \end{bmatrix} = \begin{bmatrix} 50\angle 0^\circ \\ 0 \end{bmatrix}$$

Example 7.18 *For the network shown in Fig. 7.58, write down the tieset matrix and obtain network equilibrium equation in matrix form using KVL.*

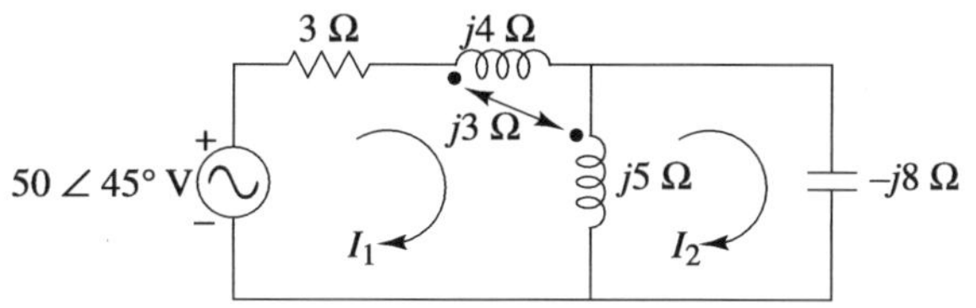

Fig. 7.58

Solution The branch currents are so chosen that they assume the direction out of the dotted terminals. Because of this choice of current direction, the mutual inductance is positive. The oriented graph and its selected tree are shown in Fig. 7.59.

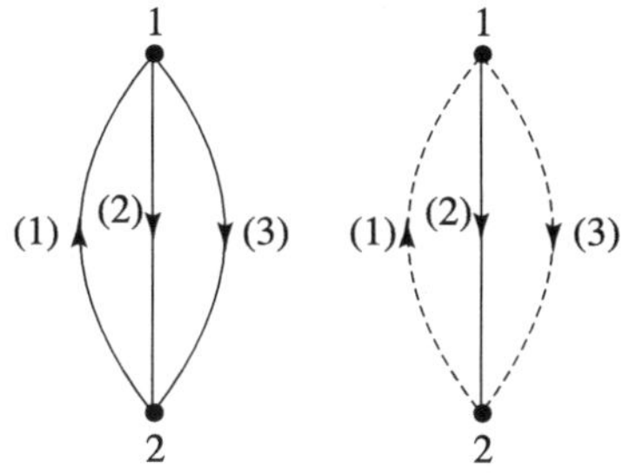

Fig. 7.59

Tieset Matrix (*B*)

$$B = \begin{array}{c} \begin{array}{ccc} 1 & 2 & 3 \end{array} \\ \begin{bmatrix} 1 & 1 & 0 \\ 0 & -1 & 1 \end{bmatrix} \end{array}$$

The KVL equation in matrix form is given by

$$B\, Z_b\, B^T\, I_l = B\, V_s - B\, Z_b\, I_s$$

Here $I_s = 0$,

$$B\, Z_b\, B^T\, I_l = B\, V_s$$

$$Z_b = \begin{bmatrix} 3+j4 & j3 & 0 \\ j3 & j5 & 0 \\ 0 & 0 & -j8 \end{bmatrix} \qquad B^T = \begin{bmatrix} 1 & 0 \\ 1 & -1 \\ 0 & 1 \end{bmatrix} \qquad V_s = \begin{bmatrix} 50\angle 45^\circ \\ 0 \\ 0 \end{bmatrix}$$

$$B\, Z_b = \begin{bmatrix} 1 & 1 & 0 \\ 0 & -1 & 1 \end{bmatrix} \begin{bmatrix} 3+j4 & j3 & 0 \\ j3 & j5 & 0 \\ 0 & 0 & -j8 \end{bmatrix} = \begin{bmatrix} 3+j7 & j8 & 0 \\ -j3 & -j5 & -j8 \end{bmatrix}$$

$$B\, Z_b\, B^T = \begin{bmatrix} 3+j7 & j8 & 0 \\ -j3 & -j5 & -j8 \end{bmatrix} \begin{bmatrix} 1 & 0 \\ 1 & -1 \\ 0 & 1 \end{bmatrix} = \begin{bmatrix} 3+j15 & -j8 \\ -j8 & -j3 \end{bmatrix}$$

$$B\, V_s = \begin{bmatrix} 1 & 1 & 0 \\ 0 & -1 & 1 \end{bmatrix} \begin{bmatrix} 50\angle 45^\circ \\ 0 \\ 0 \end{bmatrix} \begin{bmatrix} 50\angle 45^\circ \\ 0 \end{bmatrix}$$

Hence the KVL equation in matrix form is given by,

$$\begin{bmatrix} 3+j15 & -j8 \\ -j8 & -j3 \end{bmatrix} \begin{bmatrix} I_{l_1} \\ I_{l_3} \end{bmatrix} = \begin{bmatrix} 50\angle 45^\circ \\ 0 \end{bmatrix}$$

Example 7.19 *For the network shown in Fig. 7.60 ,obtain equilibrium equation on node basis.*

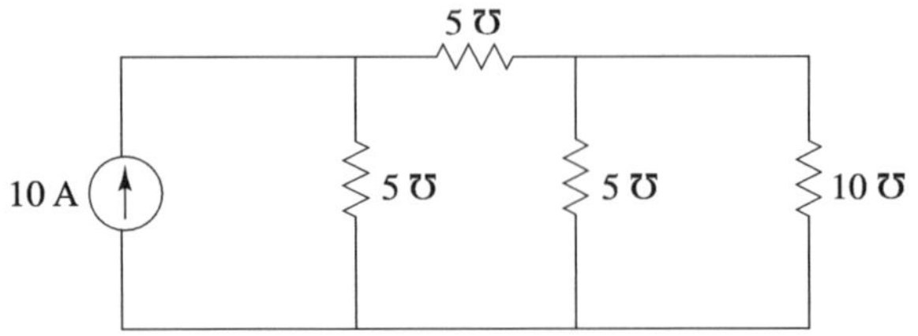

Fig. 7.60

Solution The oriented graph and one tree are shown in Fig. 7.61.

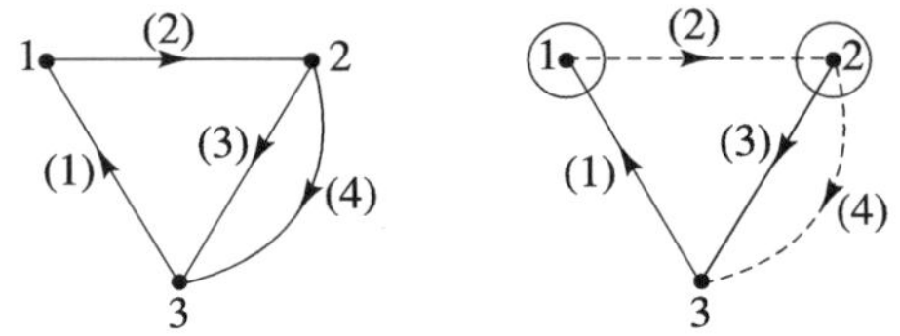

Fig. 7.61

$$Q = \begin{array}{c} \\ 1 \\ 3 \end{array}\begin{array}{c} \begin{matrix} 1 & 2 & 3 & 4 \end{matrix} \\ \begin{bmatrix} 1 & -1 & 0 & 0 \\ 0 & -1 & 1 & 1 \end{bmatrix} \end{array}$$

The KCL equation in matrix form is given by

$$Q\,Y_b\,Q^T\,V_t = Q\,I_s - Q\,Y_b\,V_s$$

Here, $V_s = 0$

$$QY_b\,Q^T\,V_t = QI_s$$

Current entering into the node is taken as negative.

$$Y_b = \begin{bmatrix} 5 & 0 & 0 & 0 \\ 0 & 5 & 0 & 0 \\ 0 & 0 & 5 & 0 \\ 0 & 0 & 0 & 10 \end{bmatrix} \qquad Q^T = \begin{bmatrix} 1 & 0 \\ -1 & -1 \\ 0 & 1 \\ 0 & 1 \end{bmatrix} \qquad I_s = \begin{bmatrix} -10 \\ 0 \\ 0 \\ 0 \end{bmatrix}$$

$$QY_b = \begin{bmatrix} 1 & -1 & 0 & 0 \\ 0 & -1 & 1 & 1 \end{bmatrix}\begin{bmatrix} 5 & 0 & 0 & 0 \\ 0 & 5 & 0 & 0 \\ 0 & 0 & 5 & 0 \\ 0 & 0 & 0 & 10 \end{bmatrix} = \begin{bmatrix} 5 & -5 & 0 & 0 \\ 0 & -5 & 5 & 10 \end{bmatrix}$$

$$QY_b\,Q^T = \begin{bmatrix} 5 & -5 & 0 & 0 \\ 0 & -5 & 5 & 10 \end{bmatrix}\begin{bmatrix} 1 & 0 \\ -1 & -1 \\ 0 & 1 \\ 0 & 1 \end{bmatrix} = \begin{bmatrix} 10 & 5 \\ 5 & 20 \end{bmatrix}$$

$$QI_s = \begin{bmatrix} 1 & -1 & 0 & 0 \\ 0 & -1 & 1 & 1 \end{bmatrix}\begin{bmatrix} -10 \\ 0 \\ 0 \\ 0 \end{bmatrix} = \begin{bmatrix} -10 \\ 0 \end{bmatrix}$$

Hence, KCL equation will be written as

$$\begin{bmatrix} 10 & 5 \\ 5 & 20 \end{bmatrix}\begin{bmatrix} v_{t_1} \\ v_{t_3} \end{bmatrix} = \begin{bmatrix} -10 \\ 0 \end{bmatrix}$$

Solving this matrix equation, we get

$$v_{t_1} = -\frac{8}{7}\ \text{V}$$

$$v_{t_3} = \frac{2}{7}\ \text{V}$$

Example 7.20 *Calculate the twig voltages using KCL equation for the network shown in Fig. 7.62.*

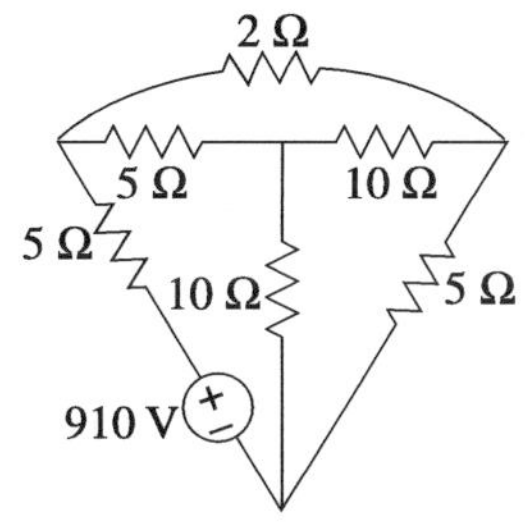

Fig. 7.62

Solution The oriented graph and one of the trees are shown in Fig. 7.63.

f-cutset 1: {1, 4, 5, 6}
f-cutset 2: {2, 4, 5}
f-cutset 3: {3, 4, 6}

$$Q = \begin{array}{c} \\ 1 \\ 2 \\ 3 \end{array}\begin{array}{c} \begin{array}{cccccc} 1 & 2 & 3 & 4 & 5 & 6 \end{array} \\ \begin{bmatrix} 1 & 0 & 0 & -1 & -1 & 1 \\ 0 & 1 & 0 & -1 & -1 & 0 \\ 0 & 0 & 1 & -1 & 0 & 1 \end{bmatrix} \end{array}$$

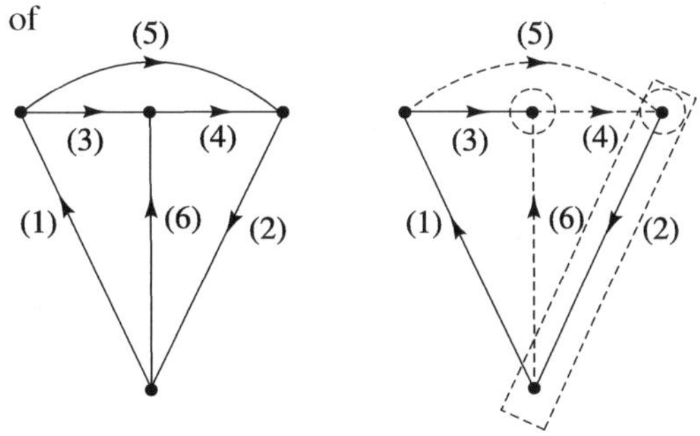

Fig. 7.63

The network equilibrium equation on node basis can be written as

$$QY_b Q^T V_t = QI_s - QY_b V_s$$

$$Y_b = \begin{bmatrix} 0.2 & 0 & 0 & 0 & 0 & 0 \\ 0 & 0.2 & 0 & 0 & 0 & 0 \\ 0 & 0 & 0.2 & 0 & 0 & 0 \\ 0 & 0 & 0 & 0.1 & 0 & 0 \\ 0 & 0 & 0 & 0 & 0.5 & 0 \\ 0 & 0 & 0 & 0 & 0 & 0.1 \end{bmatrix} \quad I_s = \begin{bmatrix} 0 \\ 0 \\ 0 \\ 0 \\ 0 \\ 0 \end{bmatrix} \quad V_s = \begin{bmatrix} 910 \\ 0 \\ 0 \\ 0 \\ 0 \\ 0 \end{bmatrix}$$

$$QY_b = \begin{bmatrix} 1 & 0 & 0 & -1 & -1 & 1 \\ 0 & 1 & 0 & -1 & -1 & 0 \\ 0 & 0 & 1 & -1 & 0 & 1 \end{bmatrix} \begin{bmatrix} 0.2 & 0 & 0 & 0 & 0 & 0 \\ 0 & 0.2 & 0 & 0 & 0 & 0 \\ 0 & 0 & 0.2 & 0 & 0 & 0 \\ 0 & 0 & 0 & 0.1 & 0 & 0 \\ 0 & 0 & 0 & 0 & 0.5 & 0 \\ 0 & 0 & 0 & 0 & 0 & 0.1 \end{bmatrix}$$

$$QY_b Q^T = \begin{bmatrix} 0.2 & 0 & 0 & -0.1 & -0.5 & 0.1 \\ 0 & 0.2 & 0 & -0.1 & -0.5 & 0 \\ 0 & 0 & 0.2 & -0.1 & 0 & 0.1 \end{bmatrix} \begin{bmatrix} 1 & 0 & 0 \\ 0 & 1 & 0 \\ 0 & 0 & 1 \\ -1 & -1 & -1 \\ -1 & -1 & 0 \\ 1 & 0 & 1 \end{bmatrix} = \begin{bmatrix} 0.9 & 0.6 & 0.2 \\ 0.6 & 0.8 & 0.1 \\ 0.2 & 0.1 & 0.3 \end{bmatrix}$$

$$QI_s = \begin{bmatrix} 1 & 0 & 0 & -1 & -1 & 1 \\ 0 & 1 & 0 & -1 & -1 & 0 \\ 0 & 0 & 1 & -1 & 0 & 1 \end{bmatrix} \begin{bmatrix} 0 \\ 0 \\ 0 \\ 0 \\ 0 \\ 0 \end{bmatrix} = \begin{bmatrix} 0 \\ 0 \\ 0 \end{bmatrix}$$

$$QY_b V_s = \begin{bmatrix} 0.2 & 0 & 0 & -0.1 & -0.5 & 0.1 \\ 0 & 0.2 & 0 & -0.1 & -0.5 & 0 \\ 0 & 0 & 0.2 & -0.1 & 0 & 0.1 \end{bmatrix} \begin{bmatrix} 910 \\ 0 \\ 0 \\ 0 \\ 0 \\ 0 \end{bmatrix} = \begin{bmatrix} 182 \\ 0 \\ 0 \end{bmatrix}$$

$$QI_s - QY_b V_s = \begin{bmatrix} -182 \\ 0 \\ 0 \end{bmatrix}$$

Hence, KCL equation can be written as,

$$QY_b Q^T V_t = QI_s - QY_b V_s$$

$$\begin{bmatrix} 0.9 & 0.6 & 0.2 \\ 0.6 & 0.8 & 0.1 \\ 0.2 & 0.1 & 0.3 \end{bmatrix} \begin{bmatrix} v_{t_1} \\ v_{t_2} \\ v_{t_3} \end{bmatrix} = \begin{bmatrix} -182 \\ 0 \\ 0 \end{bmatrix}$$

Solving this matrix equation, we get

$$v_{t_1} = -460 \text{ V}$$
$$v_{t_2} = 320 \text{ V}$$
$$v_{t_3} = 200 \text{ V}$$

Example 7.21 *For the network shown in Fig. 7.64, write down the f-cutset matrix and obtain the network equilibrium equation in matrix form using KCL and calculate v.*

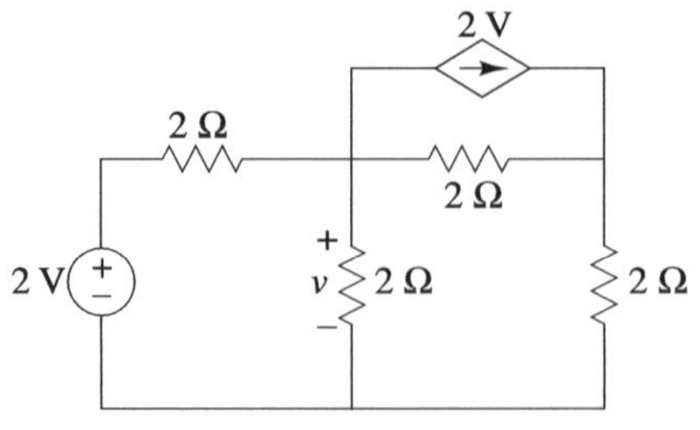

Fig. 7.64

Solution The oriented graph and its selected tree are shown in Fig. 7.65. Since voltage v is to be determined, branch 2 is chosen as twig.

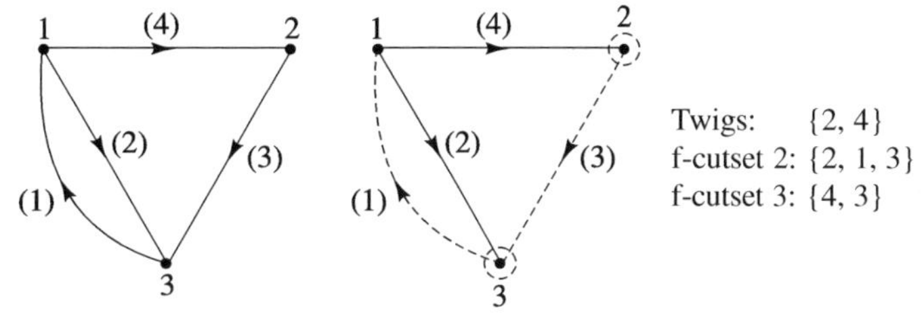

Fig. 7.65

$$Q = \begin{bmatrix} -1 & 1 & 1 & 0 \\ 0 & 0 & -1 & 1 \end{bmatrix}$$

(columns 1, 2, 3, 4)

The KCL equation in matrix form is given by

$$Q\,Y_b\,Q^T\,V_t = Q\,I_s - Q\,Y_b\,V_s$$

$$Y_b = \begin{bmatrix} 0.5 & 0 & 0 & 0 \\ 0 & 0.5 & 0 & 0 \\ 0 & 0 & 0.5 & 0 \\ 0 & 0 & 0 & 0.5 \end{bmatrix} \quad Q^T = \begin{bmatrix} -1 & 0 \\ 1 & 0 \\ 1 & -1 \\ 0 & 1 \end{bmatrix} \quad I_s = \begin{bmatrix} 0 \\ 0 \\ 0 \\ -2v \end{bmatrix} \quad V_s = \begin{bmatrix} 2 \\ 0 \\ 0 \\ 0 \end{bmatrix}$$

$$QY_b = \begin{bmatrix} -1 & 1 & 1 & 0 \\ 0 & 0 & -1 & 1 \end{bmatrix}\begin{bmatrix} 0.5 & 0 & 0 & 0 \\ 0 & 0.5 & 0 & 0 \\ 0 & 0 & 0.5 & 0 \\ 0 & 0 & 0 & 0.5 \end{bmatrix} = \begin{bmatrix} -0.5 & 0.5 & 0.5 & 0 \\ 0 & 0 & -0.5 & 0.5 \end{bmatrix}$$

$$QY_b\,Q^T = \begin{bmatrix} -0.5 & 0.5 & 0.5 & 0 \\ 0 & 0 & -0.5 & 0.5 \end{bmatrix}\begin{bmatrix} -1 & 0 \\ 1 & 0 \\ 1 & -1 \\ 0 & 1 \end{bmatrix} = \begin{bmatrix} 1.5 & -0.5 \\ -0.5 & 1 \end{bmatrix}$$

$$QI_s = \begin{bmatrix} -1 & 1 & 1 & 0 \\ 0 & 0 & -1 & 1 \end{bmatrix}\begin{bmatrix} 0 \\ 0 \\ 0 \\ -2v \end{bmatrix} = \begin{bmatrix} 0 \\ -2v \end{bmatrix}$$

$$QY_b\,V_s = \begin{bmatrix} -0.5 & 0.5 & 0.5 & 0 \\ 0 & 0 & -0.5 & 0.5 \end{bmatrix}\begin{bmatrix} 2 \\ 0 \\ 0 \\ 0 \end{bmatrix} = \begin{bmatrix} -1 \\ 0 \end{bmatrix}$$

$$QI_s - QY_bV_s = \begin{bmatrix} 1 \\ -2v \end{bmatrix}$$

Hence, the KCL equation can be written as

$$QY_b\,Q^T\,V_t = QI_s - QY_b\,V_s$$

$$\begin{bmatrix} 1.5 & -0.5 \\ -0.5 & 1 \end{bmatrix}\begin{bmatrix} v_{t_2} \\ v_{t_4} \end{bmatrix} = \begin{bmatrix} 1 \\ -2v \end{bmatrix}$$

From the figure, $v_{t_2} = v$

Solving this matrix equation, we get

$$v_{t_2} = 0.44 \text{ V}$$
$$v_{t_4} = 0.66 \text{ V}$$
$$v = v_{t_2} = 0.44 \text{ V}$$

Example 7.22 *For the network shown in Fig. 7.66, write down the f-cutset matrix and obtain the network equilibrium equation in matrix form using KCL and calculate v.*

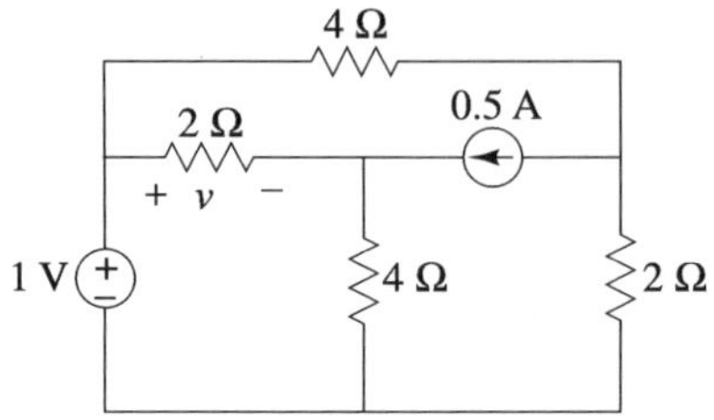

Fig. 7.66

Solution The voltage and current sources are converted into accompanied sources by source-shifting method.

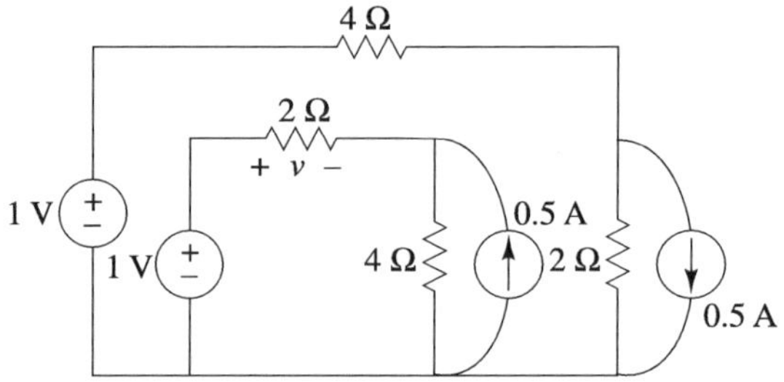

Fig. 7.67

The oriented graph and its selected tree are shown in Fig. 7.68.

$$Q = \begin{array}{c} \begin{matrix} 1 & 2 & 3 & 4 \end{matrix} \\ \begin{bmatrix} 1 & 0 & 0 & -1 \\ 0 & 1 & -1 & 0 \end{bmatrix} \end{array}$$

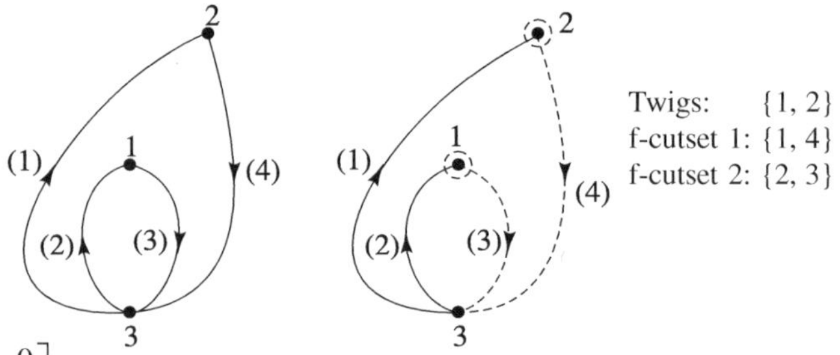

Fig. 7.68

The KCL equation in the matrix form is given by

$$QY_b Q^T V_t = QI_s - QY_b V_s$$

$$Y_b = \begin{bmatrix} 0.25 & 0 & 0 & 0 \\ 0 & 0.5 & 0 & 0 \\ 0 & 0 & 0.25 & 0 \\ 0 & 0 & 0 & 0.5 \end{bmatrix}$$

$$Q^T = \begin{bmatrix} 1 & 0 \\ 0 & 1 \\ 0 & -1 \\ 1 & 0 \end{bmatrix} \quad I_s = \begin{bmatrix} 0 \\ 0 \\ 0.5 \\ -0.5 \end{bmatrix} \quad V_s = \begin{bmatrix} 1 \\ 1 \\ 0 \\ 0 \end{bmatrix}$$

$$QY_b = \begin{bmatrix} 1 & 0 & 0 & -1 \\ 0 & 1 & -1 & 0 \end{bmatrix} \begin{bmatrix} 0.25 & 0 & 0 & 0 \\ 0 & 0.5 & 0 & 0 \\ 0 & 0 & 0.25 & 0 \\ 0 & 0 & 0 & 0.5 \end{bmatrix} = \begin{bmatrix} 0.25 & 0 & 0 & -0.5 \\ 0 & 0.5 & -0.25 & 0 \end{bmatrix}$$

$$QY_b Q^T = \begin{bmatrix} 0.25 & 0 & 0 & -0.5 \\ 0 & 0.5 & -0.25 & 0 \end{bmatrix} \begin{bmatrix} 1 & 0 \\ 0 & 1 \\ 0 & -1 \\ 1 & 0 \end{bmatrix} = \begin{bmatrix} 0.75 & 0 \\ 0 & 0.75 \end{bmatrix}$$

$$QI_s = \begin{bmatrix} 1 & 0 & 0 & -1 \\ 0 & 1 & -1 & 0 \end{bmatrix} \begin{bmatrix} 0 \\ 0 \\ 0.5 \\ -0.5 \end{bmatrix} = \begin{bmatrix} 0.5 \\ -0.5 \end{bmatrix}$$

$$QY_b V_s = \begin{bmatrix} 0.25 & 0 & 0 & -0.5 \\ 0 & 0.5 & -0.25 & 0 \end{bmatrix} \begin{bmatrix} 1 \\ 1 \\ 0 \\ 0 \end{bmatrix} = \begin{bmatrix} 0.25 \\ 0.5 \end{bmatrix}$$

$$QI_s - QY_b V_s = \begin{bmatrix} 0.25 \\ -1 \end{bmatrix}$$

Hence, the KCL equation can be written as

$$QY_b Q^T V_t = QI_s - QY_b V_s$$

$$\begin{bmatrix} 0.75 & 0 \\ 0 & 0.75 \end{bmatrix} \begin{bmatrix} v_{t_1} \\ v_{t_2} \end{bmatrix} = \begin{bmatrix} 0.25 \\ -1 \end{bmatrix}$$

Solving this matrix equation, we get

$$v_{t_1} = 0.33 \text{ V}$$

$$v_{t_2} = -1.33 \text{ V}$$

From the figure, $v = 1 + v_{t_2} = -0.33 \text{ V}$

7.13 DUALITY

Two networks are said to be the dual of each other when the mesh equations of one network are the same as the node equations of the other. Kirchhoff's voltage law and current law are same, word for word, with voltage substituted for current, independent loop for independent node pair, etc. Similarly, two graphs are said to be the dual of each other if the incidence matrix of any one of them is equal to the circuit matrix of the other. Only planar networks have duals.

Table 7.1 *Conversion for Dual Electrical Circuits*

Loop basis	*Node basis*
Current	Voltage
Resistance	Conductance
Inductance	Capacitance
Branch current	Branch voltage
Mesh	Node
Short circuit	Open circuit
Parallel path	Series path

The following steps are involved in constructing the dual of a network:

(1) Place a node inside each mesh of the given network. These internal nodes correspond to the independent nodes in the dual network.

(2) Place a node outside the given network. The external node corresponds to the datum node in the dual network.

(3) Connect all internal nodes in the adjacent mesh by dashed lines crossing the common branches. Elements which are the duals of the common branches will form the branches connecting the corresponding independent node in the dual network.

(4) Connect all internal nodes to the external node by dashed lines corresponding to all external branches. Duals of these external branches will form the branches connecting independent nodes and the datum node.

(5) A clockwise current in a mesh corresponds to a positive polarity (with respect to the datum node) at the dual independent node.

(6) A voltage rise in the direction of a clockwise mesh current corresponds to a current flowing towards the dual independent node.

Example 7.23 *Draw the dual of the network shown in Fig. 7.69.*

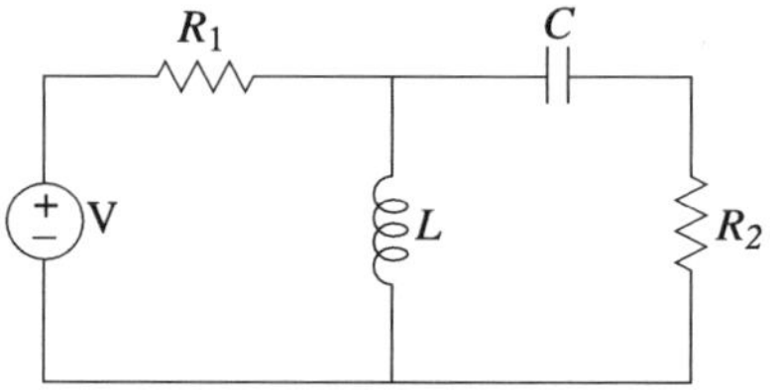

Fig. 7.69

Solution

(a) Place a node inside each mesh.

(b) Place a node outside the mesh which will correspond to the datum node.

(c) Connect two internal nodes through a dashed line. The element which is dual of the common branch (here capacitance) will form the branch connecting the corresponding independent node in the dual network.

(d) Connect all internal nodes to the external node by dashed lines crossing all the branches. The dual of these branches will form the branches connecting the independent node and datum.

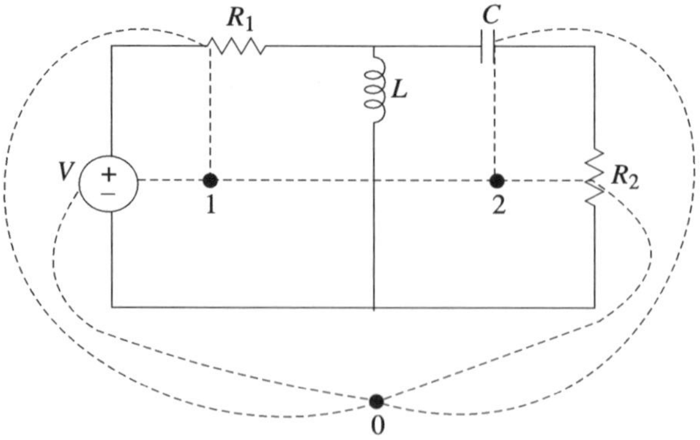

Fig. 7.70

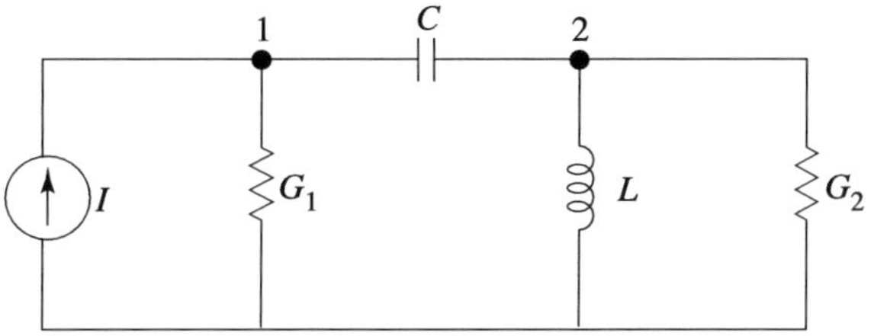

Dual network

Fig. 7.71

Example 7.24 *Draw the dual of the network of Fig. 7.72.*

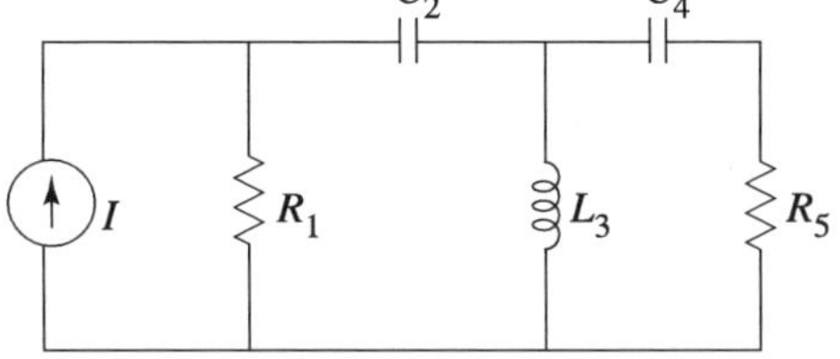

Fig. 7.72

Solution For drawing the dual network, proceed in the same way as in Example 7.23.

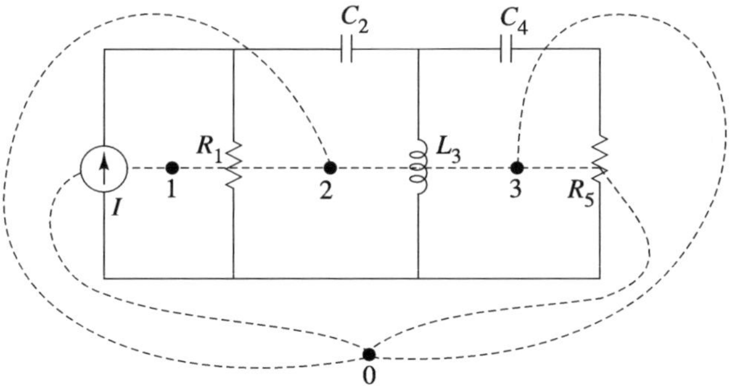

Fig. 7.73

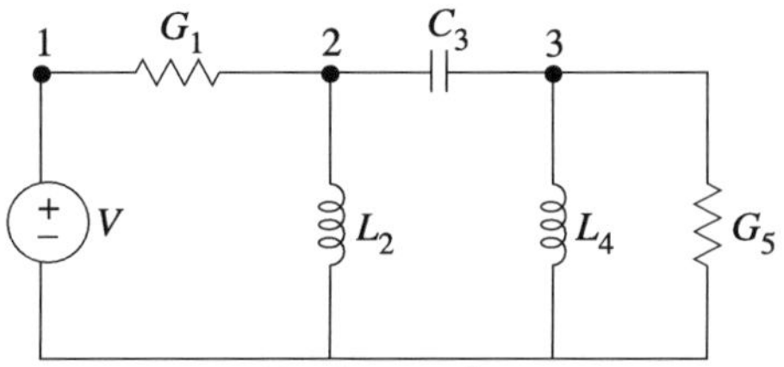

Dual network

Fig. 7.74

Example 7.25 *Draw the dual of the network shown in Fig. 7.75.*

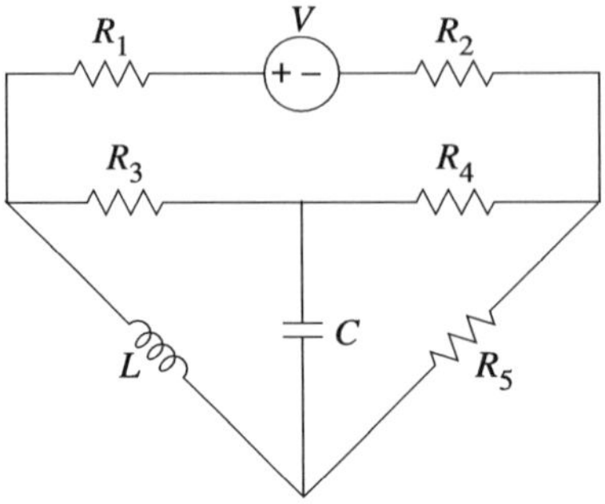

Fig. 7.75

Solution For drawing the dual network, proceed in the same way as in Example 7.23.

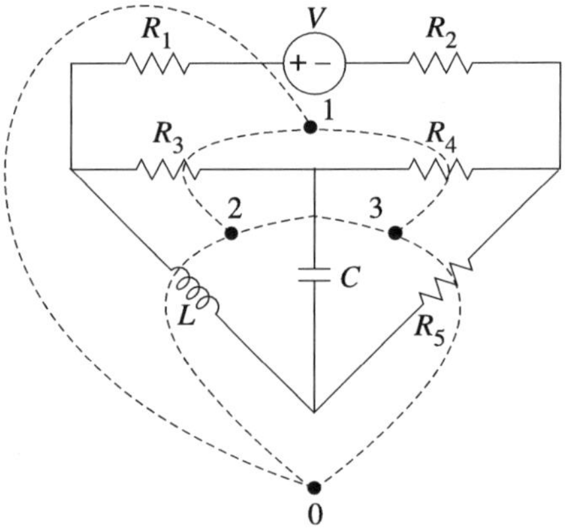

Fig. 7.76

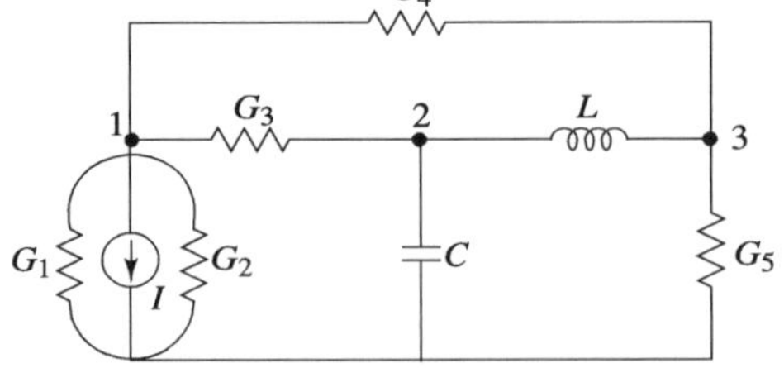

Dual network

Fig. 7.77

Exercises

1. For the networks shown, write the incidence matrix, tieset matrix and f-cutset matrix.

(i)

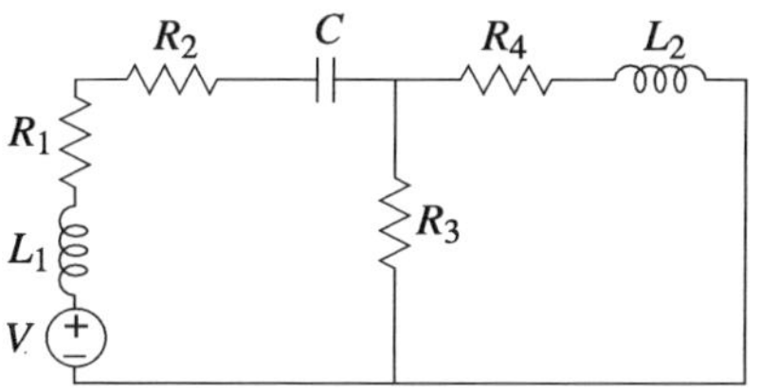

Fig. 7.78

(ii)

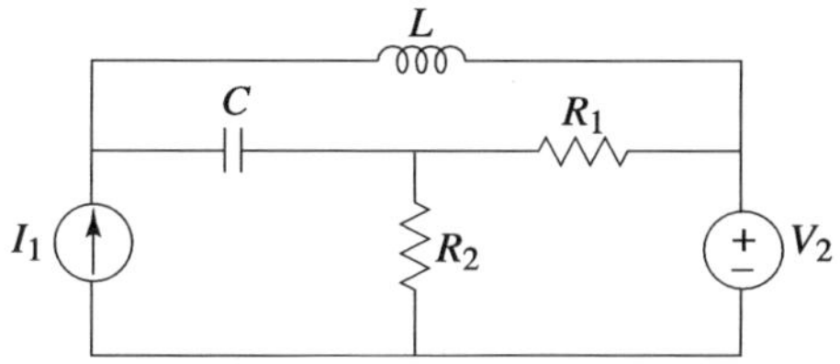

Fig. 7.79

(iii)

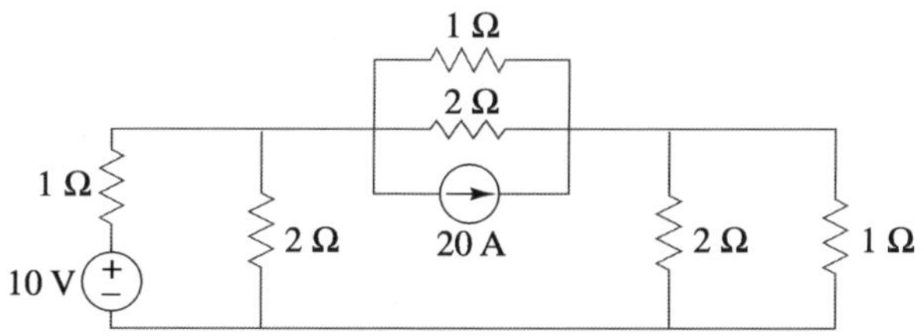

Fig. 7.80

(iv)

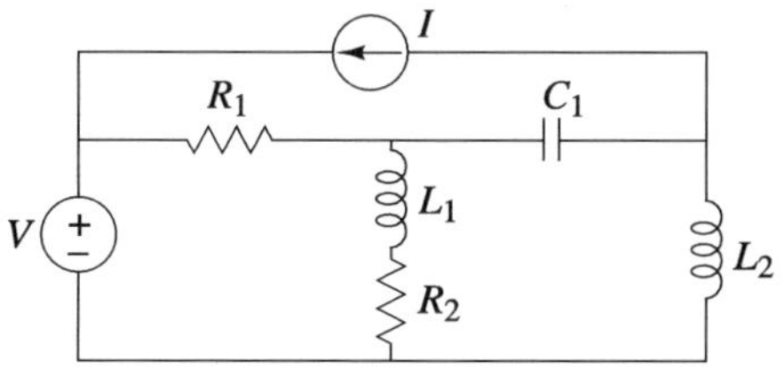

Fig. 7.81

2. For the graph shown, write the incidence matrix, tieset matrix and f-cutset matrix.

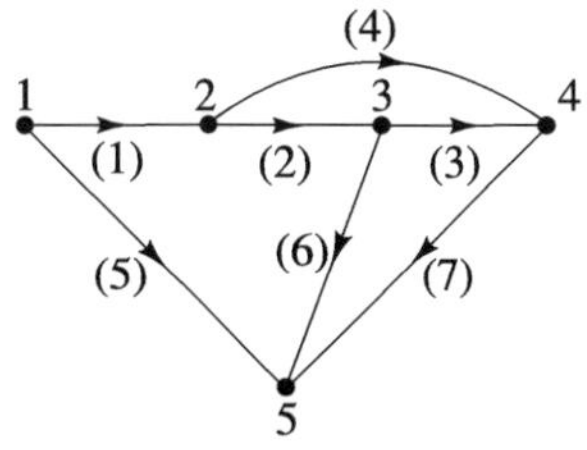

Fig. 7.82

3. The incidence matrix is given as follows:

Branches →

1	2	3	4	5	6	7	8
–1	–1	0	0	0	0	1	0
0	1	1	0	1	0	0	0
0	0	–1	–1	0	1	0	0
1	0	0	1	0	0	0	1

Draw oriented graph and write tieset matrix.

4. The incidence matrix is given below:

Branches →

	1	2	3	4	5	6	7	8	9	10
	0	0	1	1	1	1	0	1	0	0
	0	–1	–1	0	0	0	–1	0	0	–1
$A =$	–1	1	0	0	0	0	0	–1	–1	1
	1	0	0	0	–1	–1	1	0	0	0

Draw the oriented graph.

5. For the network shown in Fig. 7.83, draw the oriented graph and obtain the tieset matrix. Use this matrix to calculate the current.

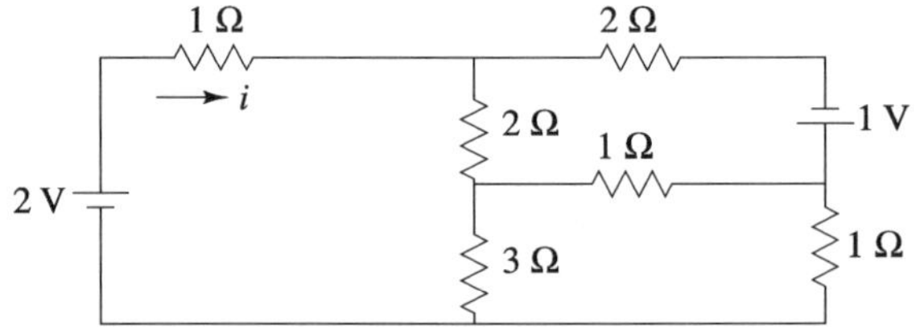

Fig. 7.83

6. Draw the dual networks for the circuits shown:

(i)

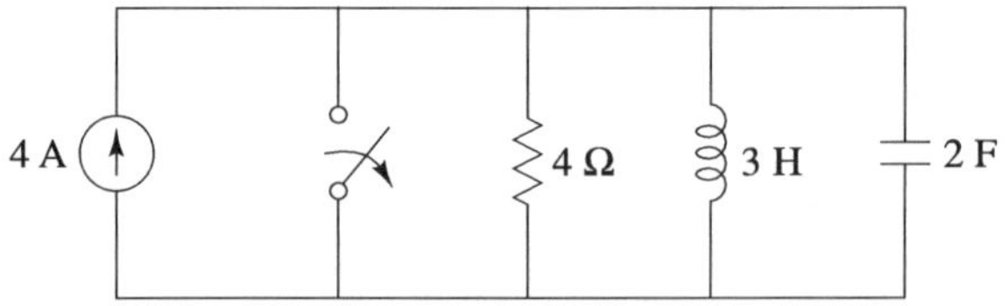

Fig. 7.84

(ii)

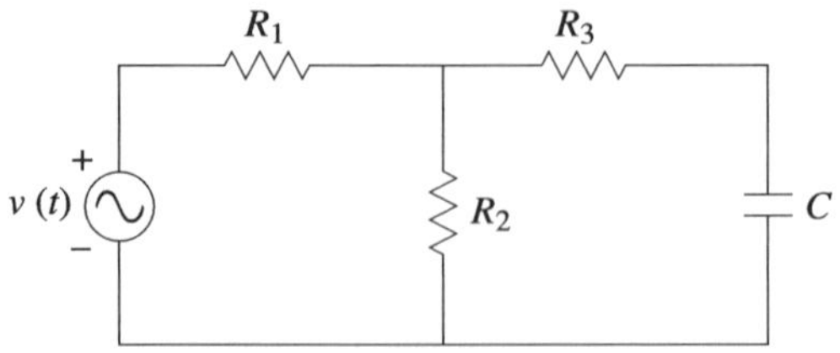

Fig. 7.85

(iii)

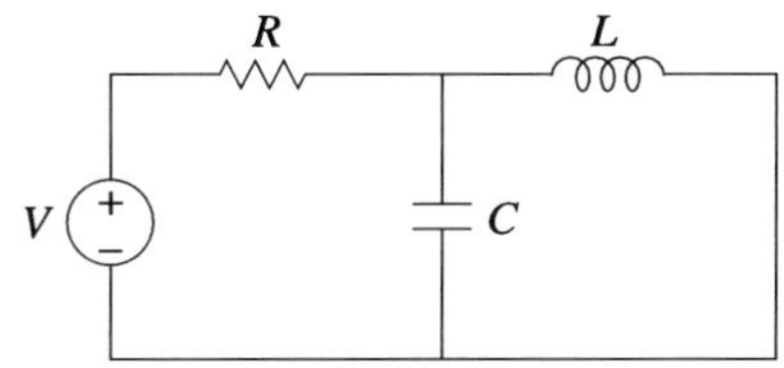

Fig. 7.86

(iv)

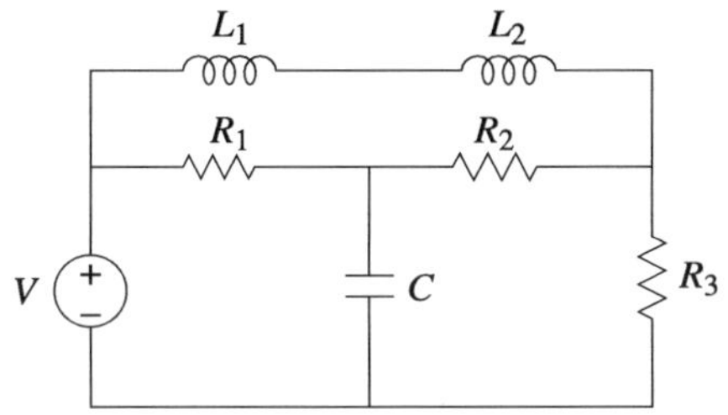

Fig. 7.87

(v)

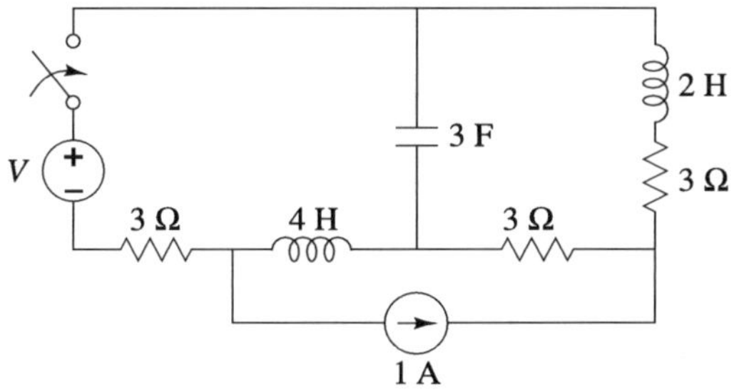

Fig. 7.88

7. Using the principles of network topology, write the loop/node equation in matrix form for the network shown in Fig. 7.89.

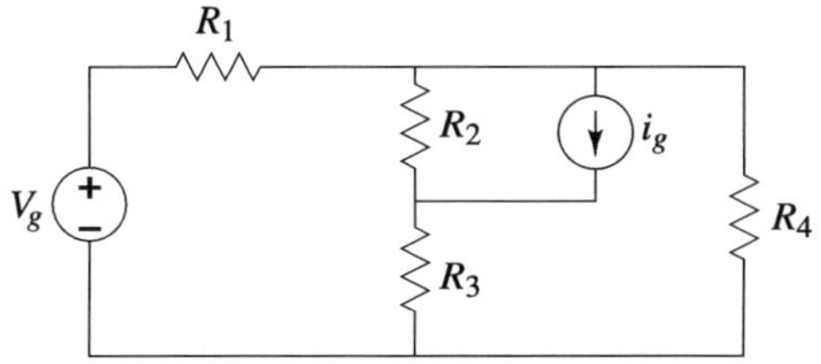

Fig. 7.89

Objective-Type Questions

1. The number of independent loops for a network with n nodes and b branches is
   (a) $n - 1$ (b) $b - n$
   (c) $b - n + 1$ (d) independent of the number of nodes
2. A network has 7 nodes and 5 independent loops. The number of branched in the network is
   (a) 13 (b) 12 (c) 11 (d) 10
3. Identify which of the following is NOT a tree of the graph shown in Fig. 7.90.
   (a) *begh* (b) *defg*
   (c) *adfg* (d) *aegh*

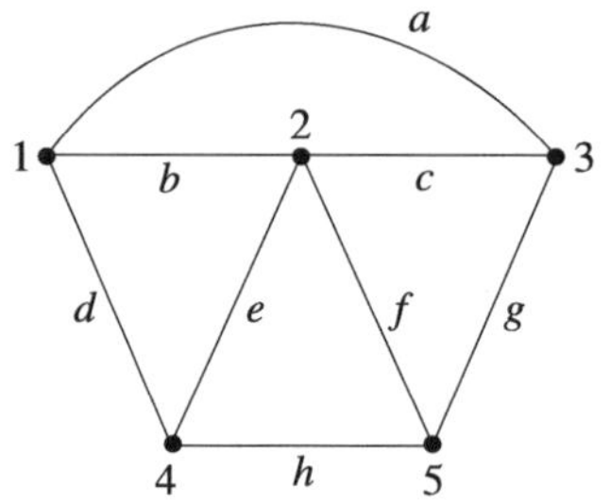

Fig. 7.90

4. The minimum number of equations required to analyze the circuit shown in Fig. 7.91 is
   (a) 3 (b) 4
   (c) 6 (d) 7

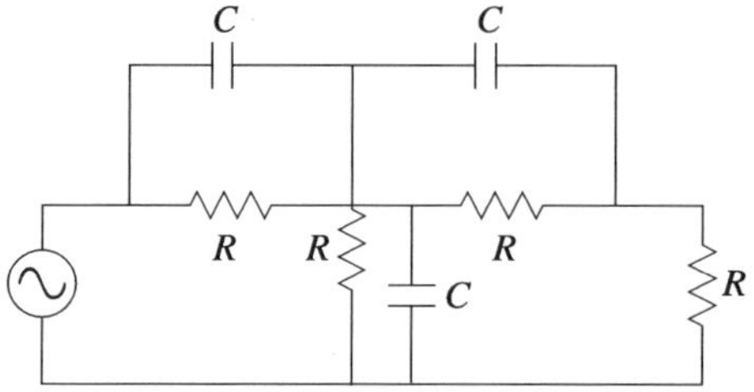

Fig. 7.91

5. Consider the network graph shown in Fig. 7.92.

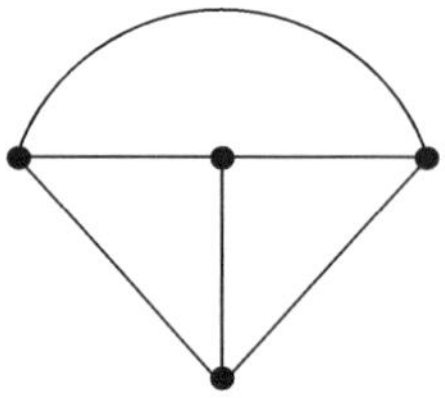
Fig. 7.92

Which one of the following is NOT a tree of this graph?

(a) 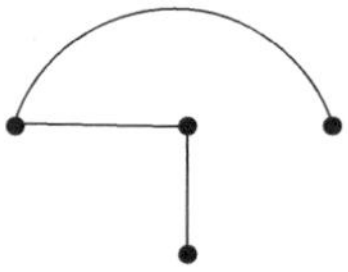
(b) 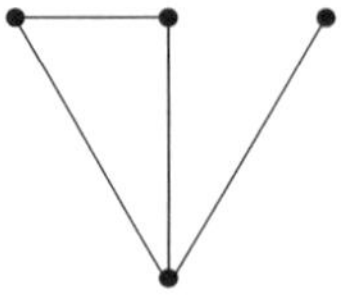
(c) 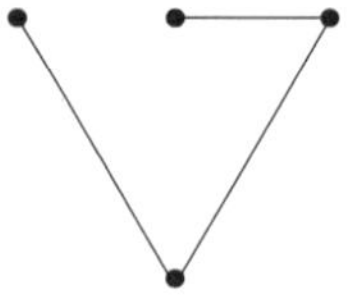
(d) 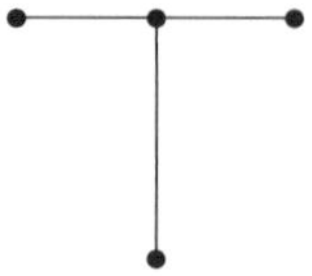

6. Figure below shows a network and its graph is drawn aside. A proper tree chosen for analysing the network will contain the edges.
(a) *ab*, *bc*, *ad* (b) *ab*, *bc*, *ca* (c) *ab*, *bd*, *cd* (d) *ac*, *bd*, *ad*

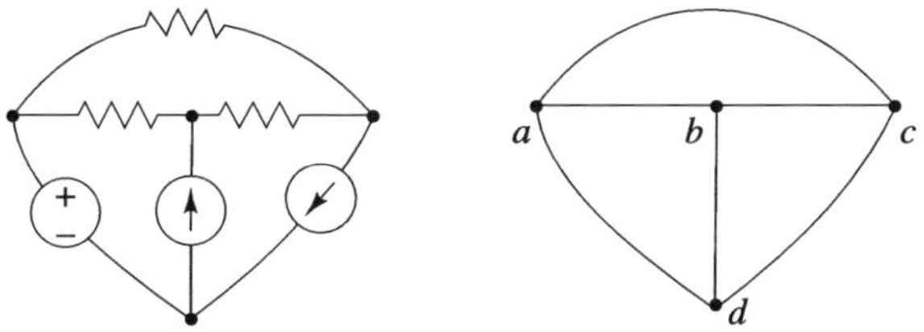

Fig. 7.93

7. The graph of an electrical network has *n* nodes and *b* branches. The number of links with respect to the choice of a tree is given by
(a) $b - n + 1$ (b) $b + n$ (c) $n - b + 1$ (d) $n - 2b - 1$

8. In the graph shown in Fig. 7.94, one possible tree is formed by the branches 4, 5, 6, 7. Then one possible fundamental cut set is
(a) 1, 2, 3, 8 (b) 1, 2, 5, 6
(c) 1, 5, 6, 8 (d) 1, 2, 3, 7, 8

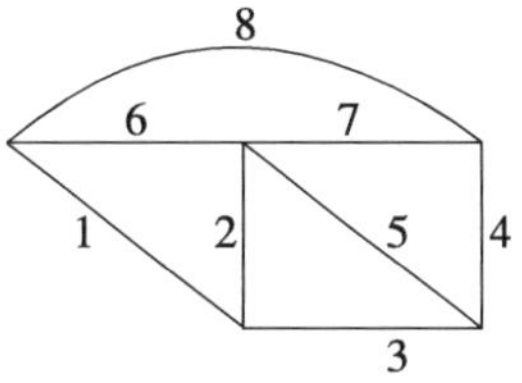

Fig. 7.94

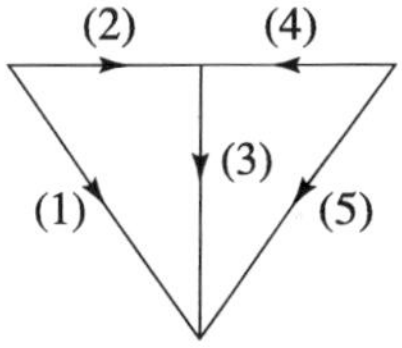

Fig. 7.95

9. Which one of the following represents the total number of trees in the graph given in Fig. 7.95?
(a) 4 (b) C
(c) 5 (d) 8

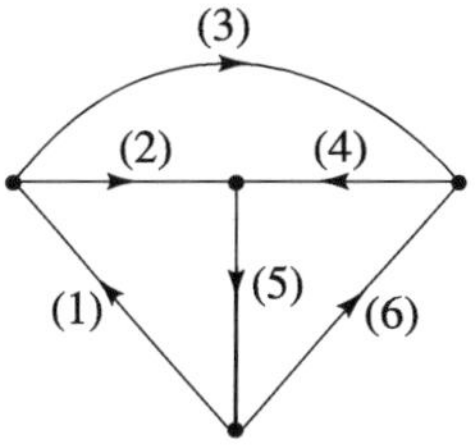

Fig. 7.96

10. Which one of the following is a cutset of the graph shown in the Fig. 7.96?
(a) 1, 2, 3 and 4 (b) 2, 3, 4 and 6
(c) 1, 4, 5 and 6 (d) 1, 2, 4 and 5

Answers to Objective-Type Questions

1. (c) **2.** (c) **3.** (c) **4.** (b) **5.** (b) **6.** (d)
7. (a) **8.** (d) **9.** (d) **10.** (d)

8 Transient Analysis

8.1 INTRODUCTION

Whenever a network containing energy storage elements such as inductor or capacitor is switched from one condition to another, either by change in applied source or change in network elements, the response current and voltage change from one state to the other state. The time taken to change from an initial steady state to the final steady state is known as the *transient period*. This response is known as *transient response* or *transients*. The response of the network after it attains a final steady value is independent of time and is called the steady-state response. The complete response of the network is determined with the help of a differential equation.

8.2 NETWORK EQUATIONS

A circuit contains resistors, inductors, capacitors and energy sources. We must first write network equations for current and voltage relationship of each circuit element in the time domain.

(1) Resistor For the resistor shown in Fig. 8.1, the v–i relationship in time domain is

$$v(t) = R\,i(t)$$

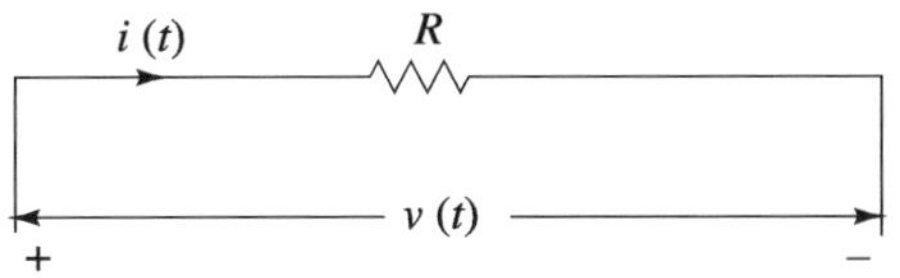

Fig. 8.1

(2) Inductor For the inductor shown in Fig. 8.2, the v–i relationships in the time domain are

$$v(t) = L\,\frac{di}{dt}$$

$$i(t) = \frac{1}{L}\int_0^t v(t)\,dt + i(0)$$

where $i(0)$ is the initial current passing through the inductor.

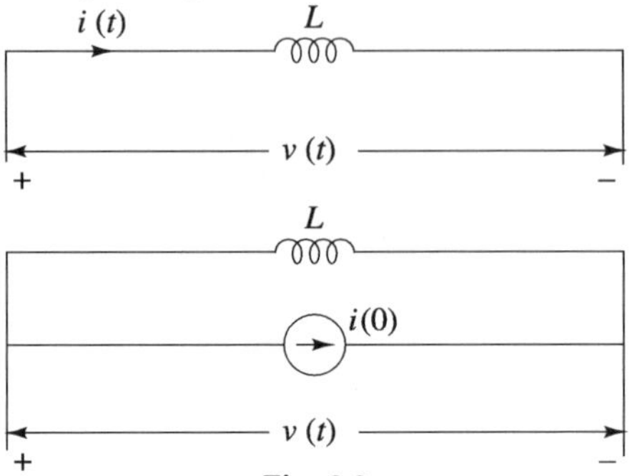

Fig. 8.2

(3) Capacitor For the capacitor shown in Fig. 8.3, the v–i relationships in the time domain are

$$i(t) = C\frac{dv(t)}{dt}$$

$$v(t) = \frac{1}{C}\int_0^t i(t)\,dt + v(0)$$

where $v(0)$ is the initial voltage on the capacitor.

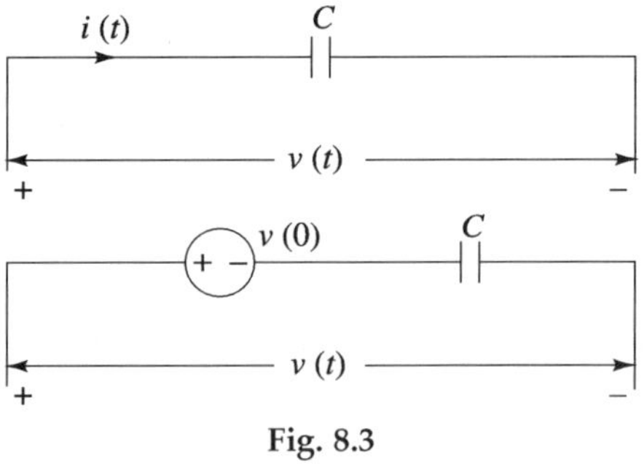

Fig. 8.3

8.3 INITIAL CONDITIONS

In solving the differential equations in the network, we get some arbitary constant. Initial conditions are used to determine these arbitrary constants. It helps us to know the behaviour of elements at the instant of switching.

To differentiate between the time immediately before and immediately after the switching, the signs '–' and '+' are used. The conditions existing just before switching are denoted as $i(0^-)$, $v(0^-)$, etc. Conditions just after switching are denoted as $i(0^+)$, $v(0^+)$.

Sometimes conditions at $t = \infty$ are used in the evaluation of arbitrary constants. These are known as final conditions.

In solving the problems for initial conditions in the network, we divide the time period in following ways:

(1) Just before switching (from $t = -\infty$ to $t = 0^-$)

(2) Just after switching (at $t = 0^+$)

(3) After switching (for $t > 0$)

If the network remains in one condition for a long time without any switching action, it is said to be under steady-state condition.

(1) Initial conditions for the resistor For a resistor, current and voltage are related by $v(t) = Ri(t)$. The current through a resistor will change instantaneously if the voltage changes instantaneously. Similarly, the voltage will change instantaneously if the current changes instantaneously.

(2) Initial conditions for the inductor Voltage across the inductor is proportional to the rate of change of current. It is impossible to change the current through an inductor by a finite amount in zero time. This requires an infinite voltage across the inductor. An inductor does not allow an abrupt change in the current through it.

If there is no current flowing through the inductor at $t = 0^-$, the inductor will act as an open circuit at $t = 0^+$. If a current of value I_0 flows through the inductor at $t = 0^-$, the inductor can be regarded as a current source of I_0 ampere at $t = 0^+$.

(3) Initial conditions for the capacitor Current through a capacitor is proportional to the rate of change of voltage. It is impossible to change the voltage across a capacitor by a finite amount in zero time. This requires an infinite current through the capacitor. A capacitor does not allow an abrupt change in voltage across it.

If there is no voltage across the capacitor at $t = 0^-$, the capacitor will act as a short circuit at $t = 0^+$. If the capacitor is charged to a voltage V_0 at $t = 0^-$, it can be regarded as a voltage source of V_0 volt at $t = 0^+$. These conditions are summarized in Fig. 8.4.

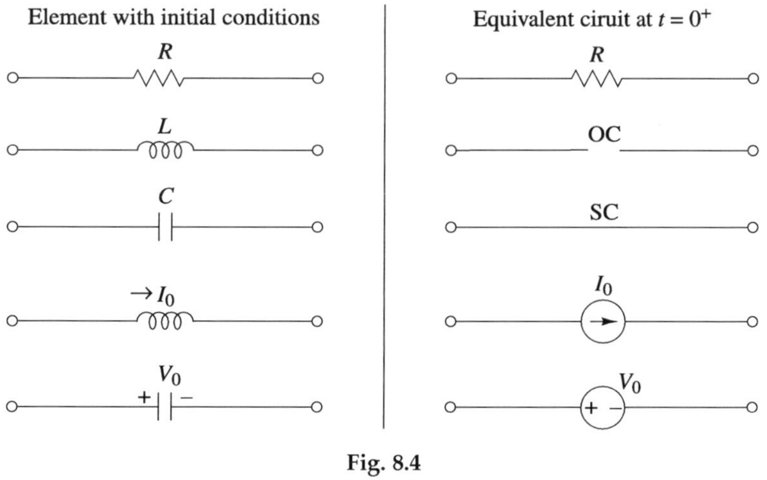

Fig. 8.4

Similarly, we can draw the chart for final conditions as shown in Fig. 8.5.

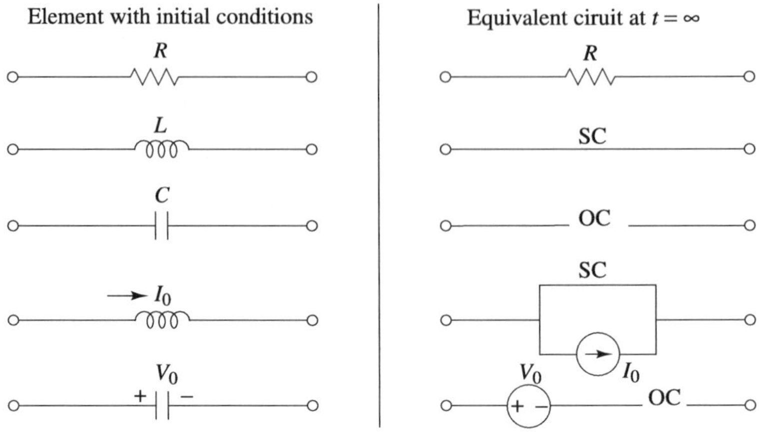

Fig. 8.5

8.4 PROCEDURE FOR EVALUATING INITIAL CONDITIONS

(1) Draw the equivalent network at $t = 0^-$. Before switching action takes place, i.e., for $t = -\infty$ to $t = 0^-$, the network is under steady-state conditions. Hence, find the current flowing through the inductors $i_L(0^-)$ and voltage across the capacitor $v_C(0^-)$.

(2) Draw the equivalent network at $t = 0^+$, i.e., immediately after switching. Replace all the inductors with open circuits or with current sources $i_L(0^+)$ and replace all capacitors by short circuits or voltage sources $v_C(0^+)$. Resistors are kept as it is in the network.

(3) Initial voltages or currents in the network are determined from the equivalent network at $t = 0^+$.

(4) Initial conditions, i.e., $\frac{di}{dt}(0^+)$, $\frac{dv}{dt}(0^+)$, $\frac{d^2i}{dt^2}(0^+)$, $\frac{d^2v}{dt^2}(0^+)$ are determined by writing integro-differential equations for the network for $t > 0$, i.e., after the switching action by making use of initial condition.

Example 8.1 *In the network of Fig. 8.6, the switch is closed at t = 0. With the capacitor uncharged, find value for i, $\frac{di}{dt}$ and $\frac{d^2i}{dt^2}$ at $t = 0^+$.*

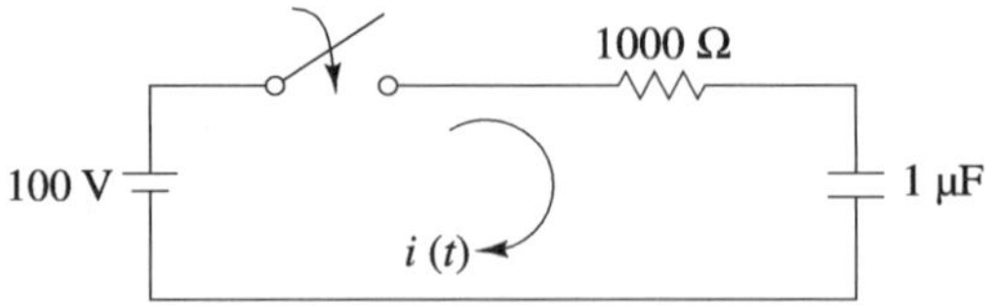

Fig. 8.6

Solution

At $t = 0^-$, the capacitor is uncharged.

$$v_C(0^-) = 0$$
$$i(0^-) = 0$$

At $t = 0^+$, the capacitor acts as a short circuit.

$$i(0^+) = \frac{100}{1000} = 0.1 \text{ A}$$

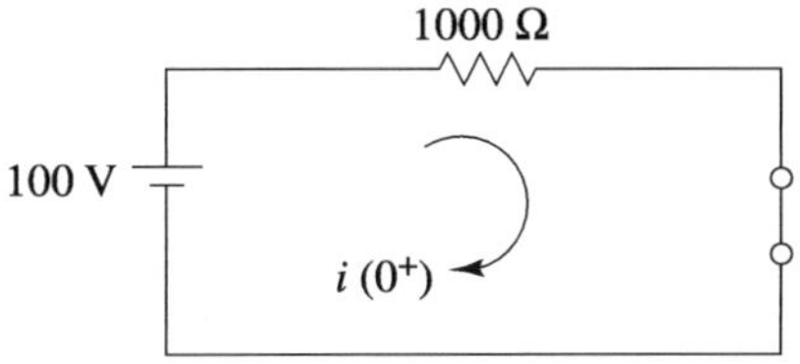

Fig. 8.7

Writing KVL equation for $t > 0$,

$$100 - 1000i - \frac{1}{1 \times 10^{-6}} \int_0^t i \, dt = 0 \qquad \ldots(i)$$

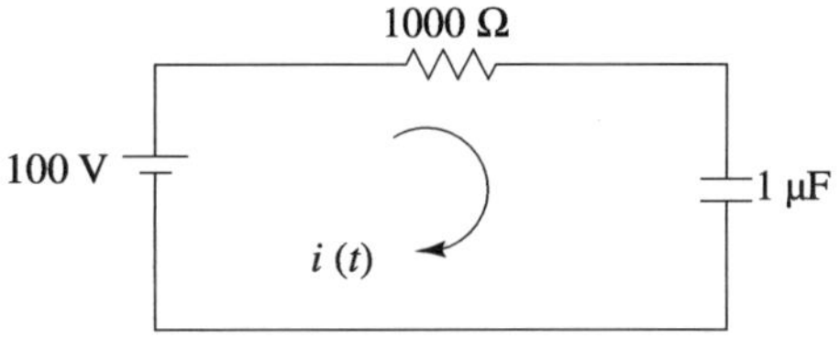

Fig. 8.8

Differentiating the Eq. (i), we get

$$-1000 \frac{di}{dt} - 10^6 i = 0$$

$$\frac{di}{dt} = -\frac{10^6}{1000} i \qquad \ldots(ii)$$

At $t = 0^+$

$$\frac{di}{dt}(0^+) = -\frac{10^6}{1000} i(0^+)$$

$$= -10^3 \times 0.1 = -100 \text{ A/s}$$

Differentiating the Eq. (ii), we get

$$\frac{d^2i}{dt^2} = -\frac{10^6}{1000} \frac{di}{dt}$$

At $t = 0^+$

$$\frac{d^2i}{dt^2}(0^+) = -\frac{10^6}{1000} \frac{di}{dt}(0^+)$$

$$= -\frac{10^6}{1000}(-100) = 10^5 \text{ A/s}^2$$

Example 8.2 *In the given network of Fig. 8.9, the switch is closed at $t = 0$. With zero current in the inductor, find i, $\frac{di}{dt}$ and $\frac{d^2i}{dt^2}$ at $t = 0^+$.*

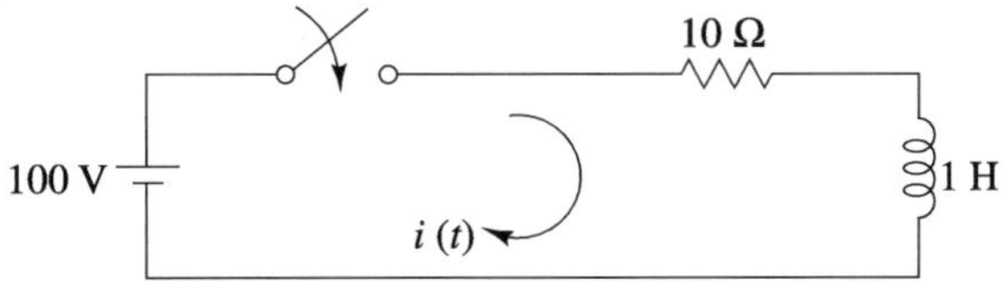

Fig. 8.9

Solution

At $t = 0^-$, no current flows through the inductor.

$$i(0^-) = 0$$

At $t = 0^+$, the inductor acts as an open circuit.

$$i(0^+) = 0$$

Writing KVL equation for $t > 0$,

$$100 - 10i - 1\frac{di}{dt} = 0 \qquad \ldots(i)$$

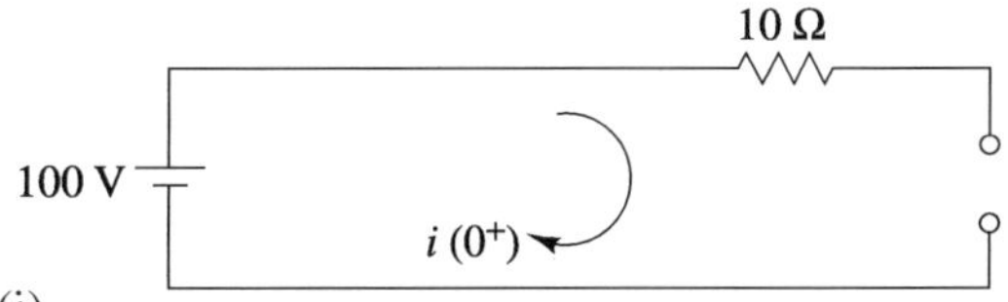

Fig. 8.10

$$\frac{di}{dt} = 100 - 10i \qquad \ldots\text{(ii)}$$

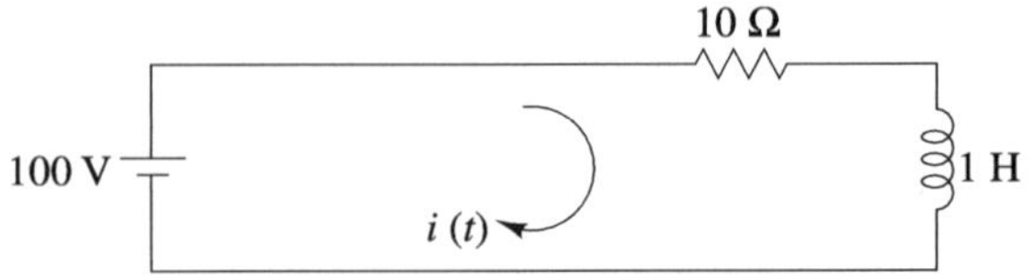

Fig. 8.11

At $t = 0^+$

$$\frac{di}{dt}(0^+) = 100 - 10i\,(0^+)$$
$$= 100 - 10\,(0) = 100 \text{ A/s}$$

Differentiating the Eq. (ii), we get

$$\frac{d^2i}{dt^2} = -10\,\frac{di}{dt}$$

At $t = 0^+$

$$\frac{d^2i}{dt^2}(0^+) = -10\,\frac{di}{dt}(0^+)$$
$$= -10\,(100) = -1000 \text{ A/s}^2$$

Example 8.3 *In the network shown in Fig. 8.12, the switch is closed. Assuming all initial conditions as zero, find i,* $\frac{di}{dt}$ *and* $\frac{d^2i}{dt^2}$ *at* $t = 0^+$.

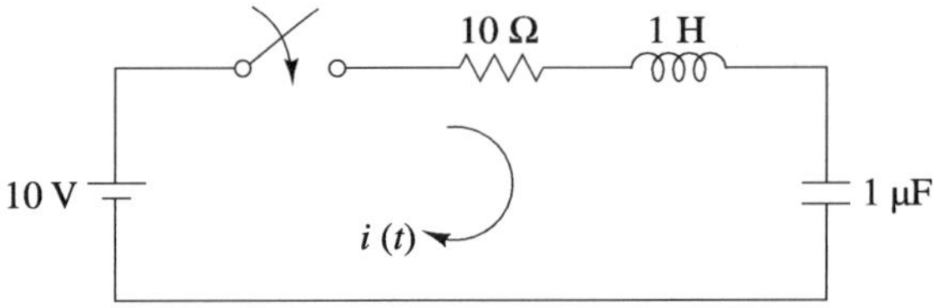

Fig. 8.12

Solution

At $t = 0^-$,

$$i\,(0^-) = 0$$
$$v_C\,(0^-) = 0$$

At $t = 0^+$, the inductor acts as an open circuit and the capacitor acts as a short circuit.

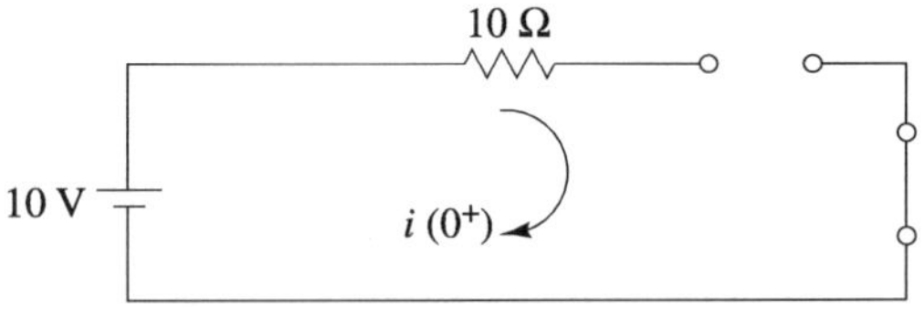

Fig. 8.13

$$i\,(0^+) = 0$$
$$v_C\,(0^+) = 0$$

Writing KVL equation for $t > 0$,

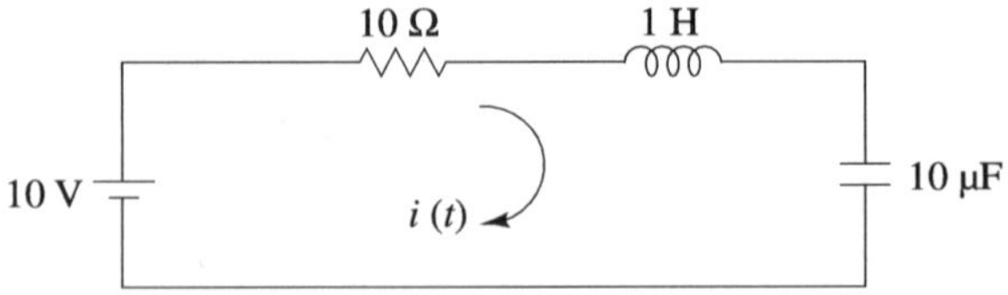

Fig. 8.14

$$10 - 10i - 1\,\frac{di}{dt} - \frac{1}{10 \times 10^{-6}}\int_0^t i\,dt = 0$$

$$10 = 10i + \frac{di}{dt} + \frac{1}{10 \times 10^{-6}}\int_0^t i\,dt \qquad \ldots\text{(i)}$$

At $t = 0^+$

$$10 = 10i\,(0^+) + \frac{di}{dt}(0^+) + 0$$

$$\frac{di}{dt}(0^+) = 10 \text{ A/s}$$

Differentiating the Eq. (i), we get

$$0 = 10\frac{di}{dt} + \frac{d^2i}{dt^2} + \frac{1}{10 \times 10^{-6}}\,i$$

At $t = 0^+$

$$0 = 10\frac{di}{dt}(0^+) + \frac{d^2i}{dt^2}(0^+) + \frac{1}{10^{-5}}\,i(0^+)$$

$$\frac{d^2i}{dt^2}(0^+) = -10 \times 10 = -100 \text{ A/s}^2$$

Example 8.4 *In the network shown in Fig. 8.15, at t = 0, the switch is opened. Calculate v,* $\frac{dv}{dt}$ *and* $\frac{d^2v}{dt^2}$ *at t = 0⁺.*

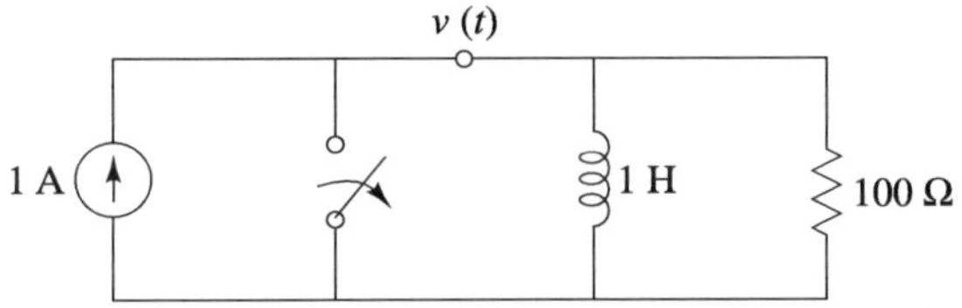

Fig. 8.15

Solution At $t = 0^-$, the switch is closed. Hence, no current flows through the inductor.

$$i_L(0^-) = 0$$

At $t = 0^+$, the inductor acts as an open circuit.

$$i_L(0^+) = 0$$

$$v(0^+) = 100 \times 1 = 100 \text{ V}$$

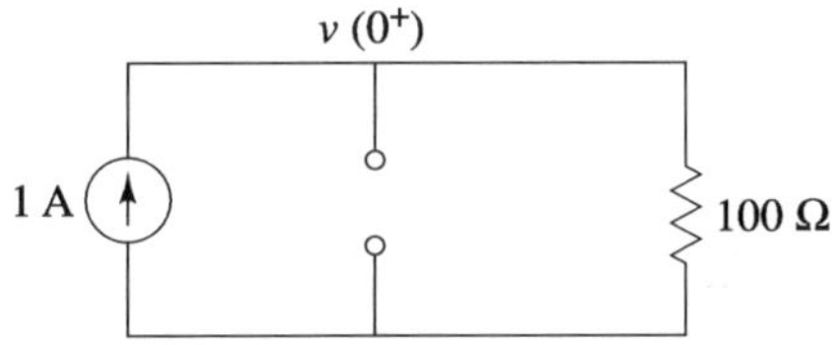

Fig. 8.16

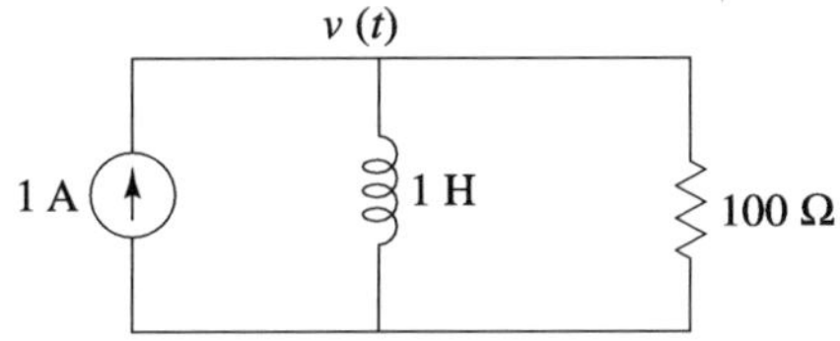

Fig. 8.17

Writing KCL equation for $t > 0$,

$$\frac{v}{100} + \frac{1}{1}\int_0^t v\,dt = 1 \qquad \ldots\text{(i)}$$

Differentiating the Eq. (i), we get

$$\frac{1}{100}\frac{dv}{dt} + v = 0 \qquad \ldots\text{(ii)}$$

At $t = 0^+$

$$\frac{dv}{dt}(0^+) = -100v\,(0^+)$$

$$= -100 \times 100 = -10000 \text{ V/s}$$

Differentiating the Eq. (ii), we get

$$\frac{1}{100}\frac{d^2v}{dt^2}+\frac{dv}{dt}=0$$

At $t = 0^+$

$$\frac{d^2v}{dt^2}(0^+) = -100\,\frac{dv}{dt}(0^+)$$

$$= -100 \times (-10^4) = 10^6 \text{ V/s}^2$$

Example 8.5 *In the given network of Fig. 8.18, the switch is opened at t = 0. Solve for v, $\frac{dv}{dt}$ and $\frac{d^2v}{dt^2}$ at t = 0^+.*

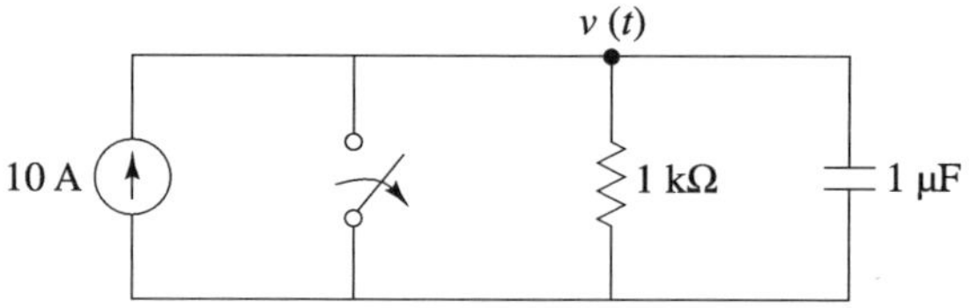

Fig. 8.18

Solution At $t = 0^-$, switch is closed. Hence, the voltage across the capacitor is zero.

$$v(0^-) = v_C(0^-) = 0$$

At $t = 0^+$, the capacitor acts as a short circuit.

$$v(0^+) = 0$$

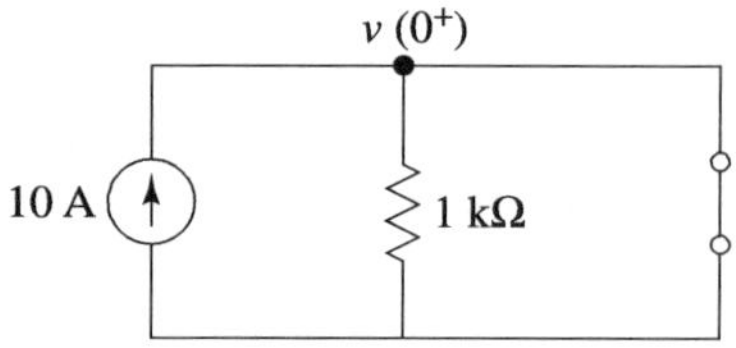

Fig. 8.19

Writing KCL equation for $t > 0$,

$$\frac{v}{1000} + 10^{-6}\frac{dv}{dt} = 10 \qquad \ldots\text{(i)}$$

At $t = 0^+$

$$\frac{v(0^+)}{1000} + 10^{-6}\frac{dv}{dt}(0^+) = 10$$

$$\frac{dv}{dt}(0^+) = \frac{10}{10^{-6}} = 10 \times 10^6 \text{ V/s}$$

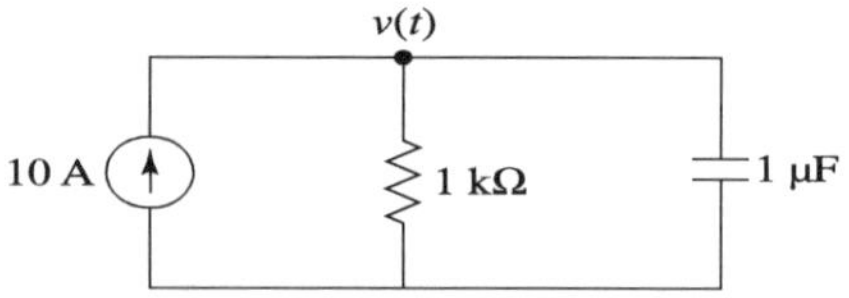

Fig. 8.20

Differentiating the Eq. (i),

$$\frac{1}{1000}\frac{dv}{dt} + 10^{-6}\frac{d^2v}{dt^2} = 0$$

At $t = 0^+$

$$\frac{1}{1000}\frac{dv}{dt}(0^+) + 10^{-6}\frac{d^2v}{dt^2}(0^+) = 0$$

$$\frac{d^2v}{dt^2}(0^+) = -\frac{1}{1000 \times 10^{-6}} \times 10 \times 10^6 = -10 \times 10^9 \text{ V/s}^2$$

Example 8.6 *For the network shown in Fig. 8.21, switch is closed at t = 0, determine v,* $\frac{dv}{dt}$ *and* $\frac{d^2v}{dt^2}$ *at* $t = 0^+$.

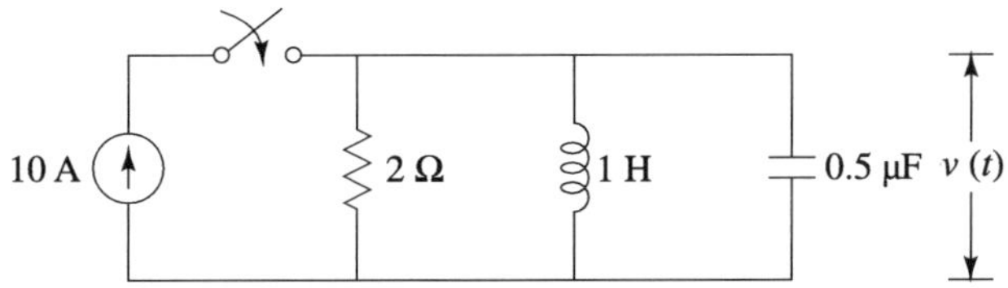

Fig. 8.21

Solution At $t = 0^-$, no current flows through the inductor and there is no voltage across the capacitor.

$$i_L(0^-) = 0$$
$$v(0^-) = v_C(0^-) = 0$$

At $t = 0^+$, the inductor acts as an open circuit and the capacitor acts as a short circuit.

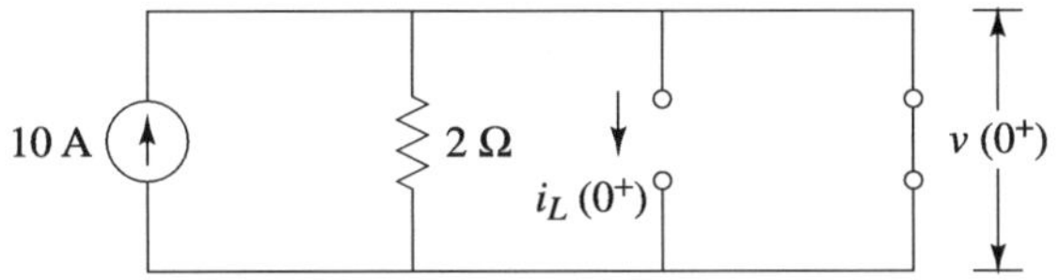

Fig. 8.22

$$i_L(0^+) = 0$$
$$v(0^+) = v_c(0^+) = 0$$

For $t > 0$,

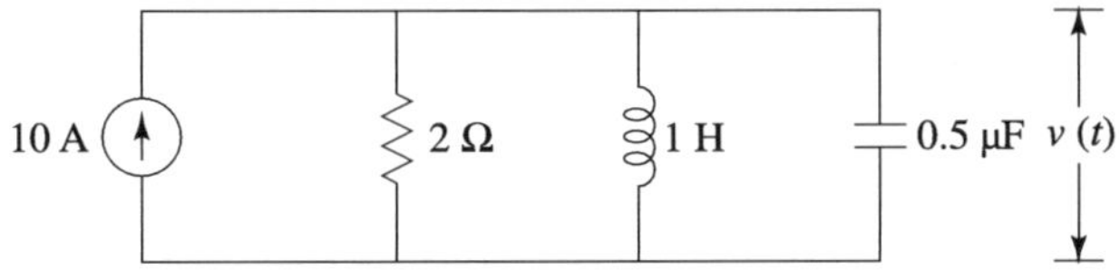

Fig. 8.23

Writing KCL equation for $t > 0$,

$$10 = \frac{v}{2} + \frac{1}{1}\int_0^t v\,dt + 0.5\times10^{-6}\frac{dv}{dt} \qquad \text{...(i)}$$

At $t = 0^+$

$$10 = \frac{v(0^+)}{2} + 0 + 0.5\times10^{-6}\frac{dv}{dt}(0^+)$$

$$\frac{dv}{dt}(0^+) = 20\times10^6 \text{ V/s}$$

Differentiating the Eq. (i), we get

$$0 = \frac{1}{2}\frac{dv}{dt} + v + 0.5\times10^{-6}\frac{d^2v}{dt^2}$$

At $t = 0^+$

$$0 = \frac{1}{2}\frac{dv}{dt}(0^+) + v(0^+) + 0.5\times10^{-6}\frac{d^2v}{dt^2}(0^+)$$

$$\frac{d^2v}{dt^2}(0^+) = -20\times10^{12} \text{ V/s}^2$$

Example 8.7 *In the network shown in Fig. 8.24, the switch is changed from the position 1 to the position 2 at t = 0, steady condition having reached before switching. Find the values of i,* $\frac{di}{dt}$ *and* $\frac{d^2i}{dt^2}$ *at* $t = 0^+$.

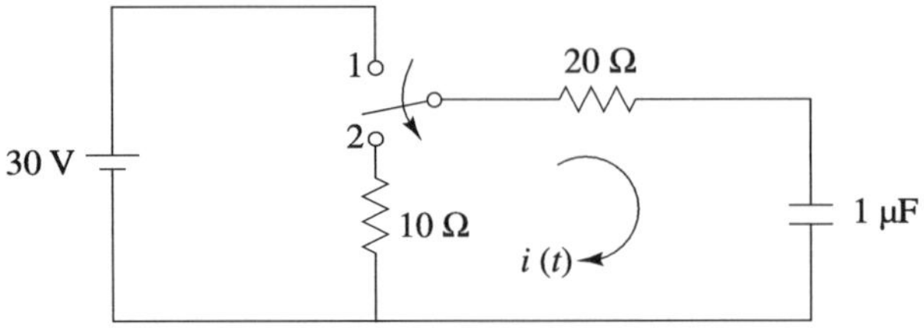

Fig. 8.24

Solution At $t = 0^-$, the network attains steady-state condition. Hence, the capacitor acts as an open circuit.

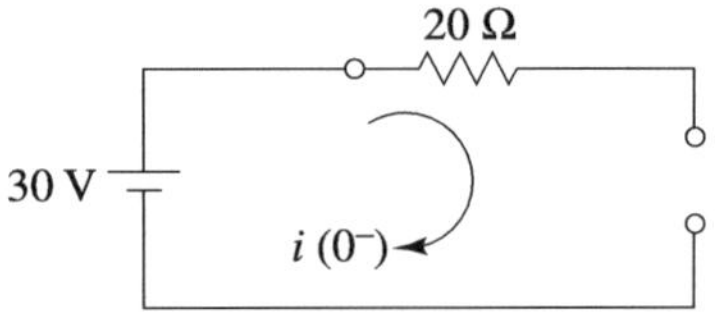

Fig. 8.25

$$v_C(0^-) = 30 \text{ V}$$

$$i(0^-) = 0$$

At $t = 0^+$, the capacitor acts as a voltage source of 30 V.

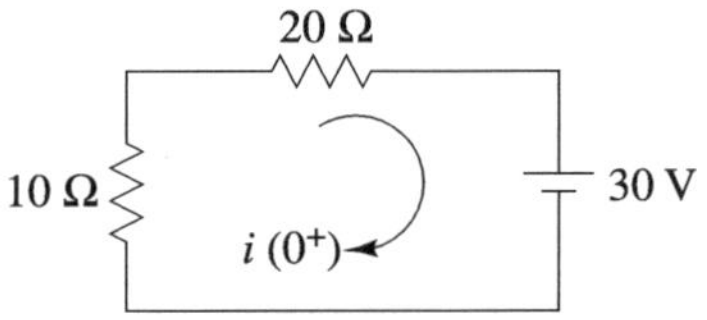

Fig. 8.26

$$i(0^+) = -\frac{30}{30} = -1 \text{ A}$$

Writing KVL equation for $t > 0$,

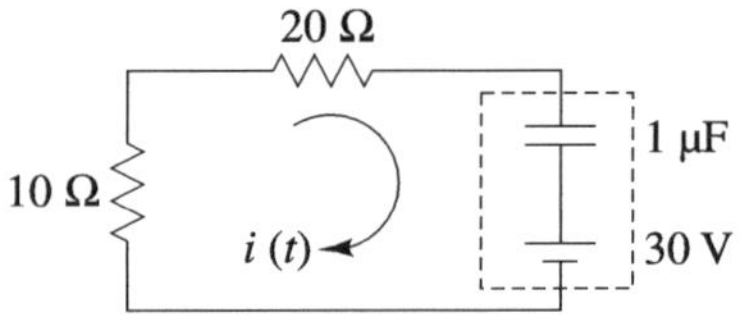

Fig. 8.27

$$-10i - 20i - \frac{1}{1 \times 10^{-6}} \int_0^t i \; dt - 30 = 0 \qquad \text{...(i)}$$

Differentating the equation (i), we get

$$-30\frac{di}{dt} - 10^6 i = 0 \qquad \text{...(ii)}$$

At $t = 0^+$

$$-30\frac{di}{dt}(0^+) - 10^6 i(0^+) = 0$$

$$\frac{di}{dt}(0^+) = -\frac{10^6(-1)}{30} = 0.33 \times 10^5 \text{ A/s}$$

Differentiating the Eq. (ii), we get

$$-30\frac{d^2i}{dt^2} - 10^6\frac{di}{dt} = 0$$

At $t = 0^+$

$$-30\frac{d^2i}{dt^2}(0^+) - 10^6\frac{di}{dt}(0^+) = 0$$

$$\frac{d^2i}{dt^3}(0^+) = -\frac{10^6 \times 0.33 \times 10^5}{30} = -1.1 \times 10^9 \text{ A/s}^2$$

Example 8.8 *In the network shown in Fig. 8.28, the switch is changed from the position 1 to the position 2 at $t = 0$, steady condition having reached before switching. Find the values of i, $\frac{di}{dt}$ and $\frac{d^2i}{dt^2}$ at $t = 0^+$.*

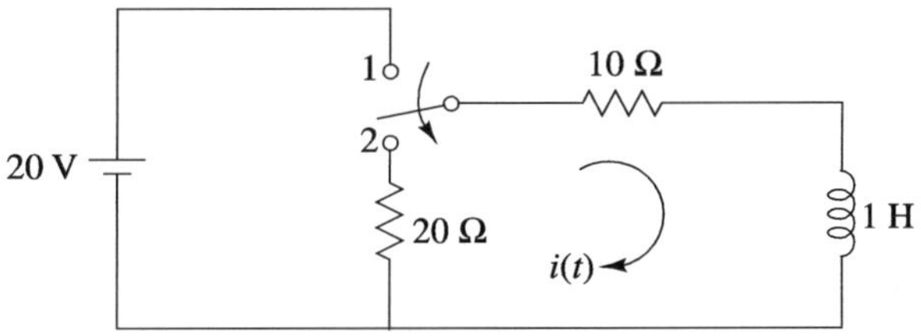

Fig. 8.28

Solution At $t = 0^-$, the network attains steady-state condition. Hence, the inductor acts as a short circuit.

$$i(0^-) = \frac{20}{10} = 2 \text{ A}$$

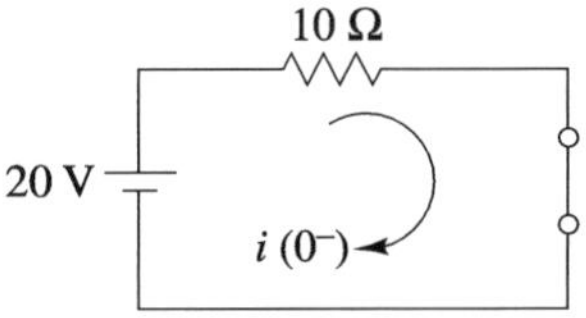

Fig. 8.29

At $t = 0^+$, the inductor acts as a current source of 2 A.

$$i(0^+) = 2 \text{ A}$$

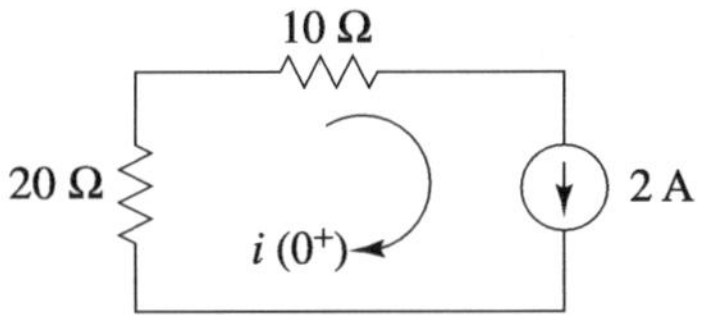

Fig. 8.30

Writing KVL equation for $t > 0$,

$$-30i - 1\frac{di}{dt} = 0 \qquad \ldots \text{(i)}$$

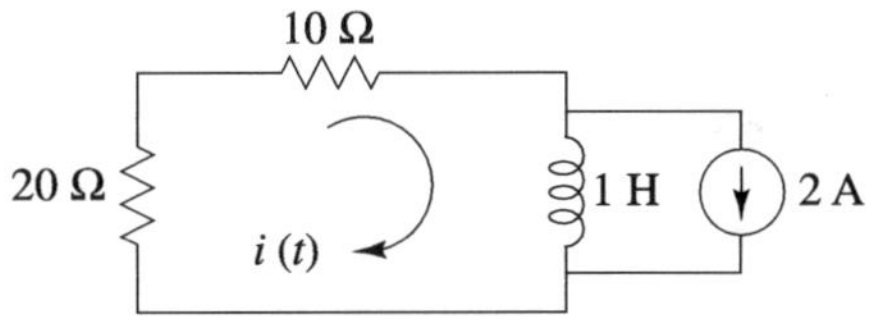

Fig. 8.31

At $t = 0^+$

$$-30\, i(0^+) - \frac{di}{dt}(0^+) = 0$$

$$\frac{di}{dt}(0^+) = -30 \times 2 = -60 \text{ A/s}$$

Differentiating the Eq. (i), we get

$$-30\frac{di}{dt} - \frac{d^2i}{dt^2} = 0$$

At $t = 0^+$

$$-30\frac{di}{dt}(0^+) - \frac{d^2i}{dt^2}(0^+) = 0$$

$$\frac{d^2i}{dt^2}(0^+) = 1800 \text{ A/s}^2$$

Example 8.9 *In the network shown in Fig. 8.32, the switch is changed from the position 1 to the position 2 at t = 0, steady condition having reached before switching. Find the values of i,* $\frac{di}{dt}$ *and* $\frac{d^2i}{dt^2}$ *at* $t = 0^+$.

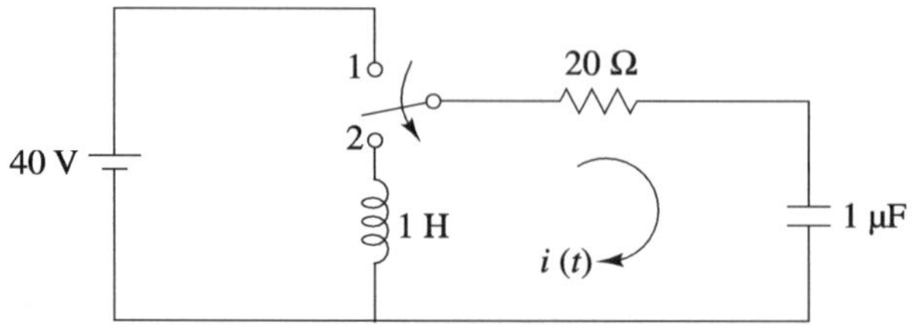

Fig. 8.32

Solution At $t = 0^-$, the network attains steady state. Hence, the capacitor acts as an open circuit.

$$v_C(0^-) = 40 \text{ V}$$
$$i(0^-) = 0$$

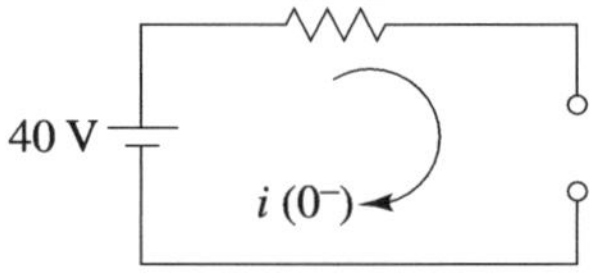

Fig. 8.33

At $t = 0^+$, the capacitor acts as a voltage source of 40 V and the inductor acts as an open circuit.

$$v_C(0^+) = 40 \text{ V}$$
$$i(0^+) = 0$$

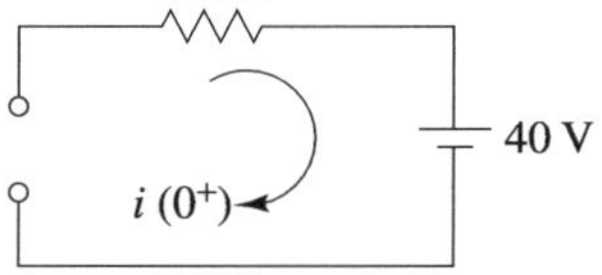

Fig. 8.34

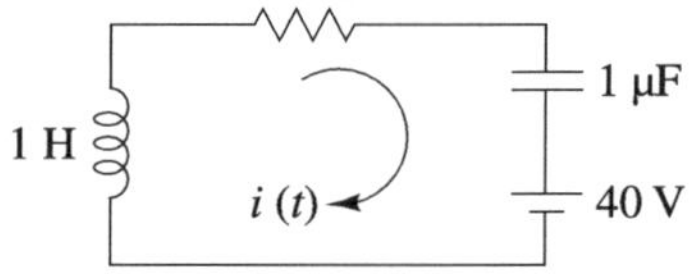

Fig. 8.35

Writing KVL equation for $t > 0$,

$$-1\frac{di}{dt} - 20i - \frac{1}{1\times10^{-6}}\int_0^t i\,dt - 40 = 0 \quad \ldots \text{(i)}$$

At $t = 0^+$

$$-\frac{di}{dt}(0^+) - 20i(0^+) - 0 - 40 = 0$$

$$\frac{di}{dt}(0^+) = -40 \text{ A/s}$$

Differentiating the equation (i), we get

$$-\frac{d^2i}{dt^2} - 20\frac{di}{dt} - 10^6\, i - 0 = 0$$

At $t = 0^+$

$$-\frac{d^2i}{dt^2}(0^+) - 20\frac{di}{dt}(0^+) - 10^6\, i(0^+) = 0$$

$$\frac{d^2i}{dt^2}(0^+) = 800 \text{ A/s}^2$$

Example 8.10 *In the network of Fig. 8.36, the switch is changed from the position 'a' to 'b' at t = 0. Solve for i,* $\frac{di}{dt}$ *and* $\frac{d^2i}{dt^2}$ *at* $t = 0^+$.

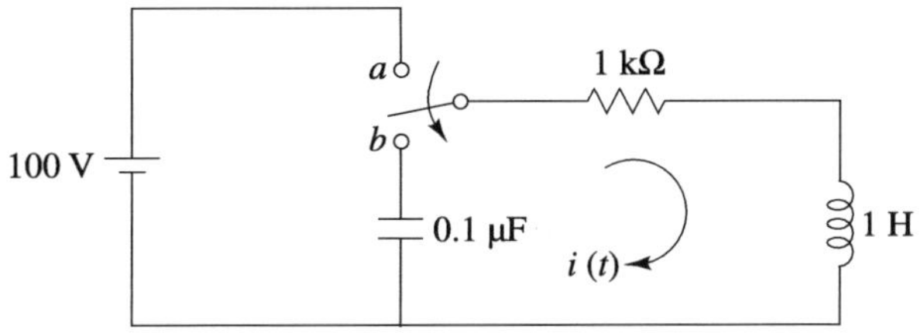

Fig. 8.36

Solution At $t = 0^-$, the network attains steady condition. Hence, the inductor acts as a short circuit.

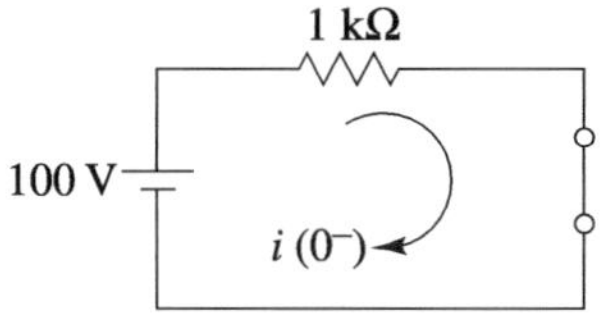

Fig. 8.37

$$i(0^-) = \frac{100}{1000} = 0.1 \text{ A}$$

At $t = 0^+$, the inductor acts as a current source of 0.1 A and the capacitor acts as a short circuit.

$$i(0^+) = 0.1 \text{ A}$$

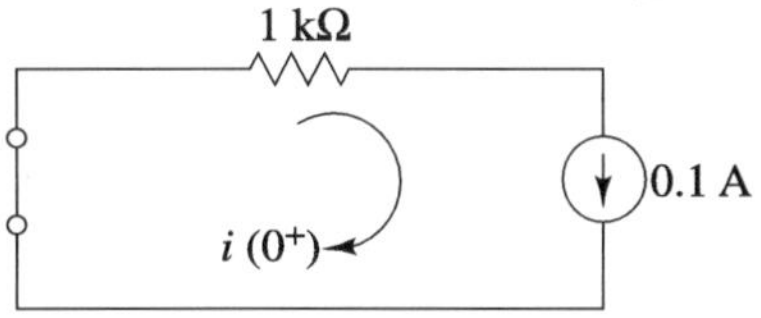

Fig. 8.38

Writing KVL equation for $t > 0$,

$$-\frac{1}{0.1 \times 10^{-6}} \int_0^t i \, dt - 1000\, i - 1 \frac{di}{dt} = 0 \qquad \text{...(i)}$$

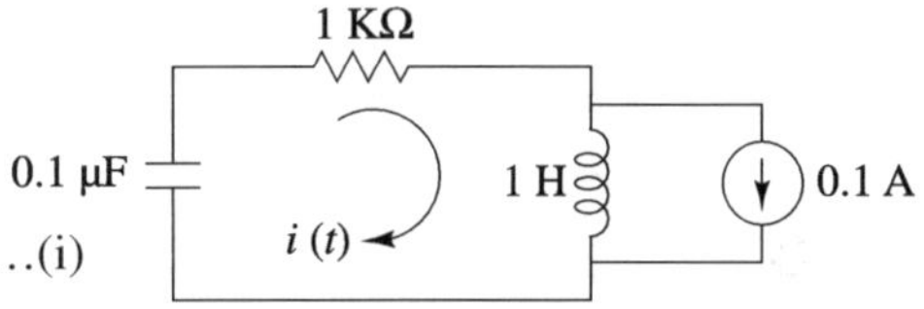

Fig. 8.39

At $t = 0^+$

$$-0 - 1000\, i(0^+) - \frac{di}{dt}(0^+) = 0$$

$$\frac{di}{dt}(0^+) = -1000\, i(0^+)$$

$$= -1000 \times 0.1 = -100 \text{ A/s}$$

Differentiating the Eq. (i), we get

$$-\frac{1}{10^{-7}} i - 1000 \frac{di}{dt} - \frac{d^2 i}{dt^2} = 0$$

At $t = 0^+$

$$-10^7\, i(0^+) - 1000 \frac{di}{dt}(0^+) - \frac{d^2 i}{dt^2}(0^+) = 0$$

$$\frac{d^2 i}{dt^2}(0^+) = -10^7 (0.1) - 1000(-100) = -9 \times 10^5 \text{ A/s}^2$$

Example 8.11 *In the network shown in Fig. 8.40, assuming all initial conditions as zero, find* $i_1(0^+)$, $i_2(0^+)$, $\frac{di_1}{dt}(0^+)$, $\frac{di_2}{dt}(0^+)$, $\frac{d^2i_1}{dt^2}(0^+)$ *and* $\frac{d^2i_2}{dt^2}(0^+)$.

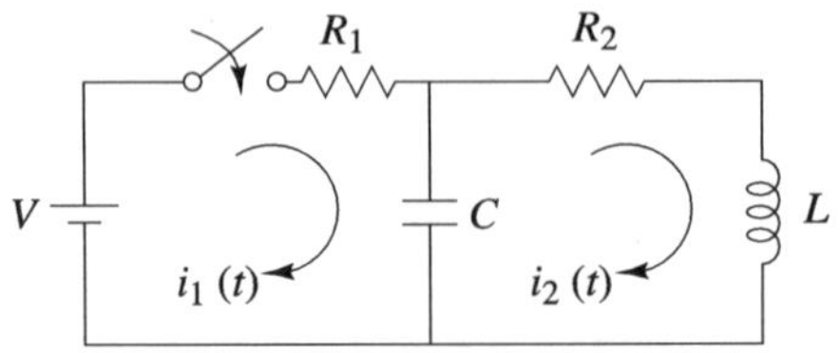

Fig. 8.40

Solution At $t = 0^-$, all initial conditions are zero.

$$v_C(0^-) = 0$$
$$i_1(0^-) = 0$$
$$i_2(0^-) = 0$$

At $t = 0^+$, the inductor acts as an open circuit and the capacitor acts as a short circuit.

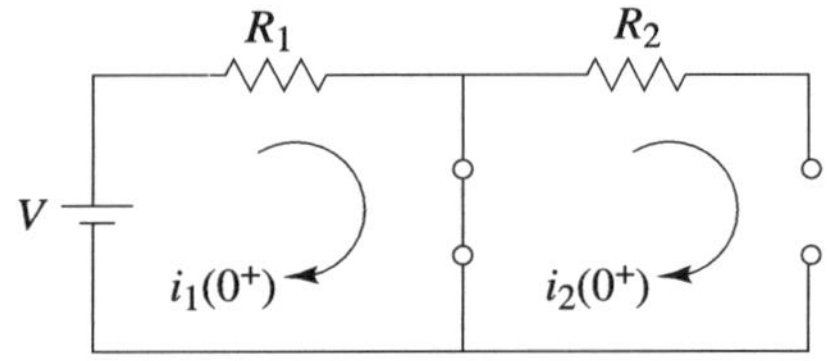

Fig. 8.41

$$i_1(0^+) = \frac{V}{R_1}$$
$$i_2(0^+) = 0$$
$$v_C(0^+) = 0$$

Writing KVL equations for two meshes for $t > 0$,

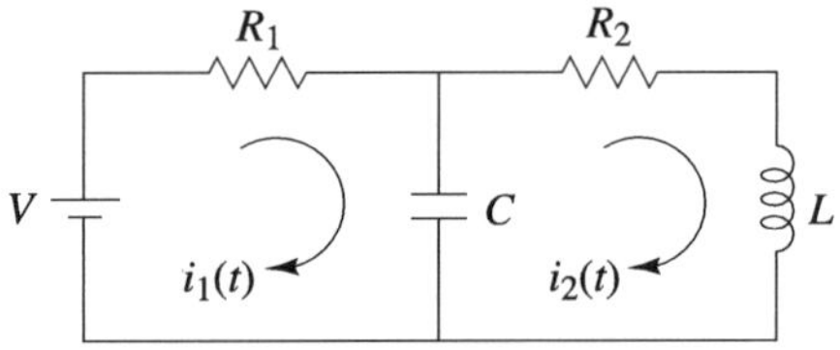

Fig. 8.42

$$R_1 i_1 + \frac{1}{C}\int_0^t (i_1 - i_2)dt = V \qquad \ldots\text{(i)}$$

and $$R_2 i_2 + L\frac{di_2}{dt} + \frac{1}{C}\int_0^t (i_2 - i_1)dt = 0 \qquad \ldots\text{(ii)}$$

From the equation (ii), at $t = 0^+$

$$R_2 i_2(0^+) + L\frac{di_2}{dt}(0^+) + \frac{1}{C}\int_0^{0^+} (i_2 - i_1)\,dt = 0$$

$$\frac{di_2}{dt}(0^+) = 0$$

Differentiating the equation (i), we get

$$R_1\frac{di_1}{dt} + \frac{1}{C}(i_1 - i_2) = 0 \qquad \ldots\text{(iii)}$$

At $t = 0^+$

$$R_1\frac{di_1}{dt}(0^+) + \frac{1}{C}i_1(0^+) - \frac{1}{C}i_2(0^+) = 0$$

$$R_1\frac{di_1}{dt}(0^+) + \frac{1}{C}\frac{V}{R_1} = 0$$

$$\frac{di_1}{dt}(0^+) = -\frac{V}{R_1^2 C}$$

Differentiating the equation (iii), we get

$$R_1 \frac{d^2 i_1}{dt^2} + \frac{1}{C}\frac{di_1}{dt} - \frac{1}{C}\frac{di_2}{dt} = 0$$

At $t = 0^+$

$$R_1 \frac{d^2 i_1}{dt^2}(0^+) + \frac{1}{C}\frac{di_1}{dt}(0^+) - \frac{1}{C}\frac{di_2}{dt}(0^+) = 0$$

$$\frac{d^2 i_1}{dt^2}(0^+) = \frac{V}{R_1^3 C^2}$$

Differentiating the Eq. (ii), we get

$$R_2 \frac{di_2}{dt} + L\frac{d^2 i_2}{dt^2} + \frac{1}{C}(i_2 - i_1) = 0$$

At $t = 0^+$

$$\frac{d^2 i_2}{dt^2}(0^+) = -\frac{R_2}{L}\frac{di_2}{dt}(0^+) - \frac{1}{LC}[i_2(0^+) - i_1(0^+)] = \frac{V}{R_1 LC}$$

Example 8.12 *In the network shown in Fig. 8.43, assuming all initial conditions as zero, find* $i_1, i_2, \frac{di_1}{dt}, \frac{di_2}{dt}, \frac{d^2 i_1}{dt^2}$ *and* $\frac{d^2 i_2}{dt^2}$ *at* $t = 0^+$.

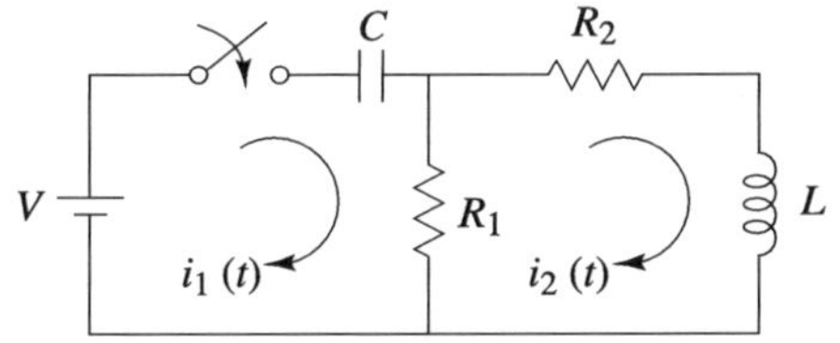

Fig. 8.43

Solution At $t = 0^-$, all initial conditions are zero.

$$v_C(0^-) = 0$$
$$i_1(0^-) = 0$$
$$i_2(0^-) = 0$$

At $t = 0^+$, the capacitor acts as a short circuit and the inductor acts as an open circuit.

$$i_1(0^+) = \frac{V}{R_1}$$
$$i_2(0^+) = 0$$
$$v_C(0^+) = 0$$

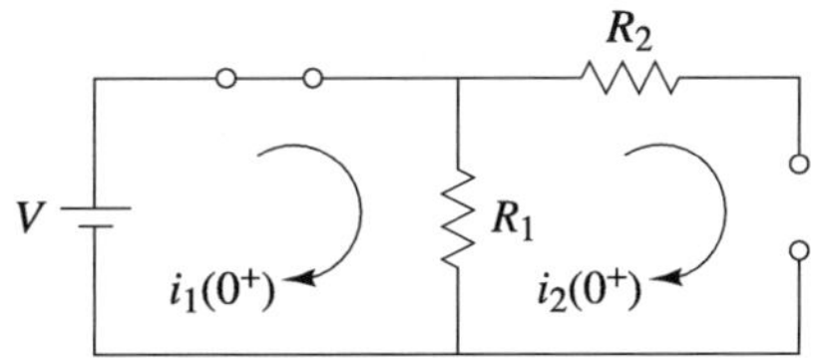

Fig. 8.44

Writing KVL equation for $t > 0$,

$$\frac{1}{C}\int_0^t i_1 \, dt + R_1(i_1 - i_2) = V \quad \ldots\text{(i)}$$

and $$R_1(i_2 - i_1) + R_2 i_2 + L\frac{di_2}{dt} = 0 \quad \ldots\text{(ii)}$$

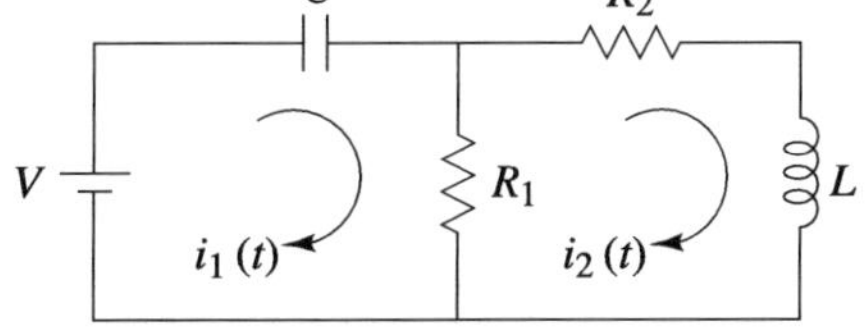

Fig. 8.45

From the Eq. (ii),

$$\frac{di_2}{dt} = \frac{1}{L}[R_1 i_1 - (R_1 + R_2) i_2] \qquad \text{...(iii)}$$

At $t = 0^+$

$$\frac{di_2}{dt}(0^+) = \frac{1}{L}[R_1 i_1 (0^+) - (R_1 + R_2) i_2 (0^+)]$$

$$= \frac{1}{L}\left[R_1 \frac{V}{R_1} - (R_1 + R_2) 0\right] = \frac{V}{L}$$

Differentiating the Eq. (i), we get

$$\frac{i_1}{C} + R_1 \frac{di_1}{dt} - R_1 \frac{di_2}{dt} = 0$$

$$\frac{di_1}{dt} = \frac{di_2}{dt} - \frac{i_1}{R_1 C} \qquad \text{...(iv)}$$

At $t = 0^+$

$$\frac{di_1}{dt}(0^+) = \frac{di_2}{dt}(0^+) - \frac{i_1(0^+)}{R_1 C}$$

$$= \frac{V}{L} - \frac{V}{R_1^2 C}$$

Differentiating the Eq. (iii), we get

$$\frac{d^2 i_2}{dt^2} = \frac{1}{L}\left[R_1 \frac{di_1}{dt} - (R_1 + R_2)\frac{di_2}{dt}\right]$$

At $t = 0^+$

$$\frac{d^2 i_2}{dt^2}(0^+) = -V\left(\frac{1}{R_1 LC} + \frac{R_2}{L^2}\right)$$

Differentiating the Eq. (iv), we get

$$\frac{d^2 i_1}{dt^2} = \frac{d^2 i_2}{dt^2} - \frac{1}{R_1 C}\frac{di_1}{dt}$$

At $t = 0^+$

$$\frac{d^2 i_1}{dt^2}(0^+) = \frac{d^2 i_2}{dt^2}(0^+) - \frac{1}{R_1 C}\frac{di_1}{dt}(0^+)$$

$$= -\frac{V}{R_1 LC} - \frac{VR_2}{L^2} - \frac{1}{R_1 C}\left(\frac{V}{L} - \frac{V}{R_1^2 C}\right)$$

$$= \frac{V}{R_1^3 C^2} - \frac{2V}{R_1 LC} - \frac{VR_2}{L^2}$$

Example 8.13 *In the network shown in Fig. 8.46, a steady state is reached with the switch open. At $t = 0$, the switch is closed. For the element values given, determine the value of $v_a(0^-)$, $v_b(0^-)$, $v_a(0^+)$ and $v_b(0^+)$.*

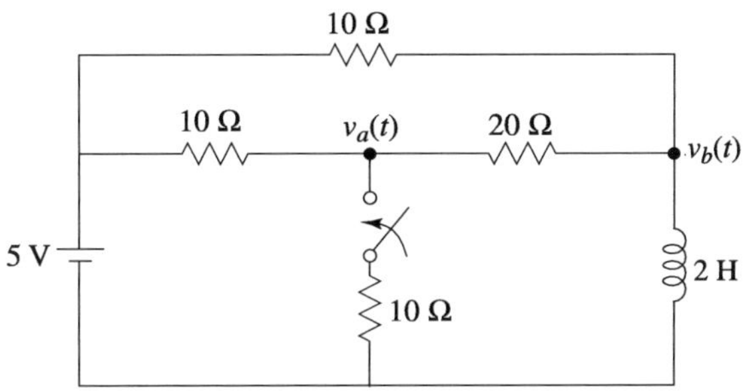

Fig. 8.46

Solution At $t = 0^-$, the network attains steady-state condition. Hence, the inductor acts as a short circuit.

$$i_L(0^-) = \frac{5}{(10 \parallel 30)} = \frac{5}{7.5} = \frac{2}{3}\text{ A}$$

$$v_b(0^-) = 0$$

$$v_a(0^-) = 5 \times \frac{20}{30} = 3.33 \text{ V}$$

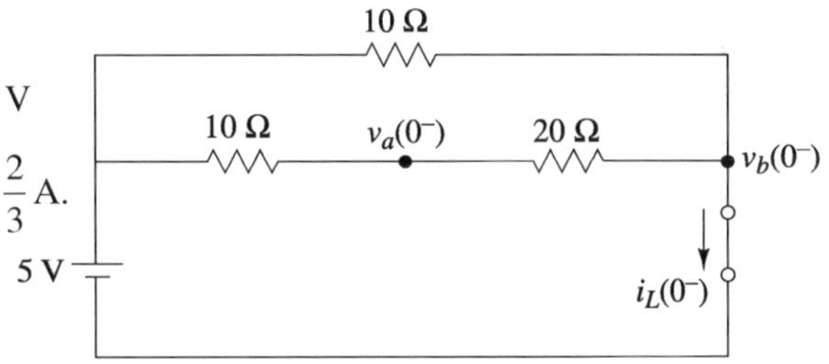

Fig. 8.47

At $t = 0^+$, the inductor acts as a current source of $\frac{2}{3}$ A.

Writing KCL equations at $t = 0^+$,

$$\frac{v_a(0^+) - 5}{10} + \frac{v_a(0^+)}{10} + \frac{v_a(0^+) - v_b(0^+)}{20} = 0$$

and
$$\frac{v_b(0^+) - v_a(0^+)}{20} + \frac{v_b(0^+) - 5}{10} + \frac{2}{3} = 0$$

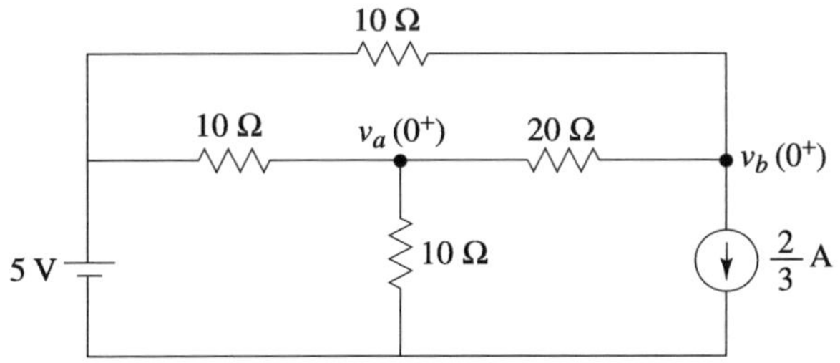

Fig. 8.48

Solving these two equations, we get

$$v_a(0^+) = 1.9 \text{ V}$$
$$v_b(0^+) = -0.477 \text{ V}$$

Example 8.14 *In the accompanying Fig. 8.49 is shown a network in which a steady state is reached with switch open. At $t = 0$, switch is closed. Determine $v_a(0^-)$, $v_a(0^+)$, $v_b(0^-)$ and $v_b(0^+)$.*

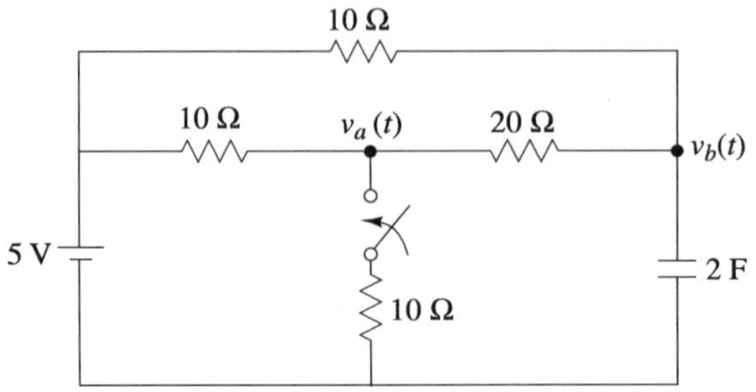

Fig. 8.49

Solution At $t = 0^-$, the network attains steady-state condition. Hence, the capacitor acts as an open circuit.

$$v_a(0^-) = 5 \text{ V}$$
$$v_b(0^-) = 5 \text{ V}$$

At $t = 0^+$, the capacitor acts as a voltage source of 5 V.

$$v_b(0^+) = 5 \text{ V}$$

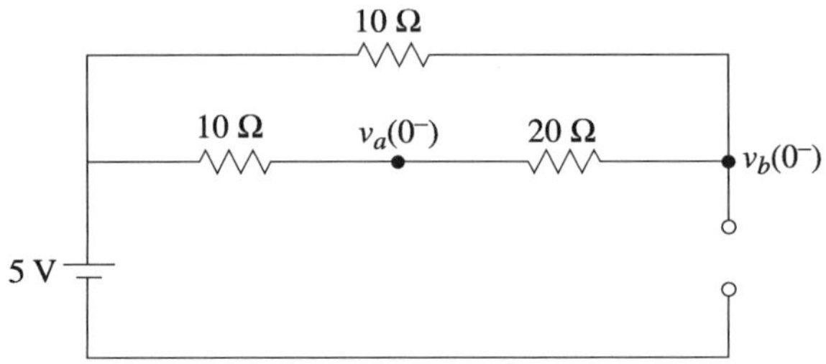

Fig. 8.50

Writing KCL equation at $t = 0^+$,

$$\frac{v_a(0^+) - 5}{10} + \frac{v_a(0^+)}{10} + \frac{v_a(0^+) - 5}{20} = 0$$

$$0.25\, v_a(0^+) = 0.75$$

$$v_a(0^+) = 3 \text{ V}$$

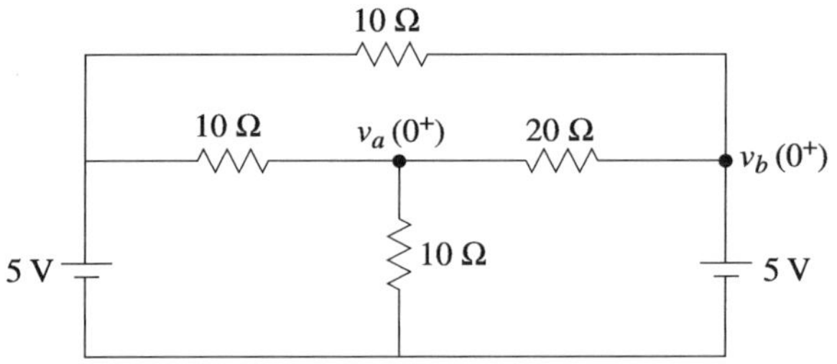

Fig. 8.51

Example 8.15 *The network shown in the Fig. 8.52 has two independent node pairs. If the switch is opened at t = 0. Find v_1, v_2, $\frac{dv_1}{dt}$ and $\frac{dv_2}{dt}$ at $t = 0^+$.*

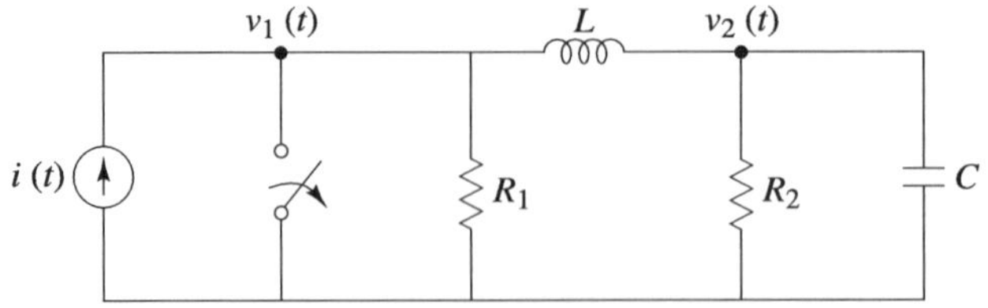

Fig. 8.52

Solution At $t = 0^-$, no current flows through the inductor and there is no voltage across the capacitor.

$$i_L(0^-) = 0$$
$$v_C(0^-) = v_2(0^-) = 0$$

At $t = 0^+$, the inductor acts as an open circuit and the capacitor acts as a short circuit.

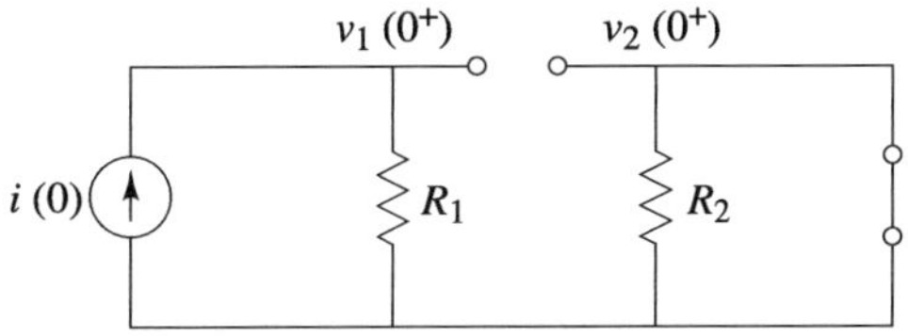

Fig. 8.53

$$i_L(0^+) = 0$$
$$v_1(0^+) = R_1\, i(0^+)$$
$$v_2(0^+) = 0$$

Writing KCL equation at Node 1 for $t > 0$,

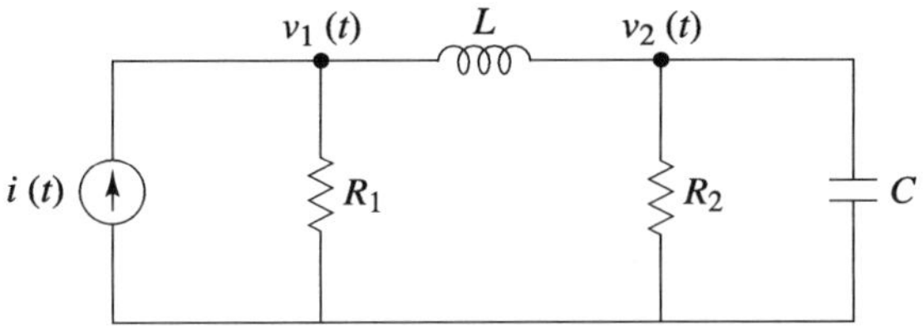

Fig. 8.54

$$i(t) = \frac{v_1}{R_1} + \frac{1}{L}\int_0^t (v_1 - v_2)dt \quad \ldots\text{(i)}$$

Differentiating the Eq. (i), we get

$$\frac{di}{dt} = \frac{1}{R_1}\frac{dv_1}{dt} + \frac{1}{L}(v_1 - v_2)$$

At $t = 0^+$

$$\frac{dv_1}{dt}(0^+) = R_1\left[\frac{di}{dt}(0^+) - \frac{1}{L}R_1 i(0^+)\right]$$

Writing KCL equation at Node 2 for $t > 0$

$$\frac{1}{L}\int_0^t (v_2 - v_1)\,dt + \frac{v_2}{R_2} + C\frac{dv_2}{dt} = 0 \quad \ldots\text{(ii)}$$

At $t = 0^+$,

$$0 + \frac{v_2(0^+)}{R_2} + C\frac{dv_2}{dt}(0^+) = 0$$

$$\frac{dv_2}{dt}(0^+) = 0$$

Example 8.16 *In the network shown in Fig. 8.55, the switch is closed at $t = 0$, with zero capacitor voltage and zero inductor current. Solve for v_1, v_2, , $\frac{dv_1}{dt}$, $\frac{dv_2}{dt}$ and $\frac{d^2v_2}{dt^2}$ at $t = 0^+$.*

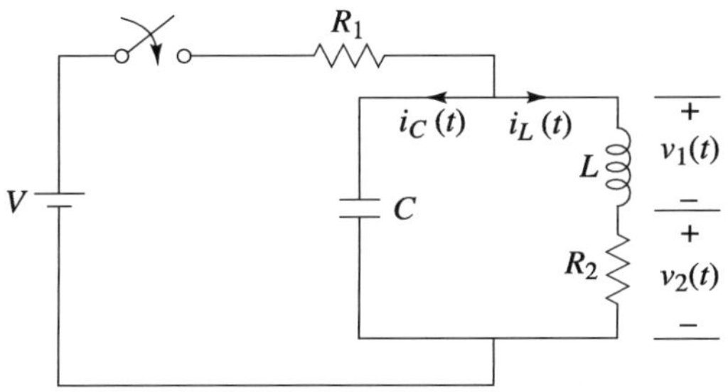

Fig. 8.55

Solution At $t = 0^-$, no current flows through the inductor and there is no voltage across the capacitor.

$$v_C(0^-) = 0$$
$$v_1(0^-) = 0$$
$$v_2(0^-) = 0$$
$$i_L(0^-) = 0$$
$$i_C(0^-) = 0$$

At $t = 0^+$, the inductor acts as an open circuit and the capacitor acts as a short circuit.

$$v_C(0^+) = 0$$
$$v_1(0^+) = 0$$
$$v_2(0^+) = 0$$
$$i_L(0^+) = 0$$
$$i_C(0^+) = \frac{V}{R_1}$$

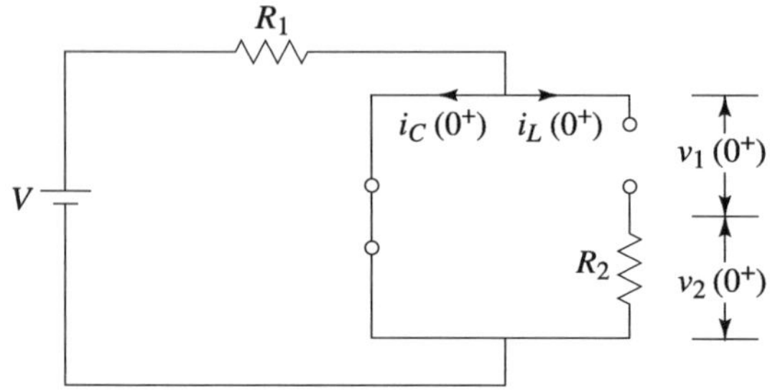

Fig. 8.56

Writing KVL equation for $t > 0$,

$$v_C(t) = v_1(t) + v_2(t) \qquad \text{...(i)}$$

Differentiating the Eq. (i), we get

$$\frac{dv_C}{dt} = \frac{dv_1}{dt} + \frac{dv_2}{dt} \qquad \text{...(ii)}$$

Now,

$$v_C = \frac{1}{C}\int_0^t i_C \, dt \qquad \text{...(iii)}$$

$$\frac{dv_C}{dt} = \frac{i_C}{C}$$

At $t = 0^+$

$$\frac{dv_C}{dt}(0^+) = \frac{i_C(0^+)}{C} = \frac{V}{R_1 C} \text{ V/s}$$

Also

$$v_1 = L\frac{di_L}{dt} \qquad \text{...(iv)}$$

$$\frac{di_L}{dt} = \frac{v_1}{L} \qquad \text{...(v)}$$

At $t = 0^+$

$$\frac{di_L}{dt}(0^+) = \frac{v_1(0^+)}{L} = 0$$

Also,

$$v_2 = R_2 \, i_L \qquad \text{... (vi)}$$

$$\frac{dv_2}{dt} = R_2\frac{di_L}{dt} \qquad \text{... (vii)}$$

At $t = 0^+$

$$\frac{dv_2}{dt}(0^+) = R_2\frac{di_L}{dt}(0^+) = 0$$

$$\frac{dv_C}{dt}(0^+) = \frac{dv_1}{dt}(0^+) + \frac{dv_2}{dt}(0^+)$$

$$\frac{dv_1}{dt}(0^+) = \frac{V}{R_1 C} \text{ V/s}$$

Differentiating the Eq. (vii), we get

$$\frac{d^2 v_2}{dt^2} = R_2 \frac{d^2 i_L}{dt^2}$$

At $t = 0^+$

$$\frac{d^2 v_2}{dt^2}(0^+) = R_2 \frac{d^2 i_L}{dt}(0^+)$$

Differentiating the Eq. (v), we get

$$\frac{d^2 i_L}{dt^2} = \frac{1}{L}\frac{dv_1}{dt}$$

At $t = 0^+$

$$\frac{d^2 i_L}{dt^2}(0^+) = \frac{1}{L}\frac{dv_1}{dt}(0^+) = \frac{1}{L}\frac{V}{R_1 C}$$

$$\frac{d^2 v_2}{dt^2}(0^+) = \frac{R_2 V}{R_1 L C} \text{ V/s}^2$$

Example 8.17 *In the network shown in Fig. 8.57, a steady state is reached with switch open. At t = 0, switch is closed. Find the three loop currents at t = 0⁺.*

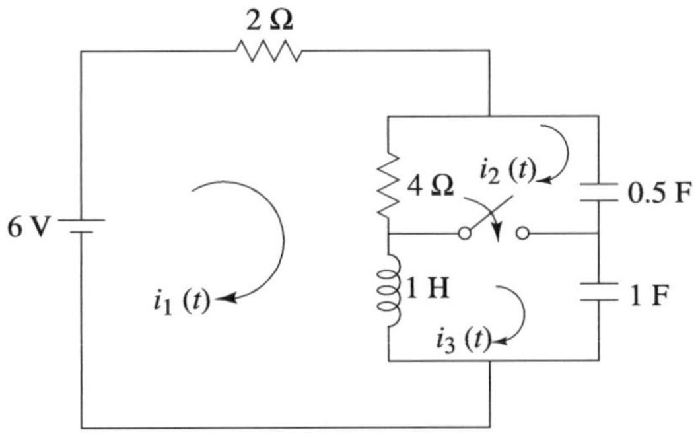

Fig. 8.57

Solution At $t = 0^-$, the network attains steady-state condition. Hence, the inductor act as a short circuit and the capacitors act as open circuits.

$$i_{4\Omega}(0^-) = i_1(0^-) = \frac{6}{6} = 1 \text{ A}$$

$$i_2(0^-) = 0$$

$$i_3(0^-) = 0$$

$$v_1(0^-) + v_2(0^-) = 6 \times \frac{4}{6} = 4 \text{ V}$$

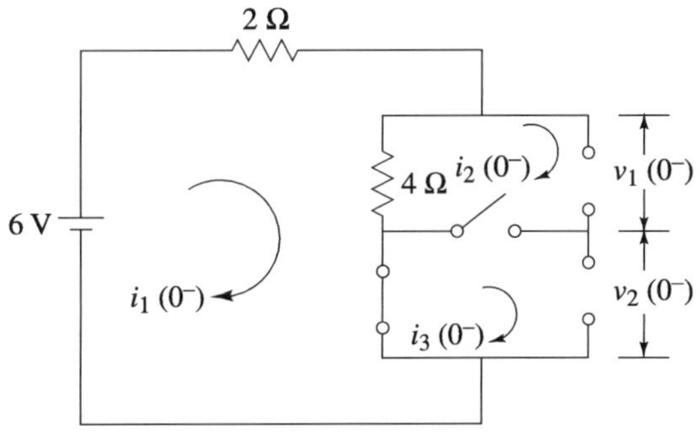

Fig. 8.58

Since the charges on capacitors are equal when connected in series,

$$Q_1 = Q_2$$
$$C_1 v_1 = C_2 v_2$$
$$\frac{v_1(0^-)}{v_2(0^-)} = \frac{C_2}{C_1} = \frac{1}{0.5} = 2$$
$$v_1(0^-) = \frac{8}{3}\text{ V}$$

and
$$v_2(0^-) = \frac{4}{3}\text{ V}$$

At $t = 0^+$, the inductor is replaced by a current source of 1 A and the capacitors are replaced by a voltage source of $\frac{8}{3}$ V and $\frac{4}{3}$ V respectively.

$$v_1(0^+) = \frac{8}{3}\text{ V}$$
$$v_2(0^+) = \frac{4}{3}\text{ V}$$

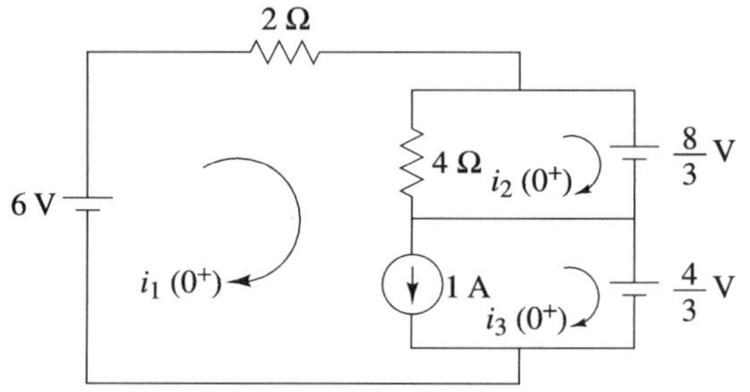

Fig. 8.59

At $t = 0^+$

$$6 - 2i_1(0^+) - \frac{8}{3} - \frac{4}{3} = 0$$
$$i_1(0^+) = 1\text{ A}$$

Now,
$$i_1(0^+) - i_3(0^+) = 1$$
$$i_3(0^+) = 0$$

Writing the KVL equation for Mesh 2,

$$-4\,[i_2(0^+) - i_1(0^+)] - \frac{8}{3} = 0$$
$$-4i_2(0^+) + 4 - \frac{8}{3} = 0$$
$$4\,i_2(0^+) = \frac{4}{3}$$
$$i_2(0^+) = \frac{1}{3}\text{ A}$$

Example 8.18 *In the network shown in Fig. 8.60, a steady state is reached with the switch open. At t = 0, the switch is closed. Determine* $v_C(0^-)$, $i_1(0^+)$, $i_2(0^+)$, $\frac{di_1}{dt}(0^+)$ *and* $\frac{di_2}{dt}(0^+)$.

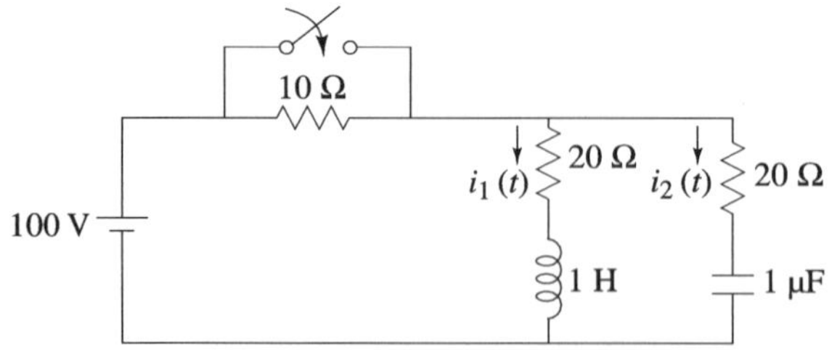

Fig. 8.60

Solution At $t = 0^-$, the network is in steady-state. Hence, the inductor acts as a short circuit and the capacitor acts as an open circuit.

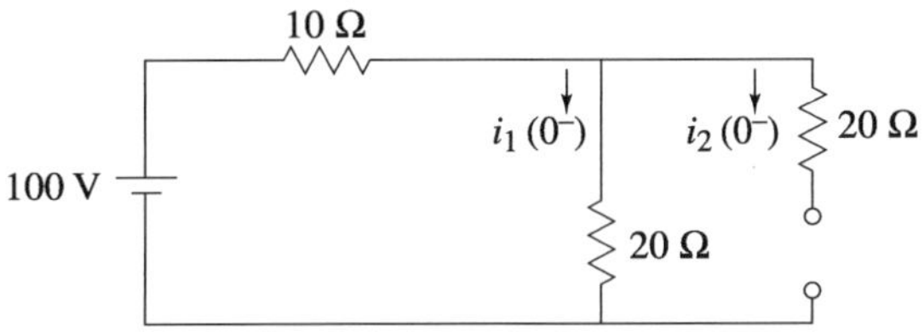

Fig. 8.61

$$v_C(0^-) = 100 \times \frac{20}{20+10} = 66.67 \text{ V}$$

$$i_1(0^-) = \frac{66.67}{20} = 3.33 \text{ A}$$

$$i_2(0^-) = 0$$

At $t = 0^+$, the inductor acts as a current source of 3.33 A and the capacitor acts as a voltage source of 66.67 V.

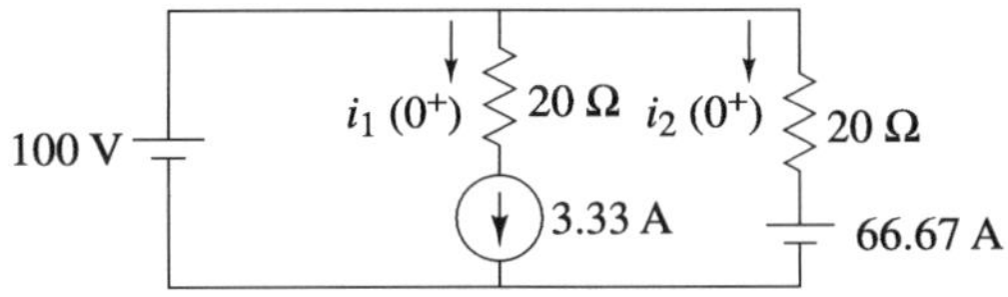

Fig. 8.62

$$v_C(0^+) = 66.67 \text{ V}$$

$$i_1(0^+) = 3.33 \text{ A}$$

$$i_2(0^+) = \frac{100 - 66.67}{20} = 1.67 \text{ A}$$

For $t > 0$,

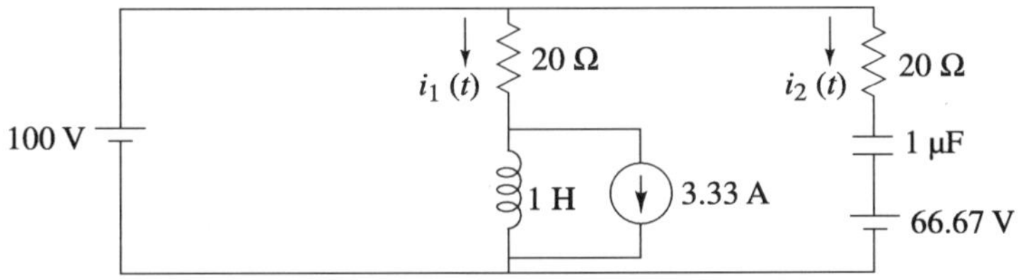

Fig. 8.63

Writing KVL equations for $t > 0$,

$$1\frac{di_1}{dt} + 20\, i_1 = 100 \qquad \text{...(i)}$$

and

$$20\, i_2 + 10^6 \int_0^t i_2\, dt = 100 - 66.67 = 33.33 \qquad \text{...(ii)}$$

At $t = 0^+$

$$\frac{di_1}{dt}(0^+) = 100 - 20\, i_1(0^+)$$

$$= 100 - 20\,(3.33) = 33.3 \text{ A/s}$$

Differentiating the Eq. (ii), we get

$$20\frac{di_2}{dt} + 10^6\, i_2 = 0$$

At $t = 0^+$

$$20\frac{di_2}{dt}(0^+) = -10^6\, i_2(0^+)$$

$$= -\frac{10^6}{20} \times 1.67 = -83500 \text{ A/s}^2$$

8.5 RESISTOR–INDUCTOR CIRCUIT

Consider a series *R-L* circuit as shown in Fig. 8.64. The switch is closed at time $t = 0$. The inductor in the circuit is initially un-energised.

Applying Kirchhoff's voltage law to the circuit for $t > 0$,

$$V = Ri + L\frac{di}{dt}$$

This is a linear differential equation of first order. It can be solved if the variables can be separated.

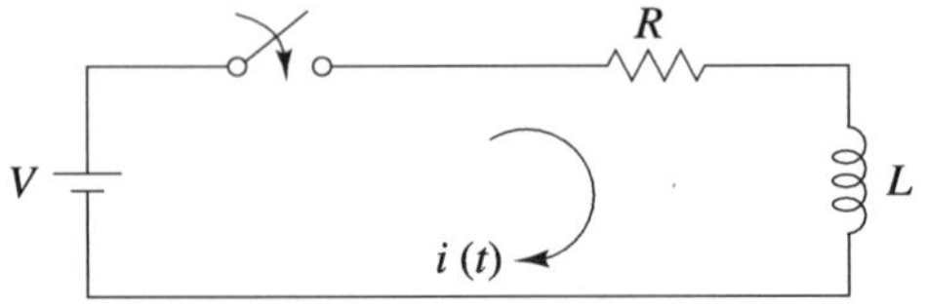

Fig. 8.64

$$(V - Ri)\,dt = L\,di$$

$$\frac{L\,di}{V - Ri} = dt$$

Integrating both the sides, we get

$$-\frac{L}{R} l_n (V - Ri) = t + K$$

where l_n designates that the logarithm is of base e and K is an arbitrary constant. K can be evaluated from the initial condition. In the circuit shown, the switch is closed at $t = 0$, i.e., just before closing the switch, the current in the inductor is zero. Since the inductor does not allow sudden change in current, at $t = 0^+$, just after the switch is closed, the current remains zero.

Setting $i = 0$ at $t = 0$, we get

$$-\frac{L}{R} l_n V = K$$

$$-\frac{L}{R} l_n (V - Ri) = t - \frac{L}{R} l_n V$$

$$-\frac{L}{R}[l_n (V - Ri) - l_n V] = t$$

$$\frac{V - Ri}{V} = e^{-\frac{R}{L}t}$$

$$i = \frac{V}{R} - \frac{V}{R} e^{-\frac{R}{L}t} \qquad \text{for } t > 0$$

The complete response is composed of two parts, the steady-state response or forced response or zero state response $\frac{V}{R}$ and transient response or natural response or zero input response $\frac{V}{R} e^{-\frac{R}{L}t}$.

The natural response is a characteristic of the circuit. Its form may be found by considering the source-free circuit. The forced response has the characteristics of forcing function, i.e., applied voltage. Thus, when the switch is closed, response reaches the steady-state value after some time interval as shown in Fig. 8.65.

Here, the transient period is defined as the time taken for the current to reach its final or steady state value from its initial value.

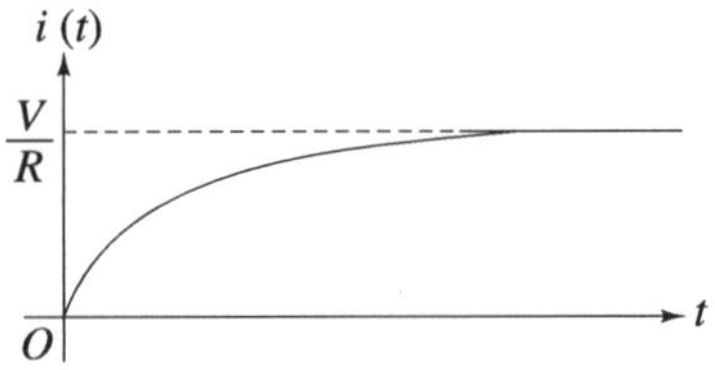

Fig. 8.65

The term $\frac{L}{R}$ is called time constant and is denoted by T.

$$T = \frac{L}{R}$$

At one time constant, the current reaches 63.2 per cent of its final value $\frac{V}{R}$.

$$i(T) = \frac{V}{R} - \frac{V}{R} e^{-\frac{1}{T}t} = \frac{V}{R} - \frac{V}{R} e^{-1} = \frac{V}{R} - 0.368\frac{V}{R} = 0.632\frac{V}{R}$$

Similarly,

$$i\,(2T) = \frac{V}{R} - \frac{V}{R}e^{-2} = \frac{V}{R} - 0.135\frac{V}{R} = 0.865\frac{V}{R}$$

$$i\,(3T) = \frac{V}{R} - \frac{V}{R}e^{-3} = \frac{V}{R} - 0.0498\frac{V}{R} = 0.950\frac{V}{R}$$

$$i\,(5T) = \frac{V}{R} - \frac{V}{R}e^{-5} = \frac{V}{R} - 0.0067\frac{V}{R} = 0.993\frac{V}{R}$$

After 5 time constants, the current reaches 99.33 per cent of its final value.

The voltage across resistor is

$$v_R = Ri = R \times \frac{V}{R}\left(1 - e^{-\frac{R}{L}t}\right)$$

$$= V\left(1 - e^{-\frac{R}{L}t}\right) \qquad \text{for } t > 0$$

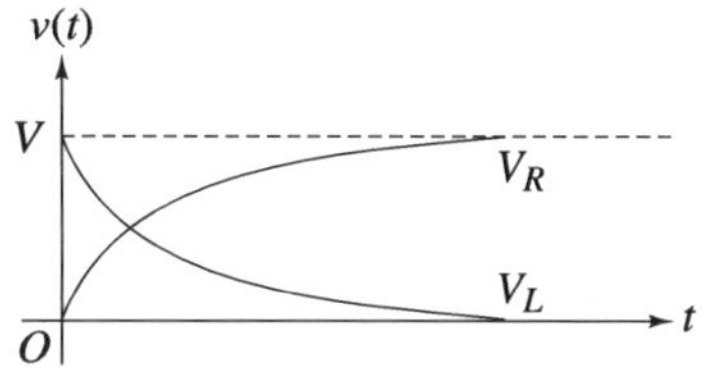

Fig. 8.66

Similarly, voltage across inductor is

$$v_L = \frac{di}{dt} = L\frac{V}{R}\frac{d}{dt}\left(1 - e^{-\frac{R}{L}t}\right)$$

$$= V\,e^{-\frac{R}{L}t} \qquad \text{for } t > 0$$

8.6 RESISTOR–CAPACITOR CIRCUIT

Consider a series R-C circuit as shown in Fig. 8.67. The switch is closed at time $t = 0$. The capacitor is initially uncharged.

Applying Kirchhoff's voltage law to the circuit for $t > 0$,

$$V = Ri + \frac{1}{C}\int_0^t i\ dt$$

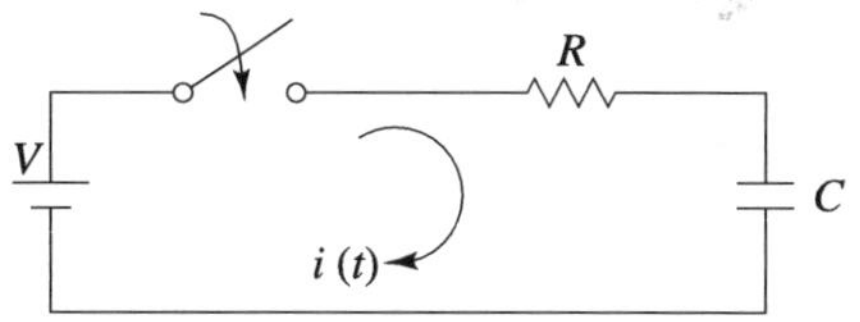

Fig. 8.67

Differentiating the above equation, we get

$$0 = R\frac{di}{dt} + \frac{i}{C}$$

$$\frac{di}{dt} + \frac{1}{RC}\,i = 0$$

This is a linear differential equation of first order. The variables may be separated to solve the equation.

$$\frac{di}{i} = -\frac{dt}{RC}$$

Integrating both the sides, we get

$$l_n\, i = -\frac{1}{RC}t + K$$

The constant K can be evaluated from initial condition. In the circuit shown, the switch is closed at $t = 0$. Since the capacitor never allows sudden change in voltage, it will act as short circuit at $t = 0^+$. Therefore current in the circuit at $t = 0^+$ is $\frac{V}{R}$.

Setting $\qquad i = \frac{V}{R}$ at $t = 0$, we get

$$l_n \frac{V}{R} = K$$

$$l_n i = -\frac{1}{RC} t + l_n \frac{V}{R}$$

$$l_n i - l_n \frac{V}{R} = -\frac{1}{RC} t$$

$$l_n \frac{i}{\left(\frac{V}{R}\right)} = -\frac{1}{RC} t$$

$$\frac{i}{V/R} = e^{-\frac{1}{RC} t}$$

$$i = \frac{V}{R} e^{-\frac{1}{RC} t} \qquad \text{for } t > 0$$

When the switch is closed, the response decays with time as shown in Fig. 8.68.

The term RC is called time constant and is denoted by T.

$$T = RC$$

After 5 time constants, the current drops to 99 per cent of its initial value.

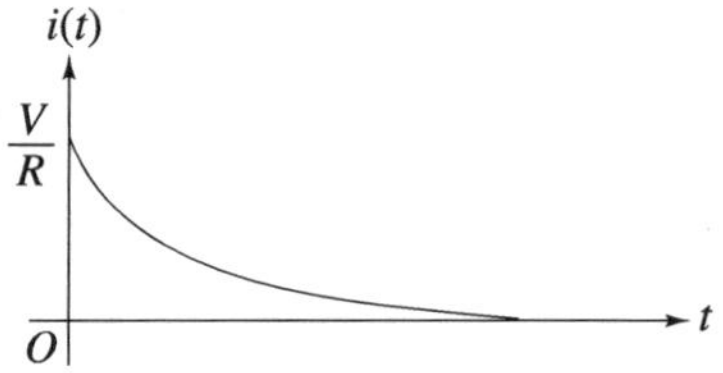

Fig. 8.68

The voltage across the resistor is

$$v_R = Ri = R \frac{V}{R} e^{-\frac{1}{RC} t}$$

$$= Ve^{-\frac{1}{RC} t} \qquad \text{for } t > 0$$

Similarly, the voltage across the capacitor is

$$v_C = \frac{1}{C} \int_0^t i \, dt$$

$$= \frac{1}{C} \int_0^t \frac{V}{R} e^{-\frac{1}{RC} t}$$

$$= -Ve^{-\frac{1}{RC} t} + K$$

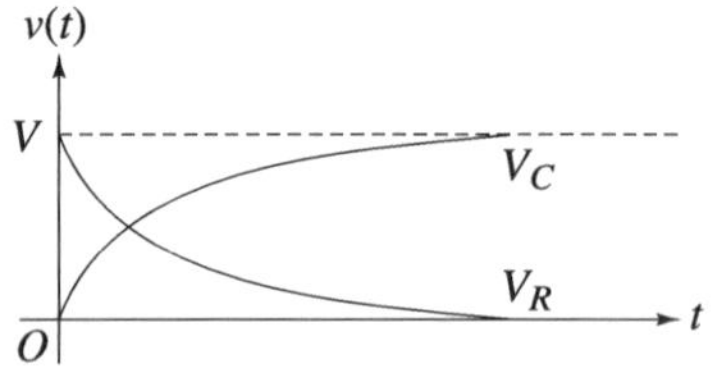

Fig. 8.69

At $t = 0$, $v_C(0) = 0$

$$K = V$$

Hence, $\qquad v_C = V\left(1 - e^{-\frac{1}{RC} t}\right) \qquad \text{for } t > 0$

8.7 RESISTOR–INDUCTOR–CAPACITOR CIRCUIT

Consider a series *R-L-C* circuit as shown in Fig. 8.70. The switch is closed at time $t = 0$. The capacitor and inductor are initially uncharged.

Applying Kirchhoff's voltage law to the circuit for $t > 0$,

$$V = Ri + L\frac{di}{dt} + \frac{1}{C}\int_0^t i\, dt$$

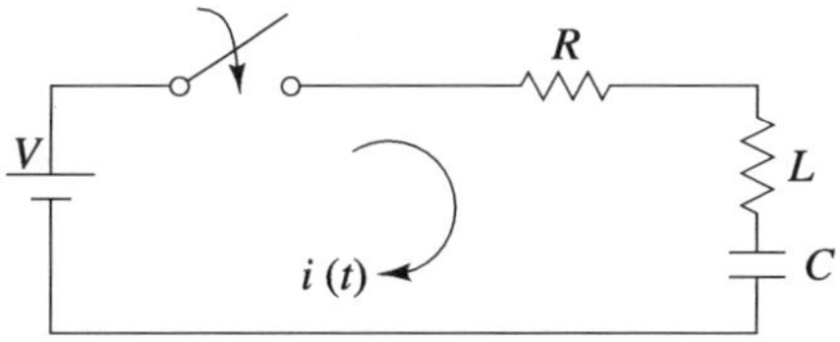

Fig. 8.70

Differentiating the above equation, we get

$$0 = R\frac{di}{dt} + L\frac{d^2i}{dt^2} + \frac{1}{C}i$$

$$\frac{d^2i}{dt^2} + \frac{R}{L}\frac{di}{dt} + \frac{1}{LC}i = 0$$

This is a second-order differential equation. The auxiliary equation or characteristic equation will be given by,

$$s^2 + \left(\frac{R}{L}\right)s + \left(\frac{1}{LC}\right) = 0$$

Let s_1 and s_2 be the roots of the equation,

$$s_1 = -\frac{R}{2L} + \sqrt{\left(\frac{R}{2L}\right)^2 - \frac{1}{LC}}$$

$$= -\alpha + \sqrt{\alpha^2 - \omega_{o^2}}$$

$$= -\alpha + \beta$$

$$s_2 = -\frac{R}{2L} - \sqrt{\left(\frac{R}{2L}\right)^2 - \frac{1}{LC}}$$

$$= -\alpha - \sqrt{\alpha^2 - \omega_o^2} = -\alpha - \beta$$

where $$\alpha = \frac{R}{2L}$$

$$\omega_o = \frac{1}{\sqrt{LC}}$$

and $$\beta = \sqrt{\alpha^2 - \omega_o^2}$$

The solution of the above second-order differential equation will be given by

$$i(t) = A_1 e^{s_1 t} + A_2 e^{s_2 t}$$

where A_1 and A_2 are constants to be determined and s_1 and s_2 are the roots of the equation.

Now depending upon the values of α and ω_o, we have 3 cases of the response.

Case I *When $\alpha > \omega_o$*

i.e., $$\frac{R}{2L} > \frac{1}{\sqrt{LC}}$$

The roots are real and unequal and it gives an overdamped response. In this case, the solution is given by

$$i = e^{-\alpha t}(A_1 e^{\beta t} + A_2 e^{-\beta t})$$

or $$i = A_1 e^{s_1 t} + A_2 e^{s_2 t} \qquad \text{for } t > 0$$

The current curve for an overdamped case is shown in Fig. 8.71.

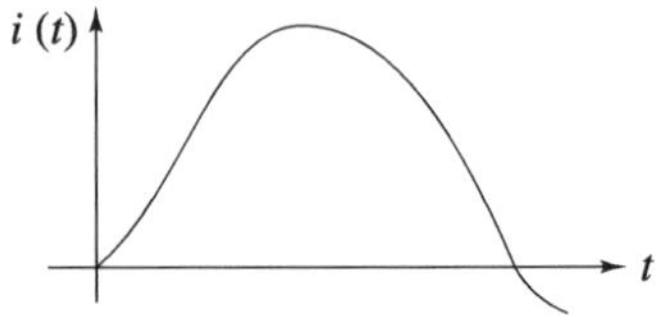

Fig. 8.71

Case II *When $\alpha = \omega_o$*

i.e., $$\frac{R}{2L} = \frac{1}{\sqrt{LC}}$$

The roots are real and equal and it gives a critically damped response. In this case the solution is given by

$$i = e^{-\alpha t}(A_1 + A_2 t) \qquad \text{for } t > 0$$

The current curve for critically damped case is shown in Fig. 8.72.

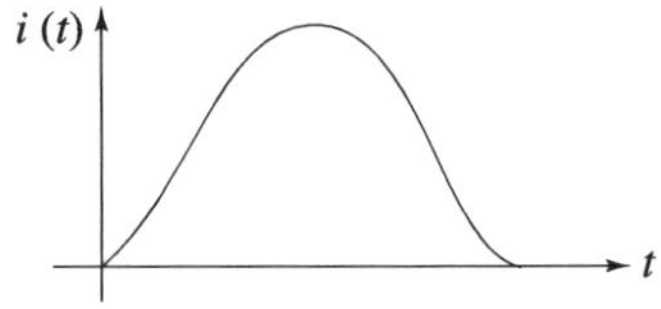

Fig. 8.72

Case III *When $\alpha < \omega_o$*

i.e., $$\frac{R}{2L} < \frac{1}{\sqrt{LC}}$$

The roots are complex conjugate and it gives an underdamped response. In this case, the solution is given by

$$i(t) = A_1 e^{s_1 t} + A_2 e^{s_2 t}$$

where $$s_{1,2} = -\alpha \pm \sqrt{\alpha^2 - \omega_o^2}$$

Let $$\sqrt{\alpha^2 - \omega_o^2} = \sqrt{-1}\sqrt{\omega_o^2 - \alpha^2}$$

$$= j\,\omega_d$$

where $$j = \sqrt{-1}$$

and $$\omega_d = \sqrt{\omega_{o^2} - \alpha^2}$$

Hence $$i(t) = e^{-\alpha t}(A_1 e^{j\omega_d t} + A_2 e^{-j\omega_d t})$$

$$= e^{-\alpha t}\left\{(A_1 + A_2)\left[\frac{e^{j\omega_d t} + e^{-j\omega_d t}}{2}\right] + j(A_1 - A_2)\left[\frac{e^{j\omega_d t} - e^{-j\omega_d t}}{j2}\right]\right\}$$

$$= e^{-\alpha t}[(A_1 + A_2)\cos \omega_d t + j(A_1 - A_2)\sin \omega_d t]$$

$$= e^{-\alpha t}(B_1 \cos \omega_d t + B_2 \sin \omega_d t) \qquad \text{for } t > 0$$

The current curve for an underdamped case is shown in Fig. 8.73.

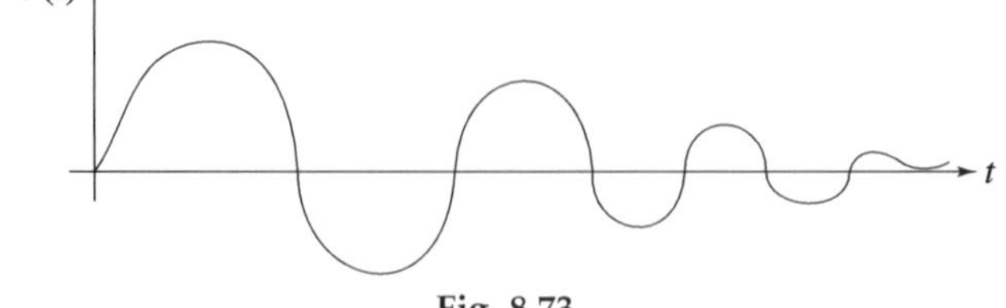

Fig. 8.73

Note:

(1) Consider a homogeneous equation,

$$\frac{di}{dt} + Pi = 0 \qquad \text{where } P \text{ is a constant.}$$

The solution of this equation will be,

$$i(t) = K e^{-Pt}$$

The value of K is obtained by putting $t = 0$ in the equation for $i(t)$.

(2) Consider a non-homogeneous equation,

$$\frac{di}{dt} + Pi = Q$$

where P is a constant and Q may be a function of the independent variable t or a constant. The solution of this equation will be

$$i(t) = e^{-Pt} \int Q\, e^{Pt}\, dt + K e^{-Pt} = \frac{Q}{P} + K e^{-Pt} = i_{ss} + i_t$$

Here, i_{ss} is the steady-state part (at $t = \infty$) and i_t is the transient part of the solution. The value of K can be obtained as follows:

$$i(t) = i(\infty) + K e^{-Pt}$$

At $t = 0$,

$$i(0) = i(\infty) + K \qquad \therefore \quad K = i(0) - i(\infty)$$

Hence $\quad i(t) = i(\infty) + [i(0) - i(\infty)]\, e^{-Pt}$

Example 8.19 *In the network of Fig. 8.74, the switch is initially at the position 1. On the steady state having reached, the switch is changed to the position 2. Find current i (t).*

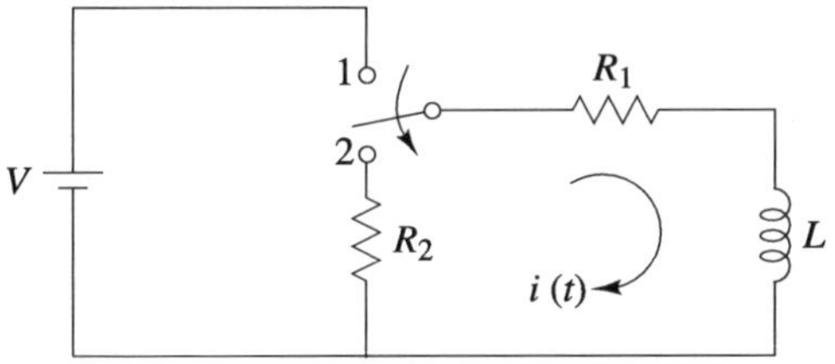

Fig. 8.74

Solution At $t = 0^-$, the network has attained steady-state condition. Hence, the inductor acts as a short circuit.

$$i(0^-) = \frac{V}{R_1}$$

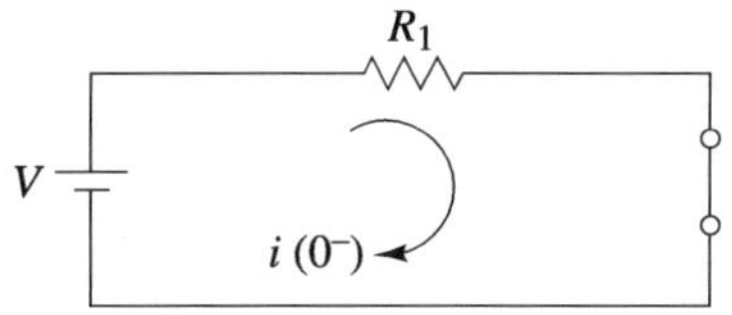

Fig. 8.75

Since the inductor does not allow sudden change in current,

$$i(0^+) = \frac{V}{R_1}$$

Writing KVL equation for $t > 0$,

$$(R_1 + R_2)\, i + L\frac{di}{dt} = 0$$

$$\frac{di}{dt} + \frac{(R_1 + R_2)}{L}\, i = 0$$

The solution of this differential equation is given by

$$i(t) = Ke^{-\left(\frac{R_1+R_2}{L}\right)t}$$

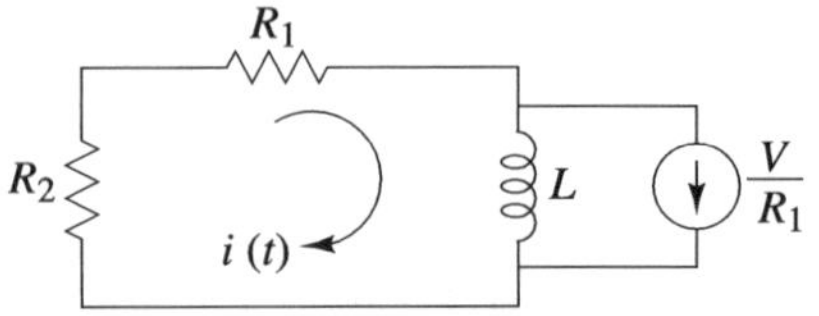

Fig. 8.76

At $t = 0$, $i(0) = \dfrac{V}{R_1}$

$$\frac{V}{R_1} = K e^0 = K$$

$$i(t) = \frac{V}{R_1} e^{-\left(\frac{R_1+R_2}{L}\right)t} \qquad \text{for } t > 0$$

Example 8.20 *In the network shown in Fig. 8.77, the switch is closed at t = 0, a steady state having previously been attained. Find the current i (t).*

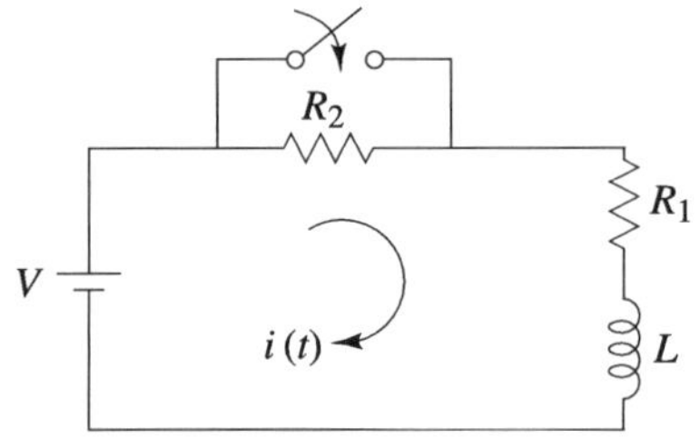

Fig. 8.77

Solution At $t = 0^-$, the network has attained steady-state condition. Hence, the inductor acts as a short circuit.

$$i(0^-) = \frac{V}{R_1 + R_2}$$

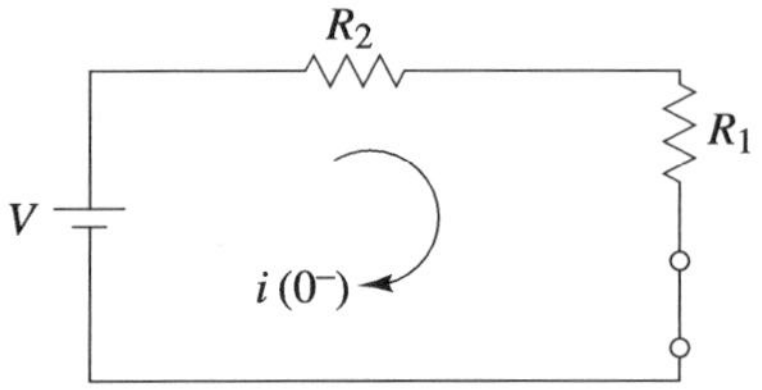

Fig. 8.78

Since the current through the inductor cannot change instantaneously,

$$i(0^+) = \frac{V}{R_1 + R_2}$$

Writing KVL equation for $t > 0$,

$$V = L\frac{di}{dt} + R_1 i$$

$$\frac{di}{dt} + \frac{R_1}{L} i = \frac{V}{L}$$

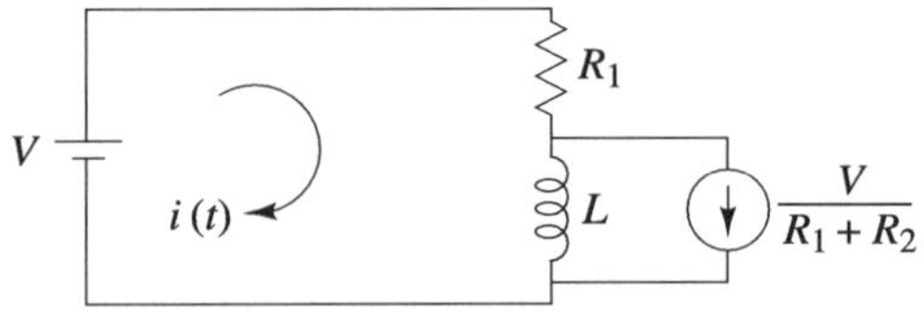

Fig. 8.79

The solution of this differential equation is given by

$$i(t) = \frac{Q}{P} + Ke^{-Pt}$$

$$= \frac{V}{R_1} + Ke^{-\frac{R_1}{L}t}$$

At $t = 0$, $i(0) = \dfrac{V}{R_1 + R_2}$

$$\frac{V}{R_1 + R_2} = \frac{V}{R_1} + K$$

$$K = -\frac{VR_2}{R_1(R_1 + R_2)}$$

$$i(t) = \frac{V}{R_1} - \frac{VR_2}{R_1(R_1 + R_2)} e^{-\frac{R_1}{L}t}$$

$$= \frac{V}{R_1}\left(1 - \frac{R_2}{R_1 + R_2} e^{-\frac{R_1}{L}t}\right) \qquad \text{for } t > 0$$

Example 8.21 *The network of Fig. 8.80 is under steady state with switch at the position 1. At t = 0, switch is moved to position 2. Find i (t).*

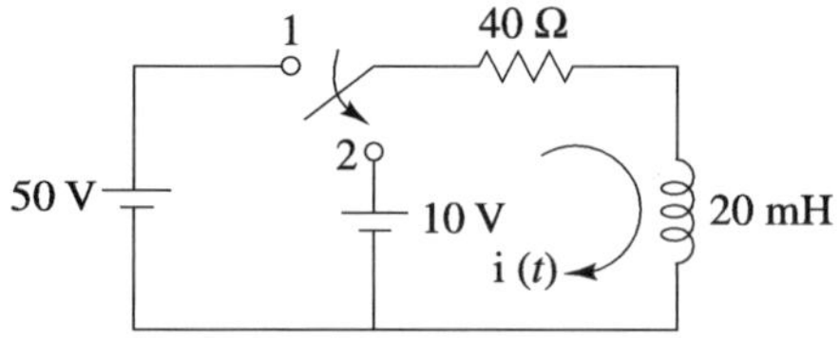

Fig. 8.80

Solution At $t = 0^-$, the network has attained steady-state condition. Hence, the inductor acts as a short circuit.

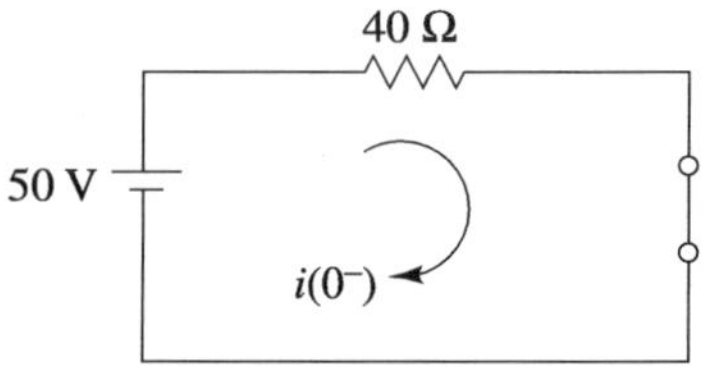

Fig. 8.81

$$i(0^-) = \frac{50}{40} = 1.25 \text{ A}$$

Since current through the inductor cannot change instantaneously,

$$i(0^+) = 1.25 \text{ A}$$

Writing KVL equation for $t > 0$,

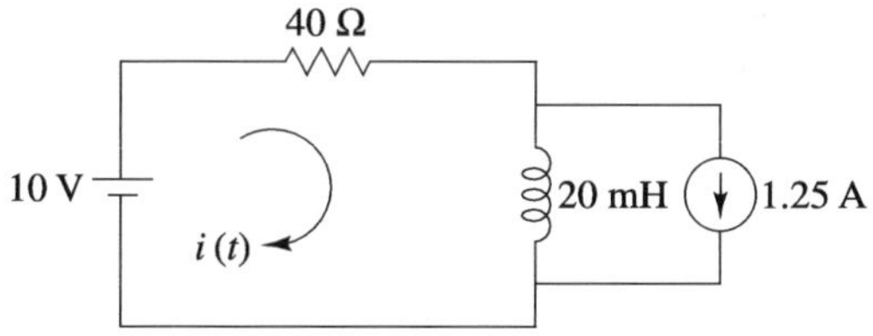

Fig. 8.82

$$20 \times 10^{-3} \frac{di}{dt} + 40\, i = 10$$

$$\frac{di}{dt} + 2 \times 10^3\, i = 500$$

Solution of this differential equation is given by

$$i(t) = \frac{Q}{P} + Ke^{-Pt}$$

$$= \frac{500}{2000} + K\, e^{-2000\, t}$$

$$= 0.25 + K\, e^{-2000\, t}$$

At $t = 0$, $i(0) = 1.25$

$$1.25 = 0.25 + K$$

$$K = 1$$

$$i(t) = 0.25 + e^{-2000\, t} \qquad \text{for } t > 0$$

Example 8.22 *The switch in the circuit of Fig. 8.83 is moved from the position 1 to 2 at t = 0. Find $v_C(t)$.*

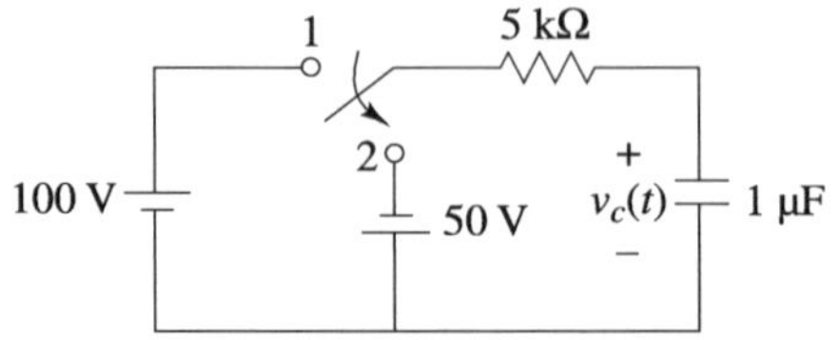

Fig. 8.83

Solution At $t = 0^-$, the network has attained steady-state condition. Hence, the capacitor acts as an open circuit.

$$v_C(0^-) = 100 \text{ V}$$

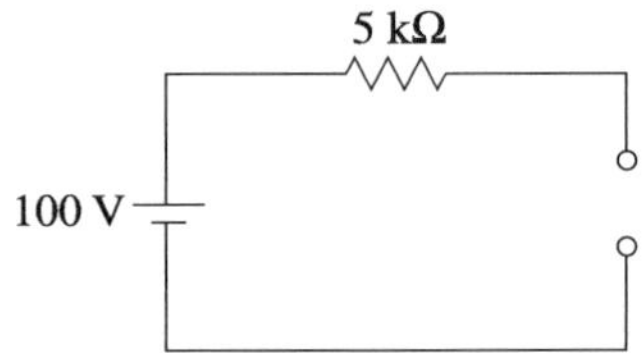

Fig. 8.84

Since the voltage across the capacitor cannot change instantaneously,

$$v_C(0^+) = 100 \text{ V}$$

Writing KCL equation for $t > 0$,

$$1 \times 10^{-6} \frac{dv_C}{dt} + \frac{v_C + 50}{5000} = 0$$

$$\frac{dv_C}{dt} + 200\, v_C = 10^4$$

The solution of this differential equation is given by

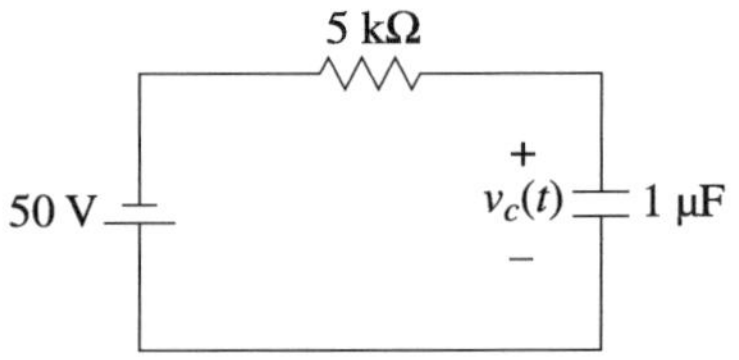

Fig. 8.85

$$v_C(t) = \frac{Q}{P} + K\, e^{-Pt}$$

$$= -50 + K\, e^{-200t}$$

At $t = 0$, $v_C(0) = 100$

$$100 = -50 + K$$

$$K = 150$$

$$v_C(t) = -50 + 150\, e^{-200t} \qquad \text{for } t > 0$$

Example 8.23 *In the network of Fig. 8.86, the switch is moved from 1 to 2 at t = 0. Determine i(t).*

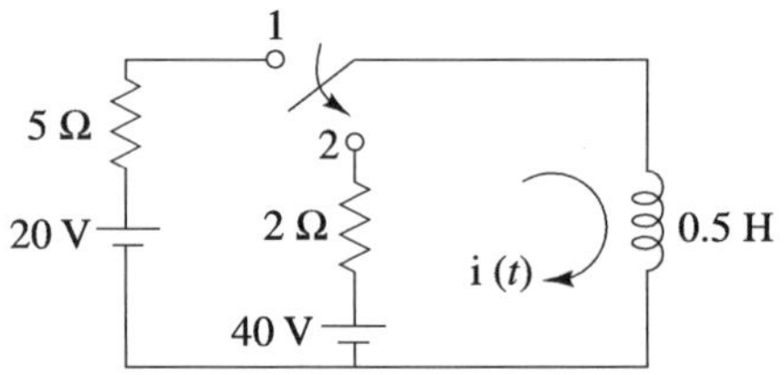

Fig. 8.86

Solution At $t = 0^-$, the network has attained steady-state condition. Hence, the inductor acts as a short circuit.

$$i(0^-) = \frac{20}{5} = 4 \text{ A}$$

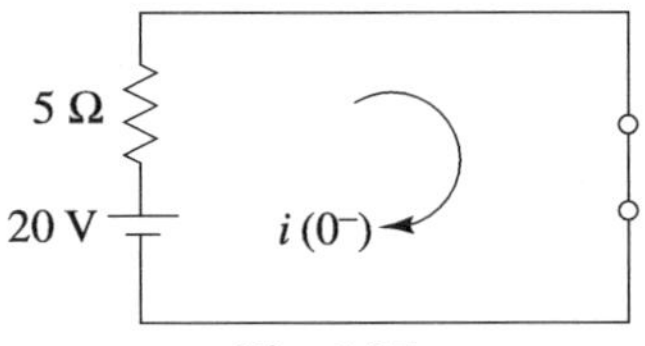

Fig. 8.87

Since the current through the inductor cannot change instantaneously,

$$i(0^+) = 4 \text{ A}$$

Writing KVL equation for $t > 0$,

$$0.5 \frac{di}{dt} + 2i = 40$$

$$\frac{di}{dt} + 4i = 80$$

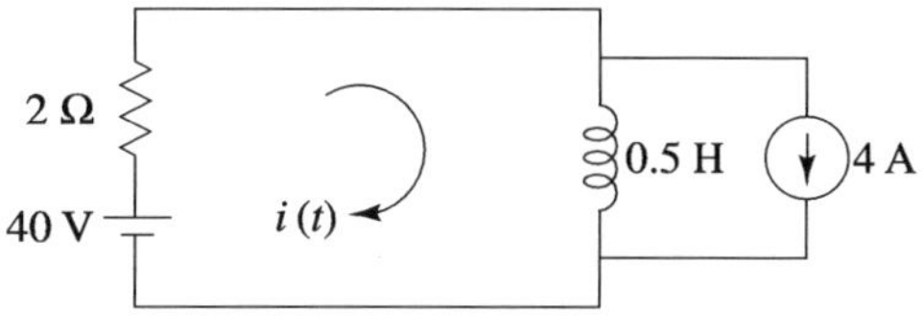

Fig. 8.88

Solution of this differential equation is given by

$$i(t) = \frac{Q}{P} + K e^{-Pt}$$

$$= \frac{80}{4} + K e^{-4t} = 20 + K e^{-4t}$$

At $t = 0$, $i(0) = 4$

$$4 = 20 + K$$

$$K = -16$$

$$i(t) = 20 - 16 e^{-4t} \qquad \text{for } t > 0$$

Example 8.24 *For the network shown in Fig. 8.89, steady state is reached with the switch closed. The switch is opened at t = 0. Obtain expressions for $i_L(t)$ and $v_L(t)$.*

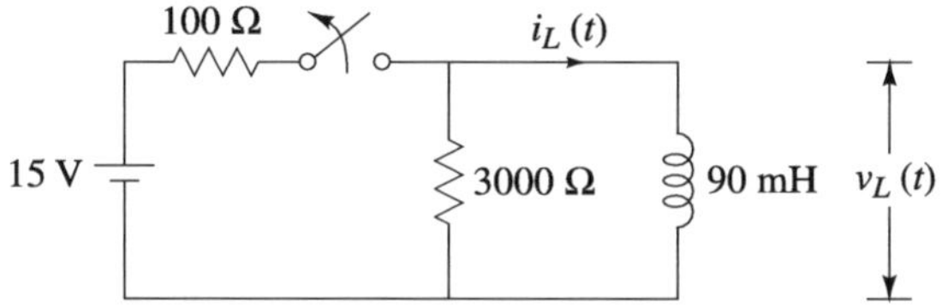

Fig. 8.89

Solution At $t = 0^-$, the network has attained steady-state condition. Hence, the inductor acts as a short circuit.

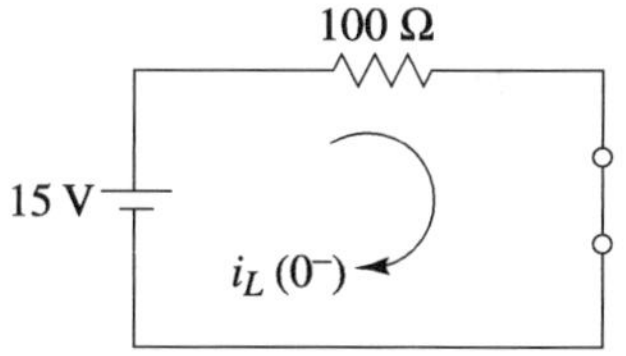

Fig. 8.90

$$i_L(0^-) = \frac{15}{100} = 0.15 \text{ A}$$

Since current through the inductor cannot change instantaneously,

$$i_L(0^+) = 0.15 \text{ A}$$

For $t > 0$,

Writing KVL equation for $t > 0$,

$$90 \times 10^{-3} \frac{di_L}{dt} + 3000\, i_L = 0$$

$$\frac{di_L}{dt} + 33.33 \times 10^3\, i_L = 0$$

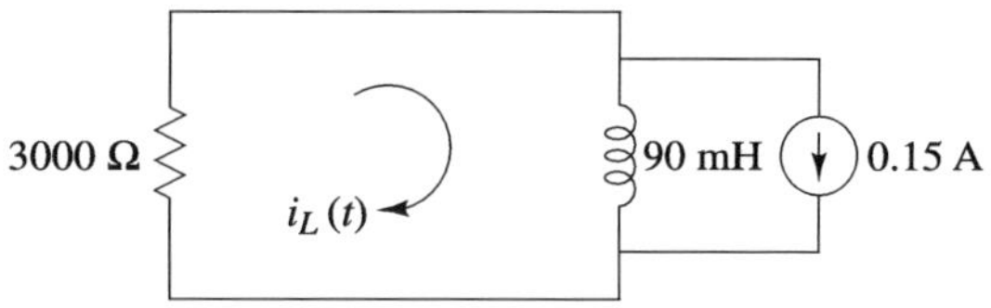

Fig. 8.91

The solution of this differential equation is given by

$$i_L(t) = K e^{-33.33 \times 10^3 t}$$

At $t = 0$, $i_L(0) = 0.15$

$$0.15 = \text{K}$$

$$i_L(t) = 0.15\, e^{-33.33 \times 10^3 t} \qquad \text{for } t > 0$$

Also, $$v_L(t) = L\frac{di_L}{dt}$$

$$= 90 \times 10^{-3}\frac{d}{dt}\left(0.15e^{-33.33\times 10^3 t}\right)$$

$$= -90 \times 10^{-3} \times 0.15 \times 33.33 \times 10^3 \times e^{-33.33\times 10^3 t}$$

$$= -450\, e^{-33.33\times 10^3 t} \qquad \text{for } t > 0$$

Example 8.25 *In the network of Fig. 8.92, the switch is open for a long time and it closes at t = 0. Find i (t).*

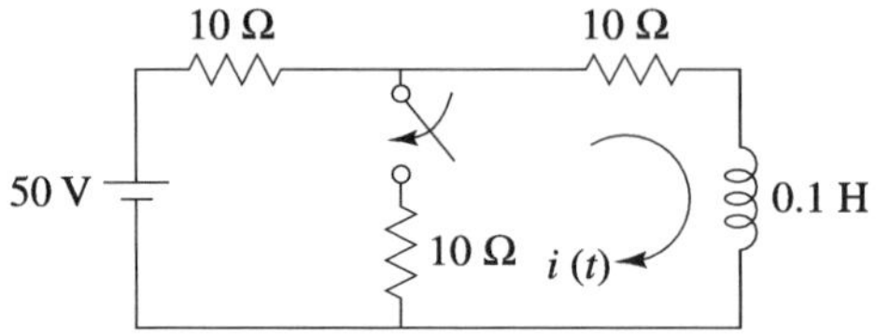

Fig. 8.92

Solution At $t = 0^-$, the network has attained steady-state condition. Hence, the inductor acts as a short circuit.

$$i(0^-) = \frac{50}{10+10} = 2.5 \text{ A}$$

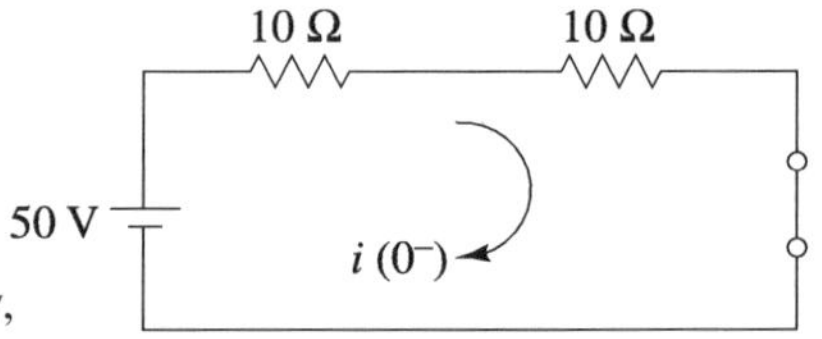

Fig. 8.93

Since current through the inductor cannot change instantaneously,

$$i(0^+) = 2.5 \text{ A}$$

For $t > 0$, representing the network to the left of the inductor by Thevenin's equivalent network,

$$V_{eq} = 50 \times \frac{10}{10+10} = 25 \text{ V}$$

$$R_{eq} = (10 \parallel 10) + 10 = 15\ \Omega$$

Writing KVL equation for $t > 0$,

$$0.1\frac{di}{dt} + 15i = 25$$

$$\frac{di}{dt} + 150i = 250$$

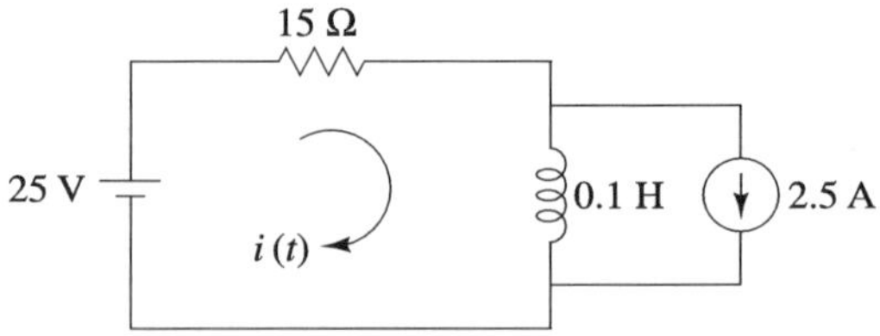

Fig. 8.94

Solution of this differential equation is given by

$$i(t) = \frac{Q}{P} + Ke^{-Pt}$$

$$= \frac{250}{150} + K\, e^{-150 t}$$

$$= 1.667 + K\, e^{-150 t}$$

At $t = 0$, $i(0) = 2.5$

$$2.5 = 1.667 + K$$

$$K = 0.833$$

$$i(t) = 1.667 + 0.833\, e^{-150 t} \qquad \text{for } t > 0$$

Example 8.26 *In the network shown in Fig. 8.95, the switch closes at $t = 0$. The capacitor is initially uncharged. Find $v_C(t)$ and $i_C(t)$.*

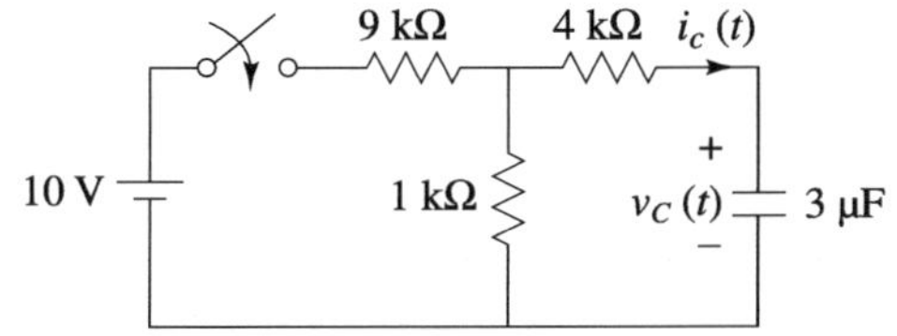

Fig. 8.95

Solution At $t = 0^-$, the capacitor is uncharged. Hence, it acts as a short circuit.

$$v_C(0^-) = 0$$
$$i_C(0^-) = 0$$

Since voltage across the capacitor cannot change instantaneously,

$$v_C(0^+) = 0$$

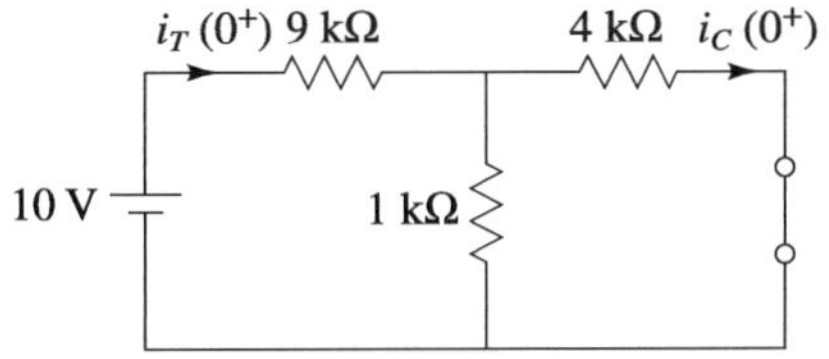

Fig. 8.96

At $t = 0^+$

$$i_T(0^+) = \left[\frac{10}{9\,\text{k} + (4\,\text{k} \parallel 1\,\text{k})}\right] = \frac{10}{9.8\,\text{k}} = 1.02 \text{ mA}$$

$$i_C(0^+) = 1.02\,\text{m} \times \frac{1\,\text{k}}{1\,\text{k} + 4\,\text{k}} = 0.204 \text{ mA}$$

For $t > 0$, representing the network to the left of the capacitor by Thevenin's equivalent network,

$$V_{eq} = 10 \times \frac{1\,\text{k}}{9\,\text{k} + 1\,\text{k}} = 1 \text{ V}$$

$$R_{eq} = (9\,\text{k} \parallel 1\,\text{k}) + 4\,\text{k} = 4.9 \text{ k}\Omega$$

Writing KCL equation for $t > 0$,

$$3 \times 10^{-6} \frac{dv_C}{dt} + \frac{v_C - 1}{4.9 \times 10^3} = 0$$

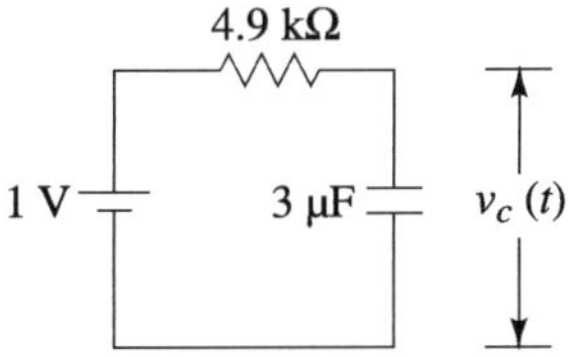

Fig. 8.97

$$\frac{dv_C}{dt} + 68.02\, v_C = 68.02$$

The solution of this differential equation is given by

$$v_C(t) = \frac{Q}{P} + Ke^{-Pt}$$
$$= 1 + K\,e^{-68.02\,t}$$

At $t = 0$, $v_C(0) = 0$

$$0 = 1 + K$$
$$K = -1$$
$$v_C(t) = 1 - e^{-68.02\,t} \qquad \text{for } t > 0$$

$$i_C(t) = C\,\frac{dv_C}{dt}$$

$$= 3 \times 10^{-6} \frac{d}{dt}(1 - e^{-68.02t})$$

$$= 3 \times 10^{-6} \times 68.02\, e^{-68.02\,t}$$

$$= 204.06 \times 10^{-6}\, e^{-68.02\,t} \qquad \text{for } t > 0$$

Example 8.27 *For the network shown in Fig. 8.98, the switch is open for a long time and closes at $t = 0$. Determine $v_C(t)$.*

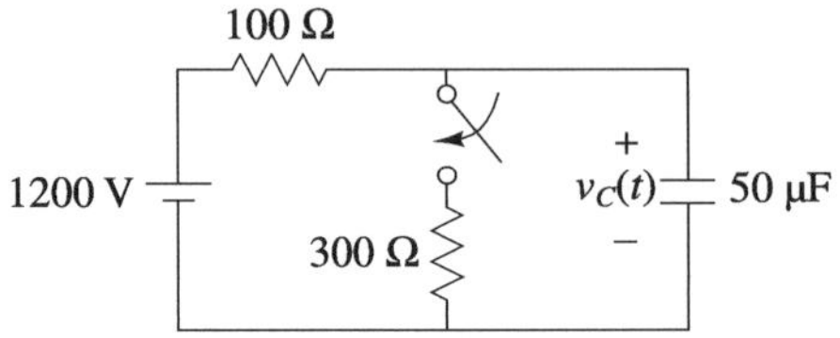

Fig. 8.98

Solution At $t = 0^-$, the network has attained steady-state condition. Hence, the capacitor acts as an open circuit.

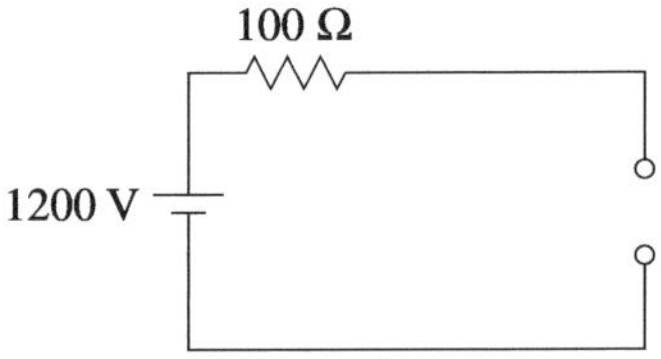

Fig. 8.99

$$v_C(0^-) = 1200 \text{ V}$$

Since the voltage across the capacitor cannot change instantaneously,

$$v_C(0^+) = 1200 \text{ V}$$

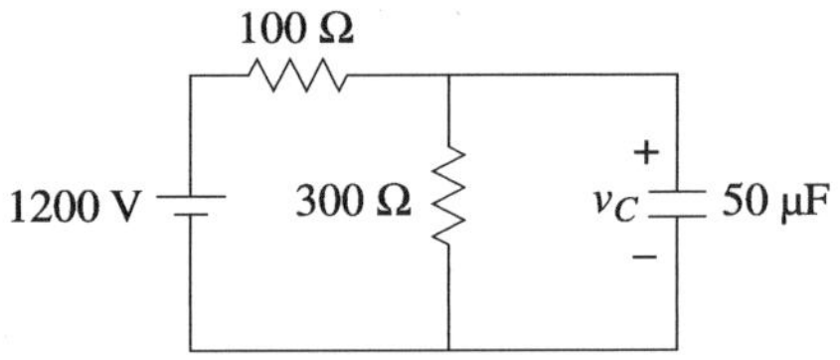

Fig. 8.100

Writing KCL equation for $t > 0$,

$$50 \times 10^{-6} \frac{dv_C}{dt} + \frac{v_C}{300} + \frac{v_C - 1200}{100} = 0$$

$$\frac{dv_C}{dt} + 266.67\, v_C = 0.24 \times 10^6$$

Solution of this differential equation is given by

$$v_C(t) = \frac{Q}{P} + Ke^{-Pt}$$

$$= \frac{0.24 \times 10^6}{266.67} + Ke^{-266.67\,t}$$

$$= 900 + Ke^{-266.67\,t}$$

At $t = 0$, $v_C(0) = 1200$

$$1200 = 900 + K$$

$$K = 300$$

$$v_C(t) = 900 + 300\, e^{-266.67\,t} \qquad \text{for } t > 0$$

Example 8.28 *In Fig. 8.101, the switch is closed at t = 0. Find i (t) for t > 0.*

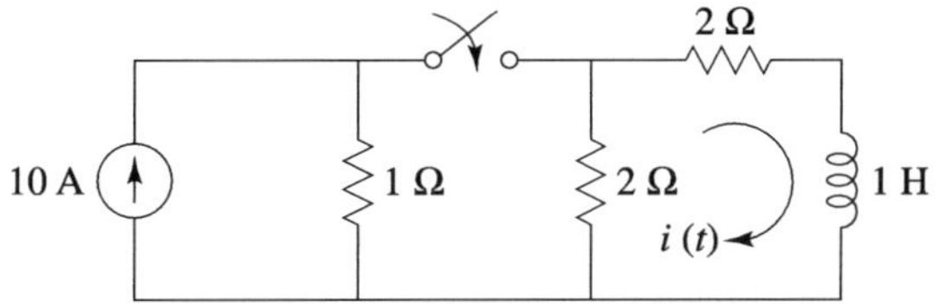

Fig. 8.101

Solution At $t = 0^-$,

$$i(0^-) = 0$$

Since current through inductor cannot change instantaneously,

$$i(0^+) = 0$$

Simplifying the network by source-transformation technique,

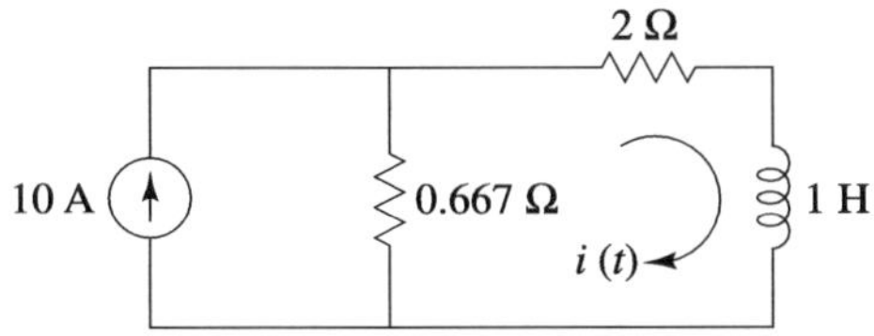

Fig. 8.102

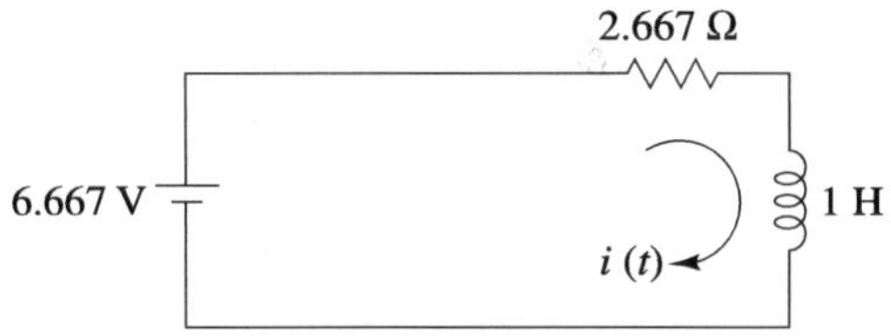

Fig. 8.103

Writing KVL equation for $t > 0$,

$$6.67 - 2.67\, i - \frac{di}{dt} = 0$$

$$\frac{di}{dt} + 2.67\, i = 6.67$$

The solution of this differential equation is given by

$$i(t) = \frac{Q}{P} + Ke^{-Pt}$$

$$= \frac{6.67}{2.67} + Ke^{-2.67t} = 2.5 + Ke^{-2.67t}$$

At $t = 0$, $i(0) = 0$

$$0 = 2.5 + K$$

$$K = -2.5$$

$$i(t) = 2.5 - 2.5\, e^{-2.67t}$$

$$= 2.5\,(1 - e^{-2.67\,t}) \qquad \text{for } t > 0$$

Example 8.29 *Find current i (t) for t > 0.*

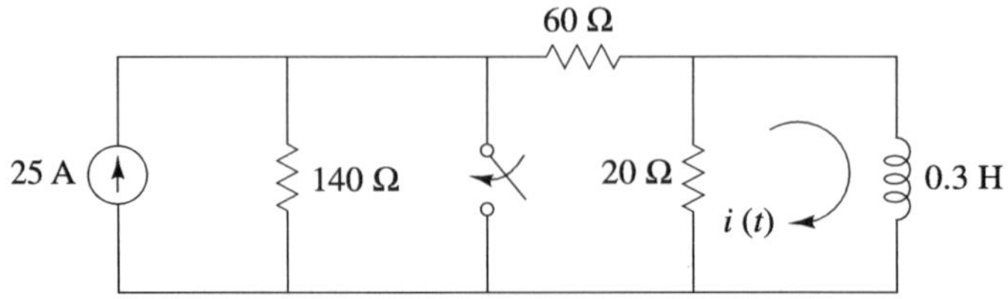

Fig. 8.104

Solution At $t = 0^-$, the inductor acts as a short circuit. Simplifying the network,

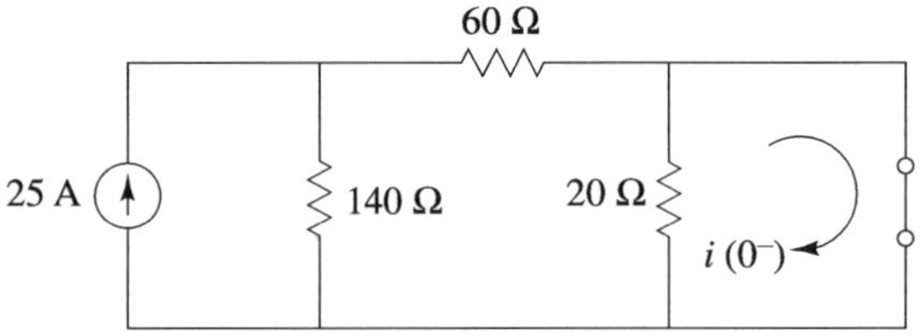

Fig. 8.105

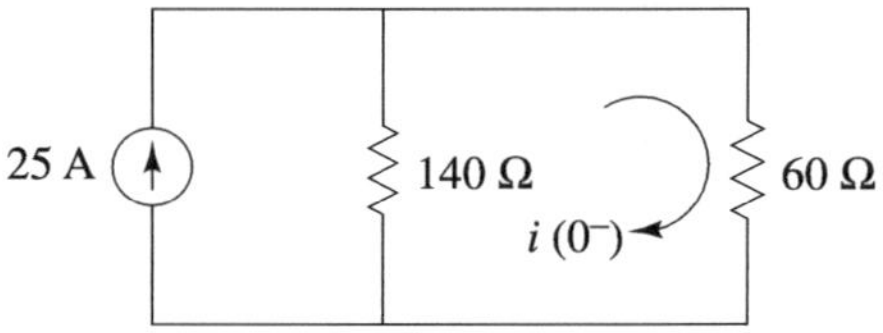

Fig. 8.106

$$i(0^-) = 25 \times \frac{140}{140 + 60} = 17.5 \text{ A}$$

Since current through the inductor cannot change instantaneously,

$$i(0^+) = 17.5 \text{ A}$$

For $t > 0$,

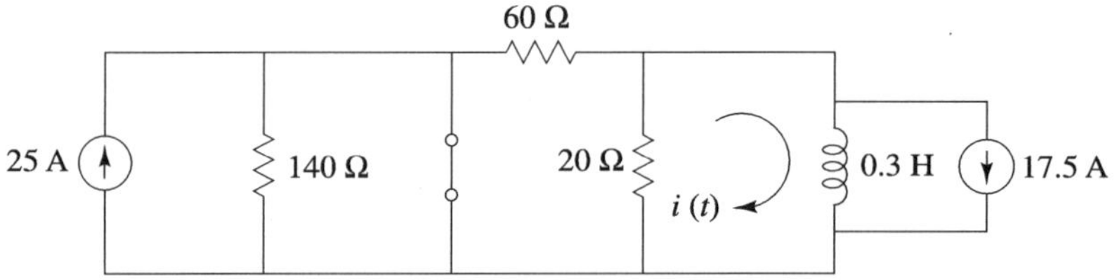

Fig. 8.107

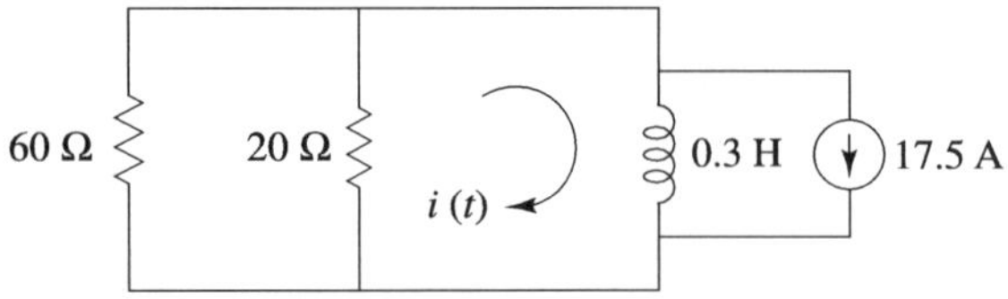

Fig. 8.108

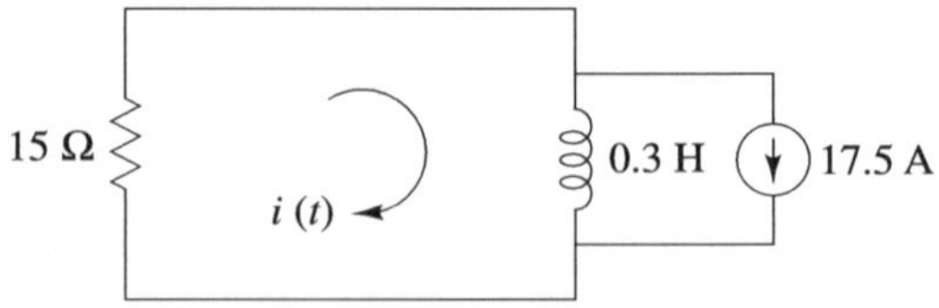

Fig. 8.109

Writing KVL equation for $t > 0$,

$$-15\, i - 0.3 \frac{di}{dt} = 0$$

$$\frac{di}{dt} + 50\,i = 0$$

Solution of this differential equation is given by,

$$i(t) = Ke^{-50\,t}$$

At $t = 0$, $i(0) = 17.5$

$$K = 17.5$$

$$i(t) = 17.5\,e^{-50\,t} \qquad \text{for } t > 0$$

Example 8.30 *In Fig. 8.110, the switch is closed at t = 0. Find $v_C(t)$ for t > 0.*

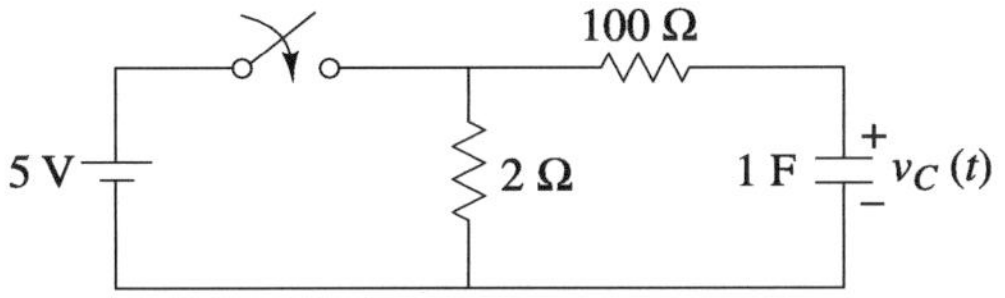

Fig. 8.110

Solution At $t = 0^-$

$$v_C(0^-) = 0$$

Since the voltage across the capacitor cannot change instantaneously,

$$v_C(0^+) = 0$$

Since the resistor of 2 Ω is connected in parallel with the voltage source of 5 V, it becomes redundant.

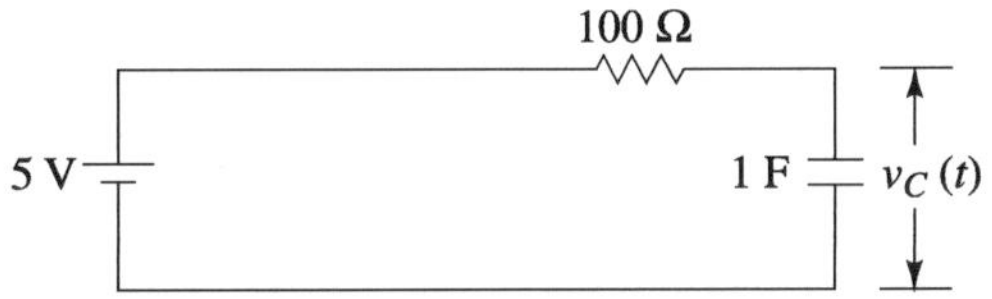

Fig. 8.111

Writing KCL equation for $t > 0$,

$$\frac{v_C - 5}{100} + 1\frac{dv_C}{dt} = 0$$

$$100\frac{dv_C}{dt} + v_C = 5$$

$$\frac{dv_C}{dt} + 0.01\,v_C = 0.05$$

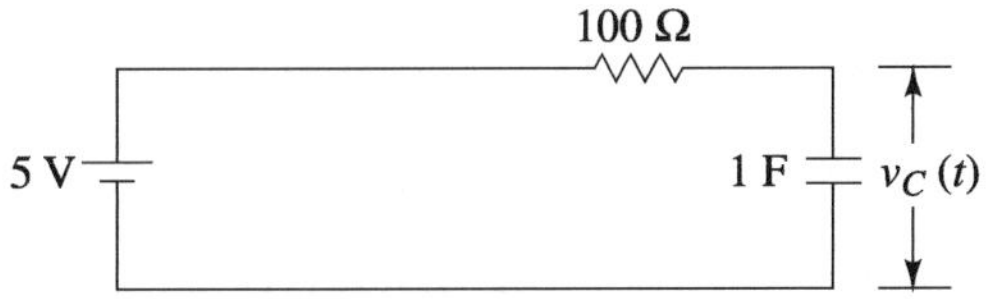

Fig. 8.112

The solution of this differential equation is given by

$$v_C(t) = \frac{Q}{P} + Ke^{-Pt}$$

$$= \frac{Q}{P} + Ke^{-0.01\,t}$$

$$= \frac{0.05}{0.01} + Ke^{-0.01\,t} = 5 + Ke^{-0.01\,t}$$

At $t = 0$, $v_C(0) = 0$

$$0 = 5 + K$$

$$K = -5$$

$$v_C(t) = 5 - 5e^{-0.01\,t} = 5\,(1 - e^{-0.01\,t}) \qquad \text{for } t > 0$$

Example 8.31 *In the network shown, the switch is shifted to position b at t = 0. Find v (t) for t > 0.*

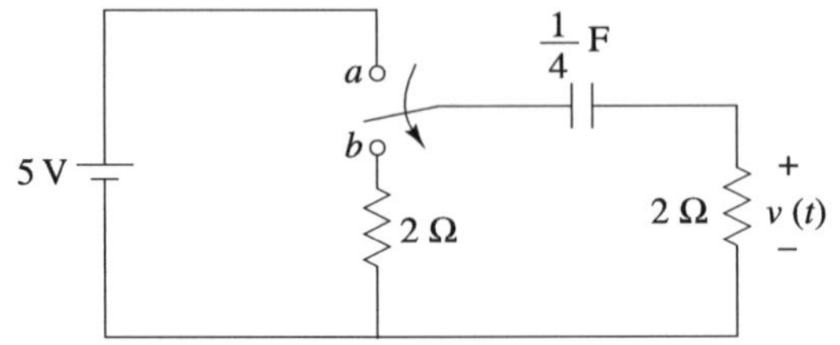

Fig. 8.113

Solution At $t = 0^-$, the network has attained steady-state condition. Hence, the capacitor acts as an open circuit.

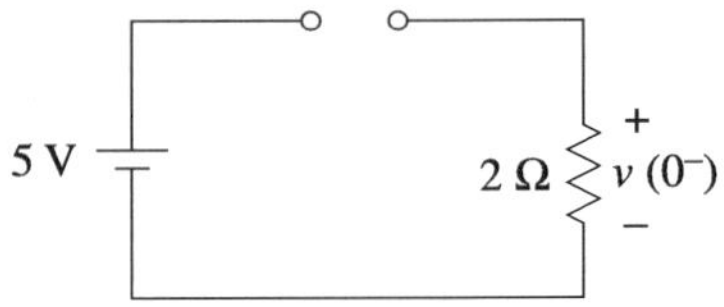

Fig. 8.114

$$v_C(0^-) = 5 \text{ V}$$
$$v(0^-) = 0$$

At $t = 0^+$, the capacitor acts as a voltage source of 5 V.

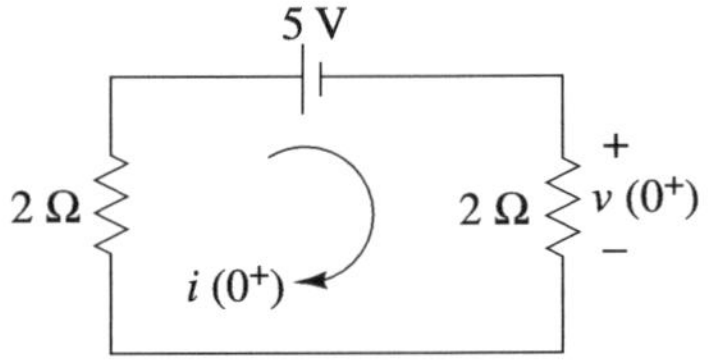

Fig. 8.115

$$i(0^+) = -\frac{5}{4} = -1.25 \text{ A}$$
$$v(0^+) = -1.25 \times 2 = -2.5 \text{ V}$$

Writing KVL equation for $t > 0$,

$$-2i - 5 - \frac{1}{1/4}\int_0^t i\,dt - 2i = 0 \qquad \ldots(1)$$

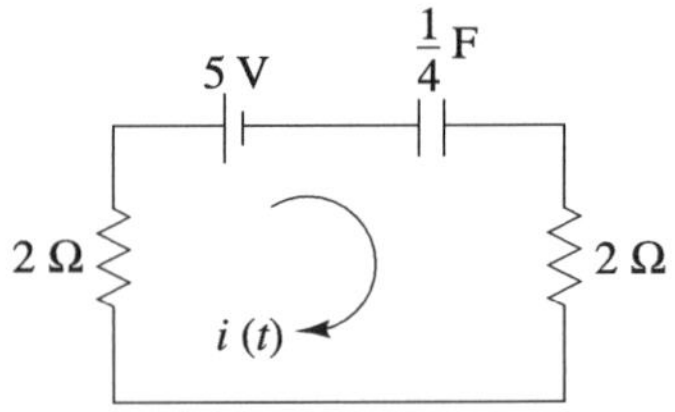

Fig. 8.116

Differentiating the Eq. (i), we get

$$-4\frac{di}{dt} - 4i = 0$$
$$\frac{di}{dt} + i = 0$$

The solution of this equation is given by

$$i(t) = Ke^{-t}$$

At $t = 0$, $i(0) = -1.25$

$$K = -1.25$$
$$i(t) = -1.25\, e^{-t} \qquad \text{for } t > 0$$
$$v(t) = 2\, i(t)$$
$$= -2.5\, e^{-t} \qquad \text{for } t > 0$$

Example 8.32 *In the network of Fig. 8.117, the switch is open for a long time and at t = 0, it is closed. Determine $v_2(t)$.*

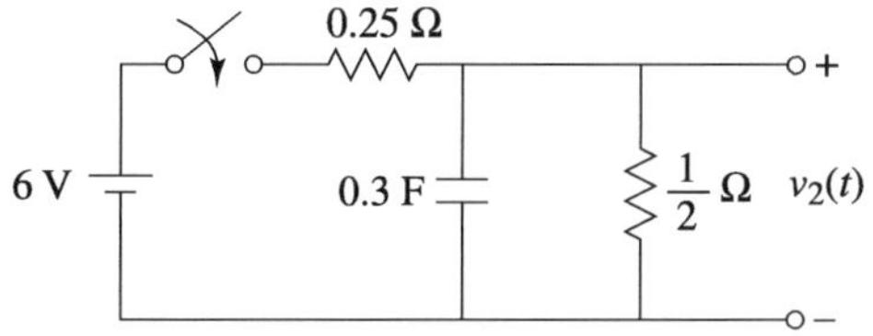

Fig. 8.117

Solution At $t = 0^-$, the switch is open.

$$v_2(0^-) = 0$$

At $t = 0^+$,

$$v_2(0^+) = 0$$

Writing KCL equation for $t > 0$,

$$\frac{v_2}{1/2} + 0.3\frac{dv_2}{dt} + \frac{v_2 - 6}{0.25} = 0$$

$$\frac{dv_2}{dt} + 20v_2 = 80$$

The solution of this differential equation is given by,

$$v_2(t) = \frac{Q}{P} + Ke^{-Pt}$$

$$= \frac{80}{20} + Ke^{-20t}$$

$$v_2(t) = 4 + K e^{-20t}$$

At $t = 0$, $v_2(0) = 0$

$$0 = 4 + K$$

$$K = -4$$

$$v_2(t) = 4 - 4e^{-20t}$$

$$= 4(1 - e^{-20t}) \qquad \text{for } t > 0$$

Example 8.33 *In the network of Fig. 8.118, the switch is in position 'a' for a long time. At t = 0, the switch is moved from a to b. Find $v_2(t)$. Assume that the initial current in the 2-H inductor is zero.*

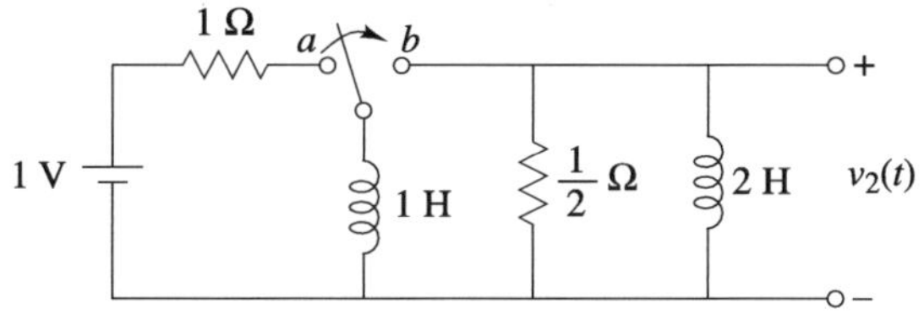

Fig. 8.118

Solution At $t = 0^-$, the switch is in the position a. The network has attained steady-state condition. Hence, the inductor acts as a short circuit.

Current through the 1-H inductor is given by

$$i(0^-) = \frac{1}{1} = 1\text{ A}$$

$$v_2(0^-) = 0$$

Since current through the inductor cannot change instantaneously,

$$i(0^+) = 1 \text{ A}$$

$$v_2(0^+) = -1 \times \frac{1}{2} = -0.5 \text{ V}$$

Writing KCL equation for $t > 0$,

$$\frac{1}{1}\int_0^t v_2 dt + i(0^+) + \frac{v_2}{1/2} + \frac{1}{2}\int_0^t v_2 dt = 0 \quad \text{...(i)}$$

Differentiating the Eq. (i), we get

$$v_2 + 2\frac{dv_2}{dt} + \frac{1}{2}v_2 = 0$$

$$\frac{dv_2}{dt} + \frac{3}{4}v_2 = 0$$

The solution of this differential equation is given by

$$v_2(t) = K e^{-(3/4)t}$$

At $t = 0$, $v_2(0) = -0.5$

$$-0.5 = K e^0$$

$$K = -0.5$$

$$v_2(t) = -0.5\, e^{-(3/4)t} \qquad \text{for } t > 0$$

Example 8.34 *In the network shown in Fig. 8.119, a steady state condition is achieved with switch open. At $t = 0$ switch is closed. Find $v_a(t)$.*

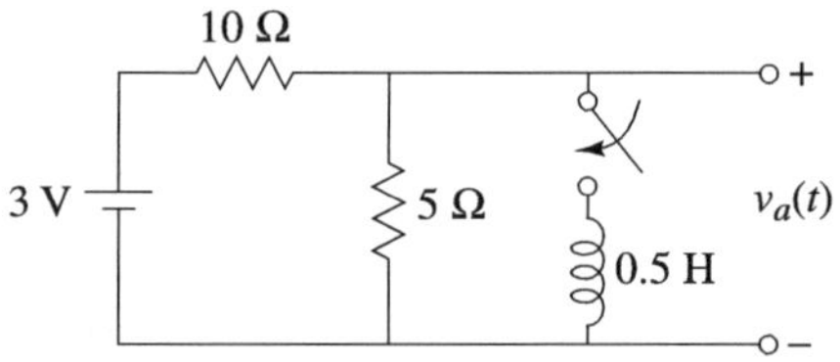

Fig. 8.119

Solution At $t = 0^-$, the network has attained steady-state condition.

$$v_a(0^-) = 3 \times \frac{5}{10+5} = 1 \text{ V}$$

At $t = 0^+$

$$v_a(0^+) = 1 \text{ V}$$

Writing KCL equation for $t > 0$,

$$\frac{1}{0.5}\int_0^t v_a\, dt + \frac{v_a}{5} + \frac{v_a - 3}{10} = 0 \quad \text{...(i)}$$

Differentiating the Eq. (i), we get

$$2v_a + 0.2\frac{dv_a}{dt} + 0.1\frac{dv_a}{dt} = 0$$

$$\frac{dv_a}{dt} + \frac{20}{3}v_a = 0$$

The solution of this differential equation is given by

$$v_a(t) = K\,e^{-(20/3)\,t}$$

At $t = 0$, $v_a(0) = 1$

$$1 = K$$

$$v_a(t) = e^{-(20/3)\,t} \qquad \text{for } t > 0$$

Example 8.35 *The switch is moved from the position a to b at t = 0, having been in the position a for a long time before t = 0. The capacitor C_2 is uncharged at t = 0. Find i (t) and v_2 (t) for t > 0.*

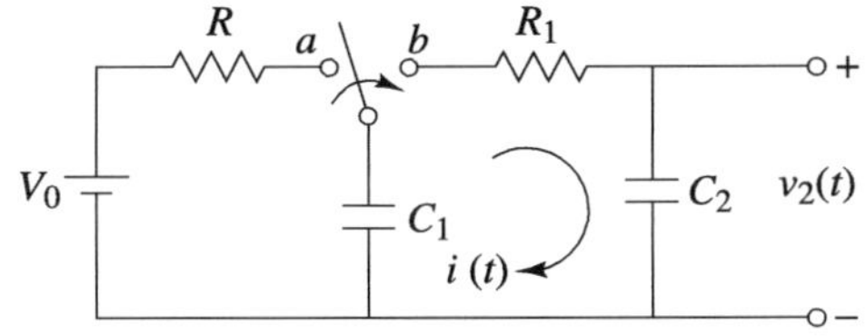

Fig. 8.120

Solution At $t = 0^-$, the network has attained steady-state condition. Hence, the capacitor acts as an open circuit and it will charge to V_0 volt.

$$v(0^-) = V_0$$

Since the voltage across the capacitor cannot change instantaneously,

$$v(0^+) = V_0$$

$$i(0^+) = \frac{V_0}{R_1}$$

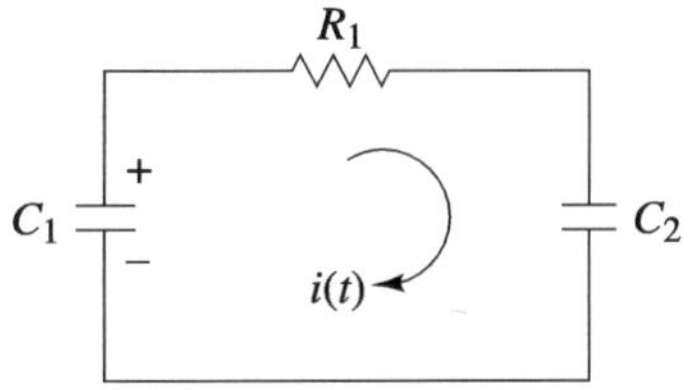

Fig. 8.121

Writing KVL equation for t > 0,

$$-\frac{1}{C_1}\int_0^t i\,dt + V_0 - R_1 i - \frac{1}{C_2}\int_0^t i\,dt = 0 \qquad \text{...(i)}$$

Differentiating the Eq. (i), we get

$$-\frac{i}{C_1} - R_1\frac{di}{dt} - \frac{i}{C_2} = 0$$

$$\frac{di}{dt} + \frac{1}{R_1}\left(\frac{C_1 + C_2}{C_1 C_2}\right) i = 0$$

The solution of this differential equation is given by

$$i(t) = K e^{-\frac{1}{R_1}\left(\frac{C_1+C_2}{C_1C_2}\right)t}$$

At $t = 0$, $i(0) = \dfrac{V_0}{R_1}$

$$K = \frac{V_0}{R_1}$$

$$i(t) = \frac{V_0}{R_1} e^{-\frac{1}{R_1}\left(\frac{C_1+C_2}{C_1C_2}\right)t}$$

$$= \frac{V_0}{R_1} e^{-\frac{1}{R_1 C} t} \qquad \text{where, } C = \frac{C_1 C_2}{C_1 + C_2}$$

$$v_2(t) = \frac{1}{C_2} \int_0^t i \, dt$$

$$= \frac{1}{C_2} \int_0^t \frac{V_0}{R_1} e^{-\frac{t}{R_1 C}} dt$$

$$= \frac{V_0}{R_1 C_2} R_1 C \left(1 - e^{-\frac{1}{R_1 C} t}\right)$$

$$= \frac{V_0}{C_2} C \left(1 - e^{-\frac{t}{R_1 C} t}\right) \qquad \text{for } t > 0$$

Example 8.36 *In the network of Fig. 8.122, determine currents $i_1(t)$ and $i_2(t)$ when the switch is closed at $t = 0$.*

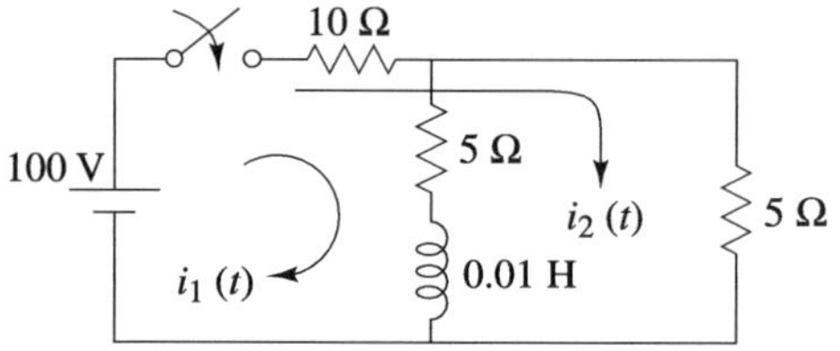

Fig. 8.122

Solution At $t = 0^-$

$$i_1(0^-) = i_2(0^-) = 0$$

At $t = 0^+$,

$$i_1(0^+) = 0$$

$$i_2(0^+) = \frac{100}{15} = 6.67 \text{ A}$$

Writing KVL equations for $t > 0$,

$$10(i_1 + i_2) + 5i_1 + 0.01 \frac{di_1}{dt} = 100 \qquad \text{...(i)}$$

$$10(i_1 + i_2) + 5i_2 = 100 \qquad \text{...(ii)}$$

From the Eq. (ii), we get

$$i_2 = \frac{100 - 10i_1}{15}$$

Substituting in the Eq. (i), we get

$$\frac{di_1}{dt} + 833\, i_1 = 3333$$

The solution of this differential equation is given by

$$i_1(t) = \frac{Q}{P} + Ke^{-Pt}$$

$$= \frac{3333}{833} + K\,e^{-833t} = 4 + K\,e^{-833t}$$

At $t = 0$, $i_1(0) = 0$

$$0 = 4 + K$$

$$K = -4$$

$$i_1(t) = 4 - 4e^{-833\,t}$$

$$= 4\left(1 - e^{-833t}\right) \qquad \text{for } t > 0$$

$$i_2(t) = \frac{100 - 10i_1}{15}$$

$$= \frac{100 - 10\left(4 - 4e^{-833t}\right)}{15}$$

$$= 4 + 2.67\,e^{-833\,t} \qquad \text{for } t > 0$$

Example 8.37 *The switch in the network of Fig. 8.123 is opened at $t = 0$. Find $i(t)$ for $t > 0$ if,*

(i) $L = \frac{1}{2}\,H$ *and* $C = 1\,F$

(ii) $L = 1\,H$ *and* $C = 1\,F$

(iii) $L = 5\,H$ *and* $C = 1\,F$

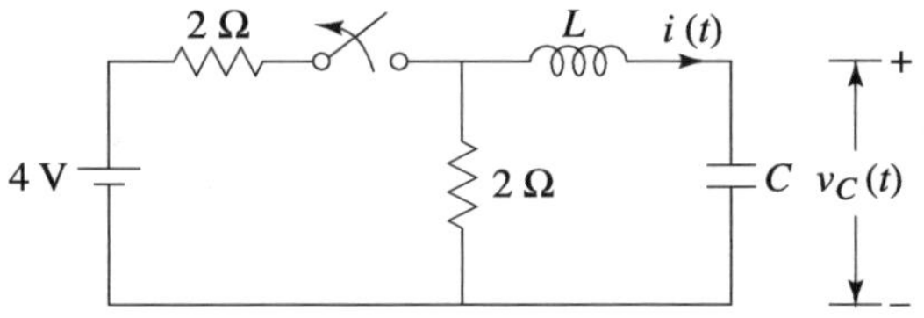

Fig. 8.123

Solution At $t = 0^-$, the network has attained steady-state condition. Hence, the inductor acts as a short circuit and the capacitor acts as an open circuit.

$$v_C(0^-) = 4 \times \frac{2}{2+2} = 2\text{ V}$$

$$i(0^-) = 0$$

Since current through the inductor and voltage across the capacitor cannot change instantaneously,

$$v_C(0^+) = 2\text{ V}$$

$$i(0^+) = 0$$

Case I *When $R = 2\,\Omega$, $L = \frac{1}{2}\,H$, $C = 1\,F$*

$$\alpha = \frac{R}{2L} = \frac{2}{2 \times \frac{1}{2}} = 2$$

$$\omega_0 = \frac{1}{\sqrt{LC}} = \frac{1}{\sqrt{\frac{1}{2} \times 1}} = \frac{1}{\sqrt{0.5}} = 1.414$$

$$\alpha > \omega_0$$

This indicates an overdamped case.

$$i(t) = A_1 e^{s_1 t} - A_2 e^{s_2 t}$$

where, $$s_1 = -\alpha - \sqrt{\alpha^2 - \omega_o^2}$$

$$= -2 - \sqrt{4-2} = -2 - \sqrt{2} = -3.414$$

and $$s_2 = -\alpha + \sqrt{\alpha^2 - \omega_o^2} = -2 + \sqrt{2} = -0.586$$

$$i(t) = A_1 e^{-3.414\,t} + A_2 e^{-0.586\,t}$$

At $t = 0$, $i\,(0) = 0$

$$A_1 + A_2 = 0 \qquad \ldots\text{(i)}$$

Also $$v_L(0^+) + v_C(0^+) + v_R(0^+) = 0$$

$$v_L(0^+) = -v_R(0^+) - v_C\,(0^+)$$

$$= -2i(0^+) - v_C\,(0^+)$$

$$= -2\text{ V} \qquad \ldots\text{(ii)}$$

$$v_L\,(0^+) = L\frac{di}{dt}(0^+)$$

$$\frac{di}{dt}(0^+) = \frac{v_L(0^+)}{L} = -\frac{2}{0.5} = -4\text{ A/s}$$

Differentiating the equation of $i\,(t)$ and putting the condition at $t = 0$, we get,

$$-3.414\,A_1 - 0.586\,A_2 = -4 \qquad \ldots\text{(iii)}$$

Solving Eqs (i) and (iii), we get

$A_1 = 1.414$ and $A_2 = -1.414$

$$i(t) = 1.414(e^{-3.414t} - e^{-0.586\,t}) \qquad \text{for } t > 0$$

Case II *When $R = 2\ \Omega$, $L = 1\ H$, $C = 1\ F$*

$$\alpha = \frac{R}{2L} = \frac{2}{2 \times 1} = \frac{2}{2} = 1$$

$$\omega_0 = \frac{1}{\sqrt{LC}} = \frac{1}{\sqrt{1}} = 1$$

$$\alpha = \omega_0$$

This indicates a critically damped case.

$$i(t) = e^{-\alpha t}\,(A_1 + A_2\,t)$$

$$= e^{-t}\,(A_1 + A_2\,t)$$

At $t = 0$, $i(0) = 0$

$$A_1 = 0$$

Also, $$v_L(0^+) = L\,\frac{di}{dt}\,(0^+)$$

$$\frac{di}{dt}(0^+) = \frac{v_L(0^+)}{L} = -\frac{2}{1} = -2\text{ A/s}$$

Differentiating the equation of $i(t)$ and putting the condition at $t = 0$, we get

$$\left.\frac{di}{dt}\right|_{t=0} = -A_1 + A_2 = -2$$

$$A_2 = -2$$

$$i(t) = -2\,t\,e^{-t} \qquad \text{for } t > 0$$

Case III *When $R = 2\ \Omega$, $L = 5$ H, $C = 1$ F*

$$\alpha = \frac{R}{2L} = \frac{2}{10} = 0.2$$

$$\omega_0 = \frac{1}{\sqrt{LC}} = \frac{1}{\sqrt{5}} = 0.447$$

$$\alpha < \omega_0$$

This indicates an underdamped case.

$$i(t) = e^{-\alpha t}(B_1 \cos \omega_d t + B_2 \sin \omega_d t)$$

where,

$$\omega_d = \sqrt{\omega_o^2 - \alpha^2}$$

$$= \sqrt{(0.447)^2 - (0.2)^2} = 0.4$$

$$s_{1,2} = -\alpha \pm j\,\omega_d$$

$$= -0.2 \pm j\,0.4$$

$$i(t) = e^{-0.2t}(B_1 \cos 0.4\,t + B_2 \sin 0.4\,t)$$

Applying the initial conditions,

$$i(0^+) = 0$$

and

$$\frac{di}{dt}(0^+) = -\frac{v_L(0^+)}{L} = -\frac{2}{5}$$

$$B_1 = i(0) = 0$$

$$B_2 = -1$$

$$i(t) = -e^{-0.2t} \sin 0.4\,t \qquad \text{for } t > 0$$

Exercises

1. The switch in Fig. 8.124 is moved from the position a to b at $t = 0$, the network having been in steady state in the position a. Determine $i_1(0^+)$, $i_2(0^+)$, $i_3(0^+)$ $\frac{di_2}{dt}(0^+)$ and $\frac{di_3}{dt}(0^+)$.

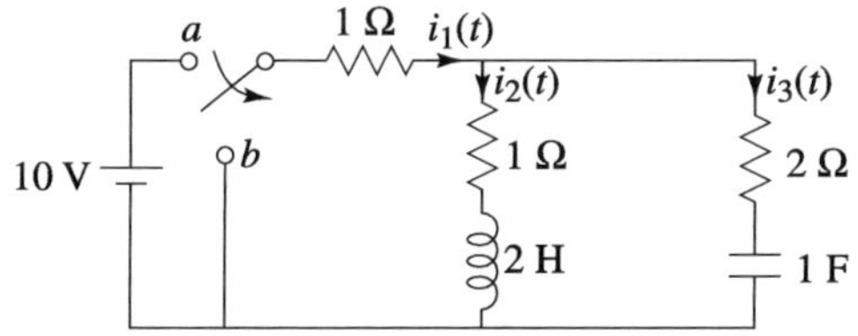

Fig. 8.124

[1.66 A, 5 A, –3.33 A, –3.33 A/s, 2.22 A/s]

2. In the network shown in Fig. 8.125, switch is closed at $t = 0$. Obtain the current $i_2(t)$.

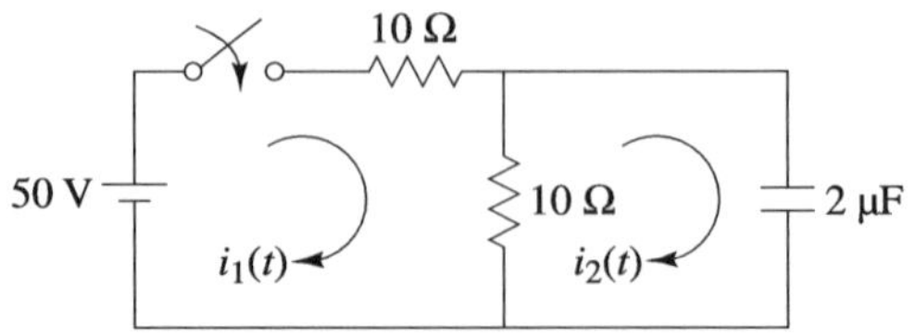

Fig. 8.125

$[i_2(t) = 5\ e^{-100000t}]$

3. The network shown in Fig. 8.126 is under steady-state when the switch is closed. At $t = 0$, it is opened. Obtain an expression for $i(t)$.

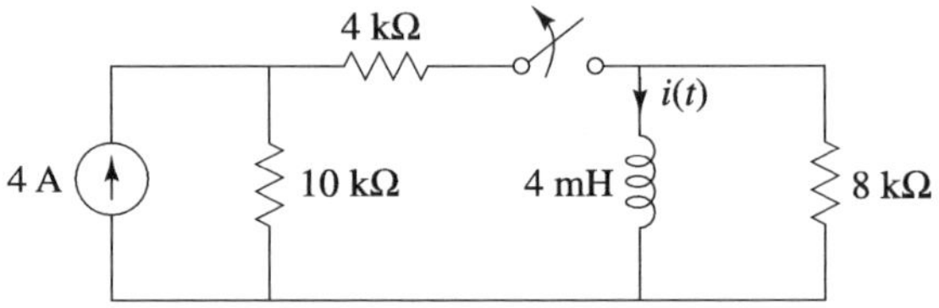

Fig. 8.126

$[i(t) = 2.857\ e^{-2 \times 10^6 t}]$

4. The switch in Fig. 8.127 is open for a long time and closes at $t = 0$. Determine $i(t)$ for $t > 0$.

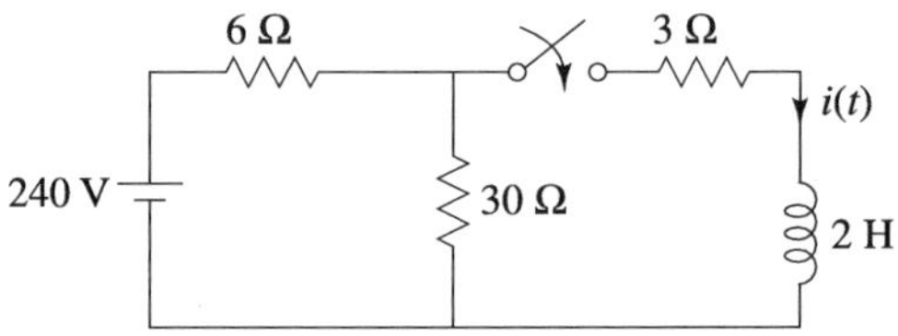

Fig. 8.127

$[i(t) = 25(1 - e^{-4t})]$

5. In the network shown, the steady state is reached with the switch open. At $t = 0$, the switch is closed. Find $v_C(t)$ for $t > 0$.

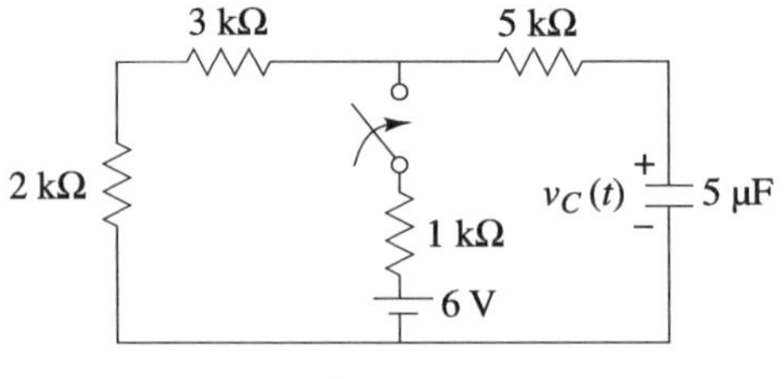

Fig. 8.128

$[v_C(t) = 5e^{-20t}]$

6. The circuit shown has acquired steady state before switching at $t = 0$.
(i) Obtain $v_C(0^+)$, $v_C(0^-)$, $i(0^+)$ and $i(0^-)$.
(ii) Obtain time constant for $t > 0$.
(iii) Find current $i(t)$ for $t > 0$.

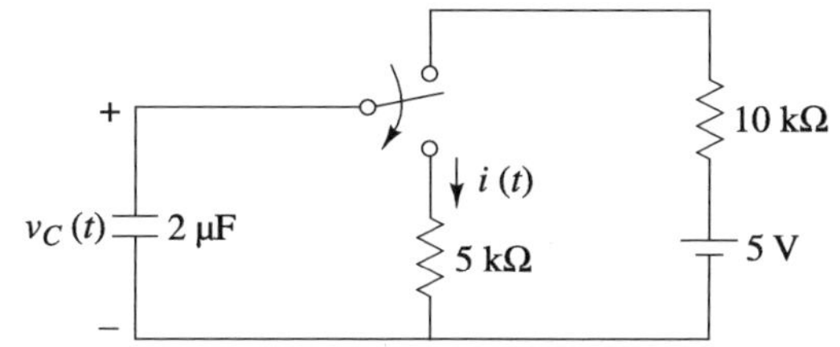

Fig. 8.129

[(i) 5 V, 5 V, 1 mA, 0, (ii) 0.01 s, (iii) e^{-100t} mA]

7. In the network shown, the switch is initially at the position 1 for a long time. At $t = 0$, the switch is changed to the position 2. Find current $i(t)$ for $t > 0$.

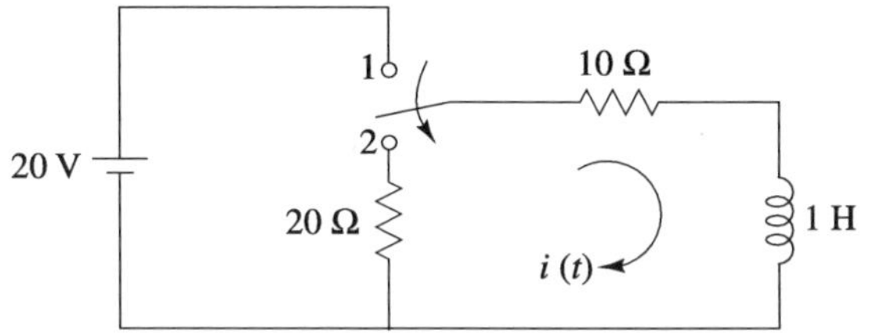

Fig. 8.130

[$i(t) = 2e^{-30t}$]

8. In the network shown, the switch is closed at $t = 0$. Find $v(t)$ for $t > 0$.

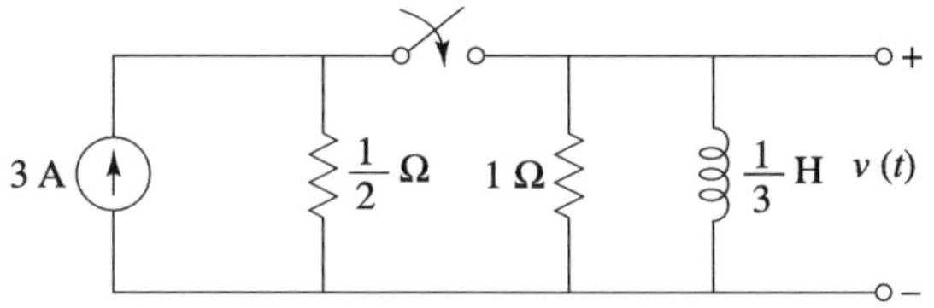

Fig. 8.131

[$v(t) = e^{-t}$]

9. In the network shown, the switch is in the position 1 for a long time and at $t = 0$, the switch is moved to the position 2. Find $v(t)$ for $t > 0$.

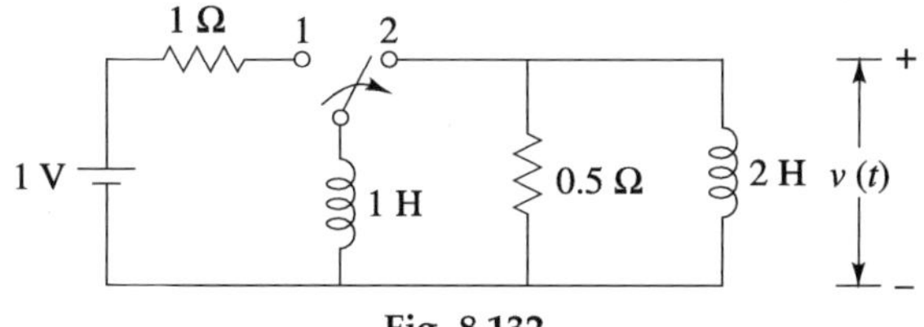

Fig. 8.132

[$v(t) = -0.5\ e^{-(3/4)t}$]

10. In Fig. 8.133, the switch is open until time t = 100 seconds and is closed for all times thereafter. Find $v(t)$ for all times greater than 100 if $v(100) = -3$.

[$v(t) = 5 - 8\ e^{-(t-100)/160}$ V]

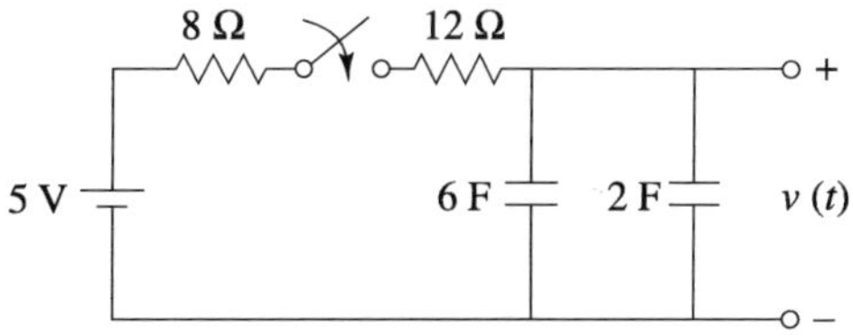

Fig. 8.133

11. A series *R-L* circuit has a constant voltage *V* applied at $t = 0$. At what time does $v_R = v_L$.

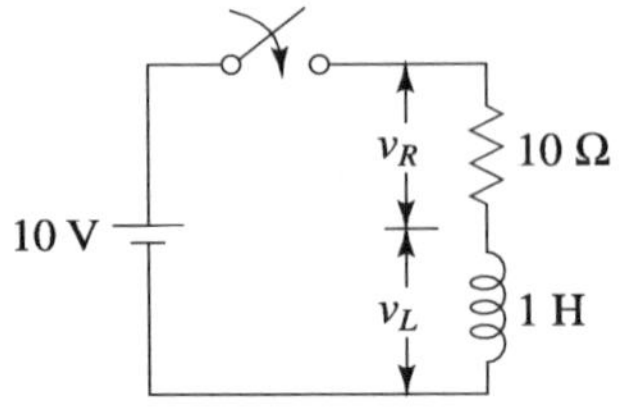

Fig. 8.134

[0.0693 s]

12. In the circuit shown, at time $t = 0$, the voltage across the capacitor is zero and the switch is moved to the position *y*. The switch is kept at position *y* for 20 seconds and then moved to position *z* and kept in that position thereafter. Find the voltage across the capacitor at $t = 30$ seconds.

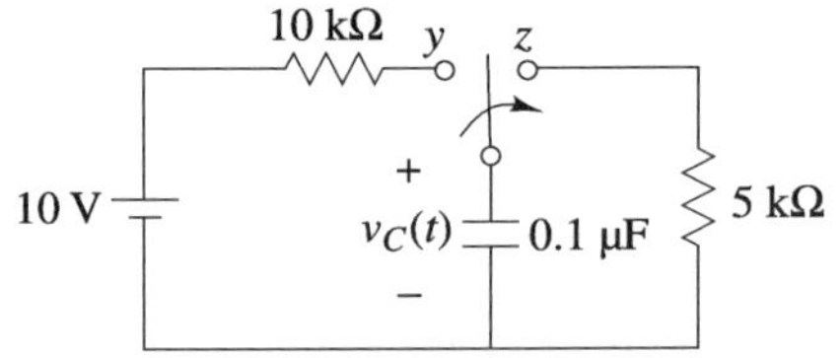

Fig. 8.135

[0]

13. Determine whether *RLC* series circuit shown in Fig 8.136 is underdamped, overdamped or critically damped. Also, find $v_L(0^+)$, $\frac{di}{dt}(0^+)$ and $i(\infty)$.

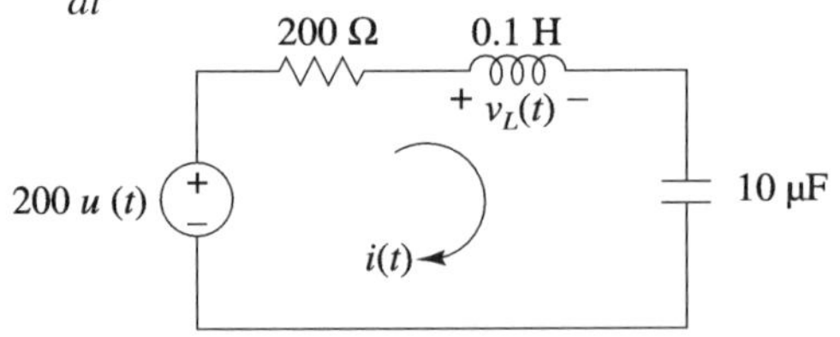

Fig. 8.136

[critically damped, 200 V, 2000 A/s, 0]

14. Determine whether *RLC* circuit of Fig 8.137 is underdamped, overdamped or critically damped. Also find $v_L(0^+)$, $\frac{di}{dt}(0^+)$, $\frac{d^2v}{dt^2}(0^+)$ if $v(t) = u(t)$.

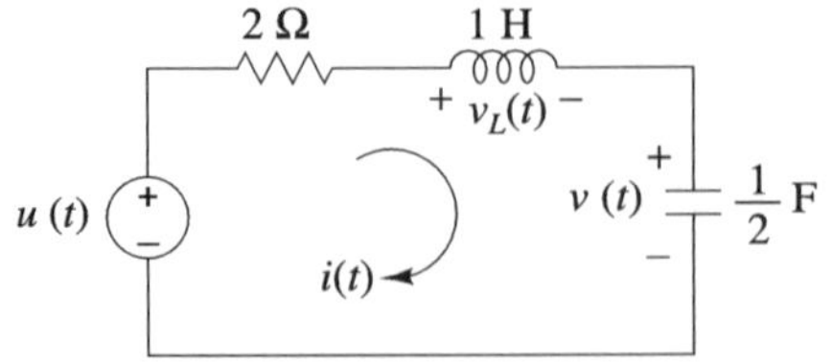

Fig. 8.137

[underdamped 1 V, 1 A/s, 2 V/s²]

Objective-Type Questions

1. The voltages v_{C_1}, v_{C_2} and v_{C_3} across the capacitors in the circuit in Fig. 8.138 under steady state are respectively

(a) 80 V, 32 V, 48 V
(b) 80 V, 48 V, 32 V
(c) 20 V, 8 V, 12 V
(d) 20 V, 12 V, 8 V

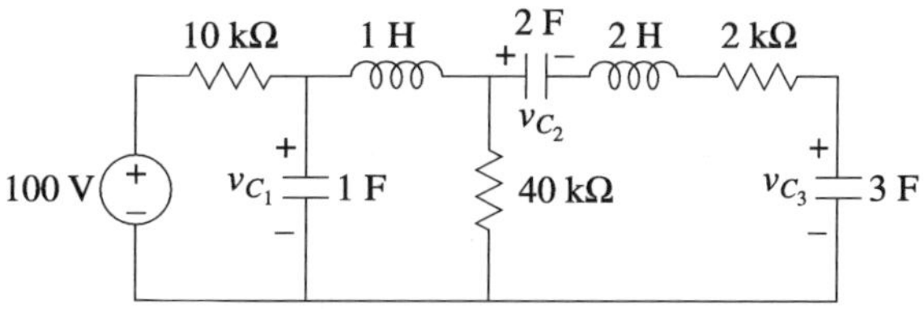

Fig. 8.138

2. In the circuit of Fig. 8.139, the voltage $v(t)$ is

(a) $e^{at} - e^{bt}$ (b) $e^{at} + e^{bt}$ (c) $ae^{dt} - be^{bt}$ (d) $ae^{at} + be^{bt}$

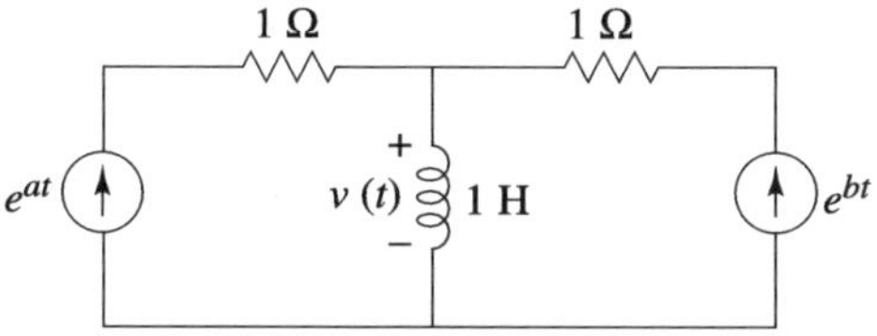

Fig. 8.139

3. The differential equation for the current $i(t)$ in the circuit of Fig. 8.140 is

(a) $2\dfrac{d^2i}{dt^2} + 2\dfrac{di}{dt} + i(t) = \sin t$

(b) $\dfrac{d^2i}{dt^2} + 2\dfrac{di}{dt} + 2i(t) = \cos t$

(c) $2\dfrac{d^2i}{dt^2} + 2\dfrac{di}{dt} + i(t) = \cos t$

(d) $\dfrac{d^2i}{dt^2} + 2\dfrac{di}{dt} + 2i(t) = \sin t$

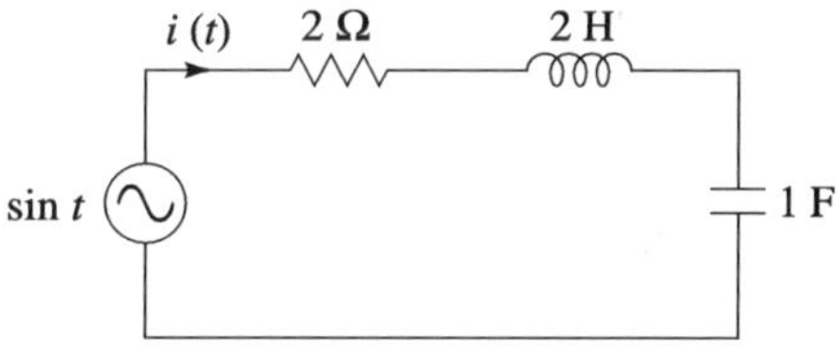

Fig. 8.140

4. At $t = 0^+$, the current i_1 is

(a) $-\dfrac{V}{2R}$ (b) $-\dfrac{V}{R}$

(c) $-\dfrac{V}{4R}$ (d) zero

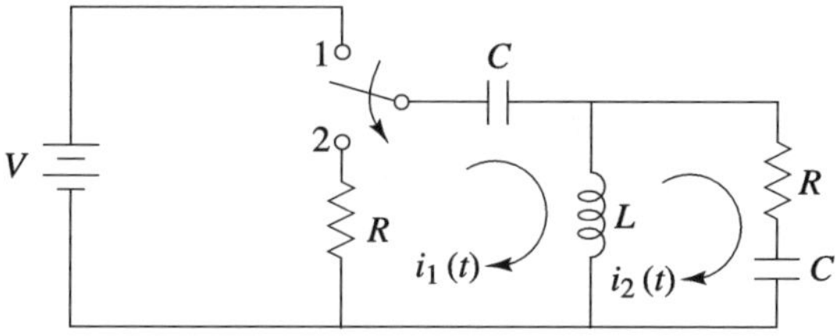

Fig. 8.141

5. For the circuit shown in Fig. 8.142, the time constant $RC = 1$ ms. The input voltage is $v_i(t) = \sqrt{2}\sin 10^3 t$. The output voltage $v_0(t)$ is equal to

(a) $\sin(10^3 - 45°)$ (b) $\sin(10^3 t + 45°)$ (c) $\sin(10^3 t - 53°)$ (d) $\sin(10^3 t + 53°)$

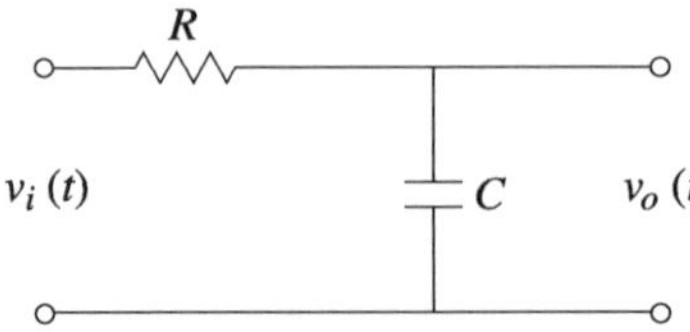

Fig. 8.142

6. For the *R-L* circuit shown in Fig. 8.143, the input voltage $v_i(t) = u(t)$. The current $i(t)$ is

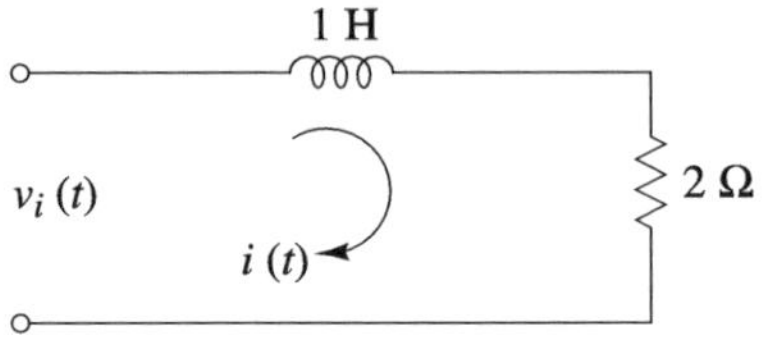

Fig. 8.143

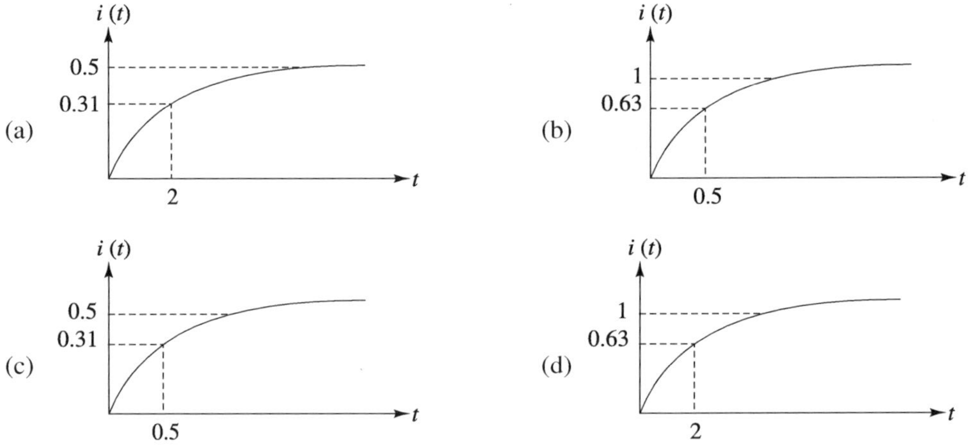

Fig. 8.144

7. The condition on *R*, *L* and *C* such that the step response $v(t)$ in Fig. 8.145 has no oscillations is

(a) $R \geq \frac{1}{2}\sqrt{\frac{L}{C}}$ (b) $R \geq \sqrt{\frac{L}{C}}$

(c) $R \geq 2\sqrt{\frac{L}{C}}$ (d) $R = \frac{1}{\sqrt{LC}}$

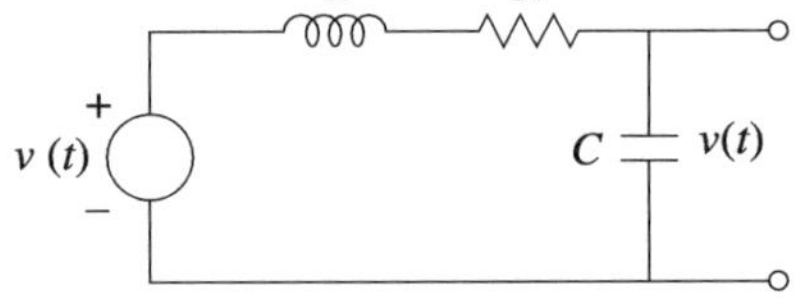

Fig. 8.145

8. The switch *S* in Fig. 8.146 closed at $t = 0$. If $v_2(0) = 10$ V and $v_g(0) = 0$ respectively, the voltages across capacitors in steady state will be

(a) $v_2(\infty) = v_1(\infty) = 0$
(b) $v_2(\infty) = 2$ V, $v_1(\infty) = 8$ V
(c) $v_2(\infty) = v_1(\infty) = 8$ V
(d) $v_2(\infty) = 8$ V, $v_1(\infty) = 2$ V

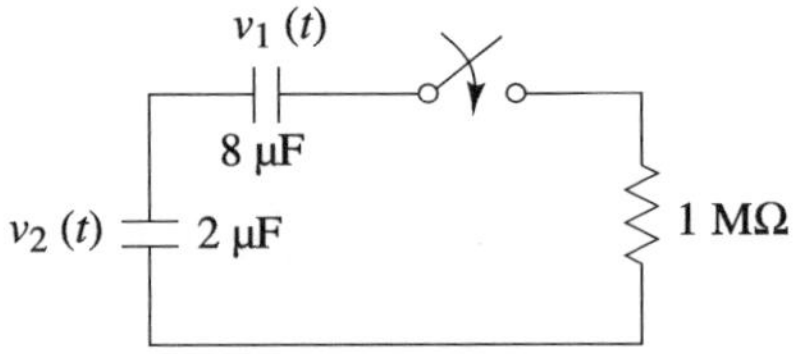

Fig. 8.146

9. The time constant of the network shown in Fig. 8.147 is

(a) 2 RC (b) 3 RC

(c) 1/2 RC (d) $\frac{2}{3}$ RC

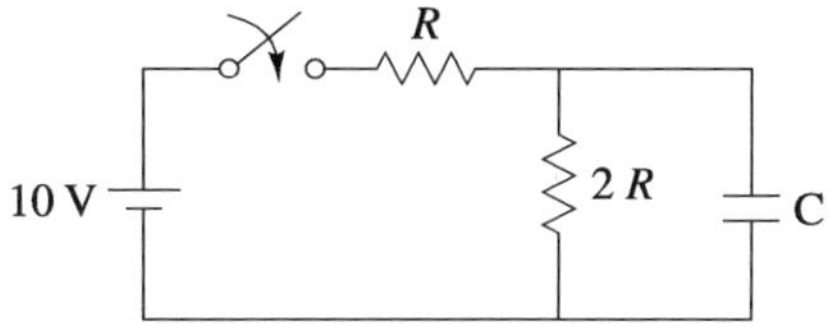

Fig. 8.147

10. In the series RC circuit shown in Fig. 8.148, the voltage across C starts increasing when the dc source is switched on. The rate of increase of voltage across C at the instant just after the switch is closed i.e., at $t = 0^+$ will be

(a) zero (b) infinity

(c) RC (d) $\frac{1}{RC}$

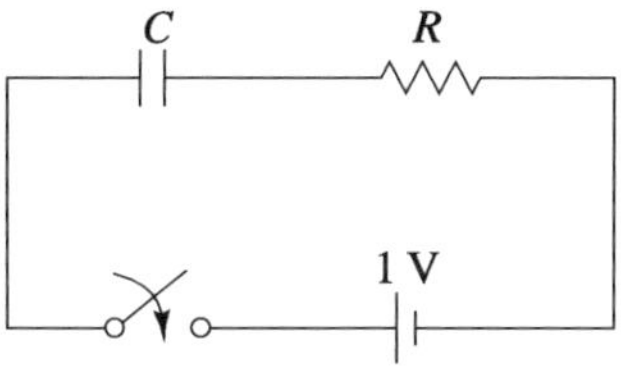

Fig. 8.148

11. The $v - i$ characteristic as seen from the terminal pair ($A - B$) of the network of Fig. 8.149(a) is shown in Fig. 8.149(b). If an inductance of value 6 mH is connected across the terminal pair, the time constant of the system will be

(a) 3 μs

(b) 12 s

(c) 32 s

(d) unknown, unless actual network is specified

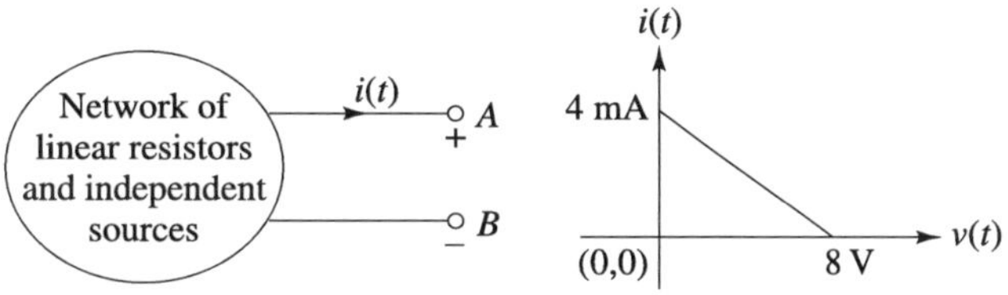

(a) (b)

Fig. 8.149

12. In the network shown in Fig. 8.150, the circuit was initially in the steady-state condition with the switch K closed. At the instant when the switch is opened, the rate of decay of current through inductance will be

(a) zero (b) 0.5 A/s

(c) 1 A/s (d) 2 A/s

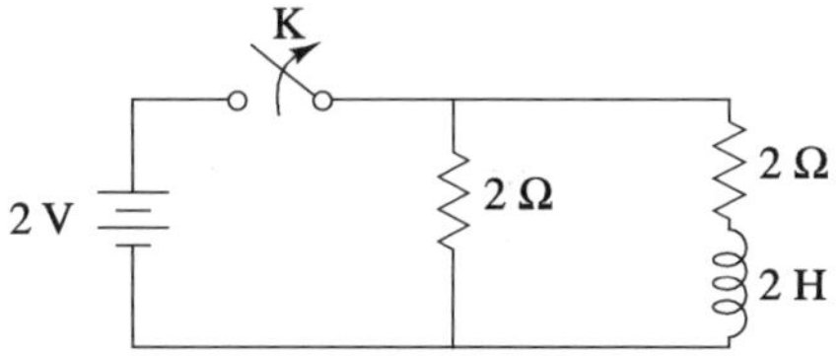

Fig. 8.150

13. A step function voltage is applied to an RLC series circuit having $R = 2\ \Omega$, $L = 1$ H and $C = 1$ F. The transient current response of the circuit would be

(a) over damped (b) critically damped (c) under damped (d) none of these

Answers to Objective-Type Questions

1. (b) **2.** (d) **3.** (c) **4.** (d) **5.** (a) **6.** (b)

7. (c) **8.** (d) **9.** (d) **10.** (d) **11.** (a) **12.** (d)

13. (b)

9 Laplace Transform and Its Application

9.1 INTRODUCTION

Time-domain analysis is the conventional method of analysing a network. For a simple network with first-order differential equation of network variable, this method is very useful. But as the order of network variable equation increases, this method of analysis becomes very tedious. For such applications, frequency domain analysis using Laplace transform is very convenient. Time-domain analysis, also known as *classical method,* is difficult to apply to a differential equation with excitation functions which contain derivatives. Laplace transform methods prove to be superior. The Laplace transform method has the following advantages:

(1) Solution of differential equations is a systematic procedure.

(2) Initial conditions are automatically incorporated.

(3) It gives the complete solution, i.e., both complementary and particular solution in one step.

9.2 LAPLACE TRANSFORMATION

The Laplace transform of a function $f(t)$ is defined as

$$F(s) = \pounds\,[f(t)] = \int_{0^-}^{\infty} f(t)\, e^{-st} dt \qquad \ldots(9.1)$$

where s is the complex frequency variable

$$s = \sigma + j\omega \qquad \ldots(9.2)$$

Here, the lower limit of integration is $t = 0^-$ instead of $t = 0^+$.

The function $f(t)$ must satisfy the following condition to possess a Laplace transform,

$$\int_{0^-}^{\infty} |\, f(t) \,|\; e^{-\sigma t}\, dt < \infty \qquad \ldots(9.3)$$

where σ is real and positive.

The inverse Laplace transform $\mathcal{L}^{-1}[F(s)]$ is

$$f(t) = \frac{1}{2\pi j} \int_{\sigma - j\infty}^{\sigma + j\infty} F(s)\, e^{st}\, ds \qquad \ldots(9.4)$$

9.3 LAPLACE TRANSFORM OF SOME IMPORTANT FUNCTIONS

9.3.1 Unit Step Function

The unit step function is defined by the equation,

$$u(t) = 1 \qquad t > 0$$
$$= 0 \qquad t < 0$$

The Laplace transform of unit step function is,

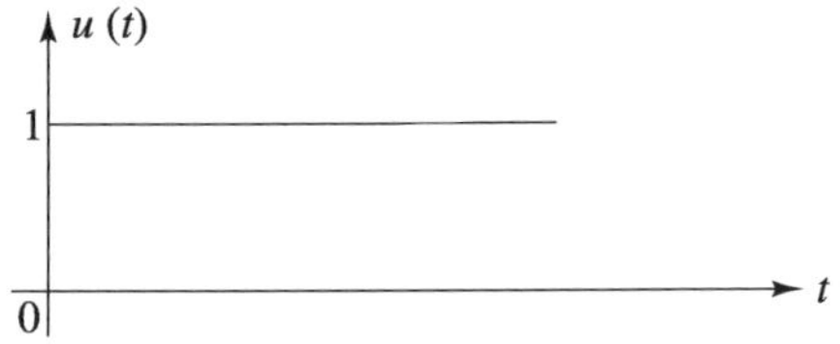

Fig. 9.1

$$\mathcal{L}[u(t)] = \int_{0^-}^{\infty} u(t)\, e^{-st}\, dt$$

$$= \int_{0^-}^{\infty} 1 \cdot e^{-st}\, dt = \left[-\frac{e^{-st}}{s} \right]_{0^-}^{\infty} = \frac{1}{s} \qquad \ldots(9.5)$$

9.3.2 Delayed or Shifted Unit Step Function

The delayed or shifted unit step function is defined by the equation

$$u(t-a) = 1 \qquad t > a$$
$$= 0 \qquad t < a$$

The Laplace transform of $u(t-a)$ is

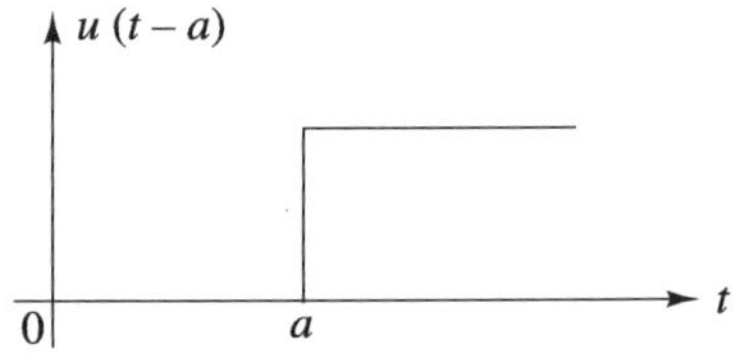

Fig. 9.2

$$\mathcal{L}[u(t-a)] = \int_{a}^{\infty} 1 . e^{-st}\, dt$$

$$= \left[-\frac{e^{-st}}{s} \right]_{a}^{\infty} = \frac{e^{-as}}{s} \qquad \ldots(9.6)$$

9.3.3 Unit Ramp Function

The unit ramp function is defined by the equation

$$r(t) = t \qquad t > 0$$
$$= 0 \qquad t < 0$$

The Laplace transform of the unit ramp function is

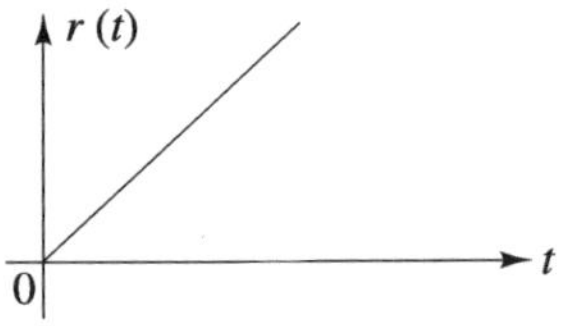

Fig. 9.3

$$\mathcal{L}[r(t)] = \int_{0^-}^{\infty} r(t)\, e^{-st}\, dt$$

$$= \int_{0^-}^{\infty} t\, e^{-st}\, dt = \frac{1}{s^2} \qquad \ldots(9.7)$$

9.3.4 Delayed Unit Ramp Function

The delayed unit ramp function is defined by the equation

$$r(t-a) = t \qquad t > a$$
$$= 0 \qquad t < a$$

The Laplace transform of $r(t-a)$ is

$$£[r(t-a)] = \int_a^{\infty} t\, e^{-st}\,dt = \frac{e^{-as}}{s^2} \qquad \ldots(9.8)$$

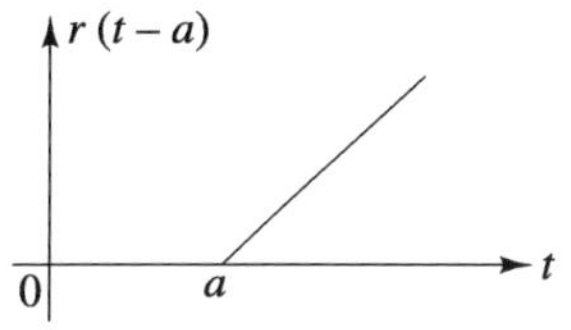

Fig. 9.4

9.3.5 Unit Impulse Function

The unit impulse function is defined by the equation

$$\delta(t) = 0 \qquad t \neq 0$$

and $$\int_{-\infty}^{\infty} \delta(t)\,dt = 1 \qquad t = 0$$

The Laplace transform of the unit impulse function is

$$£[\delta(t)] = \int_{0^-}^{\infty} \delta(t)\, e^{-st}\,dt = 1 \qquad \ldots(9.9)$$

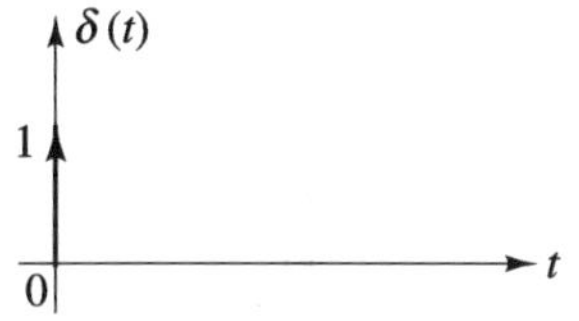

Fig. 9.5

9.3.6 Exponential Function (e^{at})

The Laplace transform of the exponential function is

$$£[e^{at}] = \int_{0^-}^{\infty} e^{at} e^{-st}\,dt$$

$$= \int_{0^-}^{\infty} e^{-(s-a)t}\,dt = \left[-\frac{e^{-(s-a)t}}{s-a}\right]_{0^-}^{\infty}$$

$$= \frac{1}{s-a} \qquad \ldots(9.10)$$

9.3.7 Sine Function

We know that $\sin \omega t = \frac{1}{2j}[e^{j\omega t} - e^{-j\omega t}]$.

The Laplace transform of sine function is

$$£[\sin \omega t] = L\left[\frac{1}{2j}(e^{j\omega t} - e^{-j\omega t})\right]$$

$$= \frac{1}{2j}\left\{£[e^{j\omega t}] - £[e^{-j\omega t}]\right\} = \frac{1}{2j}\left[\frac{1}{s-j\omega} - \frac{1}{s+j\omega}\right]$$

$$= \frac{\omega}{s^2+\omega^2} \qquad \ldots(9.11)$$

9.3.8 Cosine Function

We know that $\cos \omega t = \frac{1}{2}[e^{j\omega t} + e^{-j\omega t}]$.

The Laplace transform of cosine function is

$$£ [\cos \omega t] = £ \left[\frac{1}{2}(e^{j\omega t} + e^{-j\omega t})\right]$$

$$= \frac{1}{2}\{£ [e^{j\omega t}] + £ [e^{-j\omega t}]\} = \frac{1}{2}\left[\frac{1}{s-j\omega} + \frac{1}{s+j\omega}\right]$$

$$= \frac{s}{s^2+\omega^2} \qquad \ldots(9.12)$$

9.3.9 Exponentially Damped Function

Laplace transform of an exponentially damped function $e^{-at} f(t)$ is

$$£ [e^{-at} f(t)\, dt] = \int_{0^-}^{\infty} f(t)\, e^{-at}\, e^{-st}\, dt = \int_{0^-}^{\infty} f(t)\, e^{-(s+a)t}\, dt$$

$$= F(s+a) \qquad \ldots(9.13)$$

Thus, the transform of the function $e^{-at} f(t)$ is obtained by putting $(s + a)$ in place of s in the transform of $f(t)$.

$$£ [e^{-at} \sin \omega t] = \frac{\omega}{(s+a)^2+\omega^2} \qquad \ldots(9.14)$$

$$£ [e^{-at} \cos \omega t] = \frac{s+a}{(s+a)^2+\omega^2} \qquad \ldots(9.15)$$

Table 9.1 *Laplace Transformation*

Sr. No.	$f(t)$	$F(s)$
1	$f(t)$	$F(s) = \int_{0^-}^{\infty} f(t)\, e^{-st}\, dt$
2	$a f_1(t) + b f_2(t)$	$a F_1(s) + bF_2(s)$
3	$f(t-a)\, u(t-a)$	$e^{-as} F(s)$
4	$e^{-at} f(t)$	$F(s+a)$
5	$\frac{d}{dt} f(t)$	$sF(s) - f(0^-)$
6	$\int_{0^-}^{t} f(t)\, dt$	$\frac{1}{s} F(s)$
7	t	$\frac{1}{s^2}$
8	t^n	$\frac{n!}{s^{n+1}}$

(Contd)

Sr. No.	$f(t)$	$F(s)$
9	$t\,f(t)$	$-\dfrac{d}{ds}F(s)$
10	$\dfrac{1}{t}f(t)$	$\int_s^{\infty} F(s)\,ds$
11	$f(at)$	$\dfrac{1}{a}F\left(\dfrac{s}{a}\right)$
12	$u(t)$	$\dfrac{1}{s}$
13	$u(t-a)$	$\dfrac{e^{-as}}{s}$
14	$r(t)$	$\dfrac{1}{s^2}$
15	$r(t-a)$	$\dfrac{e^{-as}}{s^2}$
16	$\delta(t)$	1
17	$\delta(t-a)$	e^{-as}
18	e^{at}	$\dfrac{1}{s-a}$
19	$\sin \omega t$	$\dfrac{\omega}{s^2+\omega^2}$
20	$\cos \omega t$	$\dfrac{s}{s^2+\omega^2}$
21	$e^{-at}\sin \omega t$	$\dfrac{\omega}{(s+a)^2+\omega^2}$
22	$e^{-at}\cos \omega t$	$\dfrac{s+a}{(s+a)^2+\omega^2}$

9.4 THE TRANSFORMED CIRCUIT

Voltage–current relationships of network elements can also be represented in the frequency domain.

Resistor For the resistor, the v–i relationship in time domain is

$$v(t) = R\,i(t) \qquad \ldots(9.16)$$

The corresponding frequency–domain relation will be given as

$$V(s) = RI(s) \qquad \ldots(9.17)$$

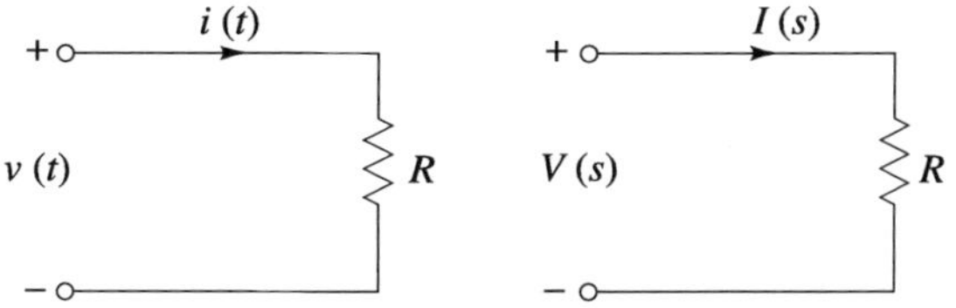

Fig. 9.6 *Resistor*

Inductor For the inductor, the v–i relationships in time domain are

$$v(t) = L\,\frac{di}{dt} \qquad \ldots(9.18)$$

$$i(t) = \frac{1}{L}\int_{0^-}^{t} v(t)\,dt + i(0^-) \qquad \ldots(9.19)$$

Transforming both the equations, we get

$$V(s) = Ls\,I(s) - L\,i(0^-) \qquad \ldots(9.20)$$

$$I(s) = \frac{1}{Ls}V(s) + \frac{i(0^-)}{s} \qquad \ldots(9.21)$$

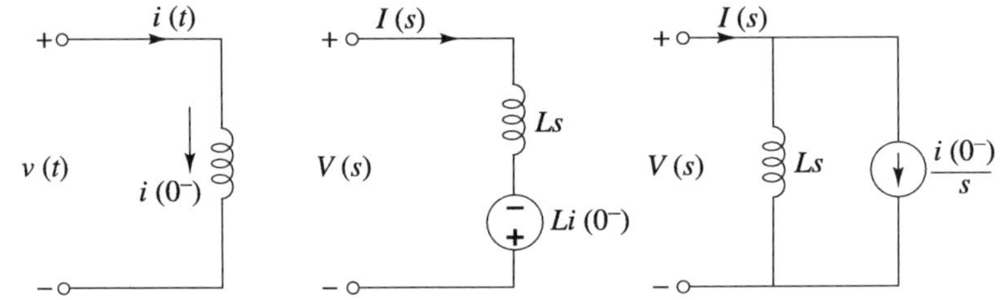

Fig. 9.7 *Inductor*

Capacitor For capacitor, the v–i relationships in time domain are

$$v(t) = \frac{1}{C}\int_{0^-}^{t} i(t)\,dt + v(0^-) \qquad \ldots(9.22)$$

$$i(t) = C\,\frac{dv}{dt} \qquad \ldots(9.23)$$

The corresponding frequency–domain relations are given as

$$V(s) = \frac{1}{Cs}I(s) + \frac{v(0^-)}{s} \qquad \ldots(9.24)$$

$$I(s) = Cs\,V(s) - Cv(0^-) \qquad \ldots(9.25)$$

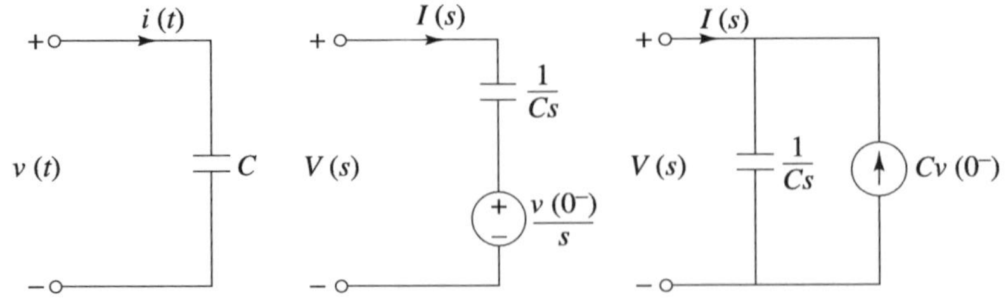

Fig. 9.8 *Capacitor*

9.5 RESISTOR–INDUCTOR CIRCUIT

Consider a series *R*-*L* circuit as shown in Fig. 9.9. The switch is closed at time $t = 0$.

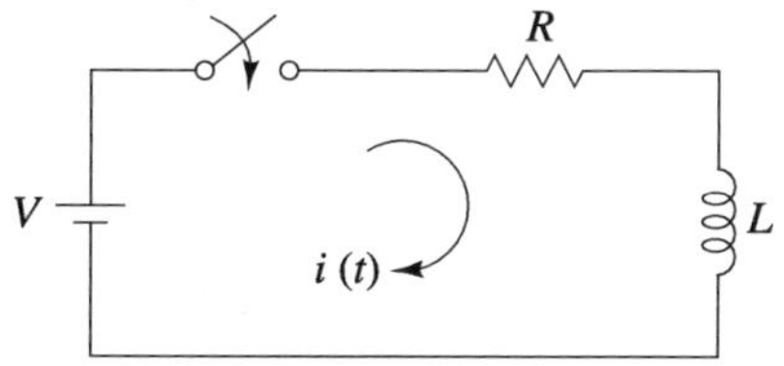

Fig. 9.9

For $t > 0$, the transformed network is shown in Fig. 9.10

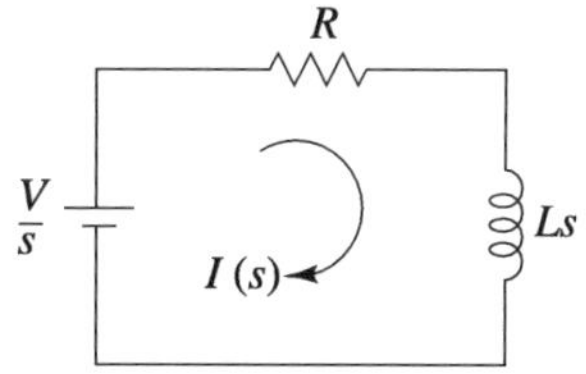

Fig. 9.10

Applying KVL to the Mesh,

$$RI(s) + LsI(s) = \frac{V}{s}$$

$$I(s) = \frac{\frac{V}{L}}{s\left(s+\frac{R}{L}\right)} \qquad \ldots(9.26)$$

By partial-fraction expansion,

$$I(s) = \frac{A}{s} + \frac{B}{s+\frac{R}{L}}$$

$$A = sI(s)\big|_{s=0} = s \times \left.\frac{\frac{V}{L}}{s\left(s+\frac{R}{L}\right)}\right|_{S=0} = \frac{V}{R}$$

$$B = \left(s+\frac{R}{L}\right)I(s)\bigg|_{s=-R/L} = \left(s+\frac{R}{L}\right) \times \left.\frac{\frac{V}{L}}{s\left(s+\frac{R}{L}\right)}\right|_{s=-R/L} = -\frac{V}{R}$$

$$I(s) = \left[\frac{\frac{V}{R}}{s} + \frac{\left(-\frac{V}{R}\right)}{s+\frac{R}{L}}\right]$$

Taking the inverse Laplace transform,

$$i(t) = \frac{V}{R} - \frac{V}{R}e^{-(R/L)t}$$

$$= \frac{V}{R}\left[1 - e^{-(R/L)t}\right] \qquad \text{for } t > 0 \qquad \ldots(9.27)$$

9.6 RESISTOR–CAPACITOR CIRCUIT

Consider a series *R*-*C* circuit as shown in Fig. 9.11. The switch is closed at time $t = 0$.

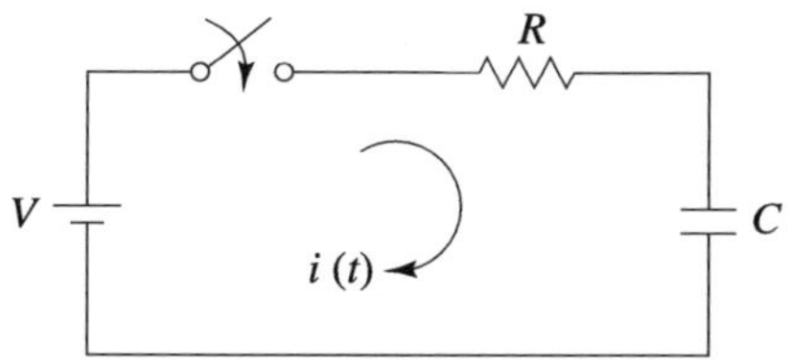

Fig. 9.11

For $t > 0$, the transformed network is shown in Fig. 9.12
Applying KVL to the Mesh,

$$R\,I(s) + \frac{1}{Cs}\,I(s) = \frac{V}{s}$$

$$I(s) = \frac{\frac{V}{R}}{s + \frac{1}{RC}}$$

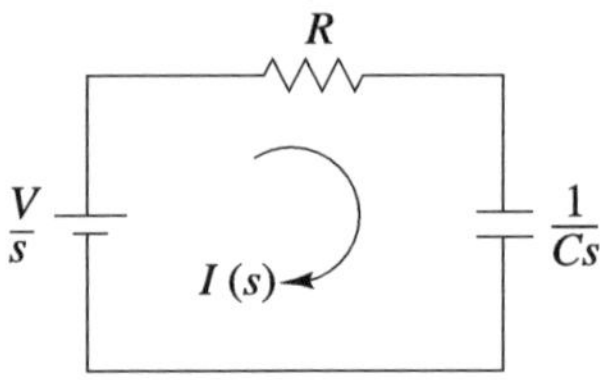

Fig. 9.12

Taking the inverse Laplace transform,

$$i(t) = \frac{V}{R}\,e^{-(1/RC)t} \qquad \text{for } t > 0 \qquad \text{...(9.28)}$$

9.7 RESISTOR–INDUCTOR–CAPACITOR CIRCUIT

Consider a series *R-L-C* circuit shown in Fig. 9.13. The switch is closed at time $t = 0$.

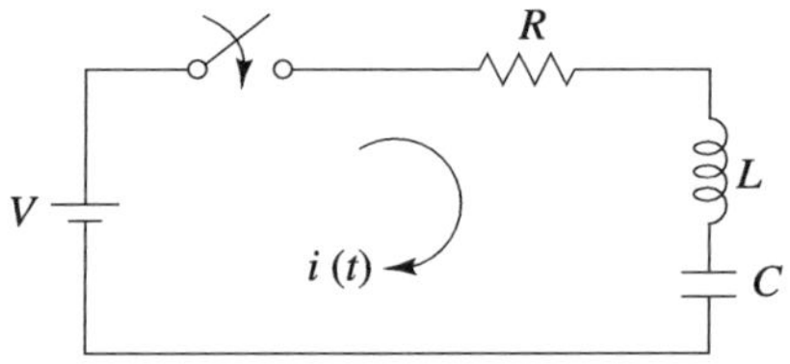

Fig. 9.13

For $t > 0$, the transformed network is shown in Fig. 9.14
Applying KVL to the Mesh,

$$RI(s) + Ls\,I(s) + \frac{1}{Cs}\,I(s) = \frac{V}{s}$$

$$I(s) = \frac{\frac{V}{L}}{s^2 + \frac{R}{L}s + \frac{1}{LC}}$$

$$= \frac{\frac{V}{L}}{(s - s_1)(s - s_2)} \qquad \text{...(9.29)}$$

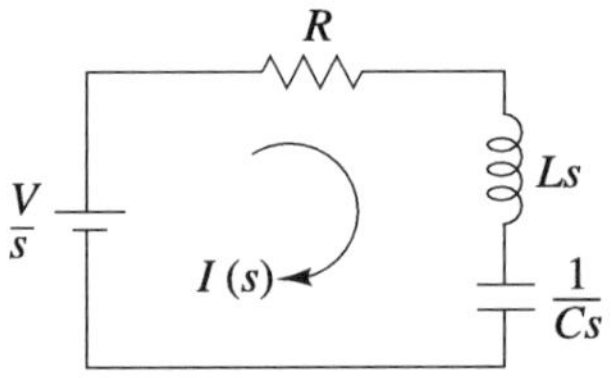

Fig. 9.14

where s_1 and s_2 are the roots of the equation $s^2 + \left(\frac{R}{L}\right)s + \left(\frac{1}{LC}\right) = 0.$

$$s_1 = -\frac{R}{2L} + \sqrt{\left(\frac{R}{2L}\right)^2 - \frac{1}{LC}}$$

$$= -\alpha + \sqrt{\alpha^2 - \omega_o^2} = -\alpha + \beta$$

$$s_2 = -\frac{R}{2L} - \sqrt{\left(\frac{R}{2L}\right)^2 - \frac{1}{LC}}$$

$$= -\alpha - \sqrt{\alpha^2 - \omega_o^2} = -\alpha - \beta$$

where $$\alpha = \frac{R}{2L}$$

$$\omega_o = \frac{1}{\sqrt{LC}}$$

and $$\beta = \sqrt{\alpha^2 - \omega_o^2}$$

By partial-fraction expansion, of $I(s)$,

$$I(s) = \frac{A}{s - s_1} + \frac{B}{s - s_2}$$

$$A = (s - s_1)\, I(s)\big|_{s=s_1}$$

$$= \frac{\frac{V}{L}}{s_1 - s_2}$$

$$B = (s - s_2)\, I(s)\big|_{s=s_2}$$

$$= \frac{\frac{V}{L}}{s_2 - s_1} = -\frac{\frac{V}{L}}{s_1 - s_2}$$

$$I(s) = \frac{V}{L(s_1 - s_2)}\left[\frac{1}{s - s_1} - \frac{1}{s - s_2}\right]$$

Taking the inverse Laplace transform,

$$i(t) = \frac{V}{L(s_1 - s_2)}\left[e^{s_1 t} - e^{s_2 t}\right] \qquad \ldots(9.30)$$

$$= A_1\, e^{s_1 t} + A_2\, e^{s_2 t}$$

where A_1 and A_2 are constants to be determined and s_1 and s_2 are the roots of the equation.

Now depending upon the values of s_1 and s_2, we have 3 cases of the response.

Case I When the roots are real and unequal, it gives an overdamped response.

$$\frac{R}{2L} > \frac{1}{\sqrt{LC}}$$

$$\alpha > \omega_o$$

In this case, the solution is given by

$$i(t) = e^{-\alpha t}\,(A_1\, e^{\beta t} + A_2\, e^{-\beta t})$$

or $$i(t) = A_1\, e^{s_1 t} + A_2\, e^{s_2 t} \qquad \text{for } t > 0$$

Case II When the roots are real and equal, it gives a critically damped response.

$$\frac{R}{2L} = \frac{1}{\sqrt{LC}}$$

$$\alpha = \omega_o$$

In this case, the solution is given by

$$i(t) = e^{-\alpha t}\,(A_1 + A_2\, t) \qquad \text{for } t > 0$$

Case III When the roots are complex conjugate, it gives an underdamped response.

$$\frac{R}{2L} < \frac{1}{\sqrt{LC}}$$

$$\alpha < \omega_o$$

In this case, the solution is given by

$$i(t) = A_1\, e^{s_1 t} + A_2\, e^{s_2 t}$$

where $s_{1,2} = -\alpha \pm \sqrt{\alpha^2 - \omega_o^2}$

Let $\sqrt{\alpha^2 - \omega_o^2} = \sqrt{-1}\sqrt{\omega_o^2 - \alpha^2}$

$$= j\,\omega_d$$

where $j = \sqrt{-1}$

and $\omega_d = \sqrt{\omega_o^2 - \alpha^2}$

Hence $i(t) = e^{-\alpha t}(A_1\, e^{j\omega_d t} + A_2\, e^{-j\omega_d t})$

$$= e^{-\alpha t}\left[(A_1 + A_2)\left\{\frac{e^{j\omega_d t} + e^{-j\omega_d t}}{2}\right\} + j(A_1 - A_2)\left\{\frac{e^{j\omega_d t} - e^{-j\omega_d t}}{2j}\right\}\right]$$

$$= e^{-\alpha t}\,[(A_1 + A_2)\cos \omega_d t + j\,(A_1 - A_2)\sin \omega_d t]$$

$$= e^{-\alpha t}\,(B_1 \cos \omega_d t + B_2 \sin \omega_d t) \qquad \text{for } t > 0$$

Example 9.1 *In the network of Fig. 9.15, the switch is moved from a to b at t = 0. Determine i (t) and v_c (t).*

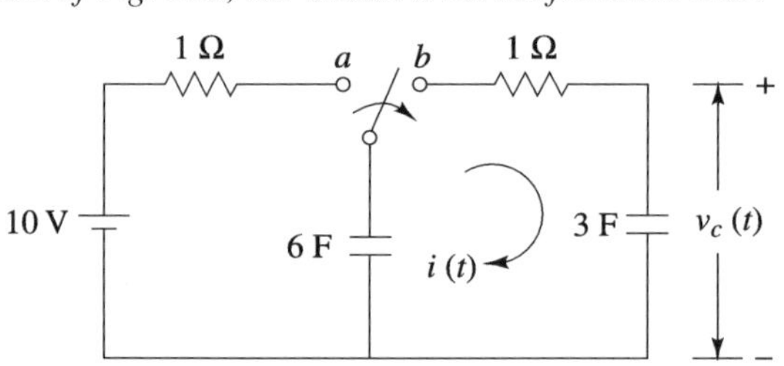

Fig. 9.15

Solution At $t = 0^-$, the network has attained steady-state condition. Hence, the capacitor of 6 F acts as an open circuit.

$$v_{6\,F}(0^-) = 10 \text{ V}$$
$$i\,(0^-) = 0$$
$$v_{3\,F}(0^-) = 0$$

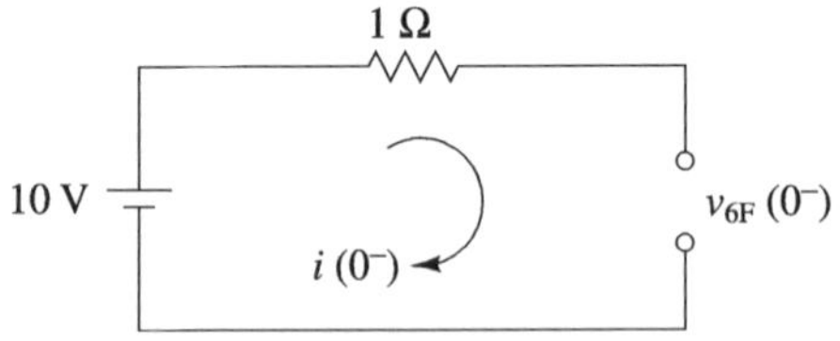

Fig. 9.16

Since voltage across the capacitor cannot change instantaneously,

$$v_{6\,F}(0^+) = 10 \text{ V}$$
$$v_{3\,F}(0^+) = 0$$

For $t > 0$, the transformed network is shown in Fig. 9.17. Applying KVL to the Mesh for $t > 0$,

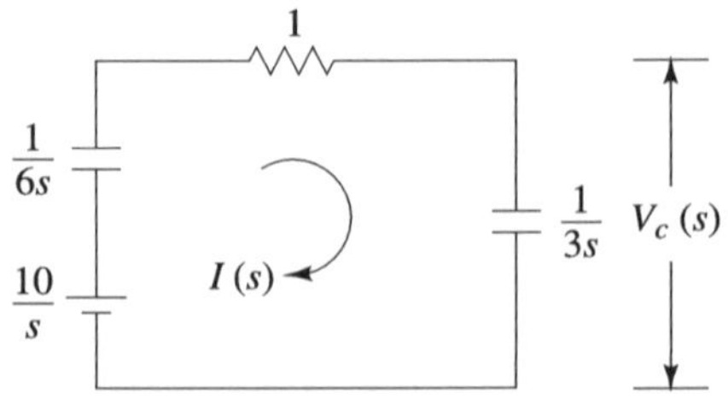

Fig. 9.17

$$\frac{10}{s} - \frac{1}{6s}\,I(s) - I\,(s) - \frac{1}{3s}\,I(s) = 0$$

$$\frac{1}{6s}\,I(s) + I(s) + \frac{1}{3s}\,I(s) = \frac{10}{s}$$

$$I(s) = \frac{10}{s\left(1+\frac{1}{6s}+\frac{1}{3s}\right)}$$

$$= \frac{60}{6s+3} = \frac{10}{s+0.5}$$

Taking the inverse Laplace transform,

$$i(t) = 10e^{-0.5t} \qquad \text{for } t > 0$$

Voltage across the 3-F capacitor is given by

$$V_c(s) = \frac{1}{3s} I(s) = \frac{10}{3s(s+0.5)}$$

By partial-fraction expansion,

$$V_c(s) = \frac{A}{s} + \frac{B}{s+0.5}$$

$$A = sV_c(s)\big|_{s=0}$$

$$= \frac{10}{3(s+0.5)}\bigg|_{s=0} = \frac{20}{3}$$

$$B = (s+0.5)V_c(s)\big|_{s=-0.5}$$

$$= \frac{10}{3s}\bigg|_{s=-0.5} = -\frac{20}{3}$$

$$V_c(s) = \frac{\frac{20}{3}}{s} - \frac{\frac{20}{3}}{s+0.5}$$

Taking the inverse Laplace transform,

$$v_c(t) = \frac{20}{3} - \frac{20}{3}e^{-0.5t}$$

$$= \frac{20}{3}(1-e^{-0.5t}) \qquad \text{for } t > 0$$

Example 9.2 *The switch in the network shown in Fig. 9.18 is closed at t = 0. Determine the voltage across the capacitor.*

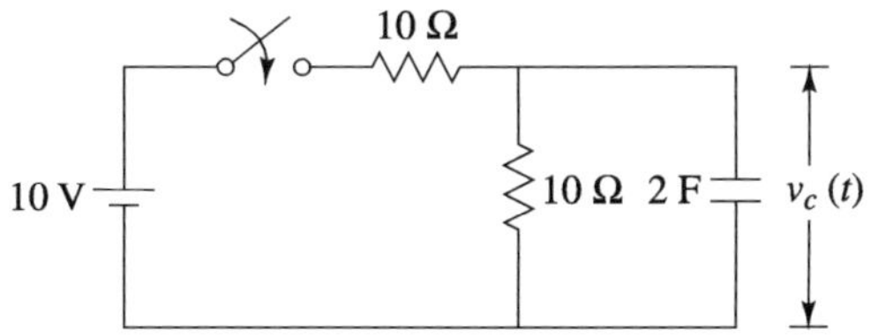

Fig. 9.18

Solution At $t = 0^-$, the capacitor is uncharged.

$$v_c(0^-) = 0$$

Since the voltage across the capacitor cannot change instantaneously,

$$v_c(0^+) = 0$$

For $t > 0$, the transformed network is shown in Fig. 9.19.

Applying KCL at Node for $t > 0$,

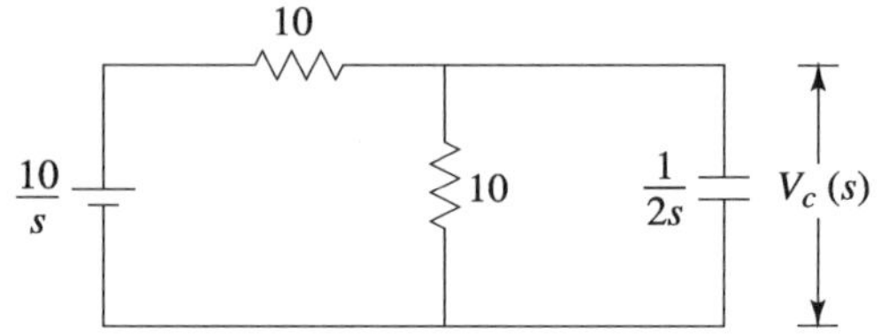

Fig. 9.19

$$\frac{V_c(s) - \frac{10}{s}}{10} + \frac{V_c(s)}{10} + \frac{V_c(s)}{\frac{1}{2s}} = 0$$

$$2s\, V_c(s) + 0.2\, V_c(s) = \frac{1}{s}$$

$$V_c(s) = \frac{1}{s(2s+0.2)} = \frac{0.5}{s(s+0.1)}$$

By partial-fraction expansion,

$$V_c(s) = \frac{A}{s} + \frac{B}{s+0.1}$$

$$A = sV_c(s)|_{s=0} = \left.\frac{0.5}{s+0.1}\right|_{s=0} = \frac{0.5}{0.1} = 5$$

$$B = (s+0.1)V_c(s)|_{s=-0.1} = \left.\frac{0.5}{s}\right|_{s=-0.1} = -\frac{0.5}{0.1} = -5$$

$$V_c(s) = \frac{5}{s} - \frac{5}{s+0.1}$$

Taking inverse Laplace transform,

$$v_c(t) = 5 - 5\, e^{-0.1t} \qquad \text{for } t > 0$$

Example 9.3 *In the network of Fig. 9.20, the switch is moved from the position 1 to 2 at t = 0, steady-state conditions having been established in the position 1. Determine i (t) for t > 0.*

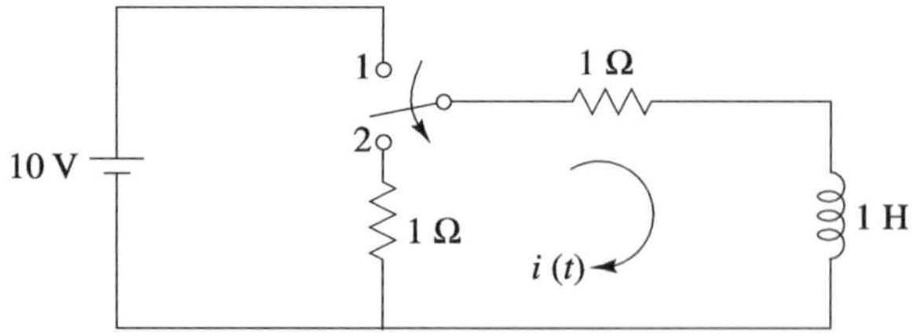

Fig. 9.20

Solution At $t = 0^-$, the network has attained steady-state condition. Hence, the inductor acts as a short circuit.

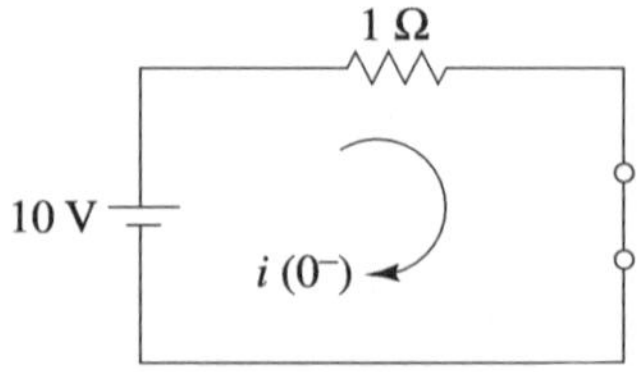

Fig. 9.21

$$i(0^-) = \frac{10}{1} = 10 \text{ A}$$

Since the current through the inductor cannot change instantaneously,

$$i(0^+) = 10 \text{ A}$$

For $t > 0$, the transformed network is shown in Fig. 9.22.

Applying KVL to the Mesh for $t > 0$,

$$-I(s) - I(s) - sI(s) + 10 = 0$$

$$I(s)(s+2) = 10$$

$$I(s) = \frac{10}{s+2}$$

Taking inverse Laplace transform,

$$i(t) = 10e^{-2t} \quad \text{for } t > 0$$

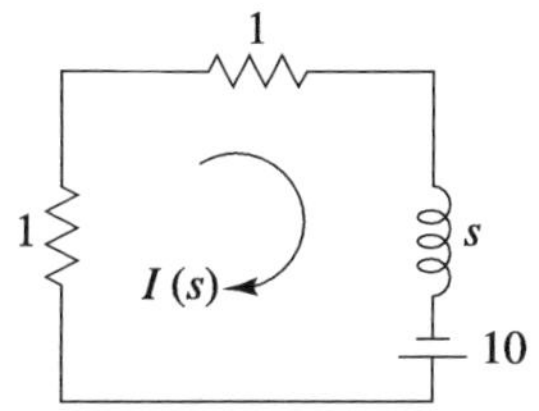

Fig. 9.22

Example 9.4 *The network of Fig. 9.23 was initially in the steady state with the switch in the position a. At t = 0, the switch goes from a to b. Find an expression for voltage v (t) for t > 0.*

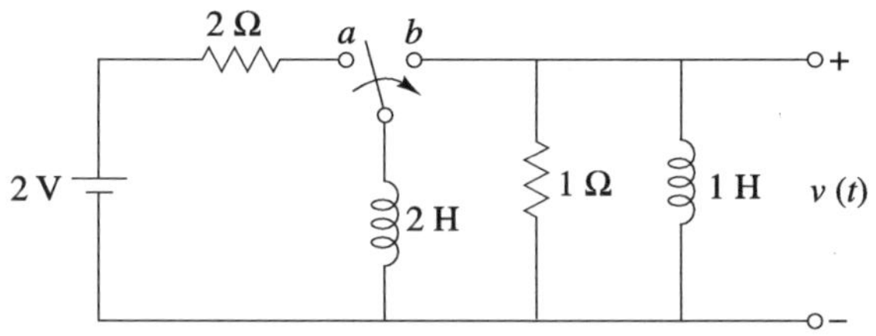

Fig. 9.23

Solution At $t = 0^-$, the network has attained steady-state condition. Hence, the inductor acts as a short circuit.

$$i(0^-) = \frac{2}{2} = 1 \text{ A}$$

Since current through the inductor cannot change instantaneously,

$$i(0^+) = 1 \text{ A}$$

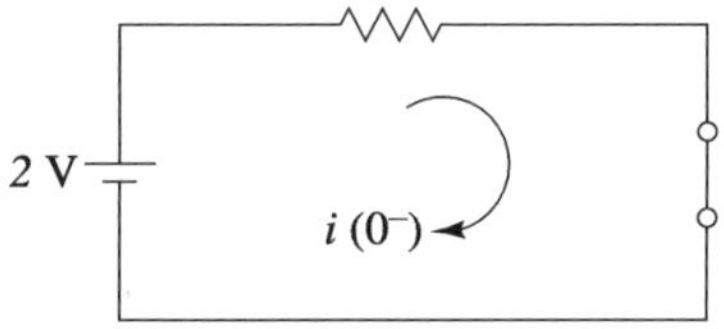

Fig. 9.24

For $t > 0$, the transformed network is shown in Fig. 9.25.

Applying KCL at Node for $t > 0$

$$\frac{V(s)+2}{2s} + \frac{V(s)}{1} + \frac{V(s)}{s} = 0$$

$$V(s)\left(1 + \frac{3}{2s}\right) = -\frac{1}{s}$$

$$V(s) = \frac{2}{2s+3} = -\frac{1}{s+1.5}$$

Taking the inverse Laplace transform,

$$v(t) = -e^{-1.5t} \quad \text{for } t > 0$$

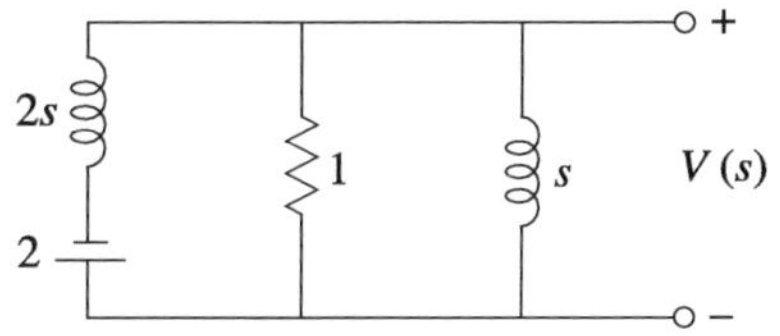

Fig. 9.25

Example 9.5 *The switch in Fig. 9.26 is opened at time t = 0. Determine the voltage v(t) for t > 0.*

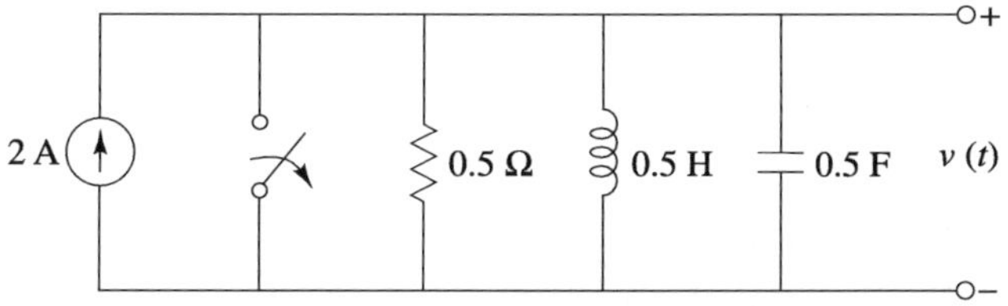

Fig. 9.26

Solution At $t = 0^-$, the network has attained steady-state condition. Hence, the inductor acts as a short circuit and the capacitor acts as an open circuit.

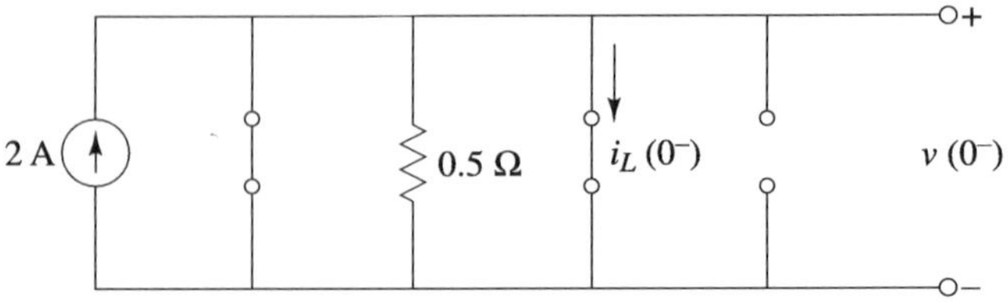

Fig. 9.27

$$i_L(0^-) = 0$$
$$v(0^-) = 0$$

Since current through the inductor and voltage across the capacitor cannot change instantaneously,

$$i_L(0^+) = 0$$
$$v(0^+) = 0$$

For $t > 0$, the transformed network is shown in Fig. 9.28.

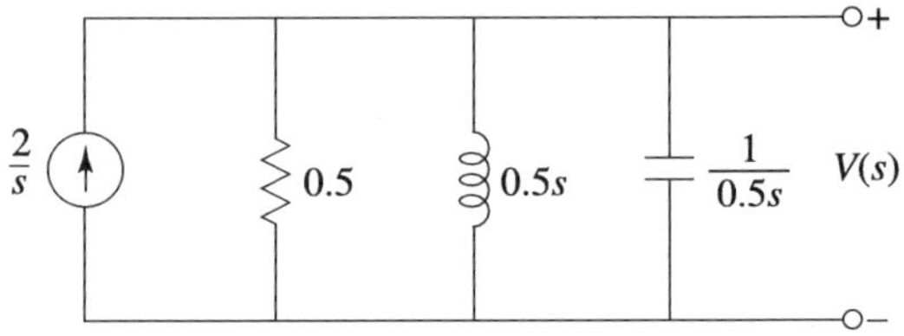

Fig. 9.28

Applying KCL at Node for $t > 0$,

$$\frac{V(s)}{0.5} + \frac{V(s)}{0.5s} + \frac{V(s)}{\frac{1}{0.5s}} = \frac{2}{s}$$

$$2V(s) + \frac{2}{s}V(s) + 0.5s\,V(s) = \frac{2}{s}$$

$$V(s) = \frac{\frac{2}{s}}{\frac{2}{s} + 0.5s + 2}$$

$$= \frac{4}{s^2 + 4s + 4} = \frac{4}{(s+2)^2}$$

Taking inverse Laplace transform,

$$v(t) = 4\,t\,e^{-2t} \qquad \text{for } t > 0$$

Example 9.6 *In the network of Fig. 9.29, the switch is closed and steady-state is attained. At t = 0, switch is opened. Determine the current through the inductor.*

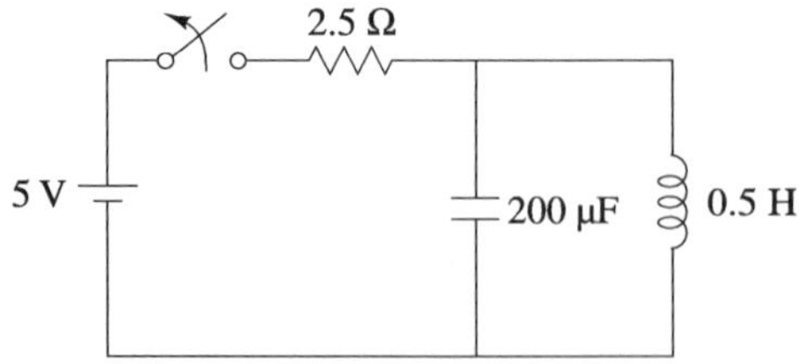

Fig. 9.29

Solution At $t = 0^-$, the switch is closed and steady-state condition is attained. Hence, the inductor acts as a short circuit and the capacitor acts as an open circuit.

Current through inductor is same as the current through the resistor.

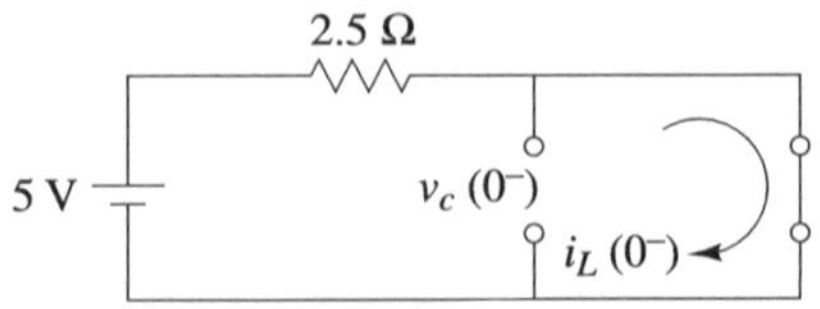

Fig. 9.30

$$i_L(0^-) = \frac{5}{2.5} = 2\text{ A}$$

Voltage across the capacitor is zero as it is connected in parallel with a short.

$$v_c(0^-) = 0$$

Since voltage across the capacitor and current through the inductor cannot change instantaneously,

$$i_L(0^+) = 2\text{ A}$$

$$v_c(0^+) = 0$$

For $t > 0$, the transformed network is shown in Fig. 9.31.

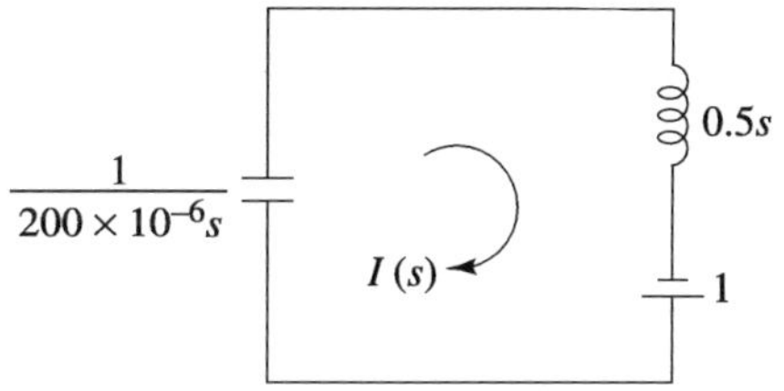

Fig. 9.31

Applying KVL to the Mesh for $t > 0$,

$$-\frac{1}{200 \times 10^{-6}\, s} I(s) - 0.5s\, I(s) + 1 = 0$$

$$0.5s\, I(s) - 1 + 5000\frac{I(s)}{s} = 0$$

$$I(s) = \frac{1}{0.5s + \frac{5000}{s}} = \frac{2s}{s^2 + 10000}$$

Taking inverse Laplace transform,

$$i(t) = 2 \cos 100\, t \qquad \text{for } t > 0$$

Example 9.7 *In the network of Fig. 9.32, the switch is closed for a long time and at t = 0, the switch is opened. Determine the current through the capacitor.*

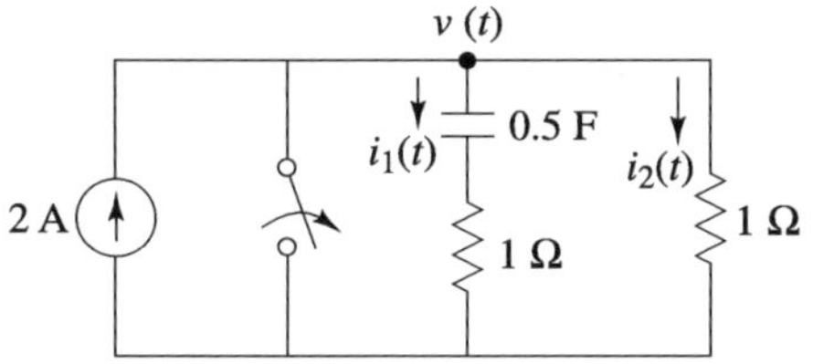

Fig. 9.32

Solution At $t = 0^-$, the switch is closed and steady-state condition is reached.

$$v_c(0^-) = 0$$

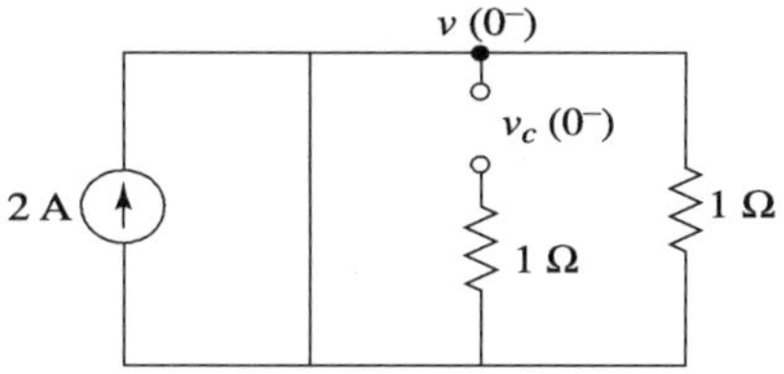

Fig. 9.33

Since voltage across the capacitor cannot change instantaneously,

$$v_c(0^+) = 0$$

For $t > 0$, the transformed network is shown in Fig. 9.34.

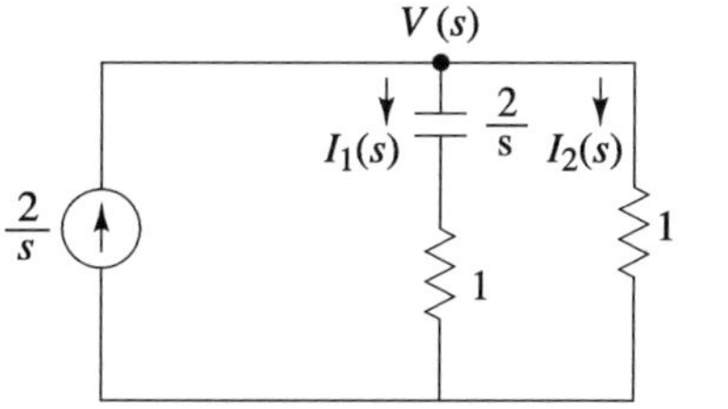

Fig. 9.34

Applying KVL to two parallel branches,

$$\frac{2}{s} I_1(s) + I_1(s) = I_2(s)$$

Applying KCL at Node for $t > 0$,

$$\frac{2}{s} = I_1(s) + I_2(s)$$

$$\frac{2}{s}I_1(s) + I_1(s) = \frac{2}{s} - I_1(s)$$

$$\frac{2}{s}I_1(s) + 2\,I_1(s) = \frac{2}{s}$$

$$I_1(s) = \frac{\frac{2}{s}}{\frac{2}{s}+2} = \frac{1}{s+1}$$

Taking the inverse Laplace transform,

$$i_1(t) = e^{-t} \qquad \text{for } t > 0$$

Example 9.8 *In the network of Fig. 9.35, the switch is moved from a to b, at t = 0. Find v(t).*

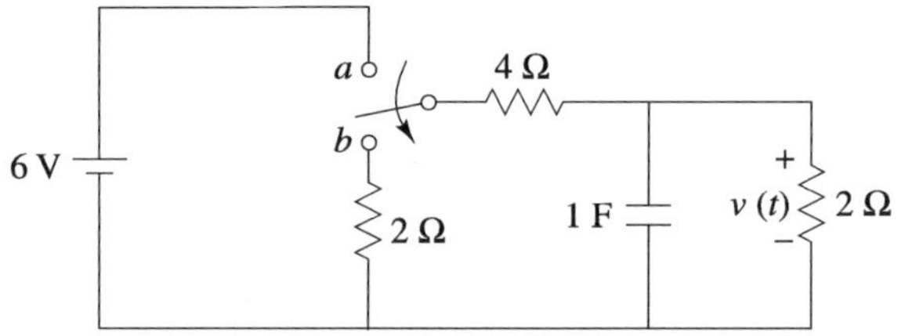

Fig. 9.35

Solution At $t = 0^-$, steady-state condition is reached. Hence, the capacitor acts as an open circuit.

$$v(0^-) = 6 \times \frac{2}{4+2} = 2\text{ V}$$

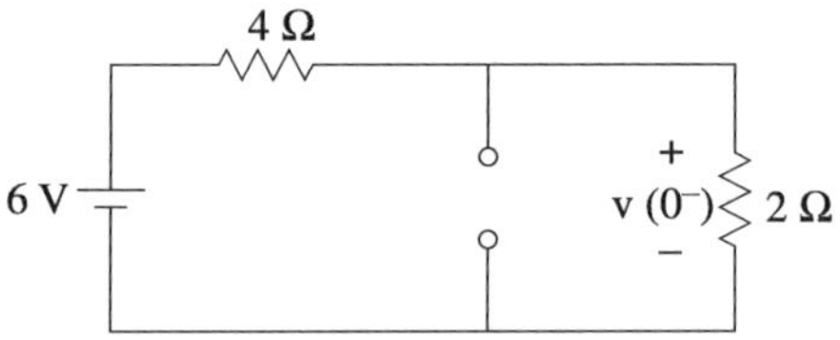

Fig. 9.36

Since voltage across the capacitor cannot change instantaneously,

$$v(0^+) = 2\text{ V}$$

For $t > 0$, the transformed network is shown in Fig. 9.37.

Applying KCL at Node for $t > 0$,

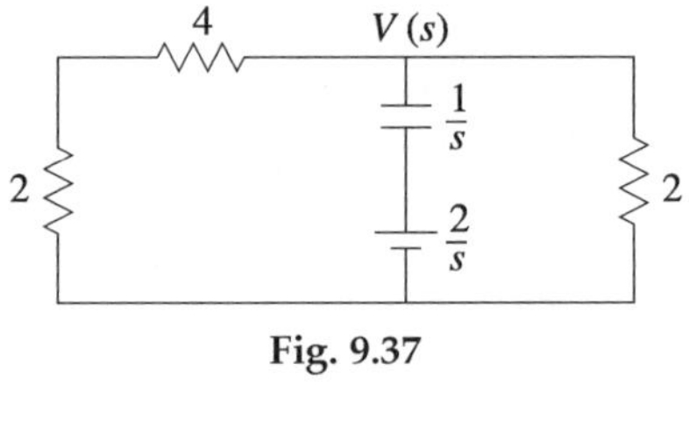

Fig. 9.37

$$\frac{V(s)}{6} + \frac{V(s) - \frac{2}{s}}{\frac{1}{s}} + \frac{V(s)}{2} = 0$$

$$V(s)\left[\frac{2}{3} + s\right] = 2$$

$$V(s) = \frac{2}{s + \frac{2}{3}}$$

Taking the inverse Laplace transform,

$$v(t) = 2\,e^{-(2/3)\,t} \qquad \text{for } t > 0$$

Example 9.9 *In the network of Fig. 9.38, the switch is opened at t = 0. Find i(t).*

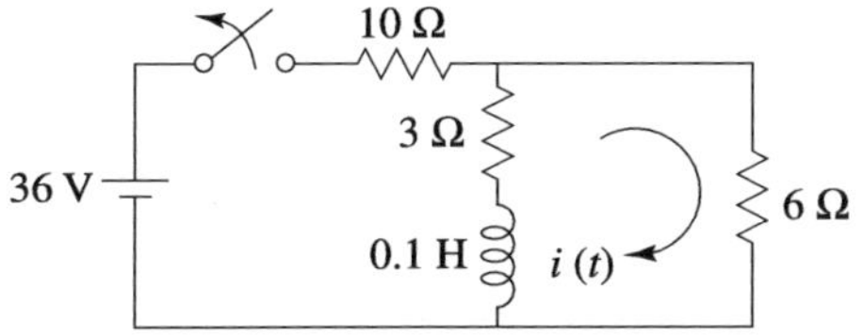

Fig. 9.38

Solution At $t = 0^-$, the switch is closed and steady-state condition is reached. Hence, the inductor acts as a short circuit.

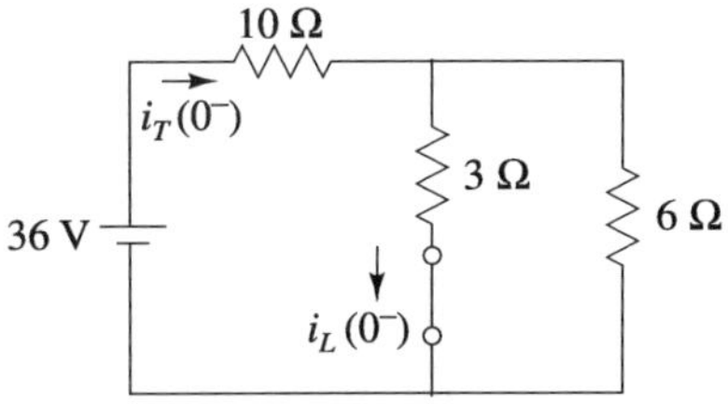

Fig. 9.39

$$i_T(0^-) = \frac{36}{10+(3 \parallel 6)} = \frac{36}{10+2} = 3 \text{ A}$$

$$i_L(0^-) = 3 \times \frac{6}{6+3} = 2 \text{ A}$$

Since current through the inductor cannot change instantaneously,

$$i_L(0^+) = 2 \text{ A}$$

For $t > 0$, the transformed network is shown in Fig. 9.40.

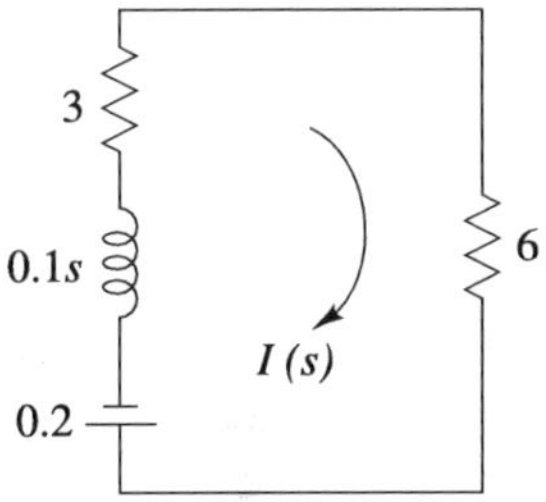

Fig. 9.40

Applying KVL to Mesh for $t > 0$,

$$-0.2 - 0.1s\, I(s) - 3I(s) - 6I(s) = 0$$

$$0.1sI(s) + 9I(s) = 0.2$$

$$I(s) = \frac{0.2}{0.1s+9} = \frac{2}{s+90}$$

Taking inverse Laplace transform,

$$i(t) = 2\, e^{-90\,t} \qquad \text{for } t > 0$$

Example 9.10 *The network shown in Fig. 9.41 has acquired steady-state with the switch closed for t < 0. At t = 0, the switch is opened. Obtain i (t) for t > 0.*

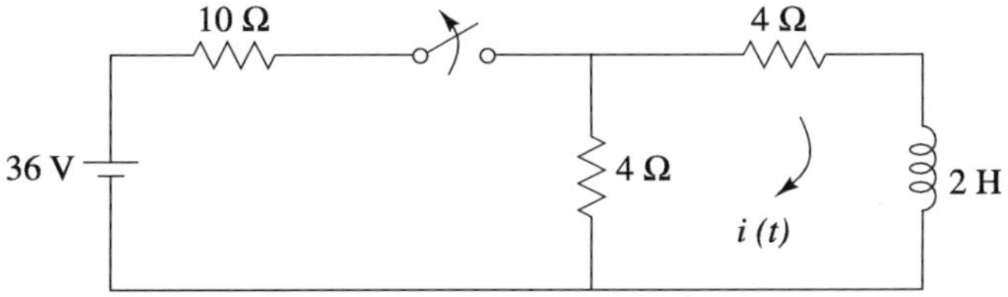

Fig. 9.41

Solution At $t = 0^-$, the switch is closed and the network has acquired steady-state. Hence, the inductor acts as a short circuit.

$$i_T(0^-) = \frac{36}{10+(4 \parallel 4)}$$

$$= \frac{36}{10+2} = 3\text{ A}$$

$$i(0^-) = 3 \times \frac{4}{4+4} = 1.5\text{ A}$$

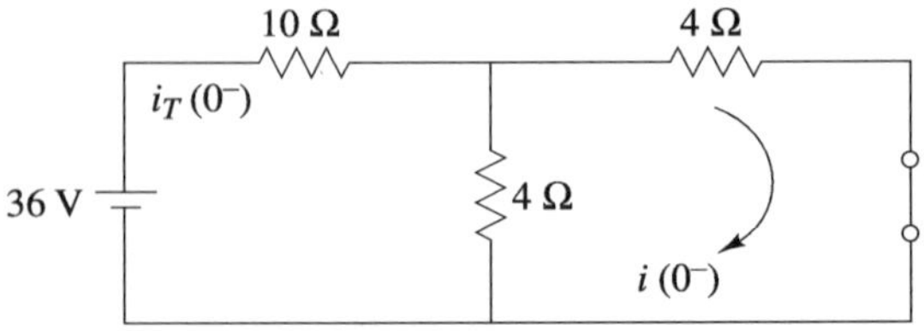

Fig. 9.42

Since current through the inductor cannot change instantaneously,

$$i(0^+) = 1.5\text{ A}$$

For $t > 0$, the transformed network is shown in Fig. 9.43.

Applying KVL to the Mesh for $t > 0$,

$$-4I(s) - 4I(s) - 2sI(s) + 3 = 0$$

$$8I(s) + 2sI(s) = 3$$

$$I(s) = \frac{3}{2s+8} = \frac{1.5}{s+4}$$

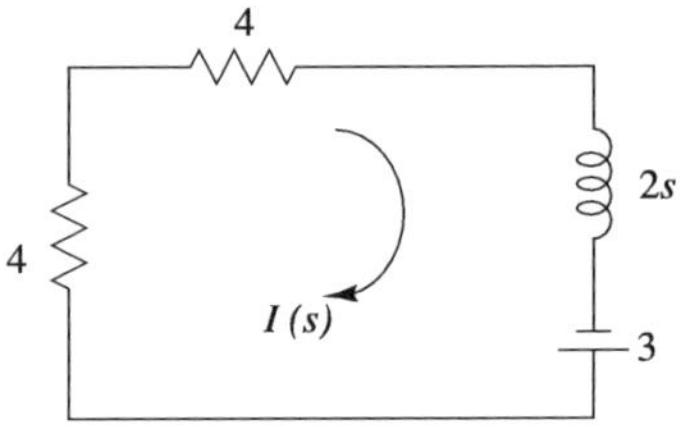

Fig. 9.43

Taking the inverse Laplace transform,

$$i(t) = 1.5\, e^{-4t} \qquad \text{for } t > 0$$

Example 9.11 *The network shown in Fig. 9.44 has acquired steady-state at t < 0 with the switch open. The switch is closed at t = 0. Determine v (t).*

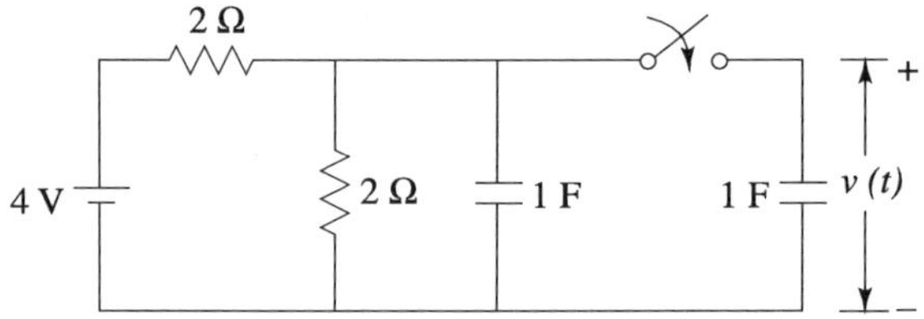

Fig. 9.44

Solution At $t = 0^-$, steady-state condition is reached. Hence, the capacitor of 1 F acts as an open circuit.

$$v(0^-) = 4 \times \frac{2}{2+2} = 2\text{ V}$$

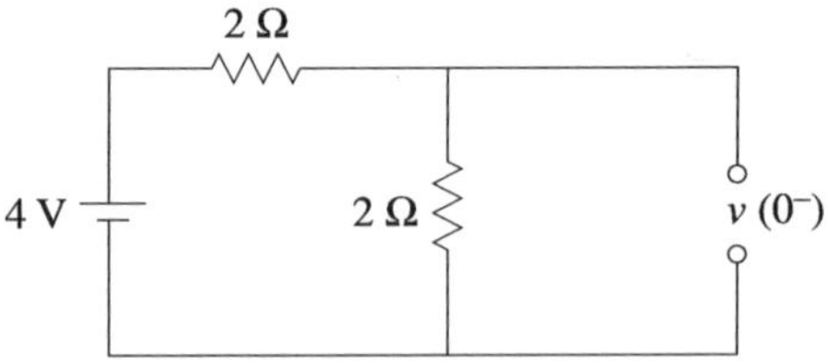

Fig. 9.45

Since voltage across the capacitor cannot change instantaneously,

$$v(0^+) = 2\text{ V}$$

For $t > 0$, the transformed network is shown in Fig. 9.46.

Applying KCL at the Node for $t > 0$,

$$\frac{V(s) - \frac{4}{s}}{2} + \frac{V(s)}{2} + \frac{V(s) - \frac{2}{s}}{\frac{1}{s}} + \frac{V(s)}{\frac{1}{s}} = 0$$

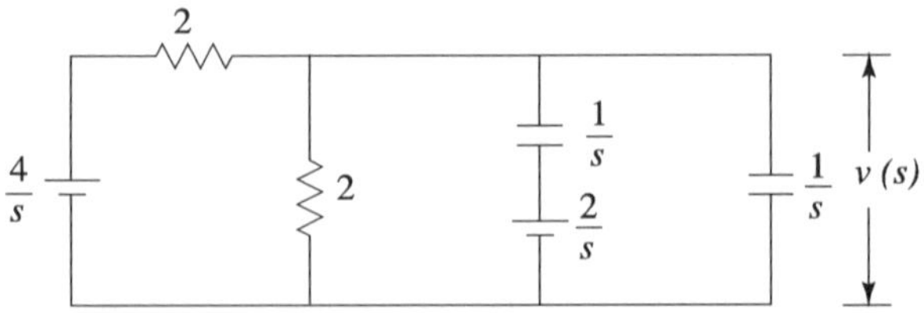

Fig. 9.46

$$2s\, V(s) + V(s) = \frac{2}{s} + 2$$

$$V(s) = \frac{\frac{2}{s}+2}{2s+1}$$

$$= \frac{2s+2}{s(2s+1)} = \frac{2}{s} - \frac{2}{2s+1} = \frac{2}{s} - \frac{1}{s+0.5}$$

Taking the inverse Laplace transform,

$$v(t) = 2 - e^{-0.5\,t} \qquad \text{for } t > 0$$

Example 9.12 *In the network shown in Fig. 9.47, the switch is opened at t = 0. Steady-state condition is achieved before t = 0. Find i (t).*

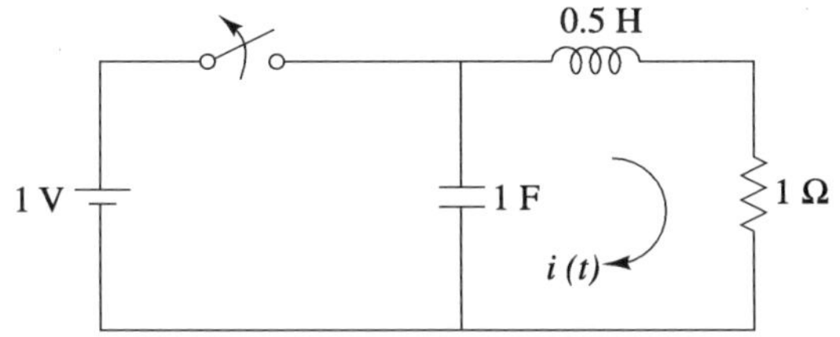

Fig. 9.47

Solution At $t = 0^-$, the switch is closed and steady-state condition is achieved. Hence, the capacitor acts as an open circuit and the inductor acts as a short circuit.

$$v_C(0^-) = 1 \text{ V}$$
$$i(0^-) = 1 \text{ A}$$

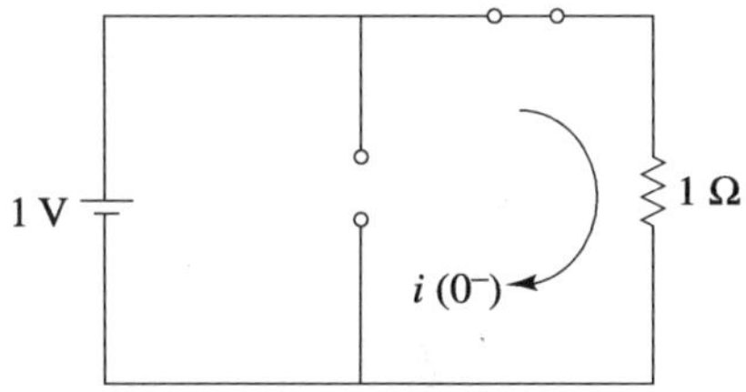

Fig. 9.48

Since current through the inductor and voltage across the capacitor cannot change instantaneously,

$$v_C(0^+) = 1 \text{ V}$$
$$i(0^+) = 1 \text{ A}$$

For $t > 0$, the transformed network is shown in Fig. 9.49.

Applying KVL to the Mesh for $t > 0$,

$$\frac{1}{s} - \frac{1}{s} I(s) - 0.5s\, I(s) + 0.5 - I(s) = 0$$

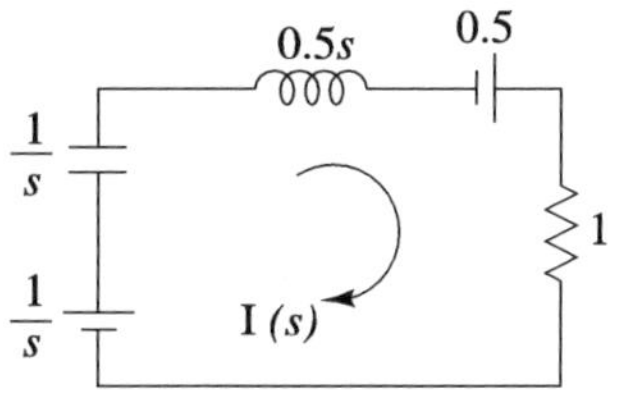

Fig. 9.49

$$0.5 + \frac{1}{s} = \frac{1}{s} I(s) + 0.5\, s\, I(s) + I(s)$$

$$I(s)\left[1 + \frac{1}{s} + 0.5s\right] = 0.5 + \frac{1}{s}$$

$$I(s) = \frac{s+2}{s^2+2s+2} = \frac{(s+1)+1}{(s+1)^2+1}$$

$$= \frac{s+1}{(s+1)^2+1} + \frac{1}{(s+1)^2+1}$$

Taking the inverse Laplace transform,

$$i(t) = e^{-t}\cos t + e^{-t}\sin t \qquad \text{for } t > 0$$

Example 9.13 *In the network shown in Fig. 9.50, the switch is closed at $t = 0$, the steady-state being reached before $t = 0$. Determine current through inductor of 3 H.*

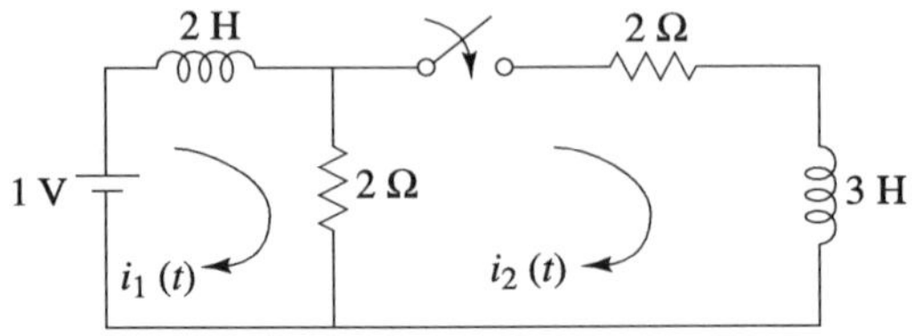

Fig. 9.50

Solution At $t = 0^-$, steady-state condition is reached. Hence, the inductor of 2 H acts as a short circuit.

$$i_1(0^-) = \frac{1}{2}\text{ A}$$

$$i_2(0^-) = 0$$

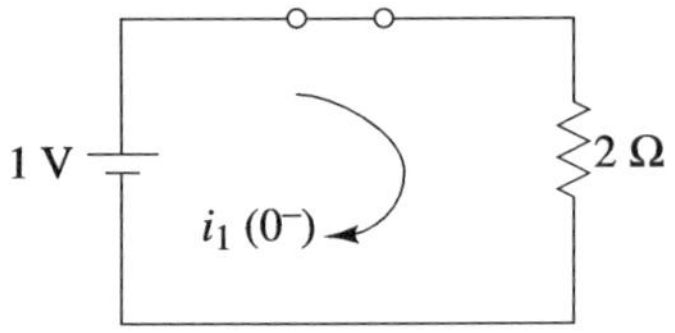

Fig. 9.51

Since current through the inductor cannot change instantaneously,

$$i_1(0^+) = \frac{1}{2}\text{ A}$$

$$i_2(0^+) = 0$$

For $t > 0$, the transformed network is shown in Fig. 9.52.

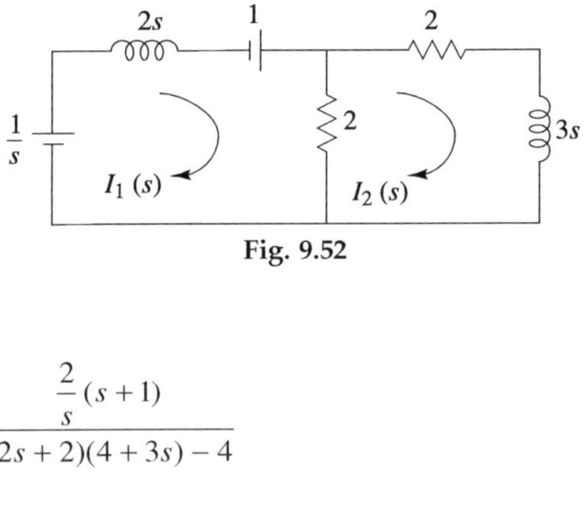

Fig. 9.52

Applying KVL to Mesh 1,

$$\frac{1}{s} - 2s\, I_1(s) + 1 - 2\,[I_1(s) - I_2(s)] = 0$$

$$(2 + 2s)\, I_1(s) - 2I_2(s) = 1 + \frac{1}{s}$$

Applying KVL to Mesh 2,

$$-2\,[I_2(s) - I_1(s)] - 2I_2(s) - 3s\, I_2(s) = 0$$

$$-2I_1(s) + (4 + 3s)\, I_2(s) = 0$$

By Cramer's Rule,

$$I_2(s) = \frac{\begin{vmatrix} 2+2s & 1+1/s \\ -2 & 0 \end{vmatrix}}{\begin{vmatrix} 2+2s & -2 \\ -2 & 4+3s \end{vmatrix}} = \frac{\frac{2}{s}(s+1)}{(2s+2)(4+3s)-4}$$

$$= \frac{s+1}{s(3s^2+7s+2)} = \frac{s+1}{3s\left(s+\frac{1}{3}\right)(s+2)} = \frac{\frac{1}{3}(s+1)}{s(s+2)\left(s+\frac{1}{3}\right)}$$

By partial-fraction expansion,

$$I_2(s) = \frac{A}{s} + \frac{B}{s+2} + \frac{C}{s+\frac{1}{3}}$$

$$A = \left.\frac{\frac{1}{3}(s+1)}{(s+2)\left(s+\frac{1}{3}\right)}\right|_{s=0} = \frac{1}{2}$$

$$B = \left.\frac{\frac{1}{3}(s+1)}{s\left(s+\frac{1}{3}\right)}\right|_{s=-2} = -\frac{1}{10}$$

$$C = \left.\frac{\frac{1}{3}(s+1)}{s(s+2)}\right|_{s=-1/3} = -\frac{2}{5}$$

$$I_2(s) = \frac{1}{2s} - \frac{1}{10(s+2)} - \frac{2}{5\left(s+\frac{1}{3}\right)}$$

Taking inverse Laplace transform,

$$i_2(t) = \frac{1}{2} - \frac{1}{10}e^{-2t} - \frac{2}{5}e^{-(1/3)t} \qquad \text{for } t > 0$$

9.8 TRANSIENT AND STEADY STATE RESPONSE OF R-L CIRCUIT TO VARIOUS FUNCTIONS

Consider a series *R-L* circuit shown in Fig. 9.53.

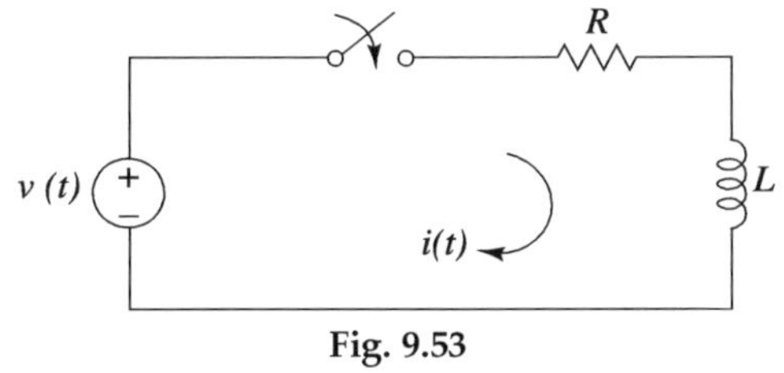

Fig. 9.53

For $t > 0$, the transformed network is shown in Fig. 9.54.

Applying KVL to the Mesh,

$$R\,I(s) + Ls\,I(s) = V(s)$$

$$I(s) = \frac{V(s)}{R + Ls}$$

$$I(s) = \frac{1}{L}\frac{V(s)}{s + \frac{R}{L}} \qquad \ldots(9.32)$$

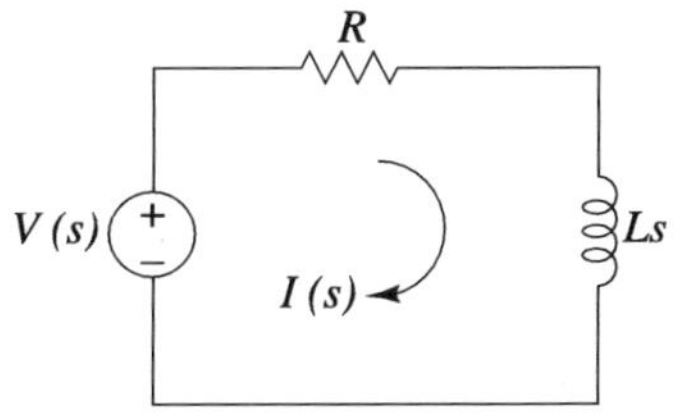

Fig. 9.54

(i) When the unit step signal is applied,

i.e., $$v(t) = u(t) \qquad \ldots(9.33)$$

Taking Laplace transform,

$$V(s) = \frac{1}{s} \qquad \ldots(9.34)$$

$$I(s) = \frac{1}{L}\frac{\frac{1}{s}}{s+\frac{R}{L}}$$

$$= \frac{1}{L}\frac{1}{s\left(s+\frac{R}{L}\right)}$$

By partial-fraction expansion,

$$I(s) = \frac{1}{L}\left(\frac{A}{s}+\frac{B}{s+\frac{R}{L}}\right)$$

$$A = s\,I(s)\big|_{s=0} = \left.\frac{1}{s+\frac{R}{L}}\right|_{s=0} = \frac{L}{R}$$

$$B = \left(s+\frac{R}{L}\right)I(s)\big|_{s=-R/L} = \left.\frac{1}{s}\right|_{s=-R/L} = -\frac{L}{R}$$

$$I(s) = \frac{1}{L}\left(\frac{L}{R}\times\frac{1}{s}-\frac{L}{R}\times\frac{1}{s+\frac{R}{L}}\right)$$

$$= \frac{1}{R}\left(\frac{1}{s}-\frac{1}{s+\frac{R}{L}}\right)$$

Taking inverse Laplace transform,

$$i(t) = \frac{1}{R}\left[1-e^{-(R/L)t}\right] \qquad \text{for } t > 0 \qquad \ldots(9.35)$$

(ii) When unit ramp signal is applied,

$$v(t) = r(t) = t \qquad \text{for } t > 0 \qquad \ldots(9.36)$$

Taking Laplace transform,

$$V(s) = \frac{1}{s^2} \qquad \ldots(9.37)$$

$$I(s) = \frac{1}{L} \times \frac{1}{s^2\left(s + \frac{R}{L}\right)}$$

By partial-fraction expansion,

$$\frac{1}{L} \times \frac{1}{s^2\left(s + \frac{R}{L}\right)} = \frac{A}{s} + \frac{B}{s^2} + \frac{C}{s + \frac{R}{L}}$$

$$\frac{1}{L} = As\left(s + \frac{R}{L}\right) + B\left(s + \frac{R}{L}\right) + Cs^2$$

Putting $s = 0$, we get

$$B = \frac{1}{R}$$

Putting $s = -R/L$, we get

$$C = \frac{L}{R^2}$$

Comparing coefficients of s^2,

$$A + C = 0$$

$$A = -C = -\frac{L}{R^2}$$

$$I(s) = -\frac{L}{R^2}\frac{1}{s} + \frac{1}{R}\frac{1}{s^2} + \frac{L}{R^2}\frac{1}{s + \frac{R}{L}}$$

Taking inverse Laplace transform,

$$i(t) = -\frac{L}{R^2} + \frac{1}{R}t + \frac{L}{R^2}e^{-(R/L)t}$$

$$= \frac{1}{R}t - \frac{L}{R^2}[1 - e^{-(R/L)t}] \qquad \text{for } t > 0 \qquad \ldots(9.38)$$

(iii) When unit impulse signal is applied.

$$v(t) = \delta(t) \qquad \ldots(9.39)$$

Taking Laplace transform,

$$V(s) = 1 \qquad \ldots(9.40)$$

$$I(s) = \frac{1}{L}\frac{1}{s + \frac{R}{L}}$$

Taking inverse Laplace transform,

$$i(t) = \frac{1}{L}e^{-(R/L)t} \qquad \text{for } t > 0 \qquad \ldots(9.41)$$

9.9 TRANSIENT AND STEADY-STATE RESPONSE OF R–C CIRCUIT TO VARIOUS FUNCTIONS

Consider a series *R-C* circuit as shown in Fig. 9.55.

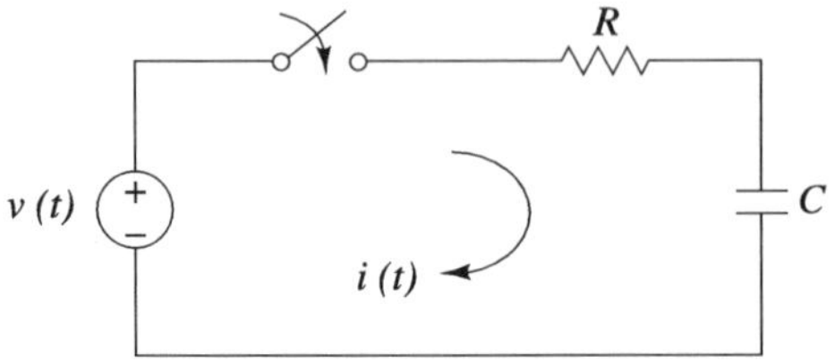

Fig. 9.55

For $t > 0$, the transformed network is shown in Fig. 9.56.

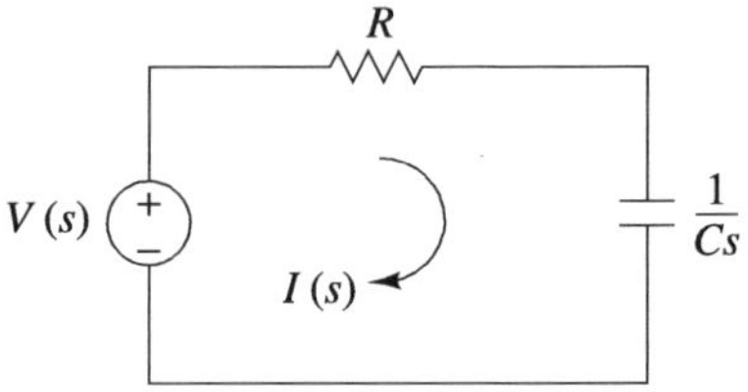

Fig. 9.56

Applying KVL to the Mesh,

$$\frac{1}{Cs} I(s) + R\, I(s) = V(s)$$

$$I(s) = \frac{V(s)}{\frac{1}{Cs} + R} = \frac{sV(s)}{R\left(s + \frac{1}{RC}\right)} \qquad \ldots(9.42)$$

(i) When unit step signal is applied.

$$v(t) = u(t)$$

Taking Laplace transform,

$$V(s) = \frac{1}{s}$$

$$I(s) = \frac{s \times \frac{1}{s}}{R\left(s + \frac{1}{RC}\right)} = \frac{1}{R\left(s + \frac{1}{RC}\right)}$$

Taking inverse Laplace transform,

$$i(t) = \frac{1}{R} e^{-(1/RC)\, t} \qquad \text{for } t > 0 \qquad \ldots(9.43)$$

(ii) When unit ramp signal is applied,

$$v(t) = r(t) = t$$

Taking Laplace transform,

$$V(s) = \frac{1}{s^2}$$

$$I(s) = \frac{s \times \frac{1}{s^2}}{R\left(s + \frac{1}{RC}\right)} = \frac{1}{Rs\left(s + \frac{1}{RC}\right)}$$

By partial-fraction expansion,

$$I(s) = \frac{A}{s} + \frac{B}{s + \frac{1}{RC}}$$

$$A = s\,I(s)\,|_{s=0}$$

$$= \left.\frac{1}{R\left(s+\frac{1}{RC}\right)}\right|_{s=0} = C$$

$$B = \left.\left(s+\frac{1}{RC}\right)I(s)\right|_{s=-1/RC}$$

$$= \left.\frac{1}{Rs}\right|_{s=-1/RC} = -\,C$$

$$I(s) = \frac{C}{s} - \frac{C}{s+\frac{1}{RC}}$$

Taking inverse Laplace transform,

$$i(t) = C - C\,e^{-(1/RC)t} \qquad \text{for } t > 0 \qquad \ldots(9.44)$$

(iii) When unit inpulse signal is applied.

$$v(t) = \delta(t)$$

Taking Laplace transform,

$$V(s) = 1$$

$$I(s) = \frac{s}{R\left(s+\frac{1}{RC}\right)}$$

$$= \frac{s+\frac{1}{RC}-\frac{1}{RC}}{R\left(s+\frac{1}{RC}\right)} = \frac{1}{R}\left(1-\frac{\frac{1}{RC}}{s+\frac{1}{RC}}\right)$$

Taking inverse Laplace transform,

$$i(t) = \frac{1}{R}\left[\delta(t) - \frac{1}{RC}e^{-(1/RC)t}\right] \quad \text{for } t > 0 \qquad \ldots(9.45)$$

Example 9.14 *At t = 0, unit pulse voltage of unit width is applied to a series RL circuit as shown in Fig. 9.57. Obtain an expression for i(t).*

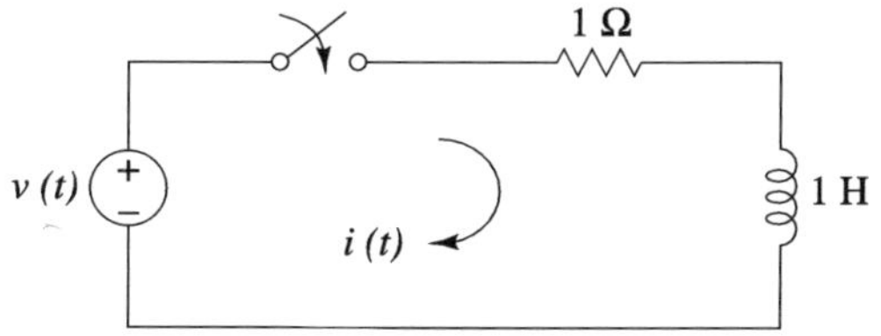

Fig. 9.57

Solution $v(t) = u(t) - u(t-1)$

$$V(s) = \frac{1}{s} - \frac{e^{-s}}{s} = \frac{1-e^{-s}}{s}$$

For $t > 0$, the transformed network is shown in Fig. 9.58.
Applying KVL to the Mesh,

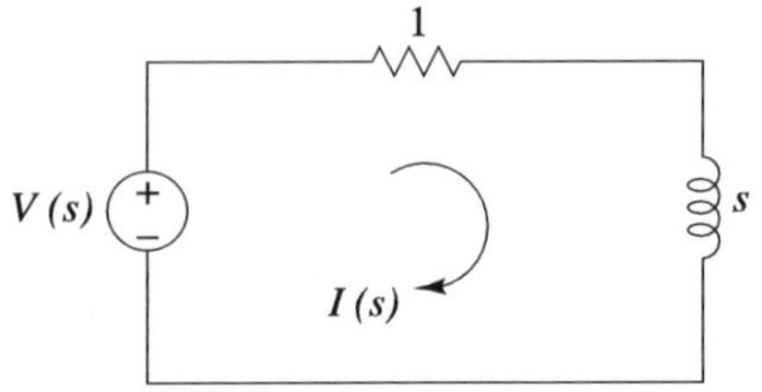

Fig. 9.58

$$V(s) = I(s) + s\,I(s) = \frac{1-e^{-s}}{s}$$

$$I(s) = \frac{1-e^{-s}}{s(s+1)}$$

$$= \frac{1}{s(s+1)} - \frac{e^{-s}}{s(s+1)}$$

$$= \frac{1}{s} - \frac{1}{s+1} - \frac{e^{-s}}{s} + \frac{e^{-s}}{s+1}$$

Taking inverse Laplace transform,

$$i(t) = u(t) - e^{-t}\,u(t) - u(t-1) + e^{-(t-1)}\,u(t-1)$$
$$= (1 - e^{-t})\,u(t) - [1 - e^{-(t-1)}]\,u(t-1) \qquad \text{for } t > 0$$

Example 9.15 *A rectangular voltage pulse of unit height and T-seconds duration is applied to a series R-C network at t = 0. Obtain the expression for the current i (t). Assume the capacitor to be initially uncharged.*

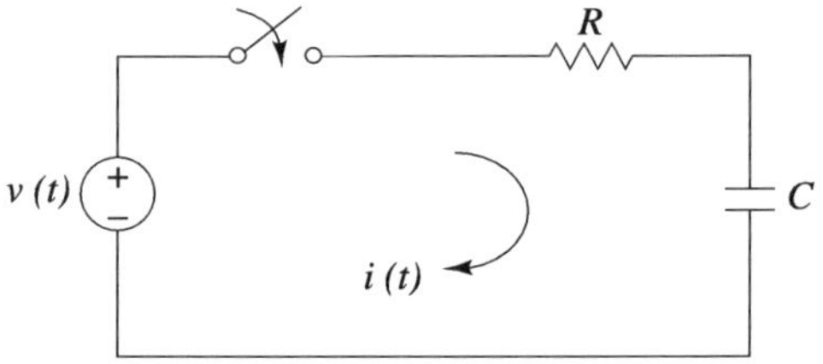

Fig. 9.59

Solution $v(t) = u(t) - u(t-T)$

$$V(s) = \frac{1}{s} - \frac{e^{-sT}}{s} = \frac{1-e^{-sT}}{s}$$

The transformed network is shown in Fig. 9.60.
Applying KVL to the Mesh for $t > 0$,

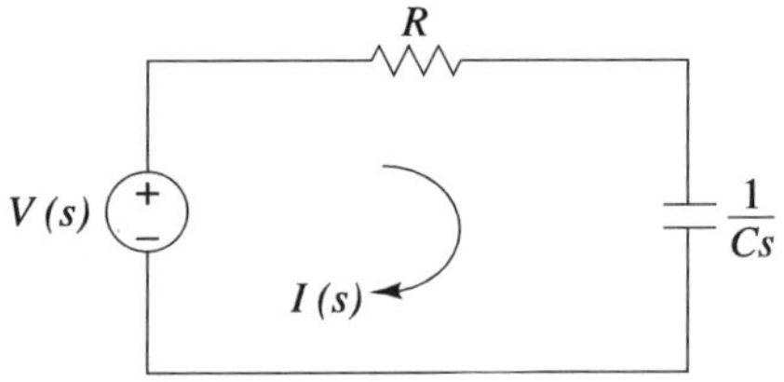

Fig. 9.60

$$V(s) = R\,I(s) + \frac{1}{Cs}I(s) = \frac{1-e^{-sT}}{s}$$

$$I(s)\left[R + \frac{1}{Cs}\right] = \frac{1}{s}(1 - e^{-sT})$$

$$I(s) = \frac{1-e^{-sT}}{R\left(s + \frac{1}{RC}\right)} = \frac{1}{R}\left[\frac{1}{s + \frac{1}{RC}} - \frac{e^{-sT}}{s + \frac{1}{RC}}\right]$$

Taking inverse Laplace transform,

$$i(t) = \frac{1}{R}\left[e^{-(1/RC)t}u(t) - e^{-(1/RC)(t-T)}u(t-T)\right] \qquad \text{for } t > 0$$

Example 9.16 *For the network shown, determine the current i (t) when the switch is closed at t = 0 with zero initial conditions.*

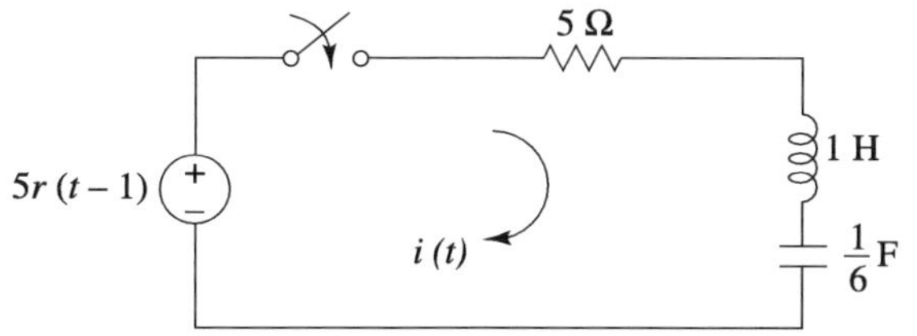

Fig. 9.61

Solution The transformed network is shown in Fig. 9.62.

Applying KVL to the Mesh for $t > 0$,

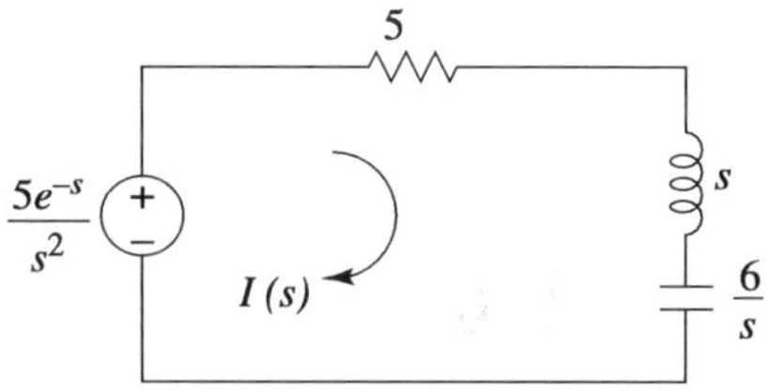

Fig. 9.62

$$\frac{5e^{-s}}{s^2} - 5I(s) - s\,I(s) - \frac{6}{s}I(s) = 0$$

$$5I(s) + s\,I(s) + \frac{6}{s}I(s) = \frac{5e^{-s}}{s^2}$$

$$I(s) = \frac{5e^{-s}}{s(s^2+5s+6)} = \frac{5e^{-s}}{s(s+3)(s+2)}$$

By partial-fraction expansion,

$$\frac{1}{s(s+3)(s+2)} = \frac{A}{s} + \frac{B}{s+3} + \frac{C}{s+2}$$

$$A = \left.\frac{1}{(s+3)(s+2)}\right|_{s=0} = \frac{1}{6}$$

$$B = \left.\frac{1}{s(s+2)}\right|_{s=-3} = \frac{1}{3}$$

$$C = \left.\frac{1}{s(s+3)}\right|_{s=-2} = -\frac{1}{2}$$

$$I(s) = 5e^{-s}\left[\frac{1}{6s} + \frac{1}{3(s+3)} - \frac{1}{2(s+2)}\right]$$

$$= \frac{5}{6}\frac{e^{-s}}{s} + \frac{5}{3}\frac{e^{-s}}{s+3} - \frac{5}{2}\frac{e^{-s}}{s+2}$$

Taking inverse Laplace transform,

$$i(t) = \frac{5}{6}u(t-1) + \frac{5}{3}e^{-3(t-1)}u(t-1) - \frac{5}{2}e^{-2(t-1)}u(t-1) \qquad \text{for } t > 0$$

Example 9.17 *For the network shown in Fig. 9.63, determine the current i (t) when the switch is closed at t = 0. Assume that initial current in the inductor is zero.*

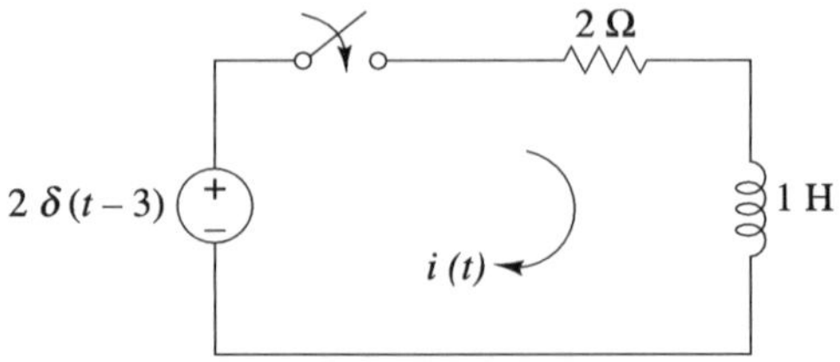

Fig. 9.63

Solution The transformed network is shown in Fig. 9.64.

Applying KVL to Mesh for $t > 0$,

$$2e^{-3s} - 2I(s) - s\,I(s) = 0$$

$$2I(s) + s\,I(s) = 2e^{-3s}$$

$$I(s) = \frac{2e^{-3s}}{s+2}$$

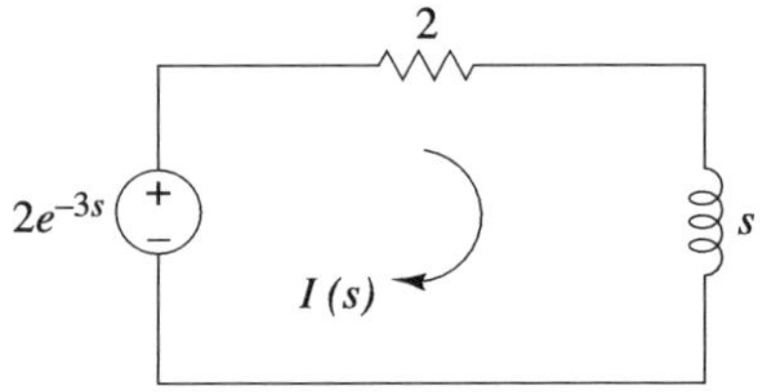

Fig. 9.64

Taking inverse Laplace transform,

$$i(t) = 2e^{-2(t-3)}\,u(t-3) \qquad \text{for } t > 0$$

Example 9.18 *Find impulse response of the current i (t) in the network shown in Fig. 9.65.*

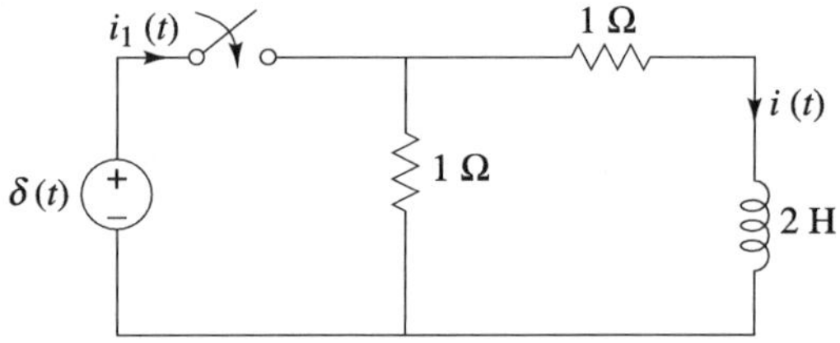

Fig. 9.65

Solution The transformed network is shown in Fig. 9.66.

Total impedance $$Z(s) = \frac{1(2s+1)}{2s+1+1} = \frac{2s+1}{2s+2}$$

$$I_1(s) = \frac{V(s)}{Z(s)}$$

$$= \frac{1}{\dfrac{2s+1}{2s+2}} = \frac{2s+2}{2s+1}$$

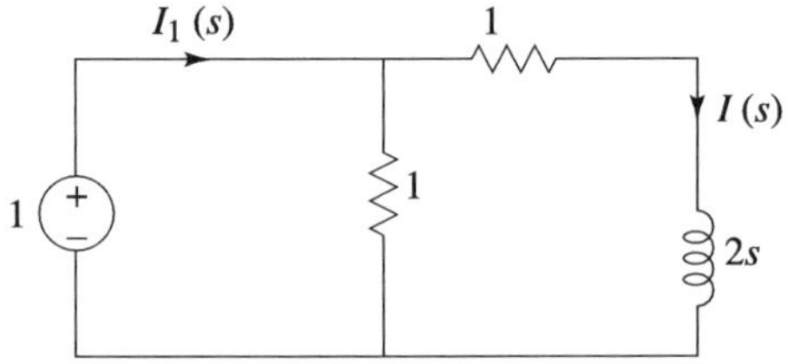

Fig. 9.66

By current-division formula,

$$I(s) = I_1(s) \times \frac{1}{2s+2}$$

$$= \frac{2s+2}{2s+1} \times \frac{1}{2s+2} = \frac{1}{2s+1} = \frac{1}{2}\,\frac{1}{s+0.5}$$

Taking inverse Laplace transform,

$$i(t) = \frac{1}{2}e^{-0.5\,t}\,u(t) \qquad \text{for } t > 0$$

Example 9.19 *Find the impulse response of the voltage across the capacitor in the network shown in Fig. 9.67. Also determine response $v_c(t)$ for step input.*

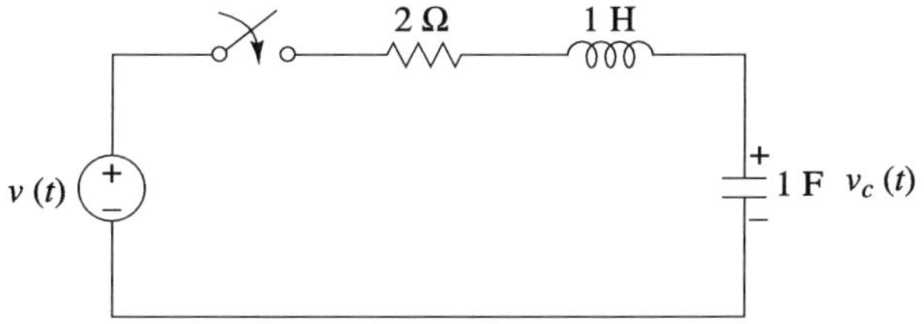

Fig. 9.67

Solution The transformed network is shown in Fig. 9.68.

By voltage division formula,

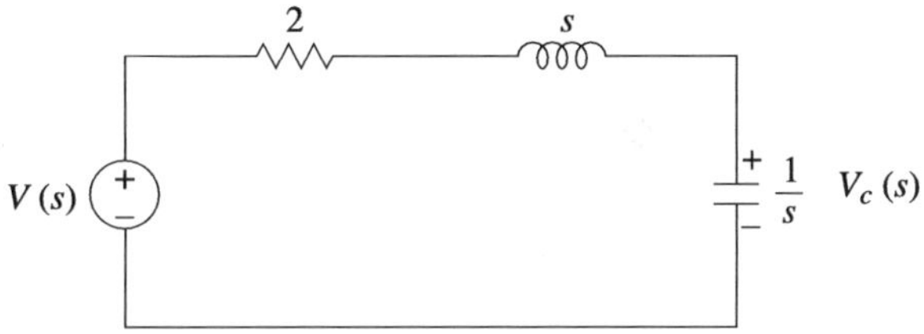

Fig. 9.68

$$V_c(s) = V(s) \times \frac{\frac{1}{s}}{2 + s + \frac{1}{s}}$$

$$= \frac{V(s)}{s^2 + 2s + 1} = \frac{V(s)}{(s+1)^2}$$

(i) For impulse input,

$$V(s) = 1$$

$$V_c(s) = \frac{1}{(s+1)^2}$$

Taking inverse Laplace transform,

Impulse response $\quad v_c(t) = t\, e^{-t}\, u(t) \quad$ for $t > 0$

(ii) For step input,

$$V(s) = \frac{1}{s}$$

$$V_c(s) = \frac{1}{s(s+1)^2}$$

By partial-fraction expansion,

$$V_c(s) = \frac{A}{s} + \frac{B}{s+1} + \frac{C}{(s+1)^2}$$

$$1 = A(s+1)^2 + Bs(s+1) + Cs$$

$$= A(s^2 + 2s + 1) + B(s^2 + s) + Cs$$

$$= s^2(A + B) + s(2A + B + C) + A$$

Comparing coefficient of s^2, s^1 and s^0, we have

$$A = 1$$

$$A + B = 0$$

$$B = -A = -1$$

$$2A + B + C = 0$$

$$C = -2A - B = -2 + 1 = -1$$

$$V_c(s) = \frac{1}{s} - \frac{1}{s+1} - \frac{1}{(s+1)^2}$$

Taking inverse Laplace transform,

$$v_c(t) = u(t) - e^{-t}\,u(t) - te^{-t}\,u(t)$$

$$= (1 - e^{-t} - t\,e^{-t})\,u(t) \qquad \text{for } t > 0$$

Example 9.20 *Determine the expression for $v_L(t)$ in the network shown in Fig. 9.69. Find $v_L(t)$ when (i) $v_s(t) = \delta(t)$, and (ii) $v_s(t) = e^{-t}\,u(t)$.*

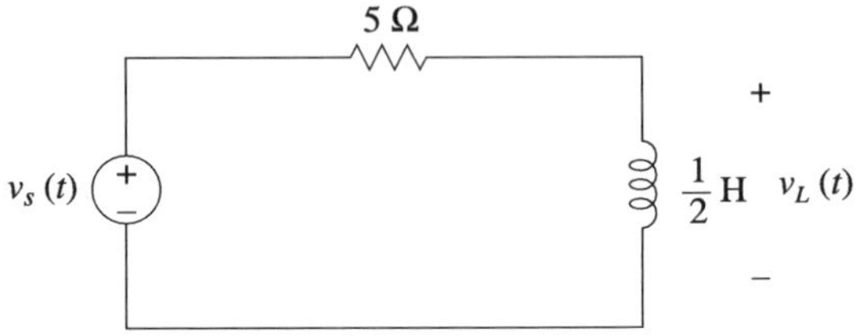

Fig. 9.69

Solution The transformed network is shown in Fig. 9.70.

By voltage-division formula,

$$V_L(s) = V_s(s) \times \frac{\frac{s}{2}}{\frac{s}{2}+5} = \frac{s}{s+10}\,V_s(s)$$

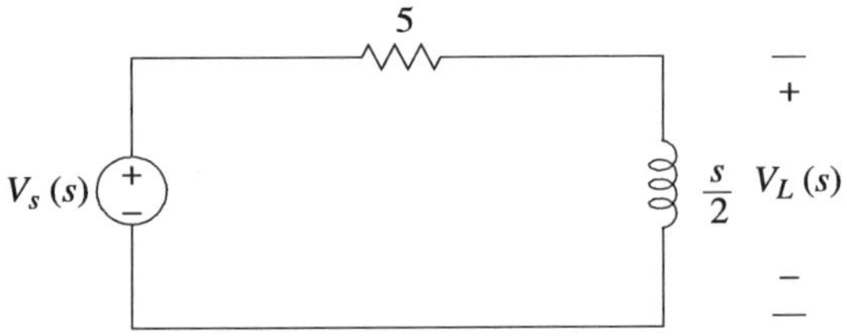

Fig. 9.70

(i) For impulse input,

$$V_s(s) = 1$$

$$V_L(s) = \frac{s}{s+10}$$

$$= \frac{s+10-10}{s+10} = 1 - \frac{10}{s+10}$$

Taking inverse Laplace transform,

$$V_L(t) = \delta(t) - 10\,e^{-10t}\,u(t) \qquad \text{for } t > 0$$

(ii) For $v_s(t) = e^{-t}\,u(t)$,

$$V_s(s) = \frac{1}{s+1}$$

$$V_L(s) = \frac{1}{s+1}\,\frac{s}{s+10} = \frac{s}{(s+10)(s+1)}$$

By partial-fraction expansion,

$$V_L(s) = \frac{A}{s+10} + \frac{B}{s+1}$$

$$A = \left.\frac{s}{s+1}\right|_{s=-10} = \frac{10}{9}$$

$$B = \left.\frac{s}{s+10}\right|_{s=-1} = -\frac{1}{9}$$

$$V_L(s) = \frac{10}{9}\frac{1}{s+10} - \frac{1}{9}\frac{1}{s+1}$$

Taking inverse Laplace transform,

$$v_L(t) = \frac{10}{9}e^{-10\,t}\,u(t) - \frac{1}{9}\,e^{-t}\,u(t)$$

$$= \left(\frac{10}{9}e^{-10t} - \frac{1}{9}e^{-t}\right)u(t) \qquad \text{for } t > 0$$

Example 9.21 *For the network shown in Fig. 9.71, find the response $v_0\,(t)$.*

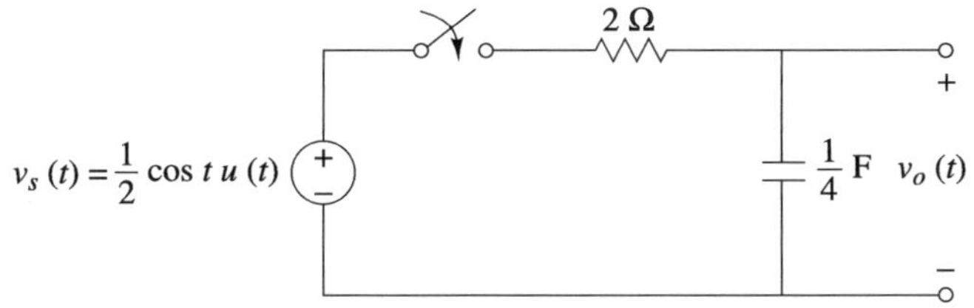

Fig. 9.71

Solution The transformed network is shown in Fig. 9.72.

$$V_s\,(s) = \frac{1}{2}\frac{s}{s^2+1}$$

By voltage-division formula,

$$V_o(s) = V_s\,(s) \times \frac{\frac{4}{s}}{2+\frac{4}{s}}$$

$$= \frac{2V_s(s)}{s+2} = \frac{s}{(s^2+1)(s+2)}$$

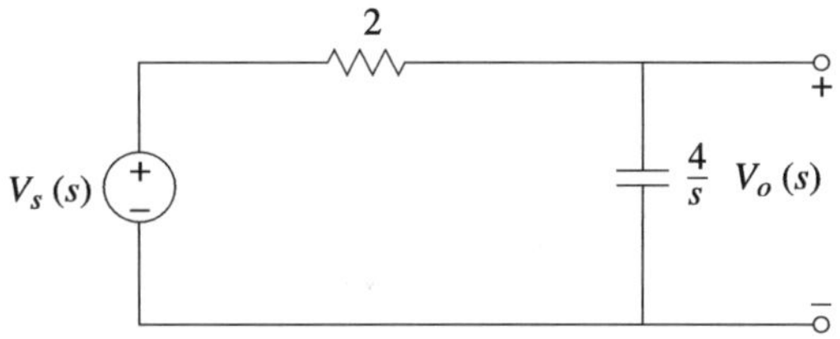

Fig. 9.72

By partial-fraction expansion,

$$V_o(s) = \frac{As+B}{s^2+1} + \frac{C}{s+2}$$

$$s = (As + B)\,(s + 2) + c(s^2 + 1)$$

$$s = s^2(A + C) + s(2A + B) + (2B + C)$$

Comparing coefficient of s^2, s and s^0, we have

$$A + C = 0$$

$$2A + B = 1$$

$$2B + C = 0$$

Solving the equations,we get

$$A = 0.4$$

$$B = 0.2$$

$$C = -0.4$$

$$V_o(s) = \frac{0.4s+0.2}{s^2+1} - \frac{0.4}{s+2} = \frac{0.4s}{s^2+1} + \frac{0.2}{s^2+1} - \frac{0.4}{s+2}$$

Taking the inverse Laplace transform,

$$i(t) = 0.4 \cos t + 0.2 \sin t - 0.4\, e^{-2t} \qquad \text{for } t > 0$$

Example 9.22 *Determine the current i (t) in the network shown in Fig. 9.73, when the switch s is closed at t = 0.*

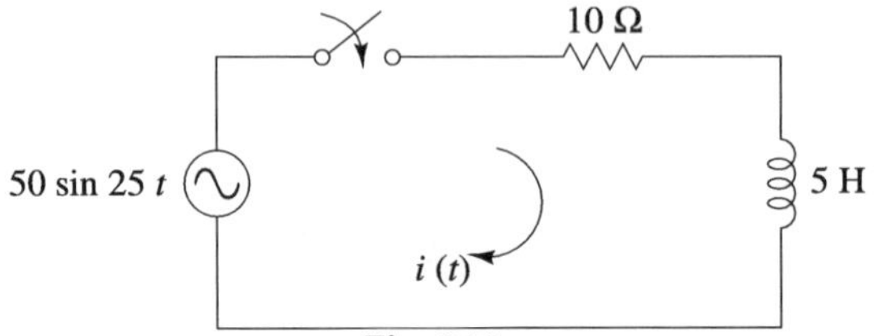

Fig. 9.73

Solution The transformed network is shown in Fig. 9.74.

Applying KVL to the Mesh for $t > 0$,

$$\frac{1250}{s^2+625} = 10\, I(s) + 5s\, I(s)$$

$$I(s) = \frac{250}{(s^2+625)(s+2)}$$

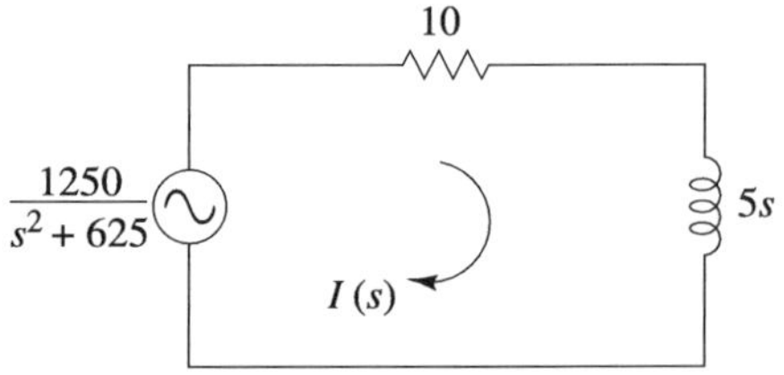

Fig. 9.74

By partial-fraction expansion,

$$I(s) = \frac{As+B}{s^2+625} + \frac{C}{s+2}$$

$$250 = (As+B)(s+2) + C(s^2+625)$$

$$= s^2(A+C) + s(2A+B) + (2B+625C)$$

Comparing coefficients, we have

$$A + C = 0$$

$$2A + B = 0$$

$$2B + 625C = 250$$

Solving the equations,we get

$$A = -0.397$$

$$B = 0.795$$

$$C = 0.397$$

$$I(s) = \frac{-0.397s + 0.795}{s^2+625} + \frac{0.397}{s+2} = -\frac{0.397s}{s^2+625} + \frac{0.795}{s^2+625} + \frac{0.397}{s+2}$$

Taking the inverse Laplace transform,

$$i(t) = -0.397 \cos 25t + 0.032 \sin 25t + 0.397\, e^{-2t} \qquad \text{for } t > 0$$

Example 9.23 *The network shown in Fig. 9.75 is at rest for t < 0. If the voltage v (t) = u (t) cos t + A δ (t) is applied to the network, determine the value of A so that there is no transient term in the current response i (t).*

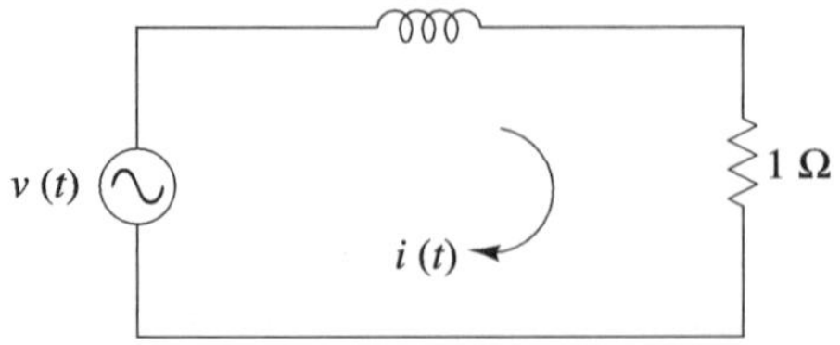

Fig. 9.75

$$v(t) = u(t)\cos t + A\delta(t)$$

$$V(s) = \frac{s}{s^2+1} + A$$

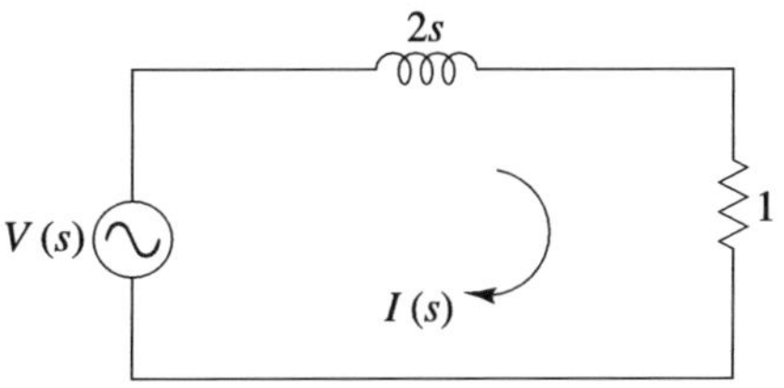

Fig. 9.76

Solution The transformed network is shown in Fig. 9.76.

Applying KVL to the Mesh for $t > 0$,

$$V(s) = 2s\,I(s) + I(s) = \frac{s}{s^2+1} + A$$

$$I(s) = \frac{s + A(s^2+1)}{2\left(s+\frac{1}{2}\right)(s^2+1)} = \frac{K_1}{s+\frac{1}{2}} + \frac{K_2 s + K_3}{s^2+1}$$

The transient part of the response is given by the first term. Hence, for the transient term to vanish, $K_1 = 0$.

$$K_1 = \left(s+\frac{1}{2}\right) I(s)\Big|_{s=-\frac{1}{2}} = \frac{\frac{-1}{2} + A\left(\frac{5}{4}\right)}{2\left(\frac{5}{4}\right)}$$

When $\quad K_1 = 0$

$$\frac{5}{4}A = \frac{1}{2}$$

$$A = \frac{2}{5} = 0.4$$

Example 9.24 *The network shown has zero initial conditions. A voltage $v_i(t) = \delta(t)$ applied to two terminal network produces voltage $v_0(t) = [e^{-2t} + e^{-3t}]\,u(t)$. What should be $v_i(t)$ to give $v_o(t) = t\,e^{-2t}\,u(t)$?*

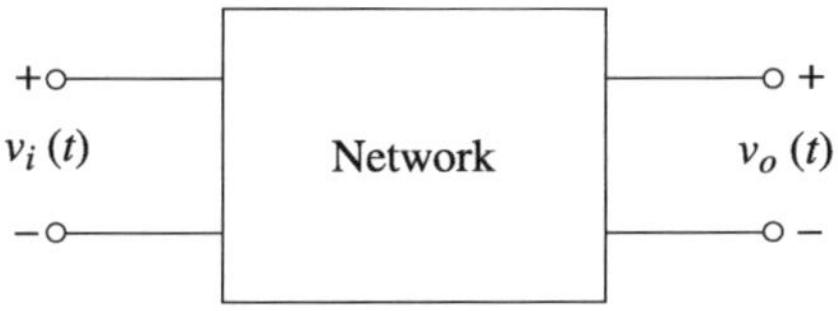

Fig. 9.77

Solution For $v_i(t) = \delta(t)$,

$$V_i(s) = 1$$

$$v_o(t) = [e^{-2t} + e^{-3t}]u(t)$$

$$V_o(s) = \frac{1}{s+2} + \frac{1}{s+3}$$

Hence system function $\quad H(s) = \dfrac{V_o(s)}{V_i(s)}$

$$= \frac{1}{s+2} + \frac{1}{s+3} = \frac{2s+5}{(s+2)(s+3)} \qquad \ldots\text{(i)}$$

For $v_o(t) = t\,e^{-2t}\,u(t)$,

$$V_o(s) = \frac{1}{(s+2)^2}$$

From the Eq. (i),

$$V_i(s) = \frac{V_o(s)}{H(s)} = \frac{1}{(s+2)^2} \times \frac{(s+2)(s+3)}{2s+5} = \frac{(s+3)}{2(s+2.5)(s+2)}$$

By partial-fraction expansion,

$$V_i(s) = \frac{A}{s+2} + \frac{B}{s+2.5}$$

$$A = 1$$

$$B = -0.5$$

$$V_i(s) = \frac{1}{s+2} - \frac{0.5}{s+2.5}$$

Taking inverse Laplace transform,

$$v_i(t) = e^{-2t} - 0.5e^{-2.5t} \qquad \text{for } t > 0$$

Example 9.25 *A unit impulse applied to two terminal black box produces a voltage $v_0(t) = 2e^{-t} - e^{-3t}$. Determine the terminal voltage when a current pulse of 1 A height and a duration of 2 seconds is applied at the terminal.*

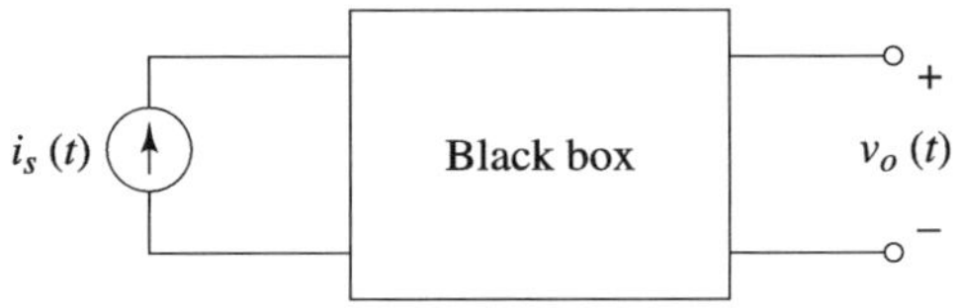

Fig. 9.78

Solution

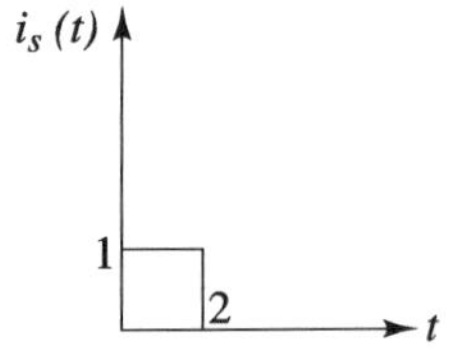

Fig. 9.79

$$v_0(t) = 2e^{-t} - e^{-3t}$$

$$V_0(s) = \frac{2}{s+1} - \frac{1}{s+3}$$

When $i_s(t) = \delta(t)$,

$$I_s(s) = 1$$

$$V_0(s) = Z(s)\ I_s(s) \qquad \ldots(i)$$

$$Z(s) = \frac{V_0(s)}{I_s(s)} = \frac{2}{s+1} - \frac{1}{s+3}$$

When $i_s(t)$ is a pulse of 1-A height and a duration of 2 seconds then,

$$i_s(t) = u(t) - u(t-2)$$

$$I_s(s) = \frac{1}{s} - \frac{e^{-2s}}{s}$$

From the Eq. (i),

$$V_0(s) = \left[\frac{2}{s+1} - \frac{1}{s+3}\right]\left[\frac{1}{s} - \frac{e^{-2s}}{s}\right]$$

$$= \frac{2}{s(s+1)} - \frac{1}{s(s+3)} - \frac{2e^{-2s}}{s(s+1)} + \frac{e^{-2s}}{s(s+3)}$$

$$= 2\left[\frac{1}{s} - \frac{1}{s+1}\right] - \frac{1}{3}\left[\frac{1}{s} - \frac{1}{s+3}\right] - 2e^{-2s}\left[\frac{1}{s} - \frac{1}{s+1}\right] + \frac{e^{-2s}}{3}\left[\frac{1}{s} - \frac{1}{s+3}\right]$$

Taking the inverse Laplace transform,

$$v(t) = 2[u(t) - e^{-t}u(t)] - \frac{1}{3}[u(t) - e^{-3t}u(t)] - 2[u(t-2) - e^{-(t-2)}u(t-2)] + \frac{1}{3}[u(t-2) - e^{-3(t-2)}u(t-2)]$$

for $t > 0$.

Exercises

1. For the network shown, the switch is closed at $t = 0$. Find the current $i_1(t)$ for $t > 0$.

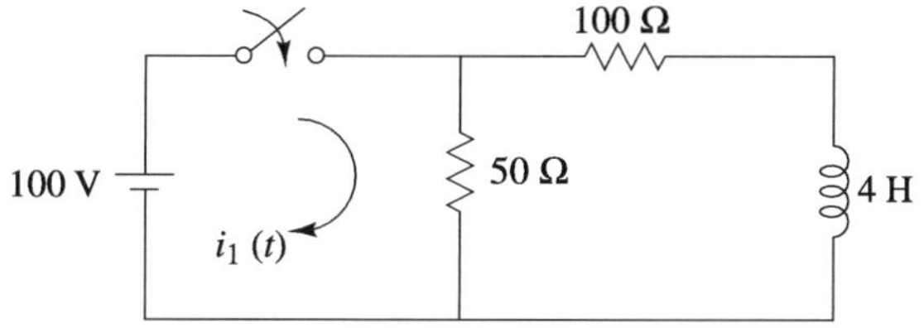

Fig. 9.80

$[i_1(t) = 3 - e^{-25t}]$

2. Determine the current $i(t)$ in the network of Fig. 9.81, when the switch is closed at $t = 0$. The inductor is initially unenergized.

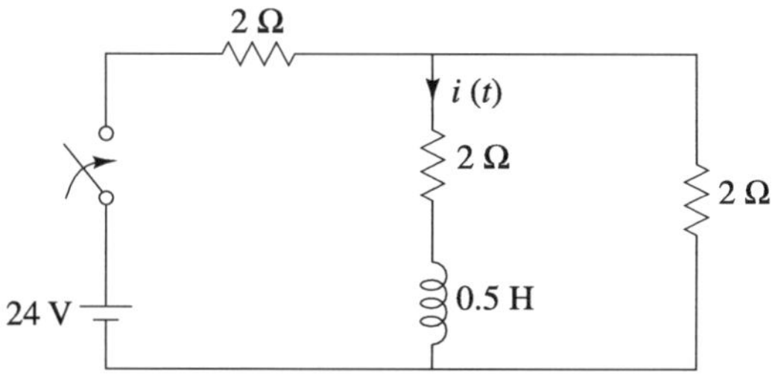

Fig. 9.81

$[i(t) = 4(1 - e^{-6t})]$

3. In the network of Fig. 9.82, after the switch has been in the open position for a long time, it is closed at $t = 0$. Find the voltage across the capacitor.

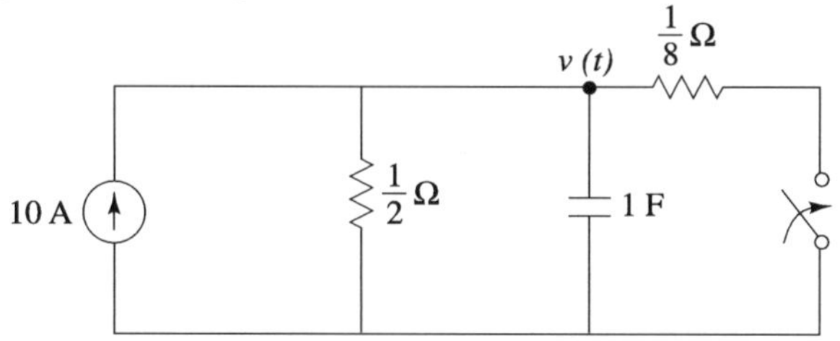

Fig. 9.82

$[v(t) = 1 + 4\,e^{-10t}]$

4. The circuit of Fig. 9.83, has been in the condition shown for a long time. At $t = 0$, switch is closed. Find $v(t)$ for $t > 0$.

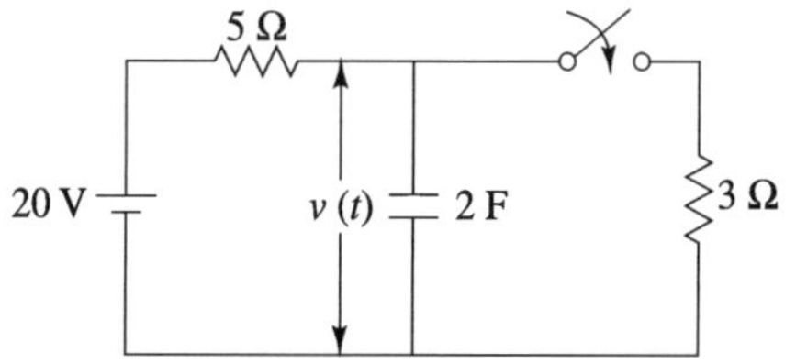

Fig. 9.83

$[v(t) = 7.5 + 12.5\,e^{-(4/15)t}]$

5. Figure 9.84 shows a circuit which is in the steady-state with the switch open. At $t = 0$, the switch is closed. Determine the current $i(t)$. Find its value at $t = 0.114\ \mu$ seconds.

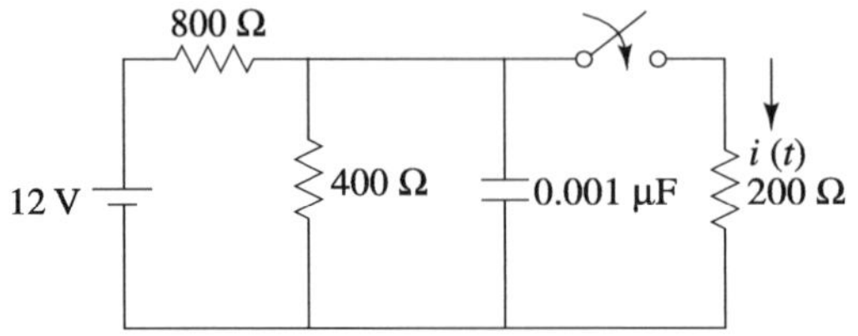

Fig. 9.84

$[i(t) = 0.00857 + 0.01143\ e^{-8.75 \times 10^6 t}, 0.013\ \text{A}]$

6. Find $i(t)$ for the network shown in Fig. 9.85.

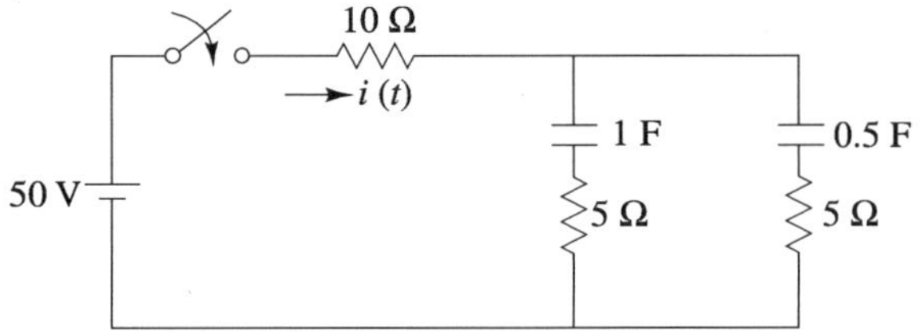

Fig. 9.85

$[i(t) = 0.125\ e^{-0.308t} + 3.875\ e^{-0.052\ t}]$

7. Determine $v(t)$ where $i_L(0^-) = 15$ A and $v_c(0^-) = 5$ V.

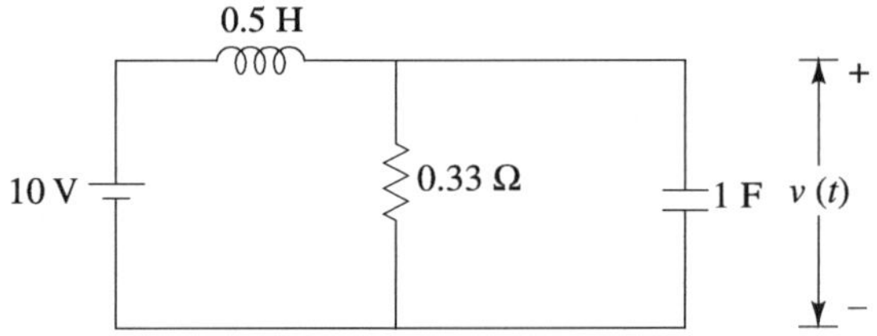

Fig. 9.86

$[v(t) = 10 - 10e^{-t} + 5e^{-2t}]$

8. The network shown has acquired steady state with the switch at position 1 for $t < 0$. At $t = 0$, the switch is thrown to the position 2. Find $v(t)$ for $t > 0$.

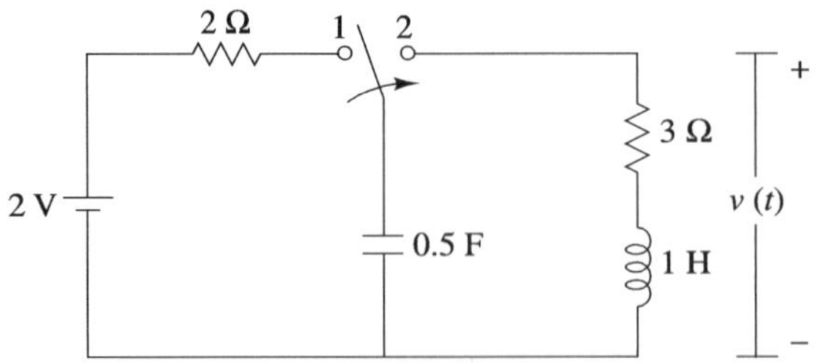

Fig. 9.87

$[v(t) = 4e^{-t} - 2\ e^{-2t}]$

9. In the network shown in Fig. 9.88, the switch is closed at $t = 0$. Find current $i_1(t)$ for $t > 0$.

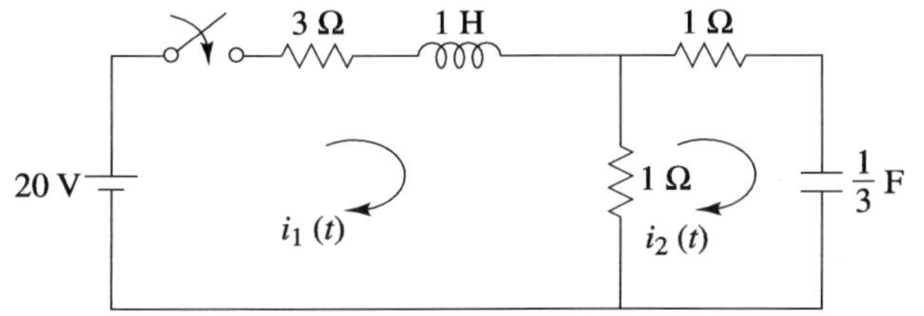

Fig. 9.88

$[i_1(t) = 5 + 5e^{-2t} - 10e^{-3t}]$

10. In the network shown in Fig. 9.89, the switch is closed at $t = 0$. Find the current through the 30-Ω resistor.

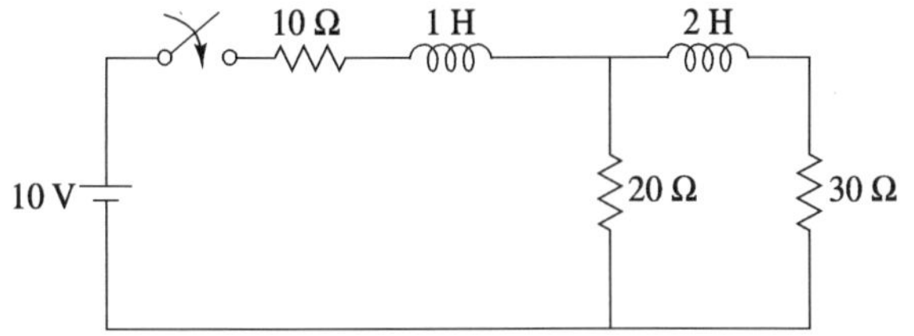

Fig. 9.89

$[i(t) = 0.1818 - 0.265\, e^{-13.14\, t} + 0.083\, e^{-41.86\, t}]$

11. The network shown in Fig. 9.90 is in steady state with s_1 closed and s_2 open. At $t = 0$, s_1 is opened and s_2 is closed. Find the current through the capacitor.

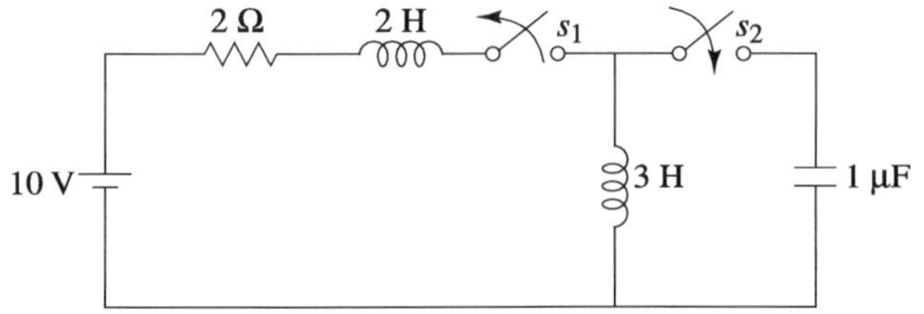

Fig. 9.90

$[i(t) = 5 \cos (0.577 \times 10^3\, t)]$

12. In the network shown in Fig. 9.91, find currents $i_1\,(t)$ and $i_2\,(t)$ for $t > 0$.

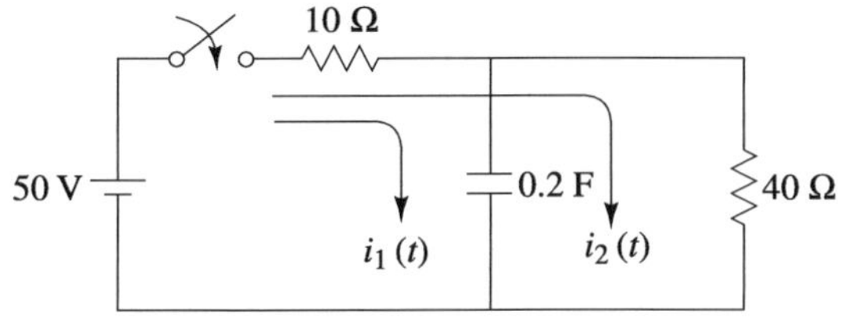

Fig. 9.91

$[i_1(t) = 5\, e^{-0.625\, t},\ i_2\,(t) = 1 - e^{-0.625\, t}]$

13. For the network shown in Fig. 9.92, find currents $i_1(t)$ and $i_2(t)$ for $t > 0$.

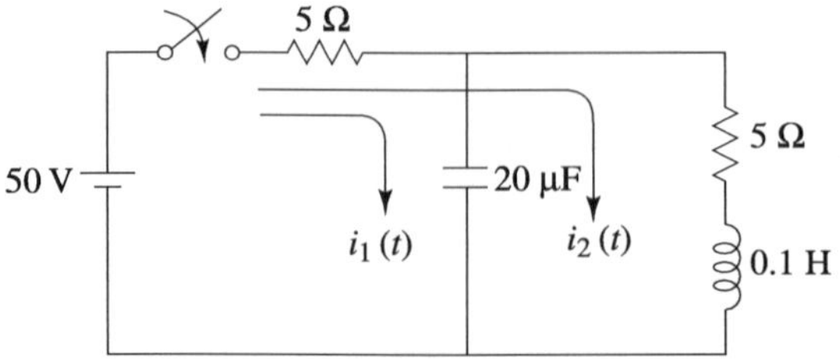

Fig. 9.92

$$\begin{bmatrix} i_1(t) = 0.101e^{-100.5t} + 10.05e^{-9949.5t} \\ i_2(t) = 5 - 5.05e^{-100.5t} + 0.05e^{-9949.5t} \end{bmatrix}$$

14. In the network shown in Fig. 9.93, the switch is opened at $t = 0$, the steady state having been established previously. Find $i(t)$ for $t > 0$.

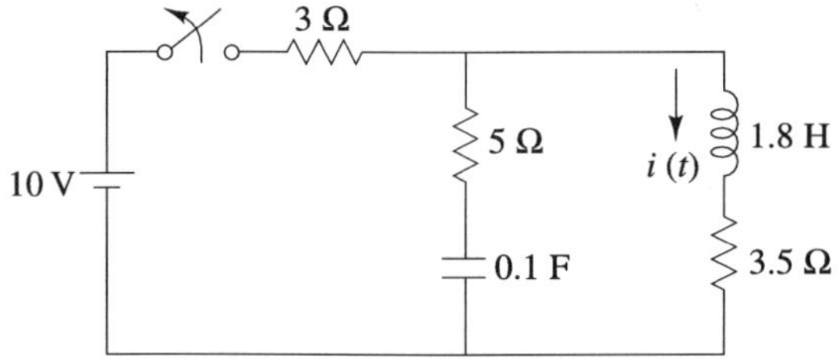

Fig. 9.93

$[i(t) = 1.5124e^{-2.22t} + 3.049e^{-2.5t}]$

15. Find the current $i(t)$, if the switch is closed at $t = 0$. Assume initial conditions to be zero.

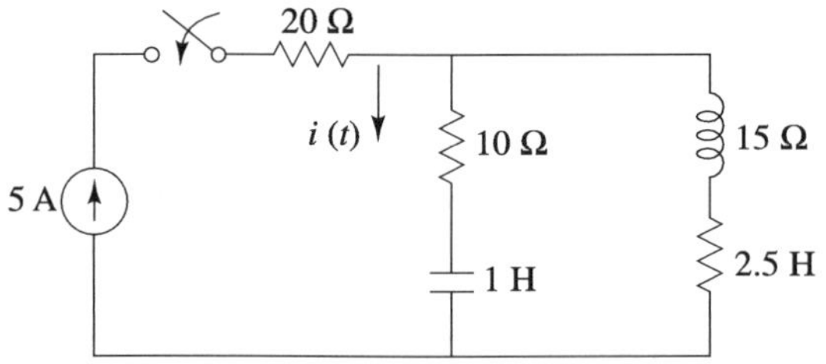

Fig. 9.94

$[i(t) = 3 + 0.57e^{-7.14t}]$

16. In the network shown in Fig. 9.95, find the voltage $v(t)$ for $t > 0$.

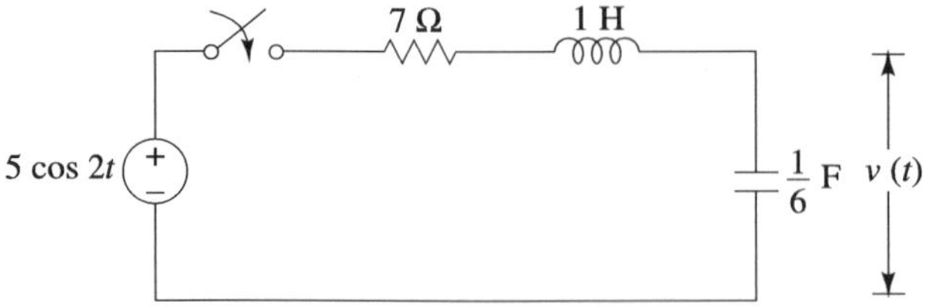

Fig. 9.95

$$\left[v(t) = -\frac{6}{5}e^{-t} + \frac{9}{10}e^{-6t} + \frac{3}{10}\cos 2t + \frac{21}{10}\sin 2t \right]$$

17. For the network shown, determine $v(t)$ when the input is

(i) an impulse function $[e^{-t}]\,u(t)$

(ii) a unit step function $[1 - e^{-t}]\,u(t)$

(iii) $i(t) = 4e^{-t}\,u(t)$ $[4t\,e^{-t}]\,u(t)$

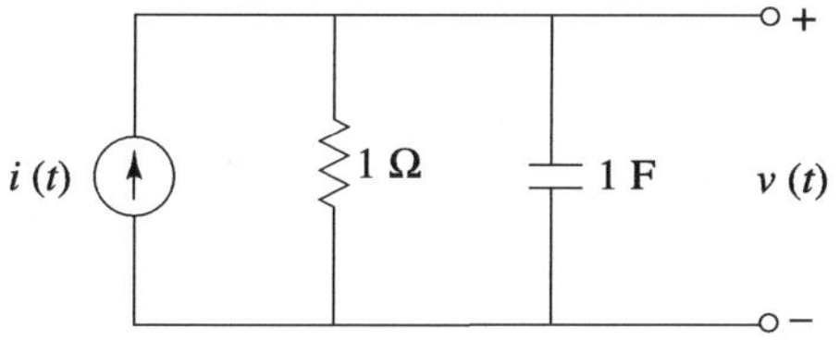

Fig. 9.96

18. For a unit ramp input, find the response $v_c(t)$ for $t > 0$.

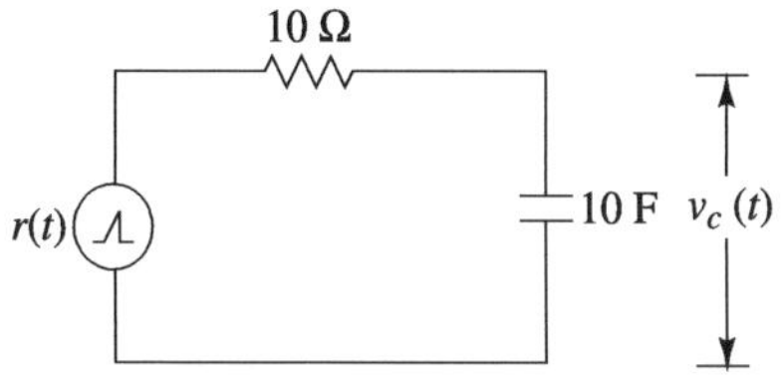

Fig. 9.97

$[v_c(t) = -100\,u(t) + 100e^{-0.01t}\,u(t) + tu(t)]$

Objective-Type Questions

1. If the Laplace transform of the voltage across a capacitor of value $\frac{1}{2}$ F is

$$V_c(s) = \frac{1}{s^2+1}$$

the value of the current through the capacitor at $t = 0^+$ is

(a) 0 (b) 2 A (c) $\frac{1}{2}$ A (d) 1 A

2. The response of an initially relaxed linear constant parameter network to a unit impulse applied at $t = 0$ is $4e^{-2t}\,u(t)$. The response of this network to a unit step function will be

(a) $2[1 - e^{-2t}]\,u(t)$ (b) $4[e^{-t} - e^{-2t}]\,u(t)$ (c) $\sin 2t$ (d) $(1 - 4\,e^{-4t})\,u(t)$

3. The Laplace transform of a unit ramp function starting at $t = a$ is

(a) $\frac{1}{(s+a)^2}$ (b) $\frac{e^{-as}}{(s+a)^2}$ (c) $\frac{e^{-as}}{s^2}$ (d) $\frac{a}{s^2}$

4. The Laplace transform of $e^{at}\cos \alpha t$ is equal to

(a) $\frac{s-\alpha}{(s-\alpha)^2+\alpha^2}$ (b) $\frac{s+\alpha}{(s-\alpha)^2+\alpha^2}$ (c) $\frac{1}{(s-\alpha)^2}$ (d) none of the above

5. The circuit shown in Fig. 9.98 has initial current $i(0^-) = 1$ A through the inductor and an initial voltage $v_c(0^-) = -1$ V across the capacitor. For input $v(t) = u(t)$, the Laplace transform of the current $i(t)$ for $t \geq 0$ is

(a) $\dfrac{s}{s^2+s+1}$ (b) $\dfrac{s+2}{s^2+s+1}$

(c) $\dfrac{s-2}{s^2+s+1}$ (d) $\dfrac{s-2}{s^2+s+1}$

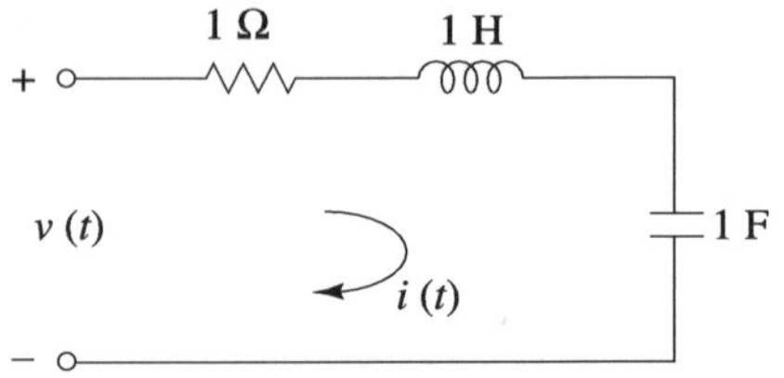

Fig. 9.98

6. A square pulse of 3 volts amplitude is applied to an *R-C* circuit shown in Fig. 9.99. The capacitor is initially uncharged. The output voltage v_0 at time $t = 2$ seconds is

(a) 3 V (b) –3 V (c) 4 V (d) –4 V

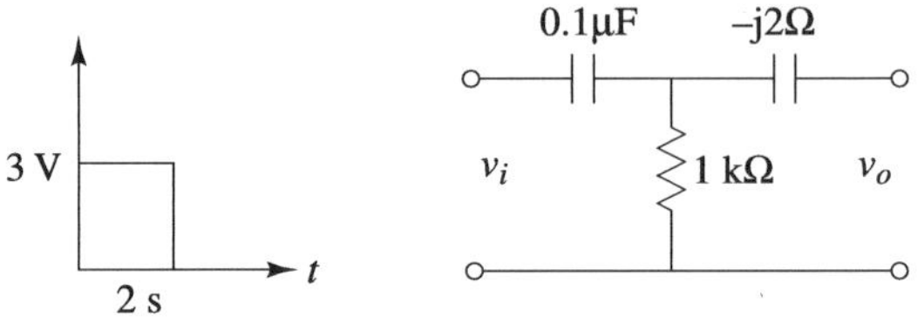

Fig. 9.99

7. A 2-mH inductor with some initial current can be represented as shown. The value of the initial current is

(a) 0.5 A (b) 2 A

(c) 1 A (d) 0

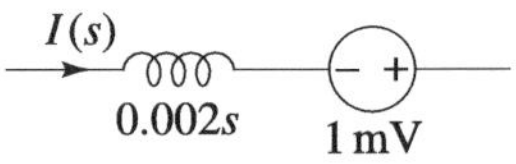

Fig. 9.100

8. A current impulse $5\,\delta(t)$ is forced through a capacitor C. The voltage $v_c(t)$ across the capacitor is given by

(a) $5\,t$ (b) $5\,u(t) - C$ (c) $\dfrac{5}{C}t$ (d) $\dfrac{5\,u(t)}{C}$

9. In the circuit shown in Fig. 9.101, it is desired to have a constant direct current $i(t)$ through the ideal inductor L. The nature of the voltage source $v(t)$ must be

(a) a constant voltage

(b) a linearly increasing voltage

(c) an ideal impulse

(d) as exponential increasing voltage

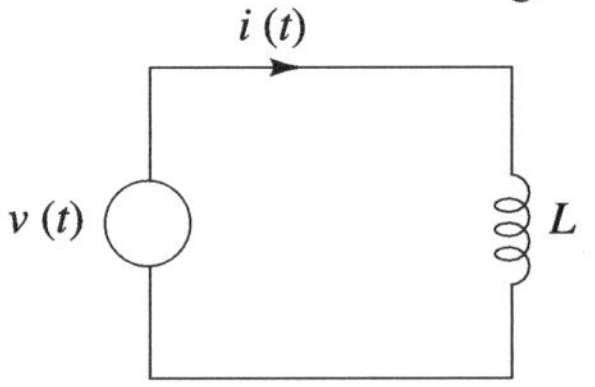

Fig. 9.101

10. When a unit impulse voltage is applied to an inductor of 1 H, the energy supplied by the source is

(a) ∞ (b) 1 J (c) $\dfrac{1}{2}$ J (d) 0

Answers to Objective-Type Questions

1. (c) **2.** (a) **3.** (c) **4.** (a) **5.** (b) **6.** (b)

7. (a) **8.** (d) **9.** (c) **10.** (c)

10 Network Functions

10.1 INTRODUCTION

A network function gives the relation between currents or voltages at different parts of the network. It is broadly classified as *driving point* and *transfer function*. It is associated with terminals and ports.

Any network may be represented schematically by a rectangular box. Terminals are needed to connect any network to any other network or for taking some measurements. Two such associated terminals are called *terminal pair* or *port*. If there is only one pair of terminals in the network, it is called a one-port network. If there are two pairs of terminals, it is called a two-port network. The port to which energy source is connected is called the *input port*. The port to which load is connected is known as the *output port*. One such network having only one pair of terminals (1 – 1′) is shown in Fig. 10.1 (a) and is called *one-port network*. Figure 10.1(b) shows a two port network with two pairs of terminals. The terminals 1 – 1′ together constitute a port. Similarly the terminals 2 – 2′ constitute another port.

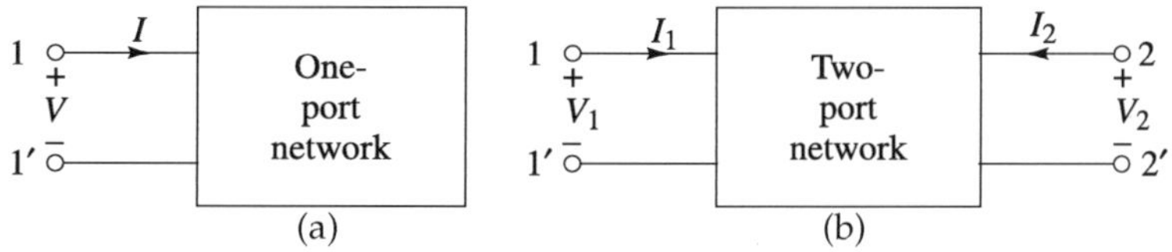

Fig. 10.1

A voltage and current are assigned to each of the two ports. V_1 and I_1 are assigned to the input port whereas V_2 and I_2 are assigned to the output port. It is also assumed that currents I_1 and I_2 are entering into the network at the upper terminals 1 and 2 respectively.

10.2 DRIVING-POINT FUNCTIONS

If excitation and response are measured at the same ports, the network function is known as the driving-point function. For a one-port network of Fig. 10.1 (a), only one voltage and current are specified and hence only one network function (and its reciprocal) can be defined.

(i) Driving-point impedance function It is defined as the ratio of the voltage transform at one port to the current transform at the same port. It is denoted by $Z(s)$.

$$Z(s) = \frac{V(s)}{I(s)} \qquad \text{...(10.1)}$$

(ii) Driving-point admittance function It is defined as the ratio of the current transform at one port to the voltage transform at the same port. It is denoted by $Y(s)$.

$$Y(s) = \frac{I(s)}{V(s)} \qquad \text{...(10.2)}$$

For a two-port network, the driving-point impedance function and driving-point admittance function at Port 1 are

$$Z_{11}(s) = \frac{V_1(s)}{I_1(s)} \qquad \text{...(10.3)}$$

$$Y_{11}(s) = \frac{I_1(s)}{V_1(s)} \qquad \text{...(10.4)}$$

Similarly, at Port 2, we have

$$Z_{22}(s) = \frac{V_2(s)}{I_2(s)} \qquad \text{...(10.5)}$$

$$Y_{22}(s) = \frac{I_2(s)}{V_2(s)} \qquad \text{...(10.6)}$$

10.3 TRANSFER FUNCTION

The transfer function is used to describe networks which have at least two ports. It relates a voltage or current at one port to the voltage or current at another port. These functions are also defined as the ratio of a response transform to an excitation transform. Thus, there are four possible forms of transfer functions.

(1) Voltage transfer function It is defined as the ratio of the voltage transform at one port to the voltage transform at another port. It is denoted by $G(s)$.

$$G_{12}(s) = \frac{V_2(s)}{V_1(s)}$$

$$G_{21}(s) = \frac{V_1(s)}{V_2(s)} \qquad \text{...(10.7)}$$

(2) Current transfer function It is defined as the ratio of the current transform at one port to the current transform at another port. It is denoted by $\alpha(s)$.

$$\alpha_{12}(s) = \frac{I_2(s)}{I_1(s)} \qquad \text{...(10.8)}$$

$$\alpha_{21}(s) = \frac{I_1(s)}{I_2(s)}$$

(3) Transfer impedance function It is defined as the ratio of voltage transform at one port to the current transform at another port. It is denoted by $Z(s)$.

$$Z_{12}(s) = \frac{V_2(s)}{I_1(s)}$$

$$Z_{21}(s) = \frac{V_1(s)}{I_2(s)} \qquad \ldots(10.9)$$

(4) Transfer admittance function It is defined as the ratio of current transform at one port to the voltage transform at another port. It denoted by $Y(s)$.

$$Y_{12}(s) = \frac{I_2(s)}{V_1(s)}$$

$$Y_{21}(s) = \frac{I_1(s)}{V_2(s)} \qquad \ldots(10.10)$$

Example 10.1 *Determine the driving-point impedance function of a one-port network shown in Fig. 10.2.*

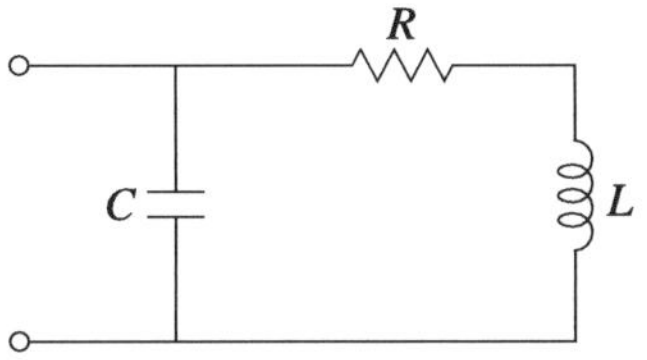

Fig. 10.2

Solution The transformed network is shown in Fig. 10.3.

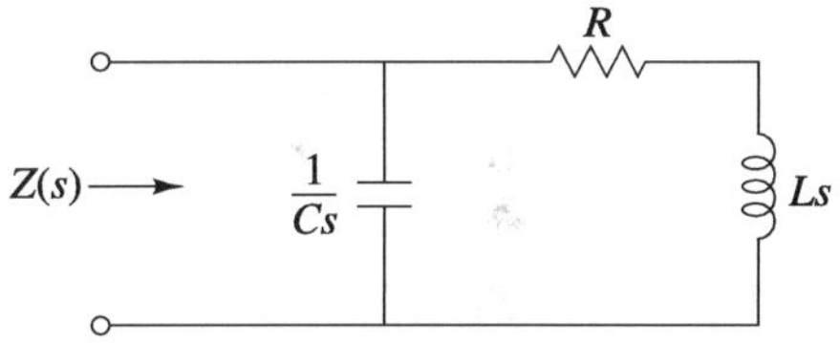

Fig. 10.3

$$Z(s) = \frac{(R + Ls)\dfrac{1}{Cs}}{\dfrac{1}{Cs} + (R + Ls)}$$

$$= \frac{R + Ls}{LCs^2 + RCs + 1} = \frac{1}{C}\,\frac{s + \dfrac{R}{L}}{s^2 + \dfrac{R}{L}s + \dfrac{1}{LC}}$$

Example 10.2 *Find voltage transfer function of the two-port network shown in Fig. 10.4.*

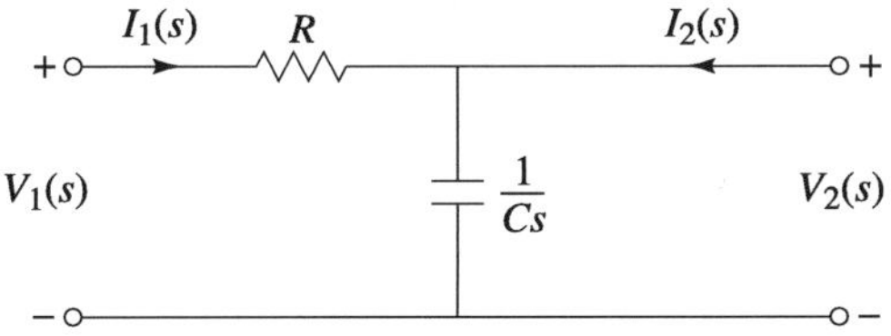

Fig. 10.4

Solution By voltage division formula,

$$V_2(s) = V_1(s) \times \frac{\frac{1}{Cs}}{R + \frac{1}{Cs}}$$

Voltage transfer function $\dfrac{V_2(s)}{V_1(s)} = \dfrac{\frac{1}{Cs}}{R + \frac{1}{Cs}} = \dfrac{\frac{1}{RC}}{s + \frac{1}{RC}}$

Example 10.3 *Determine the driving-point impedance of the network shown in Fig. 10.5.*

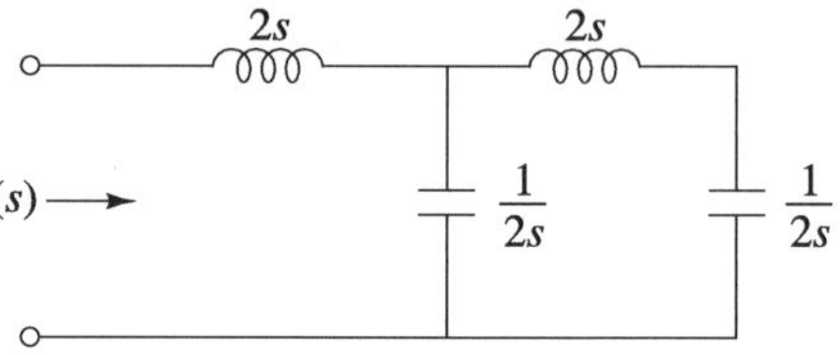

Fig. 10.5

Solution

$$Z(s) = 2s + \frac{\frac{1}{2s}\left(2s + \frac{1}{2s}\right)}{\frac{1}{2s} + 2s + \frac{1}{2s}}$$

$$= 2s + \frac{\frac{1}{2s}\left(2s + \frac{1}{2s}\right)}{\frac{2+4s^2}{2s}} = 2s + \frac{\left(2s + \frac{1}{2s}\right)}{2+4s^2}$$

$$= \frac{4s + 8s^3 + 2s + \frac{1}{2s}}{2+4s^2} = \frac{16s^4 + 12s^2 + 1}{8s^3 + 4s}$$

Example 10.4 *Determine the driving-point impedance of the network shown in Fig. 10.6.*

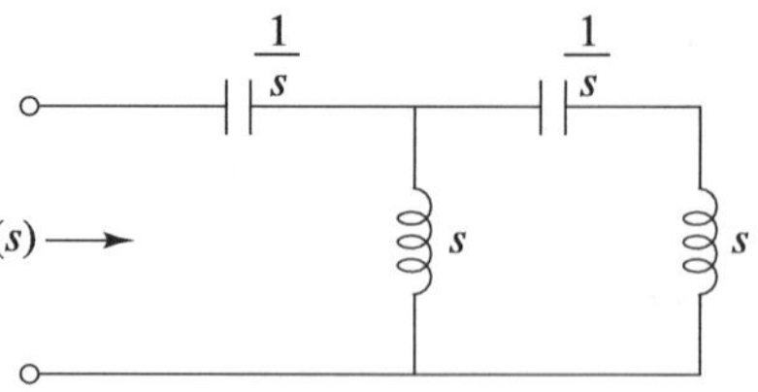

Fig. 10.6

Solution

$$Z(s) = \frac{1}{s} + \frac{s\left(\frac{1}{s} + s\right)}{s + \frac{1}{s} + s}$$

$$= \frac{1}{s} + \frac{(1+s^2)s}{2s^2+1} = \frac{1}{s} + \frac{s+s^3}{2s^2+1}$$

$$= \frac{s^4+3s^2+1}{2s^3+s}$$

Example 10.5 *Find the driving-point admittance function of the network shown in Fig. 10.7.*

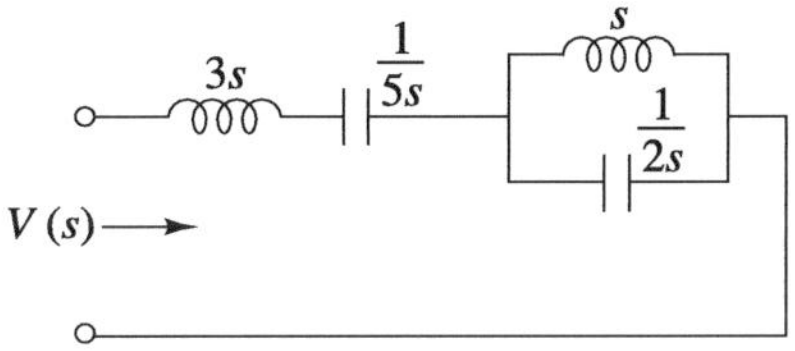

Fig. 10.7

Solution

$$Z(s) = 3s + \frac{1}{5s} + \frac{s\left(\frac{1}{2s}\right)}{s+\frac{1}{2s}} = 3s + \frac{1}{5s} + \frac{s}{2s^2+1}$$

$$= \frac{30s^4+15s^2+2s^2+1+5s^2}{5s(2s^2+1)} = \frac{30s^4+22s^2+1}{5s(2s^2+1)}$$

$$Y(s) = \frac{1}{Z(s)} = \frac{5s(2s^2+1)}{30s^4+22s^2+1}$$

10.4 ANALYSIS OF LADDER NETWORKS

The network functions of a ladder network can be obtained by a simple method. This method depends upon the relationships that exist between the branch currents and node voltages of the ladder network. Consider the network shown in Fig. 10.8 where all the impedances are connected in series branches and all the admittances are connected in parallel branches.

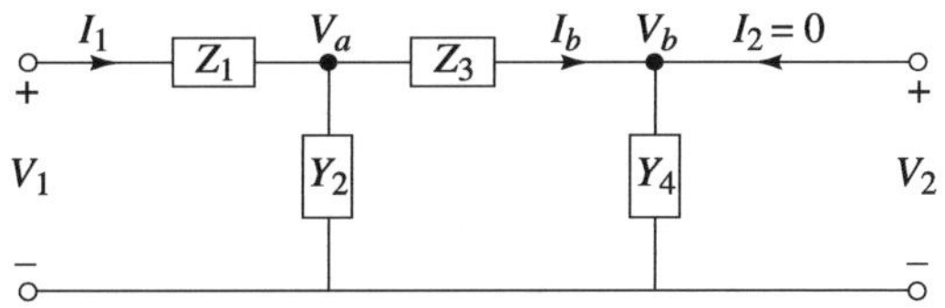

Fig. 10.8

Analysis is done by writing the set of equations. In writing these equations, we begin at the port 2 of the ladder and work towards the Port 1.

$$V_b = V_2$$
$$I_b = Y_4 V_2$$
$$V_a = Z_3 I_b + V_2 = (1 + Z_3 Y_4) V_2$$
$$I_1 = Y_2 V_a + I_b$$
$$= [Y_2 (1 + Z_3 Y_4) + Y_4] V_2$$
$$V_1 = Z_1 I_1 + V_a$$
$$= [Z_1 \{Y_2 (1 + Z_3 Y_4) + Y_4\} + (1 + Z_3 Y_4)] V_2 \quad \ldots (10.11)$$

Thus, each succeeding equation takes into account one new impedance or admittance. We observe that, except the first two equations, each subsequent equation is obtained by multiplying the equation just preceding it by imittance (either impedance or admittance) that is next down the line and then adding to this product the equation twice preceding it. After writing these equations, we can obtain any network function.

Example 10.6 *Find the network functions* $\frac{V_1}{I_1}$, $\frac{V_2}{V_1}$ *and* $\frac{V_2}{I_1}$ *for the network shown in Fig. 10.9.*

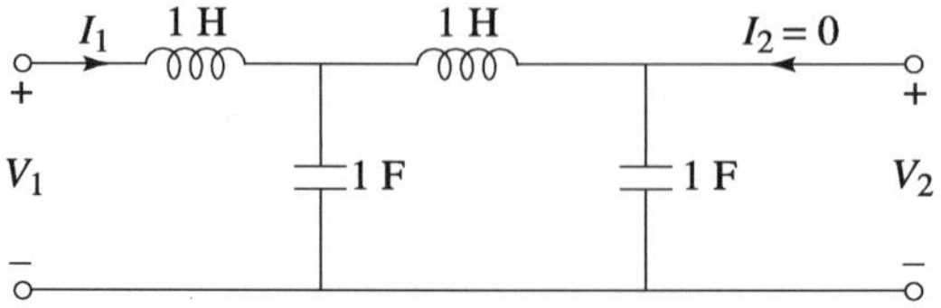

Fig. 10.9

Solution The transformed network is shown in Fig. 10.10.

$$V_b = V_2$$

$$I_b = \frac{V_2}{\frac{1}{s}} = s\,V_2$$

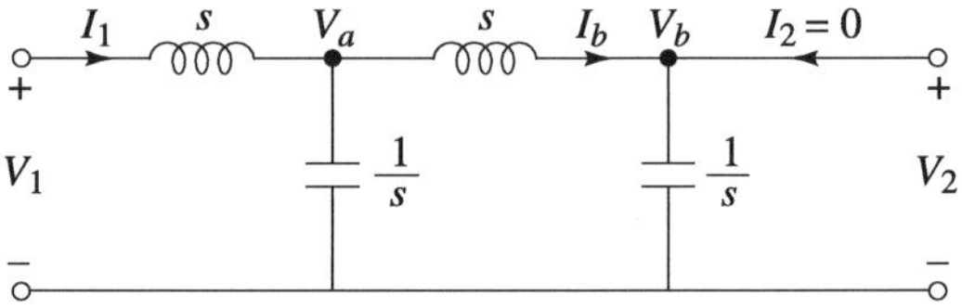

Fig. 10.10

$$V_a = s\,I_b + V_2$$
$$= s\,(s\,V_2) + V_2 = (s^2 + 1)\,V_2$$

$$I_1 = \frac{V_a}{\frac{1}{s}} + I_b = s\,V_a + I_b$$
$$= s\,(s^2 + 1)\,V_2 + s\,V_2 = (s^3 + 2s)\,V_2$$

$$V_1 = s\,I_1 + V_a$$
$$= s\,(s^3 + 2s)\,V_2 + (s^2 + 1)\,V_2$$
$$= (s^4 + 2s^2 + s^2 + 1)\,V_2$$
$$= (s^4 + 3s^2 + 1)\,V_2$$

Hence,

$$\frac{V_1}{I_1} = \frac{(s^4 + 3s^2 + 1)V_2}{(s^3 + 2s)V_2} = \frac{s^4 + 3s^2 + 1}{s^3 + 2s}$$

$$\frac{V_2}{V_1} = \frac{V_2}{(s^4 + 3s^2 + 1)V_2} = \frac{1}{s^4 + 3s^2 + 1}$$

$$\frac{V_2}{I_1} = \frac{V_2}{(s^3 + 2s)V_2} = \frac{1}{s^3 + 2s} = \frac{1}{s(s^2 + 2)}$$

Example 10.7 *Find the network functions* $\frac{V_2}{V_1}$, $\frac{V_1}{I_1}$ *and* $\frac{V_2}{I_1}$ *for the network in Fig. 10.11.*

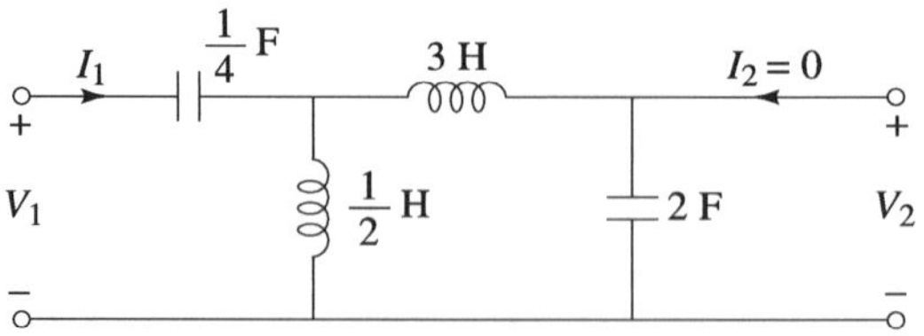

Fig. 10.11

Solution The transformed network is shown in Fig. 10.12.

$$V_b = V_2$$

$$I_b = \frac{V_2}{\frac{1}{2s}} = 2s\,V_2$$

$$V_a = 3s\,I_b + V_2$$
$$= 3s\,(2sV_2) + V_2$$
$$= (6s^2 + 1)\,V_2$$

$$I_1 = \frac{V_a}{\frac{s}{2}} + I_b$$

$$= \frac{2}{s}(6s^2 + 1)\,V_2 + 2s\,V_2 = \left(\frac{14s^2 + 2}{s}\right) V_2$$

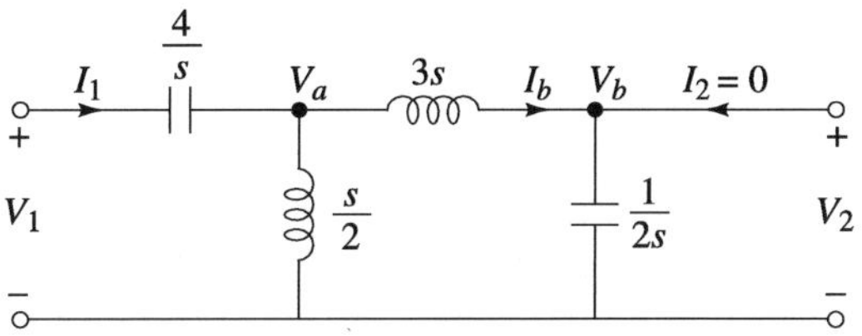

Fig. 10.12

$$V_1 = \frac{4}{s} I_1 + V_a$$

$$= \frac{4}{s}\left(\frac{14s^2 + 2}{s}\right) V_2 + (6s^2 + 1)\,V_2 = \left(\frac{6s^4 + 57s^2 + 8}{s^2}\right) V_2$$

Hence,

$$\frac{V_1}{I_1} = \frac{6s^4 + 57s^2 + 8}{14s^3 + 2s}$$

$$\frac{V_2}{V_1} = \frac{s^2}{6s^4 + 57s^2 + 8}$$

$$\frac{V_2}{I_1} = \frac{s}{14s^2 + 2}$$

Example 10.8 *For the network shown in Fig. 10.13, determine transfer function* $\frac{V_2}{V_1}$.

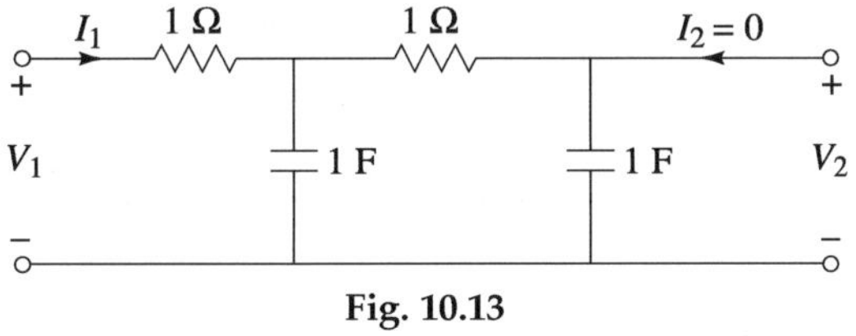

Fig. 10.13

Solution The transformed network is shown in Fig. 10.14.

$$V_b = V_2$$

$$I_b = \frac{V_2}{\frac{1}{s}} = s\,V_2$$

$$V_a = 1\,I_b + V_2 = s\,V_2 + V_2 = (s + 1)\,V_2$$

$$I_1 = \frac{V_a}{\frac{1}{s}} + I_b$$

$$= s\,V_a + I_b = s\,(s + 1)\,V_2 + s\,V_2 = (s^2 + 2s)\,V_2$$

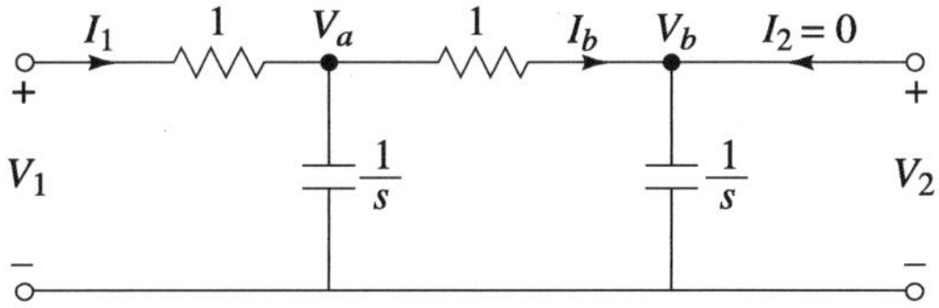

Fig. 10.14

$$V_1 = 1\,I_1 + V_a = (s^2 + 2s)\,V_2 + (s + 1)\,V_2$$
$$= (s^2 + 3s + 1)\,V_2$$

Hence, $$\frac{V_2}{V_1} = \frac{V_2}{(s^2+3s+1)V_2} = \frac{1}{s^2+3s+1}$$

Example 10.9 *For the network shown in Fig. 10.15, determine the voltage transfer function* $\frac{V_2}{V_1}$.

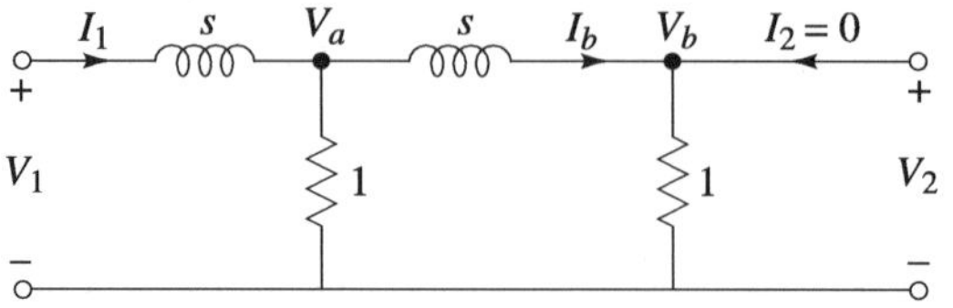

Fig. 10.15

Solution

$$V_b = V_2$$
$$I_b = \frac{V_2}{1} = V_2$$
$$V_a = s\,I_b + V_2 = s\,V_2 + V_2 = (s+1)\,V_2$$
$$I_1 = \frac{V_a}{1} + I_b$$
$$= (s+1)\,V_2 + V_2 = (s+2)\,V_2$$
$$V_1 = s\,I_1 + V_a$$
$$= s\,(s+2)\,V_2 + (s+1)\,V_2 = (s^2+3s+1)\,V_2$$

Hence, $$\frac{V_2}{V_1} = \frac{V_2}{(s^2+3s+1)V_2} = \frac{1}{s^2+3s+1}$$

Example 10.10 *For the ladder network of Fig. 10.16, find the driving point-impedance at the 1–1′ terminal with 2–2′ open.*

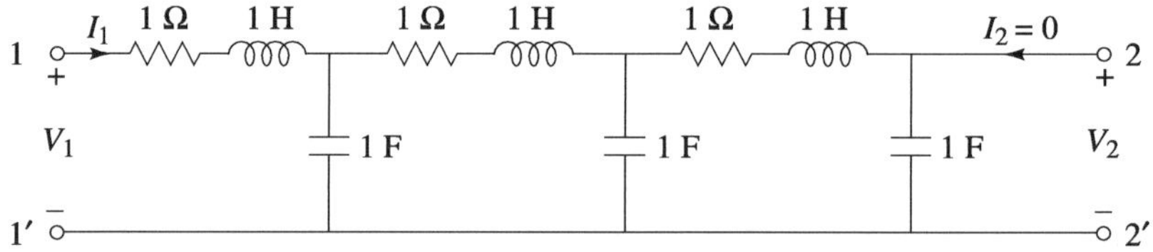

Fig. 10.16

Solution The transformed network is shown in Fig. 10.17.

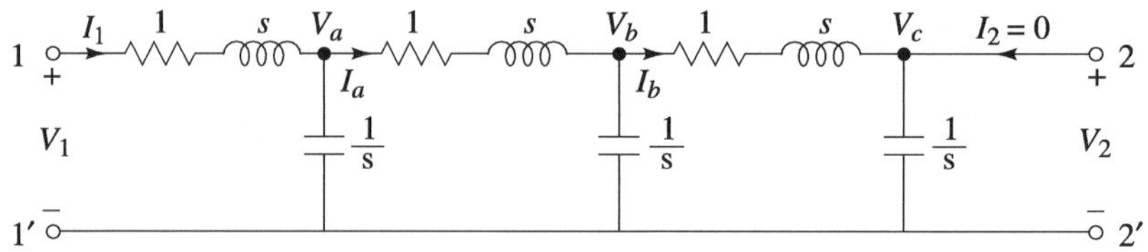

Fig. 10.17

$$V_c = V_2$$
$$I_b = \frac{V_2}{\frac{1}{s}} = s\,V_2$$

$$V_b = (s+1)\,I_b + V_2 = (s+1)\,s\,V_2 + V_2 = (s^2+s+1)\,V_2$$

$$I_a = \frac{V_b}{\frac{1}{s}} + I_b$$

$$= s\,V_b + I_b = s\,(s^2+s+1)\,V_2 + s\,V_2 = (s^3+s^2+2s)\,V_2$$

$$V_a = (s+1)\,I_a + V_b$$

$$= (s+1)\,(s^3+s^2+2s)\,V_2 + (s^2+s+1)\,V_2 = (s^4+2s^3+4s^2+3s+1)\,V_2$$

$$I_1 = \frac{V_a}{\frac{1}{s}} + I_a = s\,V_a + I_a$$

$$= s\,(s^4+2s^3+4s^2+3s+1)\,V_2 + (s^3+s^2+2s)\,V_2 = (s^5+2s^4+5s^3+4s^2+3s)\,V_2$$

$$V_1 = (s+1)\,I_1 + V_a$$

$$= (s+1)\,(s^5+2s^4+5s^3+4s^2+3s)\,V_2 + (s^4+2s^3+4s^2+3s+1)\,V_2$$

$$= (s^6+3s^5+8s^4+11s^3+11s^2+6s+1)\,V_2$$

Hence,
$$Z_{11} = \frac{V_1}{I_1} = \frac{s^6+3s^5+8s^4+11s^3+11s^2+6s+1}{s^5+2s^4+5s^3+4s^2+3s}$$

Example 10.11 *Determine the voltage transfer function* $\frac{V_2}{V_1}$ *for the network shown in Fig. 10.18.*

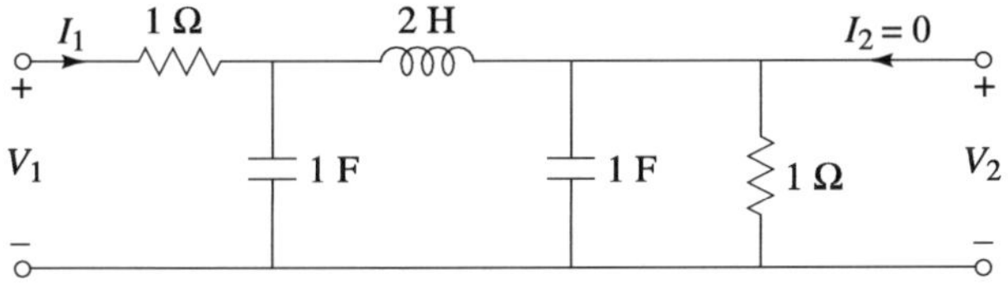

Fig. 10.18

Solution The transformed network is shown in Fig. 10.19.

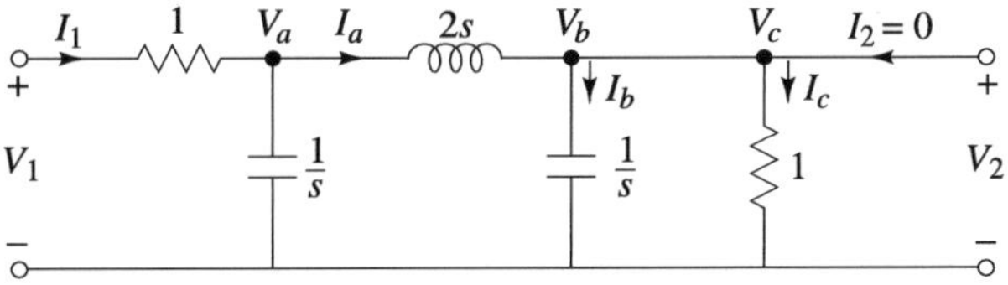

Fig. 10.19

$$V_c = V_b = V_2$$

$$I_a = I_b + I_c$$

$$= \frac{V_2}{\frac{1}{s}} + \frac{V_2}{1} = s\,V_2 + V_2$$

$$= (s+1)\,V_2$$

$$V_a = 2s\,I_a + V_2$$

$$= 2s\,(s+1)\,V_2 + V_2$$

$$= (2s^2+2s+1)\,V_2$$

$$I_1 = \frac{V_a}{\frac{1}{s}} + I_a = s\,V_a + I_a$$

$$= s\,(2s^2+2s+1)\,V_2 + (s+1)\,V_2 = (2s^3+2s^2+2s+1)\,V_2$$

$$V_1 = 1I_1 + V_a$$

$$= (2s^3+2s^2+2s+1)\,V_2 + (2s^2+2s+1)\,V_2 = (2s^3+4s^2+4s+2)\,V_2$$

Hence, $$\frac{V_2}{V_1} = \frac{1}{2s^3 + 4s^2 + 4s + 2}$$

Example 10.12 *For the network shown in Fig. 10.20, determine the transfer function* $\frac{I_2}{V_1}$.

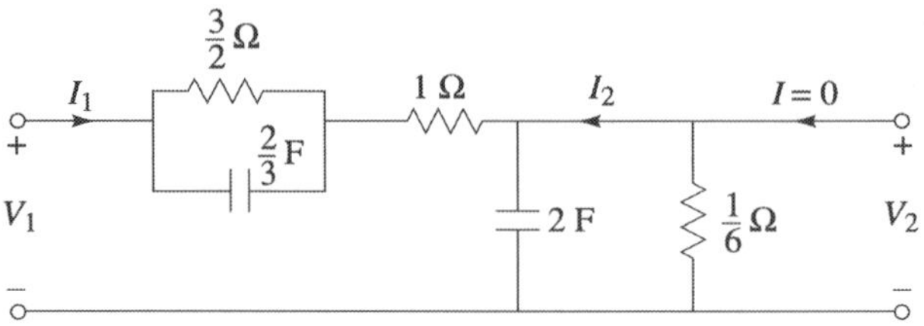

Fig. 10.20

Solution The transformed network is shown in Fig. 10.21.

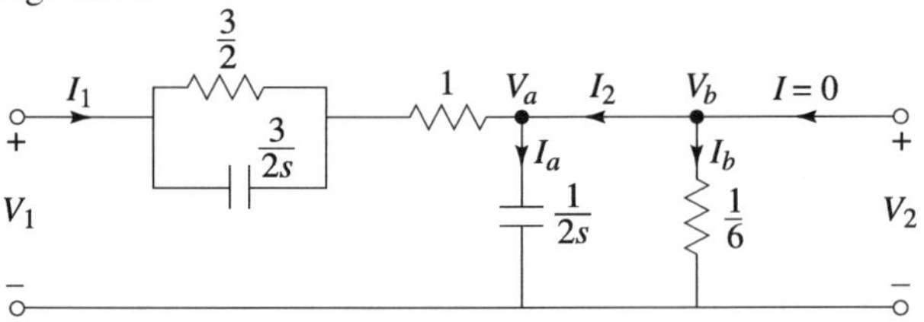

Fig. 10.21

$$V_a = V_b = V_2$$

$$I_1 = I_a + I_b$$

$$= \frac{V_2}{\frac{1}{2s}} + \frac{V_2}{\frac{1}{6}} = 2s\, V_2 + 6V_2$$

$$= (2s + 6)\, V_2$$

$$V_1 = \left[\frac{\frac{3}{2}\times\frac{3}{2s}}{\frac{3}{2}+\frac{3}{2s}} + 1\right] I_1 + V_2$$

$$= \left(\frac{9}{6s+6} + 1\right) I_1 + V_2 = \frac{(6s+15)}{6s+6}(2s+6)\, V_2 + V_2$$

$$= \left[\frac{(2s+5)(s+3)}{(s+1)} + 1\right] V_2 = \left[\frac{(2s+5)(s+3)+(s+1)}{s+1}\right] V_2$$

$$= \left(\frac{2s^2 + 6s + 5s + 15 + s + 1}{s+1}\right) V_2$$

$$= \left(\frac{2s^2 + 12s + 16}{s+1}\right) V_2 = \frac{2(s^2+6s+8)}{s+1} V_2 = \frac{2(s+4)(s+2)}{s+1} V_2$$

Also, $$I_2 = -I_b = -6V_2$$

$$\frac{I_2}{V_1} = \frac{-6V_2}{\frac{2(s+4)(s+2)}{s+1} V_2} = \frac{-3(s+1)}{(s+4)(s+2)}$$

Example 10.13 *For the network shown in Fig. 10.22, compute* $\alpha_{12}(s) = \frac{I_2(s)}{I_l(s)}$ *and* $Z_{12}(s) = \frac{V_2(s)}{I_l(s)}$.

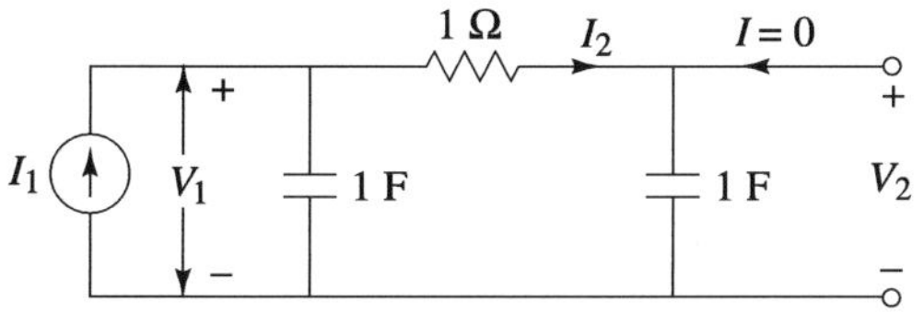

Fig. 10.22

Solution The transformed network is as shown in Fig. 10.23.

$$V_a = V_2$$

$$I_2 = \frac{V_2}{\frac{1}{s}} = sV_2$$

$$V_1 = 1\, I_2 + V_a$$
$$= sV_2 + V_2 = (s+1)\, V_2$$

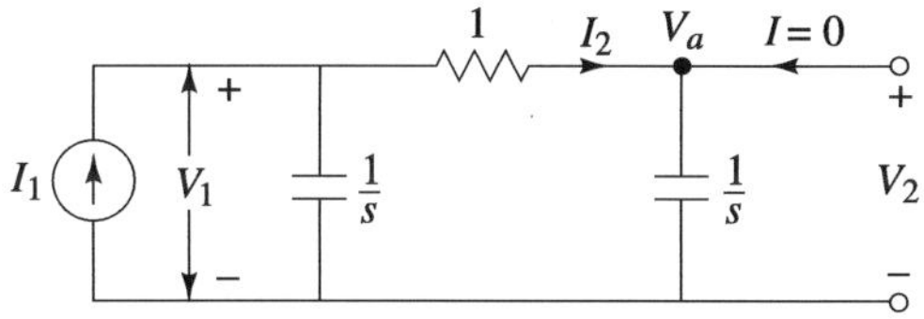

Fig. 10.23

$$I_1 = \frac{V_1}{\frac{1}{s}} + I_2$$
$$= sV_1 + I_2 = s\,(s+1)\, V_2 + sV_2$$
$$= (s^2 + 2s)\, V_2$$

Hence, $$\alpha_{12}(s) = \frac{I_2}{I_1} = \frac{1}{s+2}$$

and $$Z_{12}(s) = \frac{V_2}{I_1} = \frac{1}{s^2 + 2s}$$

Example 10.14 *Determine the voltage ratio* $\frac{V_2}{V_1}$, *current ratio* $\frac{I_2}{I_1}$, *transfer impedance* $\frac{V_2}{I_1}$ *and driving-point impedance* $\frac{V_1}{I_1}$ *for the network shown in Fig. 10.24.*

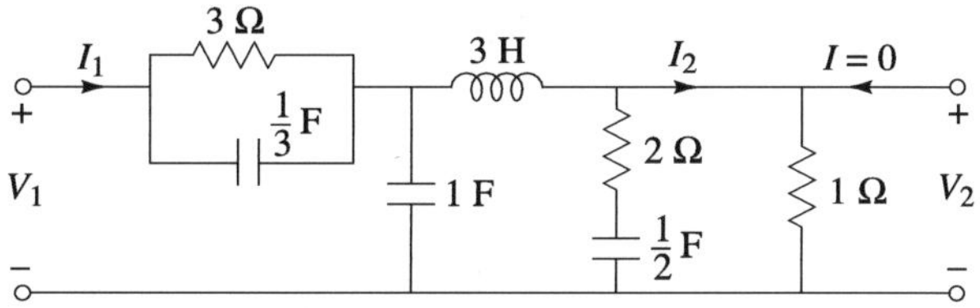

Fig. 10.24

Solution Transformed network is shown in Fig. 10.25.

$$V_c = V_b = V_2$$

$$I_2 = \frac{V_2}{1} = V_2$$

$$I_a = I_b + I_2$$

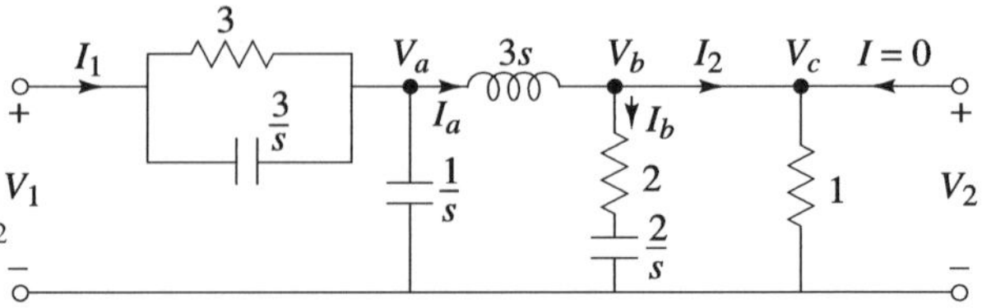

Fig. 10.25

$$= \frac{V_2}{2+\frac{2}{s}} + \frac{V_2}{1}$$

$$= \frac{s}{2s+2} V_2 + V_2 = \left(\frac{3s+2}{2s+2}\right) = V_2$$

$$V_a = 3s\, I_a + V_2$$

$$= \frac{3s(3s+2)}{2s+2} V_2 + V_2 = \left(\frac{9s^2+8s+2}{2s+2}\right) V_2$$

$$I_1 = \frac{V_a}{\frac{1}{s}} + I_a$$

$$= s\, V_a + I_a = \frac{s(9s^2+8s+2)}{2s+2} V_2 + \left(\frac{3s+2}{2s+2}\right) V_2$$

$$= \left(\frac{9s^3+8s^2+5s+2}{2s+2}\right) V_2$$

$$V_1 = \left[\frac{3\times\frac{3}{s}}{3+\frac{3}{s}}\right] I_1 + V_a = \left(\frac{3}{s+1}\right) I_1 + V_a$$

$$= \left(\frac{3}{s+1}\right)\left(\frac{9s^3+8s^2+5s+2}{2s+2}\right) V_2 + \left(\frac{9s^2+8s+2}{2s+2}\right) V_2$$

$$= \left[\frac{27s^3+24s^2+15s+6+9s^3+8s^2+2s+9s^2+8s+2}{(s+1)(2s+2)}\right] V_2$$

$$= \frac{(36s^3+41s^2+25s+8)}{(s+1)(2s+2)} V_2 = \left(\frac{36s^3+41s^2+25s+8}{2s^2+4s+2}\right) V_2$$

Hence,
$$\frac{V_2}{V_1} = \frac{2s^2+4s+2}{36s^3+41s^2+25s+8}$$

$$\frac{I_2}{I_1} = \frac{2s+2}{9s^3+8s^2+5s+2}$$

$$\frac{V_2}{I_1} = \frac{2s+2}{9s^3+8s^2+5s+2}$$

$$\frac{V_1}{I_1} = \frac{36s^3+41s^2+25s+8}{(s+1)(9s^3+8s^2+5s+2)} = \frac{36s^3+41s^2+25s+8}{9s^4+17s^3+13s^2+7s+2}$$

10.5 ANALYSIS OF NON-LADDER NETWORK

The above method is applicable for ladder networks. There are other network configurations to which the technique described is not applicable.

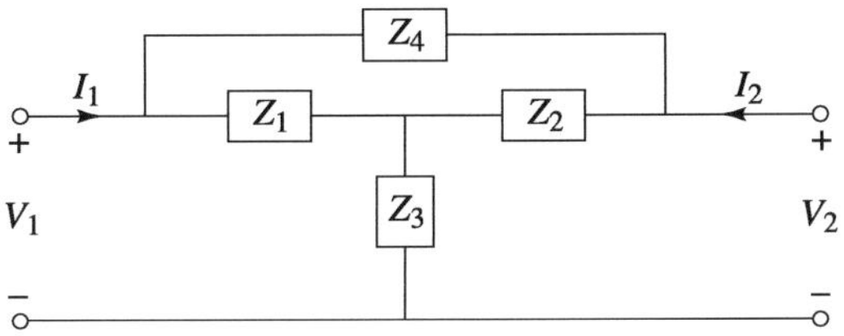

Fig. 10.26

For such a type of network, it is necessary to express the network functions as a quotient of determinants, formulated on KVL and KCL basis.

Example 10.15 *For the network shown in Fig. 10.27, determine $Z_{11}(s)$, $G_{12}(s)$ and $Z_{12}(s)$.*

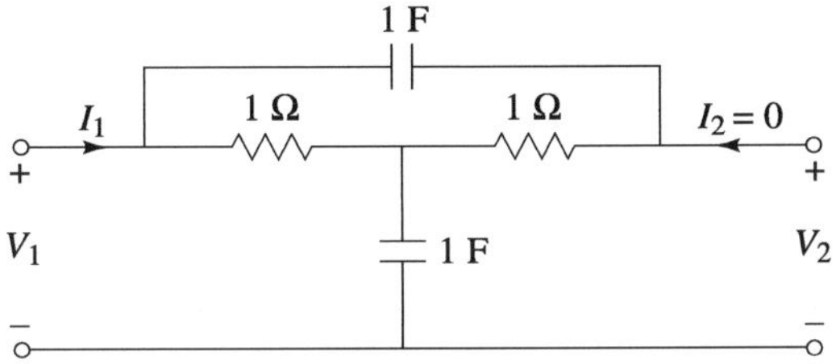

Fig. 10.27

Solution The transformed network is shown in Fig. 10.28.

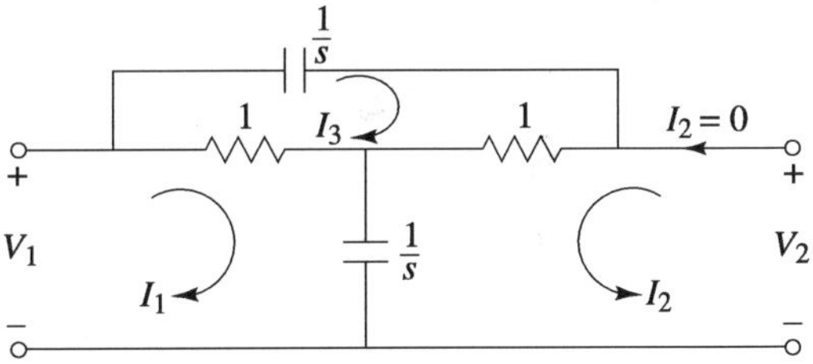

Fig. 10.28

Applying KVL to Mesh 1,

$$V_1 = \left(1+\frac{1}{s}\right)I_1 - I_3 \qquad \ldots\text{(i)}$$

Applying KVL to Mesh 2,

$$V_2 = \frac{1}{s}\,I_1 + I_3 \qquad \ldots\text{(ii)}$$

Applying KVL to Mesh 3,

$$I_1 - \left(2+\frac{1}{s}\right) I_3 = 0$$

$$I_3 = \left(\frac{s}{2s+1}\right) I_1 \qquad \ldots\text{(iii)}$$

Substituting the Eq. (iii) in Eqs (i) and (ii),

$$V_1 = \left(1+\frac{1}{s}\right)I_1 - \left(\frac{s}{2s+1}\right) I_1$$

$$= \left(\frac{s+1}{s} - \frac{s}{2s+1}\right) I_1 = \left[\frac{s^2+3s+1}{s(2s+1)}\right] I_1$$

$$V_2 = \frac{1}{s}\,\mathrm{I}_1 + \frac{s}{2s+1} I_1 = \left[\frac{s^2+2s+1}{s(2s+1)}\right] I_1$$

Hence,

$$Z_{11}(s) = \frac{V_1}{I_1} = \frac{s^2+3s+1}{s(2s+1)}$$

$$Z_{12}(s) = \frac{V_2}{I_1} = \frac{s^2+2s+1}{s(2s+1)}$$

$$G_{12}(s) = \frac{V_2}{V_1} = \frac{s^2+2s+1}{s^2+3s+1}$$

Example 10.16 *For the network shown in Fig. 10.29, find the driving-point admittance Y_{11} and transfer admittance Y_{12}.*

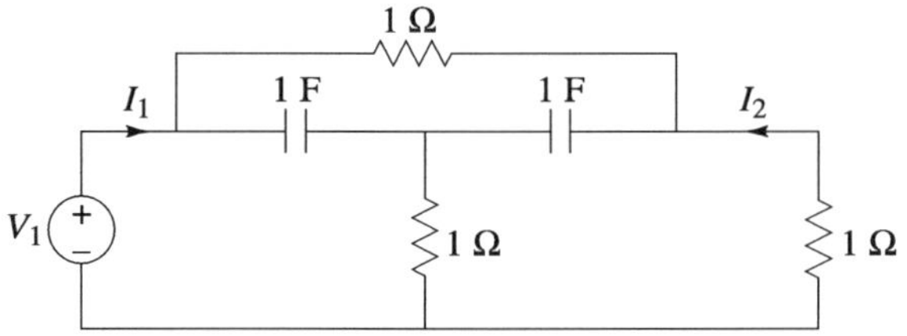

Fig. 10.29

Solution The transformed network is shown in Fig. 10.30.

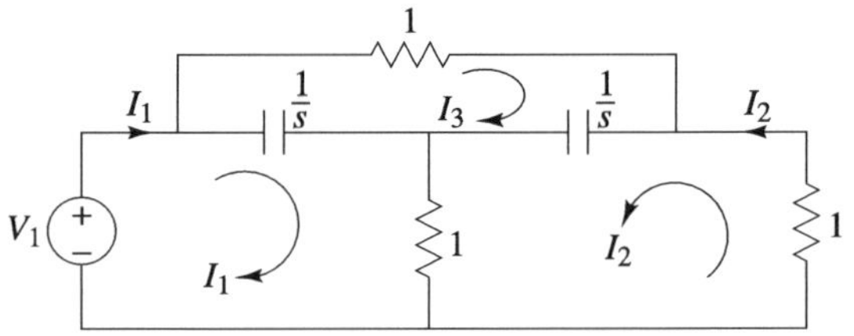

Fig. 10.30

Applying KVL to Mesh 1,

$$V_1 = \left(\frac{1}{s}+1\right) I_1 + I_2 - \frac{1}{s} I_3 \qquad \text{...(i)}$$

Applying KVL to Mesh 2,

$$0 = I_1 + \left(2+\frac{1}{s}\right) I_2 + \frac{1}{s} I_3 \qquad \text{...(ii)}$$

Applying KVL to Mesh 3,

$$0 = -\frac{1}{s} I_1 + \frac{1}{s} I_2 + \left(\frac{2}{s}+1\right) I_3 \qquad \text{...(iii)}$$

Writing these equations in matrix form,

$$\begin{bmatrix} V_1 \\ 0 \\ 0 \end{bmatrix} = \begin{bmatrix} \frac{1}{s}+1 & 1 & \frac{-1}{s} \\ 1 & 2+\frac{1}{s} & \frac{1}{s} \\ \frac{-1}{s} & \frac{1}{s} & \frac{2}{s}+1 \end{bmatrix} \begin{bmatrix} I_1 \\ I_2 \\ I_3 \end{bmatrix}$$

$$I_1 = \frac{\Delta_1}{\Delta}$$

$$\Delta = \begin{vmatrix} \frac{1}{s}+1 & 1 & \frac{-1}{s} \\ 1 & 2+\frac{1}{s} & \frac{1}{s} \\ \frac{-1}{s} & \frac{1}{s} & \frac{2}{s}+1 \end{vmatrix}$$

$$= \left(1+\frac{1}{s}\right)\left[\left(2+\frac{1}{s}\right)\left(1+\frac{2}{s}\right)-\frac{1}{s^2}\right] - 1\left[(1)\left(1+\frac{2}{s}\right)+\frac{1}{s^2}\right] - \frac{1}{s}\left[(1)\left(\frac{1}{s}\right)+\left(\frac{1}{s}\right)\left(2+\frac{1}{s}\right)\right]$$

$$= \frac{s^2+5s+2}{s^2}$$

$$\Delta_1 = \begin{vmatrix} V_1 & 1 & \frac{-1}{s} \\ 0 & 2+\frac{1}{s} & \frac{1}{s} \\ 0 & \frac{1}{s} & \frac{2}{s}+1 \end{vmatrix}$$

$$= V_1\left[\left(2+\frac{1}{s}\right)\left(1+\frac{2}{s}\right)-\frac{1}{s^2}\right] = V_1\left(\frac{2s^2+5s+1}{s^2}\right)$$

$$I_1 = V_1\left(\frac{2s^2+5s+1}{s^2+5s+2}\right)$$

Driving-point admittance $$Y_{11} = \frac{I_1}{V_1} = \frac{2s^2+5s+1}{s^2+5s+2}$$

$$I_2 = \frac{\Delta_2}{\Delta}$$

$$\Delta_2 = \begin{vmatrix} \frac{1}{s}+1 & V_1 & \frac{-1}{s} \\ 1 & 0 & \frac{1}{s} \\ \frac{-1}{s} & 0 & \frac{2}{s}+1 \end{vmatrix}$$

$$= -V_1\left[\frac{2}{s}+1+\frac{1}{s^2}\right] = -V_1\left(\frac{s^2+2s+1}{s^2}\right)$$

$$I_2 = -V_1\left(\frac{s^2+2s+1}{s^2+5s+2}\right)$$

Transfer admittance $$Y_{12} = \frac{I_2}{V_1} = -\frac{s^2+2s+1}{s^2+5s+2}$$

10.6 POLES AND ZEROS OF NETWORK FUNCTIONS

The network function $F(s)$ can be written as ratio of two polynomials.

$$F(s) = \frac{N(s)}{D(s)} = \frac{a_0 s^n + a_1 s^{n-1} + \ldots\ldots + a_{n-1}s + a_n}{b_0 s^m + b_1 s^{m-1} + \ldots\ldots + b_{m-1}s + b_m}$$

where a_0, a_1, ..., a_n and b_0, b_1, ..., b_m are the coefficients of the polynomials $N(s)$ and $D(s)$. These are real and positive for networks of passive elements. Let $N(s) = 0$ have n roots as $z_1, z_2, \ldots\ldots, z_n$ and $D(s) = 0$ have m roots as $p_1, p_2, \ldots\ldots, p_m$. Then $F(s)$ can be written as

$$F(s) = H\,\frac{(s - z_1)(s - z_2)\ldots\ldots(s - z_n)}{(s - p_1)(s - p_2)\ldots\ldots(s - p_m)}$$

where $H = a_0/b_0$ is a constant called *scale factor* and $z_1, z_2, \ldots\ldots, z_n, p_1, p_2, \ldots\ldots, p_m$ are complex frequencies. When the variable s has the values $z_1, z_2, \ldots, z_n$, the network function becomes zero; such complex frequencies are known as the zeros of the network function. When the variable s has values $p_1, p_2, \ldots\ldots, p_m$, the network function becomes infinite; such complex frequencies are known as the poles of the network function. A network function is completely specified by its poles, zeros and the scale factor.

If the poles or zeros are not repeated, then the function is said to be having simple poles or simple zeros. If the poles or zeros are repeated, then the function is said to be having multiple poles or multiple zeros. When $n > m$, then $(n - m)$ zeros are at $s = \infty$, and for $m > n$, $(m - n)$ poles are at $s = \infty$.

The total number of zeros is equal to the total number of poles. For any network function, poles and zeros at zero and infinity are taken into account in addition to finite poles and zeros.

Poles and zeros are critical frequencies. The network function becomes infinite at poles, while the network function becomes zero at zeros. The network function has a finite, non-zero value at other frequencies.

Poles and zeros provide a representation of a network function as shown in Fig. 10.31. The zeros are shown by circles and the poles by crosses. This diagram is referred to as pole-zero plot.

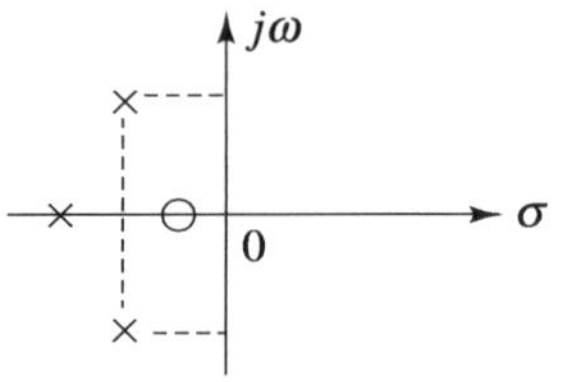

Fig. 10.31 *Pole-zero plot*

10.7 RESTRICTIONS ON POLE AND ZERO LOCATIONS FOR DRIVING- POINT FUNCTIONS [COMMON FACTORS IN *N*(*s*) AND *D*(*s*) CANCELLED]

(1) The coefficients in the polynomials $N(s)$ and $D(s)$ must be real and positive.

(2) The poles and zeros, if complex or imaginary, must occur in conjugate pairs.

(3) The real part of all poles and zeros, must be negative or zero, i.e., the poles and zeros must lie in left half of s plane.

(4) If the real part of pole or zero is zero, then that pole or zero must be simple.

(5) The polynomials $N(s)$ and $D(s)$ may not have missing terms between those of highest and lowest degree, unless all even or all odd terms are missing.

(6) The degree of $N(s)$ and $D(s)$ may differ by either zero or one only. This condition prevents multiple poles and zeros at $s = \infty$.

(7) The terms of lowest degree in $N(s)$ and $D(s)$ may differ in degree by one at most. This condition prevents multiple poles and zeros at $s = 0$.

10.8 RESTRICTIONS ON POLE AND ZERO LOCATIONS FOR TRANSFER FUNCTIONS [COMMON FACTORS IN *N* (*s*) AND *D* (*s*) CANCELLED]

(1) The coefficients in the polynomials $N(s)$ and $D(s)$ must be real, and those for $D(s)$ must be positive.
(2) The poles and zeros if complex or imaginary must occur in conjugate pairs.
(3) The real part of poles must be negative or zero. If the real part is zero then that pole must be simple.
(4) The polynomial $D(s)$ may not have any missing terms between that of highest and lowest degree, unless all even or all odd terms are missing.
(5) The polynomial $N(s)$ may have terms missing between the terms of lowest and highest degree, and some of the coefficients may be negative.
(6) The degree of $N(s)$ may be as small as zero, independent of the degree of $D(s)$.
(7) For voltage and current transfer functions, the maximum degree of $N(s)$ is the degree of $D(s)$.
(8) For transfer impedance and admittance functions, the maximum degree of $N(s)$ is the degree of $D(s)$ plus one.

Example 10.17 *Test whether the following represent driving-point immittances.*

(a) $\dfrac{5s^4 + 3s^2 - 2s + 1}{s^3 + 6s + 20}$

Solution The numerator and denominator polynomials have a missing term between those of highest and lowest degree and one of the coefficient is negative in numerator polynomial. Hence, the function does not represent driving-point immittance.

(b) $\dfrac{s^3 + s^2 + 5s + 2}{s^4 + 6s^3 + 9s^2}$

Solution The term of lowest degree in numerator and denominator polynomials differ in degree by two which is not permitted. Hence, the function does not represent driving-point immittance.

(c) $\dfrac{s^2 + 3s + 2}{s^2 + 6s + 2}$

Solution The function satisfies all the necessary conditions. Hence, it represents driving-point immittance.

Example 10.18 *Obtain the pole-zero plot of the following functions.*

(i) $$F(s) = \frac{s(s+2)}{s^2 + 2s + 2}$$

Solution $$F(s) = \frac{s(s+2)}{(s+1+j1)(s+1-j1)}$$

The function $F(s)$ has zeros at $s = 0$ and $s = -2$ and poles at $s = -1 - j1$ and $s = -1 + j1$.

The pole-zero plot is shown in Fig. 10.32.

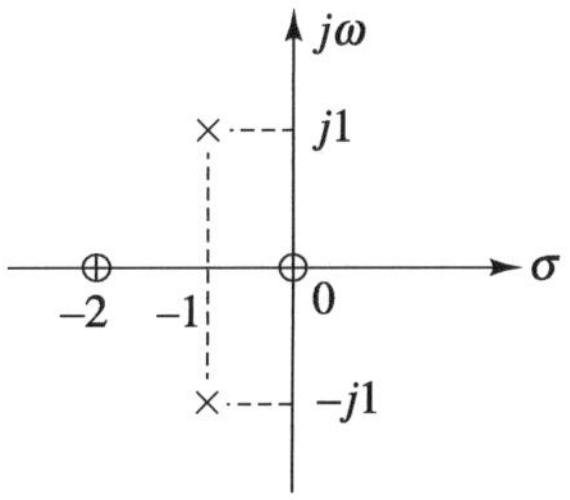

Fig. 10.32

(ii) $$F(s) = \frac{s(s+2)}{(s+1)(s+3)}$$

Solution The function $F(s)$ has zeros at $s = 0$ and $s = -2$ and poles at $s = -1$ and $s = -3$. The poles-zero plot is shown in Fig. 10.33.

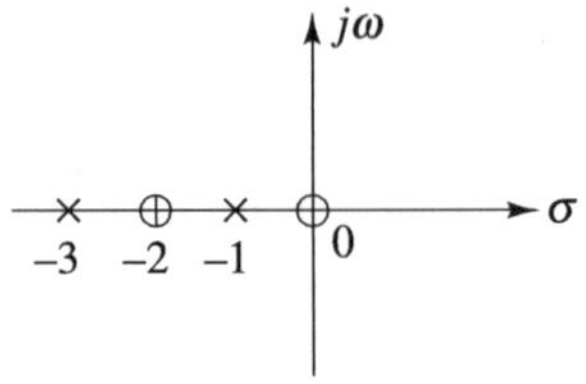

Fig. 10.33

(iii) $$F(s) = \frac{s(s+1)}{(s+2)^2(s+3)}$$

Solution The function $F(s)$ has zeros at $s = 0$ and $s = -1$ and poles at $s = -2, -2$ and $s = -3$. The pole-zero plot is shown in Fig. 10.34.

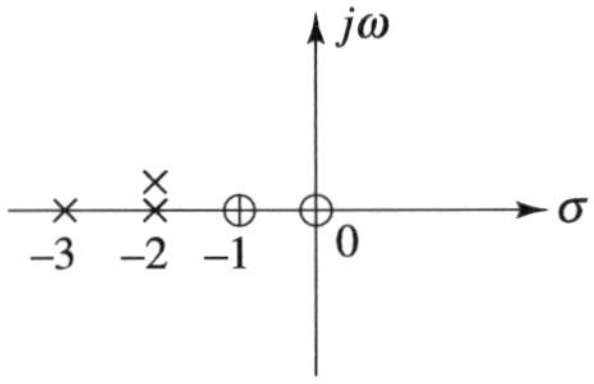

Fig. 10.34

(iv) $$F(s) = \frac{(s+1)^2(s+5)}{(s+2)(s+3+j2)(s+3-j2)}$$

Solution The function $F(s)$ has zeros at $s = -1, -1$ and $s = -5$ and poles at $s = -2$, $s = -3 \pm j2$. The pole-zero plot is shown in Fig. 10.35.

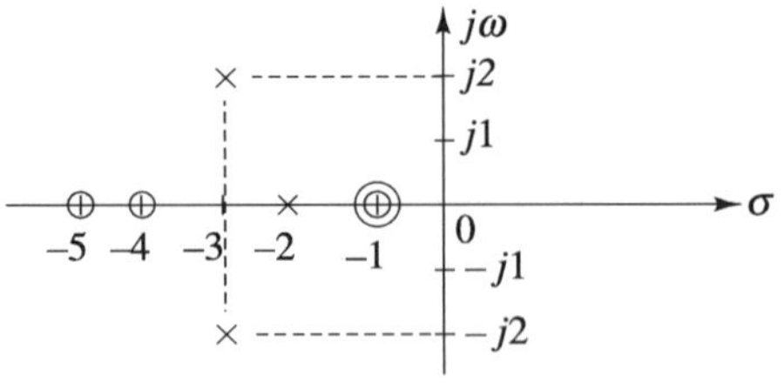

Fig. 10.35

(v) $$F(s) = \frac{s^2+4}{(s+2)(s^2+9)}$$

Solution The function $F(s)$ has zeros at $s = j2$ and $s = -j2$ and poles at $s = -2$, $s = j3$ and $s = -j3$. The pole-zero plot is shown in Fig. 10.36.

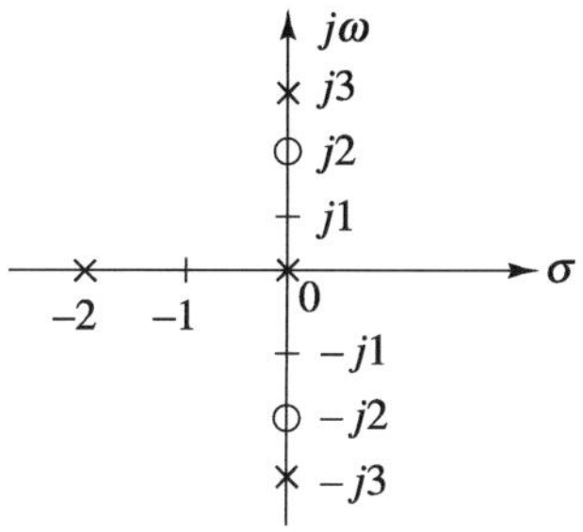

Fig. 10.36

Example 10.19 *Find poles and zeros of the impedance of the following network and plot them on the s-plane.*

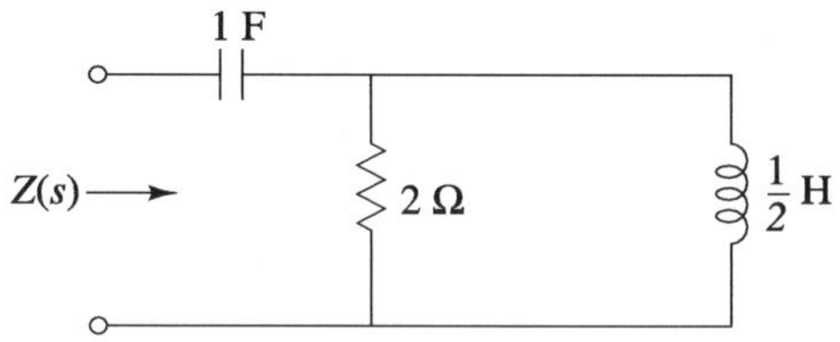

Fig. 10.37

Solution The transformed network is shown in Fig. 10.38.

$$Z(s) = \frac{1}{s} + \frac{\frac{s}{2} \times 2}{\frac{s}{2} + 2} = \frac{1}{s} + \frac{2s}{s+4}$$

$$= \frac{2s^2 + s + 4}{s(s+4)} = \frac{2(s^2 + 0.5s + 2)}{s(s+4)}$$

$$= \frac{2(s + 0.25 + j1.4)(s + 0.25 - j1.4)}{s(s+4)}$$

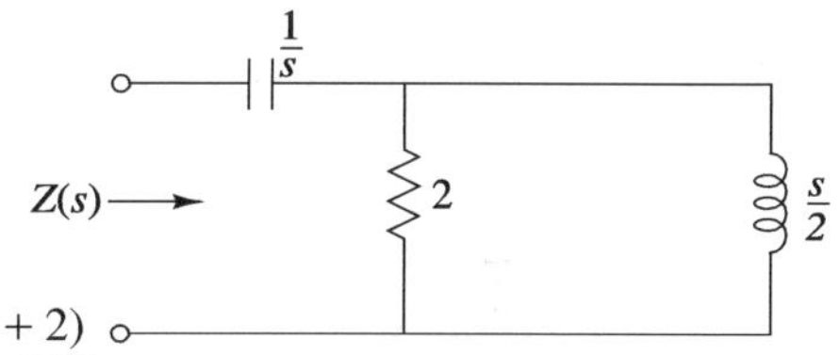

Fig. 10.38

Thus, $Z(s)$ has zeros at $s = -0.25 \pm j1.4$ and poles at $s = 0$ and $s = -4$ as shown in Fig. 10.39.

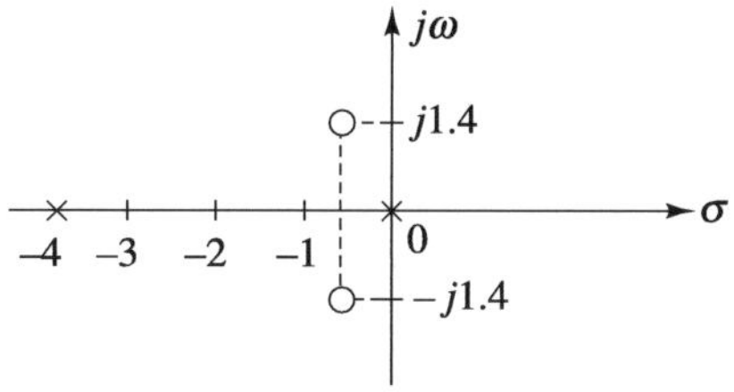

Fig. 10.39

Example 10.20 *Determine the poles and zeros of the impedance function Z (s) in the network shown in Fig. 10.40.*

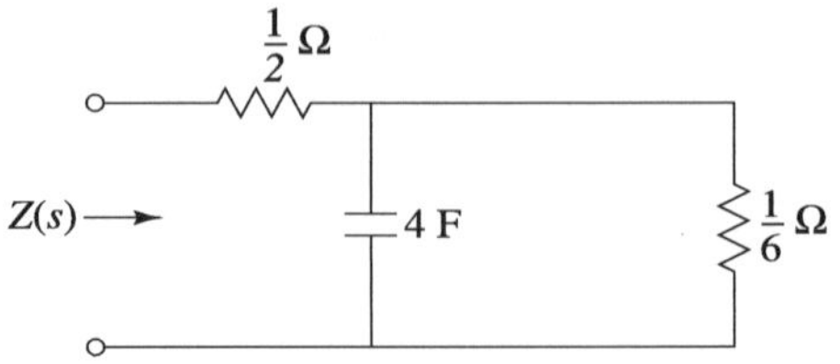

Fig. 10.40

Solution The transformed network is shown in Fig. 10.41.

$$Z(s) = \frac{1}{2} + \frac{\frac{1}{4s} \times \frac{1}{6}}{\frac{1}{4s} + \frac{1}{6}}$$

$$= \frac{1}{2} + \frac{1}{4s+6} = \frac{4s+8}{2(4s+6)}$$

$$= \frac{s+2}{2s+3} = \frac{0.5\,(s+2)}{s+1.5}$$

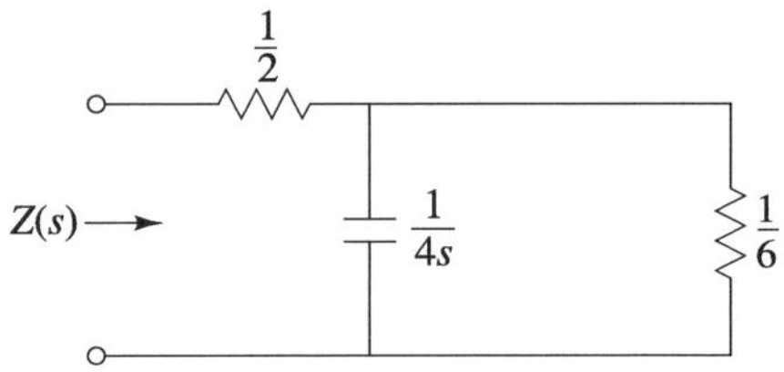

Fig. 10.41

The function $Z(s)$ has zero at $s = -2$ and pole at $s = -1.5$.

Example 10.21 *Determine Z(s) in the network shown in Fig. 10.42. Find out poles and zeros of Z (s) and plot them on s-plane.*

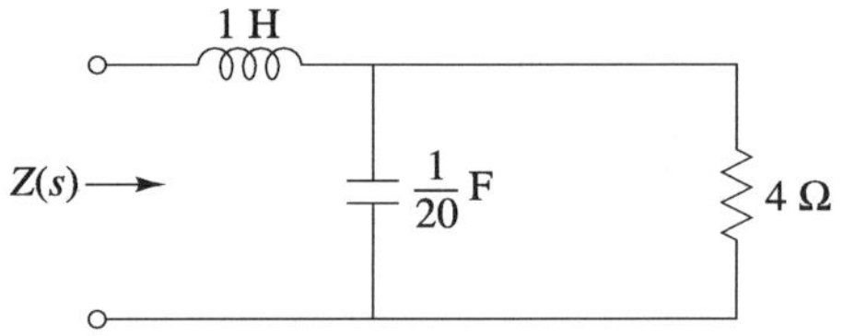

Fig. 10.42

Solution The transformed network is shown in Fig. 10.43.

$$Z(s) = s + \frac{\frac{20}{s} \times 4}{\frac{20}{s} + 4}$$

$$= s + \frac{80}{4s+20} = s + \frac{20}{s+5}$$

$$= \frac{s(s+5)+20}{s+5} = \frac{s^2+5s+20}{s+5}$$

$$= \frac{(s+2.5+j3.71)(s+2.5-j3.71)}{s+5}$$

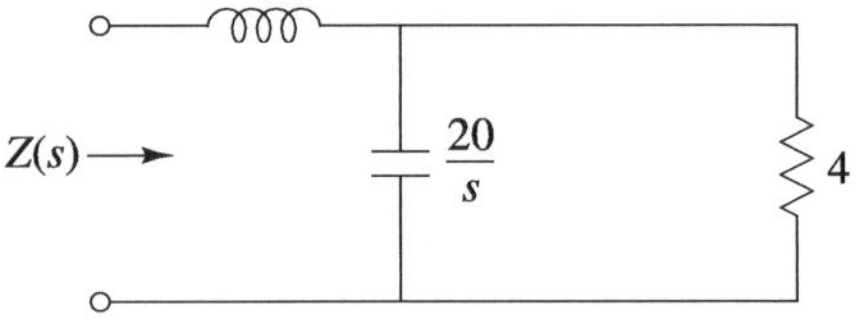

Fig. 10.43

The function $Z(s)$ has zeros at $s = -2.5 + j3.71$ and $s = -2.5 - j3.71$ and pole at $s = -5$.
The pole-zero diagram is shown in Fig. 10.44.

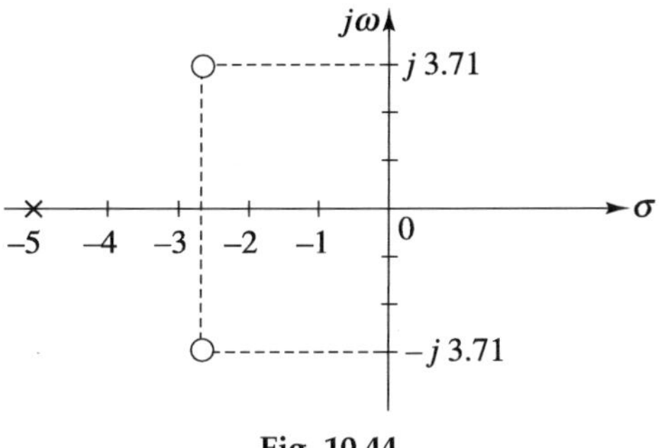

Fig. 10.44

Example 10.22 *For the network shown, plot poles and zeros of function* $\frac{I_0}{I_i}$.

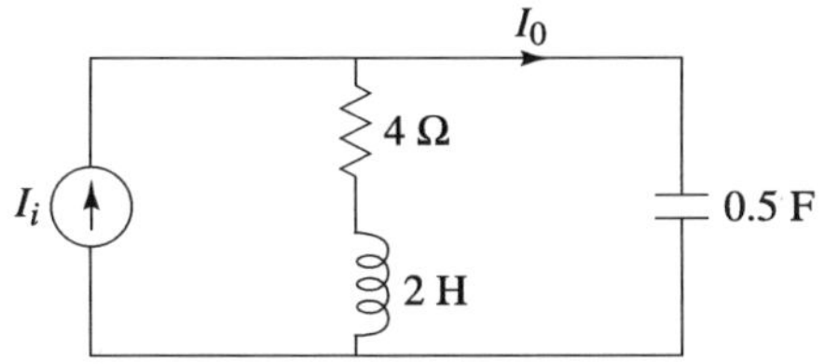

Fig. 10.45

Solution The transformed network is shown in Fig. 10.46.

By current-division formula,

$$I_0 = I_i\left(\frac{4+2s}{4+2s+\frac{2}{s}}\right)$$

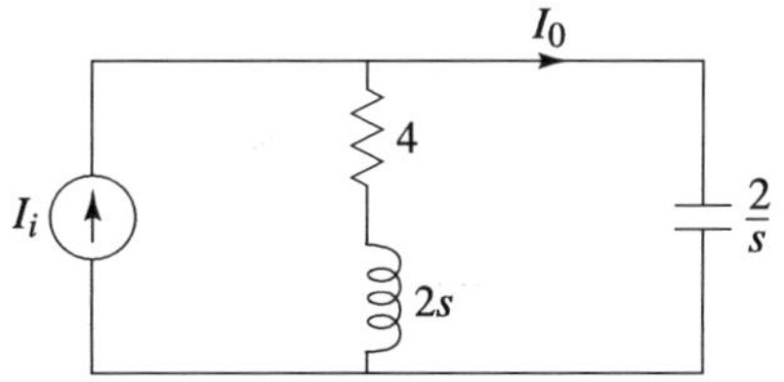

Fig. 10.46

$$\frac{I_0}{I_i} = \frac{s(4+2s)}{4s+2s^2+2}$$

$$= \frac{s(s+2)}{s^2+2s+1} = \frac{s(s+2)}{(s+1)(s+1)}$$

The function has double poles at $s = -1$ and zeros at $s = 0$ and $s = -2$.

The pole-zero diagram is shown in Fig. 10.47.

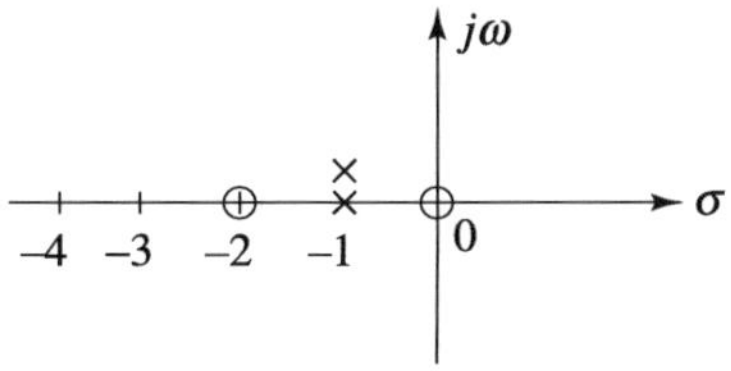

Fig. 10.47

Example 10.23 *Draw the pole-zero diagram of $\frac{I_2}{I_1}$ for the network shown.*

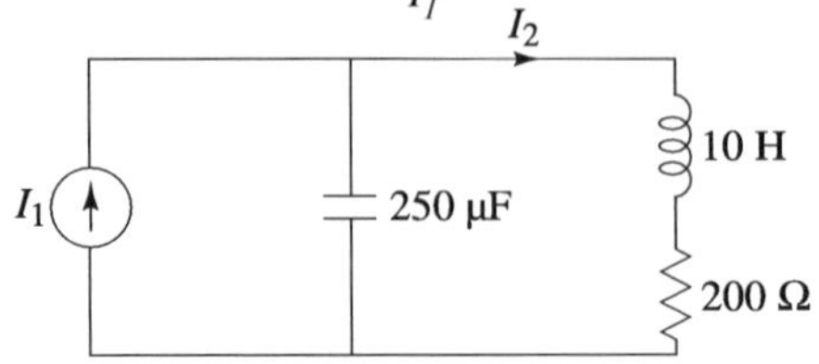

Fig. 10.48

Solution The transformed network is shown in Fig. 10.49.

By current-division formula,

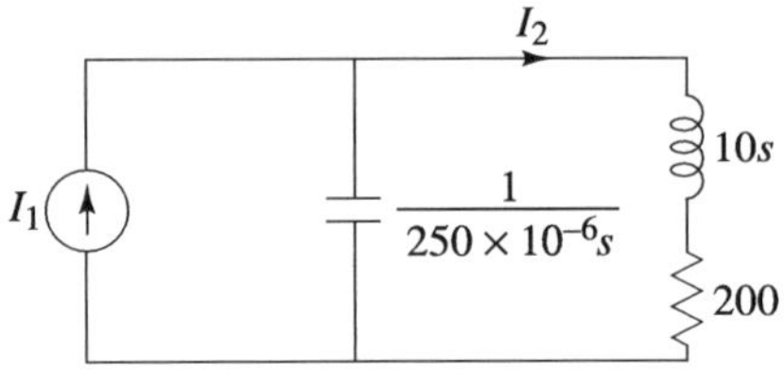

Fig. 10.49

$$I_2 = I_1 \frac{\frac{1}{250 \times 10^{-6} s}}{\frac{1}{250 \times 10^{-6} s} + 10s + 200}$$

$$\frac{I_2}{I_1} = \frac{400}{s^2 + 20s + 400}$$

The function has no zero and poles at $s = -10 \pm j17.32$.

The pole-zero diagram is shown in Fig.10.50.

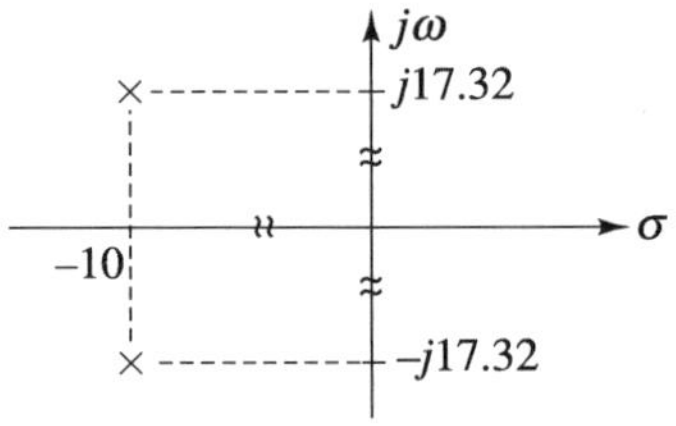

Fig. 10.50

Example 10.24 *Obtain the impedance function Z(s) for which pole-zero diagram is shown in Fig. 10.51.*

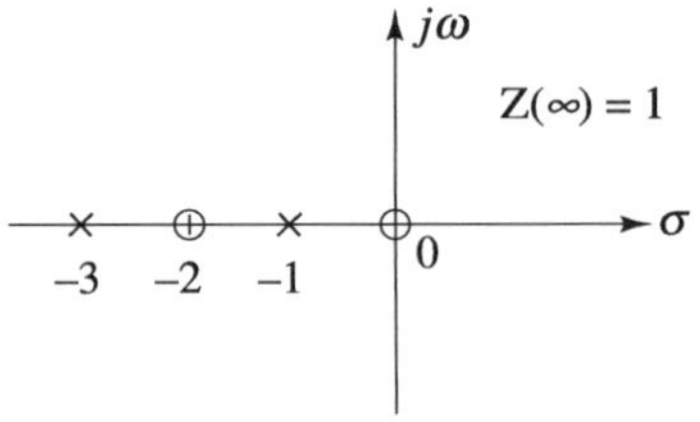

Fig. 10.51

Solution The function $Z(s)$ has poles at $s = -1$ and $s = -3$ and zeros at $s = 0$ and $s = -2$.

$$Z(s) = H \frac{s(s+2)}{(s+1)(s+3)} = H \frac{s^2\left(1+\frac{2}{s}\right)}{s^2\left(1+\frac{1}{s}\right)\left(1+\frac{3}{s}\right)}$$

For $s = \infty$,

$$Z(\infty) = H\frac{1}{(1)(1)} = H$$

When $Z(\infty) = 1$, we have

$$H = 1$$

$$Z(s) = \frac{s(s+2)}{(s+1)(s+3)}$$

Example 10.25 *Obtain the admittance function Y(s) for which the pole-zero diagram is shown in Fig. 10.52.*

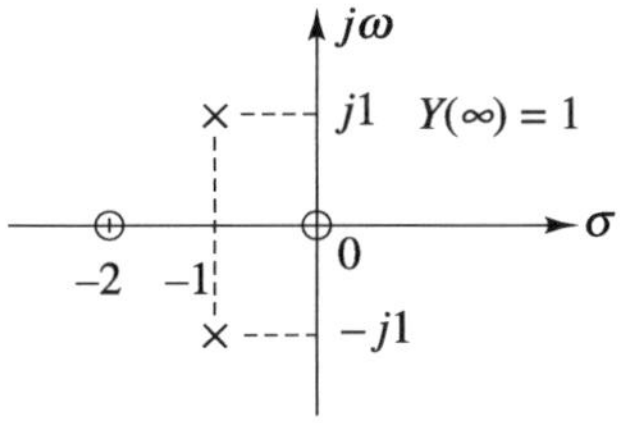

Fig. 10.52

Solution The function $Y(s)$ has poles at $s = -1 \pm j1$ and zeros at $s = 0$ and $s = -2$.

$$Y(s) = H\frac{s(s+2)}{(s+1+j1)(s+1-j1)} = H\frac{s(s+2)}{(s+1)^2+1^2}$$

$$= H\frac{s(s+2)}{s^2+2s+2} = H\frac{s^2\left(1+\frac{2}{s}\right)}{s^2\left(1+\frac{2}{s}+\frac{3}{s^2}\right)}$$

For $s = \infty$,

$$Y(\infty) = H\frac{(1)}{(1)} = H$$

When $Y(\infty) = 1$, we have

$$H = 1$$

$$Y(s) = \frac{s(s+2)}{s^2+2s+2}$$

Example 10.26 *A network and its pole-zero configuration are shown in Fig. 10.53. Determine the values of R, L and C if Z (j0) = 1.*

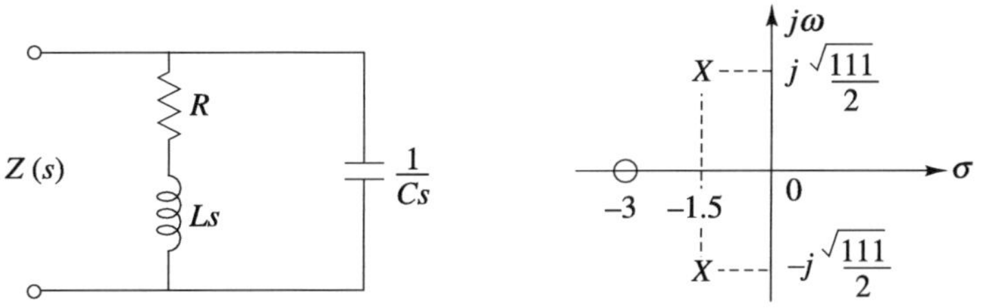

Fig. 10.53

Solution

$$Z(s) = \frac{(Ls+R)\dfrac{1}{Cs}}{(Ls+R)+\dfrac{1}{Cs}} = \frac{Ls+R}{LCs^2+RCs+1}$$

$$= \frac{\dfrac{1}{C}\left(s+\dfrac{R}{L}\right)}{s^2+\dfrac{R}{L}s+\dfrac{1}{LC}} \qquad \ldots\text{(i)}$$

From the pole-zero diagram, zero is at $s = -3$ and poles are at $s = -1.5 \pm j\,\dfrac{\sqrt{111}}{2}$.

$$Z(s) = H\,\frac{s+3}{\left(s+1.5+j\dfrac{\sqrt{111}}{2}\right)\left(s+1.5-j\dfrac{\sqrt{111}}{2}\right)}$$

$$Z(s) = H\,\frac{s+3}{\left(s+\dfrac{3}{2}\right)^2+\left(\dfrac{\sqrt{111}}{2}\right)^2} = H\,\frac{s+3}{s^2+3s+30}$$

When $Z(j0) = 1$, we have

$$1 = H\left(\frac{3}{30}\right)$$

$$H = 10$$

$$Z(s) = \frac{10(s+3)}{s^2+3s+30} \qquad \ldots\text{(ii)}$$

Comparing the Eq. (ii) with the Eq. (i), we get

$$\frac{R}{L} = 3$$

$$\frac{1}{C} = 10$$

$$\frac{1}{LC} = 30$$

Solving the above equations, we get

$$C = \frac{1}{10}F$$

$$L = \frac{1}{3}H$$

$$R = 1\,\Omega$$

Example 10.27 *A network is shown in Fig. 10.54. The poles and zeros of the driving-point function Z(s) of this network are at the following places:*

Poles at $-\dfrac{1}{2} \pm j\dfrac{\sqrt{3}}{2}$

Zero at –1
If Z (j0) = 1, determine the values of R, L and C.

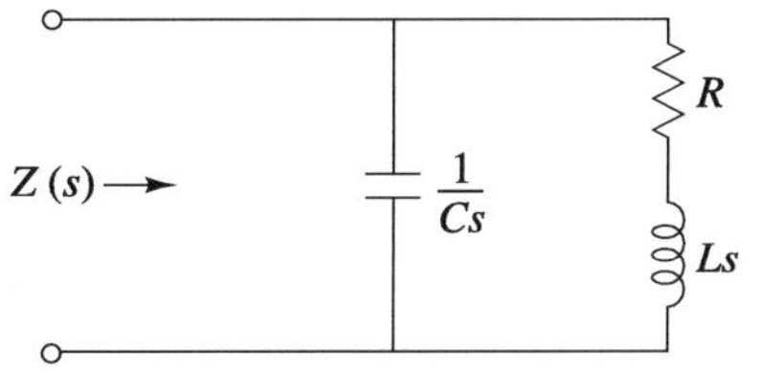

Fig. 10.54

Solution

$$Z(s) = \frac{(Ls+R)\frac{1}{Cs}}{Ls+R+\frac{1}{Cs}}$$

$$= \frac{Ls+R}{LCs^2+RCs+1} = \frac{\frac{1}{C}\left(s+\frac{R}{L}\right)}{s^2+\frac{R}{L}s+\frac{1}{LC}} \qquad \text{...(i)}$$

The poles are at $-\frac{1}{2} \pm j\frac{\sqrt{3}}{2}$ and zero is at -1.

$$Z(s) = H\frac{s+1}{\left(s+\frac{1}{2}+j\frac{\sqrt{3}}{2}\right)\left(s+\frac{1}{2}-j\frac{\sqrt{3}}{2}\right)}$$

$$= H\frac{s+1}{\left(s+\frac{1}{2}\right)^2+\left(\frac{\sqrt{3}}{2}\right)^2} = H\frac{s+1}{s^2+s+1}$$

When, $Z(j0) = 1$, we have

$$1 = H \times \frac{1}{1}$$

$$H = 1$$

$$Z(s) = \frac{s+1}{s^2+s+1} \qquad \text{...(ii)}$$

Comparing the Eq. (ii) with the Eq. (i), we get

$$C = 1$$

$$\frac{R}{L} = 1$$

$$\frac{1}{LC} = 1$$

Solving the above equations, we get

$$C = 1\text{ F}$$

$$L = 1\text{ H}$$

$$R = 1\ \Omega$$

Example 10.28 *The pole-zero diagram of the driving-point impedance function of the network of Fig. 10.55 is shown below. At dc, the input impedance is resistive and equal to 2 Ω. Determine the values of R, L and C.*

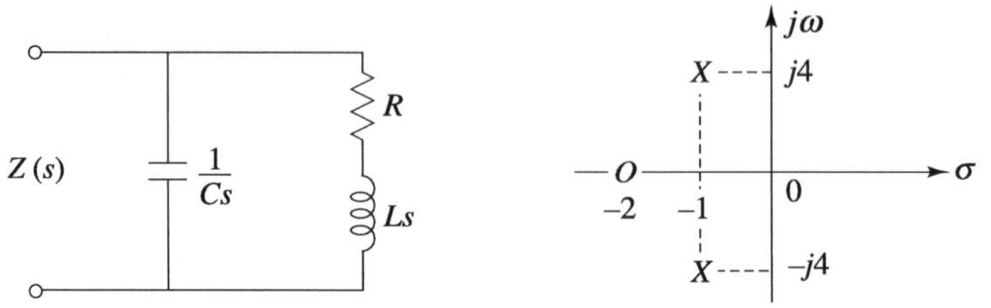

Fig. 10.55

Solution

$$Z(s) = \frac{(Ls+R)\dfrac{1}{Cs}}{Ls+R+\dfrac{1}{Cs}}$$

$$= \frac{Ls+R}{LCs^2+RCs+1} = \frac{\dfrac{1}{C}\left(s+\dfrac{R}{L}\right)}{s^2+\dfrac{R}{L}s+\dfrac{1}{LC}} \qquad \ldots\text{(i)}$$

From the pole-zero diagram, zero is at $s = -2$ and poles are at $s = -1 + j4$ and $s = -1 - j4$

$$Z(s) = H\frac{s+2}{(s+1+j4)(s+1-j4)}$$

$$= H\frac{s+2}{(s+1)^2+(4)^2} = H\frac{s+2}{s^2+2s+17}$$

At dc i.e., $\omega = 0, Z(j0) = 2$

$$2 = H\frac{2}{17}$$

$$H = 17$$

$$Z(s) = 17\frac{s+2}{s^2+2s+17} \qquad \ldots\text{(ii)}$$

Comparing the Eq. (ii) with the Eq. (i), we get

$$\frac{1}{C} = 17$$

$$\frac{R}{L} = 2$$

$$\frac{1}{LC} = 17$$

Solving the above equations, we get

$$C = \frac{1}{17}\text{F}$$

$$L = 1\text{ H}$$

$$R = 2\ \Omega$$

Example 10.29 *The network shown in Fig. 10.56 has the driving-point admittance Y (s) of the form*

$$Y(s) = H\,\frac{(s - s_1)(s - s_2)}{(s - s_3)}$$

(*a*) *Express* s_1, s_2, s_3 *in terms of R, L and C.*

(*b*) *When* $s_1 = -10 + j10^4$, $s_2 = -10 - j10^4$ *and* $Y(j0) = 10^{-1}$ *mho, find the values of R, L and C and determine the value of* s_3.

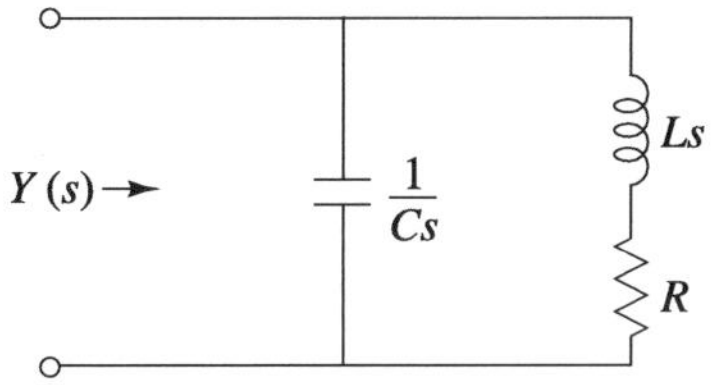

Fig. 10.56

Solution

(a)
$$Y(s) = Cs + \frac{1}{Ls + R}$$

$$= \frac{(Ls + R)Cs + 1}{Ls + R} = \frac{LCs^2 + RCs + 1}{Ls + R}$$

$$= \frac{C\left(s^2 + \frac{R}{L}s + \frac{1}{LC}\right)}{s + \frac{R}{L}} \qquad \ldots\text{(i)}$$

But
$$Y(s) = \frac{H(s - s_1)(s - s_2)}{(s - s_3)}$$

where
$$s_1, s_2 = \frac{-\frac{R}{L} \pm \sqrt{\left(\frac{R}{L}\right)^2 - \frac{4}{LC}}}{2}$$

$$= -\frac{R}{2L} \pm \sqrt{\left(\frac{R}{2L}\right)^2 - \frac{1}{LC}}$$

$$s_3 = -\frac{R}{L}$$

(b) When
$$s_1 = -10 + j10^4$$
$$s_2 = -10 - j10^4$$

$$Y(s) = H\,\frac{(s + 10 - j10^4)(s + 10 + j10^4)}{s - s_3} = H\frac{s^2 + 20s + 10^8}{s - s_3} \qquad \ldots\text{(ii)}$$

Comparing the Eq. (ii) with the Eq. (i), we get

$$\frac{R}{L} = 20$$
$$s_3 = -20$$

$$Y(s) = H\frac{(s^2 + 20s + 10^8)}{(s + 20)}$$

At $s = j0$

$$Y(j0) = H\frac{(10^8)}{20} = 10^{-1}$$
$$H = 0.02 \times 10^{-6}$$

$$Y(s) = 0.02 \times 10^{-6}\frac{(s^2 + 20s + 10^8)}{(s + 20)} \qquad \ldots \text{(iii)}$$

Comparing the Eq. (iii) with Eq. (i), we get

$$C = 0.02 \times 10^{-6}\text{ F} = 0.02\ \mu\text{F}$$

$$\frac{1}{LC} = 10^8$$

$$L = \frac{1}{2}\text{H}$$

$$\frac{R}{L} = 20$$

$$R = 10\ \Omega$$

Example 10.30 *A network and pole-zero diagram for driving-point impedance Z (s) are shown in Fig. 10.57. Calculate the values of the parameters R, L, G and C if Z (j0) = 1.*

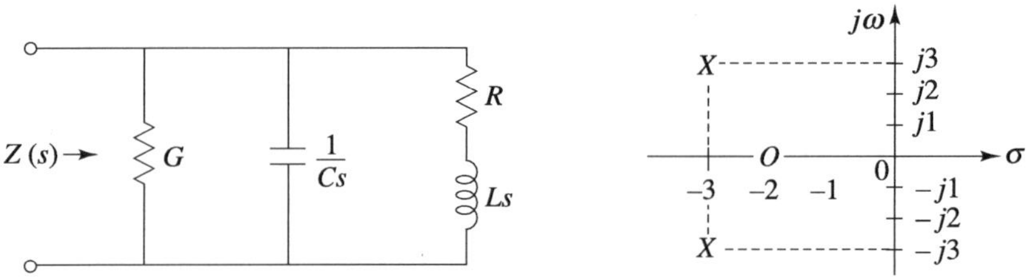

Fig. 10.57

Solution It is easier to calculate $Y(s)$ and then invert it to obtain $Z(s)$.

$$Y(s) = G + Cs + \frac{1}{Ls + R}$$

$$= \frac{(G + Cs)(Ls + R) + 1}{Ls + R} = \frac{LCs^2 + (GL + RC)s + 1 + GR}{Ls + R}$$

$$Z(s) = \frac{1}{Y(s)} = \frac{Ls + R}{LCs^2 + (GL + RC)s + 1 + GR}$$

$$= \frac{\frac{1}{C}\left(s + \frac{R}{L}\right)}{s^2 + \left(\frac{G}{C} + \frac{R}{L}\right)s + \left(\frac{1 + GR}{LC}\right)} \qquad \ldots \text{(i)}$$

From the pole-zero diagram, zero is at $s = -2$ and poles are at $s = -3 \pm j3$.

$$Z(s) = H\,\frac{(s+2)}{(s+3-j3)(s+3+j3)}$$

$$= H\,\frac{(s+2)}{(s+3)^2+9} = H\,\frac{s+2}{s^2+6s+18}$$

When $Z(j0) = 1$, we have

$$1 = H\,\frac{2}{18}$$

$$H = 9$$

$$Z(s) = \frac{9(s+2)}{(s^2+6s+18)} \qquad \ldots \text{(ii)}$$

Comparing the Eq. (ii) with the Eq. (i), we get

$$\frac{1}{C} = 9$$

$$\frac{R}{L} = 2$$

$$\frac{G}{C} + \frac{R}{L} = 6$$

$$\frac{1+GR}{LC} = 18$$

Solving the above equations, we get

$$C = \frac{1}{9}\text{ F}$$

$$L = \frac{9}{10}\text{ H}$$

$$G = \frac{4}{9}\ \mho$$

$$R = \frac{9}{5}\ \Omega$$

Example 10.31 *A series R-L-C circuit has its driving-point admittance and pole-zero diagram as shown in Fig. 10.58. Find the values of R, L and C.*

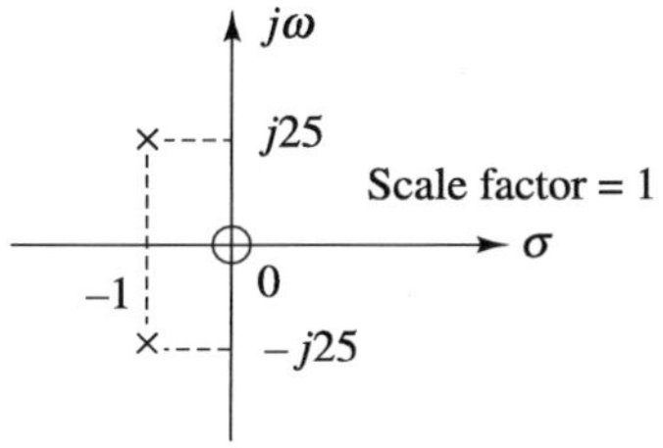

Fig. 10.58

Solution The function $Y(s)$ has poles at $s = -1 + j25$ and $s = -1 - j25$ and zero at $s = 0$.

$$Y(s) = H\frac{s}{(s+1+j25)(s+1-j25)}$$

$$= H\frac{s}{(s+1)^2 + (25)^2} = H\frac{s}{s^2+2s+626}$$

Scale factor $H = 1$

$$Y(s) = \frac{s}{s^2+2s+626} \quad \text{... (i)}$$

For a series *R-L-C* circuit,

$$Z(s) = R + Ls + \frac{1}{Cs}$$

$$= \frac{LCs^2 + RCs + 1}{Cs} = \frac{L\left(s^2 + \frac{R}{L}s + \frac{1}{LC}\right)}{s}$$

$$Y(s) = \frac{1}{Z(s)} = \frac{s}{L\left(s^2 + \frac{R}{L}s + \frac{1}{LC}\right)} \quad \text{... (ii)}$$

Comparing the Eqs (i) and (ii), we get

$$L = 1\text{ H}$$

$$\frac{1}{LC} = 626$$

$$C = \frac{1}{626}\text{ F}$$

$$\frac{R}{L} = 2$$

$$R = 2\ \Omega$$

10.9 TIME-DOMAIN BEHAVIOUR FROM THE POLE-ZERO PLOT

The time-domain behaviour of a system can be determined from the pole-zero plot. Consider a network function

$$F(s) = H\frac{(s-z_1)(s-z_2)\ldots\ldots(s-z_n)}{(s-p_1)(s-p_2)\ldots\ldots(s-p_m)}$$

The poles of this function determine the time-domain behaviour of $f(t)$. The function $f(t)$ can be determined from the knowledge of the poles, the zeros and the scale factor H. Figure 10.59 shows some pole locations and the corresponding time-domain response.

(i) When pole is at origin i.e. at $s = 0$, the function $f(t)$ represents steady-state response of the circuit i.e. dc value.

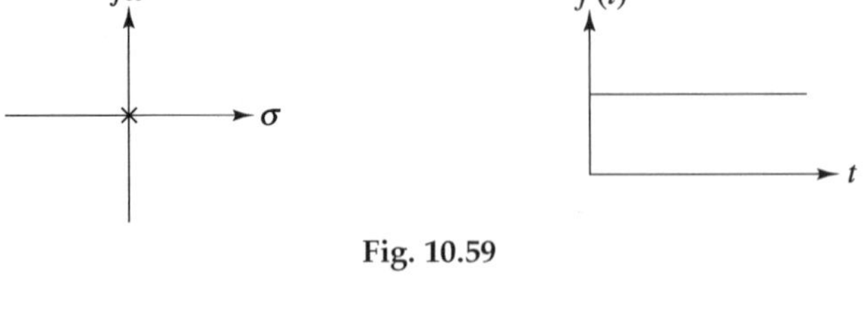

Fig. 10.59

(ii) When pole lies in the left half of the *s*-plane, the response decreases exponentially.

Fig. 10.60

(iii) When pole lies in the right half of the *s*-plane, the response increases exponentially. A pole in the right-half plane gives rise to unbounded response and unstable system.

Fig. 10.61

(iv) For $s = 0 + j\omega_n$, the response becomes $f(t) = Ae^{\pm j\omega_n t} = A(\cos\,\omega_n t \pm j\,\sin\,\omega_n t)$.The exponential response $e^{\pm j\omega_n t}$ may be interpreted as a rotating phasor of unit length. A positive sign of exponential $e^{j\omega_n t}$ indicates counterclockwise rotation, while a negative sign $e^{-j\omega_n t}$ indicates clockwise rotation. The variation of exponential function $e^{j\omega_n t}$ with time is thus sinusoidal and hence constitutes the case of sinusoidal steady state.

Fig. 10.62

(v) For $s = \sigma_n + j\omega_n$, the response becomes $f(t) = A\,e^{st} = A\,e^{(\sigma_n + j\omega_n t)} = A\,e^{\sigma_n t}\,e^{j\omega_n t}$. The resonse $e^{\sigma_n t}$ is an exponentially increasing or decreasing function. The response $e^{j\omega_n t}$ is a sinusoidal function. Hence, the response of the product of these responses will be overdamped sinusoids or underdamped sinusoids.

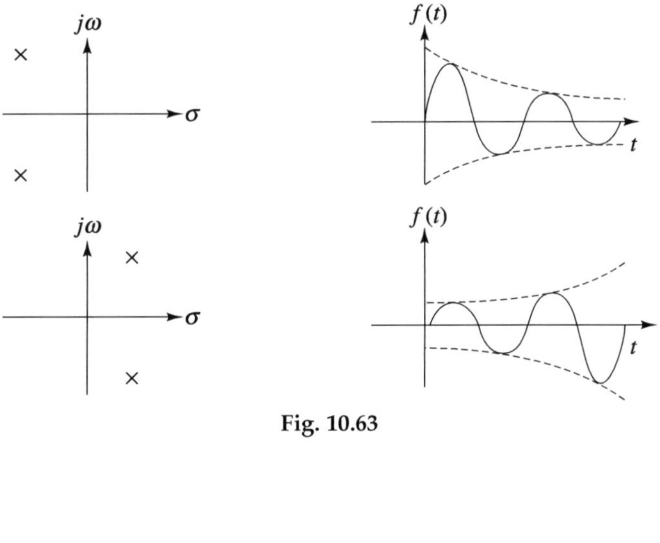

Fig. 10.63

(vi) The real part *s* of the pole is the displacement of the pole from the imaginary axis. Since σ is also the damping factor, a greater value of σ (i.e., a greater displacement of the pole from the imaginary axis) means that response decays more rapidly with time. The poles with greater displacement from the real axis correspond to higher frequency of oscillation.

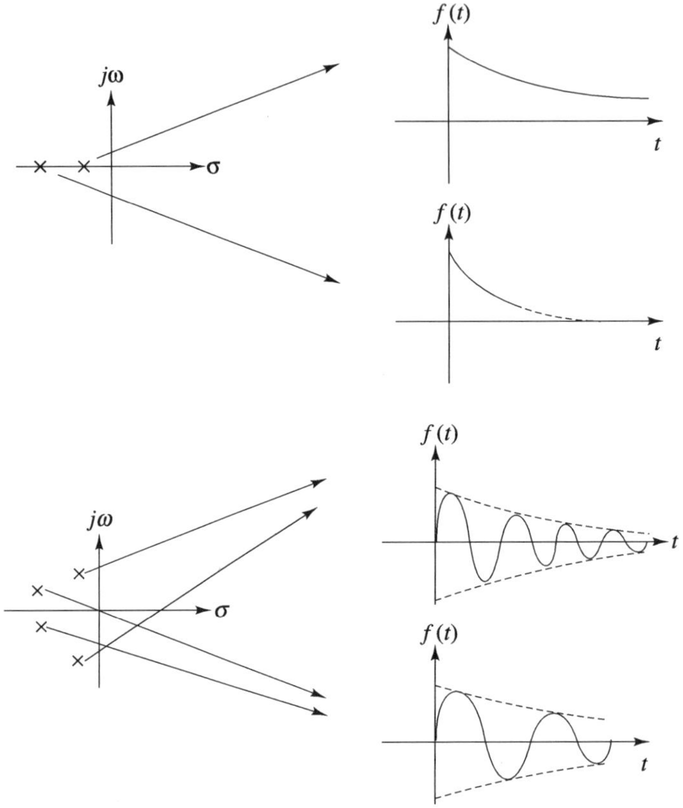

Fig. 10.64

10.10 GRAPHICAL METHOD FOR DETERMINATION OF RESIDUE

Consider a network function,

$$F(s) = H\,\frac{(s-z_1)(s-z_2)\ldots\ldots(s-z_n)}{(s-p_1)(s-p_2)\ldots\ldots(s-p_m)}$$

By partial fraction expansion,

$$F(s) = \frac{K_1}{(s-p_1)} + \frac{K_2}{(s-p_2)} + \ldots\ldots\ldots + \frac{K_m}{(s-p_m)}$$

The residue K_i is given by

$$K_i = (s-p_i)\,F(s)\,|_{s\to p_i}$$

$$= H\,\frac{(p_i-z_1)(p_i-z_2)\ldots\ldots(p_i-z_n)}{(p_i-p_1)(p_i-p_2)\ldots\ldots(p_i-p_m)}$$

Each term $(p_i - z_i)$ represents a phasor drawn from zero z_i to pole p_i.

Each term $(p_i - p_k)$, $i \neq k$, represents a phasor drawn from other poles to the pole p_i.

$$K_i = H\,\frac{\text{Product of phasors (polar form) from each zero to } p_i}{\text{Product of phasors (polar form) from other poles to } p_i}$$

The residues can be obtained by graphical method in the following way:

(1) Draw the pole-zero diagram for the given network function.

(2) Measure the distance from each of the other poles to a given pole.

(3) Measure the distance from each of the other zeros to a given pole.

(4) Measure the angle from each of the other poles to a given pole.

(5) Measure the angle from each of the other zeros to a given pole.

(6) Substitute these values in the required residue equation.

The graphical method can be used if poles are simple and complex. But it cannot be used when there are multiple poles.

Example 10.32 *The current I(s) in a network is given by* $I(s) = \dfrac{2s}{(s+1)(s+2)}$

Plot the pole-zero pattern in the s-plane and hence obtain i(t).

Solution Poles are at $s = -1$ and $s = -2$ and zero is at $s = 0$. The pole-zero plot is shown in Fig. 10.65.

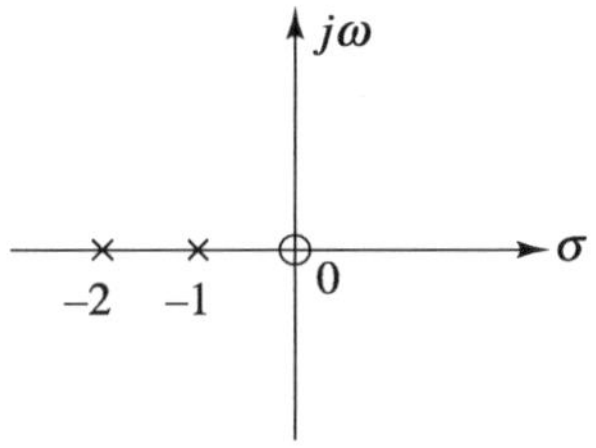

Fig. 10.65

By partial-fraction expansion,

$$I(s) = \frac{K_1}{s+1} + \frac{K_2}{s+2}$$

The coefficients K_1 and K_2, often referred as residues, can be evaluated from the pole-zero diagram.

$$K_1 = H\,\frac{\text{Phasor from zero at origin to pole at } A}{\text{Phasor from pole at } B \text{ to pole at } A}$$

$$= 2\,\frac{1\angle 180°}{1\angle 0°}$$

$$= 2\,\angle 180° = -2$$

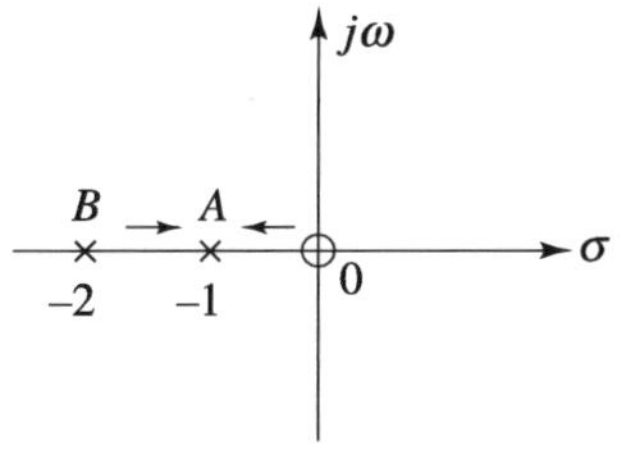

Fig. 10.66 *Evaluation of* K_1

$$K_2 = H\,\frac{\text{Phasor from zero at origin to pole at } B}{\text{Phasor from pole at } A \text{ to pole at } B}$$

$$= 2\,\frac{2\angle 180°}{1\angle 180°} = 4$$

$$I(s) = -\frac{2}{s+1} + \frac{4}{s+2}$$

Taking inverse Laplace transform,

$$i(t) = -2e^{-t} + 4e^{-2t}$$

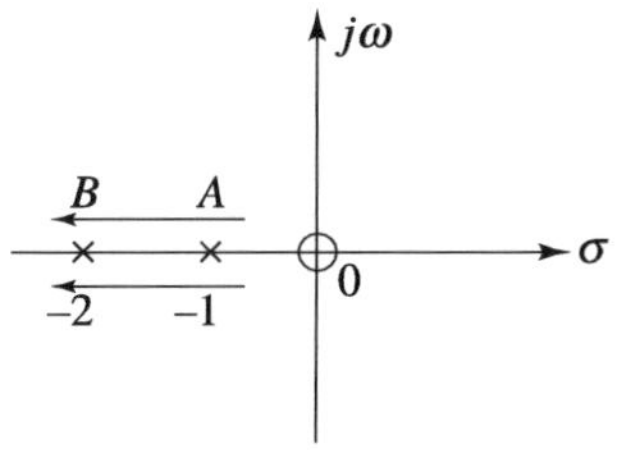

Fig. 10.67 *Evaluation of* K_2

Example 10.33 *The voltage V(s) of a network is given by*

$$V(s) = \frac{3s}{(s+2)(s^2+2s+2)}$$

Plot its pole-zero diagram and hence obtain v (t).

Solution

$$V(s) = \frac{3s}{(s+2)(s^2+2s+2)}$$

$$= \frac{3s}{(s+2)(s+1+j1)(s+1-j1)}$$

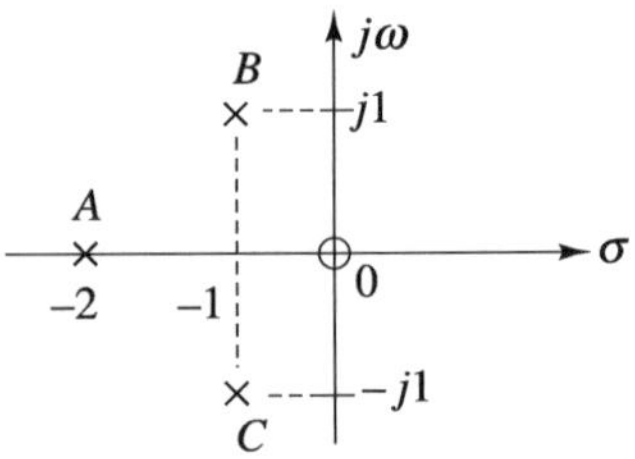

Fig. 10.68

Poles are at $s = -2$, $s = -1 \pm j1$ and zero is at $s = 0$ as shown in Fig. 10.68.

By partial-fraction expansion,

$$V(s) = \frac{K_1}{s+2} + \frac{K_2}{s+1-j1} + \frac{K_2^*}{s+1+j1}$$

The coefficients K_1, K_2 and K_2^* can be evaluated from the pole-zero diagram.

$$K_1 = \frac{3(\overrightarrow{OA})}{(\overrightarrow{BA})(\overrightarrow{CA})} = 3\frac{2\angle 180°}{(\sqrt{2}\angle -135°)(\sqrt{2}\angle 135°)}$$

$$= 3\sqrt{180°} = -3$$

$$K_2 = \frac{3(\overrightarrow{OB})}{(\overrightarrow{AB})(\overrightarrow{CB})} = 3\frac{(\sqrt{2}\angle 135^o)}{(\sqrt{2}\angle 45^o)(2\angle 90^o)} = \frac{3}{2}$$

$$K_2^* = \frac{3}{2}$$

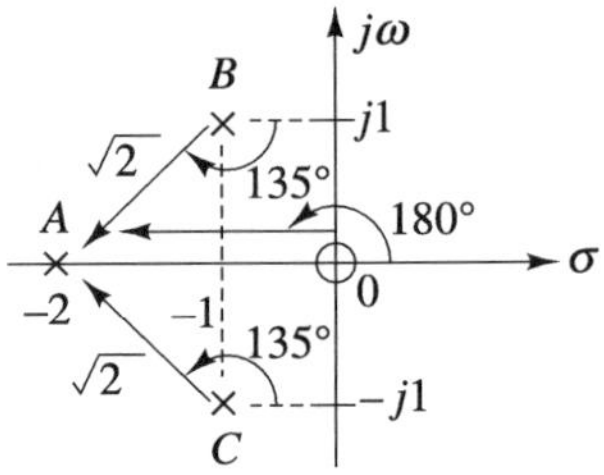

Fig. 10.69 *Evaluation of K_1*

$$V(s) = -\frac{3}{(s+2)} + \frac{\frac{3}{2}}{(s+1-j1)} + \frac{\frac{3}{2}}{(s+1+j1)}$$

Taking inverse Laplace transform,

$$v(t) = -3e^{-2t} + \frac{3}{2}\left[e^{(-1+j1)t} + e^{(-1-j1)t}\right]$$

$$= -3e^{-2t} + 2\times\frac{3}{2}e^{-t}\left(\frac{e^{j1}+e^{-j1}}{2}\right) = -3e^{-2t} + 3e^{-t}\cos t$$

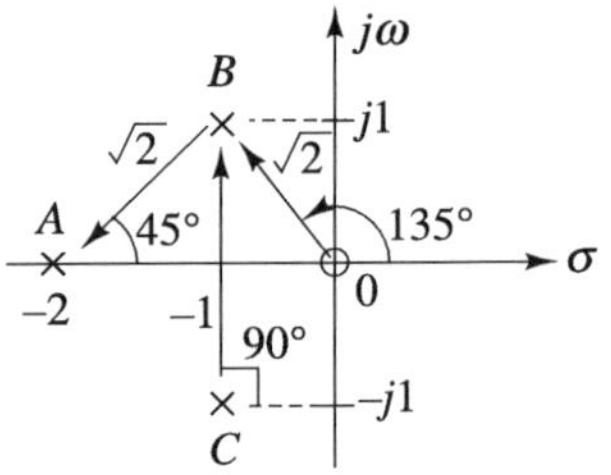

Fig. 10.70 *Evaluation of K_2*

Example 10.34 *Find the function v(t) using the pole-zero plot of following function:*

$$V(s) = \frac{(s+2)(s+6)}{(s+1)(s+5)}$$

Solution If the degree of the numerator is greater or equal to the degree of the denominator, we can divide the numerator by the denominator such that the remainder can be expanded more easily into partial fractions.

$$V(s) = \frac{s^2+8s+12}{s^2+6s+5} = 1 + \frac{2s+7}{s^2+6s+5} = 1 + \frac{2(s+3.5)}{(s+1)(s+5)}$$

By partial fraction expansion,

$$V(s) = 1 + \frac{K_1}{s+1} + \frac{K_2}{s+5}$$

K_1 and K_2 can be evaluated from the pole-zero diagram shown in Fig. 10.71 and Fig. 10.72.

$$K_1 = 2\frac{2.5\angle 0°}{4\angle 0°} = \frac{5}{4}$$

$$K_2 = 2\ \frac{1.5\angle 180°}{4\angle 180°} = \frac{3}{4}$$

$$V(s) = 1 + \frac{5}{4(s+1)} + \frac{3}{4(s+5)}$$

Taking inverse Laplace transform,

$$v(t) = \delta(t) + \frac{5}{4}e^{-t} + \frac{3}{4}e^{-5t}$$

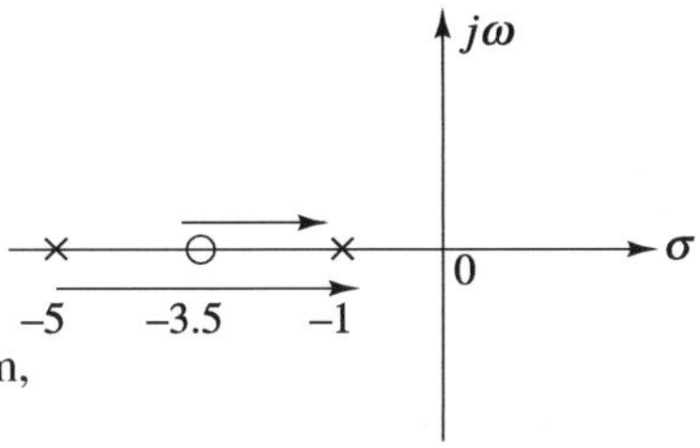

Fig. 10.71 *Evaluation of K_1*

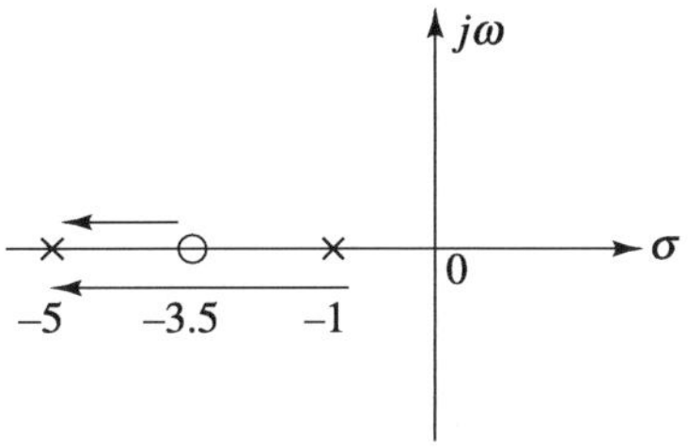

Fig. 10.72 *Evaluation of K_2*

Example 10.35 *Evaluate amplitude and phase of the network function $F(s) = \dfrac{4s}{s^2 + 2s + 2}$ from the pole- zero plot at $s = j2$.*

Solution $$F(s) = \frac{4s}{s^2+2s+2} = \frac{4s}{(s+1+j1)(s+1-j1)}$$

The pole-zero plot is shown in Fig. 10.73.

At $s = j2$, we have

$$|F(j2)| = \frac{\text{Product of phasor magnitudes from all zeros to } j2}{\text{Product of phasor magnitudes from all poles to } j2}$$

$$= \frac{2}{\left(\sqrt{2}\right)\left(\sqrt{10}\right)} = 0.447$$

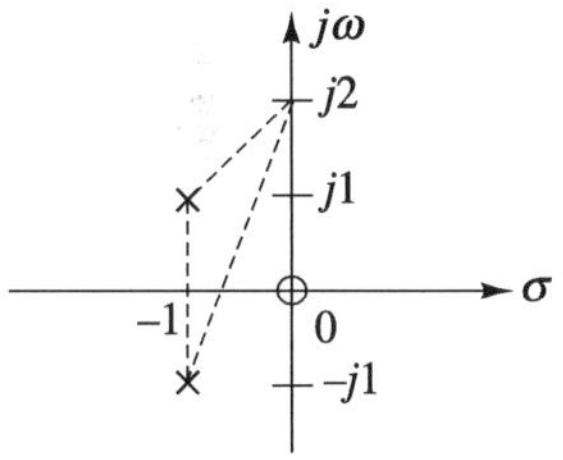

Fig. 10.73

$$f(\omega) = \tan^{-1}\left(\frac{2}{0}\right) - \tan^{-1}\left(\frac{3}{1}\right) - \tan^{-1}\left(\frac{1}{1}\right) = 90° - 71.56° - 45° = -26.56°$$

Example 10.36 *Using the pole-zero plot, find magnitude and phase of the function*

$$F(s) = \frac{(s+1)(s+3)}{s(s+2)} \quad \text{at } s = j4.$$

Solution $$F(s) = \frac{(s+1)(s+3)}{s(s+2)}$$

The pole-zero plot is shown in Fig. 10.74.

At $s = j4$, we have

$$|F(j4)| = \frac{\text{Product of vector magnitudes from all zeros to } j4}{\text{Product of vector magnitudes from all poles to } j4}$$

$$= \frac{(5)\left(\sqrt{17}\right)}{\left(\sqrt{20}\right)(4)} = 1.15$$

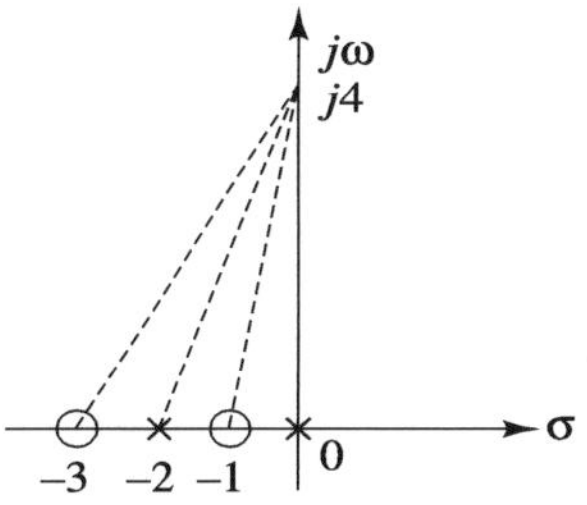

Fig. 10.74

$$\phi(\omega) = \tan^{-1}\left(\frac{4}{1}\right) + \tan^{-1}\left(\frac{4}{3}\right) - \tan^{1}\left(\frac{4}{0}\right) - \tan^{-1}\left(\frac{4}{2}\right)$$

$$= 75.96° + 53.13° - 90° - 63.43° = -24.34°$$

Example 10.37 *Plot amplitude and phase response for*

$$F(s) = \frac{s}{s+10}$$

Solution **Amplitude response**

$$F(j\omega) = \frac{j\omega}{j\omega+10}$$

$$|F(j\omega)| = \frac{\omega}{\sqrt{\omega^2+100}}$$

ω	$\lvert F(j\omega) \rvert$
0	0
10	0.707
100	0.995
1000	1

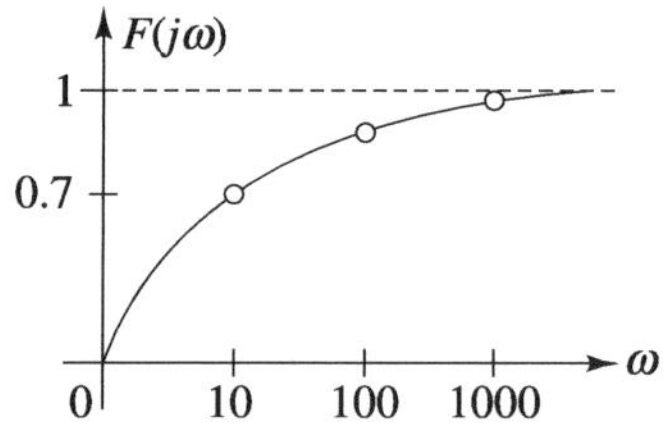

Fig. 10.75

Phase response

$$\phi(\omega) = \tan^{-1}\left(\frac{\omega}{0}\right) - \tan^{-1}\left(\frac{\omega}{10}\right) = 90° - \tan^{-1}\left(\frac{\omega}{10}\right)$$

ω	$\phi(\omega)$
0	90°
10	45°
100	5.7°
1000	0°

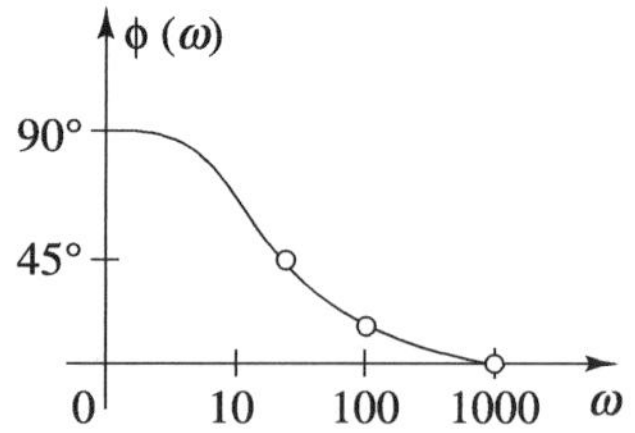

Fig. 10.76

Example 10.38 *Sketch amplitude and phase response for*

$$F(s) = \frac{s+10}{s-10}$$

Solution **Amplitude response**

$$F(j\omega) = \frac{j\omega+10}{j\omega-10}$$

$$|F(j\omega)| = \frac{\sqrt{\omega^2+100}}{\sqrt{\omega^2+100}}$$

For all ω, magnitude is unity.

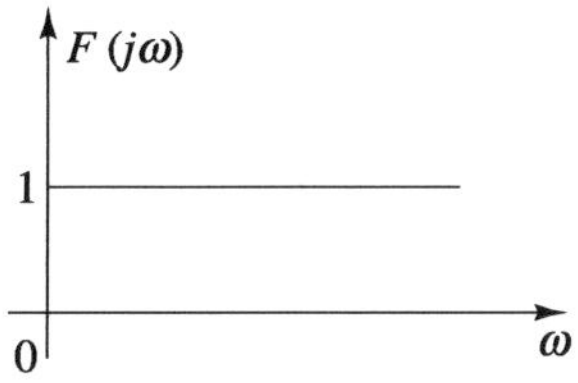

Fig. 10.77

Phase response

$$\phi(\omega) = \tan^{-1}\left(\frac{\omega}{10}\right) - \tan^{-1}\left(-\frac{\omega}{10}\right) = 2\tan^{-1}\left(\frac{\omega}{10}\right)$$

ω	$\phi(\omega)$
0	0°
10	90°
100	168.6°
1000	178.9°

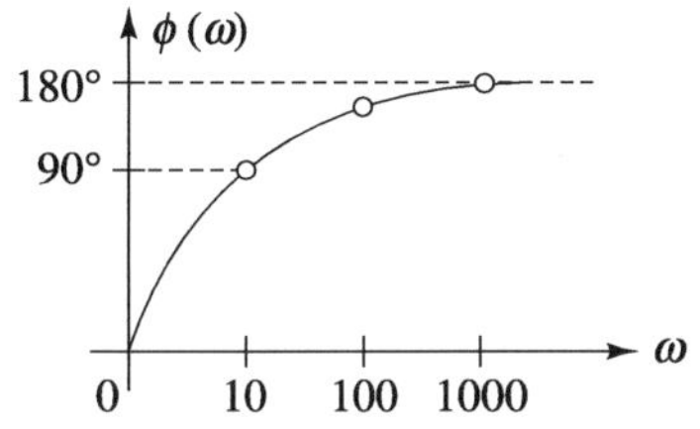

Fig. 10.78

Exercises

1. Draw the pole-zero diagram of the following network functions:

(i) $F(s) = \dfrac{s^2+4}{s^2+6s+4}$

(ii) $F(s) = \dfrac{5s-12}{s^2+4s+13}$

(iii) $F(s) = \dfrac{s+1}{(s^2+2s+2)^2}$

(iv) $F(s) = \dfrac{s(s^2+5)}{s^4+2s^2+1}$

(v) $F(s) = \dfrac{s^2+s+2}{s^4+5s^3+6s^2}$

(vi) $F(s) = \dfrac{s^2-s}{s^3+2s^2-s-2}$

(vii) $F(s) = \dfrac{s^2+3s+2}{s^2+3s}$

(viii) $F(s) = \dfrac{(s^2+4)(s+1)}{(s^2+1)(s^2+2s+5)}$

2. For the network shown in Fig. 10.79, draw the pole-zero plot of the impedance function $Z(s)$.

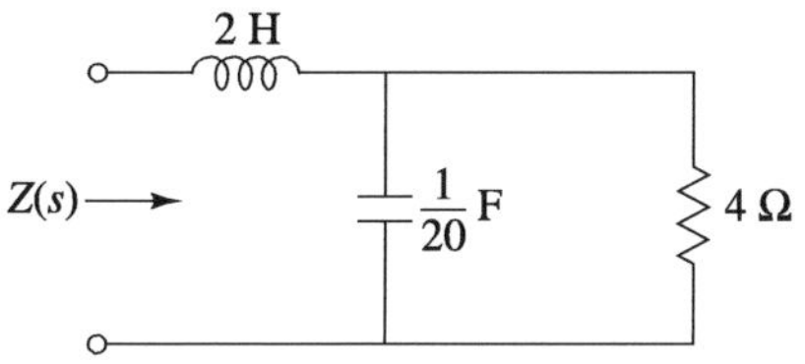

Fig. 10.79

$$\left[Z(s) = \frac{(s+2.5-j1.94)(s+2.5+j1.94)}{s+5}\right]$$

3. For the network shown in Fig. 10.80, draw the pole-zero plot of driving-point impedance function $Z(s)$.

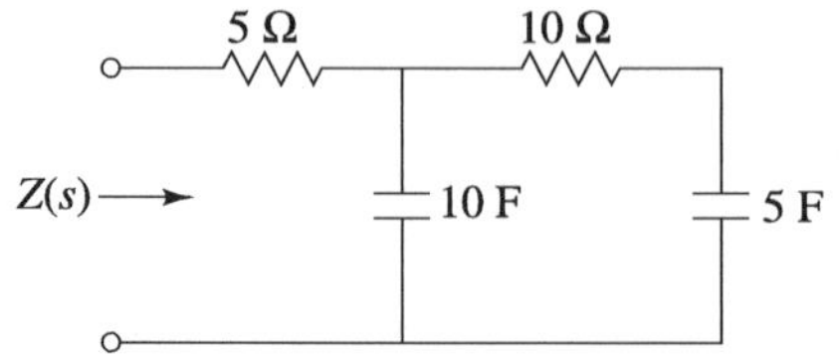

Fig. 10.80

$$\left[Z(s) = \frac{5(s+0.01)(s+0.04)}{s(s+0.03)}\right]$$

4. Find the driving-point impedance of the network shown in Fig. 10.81. Also, find poles and zeros.

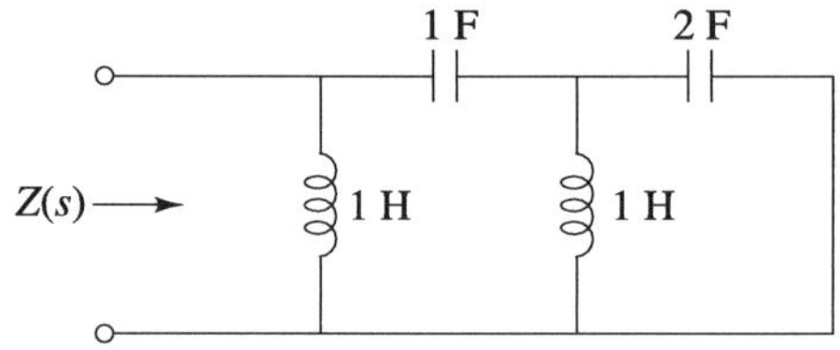

Fig. 10.81

$$\left[Z(s) = \frac{1.5s(s^2+0.33)}{(s^2+1.707)(s^2+0.293)}\right]$$

5. For the network shown in Fig. 10.82, determine $\frac{V_2}{I_g}$. Plot the pole-zero diagram of $\frac{V_2}{I_g}$.

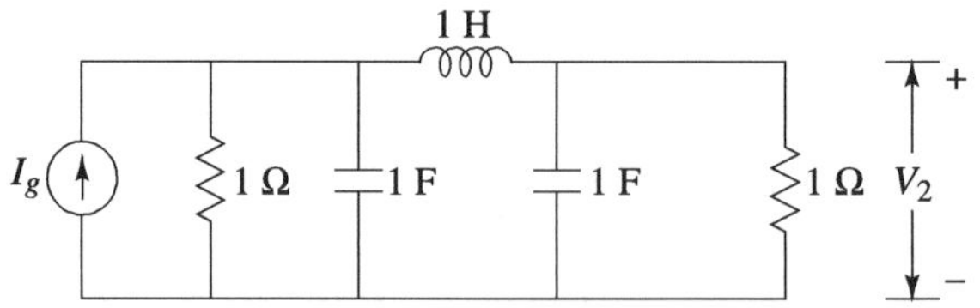

Fig. 10.82

$$\left[\frac{V_2}{I_g} = \frac{1}{s^3+2s^2+3s+2}\right]$$

6. Determine the driving-point impedance $\frac{V_1}{I_1}$, transfer impedance $\frac{V_2}{I_1}$ and voltage transfer ratio $\frac{V_2}{V_1}$ for the network shown in Fig. 10.83.

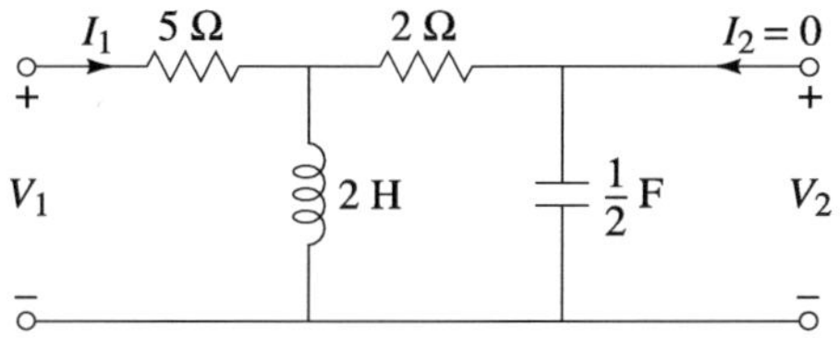

Fig. 10.83

$$\left[\frac{V_1}{I_1} = \frac{7s^2 + 7s + 5}{s^2 + s + 1}; \frac{V_2}{I_1} = \frac{2s}{s^2 + s + 1}; \frac{V_2}{V_1} = \frac{2s}{7s^2 + 7s + 5}\right]$$

7. For the network shown in Fig. 10.84, determine $\frac{V_2}{V_1}$ and $\frac{V_2}{I_1}$.

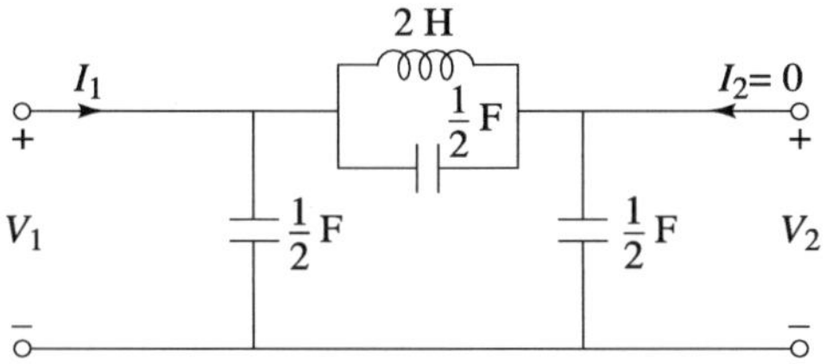

Fig. 10.84

$$\left[\frac{V_2}{V_1} = \frac{s^2 + 1}{2s^2 + 1}; \frac{V_2}{I_1} = \frac{2s^2 + 2}{s(3s^2 + 2)}\right]$$

8. Find the open-circuit transfer impedance Z_{21} and open-circuit voltage ratio G_{21} for the ladder network shown in Fig. 10.85.

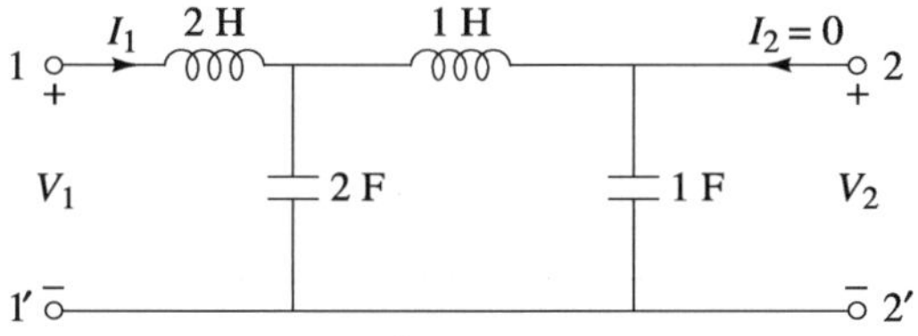

Fig. 10.85

$$\left[Z_{21} = \frac{1}{2s^3 + 3s}, G_{21} = \frac{1}{4s^4 + 7s^2 + 1}\right]$$

9. For the two-port network shown in Fig. 10.86, determine Z_{11}, Z_{21} and voltage transfer ratio $G_{21}(s)$.

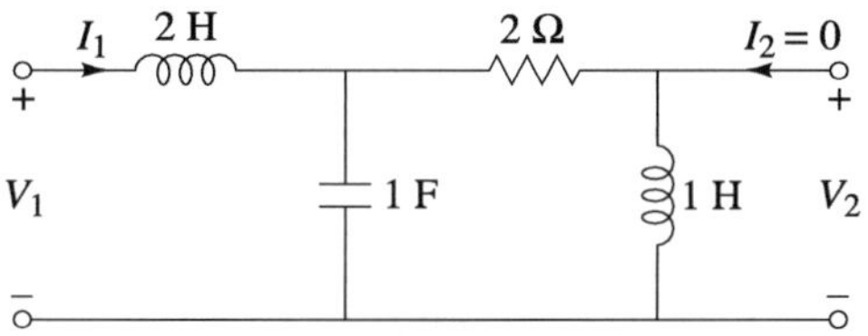

Fig. 10.86

$$\left[\frac{2s^3 + 4s^2 + 3s + 2}{s^2 + 2s + 1}, \frac{s}{s^2 + 2s + 1}, \frac{s}{2s^3 + 4s^2 + 3s + 2}\right]$$

10. Find network functions $\frac{V_2}{V_1}$ and $\frac{V_1}{I_1}$ for the circuit shown in Fig. 10.87 and plot poles and zeros of $\frac{V_2(s)}{V_1(s)}$. Find $V_2(s)$ when $V_1(s) = 1\,\angle 0°$ as a function of ω.

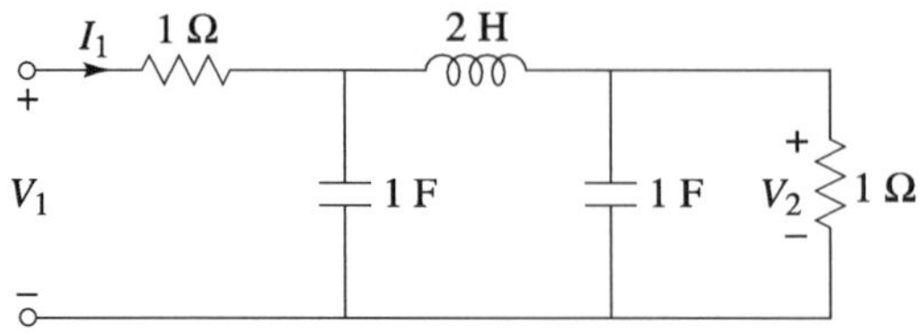

Fig. 10.87

$$\frac{V_2}{V_1} = \frac{1}{2(s^3+2s^2+2s+1)}, \frac{V_1}{I_1} = \frac{2(s^3+2s^2+2s+1)}{2s^3+2s^2+2s+1}, \frac{1}{\sqrt{(2-4\omega^2)^2+(4\omega-2\omega^3)^2}} \angle \tan^{-1}\left(\frac{4\omega-2\omega^3}{2-4\omega^2}\right)\Bigg]$$

11. For the network shown in Fig. 10.88, determine $\frac{V_1}{I_1}$ and $\frac{V_2}{I_1}$. Plot the poles and zeros of $\frac{V_2}{I_1}$ and determine magnitude and phase of $V_2(s)$, given $V_1(s) = 1\,\angle 0°$ as a function of ω.

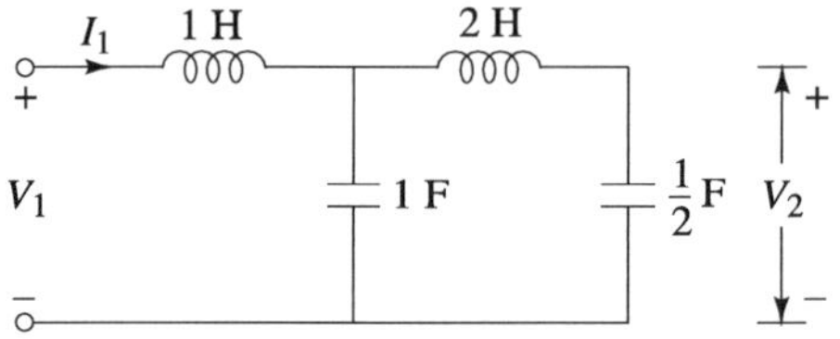

Fig. 10.88

$$\left[\frac{V_1}{I_1} = \frac{2s^4+5s^2+2}{2s^3+3s}, \frac{V_2}{I_1} = \frac{2}{2s^3+3s}, \frac{V_2}{V_1} = \frac{2}{2s^4+5s^2+2}, \frac{2}{2\omega^4-5\omega^2+2}\angle 0°\right]$$

12. For the network shown in Fig. 10.89, determine $\frac{V_1}{I_1}$ and $\frac{V_2}{V_1}$. Plot the pole and zeros for V_2 and determine magnitude and phase of $V_2(s)$, given $V_1(s) = 1\,\angle 0°$ as a function of ω.

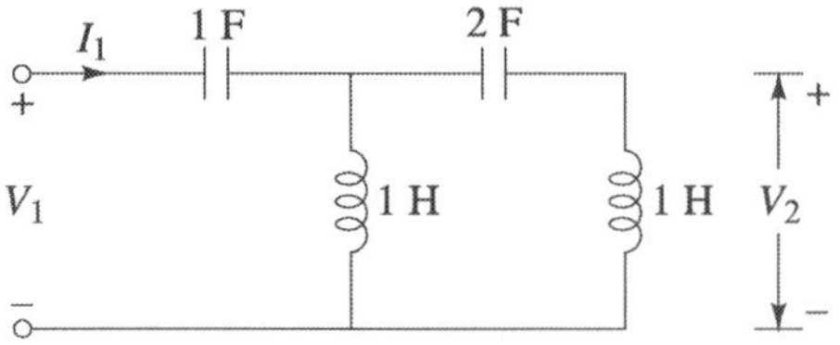

Fig. 10.89

$$\left[\frac{V_1}{I_1} = \frac{s^4+3s^2+1}{2s^3+s}, \frac{V_2}{V_1} = \frac{s^4}{s^4+3s^2+1}, \frac{\omega^4}{\omega^4-3\omega^2+1}\angle 0°\right]$$

13. For the network shown in Fig. 10.90, plot the poles and zeros of transfer impedance and determine magnitude and phase of $V_2(s)$, given $V_1(s) = 2\angle 0°$ as a function of ω. Also, find the driving-point impedance.

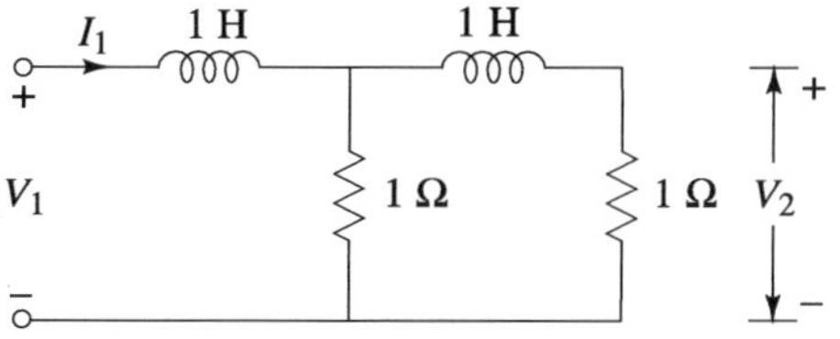

Fig. 10.90

$$\left[\frac{V_2}{I_1} = \frac{1}{s+2}, \frac{V_2}{V_1} = \frac{1}{s^2+3s+1}, \frac{2}{\sqrt{(1-\omega)^2+(3\omega)^2}} \Bigg/ \tan^{-1}\frac{3\omega}{1-\omega^2}\right]$$

14. For the network shown in Fig. 10.91, determine $\frac{V_1}{I_1}$ and $\frac{V_2}{V_1}$. Plot the poles and zeros of transfer impedance and determine magnitude and phase of $V_2(s)$, given $V_1(s) = 1\angle 0°$ as a function of ω.

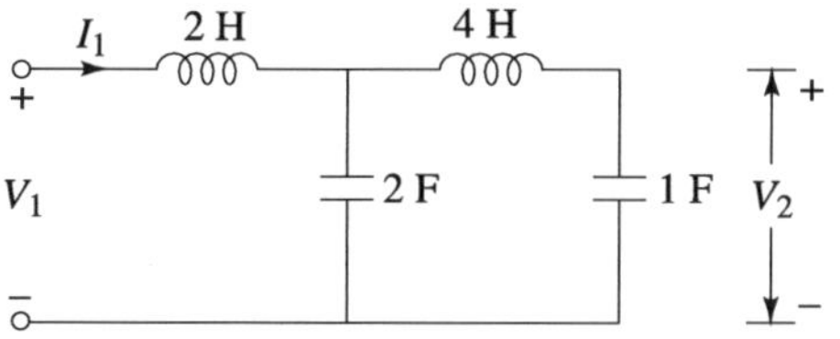

Fig. 10.91

$$\left[\frac{V_1}{I_1} = \frac{16s^4+10s^2+1}{8s^3+3s}, \frac{V_2}{I_1} = \frac{1}{8s^3+3s}, \frac{V_2}{V_1} = \frac{1}{16s^4+10s^2+1}, \frac{1}{16\omega^4-10\omega^2+1}\angle 0°\right]$$

15. For the given network function, draw the pole-zero diagram and hence obtain the time domain voltage. Verify the result analytically.

$$V(s) = \frac{5(s+5)}{(s+2)(s+7)} \qquad [v(t) = 3e^{-2t} + 2e^{-7t}]$$

16. Obtain the impedance function for which the pole-zero diagram is shown in Fig. 10.92,

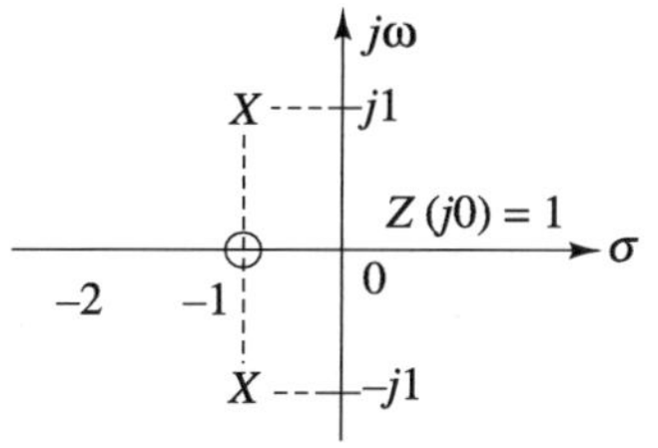

Fig. 10.92

$$\left[Z(s) = \frac{2(s+1)}{s^2+2s+2}\right]$$

17. For the network shown in Fig. 10.93, poles and zeros of driving point function $Z(s)$ are,
Poles : $(-1 \pm j4)$; zero : -2
If $Z(j0) = 1$, find the values of R, L and C.

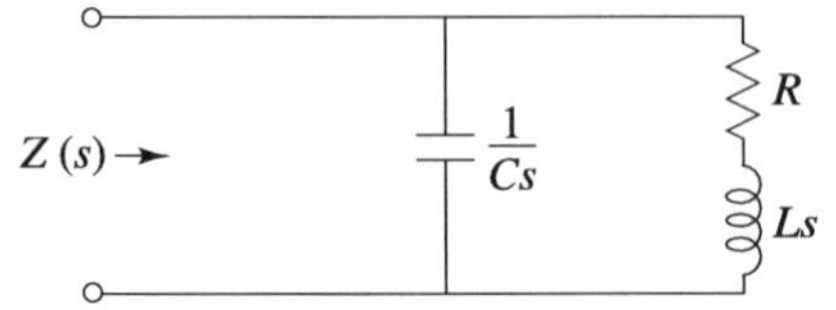

Fig. 10.93

$$\left[1\,\Omega,\ 0.5\,\text{H},\ \frac{2}{17}\,\text{F}\right]$$

18. For the two-port network shown in Fig. 10.94, find R_1, R_2 and C.

$$\frac{V_2}{V_1} = \frac{0.2}{s^2 + 3s + 2}$$

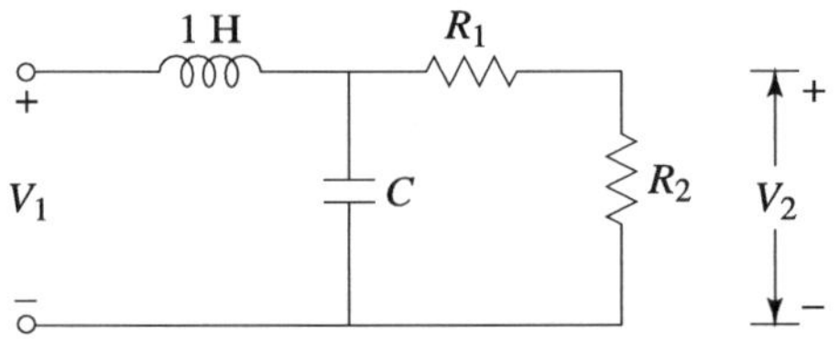

Fig. 10.94

$$\left[\frac{3}{5}\,\Omega,\ \frac{1}{15}\,\Omega,\ 0.5\,\text{F}\right]$$

Objective-Type Questions

1. Of the four networks N_1, N_2, N_3 and N_4 of Fig. 10.95 the networks having identical driving-point functions are

(a) N_1 and N_2 (b) N_2 and N_4 (c) N_1 and N_3 (d) N_1 and N_4

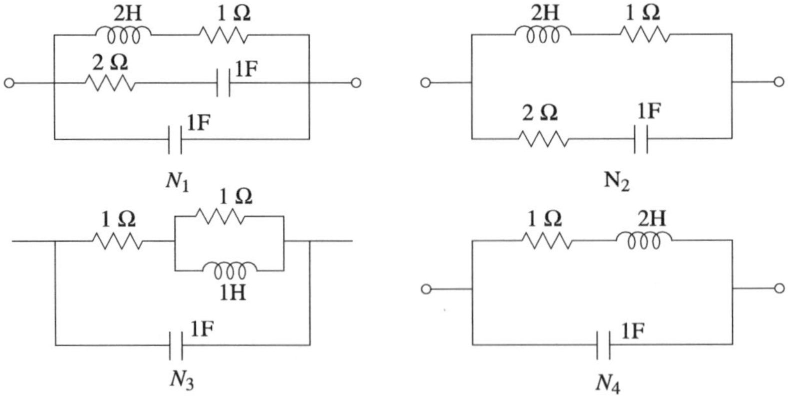

Fig. 10.95

2. The driving-point impedance $Z(s)$ of a network has the pole-zero locations as shown in Fig. 10.96. If $Z(0) = 3$, then $Z(s)$ is

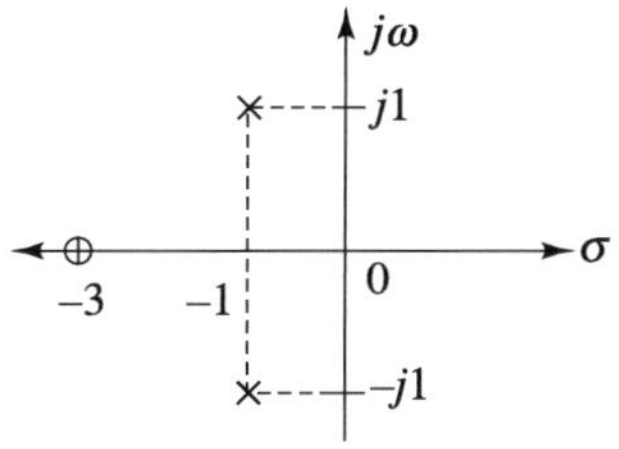

Fig. 10.96

(a) $\dfrac{3(s+3)}{s^2+2s+3}$ (b) $\dfrac{2(s+3)}{s^2+2s+2}$ (c) $\dfrac{3(s-3)}{s^2-2s-2}$ (d) $\dfrac{2(s-3)}{s^2-2s-3}$

3. For the circuit shown in Fig. 10.97 the initial conditions are zero. Its transfer function $H(s) = \dfrac{V_o(s)}{V_i(s)}$ is

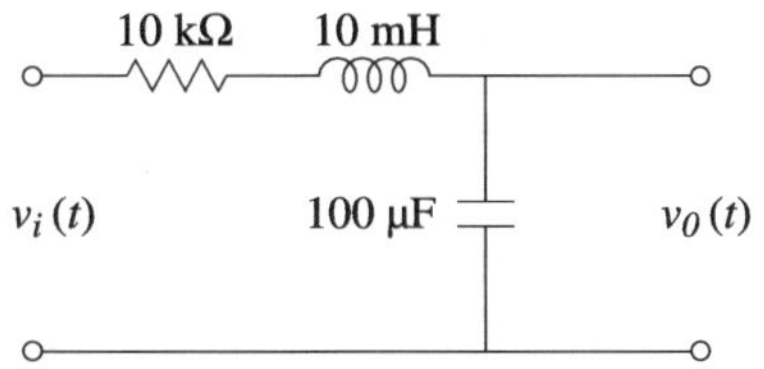

Fig. 10.97

(a) $\dfrac{1}{s^2+10^6 s+10^6}$ (b) $\dfrac{10^6}{s^2+10^3 s+10^6}$ (c) $\dfrac{10^3}{s^2+10^3 s+10^6}$ (d) $\dfrac{10^6}{s^2+10^6 s+10^6}$

4. In Fig. 10.98 shown, assume that all the capacitors are initially uncharged. If $v_i(t) = 10\,u(t)$, then $v_0(t)$ is given by

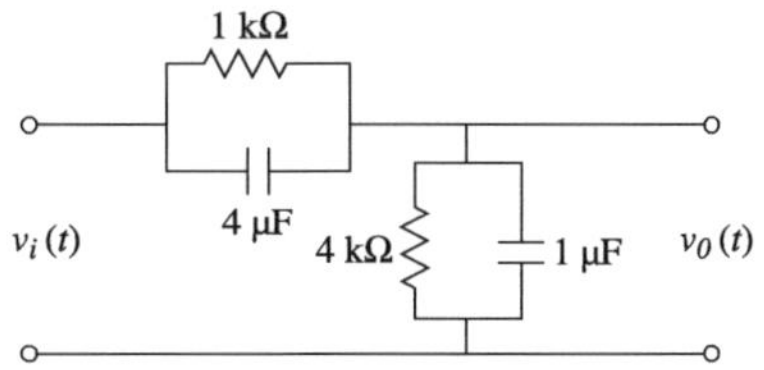

Fig. 10.98

(a) $8\,e^{-0.004\,t}$ (b) $8(1-e^{-0.004\,t})$ (c) $8\,u(t)$ (d) 8

5. A system is represented by the transfer function $\dfrac{10}{(s+1)(s+2)}$. The dc gain of this system is

(a) 1 (b) 2 (c) 5 (d) 10

6. Which one of the following is the ratio $\frac{V_{24}}{V_{13}}$ of the network shown in Fig. 10.99.

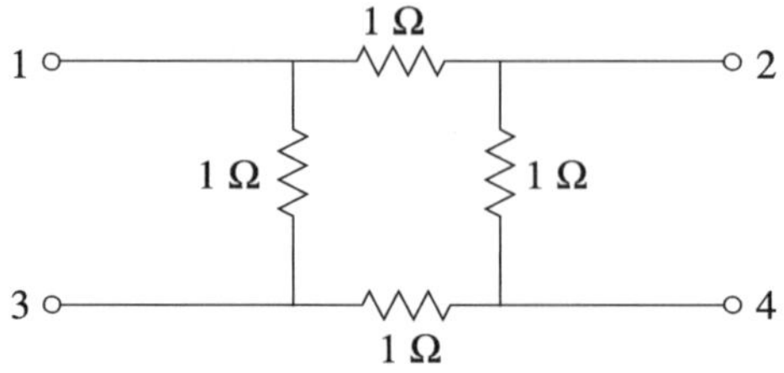

Fig. 10.99

(a) $\frac{1}{3}$ (b) $\frac{2}{3}$ (c) $\frac{3}{4}$ (d) $\frac{4}{3}$

7. A network has response with time as shown in Fig. 10.100. Which one of the following diagrams represents the location of the poles of this network?

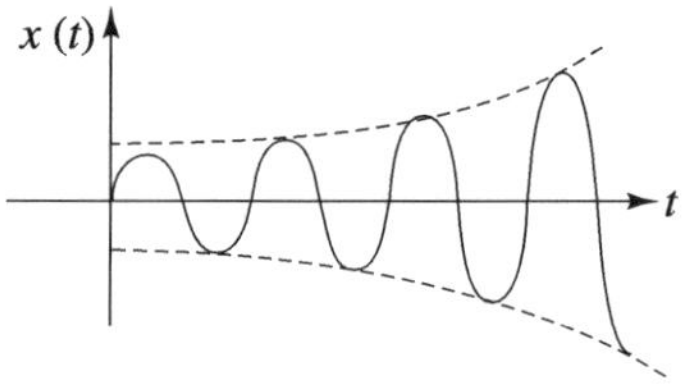

Fig. 10.100

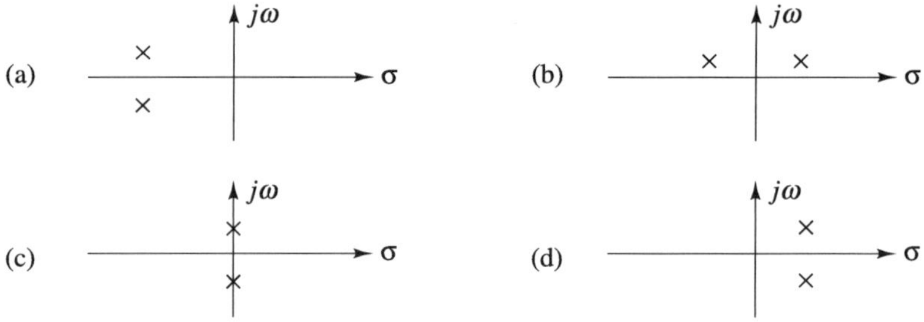

8. The transfer function of a low-pass RC network is

(a) $(RCs)(1 + RCs)$ (b) $\frac{1}{1 + RCs}$ (c) $\frac{RCs}{1 + RCs}$ (d) $\frac{s}{1 + RCs}$

9. The driving-point admittance function of the network shown in Fig. 10.101 has a

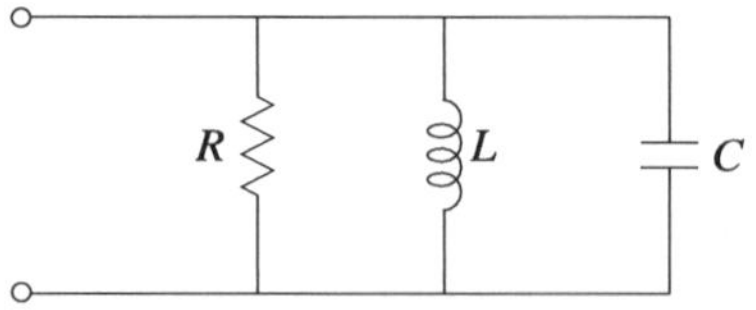

Fig. 10.101

(a) pole at $s = 0$ and zero at $s = \infty$ (b) pole at $s = 0$ and pole at $s = \infty$
(c) pole at $s = \infty$ and zero at $s = 0$ (d) pole at $s = \infty$ and zero at $s = \infty$

10. The transfer function $Y_{12}(s) = \dfrac{I_2(s)}{V_1(s)}$ for the network shown in Fig. 10.102 is

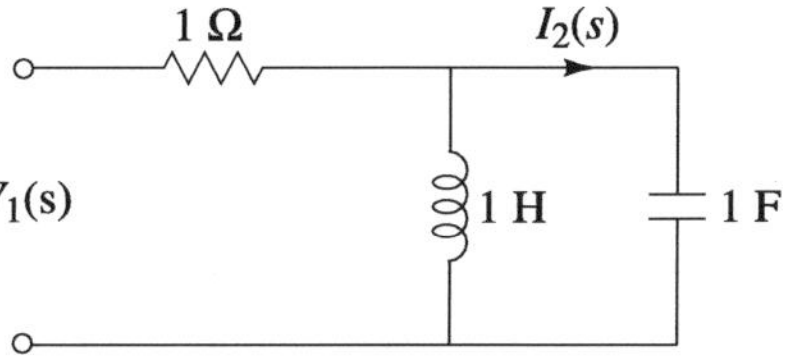

Fig. 10.102

(a) $\dfrac{s^2}{s^2+s+1}$ (b) $\dfrac{s}{s+1}$ (c) $\dfrac{1}{s+1}$ (d) $\dfrac{s+1}{s^2+1}$

11. As the poles of a network shift away from the x axis, the response

(a) remains constant (b) becomes less oscillating

(c) becomes more oscillating (d) none of these

Answers to Objective-Type Questions

1. (c) **2.** (b) **3.** (d) **4.** (c) **5.** (c) **6.** (a)

7. (d) **8.** (b) **9.** (a) **10.** (a) **11.** (c)

11 Two-Port Networks

11.1 INTRODUCTION

A two-port network has two pairs of terminals, one pair at the input known as *input port* and one pair at the output known as *output port*. There are four variables V_1, V_2, I_1 and I_2 associated with a two-port network. Two of these variables can be expressed in terms of the other two variables. Thus, there will be two dependent variables and two independent variables. The number of possible combinations generated by four variables taken two at a time is ${}^4C_2$, i.e., six. There are six possible sets of equations describing a two-port network.

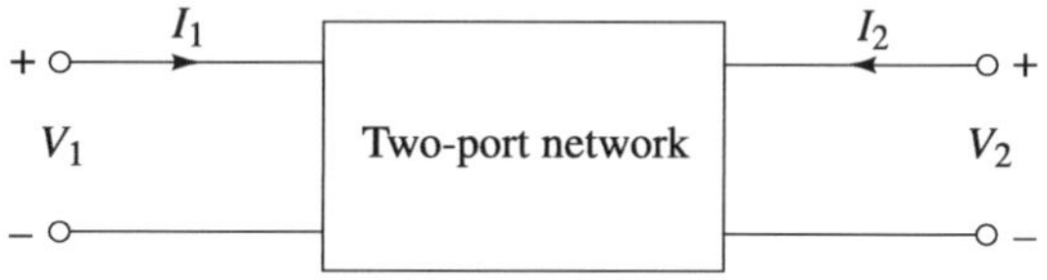

Fig. 11.1

Table 11.1 *Two-Port Parameters*

Parameter	*Variables*		*Equation*
	Express	*In terms of*	
Open-Circuit Impedance	V_1, V_2	I_1, I_2	$V_1 = Z_{11} I_1 + Z_{12} I_2$ $V_2 = Z_{21} I_1 + Z_{22} I_2$
Short-Circuit Admittance	I_1, I_2	V_1, V_2	$I_1 = Y_{11} V_1 + Y_{12} V_2$ $I_2 = Y_{21} V_1 + Y_{22} V_2$
Transmission	V_1, I_1	V_2, I_2	$V_1 = AV_2 - BI_2$ $I_1 = CV_2 - DI_2$
Inverse Transmission	V_2, I_2	V_1, I_1	$V_2 = A' V_1 - B' I_1$ $I_2 = C' V_1 - D' I_1$
Hybrid	V_1, I_2	I_1, V_2	$V_1 = h_{11} I_1 + h_{12} V_2$ $I_2 = h_{21} I_1 + h_{22} V_2$
Inverse Hybrid	I_1, V_2	V_1, I_2	$I_1 = g_{11} V_1 + g_{12} I_2$ $V_2 = g_{21} V_1 + g_{22} I_2$

11.2 OPEN-CIRCUIT IMPEDANCE PARAMETERS (Z PARAMETERS)

The Z parameters of a two-port network may be defined by expressing two-port voltages V_1 and V_2 in terms of two-port currents I_1 and I_2.

$$(V_1, V_2) = f(I_1, I_2) \quad \text{...(11.1)}$$

$$V_1 = Z_{11} I_1 + Z_{12} I_2 \quad \text{...(11.2)}$$

$$V_2 = Z_{21} I_1 + Z_{22} I_2 \quad \text{...(11.3)}$$

In matrix form, we can write

$$\begin{bmatrix} V_1 \\ V_2 \end{bmatrix} = \begin{bmatrix} Z_{11} & Z_{12} \\ Z_{21} & Z_{22} \end{bmatrix} \begin{bmatrix} I_1 \\ I_2 \end{bmatrix}$$

$$[V] = [Z]\,[I] \quad \text{...(11.4)}$$

The individual Z parameters for a given network can be defined by setting each of the port currents equal to zero.

Case 1 When the output port is open circuited,

i.e. $I_2 = 0$

$$Z_{11} = \left.\frac{V_1}{I_1}\right|_{I_2=0} \quad \text{...(11.5)}$$

where Z_{11} is the driving-point impedance with the output port open circuited. It is also called *open circuit input impedance.*

Similarly,

$$Z_{21} = \left.\frac{V_2}{I_1}\right|_{I_2=0} \quad \text{...(11.6)}$$

where Z_{21} is the transfer impedance with the output port open circuited. It is also called *open circuit forward transfer impedance.*

Case 2 When input port is open circuited,

i.e. $I_1 = 0$

$$Z_{12} = \left.\frac{V_1}{I_2}\right|_{I_1=0} \quad \text{...(11.7)}$$

where Z_{12} is the transfer impedance with the input port open circuited. It is also called *open circuit reverse transfer impedance.*

Similarly,

$$Z_{22} = \left.\frac{V_2}{I_2}\right|_{I_1=0} \quad \text{...(11.8)}$$

where Z_{22} is the open circuit driving-point impedance with the input port open circuited. It is also called *open circuit output impedance.*

As these impedance parameters are measured with either the input or output port open circuited, these are called *open circuit impedance parameters.*

The equivalent circuit of the two-port network in terms of Z parameters is as shown in Fig. 11.2.

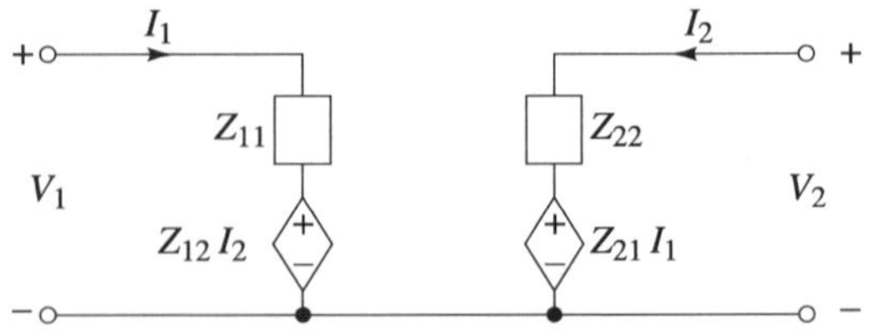

Fig. 11.2

11.2.1 Condition for Reciprocity

A network is said to be reciprocal if the ratio of excitation at one port to response at the other port is same if excitation and response are interchanged.

(a) As shown in Fig. 11.3, voltage V_s is applied at the input port with the output port short circuited.

i.e.
$$V_1 = V_s$$
$$V_2 = 0$$
$$I_2 = -I_2'$$

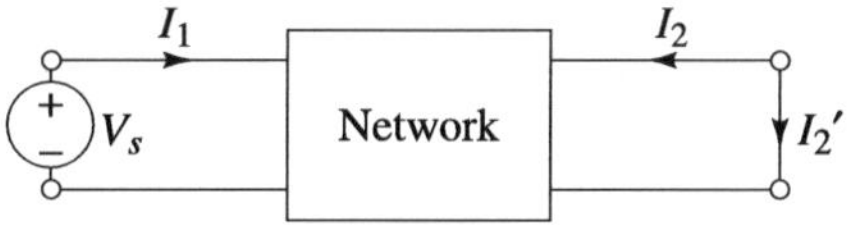

Fig. 11.3

From the Z-parameter equations,
$$V_s = Z_{11} I_1 - Z_{12} I_2'$$
$$0 = Z_{21} I_1 - Z_{22} I_2'$$
$$I_1 = \frac{Z_{22}}{Z_{21}} I_2'$$
$$V_s = Z_{11} \frac{Z_{22}}{Z_{21}} I_2' - Z_{12} I_2'$$
$$\frac{V_s}{I_2'} = \frac{Z_{11}Z_{22} - Z_{12}Z_{21}}{Z_{21}}$$

(b) Now, when voltage V_s is applied at the output port with input port short circuited.

i.e.,
$$V_2 = V_s$$
$$V_1 = 0$$
$$I_1 = -I_1'$$

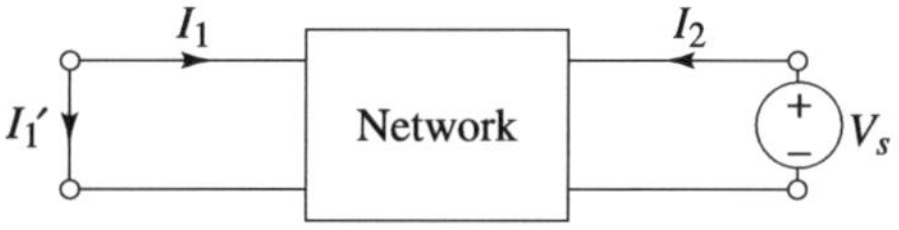

Fig. 11.4

From the Z-parameter equations,
$$0 = -Z_{11} I_1' + Z_{12} I_2$$
$$V_s = -Z_{21} I_1' + Z_{22} I_2$$
$$I_2 = \frac{Z_{11}}{Z_{12}} I_1'$$
$$V_s = -Z_{21} I_1' + Z_{22} \frac{Z_{11}}{Z_{12}} I_1'$$
$$\frac{V_s}{I_1'} = \frac{Z_{11}Z_{22} - Z_{12}Z_{21}}{Z_{12}}$$

Hence, for the network to be reciprocal,
$$\frac{V_s}{I_1'} = \frac{V_s}{I_2'}$$

i.e.
$$Z_{12} = Z_{21} \qquad \ldots(11.9)$$

11.2.2 Condition for Symmetry

For a network to be symmetrical, the voltage-to-current ratio at one port should be the same as the voltage-to-current ratio at the other port with one of the ports open circuited.

(a) When the output port is open circuited, i.e., $I_2 = 0$

From the Z-parameter equation,
$$V_s = Z_{11} I_1$$
$$\frac{V_s}{I_1} = Z_{11}$$

(b) When the input port is open circuited, i.e., $I_1 = 0$

From the Z-parameter equation,

$$V_s = Z_{22} I_2$$

$$\frac{V_s}{I_2} = Z_{22}$$

Hence, for the network to be symmetrical,

$$\frac{V_s}{I_1} = \frac{V_s}{I_2}$$

i.e.,

$$Z_{11} = Z_{22} \qquad \text{...(11.10)}$$

11.3 SHORT-CIRCUIT ADMITTANCE PARAMETERS (*Y* PARAMETERS)

The Y parameters of a two-port network may be defined by expressing the two-port currents I_1 and I_2 in terms of the two-port voltages V_1 and V_2.

$$(I_1, I_2) = f(V_1, V_2) \qquad \text{...(11.11)}$$

$$I_1 = Y_{11} V_1 + Y_{12} V_2 \qquad \text{...(11.12)}$$

$$I_2 = Y_{21} V_1 + Y_{22} V_2 \qquad \text{...(11.13)}$$

In matrix form, we can write

$$\begin{bmatrix} I_1 \\ I_2 \end{bmatrix} = \begin{bmatrix} Y_{11} & Y_{12} \\ Y_{21} & Y_{22} \end{bmatrix} \begin{bmatrix} V_1 \\ V_2 \end{bmatrix}$$

$$[I] = [Y]\,[V] \qquad \text{...(11.14)}$$

The individual Y parameters for a given network can be defined by setting each of the port voltages equal to zero.

Case 1 When the output port is short circuited,

i.e., $V_2 = 0$

$$Y_{11} = \left.\frac{I_1}{V_1}\right|_{V_2=0} \qquad \text{...(11.15)}$$

where Y_{11} is the driving-point admittance with the output port short circuited. It is also called *short-circuit input admittance.*

Similarly,

$$Y_{21} = \left.\frac{I_2}{V_1}\right|_{V_2=0} \qquad \text{...(11.16)}$$

where Y_{21} is the transfer admittance with the output port short circuited. It is also called *short-circuit forward transfer admittance.*

Case 2 When the input port is short circuited,

i.e., $V_1 = 0$

$$Y_{12} = \left.\frac{I_1}{V_2}\right|_{V_1=0} \qquad \text{...(11.17)}$$

where Y_{12} is the transfer admittance with the input port short circuited. It is also called *short-circuit reverse transfer admittance.*

Similarly,

$$Y_{22} = \left.\frac{I_2}{V_2}\right|_{V_1=0} \qquad \text{...(11.18)}$$

where Y_{22} is the short circuit driving-point admittance with the input port short circuited. It is also called the *short circuit output admittance.*

As these admittance parameters are measured with either input or output port short circuited, these are called *short circuit admittance parameters.*

The equivalent circuit of the two-port network in terms of *Y* parameters is as shown in Fig. 11.5.

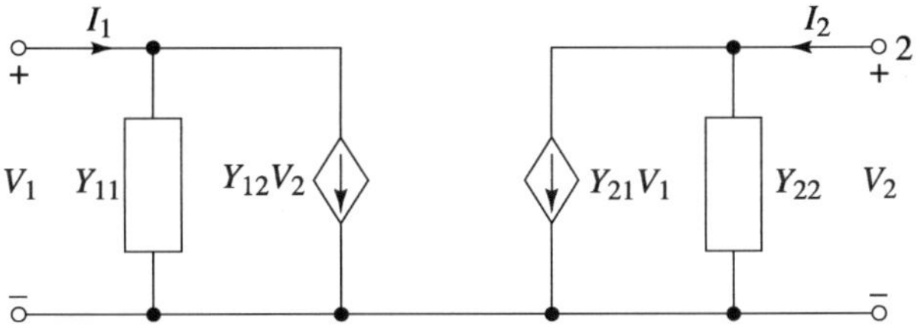

Fig. 11.5

11.3.1 Condition for Reciprocity

(a) As shown in Fig. 11.6, voltage V_s is applied at input port with the output port short circuited.

i.e.

$$V_1 = V_s$$
$$V_2 = 0$$
$$I_2 = -I_2'$$

From the *Y*-parameter equation,

$$-I_2' = Y_{21} V_s$$
$$\frac{I_2'}{V_s} = -Y_{21}$$

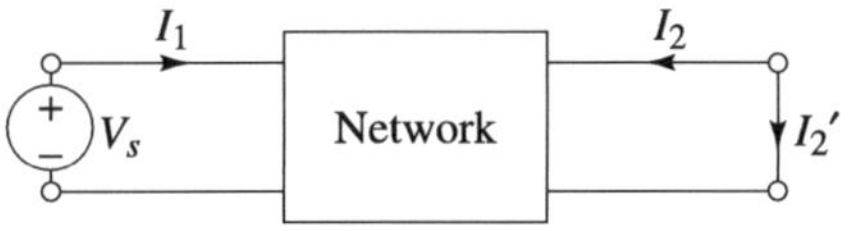

Fig. 11.6

(b) Now, when the voltage V_s is applied at output port with the input port short circuited.

i.e.

$$V_2 = V_s$$
$$V_1 = 0$$
$$I_1 = -I_1'$$

From the *Y*-parameter equation,

$$-I_1' = Y_{12} V_s$$
$$\frac{I_1'}{V_s} = -Y_{12}$$

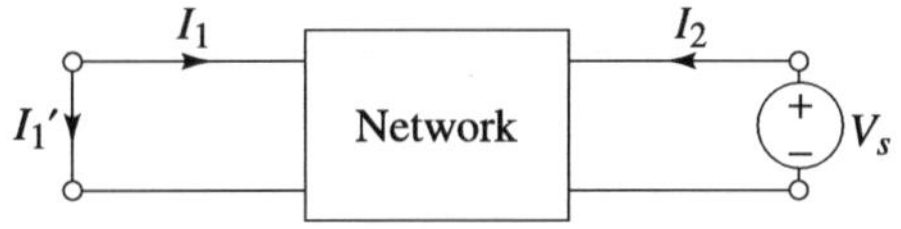

Fig. 11.7

Hence, for the network to be reciprocal,

$$\frac{I_2'}{V_s} = \frac{I_1'}{V_s}$$

i.e.,

$$Y_{12} = Y_{21} \qquad \text{...(11.19)}$$

11.3.2 Condition for Symmetry

(a) When the output port is short circuited, i.e., $V_2 = 0$.

From the *Y*-parameter equation

$$I_1 = Y_{11} V_s$$
$$\frac{V_s}{I_1} = \frac{1}{Y_{11}}$$

(b) When the input port is short circuited, i.e., $V_1 = 0$.
From the Y-parameter equation,

$$I_2 = Y_{22} V_s$$

$$\frac{V_s}{I_2} = \frac{1}{Y_{22}}$$

Hence, for the network to be symmetrical,

$$\frac{V_s}{I_1} = \frac{V_s}{I_2}$$

i.e., $$Y_{11} = Y_{22} \quad \ldots(11.20)$$

11.4 TRANSMISSION PARAMETERS (*ABCD* PARAMETERS)

The transmission parameters or chain parameters or *ABCD* parameters serve to relate the voltage and current at the input port to voltage and current at the outport port. In equation form,

$$(V_1, I_1) = f(V_2, -I_2) \quad \ldots(11.21)$$

$$V_1 = AV_2 - BI_2 \quad \ldots(11.22)$$

$$I_1 = CV_2 - DI_2 \quad \ldots(11.23)$$

Here, the negative sign is used with I_2 and not for parameters B and D. The reason the current I_2 carries a negative sign is that in transmission field, the output current is assumed to be coming out of the output port instead of going into the port.

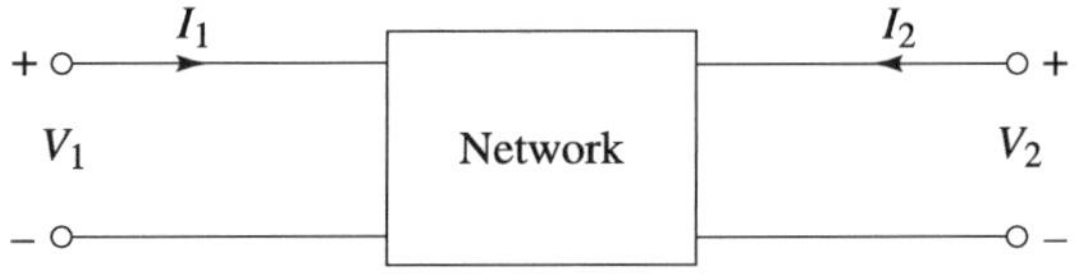

Fig. 11.8

In matrix form, we can write

$$\begin{bmatrix} V_1 \\ I_1 \end{bmatrix} = \begin{bmatrix} A & B \\ C & D \end{bmatrix} \begin{bmatrix} V_2 \\ -I_2 \end{bmatrix} \quad \ldots(11.24)$$

where matrix $\begin{bmatrix} A & B \\ C & D \end{bmatrix}$ is called transmission matrix.

For a given network, these parameters are determined as follows:

Case 1 When the output port is open circuited,

i.e., $$I_2 = 0$$

$$A = \left.\frac{V_1}{V_2}\right|_{I_2=0} \quad \ldots(11.25)$$

where A is the reverse voltage gain with the output port open circuited.

Similarly, $$C = \left.\frac{I_1}{V_2}\right|_{I_2=0} \quad \ldots(11.26)$$

where C is the transfer admittance with the output port open circuited.

Case 2 When output port is short circuited,

i.e., $$V_2 = 0$$

$$B = -\left.\frac{V_1}{I_2}\right|_{V_2=0} \quad \ldots(11.27)$$

where B is the transfer impedance with the output port short circuited.

Similarly, $$D = -\left.\frac{I_1}{I_2}\right|_{V_2=0} \quad \ldots(11.28)$$

where D is the reverse current gain with the output port short circuited.

11.4.1 Condition for Reciprocity

(a) When the voltage V_s is applied at the input port with the output port short circuited.

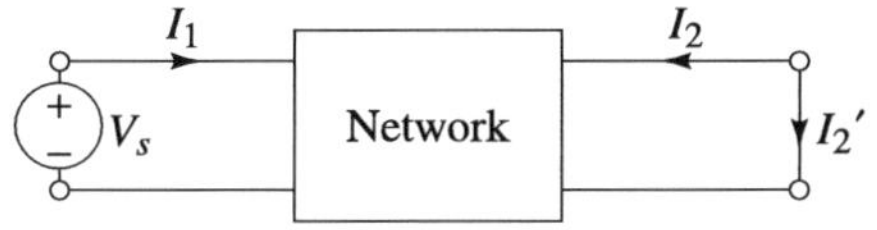

Fig. 11.9

i.e.,
$$V_1 = V_s$$
$$V_2 = 0$$
$$I_2' = -I_2$$

From the transmission parameter equation,
$$V_s = B\, I_2'$$
$$\frac{V_s}{I_2'} = B$$

(b) Now, when voltage V_s is applied at the output port with the input port short circuited.

i.e.,
$$V_2 = V_s$$
$$V_1 = 0$$
$$I_1' = -I_1$$

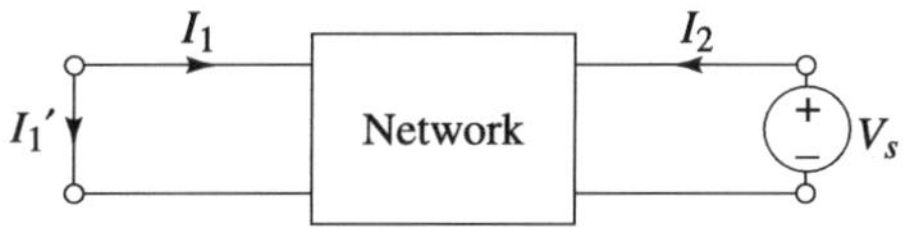

Fig. 11.10

From the transmission parameter equations,
$$0 = AV_s - BI_2$$
$$-I_1' = CV_s - DI_2$$
$$I_2 = \frac{A}{B} V_s$$
$$-I_1' = CV_s - \frac{AD}{B} V_s$$
$$\frac{V_s}{I_1'} = \frac{B}{AD - BC}$$

Hence, for the network to be reciprocal,
$$\frac{V_s}{I_2'} = \frac{V_s}{I_1'}$$

i.e., $$B = \frac{B}{AD - BC}$$

i.e., $$AD - BC = 1 \quad \ldots(11.29)$$

11.4.2 Condition for Symmetry

(a) When the output port is open circuited, i.e., $I_2 = 0$.

From the transmission-parameter equations,
$$V_s = AV_2$$

$$I_1 = CV_2$$

$$\frac{V_s}{I_1} = \frac{A}{C}$$

(b) When the input port is open circuited, i.e., $I_1 = 0$
From the transmission parameter equation,

$$CV_s = DI_2$$

$$\frac{V_s}{I_2} = \frac{D}{C}$$

Hence, for network to be symmetrical,

$$\frac{V_s}{I_1} = \frac{V_s}{I_2}$$

i.e., $$A = D \qquad \ldots(11.30)$$

11.5 INVERSE TRANSMISSION PARAMETERS (A′B′C′D′ PARAMETERS)

The inverse transmission parameters serve to relate the voltage and current at the outport port to the voltage and current at the input port. In equation form,

$$(V_2, I_2) = f(V_1, -I_1) \qquad \ldots(11.31)$$

$$V_2 = A' V_1 - B' I_1 \qquad \ldots(11.32)$$

$$I_2 = C' V_1 - D' I_1 \qquad \ldots(11.33)$$

In matrix form, we can write

$$\begin{bmatrix} V_2 \\ I_2 \end{bmatrix} = \begin{bmatrix} A' & B' \\ C' & D' \end{bmatrix} \begin{bmatrix} V_1 \\ -I_1 \end{bmatrix} \qquad \ldots(11.34)$$

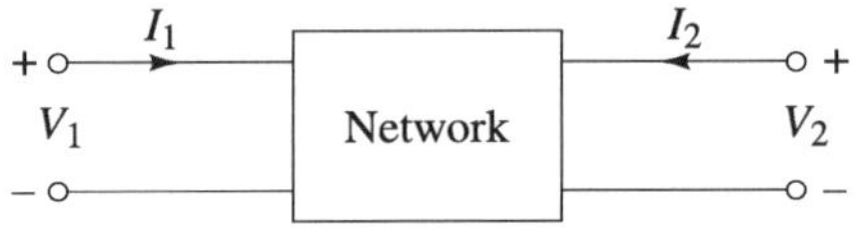

Fig. 11.11

where matrix $\begin{bmatrix} A' & B' \\ C' & D' \end{bmatrix}$ is called the *inverse transmission matrix.*

For a given network, these parameters are determined as follows:

Case 1 When the input port is open circuited,
i.e., $I_1 = 0$

$$A' = \left.\frac{V_2}{V_1}\right|_{I_1=0} \qquad \ldots(11.35)$$

where A' is the forward voltage gain with the input port open circuited.

Similarly, $$C' = \left.\frac{I_2}{V_1}\right|_{I_1=0} \qquad \ldots(11.36)$$

where C' is the transfer admittance with the input port open circuited.

Case 2 When the input port is short circuited,
i.e., $V_1 = 0$

$$B' = -\left.\frac{V_2}{I_1}\right|_{V_1=0} \qquad \ldots(11.37)$$

where B' is the transfer impedance with the input port short circuited.

Similarly, $$D' = -\left.\frac{I_2}{I_1}\right|_{V_1=0} \qquad \text{...(11.38)}$$

where D' is the forward current gain with the input port short circuited.

11.5.1 Condition for Reciprocity

(a) When a voltage V_s is applied at input port with the output port short circuited.

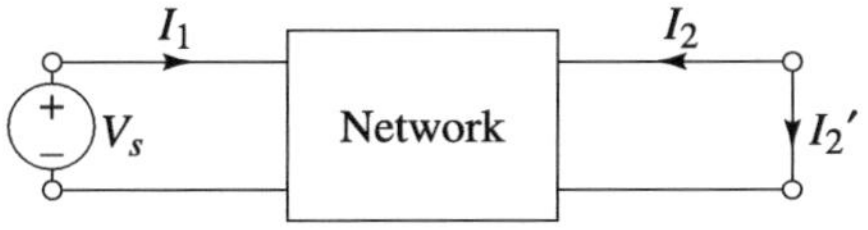

Fig. 11.12

i.e.,
$$V_1 = V_s$$
$$V_2 = 0$$
$$I_2 = -I_2'$$

From the inverse transmission parameter equations,
$$0 = A' V_s - B' I_1$$
$$-I_2' = C' V_s - D' I_1$$
$$\frac{I_2'}{V_s} = \frac{A'D' - B'C'}{B'}$$

(b) Now, when voltage V_s is applied at the output port with the input port short circuited.

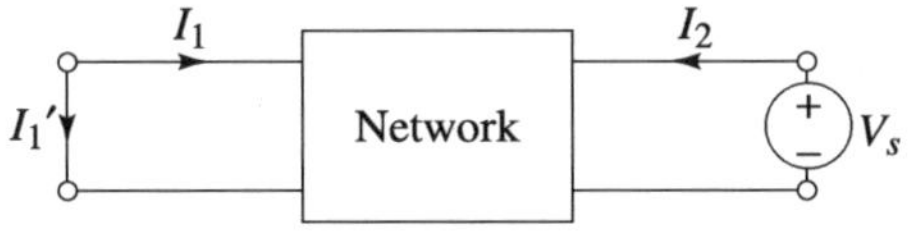

Fig. 11.13

i.e.,
$$V_2 = V_s$$
$$V_1 = 0$$
$$I_1 = -I_1'$$

From the inverse transmission parameter equations,
$$V_s = B' I_1'$$
$$I_2 = \frac{A'}{B'} V_s$$
$$\frac{I_1'}{V_s} = \frac{1}{B'}$$

Hence, for the network to be reciprocal,
$$\frac{I_2'}{V_s} = \frac{I_1'}{V_s}$$

i.e., $$A' D' - B'C' = 1 \qquad \text{...(11.39)}$$

11.5.2 Condition for Symmetry

The condition for symmetry is obtained from the Z-parameters.

$$Z_{11} = \left.\frac{V_1}{I_1}\right|_{I_2=0} = \frac{D'}{C'}$$

Similarly,

$$Z_{22} = \left.\frac{V_2}{I_2}\right|_{I_1=0} = \frac{A'}{C'}$$

For symmetrical network $Z_{11} = Z_{22}$,

$$A' = C' \qquad ...(11.40)$$

11.6 HYBRID PARAMETERS (*h* PARAMETERS)

The hybrid parameters of a two-port network may be defined by expressing the voltage of input port V_1 and current of output port I_2 in terms of current of input port I_1 and voltage of output port V_2.

$$(V_1, I_2) = f(I_1, V_2) \qquad ...(11.41)$$

$$V_1 = h_{11} I_1 + h_{12} V_2 \qquad ...(11.42)$$

$$I_2 = h_{21} I_1 + h_{22} V_2 \qquad ...(11.43)$$

In matrix form, we can write

$$\begin{bmatrix} V_1 \\ I_2 \end{bmatrix} = \begin{bmatrix} h_{11} & h_{12} \\ h_{21} & h_{22} \end{bmatrix} \begin{bmatrix} I_1 \\ V_2 \end{bmatrix} \qquad ...(11.44)$$

The individual h parameters can be defined by setting $I_1 = 0$ and $V_2 = 0$.

Case 1 When the output port is short circuited,

i.e., $V_2 = 0$

$$h_{11} = \left.\frac{V_1}{I_1}\right|_{V_2=0} \qquad ...(11.45)$$

where h_{11} is the short-circuit input impedance.

$$h_{21} = \left.\frac{I_2}{I_1}\right|_{V_2=0} \qquad ...(11.46)$$

where h_{21} is the short circuit forward current gain.

Case 2 When the input port is open circuited,

i.e., $I_1 = 0$

$$h_{12} = \left.\frac{V_1}{V_2}\right|_{I_1=0} \qquad ...(11.47)$$

where h_{12} is the open-circuit reverse voltage gain.

$$h_{22} = \left.\frac{I_2}{V_2}\right|_{I_1=0} \qquad ...(11.48)$$

where h_{22} is the open-circuit output admittance.

Since h parameters represent dimensionally an impedance, an admittance, a voltage gain and a current gain, these are called *hybrid parameters.*

The equivalent circuit of a two-port network in terms of hybrid parameters is as shown in Fig. 11.14.

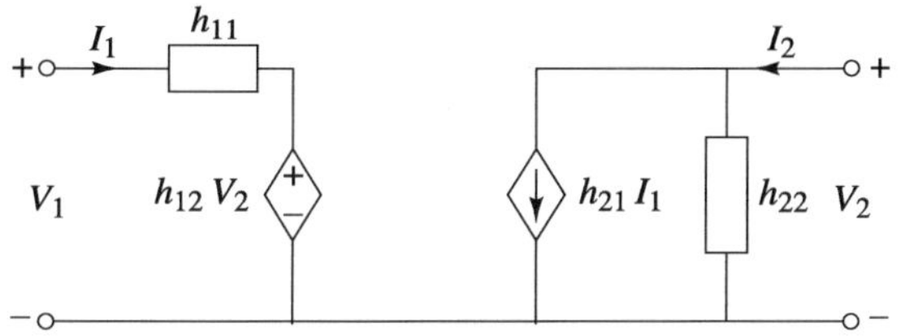

Fig. 11.14

11.6.1 Condition for Reciprocity

(a) When voltage V_s is applied at the input port and the output port is short circuited.

i.e.,
$$V_1 = V_s$$
$$V_2 = 0$$
$$I_2' = -I_2$$

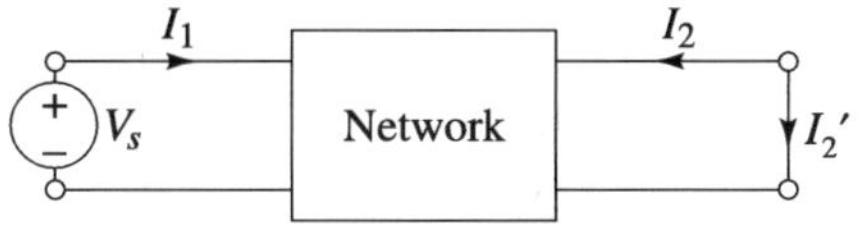

Fig. 11.15

From the h-parameter equations,
$$V_s = h_{11} I_1$$
$$-I_2' = h_{21} I_1$$
$$\frac{V_s}{I_2'} = \frac{-h_{11}}{h_{21}}$$

(b) Now, when a voltage V_s is applied at the output port with the input port short circuited.

i.e.,
$$V_1 = 0$$
$$V_2 = V_s$$
$$I_1 = -I_1'$$

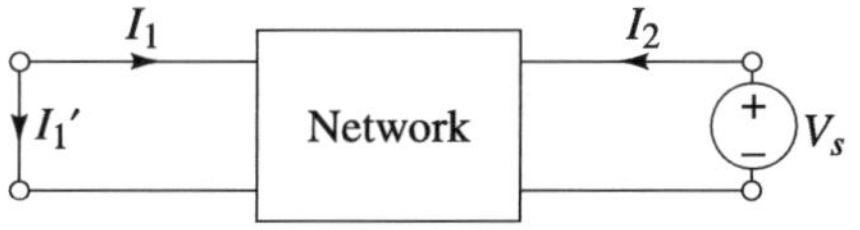

Fig. 11.16

From the h-parameter equations,
$$0 = h_{11} I_1 + h_{12} V_s$$
$$h_{12} V_s = -h_{11} I_1 = h_{11} I_1'$$
$$\frac{V_s}{I_1'} = \frac{h_{11}}{h_{12}}$$

Hence, for the network to be reciprocal,
$$\frac{V_s}{I_2'} = \frac{V_s}{I_1'}$$

i.e.,
$$h_{21} = -h_{12} \qquad \ldots(11.49)$$

11.6.2 Condition for Symmetry

The condition for symmetry is obtained from the Z-parameters.

$$Z_{11} = \left.\frac{V_1}{I_1}\right|_{I_2=0} = \left.\frac{h_{11}I_1 + h_{12}V_2}{I_1}\right|_{I_2=0}$$
$$= h_{11} + h_{12}\frac{V_2}{I_1}$$

But with $I_2 = 0$, we have

$$0 = h_{21} I_1 + h_{22} V_2$$

$$\frac{V_2}{I_1} = \frac{-h_{21}}{h_{22}}$$

$$Z_{11} = h_{11} - \frac{h_{12}h_{21}}{h_{22}} = \frac{h_{11}h_{22} - h_{12}h_{21}}{h_{22}} = \frac{\Delta h}{h_{22}}$$

where $\Delta h = h_{11} h_{22} - h_{12} h_{21}$

Similarly,

$$Z_{22} = \left.\frac{V_2}{I_2}\right|_{I_1=0}$$

With $I_1 = 0$, we have

$$I_2 = h_{22} V_2$$

$$Z_{22} = \left.\frac{V_2}{I_2}\right|_{I_1=0} = \frac{1}{h_{22}}$$

For a symmetrical network $Z_{11} = Z_{22}$

i.e., $$\frac{1}{h_{22}} = \frac{\Delta h}{h_{22}}$$

i.e., $$\Delta h = 1$$

i.e., $$h_{11} h_{22} - h_{12} h_{21} = 1 \qquad \text{...(11.50)}$$

11.7 INVERSE HYBRID PARAMETERS (*g* PARAMETERS)

The inverse hybrid parameters of a two-port network may be defined by expressing the current of the input port I_1 and voltage of the output port V_2 in terms of the voltage of the input port V_1 and the current of the output port I_2.

$$(I_1, V_2) = f(V_1, I_2) \qquad \text{...(11.51)}$$

$$I_1 = g_{11} V_1 + g_{12} I_2 \qquad \text{...(11.52)}$$

$$V_2 = g_{21} V_1 + g_{22} I_2 \qquad \text{...(11.53)}$$

In matrix form, we can write

$$\begin{bmatrix} I_1 \\ V_2 \end{bmatrix} = \begin{bmatrix} g_{11} & g_{12} \\ g_{21} & g_{22} \end{bmatrix} \begin{bmatrix} V_1 \\ I_2 \end{bmatrix} \qquad \text{...(11.54)}$$

The individual g parameters can be defined by setting $V_1 = 0$ and $I_2 = 0$.

Case I When the output port is open circuited,

i.e., $$I_2 = 0$$

$$g_{11} = \left.\frac{I_1}{V_1}\right|_{I_2=0} \qquad \text{...(11.55)}$$

where g_{11} is the open-circuit input admittance.

$$g_{21} = \left.\frac{V_2}{V_1}\right|_{I_2=0} \qquad \text{...(11.56)}$$

where g_{21} is the open-circuit forward voltage gain.

Case 2 When the input port is short circuited,

i.e., $V_1 = 0$

$$g_{12} = \left.\frac{I_1}{I_2}\right|_{V_1=0} \qquad \text{...(11.57)}$$

where g_{12} is the short-circuit reverse current gain.

$$g_{22} = \left.\frac{V_2}{I_2}\right|_{V_1=0} \qquad \text{...(11.58)}$$

where g_{22} is the short-circuit output impedance.

The equivalent circuit of a two-port network in terms of inverse hybrid parameters is as shown in Fig. 11.17.

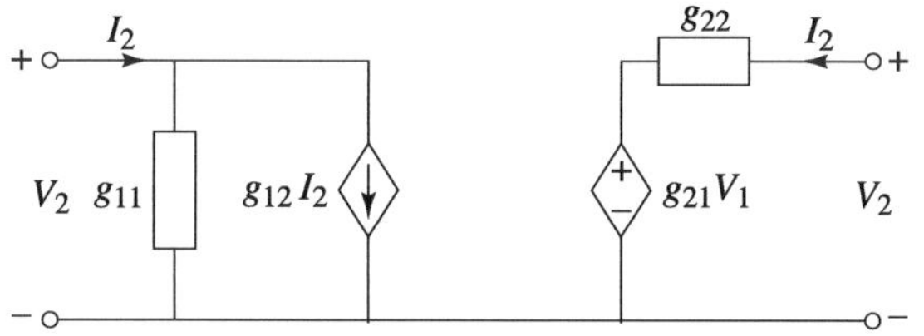

Fig. 11.17

11.7.1 Condition for Reciprocity

(a) When a voltage V_s is applied at the input port and the output port is short circuited,

i.e.,

$$V_1 = V_s$$
$$V_2 = 0$$
$$I_2' = -I_2$$

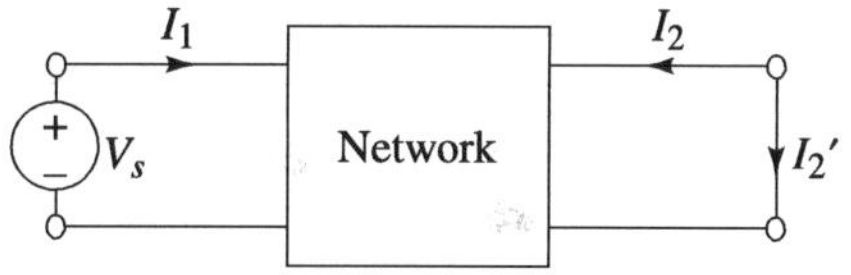

Fig. 11.18

From the g-parameter equation,

$$g_{21} V_s = g_{22} I_2'$$

$$\frac{V_s}{I_2'} = \frac{g_{22}}{g_{21}}$$

(b) Now, when the voltage V_s is applied at the output port with the input port short circuited,

i.e.,

$$V_1 = 0$$
$$V_2 = V_s$$
$$I_1 = -I_1'$$

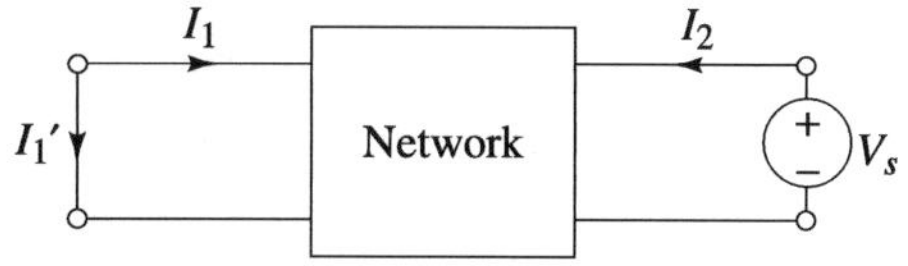

Fig. 11.19

From the g-parameter equations,

$$-I_1' = g_{12} I_2$$
$$V_s = g_{22} I_2$$

$$\frac{V_s}{I_1'} = \frac{-g_{22}}{g_{12}}$$

Hence, for the network to be reciprocal,

$$\frac{V_s}{I_2'} = \frac{V_s}{I_1'}$$

i.e., $$g_{21} = -g_{12} \qquad \text{...(11.59)}$$

11.7.2 Condition for Symmetry

The condition for symmetry is obtained from the Z-parameters.

$$Z_{11} = \left.\frac{V_1}{I_1}\right|_{I_2=0} = \frac{1}{g_{11}}$$

Similarly,

$$Z_{22} = \left.\frac{V_2}{I_2}\right|_{I_1=0}$$

$$0 = g_{11}V_1 + g_{12}I_2$$

$$V_1 = \frac{-g_{12}}{g_{11}} I_2$$

$$V_2 = g_{21}\left(\frac{-g_{12}}{g_{11}}\right) I_2 + g_{22}I_2$$

$$Z_{22} = \left.\frac{V_2}{I_2}\right|_{I_1=0}$$

$$= \frac{g_{11}g_{22} - g_{12}g_{21}}{g_{11}}$$

For a symmetrical network, $Z_{11} = Z_{22}$.

i.e.,
$$\frac{1}{g_{11}} = \frac{g_{11}g_{22} - g_{12}g_{21}}{g_{11}}$$

i.e.,
$$g_{11}g_{22} - g_{12}g_{21} = 1 \qquad \ldots(11.60)$$

Table 11.2 *Conditions for Reciprocity and Symmetry*

Parameter	*Condition for Reciprocity*	*Condition for Symmetry*
Z	$Z_{12} = Z_{21}$	$Z_{11} = Z_{22}$
Y	$Y_{12} = Y_{21}$	$Y_{11} = Y_{22}$
T	$AD - BC = 1$	$A = D$
T'	$A'D' - B'C' = 1$	$A' = D'$
h	$h_{12} = -h_{21}$	$h_{11}h_{22} - h_{12}h_{21} = 1$
g	$g_{12} = -g_{21}$	$g_{11}g_{22} - g_{12}g_{21} = 1$

Notes:

(1) To find Z-parameters of a 2-port network, apply KVL to the network.

(2) To find Y-parameters, apply KCL to the network.

(3) To find h-parameters and $ABCD$ parameters, apply KVL or KCL to the given network. Convert the equations into the standard form of Z-parameters or Y-parameters respectively and then rearrange the equations to get the standard form of h-parameters and $ABCD$-parameter equations.

Example 11.1 *Test results for a two-port network are (a) $I_1 = 0.1 \angle 0°$ A, $V_1 = 5.2 \angle 50°$ V, $V_2 = 4.1 \angle -25°$ V with Port 2 open circuited (b) $I_2 = 0.1 \angle 0°$A, $V_1 = 3.1 \angle -80°$ V, $V_2 = 4.2 \angle 60°$ V, with Port 1 open circuited. Find Z parameters.*

Solution

$$Z_{11} = \left.\frac{V_1}{I_1}\right|_{I_2=0} = \frac{5.2\angle 50^\circ}{0.1\angle 0^\circ} = 52\ \angle 50^\circ\ \Omega$$

$$Z_{21} = \left.\frac{V_2}{I_1}\right|_{I_2=0} = \frac{4.1\angle -25^\circ}{0.1\angle 0^\circ} = 41\ \angle -25^\circ\ \Omega$$

$$Z_{22} = \left.\frac{V_2}{I_2}\right|_{I_1=0} = \frac{4.2\angle 60^\circ}{0.1\angle 0^\circ} = 42\ \angle 60^\circ\ \Omega$$

$$Z_{12} = \left.\frac{V_1}{I_2}\right|_{I_1=0} = \frac{3.1\angle -80^\circ}{0.1\angle 0^\circ} = 31\ \angle -80^\circ\ \Omega$$

Hence, the Z-parameters are

$$\begin{bmatrix} Z_{11} & Z_{12} \\ Z_{21} & Z_{22} \end{bmatrix} = \begin{bmatrix} 52\angle 50^\circ & 31\angle -80^\circ \\ 41\angle -25^\circ & 42\angle 60^\circ \end{bmatrix}$$

Example 11.2 *Find the Z parameters for the network shown in Fig. 11.20.*

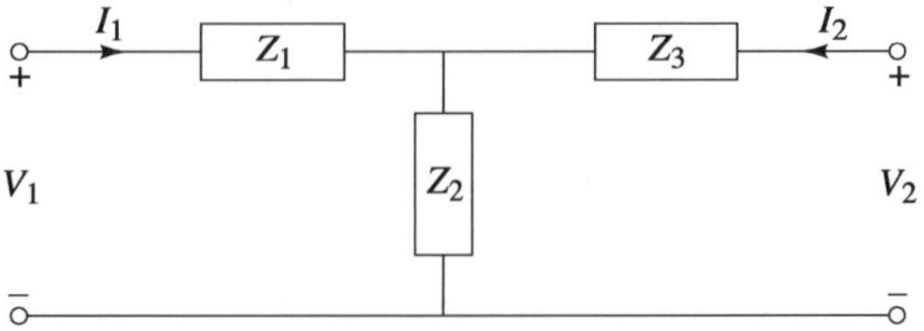

Fig. 11.20

Solution

First Method

Case 1 When the output port is open circuited, i.e., $I_2 = 0$.

Applying KVL to Mesh 1,

$$V_1 = (Z_1 + Z_2)\, I_1$$

$$Z_{11} = \left.\frac{V_1}{I_1}\right|_{I_2=0} = Z_1 + Z_2$$

Also
$$V_2 = Z_2\, I_1$$

$$Z_{21} = \left.\frac{V_2}{I_1}\right|_{I_2=0} = Z_2$$

Case 2 When the input port is open circuited, i.e., $I_1 = 0$.

Applying KVL to Mesh 2,

$$V_2 = (Z_2 + Z_3)\, I_2$$

$$Z_{22} = \left.\frac{V_2}{I_2}\right|_{I_1=0} = Z_2 + Z_3$$

Also $V_1 = Z_2 I_2$

$$Z_{12} = \left.\frac{V_1}{I_2}\right|_{I_1=0} = Z_2$$

Hence, the Z-parameters are

$$\begin{bmatrix} Z_{11} & Z_{12} \\ Z_{21} & Z_{22} \end{bmatrix} = \begin{bmatrix} Z_1 + Z_2 & Z_2 \\ Z_2 & Z_2 + Z_3 \end{bmatrix}$$

Second Method

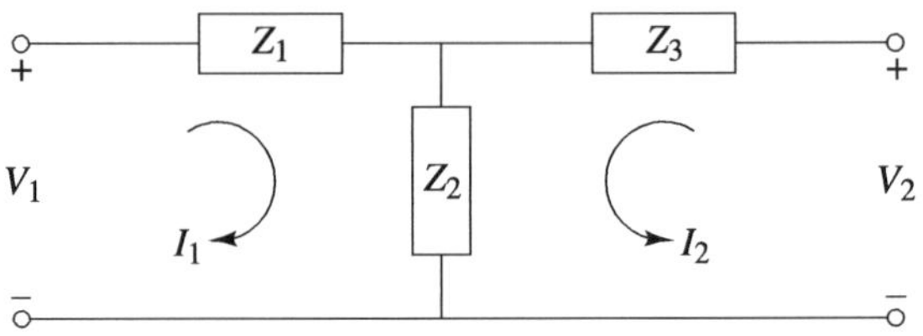

Fig. 11.21

Applying KVL to Mesh 1,

$$V_1 = Z_1 I_1 + Z_2 (I_1 + I_2)$$
$$= (Z_1 + Z_2) I_1 + Z_2 I_2 \quad \ldots\text{(i)}$$

Applying KVL to Mesh 2,

$$V_2 = Z_3 I_2 + Z_2 (I_1 + I_2)$$
$$= Z_2 I_1 + (Z_2 + Z_3) I_2 \quad \ldots\text{(ii)}$$

Comparing Eqs (i) and (ii) with Z-parameter equations, we get

$$\begin{bmatrix} Z_{11} & Z_{12} \\ Z_{21} & Z_{22} \end{bmatrix} = \begin{bmatrix} Z_1 + Z_2 & Z_2 \\ Z_2 & Z_2 + Z_3 \end{bmatrix}$$

Example 11.3 *Find Z-parameters for the network shown in Fig. 11.22.*

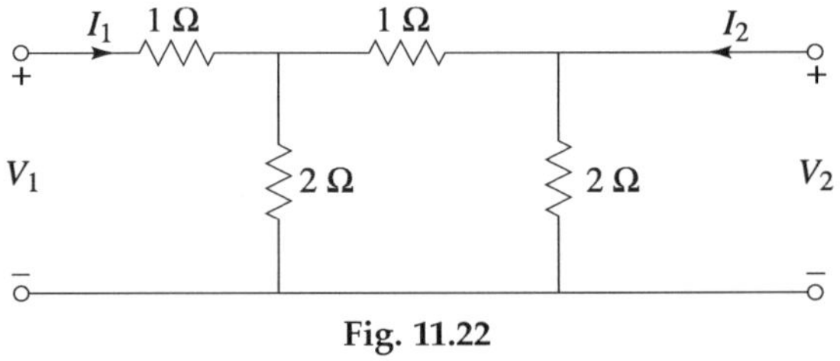

Fig. 11.22

Solution

Applying KVL to Mesh 1,

$$V_1 = 3I_1 - 2I_3 \quad \ldots\text{(i)}$$

Applying KVL to Mesh 2,

$$V_2 = 2I_2 + 2I_3 \quad \ldots\text{(ii)}$$

Applying KVL to Mesh 3,

$$-2I_1 + 2I_2 + 5I_3 = 0$$

$$I_3 = \frac{2}{5} I_1 - \frac{2}{5} I_2 \quad \ldots\text{(iii)}$$

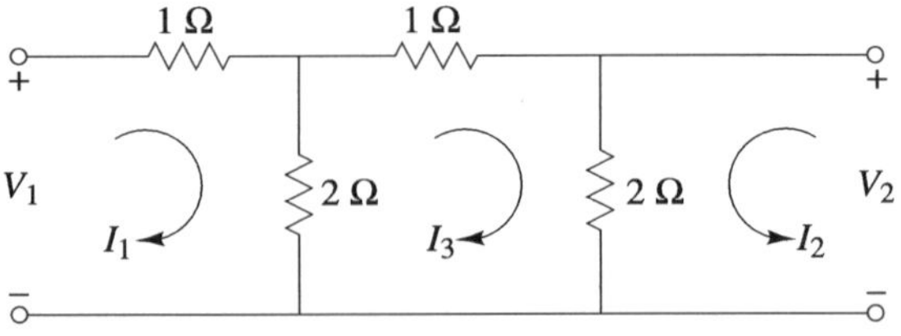

Fig. 11.23

Substituting the Eq. (iii) in the Eq. (i),

$$V_1 = 3I_1 - \frac{4}{5}I_1 + \frac{4}{5}I_2$$

$$= \frac{11}{5}I_1 + \frac{4}{5}I_2 \qquad \text{...(iv)}$$

Substituting the Eq. (iii) in the Eq. (ii),

$$V_2 = 2I_2 + \frac{4}{5}I_1 - \frac{4}{5}I_2$$

$$= \frac{4}{5}I_1 + \frac{6}{5}I_2 \qquad \text{...(v)}$$

Comparing equations (iv) and (*v*) with Z-parameter equations, we get

$$\begin{bmatrix} Z_{11} & Z_{12} \\ Z_{21} & Z_{22} \end{bmatrix} = \begin{bmatrix} 11/5 & 4/5 \\ 4/5 & 6/5 \end{bmatrix}$$

Example 11.4 *Find the Z-parameters for the network shown in Fig. 11.24.*

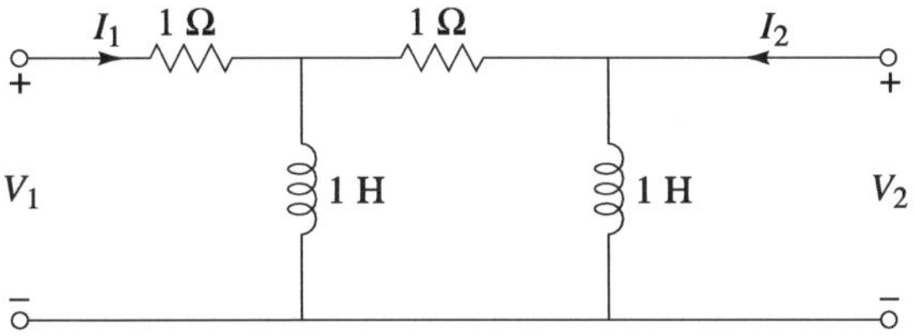

Fig. 11.24

Solution The transformed network is shown in Fig. 11.25.

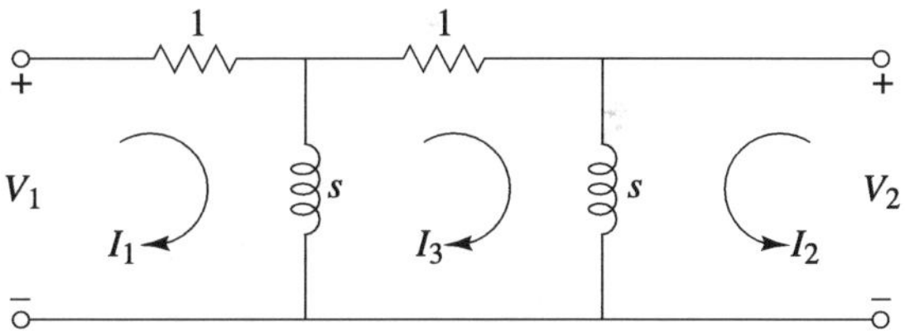

Fig. 11.25

Applying KVL to Mesh 1,

$$V_1 = (s + 1)\, I_1 - sI_3 \qquad \text{...(i)}$$

Applying KVL to Mesh 2,

$$V_2 = sI_2 + sI_3 \qquad \text{...(ii)}$$

Applying KVL to Mesh 3,

$$-sI_1 + sI_2 + (2s + 1)\, I_3 = 0$$

$$I_3 = \frac{s}{2s+1}I_1 - \frac{s}{2s+1}I_2 \qquad \text{...(iii)}$$

Substituting the Eq. (iii) in the Eq. (i),

$$V_1 = (s+1)\, I_1 - s\left(\frac{s}{2s+1}I_1 - \frac{s}{2s+1}I_2\right)$$

$$= \left(\frac{s^2 + 3s + 1}{2s+1}\right)I_1 + \left(\frac{s^2}{2s+1}\right)I_2 \qquad \text{...(iv)}$$

Substituting the Eq. (iii) in the Eq. (ii),

$$V_2 = sI_2 + s\left(\frac{s}{2s+1}I_1 - \frac{s}{2s+1}I_2\right)$$

$$= \left(\frac{s^2}{2s+1}\right)I_1 + \left(\frac{s^2 + s}{2s+1}\right)I_2 \qquad \text{...(v)}$$

Comparing the Eqs (iv) and (v) with Z-parameter equations, we get

$$\begin{bmatrix} Z_{11} & Z_{12} \\ Z_{21} & Z_{22} \end{bmatrix} = \begin{bmatrix} \dfrac{s^2+3s+1}{2s+1} & \dfrac{s^2}{2s+1} \\ \dfrac{s^2}{2s+1} & \dfrac{s^2+s}{2s+1} \end{bmatrix}$$

Example 11.5 *Find the open-circuit impedance parameters for the network shown in Fig. 11.26.*

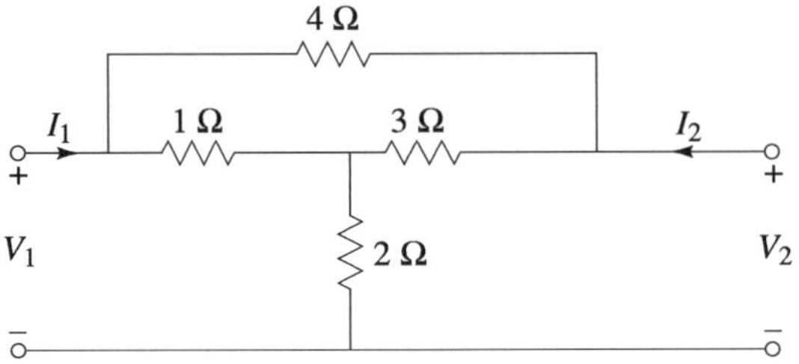

Fig. 11.26

Solution

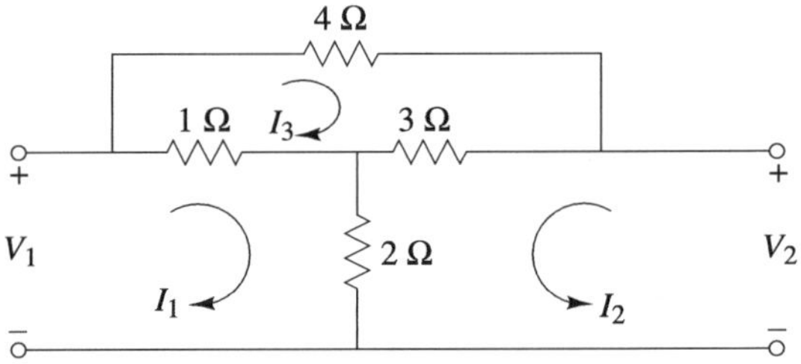

Fig. 11.27

Applying KVL to Mesh 1,

$$V_1 - (I_1 - I_3) - 2(I_1 + I_2) = 0$$

$$V_1 = 3I_1 + 2I_2 - I_3 \quad \text{...(i)}$$

Applying KVL to Mesh 2,

$$V_2 - 3(I_2 + I_3) - 2(I_1 + I_2) = 0$$

$$V_2 = 2I_1 + 5I_2 + 3I_3 \quad \text{...(ii)}$$

Applying KVL to Mesh 3,

$$4I_3 + 3(I_2 + I_3) + (I_3 - I_1) = 0$$

$$-I_1 + 3I_2 + 8I_3 = 0$$

$$I_3 = \frac{1}{8}I_1 - \frac{3}{8}I_2 \quad \text{...(iii)}$$

Substituting the Eq. (iii) in the Eq. (i),

$$V_1 = 3I_1 + 2I_2 - \left(\frac{1}{8}I_1 - \frac{3}{8}I_2\right)$$

$$= \frac{23}{8}I_1 + \frac{19}{8}I_2 \quad \text{...(iv)}$$

Substituting the Eq. (iii) in the Eq. (ii),

$$V_2 = 2I_1 + 5I_2 + 3\left(\frac{1}{8}I_1 - \frac{3}{8}I_2\right)$$

$$= \frac{19}{8}I_1 + \frac{31}{8}I_2 \quad \text{...(v)}$$

Comparing Eqs (iv) and (v) with Z-parameter equations, we get

$$\begin{bmatrix} Z_{11} & Z_{12} \\ Z_{21} & Z_{22} \end{bmatrix} = \begin{bmatrix} \dfrac{23}{8} & \dfrac{19}{8} \\ \dfrac{19}{8} & \dfrac{31}{8} \end{bmatrix}$$

Example 11.6 *Test results for a two-port network are*

(a) Port 2 short circuited : $V_1 = 50\angle 0°$ V, $I_1 = 2.1\angle -30°$ A, $I_2 = -1.1\angle -20°$ A

(b) Port 1 short circuited : $V_2 = 50\angle 0°$ V, $I_2 = 3\angle -15°$ A, $I_1 = -1.1\angle -20°$ A.

Find Y-parameters.

Solution

$$Y_{11} = \left.\frac{I_1}{V_1}\right|_{V_2=0} = \frac{2.1\angle -30°}{50\angle 0°} = 0.042\angle -30°\ \mho$$

$$Y_{21} = \left.\frac{I_2}{V_1}\right|_{V_2=0} = \frac{-1.1\angle -20°}{50\angle 0°} = -0.022\angle -20°\ \mho$$

$$Y_{12} = \left.\frac{I_1}{V_2}\right|_{V_1=0} = \frac{-1.1\angle -20°}{50\angle 0°} = -0.022\angle -20°\ \mho$$

$$Y_{22} = \left.\frac{I_2}{V_2}\right|_{V_1=0} = \frac{3\angle -15°}{50\angle 0°} = 0.06\angle -15°\ \mho$$

Hence, the Y-parameters are

$$\begin{bmatrix} Y_{11} & Y_{12} \\ Y_{21} & Y_{22} \end{bmatrix} = \begin{bmatrix} 0.042\angle -30° & -0.022\angle -20° \\ -0.022\angle -20° & 0.06\angle -15° \end{bmatrix}$$

Example 11.7 *Find Y-parameters for the network shown in Fig. 11.28.*

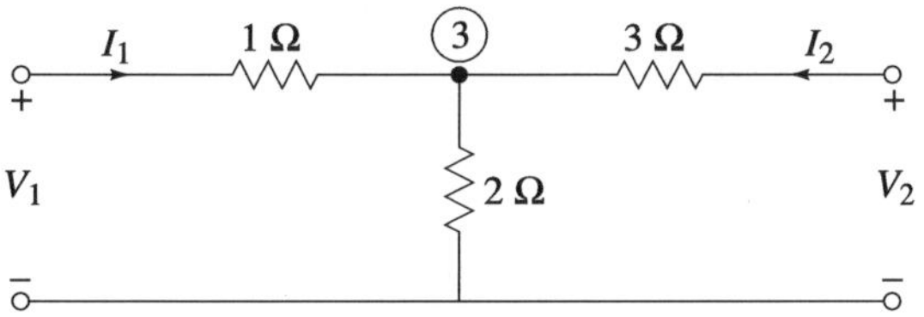

Fig. 11.28

Solution

First Method

Case I When the output port is short circuited, i.e., $V_2 = 0$.

$$R_{eq} = 1 + \frac{2\times 3}{2+3}$$

$$= 1 + \frac{6}{5} = \frac{11}{5}\ \Omega$$

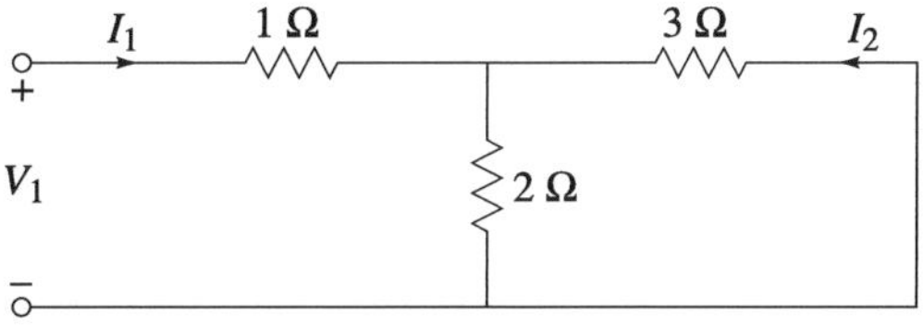

Fig. 11.29

Now
$$V_1 = \frac{11}{5} I_1$$

$$Y_{11} = \left.\frac{I_1}{V_1}\right|_{V_2=0} = \frac{5}{11}\ \mho$$

Also
$$I_2 = \frac{2}{5}(-I_1) = \frac{-2}{5} \times \frac{5}{11} V_1 = \frac{-2}{11} V_1$$

$$Y_{21} = \left.\frac{I_2}{V_1}\right|_{V_2=0} = \frac{-2}{11}\ \mho$$

Case 2 When the input port is short circuited, i.e., $V_1 = 0$.

$$R_{eq} = 3 + \frac{1\times 2}{1+2}$$

$$= 3 + \frac{2}{3} = \frac{11}{3}\ \Omega$$

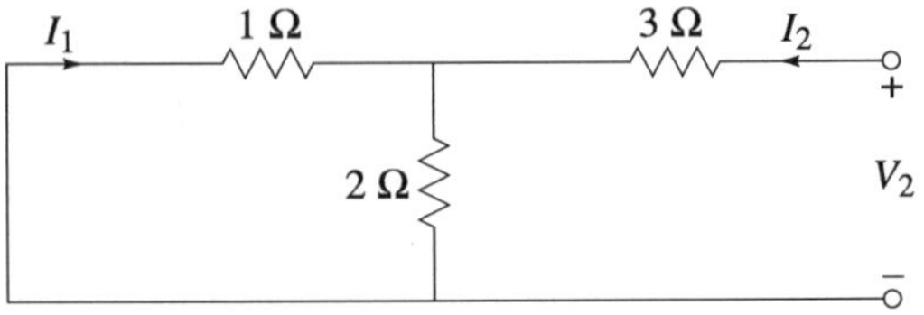

Fig. 11.30

Now
$$V_2 = \frac{11}{3} I_2$$

$$Y_{22} = \left.\frac{I_2}{V_2}\right|_{V_1=0} = \frac{3}{11}\ \mho$$

Also
$$I_1 = \frac{2}{3}(-I_2) = \frac{-2}{3}\times\frac{3}{11}V_2 = \frac{-2}{11}V_2$$

$$Y_{12} = \left.\frac{I_1}{V_2}\right|_{V_1=0} = \frac{-2}{11}\ \mho$$

Hence, the Y-parameters are

$$\begin{bmatrix} Y_{11} & Y_{12} \\ Y_{21} & Y_{22} \end{bmatrix} = \begin{bmatrix} \frac{5}{11} & \frac{-2}{11} \\ \frac{-2}{11} & \frac{3}{11} \end{bmatrix}$$

Second Method (Refer Fig. 11.28)

$$I_1 = \frac{V_1 - V_3}{1} = V_1 - V_3 \qquad \text{...(i)}$$

$$I_2 = \frac{V_2 - V_3}{3}$$

$$= \frac{V_2}{3} - \frac{V_3}{3} \qquad \text{...(ii)}$$

Applying KCL at Node 3,

$$I_1 + I_2 = \frac{V_3}{2} \qquad \text{...(iii)}$$

Substituting Eqs (i) and (ii) in the Eq. (iii),

$$V_1 - V_3 + \frac{V_2}{3} - \frac{V_3}{3} = \frac{V_3}{2}$$

$$V_1 + \frac{V_2}{3} = V_3\left(\frac{1}{2} + \frac{1}{3} + 1\right) = \frac{11}{6}V_3$$

$$V_3 = \frac{6}{11}V_1 + \frac{2}{11}V_2 \qquad \text{...(iv)}$$

Substituting the Eq. (iv) in the Eq. (i),

$$I_1 = V_1 - \frac{6}{11}V_1 - \frac{2}{11}V_2$$

$$= \frac{5}{11}V_1 - \frac{2}{11}V_2 \qquad \text{...(v)}$$

Substituting the Eq. (iv) in Eq. (ii),

$$I_2 = \frac{V_2}{3} - \frac{1}{3}\left(\frac{6}{11}V_1 + \frac{2}{11}V_2\right)$$

$$= \frac{-2}{11}V_1 + \frac{3}{11}V_2 \qquad \ldots\text{(vi)}$$

Comparing Eqs (v) and (vi) with Y-parameter equations, we get

$$\begin{bmatrix} Y_{11} & Y_{12} \\ Y_{21} & Y_{22} \end{bmatrix} = \begin{bmatrix} \frac{5}{11} & \frac{-2}{11} \\ \frac{-2}{11} & \frac{3}{11} \end{bmatrix}$$

Example 11.8 *Determine Y-parameters for the network shown in Fig. 11.31.*

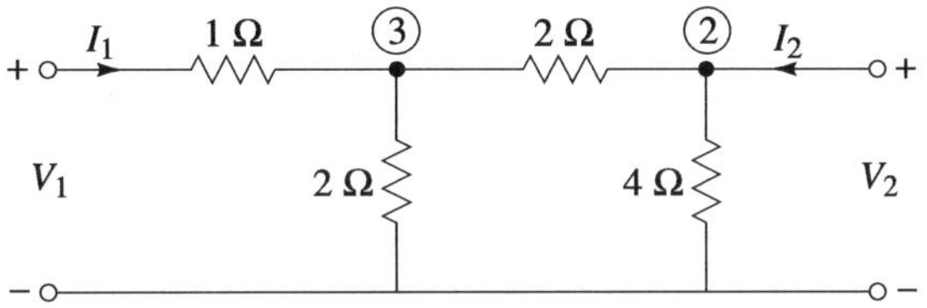

Fig. 11.31

Solution From Fig. 11.31,

$$I_1 = \frac{V_1 - V_3}{1} = V_1 - V_3 \qquad \ldots\text{(i)}$$

Applying KCL at Node 3,

$$I_1 = \frac{V_3}{2} + \frac{V_3 - V_2}{2} = V_3 - \frac{V_2}{2} \qquad \ldots\text{(ii)}$$

Applying KCL at Node 2,

$$I_2 = \frac{V_2}{4} + \frac{V_2 - V_3}{2} = \frac{3}{4}V_2 - \frac{V_3}{2} \qquad \ldots\text{(iii)}$$

Substituting the Eq. (i) in the Eq. (ii),

$$V_1 - V_3 = V_3 - \frac{V_2}{2}$$

$$V_3 = \frac{V_1}{2} + \frac{V_2}{4} \qquad \ldots\text{(iv)}$$

Substituting the Eq. (iv) in the Eq. (ii),

$$I_1 = \frac{V_1}{2} + \frac{V_2}{4} - \frac{V_2}{2} = \frac{V_1}{2} - \frac{V_2}{4} \qquad \ldots\text{(v)}$$

Substituting the Eq. (iv) in the Eq. (iii),

$$I_2 = \frac{3}{4}V_2 - \frac{1}{2}\left(\frac{V_1}{2} + \frac{V_2}{4}\right) = \frac{-V_1}{4} + \frac{5V_2}{8} \qquad \ldots\text{(vi)}$$

Comparing Eqs (v) and (vi) with Y-parameter equations, we get

$$\begin{bmatrix} Y_{11} & Y_{12} \\ Y_{21} & Y_{22} \end{bmatrix} = \begin{bmatrix} \frac{1}{2} & \frac{-1}{4} \\ \frac{-1}{4} & \frac{5}{8} \end{bmatrix}$$

Example 11.9 *Determine the short-circuit admittance parameters for the network shown in Fig. 11.32.*

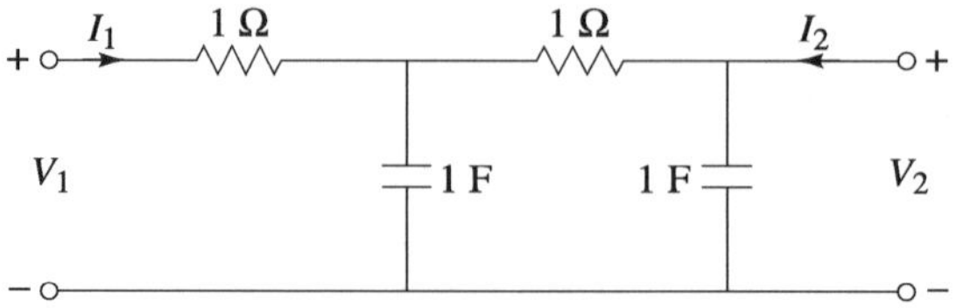

Fig. 11.32

Solution The transformed network is shown in Fig. 11.33.

From Fig. 11.33,

$$I_1 = \frac{V_1 - V_3}{1} = V_1 - V_3 \qquad \text{...(i)}$$

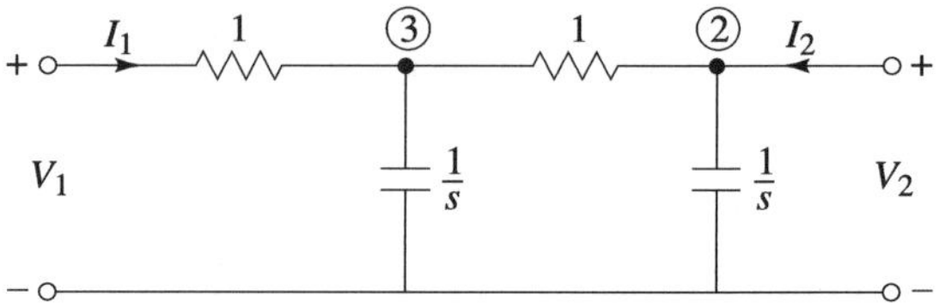

Fig. 11.33

Applying KCL at Node 3,

$$I_1 = \frac{V_3}{\frac{1}{s}} + \frac{(V_3 - V_2)}{1}$$

$$= (s+1)\,V_3 - V_2 \qquad \text{...(ii)}$$

Applying KCL at Node 2,

$$I_2 = \frac{V_2}{\frac{1}{s}} + \frac{(V_2 - V_3)}{1}$$

$$= (s+1)\,V_2 - V_3 \qquad \text{...(iii)}$$

Substituting the Eq. (i) in the Eq. (ii),

$$V_1 - V_3 = (s+1)\,V_3 - V_2$$

$$(s+2)\,V_3 = V_1 + V_2$$

$$V_3 = \frac{1}{s+2}V_1 + \frac{1}{s+2}V_2 \qquad \text{...(iv)}$$

Substituting the Eq. (iv) in the Eq. (ii),

$$I_1 = (s+1)\left(\frac{1}{s+2}V_1 + \frac{1}{s+2}V_2\right) - V_2$$

$$= \frac{s+1}{s+2}V_1 - \frac{1}{s+2}V_2 \qquad \text{...(v)}$$

Substituting the Eq. (iv) in the Eq. (iii),

$$I_2 = (s+1)\,V_2 - \left(\frac{1}{s+2}V_1 + \frac{1}{s+2}V_2\right)$$

$$= \frac{-1}{s+2}V_1 + \frac{s^2+3s+1}{s+2}V_2 \qquad \text{...(vi)}$$

Comparing Eqs (v) and (vi) with *Y*-parameter equations, we get

$$\begin{bmatrix} Y_{11} & Y_{12} \\ Y_{21} & Y_{22} \end{bmatrix} = \begin{bmatrix} \frac{s+1}{s+2} & \frac{-1}{s+2} \\ \frac{-1}{s+2} & \frac{s^2+3s+1}{s+2} \end{bmatrix}$$

Example 11.10 *Determine Y-parameters for the network shown in Fig. 11.34.*

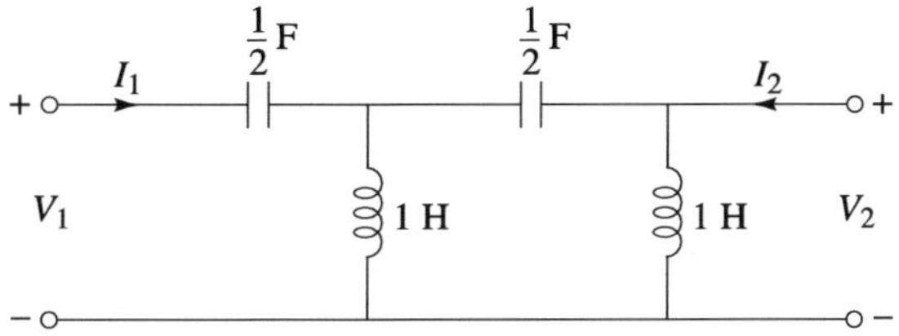

Fig. 11.34

Solution The transformed network is as shown in Fig. 11.35.

From Fig. 11.35,

$$I_1 = \frac{V_1 - V_3}{\frac{2}{s}} = \frac{s}{2}V_1 - \frac{s}{2}V_3 \qquad \ldots\text{(i)}$$

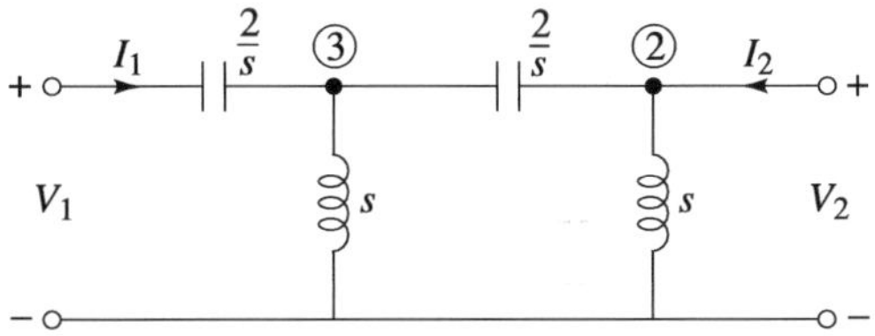

Fig. 11.35

Applying KCL at Node 3,

$$\frac{s}{2}(V_1 - V_3) = \frac{V_3}{s} + \frac{s}{2}(V_3 - V_2)$$

$$\frac{s}{2}V_3 + \frac{1}{s}V_3 + \frac{s}{2}V_3 = \frac{s}{2}V_1 + \frac{s}{2}V_2$$

$$V_3 = \frac{s^2}{2(s^2+1)}V_1 + \frac{s^2}{2(s^2+1)}V_2 \qquad \ldots\text{(ii)}$$

Substituting the Eq. (ii) in the Eq. (i),

$$I_1 = \frac{s}{2}V_1 - \frac{s}{2}\left[\frac{s^2}{2(s^2+1)}V_1 + \frac{s^2}{2(s^2+1)}V_2\right]$$

$$= \left[\frac{s}{2} - \frac{s^3}{4(s^2+1)}\right]V_1 - \frac{s^3}{4(s^2+1)}V_2$$

$$= \frac{s^3+2s}{4(s^2+1)}V_1 - \frac{s^3}{4(s^2+1)}V_2 \qquad \ldots\text{(iii)}$$

Applying KCL at Node 2,

$$I_2 = \frac{V_2}{s} + \frac{s}{2}(V_2 - V_3)$$

$$= \frac{s^2+2}{2s}V_2 - \frac{s}{2}V_3 \qquad \ldots\text{(iv)}$$

Substituting the Eq. (ii) in the Eq. (iv),

$$I_2 = \frac{s^2+2}{2s}V_2 - \frac{s}{2}\left[\frac{s^2}{2(s^2+1)}V_1 + \frac{s^2}{2(s^2+1)}V_2\right]$$

$$= \frac{-s^3}{4(s^2+1)}V_1 + \left[\frac{s^2+2}{2s} - \frac{s^3}{4(s^2+1)}\right]V_2$$

$$= \frac{-s^3}{4(s^2+1)}V_1 + \frac{s^4+6s^2+4}{4s(s^2+1)}V_2 \qquad \ldots\text{(v)}$$

Comparing the Eqs (iii) and (v) with Y-parameter equation, we get

$$\begin{bmatrix} Y_{11} & Y_{12} \\ Y_{21} & Y_{22} \end{bmatrix} = \begin{bmatrix} \frac{s^3 + 2s}{4(s^2+1)} & \frac{-s^3}{4(s^2+1)} \\ \frac{-s^3}{4(s^2+1)} & \frac{s^4 + 6s^2 + 4}{4s(s^2+1)} \end{bmatrix}$$

Example 11.11 *Obtain Y-parameters of the network shown in Fig. 11.36.*

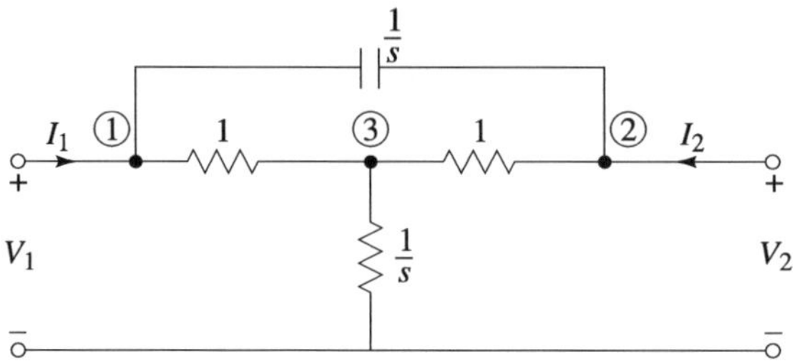

Fig. 11.36

Solution Applying KCL at Node 1,

$$I_1 = \frac{V_1 - V_3}{1} + \frac{V_1 - V_2}{\frac{1}{s}} = (s+1)\,V_1 - s\,V_2 - V_3 \qquad \text{...(i)}$$

Applying KCL at Node 2,

$$I_2 = \frac{V_2 - V_3}{1} + \frac{V_2 - V_1}{\frac{1}{s}} = (s+1)\,V_2 - s\,V_1 - V_3 \qquad \text{...(ii)}$$

Applying KCL at Node 3,

$$\frac{V_3}{\frac{1}{s}} + \frac{V_3 - V_1}{1} + \frac{V_3 - V_2}{1} = 0$$

$$(s+2)\,V_3 - V_1 - V_2 = 0$$

$$V_3 = \frac{1}{s+2} V_1 + \frac{1}{s+2} V_2 \qquad \text{...(iii)}$$

Substituting the equation (iii) in the equation (i),

$$I_1 = (s+1)\,V_1 - s\,V_2 - \left(\frac{1}{s+2} V_1 + \frac{1}{s+2} V_2\right)$$

$$= \left[\frac{(s+1)(s+2) - 1}{(s+2)}\right] V_1 - \left[\frac{s(s+2)+1}{(s+2)}\right] V_2$$

$$= \left(\frac{s^2 + 3s + 1}{s+2}\right) V_1 - \left(\frac{s^2 + 2s + 1}{s+2}\right) V_2 \qquad \text{...(iv)}$$

Substituting the Eq. (iii) in the Eq. (ii),

$$I_2 = (s+1)\,V_2 - s\,V_1 - \left(\frac{1}{s+2}V_1 + \frac{1}{s+2}V_2\right)$$

$$= -\left[\frac{s(s+2)+1}{(s+2)}\right]V_1 + \left[\frac{(s+1)(s+2)-1}{(s+2)}\right]V_2$$

$$= -\left(\frac{s^2+2s+1}{s+2}\right)V_1 + \left(\frac{s^2+3s+1}{s+2}\right)V_2 \qquad \text{...(v)}$$

Comparing Eqs (iv) and (v) with Y-parameter equations, we get

$$\begin{bmatrix} Y_{11} & Y_{12} \\ Y_{21} & Y_{22} \end{bmatrix} = \begin{bmatrix} \dfrac{s^2+3s+1}{s+2} & \dfrac{-(s^2+2s+1)}{s+2} \\ \dfrac{-(s^2+2s+1)}{s+2} & \dfrac{s^2+3s+1}{s+2} \end{bmatrix}$$

Example 11.12 *Find the transmission parameters for the network shown in Fig. 11.37.*

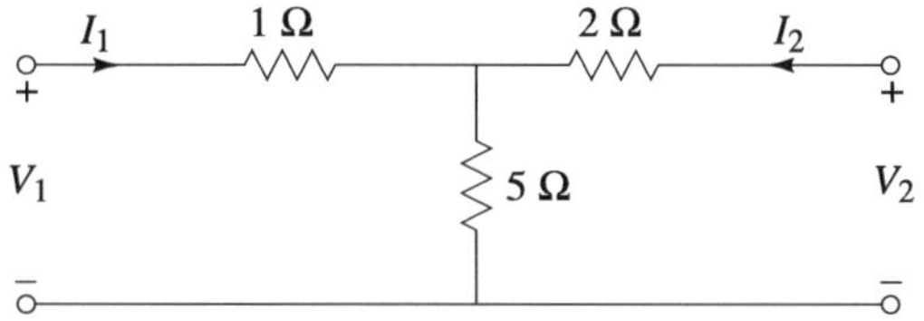

Fig. 11.37

Solution

First Method

Case 1 When the output port is open circuited, i.e., $I_2 = 0$.

$$V_1 = 6I_1$$

and

$$V_2 = 5I_1$$

$$A = \left.\frac{V_1}{V_2}\right|_{I_2=0} = \frac{6I_1}{5I_1} = \frac{6}{5}$$

$$C = \left.\frac{I_1}{V_2}\right|_{I_2=0} = \frac{1}{5}\ \text{℧}$$

Case 2 When the output port is short circuited, i.e., $V_2 = 0$.

$$R_{eq} = 1 + \frac{5\times 2}{5+2} = 1 + \frac{10}{7} = \frac{17}{7}\ \Omega$$

Now

$$V_1 = \frac{17}{7}I_1$$

and

$$I_2 = \frac{5}{7}(-I_1) = \frac{-5}{7}I_1$$

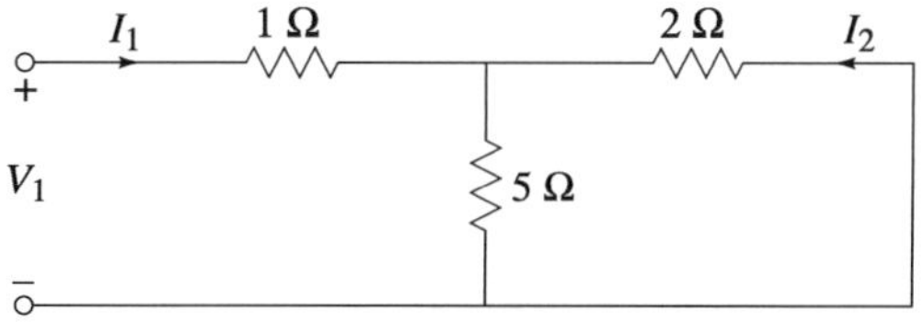

Fig. 11.38

$$B = -\left.\frac{V_1}{I_2}\right|_{V_2=0} = -\frac{\frac{17}{7}I_1}{\frac{-5}{7}I_1} = \frac{17}{5}\ \Omega$$

$$D = -\left.\frac{I_1}{I_2}\right|_{V_2=0} = \frac{7}{5}$$

Hence, the transmission parameters are

$$\begin{bmatrix} A & B \\ C & D \end{bmatrix} = \begin{bmatrix} \frac{6}{5} & \frac{17}{5} \\ \frac{1}{5} & \frac{7}{5} \end{bmatrix}$$

Second Method (Refer Fig. 11.37)

Applying KVL to Mesh 1,

$$V_1 = I_1 + 5\,(I_1 + I_2) = 6I_1 + 5I_2 \quad \text{...(i)}$$

Applying KVL to Mesh 2,

$$V_2 = 2I_2 + 5\,(I_1 + I_2) = 5I_1 + 7I_2 \quad \text{...(ii)}$$

Hence,

$$5I_1 = V_2 - 7I_2$$

$$I_1 = \frac{1}{5}V_2 - \frac{7}{5}I_2 \quad \text{...(iii)}$$

Substituting the Eq. (iii) in the Eq. (i),

$$V_1 = 6\left(\frac{1}{5}V_2 - \frac{7}{5}I_2\right) + 5I_2$$

$$= \frac{6}{5}V_2 - \frac{17}{5}I_2 \quad \text{...(iv)}$$

Comparing Eqs (iii) and (iv) with transmission parameter equations, we get

$$\begin{bmatrix} A & B \\ C & D \end{bmatrix} = \begin{bmatrix} \frac{6}{5} & \frac{17}{5} \\ \frac{1}{5} & \frac{7}{5} \end{bmatrix}$$

Example 11.13 *Obtain ABCD parameters for the network shown in Fig. 11.39.*

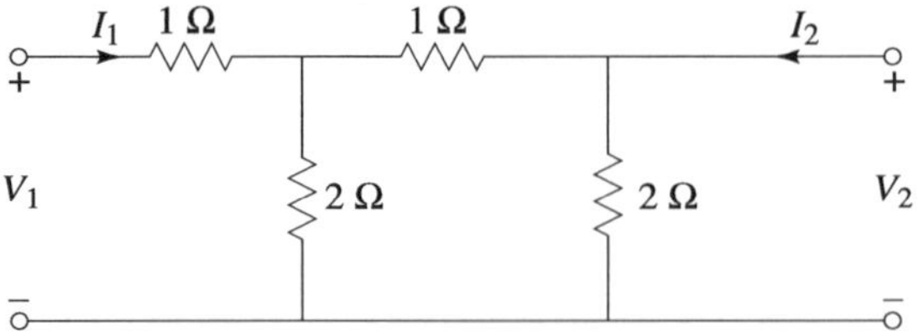

Fig. 11.39

Solution Applying KVL to Mesh 1,

$$V_1 = I_1 + 2\,(I_1 - I_3) = 3I_1 - 2I_3 \quad \text{...(i)}$$

Applying KVL to Mesh 2,

$$V_2 = 2\,(I_2 + I_3) = 2I_2 + 2I_3 \quad \text{...(ii)}$$

Applying KVL to Mesh 3,

$$2(I_3 - I_1) + I_3 + 2(I_3 + I_2) = 0$$

$$5I_3 = 2I_1 - 2I_2$$

$$I_3 = \frac{2}{5}I_1 - \frac{2}{5}I_2 \qquad \text{...(iii)}$$

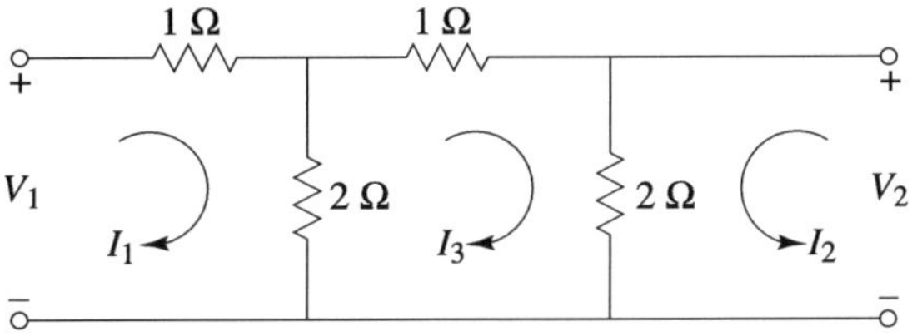

Fig. 11.40

Substituting the Eq. (iii) in the Eq. (i),

$$V_1 = 3I_1 - 2\left(\frac{2}{5}I_1 - \frac{2}{5}I_2\right)$$

$$= \frac{11}{5}I_1 + \frac{4}{5}I_2 \qquad \text{...(iv)}$$

Substituting the Eq. (iii) in the Eq. (ii),

$$V_2 = 2I_2 + 2\left(\frac{2}{5}I_1 - \frac{2}{5}I_2\right) = \frac{4}{5}I_1 + \frac{6}{5}I_2$$

$$\frac{4}{5}I_1 = V_2 - \frac{6}{5}I_2$$

$$I_1 = \frac{5}{4}V_2 - \frac{3}{2}I_2 \qquad \text{...(v)}$$

Substituting the Eq. (v) in the Eq. (iv),

$$V_1 = \frac{11}{5}\left(\frac{5}{4}V_2 - \frac{3}{2}I_2\right) + \frac{4}{5}I_2 = \frac{11}{4}V_2 - \frac{5}{2}I_2 \qquad \text{...(vi)}$$

Comparing Eqs (v) and (vi) with *ABCD* parameter equations, we get

$$\begin{bmatrix} A & B \\ C & D \end{bmatrix} = \begin{bmatrix} \frac{11}{4} & \frac{5}{2} \\ \frac{5}{4} & \frac{3}{2} \end{bmatrix}$$

Example 11.14 *Determine the transmission parameters for the network shown in Fig. 11.41.*

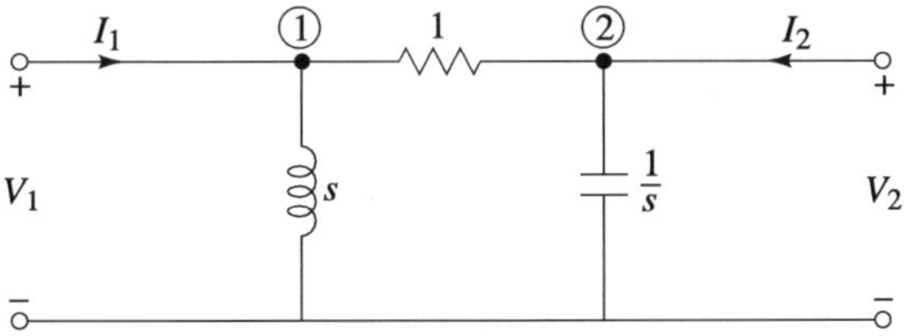

Fig. 11.41

Solution Applying KCL at Node 1,

$$I_1 = \frac{V_1}{s} + (V_1 - V_2) = \frac{s+1}{s}V_1 - V_2 \qquad \text{...(i)}$$

Applying KCL at Node 2,

$$I_2 = \frac{V_2}{\frac{1}{s}} + (V_2 - V_1)$$

$$I_2 = (s+1)V_2 - V_1$$

$$V_1 = (s+1)V_2 - I_2 \qquad \text{...(ii)}$$

Substituting the Eq. (ii) in the Eq. (i),

$$I_1 = \frac{s+1}{s}\,[(s+1)\,V_2 - I_2] - V_2$$

$$= \left[\frac{(s+1)^2}{s} - 1\right]V_2 - \frac{s+1}{s}\,I_2$$

$$= \left(\frac{s^2+s+1}{s}\right)V_2 - \left(\frac{s+1}{s}\right)I_2 \qquad \text{...(iii)}$$

Comparing Eqs (ii) and (iii) with *ABCD* parameter equations, we get

$$\begin{bmatrix} A & B \\ C & D \end{bmatrix} = \begin{bmatrix} s+1 & -1 \\ \dfrac{s^2+s+1}{s} & \dfrac{s+1}{s} \end{bmatrix}$$

Example 11.15 *Find transmission parameters for the two-port network shown in Fig. 11.42.*

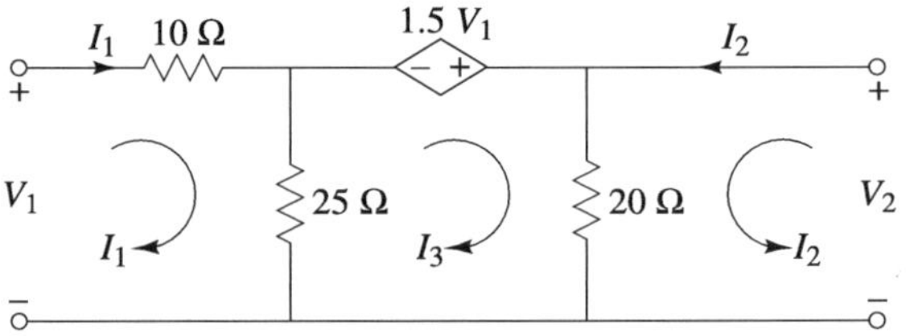

Fig. 11.42

Solution Applying KVL to Mesh 1,

$$V_1 = 10I_1 + 25(I_1 - I_3) = 35I_1 - 25I_3 \qquad \text{...(i)}$$

Applying KVL to Mesh 2,

$$V_2 = 20(I_2 + I_3) = 20I_2 + 20I_3 \qquad \text{...(ii)}$$

Applying KVL to Mesh 3,

$$25(I_3 - I_1) - 1.5V_1 + 20(I_2 + I_3) = 0$$

$$25I_3 - 25I_1 - 1.5\,(35I_1 - 25I_3) + 20I_2 + 20I_3 = 0$$

$$25I_3 - 25I_1 - 52.5I_1 + 37.5I_3 + 20I_2 + 20I_3 = 0$$

$$82.5I_3 = 77.5I_1 - 20I_2$$

$$I_3 = 0.94I_1 - 0.24I_2 \qquad \text{...(iii)}$$

Substituting the Eq. (iii) in Eq. (i),

$$V_1 = 35I_1 - 25(0.94I_1 - 0.24I_2) = 11.5I_1 + 6.1I_2 \qquad \text{...(iv)}$$

Substituting the Eq. (iii) in the Eq. (ii),

$$V_2 = 20I_2 + 20\,(0.94I_1 - 0.24I_2) = 18.8I_1 + 15.2I_2 \qquad \text{...(v)}$$

From the Eq. (v),

$$I_1 = 0.053V_2 - 0.81I_2 \qquad \text{...(vi)}$$

Substituting the Eq. (vi) in the Eq. (iv),

$$V_1 = 11.5\,(0.053V_2 - 0.81I_2) + 6.1I_2 = 0.61V_2 + 3.22I_2 \qquad \text{...(vii)}$$

Comparing Eqs (vi) and (vii) with *ABCD* parameter equations, we get

$$\begin{bmatrix} A & B \\ C & D \end{bmatrix} = \begin{bmatrix} 0.61 & 3.22 \\ 0.053 & 0.81 \end{bmatrix}$$

Example 11.16 *In the two-port network shown in Fig. 11.43, compute h-parameters from the following data:*

(a) *With the output port short-circuited :* $V_1 = 25$ V, $I_1 = 1$ A, $I_2 = 2$ A

(b) *With the input port open-circuited :* $V_1 = 10$ V, $V_2 = 50$ V, $I_2 = 2$ A

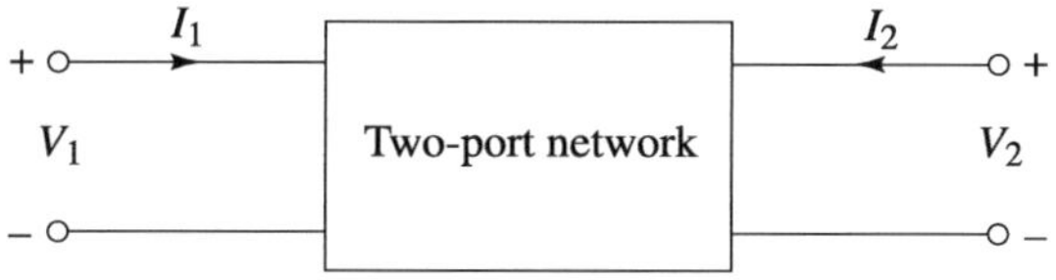

Fig. 11.43

Solution

$$h_{11} = \left.\frac{V_1}{I_1}\right|_{V_2=0} = \frac{25}{1} = 25\ \Omega$$

$$h_{21} = \left.\frac{I_2}{I_1}\right|_{V_2=0} = \frac{2}{1} = 2$$

$$h_{12} = \left.\frac{V_1}{V_2}\right|_{I_1=0} = \frac{10}{50} = 0.2$$

$$h_{22} = \left.\frac{I_2}{V_2}\right|_{I_1=0} = \frac{2}{50} = 0.04\ \mho$$

Hence, the h-parameters are

$$\begin{bmatrix} h_{11} & h_{12} \\ h_{21} & h_{22} \end{bmatrix} = \begin{bmatrix} 25 & 0.2 \\ 2 & 0.04 \end{bmatrix}$$

Example 11.17 *Determine hybrid parameters for the network of Fig. 11.44.*

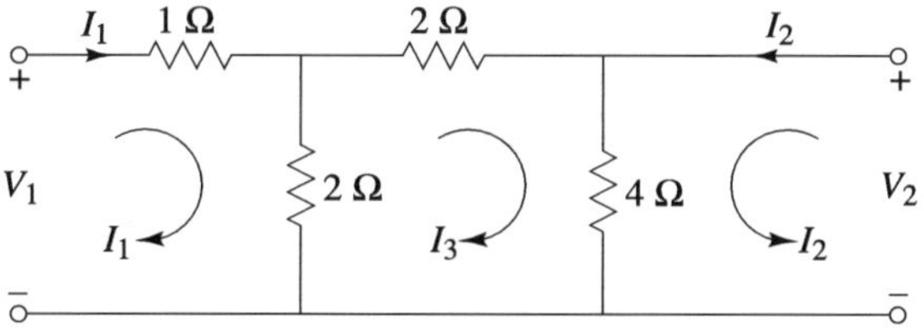

Fig. 11.44

Solution

First Method

Case I When port 2 is short circuited, i.e., $V_2 = 0$.

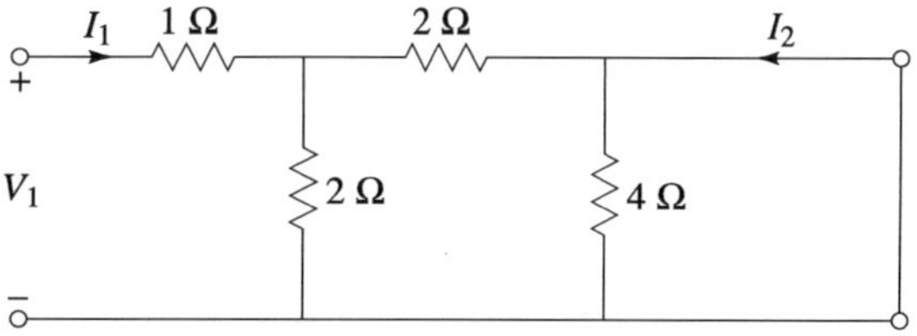

Fig. 11.45

$$R_{eq} = 1 + \frac{2 \times 2}{2 + 2} = 2\ \Omega$$

Now, $V_1 = 2I_1$

$$h_{11} = \left.\frac{V_1}{I_1}\right|_{V_2=0} = 2\ \Omega$$

Also,
$$I_2 = -I_1 \times \frac{2}{2+2} = \frac{-I_1}{2}$$
$$h_{21} = \left.\frac{I_2}{I_1}\right|_{V_2=0} = \frac{-1}{2}$$

Case 2 When Port 1 is open circuited, i.e., $I_1 = 0$.
$$R_{eq} = \frac{(2+2)\times 4}{2+2+4} = 2\ \Omega$$
$$V_1 = 2I_y$$
$$I_y = \frac{I_2}{2}$$
$$V_2 = 4I_x$$
$$I_x = \frac{I_2}{2}$$

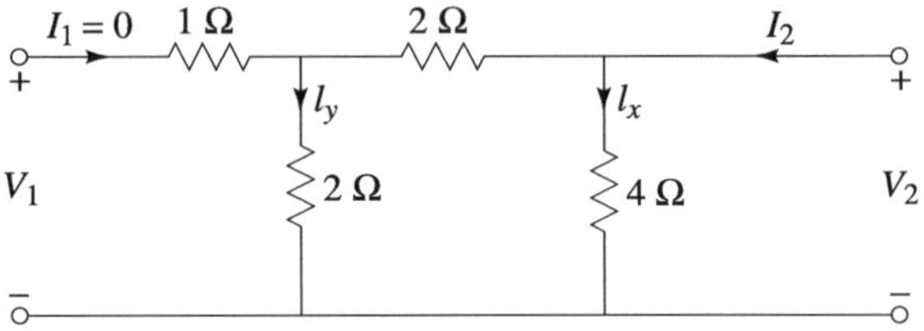

Fig. 11.46

$$h_{12} = \left.\frac{V_1}{V_2}\right|_{I_1=0} = \frac{2I_y}{4I_x} = \frac{2\times\frac{I_2}{2}}{4\times\frac{I_2}{2}} = \frac{1}{2}$$
$$h_{22} = \left.\frac{I_2}{V_2}\right|_{I_1=0} = \frac{2I_x}{4I_x} = \frac{1}{2}\ \mho$$

Hence, the h-parameters are
$$\begin{bmatrix} h_{11} & h_{12} \\ h_{21} & h_{22} \end{bmatrix} = \begin{bmatrix} 2 & \frac{1}{2} \\ \frac{-1}{2} & \frac{1}{2} \end{bmatrix}$$

Second Method Refer Fig. 11.44.

Applying KVL to Mesh 1,
$$V_1 = 3I_1 - 2I_3 \quad \ldots\text{(i)}$$
Applying KVL to Mesh 2,
$$V_2 = 4I_2 + 4I_3 \quad \ldots\text{(ii)}$$
Applying KVL to Mesh 3,
$$2(I_3 - I_1) + 2I_3 + 4(I_3 + I_2) = 0$$
$$8I_3 = 2I_1 - 4I_2$$
$$I_3 = \frac{I_1}{4} - \frac{I_2}{2} \quad \ldots\text{(iii)}$$
Substituting the Eq. (iii) in the Eq. (i),
$$V_1 = 3I_1 - 2\left(\frac{I_1}{4} - \frac{I_2}{2}\right) = \frac{5}{2}I_1 + I_2 \quad \ldots\text{(iv)}$$
Substituting the Eq. (iii) in the Eq. (ii),
$$V_2 = 4I_2 + 4\left(\frac{I_1}{4} - \frac{I_2}{2}\right)$$
$$= 4I_2 + I_1 - 2I_2 = I_1 + 2I_2$$

$$I_2 = -\frac{1}{2} I_1 + \frac{1}{2} V_2 \qquad \text{...(v)}$$

Substituting the Eq. (v) in the Eq. (iv),

$$V_1 = \frac{5}{2} I_1 - \frac{1}{2} I_1 + \frac{1}{2} V_2 = 2I_1 + \frac{1}{2} V_2 \qquad \text{...(vi)}$$

Comparing Eqs (v) and (vi) with h-parameter equations, we get

$$\begin{bmatrix} h_{11} & h_{12} \\ h_{21} & h_{22} \end{bmatrix} = \begin{bmatrix} 2 & \frac{1}{2} \\ \frac{-1}{2} & \frac{1}{2} \end{bmatrix}$$

Example 11.18 *Find h-parameters for the network shown in Fig. 11.47.*

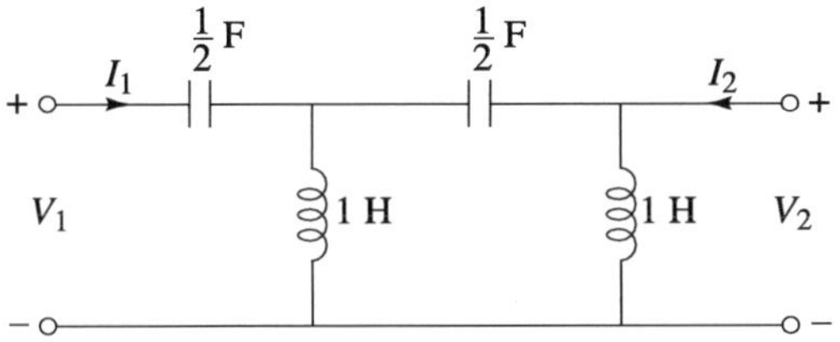

Fig. 11.47

Solution As solved in Example 11.10, derive the equations for I_1 and I_2 in terms of V_1 and V_2.

$$I_1 = \frac{s^3 + 2s}{4(s^2+1)} V_1 - \frac{s^3}{4(s^2+1)} V_2 \qquad \text{...(i)}$$

$$I_2 = \frac{-s^3}{4(s^2+1)} V_1 + \frac{s^4 + 6s^2 + 4}{4s(s^2+1)} V_2 \qquad \text{...(ii)}$$

From the Eq. (i),

$$V_1 = \frac{4(s^2+1)}{s(s^2+2)} I_1 + \frac{s^2}{s^2+2} V_2 \qquad \text{...(iii)}$$

Substituting the Eq. (iii) in the Eq. (ii),

$$I_2 = \frac{-s^3}{4(s^2+1)} \left[\frac{4(s^2+1)}{s(s^2+2)} I_1 + \frac{s^2}{s^2+2} V_2 \right] + \frac{s^4 + 6s^2 + 4}{4s(s^2+1)} V_2$$

$$= \frac{-s^2}{s^2+2} I_1 + \frac{2(s^2+1)}{s(s^2+2)} V_2 \qquad \text{...(iv)}$$

Comparing the Eqs (iii) and (iv) with h-parameter equations, we get

$$\begin{bmatrix} h_{11} & h_{12} \\ h_{21} & h_{22} \end{bmatrix} = \begin{bmatrix} \frac{4(s^2+1)}{s(s^2+2)} & \frac{s^2}{s^2+2} \\ \frac{-s^2}{s^2+2} & \frac{2(s^2+1)}{s(s^2+2)} \end{bmatrix}$$

11.8 INTERRELATIONSHIPS BETWEEN THE PARAMETERS

When it is required to find out two or more parameters of a particular network then finding each parameter will be tedious. But if we find a particular parameter then the other parameters can be found if the interrelationship between them is known.

11.8.1 Z-parameters in Terms of Other Parameters

(i) Z-parameters in terms of Y-parameters We know that

$$I_1 = Y_{11} V_1 + Y_{12} V_2$$
$$I_2 = Y_{21} V_1 + Y_{22} V_2$$

By Cramer's rule,

$$V_1 = \frac{\begin{vmatrix} I_1 & Y_{12} \\ I_2 & Y_{22} \end{vmatrix}}{\begin{vmatrix} Y_{11} & Y_{12} \\ Y_{21} & Y_{22} \end{vmatrix}} = \frac{Y_{22}I_1 - Y_{12}I_2}{Y_{11}Y_{22} - Y_{12}Y_{21}}$$

$$= \frac{Y_{22}}{\Delta Y} I_1 - \frac{Y_{12}}{\Delta Y} I_2$$

where $\Delta Y = Y_{11}Y_{22} - Y_{12}Y_{21}$

Comparing with $V_1 = Z_{11} I_1 + Z_{12} I_2$, we get

$$Z_{11} = \frac{Y_{22}}{\Delta Y} \quad \text{...(11.61)}$$

$$Z_{12} = \frac{-Y_{12}}{\Delta Y} \quad \text{...(11.62)}$$

Also,

$$V_2 = \frac{\begin{vmatrix} Y_{11} & I_1 \\ Y_{21} & I_2 \end{vmatrix}}{\Delta Y} = \frac{Y_{11}}{\Delta Y} I_2 - \frac{Y_{21}}{\Delta Y} I_1$$

Comparing with $V_2 = Z_{21} I_1 + Z_{22} I_2$, we get

$$Z_{22} = \frac{Y_{11}}{\Delta Y} \quad \text{...(11.63)}$$

$$Z_{21} = \frac{-Y_{21}}{\Delta Y} \quad \text{...(11.64)}$$

(ii) Z-parameters in terms of ABCD parameters We know that

$$V_1 = AV_2 - BI_2$$
$$I_1 = CV_2 - DI_2$$

Rewriting the second equation,

$$V_2 = \frac{1}{C} I_1 + \frac{D}{C} I_2$$

Comparing with $V_2 = Z_{21} I_1 + Z_{22} I_2$, we get

$$Z_{21} = \frac{1}{C} \quad \text{...(11.65)}$$

$$Z_{22} = \frac{D}{C} \quad \text{...(11.66)}$$

Also, $$V_1 = A\left[\frac{1}{C}I_1 + \frac{D}{C}I_2\right] - BI_2$$

$$= \frac{A}{C}I_1 + \left[\frac{AD}{C} - B\right]I_2$$

$$= \frac{A}{C}I_1 + \left[\frac{AD - BC}{C}\right]I_2$$

Comparing with $V_1 = Z_{11}I_1 + Z_{12}I_2$, we get

$$Z_{11} = \frac{A}{C} \qquad \text{...(11.67)}$$

$$Z_{12} = \frac{AD - BC}{C} \qquad \text{...(11.68)}$$

(iii) Z-parameters in terms of A′B′C′D′ parameters We know that

$$V_2 = A'V_1 - B'I_1$$
$$I_2 = C'V_1 - D'I_1$$

Rewriting the second equation,

$$V_1 = \frac{D'}{C'}I_1 + \frac{1}{C'}I_2$$

Comparing with $V_1 = Z_{11}I_1 + Z_{12}I_2$, we get

$$Z_{11} = \frac{D'}{C'} \qquad \text{...(11.69)}$$

$$Z_{12} = \frac{1}{C'} \qquad \text{...(11.70)}$$

Also, $$V_2 = A'\left[\frac{D'}{C'}I_1 + \frac{1}{C'}I_2\right] - B'I_1 = \left[\frac{A'D' - B'C'}{C'}\right]I_1 + \frac{A'}{C'}I_2$$

Comparing with $V_2 = Z_{21}I_1 + Z_{22}I_2$, we get

$$Z_{21} = \frac{A'D' - B'C'}{C'} = \frac{\Delta T'}{C'} \qquad \text{...(11.71)}$$

$$Z_{22} = \frac{A'}{C'} \qquad \text{...(11.72)}$$

(iv) Z-parameters in terms of hybrid parameters We know that

$$V_1 = h_{11}I_1 + h_{12}V_2$$
$$I_2 = h_{21}I_1 + h_{22}V_2$$

Rewriting the second equation

$$V_2 = \frac{-h_{21}}{h_{22}}I_1 + \frac{1}{h_{22}}I_2$$

Comparing with $V_2 = Z_{21}I_1 + Z_{22}I_2$, we get

$$Z_{21} = \frac{-h_{21}}{h_{22}} \qquad \text{...(11.73)}$$

$$Z_{22} = \frac{1}{h_{22}} \qquad \text{...(11.74)}$$

Also, $$V_1 = h_{11} I_1 + h_{12}\left[\frac{-h_{21}}{h_{22}} I_1 + \frac{1}{h_{22}} I_2\right]$$

$$= h_{11} I_1 + \frac{h_{12}}{h_{22}} I_2 - \frac{h_{12}h_{21}}{h_{22}} I_1$$

$$= \left[\frac{h_{11}h_{22} - h_{12}h_{21}}{h_{22}}\right] I_1 + \frac{h_{12}}{h_{22}} I_2$$

Comparing with $V_1 = Z_{11} I_1 + Z_{12} I_2$, we get

$$Z_{11} = \frac{h_{11}h_{22} - h_{12}h_{21}}{h_{22}} = \frac{\Delta h}{h_{22}} \qquad \text{...(11.75)}$$

$$Z_{12} = \frac{h_{12}}{h_{22}} \qquad \text{...(11.76)}$$

(v) Z-parameters in terms of inverse hybrid parameters We know that

$$I_1 = g_{11} V_1 + g_{12} I_2$$
$$V_2 = g_{21} V_1 + g_{22} I_2$$

Rewriting first equation,

$$V_1 = \frac{1}{g_{11}} I_1 - \frac{g_{12}}{g_{11}} I_2$$

Comparing with $V_1 = Z_{11} I_1 + Z_{12} I_2$, we get

$$Z_{11} = \frac{1}{g_{11}} \qquad \text{...(11.77)}$$

$$Z_{12} = \frac{-g_{12}}{g_{11}} \qquad \text{...(11.78)}$$

Also $$V_2 = g_{21}\left[\frac{1}{g_{11}} I_1 - \frac{g_{12}}{g_{11}} I_2\right] + g_{22} I_2$$

$$= \frac{g_{21}}{g_{11}} I_1 + \left[\frac{g_{11}g_{22} - g_{12}g_{21}}{g_{11}}\right] I_2$$

Comparing with $V_2 = Z_{21} I_1 + Z_{22} I_2$, we get

$$Z_{21} = \frac{g_{21}}{g_{11}} \qquad \text{...(11.79)}$$

$$Z_{22} = \frac{g_{11}g_{22} - g_{12}g_{21}}{g_{11}} = \frac{\Delta g}{g_{11}} \qquad \text{...(11.80)}$$

11.8.2 Y-parameters in Terms of Other Parameters

(i) Y-parameters in terms of Z-parameters We know that

$$V_1 = Z_{11} I_1 + Z_{12} I_2$$
$$V_2 = Z_{21} I_1 + Z_{22} I_2$$

By Cramer's rule,

$$I_1 = \frac{\begin{vmatrix} V_1 & Z_{12} \\ V_2 & Z_{22} \end{vmatrix}}{\begin{vmatrix} Z_{11} & Z_{12} \\ Z_{21} & Z_{22} \end{vmatrix}} = \frac{Z_{22}V_1 - Z_{12}V_2}{Z_{11}Z_{22} - Z_{12}Z_{21}}$$

$$= \frac{Z_{22}}{\Delta Z} V_1 - \frac{Z_{12}}{\Delta Z} V_2$$

where $\Delta Z = Z_{11} Z_{22} - Z_{12} Z_{21}$

Comparing with $I_1 = Y_{11} V_1 + Y_{12} V_2$, we get

$$Y_{11} = \frac{Z_{22}}{\Delta Z} \quad \text{...(11.81)}$$

$$Y_{12} = \frac{-Z_{12}}{\Delta Z} \quad \text{...(11.82)}$$

Also,

$$I_2 = \frac{\begin{vmatrix} Z_{11} & V_1 \\ Z_{21} & V_2 \end{vmatrix}}{\Delta Z}$$

$$= \frac{Z_{11}V_2 - Z_{21}V_1}{\Delta Z} = \frac{-Z_{21}}{\Delta Z} V_1 + \frac{Z_{11}}{\Delta Z} V_2$$

Comparing with $I_2 = Y_{21} V_1 + Y_{22} V_2$, we get

$$Y_{21} = \frac{-Z_{21}}{\Delta Z} \quad \text{...(11.83)}$$

$$Y_{22} = \frac{Z_{11}}{\Delta Z} \quad \text{...(11.84)}$$

(ii) Y-parameters in terms of ABCD parameters We know that

$$V_1 = AV_2 - BI_2$$

$$I_1 = CV_2 - DI_2$$

Rewriting the first equation,

$$I_2 = -\frac{1}{B} V_1 + \frac{A}{B} V_2$$

Comparing with $I_2 = Y_{21} V_1 + Y_{22} V_2$, we get

$$Y_{21} = \frac{-1}{B} \quad \text{...(11.85)}$$

$$Y_{22} = \frac{A}{B} \quad \text{...(11.86)}$$

Also,

$$I_1 = CV_2 - D\left[\frac{A}{B} V_2 - \frac{1}{B} V_1\right]$$

$$= \frac{D}{B} V_1 + \left[\frac{BC - AD}{B}\right] V_2$$

Comparing with $I_1 = Y_{11} V_1 + Y_{12} V_2$, we get

$$Y_{11} = \frac{D}{B} \quad \text{...(11.87)}$$

$$Y_{12} = \frac{BC - AD}{B} = -\frac{AD - BC}{B} = -\frac{\Delta T}{B} \qquad \text{...(11.88)}$$

(iii) Y-parameters in terms of A′B′C′D′ parameters We know that

$$V_2 = A' V_1 - B' I_1$$
$$I_2 = C' V_1 - D' I_1$$

Rewriting first equation,

$$I_1 = \frac{A'}{B'} V_1 - \frac{1}{B'} V_2$$

Comparing with $I_1 = Y_{11} V_1 + Y_{12} V_2$, we get

$$Y_{11} = \frac{A'}{B'} \qquad \text{...(11.89)}$$

$$Y_{12} = -\frac{1}{B'} \qquad \text{...(11.90)}$$

Also,

$$I_2 = C' V_1 - D' \left[\frac{A'}{B'} V_1 - \frac{1}{B'} V_2\right]$$

$$= -\left[\frac{A'D' - B'C'}{B'}\right] V_1 + \frac{D'}{B'} V_2$$

Comparing with $I_2 = Y_{21} V_1 + Y_{22} V_2$, we get

$$Y_{21} = -\left[\frac{A'D' - B'C'}{B'}\right] = -\frac{\Delta T'}{B'} \qquad \text{...(11.91)}$$

and

$$Y_{22} = \frac{D'}{B'} \qquad \text{...(11.92)}$$

(iv) Y-parameters in terms of hybrid parameters We know that

$$V_1 = h_{11} I_1 + h_{12} V_2$$
$$I_2 = h_{21} I_1 + h_{22} V_2$$

Rewriting the first equation,

$$I_1 = \frac{1}{h_{11}} V_1 - \frac{h_{12}}{h_{11}} V_2$$

Comparing with $I_1 = Y_{11} V_1 + Y_{12} V_2$, we get

$$Y_{11} = \frac{1}{h_{11}} \qquad \text{...(11.93)}$$

$$Y_{12} = \frac{-h_{12}}{h_{11}} \qquad \text{...(11.94)}$$

Also

$$I_2 = h_{22} V_2 + h_{21} \left[\frac{1}{h_{11}} V_1 - \frac{h_{12}}{h_{11}} V_2\right] = \frac{h_{21}}{h_{11}} V_1 + \left[\frac{h_{11}h_{22} - h_{12}h_{21}}{h_{11}}\right] V_2$$

Comparing with $I_2 = Y_{21} V_1 + Y_{22} V_2$, we get

$$Y_{21} = \frac{h_{21}}{h_{11}} \qquad \text{...(11.95)}$$

$$Y_{22} = \frac{\Delta h}{h_{11}} \qquad \text{...(11.96)}$$

(v) Y-parameters in terms of inverse hybrid parameters We know that

$$I_1 = g_{11} V_1 + g_{12} I_2$$
$$V_2 = g_{21} V_1 + g_{22} I_2$$

Rewriting the second equation,

$$I_2 = -\frac{g_{21}}{g_{22}} V_1 + \frac{1}{g_{22}} V_2$$

Comparing with $I_2 = Y_{21} V_1 + Y_{22} V_2$, we get

$$Y_{21} = -\frac{g_{21}}{g_{22}} \quad \text{...(11.97)}$$

$$Y_{22} = \frac{1}{g_{22}} \quad \text{...(11.98)}$$

Also,

$$I_1 = g_{11} V_1 + g_{12}\left[-\frac{g_{21}}{g_{22}} V_1 + \frac{1}{g_{22}} V_2\right]$$

$$= \left[\frac{g_{11}g_{22} - g_{12}g_{21}}{g_{22}}\right] V_1 + \frac{g_{12}}{g_{22}} V_2$$

Comparing with $I_1 = Y_{11} V_1 + Y_{12} V_2$, we get

$$Y_{11} = \frac{\Delta g}{g_{22}} \quad \text{...(11.99)}$$

$$Y_{12} = \frac{g_{12}}{g_{22}} \quad \text{...(11.100)}$$

11.8.3 *ABCD* Parameters in Terms of Other Parameters

(i) ABCD parameters in terms of Z-parameters We know that

$$V_1 = Z_{11} I_1 + Z_{12} I_2$$
$$V_2 = Z_{21} I_1 + Z_{22} I_2$$

Rewriting the second equation,

$$I_1 = \frac{1}{Z_{21}} V_2 - \frac{Z_{22}}{Z_{21}} I_2$$

Comparing with $I_1 = CV_2 - DI_2$, we get

$$C = \frac{1}{Z_{21}} \quad \text{...(11.101)}$$

$$D = \frac{Z_{22}}{Z_{21}} \quad \text{...(11.102)}$$

Also,

$$V_1 = Z_{12} I_2 + Z_{11}\left[\frac{1}{Z_{21}} V_2 - \frac{Z_{22}}{Z_{21}} I_2\right]$$

$$= Z_{12} I_2 + \frac{Z_{11}}{Z_{21}} V_2 - \frac{Z_{22}Z_{11}}{Z_{21}} I_2$$

$$= \frac{Z_{11}}{Z_{21}} V_2 - \left[\frac{Z_{11}Z_{22} - Z_{12}Z_{21}}{Z_{21}} \right] I_2$$

Comparing with $V_1 = AV_2 - BI_2$, we get

$$A = \frac{Z_{11}}{Z_{21}} \qquad \text{...(11.103)}$$

$$B = \frac{Z_{11}Z_{22} - Z_{12}Z_{21}}{Z_{21}} = \frac{\Delta Z}{Z_{21}} \qquad \text{...(11.104)}$$

(ii) ABCD parameters in terms of Y-parameters We know that

$$I_1 = Y_{11} V_1 + Y_{12} V_2$$
$$I_2 = Y_{21} V_1 + Y_{22} V_2$$

Rewriting the second equation,

$$V_1 = \frac{-Y_{22}}{Y_{21}} V_2 + \frac{1}{Y_{21}} I_2$$

Comparing with $V_1 = AV_2 - BI_2$, we get

$$A = \frac{-Y_{22}}{Y_{21}} \qquad \text{...(11.105)}$$

$$B = \frac{-1}{Y_{21}} \qquad \text{...(11.106)}$$

Also,

$$I_1 = Y_{12} V_2 + Y_{11} \left[\frac{-Y_{22}}{Y_{21}} V_2 + \frac{1}{Y_{21}} I_2 \right]$$

$$= \left[\frac{Y_{12}Y_{21} - Y_{11}Y_{22}}{Y_{21}} \right] V_2 + \frac{Y_{11}}{Y_{21}} I_2$$

Comparing with $I_1 = CV_2 - DI_2$, we get

$$C = \frac{Y_{12}Y_{21} - Y_{11}Y_{22}}{Y_{21}} = \frac{-\Delta Y}{Y_{21}} \qquad \text{...(11.107)}$$

$$D = \frac{-Y_{11}}{Y_{21}} \qquad \text{...(11.108)}$$

(iii) ABCD parameters in terms of A′B′C′D′ parameters We know that

$$V_2 = A' V_1 - B' I_1$$
$$I_2 = C' V_1 - D' I_1$$

By Cramer's rule,

$$V_1 = \frac{\begin{vmatrix} V_2 & B' \\ I_2 & D' \end{vmatrix}}{\begin{vmatrix} A' & B' \\ C' & D' \end{vmatrix}}$$

$$= \frac{D'}{\Delta T'} V_2 - \frac{B'}{\Delta T'} I_2$$

where $\Delta T' = A'D' - B'C'$

Comparing with $V_1 = AV_2 - BI_2$, we get

$$A = \frac{D'}{\Delta T'} \qquad \text{...(11.109)}$$

$$B = \frac{B'}{\Delta T'} \qquad \text{...(11.110)}$$

$$-I_1 = \frac{\begin{vmatrix} A' & V_2 \\ C' & I_2 \end{vmatrix}}{\Delta T'} = -\frac{C'}{\Delta T'} V_2 + \frac{A'}{\Delta T'} I_2$$

$$I_1 = \frac{C'}{\Delta T'} V_2 - \frac{A'}{\Delta T'} I_2$$

Comparing with $I_1 = CV_2 - DI_2$, we get

$$C = \frac{C'}{\Delta T'} \qquad \text{...(11.111)}$$

$$D = \frac{A'}{\Delta T'} \qquad \text{...(11.112)}$$

(iv) ABCD parameters in terms of hybrid parameters We know that

$$V_1 = h_{11} I_1 + h_{12} V_2$$

$$I_2 = h_{21} I_1 + h_{22} V_2$$

Rewriting the second equation,

$$I_1 = \frac{-h_{22}}{h_{21}} V_2 + \frac{1}{h_{21}} I_2$$

Comparing with $I_1 = CV_2 - DI_2$

$$D = \frac{-1}{h_{21}} \qquad \text{...(11.113)}$$

$$C = \frac{-h_{22}}{h_{21}} \qquad \text{...(11.114)}$$

Also,

$$V_1 = h_{12} V_2 + h_{11} \left[\frac{1}{h_{21}} I_2 - \frac{h_{22}}{h_{21}} V_2 \right]$$

$$= \left[\frac{h_{12}h_{21} - h_{11}h_{22}}{h_{21}} \right] V_2 + \frac{h_{11}}{h_{21}} I_2$$

Comparing with $V_1 = AV_2 - BI_2$, we get

$$A = \frac{-\Delta h}{h_{21}} \qquad \text{...(11.115)}$$

$$B = \frac{-h_{11}}{h_{21}} \qquad \text{...(11.116)}$$

(v) ABCD parameters in terms of inverse hybrid parameters We know that

$$I_1 = g_{11} V_1 + g_{12} I_2$$
$$V_2 = g_{21} V_1 + g_{22} I_2$$

Rewriting the second equation,

$$V_1 = \frac{1}{g_{21}} V_2 - \frac{g_{22}}{g_{21}} I_2$$

Comparing with $V_1 = AV_2 - BI_2$

$$A = \frac{1}{g_{21}} \quad \text{...(11.117)}$$

$$B = \frac{g_{22}}{g_{21}} \quad \text{...(11.118)}$$

Also,

$$I_1 = g_{11} \left[\frac{1}{g_{21}} V_2 - \frac{g_{22}}{g_{21}} I_2 \right] + g_{12} I_2$$

$$= \frac{g_{11}}{g_{21}} V_2 - \left[\frac{g_{11} g_{22} - g_{12} g_{21}}{g_{21}} \right] I_2$$

Comparing with $I_1 = CV_2 - DI_2$, we get

$$C = \frac{g_{11}}{g_{21}} \quad \text{...(11.119)}$$

$$D = \frac{\Delta g}{g_{21}} \quad \text{...(11.120)}$$

11.8.4 *A′B′C′D′* Parameters in Terms of Other Parameters

(i) A′B′C′D′ parameters in terms of Z-parameters We know that,

$$V_1 = Z_{11} I_1 + Z_{12} I_2$$
$$V_2 = Z_{21} I_1 + Z_{22} I_2$$

Rewriting the first equation,

$$I_2 = \frac{1}{Z_{12}} V_1 - \frac{Z_{11}}{Z_{12}} I_1$$

Comparing with $I_2 = C' V_1 - D' I_1$, we get

$$C' = \frac{1}{Z_{12}} \quad \text{...(11.121)}$$

$$D' = \frac{Z_{11}}{Z_{12}} \quad \text{...(11.122)}$$

Also,

$$V_2 = Z_{21} I_1 + Z_{22} \left[\frac{1}{Z_{12}} V_1 - \frac{Z_{11}}{Z_{12}} I_1 \right]$$

$$= Z_{12} I_1 + \frac{Z_{22}}{Z_{12}} V_1 - \frac{Z_{22} Z_{11}}{Z_{12}} I_1$$

$$= \frac{Z_{22}}{Z_{12}} V_1 - \left[\frac{Z_{11} Z_{22} - Z_{12} Z_{21}}{Z_{12}} \right] I_1$$

Comparing with $V_2 = A' V_1 - B' I_1$, we get

$$A' = \frac{Z_{22}}{Z_{12}} \quad \text{...(11.123)}$$

$$B' = \frac{Z_{11}Z_{22} - Z_{12}Z_{21}}{Z_{12}} = \frac{\Delta Z}{Z_{12}} \quad \text{...(11.124)}$$

(ii) A′B′C′D′ parameters in terms of Y-parameters We know that

$$I_1 = Y_{11} V_1 + Y_{12} V_2$$
$$I_2 = Y_{21} V_1 + Y_{22} V_2$$

Rewriting the first equation,

$$V_2 = \frac{-Y_{11}}{Y_{12}} V_1 + \frac{1}{Y_{12}} I_1$$

Comparing with $V_2 = A' V_1 - B' I_1$, we get

$$A' = \frac{-Y_{11}}{Y_{12}} \quad \text{...(11.125)}$$

$$B' = \frac{-1}{Y_{12}} \quad \text{...(11.126)}$$

Also,

$$I_2 = Y_{21} V_1 + Y_{22} \left[\frac{-Y_{11}}{Y_{12}} V_1 + \frac{1}{Y_{12}} I_1 \right]$$

$$= \left[\frac{Y_{12}Y_{21} - Y_{11}Y_{22}}{Y_{12}} \right] V_1 + \frac{Y_{22}}{Y_{12}} I_1$$

Comparing with $I_2 = C' V_1 - D' I_1$, we get

$$C' = \frac{Y_{12}Y_{21} - Y_{11}Y_{22}}{Y_{12}} = \frac{-\Delta Y}{Y_{12}} \quad \text{...(11.127)}$$

$$D' = \frac{-Y_{22}}{Y_{12}} \quad \text{...(11.128)}$$

(iii) A′B′C′D′ parameters in terms of ABCD parameters We know that

$$V_1 = AV_2 - BI_2$$
$$I_1 = CV_2 - DI_2$$

By Cramer's rule,

$$V_2 = \frac{\begin{vmatrix} V_1 & B \\ I_1 & D \end{vmatrix}}{\begin{vmatrix} A & B \\ C & D \end{vmatrix}} = \frac{D}{\Delta T} V_2 - \frac{B}{\Delta T} I_2$$

where $\Delta T = AD - BC$

Comparing with $V_2 = A' V_1 - B' I_1$, we get

$$A' = \frac{D}{\Delta T} \quad \text{...(11.129)}$$

$$B' = \frac{B}{\Delta T} \qquad \text{...(11.130)}$$

$$-I_2 = \frac{\begin{vmatrix} A & V_1 \\ C & I_1 \end{vmatrix}}{\Delta T}$$

$$= -\frac{C}{\Delta T} V_1 + \frac{A}{\Delta T} I_1$$

$$I_2 = \frac{C}{\Delta T} V_1 - \frac{A}{\Delta T} I_1$$

Comparing with $I_2 = C' V_1 - D' I_1$, we get

$$C' = \frac{C}{\Delta T} \qquad \text{...(11.131)}$$

$$D' = \frac{A}{\Delta T} \qquad \text{...(11.132)}$$

(iv) A′B′C′D′ parameters in terms of hybrid parameters We know that

$$V_1 = h_{11} I_1 + h_{12} V_2$$

$$I_2 = h_{21} I_1 + h_{22} V_2$$

Rewriting the first equation,

$$V_2 = \frac{1}{h_{12}} V_1 - \frac{h_{11}}{h_{12}} I_1$$

Comparing with $V_2 = A' V_1 - B' I_1$

$$A' = \frac{1}{h_{12}} \qquad \text{...(11.133)}$$

$$B' = \frac{h_{11}}{h_{12}} \qquad \text{...(11.134)}$$

Also,

$$I_2 = h_{21} I_1 + h_{22} \left[\frac{1}{h_{12}} V_1 - \frac{h_{11}}{h_{12}} I_1 \right]$$

$$= \frac{h_{22}}{h_{12}} V_1 + \left[\frac{h_{11}h_{22} - h_{12}h_{21}}{h_{12}} \right] I_1$$

Comparing with $I_2 = C' V_1 - D' I_1$, we get

$$C' = \frac{h_{22}}{h_{12}} \qquad \text{...(11.135)}$$

$$D' = \left[\frac{h_{11}h_{22} - h_{12}h_{21}}{h_{12}} \right] = \frac{\Delta h}{h_{12}} \qquad \text{...(11.136)}$$

(v) A′B′C′D′ parameters in terms of inverse hybrid parameters We know that

$$I_1 = g_{11} V_1 + g_{12} I_2$$

$$V_2 = g_{21} V_1 + g_{22} I_2$$

Rewriting the first equation,

$$I_2 = -\frac{g_{11}}{g_{12}} V_1 + \frac{1}{g_{12}} I_1$$

Comparing with $I_2 = C' V_1 - D' I_1$

$$C = -\frac{g_{11}}{g_{12}} \quad \text{...(11.137)}$$

$$D' = -\frac{1}{g_{12}} \quad \text{...(11.138)}$$

Also,

$$V_2 = g_{21} V_1 + g_{22}\left[\frac{-g_{11}}{g_{12}} V_1 + \frac{1}{g_{12}} I_1\right]$$

$$= -\left[\frac{g_{11}g_{22} - g_{12}g_{21}}{g_{12}}\right] V_2 + \frac{g_{22}}{g_{12}} I_1$$

Comparing with $V_2 = A' V_1 - B' I_1$, we get

$$A' = -\frac{\Delta g}{g_{12}} \quad \text{...(11.139)}$$

$$B' = \frac{g_{22}}{g_{12}} \quad \text{...(11.140)}$$

11.8.5 Hybrid Parameters in Terms of Other Parameters

(i) Hybrid parameters in terms of Z-parameters We know that

$$V_1 = Z_{11} I_1 + Z_{12} I_2$$
$$V_2 = Z_{21} I_1 + Z_{22} I_2$$

Rewriting the second equation,

$$I_2 = \frac{-Z_{21}}{Z_{22}} I_1 + \frac{1}{Z_{22}} V_2$$

Comparing with $I_2 = h_{21} I_1 + h_{22} V_2$, we get

$$h_{21} = \frac{-Z_{21}}{Z_{22}} \quad \text{...(11.141)}$$

$$h_{22} = \frac{1}{Z_{22}} \quad \text{...(11.142)}$$

Also,

$$V_1 = Z_{11} I_1 + Z_{12}\left[\frac{1}{Z_{22}} V_2 - \frac{Z_{21}}{Z_{22}} I_1\right]$$

$$= \left[\frac{Z_{11}Z_{22} - Z_{12}Z_{21}}{Z_{22}}\right] I_1 + \frac{Z_{12}}{Z_{22}} V_2$$

Comparing with $V_1 = h_{11} I_1 + h_{12} V_2$, we get

$$h_{11} = \frac{\Delta Z}{Z_{22}} \quad \text{...(11.143)}$$

$$h_{12} = \frac{Z_{12}}{Z_{22}} \quad \text{...(11.144)}$$

(ii) Hybrid parameters in terms of Y-parameters We know that

$$I_1 = Y_{11} V_1 + Y_{12} V_2$$
$$I_2 = Y_{21} V_1 + Y_{22} V_2$$

Rewriting the first equation,

$$V_1 = \frac{1}{Y_{11}} I_1 - \frac{Y_{12}}{Y_{11}} V_2$$

Comparing with $V_1 = h_{11} I_1 + h_{12} V_2$, we get

$$h_{11} = \frac{1}{Y_{11}} \qquad \text{...(11.145)}$$

$$h_{12} = \frac{-Y_{12}}{Y_{11}} \qquad \text{...(11.146)}$$

Also,

$$I_2 = Y_{22} V_2 + Y_{21} \left[\frac{1}{Y_{11}} I_1 - \frac{Y_{12}}{Y_{11}} V_2 \right]$$

$$= \left[\frac{Y_{11}Y_{22} - Y_{12}Y_{21}}{Y_{11}} \right] V_2 + \frac{Y_{21}}{Y_{11}} I_1$$

Comparing with $I_2 = h_{21} I_1 + h_{22} V_2$, we get

$$h_{21} = \frac{Y_{22}}{Y_{11}} \qquad \text{...(11.147)}$$

$$h_{22} = \frac{\Delta Y}{Y_{11}} \qquad \text{...(11.148)}$$

(iii) Hybrid parameters in terms of ABCD parameters We know that

$$V_1 = AV_2 - BI_2$$
$$I_1 = CV_2 - DI_2$$

Rewriting the second equation,

$$I_2 = \frac{-1}{D} I_1 + \frac{C}{D} V_2$$

Comparing with $I_2 = h_{21} I_1 + h_{22} V_2$, we get

$$h_{21} = \frac{-1}{D} \qquad \text{...(11.149)}$$

$$h_{22} = \frac{C}{D} \qquad \text{...(11.150)}$$

Also,

$$V_1 = AV_2 - B \left[\frac{-1}{D} I_1 + \frac{C}{D} V_2 \right]$$

$$= \frac{B}{D} I_1 + \left[\frac{AD - BC}{D} \right] V_2$$

Comparing with $V_1 = h_{11} I_1 + h_{12} V_2$, we get

$$h_{11} = \frac{B}{D} \qquad \text{...(11.151)}$$

$$h_{12} = \frac{AD - BC}{D} = \frac{\Delta T}{D} \qquad \text{...(11.152)}$$

(iv) Hybrid parameters in terms of A′B′C′D′ parameters We know that

$$V_2 = A'V_1 - B'I_1$$
$$I_2 = C'V_1 - D'I_1$$

Rewriting the first equation,

$$V_1 = \frac{B'}{A'}I_1 + \frac{1}{A'}V_2$$

Comparing with $V_1 = h_{11}I_1 + h_{12}V_2$, we get

$$h_{11} = \frac{B'}{A'} \quad \text{...(11.153)}$$

$$h_{12} = \frac{1}{A'} \quad \text{...(11.154)}$$

Also,

$$I_2 = C'\left[\frac{B'}{A'}I_1 + \frac{1}{A'}V_2\right] - D'I_1$$

$$= -\left[\frac{A'D' - B'C'}{A'}\right]I_1 + \frac{C'}{A'}V_2$$

Comparing with $I_2 = h_{21}I_1 + h_{22}V_2$, we get

$$h_{21} = -\left[\frac{A'D' - B'C'}{A'}\right] = -\frac{\Delta T'}{A'} \quad \text{...(11.155)}$$

$$h_{22} = \frac{C'}{A'} \quad \text{...(11.156)}$$

(v) Hybrid parameters in terms of inverse hybrid parameters We know that

$$I_1 = g_{11}V_1 + g_{12}I_2$$
$$V_2 = g_{21}V_1 + g_{22}I_2$$

By Cramer's rule,

$$V_1 = \frac{\begin{vmatrix} I_1 & g_{12} \\ V_2 & g_{22} \end{vmatrix}}{\begin{vmatrix} g_{11} & g_{12} \\ g_{21} & g_{22} \end{vmatrix}}$$

$$= \frac{g_{22}}{\Delta g}I_1 - \frac{g_{12}}{\Delta g}V_2$$

where $\Delta g = g_{11}g_{22} - g_{12}g_{21}$

Comparing with $V_1 = h_{11}I_1 + h_{12}V_2$, we get

$$h_{11} = \frac{g_{22}}{\Delta g} \quad \text{...(11.157)}$$

$$h_{12} = -\frac{g_{12}}{\Delta g} \quad \text{...(11.158)}$$

$$I_2 = \frac{\begin{vmatrix} g_{11} & I_1 \\ g_{21} & V_2 \end{vmatrix}}{\Delta g} = -\frac{g_{21}}{\Delta g}I_1 + \frac{g_{11}}{\Delta g}V_2$$

Comparing with $I_2 = h_{21} I_1 + h_{22} V_2$, we get

$$h_{21} = -\frac{g_{21}}{\Delta g} \quad \text{...(11.159)}$$

$$h_{22} = \frac{g_{11}}{\Delta g} \quad \text{...(11.160)}$$

11.8.6 Inverse Hybrid Parameters in Terms of Other Parameters

(i) Inverse hybrid parameters in terms of Z-parameters We know that

$$V_1 = Z_{11} I_1 + Z_{12} I_2$$
$$V_2 = Z_{21} I_1 + Z_{22} I_2$$

Rewriting the first equation,

$$I_1 = \frac{1}{Z_{11}} V_1 - \frac{Z_{12}}{Z_{11}} I_2$$

Comparing with $I_1 = g_{11} V_1 + g_{12} I_2$, we get

$$g_{11} = \frac{1}{Z_{11}} \quad \text{...(11.161)}$$

$$g_{12} = \frac{-Z_{12}}{Z_{11}} \quad \text{...(11.162)}$$

Also,

$$V_2 = Z_{21}\left[\frac{1}{Z_{11}} V_1 - \frac{Z_{12}}{Z_{11}} I_2\right] + Z_{22} I_2$$

$$= \frac{Z_{21}}{Z_{11}} V_1 + \left[\frac{Z_{11}Z_{22} - Z_{12}Z_{21}}{Z_{11}}\right] I_2$$

Comparing with $V_2 = g_{21} V_1 + g_{22} I_2$, we get

$$g_{21} = \frac{Z_{21}}{Z_{11}} \quad \text{...(11.163)}$$

$$g_{22} = \frac{\Delta Z}{Z_{11}} \quad \text{...(11.164)}$$

(ii) Inverse hybrid parameters in terms of Y-parameters We know that

$$I_1 = Y_{11} V_1 + Y_{12} V_2$$
$$I_2 = Y_{21} V_1 + Y_{22} V_2$$

Rewriting the second equation,

$$V_2 = \frac{-Y_{21}}{Y_{22}} V_1 + \frac{1}{Y_{22}} I_2$$

Comparing with $V_2 = g_{21} V_1 + g_{22} I_2$, we get

$$g_{21} = \frac{-Y_{21}}{Y_{22}} \quad \text{...(11.165)}$$

$$g_{22} = \frac{1}{Y_{22}} \quad \text{...(11.166)}$$

Also,

$$I_1 = Y_{11} V_1 + Y_{12} \left[\frac{-Y_{21}}{Y_{22}} V_1 + \frac{1}{Y_{22}} I_2 \right]$$

$$= \left[\frac{Y_{11}Y_{22} - Y_{12}Y_{21}}{Y_{22}} \right] V_1 + \frac{Y_{12}}{Y_{22}} I_2$$

Comparing with $I_1 = g_{11} V_1 + g_{12} I_2$, we get

$$g_{11} = \frac{\Delta Y}{Y_{22}} \qquad \text{...(11.167)}$$

$$g_{12} = \frac{Y_{12}}{Y_{22}} \qquad \text{...(11.168)}$$

(iii) Inverse hybrid parameters in terms of ABCD parameters We know that

$$V_1 = AV_2 - BI_2$$
$$I_1 = CV_2 - DI_2$$

Rewriting the first equation,

$$V_2 = \frac{1}{A} V_1 + \frac{B}{A} I_2$$

Comparing with $V_2 = g_{21} V_1 + g_{22} I_2$, we get

$$g_{21} = \frac{1}{A} \qquad \text{...(11.169)}$$

$$g_{22} = \frac{B}{A} \qquad \text{...(11.170)}$$

Also,

$$I_1 = C\left[\frac{1}{A} V_1 + \frac{B}{A} I_2 \right] - D\, I_2$$

$$= \frac{C}{A} I_1 - \left[\frac{AD - BC}{A} \right] I_2$$

Comparing with $I_1 = g_{11} V_1 + g_{12} I_2$, we get

$$g_{11} = \frac{C}{A} \qquad \text{...(11.171)}$$

$$g_{12} = -\left[\frac{AD - BC}{A} \right] = -\frac{\Delta T}{A} \qquad \text{...(11.172)}$$

(iv) Inverse hybrid parameters in terms of A′B′C′D′ parameters We know that

$$V_2 = A' V_1 - B' I_1$$
$$I_2 = C' V_1 - D' I_1$$

Rewriting the second equation,

$$I_1 = \frac{C'}{D'} V_1 - \frac{1}{D'} I_2$$

Comparing with $I_1 = g_{11} V_1 + g_{12} I_2$, we get

$$g_{11} = \frac{C'}{D'} \qquad \text{...(11.173)}$$

$$g_{12} = -\frac{1}{D'} \qquad \text{...(11.174)}$$

Also,
$$V_2 = A'V_1 - B'\left[\frac{C'}{D'}V_1 - \frac{1}{D'}I_2\right]$$
$$= \left[\frac{A'D' - B'C'}{D'}\right]V_1 + \frac{B'}{D'}I_2$$

Comparing with $V_2 = g_{21} V_1 + g_{22} I_2$, we get
$$g_{21} = \left[\frac{A'D' - B'C'}{D'}\right] = \frac{\Delta T'}{D'} \quad \text{...(11.175)}$$
$$g_{22} = \frac{B'}{D'} \quad \text{...(11.176)}$$

(v) Inverse hybrid parameters in terms of hybrid parameters We know that
$$V_1 = h_{11} I_1 + h_{12} V_2$$
$$I_2 = h_{21} I_1 + h_{22} V_2$$

By Cramer's rule,
$$I_1 = \frac{\begin{vmatrix} V_1 & h_{12} \\ I_2 & h_{22} \end{vmatrix}}{\begin{vmatrix} h_{11} & h_{12} \\ h_{21} & h_{22} \end{vmatrix}}$$
$$= \frac{h_{22}}{\Delta h}V_1 - \frac{h_{12}}{\Delta h}I_2$$

where $\Delta h = h_{11} h_{22} - h_{12} h_{21}$

Comparing with $I_1 = g_{11} V_1 + g_{12} I_2$, we get
$$g_{11} = \frac{h_{22}}{\Delta h} \quad \text{...(11.177)}$$
$$g_{12} = -\frac{h_{12}}{\Delta h} \quad \text{...(11.178)}$$
$$V_2 = \frac{\begin{vmatrix} h_{11} & V_1 \\ h_{21} & I_2 \end{vmatrix}}{\Delta h}$$
$$= -\frac{h_{21}}{\Delta h}V_1 + \frac{h_{11}}{\Delta h}I_2$$

Comparing with $V_2 = g_{21} V_1 + g_{22} I_2$, we get
$$g_{21} = -\frac{h_{21}}{\Delta h} \quad \text{...(11.179)}$$
$$g_{22} = \frac{h_{11}}{\Delta h} \quad \text{...(11.180)}$$

Table 11.3 *Inter-relationship between Parameters*
$\Delta X = X_{11} X_{22} - X_{12} X_{21}$

	In terms of											
	[Z]		[Y]		[T]		[T′]		[h]		[g]	
[Z]	Z_{11}	Z_{12}	$\frac{Y_{22}}{\Delta Y}$	$\frac{-Y_{12}}{\Delta Y}$	$\frac{A}{C}$	$\frac{\Delta T}{C}$	$\frac{D'}{C'}$	$\frac{1}{C'}$	$\frac{\Delta h}{h_{22}}$	$\frac{h_{12}}{h_{22}}$	$\frac{1}{g_{11}}$	$\frac{-g_{12}}{g_{11}}$
	Z_{21}	Z_{22}	$\frac{-Y_{21}}{\Delta Y}$	$\frac{Y_{11}}{\Delta Y}$	$\frac{1}{C}$	$\frac{D}{C}$	$\frac{\Delta T'}{C'}$	$\frac{A'}{C'}$	$\frac{-h_{21}}{h_{22}}$	$\frac{1}{h_{22}}$	$\frac{g_{21}}{g_{11}}$	$\frac{\Delta g}{g_{11}}$
[Y]	$\frac{Z_{22}}{\Delta Z}$	$\frac{-Z_{12}}{\Delta Z}$	Y_{11}	Y_{12}	$\frac{D}{B}$	$\frac{-\Delta T}{B}$	$\frac{A'}{B'}$	$-\frac{1}{B'}$	$\frac{1}{h_{11}}$	$\frac{-h_{12}}{h_{11}}$	$\frac{\Delta g}{g_{22}}$	$\frac{g_{12}}{g_{22}}$
	$\frac{-Z_{21}}{\Delta Z}$	$\frac{Z_{11}}{\Delta Z}$	Y_{21}	Y_{22}	$\frac{-1}{B}$	$\frac{A}{B}$	$\frac{\Delta T'}{B'}$	$\frac{D'}{B'}$	$\frac{h_{21}}{h_{11}}$	$\frac{\Delta h}{h_{11}}$	$\frac{-g_{21}}{g_{22}}$	$\frac{1}{g_{22}}$
[T]	$\frac{Z_{11}}{Z_{21}}$	$\frac{\Delta Z}{Z_{21}}$	$\frac{-Y_{22}}{Y_{21}}$	$\frac{-1}{Y_{21}}$	A	B	$\frac{D'}{\Delta T'}$	$\frac{B'}{\Delta T'}$	$\frac{-\Delta h}{h_{21}}$	$\frac{-h_{11}}{h_{21}}$	$\frac{1}{g_{21}}$	$\frac{g_{22}}{g_{21}}$
	$\frac{1}{Z_{21}}$	$\frac{Z_{22}}{Z_{21}}$	$\frac{-\Delta Y}{Y_{21}}$	$\frac{-Y_{11}}{Y_{21}}$	C	D	$\frac{C'}{\Delta T'}$	$\frac{A'}{\Delta T'}$	$\frac{-h_{22}}{h_{21}}$	$\frac{-1}{h_{21}}$	$\frac{g_{11}}{g_{21}}$	$\frac{\Delta g}{g_{21}}$
[T′]	$\frac{Z_{22}}{Z_{12}}$	$\frac{\Delta Z}{Z_{12}}$	$\frac{-Y_{11}}{Y_{12}}$	$\frac{-1}{Y_{12}}$	$\frac{D}{\Delta T}$	$\frac{B}{\Delta T}$	A'	B'	$\frac{1}{h_{12}}$	$\frac{h_{11}}{h_{12}}$	$\frac{-\Delta g}{g_{12}}$	$\frac{-g_{22}}{g_{12}}$
	$\frac{1}{Z_{12}}$	$\frac{Z_{11}}{Z_{12}}$	$\frac{-\Delta Y}{Y_{12}}$	$\frac{-Y_{22}}{Y_{12}}$	$\frac{C}{\Delta T}$	$\frac{A}{\Delta T}$	C'	D'	$\frac{h_{22}}{h_{12}}$	$\frac{\Delta h}{h_{12}}$	$\frac{-g_{11}}{g_{12}}$	$\frac{-1}{g_{12}}$
[h]	$\frac{\Delta Z}{Z_{22}}$	$\frac{Z_{12}}{Z_{22}}$	$\frac{1}{Y_{11}}$	$\frac{-Y_{12}}{Y_{11}}$	$\frac{B}{D}$	$\frac{\Delta T}{D}$	$\frac{B'}{A'}$	$\frac{1}{A'}$	h_{11}	h_{12}	$\frac{g_{22}}{\Delta g}$	$-\frac{g_{12}}{\Delta g}$
	$\frac{-Z_{21}}{Z_{22}}$	$\frac{1}{Z_{22}}$	$\frac{Y_{21}}{Y_{11}}$	$\frac{\Delta Y}{Y_{11}}$	$\frac{-1}{D}$	$\frac{C}{D}$	$\frac{-\Delta T'}{A'}$	$\frac{C'}{A'}$	h_{21}	h_{22}	$-\frac{g_{21}}{\Delta g}$	$\frac{g_{11}}{\Delta g}$
[g]	$\frac{1}{Z_{11}}$	$\frac{-Z_{12}}{Z_{11}}$	$\frac{\Delta Y}{Y_{22}}$	$\frac{Y_{12}}{Y_{22}}$	$\frac{C}{A}$	$-\frac{\Delta T}{A}$	$\frac{C'}{D'}$	$-\frac{1}{D'}$	$\frac{h_{22}}{\Delta h}$	$-\frac{h_{12}}{\Delta h}$	g_{11}	g_{12}
	$\frac{Z_{21}}{Z_{11}}$	$\frac{\Delta Z}{Z_{11}}$	$\frac{-Y_{21}}{Y_{22}}$	$\frac{1}{Y_{22}}$	$\frac{1}{A}$	$\frac{B}{A}$	$\frac{\Delta T'}{D'}$	$\frac{B'}{D'}$	$-\frac{h_{21}}{\Delta h}$	$\frac{h_{11}}{\Delta h}$	g_{21}	g_{22}

Example 11.19 *The Z parameters of a two-port network are $Z_{11} = 20\ \Omega$, $Z_{22} = 30\ \Omega$, $Z_{12} = Z_{21} = 10\ \Omega$. Find Y and ABCD parameters.*

Solution

$$\Delta Z = Z_{11}Z_{22} - Z_{12}Z_{21}$$
$$= (20)\,(30) - (10)\,(10) = 500$$

***Y*-parameters**

$$Y_{11} = \frac{Z_{22}}{\Delta Z} = \frac{30}{500} = \frac{3}{50}\ \mho$$

$$Y_{12} = \frac{-Z_{12}}{\Delta Z} = \frac{-10}{500} = \frac{-1}{50}\ \mho$$

$$Y_{21} = \frac{-Z_{21}}{\Delta Z} = \frac{-10}{500} = \frac{-1}{50}\ \mho$$

$$Y_{22} = \frac{Z_{11}}{\Delta Z} = \frac{20}{500} = \frac{2}{50}\ \mho$$

Hence, the Y-parameters are

$$\begin{bmatrix} Y_{11} & Y_{12} \\ Y_{21} & Y_{22} \end{bmatrix} = \begin{bmatrix} \frac{3}{50} & \frac{-1}{50} \\ \frac{-1}{50} & \frac{2}{50} \end{bmatrix}$$

ABCD parameters

$$A = \frac{Z_{11}}{Z_{21}} = \frac{20}{10} = 2$$

$$B = \frac{\Delta Z}{Z_{21}} = \frac{500}{10} = 50$$

$$C = \frac{1}{Z_{21}} = \frac{1}{10} = 0.1$$

$$D = \frac{Z_{22}}{Z_{21}} = \frac{30}{10} = 3$$

Hence, the $ABCD$ parameters are

$$\begin{bmatrix} A & B \\ C & D \end{bmatrix} = \begin{bmatrix} 2 & 50 \\ 0.1 & 3 \end{bmatrix}$$

Example 11.20 *Currents I_1 and I_2 entering at Port 1 and Port 2 respectively of a two-port network are given by the following equations:*

$$I_1 = 0.5\,V_1 - 0.2\,V_2$$
$$I_2 = -0.2\,V_1 + V_2$$

Find Y, Z and ABCD parameters for the network.

Solution

$$Y_{11} = \left.\frac{I_1}{V_1}\right|_{V_2=0} = 0.5\ \mho$$

$$Y_{21} = \left.\frac{I_2}{V_1}\right|_{V_2=0} = -0.2\ \mho$$

$$Y_{12} = \left.\frac{I_1}{V_2}\right|_{V_1=0} = -0.2\ \mho$$

$$Y_{22} = \left.\frac{I_2}{V_2}\right|_{V_1=0} = 1\ \mho$$

Hence, the Y-parameters are

$$\begin{bmatrix} Y_{11} & Y_{12} \\ Y_{21} & Y_{22} \end{bmatrix} = \begin{bmatrix} 0.5 & -0.2 \\ -0.2 & 1 \end{bmatrix}$$

***Z*-parameters**

$$\Delta Y = Y_{11}\,Y_{22} - Y_{12}\,Y_{21}$$
$$= (0.5)\,(1) - (-0.2)\,(-0.2) = 0.46$$

$$Z_{11} = \frac{Y_{22}}{\Delta Y} = \frac{1}{0.46} = 2.174\ \Omega$$

$$Z_{12} = \frac{-Y_{12}}{\Delta Y} = \frac{-(-0.2)}{0.46} = 0.434\ \Omega$$

$$Z_{21} = \frac{-Y_{21}}{\Delta Y} = \frac{-(-0.2)}{0.46} = 0.434\ \Omega$$

$$Z_{22} = \frac{Y_{11}}{\Delta Y} = \frac{0.5}{0.46} = 1.087\ \Omega$$

$$\begin{bmatrix} Z_{11} & Z_{12} \\ Z_{21} & Z_{22} \end{bmatrix} = \begin{bmatrix} 2.174 & 0.434 \\ 0.434 & 1.087 \end{bmatrix}$$

ABCD parameters

$$A = \frac{-Y_{22}}{Y_{21}} = \frac{-1}{-0.2} = 5$$

$$B = -\frac{1}{Y_{21}} = \frac{-1}{-0.2} = 5$$

$$C = -\frac{\Delta Y}{Y_{21}} = \frac{-0.46}{-0.2} = 2.3$$

$$D = \frac{-Y_{11}}{Y_{21}} = \frac{-0.5}{-0.2} = 2.5$$

Hence, the *ABCD* parameters are

$$\begin{bmatrix} A & B \\ C & D \end{bmatrix} = \begin{bmatrix} 5 & 5 \\ 2.3 & 2.5 \end{bmatrix}$$

Example 11.21 *Using the relation $Y = Z^{-1}$, show that* $|Z| = \frac{1}{2}\left(\frac{Z_{22}}{Y_{11}} + \frac{Z_{11}}{Y_{22}}\right)$.

Solution We know that

$$Y = Z^{-1}$$

i.e.,

$$\begin{bmatrix} Y_{11} & Y_{12} \\ Y_{21} & Y_{22} \end{bmatrix} = \begin{bmatrix} \frac{Z_{22}}{\Delta Z} & \frac{-Z_{12}}{\Delta Z} \\ \frac{-Z_{21}}{\Delta Z} & \frac{Z_{11}}{\Delta Z} \end{bmatrix}$$

$$|Z| = Z_{11}\,Z_{22} - Z_{12}\,Z_{21}$$

$$\frac{1}{2}\left(\frac{Z_{22}}{Y_{11}} + \frac{Z_{11}}{Y_{22}}\right) = \frac{1}{2}\left(\frac{Z_{22}}{\frac{Z_{22}}{\Delta Z}} + \frac{Z_{11}}{\frac{Z_{11}}{\Delta Z}}\right)$$

$$= \frac{1}{2}(\Delta Z + \Delta Z) = \frac{1}{2}(2\Delta Z) = \Delta Z = Z_{11}Z_{12} - Z_{12}Z_{21}$$

$$|Z| = \frac{1}{2}\left(\frac{Z_{22}}{Y_{11}} + \frac{Z_{11}}{Y_{22}}\right)$$

Example 11.22 *For the network shown in Fig. 11.48, find Z and Y-parameters.*

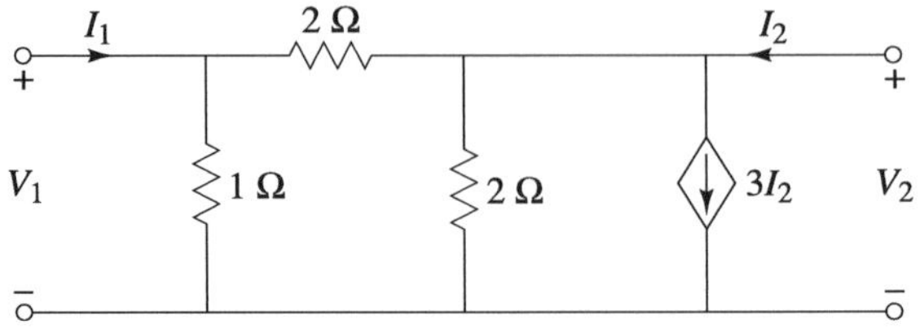

Fig. 11.48

Solution By source transformation technique,

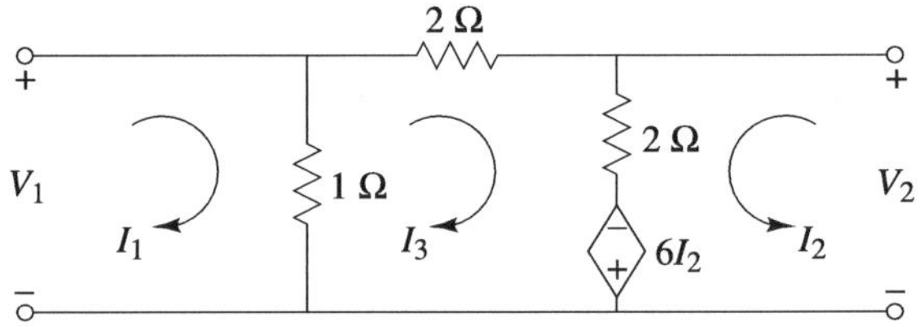

Fig. 11.49

Applying KVL to Mesh 1,

$$V_1 = I_1 - I_3 \quad \ldots\text{(i)}$$

Applying KVL to Mesh 2,

$$V_2 = 2\,(I_2 + I_3) - 6I_2 = -4I_2 + 2I_3 \quad \ldots\text{(ii)}$$

Applying KVL to Mesh 3,

$$(I_3 - I_1) + 2I_3 + 2\,(I_2 + I_3) - 6I_2 = 0$$

$$5I_3 = I_1 + 4I_2$$

$$I_3 = \frac{1}{5}I_1 + \frac{4}{5}I_2 \quad \ldots\text{(iii)}$$

Substituting the Eq. (iii) in the Eq. (i),

$$V_1 = I_1 - \frac{1}{5}I_1 - \frac{4}{5}I_2 = \frac{4}{5}\,I_1 - \frac{4}{5}\,I_2 \quad \ldots\text{(iv)}$$

Substituting the Eq. (iii) in the Eq. (ii),

$$V_2 = -4I_2 + 2\left(\frac{1}{5}I_1 + \frac{4}{5}I_2\right)$$

$$= \frac{2}{5}I_1 - \frac{12}{5}I_2 \quad \ldots\text{(v)}$$

Comparing Eqs (iv) and (v) with Z-parameter equations, we get

$$\begin{bmatrix} Z_{11} & Z_{12} \\ Z_{21} & Z_{22} \end{bmatrix} = \begin{bmatrix} \frac{4}{5} & \frac{-4}{5} \\ \frac{2}{5} & \frac{-12}{5} \end{bmatrix}$$

***Y*-parameters**

$$\Delta Z = Z_{11} Z_{22} - Z_{12} Z_{21}$$

$$= \left(\frac{4}{5}\right)\left(\frac{-12}{5}\right) - \left(\frac{-4}{5}\right)\left(\frac{2}{5}\right) = \frac{-40}{25} = \frac{-8}{5}$$

$$Y_{11} = \frac{Z_{22}}{\Delta Z} = \frac{\left(\frac{-12}{5}\right)}{\left(\frac{-8}{5}\right)} = \frac{3}{2}\ \mho$$

$$Y_{12} = \frac{-Z_{12}}{\Delta Z} = \frac{-\left(\frac{-4}{5}\right)}{\left(\frac{-8}{5}\right)} = \frac{-1}{2}\ \mho$$

$$Y_{21} = \frac{-Z_{21}}{\Delta Z} = \frac{-\left(\frac{2}{5}\right)}{\left(\frac{-8}{5}\right)} = \frac{1}{4}\ \mho$$

$$Y_{22} = \frac{Z_{11}}{\Delta Z} = \frac{\left(\frac{4}{5}\right)}{\left(\frac{-8}{5}\right)} = \frac{-1}{2}\ \mho$$

Hence, the *Y*-parameters are

$$\begin{bmatrix} Y_{11} & Y_{12} \\ Y_{21} & Y_{22} \end{bmatrix} = \begin{bmatrix} \frac{3}{2} & \frac{-1}{2} \\ \frac{1}{4} & \frac{-1}{2} \end{bmatrix}$$

Example 11.23 *Find Z and h-parameters for the network shown in Fig. 11.50.*

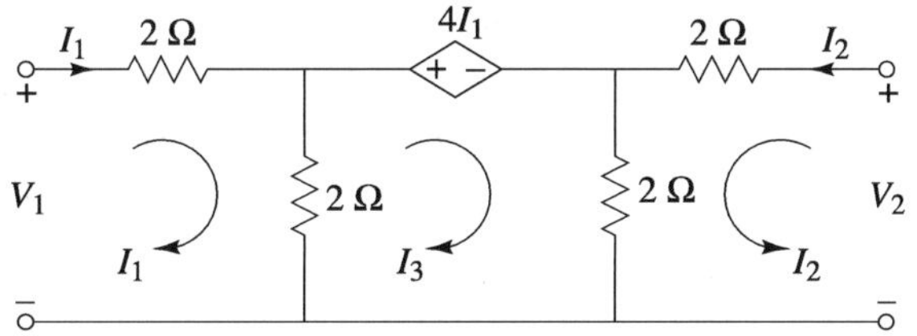

Fig. 11.50

Solution Applying KVL to Mesh 1,

$$V_1 = 2I_1 + 2(I_1 - I_3)$$
$$= 4I_1 - 2I_3 \quad \text{...(i)}$$

Applying KVL to Mesh 2,

$$V_2 = 2I_2 + 2(I_2 + I_3)$$
$$= 4I_2 + 2I_3 \quad \text{...(ii)}$$

Applying KVL to Mesh 3,

$$2\,(I_3 - I_1) + 4I_1 + 2\,(I_3 + I_2) = 0$$

$$I_1 + I_2 = -2I_3 \qquad \text{...(iii)}$$

Substituting the Eq. (iii) in the Eq. (i),

$$V_1 = 4I_1 + I_1 + I_2$$

$$= 5I_1 + I_2 \qquad \text{...(iv)}$$

Substituting the Eq. (iii) in the Eq. (ii),

$$V_2 = 4I_2 - I_1 - I_2$$

$$= -I_1 + 3I_2 \qquad \text{...(v)}$$

Comparing Eqs (iv) and (v) with Z-parameter equations, we get

$$\begin{bmatrix} Z_{11} & Z_{12} \\ Z_{21} & Z_{22} \end{bmatrix} = \begin{bmatrix} 5 & 1 \\ -1 & 3 \end{bmatrix}$$

***h*-parameters**

$$\Delta Z = Z_{11}Z_{22} - Z_{12}Z_{21}$$

$$= (5)\,(3) - (1)\,(-1) = 15 + 1 = 16$$

$$h_{11} = \frac{\Delta Z}{Z_{22}} = \frac{16}{3}\,\Omega$$

$$h_{12} = \frac{Z_{12}}{Z_{22}} = \frac{1}{3}$$

$$h_{21} = \frac{-Z_{21}}{Z_{22}} = \frac{1}{3}$$

$$h_{22} = \frac{1}{Z_{22}} = \frac{1}{3}\,℧$$

Hence, the *h*-parameters are

$$\begin{bmatrix} h_{11} & h_{12} \\ h_{22} & h_{22} \end{bmatrix} = \begin{bmatrix} \frac{16}{3} & \frac{1}{3} \\ \frac{1}{3} & \frac{1}{3} \end{bmatrix}$$

Example 11.24 *Find Y and Z-parameters for the network shown in Fig. 11.51.*

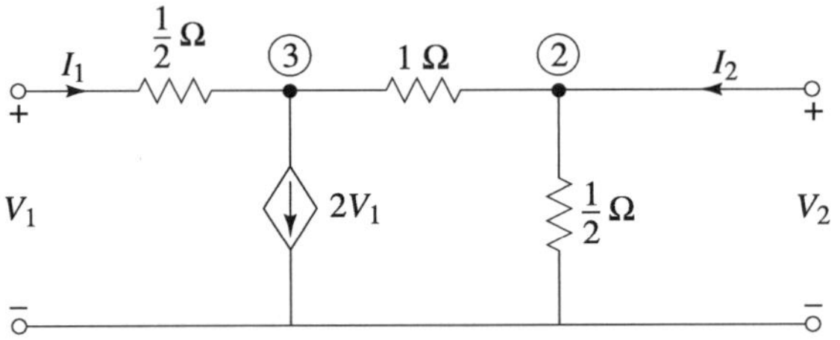

Fig. 11.51

Solution Applying KCL at Node 3,

$$2\,(V_1 - V_3) = 2V_1 + (V_3 - V_2)$$

$$V_3 = \frac{V_2}{3} \qquad \text{...(i)}$$

Now,
$$I_1 = 2V_1 + (V_3 - V_2)$$
$$= 2V_1 + \frac{V_2}{3} - V_2$$
$$= 2V_1 - \frac{2}{3} V_2 \quad \text{...(ii)}$$
$$I_2 = 2V_2 + (V_2 - V_3)$$
$$= 3V_2 - \frac{V_2}{3} = \frac{8}{3} V_2 \quad \text{...(iii)}$$

Comparing the Eqs (ii) and (iii) with Y-parameter equations, we get

$$\begin{bmatrix} Y_{11} & Y_{12} \\ Y_{21} & Y_{22} \end{bmatrix} = \begin{bmatrix} 2 & \frac{-2}{3} \\ 0 & \frac{8}{3} \end{bmatrix}$$

***Z*-parameters**

$$\Delta Y = Y_{11} Y_{22} - Y_{12} Y_{21} = (2)\left(\frac{8}{3}\right) - 0 = \frac{16}{3}$$

$$Z_{11} = \frac{Y_{22}}{\Delta Y} = \frac{\frac{8}{3}}{\frac{16}{3}} = \frac{1}{2}\,\Omega$$

$$Z_{12} = \frac{-Y_{12}}{\Delta Y} = \frac{-\left(\frac{-2}{3}\right)}{\frac{16}{3}} = \frac{1}{8}\,\Omega$$

$$Z_{21} = \frac{-Y_{21}}{\Delta Y} = \frac{0}{\frac{16}{3}} = 0$$

$$Z_{22} = \frac{Y_{11}}{\Delta Y} = \frac{2}{\frac{16}{3}} = \frac{3}{8}\,\Omega$$

Hence, the Z-parameters are

$$\begin{bmatrix} Z_{11} & Z_{12} \\ Z_{21} & Z_{22} \end{bmatrix} = \begin{bmatrix} \frac{1}{2} & \frac{1}{8} \\ 0 & \frac{3}{8} \end{bmatrix}$$

Example 11.25 *For the network shown in Fig. 11.52, find Y and Z-parameters.*

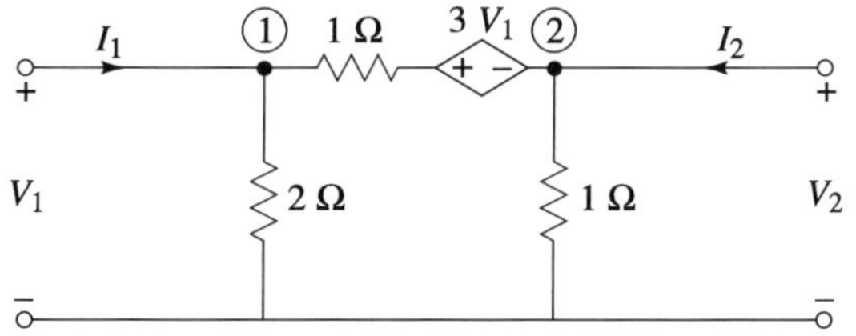

Fig. 11.52

Solution Applying KCL at Node 1,

$$I_1 = \frac{V_1}{2} + \frac{V_1 - 3V_1 - V_2}{1} = \frac{-3}{2}V_1 - V_2 \quad \text{...(i)}$$

Applying KCL at Node 2,

$$I_2 = \frac{V_2}{1} + \frac{V_2 + 3V_1 - V_1}{1}$$

$$= 2V_1 + 2V_2 \quad \text{...(ii)}$$

Comparing Eqs (i) and (ii) with Y-parameter equations, we get

$$\begin{bmatrix} Y_{11} & Y_{12} \\ Y_{21} & Y_{22} \end{bmatrix} = \begin{bmatrix} \frac{-3}{2} & -1 \\ 2 & 2 \end{bmatrix}$$

Z-parameters

$$\Delta Y = Y_{11}Y_{22} - Y_{12}Y_{21}$$

$$= \left(\frac{-3}{2}\right)(2) - (-1)(2) = -3 + 2 = -1$$

$$Z_{11} = \frac{Y_{22}}{\Delta Y} = \frac{2}{(-1)} = -2\ \Omega$$

$$Z_{12} = \frac{-Y_{12}}{\Delta Y} = -\frac{(-1)}{(-1)} = -1\ \Omega$$

$$Z_{21} = \frac{-Y_{21}}{\Delta Y} = -\frac{2}{(-1)} = 2\ \Omega$$

$$Z_{22} = \frac{Y_{11}}{\Delta Y} = \frac{\left(-\frac{3}{2}\right)}{(-1)} = \frac{3}{2}\ \Omega$$

Hence, the Z-parameters are

$$\begin{bmatrix} Z_{11} & Z_{12} \\ Z_{21} & Z_{22} \end{bmatrix} = \begin{bmatrix} -2 & -1 \\ 2 & \frac{3}{2} \end{bmatrix}$$

Example 11.26 *Find Z-parameters for the network shown in Fig. 11.53. Hence, find Y and h-parameters.*

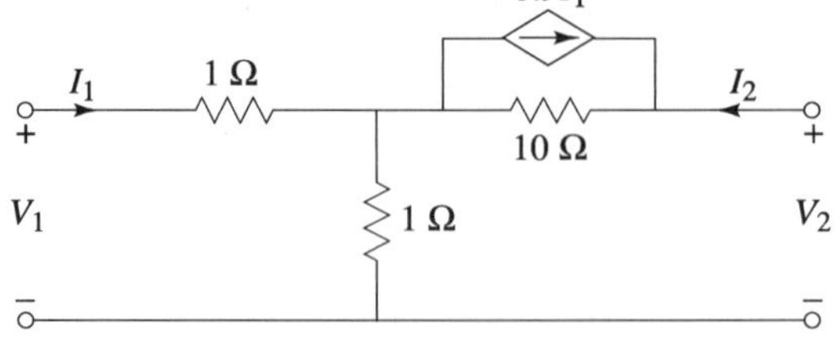

Fig. 11.53

Solution By source transformation technique,

Applying KVL to Mesh 1,

$$V_1 = 2I_1 + I_2 \quad \text{...(i)}$$

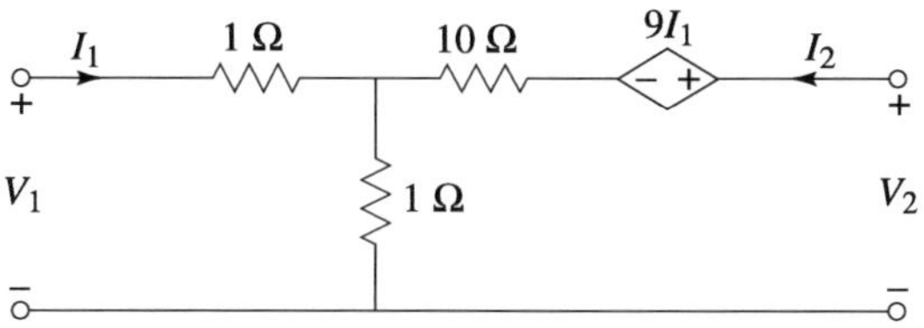

Fig. 11.54

Applying KVL to Mesh 2,

$$V_2 = 9I_1 + 10I_2 + (I_1 + I_2)$$

$$= 10I_1 + 11I_2 \qquad \ldots\text{(ii)}$$

Comparing Eqs (i) and (ii) with Z-parameter equations, we get

$$\begin{bmatrix} Z_{11} & Z_{12} \\ Z_{21} & Z_{22} \end{bmatrix} = \begin{bmatrix} 2 & 1 \\ 10 & 11 \end{bmatrix}$$

***Y*-parameters**

$$\Delta Z = Z_{11} Z_{22} - Z_{12} Z_{21}$$

$$= (2)(11) - (1)(10) = 22 - 10 = 12$$

$$Y_{11} = \frac{Z_{22}}{\Delta Z} = \frac{11}{12} \text{ ℧}$$

$$Y_{21} = \frac{-Z_{21}}{\Delta Z} = \frac{-10}{12} = \frac{-5}{6} \text{ ℧}$$

$$Y_{12} = \frac{-Z_{12}}{\Delta Z} = \frac{-1}{12} \text{ ℧}$$

$$Y_{22} = \frac{Z_{11}}{\Delta Z} = \frac{2}{12} = \frac{1}{6} \text{ ℧}$$

Hence, the *Y*-parameters are

$$\begin{bmatrix} Y_{11} & Y_{12} \\ Y_{21} & Y_{22} \end{bmatrix} = \begin{bmatrix} \frac{11}{12} & \frac{-1}{12} \\ \frac{-5}{6} & \frac{1}{6} \end{bmatrix}$$

***h*-parameters**

$$h_{11} = \frac{\Delta Z}{Z_{22}} = \frac{12}{11} \, \Omega$$

$$h_{21} = \frac{-Z_{21}}{Z_{22}} = \frac{-10}{11}$$

$$h_{12} = \frac{Z_{12}}{Z_{22}} = \frac{1}{11}$$

$$h_{22} = \frac{1}{Z_{22}} = \frac{1}{11} \text{ ℧}$$

Hence, *h*-parameters are

$$\begin{bmatrix} h_{11} & h_{12} \\ h_{21} & h_{22} \end{bmatrix} = \begin{bmatrix} \frac{12}{11} & \frac{1}{11} \\ \frac{-10}{11} & \frac{1}{11} \end{bmatrix}$$

Example 11.27 *Find Y and Z-parameters of the network shown in Fig. 11.55.*

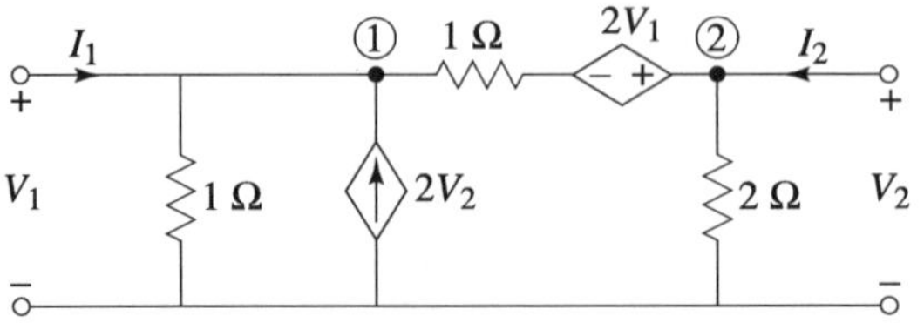

Fig. 11.55

Solution Applying KCL at Node 1,

$$I_1 + 2V_2 = \frac{V_1}{1} + \frac{3V_1 - V_2}{1}$$

$$I_1 = 4V_1 - 3V_2 \qquad \ldots\text{(i)}$$

Applying KCL at Node 2,

$$I_2 = \frac{V_2}{2} + \frac{V_2 - 2V_1 - V_1}{1}$$

$$I_2 = -3V_1 + 1.5V_2 \qquad \ldots\text{(ii)}$$

Comparing Eqs (i) and (ii) with Y-parameter equations, we get

$$\begin{bmatrix} Y_{11} & Y_{12} \\ Y_{21} & Y_{22} \end{bmatrix} = \begin{bmatrix} 4 & -3 \\ -3 & 1.5 \end{bmatrix}$$

Z-parameters

$$\Delta Y = Y_{11}Y_{22} - Y_{12}Y_{21}$$
$$= (4)(1.5) - (-3)(-3) = -3$$

$$Z_{11} = \frac{Y_{22}}{\Delta Y} = \frac{1.5}{-3} = -0.5\ \Omega$$

$$Z_{12} = \frac{-Y_{12}}{\Delta Y} = \frac{-(-3)}{-3} = -1\ \Omega$$

$$Z_{21} = \frac{-Y_{21}}{\Delta Y} = \frac{-(-3)}{-3} = -1\ \Omega$$

$$Z_{22} = \frac{Y_{11}}{\Delta Y} = \frac{4}{-3} = \frac{-4}{3}\ \Omega$$

Hence, the Z-parameters are

$$\begin{bmatrix} Z_{11} & Z_{12} \\ Z_{21} & Z_{22} \end{bmatrix} = \begin{bmatrix} -0.5 & -1 \\ -1 & \frac{-4}{3} \end{bmatrix}$$

Example 11.28 *Determine Y and Z-parameters for the network shown in Fig. 11.56.*

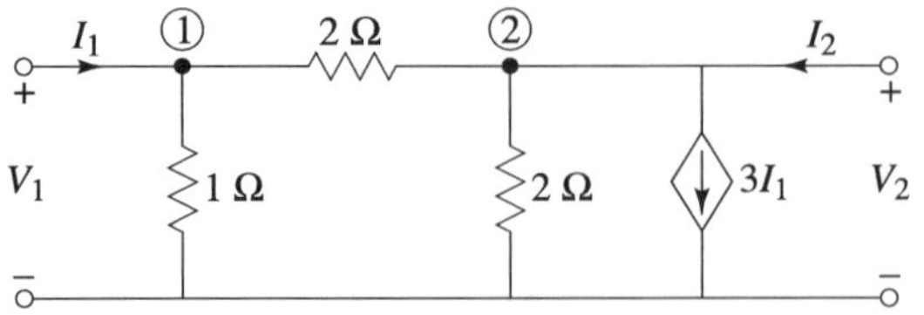

Fig. 11.56

Solution Applying KCL at Node 1,

$$I_1 = \frac{V_1}{1} + \frac{V_1 - V_2}{2}$$

$$I_1 = 1.5V_1 - 0.5V_2 \qquad \text{...(i)}$$

Applying KCL at Node 2,

$$I_2 = \frac{V_2}{2} + 3I_1 + \frac{V_2 - V_1}{2}$$

$$I_2 = \frac{V_2}{2} + \frac{V_2 - V_1}{2} + 3\,(1.5V_1 - 0.5V_2)$$

$$= 0.5V_2 + 0.5V_2 - 0.5V_1 + 4.5V_1 - 1.5V_2$$

$$= 4V_1 - 0.5V_2 \qquad \text{...(ii)}$$

Comparing Eqs (i) and (ii) with the Y-parameter equation, we get

$$\begin{bmatrix} Y_{11} & Y_{12} \\ Y_{21} & Y_{22} \end{bmatrix} = \begin{bmatrix} 1.5 & -0.5 \\ 4 & -0.5 \end{bmatrix}$$

Z-parameters

$$\Delta Y = Y_{11}Y_{22} - Y_{12}Y_{21}$$

$$= (1.5)\,(-0.5) - (-0.5)\,(4) = 1.25$$

$$Z_{11} = \frac{Y_{22}}{\Delta Y} = \frac{-0.5}{1.25} = -0.4\ \Omega$$

$$Z_{12} = \frac{-Y_{12}}{\Delta Y} = \frac{0.5}{1.25} = 0.4\ \Omega$$

$$Z_{21} = \frac{-Y_{21}}{\Delta Y} = \frac{-4}{1.25} = -3.2\ \Omega$$

$$Z_{22} = \frac{Y_{11}}{\Delta Y} = \frac{1.5}{1.25} = 1.2\ \Omega$$

Hence, the Z-parameters are

$$\begin{bmatrix} Z_{11} & Z_{12} \\ Z_{21} & Z_{22} \end{bmatrix} = \begin{bmatrix} -0.4 & 0.4 \\ -3.2 & 1.2 \end{bmatrix}$$

Example 11.29 *Determine the Y and Z-parameters for the network shown in Fig. 11.57.*

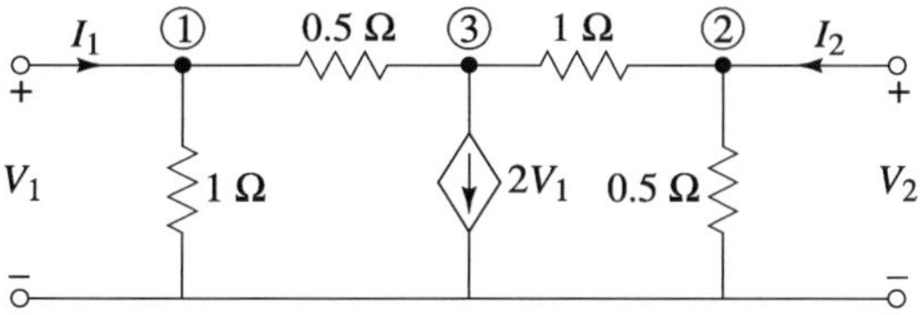

Fig. 11.57

Solution Applying KCL at Node 1,

$$I_1 = \frac{V_1}{1} + \frac{V_1 - V_3}{0.5}$$

$$I_1 = 3V_1 - 2V_3 \qquad \text{...(i)}$$

Applying KCL at Node 2,

$$I_2 = \frac{V_2}{0.5} + \frac{V_2 - V_3}{1}$$

$$I_2 = 3V_2 - V_3 \qquad \text{...(ii)}$$

Applying KCL at Node 3,

$$\frac{V_3 - V_1}{0.5} + 2V_1 + \frac{V_3 - V_2}{1} = 0$$

$$V_3 = \frac{1}{3}V_2 \qquad \text{...(iii)}$$

Substituting the Eq. (iii) in the Eqs (i) and (ii), we get

$$I_1 = 3V_1 - \frac{2}{3}V_2 \qquad \text{...(iv)}$$

$$I_2 = 0V_1 + \frac{8}{3}V_2 \qquad \text{...(v)}$$

Comparing Eqs (iv) and (v) with Y-parameter equations, we get

$$\begin{bmatrix} Y_{11} & Y_{12} \\ Y_{21} & Y_{22} \end{bmatrix} = \begin{bmatrix} 3 & \frac{-2}{3} \\ 0 & \frac{8}{3} \end{bmatrix}$$

Z-parameters

$$\Delta Y = Y_{11}Y_{22} - Y_{12}Y_{21}$$

$$= (3)\left(\frac{8}{3}\right) - 0 = 8$$

$$Z_{11} = \frac{Y_{22}}{\Delta Y} = \frac{\frac{8}{3}}{8} = \frac{1}{3}\,\Omega$$

$$Z_{12} = \frac{-Y_{12}}{\Delta Y} = \frac{\frac{2}{3}}{8} = \frac{1}{12}\,\Omega$$

$$Z_{21} = \frac{-Y_{12}}{\Delta Y}$$

$$Z_{22} = \frac{Y_{11}}{\Delta Y} = \frac{3}{8}\,\Omega$$

Hence, the Z-parameters are

$$\begin{bmatrix} Z_{11} & Z_{12} \\ Z_{21} & Z_{22} \end{bmatrix} = \begin{bmatrix} \frac{1}{3} & \frac{1}{12} \\ 0 & \frac{3}{8} \end{bmatrix}$$

Example 11.30 *Determine Z and Y-parameters of the network shown in Fig. 11.58.*

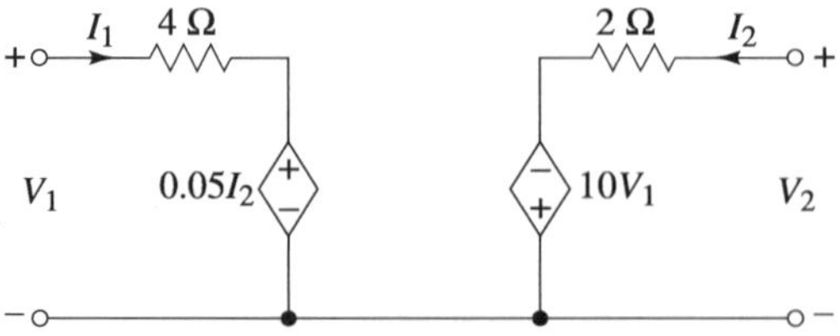

Fig. 11.58

Solution Applying KVL to Mesh 1,

$$V_1 - 4I_1 - 0.05\,I_2 = 0$$

$$V_1 = 4I_1 + 0.05I_2 \qquad \ldots\text{(i)}$$

Applying KVL to Mesh 2,

$$V_2 - 2I_2 + 10V_1 = 0$$

$$V_2 = 2I_2 - 10V_1 \qquad \ldots\text{(ii)}$$

Substituting the Eq. (i) in the Eq. (ii),

$$V_2 = 2I_2 - 40I_1 - 0.5I_2$$

$$= -40I_1 + 1.5I_2 \qquad \ldots\text{(iii)}$$

Comparing Eqs (i) and (iii) with Z-parameter equations, we get

$$\begin{bmatrix} Z_{11} & Z_{12} \\ Z_{21} & Z_{22} \end{bmatrix} = \begin{bmatrix} 4 & 0.05 \\ -40 & 1.5 \end{bmatrix}$$

***Y*-parameters**

$$\Delta Z = Z_{11}Z_{22} - Z_{12}Z_{21}$$

$$= (4)\,(1.5) - (0.05)\,(-40) = 6 + 2 = 8$$

$$Y_{11} = \frac{Z_{22}}{\Delta Z} = \frac{1.5}{8}\ \mho$$

$$Y_{12} = \frac{-Z_{12}}{\Delta Z} = \frac{-0.05}{8}\ \mho$$

$$Y_{21} = \frac{-Z_{21}}{\Delta Z} = \frac{-(-40)}{8} = \frac{40}{8}\ \mho$$

$$Y_{22} = \frac{Z_{11}}{\Delta Z} = \frac{4}{8}\ \mho$$

Hence, the *Y*-parameters are

$$\begin{bmatrix} Y_{11} & Y_{12} \\ Y_{21} & Y_{22} \end{bmatrix} = \begin{bmatrix} \frac{1.5}{8} & \frac{-0.05}{8} \\ \frac{40}{8} & \frac{4}{8} \end{bmatrix}$$

Example 11.31 *Determine Z and Y-parameters of the network shown in Fig. 11.59.*

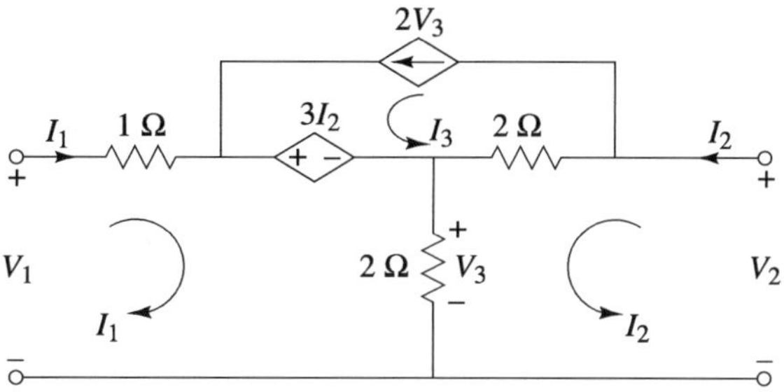

Fig. 11.59

Solution Applying KVL to Mesh 1,

$$V_1 - I_1 - 3I_2 - 2(I_1 + I_2) = 0$$

$$V_1 = 3I_1 + 5I_2 \qquad \ldots\text{(i)}$$

Applying KVL to Mesh 2,

$$V_2 - 2(I_2 - I_3) - 2(I_1 + I_2) = 0$$

$$V_2 - 2I_2 + 2I_3 - 2I_1 - 2I_2 = 0$$

$$V_2 = 2I_1 + 4I_2 - 2I_3 \qquad \ldots\text{(ii)}$$

Writing equation for Mesh 3,

$$I_3 = 2V_3 \qquad \ldots\text{(iii)}$$

From Fig. 11.59,

$$V_3 = 2\,(I_1 + I_2)$$

$$I_3 = 2V_3 = 4I_1 + 4I_2 \qquad \ldots\text{(iv)}$$

Substituting the Eq. (iv) in the Eq. (ii),

$$V_2 = -6I_1 - 4I_2 \qquad \ldots\text{(v)}$$

Comparing Eqs (i) and (v) with *Z*-parameter equations, we get

$$\begin{bmatrix} Z_{11} & Z_{12} \\ Z_{21} & Z_{22} \end{bmatrix} = \begin{bmatrix} 3 & 5 \\ -6 & -4 \end{bmatrix}$$

***Y*-parameters**

$$\Delta Z = Z_{11}Z_{22} - Z_{12}Z_{21}$$

$$= (3)\,(-4) - (5)\,(-6) = 18$$

$$Y_{11} = \frac{Z_{22}}{\Delta Z} = \frac{-4}{18} = \frac{-2}{9}\ \mho$$

$$Y_{21} = \frac{-Z_{21}}{\Delta Z} = \frac{-(-6)}{18} = \frac{1}{3}\ \mho$$

$$Y_{12} = \frac{-Z_{12}}{\Delta Z} = \frac{-5}{18}\ \mho$$

$$Y_{22} = \frac{Z_{11}}{\Delta Z} = \frac{3}{18}\ \mho$$

Hence, *Y*-parameters are

$$\begin{bmatrix} Y_{11} & Y_{12} \\ Y_{21} & Y_{22} \end{bmatrix} = \begin{bmatrix} \frac{-2}{9} & \frac{-5}{18} \\ \frac{1}{3} & \frac{3}{18} \end{bmatrix}$$

11.9 INTERCONNECTION OF TWO-PORT NETWORKS

We shall now discuss the various types of interconnections of two-port networks, namely, cascade, parallel, series, series–parallel and parallel-series. We shall derive the relation between the input and output quantities of the combined two-port networks.

11.9.1 Cascade Connection

(a) Transmission parameter representation Figure 11.60 shows two-port networks connected in cascade. In the cascade connection, the output port of the first network becomes the input port of the second network. Since it is assumed that input and output currents are positive when they enter the network, we have

$$I_1' = -I_2$$

Let A_1, B_1, C_1, D_1 be the transmission parameters of the network N_1 and A_2, B_2, C_2, D_2 be the transmission parameters of the network N_2.

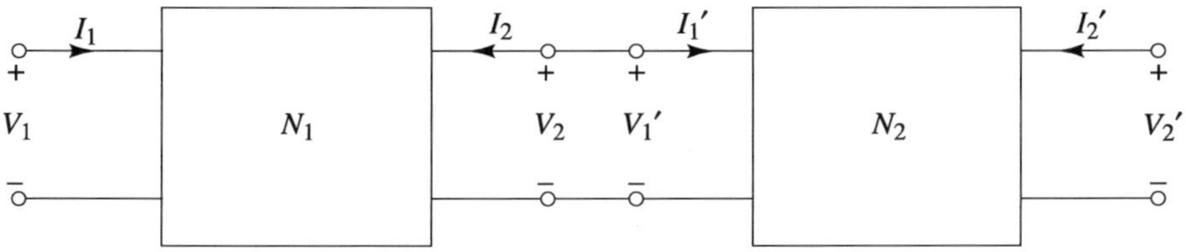

Fig. 11.60

For the network N_1,

$$\begin{bmatrix} V_1 \\ I_1 \end{bmatrix} = \begin{bmatrix} A_1 & B_1 \\ C_1 & D_1 \end{bmatrix} \begin{bmatrix} V_2 \\ -I_2 \end{bmatrix} \qquad \text{...(11.181)}$$

For the network N_2,

$$\begin{bmatrix} V_1' \\ I_1' \end{bmatrix} = \begin{bmatrix} A_2 & B_2 \\ C_2 & D_2 \end{bmatrix} \begin{bmatrix} V_2' \\ -I_2' \end{bmatrix} \qquad \text{...(11.182)}$$

Since $V_1' = V_2$ and $I_1' = -I_2$, we can write

$$\begin{bmatrix} V_2 \\ -I_2 \end{bmatrix} = \begin{bmatrix} A_2 & B_2 \\ C_2 & D_2 \end{bmatrix} \begin{bmatrix} V_2' \\ -I_2' \end{bmatrix} \qquad \text{...(11.183)}$$

Combining Eqs (11.181) and (11.183), we get

$$\begin{bmatrix} V_1 \\ I_1 \end{bmatrix} = \begin{bmatrix} A_1 & B_1 \\ C_1 & D_1 \end{bmatrix} \begin{bmatrix} A_2 & B_2 \\ C_2 & D_2 \end{bmatrix} \begin{bmatrix} V_2' \\ -I_2' \end{bmatrix} = \begin{bmatrix} A & B \\ C & D \end{bmatrix} \begin{bmatrix} V_2' \\ -I_2' \end{bmatrix} \qquad \text{...(11.184)}$$

Hence,

$$\begin{bmatrix} A & B \\ C & D \end{bmatrix} = \begin{bmatrix} A_1 & B_1 \\ C_1 & D_1 \end{bmatrix} \begin{bmatrix} A_2 & B_2 \\ C_2 & D_2 \end{bmatrix} \qquad \text{...(11.185)}$$

Equation (11.185) shows that the resultant *ABCD* matrix of the cascade connection is the product of the individual *ABCD* matrices.

(b) Inverse Transmission parameter representation Figure 11.61 shows two-port networks connected in cascade. Since it is assumed that input and output currents are positive when they enter the network, we have

$$-I_1' = I_2$$

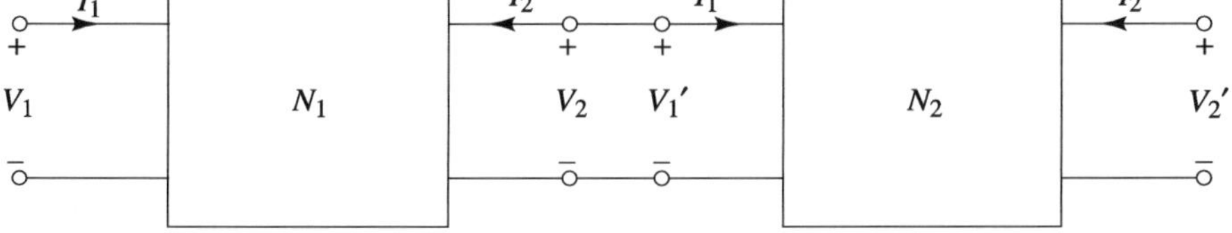

Fig. 11.61

Let A_1', B_1', C_1', D_1' be the transmission parameters of the network N_1 and A_2', B_2', C_2', D_2' be the transmission parameters of the network N_2.

For the network N_1,

$$\begin{bmatrix} V_2 \\ I_2 \end{bmatrix} = \begin{bmatrix} A_1' & B_1' \\ C_1' & D_1' \end{bmatrix} \begin{bmatrix} V_1 \\ -I_1 \end{bmatrix} \qquad \text{...(11.186)}$$

For the network N_2,

$$\begin{bmatrix} V_2' \\ I_2' \end{bmatrix} = \begin{bmatrix} A_2' & B_2' \\ C_2' & D_2' \end{bmatrix} \begin{bmatrix} V_2' \\ -I_1' \end{bmatrix} \qquad \ldots(11.187)$$

Since $V_1' = V_2$ and $-I_1' = I_2$, we can write

$$\begin{bmatrix} V_2' \\ I_2' \end{bmatrix} = \begin{bmatrix} A_2' & B_2' \\ C_2' & D_2' \end{bmatrix} \begin{bmatrix} V_2 \\ I_2 \end{bmatrix} \qquad \ldots(11.188)$$

Combining equations (11.186) and (11.188), we get

$$\begin{bmatrix} V_2' \\ I_2' \end{bmatrix} = \begin{bmatrix} A_1' & B_1' \\ C_1' & D_1' \end{bmatrix} \begin{bmatrix} A_2' & B_2' \\ C_2' & D_2' \end{bmatrix} \begin{bmatrix} V_1 \\ -I_1 \end{bmatrix} = \begin{bmatrix} A' & B' \\ C' & D' \end{bmatrix} \begin{bmatrix} V_1 \\ -I_1 \end{bmatrix} \qquad \ldots(11.189)$$

Hence,

$$\begin{bmatrix} A' & B' \\ C' & D' \end{bmatrix} = \begin{bmatrix} A_1' & B_1' \\ C_1' & D_1' \end{bmatrix} \begin{bmatrix} A_2' & B_2' \\ C_2' & D_2' \end{bmatrix} \qquad \ldots(11.190)$$

Equation (11.190) shows that the resultant $A'B'C'D'$ matrix of the cascade connection is the product of the individual $A'BC'D'$ matrices.

11.9.2 Parallel Connection

Figure 11.62 shows two-port networks connected in parallel. In the parallel connection, the two networks have the same input voltages and the same output voltages.

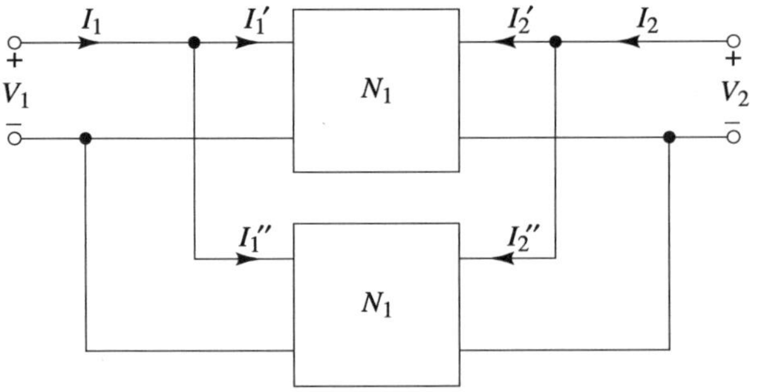

Fig. 11.62

Let Y_{11}', Y_{12}', Y_{21}', Y_{22}' be the Y-parameters of the network N_1 and Y_1'', Y_{12}'', Y_{21}'', Y_{22}'' be the Y-parameters of the network N_2.

For the network N_1,

$$\begin{bmatrix} I_1' \\ I_2' \end{bmatrix} = \begin{bmatrix} Y_{11}' & Y_{12}' \\ Y_{21}' & Y_{22}' \end{bmatrix} \begin{bmatrix} V_1 \\ V_2 \end{bmatrix} \qquad \ldots(11.191)$$

For the network N_2,

$$\begin{bmatrix} I_1'' \\ I_2'' \end{bmatrix} = \begin{bmatrix} Y_{11}'' & Y_{12}'' \\ Y_{21}'' & Y_{22}'' \end{bmatrix} \begin{bmatrix} V_1 \\ V_2 \end{bmatrix} \qquad \ldots(11.192)$$

For the combined network, $I_1 = I_1' + I_1''$ and $I_2 = I_2' + I_2''$.

Hence,
$$\begin{bmatrix} I_1 \\ I_2 \end{bmatrix} = \begin{bmatrix} I_1' + I_1'' \\ I_2' + I_2'' \end{bmatrix} = \begin{bmatrix} Y_{11}' + Y_{11}'' & Y_{12}' + Y_{12}'' \\ Y_{21}' + Y_{21}'' & Y_{22}' + Y_{22}'' \end{bmatrix} \begin{bmatrix} V_1 \\ V_2 \end{bmatrix}$$
$$= \begin{bmatrix} Y_{11} & Y_{12} \\ Y_{21} & Y_{22} \end{bmatrix} \begin{bmatrix} V_1 \\ V_2 \end{bmatrix} \qquad \text{...(11.193)}$$

Thus, the resultant Y-parameter matrix for parallel connected networks is the sum of Y matrices of each individual two-port networks.

11.9.3 Series Connection

Figure 11.63 shows two-port networks connected in series. In a series connection, both the networks carry the same input current. Their output currents are also equal.

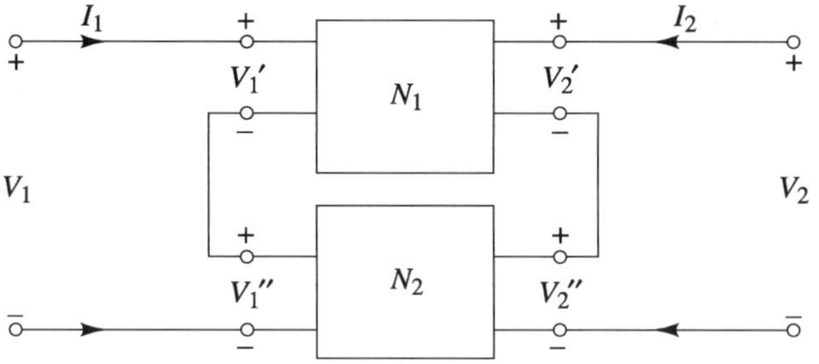

Fig. 11.63

Let $Z_{11}', Z_{12}', Z_{21}', Z_{22}'$ be the Z-parameters of the network N_1 and $Z_{11}'', Z_{12}'', Z_{21}'', Z_{22}''$ be the Z-parameters of the network N_2.

For the network N_1,

$$\begin{bmatrix} V_1' \\ V_2' \end{bmatrix} = \begin{bmatrix} Z_{11}' & Z_{12}'' \\ Z_{21}' & Z_{22}'' \end{bmatrix} \begin{bmatrix} I_1 \\ I_2 \end{bmatrix} \qquad \text{...(11.194)}$$

For the network N_2,

$$\begin{bmatrix} V_1'' \\ V_2'' \end{bmatrix} = \begin{bmatrix} Z_{11}' & Z_{12}'' \\ Z_{21}' & Z_{22}'' \end{bmatrix} \begin{bmatrix} I_1 \\ I_2 \end{bmatrix} \qquad \text{...(11.195)}$$

For the combined network $V_1 = V_1' + V_1''$ and $V_2 = V_2' + V_2''$.

Hence,
$$\begin{bmatrix} V_1 \\ V_2 \end{bmatrix} = \begin{bmatrix} V_1' + V_1'' \\ V_2' + V_2'' \end{bmatrix} = \begin{bmatrix} Z_{11}' + Z_{11}'' & Z_{12}' + Z_{12}'' \\ Z_{21}' + Z_{21}'' & Z_{22}' + Z_{22}'' \end{bmatrix} \begin{bmatrix} I_1 \\ I_2 \end{bmatrix}$$
$$= \begin{bmatrix} Z_{11} & Z_{12} \\ Z_{21} & Z_{22} \end{bmatrix} \begin{bmatrix} I_1 \\ I_2 \end{bmatrix} \qquad \text{...(11.196)}$$

Thus, the resultant Z-parameter matrix for the series-connected networks is the sum of Z matrices of each individual two-port network.

11.9.4 Series–Parallel Connection

Figure 11.64 shows two networks connected in series–parallel. Here, the input ports of two networks are connected in series and the output ports are connected in parallel.

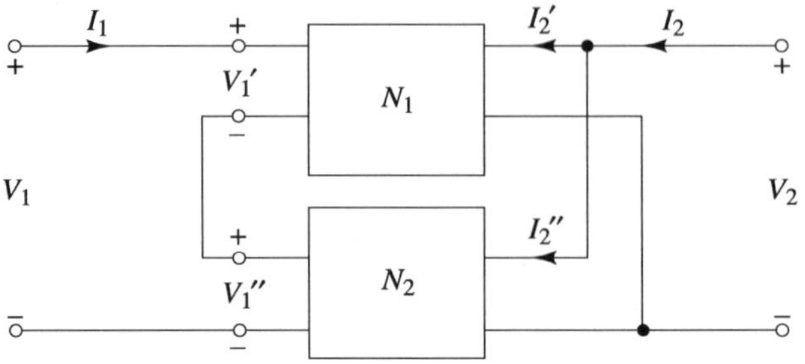

Fig. 11.64

Let $h'_{11}, h'_{12}, h'_{21}, h'_{22}$ be the h-parameters of the network N_1 and $h''_{11}, h''_{12}, h''_{21}, h''_{22}$ be the h-parameters of the network N_2.

For the network N_1,

$$\begin{bmatrix} V_1' \\ V_2' \end{bmatrix} = \begin{bmatrix} h_{11}' & h_{12}' \\ h_{21}' & h_{22}' \end{bmatrix} \begin{bmatrix} I_1 \\ V_2 \end{bmatrix} \qquad \ldots(11.197)$$

For the network N_2,

$$\begin{bmatrix} V_1'' \\ I_2'' \end{bmatrix} = \begin{bmatrix} h_{11}'' & h_{12}'' \\ h_{21}'' & h_{22}'' \end{bmatrix} \begin{bmatrix} I_1 \\ V_2 \end{bmatrix} \qquad \ldots(11.198)$$

For the combined network, $V_1 = V_1' + V_1''$ and $I_2 = I_2' + I_2''$

Hence,
$$\begin{bmatrix} V_1 \\ I_2 \end{bmatrix} = \begin{bmatrix} V_1' + V_1'' \\ I_2' + I_2'' \end{bmatrix} = \begin{bmatrix} h_{11}' + h_{11}'' & h_{12}' + h_{12}'' \\ h_{21}' + h_{21}'' & h_{22}' + h_{22}'' \end{bmatrix} \begin{bmatrix} I_1 \\ V_2 \end{bmatrix}$$

$$= \begin{bmatrix} h_{11} & h_{12} \\ h_{21} & h_{22} \end{bmatrix} \begin{bmatrix} I_1 \\ V_2 \end{bmatrix} \qquad \ldots(11.199)$$

Thus, the resultant h-parameter matrix is the sum of h-parameter matrices of each individual two-port networks.

11.9.5 Parallel–Series Connection

Figure 11.65 shows two networks connected in parallel-series. Here the input ports of two networks are connected in parallel and the output ports are connected in series.

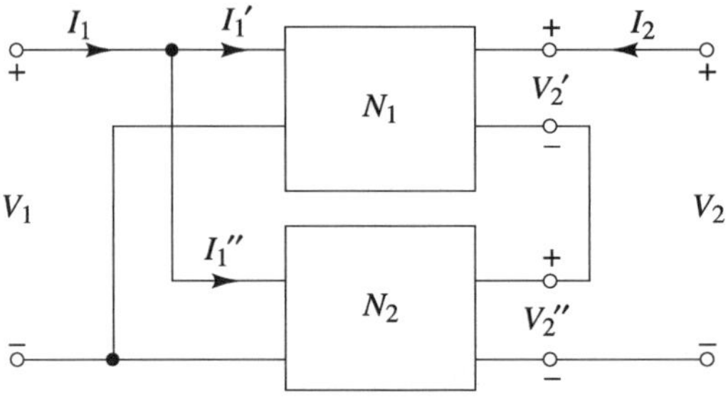

Fig. 11.65

Let g_{11}', g_{12}', g_{21}', g_{22}' be the g-parameters of the network N_1 and g_{11}'', g_{12}'', g_{21}'', g_{22}'' be the g-parameters of the network N_2.

For the network N_1,

$$\begin{bmatrix} I_1' \\ V_2' \end{bmatrix} = \begin{bmatrix} g_{11}' & g_{12}' \\ g_{21}' & g_{22}' \end{bmatrix} \begin{bmatrix} V_1 \\ I_2 \end{bmatrix} \qquad \ldots(11.200)$$

For the network N_2,

$$\begin{bmatrix} I_1'' \\ V_2'' \end{bmatrix} = \begin{bmatrix} g_{11}'' & g_{12}'' \\ g_{21}'' & g_{22}'' \end{bmatrix} \begin{bmatrix} V_1 \\ I_2 \end{bmatrix} \qquad \ldots(11.201)$$

For the combined network, $I_1 = I_1' + I_1''$ and $V_2 = V_2' + V_2''$

Hence,

$$\begin{bmatrix} I_1 \\ V_2 \end{bmatrix} = \begin{bmatrix} I_1' + I_1'' \\ V_2' + V_2'' \end{bmatrix} = \begin{bmatrix} g_{11}' + g_{11}'' & g_{12}' + g_{12}'' \\ g_{21}' + g_{21}'' & g_{22}' + g_{22}'' \end{bmatrix} \begin{bmatrix} V_1 \\ I_2 \end{bmatrix}$$

$$= \begin{bmatrix} g_{11} & g_{12} \\ g_{21} & g_{22} \end{bmatrix} \begin{bmatrix} V_1 \\ I_2 \end{bmatrix} \qquad \ldots(11.202)$$

Thus, the resultant g-parameter matrix is the sum of the g-parameter matrices of each individual two-port network.

Example 11.32 *Two identical sections of the network shown in Fig. 11.66 are connected in cascade. Obtain the transmission parameters of the overall connection.*

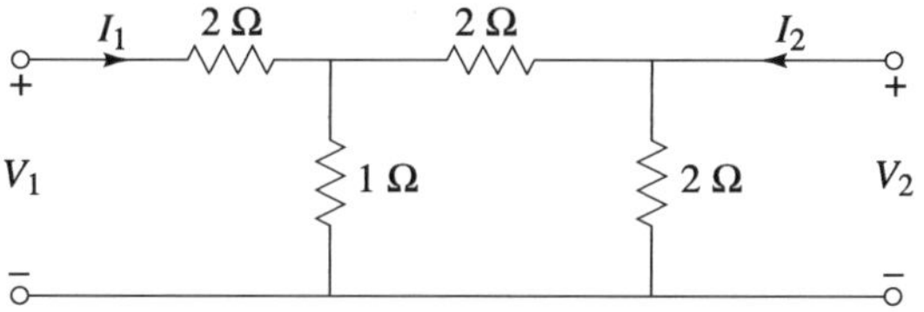

Fig. 11.66

Solution

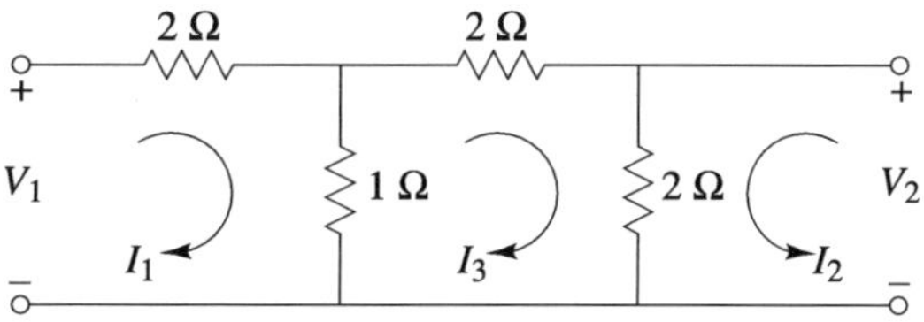

Fig. 11.67

Applying KVL to Mesh 1,

$$V_1 = 3I_1 - I_3 \qquad \ldots(\text{i})$$

Applying KVL to Mesh 2,

$$V_2 = 2I_2 + 2I_3 \qquad \ldots(\text{ii})$$

Applying KVL to Mesh 3,

$$I_1 - 2I_2 - 5I_3 = 0$$

$$I_3 = \frac{1}{5} I_1 - \frac{2}{5} I_2 \qquad \ldots(\text{iii})$$

Substituting the Eq. (iii) in the Eq. (i),

$$V_1 = 3I_1 - \left(\frac{1}{5} I_1 - \frac{2}{5} I_2\right)$$

$$= \frac{14}{5} I_1 + \frac{2}{5} I_2 \qquad \ldots(\text{iv})$$

Substituting the Eq. (iii) in the Eq. (ii),

$$V_2 = 2I_2 + 2\left(\frac{1}{5}I_1 - \frac{2}{5}I_2\right) = \frac{2}{5}I_1 + \frac{6}{5}I_2$$

$$I_1 = \frac{5}{2}V_2 - 3I_2 \qquad \text{...(v)}$$

Substituting the Eq. (v) in the Eq. (iv),

$$V_1 = \frac{14}{5}\left(\frac{5}{2}V_2 - 3I_2\right) + \frac{2}{5}I_2 = 7V_2 - 8I_2 \qquad \text{...(vi)}$$

Comparing the Eqs (vi) and (v) with *ABCD* parameter equations, we get

$$\begin{bmatrix} A_1 & B_1 \\ C_1 & D_1 \end{bmatrix} = \begin{bmatrix} 7 & 8 \\ 2.5 & 3 \end{bmatrix}$$

Hence, transmission parameters of the overall cascaded network are

$$\begin{bmatrix} A & B \\ C & D \end{bmatrix} = \begin{bmatrix} 7 & 8 \\ 2.5 & 3 \end{bmatrix}\begin{bmatrix} 7 & 8 \\ 2.5 & 3 \end{bmatrix} = \begin{bmatrix} 69 & 80 \\ 25 & 29 \end{bmatrix}$$

Example 11.33 *Determine ABCD parameters for the ladder network shown in Fig. 11.68.*

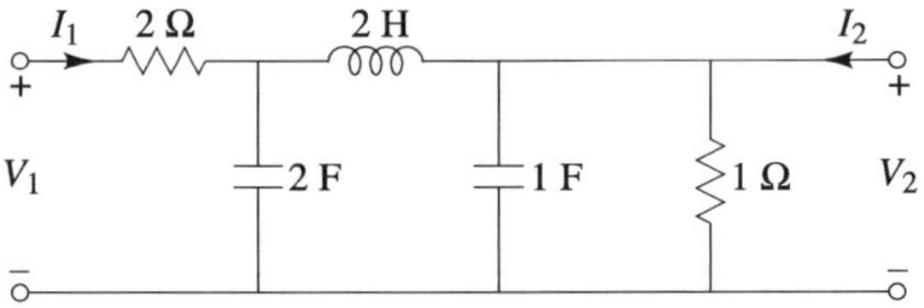

Fig. 11.68

Solution The above network can be considered as a cascade connection of two networks N_1 and N_2.

For the network N_1

Applying KVL to Mesh 1,

$$V_1 = \left(2 + \frac{1}{2s}\right)I_1 + \frac{1}{2s}I_2 \qquad \text{...(i)}$$

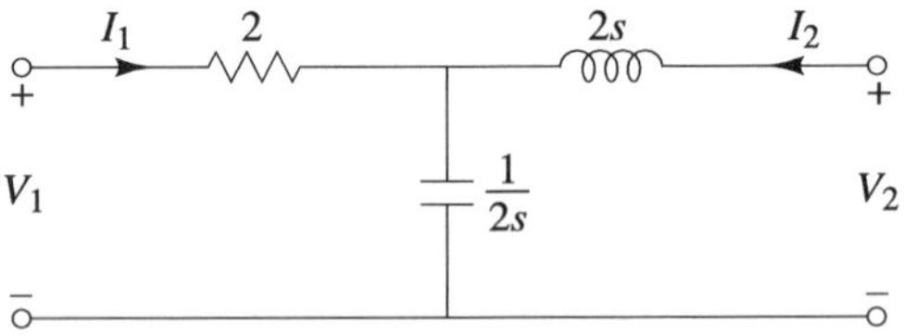

Fig. 11.69

Applying KVL to Mesh 2,

$$V_2 = \frac{1}{2s}I_1 + \left(2s + \frac{1}{2s}\right)I_2 \qquad \text{...(ii)}$$

From the Eq. (ii),

$$I_1 = 2s\,V_2 - (4s^2 + 1)\,I_2 \qquad \text{...(iii)}$$

Substituting the Eq. (iii) in the Eq. (i),

$$V_1 = \left(2 + \frac{1}{2s}\right)[2s\,V_2 - (4s^2 + 1)\,I_2] + \frac{1}{2s}I_2$$

$$= (4s + 1)\,V_2 - (8s^2 + 2s + 2)\,I_2 \qquad \text{...(iv)}$$

Comparing Eqs (iv) and (iii) with *ABCD* parameter equations, we get

$$\begin{bmatrix} A_1 & B_1 \\ C_1 & D_1 \end{bmatrix} = \begin{bmatrix} 4s+1 & 8s^2+2s+2 \\ 2s & 4s^2+1 \end{bmatrix}$$

For the network N_2

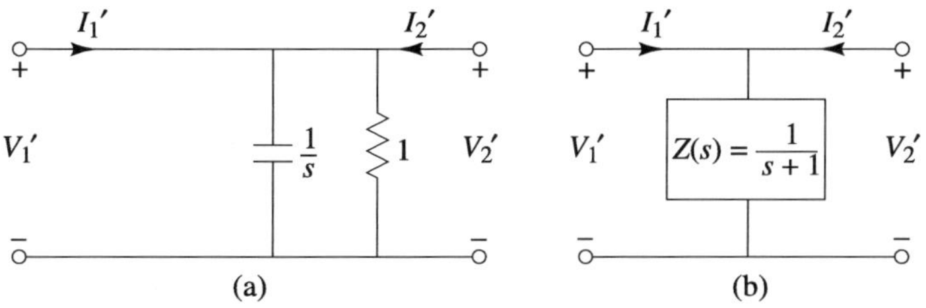

Fig. 11.70

Applying KVL to Mesh 1,

$$V_1' = \frac{1}{s+1} I_1' + \frac{1}{s+1} I_2' \qquad \text{...(i)}$$

Applying KVL to Mesh 2,

$$V_2' = \frac{1}{s+1} I_1' + \frac{1}{s+1} I_2' \qquad \text{...(ii)}$$

From the Eq. (ii),

$$I_1' = (s+1)\, V_2' - I_2' \qquad \text{...(iii)}$$

Also, $$V_1' = V_2' \qquad \text{...(iv)}$$

Comparing Eqs (iv) and (iii) with *ABCD* parameter equations, we get

$$\begin{bmatrix} A_2 & B_2 \\ C_2 & D_2 \end{bmatrix} = \begin{bmatrix} 1 & 0 \\ s+1 & 1 \end{bmatrix}$$

Hence, overall *ABCD* parameters are

$$\begin{bmatrix} A & B \\ C & D \end{bmatrix} = \begin{bmatrix} 4s+1 & 8s^2+2s+2 \\ 2s & 4s^2+1 \end{bmatrix} \begin{bmatrix} 1 & 0 \\ s+1 & 1 \end{bmatrix}$$

$$= \begin{bmatrix} 8s^3+10s^2+8s+3 & 8s^2+2s+2 \\ 4s^3+4s^2+3s+1 & 4s^2+1 \end{bmatrix}$$

Example 11.34 *Determine Y-parameters for the network shown in Fig. 11.71.*

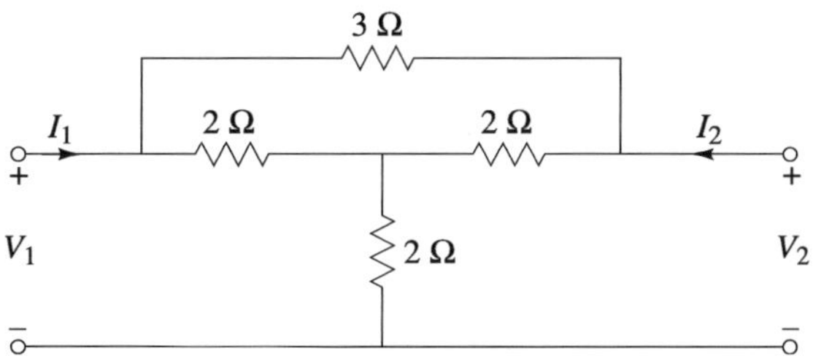

Fig. 11.71

Solution The above network can be considered as a parallel connection of two networks, N_1 and N_2.

For the network N_1

Applying KCL at Node 3,

$$I_1' + I_2' = \frac{V_3}{2} \qquad \ldots\text{(i)}$$

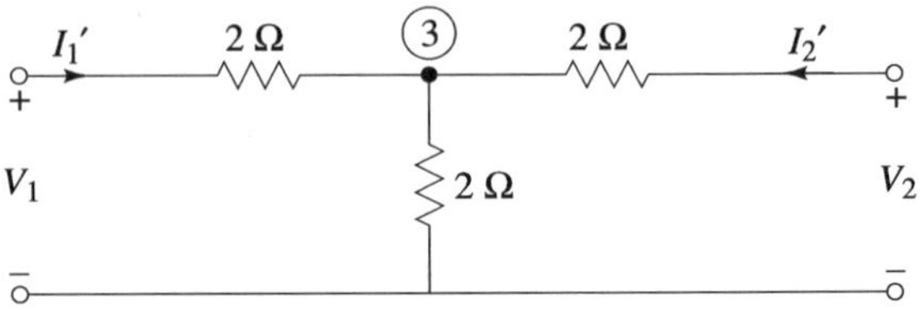

Fig. 11.72

From Fig. 11.72,

$$I_1' = \frac{V_1 - V_3}{2} \qquad \ldots\text{(ii)}$$

$$I_2' = \frac{V_2 - V_3}{2} \qquad \ldots\text{(iii)}$$

Substituting Eqs (ii) and (iii) in the Eq (i),

$$\frac{V_1 - V_3}{2} + \frac{V_2 - V_3}{2} = \frac{V_3}{2}$$

$$3V_3 = V_1 + V_2$$

$$V_3 = \frac{V_1}{3} + \frac{V_2}{3} \qquad \ldots\text{(iv)}$$

Substituting the Eq. (iv) in the Eq. (ii),

$$I_1' = \frac{V_1}{2} - \frac{1}{2}\left(\frac{V_1}{3} + \frac{V_2}{3}\right)$$

$$= \frac{1}{3}V_1 - \frac{1}{6}V_2 \qquad \ldots\text{(v)}$$

Substituting the Eq. (iv) in the Eq. (iii),

$$I_2' = \frac{V_2}{2} - \frac{1}{2}\left(\frac{V_1}{3} + \frac{V_2}{3}\right)$$

$$= -\frac{1}{6}V_1 + \frac{1}{3}V_2 \qquad \ldots\text{(vi)}$$

Comparing Eqs (v) and (vi) with Y-parameter equations, we get

$$\begin{bmatrix} Y_{11}' & Y_{12}' \\ Y_{21}' & Y_{22}' \end{bmatrix} = \begin{bmatrix} \frac{1}{3} & \frac{-1}{6} \\ \frac{-1}{6} & \frac{1}{3} \end{bmatrix}$$

For the network N_2

$$I_1'' = -I_2'' = \frac{V_1 - V_2}{3}$$

$$= \frac{1}{3}V_1 - \frac{1}{3}V_2$$

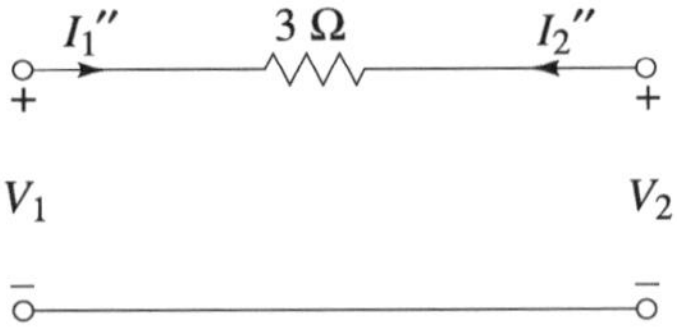

Fig. 11.73

Hence, the Y-parameters are

$$\begin{bmatrix} Y_{11}'' & Y_{12}'' \\ Y_{21}'' & Y_{22}'' \end{bmatrix} = \begin{bmatrix} \frac{1}{3} & \frac{-1}{3} \\ \frac{-1}{3} & \frac{1}{3} \end{bmatrix}$$

The overall Y-parameters of the network are

$$\begin{bmatrix} Y_{11} & Y_{12} \\ Y_{21} & Y_{22} \end{bmatrix} = \begin{bmatrix} Y_{11}' + Y_{11}'' & Y_{12}' + Y_{12}'' \\ Y_{21}' + Y_{21}'' & Y_{22}' + Y_{22}'' \end{bmatrix}$$

$$= \begin{bmatrix} \frac{1}{3}+\frac{1}{3} & \frac{-1}{6}-\frac{1}{3} \\ \frac{-1}{6}-\frac{1}{3} & \frac{1}{3}+\frac{1}{3} \end{bmatrix} = \begin{bmatrix} \frac{2}{3} & \frac{-1}{2} \\ \frac{-1}{2} & \frac{2}{3} \end{bmatrix}$$

Example 11.35 *Find Y-parameters for the network shown in Fig. 11.74.*

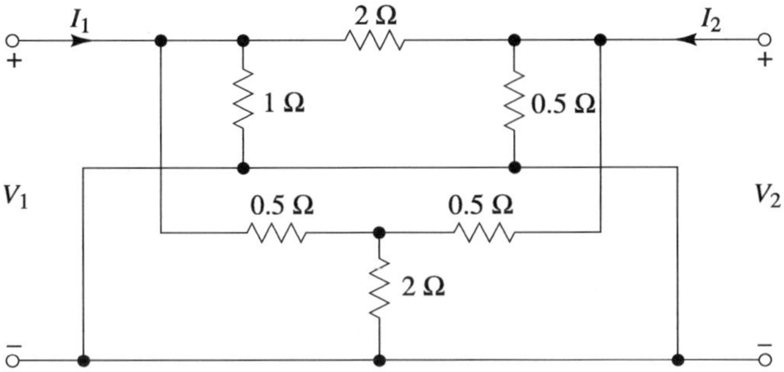

Fig. 11.74

Solution The above network can be considered as a parallel combination of two networks N_1 and N_2.

For the network N_1

Applying KCL at Node 1,

$$I_1' = \frac{V_1}{1} + \frac{V_1 - V_2}{2}$$

$$= \frac{3}{2}V_1 - \frac{1}{2}V_2 \quad \text{...(i)}$$

Applying KCL at Node 2,

$$I_2' = \frac{V_2}{0.5} + \frac{V_2 - V_1}{2}$$

$$= \frac{-1}{2}V_1 + \frac{5}{2}V_2 \quad \text{...(ii)}$$

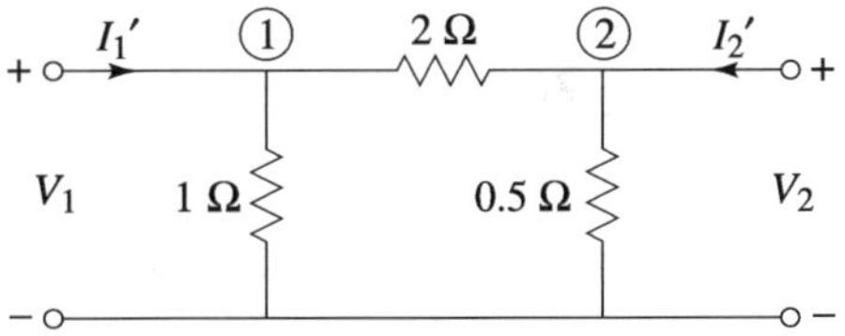

Fig. 11.75

Comparing Eqs (i) and (ii) with Y-parameter equation, we get

$$\begin{bmatrix} Y_{11}' & Y_{12}' \\ Y_{21}' & Y_{22}' \end{bmatrix} = \begin{bmatrix} \frac{3}{2} & \frac{-1}{2} \\ \frac{-1}{2} & \frac{5}{2} \end{bmatrix}$$

For the network N_2

Applying KCL at Node 3,

$$I_1'' + I_2'' = \frac{V_3}{2} \quad \text{...(i)}$$

where

$$I_1'' = \frac{V_1 - V_3}{0.5} = 2V_1 - 2V_3$$

$$I_2'' = \frac{V_2 - V_3}{0.5} = 2V_2 - 2V_3$$

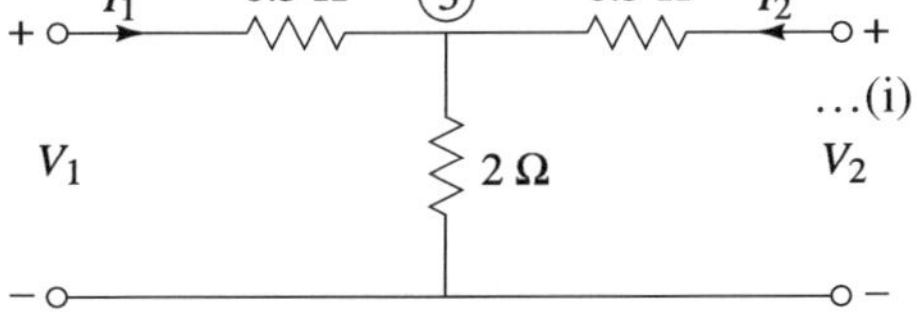

Fig. 11.76

$$2V_1 - 2V_3 + 2V_2 - 2V_3 = 0.5V_3$$

$$4.5V_3 = 2V_1 + 2V_2$$

$$V_3 = \frac{4}{9}V_1 + \frac{4}{9}V_2 \qquad \text{...(ii)}$$

$$I_1'' = 2V_1 - 2V_3 = 2V_1 - 2\left(\frac{4}{9}V_1 + \frac{4}{9}V_2\right)$$

$$= \frac{10}{9}V_1 - \frac{8}{9}V_2 \qquad \text{...(iii)}$$

and

$$I_2'' = 2V_2 - 2V_3$$

$$= 2V_2 - 2\left(\frac{4}{9}V_1 + \frac{4}{9}V_2\right)$$

$$= -\frac{8}{9}V_1 + \frac{10}{9}V_2 \qquad \text{...(iv)}$$

Comparing Eqs (iii) and (iv) with Y-parameter equations, we get

$$\begin{bmatrix} Y_{11}'' & Y_{12}'' \\ Y_{21}'' & Y_{22}'' \end{bmatrix} = \begin{bmatrix} \frac{10}{9} & \frac{-8}{9} \\ \frac{-8}{9} & \frac{10}{9} \end{bmatrix}$$

Hence, overall Y-parameters of the network are

$$\begin{bmatrix} Y_{11} & Y_{12} \\ Y_{21} & Y_{22} \end{bmatrix} = \begin{bmatrix} \frac{3}{2}+\frac{10}{9} & \frac{-1}{2}-\frac{8}{9} \\ \frac{-1}{2}-\frac{8}{9} & \frac{5}{2}+\frac{10}{9} \end{bmatrix} \begin{bmatrix} \frac{47}{18} & \frac{-25}{18} \\ \frac{-25}{18} & \frac{65}{18} \end{bmatrix}$$

Example 11.36 *Find Y-parameters for the network shown in Fig. 11.77.*

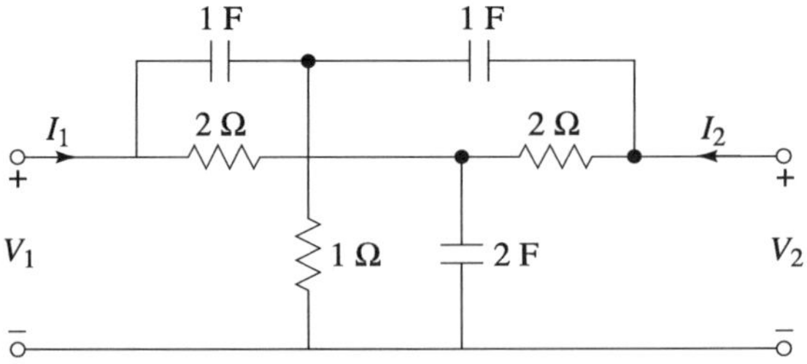

Fig. 11.77

Solution The above network can be considered as a parallel connection of two networks, N_1 and N_2.

For the network N_1

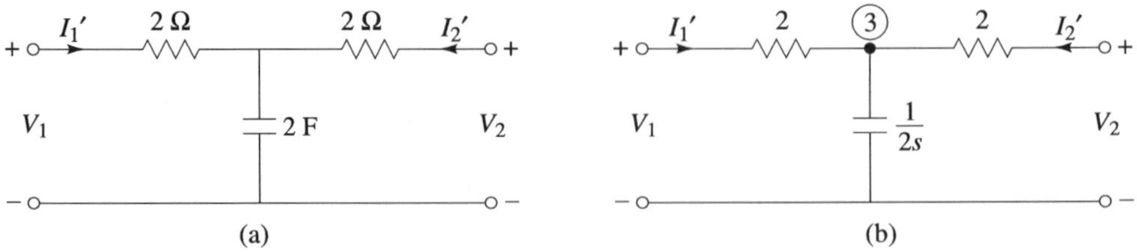

Fig. 11.78

Applying KCL at Node 3,

$$I_1' + I_2' = 2s\,(V_3) \qquad \text{...(i)}$$

From Fig. 11.78

$$I_1' = \frac{V_1 - V_3}{2} = \frac{1}{2}V_1 - \frac{1}{2}V_3 \qquad \text{...(ii)}$$

$$I_2' = \frac{V_2 - V_3}{2} = \frac{1}{2}V_2 - \frac{1}{2}V_3 \qquad \text{...(iii)}$$

Substituting the Eq. (ii) and Eq. (iii), in the Eq. (i)

$$\frac{V_1}{2} - \frac{V_3}{2} + \frac{V_2}{2} - \frac{V_3}{2} = (2s)\,V_3$$

$$(2s + 1)\,V_3 = \frac{V_1}{2} + \frac{V_2}{2}$$

$$V_3 = \frac{1}{2(2s+1)}V_1 + \frac{1}{2(2s+1)}V_2 \qquad \text{...(iv)}$$

Substituting the Eq. (iv) in the Eq. (ii),

$$I_1' = \frac{V_1}{2} - \frac{1}{2}\left[\frac{1}{2(2s+1)}V_1 + \frac{1}{2(2s+1)}V_2\right]$$

$$= \left(\frac{4s+1}{8s+4}\right)V_1 - \left(\frac{1}{8s+4}\right)V_2 \qquad \text{...(v)}$$

Substituting the Eq. (iv) in the Eq. (iii),

$$I_2' = \frac{V_2}{2} - \frac{1}{2}\left[\frac{1}{2(2s+1)}V_1 - \frac{1}{2(2s+1)}V_2\right]$$

$$= -\left(\frac{1}{8s+4}\right)V_1 + \left(\frac{4s+1}{8s+4}\right)V_2 \qquad \text{...(vi)}$$

Comparing Eqs (v) and (vi) with Y-parameter equations, we get

$$\begin{bmatrix} Y_{11}' & Y_{12}' \\ Y_{21}' & Y_{22}' \end{bmatrix} = \begin{bmatrix} \dfrac{4s+1}{8s+4} & \dfrac{-1}{8s+4} \\ \dfrac{-1}{8s+4} & \dfrac{4s+1}{8s+4} \end{bmatrix}$$

For the network N_2

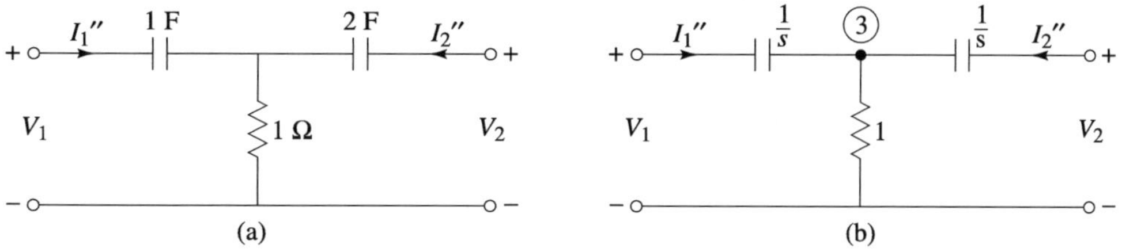

Fig. 11.79

Applying KCL at Node 3,

$$I_1'' + I_2'' = V_3 \qquad \text{...(i)}$$

From Fig. 11.79,

$$I_1'' = \frac{V_1 - V_3}{\frac{1}{s}} = s\,V_1 - s\,V_3 \qquad \text{...(ii)}$$

$$I_2' = \frac{V_2 - V_3}{\frac{1}{s}} = s\,V_2 - s\,V_3 \qquad \text{...(iii)}$$

Substituting the Eqs (ii) and (iii) in the Eq. (i),

$$s\,V_1 - s\,V_3 + s\,V_2 - s\,V_3 = V_3$$

$$(2s + 1)\,V_3 = s\,V_1 + s\,V_2$$

$$V_3 = \left(\frac{s}{2s+1}\right)V_1 + \left(\frac{s}{2s+1}\right)V_2 \qquad \text{...(iv)}$$

Substituting the Eq. (iv) in the Eq. (ii),

$$I_1'' = s\,V_1 - s\left[\left(\frac{s}{2s+1}\right)V_1 + \frac{s}{(2s+1)}V_2\right]$$

$$= \left[\frac{s(s+1)}{2s+1}\right]V_1 - \left(\frac{s^2}{2s+1}\right)V_2 \qquad \text{...(v)}$$

Substituting the Eq. (iv) in the Eq. (iii),

$$I''_2 = s\,V_2 - s\left[\left(\frac{s}{2s+1}\right)V_1 + \left(\frac{s}{2s+1}\right)V_2\right]$$

$$= -\left(\frac{s^2}{2s+1}\right)V_1 + \left[\frac{s(s+1)}{2s+1}\right]V_2 \qquad \text{...(vi)}$$

Comparing Eqs (v) and (vi) with *Y*-parameter equations, we get

$$\begin{bmatrix} Y_{11}'' & Y_{12}'' \\ Y_{21}'' & Y_{22}'' \end{bmatrix} = \begin{bmatrix} \dfrac{s(s+1)}{2s+1} & -\left(\dfrac{s^2}{2s+1}\right) \\ -\left(\dfrac{s^2}{2s+1}\right) & \dfrac{s(s+1)}{2s+1} \end{bmatrix}$$

Therefore, the overall *Y*-parameters of the network are

$$\begin{bmatrix} Y_{11} & Y_{12} \\ Y_{21}' & Y_{22} \end{bmatrix} = \begin{bmatrix} Y_{11}' + Y_{11}'' & Y_{12}' + Y_{12}'' \\ Y_{21}' + Y_{21}'' & Y_{22}' + Y_{22}'' \end{bmatrix}$$

$$= \begin{bmatrix} \dfrac{4s^2 + 8s + 1}{4(2s+1)} & \dfrac{-(4s^2+1)}{4(2s+1)} \\ \dfrac{-(4s^2+1)}{4(2s+1)} & \dfrac{4s^2 + 8s + 1}{4(2s+1)} \end{bmatrix}$$

Example 11.37 *Determine Z-parameters for the network shown in Fig. 11.80.*

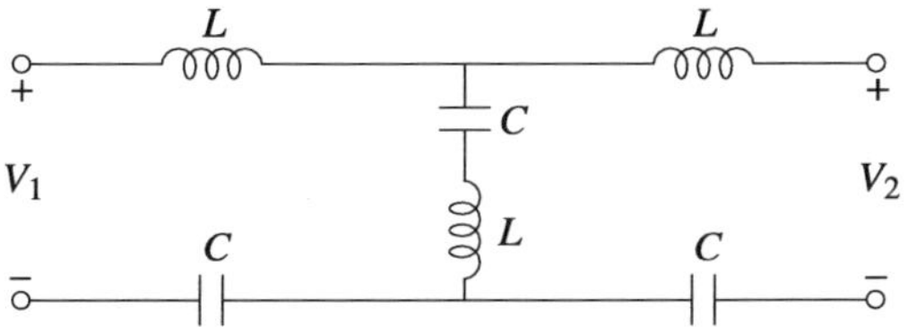

Fig. 11.80

Solution The above network can be considered as a series connection of two networks, N_1 and N_2.

For the network N_1

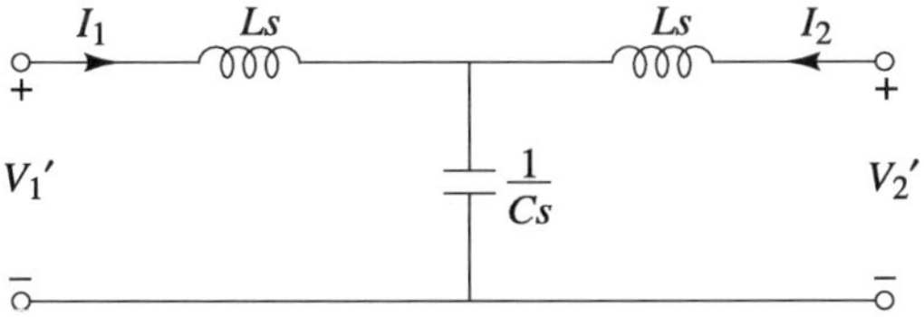

Fig. 11.81

Applying KVL to Mesh 1,

$$V_1' = \left(Ls + \frac{1}{Cs}\right) I_1 + \left(\frac{1}{Cs}\right) I_2 \qquad \ldots\text{(i)}$$

Applying KVL to Mesh 2,

$$V_2' = \left(\frac{1}{Cs}\right) I_1 + \left(Ls + \frac{1}{Cs}\right) I_2 \qquad \ldots\text{(ii)}$$

Comparing Eqs (i) and (ii) with Z-parameter equations, we get

$$\begin{bmatrix} Z_{11}' & Z_{12}' \\ Z_{21}' & Z_{22}' \end{bmatrix} = \begin{bmatrix} Ls + \dfrac{1}{Cs} & \dfrac{1}{Cs} \\ \dfrac{1}{Cs} & Ls + \dfrac{1}{Cs} \end{bmatrix}$$

For the network N_2

Applying KVL to Mesh 1,

$$V_1'' = \left(Ls + \frac{1}{Cs}\right) I_1 + (Ls)\, I_2 \qquad \ldots\text{(i)}$$

Applying KVL to Mesh 2,

$$V_2'' = (Ls)\, I_1 + \left(Ls + \frac{1}{Cs}\right) I_2 \qquad \ldots\text{(ii)}$$

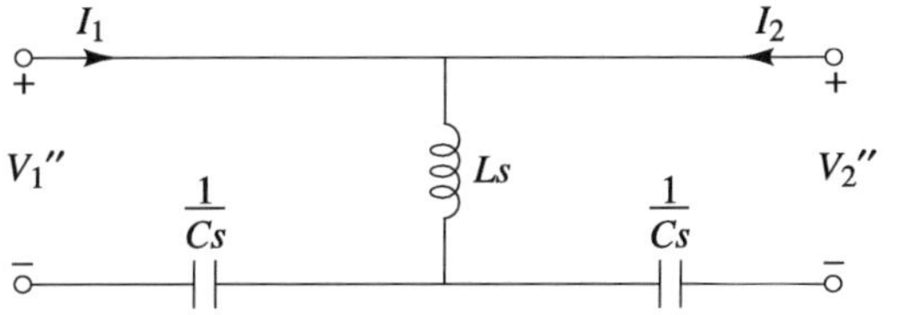

Fig. 11.82

Comparing Eqs (i) and (ii) with Z-parameter equations, we get

$$\begin{bmatrix} Z_{11}'' & Z_{12}'' \\ Z_{21}'' & Z_{22}'' \end{bmatrix} = \begin{bmatrix} Ls + \dfrac{1}{Cs} & Ls \\ Ls & Ls + \dfrac{1}{Cs} \end{bmatrix}$$

Hence, the overall Z-parameters of the network are,

$$\begin{bmatrix} Z_{11} & Z_{12} \\ Z_{21} & Z_{22} \end{bmatrix} = \begin{bmatrix} Z_{11}' + Z_{11}'' & Z_{12}' + Z_{12}'' \\ Z_{21}' + Z_{21}'' & Z_{22}' + Z_{22}'' \end{bmatrix}$$

$$= \begin{bmatrix} 2Ls + \dfrac{2}{Cs} & Ls + \dfrac{1}{Cs} \\ Ls + \dfrac{1}{Cs} & 2Ls + \dfrac{2}{Cs} \end{bmatrix} = \left(Ls + \frac{1}{Cs} \right) \begin{bmatrix} 2 & 1 \\ 1 & 2 \end{bmatrix}$$

Example 11.38 *Two identical sections of the network shown in Fig. 11.83 are connected in series. Obtain Z-parameters of the overall connection.*

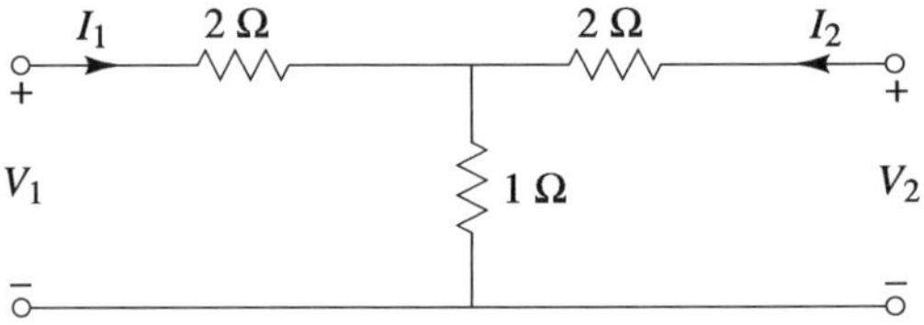

Fig. 11.83

Solution

Applying KVL to Mesh 1,

$$V_1 = 3I_1 + I_2 \qquad \ldots\text{(i)}$$

Applying KVL to Mesh 2,

$$V_2 = I_1 + 3I_2 \qquad \ldots\text{(ii)}$$

Comparing Eqs (i) and (ii) with Z-parameter equations, we get

$$\begin{bmatrix} Z_{11}'' & Z_{12}'' \\ Z_{21}'' & Z_{22}'' \end{bmatrix} = \begin{bmatrix} 3 & 1 \\ 1 & 3 \end{bmatrix}$$

Hence, Z-parameters of the overall connection are

$$\begin{bmatrix} Z_{11} & Z_{12} \\ Z_{21} & Z_{22} \end{bmatrix} = \begin{bmatrix} 3 & 1 \\ 1 & 3 \end{bmatrix} + \begin{bmatrix} 3 & 1 \\ 1 & 3 \end{bmatrix} = \begin{bmatrix} 6 & 2 \\ 2 & 6 \end{bmatrix}$$

11.10 T-NETWORK

Any two-port network can be represented by an equivalent T network as shown in Fig. 11.84.

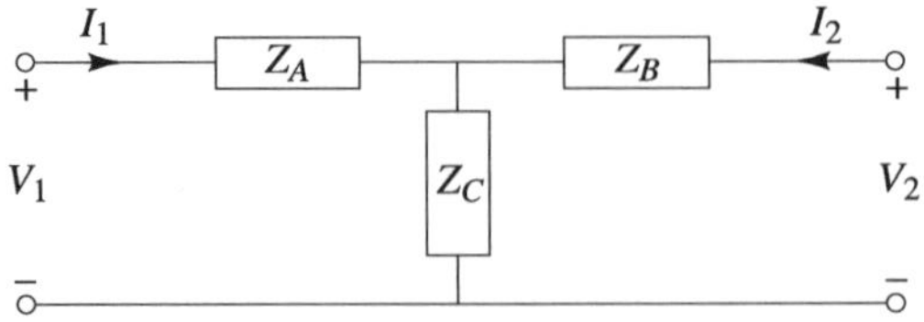

Fig. 11.84

The elements of the equivalent T network may be expressed in terms of Z-parameters.

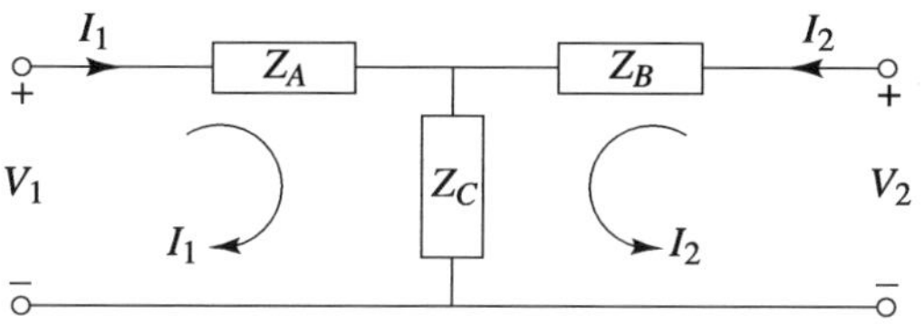

Fig. 11.85

Applying KVL to Mesh 1,

$$V_1 = Z_A I_1 + Z_C (I_1 + I_2)$$

$$V_1 = (Z_A + Z_C) I_1 + Z_C I_2 \qquad \text{...(11.203)}$$

Applying KVL to Mesh 2,

$$V_2 = Z_B I_2 + Z_C (I_2 + I_1)$$

$$V_2 = Z_C I_1 + (Z_B + Z_C) I_2 \qquad \text{...(11.204)}$$

Comparing Eqs (11.203) and (11.204) with Z-parameter equations, we get

$$Z_{11} = Z_A + Z_C$$

$$Z_{12} = Z_C$$

$$Z_{21} = Z_C$$

$$Z_{22} = Z_B + Z_C$$

Solving the above equations, we get

$$Z_A = Z_{11} - Z_{12} = Z_{11} - Z_{21} \qquad \text{...(11.205)}$$

$$Z_B = Z_{22} - Z_{21} = Z_{22} - Z_{12} \qquad \text{...(11.206)}$$

$$Z_C = Z_{12} = Z_{21} \qquad \text{...(11.207)}$$

11.11 PI (π)-NETWORK

Any two-port network can be represented by an equivalent pi (π) network as shown in Fig. 11.86.

Applying KCL at Node 1,

$$I_1 = Y_A V_1 + Y_B (V_1 - V_2)$$

$$= (Y_A + Y_B) V_1 - Y_B V_2 \qquad \text{...(11.208)}$$

Applying KCL at Node 2,

$$I_2 = Y_C V_2 + Y_B (V_2 - V_1)$$

$$= -Y_B V_1 + (Y_B + Y_C) V_2 \qquad \text{...(11.209)}$$

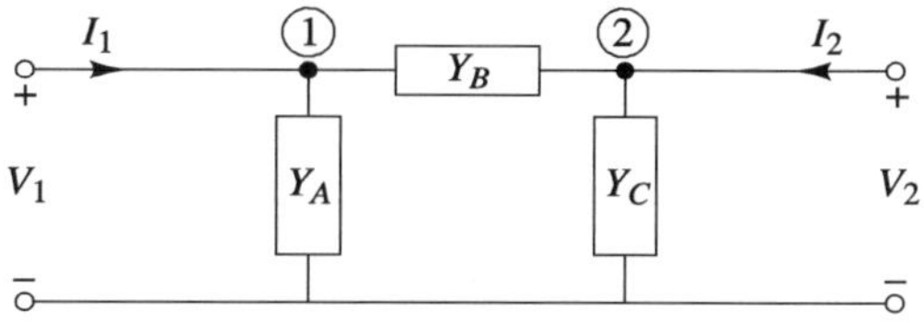

Fig. 11.86

Comparing with Y-parameter equations, we get

$$Y_{11} = Y_A + Y_B$$

$$Y_{12} = -Y_B$$

$$Y_{21} = -Y_B$$

$$Y_{22} = Y_B + Y_C$$

Solving the above equations, we get

$$Y_A = Y_{11} + Y_{12} = Y_{11} + Y_{21} \qquad \text{...(11.210)}$$

$$Y_B = -Y_{12} = -Y_{21} \qquad \text{...(11.211)}$$

$$Y_C = Y_{22} + Y_{12} = Y_{22} + Y_{21} \qquad \text{...(11.212)}$$

11.12 TERMINATED TWO-PORT NETWORKS

11.12.1 Driving-Point Impedance at Input Port

A two-port network is shown in Fig. 11.87. The output port of the network is terminated in load impedance Z_L. The input impedance of this network can be expressed in terms of parameters of two-port network.

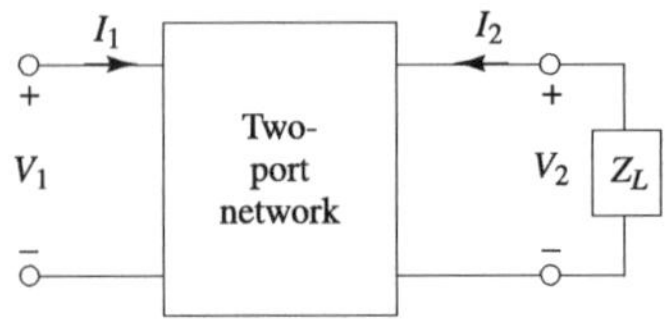

Fig. 11.87

(i) Input impedance in terms of Z-parameters We know that

$$V_1 = Z_{11} I_1 + Z_{12} I_2$$
$$V_2 = Z_{21} I_1 + Z_{22} I_2$$

From Fig. 11.87,

$$V_2 = -Z_L I_2$$
$$-I_2 Z_L = Z_{21} I_1 + Z_{22} I_2$$
$$I_2 = -\frac{Z_{21}}{Z_{22} + Z_L} I_1$$
$$Z_{in} = \frac{V_1}{I_1} = Z_{11} + Z_{12}\left(-\frac{Z_{21}}{Z_{22} + Z_L}\right)$$
$$= \frac{Z_{11}Z_{22} + Z_{11}Z_L - Z_{12}Z_{21}}{Z_{22} + Z_L}$$

If the output port is open circuited, i.e., $Z_L = \infty$,

$$Z_{in} = \lim_{Z_L \to \infty} \frac{\frac{Z_{11}Z_{22} - Z_{12}Z_{21}}{Z_L} + Z_{11}}{\frac{Z_{22}}{Z_L} + 1} = Z_{11}$$

If the output port is short circuited, i.e., $Z_L = 0$,

$$Z_{in} = \frac{Z_{11}Z_{22} - Z_{12}Z_{21}}{Z_{22}}$$

(ii) Input impedance in terms of Y-parameters We know that

$$I_1 = Y_{11}V_1 + Y_{12}V_2$$
$$I_2 = Y_{21}V_1 + Y_{22}V_2$$

From Fig. 11.87,

$$V_2 = -Z_L I_2$$
$$I_2 = -\frac{V_2}{Z_L} = -Y_L V_2 \qquad \text{where } Y_L = \frac{1}{Z_L}$$
$$-Y_L V_2 = Y_{21}V_1 + Y_{22}V_2$$
$$V_2 = \frac{-Y_{21}}{Y_{22} + Y_L} V_1$$
$$I_1 = Y_{11}V_1 + Y_{12}\left(-\frac{Y_{21}}{Y_{22} + Y_L}\right) V_1$$
$$= Y_{11} V_1 - \frac{Y_{21}Y_{12}}{Y_{22} + Y_L} V_1$$
$$= \frac{Y_{11}Y_{22} - Y_{12}Y_{21} + Y_{11}Y_L}{Y_{22} + Y_L} V_1$$
$$Z_{in} = \frac{V_1}{I_1} = \frac{Y_{22} + Y_L}{Y_{11}Y_{22} - Y_{12}Y_{21} + Y_{11}Y_L}$$

When output port is open circuited, i.e., $Y_L = 0$

$$Z_{in} = \frac{Y_{22}}{Y_{11}Y_{22} - Y_{12}Y_{21}}$$

When output port is short circuited, i.e., $Y_L = \infty$

$$Z_{in} = \lim_{Y_L \to \infty} \frac{\dfrac{Y_{22}}{Y_L} + 1}{\dfrac{Y_{11}Y_{22} - Y_{12}Y_{21}}{Y_L} + Y_{11}} = \frac{1}{Y_{11}}$$

(iii) Input impedance in terms of transmission parameters We know that

$$V_1 = AV_2 - BV_2$$
$$I_1 = CV_2 - DI_2$$

From Fig. 11.87,

$$V_2 = -Z_L I_2$$
$$I_1 = -CZ_L I_2 - DI_2 = -(CZ_L + D) I_2$$
$$I_2 = -\frac{I_1}{CZ_L + D}$$
$$V_1 = AZ_L I_2 - B = \left(-\frac{I_1}{CZ_L + D}\right) = \left(\frac{AZ_L + B}{CZ_L + D}\right) I_1$$
$$Z_{in} = \frac{V_1}{I_1} = \frac{AZ_L + B}{CZ_L + D}$$

If the output port is open circuited, i.e., $Z_L = \infty$,

$$Z_{in} = \frac{A}{C}$$

If the output port is short circuited, i.e., $Z_L = 0$,

$$Z_{in} = \frac{B}{D}$$

(iv) Input impedance in terms of hybrid parameters We know that

$$V_1 = h_{11}I_1 + h_{12}V_2$$
$$I_2 = h_{21}I_1 + h_{22}V_2$$
$$V_2 = -Z_L I_2$$
$$I_2 = h_{21}I_1 - h_{22}Z_L I_2$$
$$I_2 = \frac{h_{21}}{1 + h_{22}Z_L} I_L$$
$$V_2 = -\frac{h_{21}Z_L}{1 + h_{22}Z_L} I_1$$

Substituting the value of V_2 in V_1,

$$V_1 = h_{11}I_1 + h_{12}\left[\frac{-h_{21}Z_L}{1 + h_{22}Z_L} I_L\right]$$
$$= \left[\frac{(h_{11}h_{22} - h_{12}h_{21})Z_L + h_{11}}{1 + h_{22}Z_L}\right] I_1$$

$$Z_{in} = \frac{V_1}{I_1} = \frac{(h_{11}h_{22} - h_{12}h_{21})Z_L + h_{11}}{1 + h_{22}Z_L}$$

If the output port is open circuited, i.e., $Z_L = \infty$,

$$Z_{in} = \frac{h_{11}h_{22} - h_{12}h_{21}}{h_{22}}$$

If the output port is short circuited, i.e., $Z_L = 0$,

$$Z_{in} = h_{11}$$

11.12.2 Driving-Point Impedance at Output Port

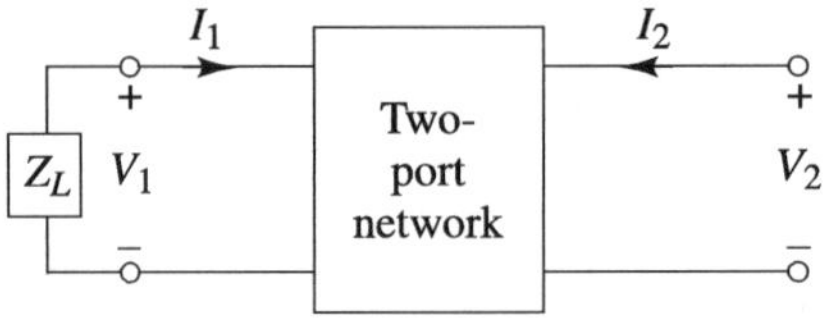

Fig. 11.88

(i) Output impedance in terms of Z-parameters We know that

$$V_1 = Z_{11}I_1 + Z_{12}I_2$$

$$V_2 = Z_{21}I_1 + Z_{22}I_2$$

From Fig. 11.88,

$$V_1 = -Z_LI_1$$

$$-I_lZ_1 = Z_{11}I_1 + Z_{12}I_2$$

$$I_1 = \left(\frac{-Z_{12}}{Z_L + Z_{11}}\right) I_2$$

$$V_2 = Z_{21}\left(\frac{-Z_{12}}{Z_L + Z_{11}}\right) I_2 + Z_{22}I_2$$

$$= I_2\left[Z_{22} - \frac{Z_{21}Z_{12}}{Z_L + Z_{11}}\right]$$

$$= \left(\frac{Z_{11}Z_{22} - Z_{12}Z_{21} + Z_{22}Z_L}{Z_{11} + Z_L}\right) I_2$$

$$Z_0 = \frac{V_2}{I_2} = \frac{Z_{11}Z_{22} - Z_{12}Z_{21} + Z_{22}Z_L}{Z_{11} + Z_L}$$

If the input port is open circuited, i.e., $Z_L = \infty$,

$$Z_0 = Z_{22}$$

If the input port is short circuited, i.e., $Z_L = 0$,

$$Z_0 = \frac{Z_{11}Z_{22} - Z_{12}Z_{21}}{Z_{11}}$$

(ii) Output impedance in terms of Y-parameters We know that

$$I_1 = Y_{11}V_1 + Y_{12}V_2$$
$$I_2 = Y_{21}V_1 + Y_{22}V_2$$

From Fig. 11.88,

$$V_1 = -Z_L I_1$$
$$I_1 = -\frac{V_1}{Z_L} = -Y_L V_1$$
$$-Y_L V_1 = Y_{11}V_1 + Y_{12}V_2$$
$$V_1 = \left(\frac{-Y_{12}}{Y_L + Y_{11}}\right) V_2$$
$$I_2 = Y_{21}\left(\frac{-Y_{12}}{Y_L + Y_{11}}\right) V_2 + Y_{22}V_2 = V_2\left[Y_{22} - \frac{Y_{21}Y_{12}}{Y_L + Y_{11}}\right]$$
$$= V_2\left[\frac{Y_{11}Y_{22} - Y_{12}Y_{21} + Y_L Y_{22}}{Y_L + Y_{11}}\right]$$
$$Z_0 = \frac{V_2}{I_2} = \frac{Y_L + Y_{11}}{Y_{11}Y_{22} - Y_{12}Y_{21} + Y_L Y_{22}}$$

If input port is open circuited, i.e., $Y_L = 0$,

$$Z_0 = \frac{Y_{11}}{Y_{11}Y_{22} - Y_{12}Y_{21}}$$

If input port is short circuited, i.e., $Y_L = \infty$,

$$Z_0 = \frac{1}{Y_{22}}$$

(iii) Output impedance in terms of ABCD parameters We know that

$$V_1 = AV_2 - BI_2$$
$$I_1 = CV_2 - DI_2$$

From Fig. 11.88,

$$V_1 = -Z_L I_1$$
$$\frac{V_1}{I_1} = -Z_L = \frac{AV_2 - BI_2}{CV_2 - DI_2}$$
$$V_2 (CZ_L + A) = I_2 (DZ_L + B)$$
$$Z_0 = \frac{V_2}{I_2} = \frac{DZ_L + B}{CZ_L + A}$$

If input port is open circuited, i.e., $Z_L = \infty$,

$$Z_0 = \frac{D}{C}$$

If input port is short circuited, i.e., $Z_L = 0$,

$$Z_0 = \frac{B}{A}$$

(iv) Output impedance in term of h-parameters We know that

$$V_1 = h_{11}I_1 + h_{12}V_{12}$$
$$I_2 = h_{21}I_1 + h_{22}V_2$$

From Fig. 11.88,

$$V_1 = -Z_L I_1$$
$$-I_1 Z_L = h_{11}I_1 + h_{22}V_2$$
$$I_1 = \left(\frac{-h_{12}}{h_{11} + Z_L}\right) V_2$$
$$I_2 = h_{21}\left(\frac{-h_{12}}{h_{11} + Z_L}\right) V_2 + h_{22}V_2$$
$$= V_2\left[\frac{h_{11}h_{22} - h_{12}h_{21} + h_{22}Z_L}{h_{11} + Z_L}\right]$$
$$Z_0 = \frac{V_2}{I_2}$$
$$= \frac{h_{11} + Z_L}{h_{11}h_{22} - h_{12}h_{21} + h_{22}Z_L}$$

If input port is open circuited, i.e., $Z_L = \infty$,

$$Z_0 = \frac{1}{h_{22}}$$

If input port is short circuited i.e., $Z_L = 0$,

$$Z_0 = \frac{h_{11}}{h_{11}h_{22} - h_{12}h_{21}}$$

Example 11.39 *The Z-parameters of a two-port network are: $Z_{11} = 10\ \Omega$, $Z_{12} = Z_{21} = 5\ \Omega$, $Z_{22} = 20\ \Omega$. Find the equivalent T-network.*

Solution

Applying KVL to Mesh 1,

$$V_1 = (Z_1 + Z_2)\, I_1 + Z_2\, I_2 \qquad \ldots\text{(i)}$$

Applying KVL to Mesh 2,

$$V_2 = Z_2\, I_1 + (Z_2 + Z_3)\, I_2 \qquad \ldots\text{(ii)}$$

Comparing Eqs (i) and (ii) with Z parameter equations, we get

$$Z_{11} = Z_1 + Z_2 = 10$$
$$Z_{12} = Z_2 = 5$$
$$Z_{21} = Z_2 = 5$$
$$Z_{22} = Z_2 + Z_3 = 20$$

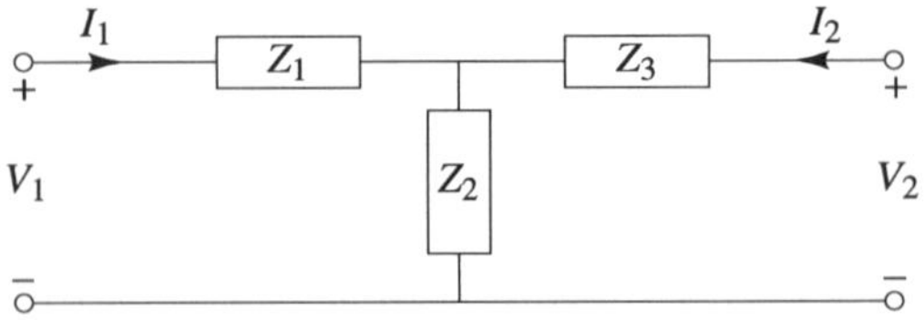

Fig. 11.89

Solving the above equations, we get

$$Z_1 = 5\ \Omega$$
$$Z_2 = 5\ \Omega$$
$$Z_3 = 15\ \Omega$$

Example 11.40 *Admittance parameters of a pi network are $Y_{11} = 0.09$ ℧, $Y_{12} = Y_{21} = -0.05$ ℧ and $Y_{22} = 0.07$ ℧. Find the values of R_a, R_b and R_c.*

Solution

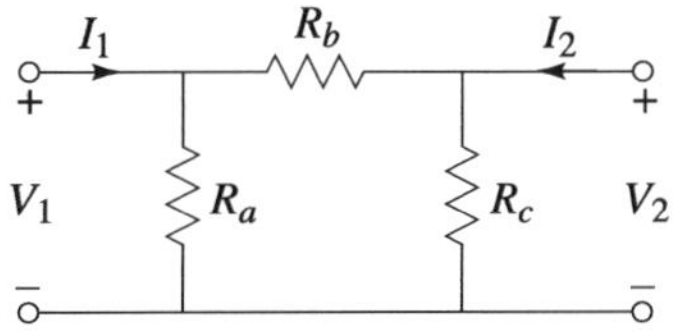

Fig. 11.90

Applying KCL at Node 1,

$$I_1 = \frac{V_1}{R_a} + \frac{V_1 - V_2}{R_b}$$

$$= \left(\frac{1}{R_a} + \frac{1}{R_b}\right) V_1 - \frac{1}{R_b} V_2 \qquad \text{...(i)}$$

Applying KCL at Node 2,

$$I_2 = \frac{V_2}{R_c} + \frac{V_2 - V_1}{R_b}$$

$$= -\frac{1}{R_b} V_1 + \left(\frac{1}{R_B} + \frac{1}{R_c}\right) V_2 \qquad \text{...(ii)}$$

Comparing Eqs (i) and (ii) with Y-parameter equations, we get

$$Y_{11} = \frac{1}{R_a} + \frac{1}{R_b} = 0.09 \ ℧$$

$$Y_{12} = -\frac{1}{R_b} = -0.05 \ ℧$$

$$Y_{21} = \frac{-1}{R_b} = -0.05 \ ℧$$

$$Y_{22} = \frac{1}{R_b} + \frac{1}{R_c} = 0.07 \ ℧$$

Solving the above equations, we get

$$R_a = 25 \ \Omega$$
$$R_b = 20 \ \Omega$$
$$R_c = 50 \ \Omega$$

Example 11.41 *Find the parameters Y_A, Y_B and Y_C of the equivalent π network as shown to represent a two-terminal pair network for which the following measurements were taken:*

(a) *With Terminal 2 short circuited, a voltage of 10 ∠0° V applied at Terminal pair 1 resulted in $I_1 = 2.5$ ∠0° A and $I_2 = -0.5$ ∠0° A.*

(b) *With Terminal 1 short circuited, the same voltage at Terminal pair 2 resulted in $I_2 = 1.5$ ∠0° A and $I_1 = -1.1$ ∠–20° A.*

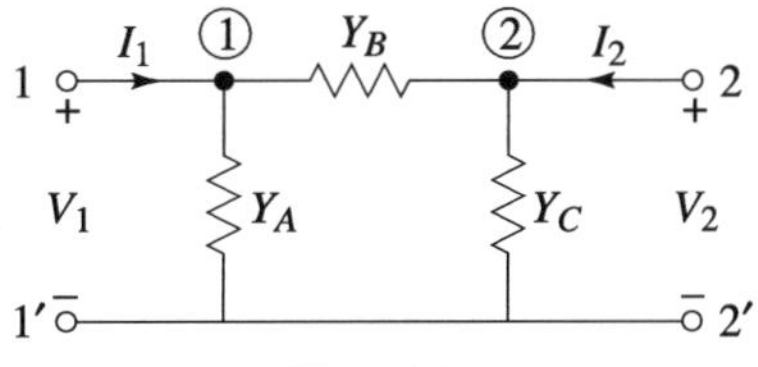

Fig. 11.91

Solution Since measurements were taken with either of the terminal pairs short circuited, we have to calculate Y-parameters first.

$$Y_{11} = \left.\frac{I_1}{V_1}\right|_{V_2=0} = \frac{2.5\angle 0^\circ}{10\angle 0^\circ} = 0.25\ \mho$$

$$Y_{21} = \left.\frac{I_2}{V_1}\right|_{V_2=0} = \frac{-0.5\angle 0^\circ}{10\angle 0^\circ} = -0.05\ \mho$$

$$Y_{22} = \left.\frac{I_2}{V_2}\right|_{V_1=0} = \frac{1.5\angle 0^\circ}{10\angle 0^\circ} = 0.15\ \mho$$

Applying KCL at Node 1,

$$I_1 = Y_A V_1 + Y_B (V_1 - V_2)$$
$$= (Y_A + Y_B) V_1 - Y_B V_2 \qquad \ldots\text{(i)}$$

Applying KCL at Node 2,

$$I_2 = Y_C V_2 + Y_B (V_2 - V_1)$$
$$= -Y_B V_1 + (Y_B + Y_C) V_2 \qquad \ldots\text{(ii)}$$

Comparing Eqs (i) and (ii) with the Y-parameter equation, we get

$$Y_{11} = Y_A + Y_B = 0.25$$
$$Y_{12} = Y_{21} = -Y_B = -0.05$$
$$Y_{22} = Y_B + Y_C = 0.15$$

Solving the above equation, we get

$$Y_A = 0.20\ \mho$$
$$Y_B = 0.05\ \mho$$
$$Y_C = 0.10\ \mho$$

Example 11.42 *A network has two input terminals (a, b) and two output terminals (c, d). The input impedance with c and d open circuited is (250 + j100) ohms and with c and d short circuited is (400 + j300) ohms. The impedance across c and d with a and b open circuited is 200 ohms. Determine the equivalent T-network parameters.*

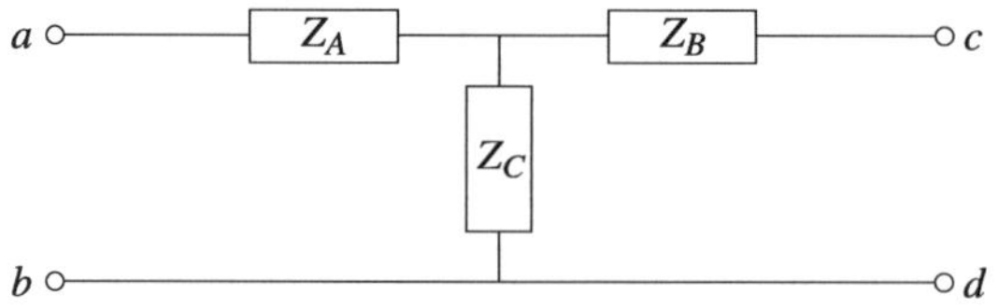

Fig. 11.92

Solution The input impedance with c and d open circuited is

$$Z_A + Z_C = 250 + j100 \qquad \ldots\text{(i)}$$

The input impedance with c and d short circuited is,

$$Z_A + \frac{Z_B Z_C}{Z_B + Z_C} = 400 + j300 \qquad \ldots\text{(ii)}$$

The impedance across c and d with a and b open circuited is

$$Z_B + Z_C = 200 \qquad \ldots\text{(iii)}$$

Subtracting the equation (i) from (ii),

$$\frac{Z_B Z_C}{Z_B + Z_C} - Z_C = 150 + j200 \quad \text{...(iv)}$$

From the equation (iii),

$$Z_B = 200 - Z_C \quad \text{...(v)}$$

Subtracting the value of Z_B in the equation (iv) and simplifying,

$$Z_C = (-100 + j200)\ \Omega \quad \text{...(vi)}$$

From Eqs (i) and (vi),

$$Z_A = (350 - j100)\ \Omega$$

From Eqs (iii) and (vi),

$$Z_B = (300 - j200)\ \Omega$$

Example 11.43 *Find the equivalent π-network for the T-network shown in Fig. 11.93.*

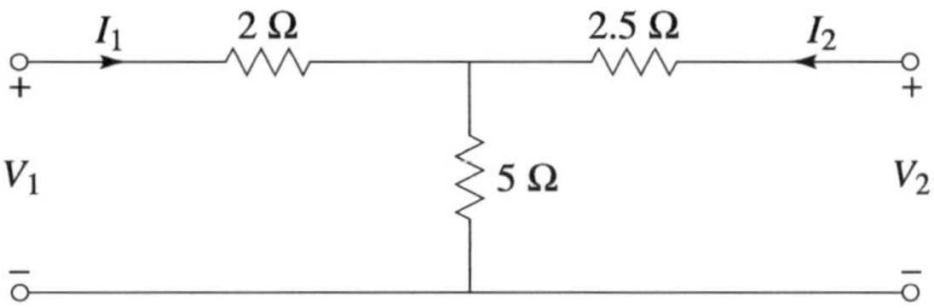

Fig. 11.93

Solution Figure 11.94 shows a *T*-network and *π*-network.

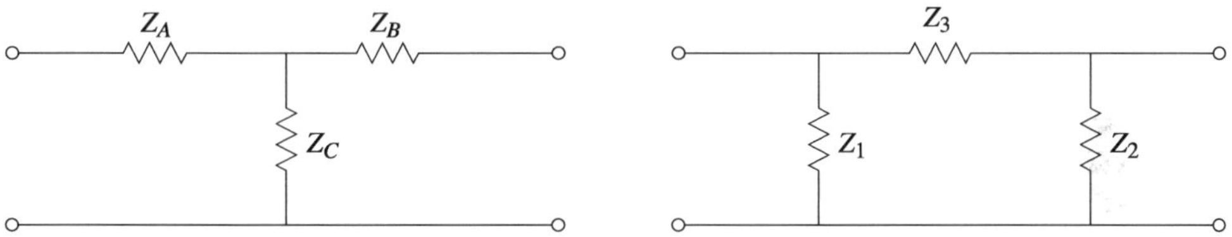

T-network (Star network) *π*-network (Delta network)

Fig. 11.94

For converting a *T*-network (star network) into an equivalent *π*-network (delta network), we can use star–delta transformation technique.

$$Z_1 = Z_A + Z_C + \frac{Z_A Z_C}{Z_B}$$

$$= 2 + 5 + \frac{2 \times 5}{2.5} = 11\ \Omega$$

$$Z_3 = Z_A + Z_B + \frac{Z_A Z_B}{Z_C}$$

$$= 2 + 2.5 + \frac{2 \times 2.5}{5} = 5.5\ \Omega$$

$$Z_2 = Z_B + Z_C + \frac{Z_B Z_C}{Z_A}$$

$$= 2.5 + 5 + \frac{2.5 \times 5}{2} = 13.75\ \Omega$$

Hence, the equivalent π-network can be shown as

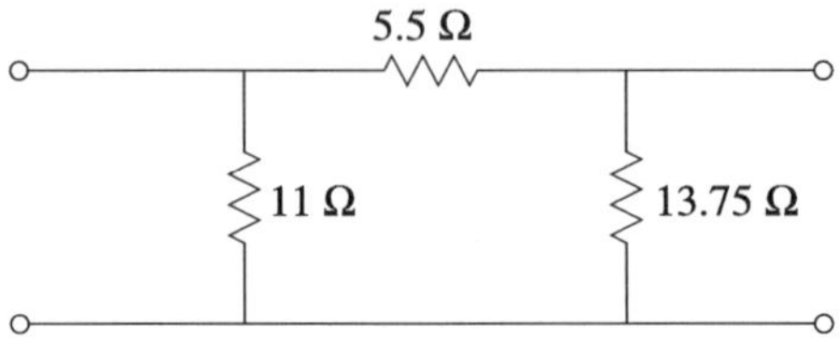

Fig. 11.95

Example 11.44 *For the network shown in Fig. 11.96, Find the equivalent T-network.*

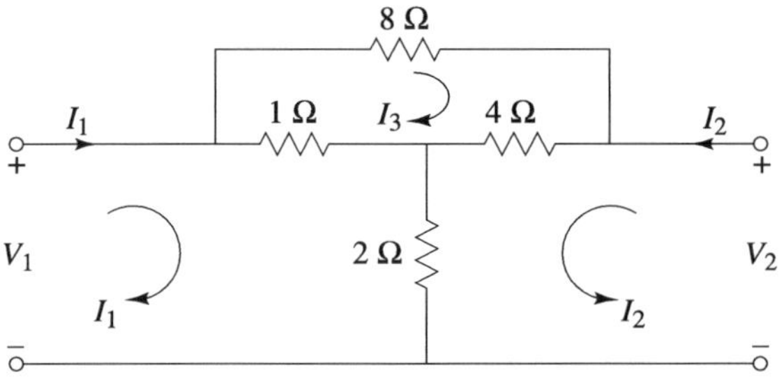

Fig. 11.96

Solution Applying KVL to Mesh 1,

$$V_1 = 3I_1 + 2I_2 - I_3 \quad \text{...(i)}$$

Applying KVL to Mesh 2,

$$V_2 = 2I_1 + 6I_2 + 4I_3 \quad \text{...(ii)}$$

Applying KVL to Mesh 3,

$$13I_3 - I_1 + 4I_2 = 0$$

$$I_3 = \frac{1}{13}I_1 - \frac{4}{13}I_2 \quad \text{...(iii)}$$

Substituting the Eq. (iii) in the Eq. (i),

$$V_1 = 3I_1 + 2I_2 - \frac{1}{13}I_1 + \frac{4}{13}I_2$$

$$= \frac{38}{13}I_1 + \frac{30}{13}I_2 \quad \text{...(iv)}$$

Substituting the Eq. (iii) in the Eq. (ii),

$$V_2 = 2I_1 + 6I_2 + 4\left(\frac{1}{13}I_1 - \frac{4}{13}I_2\right)$$

$$= \frac{30}{13}I_1 + \frac{62}{13}I_2 \quad \text{...(v)}$$

For the *T*-network,

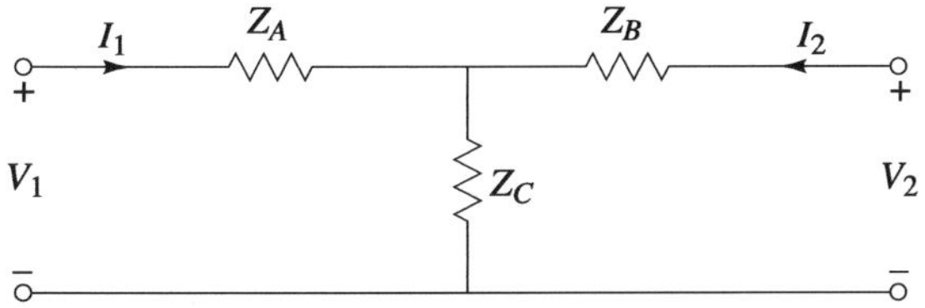

Fig. 11.97

Applying KVL to Mesh 1,

$$V_1 = (Z_A + Z_C)\, I_1 + Z_C\, I_2 \quad \text{...(vi)}$$

Applying KVL to Mesh 2,

$$V_2 = Z_C\, I_1 + (Z_B + Z_C)\, I_2 \quad \text{...(vii)}$$

Comparing Eqs (iv) and (v) with Eqs (vi) and (vii), we get

$$Z_A + Z_C = \frac{38}{13}$$

$$Z_C = \frac{30}{13}$$

$$Z_B + Z_C = \frac{62}{13}$$

Solving the above equations, we get

$$Z_A = \frac{8}{13}\,\Omega$$

$$Z_B = \frac{32}{13}\,\Omega$$

$$Z_C = \frac{30}{13}\,\Omega$$

Example 11.45 *Measurements were made on a two-terminal network shown in Fig. 11.98.*

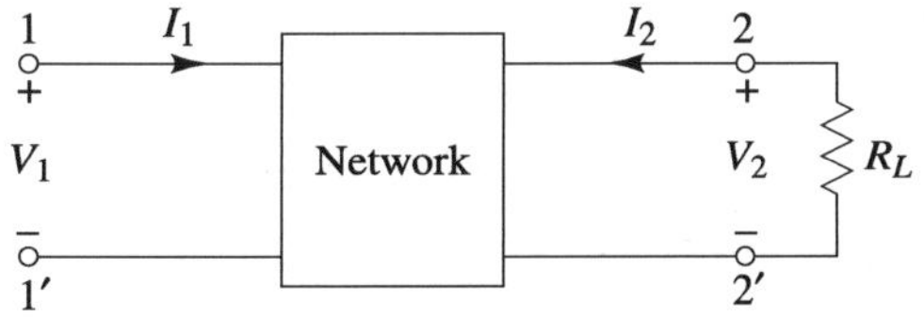

Fig. 11.98

(i) With Terminal pair 2 open, a voltage of 100 ∠0° V applied to Terminal pair 1 resulted in

$$I_1 = 10\angle 0^\circ\ A$$
$$V_2 = 25\angle 0^\circ\ V$$

(ii) With Terminal pair 1 open, the same voltage applied to Terminal pair 2 resulted in

$$I_2 = 20\angle 0^\circ\ A$$
$$V_1 = 50\angle 0^\circ\ V$$

Write mesh equations for this network. What will be the voltage across a 10-Ω resistor connected across Terminal pair 2 if a 100 ∠0° V is connected across Terminal pair 1?

Solution Since measurements were done with either of the terminal pairs open circuited, we have to calculate Z-parameters first.

$$Z_{11} = \left.\frac{V_1}{I_1}\right|_{I_2=0} = \frac{100\angle 0^\circ}{10\angle 0^\circ} = 10\ \Omega$$

$$Z_{21} = \left.\frac{V_2}{I_1}\right|_{I_2=0} = \frac{25\angle 0^\circ}{10\angle 0^\circ} = 2.5\ \Omega$$

$$Z_{22} = \left.\frac{V_2}{I_2}\right|_{I_1=0} = \frac{100\angle 0^\circ}{20\angle 0^\circ} = 5\ \Omega$$

$$Z_{12} = \left.\frac{V_1}{I_2}\right|_{I_1=0} = \frac{50\angle 0^\circ}{20\angle 0^\circ} = 2.5\ \Omega$$

Putting these values in Z-parameter equations, we get

$$V_1 = 10I_1 + 2.5I_2 \quad \ldots\text{(i)}$$

$$V_2 = 2.5I_1 + 5I_2 \quad \ldots\text{(ii)}$$

When a 10-Ω resistor is connected across Terminal pair 1,

$$V_1 = 100\angle 0^\circ \text{ V}$$

$$V_2 = -R_L I_2 = -10I_2$$

Substituting values of V_1 and V_2 in Eqs (i) and (ii),

$$100 = 10I_1 + 2.5I_2$$

and

$$-10I_2 = 2.5I_1 + 5I_2$$

$$2.5I_1 = -15I_2$$

$$I_1 = -6I_2$$

$$100 = -60I_2 + 2.5I_2$$

$$I_2 = -\frac{100}{57.5} = -1.74 \text{ A}$$

Voltage across the resistor $= -I_2 R_L = -10(-1.74) = 17.4$ V

Example 11.46 *The following equations give the voltages V_1 and V_2 at the two ports of a two-port network:*

$$V_1 = 5I_1 + 2I_2$$

$$V_2 = 2I_1 + I_2$$

A load resistor of 3 Ω is connected across Port 2. Calculate the input impedance.

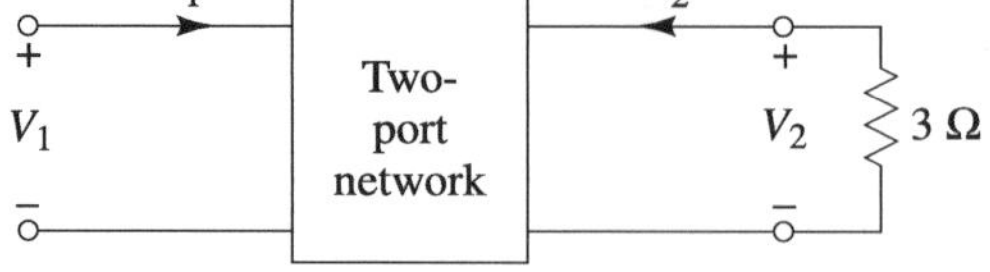

Fig. 11.99

Solution From Fig.11.99,

$$V_2 = -3I_2 \quad \ldots\text{(i)}$$

Substituting the Eq. (i) in the given equation,

$$-3I_2 = 2I_1 + I_2$$

$$I_2 = -\frac{I_1}{2} \qquad \ldots\text{(ii)}$$

Substituting the Eq. (ii) in the given equation.

$$V_1 = 5I_1 - I_1 = 4I_1$$

Input impedance $$Z_i = \frac{V_1}{I_1} = 4\ \Omega$$

Example 11.47 *The following equation gives the voltage and current at the input port of a two-port network*

$$V_1 = 5V_2 - 3I_2$$
$$I_1 = 6V_2 - 2I_2$$

A load resistance of 5 Ω is connected across the output port. Calculate the input impedance.

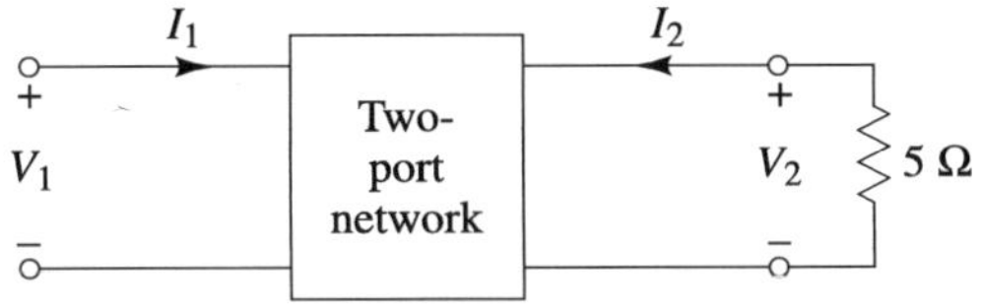

Fig. 11.100

Solution From Fig. 11.100,

$$V_2 = -5I_2$$

Substituting the value of V_2 in the given equations,

$$V_1 = 5(-5I_2) - 3I_2 = -28I_2$$
$$I_1 = 6(-5I_2) - 2I_2 = -32I_2$$

Input impedance $$Z_i = \frac{V_1}{I_1}$$

$$= \frac{-28I_2}{-32I_2} = \frac{7}{8}\ \Omega$$

Exercises

1. Determine Z-parameters for the network shown in Fig. 11.101

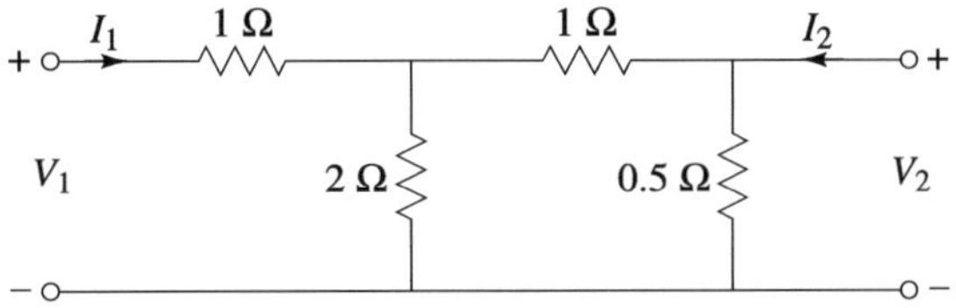

Fig. 11.101

$$Z = \begin{bmatrix} \frac{13}{7} & \frac{2}{7} \\ \frac{2}{7} & \frac{3}{7} \end{bmatrix}$$

2. Find *Z*-parameters for the network in Fig. 11.102.

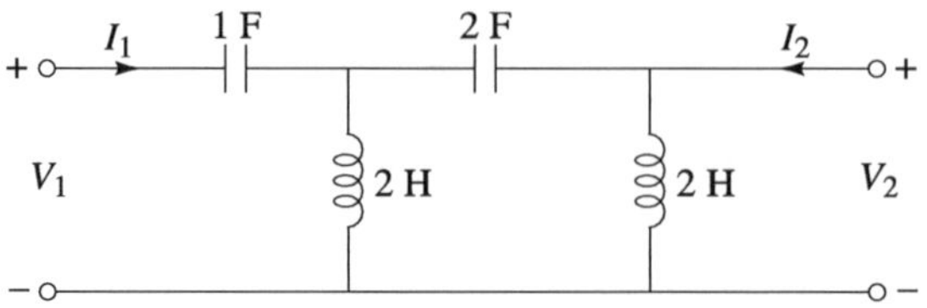

Fig. 11.102

$$Z = \begin{bmatrix} \dfrac{4s^4 + 6s^2 + 1}{4s^3 + s} & \dfrac{4s^3}{4s^2 + 1} \\ \dfrac{4s^3}{4s^2 + 1} & \dfrac{4s^3 + 2s}{4s^2 + 1} \end{bmatrix}$$

3. Find *Y*-parameters of the network shown in Fig.11.103.

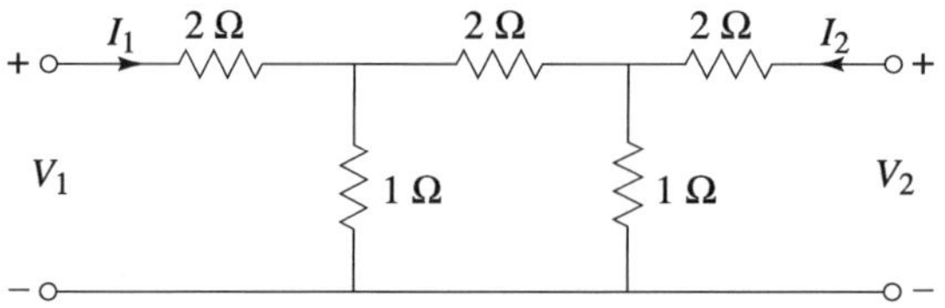

Fig. 11.103

$$Y = \begin{bmatrix} 0.36 & -0.033 \\ -0.033 & -0.36 \end{bmatrix}$$

4. Find *Y*-parameters for the network shown in Fig.11.104.

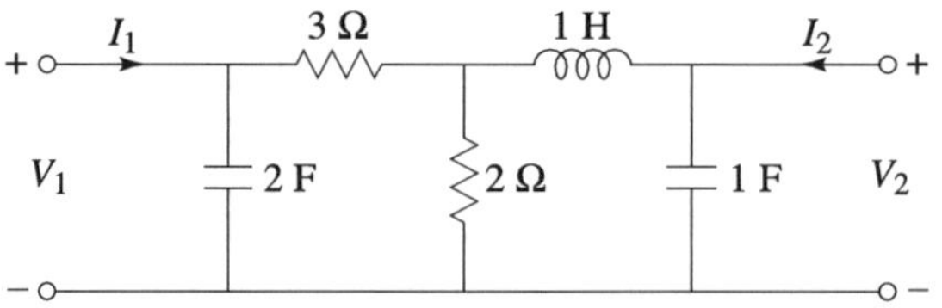

Fig. 11.104

$$Y = \begin{bmatrix} \dfrac{10s^2 + 13s + 2}{5s + 6} & \dfrac{-2}{5s + 6} \\ \dfrac{-2}{5s + 6} & \dfrac{5s^2 + 6s + 5}{5s + 6} \end{bmatrix}$$

5. Find Y-parameters for the network shown in Fig.11.105.

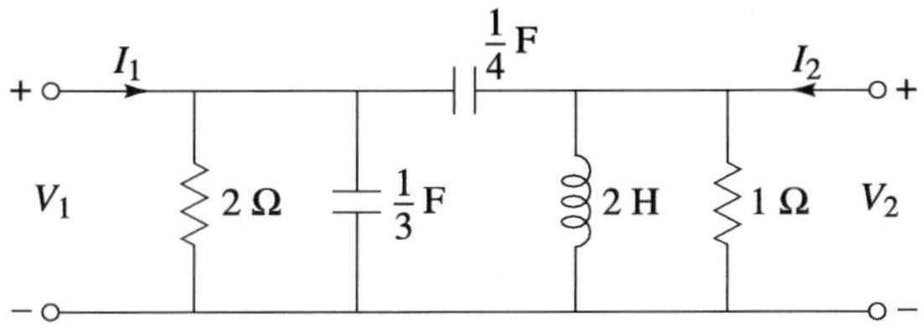

Fig. 11.105

$$Y = \begin{bmatrix} \frac{7s+6}{12} & \frac{-s}{4} \\ \frac{-s}{4} & \frac{s^2+4s+2}{4s} \end{bmatrix}$$

6. Find Y-parameters for the network shown in Fig.11.106.

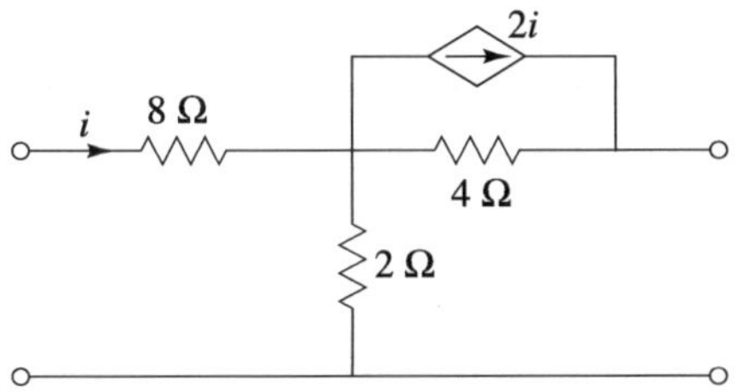

Fig. 11.106

$$Y = \begin{bmatrix} \frac{3}{20} & \frac{-1}{20} \\ \frac{-1}{4} & \frac{1}{4} \end{bmatrix}$$

7. Show the $ABCD$ parameters of the network shown in Fig.11.107.

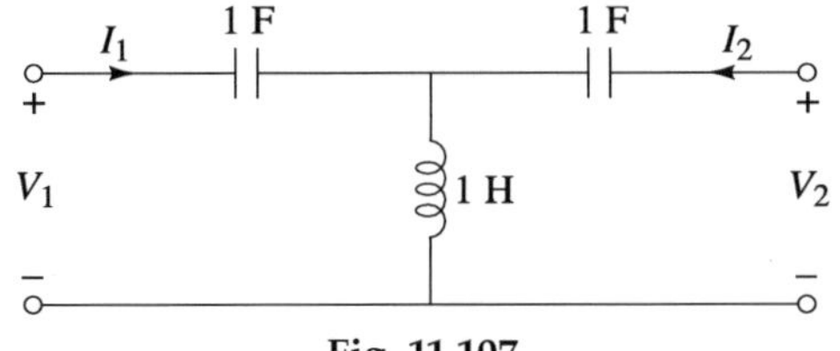

Fig. 11.107

$$\begin{bmatrix} A & B \\ C & D \end{bmatrix} = \begin{bmatrix} \frac{1+s^2}{s^2} & \frac{1+2s^2}{s^3} \\ \frac{1}{s} & \frac{1+s^2}{s^2} \end{bmatrix}$$

8. Find *ABCD* parameters for the network shown in Fig.11.108

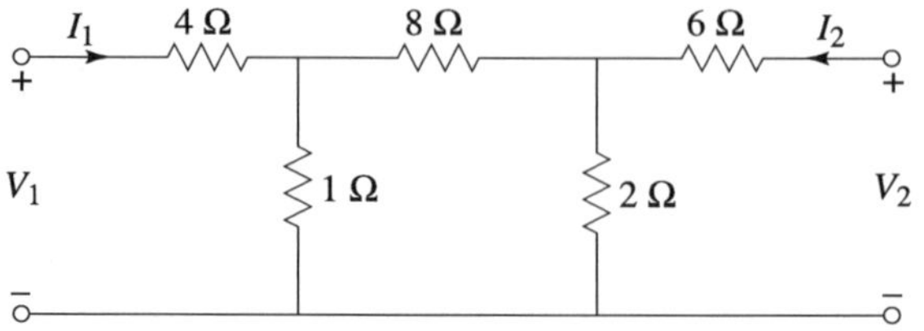

Fig. 11.108

$$\begin{bmatrix} A & B \\ C & D \end{bmatrix} = \begin{bmatrix} 27 & 206 \\ \frac{11}{2} & 42 \end{bmatrix}$$

9. For the network shown in Fig.11.109, determine parameter h_{21}.

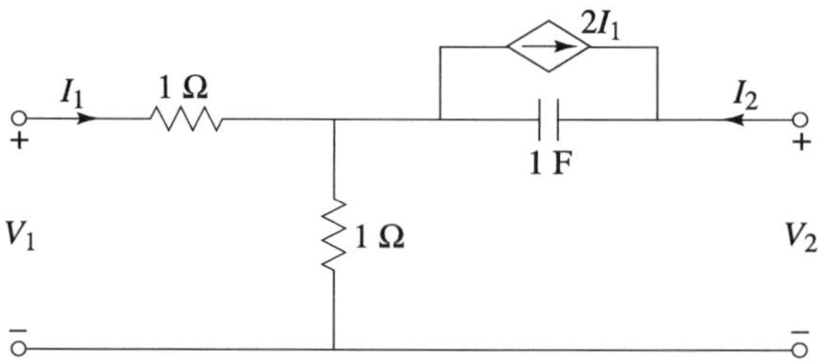

Fig. 11.109

$$h_{21} = \frac{-(2+s)}{1+s}$$

10. Determine *Y* and *Z*-parameters for the network shown in Fig. 11.110.

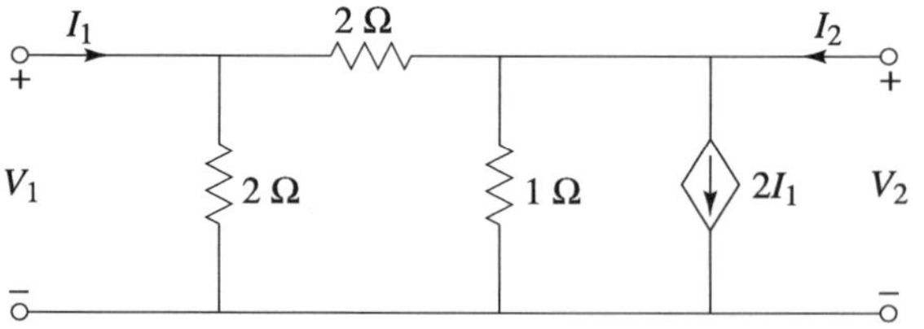

Fig. 11.110

$$\left[Y_{11} = 1 \ \mho, Y_{12} = -0.5 \ \mho, Y_{21} = 1.5 \ \mho, Y_{22} = 0.5 \ \mho \right.$$

$$\left. Z_{11} = \frac{2}{5}\,\Omega, Z_{12} = \frac{2}{5}\,\Omega, Z_{21} = -\frac{6}{5}\,\Omega, Z_{22} = \frac{4}{5}\,\Omega \right]$$

11. For the bridged T, R-C network determine Y-parameters using interconnections of two-port networks.

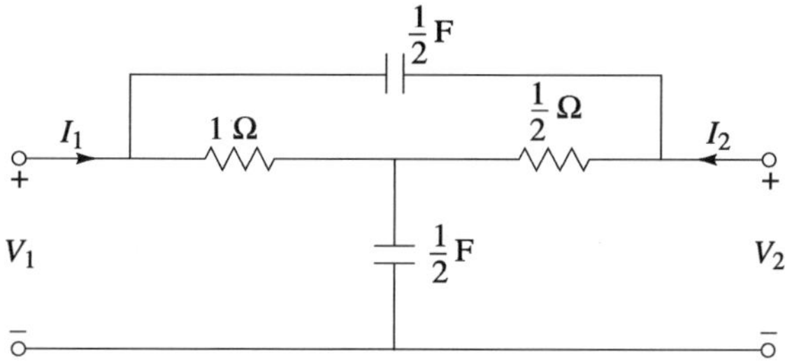

Fig. 11.111

$$Y = \begin{bmatrix} \dfrac{s^2+8s+8}{2(s+6)} & \dfrac{-(s^2+6s+8)}{2(s+6)} \\ \dfrac{-(s^2+6s+8)}{2(s+6)} & \dfrac{s^2+10s+8}{2(s+6)} \end{bmatrix}$$

12. For the network of Fig.11.112, find Y-parameters using interconnection of two-port networks.

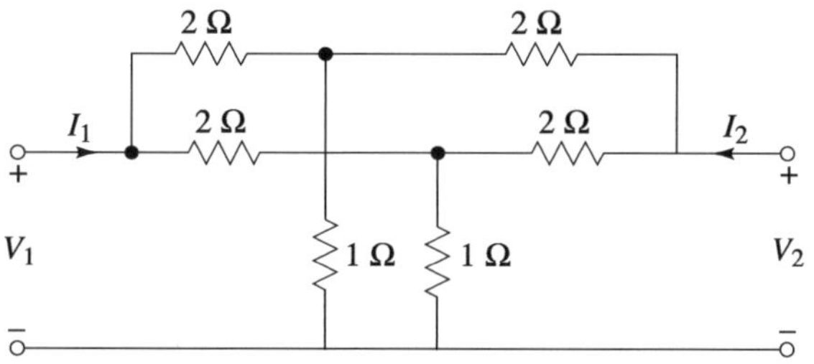

Fig. 11.112

$$Y = \begin{bmatrix} \dfrac{3}{4} & \dfrac{-1}{4} \\ \dfrac{-1}{4} & \dfrac{3}{4} \end{bmatrix}$$

13. Two identical sections of the network shown in Fig.11.113 are connected in parallel. Obtain Y-parameters of the connection.

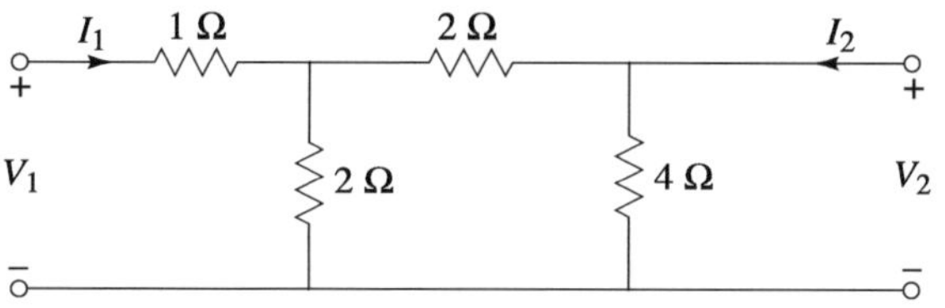

Fig. 11.113

$$Y = \begin{bmatrix} 1 & \dfrac{-1}{2} \\ \dfrac{-1}{2} & \dfrac{5}{4} \end{bmatrix}$$

14. Determine Y-parameters using interconnection of two-port networks for the network shown in Fig. 11.114.

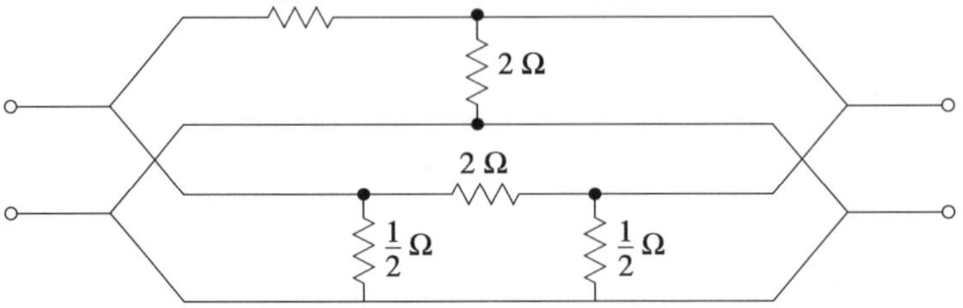

Fig. 11.114

$$Y = \begin{bmatrix} 3.1 & -0.9 \\ -0.9 & 3.1 \end{bmatrix}$$

15. If a two-port network has $Y_{11} = 1$ ℧, $Y_{12} = Y_{21} = -2$ ℧, $Y_{22} = 3$ ℧, find the equivalent pi (π) network.

Objective-Type Questions

1. The open-circuit impedance matrix of the two-port network shown in Fig. 11.115 is

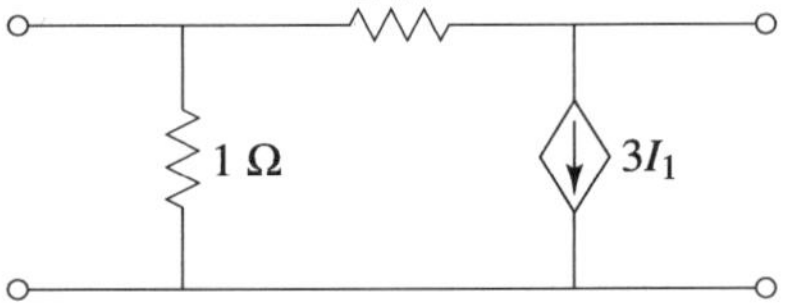

Fig. 11.115

(a) $\begin{bmatrix} -2 & 1 \\ -8 & 3 \end{bmatrix}$ (b) $\begin{bmatrix} -2 & -8 \\ 1 & 3 \end{bmatrix}$

(c) $\begin{bmatrix} 0 & 1 \\ 1 & 0 \end{bmatrix}$ (d) $\begin{bmatrix} 2 & -1 \\ -1 & 3 \end{bmatrix}$

2. Two two-port networks are connected in cascade. The combination is to be represented as a single two-port network. The parameters are obtained by multiplying the individual
(a) z-parameter matrix (b) h-parameter matrix
(c) y-parameter matrix (d) $ABCD$ parameter matrix

3. For a two-part network to be reciprocal
(a) $z_{11} = z_{22}$ (b) $y_{21} = y_{12}$ (c) $h_{21} = -h_{12}$ (d) $AD - BC = 0$

4. The short circuit admittance matrix of a two-port network is $\begin{bmatrix} 0 & -1/2 \\ 1/2 & 0 \end{bmatrix}$. The two-port network is
(a) non-reciprocal and passive (b) non-reciprocal and active
(c) reciprocal and passive (d) reciprocal and active

5. A two-port network is shown in Fig. 11.116. The parameter h_{21} for this network can be given by
(a) –1/2 (b) 1/2
(c) –3/2 (d) 3/2

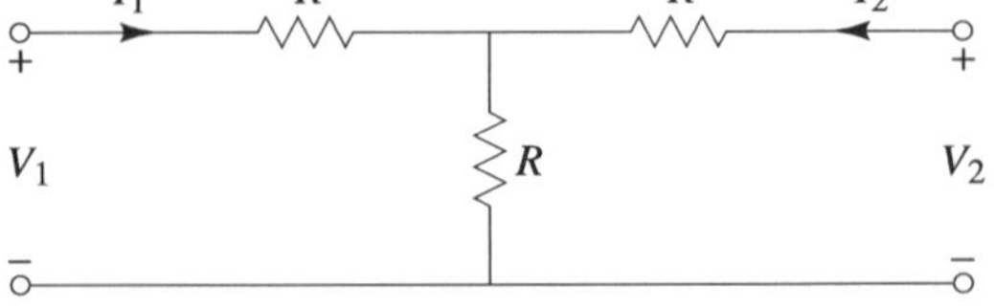

Fig. 11.116

6. The admittance parameter Y_{12} in the two-port network in Fig. 11.117 is
(a) –0.2 mho (b) 0.1 mho
(c) –0.05 mho (d) 0.05 mho

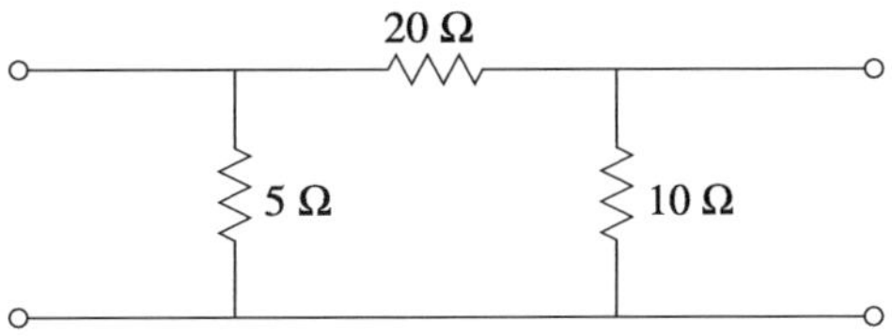

Fig. 11.117

7. The Z-parameters Z_{11} and Z_{21} for the two-port network in Fig. 11.118 are,

(a) $\frac{-6}{11}\Omega, \frac{16}{11}\Omega$ (b) $\frac{6}{11}\Omega, \frac{4}{11}\Omega$

(c) $\frac{6}{11}\Omega, \frac{-16}{11}\Omega$ (d) $\frac{4}{11}\Omega, \frac{4}{11}\Omega$

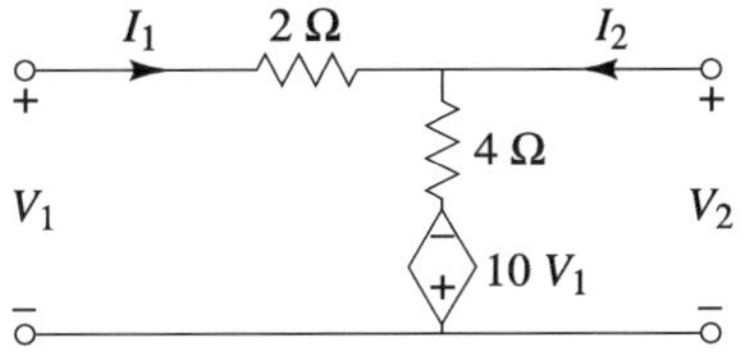

Fig. 11.118

8. The impedance parameters Z_{11} and Z_{12} of a two-port network in Fig. 11.119 are
(a) 2.75 Ω, 0.25 Ω (b) 3 Ω, 0.5 Ω
(c) 3 Ω, 0.25 Ω (d) 2.25 Ω, 0.5 Ω

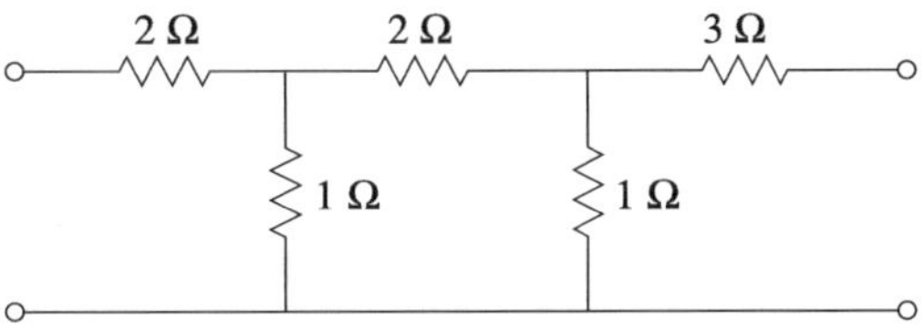

Fig. 11.119

9. The h parameters of the circuit shown in Fig. 11.120 are

(a) $\begin{bmatrix} 0.1 & 0.1 \\ -0.1 & 0.3 \end{bmatrix}$ (b) $\begin{bmatrix} 10 & -1 \\ 1 & 0.05 \end{bmatrix}$

(c) $\begin{bmatrix} 30 & 20 \\ 20 & 20 \end{bmatrix}$ (d) $\begin{bmatrix} 10 & 1 \\ -1 & 0.05 \end{bmatrix}$

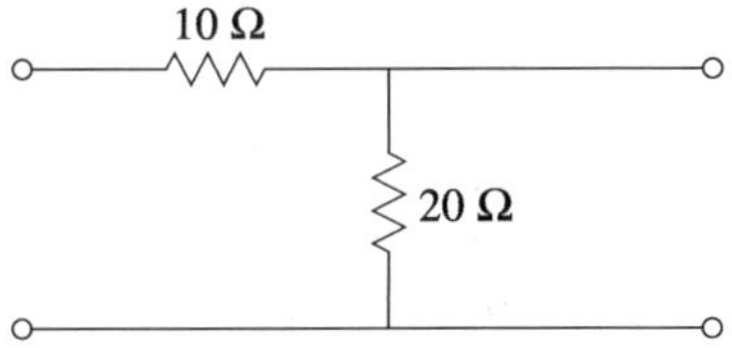

Fig. 11.120

10. A two-port network is represented by $ABCD$ parameters given by $\begin{bmatrix} V_1 \\ I_1 \end{bmatrix} = \begin{bmatrix} A & B \\ C & D \end{bmatrix}\begin{bmatrix} V_2 \\ -I_2 \end{bmatrix}$. If port 2 is terminated by R_L, then the input impedance seen at port 1 is given by

(a) $\frac{A + BR_L}{C + DR_L}$ (b) $\frac{AR_L + C}{BR_L + D}$ (c) $\frac{DR_L + A}{BR_L + C}$ (d) $\frac{B + AR_L}{D + CR_L}$

11. In the two-port network shown in Fig. 11.121, Z_{12} and Z_{21} are respectively
(a) r_e and βr_o (b) 0 and $-\beta r_o$
(c) 0 and βro (d) r_e and $-\beta r_o$

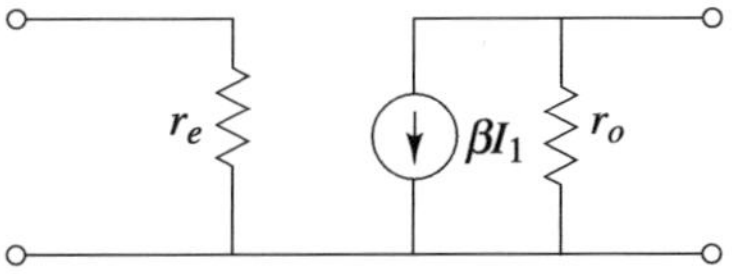

Fig. 11.121

12. If a two-port network is passive, then we have, with the usual notation, the following relationship for symmetrical network

(a) $h_{12} = h_{21}$ (b) $h_{12} = -h_{21}$
(c) $h_{11} = h_{22}$ (d) $h_{11}h_{22} - h_{12}h_{21} = 1$

13. A two-port network is defined by the following pair of equations $I_1 = 2V_1 + V_2$ and $I_2 = V_1 + V_2$. Its impedance parameters ($Z_{11}, Z_{12}, Z_{21}, Z_{22}$) are given by

(a) 2, 1, 1, 1 (b) 1, – 1, – 1, 2 (c) 1, 1, 1, 2 (d) 2, – 1, – 1, 1

14. A two-port network has transmission parameters $\begin{bmatrix} A & B \\ C & D \end{bmatrix}$. The input impedance of the network at port 1 will be

(a) $\dfrac{A}{C}$ (b) $\dfrac{AD}{BC}$ (c) $\dfrac{AB}{DC}$ (d) $\dfrac{D}{C}$

15. A two-port network is symmetrical if

(a) $Z_{11} Z_{22} - Z_{12} Z_{21} = 1$ (b) $AD - BC = 1$
(c) $h_{11} h_{22} - h_{12} h_{21} = 1$ (d) $Y_{11} Y_{22} - Y_{12} Y_{21} = 1$

16. For the network shown in Fig. 11.122, the admittance parameters are $Y_{11} = 8$ mho, $Y_{12} = Y_{21} = -6$ mho and $Y_{22} = 6$ mho. The value of Y_A, Y_B and Y_C (in mho) will be respectively

(a) 2, 6, – 6 (b) 2, 6, 0
(c) 2, 0, 6 (d) 2, 6, 8

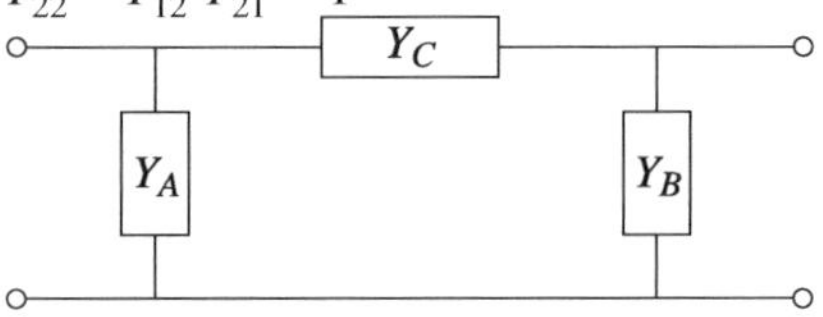

Fig. 11.122

17. The impedance matrices of two two-port networks are given by $\begin{bmatrix} 3 & 2 \\ 2 & 3 \end{bmatrix}$ and $\begin{bmatrix} 15 & 5 \\ 5 & 25 \end{bmatrix}$. If these two networks are connected in series, the impedance matrix of the resulting two-port network will be

(a) $\begin{bmatrix} 3 & 5 \\ 2 & 25 \end{bmatrix}$ (b) $\begin{bmatrix} 18 & 7 \\ 7 & 28 \end{bmatrix}$ (c) $\begin{bmatrix} 15 & 2 \\ 5 & 3 \end{bmatrix}$ (d) inderminate

18. If the π network and T network are equivalent, then the values of R_1, R_2 and R_3 will be respectively

(a) 6, 6, 6 (b) 6, 6, 9 (c) 9, 6, 9 (d) 6, 9, 6

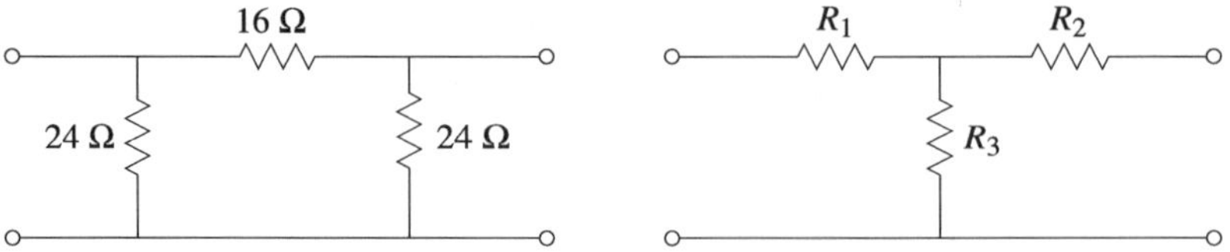

Fig. 11.123

19. For a two-port symmetrical bilateral network, if $A = 3$ and $B = 1$, the value of the parameter C will be

(a) 4 (b) 6 (c) 8 (d) 16

20. The impedance matrix for the network shown is Fig. 11.124 is

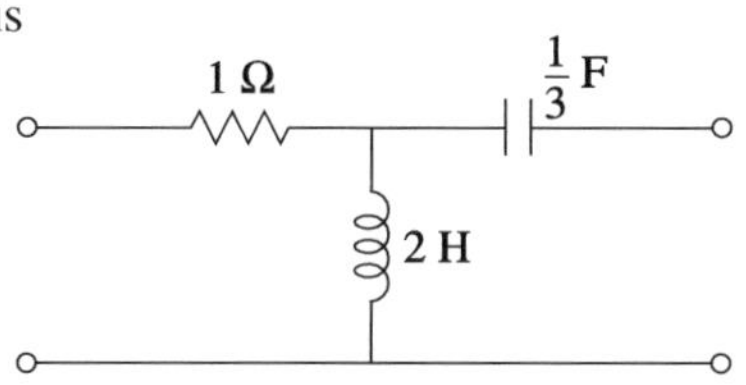

Fig. 11.124

(a) $\begin{bmatrix} 2s+1 & 2s \\ 2s & 2s+\dfrac{3}{s} \end{bmatrix}$ (b) $\begin{bmatrix} 2s+1 & -2s \\ -2s & 2s+\dfrac{3}{s} \end{bmatrix}$

(c) $\begin{bmatrix} 2s+1 & 2s \\ -2s & 2s+\dfrac{3}{s} \end{bmatrix}$ (d) $\begin{bmatrix} 2s+\dfrac{3}{2} & -2s \\ 2s & 2s+\dfrac{3}{s} \end{bmatrix}$

21. With the usual notations, a two-port resistive network satisfies the conditions

$A = D = \frac{3}{2}B = \frac{4}{3}C$. The Z_{11} of the network is

(a) $\frac{5}{3}$ (b) $\frac{4}{3}$ (c) $\frac{2}{3}$ (d) $\frac{1}{3}$

Answers to Objective-Type Questions

1. (a)	**2.** (d)	**3.** (b) & (c)	**4.** (b)	**5.** (a)	**6.** (c)
7. (c)	**8.** (a)	**9.** (d)	**10.** (d)	**11.** (b)	**12.** (d)
13. (b)	**14.** (a)	**15.** (c)	**16.** (c)	**17.** (b)	**18.** (a)
19. (c)	**20.** (a)	**21.** (b)			

12 Network Synthesis

12.1 INTRODUCTION

In the study of electrical networks, broadly there are two topics: 'Network Analysis' and 'Network Synthesis'. Any network consists of excitation, response and network function. In network analysis, network and excitation are given, whereas the response has to be determined. In network synthesis, excitation and response are given, and the network has to be determined. Thus, in network synthesis we are concerned with the realization of a network for a given excitation-response characteristic. Also, there is one major difference between analysis and synthesis. In analysis, there is a unique solution to the problem. But in synthesis, the solution is not unique and many networks can be realized.

12.2 HURWITZ POLYNOMIALS

A polynomial $P(s)$ is said to be Hurwitz if the following conditions are satisfied:

(a) $P(s)$ is real when s is real.

(b) The roots of $P(s)$ have real parts which are zero or negative.

12.3 PROPERTIES OF HURWITZ POLYNOMIALS

(1) All the coefficients in the polynomial

$$P(s) = a_n s^n + a_{n-1} s^{n-1} + \ldots\ldots + a_1 s + a_0$$

are positive. A polynomial may not have any missing terms between the highest and the lowest order unless all even or all odd terms are missing.

(2) The roots of odd and even parts of the polynomial $P(s)$ lie on the $j\omega$-axis only.

(3) If the polynomial $P(s)$ is either even or odd, the roots of polynomial $P(s)$ lie on the $j\omega$-axis only.

(4) All the quotients are positive in the continued fraction expansion of the ratio of odd to even parts or even to odd parts of the polynomial $P(s)$.

(5) If the polynomial $P(s)$ is expressed as $W(s)\, P_1(s)$, then $P(s)$ is Hurwitz if $W(s)$ and $P_1(s)$ are Hurwitz.

(6) If the ratio of the polynomial $P(s)$ and its derivative $P'(s)$ gives a continued fraction expansion with all positive coefficients then the polynomial $P(s)$ is Hurwitz.

This property helps in checking a polynomial for Hurwitz if the polynomial is an even or odd function because in such a case, it is not possible to obtain the continued fraction expansion.

Example 12.1 *State for each case, whether the polynomial is Hurwitz or not. Give reasons in each case.*

(i) $s^4 + 4s^3 + 3s + 2$

(ii) $s^6 + 5s^5 + 4s^4 - 3s^3 + 2s^2 + s + 3$

Solution (i) In the given polynomial, the term s^2 is missing and it is neither an even nor an odd polynomial. Hence, it is not Hurwitz.

(ii) Polynomial $s^6 + 5s^5 + 4s^4 - 3s^3 + 2s^2 + s + 3$ is not Hurwitz as it has a term $(-3s^3)$ which has a negative coefficient.

Example 12.2 *Test whether the polynomial* $P(s) = s^4 + s^3 + 5s^2 + 3s + 4$ *is Hurwitz.*

Solution Even part of $P(s) = m(s) = s^4 + 5s^2 + 4$

Odd part of $P(s) = n(s) = s^3 + 3s$

$$Q(s) = \frac{m(s)}{n(s)}$$

By continued fraction expansion, we have

$$
\begin{array}{l}
s^3 + 3s)\; s^4 + 5s^2 + 4 \;\;(s \\
\qquad\qquad s^4 + 3s^2 \\
\hline
\qquad\qquad 2s^2 + 4)\; s^3 + 3s \;\;\left(\frac{1}{2}s\right. \\
\qquad\qquad\qquad\qquad s^3 + 2s \\
\hline
\qquad\qquad\qquad s)\; 2s^2 + 4 \;\;(2s \\
\qquad\qquad\qquad\qquad 2s^2 \\
\hline
\qquad\qquad\qquad\qquad 4)\; s \;\;\left(\frac{s}{4}\right. \\
\qquad\qquad\qquad\qquad\quad s \\
\hline
\qquad\qquad\qquad\qquad\quad 0
\end{array}
$$

Since all the quotient terms are positive, $P(s)$ is Hurwitz.

Example 12.3 *Test whether the polynomial* $P(s) = s^3 + 4s^2 + 5s + 2$ *is Hurwitz.*

Solution Even part of $P(s) = m(s) = 4s^2 + 2$

Odd part of $P(s)\; n(s) = s^3 + 5s$

The continued fraction expansion can be obtained by dividing $n(s)$ by $m(s)$ as $n(s)$ is of higher order than $m(s)$.

$$Q(s) = \frac{n(s)}{m(s)}$$

$$4s^2 + 2\,\big)\; s^3 + 5s\; \big(\frac{s}{4}$$

$$s^3 + \frac{2s}{4}$$

$$\frac{9s}{2}\,\big)\; 4s^2 + 2\; \big(\frac{8s}{9}$$

$$4s^2$$

$$2\,\big)\; \frac{9s}{2}\; \big(\frac{9s}{4}$$

$$\frac{9s}{2}$$

$$0$$

Since all the quotient terms are positive, $P(s)$ is Hurwitz.

Example 12.4 *Test whether the polynomial $P(s) = s^4 + s^3 + 3s^2 + 2s + 12$ is Hurwitz.*

Solution Even part of $P(s) = m(s) = s^4 + 3s^2 + 12$

Odd part of $P(s) = n(s) = s^3 + 2s$

$$Q(s) = \frac{m(s)}{n(s)}$$

By continued fraction expansion, we have

$$s^3 + 2s\,)\; s^4 + 3s^2 + 12\; (s$$

$$s^4 + 2s^2$$

$$s^2 + 12\,)\; s^3 + 2s\; (s$$

$$s^3 + 12s$$

$$-10s\,\big)\; s^2 + 12\; \big(-\frac{s}{10}$$

$$s^2$$

$$12\,\big) -10s\; \big(-\frac{10}{12}s$$

$$-10s$$

$$0$$

Since two quotient terms are negative, $P(s)$ is not Hurwitz.

Example 12.5 *Prove that polynomial $P(s) = s^4 + s^3 + 2s^2 + 3s + 2$ is not Hurwitz.*

Solution Even part of $P(s) = m(s) = s^4 + 2s^2 + 2$

Odd part of $P(s) = n(s) = s^3 + 3s$

$$Q(s) = \frac{m(s)}{n(s)}$$

By continued fraction expansion, we have

$$s^3 + 3s\,)\,s^4 + 2s^2 + 2\,(\,s$$
$$\underline{s^4 + 3s^2}$$
$$-s^2 + 2\,)\,s^3 + 3s\,(-s$$
$$\underline{s^3 - 2s}$$
$$5s\,\big)\,-s^2 + 2\,\Big(-\frac{s}{5}$$
$$\underline{-s^2}$$
$$2\,\big)\,5s\,\Big(\frac{5}{2}s$$
$$\underline{5s}$$
$$0$$

Since two quotient terms are negative, $P(s)$ is not Hurwitz.

Example 12.6 *Prove that polynomial $P(s) = 2s^4 + 5s^3 + 6s^2 + 3s + 1$ is Hurwitz.*

Solution Even part of $P(s) = m(s) = 2s^4 + 6s^2 + 1$

Odd part of $P(s) = n(s) = 5s^3 + 3s$

$$Q(s) = \frac{m(s)}{n(s)}$$

By continued fraction expansion, we have

$$5s^3 + 3s\,)\,2s^4 + 6s^2 + 1\,\Big(\frac{2}{5}s$$
$$\underline{2s^4 + \frac{6}{5}s^2}$$
$$\frac{24}{5}s^2 + 1\,\Big)\,5s^3 + 3s\,\Big(\frac{25}{24}s$$
$$\underline{5s^3 + \frac{25}{24}s}$$
$$\frac{47}{24}s\,\Big)\,\frac{24}{5}s^2 + 1\,\Big(\frac{576}{235}s$$
$$\underline{\frac{24}{5}s^2}$$
$$1\,\Big)\,\frac{47}{24}s\,\Big(\frac{24}{47}s$$
$$\underline{\frac{47}{24}s}$$
$$0$$

Since all the quotient terms are positive, the polynomial $P(s)$ is Hurwitz.

Example 12.7 *Test whether the polynomial $P(s) = s^4 + 7s^3 + 6s^2 + 21s + 8$ is Hurwitz.*

Solution Even part of $P(s) = m(s) = s^4 + 6s^2 + 8$

Odd part of $P(s) = n(s) = 7s^3 + 21s$

$$Q(s) = \frac{m(s)}{n(s)}$$

By continued fraction expansion, we have

$$7s^3 + 21s \,\big)\, s^4 + 6s^2 + 8 \,\big(\frac{s}{7}$$
$$s^4 + 3s^2$$
$$3s^2 + 8 \,\big)\, 7s^3 + 21s \,\big(\frac{7}{3}s$$
$$7s^3 + \frac{56}{3}s$$
$$\frac{7}{3}s \,\big)\, 3s^2 + 8 \,\big(\frac{9}{7}s$$
$$3s^2$$
$$8 \,\big)\, \frac{7}{3}s \,\big(\frac{7}{24}s$$
$$\frac{7}{3}s$$
$$0$$

Since all the quotient terms are positive, the polynomial $P(s)$ is Hurwitz.

Example 12.8 *Check whether $P(s) = s^4 + 5s^3 + 5s^2 + 4s + 10$ is Hurwitz.*

Solution Even part of $P(s) = m(s) = s^4 + 5s^2 + 10$

Odd part of $P(s) = n(s) = 5s^3 + 4s$

$$Q(s) = \frac{m(s)}{n(s)}$$

By continued fraction expansion, we have

$$5s^3 + 4s\, s \,\big)\, s^4 + 5s^2 + 10 \,\big(\frac{s}{5}$$
$$s^4 + \frac{4}{5}s^2$$
$$\frac{21}{5}s^2 + 10 \,\big)\, 5s^3 + 4s \,\big(\frac{25}{21}s$$
$$5s^3 + \frac{250}{21}s$$
$$-\frac{166}{21}s \,\big)\, \frac{21}{5}s^2 + 10 \,\big(-\frac{441}{830}s$$
$$\frac{21}{5}s^2$$
$$10 \,\big)\, -\frac{166}{21}s \,\big(-\frac{166}{210}s$$
$$-\frac{166}{21}s$$
$$0$$

Since last two quotient terms are negative, the polynomial $P(s)$ is not Hurwitz.

Example 12.9 *Test whether the polynomial $s^5 + 3s^3 + 2s$ is Hurwitz.*

Solution Since the given polynomial contains odd functions only, it is not possible to perform a continued fraction expansion.

$$P'(s) = \frac{d}{ds} P(s) = 5s^4 + 9s^2 + 2$$

$$Q(s) = \frac{P(s)}{P'(s)}$$

By continued fraction expansion, we have

$$5s^4 + 9s^2 + 2 \,\big)\, s^5 + 3s^3 + 2s \,\big(\, \frac{s}{5}$$

$$s^5 + \frac{9}{5}s^3 + \frac{2}{5}s$$

$$\frac{6}{5}s^3 + \frac{8}{5}s \,\big)\, 5s^4 + 9s^2 + 2 \,\big(\, \frac{25}{6}s$$

$$5s^4 + \frac{20}{3}s^2$$

$$\frac{7}{3}s^2 + 2 \,\big)\, \frac{6}{5}s^3 + \frac{8}{5}s \,\big(\, \frac{18}{35}s$$

$$\frac{6}{5}s^3 + \frac{36}{35}s$$

$$\frac{20}{35}s \,\big)\, \frac{7}{3}s^2 + 2 \,\big(\, \frac{49}{12}s$$

$$\frac{7}{3}s^2$$

$$2 \,\big)\, \frac{20}{35}s \,\big(\, \frac{10}{35}s$$

$$\frac{20}{35}s$$

$$0$$

Since all the quotient terms are positive, the polynomial $P(s)$ is Hurwitz.

Example 12.10 *Test whether the polynomial $P(s)$ is Hurwitz.*

$$P(s) = s^5 + s^3 + s$$

Solution Since the given polynomial contains odd functions only, it is not possible to perform continued fraction expansion.

$$P'(s) = \frac{d}{ds} P(s) = 5s^4 + 3s^2 + 1$$

$$Q(s) = \frac{P(s)}{P'(s)}$$

By continued fraction expansion, we have

$$5s^4 + 3s^2 + 1 \,\big)\, s^5 + s^3 + s \,\big(\, \frac{s}{5}$$
$$s^5 + \frac{3}{5}s^3 + \frac{s}{5}$$
$$\frac{2}{5}s^3 + \frac{4}{5}s \,\big)\, 5s^4 + 3s^2 + 1 \,\big(\, \frac{25}{2}s$$
$$5s^4 + 10s^2$$
$$-7s^2 + 1 \,\big)\, \frac{2}{5}s^3 + \frac{4}{5}s \,\big(\, -\frac{2}{35}s$$
$$\frac{2}{5}s^3 - \frac{2}{35}s$$
$$\frac{26}{35}s \,\big)\, -7s^2 + 1 \,\big(\, -\frac{245}{26}s$$
$$-7s^2$$
$$1 \,\big)\, \frac{26}{35}s \,\big(\, \frac{26}{35}s$$
$$- \quad \frac{26}{35}s$$
$$0$$

Since the third and fourth quotient terms are negative, $P(s)$ is not Hurwitz.

Example 12.11 *Test the polynomial P(s) for Hurwitz property.*

$$P(s) = s^6 + 3s^5 + 8s^4 + 15s^3 + 17s^2 + 12s + 4$$

Solution Even part of $P(s) = m(s) = s^6 + 8s^4 + 17s^2 + 4$

Odd part of $P(s) = n(s) = 3s^5 + 15s^3 + 12s$

$$Q(s) = \frac{m(s)}{n(s)}$$

By continued fraction expansion, we have

$$3s^5 + 15s^3 + 12s \,\big)\, s^6 + 8s^4 + 17s^2 + 4 \,\big(\, \frac{1}{3}s$$
$$s^6 + 5s^4 + 4s^2$$
$$3s^4 + 13s^2 + 4 \,\big)\, 3s^5 + 15s^3 + 12s \,\big(\, s$$
$$3s^5 + 13s^3 + 4s$$
$$2s^3 + 8s \,\big)\, 3s^4 + 13s^2 + 4 \,\big(\, \frac{3}{2}s$$
$$3s^4 + 12s^2$$
$$s^2 + 4 \,\big)\, 2s^3 + 8s \,\big(\, 2s$$
$$2s^3 + 8s$$
$$0$$

The division has terminated abruptly (i.e., the number of partial quotients (that is four) is not equal to the order of polynomial (that is six) with a common factor $(s^2 + 4)$.

$$P(s) = s^6 + 3s^5 + 8s^4 + 15s^3 + 17s^2 + 12s + 4 = (s^2 + 4)(s^4 + 3s^3 + 4s^2 + 3s + 1)$$

If both the factors are Hurwitz, $P(s)$ will be Hurwitz.

Let $P_1(s) = s^2 + 4$

Since it contains only even functions, we have to find the continued fraction expansion of $\dfrac{P_1(s)}{P_1'(s)}$.

$$P_1'(s) = 2s$$

$$\frac{P_1(s)}{P_1'(s)} = \frac{s^2+4}{2s} = \frac{s^2}{2s} + \frac{4}{2s} = \frac{s}{2} + \frac{1}{s/2}$$

Since all the quotient terms are positive, $P_1(s)$ is Hurwitz.

Now, let $P_2(s) = s^4 + 3s^3 + 4s^2 + 3s + 1$

$$m_2(s) = s^4 + 4s^2 + 1$$

$$n_2(s) = 3s^3 + 3s$$

By continued fraction expansion, we have

$$\begin{array}{l}
3s^3 + 3s\,\big)\; s^4 + 4s^2 + 1 \;\big(\tfrac{1}{3}s \\
\qquad\qquad\;\; s^4 + s^2 \\
\hline
\qquad 3s^2 + 1\,\big)\; 3s^3 + 3s \;\big(s \\
\qquad\qquad\qquad\;\; 3s^3 + s \\
\hline
\qquad\qquad 2s\,\big)\; 3s^2 + 1 \;\big(\tfrac{3}{2}s \\
\qquad\qquad\qquad\; 3s^2 \\
\hline
\qquad\qquad\qquad 1\,\big)\; 2s \;\big(2s \\
\qquad\qquad\qquad\quad\; 2s \\
\hline
\qquad\qquad\qquad\quad\;\; 0
\end{array}$$

Since all the quotient terms are positive, $P_2(s)$ is Hurwitz.

Therefore $P(s) = (s^2 + 4)\,(s^4 + 3s^3 + 4s^2 + 3s + 1)$ is Hurwitz.

Example 12.12 *Test whether the polynomial* $P(s) = s^7 + 2s^6 + 2s^5 + s^4 + 4s^3 + 8s^2 + 8s + 4$ *is Hurwitz.*

Solution Even part of $P(s) = m(s) = 2s^6 + s^4 + 8s^2 + 4$

Odd part of $P(s) = n(s) = s^7 + 2s^5 + 4s^3 + 8s$

$$Q(s) = \frac{n(s)}{m(s)}$$

By continued fraction expansion, we have

$$\begin{array}{l}
2s^6 + s^4 + 8s^2 + 4\,\big)\; s^7 + 2s^5 + 4s^3 + 8s \;\big(\tfrac{s}{2} \\
\qquad\qquad\qquad\qquad\; s^7 + \tfrac{1}{2}s^5 + 4s^3 + 2s \\
\hline
\qquad\qquad \tfrac{3}{2}s^5 + 6s\,\big)\; 2s^6 + s^4 + 8s^2 + 4 \;\big(\tfrac{4}{3}s \\
\qquad\qquad\qquad\qquad\quad 2s^6 + 8s^2 \\
\hline
\qquad\qquad\qquad\qquad s^4 + 4\,\big)\; \tfrac{3}{2}s^5 + 6s \;\big(\tfrac{3}{2}s \\
\qquad\qquad\qquad\qquad\qquad\quad \tfrac{3}{2}s^5 + 6s \\
\hline
\qquad\qquad\qquad\qquad\qquad\qquad 0
\end{array}$$

Since the division has terminated abruptly it indicates a common factor $s^4 + 4$. The polynomial can be written as

$$P(s) = (s^4 + 4)\,(s^3 + 2s^2 + 2s + 1)$$

If both the factors are Hurwitz, $P(s)$ will be Hurwitz.

In the polynomial $(s^4 + 4)$, the terms s^3, s^2 and s are missing. Hence, it is not Hurwitz.

Therefore $P(s)$ is not Hurwitz.

Example 12.13 *Test whether the polynomial $2s^6 + s^5 + 13s^4 + 6s^3 + 56s^2 + 25s + 25$ is Hurwitz.*

Solution Even part of $P(s) = m(s) = 2s^6 + 13s^4 + 56s^2 + 25$

Odd part of $P(s) = n(s) = s^5 + 6s^3 + 25s$

$$Q(s) = \frac{m(s)}{n(s)}$$

By continued fraction expansion, we have

$$\begin{array}{l}
s^5 + 6s^3 + 25s\ )\ 2s^6 + 13s^4 + 56s^2 + 25\ (\ 2s \\
\qquad\qquad\qquad\quad 2s^6 + 12s^4 + 50s^2 \\
\hline
\qquad\qquad\qquad\quad s^4 + 6s^2 + 25\ )\ s^5 + 6s^3 + 25s\ (\ s \\
\qquad\qquad\qquad\qquad\qquad\qquad\qquad s^5 + 6s^3 + 25s \\
\hline
\qquad\qquad\qquad\qquad\qquad\qquad\qquad\qquad 0
\end{array}$$

The division has terminated abruptly.

$$\begin{aligned}
P(s) &= 2s^6 + s^5 + 13s^4 + 6s^3 + 56s^2 + 25s + 25 \\
&= (s^4 + 6s^2 + 25)\,(2s^2 + s + 1)
\end{aligned}$$

Let $\qquad P_1(s) = s^4 + 6s^2 + 25$

Since $P_1(s)$ contains only even functions, we have to find the continued fraction expansion of $\dfrac{P_1(s)}{P_1'(s)}$.

$$P_1'(s) = 4s^3 + 12s$$

By continued fraction expansion, we have

$$\begin{array}{l}
4s^3 + 12s\ )\ s^4 + 6s^2 + 25\ \left(\dfrac{s}{4}\right. \\
\qquad\qquad\quad s^4 + 3s^2 \\
\hline
\qquad\qquad 3s^2 + 25\ )\ 4s^3 + 12s\ \left(\dfrac{4}{3}s\right. \\
\qquad\qquad\qquad\qquad 4s^3 + \dfrac{100}{3}s \\
\hline
\qquad\qquad\qquad\qquad -\dfrac{64}{3}s\ )\ 3s^2 + 25\ \left(-\dfrac{9}{64}s\right. \\
\qquad\qquad\qquad\qquad\qquad\qquad 3s^2 \\
\hline
\qquad\qquad\qquad\qquad\qquad\qquad 25\ )\ -\dfrac{64}{3}s\ \left(-\dfrac{64}{75}s\right. \\
\qquad\qquad\qquad\qquad\qquad\qquad\qquad -\dfrac{64}{3}s \\
\hline
\qquad\qquad\qquad\qquad\qquad\qquad\qquad\quad 0
\end{array}$$

Since two of the quotient terms are negative, $P_1(s)$ is not Hurwitz.

We need not test the other factor $(2s^2 + s + 1)$ for being Hurwitz.

Hence $P(s)$ is not Hurwitz.

There is another method to test a Hurwitz polynomial. In this method, we construct the Routh–Hurwitz array for the required polynomial.

Let $$P(s) = a_n s^n + a_{n-1} s^{n-1} + a_{n-2} s^{n-2} + \ldots\ldots + a_1 s + a_0$$

The Routh–Hurwitz array is given by,

s^n	a_n	a_{n-2}	a_{n-4}	......
s^{n-1}	a_{n-1}	a_{n-3}	a_{n-5}	......
s^{n-2}	b_n	b_{n-1}	b_{n-2}	......
s^{n-3}	c_n	c_{n-1}		......
.	.			
.	.			
.	.			
.	.			
s^1	.			
s^0	.			

The coefficients of s^n and s^{n-1} rows are directly written from the given equation.

where
$$b_n = \frac{a_{n-1}a_{n-2} - a_n a_{n-3}}{a_{n-1}}$$
$$b_{n-1} = \frac{a_{n-1}a_{n-4} - a_n a_{n-5}}{a_{n-1}}$$
$$b_{n-2} = \frac{a_{n-1}a_{n-6} - a_n a_{n-7}}{a_{n-1}}$$
$$c_n = \frac{b_n a_{n-3} - a_{n-1}b_{n-1}}{b_n}$$
$$c_{n-1} = \frac{b_n a_{n-5} - a_{n-1}b_{n-2}}{b_n}$$

Hence, for polynomial $P(s)$ to be Hurwitz, there should not be any sign change in the first column of the array.

Example 12.14 *Test whether* $P(s) = s^4 + 7s^3 + 6s^2 + 21s + 8$ *is Hurwitz.*

Solution The Routh array is given by

s^4	1	6	8
s^3	7	21	
s^2	3	8	
s^1	$\frac{7}{3}$	0	
s^0	8		

As all the elements in the first column are positive, the polynomial $P(s)$ is Hurwitz.

Example 12.15 *Determine whether* $P(s) = s^4 + s^3 + 2s^2 + 3s + 2$ *is Hurwitz.*

Solution The Routh array is given by

s^4	1	2	2
s^3	1	3	
s^2	–1	2	
s^1	5	0	
s^0	2		

Since there is a sign change in the first column of the array, the polynomial $P(s)$ is not Hurwitz.

Example 12.16 *Test whether $P(s) = s^5 + 2s^4 + 4s^3 + 6s^2 + 2s + 5$ is Hurwitz.*

Solution The Routh array is given by

s^5	1	4	2
s^4	2	6	5
s^3	1	−0.5	
s^2	7	5	
s^1	−1.21		
s^0	5		

Since there is a sign change in the first column of the array, the polynomial is not Hurwitz.

Example 12.17 *Test whether the polynomial $P(s) = s^5 + s^3 + s$ is Hurwitz.*

Solution The given polynomial contains odd functions only.

$$P'(s) = 5s^4 + 3s^2 + 1$$

The Routh array is given by

s^5	1	1	1
s^4	5	3	1
s^3	0.4	0.8	
s^2	−7	1	
s^1	0.86		
s^0	1		

Since there is a sign change in the first column of the array, the polynomial is not Hurwitz.

Example 12.18 *Test whether the polynomial $P(s) = s^8 + 5s^6 + 2s^4 + 3s^2 + 1$ is Hurwitz.*

Solution The given polynomial contains even functions only.

$$P'(s) = 8s^7 + 30s^5 + 8s^3 + 6s$$

The Routh array is given by

s^8	1	5	2	3	1
s^7	8	30	8	6	0
s^6	1.25	1	2.25	1	
s^5	23.6	−6.4	−0.4	0	
s^4	1.33	2.27	1		
s^3	−46.6	−18.14	0		
s^2	1.75	1			
s^1	8.49				
s^0	1				

Since there is a sign change in the first column of the array, the polynomial is not Hurwitz.

Example 12.19 *Test whether $P(s) = s^5 + 12s^4 + 45s^3 + 60s^2 + 44s + 48$ is Hurwitz.*

Solution The Routh array is given by

s^5	1	45	44
s^4	12	60	48
s^3	40	40	
s^2	48	48	
s^1	0	0	
s^0			

Notes: When all the elements in any one row is zero, the following steps are followed :

(i) Write an auxiliary equation with the help of the coefficients of the row just above the row of zeros.

(ii) Differentiate the auxiliary equation and replace its coefficient in the row of zeros.

(iii) Proceed for the Routh test.

Auxiliary equation,

$$A(s) = 48s^2 + 48$$

$$A'(s) = 96s$$

s^5	1	45	44
s^4	12	60	48
s^3	40	40	
s^2	48	48	
s^1	96	0	
s^0	48		

Since there is no sign change in the first column of the array, the polynomial $P(s)$ is Hurwitz.

Example 12.20 *Check whether $P(s) = 2s^6 + s^5 + 13s^4 + 6s^3 + 56s^2 + 25s + 25$ is Hurwitz.*

Solution The Routh array is given by

s^6	2	13	56	25
s^5	1	6	25	
s^4	1	6	25	
s^3	0	0	0	
s^2				
s^1				
s^0				

$$A(s) = s^4 + 6s^2 + 25$$

$$A'(s) = 4s^3 + 12s$$

Now, the Routh array will be given by

s^6	2	13	56	25
s^5	1	6	25	
s^4	1	6	25	
s^3	4	12		
s^2	3	25		
s^1	–21.3			
s^0	25			

Since there is a sign change in the first column of the array, the polynomial $P(s)$ is not Hurwitz.

Example 12.21 *Determine the range of values of 'a' so that $P(s) = s^4 + s^3 + as^2 + 2s + 3$ is Hurwitz.*

Solution The Routh array is given by

$$\begin{array}{c|ccc} s^4 & 1 & a & 3 \\ s^3 & 1 & 2 & \\ s^2 & a-2 & 3 & \\ s^1 & \dfrac{2a-7}{a-2} & & \\ s^0 & 3 & & \end{array}$$

For the polynomial to be Hurwitz, all the terms in the first column of the array should be positive,

i.e.,
$$a - 2 > 0$$
$$a > 2$$

and
$$\frac{2a-7}{a-2} > 0$$
$$a > \frac{7}{2}$$

Hence, $P(s)$ will be Hurwitz when $a > \frac{7}{2}$.

Example 12.22 *Determine the range of values of K so that the polynomial $P(s) = s^3 + 3s^2 + 2s + K$ is Hurwitz.*

Solution The Routh array is given by

$$\begin{array}{c|cc} s^3 & 1 & 2 \\ s^2 & 3 & K \\ s^1 & \dfrac{6-K}{3} & 0 \\ s^0 & K & \end{array}$$

For the polynomial to be Hurwitz, all the terms in the first column of the array should be positive,

i.e.,
$$\frac{6-K}{3} > 0$$
$$6 - K > 0$$

i.e., $K < 6$ and $K > 0$

Hence, $P(s)$ will be Hurwitz for $0 < K < 6$.

12.4 POSITIVE REAL FUNCTIONS

A function $F(s)$ is positive real if the following conditions are satisfied:

(1) $F(s)$ is real for real s.

(2) The real part of $F(s)$ is greater than or equal to zero when the real part of s is greater than or equal to zero i.e.

$$\text{Re } F(s) \geq 0 \qquad \text{for Re}(s) \geq 0$$

12.4.1 Properties of Positive Real Functions

(1) If $F(s)$ is positive real then $\dfrac{1}{F(s)}$ is also positive real.
(2) The sum of two positive real functions is positive real.
(3) The poles and zeros of a positive real function cannot have positive real parts, i.e., they cannot be in the right half of the s plane.
(4) Only simple poles with real positive residues can exist on the $j\omega$-axis.
(5) The poles and zeros of a positive real function are real or occur in conjugate pairs.
(6) The highest powers of the numerator and denominator polynomials may differ at most by unity. This condition prevents the possibility of multiple poles and zeros at $s = \infty$.
(7) The lowest powers of the denominator and numerator polynomials may differ by at most unity. Hence, a positive real function has neither multiple poles nor zeros at the origin.

12.4.2 Necessary and Sufficient Conditions for Positive Real Functions

The necessary and sufficient conditions for a function with real coefficients $F(s)$ to be positive real are the following:
(1) $F(s)$ must have no poles and zeros in the right half of the s-plane.
(2) The poles of $F(s)$ on the $j\omega$-axis must be simple and the residues evaluated at these poles must be real and positive.
(3) Re $F(j\omega) \geq 0$ for all ω.

Testing of the above conditions Condition (1) requires that we test the numerator and denominator of $F(s)$ for roots in the right half of the s-plane, i.e., we must determine whether the numerator and denominator of $F(s)$ are Hurwitz. This is done through a continued fraction expansion of the odd to even or even to odd parts of the numerator and denominator.

Condition (2) is tested by making a partial-fraction expansion of $F(s)$ and checking whether the residues of the poles on the $j\omega$-axis are positive and real. Thus, if $F(s)$ has a pair of poles at $s = \pm j\omega_o$, a partial-fraction expansion gives terms of the form

$$\frac{K_1}{s - j\omega_o} + \frac{K_1^*}{s + j\omega_o}$$

Since residues of complex conjugate poles are themselves conjugate, $K_1 = K_1^*$ and should be positive and real.

Condition (3) requires that Re $F(j\omega)$ must be positive and real for all ω.

Now, to compute Re $F(j\omega)$ from $F(s)$, the numerator and denominator polynomials are separated into even and odd parts.

$$F(s) = \frac{m_1(s) + n_1(s)}{m_2(s) + n_2(s)} = \frac{m_1 + n_1}{m_2 + n_2}$$

Multiplying $N(s)$ and $D(s)$ by $m_2 - n_2$, we have

$$F(s) = \frac{m_1 + n_1}{m_2 + n_2} \frac{m_2 - n_2}{m_2 - n_2} = \frac{m_1 m_2 - n_1 n_2}{m_2^2 - n_2^2} + \frac{m_2 n_1 - m_1 n_2}{m_2^2 - n_2^2}$$

But since the product of two even functions or odd functions is itself an even function, while the product of an even and odd function is odd, we have

$$\text{Ev } F(s) = \frac{m_1 m_2 - n_1 n_2}{m_2^2 - n_2^2}$$

$$\text{Od } F(s) = \frac{m_2 n_1 - m_1 n_2}{m_2^2 - n_2^2}$$

Now, substituting $s = j\omega$ in the even polynomial gives the real part of $F(s)$ and substituting $s = j\omega$ into the odd polynomial gives imaginary part of $F(s)$.

$$\text{Ev } F(s)\,|_{s=j\omega} = \text{Re } F(j\omega)$$
$$\text{Od } F(s)\,|_{s=j\omega} = j \text{ Im } F(j\omega)$$

We have to test Re $F(j\omega) \geq 0$ for all ω.

The denominator of Re $F(j\omega)$ is always a positive quantity because

$$m_2^2 - n_2^2\Big|_{s=j\omega} \geq 0$$

Hence, the condition that Ev $F(j\omega)$ should be positive requires

$$m_1 m_2 - n_1 n_2\,|_{s=j\omega} = A(\omega^2)$$

should be positive and real for all $\omega \geq 0$.

Example 12.23 *Test whether* $F(s) = \dfrac{s+3}{s+1}$ *is a positive real function.*

Solution

(i) $$F(s) = \frac{N(s)}{D(s)} = \frac{s+3}{s+1}$$

The function $F(s)$ has pole at $s = -1$ and zero at $s = -3$.

Thus, pole and zero are in the left half of the s-plane

(ii) There is no pole on the $j\omega$ axis. Hence, the residue test is not carried out.

(iii) Even part of $N(s) = m_1 = 3$

Odd part of $N(s) = n_1 = s$

Even part of $D(s) = m_2 = 1$

Odd part of $D(s) = n_2 = s$

$$A(\omega^2) = m_1 m_2 - n_1 n_2\,|_{s=j\omega}$$
$$= (3)\,(1) - (s)\,(s)\,|_{s=j\omega}$$
$$= 3 - s^2\,|_{s=j\omega} = 3 + \omega^2$$

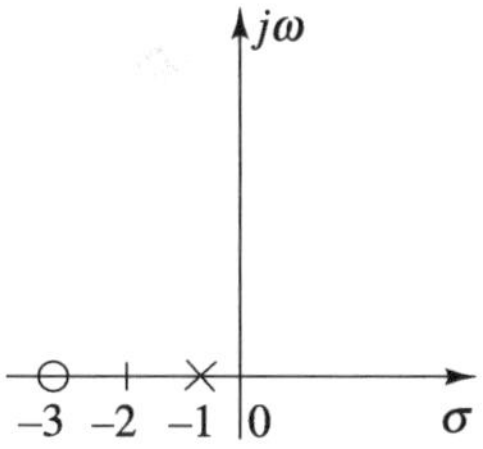

Fig. 12.1

$A(\omega^2)$ is positive for all $\omega \geq 0$

Since all the three conditions are satisfied, the function is positive real.

Example 12.24 *Test whether* $F(s) = \dfrac{s^2+6s+5}{s^2+9s+14}$ *is positive real function.*

Solution

(i) $$F(s) = \frac{N(s)}{D(s)} = \frac{s^2+6s+5}{s^2+9s+14} = \frac{(s+5)(s+1)}{(s+7)(s+2)}$$

The function $F(s)$ has poles at $s = -7$ and $s = -2$ and zeros at $s = -5$ and $s = -1$.

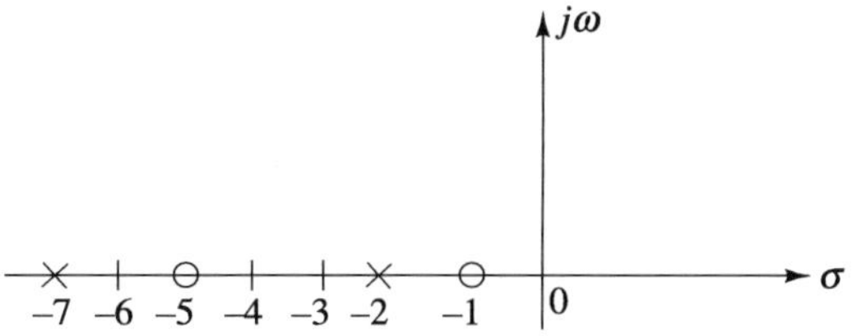

Fig. 12.2

Thus, all the poles and zeros are in the left half of the s plane.

(ii) Since there is no pole on the $j\omega$ axis, the residue test is not carried out.

(iii)

$$\text{Even part of } N(s) = m_1 = s^2 + 5$$
$$\text{Odd part of } N(s) = n_1 = 6s$$
$$\text{Even part of } D(s) = m_2 = s^2 + 14$$
$$\text{Odd part of } D(s) = n_2 = 9s$$
$$A(\omega^2) = m_1 m_2 - n_1 n_2 \,|_{s=j\omega}$$
$$= (s^2 + 5)(s^2 + 14) - (6s)(9s) \,|_{s=j\omega}$$
$$= s^4 - 35s^2 + 70 \,|_{s=j\omega} = \omega^4 + 35\omega^2 + 70$$

$A(\omega^2)$ is positive for all $\omega \geq 0$

Since all the three conditions are satisfied, the function is positive real.

Example 12.25 *Test whether* $F(s) = \dfrac{s(s+3)(s+5)}{(s+1)(s+4)}$ *is positive real function.*

Solution

(i)

$$F(s) = \frac{N(s)}{D(s)}$$
$$= \frac{s(s+3)(s+5)}{(s+1)(s+4)} = \frac{s^3 + 8s^2 + 15s}{s^2 + 5s + 4}$$

The function $F(s)$ has poles at $s = -1$ and $s = -4$ and zeros at $s = 0$, $s = -3$ and $s = -5$.

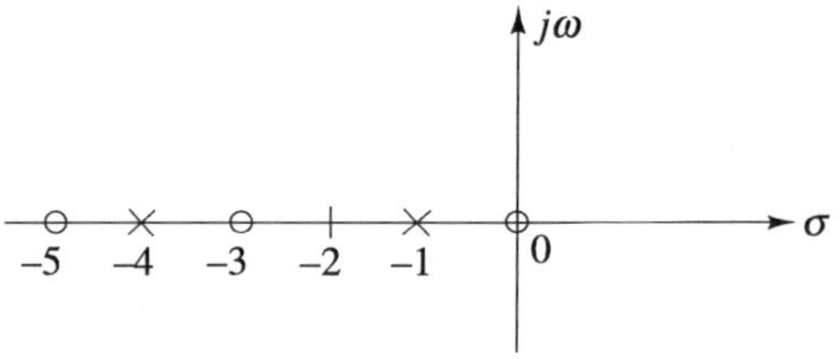

Fig. 12.3

Thus, all the poles and zeros are in the left half of the s plane.

(ii) There is no pole on the $j\omega$ axis, hence the residue test is not carried out.

(iii) Even part of $N(s) = m_1 = 8s^2$

Odd part of $N(s) = n_1 = s^3 + 15s$

Even part of $D(s) = m_2 = s^2 + 4$

Odd part of $D(s) = n_2 = 5s$

$$A(\omega^2) = m_1 m_2 - n_1 n_2 \,|_{s=j\omega}$$
$$= (8s^2)(s^2 + 4) - (s^3 + 15s)(5s) \,|_{s=j\omega}$$
$$= 3s^4 - 43s^2 \,|_{s=j\omega} = 3\omega^4 + 43\omega^2$$

$A(\omega^2)$ is positive for all $\omega \geq 0$

Since all the three conditions are satisfied, the function is positive real.

Example 12.26 *Test whether* $F(s) = \dfrac{s^2 + 1}{s^3 + 4s}$ *is positive real function.*

Solution

(i)

$$F(s) = \frac{N(s)}{D(s)} = \frac{s^2 + 1}{s^3 + 4s} = \frac{(s + j1)(s - j1)}{s(s + j2)(s - j2)}$$

The function $F(s)$ has poles at $s = 0$, $s = -j2$ and $s = j2$ and zeros at $s = -j1$ and $s = j1$. Thus, all the poles and zeros are on the $j\omega$ axis.

(ii) The poles on the $j\omega$ axis are simple. Hence, residue test is carried out.

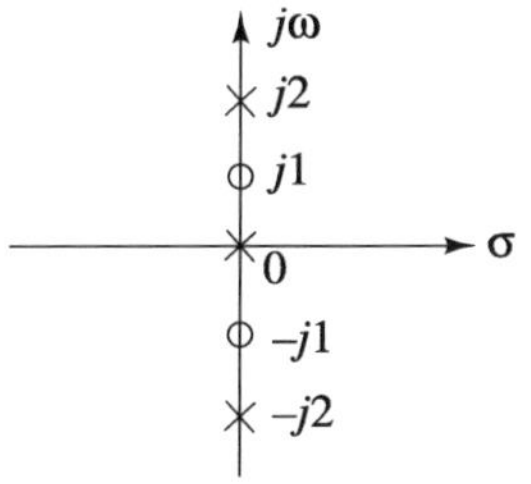

Fig. 12.4

$$F(s) = \frac{s^2+1}{s^3+4s} = \frac{s^2+1}{s(s^2+4)}$$

By partial-fraction expansion, we have

$$F(s) = \frac{K_1}{s} + \frac{K_2}{s+j2} + \frac{K_2^*}{s-j2}$$

The constants K_1, K_2 and K_2* are called residues.

$$K_1 = s\,F(s)\,|_{s=0}$$

$$= \left.\frac{s^2+1}{s^2+4}\right|_{s=0} = \frac{1}{4}$$

$$K_2 = (s+j2)\,F(s)\,|_{s=-j2}$$

$$= \left.\frac{s^2+1}{s(s-j2)}\right|_{s=-j2}$$

$$= \frac{-4+1}{(-j2)(-j2-j2)} = \frac{3}{8}$$

$$K_2^* = K_2 = \frac{3}{8}$$

Thus, residues are real and positive.

(iii) Even part of $N(s) = m_1 = s^2 + 1$

Odd part of $N(s) = n_1 = 0$

Even part of $D(s) = m_2 = 0$

Odd part of $D(s) = n_2 = s^3 + 4s$

$$A(\omega^2) = m_1 m_2 - n_1 n_2\,|_{s=j\omega}$$

$$= (s^2+1)(0) - (0)(s^3+4s)\,|_{s=j\omega} = 0$$

$A(\omega^2)$ is zero for all $\omega \geq 0$

Since all the three conditions are satisfied, the function is positive real.

Example 12.27 *Test whether* $F(s) = \dfrac{2s^3+2s^2+3s+2}{s^2+1}$ *is positive real function.*

Solution

(i)
$$F(s) = \frac{N(s)}{D(s)} = \frac{2s^3+2s^2+3s+2}{s^2+1} = \frac{2s^3+2s^2+3s+2}{(s+j1)(s-j1)}$$

Since numerator polynomial cannot be easily factorized, we will prove whether $N(s)$ is Hurwitz.

Even part of $N(s) = m(s) = 2s^2 + 2$

Odd part of $N(s) = n(s) = 2s^3 + 3s$

By continued fraction expansion, we have

$$
\begin{array}{l}
2s^2+2\,\big)\; 2s^3+3s\;\big(\,s \\
\qquad\qquad\;\; \underline{2s^3+2s} \\
\qquad\qquad\qquad s\,\big)\; 2s^2+2\;\big(\,2s \\
\qquad\qquad\qquad\quad\;\; \underline{2s^2} \\
\qquad\qquad\qquad\qquad\quad 2\,\big)\; s\;\big(\,\frac{s}{2} \\
\qquad\qquad\qquad\qquad\qquad\;\; \underline{s} \\
\qquad\qquad\qquad\qquad\qquad\;\; 0
\end{array}
$$

Since all the quotient terms are positive, $N(s)$ is Hurwitz. This indicates that zeros are in the left half of the s plane.

The function $F(s)$ has poles at $s = -j1$ and $s = j1$.

Thus, all the poles and zeros are in the left half of the s plane.

(ii) The poles on the $j\omega$ axis are simple. Hence, residue test is carried out.

$$F(s) = \frac{2s^3 + 2s^2 + 3s + 2}{s^2 + 1}$$

As the degree of the numerator is greater than that of the denominator, division is first carried out before partial-fraction expansion.

$$
\begin{array}{l}
s^2+1\,\big)\; 2s^3+2s^2+3s+2\;\big(\,2s+2 \\
\qquad\quad\;\; \underline{2s^3 \qquad\quad +2s} \\
\qquad\qquad\quad 2s^2+s+2 \\
\qquad\qquad\quad \underline{2s^2 \qquad +2} \\
\qquad\qquad\quad s
\end{array}
$$

$$F(s) = 2s + 2 + \frac{s}{s^2 + 1}$$

By partial-fraction expansion, we have

$$F(s) = 2s + 2 + \frac{K_1}{s + j1} + \frac{K_1^*}{s - j1}$$

$$K_1 = (s + j1)\, F(s)\,\big|_{s=-j1} = \frac{-j1}{-j1 - j1} = \frac{1}{2}$$

$$K_1^* = K_1 = \frac{1}{2}$$

Thus, residues are real and positive.

(iii) Even part of $N(s)$ $= m_1 = 2s^2 + 2$

Odd part of $N(s)$ $= n_1 = 2s^3 + 3s$

Even part of $D(s)$ $= m_2 = s^2 + 1$

Odd part of $D(s)$ $= n_2 = 0$

$$
\begin{aligned}
A(\omega^2) &= m_1 m_2 - n_1 n_2 \,\big|_{s=j\omega} \\
&= (2s^2 + 2)(s^2 + 1) - (2s^3 + 3s)(0)\,\big|_{s=j\omega} \\
&= 2s^4 + 4s^2 + 2\,\big|_{s=j\omega} \\
&= 2(\omega^4 - 2\omega^2 + 1) = 2(\omega^2 - 1)^2
\end{aligned}
$$

$A(\omega^2) \geq 0$ for all $\omega \geq 0$

Since all the three conditions are satisfied, the function is positive real.

Example 12.28 *Test whether* $F(s) = \dfrac{s^3 + 6s^2 + 7s + 3}{s^2 + 2s + 1}$ *is positive real function.*

Solution (i) $F(s) = \dfrac{N(s)}{D(s)} = \dfrac{s^3 + 6s^2 + 7s + 3}{s^2 + 2s + 1} = \dfrac{s^3 + 6s^2 + 7s + 3}{(s+1)(s+1)}$

Since a numerator polynomial cannot be easily factorized, we will test whether $N(s)$ is Hurwitz.

Even part of $N(s) = m(s) = 6s^2 + 3$

Odd part of $N(s) = n(s) = s^3 + 7s$

By continued fraction expansion, we have

$$
\begin{array}{l}
6s^2 + 3 \,\big)\, s^3 + 7s \,\big(\, \frac{s}{6} \\
\qquad\qquad s^3 + 0.5s \\
\hline
\qquad\qquad 6.5s \,\big)\, 6s^2 + 3 \,\big(\, 0.92s \\
\qquad\qquad\qquad\quad 6s^2 \\
\hline
\qquad\qquad\qquad\qquad 3 \,\big)\, 6.5s \,\big(\, 2.17s \\
\qquad\qquad\qquad\qquad\quad 6.5s \\
\hline
\qquad\qquad\qquad\qquad\quad 0
\end{array}
$$

Since all the quotient terms are positive, $N(s)$ is Hurwitz. This indicates that the zeros are in the left half of the s plane.

The function $F(s)$ has a double pole at $s = -1$

Thus, all the poles and zeros are in the left half of the s plane.

(ii) There is no pole on the $j\omega$ axis. Hence, the residue test is not carried out.

(iii) Even part of $N(s) = m_1 = 6s^2 + 3$

Odd part of $N(s) = n_1 = s^3 + 7s$

Even part of $D(s) = m_2 = s^2 + 1$

Odd part of $D(s) = n_2 = 2s$

$$
\begin{aligned}
A(\omega^2) &= m_1 m_2 - n_1 n_2 \big|_{s=j\omega} \\
&= (6s^2 + 3)(s^2 + 1) - (s^3 + 7s)(2s) \big|_{s=j\omega} \\
&= 4s^4 - 5s^2 + 3 \big|_{s=j\omega} = 4\omega^4 + 5\omega^2 + 3
\end{aligned}
$$

$A(\omega^2)$ is positive for all $\omega \geq 0$

Since all the three conditions are satisfied, the function is positive real.

Example 12.29 *Test whether* $F(s) = \dfrac{s^2 + s + 6}{s^2 + s + 1}$ *is a positive real function.*

Solution

(i)
$$
F(s) = \frac{N(s)}{D(s)} = \frac{s^2 + s + 6}{s^2 + s + 1}
= \frac{\left(s + \frac{1}{2} + j\frac{\sqrt{23}}{2}\right)\left(s + \frac{1}{2} - j\frac{\sqrt{23}}{2}\right)}{\left(s + \frac{1}{2} + j\frac{\sqrt{3}}{2}\right)\left(s + \frac{1}{2} - j\frac{\sqrt{3}}{2}\right)}
$$

The function $F(s)$ has zeros at $s = -\dfrac{1}{2} \pm j\dfrac{\sqrt{23}}{2}$ and poles at $s = -\dfrac{1}{2} \pm j\dfrac{\sqrt{3}}{2}$

(ii) There is no pole on the $j\omega$ axis. Hence, the residue test is not carried out.

(iii) Even part of $N(s) = m_1 = s^2 + 6$

Odd part of $N(s) = n_1 = s$

Even part of $D(s) = m_2 = s^2 + 1$

Odd part of $D(s) = n_2 = s$

$$A(\omega^2) = m_1 m_2 - n_1 n_2 \big|_{s=j\omega}$$
$$= (s^2 + 6)(s^2 + 1) - (s)(s) \big|_{s=j\omega}$$
$$= s^4 + 6s^2 + 6 \big|_{s=j\omega} = \omega^4 - 6\omega^2 + 6$$

For $\omega = 2$, $A(\omega^2) = 16 - 24 + 6 = -2$

This condition is not satisfied.

Hence, the function $F(s)$ is not positive real.

Example 12.30 *Test whether* $F(s) = \dfrac{s^2 + 4}{s^3 + 3s^2 + 3s + 1}$ *is positive real function.*

Solution

(i)
$$F(s) = \frac{N(s)}{D(s)} = \frac{s^2 + 4}{s^3 + 3s^2 + 3s + 1}$$
$$= \frac{(s + j2)(s - j2)}{(s+1)^3}$$

The function $F(s)$ has two zeros at s = $\pm j2$ and three poles at $s = -1$.

Thus, all the poles and zeros are in the left half of the s plane.

(ii) There is no pole on the $j\omega$ axis. Hence, the residue test is not carried out.

(iii) Even part of $N(s) = m_1 = s^2 + 4$

Odd part of $N(s) = n_1 = 0$

Even part of $D(s) = m_2 = 3s^2 + 1$

Odd part of $D(s) = n_2 = s^3 + 3s$

$$A(\omega^2) = m_1 m_2 - n_1 n_2 \big|_{s=j\omega} = (s^2 + 4)(3s^2 + 1) - (0)(s^3 + 3s) \big|_{s=j\omega}$$
$$= 3s^4 + 13s^2 + 4 \big|_{s=j\omega} = 3\omega^4 - 13\omega^2 + 4$$

For $\omega = 1$, $A(\omega)^2 = 3 - 13 + 4 = -6$

This condition is not satisfied.

Hence, the function $F(s)$ is not positive real.

Example 12.31 *Test whether* $F(s) = \dfrac{s^3 + 5s}{s^4 + 2s^2 + 1}$ *is positive real function.*

Solution

(i)
$$F(s) = \frac{N(s)}{D(s)} = \frac{s^3 + 5s}{s^4 + 2s^2 + 1}$$
$$= \frac{s(s^2 + 5)}{(s^2 + 1)^2} = \frac{s(s + j\sqrt{5})(s - j\sqrt{5})}{(s \pm j1)(s \pm j1)}$$

The function $F(s)$ has zeros at $s = 0$, $s = \pm j\sqrt{5}$ and two poles at $s = j1$ and two poles at $s = -j1$.

Thus, poles on the $j\omega$ axis are not simple.

Hence, the function $F(s)$ not positive real.

Example 12.32 *Test whether* $F(s) = \dfrac{s^4 + 3s^3 + s^2 + s + 2}{s^3 + s^2 + s + 1}$ *is positive real function.*

Solution

$$F(s) = \frac{N(s)}{D(s)} = \frac{s^4 + 3s^3 + s^2 + s + 2}{s^3 + s^2 + s + 1}$$

Here, it is easier to prove that $N(s)$ and $D(s)$ are Hurwitz.

By Routh array,

s^4	1	1	2
s^3	3	1	
s^2	$\frac{2}{3}$	2	
s^1	-8		
s^0	2		

Since there is a sign change in the first column of the array, $N(s)$ is not Hurwitz. Thus, all the zeros are not in the left half of the s plane. The remaining two tests need not be carried out.

Hence, the function $F(s)$ is not positive real.

12.5 ELEMENTARY SYNTHESIS CONCEPTS

We know that impedances and admittances of passive networks are positive real functions. Hence, addition of impedances of the two passive networks gives a function which is also a positive real function.

Thus, $Z(s) = Z_1(s) + Z_2(s)$ is a positive real function, if $Z_1(s)$ and $Z_2(s)$ are positive real functions.

Similarly, $Y(s) = Y_1(s) + Y_2(s)$ is a positive real function, if $Y_1(s)$ and $Y_2(s)$ are positive real functions.

There is a special terminology for synthesis procedure. We have,

$$Z(s) = Z_1(s) + Z_2(s)$$
$$Z_2(s) = Z(s) - Z_1(s)$$

Here, $Z_1(s)$ is said to have been removed from $Z(s)$ in forming the new function $Z_2(s)$. If the removed network is associated with the pole or zero of the original network impedance then that pole or zero is also said to have been removed.

There are four important removal operations.

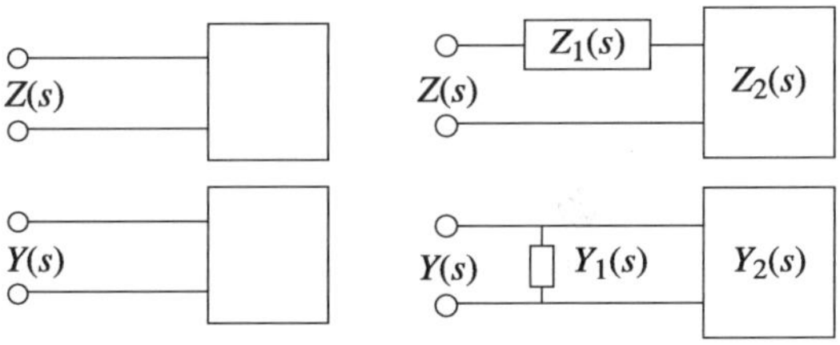

Fig. 12.5

12.5.1 Removal of a Pole at Infinity

Consider an impedance function $Z(s)$ having a pole at infinity which means that the numerator polynomial is one degree greater than the degree of the denominator polynomial.

$$Z(s) = \frac{a_{n+1}s^{n+1} + a_n s^n + \ldots + a_1 s + a_o}{b_n s^n + b_{n-1}s^{n-1} + \ldots + b_1 s + b_o}$$

$$= Hs + \frac{c_n s^n + c_{n-1}s^{n-1} + \ldots + c_1 s + c_o}{b_n s^n + b_{n-1}s^{n-1} + \ldots + b_1 s + b_o}$$

where $\quad H = \dfrac{a_{n+1}}{b_n}$

Let $\quad Z_1(s) = Hs$

and $\quad Z_2(s) = \dfrac{c_n s^n + c_{n-1}s^{n-1} + \ldots + c_1 s + c_o}{b_n s^n + b_{n-1}s^{n-1} + \ldots + b_1 s + b_o} = Z(s) - Hs$

$Z_1(s) = Hs$ represents impedance of an inductor of value H. Hence, the removal of a pole at infinity corresponds to the removal of an inductor from the network of Fig. 12.6(a).

If the given function is an admittance function $Y(s)$, then $Y_1(s) = Hs$ represents the admittance of a capacitor $Y_C(s) = Cs$. The network for $Y_1(s)$ is a capacitor of value $C = H$ as shown in Fig. 12.6(b).

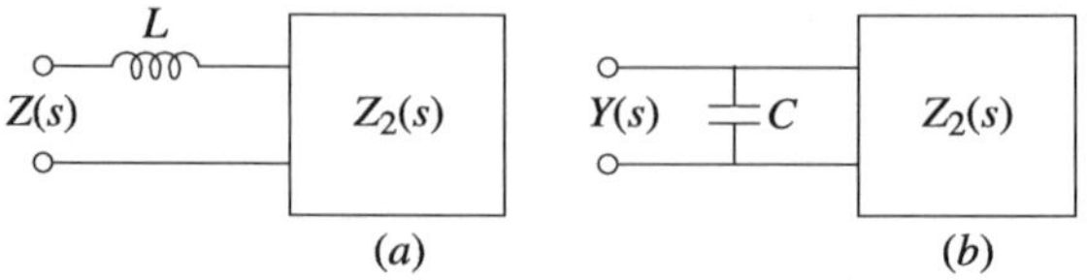

Fig. 12.6 *Network interpretation of the removal of a pole at infinity*

12.5.2 Removal of a Pole at Origin

If $Z(s)$ has a pole at the origin then it may be written as

$$Z(s) = \frac{a_o + a_1 s + \ldots + a_{n-1}s^{n-1} + a_n s^n}{b_1 s + b_2 s^2 + \ldots + b_m s^m}$$

$$= \frac{K_o}{s} + \frac{d_1 + d_2 s + \ldots + d_n s^{n-1}}{b_1 + b_2 s + \ldots + b_m s^{m-1}} = Z_1(s) + Z_2(s)$$

where $$K_o = \frac{a_o}{b_1}$$

$Z_1(s) = \dfrac{K_o}{s}$ represents the impedance of a capacitor of value $\dfrac{1}{K_o}$.

If the given function is an admittance function $Y(s)$ then removal of $Y_1(s) = \dfrac{K_o}{s}$ corresponds to an inductor of value $\dfrac{1}{K_o}$.

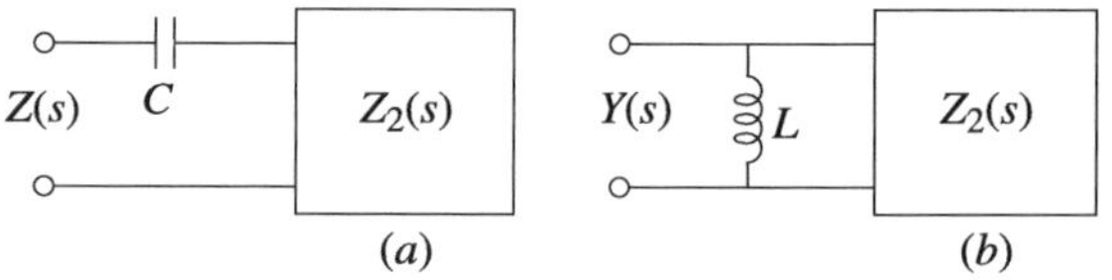

Fig. 12.7 *Network interpretation of the removal of a pole at origin*

Thus, removal of a pole from the impedance function $Z(s)$ at the origin corresponds to the removal of a capacitor, and from admittance function $Y(s)$ corresponds to removal of an inductor.

12.5.3 Removal of Conjugate Imaginary Poles

If $Z(s)$ contains poles on the imaginary axis, i.e., at $s = \pm j\omega_1$ then $Z(s)$ will have factors $(s + j\omega_1)(s - j\omega_1) = s^2 + \omega_1^2$ in the denominator polynomial

$$Z(s) = \frac{p(s)}{\left(s^2 + \omega_1^2\right) q_1(s)}$$

By partial-fraction expansion,

$$Z(s) = \frac{K_1}{s + j\omega_1} + \frac{K_1^*}{s - j\omega_1} + Z_2(s)$$

For a positive real function, $j\omega$ axis poles must themselves be conjugate and must have equal, positive and real residues.

$$K_1 = K_1^*$$

Hence, $$Z(s) = \frac{2K_1 s}{s^2 + \omega_1^2} + Z_2(s)$$

Thus, $$Z_1(s) = \frac{2K_1 s}{s^2 + \omega_1^2} = \frac{1}{\frac{s}{2K_1} + \frac{\omega_1^2}{2K_1 s}} = \frac{1}{Y_a + Y_b}$$

where $Y_a = \frac{s}{2K_1}$ is the admittance of a capacitor with $C = \frac{1}{2K_1}$

and $Y_b = \frac{\omega_1^2}{2K_1 s}$ is the admittance of an inductor with $L = \frac{2K_1}{\omega_1^2}$

If the given function is an admittance function $Y(s)$, then

$$Y_1(s) = \frac{2K_1 s}{s^2 + \omega_1^2} = \frac{1}{Z_a + Z_b} = \frac{1}{\frac{s}{2K_1} + \frac{\omega_1^2}{2K_1 s}}$$

where $Z_a = \frac{s}{2K_1}$ is the impedance of an inductor with $L = \frac{1}{2K_1}$ and $Z_b = \frac{\omega_1^2}{2K_1 s}$ is the impedance of a capacitor with $C = \frac{2K_1}{\omega_1^2}$.

Thus, removal of conjugate imaginary poles from impedance function $Z(s)$ corresponds to the removal of the parallel combination of $L - C$ and from admittance function $Y(s)$ corresponds to removal of series combination of $L - C$.

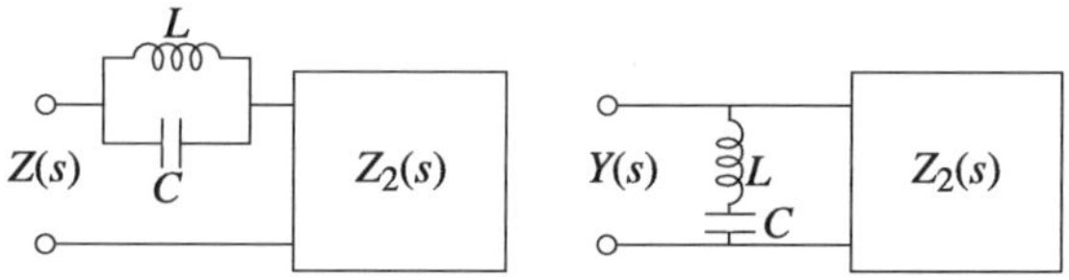

Fig. 12.8

12.5.4 Removal of a Constant

If a real number R_1 is subtracted from $Z(s)$ such that

$$Z_2(s) = Z(s) - R_1$$
$$Z(s) = R_1 + Z_2(s)$$

then R_1 represents a resistor.

If the given function is an admittance function $Y(s)$, then removal of $Y_1(s) = R_1$ represents a conductance of value R_1.

Thus, removal of a constant from impedance function $Z(s)$ corresponds to the removal of a resistance, and from admittance function $Y(s)$ corresponds to removal of a conductance.

12.6 PROPERTIES OF LC FUNCTIONS

(1) It is the ratio of odd to even or even to odd polynomials.
(2) The poles and zeros are simple and lie on the $j\omega$-axis.
(3) The poles and zeros interlace on the $j\omega$-axis.
(4) There must be either a zero or a pole at the origin and infinity.
(5) The difference between any two successive powers of numerator and denominator polynomials is at most two. There cannot be any missing terms.

(6) The highest powers of numerator and denominator polynomials must differ by unity; the lowest powers also differ by unity.

12.7 REALIZATION OF LC FUNCTIONS

There are a number of methods of realizing an LC function. But we will study only four basic forms—Foster I, Foster II, Cauer I and Cauer II forms. The Foster forms are obtained by partial-fraction expansion of $F(s)$, and the Cauer forms are obtained by continued fraction expansion of $F(s)$.

12.7.1 Foster Realization

Consider a general LC function $F(s)$ given by

$$F(s) = \frac{H(s^2+\omega_1^2)(s^2+\omega_3^2)...}{s(s^2+\omega_2^2)(s^2+\omega_4^2)...}$$

where $0 \le \omega_1^2 < \omega_2^2 < \omega_3^2...$ and H is positive.

By partial-fraction expansion of $F(s)$, we have

$$F(s) = \frac{K_0}{s} + \frac{K_2}{s+j\omega_2} + \frac{K_2}{s-j\omega_2} + \ldots + K_\infty s$$

Combining terms with conjugate poles,

$$F(s) = \frac{K_0}{s} + \frac{2K_2 s}{s^2+\omega_2^2} \ldots + K_\infty s$$

where K_0, K_i and K_∞ are the residues of $F(s)$ at the origin, at $j\omega_i$ and at infinity respectively.

These residues are given by

$$K_0 = s\,F(s)\,|_{s=0}$$

$$K_i = \left.\frac{(s^2+\omega_i^2)F(s)}{2s}\right|_{s^2=-\omega_i^2}$$

$$K_\infty = \left.\frac{F(s)}{s}\right|_{s\to\infty}$$

(1) Foster I Form If $F(s)$ represents an impedance function, it gives a series connection of impedances.

$$F(s) = Z(s) = \frac{K_0}{s} + \frac{2K_2 s}{s^2+\omega_2^2} \ldots + K_\infty s = Z_1(s) + Z_2(s) + \ldots + Z_n(s)$$

The first term $\frac{K_0}{s}$ represents the impedance of a capacitor of $\frac{1}{K_0}$ farad.

The last term $K_\infty s$ represents the impedance of an inductor of K_∞ henry.

The remaining terms, i.e., $\frac{2K_i s}{s^2+\omega_i^2}$ represent the impedance of a parallel combination of capacitor C_i and inductor L_i. For parallel combination of L_i and C_i,

$$Z(s) = \frac{1}{C_i s + \frac{1}{L_i s}}$$

$$= \frac{\left(\frac{1}{C_i}\right)s}{s^2 + \frac{1}{L_i C_i}} = \frac{2K_i s}{s^2 + \omega_i^2}$$

$$C_i = \frac{1}{2K_i} \text{ and } L_i = \frac{2K_i}{\omega_i^2}$$

Table 12.1

Impedance function	*Element*
$\frac{K_0}{s} = \frac{1}{C_0 s}$	 $C_o = \frac{1}{K_o}$
$\frac{2K_i s}{s^2 + \omega_i^2} = \frac{\left(\frac{1}{C_i}\right)s}{s^2 + \frac{1}{L_i C_i}}$	$L_i = \frac{2K_i}{\omega_i^2}$ 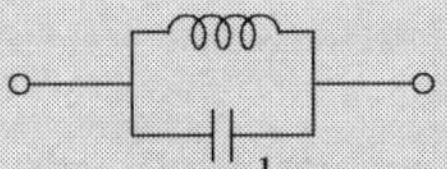 $C_i = \frac{1}{2K_i}$
$K_\infty s = Ls$	 $L_\infty = K_\infty$

The network corresponding to Foster I form is shown in Fig. 12.9.

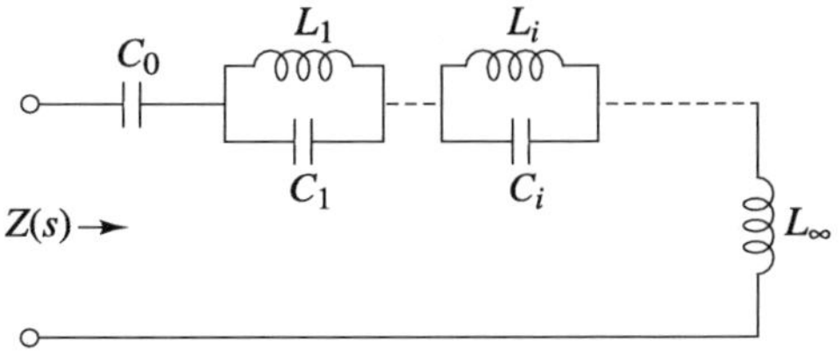

Fig. 12.9

If $Z(s)$ has no pole at the origin then capacitor C_0 is not present in the network. Similarly, if there is no pole at ∞, inductor L_∞ is not present in the network.

Foster II Form If $F(s)$ represents an admittance function, it gives the parallel combination of admittances.

$$F(s) = Y(s) = \frac{K_0}{s} + \frac{2K_2 s}{s^2 + \omega_2^2} + \ldots + K_\infty s$$
$$= Y_1(s) + Y_2(s) + \ldots Y_n(s)$$

The first term $\frac{K_0}{s}$ represents the admittance of an inductor of $\frac{1}{K_0}$ henry.

The last term $K_\infty s$ represents the admittance of a capacitor of K_∞ farad.

The remaining terms, i.e., $\frac{2K_i s}{s^2+\omega_i^2}$ represent the admittance of a series combination of an inductor L_i and a capacitor C_i. For series combination of L_i and C_i,

$$Y(s) = \frac{1}{L_i s + \frac{1}{C_i s}} = \frac{\left(\frac{1}{L_i}\right)s}{s^2 + \frac{1}{L_i C_i}} = \frac{2K_i s}{s^2+\omega_i^2}$$

$$L_i = \frac{1}{2K_i} \quad \text{and} \quad C_i = \frac{2K_i}{\omega_i^2}$$

Table 12.2

Admittance function	*Element*
$\frac{K_0}{s} = \frac{1}{L_0 s}$	 $L_o = \frac{1}{K_o}$
$\frac{2K_i s}{s^2+\omega_i^2} = \frac{\left(\frac{1}{L_i}\right)s}{s^2+\frac{1}{L_i C_i}}$	 $L_i = \frac{1}{2K_i}$, $C_i = \frac{2K_i}{\omega_i^2}$
$K_\infty s = Cs$	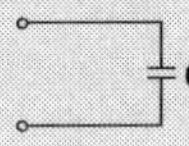 $C_\infty = K_\infty$

The network corresponding to the Foster II form is shown in Fig. 12.10.

If $Y(s)$ has no pole at the origin then inductor L_0 is not present. Similarly, if there is no pole at infinity, capacitor C_∞ is not present.

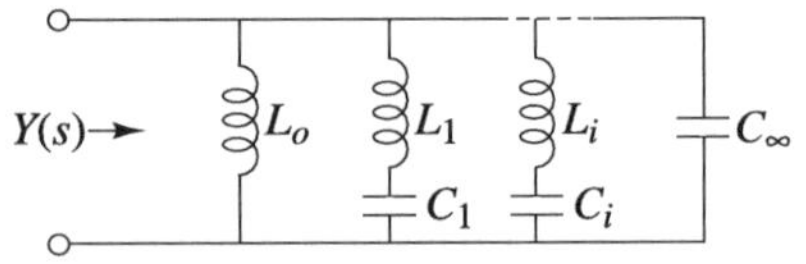

Fig. 12.10

12.7.2 Cauer Realization or Ladder Realization

(1) Cauer I Form Since the numerator and denominator polynomials of an *LC* function always differ in degrees by unity, there is always a zero or a pole at $s = \infty$. The Cauer I Form is obtained by successive removal of a pole or a zero at infinity from the function.

Consider an impedance function $Z(s)$ having a pole at infinity.

By removing the pole at infinity, we get

$$Z_2(s) = Z(s) - L_1 s$$

Now, $Z_2(s)$ has a zero at $s = \infty$. If we invert $Z_2(s)$, $Y_2(s)$ will have a pole at $s = \infty$

By removing this pole,we get

$$Y_3(s) = Y_2(s) - C_2\,s$$

Now $Y_3(s)$ has a zero at $s = \infty$, which we can invert and remove. This process continues until the remainder is zero. Each time we remove a pole, we remove an inductor or a capacitor depending on whether the function is an impedance or an admittance. The impedance $Z(s)$ can be written as a continued fraction expansion.

$$Z(s) = L_1 s + \cfrac{1}{C_2 s + \cfrac{1}{L_3 s + \cfrac{1}{C_4 s + \ldots}}}$$

Thus, the final structure is a ladder network whose series arms are inductors and shunt arms are capacitors. The Cauer I network is as shown in Fig. 12.11.

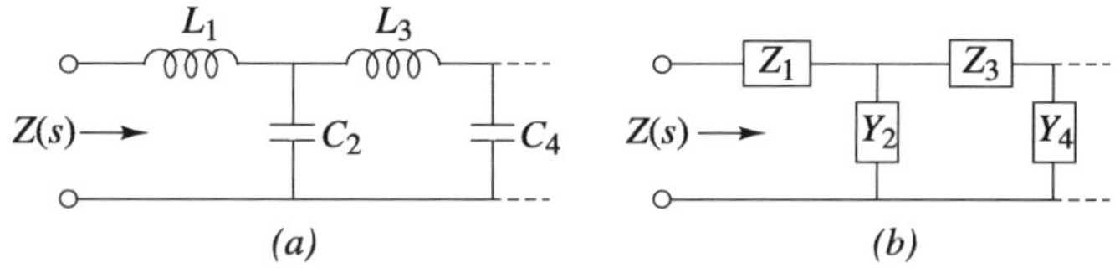

Fig. 12.11 *Cauer I Network*

If the impedance function has zero at infinity, i.e., if degree of numerator is less than that of its denominator by unity, the function is first inverted and continued fraction expansion proceeds as usual. In this case, the first element is a capacitor as shown in Fig. 12.12.

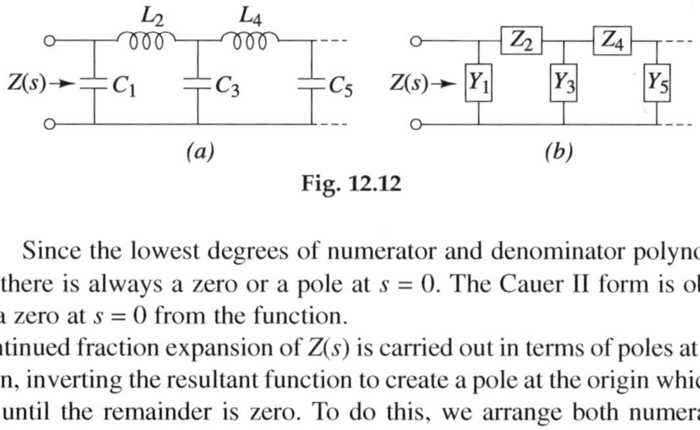

Fig. 12.12

(2) Cauer II Form Since the lowest degrees of numerator and denominator polynomials of *LC* function must differ by unity, there is always a zero or a pole at $s = 0$. The Cauer II form is obtained by successive removal of a pole or a zero at $s = 0$ from the function.

In this method, continued fraction expansion of $Z(s)$ is carried out in terms of poles at the origin by removal of the pole at the origin, inverting the resultant function to create a pole at the origin which is removed and this process is continued until the remainder is zero. To do this, we arrange both numerator and denominator polynomials in ascending order and divide the lowest power of the denominator into the lowest power of the numerator. Then we invert the remainder and divide again. The impedance $Z(s)$ can be written as a continued fraction expansion.

$$Z(s) = \frac{1}{C_1 s} + \cfrac{1}{\cfrac{1}{L_2 s} + \cfrac{1}{\cfrac{1}{C_3 s} + \cfrac{1}{\cfrac{1}{L_4 s} + \ldots}}}$$

Thus, the final structure is a ladder network whose first element is a series capacitor and second element is a shunt inductor as shown in Fig. 12.13.

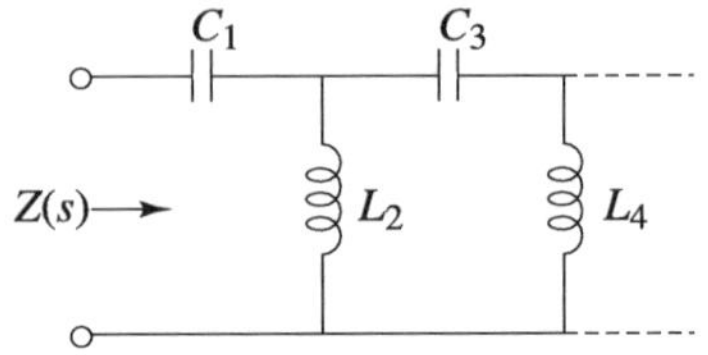

Fig. 12.13

If the impedance function has a zero at the origin then the first element is a shunt inductor and the second element is a series capacitor as shown in Fig. 12.14.

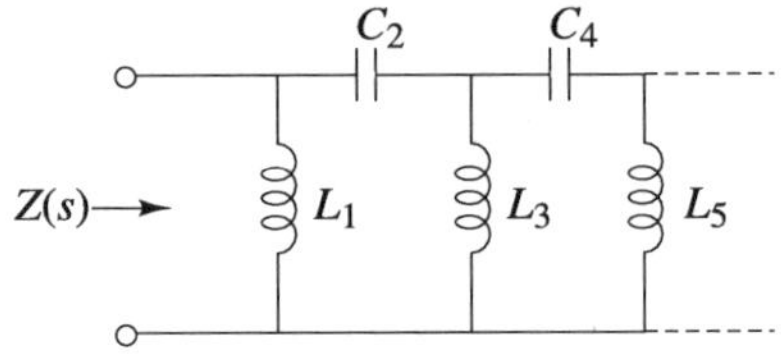

Fig. 12.14

Thus, the *LC* function $F(s)$ can be realized in four different forms. All these forms have the same number of elements and the number is equal to the number of poles and zeros of $F(s)$ including any at infinity.

Example 12.33 *Realize the Foster and Cauer forms of the following impedance function*

$$Z(s) = \frac{4(s^2+1)(s^2+9)}{s(s^2+4)}$$

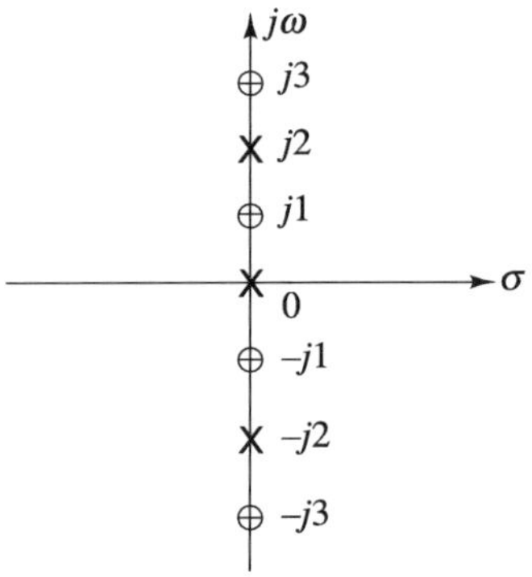

Fig. 12.15

Solution The function $Z(s)$ has poles at $s = 0$ and $s = \pm j2$ and zeros at $s = \pm j1$ and $s = \pm j3$.

From the pole-zero diagram, it is clear that poles and zeros are simple and lie on the $j\omega$ axis. Poles and zeros are interlaced. Hence, the given function is an *LC* function.

(i) Foster I Form The Foster I form is obtained by partial-fraction expansion of the impedance function $Z(s)$. But degree of numerator is greater than degree of denominator. Hence, division is first carried out.

$$Z(s) = \frac{4(s^2+1)(s^2+9)}{s(s^2+4)} = \frac{4s^4+40s^2+36}{s^3+4s}$$

$$\begin{array}{r} s^3+4s)\ 4s^4+40s^2+36\ (4s \\ 4s^4+16s^2 \\ \hline 24s^2+36 \end{array}$$

$$Z(s) = 4s + \frac{24s^2+36}{s^3+4s} = 4s + \frac{24s^2+36}{s(s^2+4)}$$

By partial-fraction expansion, we have

$$Z(s) = 4s + \frac{K_0}{s} + \frac{K_1}{s+j2} + \frac{K_1^*}{s-j2}$$

$$= 4s + \frac{K_0}{s} + \frac{2K_1 s}{s^2+4}$$

where $$K_0 = s\,Z(s)\,|_{s=0} = \frac{4(1)(9)}{4} = 9$$

$$K_1 = \left.\frac{(s^2+4)Z(s)}{2s}\right|_{s^2=-4} = \frac{4(-4+1)(-4+9)}{2(-4)} = \frac{15}{2}$$

$$Z(s) = 4s + \frac{9}{s} + \frac{15s}{s^2+4}$$

The first term represents the impedance of an inductor of 4 henry. The second term represents the impedance of a capacitor of $\frac{1}{9}$ farad. The third term represents the impedance of a parallel LC network.

For a parallel LC network,

$$Z_{LC}(s) = \frac{\left(\frac{1}{C}\right)s}{s^2 + \frac{1}{LC}}$$

By direct comparison,

$$C = \frac{1}{15}\text{ F}$$

$$L = \frac{15}{4}\text{ H}$$

The network is shown in Fig. 12.16.

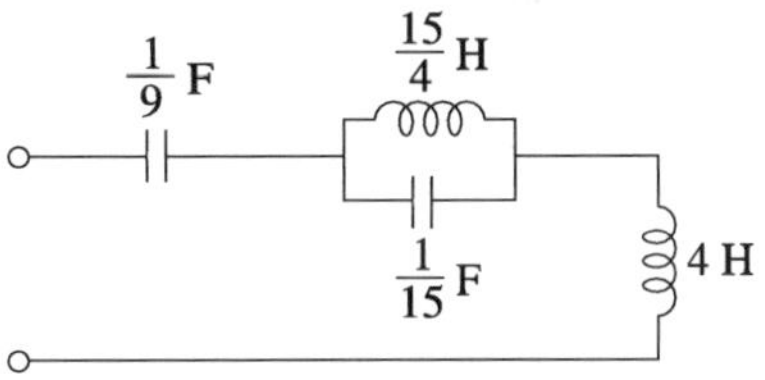

Fig. 12.16

(ii) Foster II Form The Foster II form is obtained by partial-fraction expansion of the admittance function $Y(s)$.

$$Y(s) = \frac{s(s^2+4)}{4(s^2+1)(s^2+9)}$$

By partial-fraction expansion, we have

$$Y(s) = \frac{K_1}{s+j1} + \frac{K_1^*}{s-j1} + \frac{K_2}{s+j3} + \frac{K_2^*}{s-j3}$$

$$= \frac{2K_1 s}{s^2+1} + \frac{2K_2 s}{s^2+9}$$

where $$K_1 = \left.\frac{(s^2+1)}{2s}Y(s)\right|_{s^2=-1} = \frac{(-1+4)}{8(-1+9)} = \frac{3}{64}$$

$$K_2 = \left.\frac{(s^2+9)}{2s}Y(s)\right|_{s^2=-9} = \frac{(-9+4)}{8(-9+1)} = \frac{5}{64}$$

$$Y(s) = \frac{(3/32)s}{s^2+1} + \frac{(5/32)s}{s^2+9}$$

These two terms represent admittance of a series *LC* network. For a series *LC* network,

$$Y_{LC}(s) = \frac{\left(\frac{1}{L}\right)s}{s^2 + \frac{1}{LC}}$$

By direct comparison,

$$L_1 = \frac{32}{3}\text{ H} \qquad C_1 = \frac{3}{32}\text{ F}$$

$$L_2 = \frac{32}{5}\text{ H} \qquad C_2 = \frac{5}{288}\text{ F}$$

The network is shown in Fig. 12.17.

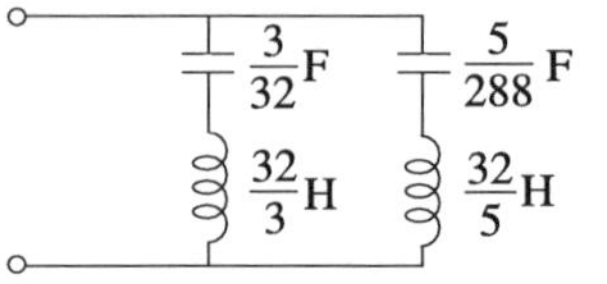

Fig. 12.17

(iii) Cauer I Form The Cauer I form is obtained from continued fraction expansion about the pole at infinity.

$$Z(s) = \frac{4s^4 + 40s^2 + 36}{s^3 + 4s}$$

Since the degree of the numerator is greater than the degree of the denominator by one, it indicates the presence of a pole at infinity.

By continued fraction expansion of $Z(s)$, we have

$s^3 + 4s\,)\,4s^4 + 40s^2 + 36\,(\,4s \leftarrow Z$

$\quad 4s^4 + 16s^2$

$\quad\overline{\qquad 24s^2 + 36\,)\,s^3 + 4s\,\left(\frac{1}{24}s \leftarrow Y\right.}$

$\qquad\qquad s^3 + \frac{3}{2}s$

$\qquad\qquad\overline{\qquad \frac{5}{2}s\,)\,24s^2 + 36\,\left(\frac{48}{5}s \leftarrow Z\right.}$

$\qquad\qquad\qquad 24s^2$

$\qquad\qquad\qquad\overline{\qquad 36\,)\,\frac{5}{2}s\,\left(\frac{5}{72}s \leftarrow Y\right.}$

$\qquad\qquad\qquad\qquad \frac{5}{2}s$

$\qquad\qquad\qquad\qquad\overline{\quad 0 \quad}$

The impedances are connected in series branches whereas the admittances are connected in parallel branches in a Cauer or ladder realisation.

The network is shown in Fig. 12.18.

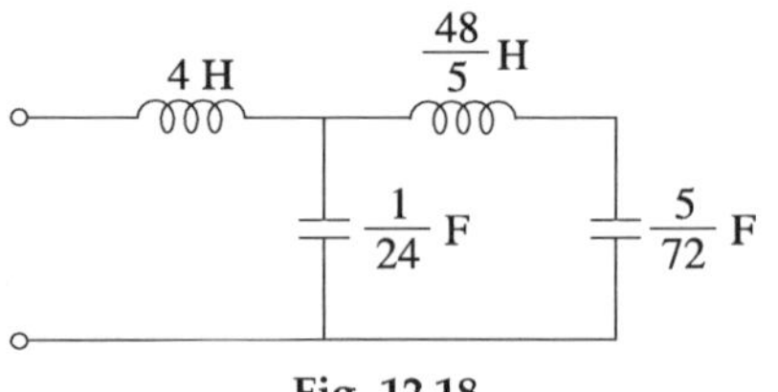

Fig. 12.18

(iv) Cauer II Form The Cauer II form is obtained from partial-fraction expansion about pole at origin.

$$Z(s) = \frac{4(s^2+1)(s^2+9)}{s(s^2+4)} = \frac{4s^4 + 40s^2 + 36}{s^3 + 4s}$$

The function $Z(s)$ has a pole at origin. Arranging the numerator and denominator polynomials in ascending order of s, we have

$$Z(s) = \frac{36 + 40s^2 + 4s^4}{4s + s^3}$$

By continued fraction expansion of $Z(s)$, we have

$$4s + s^3 \,\big)\; 36 + 40s^2 + 4s^4 \;\big(\, \frac{9}{s} \leftarrow Z$$
$$36 + 9s^2$$
$$\overline{31s^2 + 4s^4 \,\big)\; 4s + s^3 \;\big(\, \frac{4}{31s} \leftarrow Y}$$
$$4s + \frac{16}{31}s^3$$
$$\overline{\frac{15}{31}s^3 \,\big)\; 31s^2 + 4s^4 \;\big(\, \frac{961}{15s} \leftarrow Z}$$
$$31s^2$$
$$\overline{4s^4 \,\big)\; \frac{15}{31}s^3 \;\big(\, \frac{15}{124s} \leftarrow Y}$$
$$\frac{15}{31}s^3$$
$$\overline{0}$$

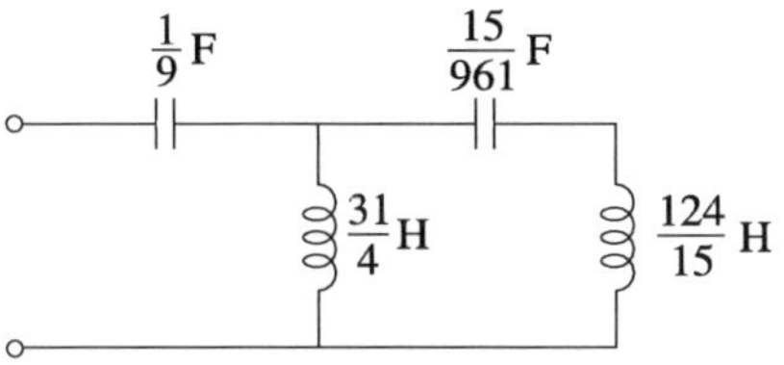

Fig. 12.19

The impedances are connected in series branches whereas the admittances are connected in parallel branches in a Cauer or ladder realisation.

The network is shown in Fig. 12.19.

Example 12.34 *Realize Foster forms of the LC impedance function*

$$Z(s) = \frac{(s^2+1)(s^2+3)}{s(s^2+2)}$$

Solution

(i) Foster I Form The Foster I form is obtained by partial-fraction expansion of the impedance function $Z(s)$. Since the degree of the numerator is greater than the degree of the denominator, division is first carried out.

$$Z(s) = \frac{(s^2+1)(s^2+3)}{s(s^2+2)} = \frac{s^4+4s^2+3}{s^3+2s}$$

$$s^3 + 2s \,)\; s^4 + 4s^2 + 3 \;(\, s$$
$$s^4 + 2s^2$$
$$\overline{2s^2 + 3}$$

$$Z(s) = s + \frac{2s^2+3}{s^3+2s} = s + \frac{2s^2+3}{s(s^2+2)}$$

By partial-fraction expansion, we have

$$Z(s) = s + \frac{K_0}{s} + \frac{K_1}{s+j2} + \frac{K_1^*}{s-j2}$$

$$= s + \frac{K_0}{s} + \frac{2K_1 s}{s^2+2}$$

where $$K_0 = s\,Z(s)\big|_{s=0} = \frac{(1)(3)}{2} = \frac{3}{2}$$

$$K_1 = \frac{(s^2+2)}{2s}Z(s)\bigg|_{s^2=-2} = \frac{(-2+1)(-2+3)}{2(-2)} = \frac{1}{4}$$

$$Z(s) = s + \frac{(3/2)}{s} + \frac{(1/2)s}{s^2+2}$$

The first term represents the impedance of an inductor of 1 henry. The second term represents the impedance of a capacitor of $\frac{2}{3}$ farad. The third term represents the impedance of a parallel LC network.

For a parallel LC network,

$$Z_{LC}(s) = \frac{\left(\frac{1}{C}\right)s}{s^2+\frac{1}{LC}}$$

By direct comparison,

$$C = 2\text{ F}$$

$$L = \frac{1}{4}\text{H}$$

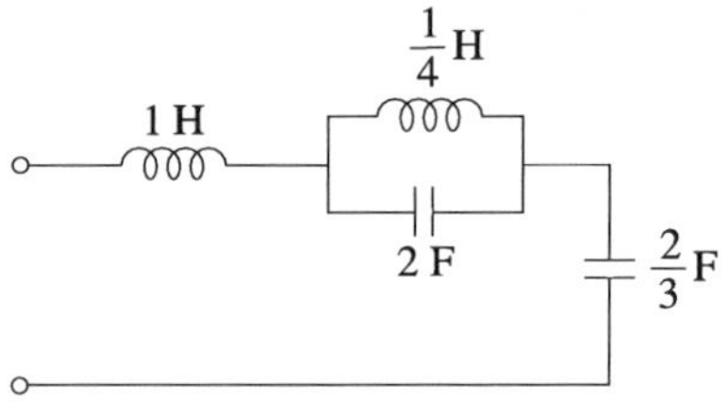

Fig. 12.20

The network is shown in Fig. 12.20.

(ii) Foster II Form The Foster II form is obtained by partial-fraction expansion of the admittance function $Y(s)$.

$$Y(s) = \frac{s(s^2+2)}{(s^2+1)(s^2+3)}$$

By partial-fraction expansion, we have

$$Y(s) = \frac{K_1}{s+j1} + \frac{K_1^*}{s-j1} + \frac{K_2}{s+j\sqrt{3}} + \frac{K_2^*}{s-j\sqrt{3}} = \frac{2K_1 s}{s^2+1} + \frac{2K_2 s}{s^2+3}$$

where $$K_1 = \frac{(s^2+1)}{2s}Y(s)\bigg|_{s^2=-1} = \frac{(-1+2)}{2(-1+3)} = \frac{1}{4}$$

$$K_2 = \frac{(s^2+3)}{2s}Y(s)\bigg|_{s^2=-3} = \frac{-3+2}{2(-3+1)} = \frac{1}{4}$$

$$Y(s) = \frac{(1/2)s}{s^2+1} + \frac{(1/2)s}{s^2+3}$$

These two terms represent admittance of a series LC network. For a series LC network,

$$Y_{LC}(s) = \frac{\left(\frac{1}{L}\right)s}{s^2+\frac{1}{LC}}$$

By direct comparison,

$$L_1 = 2 \text{ H}$$

$$C_1 = \frac{1}{2} \text{F}$$

$$L_2 = 2 \text{ H}$$

$$C_2 = \frac{1}{6} \text{ F}$$

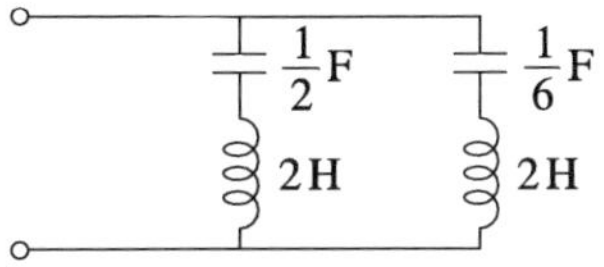

Fig. 12.21

The network is shown in Fig. 12.21.

Example 12.35 *Realize Foster forms of the following LC impedance function*

$$Z(s) = \frac{(s^2+1)(s^2+3)}{s(s^2+2)(s^2+4)}$$

Solution

(i) Foster I Form The Foster I form is obtained by partial-fraction expansion of the impedance function $Z(s)$.

By partial-fraction expansion, we have

$$Z(s) = \frac{K_0}{s} + \frac{K_1}{s+j\sqrt{2}} + \frac{K_1^*}{s-j\sqrt{2}} + \frac{K_2}{s+j2} + \frac{K_2^*}{s-j2}$$

$$= \frac{K_0}{s} + \frac{2K_1 s}{s^2+2} + \frac{2K_2 s}{s^2+4}$$

where

$$K_0 = s\,Z(s)|_{s=0} = \frac{(1)(3)}{(2)(4)} = \frac{3}{8}$$

$$K_1 = \frac{(s^2+2)}{2s} Z(s)\bigg|_{s^2=-2} = \frac{(-2+1)(-2+3)}{2(-2)(-2+4)} = \frac{1}{8}$$

$$K_2 = \frac{(s^2+4)}{2s} Z(s)\bigg|_{s^2=-4} = \frac{(-4+1)(-4+3)}{2(-4)(-4+2)} = \frac{3}{16}$$

$$Z(s) = \frac{3/8}{s} + \frac{(1/4)s}{s^2+2} + \frac{(3/8)s}{s^2+4}$$

The first term represents the impedance of a capacitor of $\frac{8}{3}$ farad. The other two terms represent the impedance of a parallel LC network.

For a parallel LC network,

$$Z_{LC}(s) = \frac{\left(\frac{1}{C}\right)s}{s^2 + \frac{1}{LC}}$$

By direct comparison,

$$C_1 = 4 \text{ F}$$

$$L_1 = \frac{1}{8} \text{H}$$

$$C_2 = \frac{8}{3}\text{ F}$$

$$L_2 = \frac{3}{32}\text{ H}$$

The network is shown in Fig. 12.22.

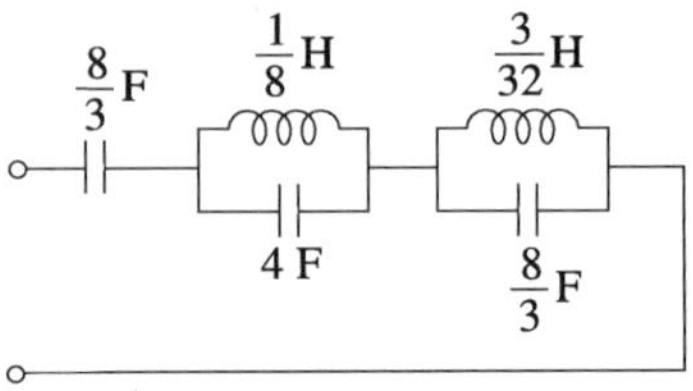

Fig. 12.22

(ii) Foster II Form The Foster II form is obtained by partial-fraction expansion of the admittance function $Y(s)$.

$$Y(s) = \frac{s(s^2+2)(s^2+4)}{(s^2+1)(s^2+3)} = \frac{s^5+6s^3+8s}{s^4+4s^2+3}$$

Since the degree of the numerator is greater than the degree of the denominator, division is first carried out.

$$\begin{array}{r} s^4+4s^2+3\;)\;s^5+6s^3+8s\;(\;s \\ s^5+4s^3+3s \\ \hline 2s^3+5s \end{array}$$

$$Y(s) = s + \frac{2s^3+5s}{s^4+4s^2+3} = s + \frac{2s^3+5s}{(s^2+1)(s^2+3)}$$

By partial-fraction expansion, we have

$$Y(s) = s + \frac{K_1}{s+j1} + \frac{K_1^*}{s-j1} + \frac{K_2}{s+j\sqrt{3}} + \frac{K_2^*}{s-j\sqrt{3}}$$

$$= s + \frac{2K_1 s}{s^2+1} + \frac{2K_2 s}{s^2+3}$$

where $$K_1 = \frac{(s^2+1)}{2s}Y(s)\bigg|_{s^2=-1} = \frac{(-1+2)(-1+4)}{2(-1+3)} = \frac{3}{4}$$

$$K_2 = \frac{(s^2+3)}{2s}Y(s)\bigg|_{s^2=-3} = \frac{(-3+2)(-3+4)}{2(-3+1)} = \frac{1}{4}$$

$$Y(s) = s + \frac{(3/2)s}{s^2+1} + \frac{(1/2)s}{s^2+3}$$

The first term represents the admittance of capacitor of 1 F. The other two terms represent admittance of a series LC network. For a series LC network,

$$Y_{LC}(s) = \frac{\left(\frac{1}{L}\right)s}{s^2+\frac{1}{LC}}$$

By direct comparison,

$$L_1 = \frac{2}{3}\text{ H}$$

$$C_1 = \frac{3}{2}\text{ F}$$

$$L_2 = 2\text{ H}$$

$$C_2 = \frac{1}{6}\text{ F}$$

The network is shown in Fig. 12.23.

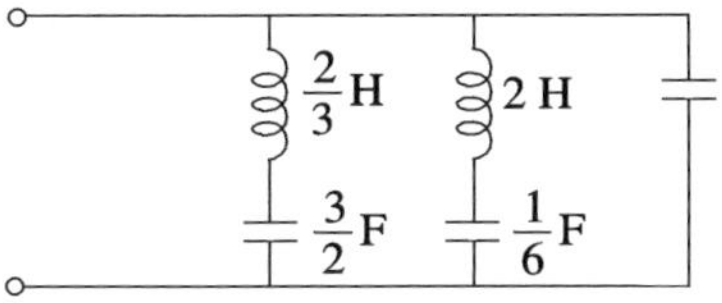

Fig. 12.23

Example 12.36 *Realize Cauer forms of following LC impedance function*

$$Z(s) = \frac{10s^4 + 12s^2 + 1}{2s^3 + 2s}$$

Solution

(i) Cauer I Form The Cauer I form is obtained from continued fraction expansion about the pole at infinity.

$$Z(s) = \frac{10s^4 + 12s^2 + 1}{2s^3 + 2s}$$

Since the degree of the numerator is greater than the degree of the denominator by one, it indicates the presence of a pole at infinity.

By continued fraction expansion of $Z(s)$, we have

$2s^3 + 2s)\ 10s^4 + 12s^2 + 1\ (5s \leftarrow Z$
$\quad 10s^4 + 10s^2$
$\quad\quad 2s^2 + 1)\ 2s^3 + 2s\ (s \leftarrow Y$
$\quad\quad\quad 2s^3 + s$
$\quad\quad\quad\quad s)\ 2s^2 + 1\ (2s \leftarrow Z$
$\quad\quad\quad\quad\quad 2s^2$
$\quad\quad\quad\quad\quad\quad 1)\ s\ (s \leftarrow Y$
$\quad\quad\quad\quad\quad\quad\quad s$
$\quad\quad\quad\quad\quad\quad\quad 0$

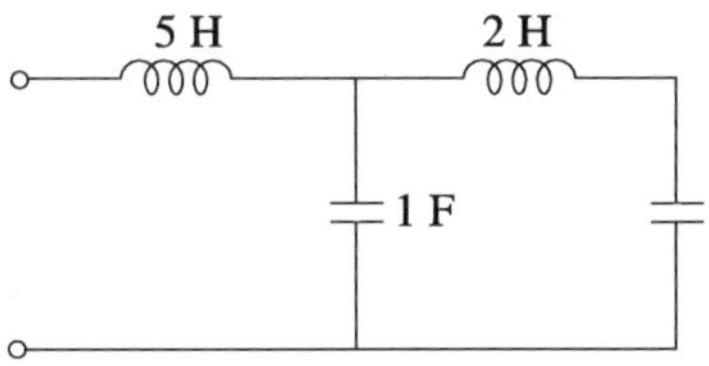

Fig. 12.24

The impedances are connected in series branches whereas the admittances are connected in parallel branches in a Cauer or ladder realisation.

The network is shown in Fig. 12.24.

(ii) Cauer II Form The Cauer II form is obtained from continued fraction expansion about the pole at the origin.

$$Z(s) = \frac{10s^4 + 12s^2 + 1}{2s^3 + 2s}$$

The function $Z(s)$ has a pole at the origin. Arranging the numerator and denominator polynomials in ascending order of s, we have

$$Z(s) = \frac{1 + 12s^2 + 10s^4}{2s + 2s^3}$$

By continued fraction expansion of $Z(s)$, we have

$$2s + 2s^3 \,\big)\, 1 + 12s^2 + 10s^4 \,\big(\frac{1}{2s} \leftarrow Z$$
$$\underline{1 + s^2}$$

$$11s^2 + 10s^4 \,\big)\, 2s + 2s^3 \,\big(\frac{2}{11s} \leftarrow Y$$
$$\underline{2s + \frac{20}{11}s^3}$$

$$\frac{2}{11}s^3 \,\big)\, 11s^2 + 10s^4 \,\big(\frac{121}{2s} \leftarrow Z$$
$$\underline{11s^2}$$

$$10s^4 \,\big)\, \frac{2}{11}s^3 \,\big(\frac{2}{110s} \leftarrow Y$$
$$\underline{\frac{2}{11}s^3}$$
$$0$$

The impedances are connected in series branches whereas the admittances are connected in parallel branches in Cauer or ladder realisation.

The network is shown in Fig. 12.25.

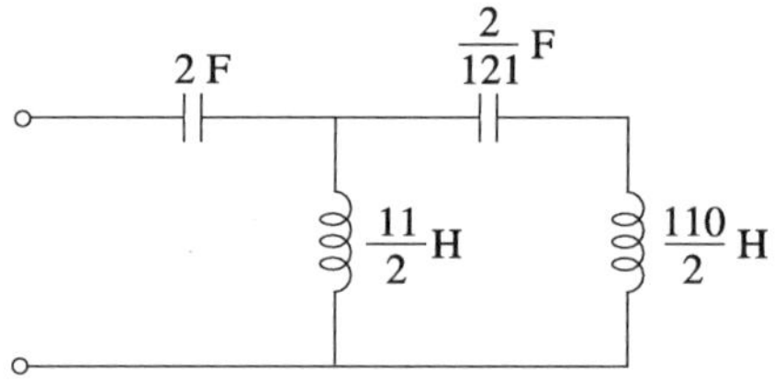

Fig. 12.25

Example 12.37 *Realize the following network function in Cauer I form*

$$Z(s) = \frac{6s^4 + 42s^2 + 48}{s^5 + 18s^3 + 48s}$$

Solution The Cauer I form is obtained by continued fraction expansion of $Z(s)$ about the pole at infinity. In the above function, the degree of the numerator is less than the degree of the denominator which indicates the presence of a zero at infinity. The admittance function $Y(s)$ has a pole at infinity. Hence, the continued fraction expansion of $Y(s)$ is carried out.

$$Y(s) = \frac{s^5 + 18s^3 + 48s}{6s^4 + 42s^2 + 48}$$

By continued fraction expansion, we have

$$6s^4 + 42s^2 + s^2 \,\big)\, s^5 + 18s^3 + 48s \,\big(\, \frac{s}{6} \leftarrow Y$$

$$s^5 + 7s^3 + 8s$$

$$11s^3 + 40s \,\big)\, 6s^4 + 42s^2 + 48 \,\big(\, \frac{6}{11} s \leftarrow Z$$

$$6s^4 + \frac{240}{11} s^2$$

$$\frac{222}{11}s^2 + 30 \,\big)\, 11s^3 + 40s \,\big(\, \frac{121}{222} s \leftarrow Y$$

$$11s^3 + \frac{5808}{222} s$$

$$\frac{3072}{222} s \,\big)\, \frac{222}{11}s^2 + 48 \,\big(\, \frac{49284}{33792} s \leftarrow Z$$

$$\frac{222}{11} s^2$$

$$48 \,\big)\, \frac{3072}{222} s \,\big(\, \frac{128}{444} s \leftarrow Y$$

$$\frac{3072}{222} s$$

$$0$$

The impedances are connected in series branches whereas the admittances are connected in parallel branches in a Cauer or ladder realisation.

The network is shown in Fig. 12.26.

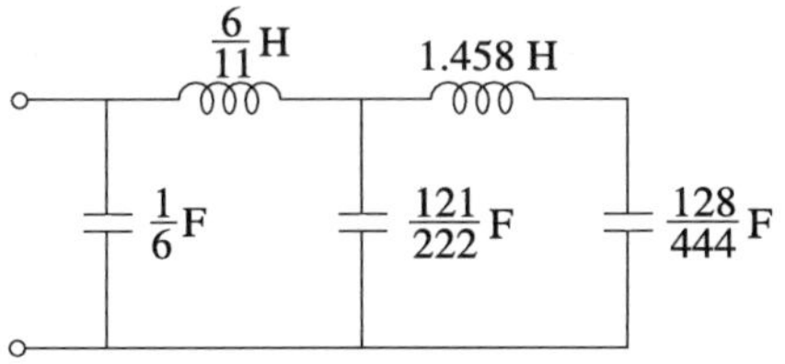

Fig. 12.26

Example 12.38 *Realize Cauer II form of the function*

$$Z_{LC}(s) = \frac{s(s^4 + 3s^2 + 1)}{3s^4 + 4s^2 + 1}$$

Solution The Cauer II form is obtained by continued fraction expansion about the pole at the origin. The given function has a zero at the origin. The admittance function $Y(s)$ has a pole at origin. Hence, the continued fraction expansion of $Y(s)$ is carried out.Arranging the polynomials in ascending order of s, we have

$$Y_{LC}(s) = \frac{3s^4 + 4s^2 + 1}{s^5 + 3s^3 + s} = \frac{1 + 4s^2 + 3s^4}{s + 3s^3 + s^5}$$

By continued fraction expansion of $Y(s)$, we have

$$s + 3s^3 + s^5 \,\Big)\, 1 + 4s^2 + 3s^4 \,\Big(\, \frac{1}{s} \leftarrow Y$$

$$\underline{1 + 3s^2 + s^4}$$

$$s^2 + 2s^4 \,\Big)\, s + 3s^3 + s^5 \,\Big(\, \frac{1}{s} \leftarrow Z$$

$$\underline{s + 2s^3}$$

$$s^3 + s^5 \,\Big)\, s^2 + 2s^4 \,\Big(\, \frac{1}{s} \leftarrow Y$$

$$\underline{s^2 + s^4}$$

$$s^4 \,\Big)\, s^3 + s^5 \,\Big(\, \frac{1}{s} \leftarrow Z$$

$$\underline{s^3}$$

$$s^5 \,\Big)\, s^4 \,\Big(\, \frac{1}{s} \leftarrow Y$$

$$\underline{s^4}$$

$$0$$

The impedances are connected in series branches whereas the admittances are connected in parallel branches in a Cauer or ladder realisation.

The network is shown in Fig. 12.27.

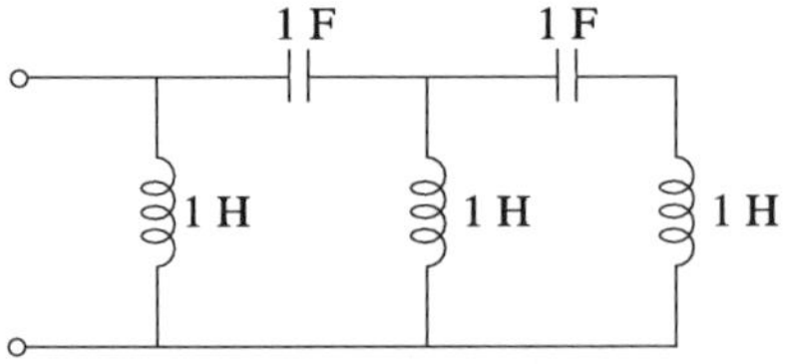

Fig. 12.27

Example 12.39 *Obtain the Cauer I form of realization for the function*

$$Z_{LC}(s) = \frac{s^5 + 7s^3 + 10s}{s^4 + 5s^2 + 4}$$

Solution The Cauer I form is obtained by continued fraction expansion of $Z_{LC}(s)$ about pole at infinity.

$$s^4 + 5s^2 + 4\ )\ s^5 + 7s^3 + 10s\ (s \leftarrow Z$$
$$s^5 + 5s^3 + 4s$$
$$\overline{\qquad\qquad\qquad}$$
$$2s^3 + 6s\ )\ s^4 + 5s^2 + 4\ \left(\frac{s}{2} \leftarrow Y\right.$$
$$s^4 + 3s^2$$
$$\overline{\qquad\qquad\qquad}$$
$$2s^2 + 4\ )\ 2s^3 + 6s\ (s \leftarrow Z$$
$$2s^3 + 4s$$
$$\overline{\qquad\qquad\qquad}$$
$$2s\ )\ 2s^2 + 4\ (s \leftarrow Y$$
$$2s^2$$
$$\overline{\qquad\qquad\qquad}$$
$$4\ )\ 2s\ \left(\frac{s}{2} \leftarrow Z\right.$$
$$2s$$
$$\overline{\qquad\qquad\qquad}$$
$$0$$

The impedances are connected in series branches whereas the admittances are connected in parallel branches in a Cauer or ladder realisation.

The network is shown in Fig. 12.28.

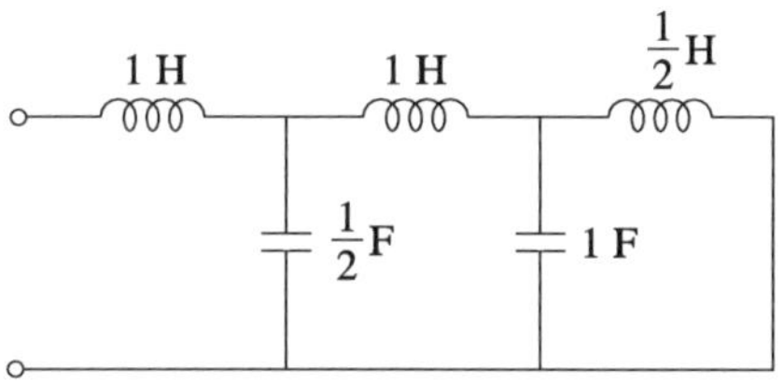

Fig. 12.28

Example 12.40 *Synthesize following LC impedance function in Cauer II form*

$$Z(s) = \frac{s^3 + 2s}{s^4 + 4s^2 + 3}$$

Solution The Cauer II form is obtained by continued fraction expansion about the pole at the origin.

$$Z(s) = \frac{s(s^2 + 2)}{(s^2 + 3)(s^2 + 1)}$$

But $Z(s)$ has a zero at the origin. Hence, the continued fraction expansion of $Y(s)$ is carried out. Arranging the polynomials in ascending order of s, we have

$$Y(s) = \frac{3 + 4s^2 + s^4}{2s + s^3}$$

By continued fraction expansion of $Y(s)$, we have

$$
\begin{array}{l}
2s + s^3 \,\big)\, 3 + 4s^2 + s^4 \,\big(\, \dfrac{3}{2s} \leftarrow Y \\
\qquad\qquad 3 + \dfrac{3}{2}s^2 \\
\hline
\qquad\qquad \dfrac{5}{2}s^2 + s^4 \,\big)\, 2s + s^3 \,\big(\, \dfrac{4}{5s} \leftarrow Z \\
\qquad\qquad\qquad\qquad 2s + \dfrac{4}{5}s^3 \\
\hline
\qquad\qquad\qquad\qquad \dfrac{s^3}{5} \,\big)\, \dfrac{5}{2}s^2 + s^4 \,\big(\, \dfrac{25}{2s} \leftarrow Y \\
\qquad\qquad\qquad\qquad\qquad \dfrac{5}{2}s^2 \\
\hline
\qquad\qquad\qquad\qquad\qquad s^4 \,\big)\, \dfrac{s^3}{5} \,\big(\, \dfrac{1}{5s} \leftarrow Z \\
\qquad\qquad\qquad\qquad\qquad\qquad \dfrac{s^3}{5} \\
\hline
\qquad\qquad\qquad\qquad\qquad\qquad 0
\end{array}
$$

The impedances are connected in series branches whereas the admittances are connected in parallel branches in a Cauer or ladder realisation.

The network is shown in Fig. 12.29.

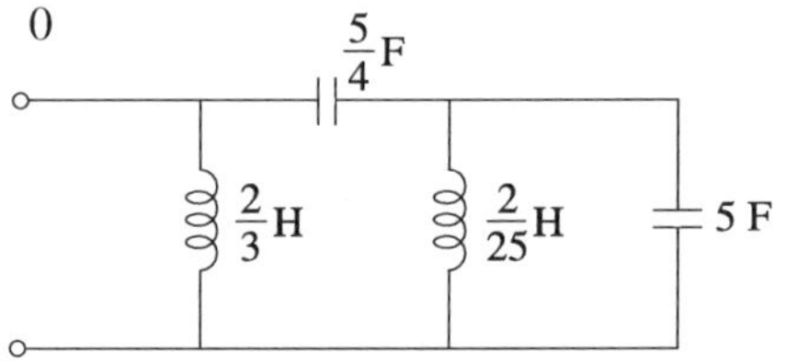

Fig. 12.29

12.8 PROPERTIES OF RC FUNCTIONS

(1) The poles and zeros are simple and are located on the negative real axis of the s plane.
(2) The poles and zeros are interlaced.
(3) The lowest critical frequency nearest to the origin is a pole.
(4) The highest critical frequency farthest to the origin is a zero.
(5) Residues evaluated at the poles of $Z_{RC}(s)$ are real and positive.
(6) The slope $\dfrac{d}{d\sigma}Z_{RC}$ is negative.
(7) $Z_{RC}(\infty) < Z_{RC}(0)$.

12.9 REALIZATION OF RC FUNCTIONS

RC functions can also be realized in four different ways. The impedance function of RC networks is given by

$$Z(s) = \frac{H(s+\sigma_1)(s+\sigma_3)\ldots}{s(s+\sigma_2)\ldots}$$

12.9.1 Foster Realization

(i) Foster I Form The Foster I form is obtained by partial-fraction expansion of $Z(s)$.

$$Z(s) = \frac{K_0}{s} + \frac{K_1}{s+\sigma_1} + \frac{K_2}{s+\sigma_2} + \ldots + K_\infty$$

where $K_0, K_1, K_2, \ldots K_\infty$ are residues of $Z(s)$.

$$K_o = s\,Z(s)\,|_{s=0}$$

$$K_i = (s+\sigma_i)\,Z(s)\,|_{s=-\sigma_i}$$

$$K_\infty = \left.\frac{Z(s)}{s}\right|_{s\to\infty}$$

The first term $\frac{K_0}{s}$ represents the impedance of a capacitor of $\frac{1}{K_o}$ farads.

The last term K_∞ represents the impedance of a resistor of K_∞ ohms.

The remaining terms, i.e., $\frac{K_i}{s+\sigma_i}$ represent the impedance of the parallel combination of resistor R_i and capacitor C_i. For parallel combination of R_i and C_i,

$$Z(s) = \frac{R_i\left(\frac{1}{C_i s}\right)}{R_i + \frac{1}{C_i s}} = \frac{K_i}{s+\sigma_i}$$

$$R_i = \frac{K_i}{\sigma_i} \text{ and } C_i = \frac{1}{K_i}$$

Table 12.3

Impedance function	*Element*
$\frac{K_0}{s} = \frac{1}{C_0 s}$	$C_o = \frac{1}{K_o}$
$\frac{K_i}{s+\sigma_i} = \frac{(R_i)\left(\frac{1}{C_i s}\right)}{R_i + \frac{1}{C_i s}}$	$R_i = \frac{K_i}{\sigma_i}$, $C_i = \frac{1}{K_i}$
$K_\infty = R_\infty$	$R_\infty = K_\infty$

The network corresponding to the Foster I form is shown in Fig. 12.30.

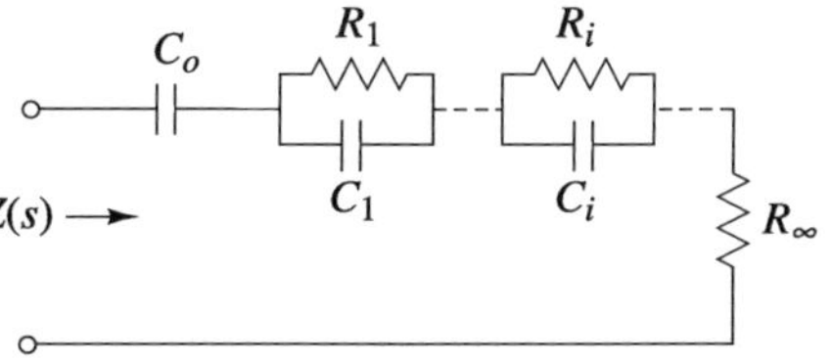

Fig. 12.30

(ii) Foster II Form The Foster II form is obtained by partial fraction expansion of $Y(s)$. Since $Y(s) = \dfrac{1}{Z(s)}$ has negative residue at its pole, Foster II form is obtained by expanding $\dfrac{Y(s)}{s}$.

$$\frac{Y(s)}{s} = \frac{K_o}{s} + \sum_{i=1}^{n} \frac{K_i}{(s+\sigma_i)} + K_\infty$$

Multiplying this equation by s,

$$Y(s) = K_o + \sum_{i=1}^{n} \frac{K_i s}{(s+\sigma_i)} + K_\infty s$$

The first term K_o represents the conductance of a resistor of $\dfrac{1}{K_o}$ ohms.

The last term $K_\infty s$ represents the admittance of a capacitor of K_∞ farads.

The remaining terms, i.e., $\dfrac{K_i s}{s+\sigma_i}$ represent the admittance of series combination of resistor R_i and capacitor C_i with $R_i = \dfrac{1}{K_i}$ ohms and $C_i = \dfrac{K_i}{\sigma_i}$ farads.

Table 12.4

Admittance function	*Element*
$K_0 = \dfrac{1}{R_0}$	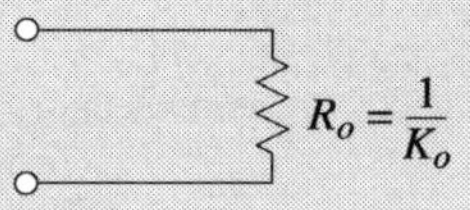
$\dfrac{K_i}{s+\omega_i} = \dfrac{\left(\dfrac{1}{R_i}\right)s}{s+\dfrac{1}{R_i C_i}}$	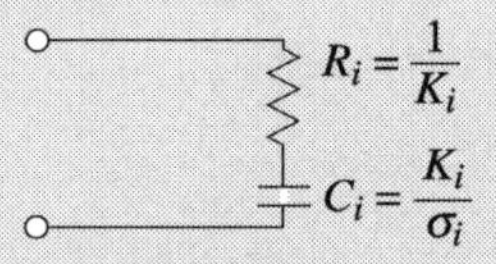
$K_\infty s = C_\infty s$	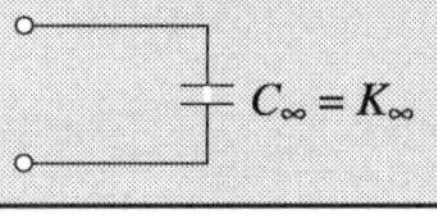

The network corresponding to the Foster II form is shown in Fig. 12.31.

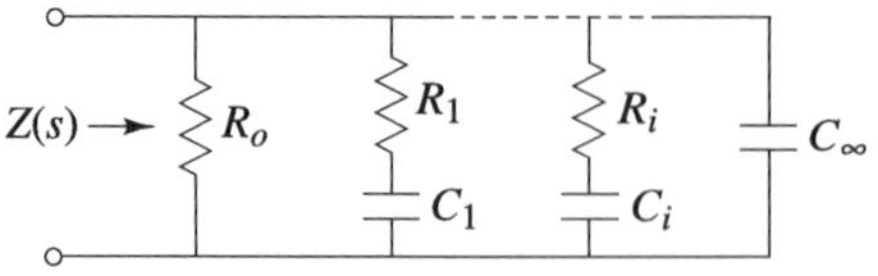

Fig. 12.31

12.9.2 Cauer Realization

(i) Cauer I Form The Cauer I form is obtained by removal of the pole from the impedance function $Z(s)$ at $s = \infty$. This is the same as a continued fraction expansion of the impedance function about infinity. The impedance $Z(s)$ can be written as a continued fraction expansion.

$$Z(s) = R_1 + \cfrac{1}{C_2 s + \cfrac{1}{R_3 + \cfrac{1}{C_4 s + \ldots}}}$$

The network is shown in Fig. 12.32.

In the network shown in Fig. 12.32, if $Z(s)$ has a zero at $s = \infty$, the first element is the capacitor C_1. If $Z(s)$ is a constant at $s = \infty$, the first element is R_1. If $Z(s)$ has a pole at $s = 0$, the last element is C_n. If $Z(s)$ is a constant at $s = 0$, the last element is R_n.

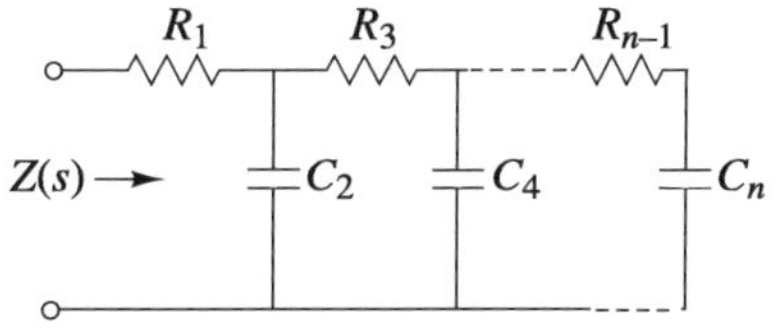

Fig. 12.32

(ii) Cauer II Form The Cauer II form is obtained by removal of the pole from the impedance function at the origin. This is the same as a continued fraction expansion of an impedance function about the origin.

If the given impedance function has a pole at the origin, it is removed as a capacitor C_1. The reciprocal of the remainder function has a minimum value at $s = 0$ which is removed as a constant of resistor R_2. If the original impedance has no pole at the origin, then the first capacitor is absent and the process is repeated with the removal of the constant corresponding to the resistor R_2.

The impedance $Z(s)$ can be written as a continued fraction expansion.

$$Z(s) = \frac{1}{C_1 s} + \cfrac{1}{\frac{1}{R_2} + \cfrac{1}{\frac{1}{C_3 s} + \cfrac{1}{\frac{1}{R_4} + \ldots}}}$$

The network is shown in Fig. 12.33.

In the network shown in Fig. 12.33, if $Z(s)$ has a pole at $s = 0$, the first element is C_1. If $Z(s)$ is a constant at $s = 0$, the first element is R_2. If $Z(s)$ has a zero at $s = \infty$, the last element is C_n. If $Z(s)$ is constant at $s = \infty$, the last element is R_n.

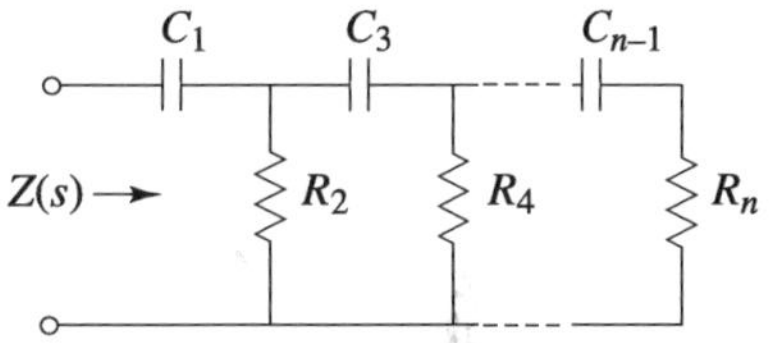

Fig. 12.33

Example 12.41 *Realize the Foster and Cauer forms of the following impedance function*

$$Z(s) = \frac{(s+1)(s+3)}{s(s+2)}$$

Solution The function $Z(s)$ has poles at $s = 0$ and $s = -2$ and zeros at $s = -1$ and $s = -3$.

From the pole-zero diagram, it is clear that poles and zeros are simple and lie on the negative real axis. The poles and zeros are interlaced and the lowest critical frequency nearest to the origin is a pole. Hence, the function $Z(s)$ is an R-C function.

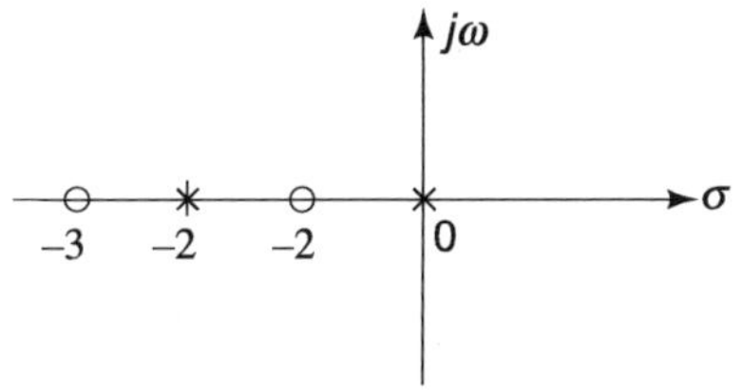

Fig. 12.34

(1) Foster I Form The Foster I form is obtained by partial fraction expansion of impedance function $Z(s)$. Since the degree of the numerator is greater than the degree of the denominator, division is first carried out

$$Z(s) = \frac{s^2 + 4s + 3}{s^2 + 2s}$$

$$\begin{array}{r} s^2+2s\ )\ s^2+4s+3\ (\ 1 \\ s^2+2s \\ \hline 2s+3 \end{array}$$

$$Z(s) = 1 + \frac{2s+3}{s^2+2s} = 1 + \frac{2s+3}{s(s+2)}$$

By partial-fraction expansion, we have

$$Z(s) = 1 + \frac{K_1}{s} + \frac{K_2}{s+2}$$

where
$$K_1 = s\,Z(s)\,|_{s=0} = \frac{(1)(3)}{2} = \frac{3}{2}$$

$$K_2 = (s+2)\,Z(s)\,|_{s=-2} = \frac{(-2+1)(-2+3)}{-2} = \frac{1}{2}$$

$$Z(s) = 1 + \frac{3/2}{s} + \frac{1/2}{s+2}$$

The first term represents the impedance of a resistor of 1 Ω. The second term represents the impedance of a capacitor of $\frac{2}{3}$ F. The third term represents the impedance of parallel RC circuit for which

$$Z_{RC}(s) = \frac{\frac{1}{C_i}}{s + \frac{1}{R_i C_i}}$$

By direct comparison,

$$R = \frac{1}{4}\,\Omega$$

$$C = 2\text{ F}$$

The network is shown in Fig. 12.35.

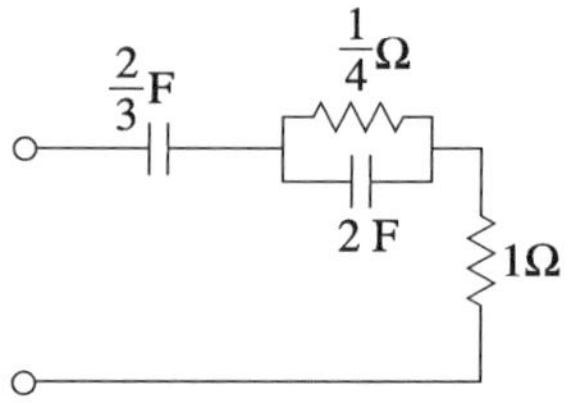

Fig. 12.35

(2) Foster II Form The Foster II form is obtained by the partial-fraction expansion of admittance function $\frac{Y(s)}{s}$.

$$Y(s) = \frac{1}{Z(s)} = \frac{s(s+2)}{(s+1)(s+3)}$$

$$\frac{Y(s)}{s} = \frac{s+2}{(s+1)(s+3)}$$

By partial-fraction expansion, we have

$$\frac{Y(s)}{s} = \frac{K_1}{s+1} + \frac{K_2}{s+3}$$

where
$$K_1 = (s+1)\frac{Y(s)}{s}\bigg|_{s=-1} = \frac{(-1+2)}{(-1+3)} = \frac{1}{2}$$

$$K_2 = (s+3)\frac{Y(s)}{s}\bigg|_{s=-3} = \frac{(-3+2)}{(-3+1)} = \frac{1}{2}$$

$$\frac{Y(s)}{s} = \frac{1/2}{s+1} + \frac{1/2}{s+3}$$

$$Y(s) = \frac{\frac{1}{2}s}{s+1} + \frac{\frac{1}{2}s}{s+3}$$

These two terms represent the admittance of a series RC circuit. For a series RC circuit.

$$Y_{RC}(s) = \frac{\left(\frac{1}{R_i}\right)s}{s + \frac{1}{R_i C_i}}$$

By direct comparison,

$$R_1 = 2\ \Omega$$
$$C_1 = \frac{1}{2}\ \text{F}$$
$$R_2 = 2\ \Omega$$
$$C_2 = \frac{1}{6}\ \text{F}$$

The network is shown in Fig. 12.36.

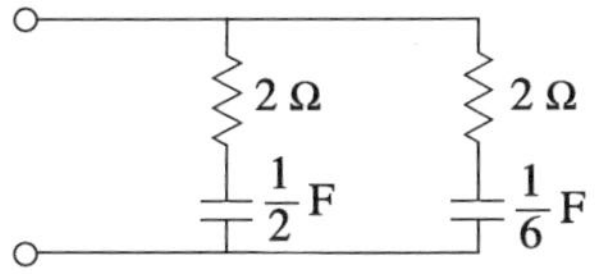

Fig. 12.36

(3) Cauer I Form The Cauer I form is obtained by continued fraction expansion about the pole at infinity.

$$Z(s) = \frac{s^2 + 4s + 3}{s^2 + 2s}$$

$$\begin{array}{l} s^2 + 2s\ )\ s^2 + 4s + 3\ (\ 1 \leftarrow Z \\ \qquad\qquad s^2 + 2s \\ \hline \qquad\qquad 2s + 3\ )\ s^2 + 2s\ (\ \frac{s}{2} \leftarrow Y \\ \qquad\qquad\qquad\quad s^2 + \frac{3s}{2} \\ \hline \qquad\qquad\qquad\qquad \frac{s}{2}\ )\ 2s + 3\ (\ 4 \leftarrow Z \\ \qquad\qquad\qquad\qquad\qquad 2s \\ \hline \qquad\qquad\qquad\qquad\qquad 3\ )\ \frac{s}{2}\ (\ \frac{s}{6} \leftarrow Y \\ \qquad\qquad\qquad\qquad\qquad\quad \frac{s}{2} \\ \hline \qquad\qquad\qquad\qquad\qquad\quad 0 \end{array}$$

The impedances are connected in series branches whereas admittances are connected in parallel branches. The network is shown in Fig. 12.37.

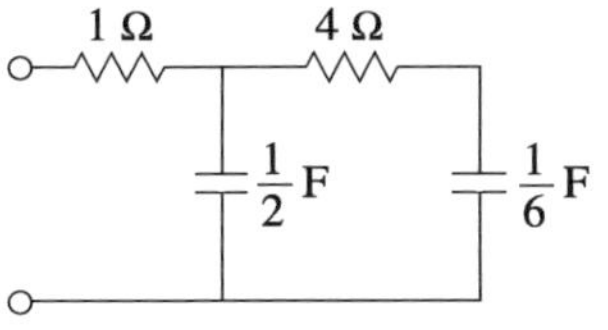

Fig. 12.37

(4) Cauer II Form The Cauer II form is obtained from continued fraction expansion about the pole at the origin. Arranging the numerator and denominator polynomials of $Z(s)$ in ascending order of s, we have

$$Z(s) = \frac{3 + 4s + s^2}{2s + s^2}$$

$$2s + s^2 \,\big)\, 3 + 4s + s^2 \,\Big(\, \frac{3}{2s} \leftarrow Z$$
$$3 + \frac{3s}{2}$$
$$\overline{\qquad\qquad\qquad\qquad}$$
$$\frac{5s}{2} + s^2 \,\big)\, 2s + s^2 \,\Big(\, \frac{4}{5} \leftarrow Y$$
$$2s + \frac{4s^2}{5}$$
$$\overline{\qquad\qquad\qquad\qquad}$$
$$\frac{s^2}{5} \,\Big)\, \frac{5s}{2} + s^2 \,\Big(\, \frac{25}{2s} \leftarrow Z$$
$$\frac{5s}{2}$$
$$\overline{\qquad\qquad\qquad\qquad}$$
$$s^2 \,\Big)\, \frac{s^2}{5} \,\Big(\, \frac{1}{5} \leftarrow Y$$
$$\frac{s^2}{5}$$
$$\overline{\qquad\qquad}$$
$$0$$

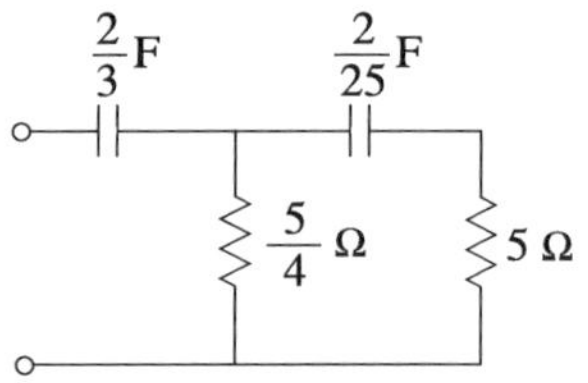

Fig. 12.38

The impedances are connected in series branches whereas admittances are connected in parallel branches. The network is shown in Fig. 12.38.

Example 12.42 *Determine the Foster form of realization of the RC impedance function.*

$$Z(s) = \frac{(s+1)(s+3)}{s(s+2)(s+4)}$$

Solution

(1) Foster I Form The Foster I form is obtained by the partial-fraction expansion of the impedance function $Z(s)$.

By partial-fraction expansion, we have

$$Z(s) = \frac{K_0}{s} + \frac{K_1}{s+2} + \frac{K_2}{s+4}$$

where

$$K_0 = s\,Z(s)\,|_{s=0} = \frac{(1)(3)}{(2)(4)} = \frac{3}{8}$$

$$K_1 = (s+2)\,Z(s)\,|_{s=-2} = \frac{(-2+1)(-2+3)}{(-2)(-2+4)} = \frac{(-1)(1)}{(-2)(2)} = \frac{1}{4}$$

$$K_2 = (s+4)\,Z(s)|_{s=-4}$$

$$= \frac{(-4+1)(-4+3)}{(-4)(-4+2)} = \frac{(-3)(-1)}{(-4)(-2)} = \frac{3}{8}$$

$$Z(s) = \frac{\frac{3}{8}}{s} + \frac{\frac{1}{4}}{s+2} + \frac{\frac{3}{8}}{s+4}$$

The first term represents the impedance of a capacitor of $\frac{8}{3}$ F. The remaining terms represent the impedance of a parallel RC circuit for which

$$Z_{RC}(s) = \frac{\frac{1}{C_i}}{s + \frac{1}{R_i C_i}}$$

By direct comparison,

$$R_1 = \frac{1}{8}\,\Omega$$

$$C_1 = 4\text{ F}$$

$$R_2 = \frac{3}{32}\,\Omega$$

$$C_2 = \frac{8}{3}\text{ F}$$

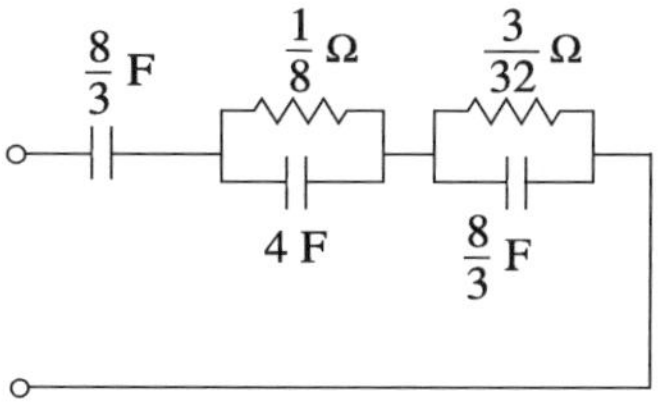

Fig. 12.39

The network is shown in Fig. 12.39.

(2) Foster II Form The Foster II form is obtained by partial-fraction expansion of admittance function $\frac{Y(s)}{s}$.

$$Y(s) = \frac{s(s+2)(s+4)}{(s+1)(s+3)}$$

$$\frac{Y(s)}{s} = \frac{(s+2)(s+4)}{(s+1)(s+3)} = \frac{s^2+6s+8}{s^2+4s+3}$$

Since the degree of the numerator is equal to the degree of the denominator, division is carried out first.

$$s^2 + 4s + 3\;)\; s^2 + 6s + 8\;(1$$
$$\underline{s^2 + 4s + 3}$$
$$2s + 5$$

$$\frac{Y(s)}{s} = 1 + \frac{2s+5}{s^2+4s+3}$$

$$= 1 + \frac{2s+5}{(s+1)(s+3)}$$

By partial-fraction expansion, we have

$$\frac{Y(s)}{s} = 1 + \frac{K_1}{s+1} + \frac{K_2}{s+3}$$

where

$$K_1 = (s+1)\left.\frac{Y(s)}{s}\right|_{s=-1}$$

$$= \frac{(-1+2)(-1+4)}{(-1+3)} = \frac{(1)(3)}{2} = \frac{3}{2}$$

$$K_2 = (s+3)\left.\frac{Y(s)}{s}\right|_{s=-3}$$

$$= \frac{(-3+2)(-3+4)}{(-3+1)} = \frac{(-1)(1)}{(-2)} = \frac{1}{2}$$

$$\frac{Y(s)}{s} = 1 + \frac{\frac{3}{2}}{s+1} + \frac{\frac{1}{2}}{s+3}$$

$$Y(s) = s + \frac{\frac{3}{2}s}{s+1} + \frac{\frac{1}{2}s}{s+3}$$

The first term represents the admittance of a capacitor of 1 F. The other two terms represent the admittance of a series *RC* circuit. For a series *RC* circuit

$$Y_{RC}(s) = \frac{\left(\frac{1}{R_i}\right)s}{s + \frac{1}{R_i C_i}}$$

By direct comparison,

$$R_1 = \frac{2}{3}\ \Omega$$

$$C_1 = \frac{3}{2}\ \text{F}$$

$$R_2 = 2\ \Omega$$

$$C_2 = \frac{1}{6}\ \text{F}$$

The network is shown in Fig. 12.40.

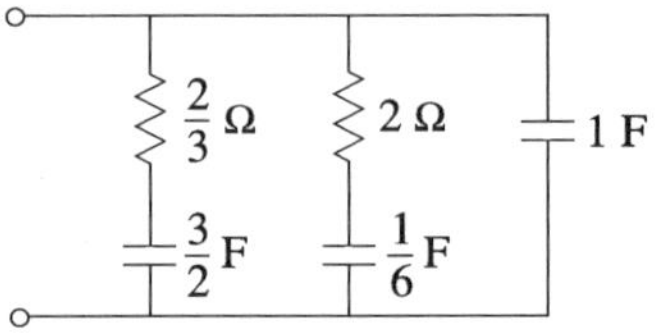

Fig. 12.40

Example 12.43 *Realize Foster forms of the following RC impedance function*

$$Z(s) = \frac{2(s+2)(s+4)}{(s+1)(s+3)}$$

Solution

(1) Foster I Form The Foster I form is obtained by the partial-fraction expansion of the impedance function $Z(s)$. Since the degree of the numerator is equal to the degree of the denominator, division is carried out first.

$$Z(s) = \frac{2s^2 + 12s + 16}{s^2 + 4s + 3}$$

$$\begin{array}{r} s^2 + 4s + 3\ )\ 2s^2 + 12s + 16\ (\ 2 \\ 2s^2 + \ \ 8s + 6 \\ \hline 4s + 10 \end{array}$$

$$Z(s) = 2 + \frac{4s+10}{s^2+4s+3} = 2 + \frac{4s+10}{(s+1)(s+3)}$$

By partial-fraction expansion, we have

$$Z(s) = 2 + \frac{K_1}{s+1} + \frac{K_2}{s+3}$$

$$K_1 = (s+1)\,Z(s)\,|_{s=-1} = \frac{2(-1+2)(-1+4)}{(-1+3)} = 3$$

$$K_2 = (s+3)\,Z(s)\,|_{s=-3} = \frac{2(-3+2)(-3+4)}{(-3+1)} = 1$$

$$Z(s) = 2 + \frac{3}{s+1} + \frac{1}{s+3}$$

The first term represents the impedance of a resistor of 2 Ω. The remaining terms represent the impedance of a parallel RC circuit for which

$$Z_{RC}(s) = \frac{\frac{1}{C_i}}{s + \frac{1}{R_i C_i}}$$

By direct comparison,

$$R_1 = 3\ \Omega$$

$$C_1 = \frac{1}{3}\ \text{F}$$

$$R_2 = \frac{1}{3}\ \Omega$$

$$C_2 = 1\ \text{F}$$

The network is shown in Fig. 12.41.

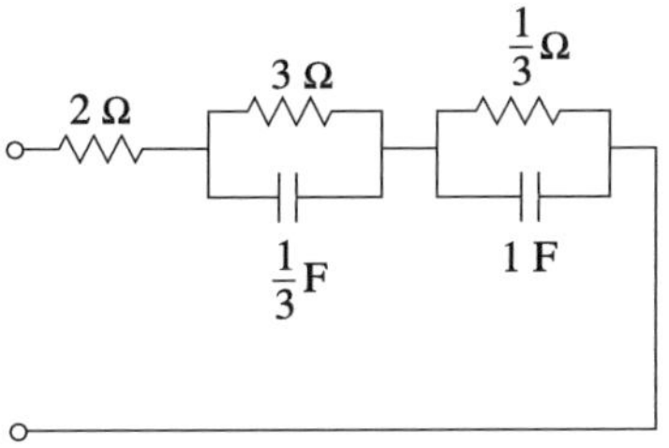

Fig. 12.41

(2) Foster II Form The Foster II form is obtained by partial-fraction expansion of admittance function $\frac{Y(s)}{s}$.

$$Y(s) = \frac{(s+1)(s+3)}{2(s+2)(s+4)}$$

$$\frac{Y(s)}{s} = \frac{(s+1)(s+3)}{2s(s+2)(s+4)}$$

By partial-fraction expansion, we have

$$\frac{Y(s)}{s} = \frac{K_0}{s} + \frac{K_1}{s+2} + \frac{K_2}{s+4}$$

where

$$K_0 = s\frac{Y(s)}{s}\bigg|_{s=0}$$

$$= \frac{(1)(3)}{(2)(2)(4)} = \frac{3}{16}$$

$$K_1 = (s+2)\frac{Y(s)}{s}\bigg|_{s=-2}$$

$$= \frac{(-2+1)(-2+3)}{2(-2)(-2+4)} = \frac{(-1)(1)}{2(-2)(2)} = \frac{1}{8}$$

$$K_2 = (s+4)\frac{Y(s)}{s}\bigg|_{s=-4}$$

$$= \frac{(-4+1)(-4+3)}{2(-4)(-4+2)} = \frac{(-3)(-1)}{2(-4)(-2)} = \frac{3}{16}$$

$$\frac{Y(s)}{s} = \frac{\frac{3}{16}}{s} + \frac{\frac{1}{8}}{s+2} + \frac{\frac{3}{16}}{s+4}$$

$$Y(s) = \frac{3}{16} + \frac{\frac{1}{8}s}{s+2} + \frac{\frac{3}{16}s}{s+4}$$

The first term represents the admittance of a resistor of $\frac{16}{3}$ Ω. The other two terms represent the admittance of a series RC circuit. For a series RC circuit.

$$Y_{RC}(s) = \frac{\left(\frac{1}{R_i}\right)s}{s + \frac{1}{R_iC_i}}$$

By direct comparison,

$$R_1 = 8\ \Omega$$

$$C_1 = \frac{1}{16}\ \text{F}$$

$$R_2 = \frac{16}{3}\ \Omega$$

$$C_2 = \frac{3}{64}\ \text{F}$$

The network is shown in Fig. 12.42.

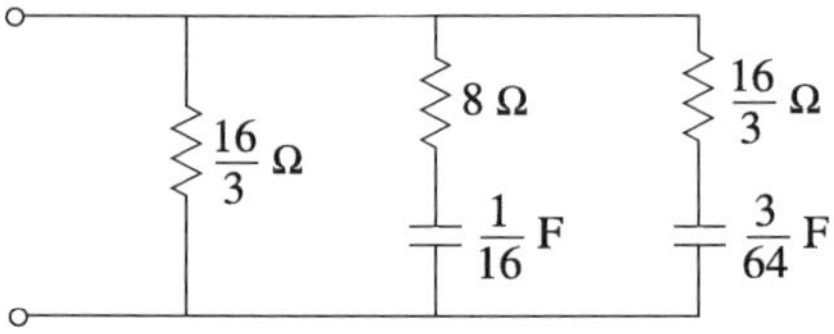

Fig. 12.42

Example 12.44 *Obtain the Cauer forms of the following RC impedance function*

$$Z(s) = \frac{(s+2)(s+6)}{2(s+1)(s+3)}$$

Solution

(1) Cauer I Form The Cauer I form is obtained by continued fraction expansion about the pole at infinity.

$$Z(s) = \frac{s^2 + 8s + 12}{2s^2 + 8s + 6}$$

By continued fraction expansion, we have

$$2s^2 + 8s + 6\ \big)\ s^2 + 8s + 12\ \big(\ \frac{1}{2} \leftarrow Z$$

$$s^2 + 4s + 3$$

$$4s + 9\ \big)\ 2s^2 + 8s + 6\ \big(\ \frac{s}{2} \leftarrow Y$$

$$2s^2 + \frac{9}{2}s$$

$$\frac{7}{2}s + 6\ \big)\ 4s + 9\ \big(\ \frac{8}{7} \leftarrow Z$$

$$4s + \frac{48}{7}$$

$$\frac{15}{7}\ \big)\ \frac{7}{2}s + 6\ \big(\ \frac{49}{30}s \leftarrow Y$$

$$\frac{7}{2}s$$

$$6\ \big)\ \frac{15}{7}\ \big(\ \frac{5}{14} \leftarrow Z$$

$$\frac{15}{7}$$

$$0$$

The impedances are connected in series branches whereas the admittances are connected in parallel branches. The network is shown in Fig. 12.43.

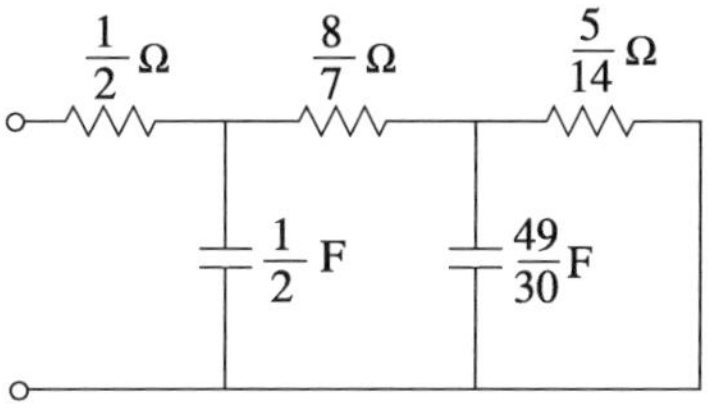

Fig. 12.43

(2) Cauer II Form The Cauer II form is obtained by continued fraction expansion about the pole at the origin. Arranging the polynomials in ascending order of s, we have

$$Z(s) = \frac{12 + 8s + s^2}{6 + 8s + 2s^2}$$

By continued fraction expansion, we have

$$\begin{array}{l} 6 + 8s + 2s^2\,)\,12 + 8s + s^2\,(\,2 \\ \qquad\qquad\quad\;\; \underline{12 + 16s + 4s^2} \\ \qquad\qquad\qquad\; -\;8s - 3s^2 \end{array}$$

Since negative term results, continued fraction expansion of $Y(s)$ is carried out.

$$Y(s) = \frac{6 + 8s + 2s^2}{12 + 8s + s^2}$$

By continued fraction expansion, we have

$$\begin{array}{l} 12 + 8s + s^2\,\big)\; 6 + 8s + 2s^2 \;\Big(\frac{1}{2} \leftarrow Y \\ \qquad\qquad\qquad\; \underline{6 + 4s + \frac{1}{2}s^2} \\ \qquad\qquad\quad 4s + \frac{3}{2}s^2\,\big)\; 12 + 8s + s^2 \;\Big(\frac{3}{s} \leftarrow Z \\ \qquad\qquad\qquad\qquad\qquad\quad \underline{12 + \frac{9}{2}s} \\ \qquad\qquad\qquad\qquad\qquad\quad \frac{7}{2}s + s^2\,\big)\; 4s + \frac{3}{2} + s^2 \;\Big(\frac{8}{7} \leftarrow Y \\ \qquad\qquad\qquad\qquad\qquad\qquad\qquad\qquad\; \underline{4s + \frac{8}{7}s^2} \\ \qquad\qquad\qquad\qquad\qquad\qquad\qquad\qquad\; \frac{5}{14}s^2\,\big)\; \frac{7}{2}s + s^2 \;\Big(\frac{49}{5s} \leftarrow Z \\ \qquad\qquad\qquad\qquad\qquad\qquad\qquad\qquad\qquad\qquad \underline{\frac{7}{2}s} \\ \qquad\qquad\qquad\qquad\qquad\qquad\qquad\qquad\qquad\qquad s^2\,\big)\; \frac{5}{14}s^2 \;\Big(\frac{5}{14} \leftarrow Y \\ \qquad\qquad\qquad\qquad\qquad\qquad\qquad\qquad\qquad\qquad\qquad \underline{\frac{5}{14}s} \\ \qquad\qquad\qquad\qquad\qquad\qquad\qquad\qquad\qquad\qquad\qquad\;\; 0 \end{array}$$

The impedances are connected in series branches, whereas the admittances are connected in parallel branches. The network is shown in Fig. 12.44.

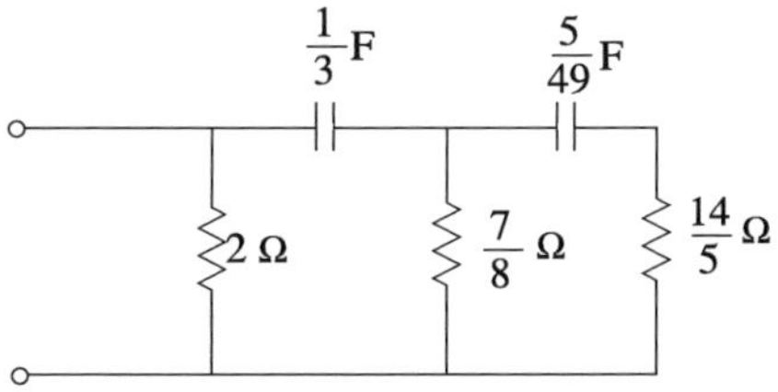

Fig. 12.44

Example 12.45 *Realize the R-C impedance in Cauer I and Foster I forms*

$$Z(s) = \frac{s+4}{(s+2)(s+6)}$$

Solution

(1) Cauer I Form The Cauer I form is obtained by continued fraction expansion of $Z(s)$ about the pole at infinity. In the above function, the degree of the numerator is less than the degree of the denominator which indicates presence of a zero at infinity. Hence, the admittance function $Y(s)$ has a pole at infinity.

$$Y(s) = \frac{s^2+8s+12}{s+4}$$

By continued fraction expansion, we have

$$s+4\,)\,s^2+8s+12\,(\,s \leftarrow Y$$
$$s^2+4s$$
$$4s+12\,\big)\,s+4\,\big(\,\frac{1}{4} \leftarrow Z$$
$$s+3$$
$$1\,\big)\,4s+12\,\big(\,4s \leftarrow Y$$
$$4s$$
$$12\,\big)\,1\,\big(\,\frac{1}{12} \leftarrow Z$$
$$1$$
$$0$$

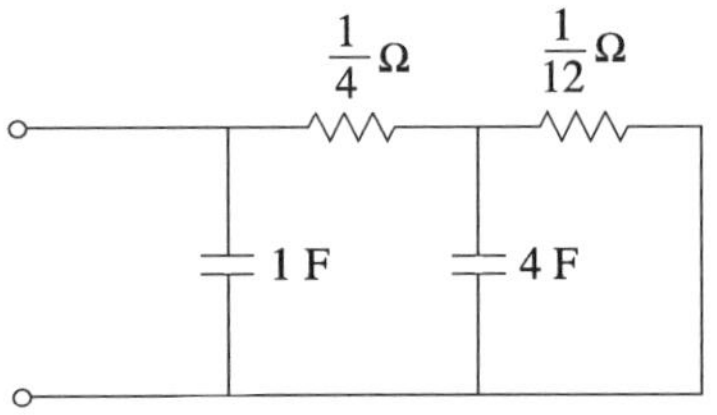

Fig. 12.45

The impedances are connected in series branches, whereas the admittances are connected in parallel branches. The network is shown in Fig. 12.45.

(2) Foster I Form The Foster I form is obtained by partial fraction expansion of $Z(s)$.

$$Z(s) = \frac{s+4}{(s+2)(s+6)}$$

By partial-fraction expansion, we have

$$Z(s) = \frac{K_1}{s+2} + \frac{K_2}{s+6}$$

where

$$K_1 = (s+2)\, Z(s)|_{s=-2}$$
$$= \frac{(-2+4)}{(-2+6)} = \frac{1}{2}$$
$$K_2 = (s+6)\, Z(s)|_{s=-6}$$
$$= \frac{(-6+4)}{(-6+2)} = \frac{1}{2}$$
$$Z(s) = \frac{1/2}{s+2} + \frac{1/2}{s+6}$$

These two terms represent the impedance of a parallel RC circuit for which

$$Z_{RC}(s) = \frac{\dfrac{1}{C_i}}{s+\dfrac{1}{R_iC_i}}$$

By direct comparison,

$$R_1 = \frac{1}{4}\ \Omega$$
$$C_1 = 2\text{ F}$$
$$R_2 = \frac{1}{12}\ \Omega$$
$$C_2 = 2\text{ F}$$

The network is shown in Fig. 12.46.

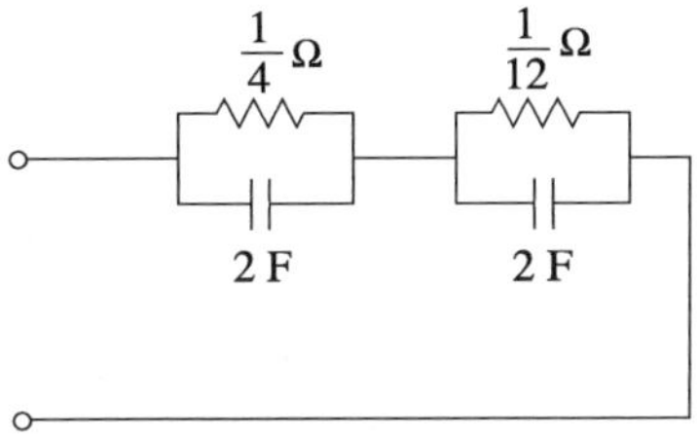

Fig. 12.46

Example 12.46 *The RC driving-point impedance function is given as*

$$Z(s) = H\,\frac{(s+1)(s+4)}{s(s+3)}$$

Realize the impedance function in the ladder form, given

$$Z(-2) = 1$$

Solution Putting $s = -2$

$$Z(-2) = H\,\frac{(-2+1)(-2+4)}{(-2)(-2+3)} = \text{H}$$
$$H = 1$$
$$Z(s) = \frac{(s+1)(s+4)}{s(s+3)}$$

(1) Cauer I Form The Cauer I form is obtained by continued fraction expansion of $Z(s)$ about the pole at infinity.

By continued fraction expansion of $Z(s)$, we have

$$\begin{array}{l}
s^2+3s\,\big)\,s^2+5s+4\,\big(\,1 \leftarrow Z \\
\qquad\qquad s^2+3s \\
\hline
\qquad\qquad 2s+4\,\big)\,s^2+3s\,\big(\,\frac{s}{2} \leftarrow Y \\
\qquad\qquad\qquad\qquad s^2+2s \\
\hline
\qquad\qquad\qquad\qquad s\,\big)\,2s+4\,\big(\,2 \leftarrow Z \\
\qquad\qquad\qquad\qquad\quad 2s \\
\hline
\qquad\qquad\qquad\qquad\quad 4\,\big)\,s\,\big(\,\frac{s}{4} \leftarrow Y \\
\qquad\qquad\qquad\qquad\qquad s \\
\hline
\qquad\qquad\qquad\qquad\qquad 0
\end{array}$$

The impedances are connected in series branches whereas admittances are connected in parallel branches. The network is shown in Fig. 12.47.

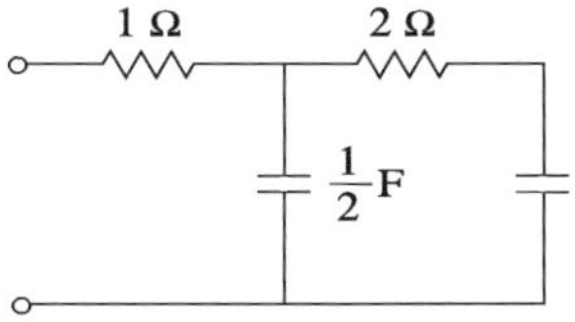

Fig. 12.47

(2) Cauer II Form The Cauer II form is obtained by continued fraction expansion about the pole at the origin. Arranging the polynomials in ascending order of s, we have

$$Z(s) = \frac{4+5s+s^2}{3s+s^2}$$

By continued fraction expansion, we have

$$
\begin{array}{l}
3s + s^2 \,\big)\; 4 + 5s + s^2 \;\Big(\dfrac{4}{3s} \leftarrow Z \\
\qquad\quad 4 + \dfrac{4s}{3} \\
\hline
\qquad\quad \dfrac{11s}{3} + s^2 \,\big)\; 3s + s^2 \;\Big(\dfrac{9}{11} \leftarrow Y \\
\qquad\qquad\qquad\quad 3s + \dfrac{9}{11}s^2 \\
\hline
\qquad\qquad\qquad\quad \dfrac{2}{11}s^2 \,\big)\; \dfrac{11}{3}s + s^2 \;\Big(\dfrac{121}{6s} \leftarrow Z \\
\qquad\qquad\qquad\qquad\qquad\quad \dfrac{11}{3}s \\
\hline
\qquad\qquad\qquad\qquad\qquad\quad s^2 \,\big)\; \dfrac{2s^2}{11} \;\Big(\dfrac{2}{11} \leftarrow Y \\
\qquad\qquad\qquad\qquad\qquad\qquad\quad \dfrac{2s^2}{11} \\
\hline
\qquad\qquad\qquad\qquad\qquad\qquad\quad 0
\end{array}
$$

The network is shown in Fig. 12.48.

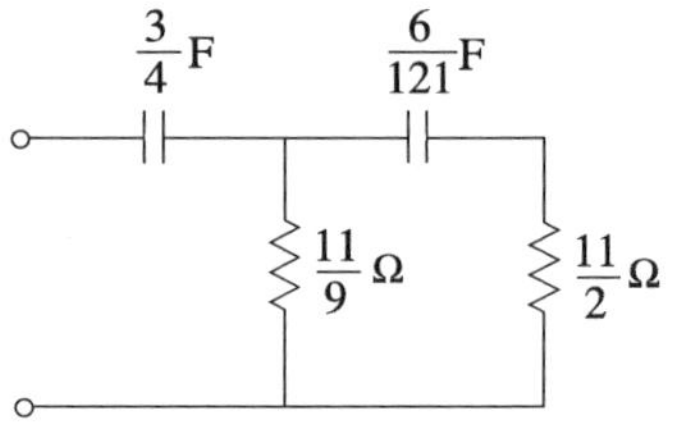

Fig. 12.48

Example 12.47 *An impedance function has the pole-zero diagram as shown. Find the impedance function such that* $Z(-4) = \dfrac{3}{4}$ *and realize in Cauer I and Foster II forms.*

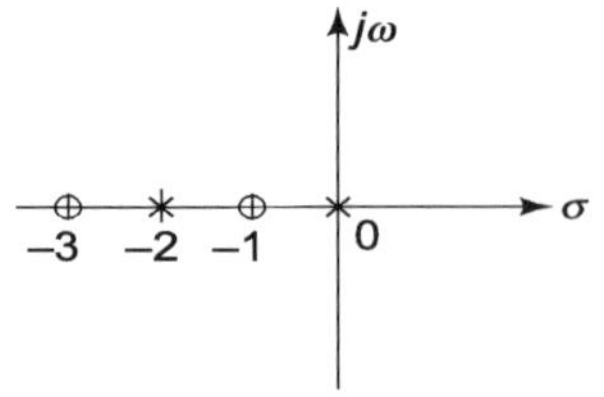

Fig. 12.49

Solution The function $Z(s)$ has poles at $s = 0$ and $s = -2$ and zeros at $s = -1$ and $s = -3$.

$$Z(s) = H\,\frac{(s+1)(s+3)}{s(s+2)}$$

Putting $\qquad s = -4$

$$Z(-4) = H\,\frac{(-4+1)(-4+3)}{(-4)(-4+2)} = H\,\frac{(-3)(-1)}{(-4)(-2)} = \frac{3}{8}\,\text{H}$$

$$\frac{3}{4} = \frac{3}{8}\,\text{H}$$

$$H = 2$$

$$Z(s) = \frac{2(s+1)(s+3)}{s(s+2)} = \frac{2s^2+8s+6}{s^2+2s}$$

(1) Cauer I Form The Cauer I form is obtained by continued fraction expansion of $Z(s)$ about the pole at infinity.

By continued fraction expansion of $Z(s)$, we have

$$\begin{array}{l}
s^2+2s\;\big)\;2s^2+8s+6\;\big(\;2 \leftarrow Z\\
\qquad\qquad\;\;\underline{2s^2+4s}\\
\qquad\qquad\qquad 4s+6\;\big)\;s^2+2s\;\big(\;\frac{s}{4} \leftarrow Y\\
\qquad\qquad\qquad\qquad\quad\;\; \underline{s^2+\frac{3}{2}s}\\
\qquad\qquad\qquad\qquad\qquad\;\; \frac{1}{2}s\;\big)\;4s+6\;\big(\;8 \leftarrow Z\\
\qquad\qquad\qquad\qquad\qquad\qquad\quad\;\underline{4s}\\
\qquad\qquad\qquad\qquad\qquad\qquad\quad\;\;6\;\big)\;\frac{1}{2}s\;\big(\;\frac{s}{12} \leftarrow Y\\
\qquad\qquad\qquad\qquad\qquad\qquad\qquad\quad\;\;\underline{\frac{1}{2}s}\\
\qquad\qquad\qquad\qquad\qquad\qquad\qquad\qquad 0
\end{array}$$

The impedances are connected in series branches whereas admittances are connected in parallel branches. The network is shown in Fig. 12.50.

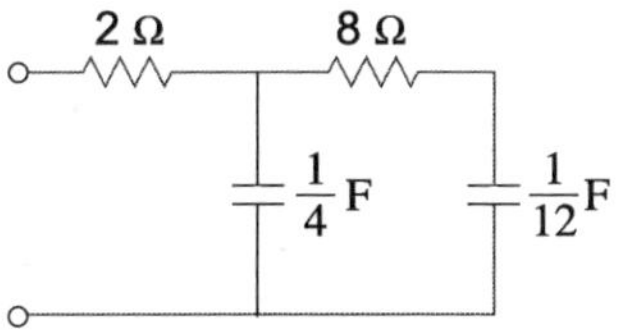

Fig. 12.50

(2) Foster II Form The Foster II form is obtained by partial fraction expansion of $\frac{Y(s)}{s}$.

$$\frac{Y(s)}{s} = \frac{s+2}{2(s+1)(s+3)}$$

$$\frac{Y(s)}{s} = \frac{K_1}{s+1} + \frac{K_2}{s+3}$$

where

$$K_1 = (s+1)\left.\frac{Y(s)}{s}\right|_{s=-1} = \frac{(-1+2)}{2(-1+3)} = \frac{1}{4}$$

$$K_2 = (s+3)\left.\frac{Y(s)}{s}\right|_{s=-3}$$

$$= \frac{(-3+2)}{2(-3+1)} = \frac{-1}{2(-2)} = \frac{1}{4}$$

$$\frac{Y(s)}{s} = \frac{\frac{1}{4}}{s+1} + \frac{\frac{1}{4}}{s+3}$$

$$Y(s) = \frac{\frac{1}{4}s}{s+1} + \frac{\frac{1}{4}s}{s+3}$$

Two terms represent the admittance of a series RC circuit. For a series RC circuit,

$$Y_{RC}(s) = \frac{\left(\frac{1}{R_i}\right)s}{s+\frac{1}{R_iC_i}}$$

By direct comparison,

$$R_1 = 4\ \Omega$$

$$C_1 = \frac{1}{4}\text{F}$$

$$R_2 = 4\Omega$$

$$C_2 = \frac{1}{12}\text{F}$$

The network is shown in Fig. 12.51.

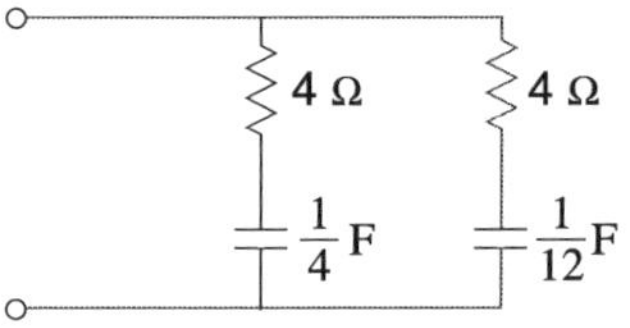

Fig. 12.51

12.10 PROPERTIES OF RL FUNCTIONS

The admittance of an inductor is similar to the impedance of a capacitor. Hence, we can conclude that the properties of an RL admittance are identical to those of an RC impedance and vice-versa, i.e.,

$$Z_{RC}(s) = Y_{RL}(s)$$
$$Z_{RL}(s) = Y_{RC}(s)$$

An RL admittance can be considered as the dual of an RC impedance and vice-versa.

12.10.1 Properties of RL Impedance Functions

(1) The poles and zeros are simple and are located on the negative real axis of the s plane.
(2) The poles and zeros are interlaced.
(3) The lowest critical frequency is a zero which may be at $s = 0$.
(4) The highest critical frequency is a pole which may be at infinity.
(5) Residues evaluated at the poles of $Z_{RL}(s)$ are real and negative while that of $\frac{Z_{RL}(s)}{s}$ are real and positive.
(6) The slope $\frac{d}{d\sigma}Z_{RL}$ is positive.
(7) $Z_{RL}(0) < Z_{RL}(\infty)$.

Example 12.48 *Indicate which of the following functions are either RL, RC or LC impedance functions.*

(i) $Z(s) = \frac{4(s+1)(s+3)}{s(s+2)}$ (ii) $Z(s) = \frac{s(s+4)(s+8)}{(s+1)(s+6)}$

(iii) $Z(s) = \frac{(s+1)(s+4)}{s(s+2)}$ (iv) $Z(s) = \frac{2(s+1)(s+3)}{(s+2)(s+16)}$

Solution (i)

$$Z(s) = \frac{4(s+1)(s+3)}{s(s+2)}$$

This is an *RC* impedance function since (i) poles and zeros are on the negative real axis, (ii) they are interlaced, and (iii) critical frequency nearest to the origin is a pole.

(ii)

$$Z(s) = \frac{s(s+4)(s+8)}{(s+1)(s+6)}$$

This is an *RL* impedance function as (i) poles and zeros are on the negative real axis, (ii) they are interlaced, and (iii) critical frequency nearest to the origin is a zero.

(iii)

$$Z(s) = \frac{(s+1)(s+4)}{s(s+2)}$$

This is an *RC* impedance function since (i) poles and zeros are on the negative real axis, (ii) they are interlaced, and (iii) critical frequency nearest to the origin is a pole.

(iv)

$$Z(s) = \frac{2(s+1)(s+3)}{(s+2)(s+6)}$$

This is an *RL* impedance function as (i) poles and zeros are on the negative real axis, (ii) they are interlaced, and (iii) critical frequency nearest to the origin is a zero.

Example 12.49 *Synthesize following RL impedance function in Foster-I and Foster-II form.*

$$Z(s) = \frac{2(s+1)(s+3)}{(s+2)(s+6)}$$

Solution

(1) Foster I Form The Foster I form is obtained by partial-fraction expansion of the impedance function $Z(s)$.

By partial-fraction expansion, we have

$$Z(s) = \frac{K_1}{s+2} + \frac{K_2}{s+6}$$

where

$$K_1 = (s+2)\,Z(s)\,|_{s=-2}$$

$$= \frac{2(-2+1)(-2+3)}{(-2+6)} = -\frac{1}{2}$$

$$K_2 = (s+6)\,Z(s)\,|_{s=-6}$$

$$= \frac{2(-6+1)(-6+3)}{(-6+2)} = -\frac{15}{2}$$

Since residues of $Z(s)$ are negative, partial fraction expansion of $\frac{Z(s)}{s}$ is carried out.

$$\frac{Z(s)}{s} = \frac{2(s+1)(s+3)}{s(s+2)(s+6)}$$

By partial fraction expansion, we have

$$\frac{Z(s)}{s} = \frac{K_0}{s} + \frac{K_1}{s+2} + \frac{K_2}{s+6}$$

where
$$K_0 = s\,\frac{Z(s)}{s}\bigg|_{s=0} = \frac{2(1)(3)}{(2)(6)} = \frac{1}{2}$$

$$K_1 = (s+2)\,\frac{Z(s)}{s}\bigg|_{s=-2}$$

$$= \frac{2(-2+1)(-2+3)}{(2)(-2+6)} = \frac{1}{4}$$

$$K_2 = (s+6)\,\frac{Z(s)}{s}\bigg|_{s=-6}$$

$$= \frac{2(-6+1)(-6+3)}{(-6)(-6+2)} = \frac{5}{4}$$

$$\frac{Z(s)}{s} = \frac{\frac{1}{2}}{s} + \frac{\frac{1}{4}}{s+2} + \frac{\frac{5}{4}}{s+6}$$

$$Z(s) = \frac{1}{2} + \frac{\frac{1}{4}s}{s+2} + \frac{\frac{5}{4}s}{s+6}$$

The first term represents the impedance of the resistor of $\frac{1}{2}\,\Omega$. The other two terms represent the impedance of the parallel *RL* circuit for which

$$Z_{RL}(s) = \frac{R_i s}{s + \frac{R_i}{L_i}}$$

By direct comparison,

$$R_1 = \frac{1}{4}\,\Omega$$

$$L_1 = \frac{1}{8}\,\text{H}$$

$$R_2 = \frac{5}{4}\,\Omega$$

$$L_2 = \frac{5}{24}\,\text{H}$$

The network is shown in Fig. 12.52.

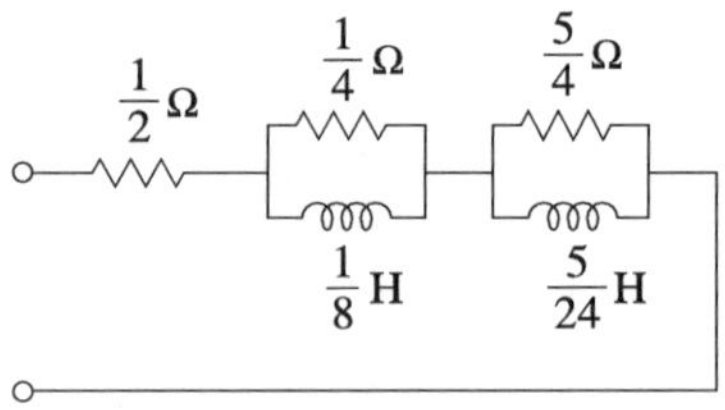

Fig. 12.52

(2) Foster II Form The Foster II form is obtained by partial fraction expansion of $Y(s)$. Since the degree of the numerator is equal to the degree of the denominator, division is first carried out.

$$Y(s) = \frac{(s+2)(s+6)}{2(s+1)(s+3)} = \frac{s^2+8s+12}{2s^2+8s+6}$$

$$2s^2 + 8s + 6 \,\big)\, s^2 + 8s + 12 \,\Big(\, \frac{1}{2}$$
$$\underline{s^2 + 4s + 3}$$
$$4s + 9$$

$$Y(s) = \frac{1}{2} + \frac{4s+9}{2s^2+8s+6}$$

$$= \frac{1}{2} + \frac{4s+9}{2(s+1)(s+3)}$$

By partial-fraction expansion, we have

$$Y_1(s) = \frac{4s+9}{2(s+1)(s+3)} = \frac{K_0}{s+1} + \frac{K_1}{s+3}$$

where

$$K_0 = (s+1)\,Y_1(s)|_{s=-1} = \frac{(-4+9)}{2(-1+3)} = \frac{5}{4}$$

$$K_1 = (s+3)\,Y_1(s)|_{s=-3} = \frac{(-12+9)}{2(-3+1)} = \frac{3}{4}$$

$$Y(s) = \frac{1}{2} + \frac{\frac{5}{4}}{s+1} + \frac{\frac{3}{4}}{s+3}$$

The first term represents the admittance of a resistor of 2 Ω. The other two terms represent the admittance of a series *RL* circuit. For a series *RL* circuit,

$$Y_{RL}(s) = \frac{\frac{1}{L_i}}{s + \frac{R_i}{L_i}}$$

By direct comparison,

$$R_1 = \frac{4}{5}\,\Omega$$

$$L_1 = \frac{4}{5}\,\text{H}$$

$$R_2 = 4\,\Omega$$

$$L_2 = \frac{4}{3}\,\text{H}$$

The network is shown in Fig. 12.53.

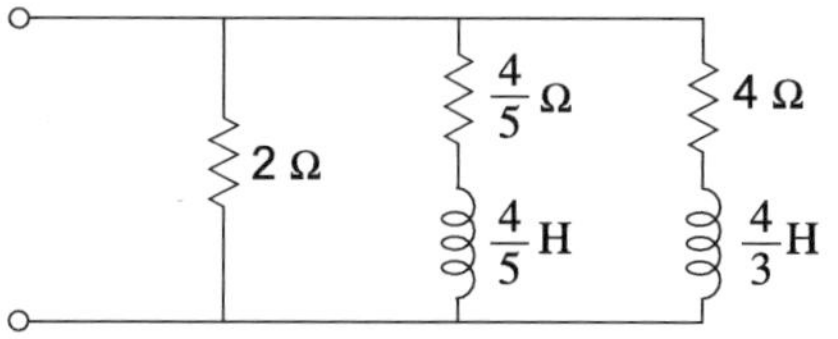

Fig. 12.53

Example 12.50 *Find the Foster forms of the following RL impedance function*

$$Z(s) = \frac{(s+1)(s+4)}{(s+5)(s+3)}$$

Solution

(1) Foster I Form The Foster I form is obtained by partial-fraction expansion of impedance function $\frac{Z(s)}{s}$.

$$\frac{Z(s)}{s} = \frac{(s+1)(s+4)}{s(s+5)(s+3)}$$

By partial-fraction expansion, we have

$$\frac{Z(s)}{s} = \frac{K_0}{s} + \frac{K_1}{s+3} + \frac{K_2}{s+5}$$

where
$$K_0 = s\left.\frac{Z(s)}{s}\right|_{s=0} = \frac{(1)(4)}{(5)(3)} = \frac{4}{15}$$

$$K_1 = (s+3)\left.\frac{Z(s)}{s}\right|_{s=-3} = \frac{(-3+1)(-3+4)}{(-3)(-3+5)} = \frac{(-2)(1)}{(-3)(2)} = \frac{1}{3}$$

$$K_2 = (s+5)\left.\frac{Z(s)}{s}\right|_{s=-5}$$

$$= \frac{(-5+1)(-5+4)}{(-5)(-5+3)} = \frac{(-4)(-1)}{(-5)(-2)} = \frac{2}{5}$$

$$\frac{Z(s)}{s} = \frac{\frac{4}{15}}{s} + \frac{\frac{1}{3}}{s+3} + \frac{\frac{2}{5}}{s+5}$$

$$Z(s) = \frac{4}{15} + \frac{\frac{1}{3}s}{s+3} + \frac{\frac{2}{5}s}{s+5}$$

The first term represents the impedance of the resistor of $\frac{4}{15}\ \Omega$. The other two terms represent the impedance of a parallel *RL* circuit for which

$$Z_{RL}(s) = \frac{R_i s}{s + \frac{R_i}{L_i}}$$

By direct comparison,

$$R_1 = \frac{1}{3}\ \Omega$$

$$L_1 = \frac{1}{9}\ \text{H}$$

$$R_2 = \frac{2}{5}\ \Omega$$

$$L_2 = \frac{2}{25}\ \text{H}$$

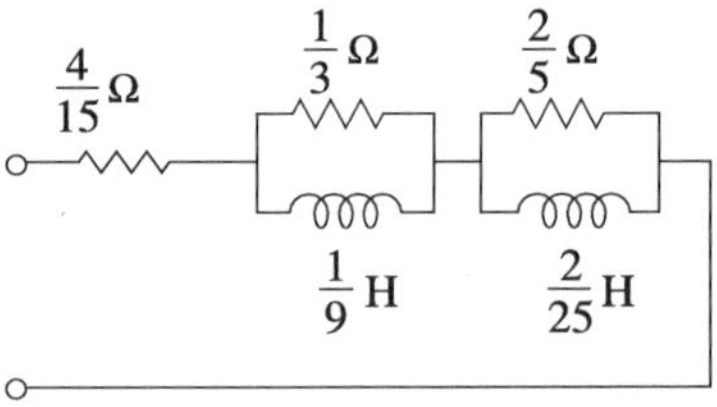

Fig. 12.54

The network is shown in Fig. 12.54.

(2) Foster II Form The Foster II form is obtained by partial fraction expansion of $Y(s)$. Since the degree of the numerator is equal to the degree of the denominator, division is first carried out.

$$Y(s) = \frac{(s+5)(s+3)}{(s+1)(s+4)} = \frac{s^2+8s+15}{s^2+5s+4}$$

$$\begin{array}{r} s^2+5s+4\ )\ s^2+8s+15\ (\ 1 \\ s^2+5s+\ \ 4 \\ \hline 3s+11 \end{array}$$

$$Y(s) = 1 + \frac{3s+11}{(s+1)(s+4)}$$

By partial-fraction expansion, we have

$$Y_1(s) = \frac{K_0}{s+1} + \frac{K_1}{s+4}$$

where $$K_0 = (s+1)\,Y_1(s)|_{s=-1} = \frac{(-3+11)}{(-1+4)} = \frac{8}{3}$$

$$K_1 = (s+4)\,Y_1(s)|_{s=-4} = \frac{(-12+11)}{(-4+1)} = \frac{1}{3}$$

$$Y_1(s) = \frac{\frac{8}{3}}{s+1} + \frac{\frac{1}{3}}{s+4}$$

$$Y(s) = 1 + \frac{\frac{8}{3}}{s+1} + \frac{\frac{1}{3}}{s+4}$$

The first term represents the admittance of a resistor of 1 Ω. The other two terms represent the admittance of a series *RL* circuit. For a series *RL* circuit,

$$Y_{RL}(s) = \frac{\frac{1}{L_i}}{s + \frac{R_i}{L_i}}$$

By direct comparison,

$$R_1 = \frac{3}{8}\ \Omega$$

$$L_1 = \frac{3}{8}\ \text{H}$$

$$R_2 = 12\ \Omega$$

$$L_2 = 3\ \text{H}$$

The network is shown in Fig. 12.55.

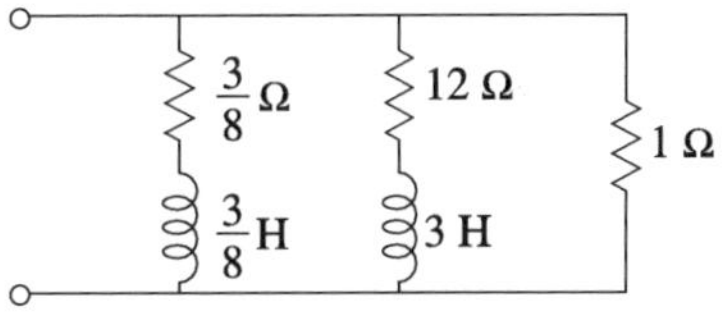

Fig. 12.55

Example 12.51 *Find the Cauer forms of the RL impedance function.*

$$Z(s) = \frac{2(s+1)(s+3)}{(s+2)(s+6)}$$

Solution

(1) Cauer I Form The Cauer I form is obtained by a continued fraction expansion of $Z(s)$ about the pole at infinity.

$$Z(s) = \frac{2(s+1)(s+3)}{(s+2)(s+6)} = \frac{2s^2+8s+6}{s^2+8s+12}$$

By continued fraction expansion, we have

$$\begin{array}{r} s^2+8s+12\ )\ 2s^2+\ 8s+\ 6\ (\ 2 \leftarrow Z \\ 2s^2+16s+24 \\ \hline -\ \ 8s-18 \end{array}$$

Since a negative term results, continued fraction expansion of $Y(s)$ is carried out.

$$Y(s) = \frac{s^2 + 8s + 12}{2s^2 + 8s + 6}$$

$$2s^2 + 8s + 6 \,\big)\, s^2 + 8s + 12 \,\big(\, \frac{1}{2} \leftarrow Y$$
$$s^2 + 4s + 3$$
$$4s + 9 \,\big)\, 2s^2 + 8s + 6 \,\big(\, \frac{1}{2} s \leftarrow Z$$
$$2s^2 + \frac{9}{2} s$$
$$\frac{7}{2} s + 6 \,\big)\, 4s + 9 \,\big(\, \frac{8}{7} \leftarrow Y$$
$$4s + \frac{48}{7}$$
$$\frac{15}{7} \,\big)\, \frac{7}{2} s + 6 \,\big(\, \frac{49}{30} s \leftarrow Z$$
$$\frac{7}{2} s$$
$$6 \,\big)\, \frac{15}{7} \,\big(\, \frac{15}{42} \leftarrow Y$$
$$\frac{15}{7}$$
$$0$$

The impedances are connected in series branches whereas the admittances are connected in parallel branches. The network is shown in Fig. 12.56.

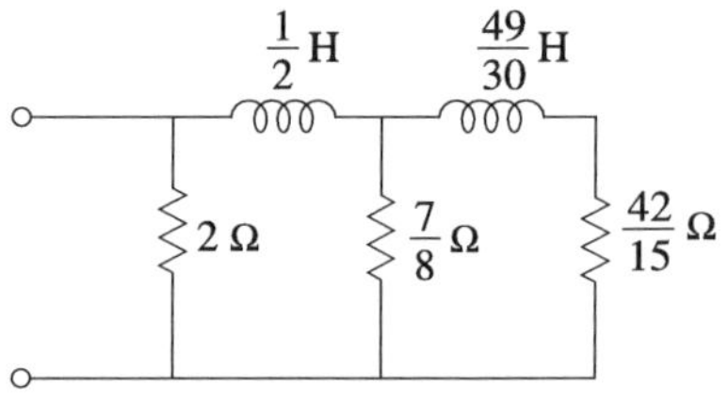

Fig. 12.56

(2) Cauer II Form The Cauer II form is obtained from a continued fraction expansion about the pole at the origin. Arranging the numerator and denominator polynomials of $Z(s)$ in ascending order of s, we have

$$Z(s) = \frac{6 + 8s + 2s^2}{12 + 8s + s^2}$$

$$12 + 8s + s^2 \,\big)\, 6 + 8s + 2s^2 \,\big(\, \frac{1}{2} \leftarrow Z$$
$$6 + 4s + \frac{1}{2} s^2$$
$$4s + \frac{3}{2} s^2 \,\big)\, 12 + 8s + s^2 \,\big(\, \frac{3}{s} \leftarrow Y$$
$$12 + \frac{9}{2} s$$
$$\frac{7}{2} s + s^2 \,\big)\, 4s + \frac{3}{2} s^2 \,\big(\, \frac{8}{7} \leftarrow Z$$
$$4s + \frac{8}{7} s^2$$

$$
\begin{array}{l}
\frac{5}{14}s^2 \,\big)\, \frac{7}{2}s + s^2 \,\big(\, \frac{98}{10s} \leftarrow Y \\
\qquad\quad \frac{7}{2}s \\
\hline
\qquad\qquad s^2 \,\big)\, \frac{5}{14}s^2 \,\big(\, \frac{5}{14} \leftarrow Z \\
\qquad\qquad\quad \frac{5}{14}s^2 \\
\hline
\qquad\qquad\qquad 0
\end{array}
$$

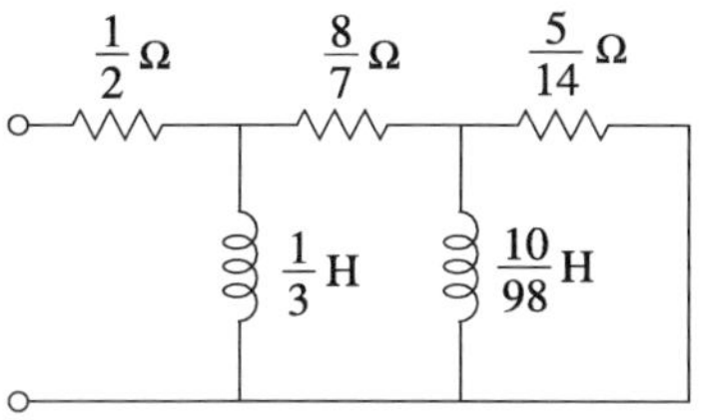

Fig. 12.57

The impedances are connected in series branches, whereas the admittances are connected in parallel branches. The network is shown in Fig. 12.57.

Example 12.52 *Obtain the Foster I and Cauer I forms of the following RL impedance function.*

$$Z(s) = \frac{s(s+4)(s+8)}{(s+1)(s+6)}$$

Solution

(1) Foster I Form The Foster I form is obtained by partial fraction expansion of $\frac{Z(s)}{s}$.

$$\frac{Z(s)}{s} = \frac{(s+4)(s+8)}{(s+1)(s+6)}$$

Since the degree of the numerator is equal to the degree of the denominator, division is first carried out.

$$
\begin{array}{l}
s^2 + 7s + 6 \,\big)\, s^2 + 12s + 32 \,\big(\, 1 \\
\qquad\qquad\quad\; s^2 + \;\;7s + \;\;6 \\
\hline
\qquad\qquad\qquad\quad 5s + 26
\end{array}
$$

$$\frac{Z(s)}{s} = 1 + \frac{5s+26}{s^2+7s+6}$$

$$= 1 + \frac{5s+26}{(s+1)(s+6)}$$

By partial-fraction expansion, we have

$$\frac{Z(s)}{s} = 1 + \frac{K_0}{s+1} + \frac{K_1}{s+6}$$

where

$$K_0 = \left.\frac{5s+26}{s+6}\right|_{s=-1}$$

$$= \frac{-5+26}{-1+6} = \frac{21}{5}$$

$$K_1 = \left.\frac{5s+26}{s+1}\right|_{s=-6}$$

$$= \frac{-30+26}{-6+1} = \frac{4}{5}$$

$$\frac{Z(s)}{s} = 1 + \frac{\frac{21}{5}}{s+1} + \frac{\frac{4}{5}}{s+6}$$

$$Z(s) = s + \frac{\frac{21}{5}s}{s+1} + \frac{\frac{4}{5}s}{s+6}$$

The first term represents the impedance of the inductor of 1 H. The other two terms represent the impedance of a parallel RL circuit for which

$$Z_{RL}(s) = \frac{R_i s}{s + \frac{R_i}{L_i}}$$

By direct comparison,

$$R_1 = \frac{21}{5}\,\Omega$$

$$L_1 = \frac{21}{5}\,\text{H}$$

$$R_2 = \frac{4}{5}\,\Omega$$

$$L_2 = \frac{4}{30}\,\text{H}$$

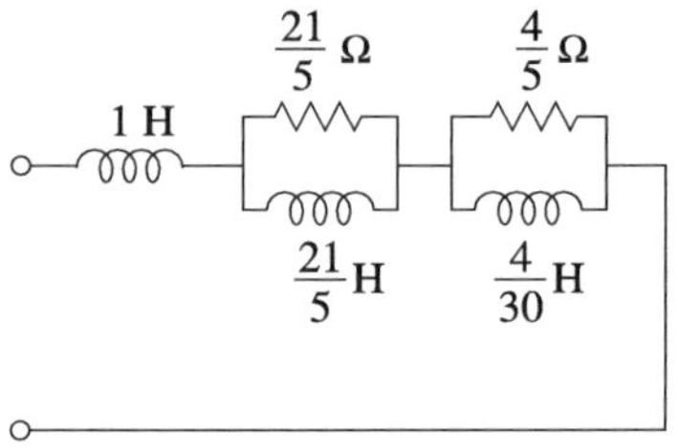

Fig. 12.58

The network is shown in Fig. 12.58.

(2) Cauer I Form The Cauer I form is obtained by continued fraction expansion of $Z(s)$ about the pole at infinity.

$$Z(s) = \frac{s^3 + 12s^2 + 32s}{s^2 + 7s + 6}$$

By continued fraction expansion, we have

$$s^2 + 7s + 6 \,\big)\, s^3 + 12s^2 + 32s \,\big(\, s \leftarrow Z$$

$$\underline{s^3 + 7s^2 + 6s}$$

$$5s^2 + 26s \,\big)\, s^2 + 7s + 6 \,\big(\, \frac{1}{5} \leftarrow Y$$

$$\underline{s^2 + \frac{26}{5}s}$$

$$\frac{9}{5}s + 6 \,\big)\, 5s^2 + 26s \,\big(\, \frac{25}{9}s \leftarrow Z$$

$$\underline{5s^2 + \frac{50}{3}s}$$

$$\frac{28}{3}s \,\big)\, \frac{9}{5}s + 6 \,\big(\, \frac{27}{140} \leftarrow Y$$

$$\underline{\frac{9}{5}s}$$

$$6 \,\big)\, \frac{28}{3}s \,\big(\, \frac{28}{18}s \leftarrow Z$$

$$\underline{\frac{28}{3}s}$$

$$0$$

The impedances are connected in series branches, whereas the admittances are connected in parallel branches. The network is shown in Fig. 12.59.

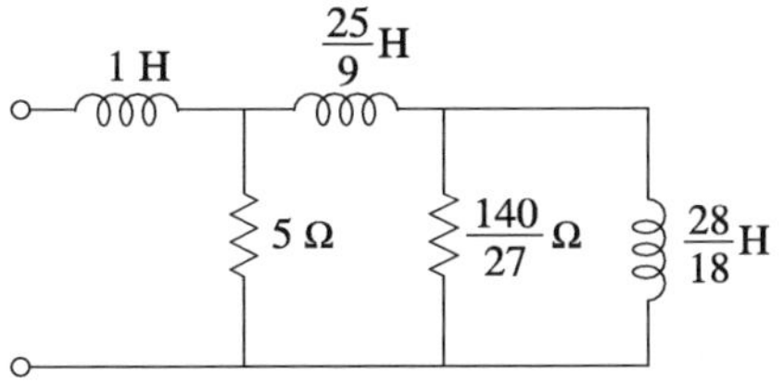

Fig. 12.59

Exercises

1. Test the following polynomials for Hurwitz property:

(i) $s^3 + s^2 + 2s + 2$ (ii) $s^4 + s^2 + s + 1$

(iii) $s^3 + 4s^2 + 5s + 2$ (iv) $s^4 + 7s^3 + 6s^2 + 21s + 8$

(v) $s^4 + s^3 + s + 1$ (vi) $s^7 + 3s^6 + 8s^5 + 15s^4 + 17s^3 + 12s^2 + 4s$

(vii) $s^7 + 2s^6 + 2s^5 + s^4 + 4s^3 + 8s^2 + 8s + 4$ (viii) $s^7 + 3s^5 + 2s^3 + s$

(ix) $s^5 + 2s^3 + s$ (x) $s^3 + 2s^2 + 4s + 2$

(xi) $s^4 + s^3 + 4s^2 + 2s + 3$ (xii) $s^5 + 8s^4 + 24s^3 + 28s^2 + 23s + 6$

(xiii) $s^7 - 2s^6 + 2s^5 + 9s^2 + 8s + 4$ (xiv) $s^7 + 3s^5 + 2s^3 + 3$

(xv) $s^5 + s^3 + s$ (xvi) $s^6 + 7s^4 + 5s^3 + s^2 + s$

(xvii) $s^4 + s^3 + 2s^2 + 3s + 2$

2. Determine whether the following functions are positive real:

(i) $\dfrac{s^3 + 5s}{s^4 + 2s^2 + 1}$ (ii) $\dfrac{s(s+3)(s+5)}{(s+1)(s+4)}$ (iii) $\dfrac{2s^2 + 2s + 1}{s^3 + 2s^2 + s + 2}$

(iv) $\dfrac{s^4 + 3s^3 + s^2 + s + 2}{s^3 + s^2 + s + 1}$ (v) $\dfrac{2s^3 + 2s^2 + 3s + 2}{s^2 + 1}$ (vi) $\dfrac{s^2 + 2s + 1}{s^2 + 4s + 4}$

(vii) $\dfrac{s^3 + 2s^2 + 3s + 1}{s^3 + 2s^2 + s + 2}$ (viii) $\dfrac{s^3 + 2s^2 + s + 1}{s^2 + s + 1}$ (ix) $\dfrac{s + 4}{s^2 + 2s + 1}$

(x) $\dfrac{s^2 + 4s + 3}{s^2 + 6s + 8}$ (xi) $\dfrac{s^2 + 1}{s^3 + 4s}$ (xii) $\dfrac{s^4 + 2s^3 + 3s^2 + 1}{s^4 + s^3 + 3s^2 + 2s + 1}$

(xiii) $\dfrac{s^2 + 2s + 4}{(s+1)(s+3)}$ (xiv) $\dfrac{2s + 4}{s + 5}$ (xv) $\dfrac{s^2 + 2s}{s^2 + 1}$

(xvi) $\dfrac{s^2 + 4}{s^3 + 3s^2 + 3s + 1}$.

3. Determine whether the following functions are *LC*, *RC* or *RL* function:

(i) $F(s) = \dfrac{2(s+1)(s+3)}{(s+2)(s+6)}$ (ii) $Z(s) = \dfrac{3(s+2)(s+4)}{s(s+3)}$ (iii) $Z(s) = \dfrac{(s+1)(s+8)}{(s+2)(s+4)}$

(iv) $Z(s) = \dfrac{Ks(s^2+4)}{(s^2+1)(s^2+3)}$ (v) $Z(s) = \dfrac{2(s^2+1)(s^2+9)}{s(s^2+2)}$ (vi) $Z(s) = \dfrac{4(s+1)(s+3)}{s(s+2)}$

(vii) $Z(s) = \dfrac{(s^2+1)(s^2+3)}{s(s^2+2)}$ (viii) $F(s) = \dfrac{(s+1)(s+2)}{s(s+3)}$ (ix) $Z(s) = \dfrac{(s+1)(s+3)}{(s+2)(s+4)}$

(x) $Z(s) = \dfrac{(s+2)(s+4)}{(s+1)}$ (xi) $Y(s) = \dfrac{4(s+3)}{(s+1)(s+5)}$ (xii) $Y(s) = \dfrac{2(s+1)(s+3)}{(s+2)(s+6)}$

(xiii) $Z(s) = \dfrac{s(s^2+4)(s^2+16)}{(s^2+9)(s^2+25)}$ (xiv) $Z(s) = \dfrac{(s^2+1)(s^2+8)}{s(s^2+4)}$.

4. Realise the following functions in Foster I form:

(i) $Z(s) = \dfrac{3(s+2)(s+4)}{s(s+3)}$ (ii) $Z(s) = \dfrac{2(s^2+1)(s^2+9)}{s(s^2+4)}$ (iii) $F(s) = \dfrac{4(s+1)(s+3)}{(s+2)(s+6)}$

(iv) $Z(s) = \dfrac{s+4}{(s+2)(s+6)}$ (v) $Z(s) = \dfrac{(s+1)(s+4)}{s(s+2)}$ (vi) $Y(s) = \dfrac{(s+2)(s+5)}{s(s+4)(s+6)}$

(vii) $Z(s) = \dfrac{s^2+2s+2}{s^2+s+1}$

5. Realise the following functions in Foster II form:

(i) $Z(s) = \dfrac{3(s+2)(s+4)}{s(s+3)}$ (ii) $Z(s) = \dfrac{2(s^2+1)(s^2+9)}{s(s^2+4)}$ (iii) $Z(s) = \dfrac{4(s+1)(s+3)}{(s+2)(s+6)}$

(iv) $Y(s) = \dfrac{4(s^2+4)(s^2+25)}{s(s^2+16)}$ (v) $Z(s) = \dfrac{(s+2)(s+5)}{s(s+4)(s+6)}$

6. Realise the following functions in Cauer I form:

(i) $F(s) = \dfrac{2(s+1)(s+3)}{(s+2)(s+6)}$ (ii) $Z(s) = \dfrac{(s^2+1)(s^2+3)}{s(s^2+2)}$ (iii) $Z(s) = \dfrac{(s+1)(s+3)}{(s+2)(s+4)}$

(iv) $Z(s) = \dfrac{2(s^2+1)(s^2+9)}{s(s^2+4)}$ (v) $F(s) = \dfrac{(s+1)(s+3)}{s(s+2)}$ (vi) $Z(s) = \dfrac{s+4}{(s+2)(s+6)}$

(vii) $Z(s) = \dfrac{6(s+2)(s+4)}{s(s+3)}$ (viii) $Z(s) = \dfrac{s^3+2s}{s^4+4s^2+3}$ (ix) $Z(s) = \dfrac{s(s^2+2)(s^2+5)}{(s^2+1)(s^2+3)}$

(x) $Z(s) = \dfrac{s^2+2s+2}{s^2+s+1}$

7. Realise the following function in Cauer II form:

(i) $F(s) = \dfrac{(s^2+1)(s^2+3)}{s(s^2+2)}$ (ii) $Z(s) = \dfrac{(s+1)(s+3)}{(s+2)(s+4)}$ (iii) $Z(s) = \dfrac{2(s^2+1)(s^2+9)}{s(s^2+4)}$

(iv) $Z(s) = \dfrac{(s+1)(s+3)}{s(s+2)}$ (v) $F(s) = \dfrac{s^3+12s^2+32s}{s^2+7s+6}$ (vi) $Z(s) = \dfrac{2(s+1)(s+3)}{(s+2)(s+6)}$

(vii) $Z(s) = \dfrac{s(s^2+2)(s^2+5)}{(s^2+1)(s^2+3)}$ (viii) $Z(s) = \dfrac{s^2+2s+2}{s^2+s+1}$

8. An impedance function has the pole-zero diagram as shown in Fig. 12.60 below. If $Z(-2) = 3$, synthesize the impedance function in Foster and Cauer forms.

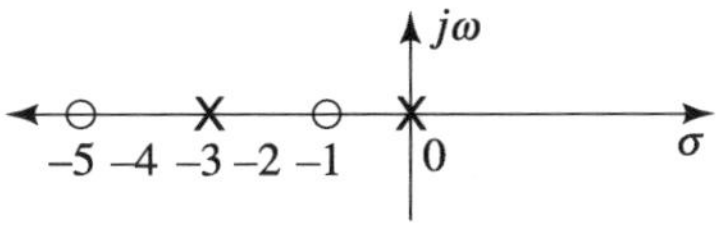

Fig. 12.60

9. An impedance function has the pole-zero diagram as shown. Find the impedance function such that $Z(-4) = \frac{8}{3}$ and realise in Cauer form.

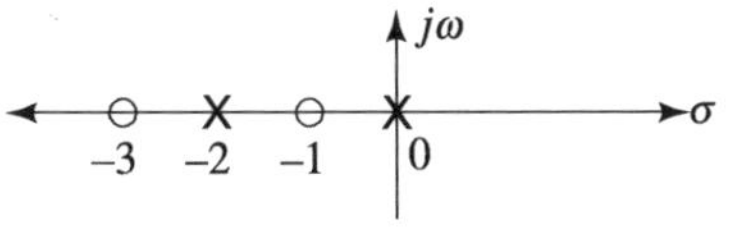

Fig. 12.61

10. For the realization of a given function $F(s)$.

$$F(s) = \frac{K_0}{s} + \sum_{i=1}^{n} \frac{sK_i}{\left(s^2 + \omega_i^2\right)} + sK_\infty$$

where K_o, K_i ($i = 1, 2, 3, \ldots n$) and K_∞ are constants.

(i) Mention the type of function (*RC*, *RL* or *LC*)

(ii) Given that $K_o = 6$, $K_1 = 8$, $\omega_1 = 4$, $K_2 = 10$, $\omega_2 = 8$, $K_\infty = 5$, find the component values of realized network for $F(s) = Z(s)$ and $F(s) = Y(s)$. Draw neat diagrams.

Objective-Type Questions

1. The necessary and sufficient condition for a rational function $F(s)$ to be the driving-point impedance of an *RC* network is that all poles and zeros should be

(a) simple and lie on the negative real axis in the s-plane
(b) complex and lie in the left half of s-plane
(c) complex and lie in the right-half of s-plane
(d) simple and lie on the positive real axis of the s-plane

2. The number of roots of $s^3 + 5s^2 + 7s + 3 = 0$ in the left half of s-plane is

(a) zero (b) one (c) two (d) three

3. The first and the last critical frequencies of a driving-point impedance function of a passive network having two kinds of elements, are a pole and a zero respectively. The above property will be satisfied by

(a) *RL* network only (b) *RC* network only
(c) *LC* network only (d) *RC* as well as *RL* network

4. The pole-zero pattern of a particular network is shown in Fig. 12.62. It is that of an

(a) *LC* network (b) *RC* network (c) *RL* network (d) none of these

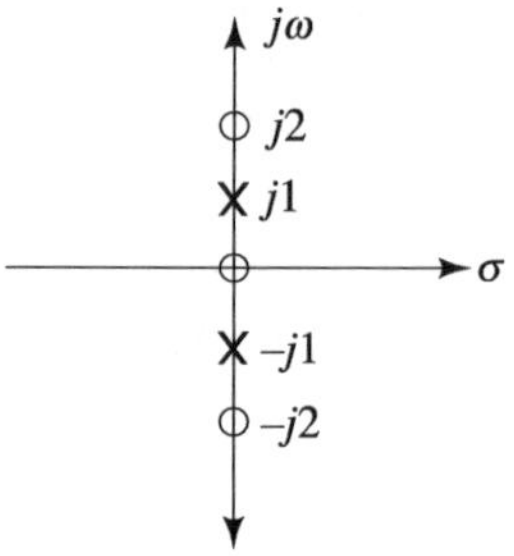

Fig. 12.62

5. The first critical frequency nearest to the origin of the complex frequency plane for an *R–L* driving-point impedance function will be
(a) a zero in the left-half plane
(b) a zero in the right-half plane
(c) a pole in the left-half plane
(d) a pole in the right-half plane

6. Consider the following polynomials:
$P_1 = s^8 + 2s^6 + 4s^4$
$P_2 = s^6 - 3s^2 + 2s^2 + 1$
$P_3 = s^4 + 3s^3 + 3s^2 + 2s + 1$
$P_4 = s^7 + 2s^6 + 2s^4 + 4s^3 + 8s^2 + 8s + 4$
which one of these polymials is not Hurwitz?
(a) P_1 (b) P_2 (c) P_3 (d) P_4

7. For very high frequencies, the driving-point admittance function, $Y(s) = \dfrac{4(s+1)(s+3)}{s(s+2)(s+4)}$ behaves as
(a) a resistance of $\dfrac{3}{2}\,\Omega$ (b) a capacitance of 4 F
(c) an inductance of $\dfrac{1}{4}$ H (d) an inductance of 4 H

8. The driving-point impedance $Z(s) = \dfrac{s+3}{s+4}$ behaves as
(a) a resistance of 0.75 Ω at low frequencies
(b) a resistance of 1 Ω at high frequencies
(c) both (a) and (b) above
(d) none of the above

9. An *RC* driving-point impedance function has zeros at $s = -2$ and $s = -5$. The admissible poles for the functions would be
(a) $s = 0, s = -6$ (b) $s = -1, s = -3$ (c) $s = 0, s = -1$ (d) $s = -3, s = -4$

10. Consider the following from the point of view of possible realisation as driving-point impedances using passive elements:

1. $\dfrac{1}{s(s+5)}$ 2. $\dfrac{s+3}{s^2(s+5)}$ 3. $\dfrac{s^2+3}{s^2(s^2+5)}$ 4. $\dfrac{s+5}{s(s+5)}$

Of these, the realisable are
(a) 1, 2 and 4 (b) 1, 2 and 3 (c) 3 and 4 (d) none of these

11. The poles and zeros of a driving-point function of a network are simple and interlace on the negative real axis with a pole closest to the origin. It can be realised
   (a) by an *LC* network
   (b) as an *RC* driving point impedance
   (c) as an *RC* driving point admittance
   (d) only by an *RLC* network

12. If $F_1(s)$ and $F_2(s)$ are two positive real functions then the function which is always positive real, is
   (a) $F_1(s)\,F_2(s)$ (b) $\dfrac{F_1(s)}{F_2(s)}$ (c) $\dfrac{F_1(s)\,F_2(s)}{F_1(s)+F_2(s)}$ (d) $F_1(s) - F_2(s)$

13. The circuit shown in Fig. 12.63 is
   (a) Cauer I form (b) Foster I form (c) Cauer II form (d) Foster II form

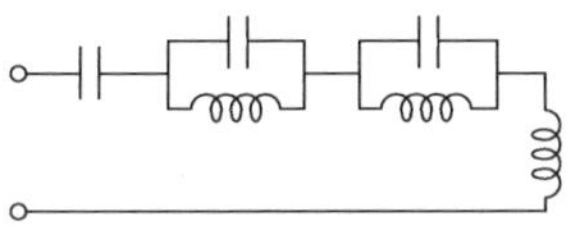

Fig. 12.63

14. For an *RC* driving-point impedance function, the poles and zeros
   (a) should alternate on the real axis
   (b) should alternate only on the negative real axis
   (c) should alternate on the imaginary axis
   (d) can lie anywhere on the left half-plane

Answers to Objective-Type Questions

1. (a)	**2.** (a)	**3.** (b)	**4.** (a)	**5.** (a)	**6.** (b)
7. (c)	**8.** (c)	**9.** (b)	**10.** (a)	**11.** (b)	**12.** (c)
13. (b)	**14.** (b)				

Appendix 1

UNITS USED IN ELECTRICITY AND MAGNETISM

Quantity	*Quantity Symbol*	*Unit*	*Unit Symbol*
Admittance	Y	siemens	S
Angular velocity	ω	radian per second	rad/s
Capacitance	C	farad	F
Charge	q	Coulomb	C
Conductance	G	siemens	S
Conductivity	σ	siemens per metre	S/m
Current-Steady or			
rms value	I	ampere	A
Instantaneous value	i	ampere	A
Maximum value	I_m	microampere	μA
Potential Difference	V	volt	V
Instantaneous value	v	volt	V
Frequency	f	hertz	Hz
Impedance	Z	ohm	Ω
Inductance, self	L	henry	H
Inductance, mutual	M	henry	H
Permeability of free space			
or magnetic constant	μ_0	henry per metre	H/m

(*Contd*)

Quantity	*Quantity Symbol*	*Unit*	*Unit Symbol*
Permeability, relative	μ_r	henry per metre	H/m
Permittivity of free space or electric constant	ε_0	farad per metre	F/m
Permittivity, relative	ε_r	farad per metre	F/m
Permittivity, absolute	ε	farad per metre	F/m
Power	P	watt	W
Reactance	X	ohm	Ω
Resistance	R	ohm	Ω
Resistivity	ρ	ohm metre	Ωm
Susceptance	B	siemens	S

Appendix 2

ABBREVIATIONS IN MULTIPLES AND SUB-MULTIPLES

Symbol	*Prefix*	*Multiplying Factor*
T	tera	10^{12}
G	giga	10^{9}
M	mega	10^{6}
k	kilo	10^{3}
d	deci	10^{-1}
c	centi	10^{-2}
m	milli	10^{-3}
μ	micro	10^{-6}
n	nano	10^{-9}
p	pico	10^{-12}

Index

A

B

C

K

L

M

N

O

P

Q

R

S

T

U

V

W

Y

Z

Made in United States
Orlando, FL
04 November 2026

90207698R00376